Cracks and Fracture

To Anne

Cracks and Fracture
K. Bertram Broberg
Department of Mathematical Physics
University College Dublin, Ireland

ACADEMIC PRESS
San Diego • London • Boston
New York • Tokyo • Toronto

This book is printed on acid-free paper

Copyright © 1999 by ACADEMIC PRESS

All Rights Reserved
No part of this publication may be reproduced or transmitted in any form or by any means electronic or mechanical, including photocopy recording, or any transformation storage and retrieval system, without permission in writing from the publisher.

Academic Press
525 B Street, Suite 1900, San Diego, California 92101-4495, USA
http://www.apnet.com

Academic Press
24-28 Oval Road, London NW1 7DX, UK
http://www.hbuk.co.uk/ap/

ISBN 0-12-134130-5

A catalogue record for this book is available from the British Library

Printed in Great Britain by Cambridge University Press, Cambridge

99 00 01 02 03 04 CUP 9 8 7 6 5 4 3 2 1

Contents

Preface vi

Introduction 1

Chapter 1 The Process Region   5
1.1   Introduction 5
1.2   Micro-separations and coalescences 5
       General considerations 5
       Dominant kernels of micro-separation 7
       The significance of an intrinsic material length parameter 8
       Micro-separations and coalescences in metallic polycrystalline materials   9
       Micro-separations and coalescences in ceramics 11
       Micro-separations and coalescences in polymers 12
       Micro-separations and coalescences in composites 13
       Necking near a crack edge in thin sheets and plates 13
1.3   The cell model of materials 15
1.4   Definition of the process region 18
1.5   Dimensions of the process region 20
1.6   Models of the process region 22
1.7   Experimental observations 25

Chapter 2 The Fracture Process   27
2.1   Introduction 27
2.2   Pre-existing cracks 27
2.3   Loading before crack growth 29
2.4   Onset of crack growth 30
2.5   Stable crack growth 31
       General properties of stable crack growth 31
       Measurement of the amount of crack growth 33
2.6   Onset of unstable crack growth 35
2.7   Unstable crack growth 36
       General considerations 36
       Crack growth direction 37
       Branching 38
2.8   The plastic region and its wake 39
       General considerations 39

	The plastic region before unstable crack growth	39
	The plastic region during crack growth	43
	The wakes of the primary plastic region and the process region; the secondary plastic region	44

Chapter 3 Basic Relations in Crack Mechanics 45
3.1 Introduction 45
    General considerations 45
    Boundary conditions 45
    The three symmetry modes 46
3.2 Path-independent integrals 48
    General considerations 48
    A path-independent integral for plates 49
    The J-integral 52
    Extension to the plastic region at a stationary crack 53
    Moving cracks: the steady state approximation 54
    Energy flow to the crack edge region 55
    Interpretation of the P- and J-integrals as energy flux at steady crack edge motion 57
    Estimates of P- or J-integrals for stationary cracks from load-displacement curves 59
    The path-independent M-integral 61
    The J-integral for anti-plane shearing and for axisymmetric cases 63
    A path-independent integral for stationary cracks in dynamic fields 66
    A path-area integral incorporating inelastic strains, inertia and body forces 67
3.3 Stress-strain field at small scale yielding 67
    General considerations 67
    Importance and dominance of inverse square-root terms 68
    The autonomy 73
    The stress intensity factor 74
    Dominating terms in modes I and II 77
    The universal character of the asymptotic stress fields 83
3.4 Cracks in non-linear elastic environments 84
3.5 Results based on energy considerations 85
    Energy flow to the crack edge 85
    Interpretation of the Nilsson integral 90
    Weight functions 90
    Weight functions for stationary cracks, subjected to dynamic loading 94
    Weight functions for moving cracks 95
3.6 The Barenblatt model 95

Chapter 4 Elastostatic Cracks 99
4.1 Introduction 99
4.2 Eigenfunction expansions 99
4.3 Complex potentials for anti-plane strain 103
    Basic equations 103
    Mode III cracks 105

	Infinite body	106
	Finite outer dimensions	110
4.4	Complex potentials for in-plane problems	111
	Basic equations	111
	An alternative formulation: the Westergaard method	114
	Westergaard method for mode I	115
	Westergaard method for mode II	121
	The basic equations for in-plane loading in polar coordinates	125
4.5	Conformal mapping	126
	General considerations	126
	Mode III crack in a long strip	127
	The elliptic hole approach	130
4.6	Some applications of complex potentials	132
	The canonical problem in the theory of cracks and fracture	132
	The mode II counterpart	135
	The mode III counterpart	135
	Energy relations	136
	Loads other than remote stresses	138
	Arrays of cracks	140
	Circular arc crack	142
4.7	Integral equations	148
	Superposition of elementary loads	148
	Arrays of dislocations	152
	Dislocation arrays in finite and semi-infinite regions	155
4.8	Dual integral equations	160
4.9	Wiener-Hopf technique	163
	General considerations	163
	A mode I edge crack in a large plate	164
	Slanting edge crack	169
4.10	Kinked and branched cracks	171
	The origin of kinked cracks	171
	The infinitesimal kink	171
	Branched cracks	173
4.11	Directional stability of cracks	174
4.12	Interface cracks	177
	General considerations	177
	The interface crack under in-plane deformation	178
	The interface crack under anti-plane strain	185
	Crack perpendicular to an interface	186
4.13	Three-dimensional crack problems	190
	Introduction	190
	The elliptic crack	191
	Cracks in curved shells	191
	Semi-infinite plane crack subjected to crack face loading	192
	Crack with slightly curved front	201
	Crack in a plate with consideration of transverse shear	205
4.14	Cracks in anisotropic linearly elastic media	206

	Introduction 206
	General relations 207
	Plane cases 210
	Basic relations for cracks in anisotropic media 215
	Cracks under anti-plane strain 215
	Cracks under generalized plane strain 216
	Interface cracks 226
	Interface crack subjected to remote loading 234
4.15	Cracks in isotropic viscoelastic media 236
	General definitions and relations 236
	Viscoelastic models 238
	The elastic-viscoelastic correspondence principle 238
	Stationary cracks 239
	Slowly propagating crack 243
	Energy flow into a cohesive region 244

Chapter 5 Elastoplastic Stresses and Strains 247

5.1	Introduction 247
5.2	Basic relations 248
5.3	Stationary mode III cracks 254
	General considerations for perfect plasticity 254
	The hodograph transform 258
	The Hult-McClintock problem 260
	Large scale yielding at a mode III crack in a finite elastic-perfectly plastic body 263
	Loss of autonomy at large scale yielding 268
	Conditions under which proportional loading prevails in the plastic region 270
	Mode III small scale yielding in strain hardening materials 271
	Strains in the crack edge vicinity at large scale yielding in strain hardening materials 273
	A note on the near edge autonomy for mode III elastic-plastic cracks . 276
	Matching solutions for elastic and plastic regions 276
	Fully plastic solutions 277
5.4	Stationary mode I cracks 278
	Cracks in perfectly plastic materials. Slip line solutions 278
	Cracks in strain hardening materials 285
	Mode I plastic region at small scale yielding 290
	The mode I Leonov-Panasyuk-Dugdale model 292
5.5	Stationary mode II cracks 296
	Cracks in perfectly plastic materials. Slip line solutions 296
	The mode II Leonov-Panasyuk-Dugdale model 297
5.6	Slowly moving mode III cracks 298
	General considerations for perfect plasticity and steady state ... 298
	Constant stress and centered fan stress sectors 299
	Stresses and strains in the primary plastic region 300
	Stresses and strains in the secondary plastic region 302

		Assembly of sectors close to the crack edge 303
		Strain hardening 307
5.7		Slowly moving mode I and II cracks 308
		General considerations 308
		Constant stress and centered fan sectors 310
		The rate of deformation tensor 311
		Asymptotic deformations in centered fan sectors 312
		The asymptotic field in elastically deforming sectors 314
		Matching conditions at sector boundaries 315
		Assembly of sectors 320
		The significance of a finite process region 326

Chapter 6 Elastodynamic Crack Mechanics — 328

- 6.1 Introduction 328
- 6.2 The Yoffe problem 330
  - Introduction 330
  - Statement of the mode II Yoffe problem 331
  - Solution for a moving line load 332
  - Solution of the mode II Yoffe problem 334
  - The crack edge vicinity in modes I and II 336
  - Crack face loads. The Craggs problem 344
- 6.3 Friction and cohesion in mode II crack propagation 345
  - Sub-Rayleigh crack velocities 345
  - Intersonic crack velocities 348
- 6.4 Steady state mode III crack propagation 356
- 6.5 Steady state crack propagation in anisotropic materials 360
- 6.6 Steady state crack propagation in viscoelastic materials 365
  - Formulation of a Wiener-Hopf equation for mode III crack propagation 365
  - Solution for the standard linear solid 367
  - Solution for models other than the standard linear solid 371
  - Energy flux into the crack edge and energy dissipation 373
- 6.7 Crack propagation in a strip 375
- 6.8 Crack propagation along an interface 377
  - Introduction 377
  - The subsonic anti-plane case 377
  - The subsonic in-plane case 379
  - The intersonic in-plane case 386
  - The subsonic/supersonic anti-plane case 390
  - The subsonic/intersonic in-plane case 394
- 6.9 Crack expanding with constant velocity 398
  - Self-similar dynamic problems 398
  - Symmetric mode I crack expansion 402
  - Symmetric and nonsymmetric mode II crack expansion 414
  - Intersonic mode II crack expansion 419
  - Symmetric mode III crack expansion 422
  - Self-similar expansion of circular and elliptic cracks 426
- 6.10 Face loaded mode III cracks 428

	A general remark 428
	A self-similar problem with wide applicability 428
	Superposition of elementary solutions............................. 434
	Moving pair of opposed concentrated crack face forces 434
	General time-independent crack face loading...................... 435
	Crack motion from an equilibrium state........................... 436
	Time-dependent crack face loads 438
6.11	Face loaded in-plane mode cracks...................................... 439
	Face loaded mode I cracks 440
	Face loaded mode II cracks....................................... 448
	Moving pair of opposed concentrated crack face forces 449
	General time-independent in-plane mode crack face loading 450
	In-plane mode crack motion from an equilibrium state 450
	Self-similar time-dependent crack face loading 451
	General time-dependent crack face loading 451
6.12	Non-constant crack velocity and crack arrest 456
	Sudden arrest of a mode III crack subjected to crack face loading 457
	Sudden arrest of in-plane mode cracks 463
	Crack growth at non-constant velocity 464
	Sudden change of crack speed 467
	Arrest from steady state crack propagation........................ 467
	Arrest of an expanding crack 468
6.13	Stress wave loading of cracks.. 469
	General considerations... 469
	Step function SH wave impinging on a crack...................... 470
	Step function SV wave impinging on a crack...................... 472
	Step function P wave impinging on a crack 473
	Crack growth after a delay time 474
	Crack arrest after stress pulse loading 481
	Non-constant crack velocity after stress pulse loading 483
	Stress wave incidence on a crack of finite length................... 486
	Stress wave incidence on a crack in a strip 488
6.14	Three-dimensional elastodynamic crack problems 491
	Oblique P wave incidence on a crack edge 491
	Three-dimensional face loading on semi-infinite cracks 499
	General three-dimensional loading on planar cracks................ 501

Chapter 7 Elastoplastic Crack Dynamics   **509**

7.1	Introduction 509
7.2	Mode III crack propagation 510
	Governing equations... 510
	Asymptotic solution 511
	Solution for the symmetry plane ahead of the crack 514
	The paradox resolved.. 519
	On the validity of asymptotic analyses. Ståhle's solution 521
7.3	Asymptotic solutions for in-plane crack propagation 522
	General relations... 522

	Asymptotic solution for mode I 525
	Asymptotic solution for mode II 531
7.4	Temperature changes at moving crack edges 536
	General considerations... 536
	Heat conduction and temperature distribution near a moving crack edge 539

Chapter 8 Physical and Engineering Aspects of Fracture   544

8.1 Introduction 544
8.2 Processes in the crack edge vicinity 545
    Void growth 545
    Internal local instability in a continuum model.................. 552
    Localization of plastic flow in shear 556
    Competition between decohesion and plastic flow 556
    Competition between cleavage and plastic flow at a crack edge 562
    Cleavage cracking in the presence of plastic flow 563
    Toughening mechanisms 565
    Fractal models for cracks.. 572
    Fracto-emission 573
8.3 Prediction of onset of crack growth 574
    General considerations. The concept of autonomy before crack growth 574
    Small scale yielding 575
    Large scale yielding 577
8.4 Prediction of fracture 579
    Global *versus* local criteria 579
    Small scale yielding 581
    Large scale yielding 587
    Dependence on plate thickness.................................... 595
8.5 Weibull's statistical theory of the strength of materials 596
    Introduction... 596
    Weibull's theory 597
    Prediction of ductile-brittle transition, using Weibull statistics 600
8.6 Crack directions... 603
    Introduction... 603
    Crack directions under small scale yielding 606
    Crack growth in compression 609
    Directional stability during wedging 613
8.7 Applications to engineering 614
    Introduction... 614
    Linear elastic fracture mechanics................................ 615
    The J-resistance curve method 615
    The J-N method 616

Chapter 9 Dynamic Processes in Fracture Mechanics   625

9.1 Dynamic crack propagation 625
    Introduction... 625
    The crack edge equation of motion 629
    A model for dynamic crack propagation............................ 630

	Simulations, using the cell model 635

        Simulations, using the cell model 635
        Other simulations of dynamic crack propagation 637
        Dynamic crack propagation in viscoplastic materials 638
        Dynamic crack propagation in modes II and III. Shear banding 642
9.2 Branching 646
        Introduction 646
        A hypothesis 646
        Branching symmetry 648
        Attempted branching 649
        Branching synchronicity 650
        Macroscopic criteria for the onset of branching 651
        Simulations of branching 652
9.3 The maximum crack velocity 653
        Mode I 653
        Modes II and III 654
9.4 Crack arrest 655
        Situations in which a running crack may be arrested 655
        Conditions for crack arrest 656
        Re-initiation 657
        The crack arrest stress intensity factor 658
        The role of the kinetic energy in crack arrest 659

Appendices   661
A1 Formulae and general relations 661
        Vector operations 661
        Displacements and stresses expressed by displacement potentials 662
        Elastic constants 664
A2 The first boundary value problem for the half-plane 665
        In-plane problem 665
        Anti-plane problem 667
A3 Some formulae related to the Airy stress function 668
        Representation of stresses 668
        Representation of strains and displacements at linear elasticity 668
        Equation and basic solutions 669
        Solutions for the vicinity of crack edges or sharp notches 669
A4 Analytic functions 670
        Definition. Cauchy-Riemann equations 670
        Singular points 671
        Power series representation. Analytic continuation 672
        Connection with Fourier series 674
        Cauchy's integral theorem 674
        Cauchy's integral formula 675
        Plemelj's formulae 677
        Liouville's theorem 678
        Decomposition of analytic functions 679
A5 Laplace transforms 680
        One-sided Laplace transforms 680

	Two-sided Laplace transforms 682
	Asymptotic relations 683
A6	Mellin transforms 684
A7	Hankel transforms 684
A8	Stresses near a sharp notch 685
A9	Basic relations for stress waves in elastic solids 687
	Basic equations for motion. The two fundamental wave types 687
	Waves from an embedded point source 690
	Wave guides 691
	Waves in plane strain or plane stress 691
	Surface waves – Rayleigh waves 694
	Waves in linearly elastic anisotropic media 696
A10	Formulae related to path-independent integrals 697
	Energy flow to the process region 697
	Relations between P- or J-integrals and load-displacement records ... 699

References 702

Index 733

Preface

"What's the title?" Valerie Hemingway asked when Anne, my wife, told her that my book was about to be completed. "*Cracks and Fracture*" said I. "Are these the two main characters?" she asked.

Indeed, they are. The term *Cracks* evokes intellectual queries about the nature of materials and their modelling, while *Fracture* evokes concern about practical realities. Each of these two characters has inspired scientific explorations, impressive engineering feats and voluminous literatures. But they belong together and they inspire one another.

The story of this book parallels the development of personal computers, because I began the writing already in 1973. At that time, very few males knew how to handle a typewriter, and I was no exception. Consequently, during the first two years, I wrote about two hundred fifty pages manually, and in this form, the incomplete manuscript was used in Sweden as text material for graduate students at the Lund Institute of Technology and (still) at the Technical University of Luleå (my handwriting is legible, although I am not proud of its aesthetic qualities).

After the first two years, I found that the writing occupied too much time, and I slowed down considerably. It was resumed after my retirement in 1990. Before that, however, the advent of personal computers forced me to learn how to handle a keyboard. I found a good program for text with mathematical formulae, Lotus Manuscript, into which I transferred my previous work. Then I became aware of LaTeX with its many convenient features, among them the excellent cross-referencing and handling of equations. With a primitive home-made program, I succeeded in translating my previous writing, about 300 pages, into LaTeX, and this has been used ever since.

For figure drawing, the progress has been equally revolutionary for authors of technical literature, although much remains to be done in terms of integrated programs. When I started writing, figure making was an almost exclusively manual affair. Now, most figures in the book are computer produced, except for a few that needed artistic skill and these were drawn manually by Ms Doris Nilsson, Lund Institute of Technology, Sweden. For the rest, I have mainly used TeXdraw and Gnuplot 3.6, but also Xfig, Coplot and Maple V. All these figures have been transferred to encapsulated postscript, those produced by TeXdraw with the aid of Ghostview. Platforms, as for the writing, have been DOS and Linux.

In retrospect, I find that this book is very different from the one I intended to write since 1973. First of all, of course, as a result of the impressive progress in the field, most of the contents relates to work done after 1973. Secondly, the original plan was to devote one half of the book to basic relations and the other half to practical applications. The methods by which most of the practical applications I had in mind were handled in the mid 70's would now be considered as either obsolete or incomplete, whereas most of the basics would still be up to date, even though many important more recent contributions would have had to be added. Therefore, I decided to concentrate on basic relations and keep the practical applications to a minor part, mainly to be found in the last two chapters.

The book presents a comprehensive and critical overview on the mechanics of cracks and fracture. Fresh perspectives are offered on these fields, highlighting central analytical methods and their applications as well as general theoretical and experimental results. Fundamental concepts are introduced, considering the specific field of cracks and fracture as well as general mathematical models and analytic methods.

General crack mechanics and fracture processes due to monotone loading or to rather few load applications constitute the main focus of this work. Much of the material included has bearings also on numerical methods, fatigue and creep, but these fields have not been explored to any depth. Material models used include isotropic and anisotropic elastic materials as well as elastic-plastic and viscoelastic materials. Composite materials are not specifically included, but stationary and dynamic interface cracks are studied.

The selection of material for the book and the way it is presented reflects my own interest and experiences. I have tried to give appropriate credit to those who have made important contributions to the field, and much work has been devoted to trace original sources, something that now and then motivated consultation of literature outside the Western tradition.

Cracks and Fracture should appeal to researchers and graduate students in engineering fields related to the sciences of mechanics and materials, in geophysics, and in mathematical physics. The engineering aspects include fracture processes in engineering structures and materials. The geophysical aspects include fracture processes in mining and in seismic events. Of interest in the field of mathematical physics are several analytical methods, such as the Wiener-Hopf technique, the Cagniard method, Hilbert problems, transform methods and complex potential methods.

It is asumed that readers are familiar with basic continuum mechanics and mathematical analysis, such as offered at university level for undergraduate students in the fields of engineering and science. Whenever possible within the constraints of space, derivation of results are pursued in sufficient detail to allow readers to follow their whole course. Analytical methods are generally introduced in the context of applications. Certain analytical tools, such as analytic functions and Laplace, Mellin and Hankel transforms, are described in appendices, which also present the basic theory for elastic wave propagation and certain formulae and general relations.

For the benefit of research workers who intend to use *Cracks and Fracture* as a reference work, there is an extensive Index. Some care has also been taken to make different parts reasonably self-contained or otherwise supplied with direct references to associated parts.

For graduate studies, parts of *Cracks and Fracture* may be chosen in different ways, if the whole book is considered too voluminous. For less mathematically demanding studies, Chapters 1, 2, 8 and 9 may be chosen. These chapters may anyway be chosen as introductory reading. For studies with the emphasis on mathematical methods, selections may be made with the help of the Index: for example, there are useful index entries on Hilbert problems and the Wiener-Hopf technique. Other selections may be made for studies in special methods, such as complex potentials and integral equations. It is also possible to select certain classes of crack problems: thus, for instance, the parts on viscoelastic materials or dynamic crack propagation may be found to be sufficiently self-contained for special studies.

During the long period of writing, many colleagues and other friends have contributed in different ways. Valuable comments on the early, manually written version were given by Drs Hans Andersson and Christer Ljung at the Lund Institute of Technology. The present version, in its entirety, was carefully read by Professor Michael Hayes, Department of Mathematical Physics, University College Dublin, Ireland, and I am deeply grateful for his many thoughtful and qualified suggestions as well as for his encouragement and support during the final years of the completion of the manuscript. Substantial parts of the manuscript were read by Dr Solveig Melin, Department of Solid Mechanics, Lund Institute of Technology, who, in particular, examined mathematical expressions. I am also very grateful to her for finding relevant literature for me during several years. Mr Neal Murphy, Department of Mechanical Engineering, University College Dublin, also examined substantial parts of the manuscript and gave valuable comments. Dr Daniel Rittel, Materal Mechanics Laboratory, Technion–Israel Institute of Technology, Haifa, gave insightful points of view on some parts of the manuscript. All manually drawn figures were skilfully made by Ms Doris Nilsson, Department of Solid Mechanics, Lund Institute of Technology.

I am also grateful for assistance with regard to computer software or hardware offered gratuitously by Dr Matti Ristinmaa, Lund Institute of Technology, and by Dr Adrian Ottewill and Dr Alun Carr, University College Dublin. I also profited from Dr Carr's considerable expertise in materials science. Several colleagues and former students in many different countries, have contributed during the years by sending me reprints of their papers or assisting me in other ways. I have also got many pieces of advice from friends, once they knew that I was involved in writing a book.

A special thanks goes to Ms Bridget Shine, Academic Press, London, for her skilful and efficient handling of matters related to the preparation of this book. From the first time I contacted Academic Press, she has taken a wholehearted interest in the work, and I am grateful for her continuous support. I am also grateful to Ms Alison Woollatt, who made the style file for this book, and to Ms Tamsin Cousins and Ms Manjula Goonawardena for their work with its production.

Above all, I express my thanks to my dear wife, Anne Buttimer, for her patience, understanding, encouragement and helpful advice during the years leading to the completion of *Cracks and Fracture*.

Introduction

What is fracture? The simplest answer would be: "the process of breaking" or "the condition of being broken". Actually, both *fracture* and *break* can be traced to the same Indo-European root, *bhreg* (to break). Each term can be either a noun or a verb.

Common-day use of the word "fracture" is not well-defined. It may refer to something that is being (or has been) broken apart, or it may indicate the presence of a crack. In the present work, "fracture" will be reserved for unstable crack growth, either all the way through a structural part or so far that the strength or stiffness of the part becomes considerably reduced.

A crack may be defined as a material separation by opening or sliding, with the separation distance substantially smaller than the separation extent – the crack length. The separation distance is often comparable to certain micro-structural length dimensions, for instance the distance between larger inhomogeneities in the material, such as inclusions. In extreme cases, the separation distance may be of the order of the atomic distance, and the crack length, while still large compared to this distance, may be smaller than some larger micro-structural dimensions, for instance a grain size. It is then appropriate to talk about a *micro-crack*. Micro-cracks play an important part in fracture processes, but so also do other types of material separation on a microscopic level, particularly internal voids (holes). The general term for a material separation on a microscopic level is a *micro-separation*.

From a practical point of view, many cracks may be considered as harmless, i.e. not leading to fracture. A big structure, such as a tanker, contains probably several thousand macroscopic cracks and several million micro-cracks. Essentially only those cracks that are situated in highly strained regions should be regarded as potential fracture initiators, and this only if they are larger than a certain size. It is a major objective of *fracture mechanics* to find out which cracks constitute an obvious risk for fracture and which do not.

Crack growth depends on loading conditions and environmental conditions. It may be extremely fast, over $1000\,\mathrm{m/s}$, and it may be extremely slow, less than $1\,\mathrm{mm/year}$.

Loading conditions include many distinct types, static, dynamic, load controlled, grip controlled, etc. An important distinction should be made between *monotone loading*, i.e. monotonically increasing load until either a certain level is reached or mechanical failure occurs, and *repeated loading*, cyclic or non-cyclic. Very few structures, such as objects loaded by their own weight, are actually exposed to monotone loading. Some structures are exposed to considerably more than ten thousand load applications. If fracture occurs after so many load applications it is referred to as *fa-*

tigue, or, more clearly, *high-cycle fatigue*. By *low-cycle fatigue* is meant fatigue after relatively few load applications, usually fewer than about ten thousand†.

Environmental conditions such as temperature and corrosive atmosphere influence crack growth. At a high temperature, usually several hundred degrees Celsius, metals show crack growth through *creep*, i.e. slow crack growth even at constant load.

The basic framework presented in this book is to a large extent common to most types of crack growth and fracture. High-cycle fatigue, creep crack growth and crack growth under corrosion will not, however, be explicitly covered.

Fracture is only one way by which mechanical failure can occur. Other types of processes leading to mechanical failure are corrosion and wear. These mechanisms do not belong to the scope of the present book. Very closely related to fracture, however, is *plastic collapse*. In a ductile tensile test piece, for instance, plastic instability precedes crack growth (at least on a macro-scale), and the fact that the final rupture occurs through fracture is uninteresting from a practical point of view: quite obviously the plastic instability should be given the blame for the failure. In other cases, the opposite order of events occurs, for instance at failure caused by bending of a ductile beam: crack growth may reduce the beam stiffness so much that plastic collapse takes over, but the failure may already be a fact when this happens, so the occurrence of plastic collapse is fairly uninteresting from a practical point of view.

In engineering structures, crack growth occurs generally through opening of a gap between the crack surfaces. This mechanism is conventionally referred to as the *opening mode* or *mode* I. In other cases, mainly in earthquakes and related events, crack growth occurs through sliding between the crack surfaces. This mode is called the *sliding mode* or *shearing mode*, and there are two varieties, *mode* II and *mode* III, depending on whether the sliding direction is normal to or parallel with the crack edge. Frequently both of these modes occur together (so called mixed mode growth), but mode I does not appear to mix readily with the other two.

Fracture mechanics is a rather young discipline. Even though the interest in fracture prediction probably is older than our civilization, the systematic approach to problems concerning growth of pre-existing macroscopic cracks, which is what fracture mechanics is about, is typically a 20th century concern. Basic mathematical tools were created by Kolosov (1909) in his doctoral thesis at University of Dorpat (present Tartu), Estonia. Inglis (1913), obviously independently, also solved a basic crack problem, and, in a discussion of Inglis' paper, B. Hopkinson (1913) suggested that nonlinear phenomena near the crack edge should be taken into account. This was finally done by Griffith (1920), but, by using energy considerations and the concept of surface energy, he avoided an analysis of the crack edge neighbourhood. Griffith's experiments with thin glass rods prompted Weibull (1939a,b) to establish a statistical theory of fracture. Orowan (1952) extended Griffith's approach to all cases of *small scale yielding* (in which plastic flow is confined to a small region near the crack edge) by inclusion of all dissipative energy, essentially the surface energy and plastic work. Irwin (1957) introduced new and expedient concepts such as the *stress intensity factor* (originally the crack driving force) and the *energy release rate*. The *critical stress intensity factor* or, equivalently, the *fracture toughness*, became concepts that laid the foundation of

---

† There is, of course, no distinct boundary between high-cycle and low-cycle fatigue – ten thousand load applications is only mentioned here as an indication.

the *linear (elastic) fracture mechanics* (LEFM). Barenblatt (1959a,b) introduced the concept of *autonomy* of the field near the crack edge, and a linearized model of the crack edge vicinity lead to his concept of *cohesion modulus*. In fact, all the different concepts used in LEFM are developed explicitly or tacitly under the assumption of autonomy, which thus provides the very basis for LEFM. Briefly expressed, autonomy implies that the processes near a crack edge are always the same in each material, regardless of body and loading geometry, under certain specified general conditions.

After the 1950s the development in the fields of crack mechanics and fracture mechanics has been quite impressive, quantitatively and qualitatively. It is not possible to describe this development in a rather limited space, but a few names will be mentioned. First to mind comes J.R. Rice, who has made outstanding contributions to virtually all fields in crack and fracture mechanics, from the mid 60s, including the introduction of the J-integral concept, a path-independent integral, for crack analysis (Rice 1968a), which laid the foundation for the nonlinear fracture mechanics, to recent contributions concerning three-dimensional dynamic crack propagation (Geubelle and Rice, 1995, Cochard and Rice, 1997, Morrisey and Rice, 1998). His impact on the whole field has been singular and enormous. In the dynamic field, the significant and pioneering contributions by B.V. Kostrov and L.B. Freund deserve particular mention. Kostrov solved several problems of importance for earthquake source physics and dynamic crack propagation in general. He was the first to solve a problem of nonconstant crack expansion (Kostrov, 1966). Among numerous contributions by Freund may be mentioned a series of four papers on crack propagation with nonconstant velocity and other dynamic problems, such as stress wave interaction with cracks (Freund 1972a,b, 1973, 1974a). Finally, T. Yokobori should be mentioned, both for his outstanding merits as a scientist and for his organizational talents: he is the Founder President of the International Congress of Fracture, which started in 1965 and has had a profound importance as a forum for exchange of ideas and experiences through their quadrennial conferences. His book on fracture mechanics, first published in Japanese 1955 and translated into English ten years later (Yokobori 1965), appears to be the first monograph in the field.

There exists a fairly large number of books and overviews related to fracture mechanics. A summing up of the state of the art at the end of the 1960s is given in a seven volume large treatise on fracture, edited by H. Liebowitz (1968-1972). Other books were written by Anderson (1995), Broek (1982), Cherepanov (1979), Freund (1990), Hahn (1976), Hellan (1984), Herzberg (1983), Kanninen and Popelar (1985), Karihaloo (1995), Knott (1979), Lawn (1993) and Yokobori (1965). There are also several books on specialized subjects, such as computer methods in fracture mechanics, fatigue crack propagation and creep crack growth. Some of these, for instance the book by Riedel (1987) on fracture at high temperatures, also give account for general properties of cracks and fracture.

This volume builds on research work in various subfields of crack and fracture mechanics from all over the world. The selection naturally reflects my own interests and experiences. The two first chapters deal with the physical processes in the vicinity of the crack edge and the development of fracture. Chapter 3 develops general basic concepts and relations in crack mechanics, such as path-independent integrals, stress intensity factors and energy flux into the process region. Chapters 4-7 deal with the analysis of elastostatic cracks, stationary or slowly moving elastic-plastic cracks, elastodynamic crack processes and elastic-plastic crack dynamics. In Chapter 8, physical and engineering aspects of the processes leading to fracture are considered, and Chapter 9 deals with dynamic fracture mechanics. The appendices include general formulae, the basic theory of analytic functions, introduction to Laplace, Mellin and Hankel transforms, and description of certain basic relations, for instance for stress waves in solids. There is an extensive bibliography, covering references to both classical and recent work.

1
The Process Region

1.1 Introduction

Regardless of the size of a structure that fails through fracture, the fracture process always takes place in a small region near the crack edge. This region, the *process region*, is subjected to very high loads. Material separations occur in this region, which is very special in the structure: whereas other parts usually may be described by appropriate constitutive equations, this is not possible for the process region. Continuum mechanics simply does not work in this context (one exception, process region in a thin sheet, will be discussed later).

The physical appearance of the process region and the physical processes going on vary between different materials. There are even variations within the same material under different environmental conditions, for instance different temperatures, and under different loading conditions, for instance different rates of loading. More or less random variations occur along the crack edge, and the properties of the process region may change during crack growth. However, irrespective of the plethora of variations, certain features are common to virtually all process regions.

The importance of the process region was recognized by, e.g., Broberg (1968), Wnuk and Knauss (1970) and Wnuk (1971, 1973, 1974). Wnuk considered in particular the structure of the process region in viscoelastic and viscoplastic materials.

1.2 Micro-separations and coalescences

General considerations

The significant process in the process region is the nucleation and growth of micro-separations, eventually culminating in coalescence with the main crack. By *micro-separation* is meant a material separation, *decohesion*, on a micro-structural level. Micro-separations may be of many types, dependent mainly on the material, but also on the local load velocity and triaxiality. The following list should be considered as a broad survey:

Micro-cracks, common in metals at low temperatures and in polycrystalline ceramics.
Voids, common in metals at room temperature.
Cavities at grain boundaries, common in metals at high temperatures.
Rupture and disentangling of molecules, common in polymers.

1. THE PROCESS REGION

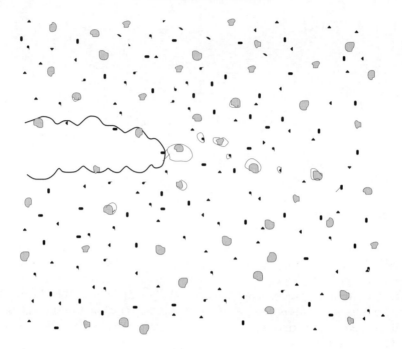

Fig. 1.2.1 Several processes are going on in the process region. The figure shows one example: void growth from two populations of particles, large and small. It is very simplified: thus, for instance, it does not show grain boundaries.

Rupture of atomic bonds, common in ceramics like glass.

Dislocation slip, "sliding off", common in thin metal sheets.

Micro-separations are nucleated at inhomogeneities in the material, for instance a particle, a hole, a flaw†, a weak grain boundary spot, a twin or a dislocation intersection. The *mean distance*, d, between dominating nucleation kernels is significant, because it controls the size of the process region at low crack velocities.

The expression "dominating nucleation kernels" needs an explanation. Often two or more populations of nucleation kernels exist side by side. Nucleation of micro-separations then usually occurs most easily at one population than at the other ones, i.e., it is dominant.

Figure 1.2.1 illustrates the typical complexity and richness of micro-structural processes in the process region. The example is schematic, but might refer to some steel at room temperature. The example assumes two kinds of particles‡ as nucleation kernels, one larger, but less frequent than the other.

---

† The difference between a hole and a flaw is essentially the shape: a hole is rounded, whereas a flaw is disc-shaped. On a micro-scale both may be very irregular.
‡ Particles are generally abundant in metals; they may be (unwanted) inclusions from the fabrication melt, or second-phase particles such as carbides. Typically, the particle size may vary between 0.01 μm and several μm.

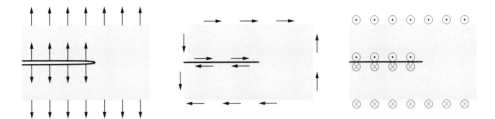

Fig. 1.2.2 Opening and sliding modes. Arrows close to the crack show displacement directions, and the other arrows show the outer loads. Arrows pointing towards the reader are symbolized by ⊙ and arrows pointing away are symbolized by ⊗.

After severe local plastic flow, stress concentrations are created at particles, due to a combination of stiffness differences and geometrical factors†. During load increase, such concentrations eventually become high enough to initiate micro-separations at particle-matrix interfaces. These micro-separations propagate along the interfaces, until the particles are effectively debonded, so that voids are created. Alternatively, a particle may fracture, especially if it is large and brittle (e.g., Broek 1972, Thomason 1990), and a void is then formed by continued plastic flow in the surrounding matrix. In general, voids are more easily created at larger than at smaller particles (Broek 1971, 1972). In some materials, particles are already debonded before loading, due to inhomogeneous thermal shrinking during fabrication of the material.

During continued loading, the voids grow due to plastic and viscoplastic flow. Eventually, coalescence between a void and the main crack occurs, by some sliding off or rupturing mechanism. Coalescences generally also take place between voids.

Micro-separations may be created by tensile forces, as in the example discussed, or by shearing forces. In general, tensile forces are the dominating cause in the *opening mode*, i.e. when the macroscopic crack is opened by the outer load; see Fig. 1.2.2. Creation by shear forces is dominating in the *sliding* (or *shearing*) *mode*, i.e. when the outer load causes the faces of the macroscopic crack to slide against each other; cf. Fig. 1.2.2. In the sliding mode, the crack face slip may be perpendicular or normal to the crack edge. Opening and shearing modes will be discussed in Chapter 3.

A thorough and very informative review on nucleation, growth and coalescence of micro-separations is given by Curran *et al.* (1987).

Dominant kernels of micro-separation

Which set of nucleation kernels in the example shown by Fig. 1.2.1 is of dominating significance? The larger particles? The smaller but more frequent ones? All particles? Still smaller inhomogeneities than particles? It is not easy or meaningful to give a precise answer. A rough rule is that the dominating kernels are those from which micro-separations are first nucleated, but if they are very infrequent, smaller but more abundant kernels might dominate the process. A working procedure may be to disre-

---

† For very small particles, this may take place on such a small scale that concepts like plastic flow and stress become inadequate.

gard one set of kernels and try to estimate the consequence of their absence for the process region.

In general, micro-separations are primarily nucleated from the large particles†, but when this is done, the loss of load carrying capacity at those particles shifts the load to the smaller particles, leading to secondary micro-separations between the primary ones. In this way, the smaller particles may contribute significantly to the coalescence between the larger particles or between the larger particles and the main crack. On the other hand, the outer load required for opening of micro-separations at the smaller particles might have been much higher in the absence of the larger particles. The dominating nucleation kernels may then be taken as the larger particles, and the intrinsic length parameter d as the average distance between them. Strictly speaking, however, d is not a pure material parameter, because the micro-separation mechanism might vary in the same material depending on the environment. Thus, in some steels, it may consist of void formation at room temperatuture, but change to micro-crack formation at sufficiently low temperatures. At very high temperatures, it may change to grain boundary cavitation. There is also a clear effect of the local load triaxiality, so that high triaxiality favours micro-crack formation, whereas low triaxiality favours void formation.

The significance of an intrinsic material length parameter

Consider a crack with characteristic length a in a large body, i.e., a body with outer dimensions so much larger than a that it may be regarded as infinite. Let the body be subjected to a slowly increasing outer load. Fracture, i.e. unstable crack growth, will occur either when the load has reached a certain critical value or not at all. In the latter case, which is very rare, the crack will become more and more rounded and will expand under increasing load without any clearly recognized moment of instability. A crack in chewing gum (after some chewing) is an example. Experience tells that, in most cases involving materials of engineering and geophysical interest, unstable crack growth occurs suddenly at some critical load, which may be given as a remote stress, σ_{crit}^{∞} (say). Furthermore, this load is known to decrease with increasing a in a given material (a large crack is more "dangerous" than a small one). Dimensional analysis then implies that the fracture load depends on a material stress parameter, σ_0 (say), and on an intrinsic length parameter, d, i.e. $\sigma_{crit}^{\infty}/\sigma_0 = f(d/a)$, where $f(\cdot)$ decreases with its argument. Thus, the existence of an intrinsic length parameter follows from experience.

The intrinsic length parameter in a material, where the dominating micro-separation mechanism is void growth from particles, may be identified with the average distance between those particles that are dominating kernels of void nucleation. In other kinds of materials, the intrinsic length parameter may be less easy to identify, but some attempts will be made to discuss its physical origin for different kinds of materials.

There are cases for which the significant length parameter needed in a dimensional analysis is not intrinsic in the material, but a characteristic body length. Thus, in sufficiently thin sheets, the sheet thickness may be more significant than any intrinsic length parameter for the material. This will be discussed later.

---

† In some materials they are nucleated from pre-existing holes or flaws, with similar micro-mechanisms as here discussed for particles.

Micro-separations and coalescences in metallic polycrystalline materials

Micro-separations in metallic polycrystalline materials are nucleated, as a rule, after severe local plastic flow. They may appear as micro-cracks, either a cleavage along a crystallographic plane or a so-called quasi-cleavage, or as voids or cavities. With some simplification it may be said that micro-cracks are generated by high stresses and voids by large strains with a strong shear component. Voids are expected to occur at a high degree of overall (macroscopic) plastic flow, and micro-cracks at a stress build-up without significant relief by plastic flow. In general, both micro-cracks and voids grow inside the grains, thus eventually producing *transgranular fracture*.

In many metals, void formation is the dominating micro-separation mechanism at temperatures around room temperature and higher, whereas micro-crack opening dominates at sufficiently low temperatures, when the resistance to slip becomes high. This resistance effect is very strong in body-centered cubic crystals. The local stress composition also influences the preference: the tendency towards micro-crack formation increases with stress triaxiality. This, indeed, seems logical because high triaxiality impedes plastic flow.

Voids may also be formed by cavitation† caused by a very high mean stress in very ductile materials. Such tension may occur in a metal under high constraint, for instance a thin metal layer, sandwiched between ceramic plates, as analysed by Varias *et al.* (1991, 1992) and by He *et al.* (1996) on the basis of experiments by Reimanis *et al.* (1991).

At very high temperatures, usually several hundred degrees, a third kind of micro-separation might dominate: cavities at grain boundary spots. In this case an *intergranular fracture* eventually results as cavities along grain boundary facets coalesce. The tendency toward intergranular fracture may be substantial even around room temperature in materials with weak grain boundaries. The presence of hydrogen atoms in steel, for instance, may lead to grain boundary embrittlement: the small hydrogen atoms move rather freely around and tend to migrate to open spaces at grain boundaries‡. Electrochemical processes may contribute to grain boundary weakening: electrical potential gradients may be created during segregation in the crystallization process, resulting in chemical differences between grain boundaries and interiors. In other contexts, electrochemical processes may increase the resistance to fracture. Weibull (1938) reports that the ultimate strength of glass plates subjected to certain electrolytes may increase more than 100 per cent.

Other cases of *environmental assisted cracking* are *stress corrosion cracking*, SCC, *liquid-metal embrittlement* and *neutron embrittlement*. In SCC, cracking occurs by tensile stresses in combination with a chemical agent that would not otherwise cause general corrosion. It is specific to certain alloy–chemical combinations. Similarly, liquid-metal embrittlement occurs for certain liquid-metal combinations. Thus, for instance, alpha-brass, which is otherwise very ductile, fails in a brittle manner and at a much reduced load under the action of mercury. Neutron irradiation may increase the nil ductility temperature (see page 15) in structural steel by as much as 200°C.

Environmental assisted cracking may reduce the strength of materials considerably, more than by a factor of two. Although very important, it falls beyond the scope of

---

† Cavitation will be further discussed on pages 549ff.
‡ Hydrogen embrittlement may also involve transgranular cleavage or even ductile fracture.

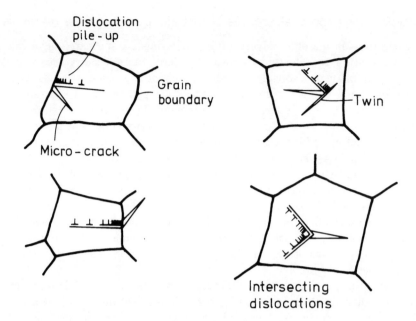

Fig. 1.2.3 Micro-cracks may be formed by different mechanisms. Here are a few examples.

the present work. Reference may be made to Thomson (1983), Fontana (1987) and Dieter (1988).

Void formation and growth was described and discussed in the example connected with Fig. 1.2.1. It may be recognized from a study of the fracture surfaces, because void coalescences leave dimples (shallow, rounded depressions), which make the surfaces appear rough, like a moon landscape filled with craters. The intrinsic length parameter, d, may be identified with the distance between the dominant particles for void nucleation.

Some mechanisms that have been suggested as initiators of micro-cracks are schematically shown in Fig. 1.2.3, but micro-crack nucleation may occur at any location of high stress concentration. Such concentration may appear as a result of elastic anisotropy mismatch or thermal expansion mismatch, particularly near grain vertices or junctions (Ghahremani et al. 1990) or at particles. It may be already present before loading, i.e., as residual stresses on a micro-scale, developed during cooling of the fabrication melt.

In a polycrystalline material, a micro-crack, once formed, may rapidly extend across the grain, but rarely into other grains, because the lattice directions change at grain boundaries. On the other hand, a micro-crack that meets a grain (or subgrain) boundary may initiate a micro-crack in the adjacent grain. Because these two micro-cracks in general have different orientations, they usually touch each other at only one point (or not at all). In this way a broken sheet of micro-cracks is created along the prospective forward growth direction of the main crack; see Fig. 1.2.4. The bridges between neighbouring micro-cracks are eventually sheared off by intense plastic flow, a process that generally requires much more energy than formation and growth of the micro-

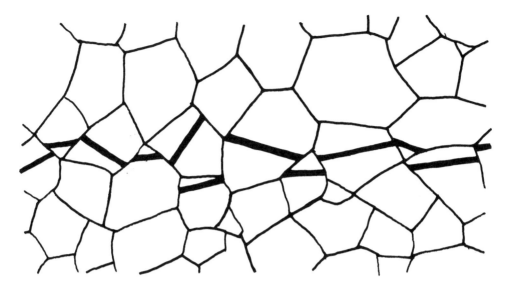

Fig. 1.2.4 Micro-cracks in different grains have different orientations and therefore their coalescence involves intense plastic flow when bridges are sheared off.

cracks. The resulting fracture surface is stepwise flat, showing glittering facets, with usually "fibrous" borders between the facets (the broken bridges). The intrinsic length parameter d is of the order of the width of the facets, which may be about equal to the grain size.

Under some conditions, for instance a high ambient pressure, *shear banding* may occur in front of a crack edge. Shear banding may be preceded by grain distortion/rotation, caused by severe straining, and appearing as a thin layer of such grains. At very high rates of strain, the localization to a layer may be due to thermal softening. On a micro-scale, shear banding may occur in the highly strained region between voids opened at two large particles. Complete shear failure occurs by coalescences in a so-called void sheet, containing voids opened at smaller particles (Tvergaard 1982a, Needleman *et al.* 1992, Faleskog and Shih 1997).

Micro-separations and coalescences in ceramics

Whereas the basic micro-structure of a metal consists of positively charged ions held together by a negatively charged "electron cloud", the basic micro-structure of a ceramic consists of atoms tied to other atoms by sharing assigned electrons. The nature of such bonds is usually either ionic or covalent. This implies a very rigid structure, and internal slip cannot occur (like dislocation slip in a metal) without permanent rupture of the bonds. The structure may be *crystalline*, i.e. possessing a high degree of regularity, like diamond or quartz, or *amorphous*, i.e. having a low degree of regularity, like glass. See further, e.g. Brook (1991).

In a pure crystalline ceramic, the micro-separation mechanism is micro-crack formation through cleavage. There might not be any larger inhomogeneities of importance than the atoms, and the intrinsic length parameter d is therefore of the order of the inter-atomic distance.

In an amorphous atomic structure there are always inhomogeneities on a scale above the atomic one. They consist of small groups of atoms, which differ from the surrounding in density and stiffness. Sometimes these groups appear as local crystallinities. Atoms of very different size are usually present in materials like glass, which also may contain comparatively large holes. Slip, in the same sense as for metals, i.e. a regular repetitive displacement along a plane, cannot occur, but plastic deformation (i.e., an irreversible deformation) nevertheless takes place on a micro-scale in amorphous atomic structures during loading. More radical distortions of the atomic structure can only take place in regions where atomic bonds have been broken.

The intrinsic length parameter d in an amorphous ceramic may be of the order of ten inter-atomic distances. However, in ceramic materials with porosities (such as pottery), the intrinsic length parameter is much larger, about equal to the average distance between the pores.

In polycrystalline ceramics both transgranular and intergranular cracking may occur. Transgranular micro-crack growth is more common for large grains than for small ones. The grain size may vary widely in the same ceramic. In igneous rocks, crystals of millimetre size or more may be found in a very fine-grained groundmass. Micro-separations are formed along facets of the larger grains, and coalescences occur when bridges of tougher, but less hard fine-grained material collapse. Sometimes a large number of micro-separations line up in a long row, with a number of unbroken bridges; this produces a narrow process region, similar to the one sometimes found in metals (see Fig. 1.2.4). In some materials, for instance concrete, such regions may be several decimetres long (Hillerborg et al. 1976), and in geological materials, the process region of a propagating earthquake slip may be even longer (Rice 1980).

In rocks under high compression, micro-separations may be produced by macroscopic shear stresses and then linked to a shear band. This is known to happen for earthquake slip, which may occur in a several millimetre thick layer, in which large and hard grains may rotate with respect to the groundmass.

Micro-separations and coalescences in polymers

In a polymer the bonds between the molecules are much weaker than the bonds between the atoms in a molecule. Plastic deformation occurs by stretching, sliding and disentangling of molecules. Several polymers fail mainly through plastic collapse, but many fail by fracture, for instance PMMA and polystyrene.

There are several possible nucleation sites for micro-separation. Depending on the polymer, these may be crystalline regions, inclusions, flaws, holes, etc. Flaws and holes may exist simply as randomly distributed empty regions on the molecular scale. In many polymers, for instance polyethylene, crystalline regions appear as thin (typically about 20 nm) lamellae separated by about equally thin layers of glassy (amorphous) polymer. The lamellae are organized in larger units (up to $1\mu m$ thick), so-called spherulites. The intrinsic length parameter d may be of the order of spherulite thickness.

In glassy (amorphous) thermoplastics, e.g. polystyrene, the process region at a crack edge is often visible as a thin layer or *craze*, with a usually lighter colour than am-

Fig. 1.2.5 Schematic view of crazing at the edge of a crack in a polymer. In the plane of the crack, the molecules form fibrils, between which voids appear. The fibrils are broken at the crack (to the left), but intact and load-carrying, apart from some broken molecules, in the process region.

bient material. In the opening mode, a craze may reach a thickness of about a tenth of a millimetre. An inspection reveals that a craze contains tiny load-bearing fibrils (typically considerably less than 1 μm in diameter), stabilized by molecular entanglements; see Fig. 1.2.5. The friction forces set up when these fibrils are drawn out are generally strong enough to cause fibril rupture before complete withdrewal. The fibrils may then have been elongated to several times their original length, mainly because of straightening of molecules. Some molecules may have been pulled out (chain disentanglement) and some may have ruptured (chain scission). Organization into fibrils appears to occur during the crazing process, and the nucleation sites may be holes. See further, e.g., McCrum *et al.* (1988).

Void formation and growth occurs in some polymers. One example is glassy polymers which have been toughened by small rubber particles. Crazing is then suppressed and holes are opened by cavitation in the rubber particles. Hole growth leads to shear banding and failure (Steenbrink *et al.* 1997).

Micro-separations and coalescences in composites

Composite materials will not be explicitly considered in the present work. The crack path may be quite complex, sometimes following interfaces between different constituents, sometimes proceeding inside a constituent. It is hardly possible to describe the process region at cracks in composites in general terms. However, the process region in composites, consisting of a matrix with fillers in the form of rounded particles, is similar to that of a steel in which voids are opened at particles, although the scale may be very different.

Necking near a crack edge in thin sheets and plates

In a thin sheet, a special kind of deformation may develop at the crack edge, viz., a necking region. No micro-separations are instrumental, and the crack edge neigh-

Fig. 1.2.6 Formation of necking in a thin plate. Note that the height of the necking region is determined by the plate thickness.

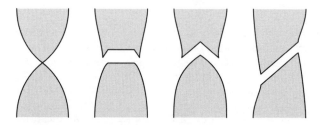

Fig. 1.2.7 Left: complete slip penetration, middle figures: shear lip formation, right figure: asymmetric cross-sectional slip.

bourhood may be treated as a continuum. Near the crack edge, the height of the necking region is generally roughly equal to the sheet thickness. Figure 1.2.6 shows schematically the formation of a necking region.

A necking region at the edge of a crack in a sufficiently thin sheet, grows continually under loading until complete sliding off occurs. Its formation and growth can, in principle, be handled by means of continuum mechanics, but the necessary three-dimensional treatment, involving large strain plasticity or viscoplasticity, may be cumbersome. The necking region may be considered as a process region, with characteristic length parameter about equal to the sheet thickness. A necking region is thus a geometry related process region, whereas the process regions previously discussed are material related.

Above a certain plate thickness, a material related process region develops in the plate interior and a geometry related process region at the surfaces. In such cases, growth starts earlier in the interior and later at the plate surfaces. This creates a curved crack edge, the so-called "tunnelling" or "thumbnail" effect.

Necking is a result of plastic flow under plane stress domination. In front of a crack edge, the maximum in-plane shear stress is comparatively small, but large shear stresses develop on planes forming about 45° angles with the plate surfaces. During continued growth, the necking region penetrates deeper towards the plate interior, in some cases ending in complete slip penetration through the plate thickness, in other cases leading to crack growth under incomplete penetration, creating *shear lips* near the plate surfaces. A few examples are shown in Fig. 1.2.7.

It may be noted that fracture may occur under conditions varying from plane strain

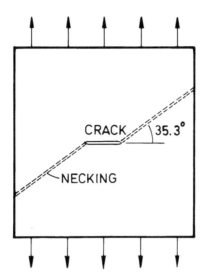

Fig. 1.2.8 Necking from a crack edge may depend on the presence of outer boundaries.

to plane stress domination, depending on the temperature. This is well known from Charpy tests or dynamic-tear (DT) tests (fracturing notched or cracked specimens by impact loading). Below a certain temperature, the *nil ductility temperature*, NDT, the impact energy needed for fracturing is roughly constant, the *lower shelf*, and the fracture surface is flat and runs straight across the specimen. Above a certain temperature, the critical impact energy is also roughly constant, the *upper shelf*, and the fracture surface is usually slanted or showing pronounced shear lips. For intermediate temperatures, the *brittle-ductile temperature transition region*, the fracture surface usually shows varying degrees of shear lips.

It is often experienced that the presence of an outer (in-plane) sheet boundary controls the formation of necking from a crack edge. Necking is then formed in the ligament between the boundary and the crack, as shown in Fig. 1.2.8. For a central crack in a sheet, such necking does not usually follow the crack direction, but forms at an angle which is about 35.3° in an isotropic (Huber-von Mises) material, according to plasticity theory (Hill 1950).

1.3 The cell model of materials

Continuum mechanics is a material model, by which the material properties may be adequately described for an arbitrarily small material volume. This model does not work for the process region. A look on Fig. 1.2.1 suggests that the smallest material unit, that can provide reasonably sufficient information about crack growth characteristics, must contain one large particle and its accompanying smaller particles. For material related process regions in general, this leads to the picture of a material composed of cells, each cell containing one micro-separation kernel of dominant type (Broberg 1979a, 1982). The cell boundaries may be taken as the midplanes between

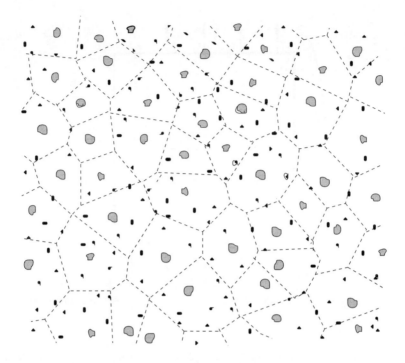

Fig. 1.3.1 A material may be envisaged as composed of cells, each cell containing one microseparation kernel of the dominant type.

the dominant kernels, creating so-called Voronoi polyhedral cells. Figure 1.3.1, derived from Fig. 1.2.1, illustrates this schematically in two dimensions.

The most important property of a cell is that of *cohesion-decohesion*. This property may be illustrated by the cohesion-decohesion curve, shown in Fig. 1.3.2 for a cubic cell, containing a central spherical hole and loaded by uniaxial straining in the vertical direction. Note, however, that decohesion also may take place as a result of shearing rather than tensile forces, and for micro-cracks rather than rounded voids. The discussion to follow may be adapted to such other cases with obvious modifications.

The cohesion-decohesion curve exhibits one ascending, cohesive, part and one descending, decohesive, part. The load is expressed by the mean stress, σ, over the upper or lower cube sides, and the displacement between these sides is expressed by δ. The maximum load, $\sigma = \sigma_D$, is reached at $\delta = \delta_D$. The curve ends abruptly by complete failure at $\delta = \delta_{max}$. If unloading takes place from any stage, the displacement decreases somewhat, as indicated by the arrow in Fig. 1.3.2.

The three parameters, σ_D, δ_D and δ_{max}, are important characteristics of cell behaviour. The significance of the displacement measures is that instability takes place under load control at δ_D and under grip control[†] at δ_{max}. Thus, it is tacitly assumed that a full cohesion-decohesion curve is obtained under grip control, but the loading

---

† Load control denotes a boundary load that does not change with changing boundary displacement; grip control, on the other hand, denotes a boundary displacement that does not change with changing boundary load.

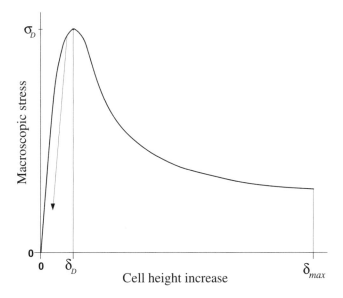

Fig. 1.3.2 The cohesion-decohesion curve: after the maximum load is reached, decohesion occurs under decreasing load.

of a cell *in situ* is generally between load and grip controlled, implying that instability occurs at a displacement between δ_D and δ_{max}. A cell *in situ* is either in the cohesive or the decohesive state. Once it has entered the decohesive state it will not normally return to the cohesive state.

During loading, the hole grows and becomes more and more flattened laterally, as indicated in Fig. 1.3.2. This behaviour was demonstrated by Andersson (1977), who used finite element methods on a cell of rigid-perfectly plastic material, similar to the cubic cell described, but with rotational symmetry. This lateral flattening is typical for cells in the process region of a crack subjected to loads which tend to cause crack opening. In other environments, such as a tensile test piece, the cell may instead be elongated in the axial direction (McClintock 1966, Rice and Tracey 1969).

A cell usually contains more micro-separation kernels than the dominating one. The cohesion-decohesion curve will then display a steeper descending part, due to activation and growth of secondary kernels, and the ratio δ_{max}/δ may then be considerably smaller. An example for a material with two populations of particles, the larger ones dominating, but less numerous, is shown in Fig. 1.3.3, which is based on work by Faleskog and Shih (1997).

Note that the cohesion-decohesion relation was introduced for a cubic cell, containing a central spherical hole and loaded by uniaxial straining. However, cell shapes and size vary, and so does the shape and size of the nucleation kernel or kernels. The cubic cell should therefore be considered as a useful model. The response to loading of a non-regular cell cannot be expressed by the cohesion-decohesion curve, but the general behaviour may be inferred from this curve. Thus, the division between cohesive and decohesive states is recognized as the instant when unstable growth would occur under load control.

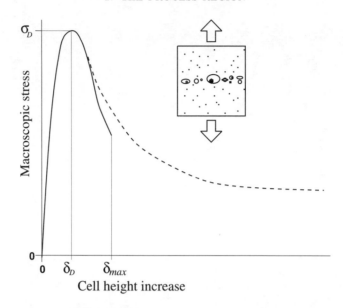

Fig. 1.3.3 The cohesion-decohesion curve for a material with two population particles. Debonding starts at the large particle, but smaller particles contribute significantly to decohesion. The dashed curve indicates decohesive behaviour for a material with only one kind of particle.

Xia and Shih (1995b) considered a cubic cell, whose surfaces remain plane with unchanged normal directions during loading. "Macroscopic principal strains", E_1, E_2, E_3, are defined for the cube as a whole by analogy with the continuum concept of strain. Xia and Shih determined cohesion-decohesion curves for different ratios E_1/E_2, from -1.0 to 0, whereas $E_3 = 0$. This corresponds reasonably well to situations for a cell in the process region of an opening mode crack with its edge in direction 3 and with straightforward direction 1. The result is sketched after their results; see Fig. 1.3.4.

Note that the maximum (macroscopic) cohesive stress depends on E_1/E_2, i.e., it is not a material property. It is however convenient to use a material parameter *cohesive strength*, or, in ambiguous cases, the "opening mode cohesive strength", based on uniaxial straining of a representative cubic cell for the material. Such a cohesive strength would not be very different from the maximum cohesive stress in the process region at a crack loaded in the opening mode.

Similarly, a "sliding mode cohesive strength" may be introduced, but crack growth in a shearing mode usually requires a high superposed compressive stress. Some combination of shear and compressive loading might therefore be considered.

1.4 Definition of the process region

Simply described, the process region consists of cells that have entered the decohesive state. However, in order to avoid inclusion of occasional isolated cells, which, for some reason, would otherwise satisfy the definition, it should also be required that the

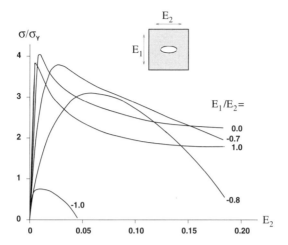

Fig. 1.3.4 Cohesion-decohesion curves for a cubic cell with one hole, subjected to biaxial macroscopic strains E_1 and E_2, whereas $E_3 = 0$. Sketched after Xia and Shih (1995b).

process region should consist of a connected region of cells terminating at the crack edge. It should also be understood that cells behind the crack edge, which are collapsed or no longer subjected to active loading, should not be considered as belonging to the process region. The process region might contain islands in the form of internal cells that are still in the cohesive state, because of random variations of cell properties.

In cases when time effects are involved the distinction between process region and other regions becomes ambiguous. The cell model may still be useful, but the cohesion-decohesion properties may depend on the rate of loading, and the concept of load control is not clearly defined for dynamic loading. In an intuitive determination of the process region at a certain instant of time, the cell boundaries may be assumed to be "frozen", i.e. subjected to instantaneous grip control. As soon as inertia movements disappear, the boundary loads are assumed to be kept constant, i.e. grip control is changed to load control. If then the displacement of some boundary facets would grow beyond any limit, the cell may be considered as belonging to the process region. Such, or similar considerations might be useful for making models of the process region.

In the absence of rate and inertia effects, the process region receives energy from adjacent cells. These cells, therefore, lose energy, but some or perhaps most of the energy that flows to the process region originates from more remote cells and even from the outer load. Figure 1.4.1 shows this schematically. The mechanical energy of the cells in the process region increases, for cells just outside it decreases. The decrease consists essentially of elastic recovery.

Note that there is a competition between decohesion and plastic flow. If the ratio σ_D/σ_Y, where σ_Y is the yield strength, overshoots a certain value, plastic flow will limit the maximum stress ahead of a crack edge to below σ_D, i.e., decohesion cannot occur. This value depends on the strain hardening and also on the so-called T-stress, the remote normal stress in the crack propagation direction. If this stress is compressive, it works together with the tensile stress normal to the crack plane to produce increased shear stresses, which implies increased plastic flow ahead of the crack edge. This tends

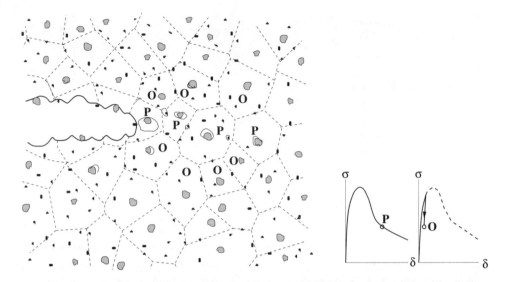

Fig. 1.4.1 Cells in the process region receive energy from adjacent cells, which actually lose energy, even though some of the energy supply to the process region is transferred from outer cells. Cells labelled P belong to the process region, cells labelled O do not. The diagrams show typical stress-displacement paths for cells inside (current position at P) and outside the process region current postion at O).

to suppress decohesion, i.e., to lower the value of σ_D/σ_Y above which no decohesion occurs. A positive T-stress implies a constraint on the plastic flow ahead of the crack edge. No plastic flow, only decohesion, results if σ_D/σ_Y falls below a certain value, dependent on the T-stress. The competition between decohesion and plastic flow will be further discussed in Section 8.2.

1.5 Dimensions of the process region

Consider a case for which rate and inertia effects are negligible. When a cell in front of the crack has reached the decohesive state, unloading occurs not only for that cell, but also for adjacent cells above and below, preventing these cells reaching the decohesive state. This indicates that the height of the process region is about equal to the cell thickness. However, because the array of cells is irregular, the average height of the process region should be somewhat larger. The intrinsic length parameter d may be taken as the height of an individual cell, so that the average height, h_p, of the process region may be somewhat larger, but hardly as large as $1.5d$.

The length of the process region, i.e. its extension in the forward direction from the crack edge, may vary widely. A study was made by Andersson and Bergkvist (1970) for an opening mode process region that may be considered as a layer of cubic cells. Their result shows that the length to height ratio depends on the ratio δ_{max}/δ_D. After taking realistic cohesion-decohesion curve shapes and distortion of cells close to the crack edge into account, the dependence may be expressed, with wide variations between different materials, by the relations

$$r_p/h_p \approx \begin{cases} 0.5\delta_{max}/\delta_D & \text{for } \delta_{max}/\delta_D > 2 \\ 1 & \text{for } \delta_{max}/\delta_D \leq 2 \end{cases} \quad (1.5.1)$$

where r_p is the length of the process region. If, for example, $\delta_{max}/\delta_D = 20$, then $r_p = 10 h_p$.

The expression (1.5.1) seems to underestimate r_p/h_p substantially (by a factor of 2–3) for a growing crack in an elastic-plastic surrounding, according to comparison with results by Tvergaard and Hutchinson (1992). However, r_p/h_p also depends on the cohesive strength, on the T-stress and on the strain hardening. It increases with increasing σ_D/σ_Y, with increasing T/σ_Y, and with the strain hardening (Xia and Shih 1995a). It also increases during the early stages of crack growth (Ståhle 1985, Xia and Shih 1995a), except for materials with high cohesive strength. In particular, if the cohesive stress approaches the maximum level for which a process region can be formed, r_p/h_p approaches unity, i.e. the process region contains only one cell in the forward direction.

In some materials the ratio r_p/h_p may be very large. One example concerns metallic polycrystalline materials in which the process region is formed by micro-cracks, as shown in Fig. 1.2.4. The length parameter d is about the average grain-size and h_p may be estimated at about $1.2d$. The micro-cracks are formed at a high overall stress but impeded overall plastic flow, as discussed in Section 1.2. The macroscopic strain δ/d in a grain may be as low as a few tenths of a per cent when micro-cracks are nucleated, but it may be ten per cent or more, cf. Fig. 1.2.4, before the bridges are broken. Thus, δ_{max}/δ_D may be of the order of a hundred, and consequently r_p/h_p may be of the order of fifty. The case shown in Fig. 1.2.4 indicates $r_p/h_p \approx 40$. Similar large r_p/h_p ratios may be found in concrete (Hillerborg *et al.* 1976). Also sliding mode process regions may show very large r_p/h_p ratios; one example is given by earthquake slip (*cf.* Rice 1980).

The length r_p of the process region in relation to the forwards extension R_p of the plastic region is of considerable interest. The embedment of the process region in the plastic region depends on several factors and will be discussed in Section 8.2.

In absolute measures, the height of the process region in the absence of rate or inertia effects may vary from a few inter-atomic distances, as in a pure crystalline ceramic, to several millimetres, as in some coarse-grained materials or materials with larger imperfections, such as porosities. A necking type region in a sheet may also be several millimetres high. As the distance between the dominant kernels of micro-separations in metallic polycrystalline materials generally is smaller than the grain size, the height of a material related process region (i.e. not a necking type) is typically smaller than 0.1 mm. In amorphous materials with strong atomic bonds it is typically smaller than 1 μm. In polymers, it might be of the order of 0.01 mm or smaller.

Viscosity effects may increase the height of the process region to more than the cell height, because each material cell may grow indefinitely in some direction under load control, even at very low loads. Thus, decreasing loads on offside cells does not necessarily lead to reverses cell growth.

Inertia, as well as viscosity, tends to increase the height of the process region (Broberg 1979a, 1985, Ravi-Chandar 1982, Ravi-Chandar and Knauss 1984c, Johnson 1992b, 1993). Information about decohesion at one central cell in the process region

Fig. 1.6.1 Schematic view of a real process region and its wake (hatched), using the cell model. In three dimensions the process region consists of a wrinkled layer of adjoining cells.

may not reach neighbouring offside cells in time to prevent them entering the decohesive state. This behaviour may be accentuated by the low velocities at which plastic deformations are propagated, compared to the velocities of elastic disturbances. For fast running cracks, they may be even lower than the crack edge velocity. A shielding effect appears, impeding the growth of central cells in the process region. See further Section 9.1.

Due to the increase of its dimensions, the process region at fast crack propagation consumes more energy than at slow crack propagation. For very slow growth the energy consumed by the process region per unit of crack growth is the minimum energy required (Broberg 1979a, 1985) for crack growth through coalescences with micro-separations. At fast crack growth the energy consumption by the process region may be much larger than the minimum energy requirement (Paxson and Lucas 1973).

1.6 Models of the process region

Fig. 1.6.1, which is derived from Fig. 1.2.1, is an attempt to show how a real process region might look like in a cross-section normal to the crack edge. The material is divided into two distinct parts: the process region and the material outside. The mechanical behaviour of the process region may be described by relations between its boundary forces and displacements. The material outside may be considered as

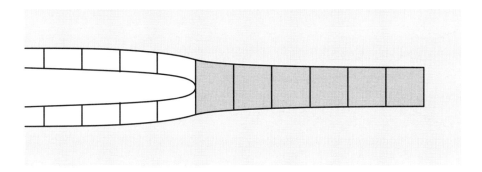

Fig. 1.6.2 Idealized model of a process region.

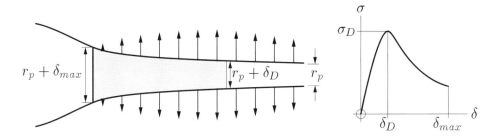

Fig. 1.6.3 The cohesion-decohesion relation for the box model consists of a σ-δ-relation.

a continuum and therefore described by means of continuum-mechanical constitutive equations.

A representative and workable model of the process region may be based on a simple shape, such as the one with originally cubic cells shown in Fig. 1.6.2. This model region should include cells rather far ahead of the crack edge, because the length of the process region is not known *a priori*. The relations between boundary forces and displacements should be taken from the appropriate cohesion-decohesion relations. Models of this kind have been successfully used in an integral equation formulation by Andersson and Bergkvist (1970), and in finite element treatments by Andersson (1975), Johnson (1992b, 1993), Shih and Xia (1994), Xia and Shih (1995a,b) and Shih *et al.* (1997). Johnson, who considered fast crack growth, used several layers of cells in the region expected to encompass the process region, whereas the others, who considered slow crack growth, used only one layer. Andersson and Bergkvist (1970) used what could be called a box model, i.e., they did not divide the layer into cells; see Fig. 1.6.3.

When the ratio r_p/h_p between length and height of the process region is large (say 4 or more) a process region model with zero height, as shown in Fig. 1.6.4, would give approximately the same results as the cell model. Even for such a reduced model it makes sense to use σ-δ-relations, but such relations cannot then include an ascending part; they would only be decohesion relations. The generic name for this kind of process regions is the *Barenblatt model* from the pioneering work by Baren-

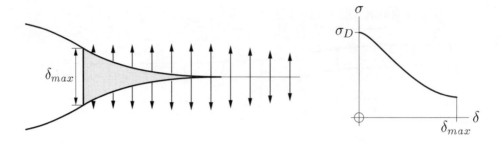

Fig. 1.6.4 A line model of the process region. The opening mode is characterized by a relation between normal stress σ and normal separation distance δ; for shearing modes there is a relation between the tangential stress τ and the tangential displacement.

blatt (1959a,c). Such models have been used in numerous analytical treatments, e.g. Leonov and Panasyuk (1959), Panasyuk (1960), Vitvisky and Leonov (1961), Atkinson (1967), Ståhle (1986). Models of this kind may also be used to simulate necking near the crack edge in a thin sheet. In some treatments, the length of the model may be prescribed and the σ-δ-relation substituted by a σ-x-relation, where x is the forwards distance from the crack edge. Such a relation is usually more convenient to handle, but might be difficult to establish from the decohesion properties, and it implies approximative treatment for growing cracks. A special version of the Barenblatt model for necking regions in thin sheets is the Leonov-Panasyuk-Dugdale model (see pages 292ff), first established by Leonov and Panasyuk (1959), Panasyuk (1960) and Dugdale (1960).

A further reduction of the cell model of the process region consists of a reduction of both height and length to zero†. This may be called a *point model* from its appearance in a cross-section normal to the crack edge. It is simply an energy sink, and it is primarily intended to be used in an elastic surrounding. It is relevant only when the total energy dissipation takes place in a region whose linear extensions from the crack edge are very small compared to the crack length, i.e. at so called *small scale yielding*. The prescribed amount of energy consumption per unit of crack growth attributed to the model includes energy dissipation in the plastic region outside the process region. The model is therefore not a true process region model, except in the rare cases when the plastic energy dissipation outside the process region is negligible compared to the energy dissipation in the process region. On the other hand, it has been the most widely used model for a near-edge region, because much work in crack and fracture mechanics has been devoted to small scale yielding.

There is one obvious disadvantage with the point model. The constitutive equations for the (linear or non-linear) elastic surrounding must be extrapolated to infinite strains. This implies uncertainties about how close to the crack edge calculated stresses and strains represent real stresses and strains sufficiently accurately.

---

† Historically, this model was not a result of a reduction from more elaborate models, but it was introduced *ab initio* by Griffith (1920) in his theory of fracture.

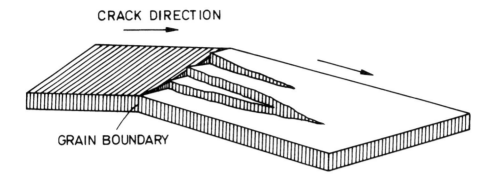

Fig. 1.7.1 The river pattern develops when cleavage propagates over grain or subgrain borders, which implies different directions of cleavage planes.

1.7 Experimental observations

Direct observation of process regions is generally very difficult. Obvious reasons are the smallness of process region, the fact that interesting processes take place inside the body, the difficulty to measure internal forces, and the vast variation and complexity of the processes going on. Attempts have been made to study the process region in thin sheets by observing deformations on the surfaces; see e.g. Bergkvist and Andersson (1972). In other attempts. the material is cut through a plane normal to the crack edge, and micro-separations may be studied this way. An excellent overview of experimental observations of the process region is given by Curran *et al.* (1987).

The most commonly employed method for getting information about the processes in the process region is *fractography*, i.e. an inspection of the fracture surfaces. When using this method, care should be taken to identify the stage in the fracture on which the inspection is focussed: is it the start of fracture or is it a later stage, perhaps one in which the fracture has been running fast? Excellent and well illustrated handbooks are available, for instance the ASM handbook on fractography (1987).

Already by the naked eye, some information may be gained about the micro-separation mechanism. In a metallic polycrystalline material the appearance of glittering facets in somewhat different orientations indicates micro-crack formation through cleavage or quasi-cleavage. Another indication of micro-crack domination is that the fracture surfaces fit well together, indicating what is often, although rather imprecisely, called "brittle fracture". A fibrous surface indicates void formation and growth until coalescences. The fracture surfaces do not fit well together, and this kind of fracture is often referred to as "ductile fracture".

With an optical microscope or a scanning electron-microscope the characteristics of the fracture surfaces can be studied in great detail. At cleavage mechanisms a "river pattern", as shown in Fig. 1.7.1, is often seen. Such a pattern develops when a cleavage propagates over a border (a grain or subgrain boundary) between different directions of crystallographic cleavage planes. The river pattern may be used to reconstruct the origin of cleavage micro-cracks. Fibrous fracture surfaces reveal dimple patterns. Sometimes the particle at which the void was formed can be seen at the dimple bottom. Fatigue fractures are usually easily recognized, even by the naked eye: the surfaces often appear to be very smooth and a closer investigation reveals striations. High crack velocities lead in general to rough fracture surfaces with the roughness increasing with the velocity (Kerkhof 1970, Ravi-Chandar 1982, Ravi-Chandar and Knauss 1984b). In glasses, for instance, the descriptive terms of "mirror", "mist" and "hackle" surfaces (Andrews 1959) are used to denote different degrees of roughness, related to the crack velocity. An excellent overview of different kinds of micro-separation in different materials is given by Curran *et al.* (1987).

2
The Fracture Process

2.1 Introduction

Fracture is often considered as a process in which increased loading suddenly causes accelerated growth of a pre-existing crack. A closer study, however, reveals three distinct phases, 1) loading without crack growth, 2) *stable crack growth*, and 3) *unstable crack growth*. Stable crack growth may, in principle, be controlled with the loading device, so that, for instance, a prescribed slow crack growth may be obtained. This is not possible for unstable crack growth, which occurs spontaneously.

Four distinctly different regions may be recognized in a crack edge vicinity. Nearest the edge is the *process region*, discussed in the previous chapter. When the crack edge advances, a *wake of the process region* is left behind. Outside the process region there is generally a *plastic region*. When the crack edge advances a *wake of the plastic region* is left behind. In this wake the material is again deforming elastically, because of unloading, but reversed plastic flow may eventually occur, so that a *secondary plastic region* appears behind the wake of the *primary plastic region*. The process region and the primary and secondary plastic regions are the *dissipative regions*. Outside them is the elastic region. In Fig. 2.1.1 the three phases of the fracture process are illustrated, together with the process region and the (primary) plastic region with its wake.

The present chapter will mainly deal with the opening mode, and some features discussed may be recognized as being specific to this mode. However, several features also carry over to the shearing modes.

2.2 Pre-existing cracks

Pre-existing cracks are very common and virtually impossible to avoid in large structures. In some solids, for instance glass, tiny surface cracks appear spontaneously as a result of chemical agents, even in seemingly neutral environments, such as air with normal humidity. In other cases, cracks are opened as a result of thermal stresses, created, for instance, after heat treatment (hardening) or welding. Cracks are also frequently formed during manufacturing of the material or as a result of mechanical processes during manufacturing or joining structural parts.

A pre-existing crack is generally not simply a sharp slot in a virgin material. Such a slot would be only a few inter-atomic distances wide, but pre-existing cracks in steels,

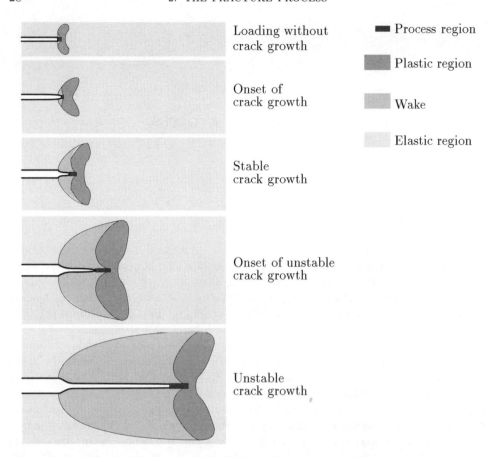

Fig. 2.1.1 The three phases of the fracture process are: 1. Loading without crack growth, but with increasing damage near the crack edge. 2. Stable crack growth. 3. Unstable crack growth. The four different regions near the crack edge are: 1. The process region. 2. The plastic region. 3. The wake of the plastic region. 4. The virgin elastic region.

for instance, may show openings of several hundred inter-atomic distances, and sometimes much more. The material state in the crack edge vicinity varies considerably, depending on the history of crack formation. This variety calls for a philosophy of handling fracture problems that does not need to consider the previous history, which, moreover, is generally poorly known. On the other hand, it is, of course, desirable to know whether the mechanism that caused cracking is still present. It is, for example, important to recognize the existence of residual stresses that have caused cracks during or after welding.

Most often, a pre-existing crack does not constitute a greater hazard than a sharp slot of the same size. Therefore, the simplifying assumption of a sharp slot, is generally appropriate for estimates on the likelihood of crack growth and fracture, and it certainly makes a mathematical analysis much more tractable.

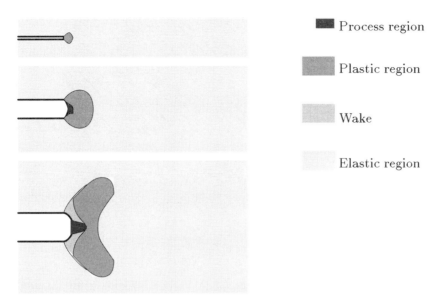

Fig. 2.3.1 During loading of an originally sharp crack, plastic flow near the crack edge causes blunting. The blunted region is sometimes referred to as the "stretched zone".

2.3 Loading before crack growth

Suppose that a crack is so oriented that the ambient stresses tend to open it. Even a small load causes a separation of the crack faces, and a strain concentration appears at the crack edge(s). In most materials, plastic flow follows, and during further loading the strains or stresses become sufficiently high to initiate micro-separations: a process region develops. Continued increase of the load causes growth of both the process region and the plastic region. Eventually, coalescences occur between micro-separations and the main crack: the crack starts growing.

The sequence of events is illustrated in Fig. 2.3.1. During loading without crack growth, the plastic flow near the crack edge and the height increase of the process region causes considerable *blunting* of the edge, forming the so-called *stretched zone*. In some materials, for instance mild steel, the blunting may be visible by the naked eye. A machine produced pre-existing crack may be substantially blunted; this may lead to considerably increased resistance to onset of crack growth.

Blunting may be studied experimentally in different ways. It can be observed optically in a cut normal to the crack edge. Another way is to pour a mould into the crack, which is withdrawn after the mould has solidified and the crack has been opened. In a less direct way, a CMOD-measurement (Crack Mouth Opening Displacement), the change in crack opening during loading is determined at the crack mouth, generally by means of a clip-gauge; see Fig. 2.3.2. This method, which is extensively used in fracture mechanics tests, is known as COD-determination (Crack Opening Displacement). It is based on some estimated relation between crack blunting and mouth opening, assuming that the crack has not grown.

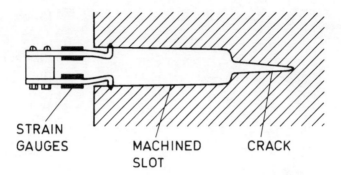

Fig. 2.3.2 Clip-gauge measurement of crack mouth opening.

2.4 Onset of crack growth

It is usually very difficult to detect when a crack starts growing. Even the very concept of incipient crack growth is difficult to define. Crack growth occurs when microseparations in front of the crack edge coalesce with the main crack, but the microseparations are unevenly spaced and of different sizes. They may even be of different types along the crack edge. Some coalescences with the main crack may therefore occur long before coalescences along the major part of the edge. Even from a macroscopic point of view, crack growth may occur early along some part of the crack edge and later at other parts, as at the "thumbnail effect".

The difficulty in identifying incipient crack growth is similar to that encountered in the determination of the "elastic limit" from a tensile test. In that case the difficulty is resolved by a convention, the idea of which is to define the yield stress as the stress when a small, but yet safely detectable permanent elongation (usually 0.2%) has occurred. The same idea, applied to crack growth, leads to definition of incipient crack growth as the state when a small, but yet safely detectable amount of crack growth has occurred. Extraordinary and sophisticated techniques should not be needed. Methods to determine the amount of stable crack growth will be discussed in the next section.

A convention that allows unambiguous determination of incipient crack growth, like the one discussed, may be needed for testing purposes. In theoretical treatments, it is nevertheless usually assumed that onset of crack growth occurs smoothly and simultaneously along the whole crack edge. Such idealizations are common in applied mechanics: in elastic-plastic theories, for instance, the state is assumed to be completely elastic until the yield condition is reached, and homogeneous plastic flow is assumed to occur immediately afterwards.

The onset of crack growth depends on several factors: material properties, body geometry (including crack geometry), load distribution, load magnitude and environmental conditions. Time effects often play a part as a result of viscoplastic flow in the process region and its vicinity. In other cases diffusion of impurity atoms towards the process region may cause delayed onset of crack growth after load application. Time effects will be more important the further the crack growth process proceeds toward unstable crack growth.

Onset of stable crack growth is governed by a *local condition*, describing when the

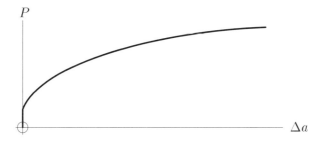

Fig. 2.5.1 Relation between stable crack growth, Δa, and outer load, P.

process region reaches a certain critical state. With experience from the development of the process region up to onset of crack growth in a certain material, an imagined observer who could overlook the whole process region and its immediate vicinity, but not necessarily other parts of the body, would be able to tell when crack growth is about to occur. In most cases of engineering interest, the development of size, shape and deformations of the process region will always be the same in the same material: the observer will not be able to see any differences, except those related to the nonregular distribution of micro-separations. This independence on body and loading geometry is what Barenblatt (1959c) called *autonomy*, a concept that will be used frequently in the present work. It has played a dominating role in fracture mechanics, although often intuitively taken for granted rather than explicitly recognized. It should, however, be remarked already here that there are several exceptions to autonomy; thus, the loading stuation (in particular whether it produces crack face opening or sliding) and environmental conditions have to be specified.

2.5 Stable crack growth

General properties of stable crack growth

Even though the total amount of crack growth during the stable phase often is much smaller than the crack length, and therefore might be difficult to detect, the outer load generally has to be increased considerably, often by a factor of two or more. Fig. 2.5.1 shows a typical relation between the amount of stable crack growth and the outer load.

In a rate-independent material, the rate of growth is directly related to the current rate of loading. If the loading device, for instance the grips in a tensile testing machine, can be manipulated, then it is possible to stop the growth or change the rate of growth as desired. This possiblility, even if it only exists in principle, may be taken to define the concept of stable crack growth. In a rate-dependent material, the concept of stable crack growth is less well defined, but is often used if the contribution to crack growth from the rate-dependence is small.

Comparatively small changes of load or grip cause substantial crack growth towards the end of the stable phase, especially at small scale yielding. Viscoplastic effects may then be significant, even if they have been negligible at earlier stages.

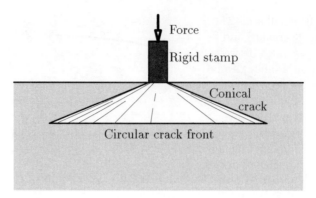

Fig. 2.5.2 Example of loading geometry suppressing unstable crack growth, thus capable of producing virtually unlimited stable crack growth.

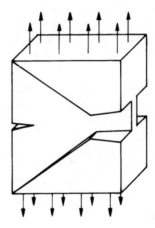

Fig. 2.5.3 Example of loading geometry suppressing stable crack growth, thus capable of producing unstable crack growth directly from a stationary crack.

The total amount of stable crack growth depends on several factors: material properties, environmental conditions, body geometry and loading geometry. In fact, virtually unlimited amounts of stable crack growth may be obtained solely by choosing a suitable loading geometry. Wedging of a plate is an obvious example, although conceptually somewhat ambiguous, because the crack often grows through jumps. Another example is shown in Fig. 2.5.2, showing conical crack surfaces†. On the other hand, body and loading geometries can be designed so as to suppress the stable phase entirely. Fig. 2.5.3 shows the idea of such designs.

In common engineering cases the total amount of stable crack growth depends very much on the embedment of the process region in the plastic region. As a result it may be either larger or smaller (but seldom much smaller) than the forward extent of the

---

† Such conical surfaces were obtained in spectacular experiments by Roesler (1956), using a cylindrical indentor on a glass block.

plastic region at small scale yielding. An illustrative example is found from numerical calculations by Tvergaard and Hutchinson (1992). A shallow embedment was obtained by choosing a low ratio between cohesive strength and yield strength; this resulted in stable crack growth about equal to the forwards extension of the plastic region. On the other hand, a deep embedment, obtained by choosing a higher ratio, resulted in stable growth amounting to several times the forwards extension of the plastic region.

Stable crack growth is governed by local conditions at the process region, but the onset of unstable growth follows as a result of a global instability. This will be discussed in Section 2.6. Stopping stable crack growth at some instant for examination, reveals that the the blunting does not accompany the moving edge; this is again sharp apart from the height extension of the process region; see Fig. 2.1.1. This reduction of blunting seems to imply that the ratio between hydrostatic and deviatoric strains and stresses in the process region vicinity increases after some amount of crack growth, which might change the micro-separation type from voids to micro-cracks.

Measurement of the amount of crack growth

Measuring the amount of stable crack growth causes in general great experimental problems. There are, however, exceptions. In thin plates, stable crack growth may amount to several times the plate thickness, and it may then easily be observed on the plate surfaces (see e.g. Broek 1968). Some uncertainty may prevail due to the possibility of a thumbnail effect. It was once described to the author that a failed ship-plate showed a crack that continued inside the plate along a distance of about 1 m, without penetrating the surfaces. This part of the crack was detected because the necking of the remaining ligaments was visible on the plate surfaces. This particular case was certainly associated with unstable and rather fast crack growth, but the observation shows that conclusions drawn from cracks visible on the surface of a body may be misleading.

Continuous recording of crack growth may be made by means of acoustical or electrical impedance methods. Such methods do not always give a satisfactory accuracy, but there has been substantial hope that recording of *acoustic emission* on engineering structures could give a warning about crack growth that might elude inspections. Similarly, tremors in the earth's crust may give a warning about an impending earthquake.

Intermittent measurements may be carried out in a number of different ways on laboratory test specimens. One simple way is to make *compliance measurements*: the stiffness of the specimen decreases with increasing crack length. The specimen is unloaded, whereby the elastic stiffness, or its inverse, the compliance, is recorded. Several determinations, by cycles of unloading, reloading and further loading, may be performed for this purpose during a test which otherwise is designed for monotone loading; see Fig. 2.5.4. This method is frequently used and is fairly accurate for deeply cracked specimens. It is not always remembered, however, that even a small number of load cycles might significantly influence the growth process by low-cycle fatigue action; cf. pages 617ff. A calibration procedure may use a machined slot, rather than the sharp crack usually required for test specimens. In this way, crack growth is suppressed, and by using slots of different lengths calibration curves can be produced.

A very accurate method of determining crack growth under laboratory conditions,

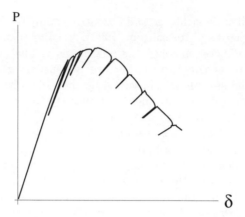

Fig. 2.5.4 Load-displacement (P-δ) record, obtained by frequent unloading and reloading. The change of slope in elastic regions gives a measure of the compliance changes during loading.

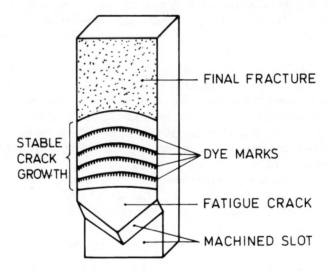

Fig. 2.5.5 A dye may be used during laboratory testing to mark the position of the crack edge at different stages of the stable phase.

consists of immersing the crack with a dye repeatedly during loading (Broberg 1975, Tryding 1991). The specimen should be oriented with the crack mouth pointing upwards. The loading is kept constant for a while until the dye has dried. Because of gravity, the dye concentrates near the crack edge, and the amount of crack growth can be seen on the fracture surfaces after completed test from the dye markings; see Fig. 2.5.5. An advantage with the method is that not only the position but also the shape of the crack edge at intervals, may be determined. An alternative, but somewhat brutal method, is to oxidize the crack surfaces by heating the crack region, for instance by means of an acetylene flame.

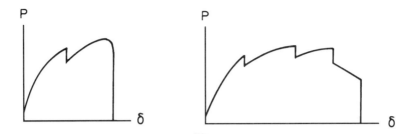

Fig. 2.6.1 Load-displacement record, showing sudden, but rather small crack advances.

2.6 Onset of unstable crack growth

A central question in fracture mechanics concerns the circumstances leading to unstable crack growth, which is usually considered as synonymous with fracture. It is the result of *global instability* (Broberg 1973a). This implies involvement of the whole body, including the loading device. This situation is well-known from the final fracturing of a tensile testpiece, which occurs sooner in a soft than in a stiff machine, and also sooner for a long testpiece than for a short. For a crack in a large plate, unstable crack growth occurs at large scale yielding (i.e., for a relatively small crack) before the $P - \Delta a$ curve, Fig. 2.5.1, is levelling out, but at small scale yielding when the curve is levelling out.

The contribution to stable crack growth by viscoplastic flow may become significant towards the end of the stable phase, as often experienced for thin sheets, for instance aluminium or celluloid sheets. It implies a conceptual difficulty, because onset of unstable crack growth is not controlled solely by the current loading situation, but also by viscoplasticity. From a practical point of view, unstable crack growth is as unwanted if caused by viscoplastic flow as if by load increase. In tests, ample time should be allowed for viscoplastic flow. Inspection of the load-displacement relation during the test will generally give information about whether the rate of loading is sufficiently low.

As well as viscoplastic flow, other time-effects may be present, for instance diffusion of impurity atoms toward the crack edge. Such effects might be very strong. An excess of hydrogen in steel may cause fracture at comparatively small loads; see e.g. Dieter (1988). Water vapour may cause fracture in glass after a few weeks at a load which is only half the load needed to cause fracture after some minutes; see e.g. Shand (1958). Several other material combinations lead to environmentally assisted cracking; cf. page 9.

Generally, a structural part or test piece is rapidly fully broken by a fast running fracture, accompanied by a sharp sound immediately after onset of unstable crack growth. Sometimes, however, some doubt may arise as to whether unstable crack growth has occurred or not. In these cases, a sudden, though very small, crack growth occurs, whereupon stable crack growth continues during further loading, until an undisputable fracture occurs. Such events appear as steps in a load versus displacement record; see Fig. 2.6.1.

The cause of such sudden crack growth steps may be of different kinds. In some cases, the origin can be traced to inhomogeneous material. One example is welded plates, where different regions possess different properties. Unstable crack growth may be initiated in the heat affected zone but be arrested in the base material. In other cases the body shape, or the load distribution, is such that a crack is soon arrested after onset of unstable growth. One example is crack growth due to thermal stresses in a plate or a hose: such stresses might vary from tensile to compressive along the prospective crack path. Frequently the explanation to a perceived sudden crack growth step is due to a so-called *pop-in* phenomenon, occurring only in the plate interior. Thus, it is a thumbnail effect. The necking regions near the plate surfaces provide higher resistance to crack growth, so that further increase of the load is needed to produce unstable growth over the whole cross-section of the plate.

2.7 Unstable crack growth

General considerations

After onset of unstable crack growth, the crack edge generally accelerates to a very high velocity, often several hundred meters per second, and sometimes to a few thousand meters per second. The energy required for conversion from a static to a dynamic state of the structure is provided by stress-strain energy† release from the body, sometimes assisted by energy supply from the loading device. At grip control, the source is stress-strain energy from the body, only. At load control, unstable crack growth occurs when the maximum load is reached, i.e. as soon as the preceding stable growth tends to occur under constant load. The stress-strain energy released, together with the energy supplied from the outer load goes to kinetic energy and to what is required by the dissipative region at the crack edge to sustain crack growth.

The general theoretical upper limit of crack propagation velocity is the Rayleigh wave velocity for mode I cracks (Barenblatt and Cherepanov 1960, Broberg 1960, Craggs 1960), the P wave velocity for mode II cracks (Freund 1979, Burridge *et al.* 1979, Broberg 1989a), and the S wave velocity for mode III cracks. For mode II crack propagation, velocities between the Rayleigh and the S wave velocities are theoretically impossible. The arguments for theoretically possible and impossible regions will be discussed in Section 6.2. The general theoretical upper limits are hardly reached in reality. For mode I cracks, velocities of about 70 per cent of the Rayleigh wave velocity have been observed (Schardin 1959), but hardly more. For each material, an upper limit seems to exist. There might be several reasons for this limitation. One reason is the increasing energy demand of the process region in the high velocity range. Another is onset of branching. These phenomena will be discussed in Sections 9.1 and 9.2.

Very often, the maximum crack edge velocity reached depends on limitations of the energy flow to the crack edge region rather than on the material. For the long strip configuration (see Fig. 2.7.1) under grip controlled conditions, the energy release from the stress-strain field per unit of crack growth is limited to some maximum value, given by the imposed grip displacement. This may be shown by means of a very simple argument, cf. Section 6.7.

---

† This energy is also called strain energy or stored energy.

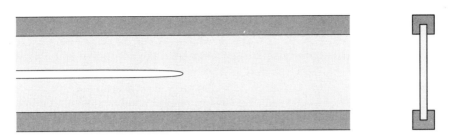

Fig. 2.7.1 Crack growth in a long strip under grip control. Because no energy disappears through the strip edges, and the only dissipative region is that at the crack edge, the energy flow to the crack edge per unit of crack advance must equal the energy accumulated in the strip per unit of length before introduction of the crack.

In the low velocity region, the energy demand for the dissipative regions may decrease somewhat with increasing crack velocity in some materials, but in other materials it increases, though at a much lower rate than at high velocities. The differences between different materials in the low velocity region may be related to material behaviour known from rapid tensile tests: the energy dissipation decreases in the low velocity region with increasing loading velocity in some materials but increases in other (Wihlborg 1976). This will be further discussed in Section 9.1.

A simple consideration shows that viscoplastic flow quite obviously plays an important part in rapid crack growth. Suppose, for instance, that the linear dimensions of the plastic region are of the order of 1 mm and that the crack edge velocity is of the order of 1 m/s. A material point close to the crack path is then being swept over by the plastic region during only 1 ms. Due to delay of the onset of plastic flow (Clark and Wood 1949) and to viscoplasticity, a time duration of this magnitude will generally impede plastic flow considerably.

Crack growth direction

A crack in an engineering structure will generally grow in a plane normal to the maximum principal stress (tacitly understood as the stress at the crack site in the absence of the crack). If the original crack orientation does not coincide with such a plane, a smooth adjustment occurs after initial kinking, i.e. a change of direction, which is abrupt at small scale of yielding; see Fig. 2.7.2. There are some exceptions to this general rule. Cracks in anisotropic materials tend to follow planes of material symmetry rather than loading symmetry. In a region of intense plastic flow, a crack may grow under in-plane shear as, for instance, in a "cup and cone" fracture of a tensile test piece. In the previous chapter, it was mentioned that a band of necking in a thin sheet may outline the path for a crack, and this path is often not normal to the overall maximum principal stress, cf. page 14.

Even at isotropy and small scale yielding, the crack may grow at approximately 45° angle to the maximum principal stress, i.e., in a direction of principal shear stress. This is extremely seldom experienced in engineering structures, except in some cases of high-cycle fatigue, but it appears as a rule in earthquakes. The reason seems to

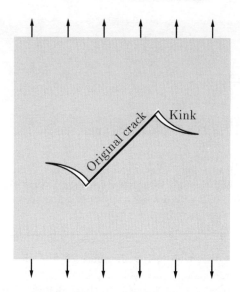

Fig. 2.7.2 Cracks in engineering structures tend to grow in a plane normal to the maximum principal stress. If the crack is not originally oriented so, a smooth adjustment towards this direction occurs after initial kinking.

Fig. 2.7.3 Examples of branching geometries.

be that the opening mode is suppressed by a high confining pressure (Nemat-Nasser and Horii 1982, Horii and Nemat-Nasser 1985,1986, Melin 1986, Nemat-Nasser 1995). By simulating such loading conditions at laboratory tests, crack propagation in the direction of in-plane shear may be provoked even in PMMA (Broberg 1987b).

Branching

Branching of a propagating crack generally occurs only at high crack edge velocities, most often larger than one fifth of the Rayleigh wave velocity. In materials like PMMA, branching generally does not occur until about 70% of the Rayleigh velocity is reached (Paxson and Lucas 1973). The phenomenon appears to be associated mainly with mode I crack growth, and it seems to be the major factor that sets an upper limit to the maximum attainable crack edge velocity in many materials. In most cases, two symmetrical crack branches are created, but sometimes there is asymmetrical branching and also branching with three or more branches. Fig. 2.7.3 shows examples of branching geometries. Very often, multiple branching occurs, i.e. each branch encounters branching and so on. This creates sometimes very regular geometrical patterns; see Fig. 2.7.4. Branching will be further discussed in Section 9.2.

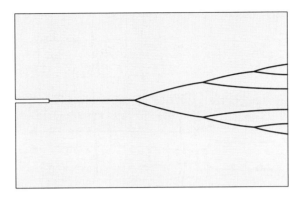

Fig. 2.7.4 Multiple branching.

2.8 The plastic region and its wake

General considerations

In contrast to the process region, the plastic region may usually be adequately analysed by continuum mechanics. Thus, the concepts of stress and strain may be used, but the occurrence of very large strains in the innermost parts of the plastic region (just outside the process region) might necessitate large strain theories and also considerations of void formation and growth, as in the Gurson model (Gurson 1977, Tvergaard 1981, 1982b, Tvergaard and Needleman 1988). It might appear contradictory to consider void growth in a continuum theory, but this can be done if the number of voids is large in each volume subjected to approximately uniform boundary conditions.

The most strained parts of the plastic region may sometimes be seen directly without any special precautions, as a roughness on the body surface or, in some transparent materials, like celluloid, as an opaque region. Certain techniques, for instance etching, may be used to improve or develop the visibility of plastic flow. Plastic strains may also be determined by using grids, as shown in Fig. 2.8.1, or moiré methods.

The plastic region before unstable crack growth

The size of the plastic region at onset of unstable crack growth, measured for instance as the largest linear extension from the crack edge, increases during loading until unstable crack growth occurs. Fig. 2.8.2 shows, schematically, how the size of the plastic region at the edge of a crack in a large plate can vary at onset of unstable crack growth, depending on the crack length. It is smallest at the largest crack lengths and largest at the smallest crack lengths.

As indicated in Fig. 2.8.2, not only the size of the plastic region but also its shape is approximately the same for different crack lengths, if these are larger than a certain minimum value. Then, autonomy prevails, not only for the process region, but also for the plastic region, i.e. it holds for the whole dissipative region at the crack edge. This is a case of small scale yielding, but the conventional definition of small scale yielding, an ASTM condition to be discussed on page 581, is generous enough to include cases

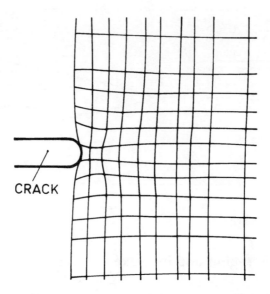

Fig. 2.8.1 Plastic strains in a celluloid sheet, revealed by a grid, applied on the sheet surfaces before loading. (Sketched from Bergkvist and Andersson 1972)

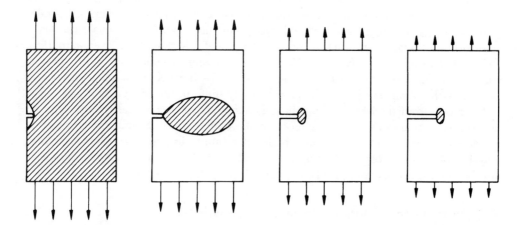

Fig. 2.8.2 Examples of how the size of the plastic region just before onset of unstable crack growth can vary with the length of a crack in a large plate.

for which the size and shape of the plastic region may vary tangibly (Larsson and Carlsson 1973).

Cases in which the plastic region is too large to qualify as small scale yielding, are referred to as *large scale yielding*. For clarity, however, the concept *net section plastic flow* will be used when a plastic region traverses a ligament from the process region to an outer boundary of the body.

Before crack growth, the strains in the plastic region at small scale yielding decrease approximately as the inverse distance to the crack edge, except near the process region.

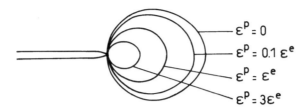

Fig. 2.8.3 In most parts of the plastic region the plastic strains are smaller than the elastic ones.

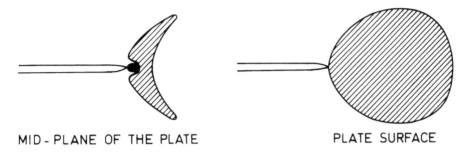

MID-PLANE OF THE PLATE   PLATE SURFACE

Fig. 2.8.4 The shape of the small scale yielding plastic region at the opening mode in the mid-plane of a plate that is much thicker than the linear extension of the plastic region, i.e. dominated by plane strain conditions.

This will be shown in Sections 5.3-5.5. A direct implication of this fact is that the strains in most parts of the plastic region are smaller than the elastic strains, and in the peripheral parts much smaller; see Fig. 2.8.3. As a consequence, the outer boundary of the plastic region is poorly defined. It is very sensitive to material inhomogeneities, and, for numerical calculations, to the numerical accuracy.

Figure 2.8.4 shows the shape of the opening mode small scale yielding plastic region in the mid-plane of a plate that is much thicker than the linear extension of the plastic region. This region is therefore subject to plane strain conditions. It has two symmetrical lobes at approximately 70° angle to the straightforward direction. Its straightforward linear extension is much smaller than the extension of the lobes, indicating by plasticity theory that the two in-plane principal stresses are approximately equal in the straight ahead of the crack. Towards the plate surfaces, however, plane stress dominates, causing a quite different shape of the plastic region, as shown in Fig. 2.8.4.

The shape of a plastic region at the edge of a crack in a thin sheet is not as well defined as in plane strain. If the sheet is so thin that no significant material related process region develops, the plastic region may consist essentially of a necking region as previously discussed; see Fig. 2.8.5. Alternatively, a split necking region may develop, if the ligament between the crack edge and the sheet boundary is short enough. In a somewhat thicker plate, a material related process region may develop, but the

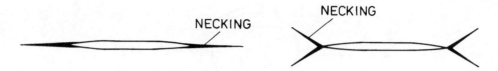

Fig. 2.8.5 Necking from a crack edge in a thin sheet. If the ligament between the crack edge and the sheet boundary is short enough, split necking occurs.

boundary of the plastic region may be situated in the plane stress region†. The shape of such a region at small scale yielding is rather rounded, as also shown in Fig. 2.8.1. However, even in thin sheets, the plastic region may be rounded. Bergkvist and Andersson (1972) performed experiments on PVC and celluloid. Tensile tests revealed almost perfectly plastic behaviour for both materials, but with a considerable difference in ultimate strain: 3.5% for a PVC sheet and 25% for a celluloid sheet. At the edge of a crack in the PVC sheet a necking region developed, but in the celluloid sheet a rounded plastic region was formed.

Very peculiar distributions of large strains have sometimes been observed in plates (Bergkvist and Andersson 1972); see Fig. 2.8.1. They seem to be caused by interaction between plane strain and plane stress plastic regions in cases when the inner part of the plastic region is subjected to plane strain and the outer parts to plane stress. The interference may cause a wavy pattern of plastic flow (Bergkvist and Andersson 1972). The plane strain region is stiffer than the plane stress region, because deformations occur under approximately equal in-plane principal stresses. It can therefore act as a hinge when the crack is being opened. In fact, compressive in-plane strains have been observed near a crack edge (Bergkvist and Andersson 1972); see Fig. 2.8.1. This indicates the possible existence of a small elastic enclave inside the plastic region.

The directions of principal strain vary largely around the crack edge. Some rotations occur during loading, but they are generally not very significant before crack growth, and therefore approximate proportional loading prevails. This implies that a total strain theory may give reasonable estimates, and in a few cases an exact description, of the stress-strain distribution in the plastic region.

In the in-plane shear mode, the shape of the small scale yielding plastic region depends on the strain hardening and on the elastic constants. One possible example is shown in Fig. 2.8.6. There are mainly tensile stresses on one side of the crack plane and mainly compressive stresses on the other side. The two sides do not necessarily have the same size and shape, because the plastic flow properties may be different for compressive and tensile stresses.

The plastic region at anti-plane shear may be analytically determined in a number of cases. In the simplest case, small scale yielding and perfect plasticity, the plastic region is circular (Hult and McClintock 1956); see Fig. 2.8.7. It is also circular at small scale yielding and power-law strain hardening (Rice 1967).

---

† Plane stress conditions dominate at distances from the crack edge that are longer than about half the plate thickness; see page 205.

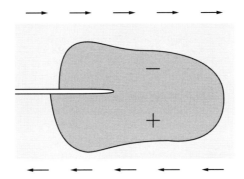

Fig. 2.8.6 The shape of the small scale yielding plastic region under in-plane shearing is not necessarily symmetric with respect to the crack plane. Plus and minus signs indicate tensile and compressive stress states, respectively.

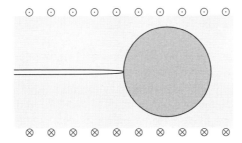

Fig. 2.8.7 Under anti-plane strain, the small scale yielding plastic region is circular. Arrows pointing toward the reader are symbolized by ⊙ and arrows pointing away are symbolized by ⊗.

The plastic region during crack growth

During crack growth, rotations of the directions of principal strains occur in the plastic region, necessitating an incremental strain theory. During the stable phase, the size of the plastic region increases, due to increased loading of the body; see Fig. 2.1.1. The shape of a small scale yielding plastic region does not change significantly during this phase, except in cases when it grows into the plane stress region during stable crack growth, after having been developed inside the plane strain region. The strain gradients are smaller than before crack growth; see Sections 5.6-5.7.

The shape and size of a small scale yielding plastic region change during unstable crack growth. The situation is highly complex. The growth of the process region with increasing crack velocity, discussed in the preceding chapter, tends to increase the size of the plastic region. The opposite trend is associated with rate dependent delayed yielding and viscoplasticity (Broberg 1979a).

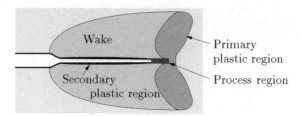

Fig. 2.8.8 Primary and secondary plastic regions.

The wakes of the primary plastic region and the process region; the secondary plastic region

During crack growth, a wake is left behind the plastic region; see Fig. 2.1.1. The wake is created through unloading, whereby deformation changes again become purely elastic and the plastic strains remain constant at each material point. Actually, a small part of the rear boundary of the plastic region starts moving forwards already before crack growth, because of unloading transferred from the process region, as shown in Fig. 2.3.1.

The process region also leaves a wake behind as it moves forward. This wake is very insignificant and may be considered as consisting of passive material with very small stiffness.

In some cases two disconnected plastic regions develop: to distinguish them the first one is called the *primary plastic region* and the second one, which develops in the wake of the primary plastic region, because of reversed plastic flow, is referred to as the *secondary plastic region*. Fig. 2.8.8 shows one example. The height of the secondary plastic region is considerably smaller than the height of the primary one and may be neglected in most contexts.

3
Basic Relations in Crack Mechanics

3.1 Introduction

General considerations

Because the process region cannot be treated as a continuum, crack and fracture problems cannot be solved simply by calculating stresses and strains in the body. On the other hand, knowledge of stresses and strains in the continuum outside the process region is essential for understanding the process of crack growth and fracture. Both analytical and numerical calculations play important roles. Analytical methods are generally based on partial differential equations or integral equations. Among numerical methods the finite element methods dominate.

Due to the complexity of real phenomena concerning cracks and fracture, analytical methods almost invariably require highly idealized models of body geometry, process region characteristics and continuum constitutive equations. Nevertheless, analytical solutions give an insight into and an understanding about basic relations that might be difficult to identify and extract from numerical treatments. Analytical solutions, particularly exact ones, are also very useful for controlling the accuracy of numerical methods, or even to test such methods for possible errors.

Boundary conditions

Most crack problems are boundary value problems. Conditions on boundaries of the continua may be divided into the following three categories:

1: Conditions on the outer boundaries of the body, including the crack faces. These conditions usually consist of specification of tractions and/or displacements.

2: Continuity conditions on the interfaces between different regions in the continuum, such as the elastic region, the plastic region and its wake or regions occupied by different materials. These conditions are automatically taken care of in some treatments, for instance finite element calculations.

3: Conditions on the boundary to the process region. These conditions depend on the response of the process region model to loads or displacements.

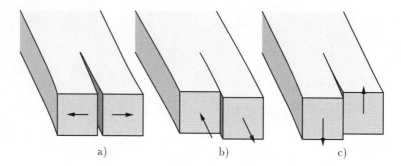

Fig. 3.1.1 The three symmetry modes. a) The in-plane opening mode. b) The in-plane shearing mode. c) The anti-plane shearing mode. Balancing forces for the shearing modes are not shown.

The three symmetry modes

In order to avoid complications, general discussions in the following will, unless otherwise stated, refer to cases in which the stress-strain field near the crack edge possesses certain symmetry properties. For linear problems, this does not impose a limitation, because each such problem may be considered as a superposition of three part-problems, each possessing symmetry properties near the crack edge, so-called symmetry *modes*. These modes were introduced by Irwin (1960), and they play an enormously important part in structuring the analysis of cracks and fracture. They are

1: *The opening mode*, explained by Fig. 3.1.1a. This mode is traditionally referred to as mode I. Two varieties, plane strain and plane stress, are of special interest.

2: *The in-plane shearing* (or sliding) mode, explained by Fig. 3.1.1b. This mode is traditionally referred to as mode II. Also for this mode the two varieties, plane strain and plane stress, are of special interest.

3: *The anti-plane shearing* (or sliding) mode, explained by Fig. 3.1.1c. This mode is traditionally referred to as mode III.

The terms "in-plane" and "anti-plane" do not imply that only plane problems are considered. The crack edge does not necessarily need to be straight. A penny-shaped (circular) crack in a large body may be loaded in the opening mode around its whole periphery or it may be loaded in mixtures of the two shearing modes. If it is subjected to arbitrary loading, it is possible to make a division into the three modes for any sufficiently close vicinity of the crack edge along the periphery. This is also possible for a non-planar crack with arbitrary edge shape.

With a coordinate system as shown in Fig. 3.1.2, and with u, v, w being the components of the displacement vector $\boldsymbol{u}$, the three modes may be specified as follows ("horizontal" refers to the x direction, "vertical" to the y direction):

3.1 INTRODUCTION

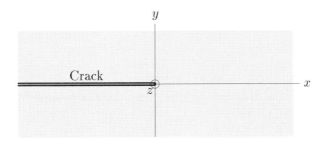

Fig. 3.1.2 Crack edge neighbourhood and coordinate system. The z-axis points toward the reader.

Mode I. Horizontal displacements are symmetric, vertical displacements anti-symmetric:

$$u(x, -y, z) = u(x, y, z), \quad v(x, -y, z) = -v(x, y, z) \tag{3.1.1}$$

$$w = 0 \quad \text{(plane strain)}, \quad \frac{\partial^2 w}{\partial z^2} = 0 \quad \text{(plane stress)} \tag{3.1.2}$$

Mode II. Horizontal displacements are anti-symmetric, vertical displacements symmetric:

$$u(x, -y, z) = -u(x, y, z), \quad v(x, -y, z) = v(x, y, z) \tag{3.1.3}$$

$$w = 0 \quad \text{(plane strain)}, \quad \frac{\partial^2 w}{\partial z^2} = 0 \quad \text{(plane stress)} \tag{3.1.4}$$

Mode III. The only non-vanishing displacement, w is anti-symmetric:

$$u = v = 0, \quad w(x, -y, z) = -w(x, y, z), \quad \frac{\partial w}{\partial z} = 0 \tag{3.1.5}$$

It is also required that the three modes possess certain stress symmetries. These are automatically satisfied for modes I and III if the displacement symmetries are satisfied and the material is isotropic. Thus:

Mode I. Normal stresses are symmetric, shear stresses anti-symmetric:

$$\sigma_x(x, -y, z) = \sigma_x(x, y, z), \quad \sigma_y(x, -y, z) = \sigma_y(x, y, z) \tag{3.1.6}$$

$$\sigma_z(x, -y, z) = \sigma_z(x, y, z), \quad \tau_{xy}(x, -y, z) = -\tau_{xy}(x, y, z) \tag{3.1.7}$$

$$\tau_{xz} = \tau_{yz} = 0 \tag{3.1.8}$$

Mode III. The only non-vanishing stresses are τ_{xz}, anti-symmetric, and τ_{yz}, symmetric:

$$\sigma_x = \sigma_y = \sigma_z = \tau_{xy} = 0 \tag{3.1.9}$$

$$\tau_{xz}(x, -y, z) = -\tau_{xz}(x, y, z), \quad \tau_{yz}(x, -y, z) = \tau_{yz}(x, y, z) \tag{3.1.10}$$

$$\frac{\partial \tau_{xz}}{\partial z} = \frac{\partial \tau_{yz}}{\partial z} = 0 \tag{3.1.11}$$

For in-plane problems there is no dependence on the anti-plane coordinate z, except for w in mode II. Thus, $u = u(x, y), \sigma_y = \sigma_y(x, y)$, etc. For non-plane problems, curvilinear rather than Cartesian coordinates may be more convenient. For the penny-shaped crack, for instance, it would be more natural to use cylindrical coordinates.

For mode II, isotropy is not sufficient for appropriate stress symmetries. However, for a material possessing both isotropy and stress-strain relations which are symmetric with respect to tension-compression, the displacement symmetry properties of mode II imply the appropriate stress symmetries. Then, normal stresses are anti-symmetric, shear stresses are symmetric:

$$\sigma_x(x,-y) = -\sigma_x(x,y), \quad \sigma_y(x,-y) = -\sigma_y(x,y) \tag{3.1.12}$$

$$\sigma_z(x,-y) = -\sigma_z(x,y), \quad \tau_{xy}(x,-y) = \tau_{xy}(x,y) \tag{3.1.13}$$

$$\tau_{xz} = \tau_{yz} = 0 \tag{3.1.14}$$

However, few, if any, stress-strain relations are symmetric with respect to tension-compression in non-linear regions. Therefore, pure mode II conditions are hardly possible to realize exactly in the continuum close to the process region, but in practical situations, such as earthquake sliding motion, they may still be assumed to prevail with sufficient accuracy.

Isotropy is not an absolute condition for symmetry mode relations; certain kinds of anisotropy may be allowed, if the crack is properly oriented with respect to the anisotropy directions. Similarly, separate treatment of the three modes is very fruitful for non-linear cases, even though the convenient principle of superposition does not hold. In this context, it should be noted that many practical situations involve pure modes, so the need for superposition of modes is generally not a problem.

In some linear cases, it is impossible to make a partition into symmetry modes. One example of considerable practical interest is the interface crack, for which the displacement pattern in the crack edge vicinity is incompatible with such a partition. Interface cracks will be discussed in Section 4.12.

3.2 Path-independent integrals

General considerations

For elastic stress-strain fields, a surface-integral representation of a property related to the presence of a defect (a singularity or an inhomogeneity) was derived by Eshelby (1957), based on an earlier work (Eshelby 1951). He noticed that the integral vanished for a closed surface, embracing a homogeneous elastic material, even with allowance for anisotropy and finite deformation. Later, Günther (1962), obviously unaware of Eshelby's results, introduced some similar surface- and line-integrals, also with conservation properties, i.e. vanishing for closed surfaces or paths. Cherepanov (1967) and Rice (1968a) introduced a path-independent integral, the J-integral, for plane elastostatic fields. It was later identified as a special case of the integral established by Eshelby, but it was mainly due to the paper by Rice (1968a) that its potential as a tool for analysis of cracks was recognized. It has been extensively applied in fracture mechanics, especially in formulations of crack growth criteria. It will be introduced in this section as a special case of a path-independent integral for plates, and its use will be discussed in later sections. Some related path-independent integrals with applications to crack problems will also be discussed.

Path-independent integrals were also discussed by Knowles and Sternberg (1972), who, like Günther (1962), made use of a theorem by Noether (1918), enabling generation of conservation laws expressed by surface- and line-integrals. In particular, they studied two integrals for plane elastostatic fields, now known as the L- and M-integrals, and previously established by Günther (1962). Interpretation of these integrals in terms of energy flux was subsequently made by Budiansky and Rice (1973). Rice (1985a) discussed several applications of the J- and M- integrals, and in a later paper, (Rice 1988a), he further discussed the usefulness of path-independent integrals for integration of the equations governing the asymptotic field near a crack edge. Some variations of previously known path-independent integrals were considered by Bui (1974), Carlsson (1974) and Eshelby (1974). Bui and Carlsson introduced integrals based on the complementary energy. Bui (1974) showed that the complementary energy flux is the same as the energy flux, even for non-linear elasticity. Thus, as remarked by Moran and Shih (1987), who performed a quite general treatment of path-independent integrals, the complementary integrals provide no new information in themselves.

The path-independent integral introduced by Cherepanov (1967) was derived for a moving crack. It also took heat flow, body forces and inertia into consideration, but the path was assumed to be drawn sufficiently close to the crack edge to avail of asymptotic properties in this vicinity.

A path-independent integral for plates

An extension of the J-integral to a path-independent integral for a plate subjected to in-plane loading was given by Broberg (1979b, 1987a). This integral may be written

$$\boxed{P = \int_\Gamma [W_{av}\, dx_2 - (n_j \sigma_{ji} \partial u_i / \partial x_1)_{av}\, ds] \quad (i = 1, 2, 3;\; j = 1, 2)} \quad (3.2.1)$$

where subscript av indicates the average over the plate thickness, Γ is a path in the mid-plane of the plate, W the stress-strain energy per unit of volume, n_j a unit normal to Γ, u_i the displacement, x_j Cartesian coordinates with x_3 in the thickness direction, and s the arc length along Γ, increasing in the direction for which n_j points to the right, seen from the positive x_3 side; see Fig. 3.2.1.

The stress σ_{ji} $(i, j = 1, 2, 3)$ is defined by the relation

$$\sigma_{ji} = \frac{\partial W}{\partial u_{i,j}} \quad (i, j = 1.2.3) \quad (3.2.2)$$

where

$$u_{i,j} = \frac{\partial u_i}{\partial x_j} \quad (3.2.3)$$

The stress tensor σ_{ij} as defined by equation (3.2.2), which is not restricted to small deformations, is not symmetric, except for $|u_{i,j}| \ll 1$, but the equilibrium equations take the simple form

$$\sigma_{ji,j} = 0 \quad (3.2.4)$$

where the summation convention is used.

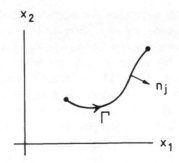

Fig. 3.2.1 The plate integral is independent of the path between two points in the x_1x_2-plane.

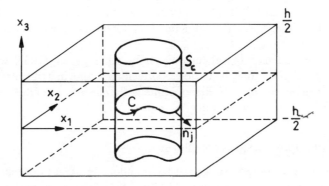

Fig. 3.2.2 Closed path C inside the plate.

The path-independence of P may be proved (cf. Broberg 1987a) by first considering the expression

$$P_1 = \frac{1}{h}\oint_S (Wn_1 - n_j\sigma_{ji}\partial u_i/\partial x_1)dS \quad (i,j=1,2,3) \tag{3.2.5}$$

where S is the total surface of a part of the plate bounded by the plate surfaces $x_3 = \pm h/2$ and by a cylindrical surface S_C, parallel to the x_3 direction and cutting the plane $x_3 = 0$ along a closed path C entirely inside the plate; see Fig. 3.2.2. n_j is the outward normal of S.

By applying Green's theorem, the area integral may be transformed to a volume integral,

$$P_1 = \frac{1}{h}\int_V \left[\frac{\partial W}{\partial x_1} - \frac{\partial(\sigma_{ji}\partial u_i/\partial x_1)}{\partial x_j}\right]dV \tag{3.2.6}$$

where V is the volume inside S. The equilibrium equations (3.2.4), together with the fact that W is function of $u_{i,j}$, imply that

$$P_1 = \frac{1}{h}\int_V \left(\frac{\partial W}{\partial u_{i,j}}\cdot\frac{\partial u_{i,j}}{\partial x_1} - \sigma_{ji}\frac{\partial^2 u_i}{\partial x_j \partial x_1}\right)dV \tag{3.2.7}$$

Thus, $P_1 = 0$, because $\sigma_{ji} = \partial W/\partial u_{i,j}$.

3.2 PATH-INDEPENDENT INTEGRALS

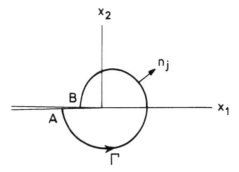

Fig. 3.2.3 Path encircling a crack edge. If the crack faces are traction free, the position of points A and B on the crack faces does not influence the value of the P-integral, provided that A and B do not lie on the wake of the dissipative region.

The intersection between S_C and the plane plate surfaces encloses a surface A_+ on $x_3 = h/2$ and A_- on $x_3 = -h/2$. Equation (3.2.5) may thus be written

$$P_1 = \int_C \left[W_{av} n_1 - \left(n_j \sigma_{ji} \frac{\partial u_i}{\partial x_1} \right)_{av} \right] ds - \frac{1}{h} \oint_{A_+ + A_-} \sigma_{3i} \frac{\partial u_i}{\partial x_1} dS \qquad (3.2.8)$$

where $j \neq 3$.

In the case of in-plane loading, $\sigma_{3i} = 0$ on the plate surfaces, and for plane strain, $\sigma_{31} = \sigma_{32} = 0$, $\partial u_3/\partial x_1 = 0$. Thus, in both cases, $\sigma_{3i} \partial u_i/\partial x_1 = 0$ on the plate surfaces. Then, the area integral vanishes, i.e.,

$$P_1 = \int_C [W_{av} n_1 - (n_j \sigma_{ji} \partial u_i/\partial x_1)_{av}] ds = 0 \quad (i = 1, 2, 3;\ j = 1, 2) \qquad (3.2.9)$$

The integral in (3.2.9) coincides with the P-integral in (3.2.1) for a closed path. This completes the proof of path-independence of the P-integral.

It is immediately evident that the vector P_k, obtained by generalizing index 1 in (3.2.9) to index $k = 1, 2, 3$, is path-independent; there is a slight difference in the proof for $k = 3$.

For a non-closed path between two fixed points A and B, the P-integral in general takes on a non-zero value. Assume that A is a point on one crack face, B a point on the opposite surface and Γ a path encircling the dissipative regions at the crack edge, as shown in Fig. 3.2.3. The crack faces are assumed to be plane and traction free. Then, if the coordinate system is oriented so that the x_1-axis is parallel to the crack faces and perpendicular to the crack edge, the points A and B may be moved along the crack faces without influencing the value of the P-integral, as long as they do not lie in the wake of the dissipative region. This follows from the fact that $dx_2 = 0$ and $n_j \sigma_{ji} = 0$ in the continuum part of the crack faces.

The fact that the P-integral takes on the same value for any path that encircles the dissipative regions at a crack edge is very useful, because this value obviously only depends on the situation inside the innermost path possible. The value is zero before loading, and during loading it provides information about the state in the neighbourhood of the crack edge. In the next subsection it will be shown that the path-independence is preserved, under certain conditions, even if the path is shrunk to go through the plastic region.

The J-integral

The J-integral may be found from the P-integral by specialization to plane stress or plane strain. For plane stress, $\sigma_{j3} = 0$ and for plane strain, $\partial u_3/\partial x_1 = 0$. In both cases, the subscript av in equation (3.2.9) may be dropped, because both terms in the integrand are independent of x_3. Thus, the integral

$$J = \int_\Gamma [W \mathrm{d}x_2 - n_j \sigma_{ji} \partial u_i/\partial x_1 \, \mathrm{d}s] = 0 \quad (i,j = 1,2) \qquad (3.2.10)$$

is path-independent. This is the J-integral (Cherepanov 1967, Rice 1968a).

An alternative way of writing the J-integral is

$$J = \int_\Gamma \left[W \mathrm{d}y - \boldsymbol{T} \frac{\partial \boldsymbol{u}}{\partial x} \mathrm{d}s \right] \qquad (3.2.11)$$

where $\boldsymbol{T}$ is the traction vector across the boundary with outward normal n, and $\boldsymbol{u}$ is the displacement vector on Γ.

Note that the derivation of the P-integral used a non-symmetric stress tensor, equation (3.2.2), which is not restricted to small strains; consequently this also applies to the J-integral. For a symmetric stress tensor, extension to finite strains leads to a somewhat different form of the J-integral than (3.2.10), cf. Broberg (1971).

Note also that the condition $\partial u_3/\partial x_1 = 0$, used to express plane strain, also covers more general states, in particular those for which u_3 is non-zero, but dependent only on x_3.

The applications of the J-integral to crack problems are similar to those of the P-integral. In many cases, approximate plane stress prevails outside the dissipative regions around the edge of a crack through the thickness of a plate, and the simpler J-integral may be used rather than the P-integral. For some other body geometries, for instance a penny-shaped crack in a large body, plane strain may be assumed for a small vicinity of a crack edge and the J-integral (but not the P-integral) may be used for paths within this vicinity.

For brevity, a path in a region of a plate where W is a single-valued function of $u_{i,j}$ will be called an "admissible path" for the P-integral and, if in addition, either plane stress or plane strain prevails in the region, the path is also an admissible path for the J-integral.

For a plate with thickness h, subjected to in-plane loading and containing a crack, $0 < x_1 < a$, $x_2 = 0$, through the whole plate thickness, the P-integral for a path remote from the crack edge $x_1 = a$ equals the P-integral for an admissible path close to the crack edge. Here, as in the following, it is tacitly understood that the path starts from one crack face and ends on the other. Far from the crack edge, approximate plane stress conditions prevail, whereas conditions close to plane strain prevail in the edge vicinity, except near the plate surfaces. Thus, if $a \gg h$, the remote path may be positioned entirely inside the plane stress region. Such a path is then an admissible path for the J-integral. If the near edge path may be positioned sufficiently close to the edge, plane strain is approached, and then also this path is an admissible path for the J-integral.

Actually, the argument used in the derivation of (3.2.10) is not applicable, because plane strain is not approached in a way that the gradient $\partial u_3/\partial x_1 \to 0$ vanishes, but it is bounded (Yang and Freund 1985), and therefore it follows from (3.2.8) that the P-integral nevertheless approaches the J-integral for the path near the crack edge. Thus, for both paths $P = J$, i.e.,

$$P = (J)_{\Gamma=\Gamma_r} = (J)_{\Gamma=\Gamma_n} \qquad (3.2.12)$$

where Γ_r denotes the remote and Γ_n the near path. Thus, if an admissible path for the J-integral can be found sufficiently close to the crack edge, the value of J for this path may be calculated by using a remote path instead. By "sufficiently close" should be understood a path that is considerably closer than about $h/2$ from the edge, cf. Yang and Freund (1985) and Rosakis and Ravi-Chandar (1986). Note that only the P-integral and not the J-integral is defined for paths in the intermediate region between plane stress and plane strain. On the other hand, because plane stress prevails on the plate surfaces, the J-integral is path-independent for paths on these surfaces, whereas the P-integral is not defined for such paths.

The requirement of approximative plane strain near the edge of a crack through the thickness in a plate generally implies rather severe restrictions on the possibility of drawing an admissible J-integral path within the distance $h/2$ from the crack edge. The crack edge must be straight† (i.e. no thumbnail effect) and the plate thickness must in general be much larger than the extension of the dissipative regions from the crack edge. These requirements are usually met only under certain laboratory conditions and then only before stable crack growth.

The fact that a remote path may be used has certain advantages. One is that stresses and strains might be less accurately known near the edge, for instance at finite element calculations. Another one is that often the J-value for a remote path can be calculated very accurately under the considerably simplifying assumption of plane stress in the whole plate, i.e. including the crack edge vicinity. A third advantage is that special paths, for instance along the outer contour of a plate, may be used.

Extension to the plastic region at a stationary crack

From the derivation of the path-independence of the P- and J-integrals it follows that all segments of the path Γ must be situated in regions where W is a single-valued function of $u_{i,j}$. This is by definition the case for an elastic (linear or non-linear) material. It is also the case for plastic deformation under proportional loading, as long as no unloading takes place. During *proportional loading*, the proportions between the components of the stress tensor remain unchanged during loading. Then, the stress and strain tensors are coaxial and the *total strain theory* of plasticity coincides with the *incremental strain theory*, cf. Section 5.2.

The fact that the P- and J-integrals are path-independent even for paths inside a plastic region during proportional loading is very important, because approximate proportional loading generally prevails in a not too large plastic region at the edge of

---

† It is, however, possible to define a vector J-integral, analogous to the vector P_k, for the immediate vicinity of a point on a curved crack edge (Kishimoto *et al.* 1980, Herrmann and Hauck 1995, 1997; see also Moran and Shih 1987).

a stationary crack, except in some vicinity of the process region where unloading has taken place. The integrals thus retain their values for paths Γ much closer to the crack edge than just outside the plastic region and consequently they sift out information about the state very close to the edge. The strains are often large in this vicinity, and therefore it is important that the path independence of the P- and J-integrals could be proved without restriction to small strains.

For increasing scale of yielding, the deviations from proportional loading become more and more pronounced, and when the plastic region at the crack edge approaches an outer boundary – net section yielding – the path-independence of the P- and J-integrals inside the plastic region becomes very poor.

Note that proportional loading does not prevail in the plastic region at the edge of a moving crack.

Moving cracks: the steady state approximation

A concept of fundamental importance in the theory of crack mechanics is that of *steady state motion*. Approximate steady state prevails in a crack edge vicinity during slow crack motion, if the mechanical state in this vicinity does not change appreciably during crack growth. During fast crack growth, also the velocity should stay reasonably constant.

Obviously there are very few situations in which steady state is realized with reasonable accuracy in a region surrounding the whole crack. In most cases the steady state region is limited to some vicinity of the crack edge and to some crack propagation phase. As an example, consider a crack in a plate, expanding in the positive x direction along the plane $y = 0$. Consider a fixed rectangular region, $\mathcal{S}\{b \leq x \leq c, |y| \leq \Delta y_0\}$, which is traversed by the crack. Then, follow another rectangular region, $\mathcal{C}$, with extension Δx_0 in the x direction and $|y| \leq \Delta y_0$ in the y direction, travelling with and containing the crack edge. It is assumed that Δx_0 is substantially smaller than $c - b$, but not necessarily smaller than $2\Delta y_0$. Then, approximately steady state conditions prevail in $\mathcal{C}$, if its mechanical state does not change appreciably during its motion through $\mathcal{S}$.

Consider two points in $\mathcal{S}$, P:(x_1, y_1) and Q:$(x_1 - \mathrm{d}x_1, y_1)$, $\mathrm{d}x_1 > 0$. During a crack advance $\mathrm{d}a = \mathrm{d}x_1$ the state in P changes to that which prevailed in Q before crack advance. Thus, the exchange

$$\frac{\partial}{\partial a} \to -\frac{\partial}{\partial x} \qquad (3.2.13)$$

may be made. Obviously, this relation holds for steady state motion in general.

The active plastic region at the edge of a crack, which is growing stably under small scale conditions, reaches approximate steady state conditions before unstable growth occurs, as the $P - \Delta a$ curve (Fig. 2.5.1) is levelling out. At large scale yielding, unstable crack growth occurs earlier, and the dissipative region does not reach steady state propagation. However, the process region may do so.

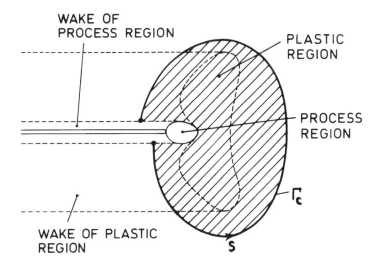

Fig. 3.2.4 Integration path Γ_c and area of integration $A - A_{pr}$ (hatched).

Energy flow to the crack edge region

Energy flux into the process region

Crack propagation occurs when the energy flow from the stress-strain field to the crack edge region is sufficient for supporting the processes leading to coalescences of micro-separations with the main crack. The analysis of the energy flow to the process region is rather complicated in the general case. Details are found in Appendix A10. For a crack in a plate, it is given by (A10.5) as

$$\left(\frac{\mathrm{d}D_{pr}}{\mathrm{d}a}\right)_{av} \approx \int_{\Gamma_c} [W_{av}\mathrm{d}x_2 + (n_j\sigma_{ji}\partial u_i/\partial a)_{av}\mathrm{d}s]$$

$$- \int_{A-A_{pr}} \left[\frac{\partial W_{av}}{\partial a} + \frac{\partial W_{av}}{\partial x_1}\right]\mathrm{d}A \quad (i=1,2,3;\ j=1,2) \quad (3.2.14)$$

where subscript av, as before, indicates the average over the plate thickness, D_{pr} is the energy dissipation in the process region per unit of plate thickness, Γ_c is a path in the mid-plane of the plate, encircling the active process region and ending on both sides of its wake (see Fig. 3.2.4), $A - A_{pr}$ is the area in the mid-plane inside Γ_c and outside the active process region and its wake, and W is the stress-strain energy (not restricted to a single-valued function of the displacement gradients). The crack length a is defined as its value in the mid-plane (the crack edge might be curved across the plate thickness).

The rather complex structure of the expression for $(\mathrm{d}D_{pr}/\mathrm{d}a)_{av}$ becomes simpler for steady state. Use of (3.2.13) implies that the area integral in (3.2.14) disappears if the path Γ_c can be positioned entirely inside the steady state region†. Thus, the average energy flow into the process region per unit of crack edge advance and per

---

† This implies that the active process region must be situated inside a steady state region. This does not necessarily require small scale yielding, as does steady state conditions for the entire dissipative region.

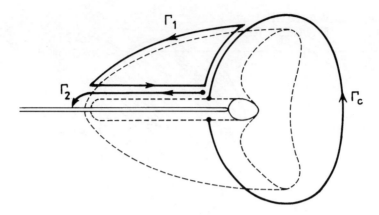

Fig. 3.2.5 The path Γ_c is drawn entirely inside an approximately steady state region outside the active dissipative regions. Γ_1 and Γ_2 are drawn inside currently elastically deforming regions. Together with Γ_c they enclose the dissipative regions and their wakes.

unit of plate thickness, in the following called the *energy flux*, is given by

$$\left(\frac{\mathrm{d}D_{pr}}{\mathrm{d}a}\right)_{av} \approx \int_{\Gamma_c} [W_{av}\mathrm{d}x_2 - (n_j\sigma_{ji}\partial u_i/\partial x_1)_{av}\mathrm{d}s] \qquad (3.2.15)$$

Because the left member of (3.2.15) is a physical quantity, independent of Γ_c, the integral is also independent of Γ_c. Note that it coincides formally with the P-integral, but the prerequisites are different. The integral in (3.2.15) requires the path to be drawn inside a steady state region, but poses no restrictions on W; the P-integral requires a path drawn in a region where W is a single-valued function of the displacement gradients. The path for the P-integral is therefore not allowed to cut through the wake of the plastic region, but the path for the path-independent integral in (3.2.15) must do so (but not traverse the wake of the process region).

Energy flux into the entire active dissipative region
Assume that the steady state region is large enough to allow Γ_c to encircle the whole active plastic region and end at the wake of the process region; see Fig. 3.2.5. The possible existence of a secondary plastic region is neglected, cf. page 44.

Writing $W = W_e + W_p$, where W_e is the recoverable (elastic) and W_p the irrecoverable (dissipated, plastic) part of the total stress-strain energy, the average energy dissipated in the plastic region per unit of crack edge advance and per unit of plate thickness is

$$\left(\frac{\mathrm{d}D_{pl}}{\mathrm{d}a}\right)_{av} = -\int_{\Gamma_c} (W_p)_{av}\mathrm{d}x_2 \qquad (3.2.16)$$

Then, from (3.2.15) it follows that the total energy dissipation per unit of crack edge advance,

$$h(\mathrm{d}D/\mathrm{d}a)_{av} = h(\mathrm{d}D_{pr}/\mathrm{d}a)_{av} + h(\mathrm{d}D_{pl}/\mathrm{d}a)_{av} \qquad (3.2.17)$$

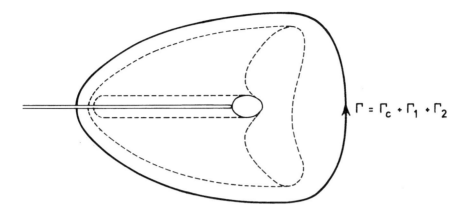

Fig. 3.2.6 Integration path Γ.

is given by the expression

$$\left(\frac{dD}{da}\right)_{av} = \int_{\Gamma_c} [(W_e)_{av} dx_2 - (n_j \sigma_{ji} \partial u_i / \partial x_1)_{av} ds] \quad (i = 1, 2, 3; \; j = 1, 2) \quad (3.2.18)$$

where Γ_c encircles the crack edge in the continuum part of the steady state region. This was noted first by Andersson (1974) under the assumption of plane stress or strain. It should be recalled that Γ_c cuts through the wake of the plastic region as a consequence of the steady state.

The integral in (3.2.15) may be generalized to cover also dynamic steady crack edge motion by exchanging D_{pr} by $D_{pr} + T_{pr}$ and W by $W + T$, where T_{pr} is kinetic energy in the process region and its wake inside S_Γ and T the kinetic energy per unit of volume. Likewise, the integral in (3.2.18) may be generalized to dynamic cases by exchanging D_{pr} by $D_{pr} + T_{pr}$ and W_e by $W_e + T$.

The integrals in (3.2.15) and (3.2.18) differ mainly through the use of W in the former and W_e in the latter, but also through the possibility of drawing the path of integration inside the plastic region (but not inside the process region) in (3.2.15). This possiblity is important when the steady state region does not cover the entire plastic region. The expression (3.2.18) is particularly useful for pronounced steady state conditions, such as crack propagation in a long strip (Fig. 2.7.1).

Interpretation of the P- and J-integrals as energy flux at steady crack edge motion

The path for the integral in (3.2.18) may be extended to encircle also the wakes of the plastic region and the process region with rather little change of the integral value. Fig. 3.2.5 shows the minimum extension needed: it consists of the paths Γ_1 and Γ_2, both divided into one upper and one lower branch. It is assumed that the path Γ_c cuts the wakes of the plastic region in the x_2 direction and is drawn as far behind the crack edge as possible without violating the conditions of approximate steady state.

In order to show that the integration path in (3.2.18) may be extended to the path

$\Gamma = \Gamma_c + \Gamma_1 + \Gamma_2$ (see Fig. 3.2.6) with relatively small change of the integral value, the displacement gradients are decomposed into an elastic and a plastic part:

$$\frac{\partial u_i}{\partial x_1} = \left(\frac{\partial u_i}{\partial x_1}\right)^e + \left(\frac{\partial u_i}{\partial x_1}\right)^p \quad (i = 1, 2, 3) \tag{3.2.19}$$

Then, since Γ_1 encircles a continuum in which only elastic deformations take place,

$$\int_{\Gamma_1} \{(W_e)_{av} dx_2 - [n_j \sigma_{ji} (\partial u_i/\partial x_1)^e]_{av} ds\} = 0 \quad (i = 1, 2, 3;\ j = 1, 2) \tag{3.2.20}$$

because the integral equals the P-integral for a closed path in an elastic region. Thus, from (3.2.18):

$$\left(\frac{dD}{da}\right)_{av} = \int_{\Gamma_c + \Gamma_1} [(W_e)_{av} dx_2 - (n_j \sigma_{ji} \partial u_i/\partial x_1)_{av}] ds$$

$$+ \int_{\Gamma_1} [n_j \sigma_{ji} (\partial u_i/\partial x_1)^p]_{av} ds \quad (i, j = 1, 2) \tag{3.2.21}$$

If the extension in the x_1 direction of the steady state region is considerably larger than the extension of the plastic region in the x_2 direction, then the stresses in the wake of the plastic region behind Γ_c are very small so that the last integral in (3.2.21) ought to be small compared to dD/da. Furthermore, if the crack edges are traction free, the term $(n_j \sigma_{ji})_{av} \approx (\sigma_{2i})_{av}$ can be neglected along Γ_2, and the differentials

$$(W_e)_{av} dx_2 - (n_1 \sigma_{1i} \partial u_i/\partial x_1)_{av} ds \tag{3.2.22}$$

give a contribution to the integral that is very small compared to $(dD_{pr}/da)_{av}$, because W_e and σ_{1i} contribute essentially only along the short part of Γ_2 that runs through a region that has remained elastic. Thus,

$$\left(\frac{dD}{da}\right)_{av} \approx \int_\Gamma [W_{av} dx_2 - (n_j \sigma_{ji} \partial u_i/\partial x_1)_{av} ds] \quad (i = 1, 2, 3;\ j = 1, 2) \tag{3.2.23}$$

where W_e could be replaced by W because the path Γ encircles all dissipative regions. Because Γ is an admissible path for the P-integral, the right member of relation (3.2.23) equals the P-integral, and then Γ may be extended to a more remote path without changing the value of the integral. Thus,

$$\boxed{\left(\frac{dD}{da}\right)_{av} \approx P} \tag{3.2.24}$$

This relation implies that the average energy flux is approximately equal to the P-integral, if approximate steady state conditions are well established in a region containing the active plastic region and a significant portion of the wake behind the active plastic region. The discussion leading to (3.2.23) indicates that this "significant portion" should extend at least as far in the x_1 direction as the extension of the plastic region in the x_2 direction. This might, however, be an unnecessarily restrictive condition, and for practical purposes it should be possible to interpret the P-integral as the *energy flux at small scale yielding shortly before onset of unstable crack growth*, but not during earlier stages. If Γ can be drawn in a region where plane stress or plane strain prevails, this interpretation is also valid for the J-integral, cf. Rice (1968a) and Cherepanov (1967).

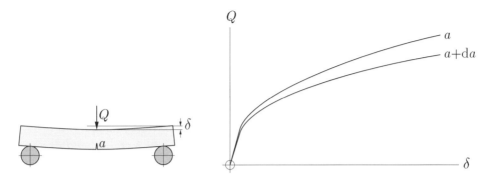

Fig. 3.2.7 Load-displacement curve for a structure with a stationary crack with crack length a or $a + \mathrm{d}a$.

Note that the integrals in (3.2.18) and (3.2.23) differ both with respect to the integrand and to the path of integration. The path Γ_c must be drawn entirely inside the steady state region, whereas Γ must encircle the entire dissipative region and its wakes. Expression (3.2.23) cannot therefore be used for determination of the energy flux into the dissipative region for crack propagation in the long strip (Fig. 2.7.1).

The energy flux $(\mathrm{d}D/\mathrm{d}a)_{av}$ is traditionally denoted by $\mathcal{G}$ and called the *energy release rate*, a concept introduced by Irwin (1957, 1960). In many cases, however, for instance, crack growth due to a moving wedge, it is difficult to associate the energy flux with an energy release, and therefore the more general term "energy flux", is used in the present work.

Estimates of P- or J-integrals for stationary cracks from load-displacement curves

Consider a plate of thickness h, containing an edge crack of length a. For simplicity it is assumed to be symmetrical with respect to the crack plane and to be subjected to loading with either mode I or mode II symmetry from a virgin state. One example could be a 3PB (three point bend) specimen for mode I loading. Loading is performed so that a displacement δ is imposed on one point of the plate boundary, whereby a force $Q = Q(\delta, a)$ is obtained at this point, see Fig. 3.2.7. Energy dissipation is assumed to be confined to a crack edge region that is small compared to a. No crack growth is assumed. This case will be referred to as case 1.

Similarly, if the length of the stationary crack had been $a + \mathrm{d}a$ rather than a, where $\mathrm{d}a \ll a$, a force $Q = Q(\delta, a + \mathrm{d}a)$, would have been obtained at the displacement δ. This case will be referred to as case 2.

An inspection of the crack edge neighbourhood in the two cases would hardly show any difference: the size, shape and loading state of the process region would be virtually the same.

Now, make the *theoretical* assumption that crack growth takes place from case 1 at fixed δ, producing an increase of the crack length from a to $a + \mathrm{d}a$ and proceeding as a pure translation of the near edge region. Steady state conditions would thus prevail in the crack edge neighbourhood during the crack growth. This case will be referred to as case 3. While such a crack growth is physically admissible, it should not be confused with most real cases of stable crack growth, during which the dissipative region in general will change during the growth in a material specific way, i.e. it would not simply be translated with the crack edge.

According to (3.2.24), the assumption of steady state, together with the fact that the dissipative region is much smaller than a, implies that the energy flow to the dissipative region during crack growth is $hP\mathrm{d}a$, where P is the value of the P-integral. If plane stress or plane strain prevails everywhere, P may be replaced by J, the value of the J-integral. Note that the transition from case 1 to case 3 takes place without energy supply from Q, because δ is constant. The energy dissipation is then provided by energy flow from the stress-strain field.

There is a difference between the situation arrived at in case 3, after the crack growth $\mathrm{d}a$, and the one in case 2, because of the trace left from the travelling dissipative region on the crack faces between a and $a+\mathrm{d}a$ in case 3. However, this trace would not cause any significant change of the stresses in the vicinity of the crack faces, because its extension is assumed to be small compared to a. Thus, the boundary condition of traction free crack faces would apply to case 3 as well as to case 2, and, consequently, the elastic stress-strain field in the two cases would be virtually the same. Thus, the elastic stress-strain energy would be about the same in cases 2 and 3, but the dissipated energy would be different. This implies that

$$\int_0^\delta [Q(\delta', a) - Q(\delta', a+\mathrm{d}a)]\mathrm{d}\delta' = hP\,\mathrm{d}a \qquad (3.2.25)$$

where the left member expresses the difference in accumulated stress-strain energy between case 1 and case 2 and the right member is the energy dissipation during crack growth from a to $a+\mathrm{d}a$ in case 3. The expression may be simplified to

$$\boxed{hP(\delta, a) = -\int_0^\delta \frac{\partial Q(\delta', a)}{\partial a}\mathrm{d}\delta'} \qquad (3.2.26)$$

In a general case, forces are given along some part of the boundary and displacements are prescribed on other parts. Then, the energy dissipation derives both from work done by the boundary forces and from release of stress-strain energy. Thus,

$$hP\mathrm{d}a = \sum \boldsymbol{Q}_i \mathrm{d}\boldsymbol{\delta}_i - \mathrm{d}U = -\mathrm{d}\Pi \qquad (3.2.27)$$

where $\boldsymbol{\delta}_i$ is the boundary displacement under the force $\boldsymbol{Q}_i$, U is the total stress-strain energy and Π is the total potential energy. Therefore, for a stationary crack and not too large scale of yielding,

$$\boxed{P = -\frac{1}{h} \cdot \frac{\mathrm{d}\Pi}{\mathrm{d}a}} \qquad (3.2.28)$$

If plane stress or plane strain prevails in the main part of the plate, then the P-integral may be substituted by the J-integral, and in this form the relation (3.2.28) was derived by Rice (1968a).

Some useful relations, similar to (3.2.26), relating P or J to quantities from load-displacement records, are given in Appendix A10. However, experimentally found relations $Q = Q(\delta, a)$ are generally not reproducible enough at large scale yielding to allow determination of $\partial Q(\delta, a)/\partial a$ or $\partial \delta(Q, a)/\partial a$ with sufficient accuracy, except if a great number of experiments are performed and the results are smoothened by statistical methods.

The difficulty of obtaining sufficiently accurate experimental relations between Q, δ and a, may be circumvented by using semi-empirical methods, avoiding differention of Q or δ with respect to the crack length a and allowing determination of J from one single experimental relation $Q = Q(\delta, a)$. One such method is described by Rice et al. (1973).

The path-independent M-integral

The path-independent M-integral, established by Günther (1962) and later by Knowles and Sternberg (1972), is

$$M = \int_\Gamma \left(W n_i x_i - n_j \sigma_{jk} \frac{\partial u_k}{\partial x_i} x_i \right) \mathrm{d}s \qquad (3.2.29)$$

where the notations are the same as used for the J-integral. Note, in particular, that n_i is the unit normal of Γ pointing to the right during travel along Γ in positive s-direction, and that it is the outward normal when Γ is a closed path. The path-independence may be shown in the same way as for the J-integral. Note, however, that even though $M = 0$ for a closed path, irrespective of the position of the coordinate origin, the value of M for a non-closed path depends on this position.

Transformation to cylindrical coordinates gives

$$M = \int_\Gamma \left(W \boldsymbol{n} \cdot \boldsymbol{r} - \boldsymbol{T} \cdot \frac{\partial \boldsymbol{u}}{\partial r} r \right) \mathrm{d}s \qquad (3.2.30)$$

where $\boldsymbol{T}$ is the traction vector across the boundary with outward normal n, and $\boldsymbol{u}$ is the displacement vector on Γ. This form shows a special feature of the M-integral, which is very useful in certain cases. Thus, the product $\boldsymbol{n} \cdot \boldsymbol{r}$ vanishes along radial lines from the coordinate origin. If in addition $\boldsymbol{T} = 0$ (as on a traction free surface) or $\partial \boldsymbol{u}/\partial r = 0$ (as on crack faces kept open by a wedge of constant thickness under full friction) along such radial lines, or else if a combination of displacements and tractions is such that the product $\boldsymbol{T} \cdot \partial \boldsymbol{u}/\partial r$ vanishes along radial lines, then the integrand of the M-integral vanishes. Furthermore, if a part of Γ is sufficiently far away from the origin to be considered as infinitely remote, then the contribution to the integral along this part may be vanishingly small – this occurs if the product of displacement gradients and stresses decreases as $r^{-\alpha}$, where $\alpha > 2$. Such features were pointed out by Eshelby (1974) and also by Freund (1978) and Ouchterlony (1978, 1980), who demonstrated the use of the M-integral for expedient determination of the energy flux into the edge of an advancing crack, and thereby the stress intensity factor (a connection which will be discussed later). See also Rice (1985a) for a discussion of the use of the M-integral in dislocation problems, and Kubo (1982) for its use in relation to interface cracks.

As an example, consider a semi-infinite crack in an infinite, linearly elastic plate. The crack faces are subjected to a pair of opposite compressive line forces Q, per unit

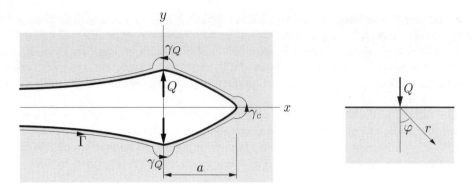

Fig. 3.2.8 Body geometry, coordinate system and integration path.

of length, acting at a distance a behind the crack edge so that mode I opening results, see Fig. 3.2.8.

A Cartesian coordinate system with origin between the application points of the load and with the crack edge at $x = a$ is introduced, together with a cylindrical coordinate system with the same origin and $\varphi = 0$ in the negative y direction. The path Γ is drawn along the crack faces, avoiding the crack edge by an infinitesimally small indentation γ_c and the load application points by infinitesimally small indentations γ_Q, and then completed to a closed path at infinity. Obviously the total stress-strain energy is finite, which implies that the energy density decreases faster than in proportion to $1/r^2$, and thus the contribution from the remote part is zero. So also is the contribution from the parts of the path following the crack edges, because $\boldsymbol{n} \cdot \boldsymbol{r} = 0$. The only contributions thus originate from the indentations, so that

$$\int_{\gamma_c} \left(W n_i x_i - n_j \sigma_{jk} \frac{\partial u_k}{\partial x_i} x_i \right) \mathrm{d}s + 2 \int_{\gamma_Q} \left(W n_i x_i - n_j \sigma_{jk} \frac{\partial u_k}{\partial x_i} x_i \right) \mathrm{d}s = 0 \quad (3.2.31)$$

In the first integral $x_1 \approx a$ and $|x_2| \ll a$, so it reduces to

$$a \int_{\gamma_c} \left(W n_1 - n_j \sigma_{jk} \frac{\partial u_k}{\partial x_1} \right) \mathrm{d}s = aJ = a\mathcal{G} \quad (3.2.32)$$

where J is the J-integral and $\mathcal{G}$ is the energy flux into the crack edge during an assumed infinitesimally small crack advance.

The second integral in (3.2.31) may be evaluated from the case of a normal compressive line force Q, per unit of length, on the surface of a semi-infinite solid; see Fig. 3.2.8. In Appendix A2 it is shown that such a force gives rise to the stresses

$$\sigma_r = -\frac{2Q \cos \varphi}{\pi r}, \qquad \sigma_\varphi = \tau_{r\varphi} = 0 \quad (3.2.33)$$

in a cylindrical coordinate system with its origin at the load application and $\varphi = 0$ pointing perpendicularly from the surface into the solid. The integration path γ_Q is taken to be semi-circular. Using the form (3.2.30) of the M-integral implies that

$\boldsymbol{n} \cdot \boldsymbol{r} = r$ and $\boldsymbol{T} \cdot \partial \boldsymbol{u}/\partial r = \sigma_r \partial u_r/\partial r$. Thus, the second integral in (3.2.31) becomes

$$\int_{\gamma_Q} \left(W \boldsymbol{n} \cdot \boldsymbol{r} - \boldsymbol{T} \cdot \frac{\partial \boldsymbol{u}}{\partial r} r \right) \mathrm{d}s = \int_\pi^{2\pi} \left(W - \sigma_r \frac{\partial u_r}{\partial r} \right) r^2 \mathrm{d}\varphi \qquad (3.2.34)$$

Furthermore, Hooke's law gives $\partial u_r/\partial r = \epsilon_r = (1-\nu)\sigma_r/2\mu$, where ν is Poisson's ratio and μ the modulus of rigidity. Thus, finally,

$$\int_{\gamma_Q} \left(W \boldsymbol{n} \cdot \boldsymbol{r} - \boldsymbol{T} \cdot \frac{\partial \boldsymbol{u}}{\partial r} r \right) \mathrm{d}s = -\frac{(1-\nu)Q^2}{2\pi\mu} \qquad (3.2.35)$$

and then, from (3.2.31) and (3.2.32),

$$\mathcal{G} = \frac{(1-\nu)Q^2}{\pi\mu a} \qquad (3.2.36)$$

The example may be considered as the special case $\alpha = \pi$ of a wedge $-\alpha \leq \varphi \leq \alpha$, provided with a symmetrically situated crack with length a from the apex and loaded by a pair of opposite forces at the apex, one on each side of the crack. The necessary relations are given in Appendix A2, also allowing for arbitrary direction of the forces (though in the plane and mutually opposing). Such and other examples are given by Freund (1978) and Ouchterlony (1978, 1980).

The J-integral for anti-plane shearing and for axisymmetric cases

The P- and J-integrals have been defined for in-plane loading and therefore they are applicable to crack modes I and II. A J-integral, J_{III}, may also be defined for anti-plane shearing, i.e. with application to mode III cracks. Using essentially the same method as for mode I, it can be shown that

$$\boxed{J_{III} = \int_\Gamma (W \mathrm{d}x_2 - n_j \sigma_{j3} \partial u_3/\partial x_1 \, \mathrm{d}s) \quad (j=1,2)} \qquad (3.2.37)$$

is path-independent for paths along which W is a single-valued function of $u_{3,j}$. The same general procedure as in the preceding subsection can be used to show that the J_{III}-integral may be interpreted approximately as the energy flux, if steady state conditions are well established in the crack edge neighbourhood and small scale yielding prevails.

The J-integral is not path-independent in the case of axial symmetry. However, in a small region near the edge of a penny-shaped crack approximate plane strain prevails. Therefore the value of the J-integral encircling the crack edge at a sufficiently close distance, in a plane through the symmetry axis, would be expected to describe the same state inside the path as in a true plane strain case.

It is possible to show that the sum of the J-integral for a path Γ and a certain area integral for the area inside Γ is independent of Γ (Broberg 1976, Bergkvist and Guex-Le 1977). Fig. 3.2.9 shows such a path and a cylindrical coordinate system r, φ, z. The path is closed, is contained in the plane $\varphi = 0$, and is assumed to be viewed in the direction of increasing φ. A Cartesian coordinate system is simultaneously introduced with the x_3-axis coinciding with the z-axis and the $x_1 x_3$ plane with the plane $\varphi = 0$.

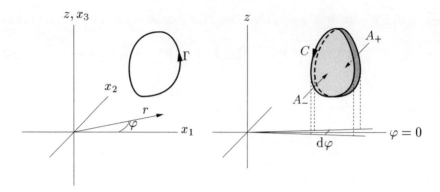

Fig. 3.2.9 Left figure: Path Γ in the $x_1 x_3$ plane ($\varphi = 0$). Right figure: A thin volume V generated by rotation of a closed curve C in the rz plane from $-\mathrm{d}\varphi/2$ to $\mathrm{d}\varphi/2$. Also shown are the two areas A_+ and A_- of the plane surfaces of the volume.

It is assumed that the strains are small compared to unity. Then, with obvious notations, the stresses and displacements are:

$$\sigma_{11} = \sigma_r(r,z)\cos^2\varphi + \sigma_\varphi(r,z)\sin^2\varphi \qquad (3.2.38)$$

$$\sigma_{22} = \sigma_r(r,z)\sin^2\varphi + \sigma_\varphi(r,z)\cos^2\varphi \qquad (3.2.39)$$

$$\sigma_{12} = [\sigma_r(r,z) - \sigma_\varphi(r,z)]\cos\varphi\sin\varphi \qquad (3.2.40)$$

$$\sigma_{33} = \sigma_z(r,z), \quad \sigma_{31} = \sigma_{zr}(r,z) \qquad (3.2.41)$$

$$u_1 = u_r(r,z)\cos\varphi, \quad u_2 = u_r(r,z)\sin\varphi, \quad u_3 = u_z(r,z) \qquad (3.2.42)$$

Thus, near $\varphi = 0$:

$$\left.\begin{array}{l}
\sigma_{11} \approx \sigma_r \\
\sigma_{22} \approx \sigma_\varphi \\
\sigma_{12} \approx (\sigma_r - \sigma_\varphi)\varphi \approx (\sigma_{11} - \sigma_{22})\varphi \\
u_1 \approx u_r, \quad u_{1,1} \approx \partial u_r/\partial r \\
u_2 \approx u_r\varphi, \quad u_{2,1} \approx (u_{1,1} - u_1/x_1)\varphi, \quad u_{2,3} \approx u_{1,3}\varphi \\
u_3 = u_z, \quad u_{3,1} \approx \partial u_z/\partial r
\end{array}\right\} \qquad (3.2.43)$$

Now, consider the thin volume V cut out by a surface generated by rotation around the z axis of a closed curve C in the r,z plane, and by the two planes $\varphi = \pm\mathrm{d}\varphi/2$, where $\mathrm{d}\varphi \ll 1$. The total surface of V is denoted by S. The expression

$$R = \oint_S \frac{1}{x_1}(W n_1 - n_j \sigma_{ji} u_{i,1})\mathrm{d}S + \int_V \left(\frac{W}{x_1^2} - \frac{\sigma_{1i} u_{i,1}}{x_1^2}\right)\mathrm{d}V \qquad (3.2.44)$$

where n_j is the outward normal of S, can be shown to vanish. In order to do so, R is first written as a volume integral, by using Green's theorem to transform the area integral. Thus,

$$R = \int_V \left[\frac{\partial}{\partial x_1}\left(\frac{W}{x_1}\right) - \frac{\partial}{\partial x_j}\left(\frac{\sigma_{ji}u_{i,1}}{x_1}\right)\right]dV + \int_V \left(\frac{W}{x_1^2} - \frac{\sigma_{1i}u_{i,1}}{x_1^2}\right)dV$$
$$= \int_V \frac{1}{x_1}\left[\frac{\partial W}{\partial x_1} - \frac{\partial}{\partial x_j}(\sigma_{ji}u_{i,1})\right]dV \quad (3.2.45)$$

The vanishing of this integral is then shown in the same way as for the integral in (3.2.6).

The integral R will now be transformed into the sum of a line integral and an area integral. Use of the notations A_+ and A_- for the plane surfaces $\varphi = \pm d\varphi/2$ of the volume V, results in the expressions

$$R \approx \int_C (Wn_1 - n_j\sigma_{ji}u_{i,1})_{i,j=1,3}\, d\varphi\, ds$$
$$+ \int_{A_++A_-} \frac{1}{x_1}(Wn_1 - n_j\sigma_{ji}u_{i,1})dA + \int_A \left(\frac{W}{x_1} - \frac{\sigma_{1i}u_{i,1}}{x_1}\right)d\varphi\, dA \quad (3.2.46)$$

where s is the arc length along C, positive in the direction for which n_j points to the right, and the notation $A = A_+ + A_-$ is used when there is no need to distinguish between the two plane surfaces.

By using Eqs (3.2.43) and observing that $n_1 \approx -d\varphi/2$ and $n_3 = 0$ on A_+ and A_-, whereas $n_2 \approx +1$ on A_+ and $n_2 \approx -1$ on A_-, the following result is obtained, after omission of second order terms:

$$R \approx \int_C (Wn_1 - n_j\sigma_{ji}u_{i,1})_{i,j=1,3}\, d\varphi\, ds$$
$$+ \int_A \left(\frac{\sigma_{22}u_1}{x_1^2} - \frac{\sigma_{11}u_{1,1} + \sigma_{13}u_{3,1}}{x_1}\right)d\varphi\, dA \quad (3.2.47)$$

Writing $R/d\varphi = J_R$, returning to cylindrical coordinates, and letting $d\varphi \to 0$, the following equalities are found:

$$J_R = \int_C \left\{W dz - \left[\sigma_r \frac{\partial u_r}{\partial r}\cos\psi - \sigma_{rz}\left(\frac{\partial u_r}{\partial r}\sin\psi - \frac{\partial u_z}{\partial r}\cos\psi\right) - \sigma_z \frac{\partial u_z}{\partial r}\sin\psi\right]ds\right\}$$
$$- \int_A \frac{1}{r}\left(\sigma_r\epsilon_r - \sigma_\varphi\epsilon_\varphi + \tau_{rz}\frac{\partial u_z}{\partial r}\right)dA = 0 \quad (3.2.48)$$

where ψ is the angle between the outward normal to Γ and the r direction. The first integral in (3.2.48) is recognized as the J-integral. This fact is easiest seen by comparison with (3.2.47).

Consider now a path Γ, encircling the edge of a penny-shaped crack $r < a$, $z = 0$, starting at a point of the lower surface and terminating at a point of the upper surface. The path is assumed not to intersect the axis $r = 0$. If the closed path C in (3.2.48) is replaced by the path Γ, J_R is obviously independent of Γ though not necessarily zero. Because its area integral is monotonically decreasing toward zero when Γ is shrunk towards the crack edge, J_R approaches J for a path near the crack edge. Thus, with the same notations as in (3.2.12),

$$J_R = J - \oint_A \frac{1}{r}\left(\sigma_r\epsilon_r - \sigma_\varphi\epsilon_\varphi + \tau_{rz}\frac{\partial u_z}{\partial r}\right)dA \approx (J)_{\Gamma_n} \quad (3.2.49)$$

Note that r here denotes the distance to the symmetry axis, not the distance to the crack edge. Equation (3.2.49) can be used to determine a value of J for a path near the crack edge from the value of the path-area integral J_R for a more remote path. This may be advantageous in finite element calculations. The contribution to the area integral from the region near the crack edge, where stresses and strains might be less accurately calculated, is generally very small compared to J, and therefore the procedure gives about the same accuracy as calculation of J for plane cases. It has been used successfully by Bergkvist (1977).

A path-independent integral for stationary cracks in dynamic fields

Nilsson (1973a) suggested a path-independent integral in the p-plane, where p is the Laplace transform variable for transformation with respect to time t. The integral can be applied to dynamic problems with stationary cracks, particularly for waves impinging on a crack. Laplace transforms are described in Appendix A5: note that a dimension-true formulation is used throughout the present work. The Laplace transformed stresses and displacements are

$$S_{ij} = \mathcal{L}_{pt}(\sigma_{ij}), \quad U_i = \mathcal{L}_{pt}(u_i) \tag{3.2.50}$$

Here, $\mathcal{L}_{pt}$ is the Laplace transform operator and the symbol $\mathcal{L}_{pt}^{-1}$ will be used to denote inverse Laplace transformation. A quadratic function $W^*(U_{i,j})$ is also introduced, such that

$$\mathcal{L}_{pt}^{-1} \frac{\partial W^*}{\partial U_{i,j}} = \sigma_{ji} \tag{3.2.51}$$

Then,

$$S_{ji} = \frac{\partial W^*}{\partial U_{i,j}} \tag{3.2.52}$$

This is analogous to (3.2.2) and the continued derivation is similar to that of the P-integral, but the equations of motion should be used rather than the equilibrium equations (3.2.4):

$$\sigma_{ji,j} = \varrho \frac{\partial^2 u_i}{\partial t^2} \tag{3.2.53}$$

Laplace transformation of these equations gives:

$$S_{ji,j} = \varrho p^2 U_i \quad \text{if } (u_i)_{t=0} = 0 \text{ and } (\partial u_i/\partial t)_{t=0} = 0 \tag{3.2.54}$$

With use of this equation and otherwise proceeding in the same way as for the P-integral in (3.2.1), path independence can be shown for the integral

$$P_N = \int_\Gamma \left[\left(W^* + \frac{1}{2} \varrho p^2 U_i U_i \right)_{av} n_1 - n_j (S_{ji} \partial U_i/\partial x_1)_{av} \right] ds \quad (j = 1, 2) \tag{3.2.55}$$

where $i = 1, 2, 3$ for modes I and II and $i = 3$ for mode III.

As for the P-integral, specialization to plane strain for modes I and II implies that subscript av may be dropped. For plane stress, however, the term $\frac{1}{2}\varrho p^2 (U_3^2)_{av}$, which obviously reflects lateral inertia, does not in general disappear. Anyway, this term is small in a thin plate, and after its omission the Nilsson integral for plane strain or plane stress is found to be

$$J_N = \int_\Gamma \left[\left(W^* + \frac{1}{2}\varrho p^2 U_i U_i \right) n_1 - n_j S_{ji} \frac{\partial U_i}{\partial x_1} \right] ds \quad (j = 1, 2) \quad (3.2.56)$$

where $i = 1, 2$ for modes I and II and $i = 3$ for mode III.

The analogy with the J-integral is somewhat formal: thus, for instance, W^* is not the Laplace transform of the stress-strain energy density W.

Of particular interest is, as Nilsson showed, that the integral also may be used for viscoelastic stress-strain fields. This may be seen as a consequence of the elastic-viscoelastic correspondence principle, page 238, which implies that the Laplace transform of a solution of a certain viscoelastic problem may be found from the Laplace transform of the corresponding elastic problem, simply by replacing the elastic moduli by their Laplace transformed viscoelastic counterparts.

The physical interpretation of the Nilsson integral follows on page 90.

A path-area integral incorporating inelastic strains, inertia and body forces

Kishimoto et al. (1980) derived the path-area integral

$$\hat{J}_k = \int_{\Gamma+\Gamma_s} \left\{ W_e(\epsilon_{ij}^e) n_k - n_j \sigma_{ji} \frac{\partial u_i}{\partial x_k} \right\} ds$$

$$+ \int_A \left\{ \sigma_{ij} \frac{\partial \epsilon_{ij}^*}{\partial x_k} + \left(\varrho \frac{\partial u_i}{\partial t^2} - F_i \right) \frac{\partial u_i}{\partial x_k} \right\} dA \quad (3.2.57)$$

where Γ is a path encircling the crack edge, Γ_s consists of paths along the crack faces†, n_i is the outward normal to Γ and Γ_s, s is the arc length along $\Gamma + \Gamma_s$, A is the area inside $\Gamma + \Gamma_s$, ϵ_{ij}^e and ϵ_{ij}^* are elastic and inelastic parts, respectively, of the deformation tensor, W_e is a stress-strain energy function, defined by $\partial W_e/\partial \epsilon_{ij} = \sigma_{ij}$, ϱ is the density and F_i the body force.

The path-area integral is independent of the choice of Γ. It is convenient for numerical computation in the same way as discussed for the the path-area integral J_R for axisymmetric cases, and it has been used, for instance, by Sakata et al. (1978) for analysis of cracked rotating disks. It may also be adapted for viscoelasticity, as shown by Kishimoto et al. (1982).

3.3 Stress-strain field at small scale yielding

General considerations

At small scale yielding, the linear size of the dissipative regions near the crack edge is much smaller than the crack length and also much smaller than the shortest ligament from the crack edge to the outer boundary of the body. It does not necessarily need to be smaller than linear body dimensions in the crack edge direction, for instance the thickness of a sheet.

---
† Γ_s should follow the crack faces from the points where Γ ends to the process region. Note that the contribution from Γ_s to the integrand of the line integral is necessarily zero only for $k = 1$.

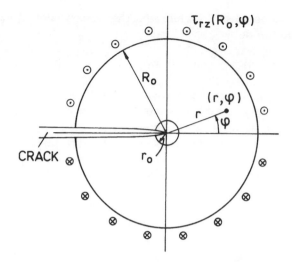

Fig. 3.3.1 Circular cylinder, subjected to anti-symmetric shear load. Symbols ⊙ and ⊗ represent arrows pointing toward and away from the reader, respectively.

In analytical as well as numerical calculations concentrating on the stress-strain field surrounding a crack at small scale of yielding, it is generally greatly simplifying to assume

a: that the linear size of the dissipative region is negligibly small compared to the crack length and other relevant linear body dimensions, and
b: that remote stresses in all directions from the crack edge are negligibly small compared to the stresses near the process region.

This limiting case will be referred to as *infinitesimally small scale yielding*.

Small scale yielding does not in general prevail for cracks in engineering structures, because efficient utilization of the material implies high overall stresses, often of the order of half the yield stress. Still, the concept of infinitesimally small scale yielding is very useful, either for approximative calculations of certain field characteristics near the crack edge or as a basis for extrapolations to larger scale of yielding.

Importance and dominance of inverse square-root terms

Stresses and displacements outside an arbitrary dissipative region
In order to calculate the stresses and strains outside the process region, a model of this region is generally needed. However, stresses and strains beyond some distance from the process region can often be calculated with reasonable accuracy even if the properties of the process region are not known. If interest is focussed on stresses and strains far away from the whole dissipative region at small scale yielding, acceptable accuracy can often be obtained without closer specification of the properties of this region. An example will be shown in order to explore this possibility.

The example concerns anti-plane shear, i.e. mode III. This mode is particularly simple to analyse, because the displacement vector is everywhere parallel to the crack edge. A crack is supposed to be situated along the symmetry plane of a circular cylinder with radius R_0, subjected to anti-symmetrically distributed loads, directed in parallel with the crack edge; see Fig. 3.3.1. A cylindrical coordinate system r, φ, z is introduced with $r = 0$ coinciding with the crack edge and $\varphi = \pm\pi$ with the crack faces. A radius r_0 is further introduced such that all dissipative regions (including wakes) are contained inside the cylinder $r = r_0$; but, apart from this requirement, r_0 should be chosen as small as possible. Linear elasticity is assumed to prevail in $r_0 < r \leq R_0$.

The displacement in the positive z direction is denoted by w, and the stresses by τ_{rz} and $\tau_{\varphi z}$. The stress $\tau_{rz}(R_0, \varphi)$ is taken as boundary traction on the cylinder studied. The condition of equilibrium is

$$\frac{\partial(\tau_{rz} r)}{\partial r} + \frac{\partial \tau_{\varphi z}}{\partial \varphi} = 0 \qquad (3.3.1)$$

Hooke's law gives

$$\tau_{rz} = \mu \frac{\partial w}{\partial r}, \qquad \tau_{\varphi z} = \mu \cdot \frac{1}{r} \cdot \frac{\partial w}{\partial \varphi} \qquad (3.3.2)$$

where μ is the modulus of rigidity. Insertion into (3.3.1) yields

$$\frac{\partial}{\partial r}\left(r \frac{\partial w}{\partial r}\right) + \frac{1}{r} \cdot \frac{\partial^2 w}{\partial \varphi^2} = 0 \qquad (3.3.3)$$

The left member of this equation is recognized as $r\Delta w$, where Δ is the Laplace operator. Thus, the equation can be written in a compact and coordinate-invariant form of the *Laplace equation*

$$\Delta w = 0 \qquad (3.3.4)$$

Solutions of the Laplace equation are known as *harmonic functions*. The equation is valid for linearly elastic and isotropic regions in all mode III cases. Specification to the current problem is made by the boundary conditions. The crack faces are assumed to be traction free, and thus

$$\frac{\partial w}{\partial \varphi} = 0 \text{ for } \varphi = \pm\pi \qquad (3.3.5)$$

At the outer boundary, the condition is

$$\frac{\partial w}{\partial r} = \frac{1}{\mu}\tau_{rz}(R_0, \varphi) \text{ for } r = R_0 \qquad (3.3.6)$$

where $\tau_{rz}(R_0, \varphi)$ is the outer load.

The condition at the remaining boundary, $r = r_0$, may be given as the stress $\tau_{rz}(r_0, \varphi)$ or as the displacement $w(r_0, \varphi)$, but neither of these functions is known. However, $w(r_0, \varphi)$ may be considered as the displacement response at $r = r_0$ when the region $r \leq r_0$ is subjected to the load $\tau_{rz}(r_0, \varphi)$. Therefore, provided that the history of loading is given, for instance as monotone loading, there exists a unique relation between $w(r_0, \varphi)$ and $\tau_{rz}(r_0, \varphi)$ for each material with time independent response to loading. This relation is obviously completely independent of the load on the mantle $r = R_0$.

Due to the prevailing anti-symmetry, $w(r_0, \varphi)$ can be expressed by the Fourier series

$$w(r_0, \varphi) = \sum_{1,3,..} \delta_n \sin \frac{n\varphi}{2} \quad (3.3.7)$$

in which each δ_n is uniquely given by $\tau_{rz}(r_0, \varphi)$, due to the unique relation between $w(r_0, \varphi)$ and $\tau_{rz}(r_0, \varphi)$. The boundary condition for $r = r_0$ is therefore given by (3.3.7).

Fundamental solutions to equation (3.3.4), satisfying the boundary condition (3.3.5) are

$$w_n = A_n r^{n/2} \sin \frac{n\varphi}{2}, \quad w_{-n} = B_n r^{-n/2} \sin \frac{n\varphi}{2} \quad (3.3.8)$$

where $n = 1, 3, 5, \ldots$, and A_n and B_n are constants. A complete general solution, satisfying the boundary condition (3.3.5), is therefore

$$w = \sum_{1,3,..} (A_n r^{n/2} + B_n r^{-n/2}) \sin \frac{n\varphi}{2} \quad (3.3.9)$$

The boundary condition (3.3.6) gives for $-\pi < \varphi < \pi$:

$$\sum_{1,3,..} \frac{n}{2} (A_n R_0^{n/2-1} - B_n R_0^{-n/2-1}) \sin \frac{n\varphi}{2} = \frac{1}{\mu} \tau_{rz}(R_0, \varphi) \quad (3.3.10)$$

Then, by Fourier analysis,

$$A_n R_0^{n/2-1} - B_n R_0^{-n/2-1} = \frac{2\gamma_n}{n} \quad (3.3.11)$$

where

$$\gamma_n = \frac{1}{\pi} \int_{-\pi}^{+\pi} \frac{1}{\mu} \tau_{rz}(R_0, \varphi) \sin \frac{n\varphi}{2} \, d\varphi \quad (3.3.12)$$

Similarly, the boundary condition (3.3.7) gives

$$A_n r_0^{n/2} + B_n r_0^{-n/2} = \delta_n \quad (3.3.13)$$

After determination of the constants A_n and B_n from (3.3.11) and (3.3.13) the displacement is obtained from (3.3.9) as

$$w = \sum_{1,3,..} \frac{1}{1+(r_0/R_0)^n} \left\{ \frac{2\gamma_n r_0}{n} \left(\frac{r_0}{R_0}\right)^{n/2-1} \left[\left(\frac{r}{r_0}\right)^{n/2} - \left(\frac{r_0}{r}\right)^{n/2}\right] \right.$$
$$\left. + \delta_n \left[\left(\frac{r_0}{r}\right)^{n/2} + \left(\frac{r_0}{R_0}\right)^n \left(\frac{r}{r_0}\right)^{n/2}\right] \right\} \sin \frac{n\varphi}{2} \quad (3.3.14)$$

and the stresses follow from (3.3.2):

$$\tau_{rz} = \mu \sum_{1,3,..} \frac{1}{1+(r_0/R_0)^n} \left\{ \gamma_n \left(\frac{r_0}{R_0}\right)^{n/2-1} \left[\left(\frac{r}{r_0}\right)^{n/2-1} + \left(\frac{r_0}{r}\right)^{n/2+1}\right] \right.$$
$$\left. - \frac{n\delta_n}{2r_0} \left[\left(\frac{r_0}{r}\right)^{n/2+1} - \left(\frac{r_0}{R_0}\right)^n \left(\frac{r}{r_0}\right)^{n/2-1}\right] \right\} \sin \frac{n\varphi}{2} \quad (3.3.15)$$

3.3 STRESS-STRAIN FIELD AT SMALL SCALE YIELDING

$$\tau_{\varphi z} = \mu \sum_{1,3,..} \frac{1}{1+(r_0/R_0)^n} \left\{ \gamma_n \left(\frac{r_0}{R_0}\right)^{n/2-1} \left[\left(\frac{r}{r_0}\right)^{n/2-1} - \left(\frac{r_0}{r}\right)^{n/2+1}\right] \right.$$
$$\left. + \frac{n\delta_n}{2r_0}\left[\left(\frac{r_0}{r}\right)^{n/2+1} + \left(\frac{r_0}{R_0}\right)^n \left(\frac{r}{r_0}\right)^{n/2-1}\right] \right\} \cos\frac{n\varphi}{2} \quad (3.3.16)$$

This result will now be used to demonstrate that the state inside the region $r = r_0$ is almost exclusively controlled by the load component containing γ_1, provided that the scale of yielding, which here may be represented by $s_y = r_0/R_0$, is sufficiently small.

Control by the inverse square-root terms of the state at the crack edge

For $r = r_0$ equation (3.3.15) leads to:

$$\tau_{rz}(r_0,\varphi) + \frac{\mu}{2r_0} \sum_{1,3,..} n\delta_n \frac{1-s_y^n}{1+s_y^n} \sin\frac{n\varphi}{2} = 2\mu \sum_{1,3,..} \gamma_n \frac{s_y^{n/2-1}}{1+s_y^n} \sin\frac{n\varphi}{2} \quad (3.3.17)$$

For $s_y \ll 1$, the second term of the left member can be written as an operation on $\tau_{rz}(r_0,\varphi)$, because the constants δ_n are uniquely given by $\tau_{rz}(r_0,\varphi)$ as described in connection with equation (3.3.7). Thus, for $s_y \ll 1$,

$$\tau_{rz}(r_0,\varphi) + T[\tau_{rz}(r_0,\varphi)] \approx 2\mu\gamma_1 \left(\frac{R_0}{r_0}\right)^{1/2} \sin\frac{\varphi}{2} \quad (3.3.18)$$

where T is an operator that depends only on φ, r_0 and the properties of the material inside $r = r_0$, i.e. it does not contain R_0 or γ_n. Then, in principle, $\tau_{rz}(r_0,\varphi)$ can be solved from (3.3.18) and the result will depend only on $\mu\gamma_1(R_0)^{1/2}$, which is the coefficient for the $r^{-1/2}$-term in the series expansions (3.3.15) and (3.3.16) for both $\tau_{rz}(r,\varphi)$ and $\tau_{\varphi z}(r,\varphi)$ when $s_y \to 0$. For non-zero s_y, the coefficient for the inverse square-root term is

$$\mu\gamma_1(R_0)^{1/2} + \mu\frac{\delta_1 r_0^{1/2}}{2R_0} \quad (3.3.19)$$

but it can be shown that the second term is much smaller than the first one if $s_y \ll 1$. To this end, the fact is used that the region inside $r = r_0$ cannot be softer than a hole. For a hole,

$$\frac{n\delta_n}{2r_0} = \frac{2\gamma_n}{1-s_y^n} s_y^{n/2-1} \quad (3.3.20)$$

as is seen by setting $\tau_{rz}(r_0,\varphi) = 0$ in (3.3.17). Thus, the ratio between the terms in expression (3.3.19) is

$$\frac{\delta_1}{2\gamma_1 R_0} \cdot \left(\frac{r_0}{R_0}\right)^{1/2} = \frac{2s_y}{1-s_y} \quad (3.3.21)$$

for a hole, and certainly considerably smaller when the region inside $r = r_0$ contains solid matter. This proves that the second term is much smaller than the first one if $s_y \ll 1$. Thus, it is demonstrated that *the state in the near-edge region $r < r_0$ is controlled exclusively by the load component containing γ_1 if the scale of yielding is sufficiently small*.

For simplicity, a circular cylinder was considered. A study of the demonstration procedure reveals that the same conclusion would have been arrived at for any other symmetrical body shape and anti-symmetric loading. From equations (3.3.2) and (3.3.9), it follows that the stresses $\tau_{rz}(r,\varphi)$ and $\tau_{\varphi z}(r,\varphi)$ can generally be represented by series expansions of the form

$$\tau_{rz} = \sum_{1,3,...} (a_n r^{n/2-1} - b_n r^{-n/2-1}) \sin\frac{n\varphi}{2} \qquad (3.3.22)$$

$$\tau_{\varphi z} = \sum_{1,3,...} (a_n r^{n/2-1} + b_n r^{-n/2-1}) \cos\frac{n\varphi}{2} \qquad (3.3.23)$$

In particular, the coefficient for the $r^{-1/2}$-term is the same for both stresses. Generalization from the case of the circular cylinder implies that in general *the coefficient of the inverse square-root terms controls the state in the near-edge region $r < r_0$ if the scale of yielding is sufficiently small.* Because information about the state inside the near-edge region is essential for predictions about crack growth and fracture, this is a very important result. A closer study of the influence of the scale of yielding is therefore called for.

The scale of yielding in the general case may be measured in essentially the same way as for the circular cylinder; it is only a question of finding the largest radius R_0 such that linear elasticity prevails in $r_0 < r \leq R_0$. Often this radius is close to the crack length a, but sometimes it is given by the shortest ligament b from the crack edge to the outer boundary of the body. Thus, the scale of yielding may be defined as $s_y = r_0/a$ if $a > b$ and otherwise as $s_y = r_0/b$.

An inspection of (3.3.15) reveals that (3.3.18) gives $\tau_{rz}(r_0,\varphi)$ approximately within a relative error s_y (i.e., the absolute error $\Delta\tau_{rz}$ is within $\pm s_y \tau_{rz}$), provided that $|\gamma_1|$ is not substantially smaller than $|\gamma_n| s_y^{(n-1)/2}$. The same applies to $\tau_{\varphi z}(r_0,\varphi)$. If "pathological" loading cases are excluded, it is only $|\gamma_3|$ that needs to be considered; it will, however, be assumed that a "normal" loading case prevails, such that $|\gamma_3|$ is smaller than, or of the same order as $|\gamma_1|$. An example is a horizontal edge crack in a large plate, whose upper and lower edges are subjected to constant shear tractions; in this case $\gamma_3 \approx 3\gamma_1/4$. The maximum stresses on the boundary of the near-edge region $r = r_0$ are thus controlled by the coefficient for the inverse square-root terms within a relative error s_y, if $s_y \ll 1$. These boundary stresses express the state inside the near-edge region.

Dominance of the inverse square-root terms
Whereas terms containing γ_n, $n \geq 3$ may be neglected in the neighbourhood of $r = r_0$, and terms containing δ_n may be neglected in the neighbourhood of $r = R_0$ at small scale yielding, *the inverse square-root terms are significant in the whole interval $r_0 \leq r \leq R_0$*. They dominate in a certain annular region, sufficiently close to the crack edge to ensure domination over terms containing γ_n, $n \geq 3$, and sufficiently far from the crack edge to ensure domination over δ_n-terms. The stress $\tau_{\varphi z}(r,\varphi)$ in this region equals approximately the $r^{-1/2}$-term, but with a relative error of about $\sqrt{s_y}$, which is considerably larger than the relative error s_y, by which this term controls the state in the near-edge region $r < r_0$. In order to show this, the annular region in which the inverse square-root term is most dominating must be found. This may be done by

first looking on the terms which compete at large r and then on those that compete at small r.

For $(r/r_0)^2 \gg 1$, $\varphi = 0$, equation (3.3.16) may be written as

$$\tau_{\varphi z} \approx \frac{\mu\gamma_1}{1+s_y}\left(\frac{r}{R_0}\right)^{-1/2} + \frac{\mu\gamma_3}{1+s_y^3}\left(\frac{r}{R_0}\right)^{1/2} + \frac{\mu\gamma_5}{1+s_y^5}\left(\frac{r}{R_0}\right)^{3/2} + \cdots$$

$$+ \delta_n\text{-terms} \quad (3.3.24)$$

Hence, if $\gamma_3 \approx \gamma_1$ and if the error contributed by the δ_n-terms is disregarded, the stress $\tau_{\varphi z}(r,0)$ is expressed by the inverse square-root term within an error of

$$p \approx r/R_0 \quad (3.3.25)$$

In order to estimate the error introduced by the δ_n-terms, it is necessary to know some representative magnitudes of these terms. A solution for the case of perfect plasticity and infinitesimally small scale yielding, obtained by letting $R_0 \to \infty$ whereas r_0 remains finite, was given by Hult and McClintock (1956); see Section 5.3. For $\varphi = 0$ it gives

$$\tau_{\varphi z} = \tau_Y \left(2\frac{r}{r_0}\right)^{-1/2}\left(1 - \frac{r_0}{2r}\right)^{-1/2}, \quad r \geq r_0 \quad (3.3.26)$$

where τ_Y is the yield stress in shear, showing that $\tau_{\varphi z}$ is expressed by the inverse square-root term within an error of

$$p \approx \frac{r_0}{4r} \quad (3.3.27)$$

if terms increasing with r are disregarded. After returning to the case of finite R_0 and r_0, it is found, by adding the errors (3.3.25) and (3.3.27) and seeking the minimum of the composed error, that *the inverse square-root term is most dominating in an annular region at about*

$$r = r_a \approx \frac{1}{2}\sqrt{r_0 R_0} \quad (3.3.28)$$

and that $(\tau_{\varphi z})_{\varphi=0}$ *then is expressed by this term within an error of*

$$p \approx \sqrt{\frac{r_0}{R_0}} \approx \sqrt{s_y} \quad (3.3.29)$$

This result should hold even without restriction to circular cylindrical bodies. If, for example, the relative error within which the coefficient of the inverse square-root terms controls the state of the dissipative region at the crack edge is $s_y = 0.01$, then $p \approx 0.1$, and the inverse square-root term is most dominating at about $r = r_a = 5r_0$. Note that these relations are very approximate and strongly influenced by body and loading geometry, but the example shows that the inverse square-root terms describe the state of the region in the crack edge vicinity very accurately, even if their dominance over other terms is not very pronounced anywhere.

Fig. 3.3.2 shows schematically $\tau_{\varphi z}(r,0)$ and some of its components.

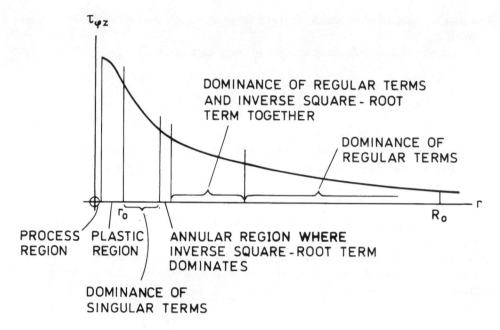

Fig. 3.3.2 The stress $\tau_{\varphi z}$ straight ahead of the crack in regions of different character. For the process region, stresses cannot be defined.

The autonomy

In the previous sub-section it was shown that the state in the dissipative region near a crack edge at small scale yielding is determined by the single load component $\mu\gamma_1$ within a relative error s_y. This result is independent on body and loading geometry as long as these comply with the requirements for mode III with exclusion of "pathological" cases. During monotone loading, the dissipative regions experience a development, which for a given material depends only on $\mu\gamma_1$, but not on body and loading geometry. For corresponding stages of this development in different bodies of the same material, the stress-strain distributions in the plastic regions are the same, and the process regions are the same, apart from statistical veriations of micro-separation sizes and distributions. The mode III case thus illustrates the concept of autonomy and shows that this concept is applicable within the class of problems that is characterized by small scale yielding†.

It is obviously possible to express the state in the dissipative region at small scale yielding by γ_1, but calculation of γ_1 is generally difficult. The stress intensity factor offers a usually far more convenient alternative.

The stress intensity factor

Assuming Hooke's law to be valid without limitations to stresses and strains, enables the dissipative region to be shrunk to the crack edge. Infinitesimally small scale

---

† This property carries over to mode II, but not in general to mode I; see page 585.

yielding, $s_y \to 0$, is then obtained. This corresponds to keeping R_0 fixed and letting $r_0/R_0 \to 0$ in equations (3.3.14)-(3.3.16). For $r/r_0 \gg 1$, which is then satisfied for arbitrarily small positive r, the following asymptotic expressions are obtained:

$$w = \sum_{1,3,\ldots} \frac{2\gamma_n R_0}{n} \left(\frac{r}{R_0}\right)^{n/2} \sin \frac{n\varphi}{2} \tag{3.3.30}$$

$$\tau_{rz} = \mu \sum_{1,3,\ldots} \gamma_n \left(\frac{r}{R_0}\right)^{n/2-1} \sin \frac{n\varphi}{2} \tag{3.3.31}$$

$$\tau_{\varphi z} = \mu \sum_{1,3,\ldots} \gamma_n \left(\frac{r}{R_0}\right)^{n/2-1} \cos \frac{n\varphi}{2} \tag{3.3.32}$$

These expressions are much simpler than (3.3.14)-(3.3.16) and, apart from the inverse square-root term for the stresses, they do not contain negative powers of r. A comparison shows that the relative difference between inverse square-root terms in (3.3.31)-(3.3.32) and corresponding terms in (3.3.15)-(3.3.16) is approximately equal to s_y. The simpler expressions (3.3.31)-(3.3.32) can therefore be used to find the coefficient for the important inverse square-root terms, if a relative error of the order of s_y can be tolerated. This coefficient is conventionally written as $K_{III}/\sqrt{2\pi}$, and K_{III} is called *the stress intensity factor*, a concept introduced by Irwin (1957, 1960). The factor $1/\sqrt{2\pi}$ reflects a desire to avoid a factor 2π in an expression for the energy flux into an advancing crack edge. The appropriate definition for the mode III stress intensity factor, covering general mode III cases, is

$$\boxed{K_{III} = \lim_{r \to 0} \left[\sqrt{2\pi r}\, \tau_{\varphi z}(r, 0)\right]} \tag{3.3.33}$$

and it may be positive or negative: note that the definition implies that *it is positive if the shear stresses in front of the crack edge tend to rotate matter in the clockwise direction as seen from inside the crack.*

Equations (3.3.30)-(3.3.32) could, of course have been obtained directly from equation (3.3.4) and the boundary conditions (3.3.5)-(3.3.6), i.e. by leaving out the boundary condition (3.3.7), leading to equation (3.3.13). This, however, creates a problem: a unique solution is not obtained, because only equation (3.3.11) relates A_n and B_n. One condition is clearly missing, but which? The most safe way to establish one further condition is to require *bounded stress-strain energy*: all power plants in the world together cannot produce energy above any predetermined amount. The stress-strain energy inside a region $r < R$ is expressed by the integral

$$\int_{-\pi}^{\pi} \int_0^R \frac{1}{2\mu}(\tau_{rz}^2 + \tau_{\varphi z}^2) r \, dr \, d\varphi \tag{3.3.34}$$

implying that τ_{rz} and $\tau_{\varphi z}$ must not contain powers $r^{-\alpha}$, $\alpha \geq 1$. Thus, $B_n = 0$ and equations (3.3.30)-(3.3.32) are obtained.

Note that the expression "bounded stress-strain energy" tacitly assumes that a finite volume is considered. Unboundedness introduced by idealizations to infinite geometries, should be disregarded, of course. Note further that unbounded *stress-strain energy density*, i.e. energy per unit of volume, is not forbidden by the boundedness condition, but might appear at a point or a line in a continuum idealization. This actually occurs for $r = 0$, i.e. the crack edge, in the idealized case leading to (3.3.30)-(3.3.32). The line $r = 0$ displays *singularities in energy density and stresses*, whereas the displacement is regular.

While the inverse square-root stress terms in the expression for the stresses are dominant within some annular region around the crack edge in a real case of small scale yielding, they are dominant in a *crack edge vicinity* $0 < r < r_a$ at infinitesimally small scale yielding, if r_a is appropriately chosen with respect to the accuracy specified for domination. Obviously this accuracy may be arbitrarily high: there is always a crack edge vicinity in which the inverse square-root term dominates in this idealization. In this vicinity the asymptotic stresses are expressed by the relations

$$\tau_{rz} = \frac{K_{III}}{\sqrt{2\pi r}} \sin \frac{\varphi}{2}, \quad \tau_{\varphi z} = \frac{K_{III}}{\sqrt{2\pi r}} \cos \frac{\varphi}{2} \qquad (3.3.35)$$

or, in Cartesian components, as

$$\tau_{xz} = -\frac{K_{III}}{\sqrt{2\pi r}} \sin \frac{\varphi}{2}, \quad \tau_{yz} = \frac{K_{III}}{\sqrt{2\pi r}} \cos \frac{\varphi}{2} \qquad (3.3.36)$$

The reason for the name "stress intensity factor" is now obvious. It is not possible to determine the stress $\tau_{\varphi z}(r, 0)$ at the crack edge in this idealization, but it is possible to express the strength or "intensity" of the singularity. The stress intensity factor controls the state in the dissipative region near the crack edge at small scale yielding; in particular, mode III crack growth occurs when a certain critical value of the stress intensity factor, dependent only on the material and environmental conditions, is reached.

The inverse square-root term is now the only term that is singular at $r = 0$; the next strongest term is the $r^{1/2}$-term, which is regular and even vanishing at $r = 0$. It will later be shown that mode I does not exhibit a similarly strong dominance of the inverse square-root term: a constant term, i.e. an r^0-term appears in general.

The importance of the inverse square-root term, and its dominance in the crack edge vicinity in the idealized case of infinitesimally small scale yielding, suggests a study of this vicinity, only, i.e. the condition at the outer boundary is left out, and only the dominating term in a sufficiently close crack tip vicinity is retained. This implies use only of equations (3.3.4)-(3.3.5); the stresses are then obtained as in equations (3.3.35) and the displacement as

$$w = \frac{K_{III}}{\mu} \sqrt{\frac{2r}{\pi}} \sin \frac{\varphi}{2} \qquad (3.3.37)$$

The crack face slip, i.e. the distance by which two originally adjacent points, one on each crack face, are separated, is given as

$$\delta_0 = w(r, \pi) - w(r, -\pi) = 2\frac{K_{III}}{\mu} \sqrt{\frac{2r}{\pi}} \qquad (3.3.38)$$

For small scale yielding there exists a unique relation between the J-integral and the stress intensity factor. From (3.2.37), it follows that

$$J_{III} = \int_{-\pi}^{\pi} \left[Wr\cos\varphi - \tau_{rz}\left(\frac{\partial w}{\partial r}\cos\varphi - \frac{1}{r}\frac{\partial w}{\partial \varphi}\sin\varphi\right) r \right] d\varphi$$

$$= \int_{-\pi}^{\pi} \left[\frac{1}{2\mu}(\tau_{rz}^2 + \tau_{\varphi z}^2)\cos\varphi - \frac{1}{\mu}\tau_{rz}^2\cos\varphi + \frac{1}{\mu}\tau_{rz}\tau_{\varphi z}\sin\varphi \right] r\, d\varphi$$

$$= \frac{1}{2\mu} \int_{-\pi}^{\pi} [(\tau_{\varphi z}^2 - \tau_{rz}^2)\cos\varphi + 2\tau_{rz}\tau_{\varphi z}\sin\varphi] r\, d\varphi \quad (3.3.39)$$

Insertion of expressions (3.3.35) for the shear stresses gives the sought relation:

$$\boxed{J_{III} = \frac{K_{III}^2}{2\mu}} \quad (3.3.40)$$

Dominating terms in modes I and II

Asymptotic stresses for mode I
Expressions for dominating stresses and displacements in the crack edge vicinity at infinitesimally small scale yielding in modes I and II, analogous to those obtained in (3.3.35)-(3.3.37) for mode III, may be found in essentially the same way, although different methods for solution of the mathematical problem exist. Several methods will be presented later on; here use of the Airy stress function (see Appendix A3) will be demonstrated for mode I. A similar procedure can be used for the more general case of a notch rather than a sharp crack; see Appendix A8.

With the Airy stress function, Φ, the stresses are represented by the expressions

$$\sigma_r = \frac{1}{r} \cdot \frac{\partial \Phi}{\partial r} + \frac{1}{r^2} \cdot \frac{\partial^2 \Phi}{\partial \varphi^2}, \quad \sigma_\varphi = \frac{\partial^2 \Phi}{\partial r^2} \quad (3.3.41)$$

$$\tau_{r\varphi} = -\frac{\partial}{\partial r}\left(\frac{1}{r} \cdot \frac{\partial \Phi}{\partial \varphi}\right) \quad (3.3.42)$$

For an isotropic linearly elastic medium, Φ satisfies the biharmonic equation

$$\Delta\Delta\Phi = 0 \quad (3.3.43)$$

Consider now the vicinity of a crack edge, $r = 0$, with crack faces at $\varphi = \pm\pi$. Plane strain or plane stress is assumed. The boundary conditions are

$$\sigma_\varphi = \tau_{r\varphi} = 0 \text{ for } \varphi = \pm\pi \quad (3.3.44)$$

because the crack faces are traction free.

Some solutions to the biharmonic equation, together with expressions for stresses and displacement gradients are given in Appendix A3. For crack problems, the condition of bounded stress-strain energy restricts these solutions to those for which $\Phi/r \to 0$ as $r \to 0$. This excludes, for instance, $r\ln r$, $r\varphi\sin\varphi$ and $\Phi = r\cos\varphi$. A complete, general solution of (3.3.43), symmetrical in φ to satisfy mode I conditions, will be of the form

$$\Phi = A(s)r^{s+2}\cos s\varphi + B(s)r^{s+2}\cos(s\varphi + 2\varphi) \quad (3.3.45)$$

where $s > -1$. This gives

$$\sigma_\varphi = (s+1)(s+2)r^s[A(s)\cos s\varphi + B(s)\cos(s\varphi + 2\varphi)] \qquad (3.3.46)$$
$$\tau_{r\varphi} = (s+1)r^s[sA(s)\sin s\varphi + (s+2)B(s)\sin(s\varphi + 2\varphi)] \qquad (3.3.47)$$

whereupon insertion into the boundary conditions (3.3.44) leads to

$$[A(s) + B(s)]\cos \pi s = 0, \qquad [sA(s) + (s+2)B(s)]\sin \pi s = 0 \qquad (3.3.48)$$

The first equation is satisfied by $s = n/2$, where $n = -1, 1, 3, 5, \cdots$, and then the second equation gives

$$\frac{n}{2}A_n + \left(\frac{n}{2} + 2\right)B_n = 0, \qquad n = -1, 1, 3, 5, \cdots \qquad (3.3.49)$$

where $A_n = A(n/2)$ and $B_n = B(n/2)$. The second equation is satisfied by $s = n/2$, where $n = 0, 2, 4, 6, \cdots$, and then the first equation gives

$$A_n + B_n = 0, \qquad n = 0, 2, 4, 6, \cdots \qquad (3.3.50)$$

Thus,

$$\Phi = A_{-1}r^{3/2}\left[\cos\frac{\varphi}{2} + \frac{1}{3}\cos\frac{3\varphi}{2}\right]$$
$$+ \sum_{1,3,5,\cdots} A_n r^{n/2+2}\left[\cos\frac{n\varphi}{2} - \frac{n}{n+4}\cos\left(\frac{n\varphi}{2} + 2\varphi\right)\right]$$
$$+ \sum_{0,2,4,\cdots} A_n r^{n/2+2}\left[\cos\frac{n\varphi}{2} - \cos\left(\frac{n\varphi}{2} + 2\varphi\right)\right] \qquad (3.3.51)$$

The stresses are then found from (3.3.41)-(3.3.42). The first term in the expression for Φ is the dominant solution in the crack edge vicinity. For $\varphi = 0$,

$$\sigma_\varphi = A_{-1}r^{-1/2} = \frac{K_I}{\sqrt{2\pi r}} \qquad (3.3.52)$$

where use has been made of the definition for the stress intensity factor at mode I:

$$\boxed{K_I = \lim_{r \to 0}\left[\sqrt{2\pi r}\,\sigma_\varphi(r, 0)\right]} \qquad (3.3.53)$$

This stress intensity factor is positive if the in-plane stresses in front of the crack tend to dilate matter, i.e., if crack opening is produced near the edge considered. Thus, the resulting mode I stress intensity factor is always non-negative. However, negative mode I stress intensity factors may sometimes appear as temporary results during calculations involving superposition of different load cases, but the net result must be non-negative. A negative net result indicates crack closure near the edge, which necessitates changes of the original boundary conditions.

The singular stresses at a mode I crack edge at infinitesimally small scale yielding may now be written in the form

$$\sigma_r = \frac{K_I}{\sqrt{2\pi r}} \cos\frac{\varphi}{2} \cdot \frac{3-\cos\varphi}{2} = \frac{K_I}{\sqrt{2\pi r}} \cos\frac{\varphi}{2}\left(1+\sin^2\frac{\varphi}{2}\right) \quad (3.3.54)$$

$$\sigma_\varphi = \frac{K_I}{\sqrt{2\pi r}} \cos\frac{\varphi}{2} \cdot \frac{1+\cos\varphi}{2} = \frac{K_I}{\sqrt{2\pi r}} \cos^3\frac{\varphi}{2} \quad (3.3.55)$$

$$\tau_{r\varphi} = \frac{K_I}{\sqrt{2\pi r}} \sin\frac{\varphi}{2} \cdot \frac{1+\cos\varphi}{2} = \frac{K_I}{\sqrt{2\pi r}} \sin\frac{\varphi}{2}\cos^2\frac{\varphi}{2} \quad (3.3.56)$$

$$\sigma_z = \begin{cases} \nu(\sigma_r+\sigma_\varphi) & \text{for plane strain} \\ 0 & \text{for plane stress} \end{cases} \quad (3.3.57)$$

By using Hooke's law, the strains are found to be

$$\epsilon_r = \frac{1}{4(1-k^2)\mu}[\sigma_r - (1-2k^2)\sigma_\varphi] \quad (3.3.58)$$

$$\epsilon_\varphi = \frac{1}{4(1-k^2)\mu}[\sigma_\varphi - (1-2k^2)\sigma_r] \quad (3.3.59)$$

$$\gamma_{r\varphi} = \frac{1}{\mu}\tau_{r\varphi} \quad (3.3.60)$$

where k is related to Poisson's ratio ν by the equations

$$k^2 = \begin{cases} \dfrac{1-2\nu}{2(1-\nu)} & \text{for plane strain} \\ \dfrac{1-\nu}{2} & \text{for plane stress} \end{cases} \quad (3.3.61)$$

The use of k together with μ is, of course, motivated by the possibility of expressing both plane stress and plane strain cases simultaneously. This may also be done in other ways, but there is one additional advantage with the choice made here, viz. that k has a definite physical significance. In static cases, the ratio k^2/μ expresses the compliance during in-plane unidirectional straining: if the in-plane strain $\epsilon_x = 0$, then an in-plane stress σ_y causes the strain $\epsilon_y = k^2\sigma_y/\mu$ either for plane strain ($\epsilon_z = 0$) or for plane stress ($\sigma_z = 0$), with the appropriate value of k. In dynamic cases, k is the ratio c_S/c_P between the propagation velocity of S and P waves, either in plane strain or in plane stress, assuming the latter to be realized by an infinitesimally thin plate. The constant k is associated with the constant κ, used by e.g. Muskhelishvili (1953a,b), through the relation $\kappa = (1+k^2)/(1-k^2)$. (Note that the symbol κ in the present work denotes the compression modulus.)

From the strains, the displacements u_r and u_φ can be obtained to within a rigid-body movement by integrations, using the first two of the expressions

$$\frac{\partial u_r}{\partial r} = \epsilon_r, \quad \frac{1}{r}\cdot\frac{\partial u_\varphi}{\partial \varphi} + \frac{u_r}{r} = \epsilon_\varphi, \quad \frac{\partial u_\varphi}{\partial r} - \frac{u_\varphi}{r} + \frac{1}{r}\cdot\frac{\partial u_r}{\partial \varphi} = \gamma_{r\varphi} \quad (3.3.62)$$

In particular the *crack face opening*, i.e. the distance by which the crack faces are separated, is found to be

$$\delta_0 = u_\varphi(r,-\pi) - u_\varphi(r,\pi) = \frac{K_I}{(1-k^2)\mu}\sqrt{\frac{2r}{\pi}} \quad (3.3.63)$$

The term for $n = 0$ in the expression for Φ corresponds to constant uniaxial stress, $\sigma_x = 4A_0$, $\sigma_y = \tau_{xy} = 0$. (The positive x axis coincides with $\varphi = 0$.)

Asymptotic stresses for mode II
A general experience is that *once a mode I problem has been solved, the solution of the corresponding mode II problem is obtained through the same procedure with simple and apparent modifications.* Here, such modifications, essentially choosing an Airy stress function which is anti-symmetric in φ, result in

$$\Phi = -A_{-1}r^{3/2}\left[\sin\frac{\varphi}{2} + \sin\frac{3\varphi}{2}\right]$$
$$+ \sum_{1,3,5,\cdots} A_n r^{n/2+2}\left[\sin\frac{n\varphi}{2} - \sin\left(\frac{n\varphi}{2} + 2\varphi\right)\right]$$
$$+ \sum_{2,4,6,\cdots} A_n r^{n/2+2}\left[\sin\frac{n\varphi}{2} - \frac{n}{n+4}\sin\left(\frac{n\varphi}{2} + 2\varphi\right)\right] \quad (3.3.64)$$

Note that there is no term for $n = 0$. Use of (3.3.41)-(3.3.42) leads to the mode II expressions for the stress field in the vicinity of the crack edge:

$$\sigma_r = \frac{K_{II}}{\sqrt{2\pi r}}\sin\frac{\varphi}{2}\cdot\frac{3\cos\varphi - 1}{2} = \frac{K_{II}}{\sqrt{2\pi r}}\sin\frac{\varphi}{2}\left(1 - 3\sin^2\frac{\varphi}{2}\right) \quad (3.3.65)$$

$$\sigma_\varphi = -3\frac{K_{II}}{\sqrt{2\pi r}}\sin\frac{\varphi}{2}\cdot\frac{1+\cos\varphi}{2} = -\frac{3K_{II}}{\sqrt{2\pi r}}\sin\frac{\varphi}{2}\cos^2\frac{\varphi}{2} \quad (3.3.66)$$

$$\tau_{r\varphi} = \frac{K_{II}}{\sqrt{2\pi r}}\cos\frac{\varphi}{2}\cdot\frac{3\cos\varphi-1}{2} = \frac{K_{II}}{\sqrt{2\pi r}}\cos\frac{\varphi}{2}\left(1 - 3\sin^2\frac{\varphi}{2}\right) \quad (3.3.67)$$

$$\sigma_z = \begin{cases} \nu(\sigma_r + \sigma_\varphi) & \text{for plane strain} \\ 0 & \text{for plane stress} \end{cases} \quad (3.3.68)$$

and the crack face slip is

$$\delta_0 = u_r(r,\pi) - u_r(r,-\pi) = \frac{K_{II}}{(1-k^2)\mu}\sqrt{\frac{2r}{\pi}} \quad (3.3.69)$$

Here, the mode II stress intensity factor K_{II} has been used; it is defined as

$$\boxed{K_{II} = \lim_{r \to 0}\left[\sqrt{2\pi r}\,\tau_{r\varphi}(r,0)\right]} \quad (3.3.70)$$

and it may be positive or negative. A sign convention can be made only with reference to a coordinate direction. Let the z axis be parallel with the crack edge. Then, *the stress intensity factor is positive if the shear stresses in front of the crack edge tend to rotate matter in the clockwise direction when looking in the negative z direction.* This is of course the usual view.

Higher order terms in modes I and II. The T-stress
Whereas mode III in general exhibits stresses proportional to $r^{-1/2}$, $r^{1/2}$, $r^{3/2}$, etc., the general mode I stress field contains terms proportional to $r^{-1/2}$, r^0, $r^{1/2}$, r, etc. (Williams 1952). The asymptotic r^0-term, often referred to as the T-stress, is absent in mode II, which otherwise, in general, contains terms with the same powers in r as mode

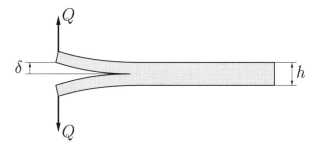

Fig. 3.3.3 A cracked structure in the shape of a double cantilever, subjected to mode I loading.

I. The T-stress consists of a constant contribution to σ_x, and it equals σ_x on the crack faces immediately behind the crack edge. The magnitude of the T-stress is not uniquely related to the magnitude of K_I; it is thus not determinable by asymptotic analysis. Its contribution to the stress field near the process region may be considerable, and it may be positive or negative. In small scale yielding, it might even be larger than the inverse square-root stresses at the plastic region boundary. A simple example is given by an approximate analysis of a double cantilever; see Fig. 3.3.3.

Elementary beam theory gives

$$\delta = Qa^3/(3EI), \quad I = bh^3/12, \quad bJ = \mathrm{d}(Q\delta)/\mathrm{d}a,$$
$$K_I^2 = EJ, \quad \sigma_b = Qa/W, \quad W = bh^2/6 \quad (3.3.71)$$

where δ is half the crack opening under the forces Q, σ_b is the bending stress on the crack faces near the edge, and h is the height of each cantilever. The expression for J follows from the fact that J in this case equals the energy flux and there is an energy release $[\mathrm{d}(Q\delta)/\mathrm{d}a]\mathrm{d}a$ from the cantilever during a crack advance $\mathrm{d}a$ under fixed δ. The equations give

$$\sigma_b = \frac{6Qa}{bh^2}, \quad K_I = \frac{2\sqrt{3}Qa}{bh\sqrt{h}} \quad (3.3.72)$$

The second expression shows that the inverse square-root term contributes

$$\sigma_i = \frac{K_I}{\sqrt{2\pi r}} = \sqrt{\frac{6}{\pi}} \cdot \frac{Qa}{bh\sqrt{hr_0}} \quad (3.3.73)$$

or less to the stresses near a boundary $r = r_0$, encircling the dissipative region. The contribution from σ_b, the T-stress, is obviously of the order of $4\sqrt{r_0/h}\,\sigma_i$. Because h is the shortest ligament, the scale of yielding may be taken as $s_y = r_0/h$; see page 72. This implies that the inverse square-root term controls the dissipative region to within an error of about $4\sqrt{s_y}$, i.e., the control is considerably less strong than in mode III cases. The scale of yielding has to be very small, indeed, in order to ensure autonomy of the dissipative region for mode I cases in general.

Note that the ratio

$$\frac{T}{K_I} = \frac{1}{\sqrt{h}} \quad (3.3.74)$$

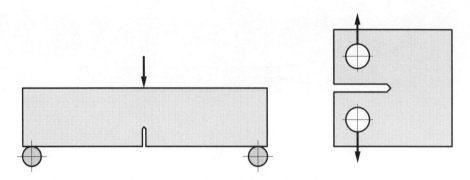

Fig. 3.3.4 Left figure: The SENB (Single Edge Notched Bend) or 3PB (3 Point Bend) specimen. Right figure: CT (Compact Tension) specimen.

for the double cantilever, so that T may be expressed in terms of the stress intensity factor multiplied by a geometry dependent factor, here $1/\sqrt{h}$. Usually a crack length parameter, a, is a more appropriate geometric characteristic, and the geometry factor is then, conventionally, taken as $B/\sqrt{\pi a}$, where the non-dimensional constant B is the *biaxiality parameter*, introduced by Leevers and Radon (1982). Thus,

$$B = \frac{T\sqrt{\pi a}}{K_I} \qquad (3.3.75)$$

If, for instance, $a = 10h$ for the double cantilever, then the biaxiality parameter $B = \sqrt{10\pi}$. For a small crack with length $2a$ in a large plate subjected to unidirectional remote mode I loading, $B = -1$; this simple result explains the normalization of B. For a SENB (Single Edge Notched bar in Bending) specimen (also called 3PB (3 Point Bend) specimen (see Fig. 3.3.4) with height W, B has been computed by Al-Ani and Hancock (1991); their results were fitted approximately to the expression

$$B = -0.46(1 - \frac{a}{W}) + 2.5\left(\frac{a}{W}\right)^2 \qquad (3.3.76)$$

by Kirk *et al.* (1993). Thus, B is negative when a/W is smaller than about 0.34. For the same specimen in tension, the SEN (single edge notched) specimen, B is lower than for the SENB specimen (except in the limit $a/W \to 0$) and is negative for a/W less than about 0.6. For a CT (Compact Tension) specimen (see Fig. 3.3.4) B is positive and comparative large (depending on the crack length). Some analytical methods for calculating stress intensity factors and T-stresses will be presented in the next chapter. Among finite element methods for calculating these quantities, the one developed by Sham (1991) may be mentioned.

For pure mode II, by definition, there cannot be any non-zero normal stresses σ_x or σ_y acting on the symmetry plane. Superposition of such stresses is, of course, possible, but strictly they should be classified as belonging to mode I.

It follows from the definition of mode II that the hydrostatic component of the stress field must be anti-symmetric with respect to the crack plane. This can, however, be realized only if the material response is everywhere symmetric with respect to compression and tension. But this is normally the case only in linearly elastic regions, and significant deviations from mode II symmetry may be expected for the dissipative region at the crack edge and the immediate elastic neighbourhood, even if body and loading geometry conform with the requirements for mode II.

The universal character of the asymptotic stress fields

The idealization to infinitesimally small scale yielding may be arrived at from a real case with a planar crack, either by retaining the real size of the dissipative region at the crack edge and assuming infinite size of both body and crack length, or by retaining the real geometry of the body but assuming vanishing size of the dissipative region near the crack edge. The former case is useful for studies of the dissipative region in plane cases; the remote outer load is then taken from (3.3.35), (3.3.54), (3.3.56), (3.3.65) and (3.3.67)†. In the latter case, the real outer loads are retained, but the elastic constitutive equations must be extrapolated towards infinite strains. The dominating stress and displacement terms then express the stress and displacement fields *asymptotically* as the crack edge is approached.

From equations (3.3.35), (3.3.54)-(3.3.57) and (3.3.65)-(3.3.68) it follows that the asymptotic stress field may be written in a compact form:

$$\sigma_{ij} = \frac{K_I}{\sqrt{2\pi r}} f_{ij}^I(\varphi) + \frac{K_{II}}{\sqrt{2\pi r}} f_{ij}^{II}(\varphi) + \frac{K_{III}}{\sqrt{2\pi r}} f_{ij}^{III}(\varphi) \qquad (3.3.77)$$

where the functions $f_{ij}(\varphi)$, with two exceptions, are independent of the material, i.e. they are *universal functions*. The exceptions are f_{33}^I and f_{33}^{II} for plane strain: they are proportional to Poisson's ratio. Due to the definition of the stress intensity factor, cf. (3.3.33), (3.3.53) and (3.3.70),

$$f_{22}^I(0) = f_{21}^{II}(0) = f_{23}^{III}(0) = 1 \qquad (3.3.78)$$

The strain and displacement fields may also be expressed in compact forms, but the angular dependence is not universal, although the material dependence, which is manifested through Poisson's ratio, is generally rather weak. In particular the asymptotic displacement of the crack face $\varphi = \pi$ may be written as

$$u_i = \frac{K_I}{(1-k^2)\mu} \sqrt{\frac{r}{2\pi}} \delta_{2i} + \frac{K_{II}}{(1-k^2)\mu} \sqrt{\frac{r}{2\pi}} \delta_{1i} + \frac{K_{III}}{\mu} \sqrt{\frac{2r}{\pi}} \delta_{3i} \qquad (3.3.79)$$

where δ_{ij} is Kronecker's delta, which equals unity if $i = j$ and zero if $i \neq j$.

The asymptotic in-plane stress fields are identical for plane stress and plane strain, but the difference in σ_z gives rise to quite dissimilar effective stress fields in mode I. Straight ahead of the crack the two in-plane principal stresses are equal, which implies that the maximum shear stress in plane strain is only the fraction $(1 - 2\nu)$ of that in plane stress. The Huber-von Mises effective stress in the mode I crack growth direction decays considerably faster in plane strain than in plane stress, except for very low values of ν. In other directions the difference is less conspicuous. As a consequence of the effective stress dissimilarities between the two cases, the appearance of the plastic region is also quite different, as will be discussed later.

---

† This is the so-called boundary layer approach, which will be described on pages 119ff.

Sometimes it is more convenient to calculate a stress intensity factor from the crack face displacement than from the stress ahead of the crack edge:

$$K_I = (1 - k^2)\mu \lim_{r \to 0}\left[\sqrt{\frac{2\pi}{r}}(v)_{\varphi=\pi}\right] \qquad (3.3.80)$$

$$K_{II} = (1 - k^2)\mu \lim_{r \to 0}\left[\sqrt{\frac{2\pi}{r}}(u)_{\varphi=\pi}\right] \qquad (3.3.81)$$

$$K_{III} = \mu \lim_{r \to 0}\left[\sqrt{\frac{\pi}{2r}}(w)_{\varphi=\pi}\right] \qquad (3.3.82)$$

3.4 Cracks in non-linear elastic environments

It is generally tacitly understood that "small scale yielding" refers to a material that is linearly elastic outside the dissipative region at a crack edge. This situation is well approximated for important engineering and geological materials, but it is not satisfactory for materials like rubber, certain plastics (particularly in thin sheets), etc. Several results exist about the asymptotic stress and deformation field at the edge of cracks in non-linearly elastic materials. Different material models have been used, all of them hyperelastic, i.e., the existence of an energy function is assumed. This energy function defines different classes of materials, such as neo-Hookean materials (Rivlin 1948), generalized neo-Hookean materials (Knowles 1977), which allow different degrees of "hardening", Mooney-Rivlin materials (Mooney 1940, Rivlin 1947), Hadamard materials (Hadamard 1903) and Ogden-Ball materials (Ogden 1972a,b, Ball 1977). These materials can be subjected to internal constraints, for instance incompressibility or the Bell constraint (Bell 1985, 1989). The latter implies that the sum of the three principal stretches, λ_1, λ_2 and λ_3, should equal 3 (the stretch λ is the ratio ds/dS, where dS and ds are the distances between two neighbouring material points in the undeformed and the deformed configuration, respectively).

An incompressible neo-Hookean material was used by Wong and Shield (1969), who found an approximate solution of the global problem of a finite crack in an infinite thin sheet, subjected to mode I (actually biaxial) loading. Knowles and Sternberg (1973, 1974) determined the asymptotic field at the edge of a crack in compressible materials, with different degrees of "hardening", under plane strain mode I loading, and Knowles (1977) solved the mode III problem for a generalized neo-Hookean material. These solutions showed results somewhat similar to those for linear elasticity: the asymptotic crack profile is parabolic or parabolic-like, and there are singularities in displacement gradients and stresses at the crack edge. However, for certain special materials bounded stresses may appear at the crack edge, even though there are unbounded displacement gradients (Knowles 1977).

Mixed mode (I and II), or modes I and II separately, were considered by Stephenson (1982) for an incompressible Mooney-Rivlin material, by Knowles (1981) for rather general classes of hyperelastic materials, by Le (1992) for a compressible Hadamard material, and by Le and Stumpf (1993) for an Ogden-Ball compressible (rubberlike) material, all assuming plane strain conditions. Geubelle and Knauss (1994a) considered mixed mode loading of incompressible generalized neo-Hookean materials under plane stress. Surprisingly, in all these cases, mode I symmetric asymptotic displacement fields occurred for all loading mode mixtures, although, in a global context, these fields would be rotated with respect to the original crack plane. Thus, even for purely anti-symmetric in-plane loading (mode II), the crack may open symmetrically (after appropriate adjustment for rigid body rotation), and anti-symmetric displacement asymptotic field components are not admitted. For some materials, apparent interpenetration results, rather than opening, implying, of course, that the solution is physically impossible (Knowles 1981). There may be more than one stress singularity (Geubelle and Knauss 1994a), and the dominating singularity is then stronger than at linear elasticity.

A totally different asymptotic field was obtained by Tarantino (1997), who considered plane stress mixed mode in-plane loading for a special model of a Bell material – that is an isotropic elastic material subject to the internal constraint of Bell. The Bell constraint is interesting, because, as found experimentally, it applies with good accuracy to a variety of metals subjected to large deformations (Bell 1985, 1989). Although Bell's observations did not include the unloading phase, a consistent mathematical formulation for hyperelastic materials under the Bell constraint is possible (Beatty and Hayes 1992a,b). A consequence of the Bell constraint is that the material volume decreases during every deformation from the undeformed state.

In agreement with previous results for other hyperelastic materials, it was found by Tarantino (1997) that the asymptotic field is mode I symmetric. However, it is non-singular! Thus, the stresses are finite and the crack closes smoothly at the crack edge. The asymptotic crack opening was found to be proportional to $r^{3/2}$, where r is the distance from the crack edge. The possibility of smooth closing for certain non-linear elastic materials appears to have been previously anticipated by Herrmann (1989).

Consideration of non-linear elasticity has also played an important role in clarifying certain features of the asymptotic field at the edge of a crack along the interface between two dissimilar linearly elastic materials. This will be discussed in the context of interface cracks, pages 177ff.

3.5 Results based on energy considerations

Energy flow to the crack edge

Crack growth requires energy dissipation in the crack edge region. The energy is provided by the elastic stress-strain field through energy flow towards the advancing crack edge. At small scale yielding, the energy flow is rather accurately controlled by the stress intensity factor, i.e. it is fairly independent of the size of the dissipative region. It may therefore be calculated under the simplifying assumption of infinitesimally small scale yielding. However, a non-zero stress intensity factor exists only when the crack edge region is controlled by an inverse square-root stress component, and this is not always the case. One counterexample is intersonic mode II crack propagation. In such cases the size of the dissipative region is crucial, see pages 353ff., so that the simplification of infinitesimally small scale yielding cannot be assumed for determination of the energy flow into the edge region. Therefore, even though a non-zero stress intensity factor is assumed here, a rather generally applicable method to calculate the energy flow will be used.

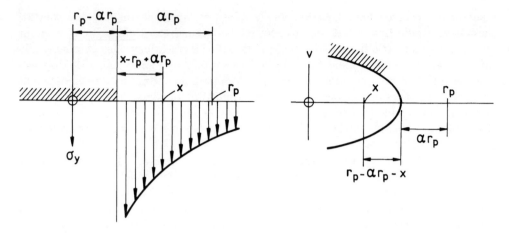

Fig. 3.5.1 Left figure: One load element to be used in a superposition scheme. Right figure: The corresponding displacement element.

Consider first a case of infinitesimally small scale yielding. In the vicinity of the edge of a mode I crack, $x < 0$, $y = \pm 0$, in plane stress or plane strain, the stress

$$\sigma_y(x,0) = \frac{K_I}{\sqrt{2\pi x}} \cdot U(x) \qquad (3.5.1)$$

and the displacement of the upper crack face, which equals half the crack face opening, given by (3.3.63), is

$$v_+(x,0) = \frac{K_I}{(1-k^2)\mu} \sqrt{\frac{-x}{2\pi}} U(-x) \qquad (3.5.2)$$

In these expressions, $U(x)$ is the *Unit step function*, defined by the relation

$$U(x) = \begin{cases} 0 & \text{for } x < 0 \\ 1 & \text{for } x > 0 \end{cases} \qquad (3.5.3)$$

When the crack edge is advancing, it acts as an energy sink. The energy consumption of the sink cannot be directly calculated by multiplication of the stress (3.5.1) and the displacement (3.5.2), because these expressions do not share a common x-interval of validity. This may, however, be arranged by removal of the singularity. To this end, a Barenblatt process region, cf. Section 1.6, will be constructed in the interval $0 < x < r_p$, by first considering the semi-infinite solid $y \geq 0$ and superposing loads of the form

$$d\sigma_y(x,0) = dK \cdot \frac{U(x - r_p + \alpha r_p)}{\sqrt{2\pi(x - r_p + \alpha r_p)}}, \quad 0 \leq \alpha \leq 1, \quad K = K(\alpha) \qquad (3.5.4)$$

on $y = 0$ (see Fig. 3.5.1), and then proceed in the same way for $y \leq 0$.

According to equations (3.3.77)-(3.3.79), each load $d\sigma_y(x,0)$ gives rise to displacements, which on $y = +0$ are

$$dv_+(x,0) = \frac{dK}{(1-k^2)\mu} \sqrt{2\pi(-x + r_p - \alpha r_p)} U(-x + r_p - \alpha r_p) \qquad (3.5.5)$$

Superposition gives:

$$\sigma_y(x,0) = \int_0^1 \frac{K'(\alpha)U(x - r_p + \alpha r_p)}{\sqrt{2\pi(x - r_p + \alpha r_p)}} d\alpha \qquad (3.5.6)$$

$$v_+(x,0) = \frac{1}{(1-k^2)\mu} \int_0^1 \frac{K'(\alpha)\sqrt{-x + r_p - \alpha r_p}}{\sqrt{2\pi}} U(-x + r_p - \alpha r_p) d\alpha \qquad (3.5.7)$$

On the surface $0 < x < r_p$, $y = 0$, neither the normal stress nor the normal displacement is zero, whereas the normal stress is zero for $x < 0$, $y = 0$, and the normal displacement is zero for $x > r_p$, $y = 0$, i.e. the result is a Barenblatt region, with a profile that depends on $K(\alpha)$. For $x \gg r_p$ integration of the right member of (3.5.6) gives the approximative relation

$$\sigma_y(x,0) \approx \frac{1}{\sqrt{2\pi x}} \int_0^1 K'(\alpha) d\alpha = \frac{1}{\sqrt{2\pi x}} [K(1) - K(0)] \qquad (3.5.8)$$

which becomes exact in the limit $r_p/a \to 0$, where a is any finite length dimension in the body, for instance the crack length. The stress σ_y should equal the expression (3.5.1) for $x \gg r_p$, and thus,

$$K(1) - K(0) = K_I \qquad (3.5.9)$$

The energy flow from the stress-strain field to the Barenblatt region, per unit of steady state advance of the crack and per unit of length along the edge, i.e. the energy flux, is

$$\mathcal{G} = -2 \int_0^{r_p} \sigma_y \frac{\partial v_+}{\partial x} dx \qquad (3.5.10)$$

where factor 2 accounts for the two crack faces and the minus sign is taken because v_+ is the displacement on the upper crack face. Insertion from equations (3.5.6)-(3.5.7) gives

$$\mathcal{G} \approx \frac{1}{2\pi(1-k^2)\mu} \int_0^{r_p} \int_0^1 \int_0^1 K'(\alpha)K'(\beta)$$
$$\times \frac{U(x - r_p + \alpha r_p)U(-x + r_p - \beta r_p)}{\sqrt{(x - r_p + \alpha r_p)(-x + r_p - \beta r_p)}} d\alpha \, d\beta \, dx \qquad (3.5.11)$$

Change of the order of integration leads to the integral

$$\int_0^{r_p} \frac{U(x - r_p + \alpha r_p)U(-x + r_p - \beta r_p)}{\sqrt{(x - r_p + \alpha r_p)(-x + r_p - \beta r_p)}} dx$$
$$= \int_{r_p - \alpha r_p}^{r_p - \beta r_p} \frac{U(\alpha - \beta) dx}{\sqrt{(x - r_p + \alpha r_p)(-x + r_p - \beta r_p)}} = \pi U(\alpha - \beta) \qquad (3.5.12)$$

whereupon the energy flux is found:

$$\mathcal{G} \approx \frac{1}{2(1-k^2)\mu} \int_0^1 \int_0^1 K'(\alpha) K'(\beta) U(\alpha - \beta) \mathrm{d}\alpha \, \mathrm{d}\beta$$

$$= \frac{1}{2(1-k^2)\mu} \int_0^1 K'(\alpha) \int_0^\alpha K'(\beta) \mathrm{d}\beta \, \mathrm{d}\alpha$$

$$= \frac{1}{2(1-k^2)\mu} \int_0^1 [K(\alpha) - K(0)] K'(\alpha) \mathrm{d}\alpha$$

$$= \frac{1}{4(1-k^2)\mu} [K(1) - K(0)]^2 = \frac{K_I^2}{4(1-k^2)\mu} \quad (3.5.13)$$

Note that the result does not contain r_p and that it becomes exact in the limit $r_p/a \to 0$. It can be shown that the relative deviation from the limit is of the order of r_p/a if $r_p/a \ll 1$. The limit, the energy flux at infinitesimally small scale yielding, may also be written in the form

$$\mathcal{G} = J = \frac{K_I^2}{4(1-k^2)\mu} = \begin{cases} \dfrac{K_I^2}{E} & \text{for plane stress} \\ \dfrac{(1-\nu^2) K_I^2}{E} & \text{for plane strain} \end{cases} \quad (3.5.14)$$

because the energy flux equals the J-integral for infinitesimally small scale yielding and steady state crack advance, cf. page 58. E is the modulus of elasticity (Young's modulus).

The energy flux for mode II is obtained in the same way:

$$\mathcal{G} = J = \frac{K_{II}^2}{4(1-k^2)\mu} = \begin{cases} \dfrac{K_{II}^2}{E} & \text{for plane stress} \\ \dfrac{(1-\nu^2) K_{II}^2}{E} & \text{for plane strain} \end{cases} \quad (3.5.15)$$

Note that the corresponding expression for mode III has already been given by equation (3.3.40):

$$\mathcal{G} = J_{III} = \frac{K_{III}^2}{2\mu} \quad (3.5.16)$$

If the crack is subjected to *mixed mode loading* and is moving straightforward (which appears to be an extremely rare event if mode I is involved), the energy flux is the sum of the contributions from each mode:

$$\mathcal{G} = \frac{K_I^2}{4(1-k^2)\mu} + \frac{K_{II}^2}{4(1-k^2)\mu} + \frac{K_{III}^2}{2\mu} \quad (3.5.17)$$

This is obvious from the derivation of result (3.5.13), showing that energy is released in mode I by the action of σ_y on displacement v: superposition of a mode II or III stress-strain state provides displacements u or w which do not produce energy together with σ_y.

The result (3.5.14) was obtained by Irwin (1957), who used the following simple

3.5 RESULTS BASED ON ENERGY CONSIDERATIONS

argument: Suppose that a cut of length Δa is made straight ahead of the crack, but that the crack is prevented from opening up by surface tractions with components equal to the stresses τ_{xy}, σ_y and τ_{yz} before the cut was made. These stresses can be written as $\sigma_{2i}(r)$, $(i=1,2,3)$, where r is the distance from the crack edge, $0 < r < \Delta a$. Now, let the crack open up slowly by relaxing the tractions along the cut. During relaxation, the force $\sigma_{2i}(r)\mathrm{d}r$ on an element $\mathrm{d}r$ of the upper cut face performs a negative work under increasing displacement u_i^+. After completed relaxation, this displacement equals its value, before the cut was made, at the distance $\Delta a - r$ behind the original position of the crack edge. Thus, the total work, considering both cut faces, amounts to

$$-\frac{1}{2}\int_0^{\Delta a} \sigma_{2i}(r) \frac{\mathrm{d}[u_i^+(\Delta a - r) - u_i^-(\Delta a - r)]}{\mathrm{d}r} \mathrm{d}r \qquad (3.5.18)$$

and thus the energy flux into the crack edge at infinitesimally small scale yielding is

$$\mathcal{G} = -\lim_{\Delta a \to 0} \left\{ \frac{1}{2}\int_0^{\Delta a} \sigma_{2i}(r) \frac{\mathrm{d}[u_i^+(\Delta a - r) - u_i^-(\Delta a - r)]}{\mathrm{d}r} \mathrm{d}r \right\} \qquad (3.5.19)$$

assuming a finite limit exists. For inverse square-root singularities the result may be written as

$$\mathcal{G} = \lim_{r \to 0}\left\{\sqrt{r}\sigma_{2i}(r) \cdot \sqrt{r}\frac{\mathrm{d}[u_i^+(r) - u_i^-(r)]}{\mathrm{d}r}\right\}$$
$$\times \lim_{\Delta a \to 0}\left(\frac{1}{2}\int_0^{\Delta a}\frac{\mathrm{d}r}{\sqrt{r}\cdot\sqrt{\Delta a - r}}\right) = \frac{\pi}{2}AB \qquad (3.5.20)$$

where

$$A = \lim_{r \to 0}[\sqrt{r}\sigma_{2i}(r)], \quad B = \lim_{r \to 0}\left\{\sqrt{r}\frac{\mathrm{d}[u_i^+(r) - u_i^-(r)]}{\mathrm{d}r}\right\} \qquad (3.5.21)$$

Attempts to insert expressions (3.5.1)-(3.5.2) directly into an integral of type (3.5.10) are sometimes found in the literature. This results in an integrand proportional to

$$\frac{U(x)}{\sqrt{x}} \cdot \frac{U(-x)}{\sqrt{-x}} \qquad (3.5.22)$$

which is claimed to equal $(\pi/2)\delta(x)$, where $\delta(x)$ is the Dirac delta function, which satisfies the conditions

$$\delta(x) = 0 \text{ for } x \neq 0, \quad \int_{-\infty}^{+\infty} \delta(x)\, \mathrm{d}x = 1 \qquad (3.5.23)$$

However, the integrand suggested is not uniquely integrable. By choosing, for instance,

$$U(x) = \lim_{\epsilon \to 0} \frac{\frac{1}{2}[1 + x/\sqrt{x^2 + \epsilon^2}]}{(1 + \alpha^2\epsilon^2/x^2)^{1/4}} \qquad (3.5.24)$$

the integral is

$$\int_{-\infty}^{+\infty} \frac{U(x)}{\sqrt{x}} \cdot \frac{U(-x)}{\sqrt{-x}} \mathrm{d}x = \int_{-\infty}^{+\infty}\frac{\mathrm{d}x}{4(x^2+1)\sqrt{x^2+\alpha^2}} \qquad (3.5.25)$$

which equals $\pi/2$ for $\alpha = 2$, but else can take on any other positive value.

Interpretation of the Nilsson integral

By shrinking the path Γ for the Nilsson integral J_N, defined by (3.2.56), to a near-edge path γ, use can be made of the fact that stresses and displacements depend on time t only through the stress intensity factor $K(t)$ in the vicinity of the crack edge, that is, for instance, $\sigma_{ji}/K(t)$ is independent of t. Thus,

$$S_{ji} = \frac{K(p)}{K(t)}\sigma_{ji}(t), \quad U_i = \frac{K(p)}{K(t)}u_i(t), \quad W^* = \left[\frac{K(p)}{K(t)}\right]^2 W(t) \qquad (3.5.26)$$

and therefore

$$J_N = \left[\frac{K(p)}{K(t)}\right]^2 \int_\gamma \left[\left(W + \frac{1}{2}\rho p^2 u_i u_i\right)n_1 - n_j\sigma_{ji}\frac{\partial u_i}{\partial x_1}\right]\mathrm{d}s \quad (j=1,2) \qquad (3.5.27)$$

Now, p may be considered as finite and positive, cf. Appendix A5, and then the second term in the integrand vanishes as γ approaches the crack edge, because u_i vanishes. The integral then equals the J-integral, which may be written as $K^2/[4(1-k^2)\mu]$, according to (3.5.14)-(3.5.15). Thus, from (3.5.27):

$$J_N = \frac{[K(p)]^2}{4(1-k^2)\mu} \qquad (3.5.28)$$

Weight functions

A familiar experience from finite element calculations is that the solution of a linear elastic problem, involving one specific loading geometry, provides easy access to solutions of problems for the same body geometry but other loading geometries. A stress intensity factor K, for instance, may be expressed by an integral of the general type

$$K = \int_\Gamma h_i(s) n_j \sigma_{ji}(s)\,\mathrm{d}s \qquad (3.5.29)$$

so that knowledge of the *weight function* $h_i(s)$ enables determination of K under the action of tractions $n_j\sigma_{ji}(s)$ on the outer boundary Γ of the body. s is an arc length coordinate along the boundary. The weight function is frequently also called the *influence function*. For modes I and II, $i,j = 1,2$, and for mode III, $i=3; j=1,2$. The expression is easily interpreted as a result of superposition of contributions from load elements $n_j\sigma_{ji}(s)\mathrm{d}s$. It is understood that the boundary load conforms with the requirements for the mode in question, without pressing the crack faces or parts of them together.

It will now be shown that the weight function can be found, if the change of boundary stresses and strains during an infinitesimal crack advance is known for any particular loading geometry. To this end, the area A between the boundary Γ and a near-edge path γ, starting on one crack face, encircling the edge, and ending on the other crack face, is studied, see Fig. 3.5.2. A pure mode, I, II or III, is assumed at the edge considered. The path Γ should include loaded parts of the crack faces, but it is assumed that these are not loaded near the edge. A positive direction of arc length s is chosen so that A is to the left of Γ and to the right of γ. The energy balance for the region

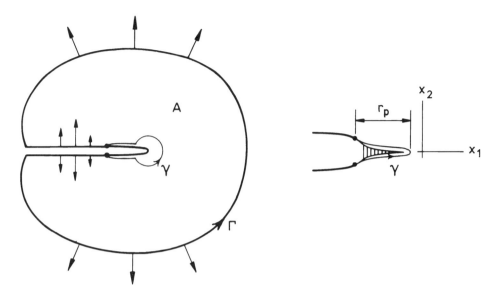

Fig. 3.5.2 The path Γ along the plate boundary, including loaded parts of the crack faces, and the path γ around the crack edge. For energy considerations it is convenient to assume γ to be drawn just outside a Barenblatt region, as shown to the right in the figure. Pure mode I, II or III conditions are assumed to prevail near the crack edge, but otherwise no symmetry conditions are invoked.

A is given by the equation

$$\frac{d}{da}\int_A \frac{1}{2}\sigma_{ji}u_{i,j}\,dA - \int_\Gamma n_j\sigma_{ji}\frac{\partial u_i}{\partial a}\,ds + \int_\gamma n_j\sigma_{ji}\frac{\partial u_i}{\partial a}\,ds = 0 \qquad (3.5.30)$$

The area integral expresses the increase of stress-strain energy in A and the line integrals express the energy flow from the area. In view of the equilibrium condition

$$\frac{\partial \sigma_{ji}}{\partial x_j} = 0 \qquad (3.5.31)$$

the area integral may be written as

$$\int_A \frac{1}{2}\frac{\partial}{\partial x_j}(\sigma_{ji}u_i)\,dA \qquad (3.5.32)$$

and then transformed to two line integrals by application of Green's theorem. Thus,

$$\frac{d}{da}\int_\Gamma \frac{1}{2}n_j\sigma_{ji}u_i\,ds - \frac{d}{da}\int_\gamma \frac{1}{2}n_j\sigma_{ji}u_i\,ds - \int_\Gamma n_j\sigma_{ji}\frac{\partial u_i}{\partial a}\,ds$$
$$+ \int_\gamma n_j\sigma_{ji}\frac{\partial u_i}{\partial a}\,ds = 0 \quad (3.5.33)$$

which is simplified to

$$\int_\Gamma \frac{1}{2}n_j\frac{\partial \sigma_{ji}}{\partial a}u_i\,ds - \int_\Gamma \frac{1}{2}n_j\sigma_{ji}\frac{\partial u_i}{\partial a}\,ds = \int_\gamma \frac{1}{2}n_j\frac{\partial \sigma_{ji}}{\partial a}u_i\,ds - \int_\gamma \frac{1}{2}n_j\sigma_{ji}\frac{\partial u_i}{\partial a}\,ds \quad (3.5.34)$$

A Barenblatt region of length $r_p \gg \mathrm{d}a$, but also infinitesimally small and much smaller than the distance from the edge to any loaded part of the crack faces (which distance therefore also may be infinitesimally small), is now introduced. The path γ is drawn along the boundary of this region; see Fig. 3.5.2. Considering that steady state conditions prevail in a close neighbourhood of the crack edge, the right member of (3.5.34) may be written as

$$\int_0^{r_p} \sigma_{2i} \frac{\partial u_i}{\partial x_1} \mathrm{d}x_1 - \int_0^{r_p} \frac{\partial \sigma_{2i}}{\partial x_1} u_i \, \mathrm{d}x_1 = -[\sigma_{2i} u_i]_0^{r_p} + 2 \int_0^{r_p} \sigma_{2i} \frac{\partial u_i}{\partial x_1} \mathrm{d}x_1$$

$$= -\int_\gamma \left(W \, \mathrm{d}x_2 - n_j \sigma_{ji} \frac{\partial u_i}{\partial x_1} \right) \mathrm{d}s = -J \quad (3.5.35)$$

where, in order to obtain formal agreement with the J-integral, $W \, \mathrm{d}x_2$ and index j were incorporated; note that both $\mathrm{d}x_2$ and n_1 vanish along γ. The traction $n_2 \sigma_{2i} = 0$, except for $i = 1$ in mode I, for $i = 2$ in mode II, and for $i = 3$ in mode III. Now, from (3.5.15)-(3.5.17) it follows that J may be written in the form $J = K^2/H$, where K is the stress intensity factor and

$$H = \begin{cases} 4(1 - k^2)\mu & \text{for modes I and II} \\ 2\mu & \text{for mode III} \end{cases} \quad (3.5.36)$$

Consider now two different loading geometries, $\sigma_{ji}^{(k)}(s)$, where $k = 1, 2$, acting on Γ and leading to the same pure mode in the crack edge vicinity. It is temporarily assumed that these loads are kept fixed everywhere on Γ during the crack advance $\mathrm{d}a$: this implies $\partial \sigma_{ji}/\partial a = 0$ on Γ. If the two loading systems act separately, the stress intensity factors will be given by the expression

$$\frac{(K^{(k)})^2}{H} = \int_\Gamma \frac{1}{2} n_j \sigma_{ji}^{(k)} \frac{\partial u_i^{(k)}}{\partial a} \mathrm{d}s \quad (3.5.37)$$

according to (3.5.34)-(3.5.35). If both systems act simultaneously, stress intensity factors, stresses and displacements will be linearly superposed:

$$\frac{(K^{(1)} + K^{(2)})^2}{H} = \int_\Gamma \frac{1}{2} n_j (\sigma_{ji}^{(1)} + \sigma_{ji}^{(2)}) \cdot \left(\frac{\partial u_i^{(1)}}{\partial a} + \frac{\partial u_i^{(2)}}{\partial a} \right) \mathrm{d}s \quad (3.5.38)$$

and hence, by use of (3.5.37):

$$2 \frac{K^{(1)} K^{(2)}}{H} = \int_\Gamma \frac{1}{2} n_j \left(\sigma_{ji}^{(1)} \frac{\partial u_i^{(2)}}{\partial a} + \sigma_{ji}^{(2)} \frac{\partial u_i^{(1)}}{\partial a} \right) \mathrm{d}s \quad (3.5.39)$$

By the Maxwell-Betti reprocity theorem†, the two terms within ( ) in the integrand are equal. The result may therefore be written in the form

$$K^{(2)} = \int_\Gamma n_j \sigma_{ji}^{(2)} \cdot \frac{H}{2K^{(1)}} \cdot \frac{\partial u_i^{(1)}}{\partial a} \mathrm{d}s = \int_\Gamma n_j \sigma_{ji}^{(2)} h_i(s) \, \mathrm{d}s \quad (3.5.40)$$

where $u_i^{(1)} = u_i^{(1)}(s, a)$ and

$$h_i(s) = \frac{H}{2K^{(1)}} \cdot \frac{\partial u_i^{(1)}}{\partial a} \quad (3.5.41)$$

---

† See, e.g., Bishlinghoff et al. (1990).

Because $\sigma_{ji}^{(1)}(s)$ may be arbitrarily chosen within the restriction regarding loading mode, it follows that $h_i(s)$ depends only on the body geometry and on the mode of loading. It is, of course, the weight function, cf. (3.5.29).

It was temporarily assumed that the stresses $\sigma_{ji}(s)$ were kept fixed along the whole boundary Γ during the crack advance da. By using (3.5.34) without this restriction, a generalization follows immediately to the case when tractions are prescribed on one part, Γ_σ of Γ and displacements on the remaining part, Γ_u. The influence of body forces may also be included, by adding an area integral

$$-\int_A \rho f_i \frac{\partial u_i}{\partial a} \mathrm{d}A \tag{3.5.42}$$

to the left member of (3.5.30) and changing the equilibrium condition (3.5.31) to

$$\frac{\partial \sigma_{ji}}{\partial x_j} + \rho f_i(x_1, x_2) = 0 \tag{3.5.43}$$

where ρ is the density and f_i is body force (force per unit mass). A general expression for the stress intensity factor is then found to be

$$\boxed{K = \int_{\Gamma_\sigma} h_i(s) n_j \sigma_{ji}(s) \,\mathrm{d}s + \int_{\Gamma_u} g_i(s) u_i(s) \,\mathrm{d}s + \int_A h_i^*(x_1, x_2) \rho f_i \,\mathrm{d}s} \tag{3.5.44}$$

where

$$g_i(s) = \frac{H}{2K^{(1)}} n_j \frac{\partial \sigma_{ji}^{(1)}(s,a)}{\partial a}, \quad s \in \Gamma_u \tag{3.5.45}$$

$$h_i^*(x_1, x_2) = \frac{H}{2K^{(1)}} \cdot \frac{\partial u_i^{(1)}(x_1, x_2, a)}{\partial a} \tag{3.5.46}$$

are the weight functions for prescribed boundary displacements and for body forces, respectively. The weight function $h_i(s)$ is still defined by (3.5.41), but only for Γ_σ.

A body force on the boundary is equivalent to a boundary traction; hence, on each point of the boundary the weight function $h_i^*(x_1, x_2)$ for body forces has the same value as the weight function $h_i(s)$ for boundary tractions.

Note that expressions (3.5.45)-(3.5.46) contain partial derivatives with respect to advance of one crack edge. Care should therefore be exercised when, for instance, the crack position is given as $|x| < a$, $y = 0$. The simplest way is to temporarily introduce $b = -a$ and $c = a$, whereupon partial differentiation is made with respect to the crack edge considered ($x = b$ or $x = c$).

Elastostatic weight functions for cracked bodies were introduced by Bueckner (1970) and the theory was further developed by Rice (1972), who obtained the results shown here. The weight function theory for cracks was later advanced towards three-dimensional cases, for instance planar cracks with curved fronts or crack interaction with inhomogeneities such as particles and dislocations. A comprehensive theory and a review of earlier work is given in Rice (1989). So-called second order weight functions for determination of the T-stress have been developed by Sham (1991) for use in finite element calculations. Weight functions for interface cracks were developed by H. Gao (1992), Banks-Sills (1993) and Fett et al. (1997).

Weight functions for stationary cracks, subjected to dynamic loading

Freund and Rice (1974) used Laplace transformed stresses and displacements for a body containing a stationary crack, subjected to dynamic loading. They showed the existence of a Laplace transformed weight function, related to Laplace transformed displacements in formally the same way as for the static case. This will be demonstrated here, employing essentially the same method as for the static case. However, the Laplace transforms involved will concern stresses and displacements and their derivatives, but not energy quantities, except formally as products of Laplace transformed stresses and displacement gradients. Therefore, the energy argument leading to (3.5.30) cannot be used, but the formal similarity may be exploited. In fact, (3.5.30) depends on a property of the energy density W, that can be described without reference to any physical interpretation of W. Thus, if a function $L(u_i, u_{i,j})$ satisfies the equality

$$\frac{\partial L}{\partial u_i} - \frac{\partial}{\partial x_j}\left(\frac{\partial L}{\partial u_{i,j}}\right) = 0 \qquad (3.5.47)$$

then a variation δu_i of u_i, which also causes a variation of $u_{i,j}$, leads to the equality

$$\delta \int_A L(u_i, u_{i,j}) \mathrm{d}A - \int_\Gamma n_j \frac{\partial L}{\partial u_{i,j}} \delta u_i \, \mathrm{d}s = 0 \qquad (3.5.48)$$

where A is a region in the (x_1, x_2)-plane, Γ its boundary and s an arc length, with positive direction such that A is to the left of Γ. The proof is simple, it makes use of Green's theorem, leading to

$$\delta \int_A L(u_i, u_{i,j}) \mathrm{d}A = \int_A \left(\frac{\partial L}{\partial u_i} \delta u_i + \frac{\partial L}{\partial u_{i,j}} \delta u_{i,j}\right) \mathrm{d}A$$
$$= \int_A \left[\frac{\partial L}{\partial u_i} \delta u_i - \frac{\partial}{\partial x_j}\left(\frac{\partial L}{\partial u_{i,j}}\right) \delta u_i\right] \mathrm{d}A$$
$$+ \int_A \frac{\partial}{\partial x_j}\left(\frac{\partial L}{\partial u_{i,j}} \delta u_i\right) \mathrm{d}A = \int_\Gamma n_j \frac{\partial L}{\partial u_{i,j}} \delta u_i \, \mathrm{d}s \qquad (3.5.49)$$

where n_i is the outward normal of Γ.

By writing $\sigma_{ji} = C_{ijkm} u_{k,m}$ so that $W = \frac{1}{2} C_{ijkm} u_{k,m} u_{i,j}$, it is easy to verify that W satisfies the equality (3.5.47) for L: this becomes in fact the equilibrium condition. Note also that substitution of L by W in equation (3.5.48) leads to equation (3.5.30) if the variation δ refers to a crack advance.

In the dynamic case, the momentum balance equation reads

$$\frac{\partial \sigma_{ji}}{\partial x_j} + \rho f_i = \rho \frac{\partial^2 u_i}{\partial t^2} \qquad (3.5.50)$$

so that

$$\frac{\partial S_{ji}}{\partial x_j} + \rho F_i = \rho p^2 U_i \qquad (3.5.51)$$

where S_{ji}, U_i and F_i are the Laplace transforms of σ_{ji}, u_i and f_i, respectively. It is assumed that the body is at rest and without dynamic forces at $t = 0$, i.e. $u_{i,t} = 0$ and $u_{i,tt} = 0$ for $t = 0$.

Now, by choosing
$$L = \frac{1}{2}(S_{ji}U_{i,j} + \rho p^2 U_i U_i) - \rho F_i U_i \qquad (3.5.52)$$

which satisfies (3.5.47), the procedure used in the static case can be followed step by step, starting by insertion of (3.5.52) into (3.5.48) in order to get the dynamic version of equation (3.5.30), with inclusion of body forces. The result is found to be expressible by equations (3.5.44)-(3.5.46) after exchanging stresses, displacements and weight functions by the Laplace transforms of their dynamic counterparts.

Weight functions for moving cracks

A two-dimensional weight function was constructed by Burridge (1976) for a semi-infinite crack, moving at non-constant speed. Later, Willis and Movchan (1995, 1997) and Movchan and Willis (1995) constructed three-dimensional weight functions for a semi-infinite crack, moving at constant speed. Their results will be discussed later (page 500).

3.6 The Barenblatt model

The Barenblatt model of a process region (Barenblatt 1959a,b,c) was used in Section 3.5, where expressions for the normal stress σ_y and the normal displacement v were given by equations (3.5.6) and (3.5.7). Here, some features of the Barenblatt region will be studied, particularly the stress and displacement distributions and the conditions for smooth opening. It is assumed that the Barenblatt region develops in an elastic surrounding.

Introduce the non-dimensional coordinate
$$\epsilon = \frac{r_p - x}{r_p} \qquad (3.6.1)$$

which equals zero at the front of the process region and unity at the edge of the crack (the rear end of the process region), the non-dimensional cohesive stress
$$g(\epsilon) = \frac{\sqrt{2\pi r_p}}{K}\sigma_y(x) \text{ for } 0 \leq \epsilon \leq 1 \qquad (3.6.2)$$

and the non-dimensional cohesive displacement
$$h(\epsilon) = \frac{(1-k^2)\mu}{K} \cdot \sqrt{\frac{2\pi}{r_p}} v_+(x) \text{ for } 0 \leq \epsilon \leq 1 \qquad (3.6.3)$$

which equals half the opening of the process region. Then, equations (3.5.6) and (3.5.7) take the non-dimensional forms
$$g(\epsilon) = \int_\epsilon^1 \frac{f'(\alpha)}{\sqrt{\alpha - \epsilon}}\,d\alpha \implies g(1) = 0 \qquad (3.6.4)$$
$$h(\epsilon) = \int_0^\epsilon f'(\alpha)\sqrt{\epsilon - \alpha}\,d\alpha \implies h(0) = 0 \qquad (3.6.5)$$

where

$$f'(\alpha) = \frac{K'(\alpha)}{K} \qquad (3.6.6)$$

and thus $f(1) = 1$, if (arbitrarily) $f(0) = 0$.

The problem now consists of determining the cohesive displacement, $h(\epsilon)$, when the cohesive stress, $g(\epsilon)$, is known and *vice versa*. First, however, note that $g(\epsilon)$ and $h(\epsilon)$ are subject to the following conditions:

$$\int_0^1 \frac{g(\epsilon)}{\sqrt{\epsilon}} d\epsilon = \int_0^1 \int_0^1 \frac{f'(\alpha) U(\alpha - \epsilon)}{\sqrt{\epsilon(\alpha - \epsilon)}} d\alpha\, d\epsilon$$

$$= \int_0^1 f'(\alpha) \int_0^\alpha \frac{d\epsilon}{\sqrt{\epsilon(\alpha - \epsilon)}} d\alpha = \pi \qquad (3.6.7)$$

$$\int_0^1 \frac{h'(\epsilon)}{\sqrt{1-\epsilon}} d\epsilon = \int_0^1 \int_0^1 \frac{f'(\alpha) U(\epsilon - \alpha)}{\sqrt{(1-\epsilon)(\epsilon - \alpha)}} d\alpha\, d\epsilon$$

$$= \int_0^1 f'(\alpha) \int_\alpha^1 \frac{d\epsilon}{\sqrt{(1-\epsilon)(\epsilon - \alpha)}} d\alpha = \pi \qquad (3.6.8)$$

where $U(\cdot)$ is the unit step function. Further, it is assumed that $\int_0^1 g'(\gamma)/\sqrt{\gamma}\, d\gamma$ is convergent, which is also a reasonable physical condition. Condition (3.6.7) is a smooth closing condition, as will later be shown; cf. (4.4.46).

The problem of finding $h(\epsilon)$ when $g(\epsilon)$ is known will now be considered. The solution of the Abelian integral equation (3.6.4) is (see *e.g.* Kanwal 1971)

$$f'(\alpha) = -\frac{1}{\pi} \int_\alpha^1 \frac{g'(\epsilon)}{\sqrt{\epsilon - \alpha}} d\epsilon \qquad (3.6.9)$$

and thus, from (3.6.5):

$$h(\epsilon) = -\frac{1}{\pi} \int_0^\epsilon \sqrt{\epsilon - \alpha} \int_\alpha^1 \frac{g'(\gamma)}{\sqrt{\gamma - \alpha}} d\gamma\, d\alpha$$

$$= -\frac{1}{\pi} \int_0^1 g'(\gamma) \int_0^\epsilon \sqrt{\frac{\epsilon - \alpha}{\gamma - \alpha}} U(\gamma - \alpha) d\alpha\, d\gamma$$

$$= -\frac{1}{\pi} \int_0^\epsilon g'(\gamma) \int_0^\gamma \sqrt{\frac{\epsilon - \alpha}{\gamma - \alpha}} d\alpha\, d\gamma - \frac{1}{\pi} \int_\epsilon^1 g'(\gamma) \int_0^\epsilon \sqrt{\frac{\epsilon - \alpha}{\gamma - \alpha}} d\alpha\, d\gamma \qquad (3.6.10)$$

Elementary integrations, including a partial integration, give

$$h(\epsilon) = \frac{\sqrt{\epsilon}}{\pi} \int_0^1 \frac{g(\gamma)}{\sqrt{\gamma}} d\gamma - \frac{1}{2\pi} \int_0^1 g(\gamma) \ln\left|\frac{\sqrt{\epsilon} + \sqrt{\gamma}}{\sqrt{\epsilon} - \sqrt{\gamma}}\right| d\gamma \qquad (3.6.11)$$

and thus, in view of (3.6.7):

$$\boxed{h(\epsilon) = \sqrt{\epsilon} - \frac{1}{2\pi} \int_0^1 g(\gamma) \ln\left|\frac{\sqrt{\epsilon} + \sqrt{\gamma}}{\sqrt{\epsilon} - \sqrt{\gamma}}\right| d\gamma} \qquad (3.6.12)$$

3.6 THE BARENBLATT MODEL

This expression gives the half opening, $h(\epsilon)$, of the process region when the cohesive stress distribution, $g(\epsilon)$, is known. For $\epsilon \ll 1$ the result is

$$h(\epsilon) \approx -\frac{2}{3\pi} \int_0^1 \frac{g'(\gamma)}{\sqrt{\gamma}} \mathrm{d}\gamma \cdot \epsilon^{3/2} \tag{3.6.13}$$

showing that the process region opens smoothly and that the condition

$$\int_0^1 \frac{g'(\gamma)}{\sqrt{\gamma}} \mathrm{d}\gamma < 0 \tag{3.6.14}$$

gives a necessary restriction on $g(\epsilon)$. The physical condition $h(\epsilon) > 0$ for $\epsilon > 0$ is obviously satisfied if $g'(\epsilon)$ is non-positive everywhere, although this is not necessary. An example where $g'(\epsilon)$ is negative in some interval is given by a cohesion-decohesion curve with two peaks (page 572).

In a similar way, assuming the half opening, $h(\epsilon)$, of the process region to be known, the cohesive stress distribution $g(\epsilon)$ is found to be

$$g(\epsilon) = \frac{2\sqrt{1-\epsilon}}{\pi} \fint_0^1 \frac{h'(\gamma)}{(\gamma-\epsilon)\sqrt{1-\gamma}} \mathrm{d}\gamma \tag{3.6.15}$$

where the integral is to be taken in the sense of the Cauchy principal value.

In dimensional form, conditions (3.6.7) and (3.6.8) read:

$$\int_0^{r_p} \frac{\sigma_y(s)\,\mathrm{d}s}{\sqrt{s}} = \sqrt{\frac{\pi}{2}} K \tag{3.6.16}$$

$$\int_0^{r_p} \frac{v'_+(s)\,\mathrm{d}s}{\sqrt{r_p-s}} = \frac{\sqrt{2\pi}K}{4(1-k^2)\mu} \tag{3.6.17}$$

where σ_y and v are written as functions of $s = r_p - x$, which is the distance from the front end of the process region. The right members of these conditions contain quantities determined by properties of the continuum outside the process region. The left members are given by quantities characteristic of the (autonomous) process region. Thus, the equations express how the continuum field is related to the properties of the process region. Note that (3.6.16), as (3.6.7), is a condition for smooth opening (or closing).

The left member of (3.6.16) consists of the quantity

$$T_I = \int_0^{r_p} \frac{\sigma_y(s)\,\mathrm{d}s}{\sqrt{s}} \tag{3.6.18}$$

which Barenblatt (1959a,b,c) called the *cohesion modulus*.

Due to the autonomy of the process region, connections other than those given by (3.6.16) and (3.6.17) can be found. One such connection is based on the observation that the J-integral for a path Γ just outside the process region equals

$$J = -\int_\Gamma n_j \sigma_{ji} \frac{\partial u_i}{\partial x_1} \mathrm{d}s = -2\int_0^{r_p} \sigma_y(x) \frac{\partial v_+}{\partial x} \mathrm{d}x \tag{3.6.19}$$

Thus,

$$J = 2 \int_0^{\delta_0(0)/2} \sigma_y(v_+) \, dv_+ \qquad (3.6.20)$$

where $\delta_0(0)$ is the current opening of the process region at $x = 0$. If the crack length a is much larger than the length r_p of the process region, then onset of crack growth occurs when the crack opening $\delta_0(0)$ approaches δ_{max}, the maximum separation distance in the decohesion relation. Then, the equality expresses the total area under the decohesion curve, the σ_y-δ curve. However, if a is not much larger than r_p, then crack growth starts under a crack opening $\delta_0(0)$ that may be substantially smaller than δ_{max}, cf. Ståhle (1983).

If the decohesion relation between σ_y and $\delta = 2v$ were known, say

$$\frac{\sigma_y}{\sigma_D} = F\!\left(\frac{\delta}{\delta_{max}}\right) = F\!\left(\frac{v_+}{v_{max}}\right) \qquad (3.6.21)$$

where σ_D is the cohesive strength, then the distribution $\sigma_y(x)$ may be found from (3.6.15):

$$g_D F\!\left(\frac{h(\epsilon)}{h_D}\right) = \frac{2\sqrt{1-\epsilon}}{\pi} \int_0^1 \frac{h'(\gamma)}{(\gamma-\epsilon)\sqrt{1-\gamma}} \, d\gamma \qquad (3.6.22)$$

where

$$g_D = \frac{\sigma_D \sqrt{2\pi r_p}}{K}, \qquad h_D = \frac{\sqrt{2\pi}(1-k^2)\mu v_{max}}{K\sqrt{r_p}} \qquad (3.6.23)$$

Thus, to obtain the stress distribution when the decohesion relation is known, the integral equation (3.6.22) has to be solved, which in general requires numerical methods.

Note that the decohesion curve for the Barenblatt model gives a somewhat incomplete description of the material, because it does not contain the ascending part known from cohesion-decohesion curves for real process regions. For applications in which the ascending part is important, a cell model or a box model may be used, cf. Section 1.6.

4
Elastostatic Cracks

4.1 Introduction

Linear elasticity and the idealization of crack problems to infinitesimally small scale yielding is assumed in the present chapter. This requires use of Hooke's law without limitation of stress or strain magnitudes. Then, the stress-strain field in the crack edge vicinity is uniquely determined for each mode by the stress intensity factor. Finding this factor will be the main objective here, but other field characteristics, for instance the crack opening, will occasionally also be explored.

The term "infinitesimally small scale yielding" is used in two slightly different meanings in the present work, depending on how real cases of small scale yielding are idealized. In one sense, the dissipative region is considered as infinitesimally small and the crack length finite, whereas in another sense, the dissipative region is considered as finite whereas the crack length is infinite. The former idealization, which will be used in the present chapter, is suitable for studies of the elastic stress-strain field in cracked bodies. The latter idealization is suitable for investigation of plastic regions at crack edges, and this will be used in Chapters 5 and 7.

The present chapter concentrates on analytical methods, with only a short discussion of numerical methods. This does not reflect any opinion about the relative importance of these two methods. A quick inspection would rather suggest the reverse emphasis: crack problems that can be attacked by analytical methods constitute a very restricted subset of the enormous set of important problems in engineering, geophysics and related fields. However, experience shows that numerical methods can be properly exploited only with a thorough knowledge of analytical methods and results, and these in turn are also frequently used for controlling the accuracy of numerical methods. Moreover, analytical methods often lead to better insight into general properties of the phenomena studied.

4.2 Eigenfunction expansions

The method of eigenfunction expansions was used in Section 3.3. Here, a more advanced example will be given.

Consider a twisted bar with circular cross-section and with a longitudinal crack that penetrates symmetrically from the surface to a depth $a < R$, where R is the radius of the bar. The length of the bar and the extension of the crack along the bar is much

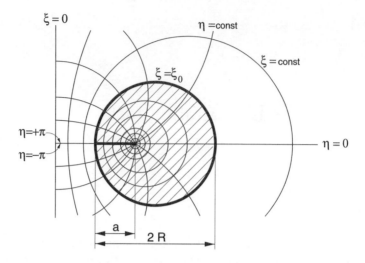

Fig. 4.2.1 Cross-section of cracked cylinder, showing coordinate orientations. Along the crack edge, $\xi = 0$.

larger than R. Torques are applied at the bar ends, producing a twist θ per unit of length along the bar. Sufficiently far from the ends of the bar, θ may be considered as constant. This portion of the bar is studied. No load acts on the lateral boundaries. Due to cross-section warping, slip parallel to the bar axis occurs, creating a mode III stress intensity near the crack edge, but it should be noted that overall deformations do not occur under clean mode III conditions.

Bipolar coordinates ξ, η, z offer a simple way to specify the boundaries. They are connected to Cartesian coordinates x, y, z by the transformations

$$x = \frac{c \sinh \xi}{N}, \quad y = \frac{c \sin \eta}{N}, \quad z = z \qquad (4.2.1)$$

where $N = \cosh \xi - \cos \eta$ and c is a constant of dimension length. Choosing

$$c = \frac{a(2R - a)}{2(R - a)} \qquad (4.2.2)$$

and a value ξ_0 given by

$$e^{-\xi_0} = 1 - \frac{a}{R} \qquad (4.2.3)$$

implies that the bipolar coordinate system may be so oriented that the line $\xi = \infty$ coincides with the crack edge, the cylinder $\xi = \xi_0$ coincides with the cylindrical boundary and the planes $\eta = \pm\pi$, $\xi_0 < \xi < \infty$, coincide with the crack faces; see Fig. 4.2.1.

It is convenient also to introduce cylindrical coordinates r, φ, z, with $r = 0$ coinciding with the crack edge and $\varphi = \pm\pi$ with the crack faces. Their relation to the bipolar coordinates for positive ξ-values is given by the equations

$$r^2 = 2c^2 \frac{e^{-\xi}}{N}, \quad \sin^2 \varphi = \frac{e^\xi \sin^2 \eta}{2N}, \quad z = z \qquad (4.2.4)$$

4.2 EIGENFUNCTION EXPANSIONS

From (3.3.1) it follows that the stresses τ_{rz} and $\tau_{\varphi z}$ can be represented by a stress function:

$$\tau_{rz} = \mu\theta \cdot \frac{1}{r} \cdot \frac{\partial \Phi}{\partial \varphi}, \quad \tau_{\varphi z} = -\mu\theta \cdot \frac{\partial \Phi}{\partial r} \tag{4.2.5}$$

Here, the factor $\mu\theta$ has been incorporated for later convenience. According to the St Venant theory of elastic torsion,

$$\Delta \Phi = -2 \tag{4.2.6}$$

with the prescription that Φ is a constant on the boundary, say

$$\Phi = 0 \text{ for } \xi = \xi_0 \text{ and for } \eta = \pm\pi \tag{4.2.7}$$

Because (4.2.6) is written in invariant form, i.e. in a form independent of coordinate systems, it also holds for a stress function Φ of ξ and η. The shear stresses, expressed in these coordinates, are found to be

$$\tau_{\xi z} = \frac{1}{c}\mu\theta(\cosh\xi - \cos\eta)\frac{\partial \Phi}{\partial \eta} \tag{4.2.8}$$

$$\tau_{\eta z} = -\frac{1}{c}\mu\theta(\cosh\xi - \cos\eta)\frac{\partial \Phi}{\partial \xi} \tag{4.2.9}$$

and the St Venant equation (4.2.6) transforms to

$$\frac{\partial^2 \Phi}{\partial \xi^2} + \frac{\partial^2 \Phi}{\partial \eta^2} = -\frac{2c^2}{N^2} \tag{4.2.10}$$

A complete general solution, symmetric in η and satisfying the condition of bounded stress-strain energy, is found in the form of an eigenfunction expansion,

$$\Phi = -\frac{c^2 \cos\eta}{N} + \sum_0^\infty A_n e^{-n\xi/2} \cos\frac{n\eta}{2} \tag{4.2.11}$$

where the eigenvalues n are integers.
For $\eta = \pm\pi$, the condition (4.2.7) gives

$$\sum_0^\infty A_n e^{-n\xi/2} \cos\frac{n\eta}{2} = -\frac{c^2}{\cosh\xi + 1} = c^2(-2e^{-\xi} + 4e^{-2\xi} - 6e^{-3\xi} + \cdots) \tag{4.2.12}$$

Thus,

$$A_n = nc^2, \quad n \text{ even} \tag{4.2.13}$$

For $\xi = \xi_0$, the condition (4.2.7) gives:

$$\sum_{1,3,\ldots} A_n e^{-n\xi_0/2} \cos\frac{n\eta}{2} = \frac{c^2 \cos\eta}{\cosh\xi_0 - \cos\eta} - c^2 \sum_{2,4,\ldots} n e^{-n\xi_0/2} \cos\frac{n\eta}{2} \tag{4.2.14}$$

Now, it is found that

$$\sum_{2,4,\ldots} n e^{-n\xi_0/2} \cos\frac{n\eta}{2} = \frac{\cosh\xi_0 \cos\eta - 1}{(\cosh\xi_0 - \cos\eta)^2} \tag{4.2.15}$$

by, for instance, first integrating the left member over ξ_0, then writing

$$\cos\frac{n\eta}{2} = \frac{1}{2}[e^{in\eta/2} + e^{-in\eta/2}] \qquad (4.2.16)$$

summing the geometrical series, and finally differentiating with respect to ξ_0. Thus,

$$\sum_{1,3,\ldots} A_n e^{-n\xi_0/2} \cos\frac{n\eta}{2} = \frac{c^2 \sin^2\eta}{(\cosh\xi_0 - \cos\eta)^2} \qquad (4.2.17)$$

Fourier analysis yields

$$A_n e^{-n\xi_0/2} = \frac{c^2}{\pi}\int_{-\pi}^{\pi}\frac{\sin^2\eta\cos(n\eta/2)}{(\cosh\xi_0 - \cos\eta)^2}d\eta = \frac{c^2}{\pi}I_n(\xi_0), \quad n \text{ odd} \qquad (4.2.18)$$

and insertion of A_n into (4.2.11) leads to the stress function

$$\Phi = -\frac{c^2\cos\eta}{N} + \frac{c^2}{\pi}\sum_{1,3,\ldots}I_n(\xi_0)e^{-n(\xi-\xi_0)/2}\cos\frac{n\eta}{2} \qquad (4.2.19)$$

Now, the principal shear stresses may be obtained from (4.2.8)-(4.2.9). The stress intensity factor is found from the shear stress

$$(\tau_{\eta z})_{\eta=0} = \frac{c}{2\pi}\mu\theta(\cosh\xi - 1)\sum_{1,3,\ldots}I_n(\xi_0)e^{-n(\xi-\xi_0)/2} \qquad (4.2.20)$$

whereupon the definition (3.3.33), together with the relation (4.2.4) gives

$$K_{III} = \lim_{r\to 0}\sqrt{2\pi r}(\tau_{\eta z})_{\eta=0} = \lim_{\xi\to\infty}2\sqrt{\pi}e^{-\xi/2}(\tau_{\eta z})_{\eta=0} = \frac{c\sqrt{c}}{\sqrt{\pi}}\mu\theta I_1(\xi_0)e^{\xi_0/2} \qquad (4.2.21)$$

The integral $I_1(\xi_0)$ is elementary:

$$I_1(\xi_0) = 2\left[\frac{1 + 3\sinh^2(\xi_0/2)}{\sinh(\xi_0/2)}\mathrm{atan}\frac{1}{\sinh(\xi_0/2)} - 3\right] \qquad (4.2.22)$$

After returning to original quantities, a and R, with the aid of (4.2.2)-(4.2.3), the stress intensity factor is found to be

$$K_{III} = \frac{\mu\theta a(2R-a)^2}{4\pi(R-a)^2}\left[\frac{4R(R-a)+3a^2}{2a\sqrt{R(R-a)}}\mathrm{atan}\frac{2\sqrt{R(R-a)}}{a} - 3\right]\sqrt{\frac{2\pi aR}{2R-a}} \qquad (4.2.23)$$

Two special cases will be considered.
1) $a \to R$:

$$K_{III} \to \frac{16\mu\theta R}{15}\sqrt{\frac{2R}{\pi}} \qquad (4.2.24)$$

2) $a/R \ll 1$:

$$K_{III} \approx \mu\theta R\sqrt{\pi a}\left[1 - \left(\frac{4}{\pi} - \frac{3}{4}\right)\frac{a}{R}\right] \qquad (4.2.25)$$

When $a/R \to 0$, the stress $\mu\theta R \to \tau_\infty$, the shear stress $\tau_{\eta z}$ at the cylindrical boundary, except in the neighbourhood of the crack. Thus,

$$K_{III} \to \tau_\infty\sqrt{\pi a} \text{ as } a/R \to 0 \qquad (4.2.26)$$

4.3 Complex potentials for anti-plane strain

Basic equations

Analytic functions play an important role for solution of two-dimensional problems in mathematical physics. In anti-plane or in-plane crack problems, displacements and stresses may be written as functions of complex potentials. Use may then be made of the basic theorems for analytic functions, powerful because of their simplicity. An introduction to the theory of analytic functions is given in Appendix A4.

In the anti-plane strain case, the only non-vanishing displacement component, w, normal to the xy-plane, satisfies the Laplace equation, cf. (3.3.4):

$$\Delta w = 0 \qquad (4.3.1)$$

Hooke's law gives

$$\tau_{xz} = \mu \frac{\partial w}{\partial x}, \quad \tau_{yz} = \mu \frac{\partial w}{\partial y} \qquad (4.3.2)$$

The real and imaginary parts of an analytic function satisfy Laplace's equation. The converse is also true: any function that satisfies Laplace's equation, i.e. any harmonic function, can be written as either the real or the imaginary part of an analytic function. Thus, if

$$f(z) = \varphi(x,y) + i\mu w(x,y), \quad z = x + iy \qquad (4.3.3)$$

is an analytic function, then equation (4.3.1) is satisfied. Now, for any analytic function the Cauchy-Riemann equations are satisfied. Thus,

$$\mu \frac{\partial w}{\partial x} = -\frac{\partial \varphi}{\partial y}, \quad \mu \frac{\partial w}{\partial y} = \frac{\partial \varphi}{\partial x} \qquad (4.3.4)$$

Inspection shows that the left members equal the shear stresses given by equations (4.3.2). Thus,

$$\boxed{\mu w = \Im f(z)} \qquad (4.3.5)$$

$$\boxed{\tau_{yz} + i\tau_{xz} = f'(z)} \qquad (4.3.6)$$

where $\Im$ denotes the imaginary part. (Note that z as a subscript and z as a complex variable are two different, easily distinguishable symbols.) Obviously, φ may be considered as a stress potential, w as a strain potential and $f(z)$ as a complex potential function for stresses and strains.

Equations (4.3.5)-(4.3.6) are valid in the region occupied by the body. Any anti-plane strain problem is thus reduced to the problem of finding a function $f(z)$ that is analytic in this region and satisfies the appropriate boundary conditions. Equation (4.3.5) is used when these conditions are given in terms of displacements and equation

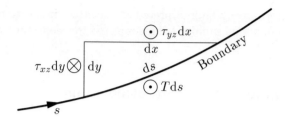

Fig. 4.3.1 Shear force T and its relation to the stresses τ_{xz} and τ_{yz}.

(4.3.6) when they are given in terms of tractions. The latter type of condition may alternatively be expressed in integrated form. If the shear force exerted on a boundary element ds, where s is the arc length along the boundary, is Tds (see Fig. 4.3.1), then

$$T\mathrm{d}s = \tau_{xz}\mathrm{d}y - \tau_{yz}\mathrm{d}x \qquad (4.3.7)$$

if positive s direction is related to the outward normal as the y to the x direction, i.e., so that the outward normal points to the right when walking in the positive s direction.

Now, it is possible to write

$$T\mathrm{d}s = -\frac{1}{2}[(\tau_{yz} + i\tau_{xz})\mathrm{d}z + (\tau_{yz} - i\tau_{xz})\mathrm{d}\bar{z}] \qquad (4.3.8)$$

where a bar denotes complex conjugation. Use of equation (4.3.6) then results in the representation

$$T\mathrm{d}s = -\frac{1}{2}\mathrm{d}[f(z) + \overline{f(z)}] = -\mathrm{d}[\Re f(z)] \qquad (4.3.9)$$

where $\Re$ denotes the real part. The left member can be integrated along the arc to give the boundary condition

$$\boxed{\Re f(z) = -\int_0^s T\mathrm{d}s} \qquad (4.3.10)$$

The point $s = 0$ may be chosen arbitrarily, because addition of a real constant to the complex potential does not influence stresses or displacements. The relation (4.3.10) is a special form of boundary conditions when tractions are given on the boundary. Another form is given by (4.3.6). When displacements are given on the boundary, (4.3.5) is to be used.

A major advantage with complex potentials lies in the fact that conformal mapping may be used when the body contour cannot be simply fitted into any of the basic orthogonal coordinate systems used in mathematical physics. Conformal mapping is particularly convenient in the case of edge cracks, thus when the region of analyticity is simply connected. Then, for instance, mapping onto a circle is possible. This may also be done for the region outside an interior crack, if the body is large enough in comparison with the crack, to be considered as infinite.

The complex potential method is not restricted to Cartesian coordinates. Stresses

4.3 COMPLEX POTENTIALS FOR ANTI-PLANE STRAIN

in polar coordinates are obtained from Cartesian stresses through the transformation relations

$$\left.\begin{array}{l}\tau_{\varphi z} = \tau_{yz}\cos\varphi - \tau_{xz}\sin\varphi \\ \tau_{rz} = \tau_{yz}\sin\varphi + \tau_{xz}\cos\varphi\end{array}\right\} \qquad (4.3.11)$$

Thus, for polar coordinates equations (4.3.5)-(4.3.6) should be modified by the substitutions

$$z = re^{i\varphi}, \quad w = w, \quad \tau_{yz} + i\tau_{xz} = (\tau_{\varphi z} + i\tau_{rz})e^{-i\varphi} \qquad (4.3.12)$$

whereas (4.3.10) takes the form

$$\Re f(z) = -\int_0^\varphi R(\varphi) T \, d\varphi \qquad (4.3.13)$$

where $R(\varphi)$ is the radius vector from the origin to the boundary.

Mode III cracks

So far no assumption about symmetry has been made, except that the only non-vanishing displacement component is w. It will now be assumed that mode III symmetry prevails so that the body is symmetric and the loading anti-symmetric with respect to the x axis. It is also assumed that there are no cracks outside the x axis. Then, w is anti-symmetric, which implies that

$$\Im f(z) = -\Im f(\bar{z}) = \Im \overline{f(\bar{z})} = \Im \overline{f}(z) \qquad (4.3.14)$$

The left and the right members give

$$f(z) - \overline{f(z)} = \overline{f}(z) - f(\bar{z}) \qquad (4.3.15)$$

which may be rearranged to

$$f(z) - \overline{f}(z) = \overline{f(z) - \overline{f}(z)} \qquad (4.3.16)$$

Now, if $f(z)$ is analytic on the lower half of the body, then $\overline{f}(z)$ is analytic on the upper half. This may be shown, for instance, by using the Cauchy-Riemann equations or by using power series representation as in Appendix A2. Then, if $f(z)$ is analytic on both the upper and the lower half, though not necessarily on the x axis, the function $\overline{f}(z)$ is also analytic on both the upper and the lower half of the region studied. Thus, the left member of (4.3.16) is analytic, except, perhaps, on the x axis. Because it equals its complex conjugated part, it must be a real function. But, by the Cauchy-Riemann equations, a real analytic function is a constant. Hence,

$$\overline{f}(z) = f(z) + C \qquad (4.3.17)$$

where C is a real constant. Complex conjugation of this expression, followed by complex conjugation of z, shows that $C = 0$. Thus, for mode III symmetry in a symmetric body:

$$\overline{f}(z) = f(z) \qquad (4.3.18)$$

Infinite body

General solutions can be given for an infinite body. The abstraction to an infinite body is useful when the shortest ligament from the crack edges is much longer than the crack length. Suppose that such a body contains a single crack, $y = 0$, $b < x < c$, and that it is subjected to the remote load $\tau_{yz} = \tau_{yz}^\infty$. The crack faces are assumed to be subjected to tractions, $\tau_{yz} = \tau_{yz}^0(x)$. Because $f(z)$ is not necessarily analytic on the x axis, it is advantageous to use separate notations, $f_+(z)$ and $f_-(z)$, for its parts on the upper and lower half-planes, respectively. Subscripts plus and minus will be used extensively for the same purpose in the continuation. Now, from (4.3.5):

$$w_+(x,0) = \frac{1}{2i\mu}[f(x+i0) - \overline{f(x+i0)}] = \frac{1}{2i\mu}[f(x+i0) - \overline{f}(x-i0)]$$
$$= \frac{1}{2i\mu}[f(x+i0) - f(x-i0)] = \frac{1}{2i\mu}[f_+(x) - f_-(x)] \quad (4.3.19)$$

where w_+ is the displacement on the upper side of the x axis. Anti-symmetry implies that $w_- = -w_+$.

It is not self-evident that w is zero on the x axis outside $b < x < c$. On the contrary, w is usually non-zero in mode III earthquake sliding motion and similar events, for which it might be more appropriate to talk about a *slipping region* than a crack. Slip from a previous event might have been deposited along the fault (the x axis) outside the current active slipping region, $b < x < c$, with subsequent consolidation, sufficiently firm to sustain shear stresses τ_{yz} without further slip during the prevailing load. For generality it is therefore assumed that

$$w_+ - w_- = 2w_0(x) \text{ for } y = 0, \quad x < b, \; x > c \quad (4.3.20)$$

Obviously, a mixed boundary value problem is posed, because stresses are given on one part of the plane $y = 0$, and displacements on another part. The remaining boundary condition, specifying the remote shear stress, will be considered later on. From equations (4.3.19)-(4.3.20) follows the boundary condition

$$f_+(x) - f_-(x) = 2i\mu w_0(x), \quad x < b, \; x > c \quad (4.3.21)$$

In view of equation (4.3.6) the tractions on the crack faces give the boundary condition

$$\tau_{yz} = \frac{1}{2}[f'_+(x) + f'_-(x)] = \tau_{yz}^0(x), \quad b < x < c \quad (4.3.22)$$

Equations (4.3.21)-(4.3.22) may be written in the form

$$f'_+(x) - f'_-(x) = 2i\mu w'_0(x), \quad x < b, \; x > c \quad (4.3.23)$$
$$f'_+(x) + f'_-(x) = 2\tau_{yz}^0(x), \quad b < x < c \quad (4.3.24)$$

These two equations constitute a Hilbert problem. The solution of such problems uses the fact that the two equations can be brought to the same form, by introducing an auxiliary function, in this case the function

$$G(z) = (z-b)^{1/2}(z-c)^{1/2} \quad (4.3.25)$$

with a branch cut along $y = 0$, $b < x < c$ and with the choice of the branch for which

4.3 COMPLEX POTENTIALS FOR ANTI-PLANE STRAIN

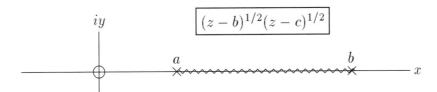

Fig. 4.3.2 Branch cuts for function definition.

$G(z) \to z$ as $z \to \infty$; see Fig. 4.3.2. Then,

$$\frac{G_-(x)}{G_+(x)} = \begin{cases} +1 & \text{for } x < b,\ x > c \\ -1 & \text{for } b < x < c \end{cases} \qquad (4.3.26)$$

and (4.3.22)-(4.3.23) may be written as one single equation,

$$f'_+(x) - \frac{G_-(x)}{G_+(x)} f'_-(x) = 2i\mu w'_0(x)[U(b-x) + U(x-c)]$$
$$+ 2\tau^0_{yz}(x)[U(x-b) - U(c-x)] = s_0(x) \qquad (4.3.27)$$

where $U(\cdot)$ is the unit step function. Multiplication of both sides with $G_+(x)$ results in

$$G_+(x)f'_+(x) - G_-(x)f'_-(x) = G_+(x)s_0(x) \qquad (4.3.28)$$

The left member contains the difference between the values on the upper and the lower side of the x axis of a function

$$G(z)f'(z) \qquad (4.3.29)$$

that is analytic in both the upper and the lower half-plane, though not everywhere on the x axis. Plemelj's formulae, given in Appendix A4, then enable a determination of this function:

$$G(z)f'(z) = \frac{1}{2\pi i} \int_{-\infty}^{\infty} \frac{G_+(\xi)s_0(\xi)}{\xi - z} d\xi \qquad (4.3.30)$$

However, this solution of the Hilbert problem (4.3.28) is not complete: a function that is analytic in the whole plane may be added, because the left member of (4.3.28) vanishes for such a function. This is equivalent to adding the general solution of the homogeneous Hilbert problem, obtained by putting the right member of (4.3.28) equal to zero. Hence,

$$f'(z) = \frac{1}{2\pi i G(z)} \int_{-\infty}^{\infty} \frac{G_+(\xi)s_0(\xi)}{\xi - z} d\xi + \frac{S_0(z)}{G(z)} \qquad (4.3.31)$$

where $S_0(z)$ is analytic in the whole plane, and thus, by Liouville's theorem a polynomial of finite degree; see Appendix A4. Thus,

$$S_0(z) = \sum_{n=0}^{N} a_n z^n \qquad (4.3.32)$$

where the coefficients a_n are real, because of the condition (4.3.18).

There is now time to consider the remaining boundary condition, which may be written as

$$\frac{1}{2}[f'(z) + \overline{f'(z)}] \to \tau_{yz}^\infty \text{ as } |z| \to \infty \quad (4.3.33)$$

Use of (4.3.31) then gives

$$\frac{1}{2}\left[\frac{a_N z^N}{z} + \frac{a_N \bar{z}^N}{\bar{z}}\right] \to \tau_{yz}^\infty \text{ as } |z| \to \infty \quad (4.3.34)$$

which obviously requires that $N = 1$ and $a_1 = \tau_{yz}^\infty$.

It remains to determine the coefficient a_0. To this end it should be recalled that the boundary condition for w was replaced by a boundary condition for w'. According to (4.3.5) integration of $f'(z)$ is needed in order to obtain w. The difference of the deposited slip between the ends of the crack is (note that $G_+(x)s_0(x)$ is imaginary):

$$w_+(c) - w_+(b) = \frac{1}{\mu}\Im[f_+(c) - f_+(b)] = -\frac{1}{\mu}\int_b^c \frac{a_1 x + a_0}{\sqrt{(x-b)(c-x)}}dx$$

$$= -\frac{\pi}{\mu}\left[a_1 \frac{b+c}{2} + a_0\right] \quad (4.3.35)$$

which gives

$$a_0 = -\frac{\mu}{\pi}[w_+(c) - w_+(b)] - \tau_{yz}^\infty \frac{b+c}{2} \quad (4.3.36)$$

and the integration constant available at integration of $f'(z)$ may, for instance, be used to fix w_+ at $z = -\infty$. As an example, the case $w_+(c) = w_+(b)$, $\tau_{yz}^0 = 0$, gives $a_0 = a_1 = 0$, and thus, from (4.3.31), the stress on the symmetry plane ahead of the crack is

$$\tau_{yz} = f'(x) = \frac{1}{\pi\sqrt{(x-b)(x-c)}}\int_b^c \frac{\tau_{yz}^0(\xi)\sqrt{(\xi-b)(c-\xi)}}{\xi - x}d\xi, \quad x > c \quad (4.3.37)$$

The solution of the general problem of a mode III crack in a large body is now complete, making it possible to calculate interesting quantities, particularly the stress intensity factor. The stress intensity factor for the edge $z = b$, following the sign convention discussed in connection with (3.3.33), is

$$K_{III} = -\lim_{x \to b-0}\left[\sqrt{2\pi(b-x)}\tau_{yz}(x)\right]$$

$$= \frac{1}{\sqrt{\pi a}}\left[\mu\int_{-\infty}^b w_0'(\xi)\sqrt{\frac{c-\xi}{b-\xi}}d\xi + \int_b^c \tau_{yz}^0(\xi)\sqrt{\frac{c-\xi}{\xi-b}}d\xi\right.$$

$$\left. + \mu\int_c^\infty w_0'(\xi)\sqrt{\frac{\xi-c}{\xi-b}}d\xi\right] - \tau_{yz}^\infty\sqrt{\pi a} - \frac{\mu}{\sqrt{\pi a}}[w_+(c) - w_+(b)] \quad (4.3.38)$$

where $a = (c-b)/2$ is the half-length of the crack. Note that the stress intensity factor for the edge $z = c$ is found after change of sign and inversion of the square roots in the integrands. Thus, for instance, for a crack face loaded semi-infinite crack,

$a \to -b/2 \to \infty$, the stress intensity factor at the edge $z = c$ is

$$K_{III} = -\sqrt{\frac{2}{\pi}} \int_{-\infty}^{c} \frac{\tau_{yz}^0(\xi)}{\sqrt{c-\xi}} d\xi \qquad (4.3.39)$$

Two special cases of particular interest will now be studied, after placing the coordinate origin so that $c = -b = a$. In the first case, $s_0(x)$ vanishes everywhere and $w_+(-a) = w_+(a)$. This corresponds to a pre-existing crack with length $2a$ in an infinite solid, which is stress free before the remote load is applied. Then,

$$K_{III} = \pm \tau_{yz}^\infty \sqrt{\pi a} \text{ for } z = \pm a \qquad (4.3.40)$$

Another interpretation of this special case follows from the observation that $\tau_{xz}(0, y) = 0$, because $f'(z)$ is real for $x = 0$. The result is therefore applicable also to a semi-infinite solid, $x \geq 0$, with an *edge crack* $y = 0$, $0 \leq x < a$, and subjected to a constant remote shear stress $\tau_{yz} = \tau_{yz}^\infty$. This case coincides with a special case of the cylindrical bar problem treated in Section 4.2: the result (4.3.40) agrees with (4.2.26).

In the other special case, $s_0(x)$ vanishes everywhere and $w_+(a) = 0$. In general, this corresponds to a constant slip deposited on $y = 0$, $x < -a$. The slip between the upper and lower part in this region is denoted by 2Δ, i.e.

$$w_+(-a) - w_-(-a) = 2\Delta \qquad (4.3.41)$$

Equation (4.3.38) now gives:

$$K_{III} = \frac{\mu \Delta}{\sqrt{\pi a}} - \tau_{yz}^\infty \sqrt{\pi a} \qquad (4.3.42)$$

for the edge $z = -a$, and, obviously

$$K_{III} = \frac{\mu \Delta}{\sqrt{\pi a}} + \tau_{yz}^\infty \sqrt{\pi a} \qquad (4.3.43)$$

for the edge $x = a$, $y = 0$. This case may be considered as having been produced by slow propagation of a friction free slipping region of constant length $2a$, from $x = -\infty$ to its current location, $y = 0$, $-a < x < a$, under constant remote stress $\tau_{yz} = \tau_{yz}^\infty$, and leaving behind a constant slip 2Δ between the two sides of the fault. Such a process involves *tearing* at $x = a$, $y = 0$, and *healing* at $x = -a$, $y = 0$. Tearing requires a certain amount of energy supply per unit of crack propagation, whereas no energy is needed for healing. On the contrary, healing might deliver energy to the stress-strain field, albeit a much smaller amount, probably in most cases negligible, in comparison with the energy required for tearing. Thus, if it is assumed that healing occurs without any energy exchange at all (energy-neutral healing), i.e. $K_{III} = 0$ at $x = -a$, $y = 0$, then

$$\Delta = \frac{\pi \tau_{yz}^\infty a}{\mu} \text{ and } K_{III} = \frac{2\mu\Delta}{\sqrt{\pi a}} \text{ at } x = -a, y = 0 \qquad (4.3.44)$$

$K_{III} = 0$ implies *smooth closing*, i.e. there is no discontinuity in $\partial w/\partial x$ at $x = -a$, $y = 0$. The shear stress on $y = 0$ for $|x| > a$ is

$$\tau_{yz} = \tau_{yz}^\infty \sqrt{\frac{x+a}{x-a}} \qquad (4.3.45)$$

and the slip distribution on the upper side of the slipping region is found by an obvious generalization of (4.3.35):

$$w_+(x) - w_+(a) = \int_x^a \frac{a_1\xi + a_0}{\sqrt{a^2 - \xi^2}} d\xi = a_1\sqrt{a^2 - x^2} + a_0 \cdot \mathrm{acos}\frac{x}{a} \text{ for } |x| < a \quad (4.3.46)$$

In the present case this reduces to

$$w_+(x) = \frac{\tau_{yz}^\infty}{\mu}\left[\sqrt{a^2 - x^2} + a \cdot \mathrm{acos}\frac{x}{a}\right] \text{ for } |x| < a \quad (4.3.47)$$

For $y = 0$, $r = a - x \ll a$,

$$w_+(x) \approx \frac{2\sqrt{2}\tau_{yz}^\infty a}{\mu}\sqrt{\frac{r}{a}} \quad (4.3.48)$$

and for $y = 0$, $r = x + a \ll a$,

$$w_+(x) \approx \frac{\pi\tau_{yz}^\infty a}{\mu}\left[1 - \frac{\sqrt{2}}{3\pi}\left(\frac{r}{a}\right)^{3/2}\right] \quad (4.3.49)$$

showing smooth closing. In agreement with this result the stress τ_{yz} is continuous at $z = -a$.

An array of collinear cracks in an infinite body is treated in essentially the same way: this will be shown for mode I.

Finite outer dimensions

For a body of finite outer dimensions, symmetric with respect to the x axis and containing a traction free crack, $y = 0$, $b < x < c$, subjected to mode III loading, the function $f'(z)$ may be written as

$$f'(z) = \frac{S(z)}{(z-b)^{1/2}(z-c)^{1/2}} \quad (4.3.50)$$

where $S(z)$ is a function that is analytic inside the outer body contour, i.e. even across the crack line. Equation (4.3.18) shows that

$$\overline{S}(z) = S(z) \quad (4.3.51)$$

If no slip has been deposited on $y = 0$, $x < b$, then $S(z)$ must satisfy the condition

$$\int_b^c \frac{S(x)}{\sqrt{(x-b)(x-c)}} dx = 0 \quad (4.3.52)$$

and, furthermore, $S(z)$ should, of course, be chosen so that the boundary conditions of the problem are satisfied. For construction of a suitable ansatz for $S(z)$, it might be helpful to use the fact that a function that is regular in a certain region can always be represented by a polynomial in the neighbourhood of any point inside the region; see Appendix A4. In special cases, this polynomial is of finite degree. Then, it represents the function in the whole region. In other cases, it is of infinite degree and then it represents the function within a certain distance from the point chosen. If no concentrated forces are acting in the body interior, this distance is at least equal to the distance to the nearest point of the boundary of the region.

In the neighbourhood of a chosen point an analytic function may be approximated by a polynomial of finite degree. Thus, for $S(z)$ the approach

$$S(z) = \sum_0^N A_n z^n \qquad (4.3.53)$$

may prove to be fruitful. Here, A_n are real constants. This approach is essential in the *boundary collocation* method. In this method, the conditions at the outer boundary of the body are satisfied for a number of selected points. Together with condition (4.3.52) this gives a set of linear equations for determination of the constants A_n. Approximations of $S(z)$ other than polynomials may, of course, also be used. Some practical hints are given by Tada, Paris and Irwin (1985)†.

The method of boundary collocation is attractive because of its simplicity. On the other hand, it might be difficult to check the validity of the solution obtained. The approximation of an analytic function in a given region by a polynomial of finite degree is not always a good choice. It would, of course, not be meaningful to increase N in cases where the function cannot be expressed by a single polynomial of infinite degree in the whole region.

4.4 Complex potentials for in-plane problems

Basic equations

For in-plane problems two complex potentials are needed. The basic relations may be established from the Papkovich representation of the displacement field (Papkovich 1932),

$$2\mu \boldsymbol{u} = 4(1-\nu)\boldsymbol{\Psi} - \text{grad}\,(\varphi + \boldsymbol{r}\cdot\boldsymbol{\Psi}) \qquad (4.4.1)$$

where, with unit vectors $\hat{\boldsymbol{x}}, \hat{\boldsymbol{y}}, \hat{\boldsymbol{z}}$ in the x, y and z directions, $\boldsymbol{u} = u\hat{\boldsymbol{x}} + v\hat{\boldsymbol{y}} + w\hat{\boldsymbol{z}}$ is the displacement vector, $\boldsymbol{r} = x\hat{\boldsymbol{x}} + y\hat{\boldsymbol{y}} + z\hat{\boldsymbol{z}}$ is the position vector and φ and $\boldsymbol{\Psi} = \psi_x\hat{\boldsymbol{x}} + \psi_y\hat{\boldsymbol{y}} + \psi_z\hat{\boldsymbol{z}}$ are scalar and vector potentials, which satisfy the equations

$$\Delta\varphi = 0, \qquad \Delta\boldsymbol{\Psi} = 0 \qquad (4.4.2)$$

Insertion shows that the Papkovich representation satisfies the equations of equilibrium, given by (A9.7) after putting the inertia term equal to zero.

As shown by Eubanks and Sternberg (1956), one of the four potential functions may be put equal to zero under certain rather general conditions. This fact is not quite surprising, because the four functions represent only three displacement components; see (4.4.1).

Specialization to plane strain, $u = u(x.y)$, $v = v(x,y)$, $w = 0$, implies that the potentials are functions of x and y, only, and then, according to (4.4.1):

$$\psi_z = 0 \qquad (4.4.3)$$

---

† This volume, *The Stress Analysis of Cracks Handbook* contains a wealth of data and other information, particularly stress intensity factors for different geometrical and loading configurations.

Because φ, ψ_x and ψ_y are harmonic functions, they may be represented by either the real or the imaginary part of an analytic function. The choice here is

$$\psi_x = \Re f(z), \quad \psi_y = \Im f(z), \quad \varphi = \Re g(z) \tag{4.4.4}$$

where $f(z)$ and $g(z)$ are analytic functions in the region occupied by the body. The interdependence of ψ_x and ψ_y is justified by the fact that essentially only two potentials would be needed to represent the two non-vanishing displacement components.

Insertion of (4.4.4) into (4.4.1) yields

$$2\mu(u+iv) = 4(1-\nu)f(z) - \left(\frac{\partial}{\partial x} + i\frac{\partial}{\partial y}\right)\{\Re g(z) + \Re[\bar{z}f(z)]\}$$

$$= 4(1-\nu)f(z) - \frac{\partial}{\partial \bar{z}}[g(z) + \overline{g(z)} + \bar{z}f(z) + z\overline{f(z)}]$$

$$= (3-4\nu)f(z) - z\overline{f'(z)} - \overline{g'(z)} \tag{4.4.5}$$

The operator $\partial/\partial\bar{z}$ refers to partial differentiation with z and $\bar{z}$ as independent variables.

Hooke's law for plane strain gives

$$\sigma_x + \sigma_y = \frac{2\mu}{1-2\nu}\left(\frac{\partial u}{\partial x} + \frac{\partial v}{\partial y}\right) = \frac{4\mu}{1-2\nu}\Re\left[\frac{\partial}{\partial z}(u+iv)\right] \tag{4.4.6}$$

$$\sigma_x - \sigma_y = 2\mu\left(\frac{\partial u}{\partial x} - \frac{\partial v}{\partial y}\right) = 4\mu\Re\left[\frac{\partial}{\partial \bar{z}}(u-iv)\right] \tag{4.4.7}$$

$$\tau_{xy} = \mu\left(\frac{\partial u}{\partial y} + \frac{\partial v}{\partial x}\right) = -2\mu\Im\left[\frac{\partial}{\partial \bar{z}}(u-iv)\right] \tag{4.4.8}$$

Thus, the stresses are represented by complex potentials:

$$\boxed{\sigma_x + \sigma_y = 4\Re[f'(z)]} \tag{4.4.9}$$

$$\boxed{\sigma_y - \sigma_x + 2i\tau_{xy} = 2[\bar{z}f''(z) + g''(z)]} \tag{4.4.10}$$

With elastic constants μ, ν, the plane stress solution is found from the plane strain solution by changing ν to $\nu/(1+\nu)$; this follows from comparing equation (4.4.6) with the corresponding equation for plane stress. Writing

$$\nu = \frac{1-2k^2}{2(1-k^2)} \text{ for plane strain}, \quad \nu = 1-2k^2 \text{ for plane stress} \tag{4.4.11}$$

which captures the conversion $\nu \to \nu/(1+\nu)$ during transition from plane strain to plane stress, results in the representation

$$\boxed{2\mu(u+iv) = \frac{1+k^2}{1-k^2}f(z) - z\overline{f'(z)} - \overline{g'(z)}} \tag{4.4.12}$$

for the displacements. It is essential to use μ together with k in order to obtain expressions valid for both plane strain and plane stress. The constant k was already introduced through (3.3.61) and has, as previously explained (see page 79), the physical significance of the ratio between the propagation velocity of S-waves and P-waves and, in the context k^2/μ, of the compliance during in-plane unidirectional straining,

4.4 COMPLEX POTENTIALS FOR IN-PLANE PROBLEMS

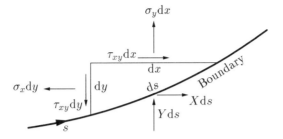

Fig. 4.4.1 Cartesian boundary forces X and Y, and their relation to the stresses σ_y and τ_{xy}

either under plane strain or plane stress. Plane stress is here assumed to be approximately realized by a thin plate, subjected to in-plane loading. The advantage with this particular choice is, of course, most apparent for dynamic cases; however, comparisons between dynamic and static solutions are often made, and they are somewhat frustrating when different elastic constants are used.

The stress σ_z in plane strain and the strain ϵ_z in plane stress must, of course, be treated separately: they follow from the relations

$$\sigma_z = \nu(\sigma_x + \sigma_y) = \frac{1 - 2k^2}{2(1 - k^2)}(\sigma_x + \sigma_y) \text{ for plane strain} \qquad (4.4.13)$$

$$\epsilon_z = -\frac{\nu}{1 - \nu}(\epsilon_x + \epsilon_y) = -\frac{1 - 2k^2}{2k^2}(\epsilon_x + \epsilon_y) \text{ for plane stress} \qquad (4.4.14)$$

Note that a rigid body displacement is represented by

$$f(z) = A + Bz, \quad g(z) = Cz \qquad (4.4.15)$$

where A and C are complex constants, which determine a translation and B is an imaginary constant, which determines a rotation. Obviously it is always possible to take $g(0) = 0$.

The representations (4.4.9)-(4.4.10) for stresses and (4.4.12) for displacements were originally established by Kolosov (1909). Their usefulness was explored by Muskhelishvili in a famous treatise (Muskhelishvili 1953a).

If the components of the force exerted on a boundary element ds, where s is the arc length along the boundary, are Xds and Ydy in the x and y directions (see Fig. 4.4.1), then

$$X\,\mathrm{d}s = \sigma_x \mathrm{d}y - \tau_{xy}\mathrm{d}x, \quad Y\,\mathrm{d}s = \tau_{xy}\mathrm{d}y - \sigma_y \mathrm{d}x \qquad (4.4.16)$$

if positive s direction is related to the outward normal as the y to the x direction, i.e. so that the outward normal points to the right when walking in the positive s direction.

Now, it is possible to write

$$(X + iY)\mathrm{d}s = -\frac{i}{2}[(\sigma_y - \sigma_x - 2i\tau_{xy})\mathrm{d}\bar{z} + (\sigma_y + \sigma_x)\mathrm{d}z] \qquad (4.4.17)$$

Use of (4.4.9)-(4.4.10) then results in the representation

$$(X + iY)\mathrm{d}s = -i\,\mathrm{d}\bigl[f(z) + z\overline{f'(z)} + \overline{g'(z)}\bigr] \qquad (4.4.18)$$

which can be integrated along the arc to give the boundary condition

$$f(z) + z\overline{f'(z)} + \overline{g'(z)} = \int_0^s (iX - Y)ds \qquad (4.4.19)$$

The point $s = 0$ may be chosen arbitrarily. Different choices may appear to correspond to different rigid-body translations, but, in fact, the rigid-body translation is controlled by the choice of $f(z_0)$ or $g'(z_0)$, where z_0 is any suitable point inside the body contour: note that equation (4.4.12) may be written in the form

$$2\mu(u + iv) = \frac{2}{1 - k^2} f(z) - \left[f(z) + z\overline{f'(z)} + \overline{g'(z)} \right] \qquad (4.4.20)$$

where the expression inside the brackets is recognized as the left member of (4.4.19).

When boundary conditions are given in terms of tractions, either equation (4.4.19) or equations (4.4.9)-(4.4.10) may be used. When they are given in terms of displacements, (4.4.9)-(4.4.10) are to be used.

An alternative formulation: the Westergaard method

So far no specific attention has been paid to crack problems in the establishment of the basic equations. Certain modifications will make the complex potential method more adapted to such problems, particularly for symmetric bodies with cracks in the symmetry plane. This approach is due to Westergaard (1939)†. It will be introduced in a more general way than the original Westergaard method‡. This is achieved by the substitutions

$$f(z) = p(z) - s(z) \qquad (4.4.21)$$
$$g'(z) = p(z) + s(z) - z[p'(z) - s'(z)] \qquad (4.4.22)$$

where $p = (g' + f + zf')/2$ and $s = (g' - f + zf')/2$ obviously are analytic inside the body. Then, equations (4.4.9)-(4.4.10) and (4.4.12) take the form

$$\sigma_x + \sigma_y = 4\Re[p'(z) - s'(z)] \qquad (4.4.23)$$

$$\sigma_y - \sigma_x + 2i\tau_{xy} = 4s'(z) - 2(z - \bar{z})[p''(z) - s''(z)] \qquad (4.4.24)$$

$$2\mu(u + iv) = \frac{1 + k^2}{1 - k^2}[p(z) - s(z)] - \overline{p(z)} - \overline{s(z)} - (z - \bar{z})\left[\overline{p'(z)} - \overline{s'(z)}\right] \qquad (4.4.25)$$

and the boundary condition (4.4.19) changes to

$$p(z) - s(z) + \overline{p(z)} + \overline{s(z)} + (z - \bar{z})\left[\overline{p'(z)} - \overline{s'(z)}\right] = \int_0^s (iX - Y)ds \qquad (4.4.26)$$

---

† Essentially the same approach was established by Carothers (1920) for "the case of a solid having an unlimited upper boundary, given by $y = 0$", subjected to normal tractions. However, it was not applied to mixed boundary value problems.
‡ Westergaard (1939) stated that the "limitation of this type of solution" consists of the requirement that "$\sigma_x = \sigma_y$ and $\tau_{xy} = 0$ at $y = 0$".

When boundary conditions are given in terms of tractions, either equation (4.4.26) or equations (4.4.23)-(4.4.24) may be used.

Equations (4.4.15) imply that the set $p(z) = A + Bz$, $s(z) = C$, where A and C are complex constants and B an imaginary constant, represents a rigid body displacement: a translation determined by A and C, and a rotation determined by B.

Westergaard method for mode I

General relations

Suppose that the body and loading geometry are symmetric with respect to the plane $y = 0$ and that mode I loading prevails. Consider two symmetrically situated points z_0 and $\bar{z}_0$. If the complex displacement is $u_0 + iv_0$ at z_0, then it is $u_0 - iv_0$ at $\bar{z}_0$. With subscripts plus and minus for functions of an argument in the upper and the lower half-plane, respectively, these displacements are

$$2\mu(u_0 + iv_0) = \frac{1+k^2}{1-k^2}[p_+(z_0) - s_+(z_0)] - \overline{p_+(z_0)} - \overline{s_+(z_0)}$$
$$- 2iy_0\left[\overline{p'_+(z_0)} - \overline{s'_+(z_0)}\right] \quad (4.4.27)$$

$$2\mu(u_0 - iv_0) = \frac{1+k^2}{1-k^2}[p_-(\bar{z}_0) - s_-(\bar{z}_0)] - \overline{p_-(\bar{z}_0)} - \overline{s_-(\bar{z}_0)}$$
$$+ 2iy_0\left[\overline{p'_-(\bar{z}_0)} - \overline{s'_-(\bar{z}_0)}\right] \quad (4.4.28)$$

Complex conjugation of the last equation gives

$$2\mu(u_0 + iv_0) = \frac{1+k^2}{1-k^2}\left[\overline{p_-(\bar{z}_0)} - \overline{s_-(\bar{z}_0)}\right] - p_-(\bar{z}_0) - s_-(\bar{z}_0)$$
$$- 2iy_0[p'_-(\bar{z}_0) - s'_-(\bar{z}_0)] \quad (4.4.29)$$

Comparison with (4.4.27) shows that

$$p_-(\bar{z}) = \overline{p_+(z)}, \quad s_-(\bar{z}) = \overline{s_+(z)} \quad (4.4.30)$$

Suppose further that the body contains one single crack, $y = 0$, $b < x < c$. Mode I loading may include prescribed tractions $(\sigma_y)_+ = (\sigma_y)_- = \sigma_y^0(x)$ and $(\tau_{xy})_+ = -(\tau_{xy})_- = \tau_{xy}^0(x)$ on the crack faces, or, alternatively, prescribed tractions $(\tau_{xy})_+ = -(\tau_{xy})_- = \tau_{xy}^0(x)$ and prescribed displacements $v_+ = -v_- = v_0(x)$ on the crack faces. A third possibility consists of prescribed displacements $v_+ = -v_- = v_0(x)$ and $u_+ = u_- = u_0(x)$ on the crack faces. Mathematically, these three possibilities turn out to be somewhat different: the condition $\tau_{xy} = \pm\tau_{xy}^0(x)$ implies that τ_{xy} is known along the whole boundary $y = 0$ of each half-plane, and the same reduction to a simple boundary value problem applies to the condition $v_\pm = \pm v_0(x)$. By far the most common condition is $(\sigma_y)_+ = (\sigma_y)_- = \sigma_y^0(x)$ together with $\tau_{xy}^0(x) = 0$.

For brevity, cases in which u and v are prescribed on the crack faces are not considered here – they are treated in essentially the same way as cases for which σ_y and τ_{xy} are prescribed. Due to the somewhat different mathematical character of different boundary conditions, it is convenient to start with the assumption that $\tau_{xy}^0(x) = 0$.

Then, it is possible to write

$$(\sigma_y - \sigma_x + 2i\tau_{xy})_- = (\sigma_y - \sigma_x + 2i\tau_{xy})_+ \text{ for } y = 0 \quad (4.4.31)$$

because σ_x and σ_y are symmetrical with respect to y and τ_{xy} vanishes for $y = 0$. In view of (4.4.24) the relation implies that

$$s'_-(x) = s'_+(x) \text{ for } y = 0 \quad (4.4.32)$$

i.e. no branch cut has to be made for $s'(z)$ along the crack line. But then $s'(z)$ is analytic on the whole line $y = 0$, because the points $z = b$ and $z = c$ cannot be poles; this follows from the restriction to bounded stress-strain energy. Hence, $s'(z)$, and then also $s(z)$, are analytic in the whole region inside the outer contour of the body, including the crack line. Clearly, the same conclusion would have been arrived at also for an array of cracks, provided that mode I conditions prevail for each crack and that no shear stresses act on the crack faces. For infinite regions, Liouville's theorem shows that $s'(z)$ is a polynomial of finite degree, and, in fact, according to (4.4.24), it reduces to a constant because the remote stresses are finite. Then, only one complex potential function, $p(z)$, is needed. This was noticed by Westergaard (1939), whose formulation of the basic equations for infinite regions and mode I loading without shear stresses on the crack faces agrees with equations (4.4.23)-(4.4.25), after $s'(z)$ has been put equal to a constant. He actually introduced the further simplification of putting this constant equal to zero: as will be evident later on, this is equivalent to a superposition of a constant σ_x-stress. Such a stress does not influence the stress intensity factor for cracks with plane surfaces in the symmetry-plane, and it may always be subtracted after the calculation, if stresses and strains are required.

Equations (4.4.23)-(4.3.6) show that for $y = 0$,

$$(\sigma_y)_+ = 2\Re p'_+(x) = p'(x+i0) + \overline{p'(x+i0)} = p'_+(x+i0) + p'_-(x-i0)$$
$$= p'_+(x) + p'_-(x) = \sigma_y^0(x) \text{ for } b < x < c \quad (4.4.33)$$

where use was made of (4.4.30). For $x < b$ and $x > c$ equation (4.4.25) gives

$$2\mu i \frac{\partial v_+}{\partial x} = \frac{1}{1-k^2}[p'_+(x) - p'_-(x)] - \frac{k^2}{1-k^2}[s'_+(x) - s'_-(x)] = 0 \quad (4.4.34)$$

where the last term vanishes because $s'(z)$ is analytic on $y = 0$. Thus,

$$p'_+(x) - p'_-(x) = 0 \text{ for } x < b, \, x > c \quad (4.4.35)$$

The results may immediately be generalized to an array of collinear cracks on $y = 0$, say on the portions L of the x axis. Then,

$$\boxed{\begin{aligned} p'_+(x) + p'_-(x) &= \sigma_y^0(x) && \text{for } x \in L \\ p'_+(x) - p'_-(x) &= 0 && \text{for } x \notin L \end{aligned}} \quad (4.4.36)$$

Infinite regions
Cases without shear stresses on the crack faces. An infinite region with remote stresses $\sigma_x = \sigma_x^\infty$ and σ_y^∞ is now assumed. Equations (4.4.33) and (4.4.35) then constitute a Hilbert problem, the solution of which is

$$p'(z) = \frac{1}{2\pi i G(z)} \int_b^c \frac{\sigma_y^0(\xi) G_+(\xi)}{\xi - z} d\xi + \frac{P(z)}{G(z)} \quad (4.4.37)$$

4.4 COMPLEX POTENTIALS FOR IN-PLANE PROBLEMS

where $G(z)$ is given by (4.3.25) and $P(z)$ is a polynomial of finite degree, the coefficients of which, due to (4.4.30) are real. The boundary conditions at infinity give

$$4\Re[p'(z) - s'(z)]_{|z|\to\infty} = \sigma_y^\infty + \sigma_x^\infty \qquad (4.4.38)$$
$$4\Re[s'(z)]_{|z|\to\infty} = \sigma_y^\infty - \sigma_x^\infty \qquad (4.4.39)$$

These equations imply that

$$P(z) = \frac{\sigma_y^\infty}{2}z + a_0, \quad s'(z) = \frac{\sigma_y^\infty - \sigma_x^\infty}{4} \qquad (4.4.40)$$

where a_0 is to be determined from the condition that the crack is closed at both ends:

$$2\mu\left[v_+(c) - v_+(b)\right] = \left[\frac{2}{1-k^2}\Im p_+(x) - \frac{2k^2}{1-k^2}\Im s(x)\right]_b^c$$

$$= \frac{2}{1-k^2}\int_b^c \frac{\sigma_y^\infty x/2 + a_0}{\sqrt{(x-b)(c-x)}} dx = 0 \qquad (4.4.41)$$

This gives

$$a_0 = -\frac{b+c}{4}\sigma_y^\infty \qquad (4.4.42)$$

The solution of this mode I problem for an infinite body and no shear stresses on the crack faces is now complete. The normal stress and the displacement gradient on the symmetry plane outside the crack are found to be

$$\sigma_y = 2\Re p'(x) = \pm \frac{1}{\sqrt{(x-b)(x-c)}}\left\{\frac{1}{\pi}\int_b^c \frac{\sigma_y^0(\xi)\sqrt{(\xi-b)(c-\xi)}}{\xi - x}d\xi \right.$$
$$\left. + \sigma_y^\infty[x - (b+c)/2]\right\} \qquad (4.4.43)$$

$$\frac{\partial v_+}{\partial x} = -\frac{1}{2(1-k^2)\mu\sqrt{(x-b)(c-x)}}\left\{\frac{1}{\pi}\oint_b^c \frac{\sigma_y^0(\xi)\sqrt{(\xi-b)(c-\xi)}}{\xi - x}d\xi\right.$$
$$\left. + \sigma_y^\infty[x - (b+c)/2]\right\} \qquad (4.4.44)$$

with the upper sign for $x > c$, lower sign for $x < b$, and with the last integral taken in the sense of the Cauchy principal value. The stress intensity factor at $x = b$ is

$$K_I^{(b)} = \lim_{x\to b-0}\left[\sqrt{2\pi(b-x)}\sigma_y(x)\right] = \sigma_y^\infty\sqrt{\pi a} - \frac{1}{\sqrt{\pi a}}\int_b^c \sigma_y^0(\xi)\sqrt{\frac{c-\xi}{\xi-b}}d\xi \qquad (4.4.45)$$

where $a = (c-b)/2$ is the half-length of the crack. Note the similarity with the corresponding mode III expression, (4.3.38).

The expression for $K_I^{(b)}$ shows that the singularity at $x = b$ disappears if

$$\frac{1}{\pi a}\int_b^c \sigma_y^0(\xi)\sqrt{\frac{c-\xi}{\xi-b}}d\xi = \sigma_y^\infty \qquad (4.4.46)$$

which may be thus considered as a condition for stress continuity at $x = b$. At the same time it is a condition for smooth closing, $\partial v_+/\partial x = 0$ at $x = b$, as found from (4.4.44).

It is equivalent to the non-dimensional criterion (3.6.7), although not specialized to crack face loading at an edge vicinity.

The result (4.4.45) does not hold if the crack faces make any firm contact with each other. One illustrative example is given by $\sigma_y^0(x) = -p_0 x/d$, $p_0 > 0$, $d > 0$. Then,

$$K_I^{(b)} = \sqrt{\pi a}\left[\sigma_y^\infty + \frac{p_0(c+3b)}{4d}\right] \qquad (4.4.47)$$

The condition that $K_I^{(b)}$ must be non-negative gives

$$b \geq b_0 = -\frac{4\sigma_y^\infty d + p_0 c}{3p_0} \qquad (4.4.48)$$

If $b < b_0$, then the crack is partially closed along the portion $b < x < b_0$. The left stress intensity factor is then zero, implying smooth closing, and the right stress intensity factor equals

$$K_I^{(c)} = \sqrt{\frac{\pi}{2}(c-b_0)}\left[\sigma_y^\infty + \frac{p_0(3c+b_0)}{4d}\right] \qquad (4.4.49)$$

Semi-infinite crack. After the modification that $G(z) = z^{1/2}$, $G(z) = \sqrt{x}$ for $x > 0$, $y = 0$, the special case of a semi-infinite crack, $c = 0$, $b \to -\infty$, is treated in the same way, noting, however, that a finite remote stress σ_y^∞ cannot be allowed, because it would cause an infinite stress intensity factor. There may, however, be other loads than $\sigma_y^0(x)$, for instance loads acting on points outside the crack plane, but it is first assumed that the crack face load $\sigma_y^0(x)$ is the only load. Equation (4.4.37) then changes to

$$p'(z) = \frac{1}{2\pi z^{1/2}}\int_{-\infty}^0 \frac{\sqrt{-\xi}\sigma_y^0(\xi)}{\xi - z}d\xi + \frac{P(z)}{z^{1/2}} \qquad (4.4.50)$$

Assume that $\sigma_y^0(x)$ is predominantly negative, so that crack opening is ensured. Then, $P(z) = 0$, because otherwise stresses and strains would decrease in proportion to $|z|^{-1/2}$ or slower as $|z| \to \infty$, implying infinite stress-strain energy. This leads to

$$\sigma_y(x) = \frac{1}{\pi\sqrt{x}}\int_{-\infty}^0 \frac{\sqrt{-\xi}\sigma_y^0(\xi)}{\xi - x}d\xi \text{ for } x > 0 \qquad (4.4.51)$$

$$\frac{\partial v_+}{\partial x} = -\frac{1}{2\pi(1-k^2)\mu\sqrt{-x}}\oint_{-\infty}^0 \frac{\sqrt{-\xi}\sigma_y^0(\xi)}{\xi - x}d\xi \text{ for } x < 0 \qquad (4.4.52)$$

where the integral in the second expression is taken in the sense of the Cauchy principal value. The stress intensity factor is

$$K_I = -\sqrt{\frac{2}{\pi}}\int_{-\infty}^0 \frac{\sigma_y^0(\xi)}{\sqrt{-\xi}}d\xi \qquad (4.4.53)$$

Why cannot infinite stress-strain energy be tolerated in this case? Recall that infinite regions are models of bodies that in some respect are considered as large. Whether this idealization should allow infinite stress-strain energy or not, depends on the original real problem. A crack in a large plate, subjected to constant remote stresses, may be idealized to a crack in an infinite plate, subjected to the same remote stresses, thus leading to infinite stress-strain energy. On the other hand, the case just studied concerned crack face loading, only. In the original real case the loads are distributed along a portion of the crack faces, and they should remain the same even in the idealization to an infinite body. Such loads cannot produce infinite energy.

Boundary layer approach. Barenblatt region. The Barenblatt model of the process region provides an example in which the crack face load $\sigma_y^0(x)$ is non-negative, so that crack opening occurs as a result of some other load. From a physical point of view, the cohesive stress $\sigma_y^0(x)$ should perhaps be considered as a reaction to the other load rather than as an active load itself, but mathematically such a distinction is not necessary. Thus, the crack face traction is $\sigma_y = \sigma_y^0(x)$ on $-r_p < x \leq 0$, and it is zero on $x < -r_p$.

Consider a plate containing a crack that is much longer than r_p, and so are the ligaments from the crack edge to the plate boundaries. Mode I loads are acting on the plate boundaries, producing a stress field, that is approximately given by the relations

$$\sigma_r(R, \varphi) = \frac{K_I}{\sqrt{2\pi R}} f_r(\varphi), \qquad \tau_{r\varphi}(R, \varphi) = \frac{K_I}{\sqrt{2\pi R}} f_{r\varphi}(\varphi) \qquad (4.4.54)$$

at distances $r = R$ from the crack edge, that are much larger than r_p, but may still be much smaller than the plate dimensions, i.e. the typical small scale yielding situation. Thus, an inverse square root field prevails near the crack edge, except in the neighbourhood of the process region.

Here, the *boundary layer* approach is convenient, so called by analogy with the common practice in hydrodynamics of stretching a boundary region to infinity. The remote stress field may be – and generally is – quite different from the inverse square root field, but it is obvious that the field inside $r = R$ will prevail, approximately, even if the field (4.4.54) is stretched by the continuation $\sigma_r = K_I f_r(\varphi)/\sqrt{2\pi r}$, $\tau_{r\varphi}(\varphi)/\sqrt{2\pi r}$, $r \to \infty$, disregarding the fact that the real body geometry may be finite. The solution (4.4.50) applies, but here $P(z)$ is a finite constant, so that

$$p'(z) = \frac{1}{2\pi i G(z)} \int_{-r_p}^{0} \frac{\sigma_y^0(\xi) G_+(\xi)}{\xi - z} d\xi + \frac{P(0)}{G(z)}, \qquad G(z) = z^{1/2} \qquad (4.4.55)$$

$$\sigma_y(x) = \frac{1}{\pi\sqrt{x}} \left[\int_{-r_p}^{0} \frac{\sqrt{-\xi}\,\sigma_y^0(\xi)}{\xi - x} d\xi + 2\pi P(0) \right] \text{ for } x > 0 \qquad (4.4.56)$$

$$\frac{\partial v_+}{\partial x} = -\frac{1}{2\pi(1-k^2)\mu\sqrt{-x}} \left[\oint_{-r_p}^{0} \frac{\sqrt{-\xi}\,\sigma_y^0(\xi)}{\xi - x} d\xi + 2\pi P(0) \right] \text{ for } x < 0 \qquad (4.4.57)$$

Letting $x \to \infty$ gives $\sigma_y(x) = K_I/\sqrt{2\pi x}$, so that

$$P(0) = \frac{K_I}{2\sqrt{2\pi}} \qquad (4.4.58)$$

Insertion into (4.4.57) and integration gives

$$v_+ = \frac{K_I \sqrt{-x}}{\sqrt{2\pi}(1-k^2)\mu} - \frac{1}{2\pi(1-k^2)\mu} \oint_{-r_p}^{0} \sigma_y^0(\xi) \ln \frac{\sqrt{-x} + \sqrt{-\xi}}{|\sqrt{-x} - \sqrt{-\xi}|} d\xi \qquad (4.4.59)$$

This result was previously obtained in non-dimensional form, (3.6.12), by another method.

The integral in (4.4.56) may be evaluated for $x \to +0$ by decomposing $\sigma_y^0(x)$ into two parts, one, $\sigma_y^0(x) - \sigma_y^0(0)$, that is zero for $x = 0$, and one, $\sigma_y^0(0)$, for which exact integration can be performed. The result is

$$\sigma_y(x) \to -\frac{1}{\pi\sqrt{x}} \int_{-r_p}^0 \frac{\sigma_y^0(\xi)}{\sqrt{-\xi}} d\xi + \sigma_y^0(0) + \frac{K_I}{\sqrt{2\pi x}} \quad \text{as } x \to +0 \quad (4.4.60)$$

Then, the condition for stress continuity (and thereby also smooth closing, because displacement gradients are continuous if the stresses are) gives

$$K_I = \sqrt{\frac{2}{\pi}} \int_{-r_p}^0 \frac{\sigma_y^0(\xi)}{\sqrt{-\xi}} d\xi \quad (4.4.61)$$

in agreement with the non-dimensional criterion (3.6.7). Insertion into (4.4.57) gives

$$\frac{\partial v_+}{\partial x} = -\frac{1}{2\pi(1-k^2)\mu\sqrt{-x}} \left[\oint_{-r_p}^0 \frac{\sqrt{-\xi}\sigma_y^0(\xi)}{\xi - x} d\xi + \sqrt{\frac{\pi}{2}} K_I \right]$$

$$= -\frac{\sqrt{-x}}{2\pi(1-k^2)\mu} \oint_{-r_p}^0 \frac{\sigma_y^0(\xi)}{\sqrt{-\xi}(\xi - x)} d\xi \quad \text{for } x < 0 \quad (4.4.62)$$

For later reference, relation (4.4.55) will be given after insertion of the value of $P(0)$:

$$p'(z) = -\frac{z^{1/2}}{2\pi} \int_{-r_p}^0 \frac{\sigma_y^0(\xi)}{\sqrt{-\xi}(\xi - z)} d\xi \quad (4.4.63)$$

In the so-called *modified boundary layer approach*, consideration is also taken of the T-stress. This may be achieved by adding $T\cos^2\varphi$ and $-T\cos\varphi\sin\varphi$ to the remote stresses σ_r and $\tau_{r\varphi}$, respectively, and then the complex potential

$$s'(z) = -\frac{T}{4} \quad (4.4.64)$$

However, the linear case is trivial, because the T-stress may alternatively be added as $\sigma_x = T$ to the solution for the remote inverse square root stresses.

Mode I shear stresses on the crack faces. Mode I shear stresses, $\pm\tau_{xy}^0(x)$ on the crack faces $b < x < c$, $y = \pm 0$ were temporarily left out from consideration. Note that such stresses, in contrast to mode II shear stresses, are of opposite sign on the two crack faces. They may appear when a wedge is forced toward a crack edge. Because linear superposition is possible, their contribution to the potentials $p'(z)$ and $s'(z)$ can be determined separately. The treatment is straightforward, and only the resulting contribution to the stress intensity factor is given:

$$K_I = \pm\frac{k^2}{2\sqrt{\pi a}} \int_b^c \tau_{xy}^0(x) dx \quad (4.4.65)$$

Upper sign refers to $z = b$ and lower sign to $z = c$. This stress intensity factor is to be added to the stress intensity factor given by (4.4.45). For superpositions like this it might happen that some of the stress intensity factors involved is negative: obviously this fact *per se* is not disqualifying, as long as the resulting mode I stress intensity factor is non-negative. However, it should also be investigated whether the crack opening is non-negative everywhere. Negative stress intensity factors and negative crack opening displacements may appear if the boundary value problem is posed for only half the body, after a non-physical boundary, $y = 0$, $x < b$, $x > c$ has been introduced.

Finite regions

For a body of finite outer dimensions, symmetric with respect to the x axis and containing a traction free crack, $y = 0$, $b < x < c$, subjected to mode I loading, the functions $p'(z)$ and $s'(z)$ may be written as

$$p'(z) = \frac{P(z)}{(z-b)^{1/2}(z-c)^{1/2}} + P'_0(z) \qquad (4.4.66)$$

$$s'(z) = S'(z) \qquad (4.4.67)$$

where $P(z)$, $P_0(z)$ and $S(z)$ are analytic in the whole region inside the outer contour of the body. Equations (4.4.30) imply that

$$P(\bar{z}) = \overline{P(z)}, \quad P'_0(\bar{z}) = \overline{P'_0(z)}, \quad S(\bar{z}) = \overline{S(z)} \qquad (4.4.68)$$

Then, from (4.4.23)-(4.4.24) and the condition of traction free crack faces, it follows that

$$\sigma_y = 2\Re p'(z) = P'_0(z) + \overline{P'_0(z)} = 2P'_0(x) = 0 \text{ for } y = 0, b < x < c \qquad (4.4.69)$$

But if an analytic function vanishes along a line segment, it vanishes in the whole region of analyticity. Thus,

$$P_0(z) = constant = C \qquad (4.4.70)$$

where, because of (4.4.30), C is a real constant. Thus:

$$p'(z) = \frac{P(z)}{(z-b)^{1/2}(z-c)^{1/2}} \qquad (4.4.71)$$

$$s'(z) = S'(z), \quad P(\bar{z}) = \overline{P(z)}, \quad S(\bar{z}) = \overline{S(z)} \qquad (4.4.72)$$

$P(z)$ and $S'(z)$ are to be determined so that the conditions at the outer boundary of the body are satisfied. However, one further condition applies to the analytic function $P(z)$, viz. that the crack closes at both edges, which it does, according to (4.4.25), if

$$\Im p(b) = \Im p(c) \qquad (4.4.73)$$

For the case of an edge crack, $y = 0$, $b \leq x < c$, where $x = b$, $y = 0$, is a point on the outer boundary of the body, equation (4.4.66) should be modified by changing the factor $(z-b)^{1/2}$ to, for instance, $(z+c-2b)^{1/2}$.

Westergaard method for mode II

General relations

For body geometry which is symmetric and loading geometry which is anti-symmetric with respect to the plane $y = 0$, i.e. mode II loading, the relations

$$p_-(\bar{z}) = -\overline{p_+(z)}, \quad s_-(\bar{z}) = -\overline{s_+(z)} \qquad (4.4.74)$$

are obtained, instead of (4.4.30). Suppose that the body contains one single crack, $y = 0$, $b < x < c$. Then, mode II conditions could include prescribed tractions $(\tau_{xy})_+ = (\tau_{xy})_- = \tau^0_{xy}(x)$ and $(\sigma_y)_+ = -(\sigma_y)_- = \sigma^0_y(x)$ on the crack faces, or, alternatively, prescribed tractions $(\sigma_y)_+ = -(\sigma_y)_- = \sigma^0_y(x)$ and prescribed displacements $u_+ = -u_- = u_0(x)$ on the crack faces. A third possibility consists of prescribed tangential displacements $u_+ = -u_- = u_0(x)$ together with normal displacements $v_+ = v_- = v_0(x)$ on the crack faces. Moreover, as in the mode III case, slip along $y = 0$, generated by previous mode II sliding events, may have been deposited, i.e. displacements $u_+ = -u_- = u_0(x)$ may exist on $y = 0$, $x < b$, $x > c$.

As in the treatment of the mode I case, prescribed displacements on the crack faces will be left out of the discussion. Consider first the case $\sigma_y^0(x) = 0$. Then, the conditions on $y = 0$ are:

$$(\sigma_y)_- = (\sigma_y)_+ = 0 \text{ for all } x \tag{4.4.75}$$

$$\tau_{xy} = \tau_{xy}^0(x) \text{ for } b < x < c \tag{4.4.76}$$

$$u_+ = u_0(x) \text{ for } x < b, x > c \tag{4.4.77}$$

According to equations (4.4.23) and (4.4.24),

$$\sigma_y + i\tau_{xy} = 2\Re p'(z) + 2i\Im s'(z) \text{ for } y = 0 \tag{4.4.78}$$

so that condition (4.4.75) gives

$$p'_+(x) + \overline{p'_+(x)} = p'_+(x) - p'_-(\bar{x}) = p'_+(x) - p'_-(x) = 0 \tag{4.4.79}$$

Thus, $p'(z)$ is analytic in the whole region inside the outer boundary. This result would have been found also for an array of collinear cracks, without normal tractions on the crack faces.

Condition (4.4.77) together with equation (4.4.78) implies:

$$s'_+(x) - \overline{s'_+(x)} = s'_+(x) + s'_-(x) = i\tau_{xy}^0(x) \text{ for } b < x < c \tag{4.4.80}$$

and condition (4.4.77), with use of (4.4.25), leads to:

$$\frac{k^2}{1-k^2}[p'_+(x) - p'_-(x)] - \frac{1}{1-k^2}[s'_+(x) - s'_-(x)] = 2\mu u'_0(x) \tag{4.4.81}$$

for $x < b$, $x > c$. In view of (4.4.79), this reduces to:

$$s'_+(x) - s'_-(x) = -2(1-k^2)\mu u'_0(x) \text{ for } x < b, x > c \tag{4.4.82}$$

The results may immediately be generalized to an array of collinear cracks on $y = 0$, say on the portions L of the x axis. Then,

$$\boxed{\begin{aligned} s'_+(x) + s'_-(x) &= i\tau_{xy}^0(x) & \text{for } x \in L \\ s'_+(x) - s'_-(x) &= -2(1-k^2)\mu u'_0(x) & \text{for } x \notin L \end{aligned}} \tag{4.4.83}$$

Infinite regions
First, specialization to an infinite region will be made. Then, $p'(z)$ is a constant, which may be put equal to zero, because it corresponds to a rigid body motion. Thus, only one complex potential is needed, as in the corresponding mode I case, although it is now $s(x)$ rather than $p(x)$.

Equations (4.4.80) and (4.4.82) constitute a Hilbert problem, the solution of which is

$$s'(z) = \frac{1}{2\pi i G(z)} \left[-2(1-k^2)\mu \int_{-\infty}^{b} \frac{u_0'(\xi) G_+(\xi)}{\xi - z} d\xi + i \int_{b}^{c} \frac{\tau_{xy}^0(\xi) G_+(\xi)}{\xi - z} d\xi \right.$$
$$\left. - 2(1-k^2)\mu \int_{c}^{\infty} \frac{u_0'(\xi) G_+(\xi)}{\xi - z} d\xi \right] + \frac{S(z)}{G(z)} \quad (4.4.84)$$

where $G(z) = (z-b)^{1/2}(z-c)^{1/2}$ as given by (4.3.25), and $S(z)$ is a polynomial of finite degree, the coefficients of which, due to (4.4.74), are imaginary. If the remote stress is $\tau_{xy} = \tau_{xy}^\infty$, then

$$2\Im \frac{S(z)}{G(z)} \to \tau_{xy}^\infty \quad \text{as } z \to \infty \quad (4.4.85)$$

which implies

$$S(z) = i\frac{\tau_{xy}^\infty}{2} z + ic_0 \quad (4.4.86)$$

where c_0 is a real constant to be determined from the known amount of slip difference between the ends of the slipping region (the crack). If the crack faces are traction free and the deposited slip is constant (or zero), both in front of and behind the crack, i.e. $u_0' = 0$, then

$$2\mu[u_+(c) - u_+(b)] = \left[-\frac{2}{1-k^2} \Re s_+(x) \right]_b^c$$
$$= -\frac{2}{1-k^2} \int_b^c \frac{\tau_{xy}^\infty x/2 + c_0}{\sqrt{(x-b)(c-x)}} dx = 2\mu[u_0(c) - u_0(b)] \quad (4.4.87)$$

which gives

$$c_0 = -\frac{(1-k^2)\mu}{\pi}[u_0(c) - u_0(b)] - \tau_{xy}^\infty \frac{b+c}{4} \quad (4.4.88)$$

and the integration constant available during integration of $s'(z)$ may, for instance, be used to fix u_+ at $z = -\infty$.

The solution of this mode II problem for an infinite body and no normal tractions on the crack faces is now complete. The stress intensity factor at $x = b$ is found to be:

$$K_{II} = \lim_{x \to b-0} \left[\sqrt{2\pi(b-x)} \tau_{xy}(x) \right] = \lim_{x \to b-0} \left[\sqrt{2\pi(b-x)} \, 2\Im s'(x) \right]$$
$$= -\frac{1}{\sqrt{\pi a}} \left[(1-k^2)\mu \int_{-\infty}^{b} u_0'(\xi) \sqrt{\frac{c-\xi}{b-\xi}} d\xi \right.$$
$$\left. + \int_b^c \tau_{xy}^0(\xi) \sqrt{\frac{c-\xi}{\xi-b}} d\xi + (1-k^2)\mu \int_c^\infty u_0'(\xi) \sqrt{\frac{\xi-c}{\xi-b}} d\xi \right]$$
$$+ \tau_{xy}^\infty \sqrt{\pi a} + \frac{(1-k^2)\mu}{\sqrt{\pi a}}[u_+(c) - u_+(b)] \quad (4.4.89)$$

where $a = (c-b)/2$ is the half-length of the crack. Note the similarity with the corresponding modes I and III expressions, (4.4.45) and (4.3.38).

Mode II normal tractions on the crack faces were temporarily left out from consideration. Note that such tractions are different from mode I tractions, because they are tensile on one crack face and compressive on the other, due to the mode II symmetry properties. The treatment is straightforward, but, because these kinds of traction are hardly ever realized, only the resulting contribution to the stress intensity factors at the crack edges is given:

$$K_{II} = \mp \frac{k^2}{2\sqrt{\pi a}} \int_b^c \sigma_y^0(x) dx \tag{4.4.90}$$

with upper sign for $z = b$, lower sign for $z = c$.

Semi-infinite cracks and incorporation of a Barenblatt process region model are treated in analogy with the corresponding mode I cases, pages 118ff.

Constant deposited slip behind the crack

By analogy with the discussion about mode III, the special case $u_0'(x) = 0$ on $x < -a$, $x > a$ and $\tau_{xy}^0(x) = 0$ on $-a < x < a$, will be studied, assuming that $u_+(a) = 0$. This case may have been generated from the propagation of a slipping region of constant length from $x = -\infty$ to its present location Then, with

$$u_+(-a) - u_-(-a) = 2\Delta \tag{4.4.91}$$

equation (4.4.89) gives

$$K_{II} = \tau_{xy}^\infty \sqrt{\pi a} \pm \frac{2(1-k^2)\mu}{\sqrt{\pi a}} \Delta \tag{4.4.92}$$

with upper sign for the edge $x = a$, lower sign for $x = -a$.

It is now assumed that no energy is associated with the healing process (energy-neutral healing). This implies that the stress intensity factor at the trailing end is zero. Then,

$$\Delta = \frac{\pi \tau_{xy}^\infty a}{2(1-k^2)\mu} \tag{4.4.93}$$

and $K_{II} = 2\tau_{xy}^\infty \sqrt{\pi a}$ for the edge $z = a$. In fact, use of the relation $K_{II} = 0$ for $z = -a$ as a criterion to characterize this special case, enables immediate determination of c_0 and then of τ_{xy} and $\partial u/\partial x$. From (4.4.87)-(4.4.88) it follows that

$$c_0 = \frac{(1-k^2)\mu}{\pi} u_+(-a) = \frac{\tau_{xy}^\infty a}{2} \tag{4.4.94}$$

and then, for $y = 0$, from (4.4.24)-(4.4.25):

$$\tau_{xy} = 2\Im s'(x) = \frac{\tau_{xy}^\infty x + 2c_0}{\sqrt{x^2 - a^2}} = \tau_{xy}^\infty \sqrt{\frac{x+a}{x-a}} \text{ for } |x| > a \tag{4.4.95}$$

$$\frac{\partial u_+}{\partial x} = -\frac{1}{(1-k^2)\mu} \Re s'(x) = -\frac{\tau_{xy}^\infty x + c_0}{2(1-k^2)\mu \sqrt{a^2 - x^2}}$$

$$= -\frac{\tau_{xy}^\infty}{2(1-k^2)\mu} \sqrt{\frac{a+x}{a-x}} \text{ for } |x| < a \tag{4.4.96}$$

4.4 COMPLEX POTENTIALS FOR IN-PLANE PROBLEMS

Note that this is also equivalent to smooth closing at $z = -a$, so that, for $0 < r = x + a \ll a$,

$$u_+(x) \approx \Delta - \frac{\tau_{xy}^\infty a}{3\sqrt{2}(1-k^2)\mu} \cdot \left(\frac{r}{a}\right)^{3/2} = \frac{\pi \tau_{xy}^\infty a}{2(1-k^2)\mu}\left[1 - \frac{\sqrt{2}}{3\pi}\left(\frac{r}{a}\right)^{3/2}\right] \qquad (4.4.97)$$

and consequently, τ_{xy} is continuous at $z = -a$.

Finite regions

For a body of finite outer dimensions, symmetric with respect to the x axis and containing a traction free crack, $y = 0$, $b < x < c$, subjected to mode II loading, the functions $p'(z)$ and $s'(z)$ may be written as

$$p'(z) = P'(z), \quad s'(z) = \frac{S(z)}{(z-b)^{1/2}(z-c)^{1/2}} + S_0'(z) \qquad (4.4.98)$$

where $P'(z)$, $S(z)$ and $S_0'(z)$ are analytic in the whole region inside the outer contour of the body. Equations (4.4.74) imply that

$$P'(\bar{z}) = -\overline{P'(z)}, \quad S(\bar{z}) = \overline{S(z)}, \quad S_0'(\bar{z}) = \overline{S_0'(z)} \qquad (4.4.99)$$

Then, from (4.4.23)-(4.4.24) and the condition of traction free crack faces, it follows that

$$\tau_{xy} = 2\Im s'(z) = S_0'(z) - \overline{S_0'(z)} = 2S_0'(x) = 0 \text{ for } y = 0, \ b < x < c \qquad (4.4.100)$$

But if an analytic function vanishes along a line segment, it vanishes in the whole region of analyticity. Thus,

$$S_0(z) = constant = C \qquad (4.4.101)$$

where, because of the second of equations (4.4.74), C is an imaginary constant. Thus:

$$p'(z) = P'(z), \quad s'(z) = \frac{S(z)}{(z-b)^{1/2}(z-c)^{1/2}} \qquad (4.4.102)$$

$$P'(\bar{z}) = -\overline{P'(z)}, \quad S(\bar{z}) = -\overline{S(z)} \qquad (4.4.103)$$

$P'(z)$ and $S(z)$ are to be determined so that the conditions at the outer boundary of the body are satisfied. However, one further condition applies to the analytic function $S(z)$, viz. that the crack closes at both edges, which it does, according to (4.4.25), if

$$\Re s(b) = \Re s(c) \qquad (4.4.104)$$

For the case of an edge crack, $y = 0$, $b \leq x < c$, where $z = b$ is a point on the outer boundary of the body, the second of equations (4.4.102) should be modified by changing the factor $(z-b)^{1/2}$ to, for instance, $(z+c-2b)^{1/2}$.

The basic equations for in-plane loading in polar coordinates

The use of polar coordinates in connection with complex potentials was briefly discussed for mode III; see (4.3.11)-(4.3.13). For in-plane loading,

$$\sigma_x = \sigma_r \cos^2\varphi + \sigma_\varphi \sin^2\varphi - 2\tau_{r\varphi}\cos\varphi\sin\varphi \qquad (4.4.105)$$

$$\sigma_y = \sigma_r \sin^2\varphi + \sigma_\varphi \cos^2\varphi + 2\tau_{r\varphi}\cos\varphi\sin\varphi \qquad (4.4.106)$$

$$\tau_{xy} = (\sigma_\varphi - \sigma_r)\cos\varphi\sin\varphi + \tau_{r\varphi}(\cos^2\varphi - \sin^2\varphi) \qquad (4.4.107)$$

$$u = u_r \cos\varphi - u_\varphi \sin\varphi \qquad (4.4.108)$$
$$v = u_r \sin\varphi + u_\varphi \cos\varphi \qquad (4.4.109)$$

giving

$$\sigma_y + \sigma_x = \sigma_\varphi + \sigma_r \qquad (4.4.110)$$
$$\sigma_y - \sigma_x + 2i\tau_{xy} = [\sigma_\varphi - \sigma_x + 2i\tau_{r\varphi}]e^{-2i\varphi} \qquad (4.4.111)$$
$$u + iv = [u_r + iu_\varphi]e^{i\varphi} \qquad (4.4.112)$$

Insertion into equations (4.4.9)-(4.4.10) and (4.4.12) yields

$$\boxed{\sigma_\varphi + \sigma_r = 4\Re[f'(z)]} \qquad (4.4.113)$$

$$\boxed{\sigma_\varphi - \sigma_r + 2i\tau_{r\varphi} = 2[\bar{z}f''(z) + g''(z)]e^{2i\varphi}} \qquad (4.4.114)$$

$$\boxed{2\mu(u_r + iu_\varphi) = \left[\frac{1+k^2}{1-k^2}f(z) - z\overline{f'(z)} - \overline{g'(z)}\right]e^{-i\varphi}} \qquad (4.4.115)$$

where $z = re^{i\varphi}$. Insertion into the Westergaard formulations (4.4.23)-(4.4.25) results in

$$\boxed{\sigma_\varphi + \sigma_r = 4\Re[p'(z) - s'(z)]} \qquad (4.4.116)$$

$$\boxed{\sigma_\varphi - \sigma_r + 2i\tau_{r\varphi} = \{4s'(z) - 2(z-\bar{z})[p''(z) - s''(z)]\}e^{2i\varphi}} \qquad (4.4.117)$$

$$2\mu(u_r + iu_\varphi) = \left\{\frac{1+k^2}{1-k^2}[p(z) - s(z)]\right.$$
$$\left. - \overline{p(z)} - \overline{s(z)} - (z-\bar{z})\left[\overline{p'(z)} - \overline{s'(z)}\right]\right\}e^{-i\varphi} \qquad (4.4.118)$$

The specialized equations (4.4.66)-(4.4.68) and (4.4.102)-(4.4.104) for traction free mode I and mode II cracks remain valid.

4.5 Conformal mapping

General considerations

The region inside the body contour in the z-plane may be mapped onto a region in the ζ-plane, where $\zeta = \xi + i\eta$. If the mapping function

$$z = \omega(\zeta) \qquad (4.5.1)$$

is analytic, the mapping is conformal, i.e. a small square is mapped as another small square. Suppose that one corner of the square is $\zeta = \zeta_0$ and a neighbouring corner is $\zeta = \zeta_0 + d\zeta$ in the ζ-plane, then these corners are $z = \omega(\zeta_0)$ and $z = \omega(\zeta_0) + \omega'(\zeta_0)d\zeta = \omega(\zeta_0) + |\omega'(\zeta_0)|\exp[i \arg \omega'(\zeta_0)]d\zeta$ in the z-plane. Thus, the sides of the square are enlarged by a factor $|\omega'(\zeta_0)|$ and rotated through an angle $\arg \omega'(\zeta_0)$ when going from the ζ- to the z-plane. Note that the mapping is conformal as long as

$$\omega'(\zeta_0) \neq \infty, \qquad \omega'(\zeta_0) \neq 0 \tag{4.5.2}$$

Mapping from the z-plane to the ζ-plane results in particularly simple expressions for mode III. With the notation $f(z) = F(\zeta)$, equations (4.3.5), (4.3.6) and (4.3.10) take the forms

$$\mu w = \Im F(\zeta), \qquad \tau_{yz} + i\tau_{xz} = \frac{F'(\zeta)}{\omega'(\zeta)}, \qquad \Re F(\zeta) = -\int_0^s T \, ds \tag{4.5.3}$$

where $s = s(x,y) = s[x(\xi,\eta), y(\xi,\eta)]$.

For modes I and II, the expressions become less simple. With the notations

$$p(z) = \Omega(\zeta), \qquad s(z) = \Sigma(\zeta) \tag{4.5.4}$$

$$p'(z) = \frac{\Omega'(\zeta)}{\omega'(\zeta)} = \Phi(\zeta), \qquad s'(z) = \frac{\Sigma'(\zeta)}{\omega'(\zeta)} = \Psi(\zeta) \tag{4.5.5}$$

equations (4.4.23)-(4.4.26) take the forms

$$\sigma_x + \sigma_y = 4\Re[\Phi(\zeta) - \Psi(\zeta)] \tag{4.5.6}$$

$$\sigma_y - \sigma_x + 2i\tau_{xy} = 4\Psi(\zeta) - 2\frac{\omega(\zeta) - \overline{\omega(\zeta)}}{\omega'(\zeta)}[\overline{\Phi'(\zeta)} - \overline{\Psi'(\zeta)}] \tag{4.5.7}$$

$$2\mu(u + iv) = \frac{1+k^2}{1-k^2}[\Omega(\zeta) - \Sigma(\zeta)] - \overline{\Omega(\zeta)} - \overline{\Sigma(\zeta)}$$
$$- \frac{\omega(\zeta) - \overline{\omega(\zeta)}}{\overline{\omega'(\zeta)}}[\overline{\Omega'(\zeta)} - \overline{\Sigma'(\zeta)}] \tag{4.5.8}$$

$$\Omega(\zeta) - \Sigma(\zeta) + \overline{\Omega(\zeta)} + \overline{\Sigma(\zeta)} + \frac{\omega(\zeta) - \overline{\omega(\zeta)}}{\overline{\omega'(\zeta)}}[\overline{\Omega'(\zeta)} - \overline{\Sigma'(\zeta)}] = \int_0^s (iX - Y) \, ds \tag{4.5.9}$$

where $s = s(x,y) = s[x(\xi,\eta), y(\xi,\eta)]$.

Mapping of a region onto a circle or a half-plane is often desirable, because it generally leads to simple expressions of the boundary conditions. Mapping onto a circle enables use of analytic potentials expressed by power series; this is not always possible for other regions.

The literature offers a multitude of examples of both analytical and numerical solutions to crack problems, using complex potentials and conformal mapping. As a demonstration of the technique for mode III, a problem involving an infinite strip with a semi-infinite crack (Field and Baker 1962) will be studied, and to demonstrate the technique for mode I or II problems, one simple and classical example, the elliptic hole problem (see e.g. Muskhelishvili 1953a) will be discussed.

Mode III crack in a long strip

Consider a long strip with its long edges subjected to fixed grip and with a crack extending along the symmetry plane from one end of the strip. For analysis, the strip

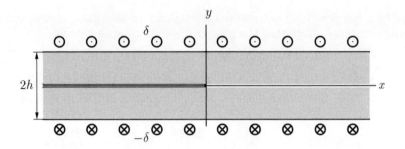

Fig. 4.5.1 Semi-infinite crack in an infinite strip. The strip edges are subjected to fixed grip, displaced perpendicularly to the strip plane the distance 2δ from each other. Thus, the upper edge is moved the distance δ toward the viewer and the lower edge the same distance from the viewer.

is assumed to be infinitely long and the crack semi-infinite; see Fig. 4.5.1. The strip thickness is assumed to be sufficiently large in comparison with its height that mode III conditions may be assumed. The upper and lower strip edges are displaced a distance 2δ from each other, in a direction perpendicular to the strip plane.

Consider now the upper half of the strip, $0 \leq y \leq h$. The boundary conditions are

$$\tau_{yz} = \mu \frac{\partial w}{\partial y} = 0 \text{ for } x < 0,\, y = 0 \tag{4.5.10}$$

$$w = 0 \text{ for } x > 0,\, y = 0, \quad w = \delta \text{ for } y = h \tag{4.5.11}$$

The two last conditions are temporarily replaced by conditions for $\partial w/\partial x$. Then, using the representation $\mu w = \Im f(z)$, the boundary conditions (4.3.5)-(4.3.6) may be written as

$$f'(x) + \overline{f'(x)} = 0 \text{ for } x < 0,\, y = 0 \tag{4.5.12}$$
$$f'(x) - \overline{f'(x)} = 0 \text{ for } x > 0,\, y = 0 \tag{4.5.13}$$
$$f'(x) - \overline{f'(x)} = 0 \text{ for } y = h \tag{4.5.14}$$

Consider now $f(z)$ as a sectionally analytic function, denoted by $f_+(z)$ for $\Im(z) > 0$ and defined as

$$f_-(z) = \overline{f_+(z)} \text{ for } \Im(z) < 0 \tag{4.5.15}$$

This results in the Hilbert problem

$$f'_+(x) + f'_-(x) = 0 \text{ for } x < 0 \tag{4.5.16}$$
$$f'_+(x) - f'_-(x) = 0 \text{ for } x > 0 \tag{4.5.17}$$
$$f'_+(x + ih) - f'_-(x - ih) = 0 \text{ for all } x \tag{4.5.18}$$

Because infinite regions are easier to handle than strips, the upper half of the strip is mapped onto the the upper half of the ζ-plane, where $\zeta = \xi + i\eta$. The mapping function is

$$z = \omega(\zeta) = -\frac{h}{\pi} \ln(1 - \zeta), \quad \Im(\zeta) \geq 0 \tag{4.5.19}$$

This mapping is shown in Fig. 4.5.2.

4.5 CONFORMAL MAPPING

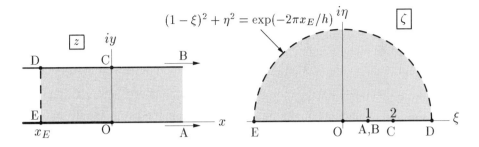

Fig. 4.5.2 Mapping of the strip $0 \leq \Im(z) \leq h$ onto $0 \leq \Im(\zeta)$.

Let $F(\zeta)$ be a sectionally analytic function in the ζ-plane, such that $F_+(\zeta) = f'_+(z)$ and $F_-(\zeta) = f'_-(z)$. Then, the Hilbert problem becomes formulated in the ζ-plane as

$$F_+(\xi) + F_-(\xi) = 0 \text{ for } \xi < 0, \, \eta = 0 \qquad (4.5.20)$$
$$F_+(\xi) - F_-(\xi) = 0 \text{ for } \xi > 0, \, \eta = 0 \qquad (4.5.21)$$

with the solution

$$F(\zeta) = \frac{P(\zeta)}{\zeta^{1/2}} \qquad (4.5.22)$$

where $P(\zeta)$ is a finite polynomial, and the square root is defined by a branch cut along the negative real axis, with the branch chosen so that the square root is real and positive for $\xi > 0$, $\eta = 0$. Thus, for instance,

$$F_+(\xi) - F_-(\xi) = -\frac{2iP(\xi)}{\sqrt{-\xi}} \text{ for } \xi < 0 \qquad (4.5.23)$$

Because $\zeta = 1 - \exp(-\pi z/h)$, the function

$$f'(z) = \frac{P(\zeta)}{\left[1 - \exp\left(\frac{-\pi z}{h}\right)\right]^{1/2}} \qquad (4.5.24)$$

Here, the branch cut in the z-plane is along the negative real axis, and the square root is real and positive along the positive real axis. It is obvious that $P(\zeta)$ is of degree zero, because the stresses are bounded as $x \to \infty$. This constant is to be determined from the condition $w(x, 0) \to \delta$ as $x \to -\infty$, which is equivalent with the original condition $w(x, h) = \delta$. Thus, cf. (4.5.23), δ is determined by integration of the expression

$$\mu \frac{\partial w}{\partial x} = \frac{f'_+(x) - f'_-(x)}{2i} = -\frac{P(0)}{\sqrt{\exp\left(\frac{-\pi x}{h}\right) - 1}} \text{ for } x < 0 \qquad (4.5.25)$$

This leads to

$$\mu\delta = \mu w(-\infty, 0) = -P(0) \int_0^{-\infty} \frac{dx}{\sqrt{\exp\left(\frac{-\pi x}{h}\right) - 1}}$$

$$= \frac{2P(0)h}{\pi} \left[\operatorname{atan}\sqrt{\exp\left(\frac{-\pi x}{h}\right) - 1} \right]_0^{-\infty} = P(0)h \quad (4.5.26)$$

which determines $P(0)$ to $\mu\delta/h$, so that

$$\tau_{yz} + i\tau_{xz} = f'(z) = \frac{\mu\delta}{h\left[1 - \exp\left(\frac{-\pi z}{h}\right)\right]^{1/2}} \quad (4.5.27)$$

Thus, for instance, the stress on the symmetry plane and the upper crack face displacement are

$$\tau_{yz} = \frac{\mu\delta}{h\sqrt{1 - \exp\left(\frac{-\pi x}{h}\right)}} \quad \text{for } x > 0 \quad (4.5.28)$$

$$w_+ = \frac{2P(0)h}{\pi} \operatorname{atan}\sqrt{\exp\left(\frac{-\pi x}{h}\right) - 1} \quad \text{for } x < 0 \quad (4.5.29)$$

and the stress intensity factor is

$$K_{III} = \frac{\sqrt{2}\mu\delta}{\sqrt{h}} \quad (4.5.30)$$

The elliptic hole approach

A sharp crack may be considered as the limiting configuration of elliptic holes, when the shorter half-axis approaches zero. The mapping function

$$\omega(\zeta) = A\left(\zeta + \frac{c}{\zeta}\right) \quad (4.5.31)$$

where

$$A = \frac{a+b}{2}, \quad c = \frac{a+b}{a-b} \quad (4.5.32)$$

maps the region outside an elliptic hole with its centre at $x = y = 0$ and half-axes a (the longer one, along the x axis) and b onto either the interior of the unit circle, by choosing $|\zeta| < 1$, or the exterior, by choosing $|\zeta| > 1$. The hole boundary is mapped onto the circle periphery $|\zeta| = 1$.

By choosing $|\zeta| \leq 1$, $\zeta \neq \pm 1$, $b = 0$, i.e. $A = a/2$ and $c = 1$, the region outside the crack $-a < x < a$, $y = 0$ is mapped onto the interior of the unit circle and the crack faces are mapped onto $|\zeta| = 1$, $\zeta \neq \pm 1$. The mapping is not conformal at the points $\zeta = \pm 1$, which correspond to the crack edges, and not at $\zeta = 0$, which corresponds to

4.5 CONFORMAL MAPPING

infinity in the z-plane. The choice of mapping function implies that

$$w(\zeta) = \frac{a}{2}\left(\zeta + \frac{1}{\zeta}\right), \quad |\zeta| \leq 1 \tag{4.5.33}$$

$$w'(\zeta) = \frac{a}{2}\left(1 - \frac{1}{\zeta^2}\right), \quad |\zeta| \leq 1, \quad \zeta \neq \pm 1, \quad \zeta \neq 0 \tag{4.5.34}$$

Then, from $w(\zeta) = z$,

$$\zeta = \frac{z - (z^2 - a^2)^{1/2}}{a} \tag{4.5.35}$$

where $(z^2 - a^2)^{1/2}$ is defined by a branch cut along $-a < x < a$, $y = 0$, with the branch chosen so that $(z^2 - a^2)^{1/2} = \sqrt{x^2 - a^2}$ for $x > a$, $y = 0$ to satisfy the condition $|\zeta| \leq 1$.

Assume that the body is subjected to remote loading, given by

$$\sigma_y \to \sigma_y^\infty, \quad \sigma_x \to 0, \quad \tau_{xy} \to 0 \text{ as } |\zeta| \to \infty \tag{4.5.36}$$

and that the crack faces are traction free. It is advantageous to superimpose a homogeneous stress state that cancels the stresses at infinity. Then, the crack faces become loaded with a negative normal stress

$$\sigma_y = -\sigma_y^\infty \text{ for } y = \pm 0 \tag{4.5.37}$$

Equations (4.5.6)-(4.5.7) will now be used, and, because $p(z)$ and $s(z)$ are analytic outside the crack, $\Omega(\zeta)$ and $\Sigma(\zeta)$ are analytic inside the unit circle, with the possible exception $\zeta = 0$, because $w(\zeta)$ is singular at this point. Assuming that they are analytic even for $\zeta = 0$, they can be expressed by the power series

$$\Omega(\zeta) = \sum_{n=1}^{\infty} a_n \zeta^n, \quad \Sigma(\zeta) = \sum_{n=1}^{\infty} b_n \zeta^n \tag{4.5.38}$$

According to (4.5.5), this gives

$$\Phi(\zeta) = \frac{2}{a} \cdot \frac{\sum_{n=1}^{\infty} n a_n \zeta^{n-1}}{1 - 1/\zeta^2} = \frac{2}{a} \cdot \frac{\sum_{n=1}^{\infty} n a_n \zeta^{n+1}}{\zeta^2 - 1} \tag{4.5.39}$$

and a similar expression for $\Psi(\zeta)$. Equations (4.5.6)-(4.5.7) yield

$$\sigma_y + i\tau_{xy} = 2\Re\Phi(\zeta) + 2i\Im\Psi(\zeta) + \cdots \tag{4.5.40}$$

where the ellipsis marks a term that vanishes both for $\zeta = 0$ and $|\zeta| = 1$, $\zeta \neq \pm 1$, corresponding to infinity and the crack faces in the z-plane.

The boundary conditions for σ_y and τ_{xy} are

$$\sigma_y + i\tau_{xy} = \begin{cases} 0 & \text{for } \zeta = 0 \\ -\sigma_y^\infty & \text{for } |\zeta| = 1 \end{cases} \tag{4.5.41}$$

The first one is automatically satisfied, and the second one gives (after noticing that $\zeta = e^{i\varphi} \to \bar{\zeta} = e^{-i\varphi} = 1/\zeta$):

$$2\Re\Phi(\zeta) = \frac{2}{a} \cdot \frac{\sum_{n=1}^{\infty} n[a_n \zeta^{n+1} - \bar{a}_n \zeta^{-n+1}]}{\zeta^2 - 1} = -\sigma_y^\infty \tag{4.5.42}$$

i.e.
$$\sum_{n=1}^{\infty} n[a_n \zeta^{n+1} - \bar{a}_n \zeta^{-n+1}] = \frac{\sigma_y^\infty a}{2}(1 - \zeta^2) \tag{4.5.43}$$

This is satisfied by
$$a_1 = -\frac{\sigma_y^\infty a}{2}, \quad a_n = 0 \quad \text{for } n > 0 \tag{4.5.44}$$

Use of the imaginary part of the second boundary condition (4.5.41) and the condition
$$\sigma_x = 0 \text{ for } \zeta = 0 \tag{4.5.45}$$

shows that $b_n = 0$. Then, both $\Omega(\zeta)$ and $\Sigma(\zeta)$ are determined and the problem is solved after resuperposing the stress $\sigma_y = \sigma_y^\infty$ everywhere. Thus, for instance,

$$(\sigma_y)_{y=0} = 2\Re\Phi(\zeta) + \sigma_y^\infty = -2\sigma_y^\infty \Re\left(\frac{\zeta}{\zeta - 1/\zeta}\right) + \sigma_y^\infty$$
$$= \sigma_y^\infty \frac{|x|}{\sqrt{x^2 - a^2}} \quad \text{for } |x| > a \tag{4.5.46}$$

$$(\sigma_y)_{y=0} = 0 \text{ for } |x| < a \tag{4.5.47}$$

The fact that a solution was obtained shows that the assumption that $\Omega(\zeta)$ and $\Sigma(\zeta)$ are analytic for $\zeta = 0$ was correct. However, if no superposition had been made to cancel the stresses at infinity, it would have been necessary to add a term proportional to $1/\zeta$ to these functions.

The success in obtaining a solution was to some extent dependent on the simple form of the mapping function. In fact, as shown by Muskhelishvili (1953a) for general problems involving complex potentials and conformal mapping, it is possible to determine the unknown coefficients (here a_n and b_n) from a finite set of linear equations, *provided that the mapping function can be expressed as a rational function – the ratio between two finite polynomials*. This is not always possible, and complex potential methods are then not suitable for solving related problems: this excludes, for instance, use of the Schwartz-Christoffel transformation. One example of a configuration, which cannot be mapped onto the unit circle by a rational function, is the asymmetrical star-shaped crack.

The choice of mapping the region outside the crack onto the interior of the unit circle is not essential. A solution may be obtained in basically the same way by mapping the region outside the crack onto the exterior of the unit circle.

4.6 Some applications of complex potentials

The canonical problem in the theory of cracks and fracture

In the preceding section some solutions were given to problems involving a crack in a large plate, subjected to remote loading. A "large plate" refers to one in which the shortest ligament from the crack to the outer plate boundary is much longer than the

4.6 SOME APPLICATIONS OF COMPLEX POTENTIALS

crack. By considering the plate as infinite, a very expedient solution procedure could be applied and rather general problems, involving non-uniform tractions on the crack faces, could be handled. However, most common mode I applications do not involve loading on the crack faces. One such problem, concerning an internal crack, will be discussed to some detail.

For a crack of length $2a$ in a large plate, subjected to remote loads $\sigma_y = \sigma_y^\infty$, $\sigma_x = \sigma_x^\infty$ and traction free crack faces, the stress intensity factor, according to (4.4.45), is

$$\boxed{K_I = \sigma_y^\infty \sqrt{\pi a}} \qquad (4.6.1)$$

Furthermore, with the coordinate origin placed at the center of the crack,

$$p'(z) = \frac{\sigma_y^\infty z}{2(z^2 - a^2)^{1/2}}, \quad s'(z) = \frac{\sigma_y^\infty - \sigma_x^\infty}{4} \qquad (4.6.2)$$

with branch cut along $-a < x < a$, $y = 0$, and the branch chosen so that $(z^2 - a^2)^{1/2} = \sqrt{x^2 - a^2}$ for $x > a$, $y = 0$. This gives

$$\sigma_x = \sigma_x^\infty - \sigma_y^\infty + \sigma_y^\infty \Re\left[\frac{z}{(z^2 - a^2)^{1/2}} - \frac{iya^2}{(z^2 - a^2)^{3/2}}\right] \qquad (4.6.3)$$

$$\sigma_y = \sigma_y^\infty \Re\left[\frac{z}{(z^2 - a^2)^{1/2}} + \frac{iya^2}{(z^2 - a^2)^{3/2}}\right] \qquad (4.6.4)$$

$$\tau_{xy} = -2\Re \frac{ya^2}{(z^2 - a^2)^{3/2}} \qquad (4.6.5)$$

$$u = \frac{\sigma_y^\infty}{2\mu}\Re\left[\frac{k^2}{1-k^2}(z^2 - a^2)^{1/2} - \frac{iyz}{(z^2 - a^2)^{1/2}}\right] - \frac{\sigma_y^\infty - \sigma_x^\infty}{4(1-k^2)\mu}x \qquad (4.6.6)$$

$$v = \frac{\sigma_y^\infty}{2\mu}\Im\left[\frac{1}{1-k^2}(z^2 - a^2)^{1/2} - \frac{iyz}{(z^2 - a^2)^{1/2}}\right] + \frac{\sigma_y^\infty - \sigma_x^\infty}{4\mu}y \qquad (4.6.7)$$

where the integration constants have been chosen so that $v = 0$ for $y = 0$, $|x| > a$ and $u = 0$ for $z = 0$. This was achieved by putting

$$p(z) = \frac{\sigma_y^\infty}{2}(z^2 - a^2)^{1/2}, \quad s(z) = \frac{\sigma_y^\infty - \sigma_x^\infty}{4}z \qquad (4.6.8)$$

This particular problem may be considered as the canonical problem in the theory of cracks and fracture. It contains almost all features of the stress-strain field of interest in the linear elastic fracture mechanics, usually written with the acronym LEFM. It was, in principle, solved by Kolosov (1909) and by Inglis (1913). The solution was used by Griffith (1920) in his famous theory of crack instability (see pages 581ff). It is still used in numerous contexts when a simple and widely recognized example with a wide range of applicability is needed.

The normal stress on the symmetry plane and the crack opening are of particular interest. For $y = 0$, $|x| > 0$, the normal stress is

$$\sigma_y = \frac{\sigma_y^\infty |x|}{\sqrt{x^2 - a^2}} = \frac{K_I |x|}{\sqrt{\pi a(x^2 - a^2)}} \qquad (4.6.9)$$

For $y = +0$, $-a < x < a$, the displacements are

$$u_+ = -\frac{\sigma_y^\infty - \sigma_x^\infty}{4(1-k^2)\mu}x \qquad (4.6.10)$$

$$v_+ = \frac{\sigma_y^\infty}{2(1-k^2)\mu}\sqrt{a^2-x^2} = \frac{K_I}{2(1-k^2)\mu}\cdot\sqrt{\frac{a^2-x^2}{\pi a}} \qquad (4.6.11)$$

The result is remarkably simple: the crack shape is elliptic and the strain ϵ_x is constant on the crack faces. A third noteworthy feature is that

$$\sigma_x = \sigma_x^\infty + \sigma_y - \sigma_y^\infty \text{ for } y = 0 \qquad (4.6.12)$$

which follows from (4.6.3)-(4.5.5). On the crack faces, $\sigma_x = -(\sigma_y^\infty - \sigma_x^\infty)$, the T-stress discussed on pages 80ff. If $\sigma_y^\infty > \sigma_x^\infty$ the stress on the crack faces is compressive; this has been experienced to cause lateral buckling in the neighbourhood of internal cracks in thin plates.

The results given by (4.6.3)-(4.6.7) are valid for either plane stress or plane strain. Real cases involving plates are, however, mixed, with plane strain appearing close to the crack edges. If the extension of the non-linear region at the edges is significantly smaller than the half-thickness, $h/2$, of the plate, then plane strain dominates in the vicinity of the edges and plane stress for distances larger than $h/2$ from the edges, cf. Yang and Freund (1985) and Rosakis and Ravi-Chandar (1986). The non-linear region is then controlled by a stress intensity factor for plane strain. This factor may be obtained with good approximation by making use of results from Section 3.2. Because the whole plate, apart from a very small region near the crack edges is subjected to plane stress, the stresses outside this small region ought to be approximately the same as in a hypothetical case where the plane stress region extends all the way to the crack edges. Equation (4.6.1) then gives the stress intensity factor, and the J-integral is found from the plane stress part of equation (3.5.14):

$$J_I = \pi(\sigma_y^\infty)^2 a/E \qquad (4.6.13)$$

This expression holds for the whole plate in the hypothetical case and with good approximation sufficiently far away from the crack edges in the real case. Equation (3.2.12), telling that the J-integral for a path near an edge of a crack in a plate equals the remote J-integral, now shows that (4.6.13) also holds for a path near the edge. Because plane strain prevails close to the edge, the plane strain part of (3.5.14) gives the stress intensity factor controlling the non-linear region,

$$K_I \approx \frac{\sigma_y^\infty \sqrt{\pi a}}{\sqrt{1-\nu^2}} \qquad (4.6.14)$$

Even if the scale of yielding is not small enough to be considered as infinitesimally small, equation (4.6.14) gives the stress intensity factor controlling the non-linear region with good approximation, if the extension of this region is significantly smaller than $h/2$ and if the plate thickness is considerably smaller than the crack length. How the size of the non-linear region may be estimated will be discussed later, as well as the

4.6 SOME APPLICATIONS OF COMPLEX POTENTIALS

question about how small the scale of yielding should be to allow reasonable accuracy by assuming infinitesimally small scale yielding.

If, on the other hand, the non-linear region extends further out from the crack edges than about $h/2$, then the whole elastic region is dominated by plane stress, and the non-linear region is controlled by the stress intensity factor for plane stress,

$$K_I = \sigma_y^\infty \sqrt{\pi a} \tag{4.6.15}$$

with reasonable accuracy if the scale of yielding is sufficiently small.

The mode II counterpart

In the mode II counterpart to the canonical mode I problem, the remote load is given by a stress $\tau_{xy} = \tau_{xy}^\infty$. Because mode II often is associated with sliding motion, slip may be deposited behind the trailing edge as described in connection with equations (4.4.93)-(4.4.96), which give stresses, displacements and deposited slip on $y = 0$ for energy-neutral healing.

For the direct counterpart to the canonical mode I problem (no slip along the symmetry plane outside the crack), the stress intensity factor is

$$K_{II} = \tau_{xy}^\infty \sqrt{\pi a} \tag{4.6.16}$$

for either plane stress or plane strain, assuming friction-free crack faces. For $y = 0$, $|x| > 0$, the tangential stress is

$$\tau_{xy} = \frac{\tau_{xy}^\infty |x|}{\sqrt{x^2 - a^2}} = \frac{K_{II}|x|}{\sqrt{\pi a(x^2 - a^2)}} \tag{4.6.17}$$

and the displacements on the upper crack face are

$$u_+ = \frac{\tau_{xy}^\infty}{2(1-k^2)\mu} \sqrt{a^2 - x^2} \tag{4.6.18}$$

$$v_+ = \frac{\tau_{xy}^\infty}{2(1-k^2)\mu} x = \frac{K_{II}}{2(1-k^2)\mu} \cdot \sqrt{\frac{a^2 - x^2}{\pi a}} \tag{4.6.19}$$

if rigid-body motion is fixed so that $\partial v/\partial x \to 0$ as $x/a \to \pm\infty$. With the aid of these expressions, weight functions can be constructed and used in the same way as described for mode I.

Mode II cracks exhibit normal stresses parallel to the crack faces. If no slip has occurred along the symmetry plane outside the crack, they are

$$\sigma_x = \mp 2\tau_{xy}^\infty \frac{x}{\sqrt{a^2 - x^2}} \quad \text{for } y = \pm 0 \tag{4.6.20}$$

Note that these stresses are singular at the crack edges.

The mode III counterpart

In the mode III counterpart to the canonical mode I problem, the remote load is given by a stress $\tau_{yz} = \tau_{yz}^\infty$. As in the mode II case, slip may be deposited behind the trailing edge. Equations (4.3.45)-(4.3.47) give stresses and displacements on $y = 0$ for such a case, assuming energy-neutral healing.

For the direct counterpart to the canonical mode I problem (no slip has occurred along the symmetry plane outside the crack), the full solution is obtained from (4.3.31) after putting $s_0(x) = 0$:

$$f'(z) = \tau_{yz}^\infty \frac{z}{(z^2 - a^2)^{1/2}}, \quad \tau_{yz} = \tau_{yz}^\infty \Re \frac{z}{(z^2 - a^2)^{1/2}}, \quad (4.6.21)$$

$$\tau_{xz} = \tau_{yz}^\infty \Im \frac{z}{(z^2 - a^2)^{1/2}}, \quad w_+ = \frac{\tau_{yz}^\infty}{\mu} \Im(z^2 - a^2)^{1/2} \quad (4.6.22)$$

The stress intensity factor is given by (4.3.40) as

$$K_{III} = \tau_{yz}^\infty \sqrt{\pi a} \quad (4.6.23)$$

and the stresses and displacements on the symmetry plane are

$$\tau_{yz} = \tau_{yz}^\infty \frac{|x|}{\sqrt{x^2 - a^2}} = \frac{K_{III}|x|}{\sqrt{\pi a(x^2 - a^2)}} \quad (4.6.24)$$

$$w_+ = \frac{\tau_{yz}^\infty}{\mu} \sqrt{a^2 - x^2} = \frac{K_{III}}{\mu} \sqrt{\frac{a^2 - x^2}{\pi a}} \quad (4.6.25)$$

On page 134, it was pointed out that mode I plane strain conditions prevail only close to the edge of a crack in a plate. The situation is similar for mode III: anti-plane strain conditions can be realized only close to the edge of a crack in a plate with traction free surfaces.

Energy relations

Consider a central crack $|x| < a$, $y = 0$ in a square body $|x|, |y| < L$, $L \gg a$. Eventually, L will be assumed to increase to infinity. Three different cases of loading will be studied; see Fig. 4.6.1. In case A the body is subjected to a constant traction $\sigma_y = \sigma_y^\infty$ at $|y| = L$, and the crack is closed by the action of crack face tractions $\sigma_y = \sigma_y^\infty$. In case B the remote stress is zero and the stresses $\sigma_y = -\sigma_y^\infty$ are applied to the crack faces, producing the displacements $v(x, 0) = v_C(x)$ on $y = +0$, $|x| < a$ and $v(x, L) = v_\infty(x)$ on $y = L$. Case C is a superposition of A and B. Then, the remote stress is $\sigma_y = \sigma_y^\infty$ and the crack faces are traction free.

Compare the change from case A to case C. A displacement change $v(x, 0) = v_C(x)$ has occurred at $y = +0$, $|x| < a$, and at $y = L$ the displacement change is $v(x, L) = v_\infty(x)$. Note that application of the crack face load, only case B, produces the remote displacement $v_\infty(x)$, and that application of the remote load, only case C, produces the crack face displacement $v_C(x)$. Betti's law[†] then tells that

$$\int_{-a}^{a} \sigma_y^\infty v_C(x) \mathrm{d}x = \int_{-\infty}^{\infty} \sigma_y^\infty v_\infty(x) \mathrm{d}x \quad (4.6.26)$$

The right member equals the stress-strain energy added to the upper half of the body by the action of the remote load during the transition from A to C. The left member equals twice the amount of stress-strain energy removed from half the body by the relaxation of the crack face load during the same transition. Thus, the net result is an

---

[†] See, e.g. Bishlinghoff et al. (1990).

4.6 SOME APPLICATIONS OF COMPLEX POTENTIALS

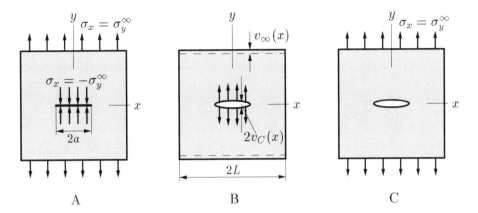

Fig. 4.6.1 Different loading cases

increase of the stress-strain energy in the upper half, equal to half the left member. Let now $L \to \infty$. Then, $v_C(x)$ is known – it is given by (4.6.11) as

$$v_C(x) = \frac{\sigma_y^\infty}{2(1-k^2)\mu} \sqrt{a^2 - x^2} \tag{4.6.27}$$

Insertion into the left integral of (4.6.26) gives the stress-strain energy increase ΔU in the whole body after the transition from case A to case C:

$$\Delta U = \frac{\pi (\sigma_y^\infty)^2 a^2}{4(1-k^2)\mu} \tag{4.6.28}$$

This may also be described as the stress-strain increase in an infinite body subjected to a remote load $\sigma_y = \sigma_y^\infty$ during opening of a crack with length $2a$ normal to direction of the load. Note that the change of potential energy, Π, is a decrease of the same size, due to the motion $v_\infty(x)$ of the remote load in its own direction:

$$\Delta \Pi = -\frac{\pi (\sigma_y^\infty)^2 a^2}{4(1-k^2)\mu} \tag{4.6.29}$$

Return now to case A and assume that the boundaries $|y| = L$ are clamped, instead of free. Relaxation of the crack face load then implies a decrease of stress-strain energy, given by

$$\Delta U = -\frac{\pi (\sigma_y^\infty)^2 a^2}{4(1-k^2)\mu} \tag{4.6.30}$$

Now, the remote load does not contribute to the potential energy, because it acts on fixed boundaries, and therefore the change of potential energy is again given by (4.6.29). Thus, the potential energy decreases by the same amount when a crack is opened in a homogeneous stress field in a very large body, regardless of whether the remote boundaries are free or clamped. The change of stress-strain energy, on the other hand, is different, being an increase if the boundaries are free, and a decrease of the same amount, if the boundaries are clamped.

Note that the crack was assumed to be pre-existing. Assume instead that the crack is opened in an originally stress free body by slow growth from zero length to length $2a$ under the action of internal pressure $p = \sigma_y^\infty$. This case, which is called case D, differs from case B in the respect that the crack is not pre-existing. During crack length increase from $2(a - da)$ to $2a$ the work performed by the pressure equals

$$dW = 2\int_{-a}^{a} \sigma_y^\infty \frac{\partial v(x,a)}{\partial a} dx \cdot da = \frac{\pi(\sigma_y^\infty)^2 a}{(1-k^2)\mu} da \qquad (4.6.31)$$

and, by integration, the total work performed is found to be equal to

$$W = \frac{\pi(\sigma_y^\infty)^2 a^2}{2(1-k^2)\mu} \qquad (4.6.32)$$

This work equals the accumulated stress-strain energy plus the energy flow into the two moving crack edges. The latter part is found by integration of the expression for the energy flux $\mathcal{G}$, given by (3.5.13). The result shows that exactly half the work performed by the pressure has been disposed of at the crack edges. The remaining half has produced stress-strain energy, so that

$$dD = dU = \frac{dW}{2} \qquad (4.6.33)$$

where D is the energy dissipated at the crack edges. Thus, the increase in stress-strain energy equals the one given by (4.6.28), and an equal amount of energy is dissipated along the crack faces.

Again, the change of potential energy is the same whether or not the outer boundaries are clamped or free, so that

$$dD = -d\Pi = \frac{\pi(\sigma_y^\infty)^2 a}{2(1-k^2)\mu} da \qquad (4.6.34)$$

It is quite another question whether it is possible in a real case to open a crack from zero length in the manner described – this would imply the unlikely coincidence that the energy flux into the process region, required for slow crack growth, increases in proportion to the crack length. However, the results (4.6.31) and (4.6.33), which are summarized in the general result (4.6.34), would still describe the energy relations at an infinitesimal crack extension, provided that σ_y^∞ is properly chosen as the material dependent stress needed for such an extension.

Loads other than remote stresses

The solution (4.6.11) may be used to calculate the stress intensity factor for loads other than a constant remote load, by using weight functions according to (3.5.44)-(3.5.46). By first superposing a constant stress $\sigma_y = -\sigma_y^\infty$, the problem is changed to one with the load acting on the crack faces, without changing the stress intensity factor. In order to avoid misunderstanding about which crack edge is to be considered, the crack edges are labelled $(b,0)$ and $(c,0)$, even if $b = -c$. The weight function (3.5.46) is then found from (4.6.11) after calculating $\partial v/\partial c$:

$$h_2(x, \pm 0) = \pm \frac{\sqrt{x-b}}{\sqrt{2\pi(c-b)(c-x)}} \quad \text{for } b < x < c \qquad (4.6.35)$$

4.6 SOME APPLICATIONS OF COMPLEX POTENTIALS

This weight function may be used for calculation of the stress intensity factor $K_I^{(c)}$ at the edge $(c,0)$, under the action of an arbitrarily distributed load $\sigma_y = \sigma_y^0(x)$ on the crack faces. Thus, by (3.5.44) it follows that

$$K_I^{(c)} = -\frac{1}{\sqrt{\pi a}} \int_b^c \sigma_y^0(\xi) \sqrt{\frac{\xi - b}{c - \xi}} d\xi \qquad (4.6.36)$$

where $a = (c-b)/2$. Note that equation (3.5.44) implies integration around the crack, i.e. from b to c along $y = +0$ and from c to b along $y = -0$, but the symmetry properties enable direct expression of K_I in the form (4.6.36). The result agrees with (4.4.45). It has to be controlled that it does not violate the condition that the crack faces must not make any firm contact with each other; see page 118.

The result (4.6.36), in turn, may be used to calculate the stress intensity factor for loads other than those restricted to the crack faces. In order to illustrate the procedure, the case of two concentrated body forces $P\hat{y}$ at $x = 0$, $y = h$ and $-P\hat{y}$ at $x = 0$, $y = -h$ is chosen. Note that the plate thickness is now denoted by $2h$. $\hat{y}$ is the unit vector in the positive y direction. For simplicity, it is assumed that $c = -b = a$.

A concentrated force $P\hat{y}$ at the point $x = 0$, $y = 0$ of an infinite plate without a crack, produces the stress

$$\sigma_y = -\frac{P}{4\pi} \cdot \frac{y[(3+k^2)x^2 + (1-k^2)y^2]}{(x^2+y^2)^2} \qquad (4.6.37)$$

This relation may be found by using (4.4.9)-(4.4.10), (4.4.12) and the ansatz $f(z) = A\ln z$, $g'(z) = B\ln z$, noting that uniqueness of $u + iv$ demands that $(1+k^2)A + (1-k^2)\overline{B} = 0$. For the two concentrated forces at $x = 0$, $y = \pm h$ it follows that the normal stress on the plane $y = 0$ (still in the absence of a crack) will be

$$\sigma_y = \frac{Ph}{2\pi} \cdot \frac{(3+k^2)x^2 + (1-k^2)h^2}{(x^2+h^2)^2} = q(x) \qquad (4.6.38)$$

This result implies that the load $\sigma_y = -q(x)$ on the crack faces gives the same stress intensity factor as the two concentrated forces at $x = 0$, $y = \pm h$, because superposition of the load $\sigma_y = +q(x)$ on the crack faces in the latter case just closes the crack. Consequently, equation (4.6.36) can be used and leads to:

$$K_I = \frac{Ph}{2\pi\sqrt{\pi a}} \int_{-a}^{a} \frac{(3+k^2)x^2 + (1-k^2)h^2}{(x^2+h^2)^2} \sqrt{\frac{a+x}{a-x}} dx \qquad (4.6.39)$$

The integral is conveniently calculated by means of residue calculus, using a closed path of integration, surrounding the two branch points $x = \pm a$, $y = 0$, but otherwise approaching the x axis between these points. Due to the symmetry, the integral equals the real part of the residue at $(0, ih)$. The result turns out to be:

$$K_I = P\sqrt{\frac{a}{\pi}} \cdot \frac{2a^2 + (1-k^2)h^2}{2(a^2+h^2)^{3/2}} \qquad (4.6.40)$$

Note that this expression reduces to

$$K_I = \frac{P}{\sqrt{\pi a}} \qquad (4.6.41)$$

when $h/a \to 0$ in agreement with (4.6.36) after putting $p(x) = -P\delta[x - (b+c)/2]$, where $\delta(x)$ is the Dirac delta function, and that it reduces to

$$K_I = \frac{(1-k^2)P}{2h}\sqrt{\frac{a}{\pi}} \tag{4.6.42}$$

when $h/a \to \infty$. This is consistent with the result, found from (4.6.38), that

$$q(x) \approx \frac{(1-k^2)P}{2\pi h} \quad \text{for } |x| \ll h \tag{4.6.43}$$

For mode I type shear tractions on the crack faces, the appropriate weight function is found after calculating $\partial u/\partial c$ from (4.6.10).

Arrays of cracks

In some cases, a number of tiny cracks appear together, for instance near the edge of a larger crack in certain ceramics. It is therefore of interest to investigate how cracks interact. Solving problems concerning multiple cracks may often proceed along the same lines as for one crack. In fact, the solutions obtained earlier for infinite regions (pages 116ff.) are still valid for an array of N collinear† cracks, $y = 0$, $b_n < x < c_n$, if the definition of the function $G(z)$ is changed to

$$G(z) = \prod_{n=1}^{N} [(z-b_n)^{1/2}(z-c_n)^{1/2}] \tag{4.6.44}$$

and the conditions of prescribed differences of deposited slip between the crack ends or, for mode I, the conditions of closed ends, are observed for each crack. This leads to determination of polynomial coefficients from equations of the type

$$\int_{b_n}^{c_n} \frac{P(x)}{G(x)} dx = H\Delta_n \tag{4.6.45}$$

where $P(x)$ is a polynomial of degree N, H is an elastic constant and Δ is a slip difference, or, for mode I, simply zero. Note that the integral is in general elementary only for $N = 1$; for $N = 2$ an elliptic integral is encountered (see page 155).

The function $G(z)$ is defined by cuts along the real axis along $b_n < x < c_n$ with the branch chosen so that $G(x) = \prod_{n=1}^{N} \sqrt{(x-b_n)(x-c_n)}$ for $x > c_N$.

Comparatively simple are *periodic arrays*, for instance an infinite row of collinear traction free cracks of equal length, uniformly spaced in an infinite region and subjected to a constant stress at infinity; see Fig. 4.6.2. Mode I is assumed, but the treatment for mode II is very similar. Because, according to (4.6.44), $G(z)$ would be infinite for $N = \infty$, a finite periodic array is first considered, with $2N + 1$ cracks with the edges at

$$z = \pm a \pm n \cdot 2d, \quad d > a, \; n = 1, 2, 3, \ldots, N \tag{4.6.46}$$

---

† The term collinear is traditionally used, rather than coplanar, because of the usual way of drawing a figure on the section $z = \text{constant}$.

4.6 SOME APPLICATIONS OF COMPLEX POTENTIALS

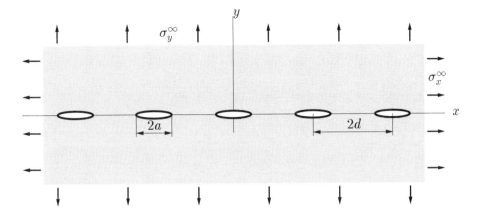

Fig. 4.6.2 Collinear periodic array of cracks subjected to remote mode I loading.

together with the edges $z = \pm a$ of the central crack. Then, after some rearrangement,

$$G(z) = (z^2 - a^2)^{1/2} \prod_1^N [(z - 2nd)^2 - a^2]^{1/2}[(z + 2nd)^2 - a^2]^{1/2} \quad (4.6.47)$$

Now, because the remote stress is finite, $P(z)/G(z)$ must be finite as $|z| \to \infty$, so that

$$P(z) + O(1/z) \to CG(z) \text{ as } |z| \to \infty \quad (4.6.48)$$

where C is constant. Because

$$G(z) \to z \prod_1^N [z^2 - (2nd)^2] \text{ as } |z| \to \infty \quad (4.6.49)$$

the result for $p'(z) = P(z)/G(z)$ may be written in a form that allows $N \to \infty$:

$$p'(z) = \frac{C}{(1 - a^2/z^2)^{1/2} \prod_1^\infty [1 - a^2/(2nd - z)^2]^{1/2}[1 - a^2/(2nd + z)^2]^{1/2}} \quad (4.6.50)$$

Regardless of the number of cracks, $s'(z)$ is constant, say

$$s'(z) = D \quad (4.6.51)$$

Branch cuts are made along the cracks, and the branch is chosen so that $p'(z) \to C$ as $z \to \infty$. The infinite product is recognized as a product representation of trigonometric functions; see e.g. Gradshteyn and Ryzhik (1980). Thus,

$$p'(z) = C \frac{\sin(\pi z/2d)}{[\cos^2(\pi z/2d) - \cos^2(\pi a/2d)]^{1/2}} \quad (4.6.52)$$

Assume remote stresses $\sigma_y = \sigma_y^\infty$, $\sigma_x = \sigma_x^\infty$. Addition of equations (4.4.23) and (4.4.24) with specialization to $y = 0$ results in

$$\sigma_y + i\tau_{xy} = 2\Re p'(x) + 2i\Im s'(x) \quad (4.6.53)$$

which shows that D must be real, because τ_{xy} vanishes for all x, and that C must be purely imaginary, because $\sigma_y = 0$ on the crack faces. For $|y| \to \infty$,

$$p'(z) \to -iC \implies C = i\sigma_y^\infty/2, \quad D = (\sigma_y^\infty - \sigma_x^\infty)/4 \qquad (4.6.54)$$

It is immediately obvious that equation (4.6.52) is consistent with the condition of closed ends of each crack. For the central crack this follows because $p'(x)$ is antisymmetric, and any crack may be considered as the central crack if the coordinate origin is placed at its centre. This would not be the case for an array in which the central crack were longer or shorter than the other cracks: attempts to generalize $p'(z)$ to such cases by modification of the factor in front of the infinite product in (4.6.50) are therefore not advisable.

Integration of (4.6.52) yields

$$p(z) = \frac{2dC}{\pi} \ln \frac{\cos(\pi z/2d) - [\cos^2(\pi z/2d) - \cos^2(\pi a/2d)]^{1/2}}{\cos(\pi a/2d)} \qquad (4.6.55)$$

$$s(z) = Dz \qquad (4.6.56)$$

where the integration constants were chosen so that $p(z) = 0$ at the edges of the cracks and $s(z) = 0$ for $z = 0$. This ensures that $u = 0$ for $z = 0$, and that $v = 0$ for all parts of the x axis that do not contain cracks.

It remains now to show that $p(z)$ is analytic in the region occupied by the body, with proper singularities at the branch points. Note that the argument of the logarithm does not vanish for any finite value of z. The only singularities, therefore, are the branch-points at the crack edges, given by (4.6.46). In the neighbourhood of these points,

$$p(z) \propto (z \mp a - 2nd)^{1/2} \qquad (4.6.57)$$

Thus, $p(z)$ is a single-valued function outside the branch cuts and the crack opening displacement is zero at the branch points. It is therefore the function sought.

The stress intensity factor is now found as

$$K_I = \lim_{r \to 0} \left[\sqrt{2\pi r} \sigma_y \right] = \lim_{r \to 0} \left[\sqrt{2\pi r} \cdot 2\Re p'(a+r) \right] = \sigma_y^\infty \sqrt{\pi a} \sqrt{\frac{2d}{\pi a} \tan \frac{\pi a}{2d}} \qquad (4.6.58)$$

The solution of this problem was obtained by Westergaard (1939) and by Koiter (1959). The stress intensity factor is plotted in Fig. 4.6.3 as a function of $(d-a)/a$. Note that the interaction between the cracks is significant only when the distance between neighbouring cracks is of the same order as the length of the cracks or smaller. The stress intensity factor is much larger than that for a single crack only when the distance between the cracks is considerably smaller than the length of the cracks. The implication for the risk of coalescence between the cracks is, however, somewhat ambiguous: the cracks do not grow straightforwards towards each other, but lose directional stability (Melin 1983), as will be discussed in Section 4.11.

The case of two collinear cracks will be treated later, using integral equations.

Circular arc crack

Muskhelishvili (1953a) studied various problems concerned with circular regions and infinite regions cut along circular arcs. Here, the stress intensity factors for a circular

4.6 SOME APPLICATIONS OF COMPLEX POTENTIALS

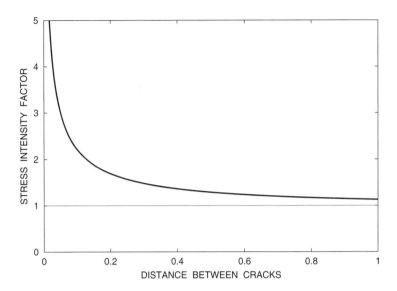

Fig. 4.6.3 Stress intensity factor for cracks in a collinear periodic array as a function of the distance between cracks. The stress intensity factor is normalized with respect to the stress intensity factor for a single crack and the distance between the cracks (from the edge of one crack to the edge of the next) is normalized with respect to the length of an individual crack, i.e., it is given as $(d-a)/a$.

arc crack in an infinite plate subjected to remote loading will be determined, essentially following the Muskhelishvili approach; see also Atluri et al. (1975). For simplicity, the radius of the arc is taken to be unity during the calculations. The crack is assumed to be situated between $z = \exp(i\theta)$ and $z = \exp(-i\theta)$ in the z-plane and to cross the positive x axis; see Fig. 4.6.4. The remote load is assumed to consist of uniaxial tension σ^∞ in a direction forming the angle α to the x axis. This choice enables consideration of any remote load simply by superposition.

Equations (4.4.113)-(4.4.114) will be used. In particular,

$$\sigma_r + i\tau_{r\varphi} = f'(z) + \overline{f'(z)} - \bar{z}\,\overline{f''(z)} - \frac{\bar{z}}{z}\,\overline{g''(z)} \qquad (4.6.59)$$

Here, $f(z)$ and $g(z)$ are analytic in the whole plane cut along the circular arc. For brevity, this cut will be called L and the region inside the unit circle will be called S_+ and the region outside S_-. Because the crack surfaces are traction free, the left member of (4.6.59) vanishes on L.

Direct use of (4.6.59) as a boundary condition would lead to difficulties, because the last term is neither analytic nor the complex conjugate of an analytic function. Muskhelishvili observed, however, that by writing the right member in the form

$$f'(z) + \overline{f'(z)} - \bar{z}\,\overline{f''(z)} - \bar{z}^2\,\overline{g''(z)} + \bar{z}\left(\bar{z} - \frac{1}{z}\right)\overline{g''(z)} \qquad (4.6.60)$$

the last term vanishes on L (because $z = \exp(i\varphi) = 1/\bar{z}$), and the three preceding terms are complex conjugates of analytic functions. These terms are now written as a

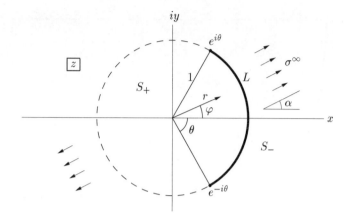

Fig. 4.6.4 Contour L of the circular arc crack in the z plane.

function $q(1/\bar{z})$, with

$$q(z) = \overline{f'}(1/z) - \frac{1}{z}\overline{f''}(1/z) - \frac{1}{z^2}\overline{g''}(1/z) \tag{4.6.61}$$

and it is possible to show that $q(z)$ is analytic in the same region as $f(z)$ and $g(z)$, except possibly at $z = 0$, where $q(z)$, as seen from (4.6.61), may have a pole of at most second order. In fact, if $F(z)$ is analytic inside (or outside) the unit circle, then $\overline{F}(1/z)$ is analytic outside (or inside) the unit circle. The proof may be made by verifying that the functions

$$\Phi = \frac{1}{2}\left[\overline{F(e^{-i\varphi}/r)} + F(e^{-i\varphi}/r)\right], \quad \Psi = \frac{-i}{2}\left[\overline{F(e^{-i\varphi}/r)} - F(e^{-i\varphi}/r)\right] \tag{4.6.62}$$

satisfy the Cauchy-Riemann equations in polar coordinates,

$$\frac{\partial \Phi}{\partial r} = \frac{1}{r}\frac{\partial \Psi}{\partial \varphi} \tag{4.6.63}$$

$$\frac{1}{r}\frac{\partial \Phi}{\partial \varphi} = -\frac{\partial \Psi}{\partial r} \tag{4.6.64}$$

The function $q(z)$ is also analytic on the circle periphery outside L, because $f(z)$ and $g(z)$ are continuous across this arc. Furthermore,

$$q(\infty) = \overline{f'}(0) \tag{4.6.65}$$

so that $g(z)$ is bounded at infinity, and

$$g''(z) = \frac{1}{z^2}f'(z) - \frac{1}{z}\overline{f''}(1/z) - \frac{1}{z^2}\overline{q}(1/z) \tag{4.6.66}$$

Equation (4.6.59) may now be written in the form

$$\sigma_r + i\tau_{r\varphi} = f'(z) + q(1/\bar{z}) + \text{ term vanishing when } |z| \to 1 \tag{4.6.67}$$

Let t be a point on L. When $z \to t$ from S_+, then $1/\bar{z} \to t$ from S_-, i.e.

$$f'(z) \to [f'(t)]_+, \quad q(1/\bar{z}) \to [q(t)]_- \tag{4.6.68}$$

4.6 SOME APPLICATIONS OF COMPLEX POTENTIALS

and hence

$$[f'(t)]_+ + [q(t)]_- = 0 \text{ on } L, \quad [f'(t)]_- + [q(t)]_+ = 0 \text{ on } L \quad (4.6.69)$$

Addition and subtraction gives a Hilbert problem,

$$[t^2 f'(t) + t^2 q(t)]_+ + [t^2 f'(t) + t^2 q(t)]_- = 0 \text{ on } L \quad (4.6.70)$$
$$[t^2 f'(t) - t^2 q(t)]_+ - [t^2 f'(t) - t^2 q(t)]_- = 0 \text{ on } L \quad (4.6.71)$$

where the factor t^2 has been introduced to ensure that the functions inside the brackets are analytic (recall that $q(z)$ might have a second order pole at $z = 0$).

Equation (4.6.71) tells that the function inside the brackets is analytic on L, and thus in the whole plane, i.e. it equals a polynomial of finite degree:

$$f'(z) - q(z) = 2 \sum_{n=0}^{N} a_n z^{n-2} \quad (4.6.72)$$

Equation (4.6.70) is solved by introducing the function

$$G(z) = (z - e^{i\theta})^{1/2}(z - e^{-i\theta})^{1/2} = (z^2 - 2z\cos\theta + 1)^{1/2} \quad (4.6.73)$$

defined so that $G(z) \to z$ as $z \to \infty$. Then, $G(0) = -1$, because $G(z) > 0$ for $z = x > 1$ and $G_+(t)/G_-(t) = -1$ on L. Multiplication with the latter factor gives

$$\{t^2 G(t)[f'(t) + q(t)]\}_+ - \{t^2 G(t)[f'(t) + q(t)]\}_- = 0 \text{ on } L \quad (4.6.74)$$

and hence

$$f'(z) + q(z) = \frac{1}{G(z)} \sum_{n=0}^{N} b_n z^{n-2} \quad (4.6.75)$$

where the polynomial has finite degree. The functions $f'(z)$ and $q(z)$ can now be determined separately:

$$f'(z) = \frac{1}{G(z)} \sum_{n=0}^{N} b_n z^{n-2} + \sum_{n=0}^{N} a_n z^{n-2} \quad (4.6.76)$$

$$g(z) = \frac{1}{G(z)} \sum_{n=0}^{N} b_n z^{n-2} - \sum_{n=0}^{N} a_n z^{n-2} \quad (4.6.77)$$

Coefficients a_n and b_n are determined by using the boundary conditions at infinity, including (4.6.65), and by ensuring that $f(z)$ and $g(z)$ are regular at $z = 0$. In order to specify boundary conditions at infinity, it is helpful to consider the homogeneous stress state which coincides at infinity with the stress state given for the problem:

$$\sigma_x + \sigma_y = \sigma^\infty, \quad \sigma_y - \sigma_x + 2i\tau_{xy} = -\sigma^\infty e^{-2i\alpha} \quad (4.6.78)$$

This stress state is represented by the analytic functions

$$f(z) = \frac{\sigma^\infty}{4} z, \quad g(z) = -\frac{\sigma^\infty}{4} e^{-2i\alpha} z^2 \quad (4.6.79)$$

where constants representing rigid body motion have been put equal to zero. Thus, turning back to the original problem,

$$f(z) \to \frac{\sigma^\infty}{4} z + \frac{A_1}{z} + \cdots \text{ as } |z| \to \infty \qquad (4.6.80)$$

$$g(z) \to -\frac{\sigma^\infty}{4} e^{-2i\alpha} z^2 + B_1 z + C_0 + \frac{C_1}{z} + \cdots \text{ as } |z| \to \infty \qquad (4.6.81)$$

i.e.

$$f'(z) \to \frac{\sigma^\infty}{4} - \frac{A_1}{z^2} + \cdots \text{ as } |z| \to \infty \qquad (4.6.82)$$

$$g''(z) \to -\frac{\sigma^\infty}{2} e^{-2i\alpha} - 2\frac{C_1}{z^3} + \cdots \text{ as } |z| \to \infty \qquad (4.6.83)$$

Note that these two functions do not contain any $1/z$-term. Equation (4.6.61) now gives:

$$q(z) \to -\frac{\sigma^\infty}{2z^2} e^{2i\alpha} + regular\, function \text{ as } |z| \to 0 \qquad (4.6.84)$$

Summing up, the remaining boundary conditions to be used are given by (4.6.65), (4.6.82)-(4.6.84) and the condition that $f'(z)$ must be bounded at $z = 0$. In order to investigate $f'(z)$ and $q(z)$ for $z \to \infty$ and $z \to 0$, the function $1/G(z)$ is expanded into power series':

$$\frac{1}{G(z)} \to \left(\frac{1}{z} + \frac{\cos\theta}{z^2} + \cdots \right) \text{ as } z \to \infty \qquad (4.6.85)$$

$$\frac{1}{G(z)} \to -\left(1 + \cos\theta \cdot z + \frac{3\cos^2\theta - 1}{2} z^2 + \cdots \right) \text{ as } z \to 0 \qquad (4.6.86)$$

The determination of the unknown coefficients is then straightforward and yields:

$$a_n = 0 \text{ for } n > 2, \quad b_n = 0 \text{ for } n > 3$$

$$a_0 = b_0 = -\frac{\sigma^\infty}{4} e^{2i\alpha}, \quad a_1 = 0, \quad b_1 = -b_0 \cos\theta$$

$$b_3 = \frac{\sigma^\infty}{4} \left[\frac{1 - \sin^2(\theta/2)\cos^2(\theta/2)\cos 2\alpha}{1 + \sin^2(\theta/2)} + i \sin^2(\theta/2) \sin 2\alpha \right]$$

$$a_2 = \frac{\sigma^\infty}{2} - b_3, \quad b_2 = -b_3 \cos\theta$$

It is now possible to determine, for instance, the stress intensity factors at the edge $z = \exp(i\theta)$:

$$K_I - iK_{II} = \lim_{\varphi \to \theta+0} \left[\sqrt{2\pi(\varphi - \theta)} (\sigma_r + i\tau_{r\varphi})_{z=\exp(i\varphi)} \right]$$

$$= \lim_{\varphi \to \theta+0} \left\{ \sqrt{2\pi(\varphi - \theta)} [f'(e^{i\varphi}) + q(e^{i\varphi})] \right\} \qquad (4.6.87)$$

where the minus-sign for K_{II} is explained by the convention mentioned in connection with definition (3.3.70). Because

$$f'(z) + q(z) = \frac{2}{G(z)} \left[\frac{b_0}{z^2} (1 - z\cos\theta) + b_3(z - \cos\theta) \right] \qquad (4.6.88)$$

4.6 SOME APPLICATIONS OF COMPLEX POTENTIALS

and

$$G(e^{i\varphi}) = (e^{i\varphi} - e^{i\theta})^{1/2}(e^{i\varphi} - e^{-i\theta})^{1/2} \to i\sqrt{\varphi - \theta}\sqrt{2\sin\theta}e^{-i\theta/2}$$

$$\text{as } \varphi \to \theta + 0 \quad (4.6.89)$$

the stress intensity factors are found from the expression

$$K_I - iK_{II} = \frac{2\sqrt{\pi}}{i\sqrt{\sin\theta}}[b_0 e^{-2i\theta}(1 - e^{i\theta}\cos\theta) + b_3(e^{i\theta} - \cos\theta)]e^{-i\theta/2}$$

$$= 2\sqrt{\pi\sin\theta}(-b_0 e^{-i\theta} + b_3)e^{-i\theta/2} \quad (4.6.90)$$

Recalling that the radius of the arc was previously taken as unity, the stress intensity factor for a radius R is

$$K_I - iK_{II} = \sigma_\infty \sqrt{\pi R \sin\theta}\,(k_I + ik_{II}) \quad (4.6.91)$$

where, as found after insertion of the expressions for b_0 and b_3:

$$k_I = \frac{1}{2}\cos(\theta/2)\left\{\frac{1}{1+\sin^2(\theta/2)} + \cos 2\alpha\left[2\cos\theta - 1 - \frac{\sin^2\theta}{4[1+\sin^2(\theta/2)]}\right]\right.$$

$$\left. + \frac{3}{2}\sin 2\alpha \cdot \sin\theta\right\} \quad (4.6.92)$$

$$k_{II} = \frac{1}{2}\sin(\theta/2)\left\{\frac{1}{1+\sin^2(\theta/2)} + \cos 2\alpha\left[2\cos\theta + 1 - \frac{\sin^2\theta}{4[1+\sin^2(\theta/2)]}\right]\right.$$

$$\left. + \sin 2\alpha \cdot \sin\theta\right\} - \frac{1}{2}\cos^3(\theta/2)\sin 2\alpha \quad (4.6.93)$$

Specialization to $\alpha = 0$ and to $\alpha = \pi/2$ gives the stress intensity factors for the remote loads $\sigma_x = \sigma^\infty = \sigma_x^\infty$ and $\sigma_y = \sigma^\infty = \sigma_y^\infty$, respectively. The stress intensity factors for a remote shear stress $\tau_{xy} = \sigma^\infty = \tau_{xy}^\infty$ are obtained by superposition of the cases $\alpha = \pi/4$, $\sigma^\infty = \tau_{xy}^\infty$ and $\alpha = -\pi/4$, $\sigma^\infty = -\tau_{xy}^\infty$. Such specializations lead to an alternative way to express the result (4.6.92)-(4.6.93):

$$\frac{K_I}{\sqrt{\pi R\sin\theta}} = \frac{\sigma_x^\infty}{2}\cos(\theta/2)\left[\frac{1-\frac{1}{4}\sin^2\theta}{1+\sin^2(\theta/2)} + 2\cos\theta - 1\right]$$

$$+ \frac{\sigma_y^\infty}{2}\cos(\theta/2)\left[\frac{1+\frac{1}{4}\sin^2\theta}{1+\sin^2(\theta/2)} - 2\cos\theta + 1\right]$$

$$+ \frac{3\tau_{xy}^\infty}{2}\sin(\theta/2)(\cos\theta + 1) \quad (4.6.94)$$

$$\frac{K_{II}}{\sqrt{\pi R\sin\theta}} = \frac{\sigma_x^\infty}{2}\sin(\theta/2)\left[\frac{1-\frac{1}{4}\sin^2\theta}{1+\sin^2(\theta/2)} + 2\cos\theta + 1\right]$$

$$+ \frac{\sigma_y^\infty}{2}\sin(\theta/2)\left[\frac{1+\frac{1}{4}\sin^2\theta}{1+\sin^2(\theta/2)} - 2\cos\theta - 1\right]$$

$$- \frac{\tau_{xy}^\infty}{2}\cos(\theta/2)(3\cos\theta - 1) \quad (4.6.95)$$

The stress intensity factors at the edge $z = \exp(-i\theta)$ are obtained by changing θ to $-\theta$ in (4.6.92)-(4.6.95). Note the crack orientation: for an orientation symmetrical with respect to the y axis, expressions (4.6.94)-(4.6.95) are valid after the substitutions $\sigma_x \to \sigma_y$, $\sigma_y \to \sigma_x$ and $\tau_{xy} \to -\tau_{xy}$.

Note that the results obtained for a remote shear load $\tau_{xy} = \tau_{xy}^\infty$, only, show a mode I stress intensity factor that is negative at one crack edge, and consequently the assumption of traction free crack faces is violated. Superposition of, for instance, a sufficiently high hydrostatic stress ($\sigma_x^\infty = \sigma_y^\infty$) will lead to non-negative mode I stress intensity factors, and, in general, it is necessary to ensure that the crack faces are traction free for the remote load chosen. Partial closure of the crack presents a difficult analytical problem, except in cases when the friction is high enough to prevent slip completely. In these cases, the open part of the circular arc can be found from the condition that the mode I stress intensity factor vanishes at the edge towards the closed part. To use this condition, the other edge is assumed to be fixed, whereas the length of the (open part of the) crack is varied and the coordinate axes chosen so that the current crack becomes symmetrical with respect to the new x axis. Then, K_I may be calculated for the edge towards the closed part as a function of the length of the open part of the crack.

Under conditions of small scale yielding, a crack in a plate almost invariably seems to grow with local mode I symmetry at the crack edge. The condition for incipient straightforward growth at one edge is therefore a remote stress orientation α such that $K_{II} = 0$ at this edge and a sufficiently high value of K_I. For another orientation the crack grows initially by kinking.

4.7 Integral equations

Instead of boundary collocation methods for solving problems formulated with the aid of complex potentials, *integral equations* may be used, cf. Muskhelishvili (1953b). Such equations may also be formulated directly, without complex potential representation for stresses and displacements. In numerical treatments, the integral equation (or equations) is reduced to a system of algebraic equations; see e.g. Erdogan and Gupta (1972). Integral equations for crack problems may also be formulated through superposition either of elementary loads or of dislocations, and they may also appear as a result of integral transformations.

Superposition of elementary loads

A simple example will show how an integral equation arises in a crack problem. Consider a central crack with length $2a$, situated in a large plate. A coordinate system x, y is introduced, such that the crack edges are $y = 0$, $x = \pm a$. The plate is subjected to a remote load $\sigma_y = \sigma_y^\infty$. Plane stress or plane strain is assumed. In comparison with the crack, the plate is assumed to be large enough to be considered as infinite. (This is the canonical mode I problem.)

Because of the symmetry, the following formulation can be given: a semi-infinite plate, $y \leq 0$, is subjected to a stress $\sigma_y = \sigma_y^\infty$ at infinity. The shear stress τ_{xy} vanishes on the whole boundary $y = 0$. The normal stress σ_y vanishes on the segment $y = 0$,

$|x| < a$. The normal displacement v at the boundary $y = 0$ vanishes on $|x| > a$. Thus, a mixed boundary value problem is formulated.

The solution of the problem proceeds in two steps. In the first one, the normal displacement v on the boundary $y = 0$ is determined for an elementary load, a concentrated normal line force, $P\hat{y}$ per unit of length, acting on the line $x = y = 0$. This well-known problem (the Boussinesq problem) has the solution

$$\frac{\partial v_-}{\partial x} = -\frac{P}{2\pi(1-k^2)\mu} \cdot \frac{1}{x}, \quad x \neq 0, y = 0 \tag{4.7.1}$$

which, for instance, may be derived with the aid of (4.4.19)-(4.4.20) by choosing $f(z) = A \ln z + B$, $g'(z) = C \ln z$. Integration of the expression is not necessary, because the original boundary condition of vanishing v on $y = 0$ may be replaced by $\partial v/\partial x = 0$.

The stress corresponding to the line force P is

$$\sigma_y = P\delta(x) \tag{4.7.2}$$

where $\delta(x)$ is Dirac's delta function, defined so that

$$\delta(x) = 0 \text{ for } x \neq 0, \quad \int_{-\infty}^{+\infty} \delta(x)\,dx = 1 \tag{4.7.3}$$

For a stress

$$d\sigma_y = q(\xi)d\xi\delta(x - \xi) \tag{4.7.4}$$

on the boundary $y = 0$, equation (4.7.1) gives

$$d\left(\frac{\partial v_-}{\partial x}\right) = -\frac{q(\xi)d\xi}{2\pi(1-k^2)\mu} \cdot \frac{1}{x-\xi} \text{ for } y = 0,\, x \neq \xi \tag{4.7.5}$$

Such displacement derivatives are superposed in the second step of the procedure, with $q(\xi)$ chosen so that the boundary conditions on $y = 0$, $x \geq 0$ for $\partial v_-/\partial x$ and σ_y are satisfied, i.e.,

$$\oint_0^\infty \frac{q(\xi)d\xi}{x - \xi} = 0 \text{ for } x > a, \quad q(x) = 0 \text{ for } 0 \leq x < a \tag{4.7.6}$$

Here, the integral should be taken in the sense of the Cauchy principal value in order to avoid the point $\xi = x$ and to ensure that finite stresses in a vicinity of this point do not give rise to any displacement when this vicinity shrinks indefinitely. Thus, a homogeneous integral equation of the first kind is obtained:

$$\oint_a^\infty \frac{q(\xi)d\xi}{\xi - x} = 0 \text{ for } x > a \tag{4.7.7}$$

The solution is

$$q(x) = \frac{Ax}{\sqrt{x^2 - a^2}} \text{ for } x > a \tag{4.7.8}$$

where A is a constant. Because of the symmetry, x can be replaced by $|x|$. The result may, for instance, be obtained by introducing a function

$$F(z) = \frac{1}{2\pi i} \int_{-\infty}^{+\infty} \frac{q(\xi)d\xi}{\xi - z} \tag{4.7.9}$$

which is analytic everywhere except on the segment $|x| < a$ of the real axis, and for which the Plemelj formulae give:

$$F_+(x) + F_-(x) = \frac{1}{\pi i} \fint_{-\infty}^{+\infty} \frac{q(\xi)\mathrm{d}\xi}{\xi - x} = 0 \quad \text{for } |x| > a \qquad (4.7.10)$$

$$F_+(x) - F_-(x) = q(x) = 0 \quad \text{for } |x| < a \qquad (4.7.11)$$

From the Hilbert problem so formulated, the solution (4.7.8) is directly obtained.
The constant A is determined to be σ_y^∞ from the condition $\sigma_y = \sigma_y^\infty$ at infinity. Thus, finally:

$$\sigma_y = \frac{\sigma_y^\infty |x|}{\sqrt{x^2 - a^2}} \quad \text{for } y = 0, |x| > a \qquad (4.7.12)$$

By using (4.7.5), the gradient of the displacement normal to the crack face is found to be

$$\frac{\partial v_-}{\partial x} = -\frac{1}{2\pi(1-k^2)\mu} \int_{-\infty}^{+\infty} \frac{q(\xi)\mathrm{d}\xi}{x - \xi}$$

$$= \frac{\sigma_y^\infty |x|}{2\pi(1-k^2)\mu} \Re \int_{-\infty}^{+\infty} \frac{\mathrm{d}\xi}{(\xi^2 - a^2)^{1/2}(\xi - x)}$$

$$= \frac{\sigma_y^\infty}{2(1-k^2)\mu} \cdot \frac{x}{\sqrt{a^2 - x^2}} \quad \text{for } |x| < a \qquad (4.7.13)$$

where the integral could be directly obtained as a residue. Integration then gives

$$v_- = -\frac{\sigma_y^\infty}{2(1-k^2)\mu} \sqrt{a^2 - x^2} \quad \text{for } |x| < a \qquad (4.7.14)$$

A more advanced example involves incorporation of a Barenblatt region at each crack end. Slow mode II crack motion in the positive x direction will be considered, for simplicity without friction along the sliding region, except in the Barenblatt regions. (Consideration of friction will be made for the corresponding dynamic case, Section 6.3.) The problem is otherwise similar to the one previously treated, after replacing the remote load by $\tau_{xy} = \tau_{xy}^\infty$, the elementary load by $T\hat{x}$, the displacement considered by u_- and the traction on the crack faces by a traction $\tau_{xy} = \tau_{xy}^0(x)$, such that

$$\tau_{xy}^0(x) = 0 \quad \text{for } -a + r_H < x < a - r_T \qquad (4.7.15)$$

where $r_H \ll a$ and $r_T \ll a$ are the lengths of the Barenblatt regions. Index H refers to "Healing" and index T to "Tearing".

Instead of equations (4.7.10)-(4.7.11), the Hilbert problem now results in

$$F_+(x) + F_-(x) = \frac{1}{\pi i} \fint_{-\infty}^{+\infty} \frac{q(\xi)\mathrm{d}\xi}{\xi - x} = 0 \quad \text{for } |x| > a \qquad (4.7.16)$$

$$F_+(x) - F_-(x) = \tau_{xy}^0(x) \quad \text{for } |x| < a \qquad (4.7.17)$$

The solution is

$$F(z) = \frac{1}{2\pi i G(z)} \int_{-a}^{a} \frac{G_+(\xi)\tau_{xy}^0(\xi)}{\xi - z} \mathrm{d}\xi + \frac{P(z)}{G(z)} \qquad (4.7.18)$$

where $P(z)$ is a finite degree polynomial and the sectionally analytic function $G(z)$ is such that $G_-(x)/G_+(x) = -1$ for $|x| > a$, $+1$ for $|x| < a$. Because no singularity should appear, a convenient choice is

$$G(z) = (a^2 - z^2)^{-1/2} \tag{4.7.19}$$

defined by branch cuts along $|x| > a$, $y = 0$, with the branch chosen so that $G(x) = 1/a$ for $x = 0$. Then,

$$F_+(x) = -F_-(x) = -\frac{\sqrt{x^2 - a^2}}{2\pi}\left[\int_{-a}^{a} \frac{\tau_{xy}^0(\xi)d\xi}{\sqrt{a^2 - \xi^2}(\xi - x)} + 2\pi i P(x)\right] \tag{4.7.20}$$

where, because $\tau_{xy} = q(x) = 2F_+(x) \to \tau_{xy}^\infty$ as $|x| \to \infty$, the polynomial $P(x) = 0$, and

$$\frac{1}{\pi}\int_{-a}^{a} \frac{\tau_{xy}^0(\xi)d\xi}{\sqrt{a^2 - \xi^2}} = \tau_{xy}^\infty \tag{4.7.21}$$

This equation imposes a condition on τ_{xy}^0 so that smooth closing and stress continuity is obtained, cf. (3.6.16).

For $|x| > a$, $y = 0$,

$$\tau_{xy} = \frac{\sqrt{x^2 - a^2}}{\pi}\int_{-a}^{a} \frac{\tau_{xy}^0(\xi)d\xi}{\sqrt{a^2 - \xi^2}(x - \xi)} \tag{4.7.22}$$

By letting $x \to a + 0$ or $x \to -a - 0$ the right hand side reaches the limits $\tau_{xy}^0(a)$ and $\tau_{xy}^0(-a)$, respectively, showing stress continuity. For $(x+a)^2 \gg r_H^2$ and $(x-a)^2 \gg r_T^2$, the relation may be written as

$$\tau_{xy} \approx \frac{1}{\pi\sqrt{2a}}\left[T_{II}\sqrt{\frac{x+a}{x-a}} - H_{II}\sqrt{\frac{x-a}{x+a}}\right] \tag{4.7.23}$$

where

$$T_{II} = \int_{a-r_T}^{a} \frac{\tau_{xy}^0(\xi)}{\sqrt{a-\xi}}d\xi, \quad H_{II} = -\int_{-a}^{-a+r_H} \frac{\tau_{xy}^0(\xi)}{\sqrt{a+\xi}}d\xi \tag{4.7.24}$$

T_{II} is obviously positive, and it can be shown that H_{II} is non-negative by considering

$$\frac{\partial u_-}{\partial x} = -\frac{i}{2(1-k^2)\mu}[F_+(x) + F_-(x)] \text{ for } |x| < a \tag{4.7.25}$$

where $F_+(x)$ is obtained for $|x| < a$ from (4.7.18) by making an indentation below the pole $\xi = x$ in the integrand:

$$F_+(x) = \frac{\sqrt{a^2 - x^2}}{2\pi i}\oint_{-a}^{a} \frac{\tau_{xy}^0(\xi)d\xi}{\sqrt{a^2 - \xi^2}(\xi - x)} + \frac{1}{2}\tau_{xy}^0(x) \tag{4.7.26}$$

where C on the integral sign denotes the Cauchy principal value†. The expression for $F_-(x)$ is similar, but with negative sign of the last term, so that

$$\frac{\partial u_-}{\partial x} = -\frac{\sqrt{a^2 - x^2}}{2\pi(1-k^2)\mu}\oint_{-a}^{a} \frac{\tau_{xy}^0(\xi)d\xi}{\sqrt{a^2 - \xi^2}(\xi - x)} \tag{4.7.27}$$

---

† The Cauchy principal value is the integral for the straight parts of the integration path that contains an infinitesimally small semi-circular indentation around the pole $\xi = x$.

which for $(x+a)^2 \gg r_H^2$ and $(a-x)^2 \gg r_T^2$ may be written as

$$\frac{\partial u_-}{\partial x} \approx \frac{1}{2\pi(1-k^2)\mu\sqrt{2a}}\left[T_{II}\sqrt{\frac{a+x}{a-x}} + H_{II}\sqrt{\frac{a-x}{a+x}}\right] \qquad (4.7.28)$$

Thus, if H_{II} were negative, slipping would vanish somewhere along the slipping region and change sign, which is physically unacceptable.

The constant T_{II} is recognized as the Barenblatt cohesion modulus, given earlier for mode I, (3.6.18), and, in the limit $r_T/a \to 0$, equilibrium demands that $T_{II} = \sqrt{\pi/2}K_{II}$, cf. (3.6.16). H_{II} is a similar modulus for the healing process. Usually, healing does not restore the cohesive bonds to anything comparable with the bonds before tearing, especially if the crack is propagating through a homogeneous material rather than along a weak interface. In such cases, H_{II} may be neglected. Such energy-neutral healing also implies that the integration interval in (4.7.22) and (4.7.27) is reduced to $(a-r_T, a)$. Then, for $x > -a$, $x+a \ll a$, relation (4.7.27) gives

$$\frac{\partial u_-}{\partial x} \approx \frac{T_{II}\sqrt{a+x}}{4\pi(1-k^2)\mu a} \qquad (4.7.29)$$

which integrates to

$$u_- \approx \frac{T_{II}(a+x)^{3/2}}{6\pi(1-k^2)\mu a} \qquad (4.7.30)$$

This shows smooth closing at the trailing end. Similarly, for $x < -a$, $|x+a| \ll a$, relation (4.7.23) gives

$$\tau_{xy} \approx \frac{T_{II}}{2\pi a}\sqrt{|x+a|} \qquad (4.7.31)$$

showing stress continuity at the trailing end.

The amount of slip deposited behind the crack, on each side, is

$$\Delta = \int_a^{-a} \frac{\partial u_+}{\partial x}\,dx = \int_{a-r_T}^a \frac{\tau_{xy}^0(\xi)}{\sqrt{a^2-\xi^2}} \oint_{-a}^a \frac{\sqrt{a^2-x^2}}{x-\xi}\,dx\,d\xi \qquad (4.7.32)$$

where a change of integration order was made. The inner integral equals $\pi\xi \approx \pi a$, and the result is

$$\Delta \approx \frac{T_{II}}{2(1-k^2)\mu}\sqrt{\frac{a}{2}} \qquad (4.7.33)$$

which in the limit $r_T/a \to 0$ becomes exact, in agreement with the result (4.4.93).

Note that the smooth closing condition (4.7.21) is a condition for $\tau_{xy}^0(x)$ when τ_{xy}^∞ is given. If the remote load is slowly increased, $\tau_{xy}^0(x)$ increases until the maximum capacity of the process region of preventing crack growth is reached.

Arrays of dislocations

In the preceding subsection, elementary loads were superposed so as to give required displacements on the boundary. Here, elementary displacements, in the form of dislocations, will be superposed so as to give required stresses on the boundary. This method was used by Yokobori et al. (1965a,b), Yokobori and Ichikawa (1965), Yokobori (1968), and Yokobori et al. (1971) for interaction between cracks of different lengths and orientations and for interaction between cracks and dislocations or slip bands. See also Kamei and Yokobori (1974) for a range of similar problems. Yokobori et al. (1965b) solved the mode I problem of two coplanar cracks of unequal length in a body large enough with respect to the cracks and the ligaments to be considered as infinite. This problem will be used to demonstrate the procedure.

Under plane strain or plane stress, the stress field around a dislocation situated at $x = y = 0$ is

$$\sigma_x = \frac{\tau_0}{(x^2 + y^2)^2}[-b_x y(3x^2 + y^2) + b_y x(x^2 - y^2)] \quad (4.7.34)$$

$$\sigma_y = \frac{\tau_0}{(x^2 + y^2)^2}[b_x y(x^2 - y^2) + b_y x(x^2 + 3y^2)] \quad (4.7.35)$$

$$\tau_{xy} = \frac{\tau_0}{(x^2 + y^2)^2}[b_x x(x^2 - y^2) + b_y y(x^2 - y^2)] \quad (4.7.36)$$

where

$$\tau_0 = \frac{1 - k^2}{\pi}\mu \quad (4.7.37)$$

and b_x and b_y are the x and y components of the Burger's vector. These components are related to a displacement discontinuity such that the resulting displacement steps are

$$b_x = \oint \frac{\partial u}{\partial s} ds \text{ and } b_y = \oint \frac{\partial v}{\partial s} ds \quad (4.7.38)$$

where the integration is taken counterclockwise around the dislocation line (here $x = y = 0$) and s is the arc length.

Expressions (4.7.34)-(4.7.36) may, for instance, be derived from (4.4.9)-(4.4.10), using the potentials

$$f(z) = -\frac{i\tau_0}{2}(b_x + ib_y)\ln z, \quad g'(z) = \frac{i\tau_0}{2}(b_x - ib_y)\ln z \quad (4.7.39)$$

which, inserted into (4.4.12) give u and v that satisfy (4.7.38).

An open mode I crack along the x axis may be considered as a continuous array of dislocations, each with Burger's vector in the y direction, implying discontinuity in the displacement v. The stress σ_y on the x axis due to one dislocation, with Burger's vector $b_y \hat{y}$, is, according to (4.7.35),

$$\sigma_y = \frac{P}{x} \quad (4.7.40)$$

where $P = \tau_0 b_y$ will be called the *strength* of the dislocation. Thus, the contribution to the stress σ_y on the x axis from a dislocation element with strength $f(\xi)d\xi$ in a continuous array of dislocation elements is

$$d\sigma_y = \frac{f(\xi)d\xi}{x - \xi} \quad (4.7.41)$$

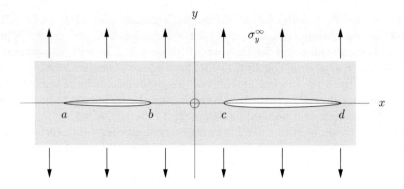

Fig. 4.7.1 Two collinear cracks subjected to remote mode I loading.

Now, suppose the two coplanar cracks to be situated along the portions L of the x axis, consisting of the segments $a < x < b$ and $c < x < d$, where $b < c$; see Fig. 4.7.1. Considering each crack as a continuous distribution of dislocations, the stress on the x axis is found to be

$$\sigma_y = \oint_L \frac{f(\xi)d\xi}{x-\xi} + \sigma_y^\infty \tag{4.7.42}$$

where σ_y^∞ is the remote stress and the integral is to be taken in the sense of the Cauchy principal value. On the crack faces, $\sigma_y = 0$ and thus,

$$\int_L \frac{f(\xi)d\xi}{\xi - x} = \sigma_y^\infty, \quad x \in L, \quad f(x) = 0, \quad x \notin L \tag{4.7.43}$$

To solve this equation, put

$$2\pi i f(x) = \varphi(x) \tag{4.7.44}$$

where $\varphi(z)$ is a complex function that vanishes as $z \to \infty$. Introduce now the sectionally analytic function

$$F(z) = \frac{1}{2\pi i} \int_\Gamma \frac{\varphi(\xi)d\xi}{\xi - z} \tag{4.7.45}$$

where Γ is the closed path containing the real axis and infinity on the upper half plane. Note that $F(z) \to 0$ as $z \to \infty$. Thus, the Plemelj formulae yield the Hilbert problem

$$F_+(x) + F_-(x) = \frac{1}{\pi i}\oint_L \frac{\varphi(\xi)d\xi}{\xi - x} = 2\sigma_y^\infty \text{ for } x \in L \tag{4.7.46}$$

$$F_+(x) - F_-(x) = \varphi(x) = 0 \text{ for } x \ni L \tag{4.7.47}$$

The solution is found after using the condition that $F(z)$ vanishes at infinity:

$$f(x) = -\frac{\sigma_y^\infty}{\pi^2 G_+(x)} \int_L \frac{G_+(\xi)d\xi}{\xi - x} + \frac{Ax + B}{\pi i G_+(x)}, \quad x \in L \tag{4.7.48}$$

where A and B are constants and

$$G(z) = [(z-a)(z-b)(z-c)(z-d)]^{1/2} \tag{4.7.49}$$

is defined through branch cuts along L and the choice that $G(z) \to z^2$ as $z \to \infty$. This makes $G_+(x) = i|G(x)|$ for $c < x < d$ and $G_+(x) = -i|G(x)|$ for $a < x < b$.

Constants A and B are determined from the conditions that each crack must close at its ends, i.e. the net Burger's vector for each crack must vanish:

$$\int_a^b f(\xi)\mathrm{d}\xi = 0, \qquad \int_c^d f(\xi)\mathrm{d}\xi = 0 \qquad (4.7.50)$$

The stress $(\sigma_y)_{y=0}$ is found after determination of A and B:

$$\sigma_y = -\oint_L \frac{f(\xi)\mathrm{d}\xi}{\xi - x} + \sigma_y^\infty \qquad (4.7.51)$$

The integral can be expressed in terms of the complete elliptic integrals K and E of the first and second kind, respectively. After lengthy calculations, the result is found to be

$$\sigma_y = \frac{\sigma_y^\infty}{2G(x)}\left[2x^2 - (a+b+c+d)x + ab + cd - (d-b)(c-a)\frac{E(m)}{K(m)}\right] \qquad (4.7.52)$$

for $x \notin L$, $y = 0$. Here,

$$m = \sqrt{\frac{(d-c)(b-a)}{(d-b)(c-a)}} \qquad (4.7.53)$$

The stress intensity factor at, for instance, crack edge $x = b$ is then found from

$$K_I^{(b)} = \lim_{x \to b+0}\left[\sqrt{2\pi(x-b)}\sigma_y(x,0)\right] \qquad (4.7.54)$$

Dislocation arrays in finite and semi-infinite regions

Another example will be used to show how the procedure works in the case of finite or semi-infinite plates with interior cracks or edge cracks. The example concerns a crack $0 \leq x < a$, $y = 0$, in a plate $x \geq 0$, i.e. an edge crack in a plate, large enough to be considered as semi-infinite. The plate is subjected to a remote stress $\sigma_y = \sigma_y^\infty$. The problem of finding the stress intensity factor will be treated in three consecutive steps. In the first step the stress $\sigma_y = \sigma_y(x)$ on $y = 0$ will be sought for a symmetrical normal traction $\sigma_x = \sigma_x^0(y)$ and a symmetrical shear traction $\tau_{xy} = \tau_{xy}^0(y)$ on $x = 0$, assuming a crack free plate. In the second step, still with a crack free plate, the result obtained in the first step will be used to find the stress $\sigma_y = \sigma_y(x)$ on $y = 0$ from a dislocation along a line $x = \xi$, $y = 0$. In the third step an integral equation is formed by requiring that the stress $\sigma_y = \sigma_y(x)$ on $y = 0$ resulting from a continuous array of dislocations on $0 \leq x < a$, $y = 0$, equals $-\sigma_y^\infty$. This procedure is comparatively simple, because the first two steps deal with a crack free plate, and the third step consists simply of superposing previously obtained solutions. The first two steps are illustrated in Fig. 4.7.2.

The first step is straightforward, and consists of a well-known solution of a simple boundary value problem for the half-plane (see Appendix A2), where, however, coordinate axes are different so that y in the Appendix shall be replaced by x, x by $-y$,

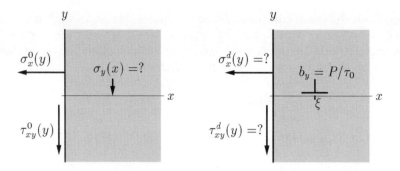

Fig. 4.7.2 Illustration of the first two steps in the procedure. In the first step the normal stress along $y=0$ is sought for arbitrary $\sigma_x^0(y)$ and $\tau_{xy}^0(y)$. In the second step, the normal stress along $y=0$ is sought for a half plane containing an edge dislocation. This is achieved by first determining $\sigma_x^d(y)$ and $\tau_{xy}^d(y)$ at $x=0$ for an infinite plate and then superposing the response found from the first step, after putting $\sigma_x^0(y) = -\sigma_x^d(y)$ and $\tau_{xy}^0(y) = -\tau_{xy}^d(y)$.

σ_x by σ_y, σ_y by σ_x and τ_{xy} by $-\tau_{xy}$. Thus,

$$\sigma_y = \frac{2}{\pi}\int_{-\infty}^{+\infty}\frac{[\sigma_x^0(s)x - \tau_{xy}^0(s)s]s^2}{(x^2+s^2)^2}\,ds \text{ for } x>0,\, y=0 \quad (4.7.55)$$

In step number two, the stress σ_y on $y=0$, caused by an edge dislocation with Burger's vector $b_y = P/\tau_0$ and located at $x=\xi$, $y=0$, is sought. If the plate were infinite, the stresses σ_x and σ_y on the plane $x=0$ would be, cf. (4.7.34) and (4.7.36),

$$\sigma_x^d = -\frac{P\xi(\xi^2-y^2)}{(\xi^2+y^2)^2},\quad \tau_{xy}^d = \frac{Py(\xi^2-y^2)}{(\xi^2+y^2)^2} \quad (4.7.56)$$

and the normal stress on $y=0$ would be (4.7.35)

$$\sigma_y = \frac{P}{x-\xi} \quad (4.7.57)$$

Removal of the left half-plane and superposition of the tractions $-\sigma_x^d$ and $-\tau_{xy}^d$ on the edge $x=0$ will result in equilibrium for the half-plane $x\geq 0$ with traction free edge $x=0$ and containing an edge dislocation with Burger's vector $b_y=P/\tau_0$, located at $x=\xi$, $y=0$. The normal stress on $y=0$ is found from the contribution (4.7.57) together with the contribution (4.7.55) after replacing σ_x^0 by $-\sigma_x^d$ and τ_{xy}^0 by $-\tau_{xy}^d$. This results in

$$\sigma_y = \frac{P}{x-\xi} + \frac{2P}{\pi}\int_{-\infty}^{+\infty}\frac{(x\xi+s^2)(\xi^2-s^2)s^2}{(\xi^2+s^2)^2(x^2+s^2)^2}\,ds \quad (4.7.58)$$

on $y=0$. The integral is elementary, and the expression reduces to

$$\sigma_y = \frac{P}{x-\xi} + Ph(x,\xi),\quad h(x,\xi)=\frac{\xi^2-4x\xi-x^2}{(x+\xi)^3} \quad (4.7.59)$$

Proceeding now to step three, letting

$$P \to \sigma_y^\infty f(\xi)d\xi \text{ and } \sigma_y \to d\sigma_y \quad (4.7.60)$$

in (4.7.59) and integrating over the interval $0 < x < a$, leads to the expression

$$\sigma_y = -\sigma_y^\infty \oint_0^a \frac{f(\xi) d\xi}{\xi - x} + \sigma_y^\infty \int_0^a h(x, \xi) f(\xi) d\xi \quad \text{for } x > 0 \qquad (4.7.61)$$

Equating this stress to $-\sigma_y^\infty$ results in the integral equation

$$\oint_0^a \frac{f(\xi) d\xi}{\xi - x} - \int_0^a h(x, \xi) f(\xi) d\xi = 1 \quad \text{for } 0 < x < a \qquad (4.7.62)$$

After this equation is solved, the stress intensity factor at $x = a$ is obtained from (4.7.61) as

$$K_I = \lim_{x \to a+0} \left[\sqrt{2\pi(x - a)} \sigma_y \right] \qquad (4.7.63)$$

It is possible to reduce the singular integral equation (4.7.62) to a Fredholm equation. To this end, the equation is written as

$$\int_0^1 \frac{f(\xi) d\xi}{\xi - x} = 1 + \int_0^1 h(x, \xi) f(\xi) d\xi = g(x), \quad 0 < x < 1 \qquad (4.7.64)$$

where, temporarily, $g(x)$ is considered as a known function, and a is put equal to unity (which may be considered as a temporary change to dimensionless coordinates x/a, ξ/a). After extending the definition of $f(x)$ to $f(x) = 0$ for $x < 0$ and $x > 1$, and introducing the sectionally analytic function

$$F(z) = \frac{1}{2\pi i} \int_{-\infty}^{+\infty} \frac{f(\xi) d\xi}{\xi - z} \qquad (4.7.65)$$

the Plemelj formulae yield:

$$F_+(x) + F_-(x) = f(x) = 0 \quad \text{for } x < 0 \text{ and } x > 1 \qquad (4.7.66)$$

$$F_+(x) - F_-(x) = \frac{1}{\pi i} \int_{-\infty}^{+\infty} \frac{f(\xi) d\xi}{\xi - x} = \frac{1}{\pi i} g(x) \quad \text{for } 0 < x < 1 \qquad (4.7.67)$$

This constitutes a Hilbert problem, the solution of which is

$$F(z) = -\frac{1}{2\pi^2 G(z)} \int_0^1 \frac{G_+(\xi) g(\xi)}{\xi - z} d\xi + \frac{P(z)}{G(z)} \qquad (4.7.68)$$

where $G(z) = [z(z-1)]^{1/2}$ with a branch cut from $z = 0$ to $z = 1$ and the branch chosen so that $G(z) \to z$ as $z \to \infty$. The polynomial $P(z)$ reduces to a constant, because the function $F(z) \to O(1/z)$ as $z \to \infty$, according to (4.7.65). Thus,

$$f(x) = F_+(x) - F_-(x) = -\frac{1}{\pi^2 G_+(x)} \oint_0^1 \frac{G_+(\xi) g(\xi)}{\xi - x} d\xi + \frac{2P(0)}{G_+(x)}$$
$$\text{for } 0 < x < 1 \quad (4.7.69)$$

The constant $P(0)$ shall be chosen to satisfy the condition that no stress singularity prevails at the outward corner $x = y = 0$. This leads to:

$$f(x) = -\frac{1}{\pi^2 G_+(x)} \int_0^1 G_+(\xi) g(\xi) \left[\frac{1}{\xi - x} - \frac{1}{\xi}\right] d\xi$$

$$= -\frac{1}{\pi^2} \sqrt{\frac{x}{1-x}} \int_0^1 \sqrt{\frac{1-\xi}{\xi}} \cdot \frac{g(\xi) d\xi}{\xi - x} \quad (4.7.70)$$

The integral in the last member equals, see (4.7.64),

$$-\pi - \int_0^1 f(s) \int_0^1 \sqrt{\frac{1-\xi}{\xi}} \cdot \frac{h(\xi, s)}{\xi - x} d\xi\, ds$$

$$= -\pi - \int_0^1 f(s) \int_0^1 \sqrt{\frac{1-\xi}{\xi}} \cdot \frac{s^2 - 4\xi s - \xi^2}{(s+\xi)^3} \cdot \frac{d\xi}{\xi - x} ds$$

$$= -\pi - \pi \int_0^1 a(s, x) f(s)\, ds \quad (4.7.71)$$

where

$$a(s, x) = \frac{s^2(2s^2 + 10s + 7) - 2xs(4s^2 + 4s + 1) - x^2(2s^2 + 2s + 1)}{2s^{1/2}(s+1)^{3/2}(s+x)^3} \quad (4.7.72)$$

Insertion into (4.7.70) gives

$$\pi \sqrt{\frac{1-x}{x}} f(x) = 1 + \int_0^1 a(s, x) f(s) ds \quad (4.7.73)$$

whereupon the substitution

$$\pi \sqrt{\frac{1-x}{x}} f(x) = \Phi(x) \quad (4.7.74)$$

leads to the Fredholm equation

$$\Phi(x) - \frac{1}{\pi} \int_0^1 K(s, x) \Phi(s) ds = 1 \quad (4.7.75)$$

where

$$K(s, x) = a(s, x) \sqrt{\frac{s}{1-s}} \quad (4.7.76)$$

After solution of the equation, the stress intensity factor is found by using (4.7.61), in which only the first integral is singular. Because values of ξ close to 1 dominate the integral, this singularity is found to be:

$$\sigma_y \to -\frac{\sigma_y^\infty}{\pi}\int_0^1 \sqrt{\frac{\xi}{1-\xi}} \cdot \frac{\Phi(\xi)\mathrm{d}\xi}{\xi - x}$$

$$\to -\frac{\sigma_y^\infty}{\pi}\Phi(1)\int_0^1 \sqrt{\frac{\xi}{1-\xi}} \cdot \frac{\mathrm{d}\xi}{\xi - x}$$

$$\to \sigma_y^\infty \frac{\Phi(1)}{\sqrt{x-1}} \quad \text{as } x \to 1+0 \quad (4.7.77)$$

which gives the stress intensity factor, (4.7.63), as

$$K_I = \sqrt{2}\Phi(1)\sigma_y^\infty\sqrt{\pi a} \quad (4.7.78)$$

after reinstatement of crack length a.

Numerical solution of (4.7.75) may be made by the approximate transformation of the integral to a sum, using a Gauss integration formula of the type

$$\int_0^1 \frac{F(x)}{\sqrt{1-x}}\mathrm{d}x \approx \sum_{n=1}^N w_n F(x_n) \quad (4.7.79)$$

where $x_n = 1 - \xi_n^2$, ξ_n is the nth positive zero of the Legendre function $P_{2N}(\xi)$, and

$$w_n = \frac{2}{(1-\xi_n^2)[P'_{2N}(\xi_n)]^2} \quad (4.7.80)$$

Solution of the resulting equation system,

$$\sum_{n=1}^N X(s_n, x_i) = \Phi(x_i) - 1$$

using $N = 22$ collocation points x_i (equal to the integration point set s_n), or more, gives, with a 3 point Lagrangian extrapolation formula for obtaining $\Phi(1)$, the stress intensity factor

$$K_I \approx 1.1215\,\sigma_y^\infty\sqrt{\pi a} \quad (4.7.81)$$

The result $K_I \approx 1.12\,\sigma_y^\infty\sqrt{\pi a}$, which is sufficiently accurate for virtually all engineering applications, is obtained by using only 6 collocation points.

Note that the method works well, even for finite plates if a solution can be found to the simple boundary value problem of prescribed tractions along the edges of the crack free plate. Naturally, the functions $h(x,\xi)$ and

$$a(s,x) = \int_0^1 \sqrt{\frac{1-\xi}{\xi}} \cdot \frac{h(\xi,s)}{\xi - x}\mathrm{d}\xi \quad (4.7.82)$$

cannot always be expressed in closed form, but $h(x,\xi)$, which represents the solution for a boundary value problem for the crack free plate, can generally be calculated numerically rather accurately, for instance by means of complex potential methods, and the ensuing numerical integration to find $a(s,x)$ should not pose any difficulties. The case of an internal crack is treated in essentially the same way, but the condition

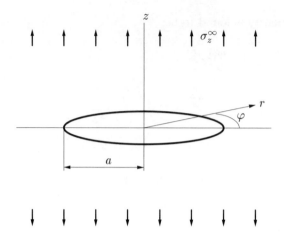

Fig. 4.8.1 Penny-shaped crack.

that no stress singularity prevails at the outward corner $x = y = 0$ has to be replaced by the condition that the crack closes at the ends.

The method has been found to yield satisfactory results for a number of symmetrical problems. One further example is given by symmetrical cracking from a circular hole in a plate (Atkinson and Thiercelin 1993). However, the method is not confined to symmetrical cases: it works, for instance, also for curved cracks; see e.g. Gol'dshtein and Salganik (1970, 1974) and Gol'dshtein and Savova (1972). One example will be given in Section 4.11 in connection with directional stability of crack paths.

4.8 Dual integral equations

The integral equations obtained in the previous section were based on the possibility of obtaining a fairly simple closed form solution for the response to an elementary load or to an elementary displacement step. An integral equation formulation of an elastostatic crack problem may also be obtained by using integral transforms. This will be demonstrated for the case of a "penny-shaped" crack in an infinite solid, a problem solved by Sneddon (1946) and, in a somewhat more general form, by Green and Zerna (1954).

A cylindrical coordinate system r, z, φ is introduced. The crack occupies the plane $z = 0$, $r < a$; see Fig. 4.8.1. The body is subjected to a remote load $\sigma_z = \sigma_z^\infty$. The half-plane $z \geq 0$ is considered. For convenience, the stress $\sigma_z = -\sigma_z^\infty$ is superposed everywhere. Then, the boundary conditions will read:

$$z = 0: \quad u_z = 0 \text{ for } r \geq a \tag{4.8.1}$$
$$\sigma_z = -\sigma_z^\infty \text{ for } r < a \tag{4.8.2}$$
$$\tau_{rz} = 0 \text{ for } 0 \leq r < \infty \tag{4.8.3}$$
$$z \to \infty: \quad \sigma_z \to 0 \tag{4.8.4}$$

4.8 DUAL INTEGRAL EQUATIONS

Because of the rotational symmetry, the Papkovich solution (4.4.1) can be written with only two potential functions, Φ and Ψ_z. Thus,

$$2\mu u_r = -z\frac{\partial \Psi_z}{\partial r} - \frac{\partial \Phi}{\partial r}, \quad 2\mu u_z = (3-4\nu)\Psi_z - z\frac{\partial \Psi_z}{\partial z} - \frac{\partial \Phi}{\partial z} \quad (4.8.5)$$

where u_r and u_z are the displacements in the r and z directions. The stresses are found from Hooke's law with use of relations (4.4.2):

$$\sigma_z = 2(1-\nu)\frac{\partial \Psi_z}{\partial z} - z\frac{\partial^2 \Psi_z}{\partial z^2} - \frac{\partial^2 \Phi}{\partial z^2} \quad (4.8.6)$$

$$\tau_{rz} = (1-2\nu)\frac{\partial \Psi_z}{\partial r} - z\frac{\partial^2 \Psi_z}{\partial r \partial z} - \frac{\partial \Phi}{\partial r \partial z} \quad (4.8.7)$$

Introduction of the Hankel transforms (see Appendix A7),

$$F = \int_0^\infty r\Phi(r,z) J_0(\xi r) dr, \quad S = \int_0^\infty r\Psi_z(r,z) J_0(\xi r) dr \quad (4.8.8)$$

transforms the partial differential equations (4.4.2) into the ordinary differential equations

$$\frac{d^2 F}{dz^2} = \xi^2 F, \quad \frac{d^2 S}{dz^2} = \xi^2 S \quad (4.8.9)$$

Making use of boundary condition (4.8.4), the solution of these equations is

$$F = A(\xi)e^{-\xi z}, \quad S = C(\xi)e^{-\xi z} \quad (4.8.10)$$

Inversion gives:

$$\Phi = \int_0^\infty \xi A(\xi)e^{-\xi z} J_0(\xi r) d\xi, \quad \Psi_z = \int_0^\infty \xi B(\xi)e^{-\xi z} J_0(\xi r) d\xi \quad (4.8.11)$$

The boundary condition (4.8.3) yields

$$B(\xi) = -\frac{\xi A(\xi)}{1-2\nu} \quad (4.8.12)$$

Thus, for $z = 0$:

$$\mu u_z = -\frac{1-\nu}{1-2\nu}\int_0^\infty \xi^2 A(\xi) J_0(\xi r) d\xi \quad (4.8.13)$$

$$\sigma_z = \frac{1}{1-2\nu}\int_0^\infty \xi^3 A(\xi) J_0(\xi r) d\xi \quad (4.8.14)$$

Combined with the boundary conditions (4.8.1) and (4.8.2), these equations become dual integral equations for solving the unknown function $A(\xi)$. The first one is satisfied by taking

$$\xi^2 A(\xi) = \int_0^a g(x) \sin(\xi x) dx \quad (4.8.15)$$

because, as may be found from integral tables, for instance Gradshteyn and Ryzhik

(1980),

$$\mu u_z = -\frac{1-\nu}{1-2\nu}\int_0^a g(x)\int_0^\infty \sin(\xi x)J_0(\xi r)\,\mathrm{d}\xi\,\mathrm{d}x$$

$$= \begin{cases} -\dfrac{1-\nu}{1-2\nu}\displaystyle\int_r^a \dfrac{g(x)}{\sqrt{x^2-r^2}}\,\mathrm{d}x & \text{for } r<a \\ 0 & \text{for } r>a \end{cases} \quad (4.8.16)$$

The stress σ_z can now be expressed with $g(x)$:

$$\sigma_z = \frac{1}{1-2\nu}\int_0^\infty\int_0^a g(x)\xi\sin(\xi x)J_0(r\xi)\,\mathrm{d}x\,\mathrm{d}\xi \qquad (4.8.17)$$

After the partial integration

$$\int_0^a g(x)\sin(\xi x)\,\mathrm{d}x = \left[-g(x)\cdot\frac{1}{\xi}\cos(\xi x)\right]_0^a + \frac{1}{\xi}\int_0^a g'(x)\cos(\xi x)\,\mathrm{d}x$$

$$= \frac{g(0)}{\xi} - \frac{g(a)}{\xi}\cos(\xi a) + \frac{1}{\xi}\int_0^a g'(x)\cos(\xi x)\,\mathrm{d}x \qquad (4.8.18)$$

further consultation of integral tables yields

$$\sigma_z = \begin{cases} \dfrac{1}{1-2\nu}\left[\dfrac{g(0)}{r}+\displaystyle\int_0^r \dfrac{g'(x)}{\sqrt{r^2-x^2}}\,\mathrm{d}x\right] & \text{for } r<a \\ \dfrac{1}{1-2\nu}\left[\dfrac{g(0)}{r}-\dfrac{g(a)}{\sqrt{r^2-a^2}}+\displaystyle\int_0^a \dfrac{g'(x)}{\sqrt{r^2-x^2}}\,\mathrm{d}x\right] & \text{for } r>a \end{cases} \quad (4.8.19)$$

Condition (4.8.2) then gives

$$g(0) = 0 \qquad (4.8.20)$$

$$\int_0^r \frac{g'(x)}{\sqrt{r^2-x^2}}\,\mathrm{d}x = -(1-2\nu)\sigma_z^\infty \qquad (4.8.21)$$

This Abelian integral equation has the solution (see e.g. Kanwal (1971))

$$g'(x) = -\frac{2}{\pi}(1-2\nu)\sigma_z^\infty \qquad (4.8.22)$$

and thus, with consideration of (4.8.20),

$$g(x) = -\frac{2}{\pi}(1-2\nu)\sigma_z^\infty x \qquad (4.8.23)$$

Now, from (4.8.16) and (4.8.19), the displacement $(u_z)_+$ on the upper crack face, and the stress σ_z on the plane $z=a$ are obtained:

$$\mu(u_z)_+ = \frac{2(1-\nu)}{\pi}\sigma_z^\infty\sqrt{a^2-r^2} \quad \text{for } r<a \qquad (4.8.24)$$

$$\sigma_z = \frac{2}{\pi}\sigma_z^\infty\left[\frac{a}{\sqrt{r^2-a^2}} - \operatorname{asin}\frac{a}{r}\right] \quad \text{for } r>a \qquad (4.8.25)$$

Returning to the original problem, the stress $\sigma_z = \sigma_z^\infty$ should be superposed, giving

$$\sigma_z = \frac{2}{\pi}\sigma_z^\infty\left[\frac{a}{\sqrt{r^2-a^2}} + \operatorname{acos}\frac{a}{r}\right] \quad \text{for } r>a \qquad (4.8.26)$$

and the stress intensity factor is found to be

$$K_I = \lim_{r \to a}\left[\sqrt{2\pi(r-a)}\sigma_z\right] = 2\sigma_z^\infty \sqrt{\frac{a}{\pi}} \qquad (4.8.27)$$

The functions $A(\xi)$ and $B(\xi)$ follow from (4.8.15) and (4.8.12):

$$A(\xi) = -\frac{2(1-2\nu)}{\pi}\sigma_z^\infty \cdot \frac{1}{\xi^4}[\sin(\xi a) - \xi a \cos(\xi a)] \qquad (4.8.28)$$

$$B(\xi) = \frac{2}{\pi}\sigma_z^\infty \cdot \frac{1}{\xi^3}[\sin(\xi a) - \xi a \cos(\xi a)] \qquad (4.8.29)$$

Then, Φ and Ψ_z are obtained from (4.8.11), so that expressions may be given for stresses and strains everywhere. For example, Hooke's law and relations (4.8.5) yield

$$\sigma_r = \frac{2\mu}{1-2\nu}\left[(1-\nu)\frac{\partial u_r}{\partial r} + \nu\left(\frac{u_r}{r} + \frac{\partial u_z}{\partial z}\right)\right] = 2\nu\frac{\partial \Psi_z}{\partial z} - z\frac{\partial^2 \Psi_z}{\partial r^2} - \frac{\partial^2 \Phi}{\partial r^2} \qquad (4.8.30)$$

For $z = 0$, the result is

$$\sigma_r = \begin{cases} -\dfrac{1+2\nu}{2}\sigma_z^\infty & \text{for } r < a \\[2mm] \dfrac{\sigma_z^\infty}{\pi}\left[\dfrac{2a}{\sqrt{r^2-a^2}} - (1+2\nu)\operatorname{asin}\dfrac{a}{r} - (1+2\nu)\dfrac{a\sqrt{r^2-a^2}}{r^2}\right] & \text{for } r > a \end{cases} \qquad (4.8.31)$$

Thus, the T-stress is

$$T = -\frac{1+2\nu}{2}\sigma_z^\infty \qquad (4.8.32)$$

The generalization to an axisymmetric load $\sigma_z = \sigma_z^0(r)$ on the crack faces in addition to the remote load (Green and Zerna 1954) is obtained by replacing the right member of (4.8.21) by $(1-2\nu)[\sigma_z^0(r) - \sigma_z^\infty]$. This leads to the stress intensity factor

$$K_I = \frac{2}{\sqrt{\pi a}}\int_0^a \frac{r[\sigma_z^\infty - \sigma_z^0(r)]}{\sqrt{a^2 - r^2}}dr \qquad (4.8.33)$$

4.9 Wiener-Hopf technique

General considerations

The Wiener-Hopf technique (Wiener and Hopf 1931, Fok 1944, Noble 1958) is a method for dealing with mixed boundary value problems or with certain integral equations, so-called Wiener-Hopf equations. Characteristic is the determination of two unknown analytic functions from one single equation. The analytic functions are generally integral transforms, for instance Laplace, Fourier or Mellin transforms, and the Wiener-Hopf technique rests heavily on the possibility of determining the asymptotic character of these transforms when the transform variable approaches certain points in the complex plane. The procedure is perhaps best introduced by means of an example: the edge crack problem treated in Section 4.7 will be used for that purpose.

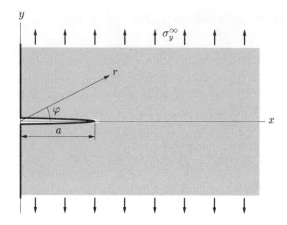

Fig. 4.9.1 Edge crack in a semi-infinite plate.

A mode I edge crack in a large plate

Consider a semi-infinite plate, $x \geq 0$, containing a crack, $0 \leq x < a$, $y = 0$, and subjected to a remote load, $\sigma_y = \sigma_y^\infty$. Alongside the Cartesian coordinate system, a polar coordinate system (r, φ) is introduced, with the same origin and such that $\varphi = 0$ coincides with the positive x axis; see Fig. 4.9.1.

The following treatment is essentially due to Koiter (1956a,b, 1965). The stress intensity factor may be found from the somewhat more convenient problem with zero remote load and the traction $\sigma_y = -\sigma_y^\infty$ on the crack faces. It is also convenient to use dimension-less coordinates $(r/a, \varphi)$, but for simplicity no new notations will be introduced: a may be envisaged as being set to unity.

Because of the symmetry it is sufficient to consider the sector $0 \leq \varphi \leq \pi/2$. The equilibrium conditions and Hooke's law read:

$$r\frac{\partial \sigma_r}{\partial r} + \frac{\partial \tau_{r\varphi}}{\partial \varphi} + \sigma_r - \sigma_\varphi = 0 \tag{4.9.1}$$

$$\frac{\partial \sigma_\varphi}{\partial \varphi} + r\frac{\partial \tau_{r\varphi}}{\partial r} + 2\tau_{r\varphi} = 0 \tag{4.9.2}$$

$$4(1-k^2)\mu \cdot \frac{\partial u_\varphi}{\partial r} = \sigma_r - (1-2k^2)\sigma_\varphi \tag{4.9.3}$$

$$4(1-k^2)\mu \cdot \frac{1}{r}\left(\frac{\partial u_\varphi}{\partial \varphi} + u_r\right) = \sigma_\varphi - (1-2k^2)\sigma_r \tag{4.9.4}$$

$$\mu\left[\frac{\partial u_r}{\partial r} - \frac{1}{r}\left(u_\varphi - \frac{\partial u_r}{\partial \varphi}\right)\right] = \tau_{r\varphi} \tag{4.9.5}$$

The following Mellin transforms (see Appendix A6) are introduced:

$$S(s,\varphi) = \int_0^\infty r^s \sigma_\varphi(r,\varphi)dr, \quad T(s,\varphi) = \int_0^\infty r^s \tau_{r\varphi}(r,\varphi)dr \tag{4.9.6}$$

$$V(s,\varphi) = \int_0^\infty r^s \frac{\partial u_\varphi}{\partial r}dr \tag{4.9.7}$$

From (4.9.1)-(4.9.5), after elimination of u_r and σ_r, the following expressions are obtained for Mellin transformed field variables:

$$4(1-k^2)\mu V = \frac{1}{s(s^2-1)}$$
$$\times \left\{ \frac{\partial^3 S}{\partial \varphi^3} + \left[4(1-k^2)s^2 - (s-1)[1+(1-2k^2)s] \right] \frac{\partial S}{\partial \varphi} \right\} \quad (4.9.8)$$

$$T = \frac{1}{s-1} \cdot \frac{\partial S}{\partial \varphi} \quad (4.9.9)$$

$$\frac{\partial^4 S}{\partial \varphi^4} + 2(s^2+1)\frac{\partial^2 S}{\partial \varphi^2} + (s^2-1)^2 S = 0 \quad (4.9.10)$$

These expressions were derived after suitable partial integrations, for instance in the derivation of (4.9.9) from (4.9.2).

The boundary conditions for the quarter-infinite plate considered, consist of traction free plate edge, $\varphi = \pi/2$, zero shear traction on $\varphi = 0$, prescribed normal traction, $-\sigma_y^\infty$, on the crack face $\varphi = 0, r < 1$ and vanishing displacement gradient, $\partial u_\varphi/\partial r = 0$, on $\varphi = 0, r > 1$. Because two boundary conditions on $\varphi = 0$ are given for only one part ($r < 1$ or $r > 1$) of this boundary, the corresponding Mellin transforms are decomposed into two parts:

$$S_+ = \int_0^1 r^s \sigma_\varphi(r,0) dr, \qquad S_- = \int_1^\infty r^s \sigma_\varphi(r,0) dr \quad (4.9.11)$$

$$V_+ = \int_0^1 r^s \left(\frac{\partial u_\varphi}{\partial r}\right)_{\varphi=0} dr, \qquad V_- = \int_1^\infty r^s \left(\frac{\partial u_\varphi}{\partial r}\right)_{\varphi=0} dr \quad (4.9.12)$$

The boundary conditions give, for $\varphi = \pi/2$,

$$(S)_{\varphi=\pi/2} = 0, \qquad (T)_{\varphi=\pi/2} = 0 \quad (4.9.13)$$

and, for $\varphi = 0$,

$$(T)_{\varphi=0} = 0, \quad S_+ = -\frac{\sigma_y^\infty}{s+1}, \quad \Re(s) > -1, \quad V_- = 0 \quad (4.9.14)$$

Because equilibrium conditions imply that the normal stress on $\varphi = 0, r > 0$ is integrable ($=\sigma_y^\infty$), it is obvious that the unknown quantity S_- is convergent for $\Re(s) \leq 0$. The other unknown boundary quantity, V_+, is certainly convergent for $\Re(s) > -1$. The strip $-1 < \Re(s) < 0$ is thus a common strip of convergence for all Mellin transforms introduced.

The general solution of (4.9.10) is

$$S = A(s)\cos[(s+1)\varphi] + B(s)\cos[(s-1)\varphi]$$
$$+ C(s)\sin[(s+1)\varphi] + D(s)\sin[(s-1)\varphi] \quad (4.9.15)$$

whereupon T and V are found from (4.9.9) and (4.9.8).
The boundary conditions (4.9.13) and the first of (4.9.14) give:

$$(S)_{\varphi=\pi/2} = -A\sin\frac{\pi s}{2} + B\sin\frac{\pi s}{2} + C\cos\frac{\pi s}{2} - D\cos\frac{\pi s}{2} = 0 \quad (4.9.16)$$

$$(s-1)(T)_{\varphi=\pi/2} = -(s+1)A\cos\frac{\pi s}{2} + (s-1)B\cos\frac{\pi s}{2}$$
$$- (s+1)C\sin\frac{\pi s}{2} + (s-1)D\sin\frac{\pi s}{2} = 0 \quad (4.9.17)$$

$$(s-1)(T)_{\varphi=0} = (s+1)C + (s-1)D = 0 \quad (4.9.18)$$

The two remaining boundary conditions, the last two of (4.9.14), are now substituted by conditions for S and V in the strip $-1 < \Re(s) < 0$, thus as if S_- and V_+ were known:

$$(S)_{\varphi=0} = A + B = S_- - \frac{\sigma_y^\infty}{s+1} \quad (4.9.19)$$

$$(V)_{\varphi=0} = \frac{1}{2(s-1)\mu}\left[\left(s - \frac{1+k^2}{1-k^2}\right)C + (s-1)D\right] = V_+ \quad (4.9.20)$$

Elimination of A, B, C and D gives a relation between S_- and V_+,

$$\frac{\sin(\pi s)}{\cos(\pi s) + 2s^2 - 1} \cdot [(s+1)S_- - \sigma_y^\infty] = 2\mu(1-k^2)(s+1)V_+ \quad (4.9.21)$$

valid in the strip $-1 < \Re(s) < 0$. This is the Wiener-Hopf equation, and it contains two unknown functions, S_- and V_+. The essential element in the solution, i.c. in the determination of both unknown functions, consists of writing the equation so that the left member is regular in the half-plane $s < 0$ and the right member is regular in the half-plane $s > -1$. To this end, a factorization (see page 679) of the first factor in the left member is necessary – the other factors and terms are regular in the appropriate half-planes. Factorization of an expression may be achieved by first decomposing its logarithm into a sum of two parts, regular in each half-plane, using the Cauchy integral. However, this integral is convergent only if the argument of the logarithm approaches unity at infinity (inside the strip). Here, it does not:

$$\frac{\sin(\pi s)}{\cos(\pi s) + 2s^2 - 1} \to \begin{cases} i & \text{as } s \to c + i\infty \\ -i & \text{as } s \to c - i\infty \end{cases} \quad (4.9.22)$$

where c is a real constant inside the strip. However, the desired behaviour can be obtained by multiplication with the function

$$\frac{s}{(B^2 - s^2)^{1/2}} \quad (4.9.23)$$

where B is a constant such that $\pm B$ is not inside the strip: it is here taken as real and larger than unity. Then,

$$H(s) = -\frac{s}{(B^2 - s^2)^{1/2}} \cdot \frac{\sin(\pi s)}{\cos(\pi s) + 2s^2 - 1} \to 1 \text{ as } s \to c \pm i\infty \quad (4.9.24)$$

and after setting $H(s) = H_+(s)H_-(s)$, where the subscripts indicate the regions of regularity, the decomposition formula (A4.51) yields

$$\ln H_-(s) = \frac{1}{2\pi i} \int_{c_2-i\infty}^{c_2+i\infty} \frac{1}{z-s} \ln \frac{-z\sin(\pi z)}{(B^2-z^2)^{1/2}[\cos(\pi z)+2z^2-1]} dz$$
$$\Re(s) < c_2, \quad c_2 \text{ inside the strip} \quad (4.9.25)$$

The substitution of c_2 by c_1 gives the expression for $-\ln H_+(s)$, regular for $\Re(s) > c_1$. Equation (4.9.21) may then be written in the form

$$-\frac{(B-s)^{1/2}}{H_-(s)}[(s+1)S_- - \sigma_y^\infty] = 2\mu(1-k^2)s(s+1)\frac{H_+(s)}{(B+s)^{1/2}}V_+$$

$$c_1 < \Re(s) < c_2 \quad (4.9.26)$$

The left member is regular for $\Re(s) < c_2$ and the right member is regular for $\Re(s) > c_1$. But then, because the two members are equal in the strip $c_1 < \Re(s) < c_2$, they define together a function that is analytic in the whole plane, possibly excluding infinity, but if so, behaving algebraically there, because physical quantities or at least their higher derivatives must be bounded everywhere. By Liouville's theorem, such a function is a finite polynomial, $P(s)$, so that

$$S_- = \frac{\sigma_y^\infty}{s+1} - \frac{P(s)H_-(s)}{(B-s)^{1/2}(s+1)}, \quad \Re(s) < c_2 \quad (4.9.27)$$

$$V_+ = \frac{1}{2(1-k^2)\mu} \cdot \frac{(B+s)^{1/2}P(s)}{s(s+1)H_+(s)}, \quad \Re(s) > c_1 \quad (4.9.28)$$

Because S_- cannot possess a singularity at $s = -1$, it follows from (4.9.27) that

$$\frac{P(-1)H_-(-1)}{(B+1)^{1/2}} = \sigma_y^\infty \quad (4.9.29)$$

and, because V_+ cannot possess a singularity at $s = 0$, it follows from (4.9.28) that

$$P(0) = 0 \quad (4.9.30)$$

The remaining part of the polynomial $P(s)$ can now be determined after studying V_+ as $s \to \infty$. The displacement gradient $(\partial u_\varphi/\partial r)_{\varphi=0}$ possesses an inverse square root singularity at $r = 1$, and it may therefore be written as

$$\left(\frac{\partial u_\varphi}{\partial r}\right)_{\varphi=0} = \frac{a(r)}{\sqrt{1-r}} \quad (4.9.31)$$

where $a(r)$ is bounded for $0 \le r \le 1$. Insertion into (4.9.12) gives

$$V_+ = \int_0^1 r^s \cdot \frac{a(r)}{\sqrt{1-r}} dr = a(\theta) \int_0^1 \frac{r^s}{\sqrt{1-r}} dr$$

$$= a(\theta)\frac{\Gamma(s+1)\Gamma(1/2)}{\Gamma(s+3/2)} \to \text{constant} \times s^{-1/2} \text{ as } s \to \infty \quad (4.9.32)$$

where θ is a value in the interval $0 \le \theta \le 1$. Then, it follows from (4.9.28) that $P(s)$ is a polynomial of first degree, and hence, according to (4.9.29):

$$P(s) = -\frac{(B+1)^{1/2}}{H_-(-1)}\sigma_y^\infty s \quad (4.9.33)$$

so that

$$S_- = \frac{\sigma_y^\infty}{s+1}\left[1 + \frac{(B+1)^{1/2}}{H_-(-1)} \cdot \frac{sH_-(s)}{(B-s)^{1/2}(s+1)}\right], \quad \Re(s) < c_2 \quad (4.9.34)$$

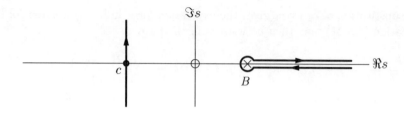

Fig. 4.9.2 Deformation of integration path from a vertical line through $s = c$ to a line folded around $s = B$. Note that the integrand vanishes for $|s| \to \infty$, $\Re(s) > c$.

$$(S)_{\varphi=0} = S_- + S_+ = \sigma_y^\infty \cdot \frac{(B+1)^{1/2}}{H_-(-1)} \cdot \frac{sH_-(s)}{(B-s)^{1/2}(s+1)}, \quad c_1 < \Re(s) < c_2 \quad (4.9.35)$$

Inversion gives:

$$\sigma_y = \frac{\sigma_y^\infty}{2\pi i} \cdot \frac{(B+1)^{1/2}}{H_-(-1)} \int_{c-i\infty}^{c+i\infty} \frac{sH_-(s)}{(B-s)^{1/2}(s+1)} r^{-s-1} \, ds \quad (4.9.36)$$

for $r > 1$. The constant c should be inside the strip of convergence, which can be taken as $-1 < \Re(s) < 0$, because c_1 and c_2 can be arbitrarily close to the strip edges. The integral is not convergent for $r = 1$, and it should not be, because a stress singularity is expected at the crack edge. The integral may be decomposed into one part, that is convergent for $r \geq 1$,

$$\int_{c-i\infty}^{c+i\infty} \left[\frac{sH_-(s)}{s+1} - 1 \right] \frac{r^{-s-1}}{(B-s)^{1/2}} \, ds \quad (4.9.37)$$

and one part, that is singular for $r = 1$:

$$\int_{c-i\infty}^{c+i\infty} \frac{r^{-s-1}}{(B-s)^{1/2}} \, ds \quad (4.9.38)$$

From this second part, the stress intensity factor can be found. The path of integration may be deformed, first to the imaginary axis and then to a part of the real axis, as shown in Fig. 4.9.2, so that

$$\int_{c-i\infty}^{c+i\infty} \frac{r^{-s-1}}{(B-s)^{1/2}} \, ds = 2 \int_B^\infty \frac{r^{-s-1}}{-i\sqrt{s-B}} \, ds \quad (4.9.39)$$

By means of the substitution $s - B = u$ the right member takes the form

$$2i \int_0^\infty \frac{e^{-(u+B+1)\ln r}}{\sqrt{u}} \, du = 2i\sqrt{\pi} e^{-(B+1)\ln r} \cdot \frac{1}{\sqrt{\ln r}} \quad (4.9.40)$$

so that in the vicinity of $r = 1 + 0$:

$$\sigma_y = \frac{\sigma_y^\infty}{\sqrt{\pi}} \cdot \frac{\sqrt{B+1}}{H_-(-1)} \cdot \frac{r^{-B-1}}{\sqrt{\ln r}} \to \frac{\sigma_y^\infty}{\sqrt{\pi}} \cdot \frac{\sqrt{B+1}}{H_-(-1)} \cdot \frac{1}{\sqrt{r-1}} \quad \text{as } r \to 1+0 \quad (4.9.41)$$

Then, reinstating the crack length a, the stress intensity factor is obtained as

$$K_I = \lim_{r \to 1+0} [\sqrt{2\pi(r-1)a}\,\sigma_y] = \frac{\sqrt{2(B+1)}}{\sqrt{\pi}H_-(-1)}\sigma_y^\infty \sqrt{\pi a} \qquad (4.9.42)$$

Thus, calculating the stress intensity factor is essentially a question of calculating $H_-(-1)$ from the integral expression for $H_-(s)$ in (4.9.25). After putting $s = -1$, the integrand is regular for $z = 0$ and the path of integration may be deformed to the imaginary axis. The substitution $y = -iz$ then leads to the result

$$\ln H_-(-1) = -\frac{1}{\pi}\int_0^\infty \frac{1}{1+y^2}\ln \frac{y\sinh(\pi y)}{\sqrt{B^2+y^2}[\cosh(\pi y) - 2y^2 - 1]}\,dy$$

$$= \ln(B+1)^{1/2} - \frac{1}{\pi}\int_0^\infty \frac{1}{1+y^2}\ln \frac{y\sinh(\pi y)}{\cosh(\pi y) - 2y^2 - 1}\,dy \qquad (4.9.43)$$

and thus

$$\frac{K_I}{\sigma_y^\infty \sqrt{\pi a}} = \sqrt{\frac{2}{\pi}}\exp\left\{\frac{1}{\pi}\int_0^\infty \frac{1}{1+y^2}\ln \frac{y\sinh(\pi y)}{\cosh(\pi y) - 2y^2 - 1}\,dy\right\} \qquad (4.9.44)$$

For numerical calculation, the substitution $y = \tan u$ is suitable, giving

$$\int_0^\infty \frac{1}{1+y^2}\ln \frac{y\sinh(\pi y)}{\cosh(\pi y) - 2y^2 - 1}\,dy$$

$$= \int_0^{\pi/2}\left[\ln \frac{y\sinh(\pi y)}{\cosh(\pi y) - 2y^2 - 1}\right]_{y=\tan u} du \qquad (4.9.45)$$

However, more rapid numerical convergence can actually be obtained by retaining B in (4.9.43) so that the stress intensity factor is found from the expression

$$\frac{K_I}{\sigma_y^\infty \sqrt{\pi a}} = \sqrt{\frac{2(B+1)}{\pi}}$$

$$\times \exp\left\{\frac{1}{\pi}\int_0^{\pi/2}\left[\ln \frac{y\sinh(\pi y)}{\sqrt{B^2+y^2}[\cosh(\pi y) - 2y^2 - 1]}\right]_{y=\tan u} du\right\} \qquad (4.9.46)$$

Using the observation from (4.9.43)-(4.9.44) that any non-negative value of B results in the same stress intensity factor, the restriction to $B > 1$ does not need to be observed during the numerical calculation. A value of B around 0.05 leads to particularly rapid convergence, giving

$$\frac{K_I}{\sigma_y^\infty \sqrt{\pi a}} = 1.1215222552... \qquad (4.9.47)$$

which, of course, far exceeds the accuracy needed in engineering applications, but might be useful for controlling numerical methods, tested on the edge crack problem.

Slanting edge crack

For the problem of an edge crack that is not normal to the plate edge, and for related problems, a method of solution was developed by Khrapkov (1971). It is related to the method outlined for the normal edge crack in the preceding subsection, although it is much more elaborate, because the geometry is no longer symmetric. Suppose that a crack forms an angle θ to the plate edge normal. Then the two sectors $\theta \leq \varphi \leq \pi/2$ and $-\pi/2 \leq \varphi \leq \theta$ must be considered separately, and the boundary conditions for each sector involve traction free plate edge, and prescribed normal and tangential tractions on the crack faces. These tractions are found after superposition of the negated remote stress everywhere, as in the case for the normal edge crack. In addition, there are coupling conditions for stresses and displacement gradients on the contact faces, i.e., in the prolongation of the crack. Two coupled Wiener-Hopf equations result, instead of one. These equations may be written in matrix form:

$$X_+(s) = A(s)X_-(s) + A(s)Q_+(s) \tag{4.9.48}$$

where

$$X_+(s) = 2(1-k^2)\mu \begin{bmatrix} V_{\varphi+}(s) \\ V_{r+}(s) \end{bmatrix}, \quad X_-(s) = \begin{bmatrix} S_-(s) \\ T_-(s) \end{bmatrix}$$

$$Q_+(s) = \begin{bmatrix} S_+(s) \\ T_+(s) \end{bmatrix}, \quad A(s) = \begin{bmatrix} a_{11}(s) & a_{12}(s) \\ a_{21}(s) & a_{22}(s) \end{bmatrix}$$

$V_{\varphi+}(s)$ and $V_{r+}(s)$ are unknown Mellin transformed displacement gradients on the crack faces, and $S_-(s)$ and $T_-(s)$ are unknown Mellin transformed normal and tangential stresses along the prolongation of the crack. The elements of $Q_+(s)$ are known Mellin transformed normal and tangential stresses on the crack faces, and the elements of $A(s)$ are known functions of s and θ.

The procedure is formally somewhat similar to the procedure when scalar instead of matrix functions are involved. Thus, the solution, after consideration of physically acceptable results, may be written in the form

$$X(s) = \frac{1}{2\pi i} H(s) \int_{c-i\infty}^{c+i\infty} \frac{H_-^{-1}(z)Q_+(z)}{z-s} dz \tag{4.9.49}$$

where $H(s)$ is a sectionally analytic matrix function, equal to $H_+(s)$ in $\Re(s) \geq 0$ and to $H_-(s)$ in $\Re(s) \leq 0$. $H_-^{-1}(s)$ is the inverse of $H_-(s)$. $H_+(s)$ and $H_-^{-1}(s)$ are found from the factorization

$$A(s) = H_+(s)H_-^{-1}(s) \tag{4.9.50}$$

Factorization of a matrix function is rather complicated, but has been described in detail by Khrapkov (1971).

After $X(s)$ has been found, the stress intensity factors K_I and K_{II} can be found in essentially the same way as the stress intensity factor for an edge crack in the preceding subsection, i.e., without inverting the Mellin transforms:

$$\begin{bmatrix} K_I \\ K_{II} \end{bmatrix} = \sqrt{2\pi} \lim_{r \to 1+0} \left\{ \sqrt{r-1} \begin{bmatrix} \sigma_\varphi(r) \\ \tau_{r\varphi}(r) \end{bmatrix} \right\}$$

$$= i\sqrt{2} \lim_{s \to -\infty} \left\{ \sqrt{s} \begin{bmatrix} S_-(s) \\ T_-(s) \end{bmatrix} \right\} = i\sqrt{2} \lim_{s \to -\infty} [\sqrt{s}X_-(s)] \tag{4.9.51}$$

With the method developed by Khrapkov, numerical results for slanting edge cracks can be obtained within any desired accuracy, because the approximations involved are

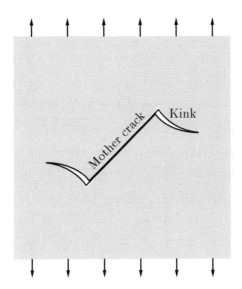

Fig. 4.10.1 Mother crack and its kinks. Kinks, propagating under mode I conditions, usually develop even if the edges of the mother crack are subjected to pure mode II loading.

limited to the calculation of a few integrals. It is also possible to find accurate solutions for the stress intensity factors at the edge of a small crack, extending from the apex of a V-shaped notch (Gallagher 1995).

4.10 Kinked and branched cracks

The origin of kinked cracks

In Section 4.6 an analytical method for studying smoothly curved cracks was demonstrated. Here, kinked cracks will be considered. Kinking is a result of incipient crack growth under mixed mode loading. Under small scale yielding conditions, cracks in plates appear to grow with local mode I symmetry at the crack edge, except when mode I growth is suppressed by a sufficiently high superposed pressure (Nemat-Nasser and Horii 1982, Horii and Nemat-Nasser 1985,1986, Melin 1986, Broberg 1987b, Nemat-Nasser 1995). After kinking, the crack continues to grow along a smooth curve. Fig. 4.10.1 shows a typical example.

The infinitesimal kink

A solution can be obtained within any desired accuracy for the problem of an infinitesimal kink at the end of a straight crack, subjected to remote loading. The method is almost identical with the one described for a slanting edge crack in Section 4.9, although the superposition procedure is slightly different. It is explained by Fig. 4.10.2, where two different problems are shown. Superposition of the cases shown in Figs 4.10.2a and b results in the original problem, i.e. remote mixed mode loading and

172   4. ELASTOSTATIC CRACKS

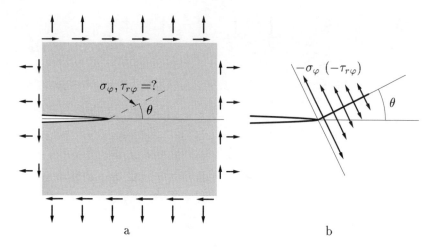

Fig. 4.10.2 a) Mother crack, subjected to remote mixed mode loading. The stresses σ_θ and $\tau_{r\theta}$ in a plane making the angle θ with the mother crack, are sought. b) A kink making the angle θ with the mother crack. The kink faces are loaded by the negated stresses $-\sigma_\theta$ and $-\tau_{r\theta}$, but there is no remote loading.

traction free faces of both the mother crack and its kink. Obviously, the stress intensity factor at the kink edge in the original problem equals the one found from the case in Fig. 4.10.2b.

An infinitesimal kink is considered. The mixed boundary loading can be expressed by the stress intensity factors K_I and K_{II} before the kink appearance. Thus, the solution of the first problem, shown by Fig. 4.10.2a, is trivial; it is obtained from (3.3.55)-(3.3.56) and (3.3.66)-(3.3.67):

$$\sigma_\varphi(r) = \frac{\cos(\theta/2)}{2\sqrt{2\pi r}}[(1+\cos\theta)K_I - 3\sin\theta K_{II}] \qquad (4.10.1)$$

$$\tau_{r\varphi}(r) = \frac{\cos(\theta/2)}{2\sqrt{2\pi r}}[\sin\theta K_I + (3\cos\theta - 1)K_{II}] \qquad (4.10.2)$$

The second problem is then properly posed, after the length of the main crack is taken as infinite, whereas the kink length is finite. The rather insignificant difference between this problem and the one for a slanting edge crack consists of different sector boundaries: here, assuming the main crack to be located along $\varphi = \pm\pi$, the sectors are $\theta \leq \varphi \leq \pi$, $-\pi \leq \varphi \leq \theta$.

Numerical results have been obtained by Bilby and Cardew (1975) and by Melin (1994). The results may be given in the form

$$\left.\begin{array}{l} k_I = R_{11}(\theta)K_I + R_{12}(\theta)K_{II} \\ k_{II} = R_{21}(\theta)K_I + R_{22}(\theta)K_{II} \end{array}\right\} \qquad (4.10.3)$$

where k_I and k_{II} are the stress intensity factors at the kink edge. The functions $R_{11}(\theta)$, $R_{12}(\theta)$, $R_{21}(\theta)$ and $R_{22}(\theta)$ are shown in Fig. 4.10.3. $R_{11}(\theta)$ and $R_{22}(\theta)$ are even, whereas $R_{12}(\theta)$ and $R_{21}(\theta)$ are odd.

If the condition for the direction of incipient kink growth in the case shown in Fig.

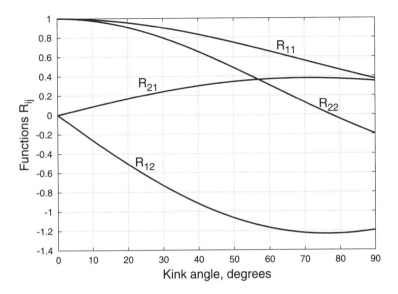

Fig. 4.10.3 Functions $R_{ij}(\theta)$ relating the stress intensity factors at the kink edge to those at the edge of the mother crack.

4.10.1 is $k_{II} = 0$, then the kink angle is given by the equation

$$\frac{K_I}{K_{II}} = -\frac{R_{22}(\theta)}{R_{21}(\theta)} \qquad (4.10.4)$$

In the limiting case when the main crack is vertical, the function $R_{22}(\theta)$ vanishes, which implies $\theta \approx -77.33^0$ if $K_{II} > 0$ (the positive root would make $k_I < 0$). The kink angle can, in principle, approach this value.

Branched cracks

Branched cracks present more difficult analytical problems than kinked cracks. However, more or less direct numerical treatments exist, that can be pursued to satisfactory accuracy for engineering purposes, e.g. Kalthoff (1973), Theocaris (1977), Vitek (1977), Lo (1978), Pärletun (1979), Hayashi and Nemat-Nasser (1981), Isida and Noguchi (1992). One interesting question concerns which angle gives pure mode I stress intensity ($K_{II} = 0$) during symmetrical branching from a straight crack, subjected to mode I loading. The result for straight branches (Isida and Noguchi 1992) appears to be that the half branching angle is about 27.3° for infinitesimal branch length, and that it decreases with increasing branch length, initially very rapidly; see Fig. 4.10.4. Obviously, branching according to the condition of pure mode I stress intensity at the branch edges would follow a curved path with branch directions changing towards the direction of the mother crack. Material inhomogeneities would obscure the direction of initial branching during experimental observations, which generally report about 15° - 20° branch directions from the symmetry plane. Note, however, that such observations generally concern branching from rapidly running cracks.

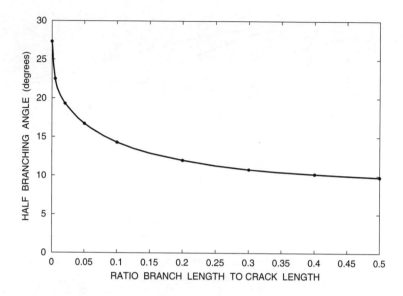

Fig. 4.10.4 The half branching angle that produces straight branches with pure mode I stress intensity at their edges. Data, shown as filled circles, after Isida and Noguchi (1992).

4.11 Directional stability of cracks

It is experienced that cracks in engineering structures generally propagate under mode I conditions in the crack edge vicinity, if small scale yielding prevails. Crack growth under mode II or III conditions seems to require an ambient pressure high enough to suppress mode I (Nemat-Nasser and Horii 1982, Horii and Nemat-Nasser 1985,1986, Melin 1986, Broberg 1987b, Nemat-Nasser 1995). They appear in earthquakes, but can also be provoked under laboratory conditions (Broberg 1987b). However, in most situations of practical interest, the crack growth direction appears to be governed by the condition of maximum mode I stress intensity factor, or, equivalently, vanishing mode II stress intensity factor. This condition has been extensively used in numerical simulations, for instance those mentioned in the context of crack branching. It is, in general less reliable when small scale yielding cannot be assumed.

The directional stability of an originally straight crack in a large plate, slowly growing under remote mode I loading, will be analysed here. Further analysis and discussion of crack paths and directional stability will be made in Section 8.6.

Consider a large plate, containing a central crack, and subjected to biaxial remote loading, $\sigma_x = \sigma_x^\infty$, $\sigma_y = \sigma_y^\infty$. In order to investigate its directional stability under slow growth, the crack is assumed to possess one straight part along $y = 0$, but deviating slightly from $y = 0$ near its ends, so that $|y'(x)|$ is infinitesimally small. The ends are situated at $x = -a_0$ and $x = a_0$ before crack growth and at $x = a_m - a$ and $a_m + a$ during crack growth, so that $x = a_m$ and $2a$ indicate the middle position and the crack length, respectively. It is assumed that $|y'(x)| \ll 1$ during crack growth. Note that symmetry with respect to the y axis is not assumed for the initial disturbance (deviation from a straight line at the ends) and the continued growth.

4.11 DIRECTIONAL STABILITY OF CRACKS

The problem about directional stability was considered by Melin (1983, 1992). The present treatment follows essentially her 1992 paper.

The condition $K_{II} = 0$ for crack growth direction will be used. From Gol'dstein and Salganik (1974)† it follows that

$$K_{II}(a_m + a) = \frac{1}{2}\sigma_y^\infty \sqrt{\pi a}\left\{y'(a_m + a) - \frac{y(a_m + a) - y(a_m - a)}{2a} \right.$$
$$\left. + \frac{2}{\pi a}\left(1 - \frac{\sigma_x^\infty}{\sigma_y^\infty}\right) \int_{-a}^{a} \sqrt{\frac{\xi + a}{a - \xi}} y'(a_m + \xi) d\xi \right\} \quad (4.11.1)$$

for the right crack edge, and the same expression after letting $a \to -a$ for the left edge. Introduce the displacement measure

$$Y(x) = [y(a_m + x) - y(a_m - x)]/2, \quad Y'(x) = [y'(a_m + x) + y'(a_m - x)] \quad (4.11.2)$$

Then,

$$K_{II}(a_m + a) + K_{II}(a_m - a)$$
$$= \frac{1}{2}\sigma_y^\infty \sqrt{\pi a}\left\{Y'(a) - \frac{Y(a)}{a} + \lambda \int_0^a \frac{Y'(\xi) d\xi}{\sqrt{a^2 - \xi^2}}\right\} \quad (4.11.3)$$

where $\lambda = 4(1 - \sigma_x^\infty/\sigma_y^\infty)/\pi$. The condition $K_{II} = 0$ gives

$$Y'(a) = \frac{Y(a)}{a} - \lambda \int_0^a \frac{Y'(\xi) d\xi}{\sqrt{a^2 - \xi^2}} \quad (4.11.4)$$

which is an integral equation for investigation of directional stability. First, however, directional stability has to be defined.

It turns out to be essential to define directional stability by considering the deviation from crack straightness as the slope $D(a) = Y(a)/a$ of the line drawn between the crack ends. Thus, the change of the deviation from straightness from the original deviation is defined by $D(a)/D(a_0)$. Directional stability is said to prevail if $D(a)/D(a_0) \to 0$ as $a/a_0 \to \infty$, whereas instability prevails if $D(a)/D(a_0)$ exceeds any predetermined value if a/a_0 is sufficiently large.

As $a/a_0 \to \infty$, $Y'(a)$ tends towards a constant or towards infinity. Assume first that it tends towards a constant, Y'_∞. Then,

$$D(a) = \frac{Y(a)}{a} \to Y'_\infty + \lambda Y'_\infty \int_0^\infty \frac{d\xi}{\sqrt{a^2 - \xi^2}} = (1 + \pi\lambda)Y'_\infty \quad (4.11.5)$$

so that $Y(a) \to (1 + \pi\lambda)Y'_\infty a$, which implies that $Y'(a) \to (1 + \pi\lambda)Y'_\infty$. Thus, either $\lambda = 0$ ($\sigma_x^\infty/\sigma_y^\infty = 1$) and $Y'_\infty \neq 0$, or $Y'_\infty = 0$.

In the former case, insertion into (4.11.4) gives

$$\frac{Y(a)}{a} = \frac{Y(a_0)}{a_0} \quad (4.11.6)$$

i.e., the crack continues to grow from $a = \pm a_0$ along a straight line. Thus, the deviation

---

† This paper contains an excellent analysis of curved cracks, based on integral equations, established from dislocation arrays.

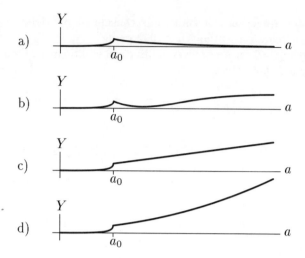

Fig. 4.11.1 Half the vertical distance between the crack edges, $Y(a)$, for a) $\sigma_x^\infty/\sigma_y^\infty < 1 - \pi/4$, b) $1 - \pi/4 < \sigma_x^\infty/\sigma_y^\infty < 1$, c) $\sigma_x^\infty/\sigma_y^\infty = 1$, and d) $\sigma_x^\infty/\sigma_y^\infty > 1$.

from straightness remains unchanged during crack growth if $\sigma_x^\infty = \sigma_y^\infty$, and neither directional stability nor instability prevails.

In the latter case, $Y_\infty' = 0$, (4.11.4) shows that

$$Y'(a) \to \frac{Y(a)}{a} - \frac{\lambda}{a}\int_0^a Y'(\xi)d\xi = (1-\lambda)\frac{Y(a)}{a} \qquad (4.11.7)$$

Hence,

$$Y(a) \to Y_\infty\left(\frac{a}{a_0}\right)^{1-\lambda}, \quad Y'(a) \to (1-\lambda)\frac{Y_\infty}{a_0}\left(\frac{a}{a_0}\right)^{-\lambda} \qquad (4.11.8)$$

where Y_∞ is a constant. Because

$$D(a) = \frac{Y(a)}{a} \to \frac{Y_\infty}{a_0}\left(\frac{a}{a_0}\right)^{-\lambda} \qquad (4.11.9)$$

the condition for directional stability, $D(a)/D(a_0) \to 0$ as $a/a_0 \to \infty$, is satisfied for $\lambda > 0$, i.e., $\sigma_x^\infty < \sigma_y^\infty$.

Note that the condition $D(a)/D(a_0) \to 0$ does not necessarily imply that $Y(a) \to 0$ as $a/a_0 \to \infty$. It does so for $\lambda > 1$, i.e., $\sigma_x^\infty/\sigma_y^\infty < 1 - \pi/4 \approx 0.2146$, but for $\lambda = 1$, i.e. $\sigma_x^\infty/\sigma_y^\infty = 1 - \pi/4$, it reaches a finite value asymptotically, and for $0 < \lambda < 1$, i.e. $1 - \pi/4 < \sigma_x^\infty/\sigma_y^\infty < 1$, $Y(a)$ increases indefinitely. Thus, in the latter case, the crack ends move away more and more from $y = 0$, but the deviation from straightness decreases more and more. These relations are illustrated in Fig. 4.11.1.

The main result, that directional instability occurs for $\sigma_x^\infty/\sigma_y^\infty > 1$, appears to be obvious, because a crack tends to propagate in a plane normal to the direction of the highest principal stress. However, it should be noted that the crack edges actually move away more and more from the plane $y = 0$ if $\sigma_x^\infty/\sigma_y^\infty \geq 1 - \pi/4 \approx 0.2146$, irrespective of the crack length. Therefore, the question of directional stability cannot be answered by considering only a crack edge vicinity, for instance by analysing the effect of an initial disturbance near the edge of a semi-infinite crack, using the stability criterion that the crack edge should move towards $y = 0$. This was done by Cotterell and Rice (1980), but their analysis needs some corrections and would then not show stability for the whole range $\sigma_x^\infty < \sigma_y^\infty$, as they reported.

Finally, it should be remarked that a positive T-stress (which for a mode I crack in a large plate implies that $\sigma_x^\infty/\sigma_y^\infty > 1$) cannot be a criterion for directional instability, as sometimes suggested. A counterexample is provided by the solution for a periodic array of cracks by Melin (1983), showing directional instability for all values of $\sigma_x^\infty/\sigma_y^\infty$; cf. pages 606ff.

4.12 Interface cracks

General considerations

Interface cracks appear along bonds between two different materials, for instance in lamellar composites, in fibre reinforced materials, in film-substrate combinations, in particle-matrix boundaries in metallic materials and in human joint replacements. They are consequently of considerable practical interest. The analysis, however, is far from simple. For cases in which infinitesimally small scale yielding and linear elasticity is assumed, an analysis indicates the existence of contact regions near crack edges and some slip in these regions. A history-independent solution is therefore not possible. On the other hand, the contact regions are extremely small in most cases of practical interest, and an approximative analysis, albeit improper in the extreme vicinity of the crack end(s), is possible. Such an analysis may be performed by treating the two material parts separately and then using coupling conditions. The inappropriateness consists of interpenetration of the two materials near the crack end(s) in the mathematical solution. Evidently, this is a consequence of the idealization to linearity: in a real case, a somewhat blunted crack end and a reasonably smoothly strained non-linear region in front of the crack would appear. Actually, no interpenetration region is found, even mathematically, if non-linear elasticity, allowing large strains, is considered, at least for some constitutive relations, cf. Knowles and Sternberg (1983), Herrmann (1989, 1992), Geubelle and Knauss (1994b,c) and Gao and Shi (1995). It should also be mentioned that an interpenetration region is not always found by linear analysis of cracks between anisotropic materials; necessary and sufficient conditions for this are given by Qu and Bassani (1989) and Bassani and Qu (1989). Their investigations included tilted orientations, occurring, for instance, at the interface between single crystals.

Cracks sometimes propagate out from an interface and continue to travel parallel with the interface. Whereas this may be due to elastic anisotropy mismatch, directing a crack out from the interface by kinking or bending, cf. He and Hutchinson (1989a), Hutchinson *et al.* (1987), it is perhaps most often caused by thermal expansion mismatch, resulting in residual normal stresses in planes normal to the interface. This phenomenon is also known to occur along the interface between two identical materials, for instance near a welded joint. Tensile residual stresses may appear close to the joint, balanced by compressive stresses further away. Such cases are not included in the following treatment.

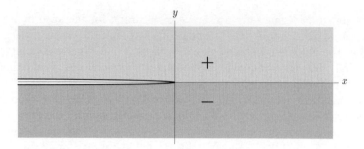

Fig. 4.12.1 A semi-infinite interface crack.

The interface crack under in-plane deformation

The crack edge vicinity

Consider a plate of two different materials, with the interface $y = 0$ and with a crack along a portion of this interface. Both materials are assumed to be isotropic and linearly elastic. Consider first the vicinity of the right crack edge, $x = 0$, by studying the idealized problem of a semi-infinite crack between two semi-infinite plates; see Fig. 4.12.1. Quantities referring to the upper plate, $y \geq 0$, will be assigned subscript + (plus) and quantities referring to the lower plate, $y \leq 0$, will be assigned subscript − (minus).

The two plates will first be treated separately. The tractions on the plate edges, $y = 0$, are denoted by $\sigma_y^0(x)$ and $\tau_{xy}^0(x)$, both vanishing for $x < 0$. According to Appendix A2, relation (A2.22), the expressions for the displacement gradients on $y = 0$ may be written as

$$\left(\frac{\partial u}{\partial x} + i\frac{\partial v}{\partial x}\right)_- = \frac{i}{2\pi(1-k_-^2)\mu_-} \fint_{-\infty}^{+\infty} \frac{\overline{s(\xi)}}{\xi - x} d\xi - \frac{k_-^2}{2(1-k_-^2)\mu_-}\overline{s(x)} \quad (4.12.1)$$

$$\left(\frac{\partial u}{\partial x} + i\frac{\partial v}{\partial x}\right)_+ = \frac{-i}{2\pi(1-k_+^2)\mu_+} \fint_{-\infty}^{+\infty} \frac{s(\xi)}{\xi - x} d\xi - \frac{k_+^2}{2(1-k_+^2)\mu_+} s(x) \quad (4.12.2)$$

where $s(x) = \sigma_y^0(x) + i\tau_{xy}^0(x)$ and the integrals are taken in the sense of the Cauchy principal value. The last term in these equations is due to the residue at $\xi = x$. Introduce the complex separation distance $\delta_c(x) = u_+ - u_- + i(v_+ - v_-)$ between points originally in contact:

$$\frac{d\delta_c(x)}{dx} = \frac{1}{2}\left[\frac{k_-^2}{(1-k_-^2)\mu_-} - \frac{k_+^2}{(1-k_+^2)\mu_+}\right]\left[\overline{s(x)} - \frac{i}{\pi\beta}\fint_{-\infty}^{+\infty}\frac{s(\xi)}{\xi - x}d\xi\right] \quad (4.12.3)$$

where

$$\beta = \frac{k_-^2[(1-k_-^2)\mu_-]^{-1} - k_+^2[(1-k_+^2)\mu_+]^{-1}}{[(1-k_-^2)\mu_-]^{-1} + [(1-k_+^2)\mu_+]^{-1}} \quad (4.12.4)$$

Use of the boundary condition that $s(x) = 0$ for $x < 0$ and the matching condition

4.12 INTERFACE CRACKS

that $\delta_c(x) = 0$ for $x > 0$, gives

$$\overline{s(x)} = 0 \text{ for } x < 0 \qquad (4.12.5)$$

$$\overline{s(x)} - \frac{i}{\pi\beta} \fint_0^{+\infty} \frac{\overline{s(\xi)}}{\xi - x} d\xi = 0 \text{ for } x > 0 \qquad (4.12.6)$$

In order to solve the equation system (4.12.5)-(4.12.6), the sectionally analytic function

$$F(z) = \frac{1}{2\pi i} \int_{-\infty}^{+\infty} \frac{\overline{s(\xi)}}{\xi - z} d\xi \qquad (4.12.7)$$

is introduced. Then, by the Plemelj formulae,

$$F_+(x) + F_-(x) = \frac{1}{\pi i} \fint_{-\infty}^{+\infty} \frac{\overline{s(\xi)}}{\xi - x} d\xi \qquad (4.12.8)$$

$$F_+(x) - F_-(x) = \overline{s(x)} \qquad (4.12.9)$$

where the integral is taken in the sense of the Cauchy principal value. Use of expressions (4.12.5)-(4.12.6) then leads to the Hilbert problem

$$F_+(x) - F_-(x) = 0 \text{ for } x < 0 \qquad (4.12.10)$$

$$F_+(x) - F_-(x) + \frac{1}{\beta}[F_+(x) + F_-(x)] = 0 \text{ for } x > 0 \qquad (4.12.11)$$

and, by writing the second equation in the form

$$F_+(x) + \frac{1-\beta}{1+\beta} F_-(x) = 0 \text{ for } x > 0 \qquad (4.12.12)$$

the equation system (4.12.10)-(4.12.11) may be written as a single equation,

$$F_+(x) - \frac{G_-(x)}{G_+(x)} F_-(x) = 0 \qquad (4.12.13)$$

where $G(z)$ is analytic in the plane with a branch cut along the positive real axis and

$$\frac{G_-(x)}{G_+(x)} = \begin{cases} 1 & \text{for } x < 0 \\ -\dfrac{1-\beta}{1+\beta} & \text{for } x > 0 \end{cases} \qquad (4.12.14)$$

Obviously, the function

$$G(z) = z^\gamma \qquad (4.12.15)$$

with a branch cut along $x > 0$, $y = 0$ and the branch chosen so that $\ln G(x) = \gamma \ln|x|$ for $y = 0$, $x < 0$, satisfies the requirements, provided that γ is appropriately chosen. For $x > 0$, the ratio

$$\frac{G_-(x)}{G_+(x)} = e^{2\pi i \gamma} = -\frac{1-\beta}{1+\beta} \implies \gamma = i\epsilon + \frac{1}{2} \qquad (4.12.16)$$

where

$$\epsilon = \frac{1}{2\pi} \ln \frac{1+\beta}{1-\beta} \qquad (4.12.17)$$

Thus, $\epsilon > 0$ if $\beta \neq 0$ and the coordinate system is oriented so that $\beta > 0$. Usually, $|\beta|$ is substantially smaller than unity, and ϵ is thus approximately equal to $\beta/3$. Note that the reverse sign for parameter ϵ is often used in the literature. Note also that any integer could have been added to γ: the choice here ensures bounded stress-strain energy.

Equation (4.12.13), written in the form

$$G_+(x)F_+(x) - G_-(x)F_-(x) = 0 \tag{4.12.18}$$

shows that the function $G(z)F(z)$ is analytic in the whole z-plane, and, by Liouville's theorem, it must equal a polynomial of finite degree. This polynomial reduces to a constant, P_0, because only the dominating solution is required. Thus,

$$F(z) = P_0 z^{-\frac{1}{2} - i\epsilon} \tag{4.12.19}$$

and, for $y = 0$, $x > 0$,

$$F_-(x) = -\frac{iP_0}{\sqrt{x}} x^{-i\epsilon} e^{\pi \epsilon}, \quad F_+(x) = \frac{iP_0}{\sqrt{x}} x^{-i\epsilon} e^{-\pi \epsilon} \tag{4.12.20}$$

Insertion into (4.12.9) gives the normal and tangential stresses on $y = 0$, $x > 0$:

$$s(x) = -\frac{2i\overline{P_0}}{\sqrt{x}} \cosh \pi\epsilon \cdot x^{i\epsilon} = \frac{K}{\sqrt{2\pi x}} \cdot \left(\frac{x}{r_0}\right)^{i\epsilon} \tag{4.12.21}$$

where a complex stress intensity factor $K = K_1 + iK_2$ and an arbitrary length parameter r_0 are introduced. Note that it is not possible to define mode I and II symmetries for the interface crack; the symbols K_1 and K_2 should therefore not be confused with the mode-distinguishing symbols K_I and K_{II}. Note also that the values of K_1 and K_2 depend on the value chosen for r_0, but the absolute value of K, i.e. $\sqrt{K\overline{K}}$, is independent of this choice, because the absolute value of $(x/r_0)^{i\epsilon}$ equals unity.

The dependence of K_1 and K_2 on the choice of r_0 may be illustrated by making two different choices, $r_0^{(1)}$ and $r_0^{(2)}$, leading to the relations

$$K_1^{(2)} = K_1^{(1)} \cos\left(\epsilon \ln \frac{r_0^{(2)}}{r_0^{(1)}}\right) - K_2^{(1)} \sin\left(\epsilon \ln \frac{r_0^{(2)}}{r_0^{(1)}}\right) \tag{4.12.22}$$

$$K_2^{(2)} = K_1^{(1)} \sin\left(\epsilon \ln \frac{r_0^{(2)}}{r_0^{(1)}}\right) + K_2^{(1)} \cos\left(\epsilon \ln \frac{r_0^{(2)}}{r_0^{(1)}}\right) \tag{4.12.23}$$

with obvious notations. Thus, for instance, the ratio K_2/K_1 has no physical significance such as that of K_{II}/K_I in the homogeneous case, which may be taken as a measure of the mode mixity.

Now, knowledge of $s(x)$ along the whole x axis, enables determination of the asymptotic stresses and strains in each medium even outside the interface, for instance by using the Airy stress function or directly by insertion into formulae (A2.18)-(A2.20) in Appendix A2, identifying σ_y^0 with $[s(x) + \overline{s(x)}]/2$ and τ_{xy}^0 with $[s(x) - \overline{s(x)}]/(2i)$. Some results are given by Sun and Jih (1987).

$F(z)$ is continuous across the real negative axis, so that for $y = 0$, $x < 0$,

$$F_+(x) = F_-(x) = \frac{P_0}{\sqrt{|x|}} |x|^{-i\epsilon} = -\frac{i\overline{K}}{2\sqrt{2\pi|x|} \cosh \pi\epsilon} \cdot \left(\frac{|x|}{r_0}\right)^{-i\epsilon} \tag{4.12.24}$$

and hence, the crack face separation can be found from its gradient, given in (4.12.3) as

$$\frac{d\delta_c(x)}{dx} = \frac{1}{2}\left[\frac{k_-^2}{(1-k_-^2)\mu_-} - \frac{k_+^2}{(1-k_+^2)\mu_+}\right] \cdot \frac{1}{\beta} \cdot [F_-(x) + F_+(x)]$$

$$= \frac{-iC_m}{2\cosh\pi\epsilon} \cdot \frac{\overline{K}}{\sqrt{2\pi|x|}}\left(\frac{|x|}{r_0}\right)^{-i\epsilon} \quad (4.12.25)$$

where

$$C_m = \frac{1}{(1-k_-^2)\mu_-} + \frac{1}{(1-k_+^2)\mu_+} \quad (4.12.26)$$

Integration gives

$$\delta_c(x) = \frac{iC_m}{(1-2i\epsilon)\cosh\pi\epsilon} \cdot \frac{\overline{K}}{\sqrt{2\pi}}\sqrt{|x|}\left(\frac{|x|}{r_0}\right)^{-i\epsilon} \quad (4.12.27)$$

By writing

$$\left(\frac{|x|}{r_0}\right)^{-i\epsilon} = e^{-i\epsilon \ln|x|/r_0} = \cos\left(\epsilon \ln \frac{|x|}{r_0}\right) - i\sin\left(\epsilon \ln \frac{|x|}{r_0}\right) \quad (4.12.28)$$

it can clearly be observed how oscillations appear in both interfacial stresses and crack face displacements, with increasing frequency toward $x = 0$, but with decreasing amplitude. Whereas this rather awkward fact does not directly disqualify the solution for the stresses, it does so for the displacements, because interpenetration results in the vicinity of $x = 0$.

A curiosity is that oscillations can be absent even for cracks along the interface between materials with different elastic constants. This requires, of course, $\epsilon = 0$, which occurs if $k_-^2/[(1-k_-^2)\mu_-] = k_+^2/[(1-k_+^2)\mu_+]$.

The energy flux is found in the same way as in Section 3.5. Equation (3.5.10) will be modified to

$$\mathcal{G} = -\int_0^{r_p}\left[\left(\sigma_y\frac{\partial v}{\partial x} + \tau_{xy}\frac{\partial u}{\partial x}\right)_+ - \left(\sigma_y\frac{\partial v}{\partial x} + \tau_{xy}\frac{\partial u}{\partial x}\right)_-\right]dx$$

$$= -\int_0^{r_p}\Im\left[s(x)\frac{d\delta_c(x)}{dx}\right]dx \quad (4.12.29)$$

which leads to the following modification of (3.5.11):

$$\mathcal{G} = \frac{1}{2\pi}\Im\left\{\frac{iC_m}{2\cosh\pi\epsilon}\int_0^{r_p}\int_0^1\int_0^1 K'(\alpha)\overline{K'(\beta)}\right.$$
$$\times (x - r_p + \alpha r_p)^{-\frac{1}{2}-i\epsilon}(-x + r_p - \beta r_p)^{-\frac{1}{2}+i\epsilon}U(x - r_p + \alpha r_p)$$
$$\left.\times U(-x + r_p - \beta r_p)d\alpha\,d\beta\,dx\right\} \quad (4.12.30)$$

Note that, as in (3.5.11), α and β are integration variables, the second one without relation to parameter β used earlier in the present section. After changing the order of integration, the integral over x, which is conveniently obtained by residue calculus, is found to be

$$\frac{\pi}{\cosh\pi\epsilon}U(\alpha - \beta) \quad (4.12.31)$$

which leads to the energy flux

$$\mathcal{G} = \frac{[(1-k_-^2)\mu_-]^{-1} + [(1-k_+^2)\mu_+]^{-1}}{8\cosh^2 \pi\epsilon} K\overline{K} \qquad (4.12.32)$$

Note that the result is independent of the choice of r_0.

A study of the operations leading from (4.12.29) to (4.12.32) shows that the following sometimes useful formula is valid:

$$\mathcal{G} = \lim_{r \to 0}\left\{\frac{\pi r}{2\cosh\pi\epsilon}\Im\left[s(r)\cdot\frac{\delta_c(r)}{dr}\right]\right\} = \lim_{r \to 0}\left\{\frac{\pi r t}{2\cosh\pi\epsilon}\cdot\frac{d(\mathbf{u}_+ - \mathbf{u}_-)}{dr}\right\} \qquad (4.12.33)$$

Here, r is the distance to the crack edge, $\mathbf{t}$ the traction vector on the cut surface $x_2 = 0$ of the lower half of the body, and $\mathbf{u}_\pm$ the crack face displacements.

Finite cracks

In order to estimate the extension of the interpenetration region, a problem involving a length parameter is needed. The obvious elementary problem concerns a crack of finite length $2a$ (including the interpenetration lengths) between two semi-infinite plates, subjected to remote loading $\sigma_y = \sigma_y^\infty$, $\tau_{xy} = \tau_{xy}^\infty$. The analysis proceeds in essentially the same way as for the semi-infinite crack; the regions of validity of (4.12.10) and (4.12.11) are changed to $|x| < a$ and $|x| > a$, respectively, and the function $G(z)$ is chosen as

$$G(z) = (z^2 - a^2)^{1/2}\left(\frac{z-a}{z+a}\right)^{i\epsilon} \qquad (4.12.34)$$

with branch cuts along $y = 0$, $|x| > a$, and the branch chosen so that $G(x) = \sqrt{a^2 - x^2}[(a-x)/(a+x)]^{i\epsilon}$ for $y = 0$, $x < a$. Then, by symmetry and the requirement of bounded stress-strain energy,

$$F(z) = \frac{P_0 + P_1 z}{G(z)} \qquad (4.12.35)$$

where P_0 and P_1 are constants. Insertion into (4.12.9) and (4.12.25) and use of the condition that $s(x) \to \sigma_y^\infty + i\tau_{xy}^\infty$ as $|x| \to \infty$ leads to

$$s(x) = \pm\frac{(\sigma_y^\infty + i\tau_{xy}^\infty)x - 2i\overline{P}_0\cosh\pi\epsilon}{\sqrt{x^2 - a^2}}\left(\frac{x-a}{x+a}\right)^{i\epsilon} \quad \text{for } |x| > a \qquad (4.12.36)$$

$$\frac{d\delta_c(x)}{dx} = -\frac{C_m}{2\cosh\pi\epsilon}\cdot\frac{(i\sigma_y^\infty + \tau_{xy}^\infty)x - 2P_0\cosh\pi\epsilon}{\sqrt{a^2 - x^2}}\cdot\left(\frac{a+x}{a-x}\right)^{i\epsilon} \quad \text{for } |x| < a \qquad (4.12.37)$$

where the upper sign refers to $x > a$ and the lower to $x < -a$.

P_0 is now to be determined so that $\delta_c(x) = 0$ for $|x| = a$, if no slip has taken place on $|x| > a$, cf. (4.4.86) and the following discussion. After integration of (4.12.37) over the interval $(-a, +a)$, the relation

$$2P_0\cosh\pi\epsilon = 2i\epsilon a(i\sigma_y^\infty + \tau_{xy}^\infty) \qquad (4.12.38)$$

is found, and thus, with upper sign referring to $x > a$, lower sign to $x < -a$:

$$s(x) = \pm\frac{(\sigma_y^\infty + i\tau_{xy}^\infty)(x + 2i\epsilon a)}{\sqrt{x^2 - a^2}}\left(\frac{x-a}{x+a}\right)^{i\epsilon} \quad \text{for } |x| > a \qquad (4.12.39)$$

$$\frac{d\delta_c(x)}{dx} = -\frac{(i\sigma_y^\infty + \tau_{xy}^\infty)C_m}{2\cosh\pi\epsilon}\cdot\frac{x - 2i\epsilon a}{\sqrt{a^2 - x^2}}\left(\frac{a+x}{a-x}\right)^{i\epsilon} \quad \text{for } |x| < a \qquad (4.12.40)$$

With consideration of (4.12.21), the complex stress intensity factor is defined as

$$K = \lim_{r \to 0}\left[\sqrt{2\pi r}\left(\frac{r}{r_0}\right)^{\mp i\epsilon} s(r)\right] \qquad (4.12.41)$$

where r is the distance ahead of or behind the crack edge and upper sign refers to the right, lower sign to the left crack edge. Note that different signs are logical, because ϵ changes sign if the assembly is turned upside down. Thus, for the right crack edge,

$$K = (\sigma_y^\infty + i\tau_{xy}^\infty)\sqrt{\pi a}(1 + 2i\epsilon)\left(\frac{r_0}{2a}\right)^{i\epsilon} \qquad (4.12.42)$$

From (4.12.39), writing $[(x-a)/(x+a)]^{i\epsilon} = \exp\{i\epsilon \ln[(x-a)/(x+a)]\}$, it is seen that no oscillations take place in the region where the argument of the cosine or sine is smaller than $\pi/2$, which implies

$$\left|\epsilon \ln \frac{a+x}{a-x}\right| < \frac{\pi}{2} \implies -[1 - e^{-\pi/(2|\epsilon|)}]a < x < [1 - e^{-\pi/(2|\epsilon|)}]a \qquad (4.12.43)$$

leaving oscillatory regions of the order of

$$e^{-\pi/(2|\epsilon|)} a \qquad (4.12.44)$$

at each crack edge. Because $|\beta| < 1/2$, $|\epsilon| < \ln 3/(2\pi) \approx 0.1748$, each oscillatory region is smaller than $2\exp[-\pi^2/\ln 3]a \approx 1.25 \times 10^{-4}a$, in agreement with estimates by England (1965) and obviously predicting an extremely small oscillatory region. In a real material, therefore, non-linear effects would take care of the oscillations, as shown for certain non-linear elastic cases by Knowles and Sternberg (1983), Herrmann (1989, 1992), Geubelle and Knauss (1994b,c) and by Gao and Shi (1995). In any event, in cases of essentially tensional loading ($\sigma_y^\infty \gg |\tau_{xy}^\infty|$), the solutions (4.12.39)-(4.12.40) would be valid with good accuracy outside such small regions near the crack edges.

Knowles and Sternberg (1983) considered the plane stress approximation (thin sheets) for an incompressible neo-Hookean material. They found an asymptotic crack opening consisting of two different parabola branches, one in each solid, and, somewhat surprisingly, these branches share a common tangent where they meet, although this tangent in general is not normal to the interface. Thus, the crack profile forms a smooth curve, but its curvature is in general discontinuous at the interface crossing. This result was corroborated for plane strain and a compressible hyperelastic material by Herrmann (1989), for plane stress and generalized neo-Hookean materials by Geubelle and Knauss (1993b,c) and for plane strain and a compressible hyperelastic material, containing five material parameters, by Gao and Shi (1995).

By introducing a contact region at each crack edge and assuming frictionless sliding under compressive normal stresses to take place in these regions, it is possible to arrive at a solution without the annoying interpenetrations. This field was first explored by Comninou (1977, 1978) and by Dunders and Comninou (1979). In her pioneering work, Comninou reduced the problem to a singular integral equation, which she solved numerically. It was later solved analytically by Gautesen and Dunders (1987a,b), thus avoiding the numerical difficulties associated with the smallness of the contact regions. For essentially tensile loading, the assumption of frictionless crack faces has been used to calculate contact region lengths very accurately by, for instance, Gautesen (1992). These lengths are extremely small and agree closely with the estimates by (4.12.44) for the lengths of oscillatory regions.

If the remote shear loading τ_{xy}^∞ is substantial, the solution (4.12.40) would imply a non-oscillatory interpenetration in addition to the small and insignificant oscillatory ones. This may be understood by examining the crack opening displacement component $\delta_2(x) = \Im[\delta_c(x)]$ near $x = \pm a$. Integration of (4.12.37) gives

$$\delta_c(x) \approx \frac{a(i\sigma_y^\infty + \tau_{xy}^\infty)C_m}{\cosh \pi\epsilon} \sqrt{\frac{r}{2a}} \left[\cos\left(\epsilon \ln \frac{2a}{r}\right) \pm i\sin\left(\epsilon \ln \frac{2a}{r}\right)\right] \quad (4.12.45)$$

where r is the distance from the crack edge. Thus,

$$\delta_2(x) \approx \frac{C_m}{\cosh \pi\epsilon} \sqrt{\frac{r}{2a}} \left[\sigma_y^\infty \cos\left(\epsilon \ln \frac{2a}{r}\right) \pm \tau_{xy}^\infty \sin\left(\epsilon \ln \frac{2a}{r}\right)\right] \quad (4.12.46)$$

This shows that interpenetration can occur near $x = -a$, even where r/a is not small, if $\tau_{xy}^\infty > 0$ and σ_y^∞ is rather small, although not necessarily much smaller than τ_{xy}^∞. This inadmissible interpenetration can extend over a quite substantial part of the crack, especially if $\sigma_y^\infty < 0$ (compressive loading), and it is then not simply absorbed by non-linear deformations in a real material. Rather it indicates the development of a contact region, in which some sliding is taking place. Note that the common situation for substantial shear loading involves one comparatively large contact region near one crack edge with a small contact (or non-linear) region near the other edge.

The necessity of incorporating interface slip along contact regions implies that a unique solution cannot be obtained: solutions would be history dependent. On the other hand, in the idealized case of vanishing friction, a solution is possible (Comninou 1977, 1978, Dunders and Comninou 1979, Gautesen and Dunders 1987a,b). An approximate, but apparently very accurate method, was devised by Dunders and Gautesen (1988) for cases when one contact region is much larger than the other one. They neglected the small and insignificant oscillatory region near one crack edge.

Some calculations of the length of the contact region were made by Gautesen (1993); they show, for instance, that the contact region is only about one per cent of the half crack length or less, if $\beta = 1/4$ and $\sigma_y^\infty > 0.4|\tau_{xy}^\infty|$, but it can extend over about one third of the crack length if $\beta = 1/4$ and $\sigma_y^\infty = 0$. Note that a contact region of this kind is not an exclusive property of interface cracks: pure shear loading, for instance, produces contact along the whole crack in the homogeneous case (identical materials), and if a suitably chosen excentric normal load were superposed, crack closure could occur along some portion of the crack and crack opening along another one. Other examples were discussed on pages 118 and 148.

The interface stresses near the crack edges are of special interest. As for cracks in homogeneous materials under small scale yielding, the state of the dissipative region ought to be controlled by the dominating singular terms found by assuming infinitesimally small scale yielding, cf. Section 3.3, although the character of these terms is more complicated for interface cracks. For $r \ll a$, where $r = |x| - a$, these terms are found from (4.12.39) to be

$$\sigma_y^0(r) = \sqrt{1 + 4\epsilon^2} \sqrt{\frac{a}{2r}} [\sigma_y^\infty \cos(\psi + \psi_0) \mp \tau_{xy}^\infty \sin(\psi + \psi_0)] \quad (4.12.47)$$

$$\tau_{xy}^0(r) = \sqrt{1 + 4\epsilon^2} \sqrt{\frac{a}{2r}} [\tau_{xy}^\infty \cos(\psi + \psi_0) \pm \sigma_y^\infty \sin(\psi + \psi_0)] \quad (4.12.48)$$

where the upper sign refers to $x > a$ and the lower sign to $x < -a$, and

$$\psi = \epsilon \ln \frac{2a}{r}, \qquad \psi_0 = \operatorname{atan}\frac{2\epsilon}{\sqrt{1+4\epsilon^2}} \qquad (4.12.49)$$

Note that ψ is not always negligible compared to unity in any crack edge vicinity, and therefore the strength of the dominating term cannot be described in the same simple way as the concept of stress intensity factor offers for cracks in homogeneous materials. How, then, can the state of the region in the crack edge vicinity be expressed?

Quite clearly, three parameters are needed to describe the state in the immediate vicinity of the edge of an interface crack. These parameters may be chosen as K_1, K_2 and r_0. The stress intensity factor K, given by the components K_1 and K_2, depends on the choice of r_0. The parameter r_0 can, in principle, be specified arbitrarily, but it has to be specified. Agreement about the choice is necessary for interpretation of a communicated value of K. Two ways to reach agreement are possible: either to choose a fix length, for instance $r_0 = 1\,\text{mm}$ for all material combinations, or to make a suitable choice for each bimaterial combination. See also Rice (1988b) for discussion on this matter.

Note that only two parameters, K_1 and K_2, are involved once r_0 has been chosen. This is the same as for a crack in a homogeneous material, for which K_I and K_{II} are needed. However, in the homogeneous case, the interpretations of K_I and K_{II} may be made separately, and they may be determined separately. Moreover, cracks are often subjected to pure mode I loading. For mixed mode loading, K_I and K_{II} might be used for predictions of the crack growth direction. For the interface crack an at least equally important question concerns the possibility of crack growth along the interface. As in the homogeneous case, a further complication is related to the T-stress, which also should be taken into consideration. Then, three parameters are involved.

The analysis of interfacial cracks was first given by Williams (1959), Cherepanov (1962), Erdogan (1963, 1965), England (1965), Malyushev and Salganik (1965) and Rice and Sih (1965). Williams (1959) described the oscillatory stress singularity, Rice and Sih (1965) introduced the convenient complex stress intensity factor, and Malyushev and Salganik (1965) determined the energy flux into the crack edge.

The interface crack under anti-plane strain

The difference between the properties of an interface crack under anti-plane strain and a mode III crack in a homogeneous medium is quite modest. Consider two mode III cracks in separate homogeneous bodies of different materials, but with the same symmetric geometry, the same array of cracks in the symmetry plane and with the same boundary loads. Then, as evident from the analysis in Section 4.3, the shear stresses on the symmetry plane are the same in both bodies, and the displacement is zero on the symmetry plane outside the cracks. Thus, it is possible to combine the lower half of one body with the upper half of the other body to obtain equilibrium, without changing displacements or stresses in the two halves. A stress intensity factor exists and is the same as for the homogeneous cases. Stresses exhibit mode III symmetry, but displacements do not. Crack sliding profiles are not symmetric. For example, the crack sliding profile consists of two half-elliptic parts in the case of a single crack in the interface between two semi-infinite media subjected to remote loading.

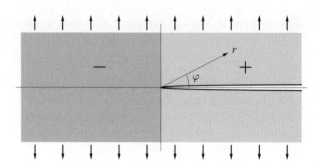

Fig. 4.12.2 Crack perpendicular to an interface.

Note the difference between interface cracks under in-plane and anti-plane deformation. Under in-plane deformation of homogeneous, symmetric bodies under mode I loading, the normal displacement on the symmetry plane outside the cracks is zero, but the tangential displacement is not. It depends on the material and it is thus not possible, as for anti-plane strain, to combine halves from two bodies of different materials without changing stresses and displacements.

If displacements, rather than stresses, are given on the boundaries, the situation becomes more complicated, unless certain specific symmetry relations prevail. Similar complications occur even for homogeneous anti-plane cases in the absence of mode III displacement anti-symmetry with respect to the crack plane. However, asymptotically, near the edge of an interface crack subjected to arbitrary anti-plane loading, a mode III stress intensity factor still exists. This follows from the continuity of the shear stress acting on the interface. An asymptotic analysis will be performed for the dynamic case, pages 377ff., also showing that the asymptotic field may be obtained by joining two halves of mode III asymptotic fields from different media.

Crack perpendicular to an interface

Zak and Williams (1963) determined the stress field in the vicinity of the edge of a crack perpendicular to the interface between two linearly elastic media and with the edge along the interface. They considered mode I loading and found a stress singularity different from the inverse square root, except for some specific material combinations, including, of course, two identical materials. Further studies were reported by Cook and Erdogan (1972).

Consider a cylindrical coordinate system, oriented so that the interface between the two materials is along $\varphi = \pm\pi/2$ and the crack faces are at $\varphi = \pm 0$. The minus subscript is used for the material at the left, i.e. at $\pi/2 \leq \varphi \leq 3\pi/2$, and the plus subscript for the material at the right; see Fig. 4.12.2. Remote loading in pure mode I is assumed, so that the crack faces are traction free.

One Airy stress function, Φ_-, is introduced for the material at the left and another one, Φ_+, for the material at the right above the crack, i.e. the quarter-plane $0 \leq \varphi \leq \pi/2$. Due to the symmetry, the problem may be solved without introducing an Airy stress function for $3\pi/2 \leq \varphi \leq 2\pi$.

4.12 INTERFACE CRACKS

The Airy stress function may be written as a series of terms of ascending powers of r. The dominant term in a crack edge vicinity is written as

$$\Phi_- = r^{s+2} F_-(\varphi) \text{ for } \pi/2 \leq \varphi \leq 3\pi/2 \tag{4.12.50}$$
$$\Phi_+ = r^{s+2} F_+(\varphi) \text{ for } \pi/2 \leq \varphi \leq 3\pi/2 \tag{4.12.51}$$

so that stresses and strains are singular as r^s. With regard to the mode I symmetry, the biharmonic equation for Φ_- yields, cf. Appendix A3,

$$F_- = A \cos[s(\varphi - \pi)] + B \cos[(s+2)(\varphi - \pi)] \tag{4.12.52}$$

whereas the full solution must be used for F_+:

$$F_+ = C \cos(s\varphi) + D \cos[(s+2)\varphi] + G \sin(s\varphi) + H \sin[(s+2)\varphi] \tag{4.12.53}$$

The boundary and matching conditions are:

$$\sigma_\varphi = 0 \text{ for } \varphi = 0, \quad \tau_{r\varphi} = 0 \text{ for } \varphi = 0 \tag{4.12.54}$$

$$(\sigma_\varphi)_- = (\sigma_\varphi)_+ \text{ for } \varphi = \frac{\pi}{2}, \quad (\tau_{r\varphi})_- = (\tau_{r\varphi})_+ \text{ for } \varphi = \frac{\pi}{2} \tag{4.12.55}$$

$$\left(\frac{\partial u_r}{\partial r}\right)_- = \left(\frac{\partial u_r}{\partial r}\right)_+ \text{ for } \varphi = \frac{\pi}{2}, \quad \left(\frac{\partial u_\varphi}{\partial r}\right)_- = \left(\frac{\partial u_\varphi}{\partial r}\right)_+ \text{ for } \varphi = \frac{\pi}{2} \tag{4.12.56}$$

Expressions in terms of F_- and F_+ of the field quantities appearing in these equations are found from Appendix A3. After some slight modifications, the following set of homogeneous equations is obtained:

$$C + D = 0, \quad sG + (s+2)H = 0, \quad A - B - 2C - 2\frac{s+1}{s+2}GT = 0 \tag{4.12.57}$$

$$sAT - (s+2)BT + 2(s+1)CT - 2sG = 0 \tag{4.12.58}$$

$$(1+\alpha)[k_-^2(s+2) - s]A - (k_-^2 - 1)(s+2)B$$
$$- 2(1-\alpha)\{[k_+^2(s+2) - s - 1]C + [k_+^2(s+1) - s]GT\} = 0 \tag{4.12.59}$$

$$(1+\alpha)[k_-^2 s - s - 2]AT - (k_-^2 - 1)(s+2)BT$$
$$+ 2(1-\alpha)\{[k_+^2(s+1) - s - 2]CT - [k_+^2 s - s - 1]G\} = 0 \tag{4.12.60}$$

Here, the first two equations have been used for elimination of D and H in the following four, $T = \tan(\pi s/2)$, and

$$\alpha = \frac{[(1-k_-^2)\mu_-]^{-1} - [(1-k_+^2)\mu_+]^{-1}}{[(1-k_-^2)\mu_-]^{-1} + [(1-k_+^2)\mu_+]^{-1}} \tag{4.12.61}$$

Vanishing determinant, the condition for non-trivial solutions, gives

$$\cos(\pi s) = 2\frac{\alpha - \beta}{1 - \beta}(s+1)^2 + \frac{\beta^2 - \alpha}{1 - \beta^2} \tag{4.12.62}$$

Parameter α and parameter β, previously defined in (4.12.4), are the two *Dunders' parameters* (Dundurs 1969), which provide a convenient and widely accepted way to specify interface properties. They are designed so that interchange of the two materials implies only a change of sign, and they vanish when the two materials have identical

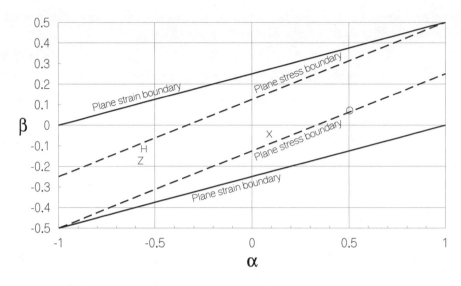

H: Al$_2$O$_3$/Ti, Plane strain, Z: Al$_2$O$_3$/Ti, Plane stress
X: Cu/Si, Plane strain
O: Cu/Al$_2$O$_3$, Plane strain

Fig. 4.12.3 Possible ranges of the Dunders parameters for plane strain and plane stress, with examples of material combinations (material with the minus subscript listed first).

elastic constants. The present definitions agree with those by Dunders if the minus subscript is identified with his subscript 1 and the plus subscript plus with his subscript 2. Note that in the literature, the roles of subscripts 1 and 2 are sometimes switched from Dunders' definitions.

The singularity exponent ϵ, defined by (4.12.17), is uniquely related to β. It will be shown that the stress singularity for a crack perpendicular to an interface may be given in terms of α and β, only. The two non-dimensional parameters designed by Dunders are thus sufficient for describing essential properties of cracks along interfaces, in spite of the fact that three non-dimensional parameters result from combinations of the four elastic constants.

The possible ranges of the Dunders' parameters are conveniently displayed in a diagram, Fig. 4.12.3, from which it is seen that the range is more narrow for plane stress than for plane strain. Some examples of material combinations, the material given the minus subscript listed first, are shown in the diagram. Interchange of the two materials corresponds to a symmetric point in the diagram, i.e., $\alpha \to -\alpha$ and $\beta \to -\beta$.

The solutions of (4.12.62) are found numerically. When $\beta = \alpha = 0$, identical material parameters, the dominating solution satisfying the requirement of bounded stress-strain energy is, of course, $s = -1/2$. Values of exponent s range from -0 at $\alpha = -1$, $\beta = -0.5$, to any arbitrarily close neighbourhood of -1 at $\alpha = 1$. Curves of constant s are shown in Fig. 4.12.4. These curves were obtained by using (4.12.62), written in

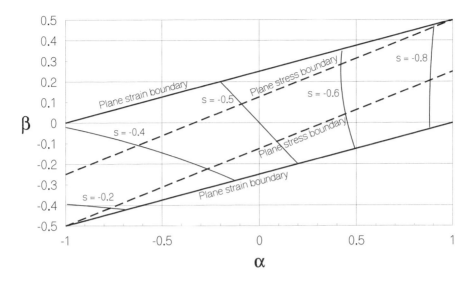

Fig. 4.12.4 Singularity exponent s for stresses and strains at the edge of a crack perpendicular to an interface.

the form

$$\alpha = \frac{(1-\beta^2)\cos\pi s + \beta[2(1+\beta)(s+1)^2 - \beta]}{2(1+\beta)(s+1)^2 - 1} \tag{4.12.63}$$

From the homogeneous equation system (4.12.57)-(4.12.60), five of the six unknowns may be expressed in terms of the sixth one, i.e. stresses and displacements may be determined to within a proportionality factor. Alternatively, they may be related to, for instance, the hoop stress straight ahead of the crack, $\sigma_\varphi(r,\pi)$. In this way it is found that the normal stress and the shear stress on the interface, normalized with respect to $\sigma_\varphi(r,\pi)$, are

$$\frac{\sigma_\varphi(r,\pi/2)}{\sigma_\varphi(r,\pi)} = \frac{(s+1)^2 + \alpha s(s+2) - \cos\pi s}{D_1} \cos\frac{\pi s}{2} \tag{4.12.64}$$

$$\frac{\tau_{r\varphi}(r,\pi/2)}{\sigma_\varphi(r,\pi)} = \frac{(1-\alpha)s(s+1) + 2\beta(s+1)^2 - \beta(1+\cos\pi s)}{D_1} \sin\frac{\pi s}{2} \tag{4.12.65}$$

where

$$D_1 = (s+1)(1-\cos\pi s) + 2\alpha s(s+1)(s+2)$$
$$- 2\beta s(s+2)^2 - \beta(s+2)(1-\cos\pi s) \tag{4.12.66}$$

Table 4.12.1 shows interface stresses for different material combinations. The crack is in the last mentioned material, e.g. in the titanium for Al_2O_3/Ti. The stresses are normalized, with the interface normal stress given as $\lim_{r\to 0}[\sigma_\varphi(r,\pi/2)/\sigma_\varphi(r,\pi)]$ and the interface shear stress as $\lim_{r\to 0}[\tau_{r\varphi}(r,\pi/2)/\sigma_\varphi(r,\pi)]$. The combinations "cork/steel"

Table 4.12.1 Normalized interface stresses for different material combinations. The crack is perpendicular to the interface. The combinations "cork/steel" and "steel/cork" represent the two extreme material combinations.

Material combination (crack-free/cracked)	Dunders' parameters α	β	Interface normal stress	Interface tangential stress
"cork/steel"	1	0.5	0	1
Ti/Al$_2$O$_3$	0.56	0.12	0.17	0.55
Homogeneous	0	0	$1/(2\sqrt{2}) \approx 0.354$	$1/(2\sqrt{2}) \approx 0.354$
Al$_2$O$_3$/Ti	-0.56	-0.12	0.45	0.24
"steel/cork"	-1	-0.5	0.5	0

and "steel/cork" are only symbolic: they represent two theoretical material combinations for which Dunders' parameters take on the extreme values $\alpha = 1$, $\beta = 0.5$ and $\alpha = -1$, $\beta = -0.5$, respectively. However, real combinations cork/steel and steel/cork are very close to the extremes: Poisson's ratio for cork is often set to zero, and its modulus of rigidity is certainly very much smaller than that for steel.

Cases when a crack reaches an interface under an oblique angle will not be discussed in detail here. Reference may be made to Erdogan and Arin (1975) and to He and Hutchinson (1989b). Solutions show that the dominating term near the crack edge possesses fixed, mixed mode angular variation for stresses and strains. However, another term, only slightly less singular and with different angular variations, also appears, and therefore the region of dominance of the first term is usually very small, and may not be significant in a real material, due to non-linear behaviour near the crack edge. The sum of the two terms might be more significant for control of the region near the crack edge. The two terms merge, because their singularities converge, in the extreme cases of interface crack and crack perpendicular to the interface.

Cases concerning cracks penetrating or kinking out of an interface have been studied by Erdogan and Biricikoglu (1973), Lu and Erdogan (1983a,b), He and Hutchinson (1989a,b) and He *et al.* (1991).

4.13 Three-dimensional crack problems

Introduction

Very few three-dimensional crack problems can be solved analytically. One exception is the plane elliptic crack in an infinite solid. Solution of this problem was given by Green and Sneddon (1950) for mode I loading and by Kassir and Sih (1966) for mixed mode loading. Stresses and displacements around an elliptic crack in an anisotropic elastic solid subjected to general remote loading were determined by Willis (1968). Further results were presented by Fabrikant (1989) for rather arbitrary types of loading. Fabrikant, who used potential theory, also developed approximate analytical solutions for planar cracks of arbitrary shape and considered interaction of coplanar cracks.

Face loaded semi-infinite cracks were analysed by Uflyand (1965), Kassir and Sih (1973, 1975), Meade and Keer (1984) and Kuo (1993). Rice (1985b, 1989) and Gao and Rice (1989) studied slightly curved cracks. Cracks in curved shells were analysed by Folias (1965a,b) and by Erdogan and Kibler (1969). The three-dimensional problem of a straight crack in a plate subjected to remote loading was analysed by Yang and Freund (1985). At the intersection between a crack edge and a free surface, for instance for a crack in a plate, the curious result is obtained that the stress singularity along the crack edge changes from the inverse square root to some other singularity when the free surface is reached (Bentham 1977, 1980, Bažant and Estenssoro 1979, 1980, Burton et al. 1984, Gudmundson and Östlund 1990).

The elliptic crack

For mode I loading, the stress intensity factor at an angle φ from the larger half-axis is (Green and Sneddon 1950)

$$K_I = \frac{\sigma^\infty \sqrt{\pi b}}{\boldsymbol{E}(m)}(1 - m^2 \cos^2 \varphi)^{1/4} \qquad (4.13.1)$$

where σ_∞ is the remote stress, a and b are the half-axes ($a > b$), $m = \sqrt{1 - b^2/a^2}$ and $\boldsymbol{E}(m)$ is the complete elliptic integral of the second kind. The maximum stress intensity factor obviously occurs at the smaller half-axis, and is

$$K_I = \frac{\sigma^\infty \sqrt{\pi b}}{\boldsymbol{E}(m)} \qquad (4.13.2)$$

This problem touches the very important problem of a semi-elliptic *surface crack* under mode I loading. Such cracks are very common in engineering structures. Usually the shorter semi-axis is perpendicular to the body surface, and the axis ratio is often about 3:4, but large variations can occur. An estimate of the stress intensity factor at the root of the crack may be found by applying the factor 1.12 to (4.13.2) to consider the edge crack effect: 1.12 is the approximate ratio between the stress intensity factors for an edge crack and a central crack, cf. (4.7.81). However, if, for instance, the crack is situated at the surface of a plate, and if its depth is not very small compared to the plate thickness, this might be a too crude approximation. Numerous numerical results exist for surface cracks in bodies of different geometries and loading conditions; see e.g. Isida et al. (1984).

Cracks in curved shells

For cracks in curved shells, the stress intensity factor varies along the edge of the crack. The pioneering work was made by Folias (1965a,b), who analysed several different cases of cracks in cylindrical, spherical and other types of shells. One very important case concerns a crack with length $2a$ in the axial direction of a cylindrical shell with the main radius R and the wall thickness h. If the shell is subjected to an internal overpressure p, the stress intensity factor increases towards the outer surface. The maximum stress intensity factor can be approximately given by the expression

$$K_I \approx \frac{pR}{h}\sqrt{\pi a(1 + 0.5\lambda^2)}, \quad \text{for } \lambda < 5 \qquad (4.13.3)$$

Semi-infinite plane crack subjected to crack face loading

where $\lambda = a\sqrt{[12(1-\nu^2)]/(Rh)}$. More accurate estimates, up to $\lambda = 8$ are given by Erdogan and Kibler (1969), who, like Folias (1965a,b), also presented results for other types of shells, crack orientations and load conditions.

The fundamental problem for crack face loaded semi-infinite plane cracks concerns the response to opposed concentrated forces, normal or tangential to the crack plane. The earliest solution seems to have appeared in a report by Uflyand (1965), who used the Papkovich representation (4.4.2) and considered normal forces and tangential forces normal to the crack edge. As shown by Eubanks and Sternberg (1956), one of the four Papkovich potentials may generally, though not always, be discarded, because there are only three displacement components. Consequently, Uflyuand used only three potentials, and was able to obtain a solution, except for the case of tangential forces parallel to the crack edge, which requires all four potentials, as shown by Kassir and Sih (1973, 1975). Later Meade and Keer (1984) found a simpler solution procedure by making use of a displacement field representation similar to but distinct from the one by Papkovich†. They also considered symmetric (not opposed) tangential crack face forces.

Meade and Keer (1984) used the displacement representation

$$2\mu \mathbf{u} = \operatorname{grad} F - (3-4\nu)Y\hat{\mathbf{y}} + 2\operatorname{curl}(\psi_x \hat{\mathbf{x}} + \psi_z \hat{\mathbf{z}}) \tag{4.13.4}$$

where F, Y, ψ_x and ψ_z are harmonic functions. In component form this reads

$$2\mu u = \frac{\partial F}{\partial x} + y\frac{\partial Y}{\partial x} + 2\frac{\partial \psi_z}{\partial y} \tag{4.13.5}$$

$$2\mu v = \frac{\partial F}{\partial y} - (3-4\nu)Y + 2(\frac{\partial \psi_x}{\partial z} - \frac{\partial \psi_z}{\partial x}) \tag{4.13.6}$$

$$2\mu w = \frac{\partial F}{\partial z} + y\frac{\partial Y}{\partial z} - 2\frac{\partial \psi_x}{\partial y} \tag{4.13.7}$$

Use of Hooke's law gives the stresses needed for boundary conditions:

$$\sigma_y = \frac{\partial^2 F}{\partial y^2} - 2(1-\nu)\frac{\partial Y}{\partial y} + y\frac{\partial^2 Y}{\partial y^2} + 2\frac{\partial^2 \psi_x}{\partial z \partial y} - 2\frac{\partial^2 \psi_z}{\partial x \partial y} \tag{4.13.8}$$

$$\tau_{xy} = \frac{\partial^2 F}{\partial x \partial y} - (1-2\nu)\frac{\partial Y}{\partial x} + y\frac{\partial^2 Y}{\partial x \partial y} + \frac{\partial^2 \psi_x}{\partial x \partial z} + \frac{\partial^2 \psi_z}{\partial y^2} - \frac{\partial^2 \psi_z}{\partial x^2} \tag{4.13.9}$$

$$\tau_{yz} = \frac{\partial^2 F}{\partial y \partial z} - (1-2\nu)\frac{\partial Y}{\partial z} + y\frac{\partial^2 Y}{\partial y \partial z} + \frac{\partial^2 \psi_x}{\partial z^2} - \frac{\partial^2 \psi_x}{\partial y^2} - \frac{\partial^2 \psi_z}{\partial x \partial z} \tag{4.13.10}$$

Consider a crack face loaded semi-infinite crack $y = 0$, $x < 0$. By choosing the loads as concentrated forces, fundamental solutions can be obtained, leading to solutions for general crack face loading by integration. The concentrated forces may be applied at opposite points, $x = -L$, $y = \pm 0$. They may be symmetric with respect to the crack plane, so as to cause crack opening, giving rise only to mode I stress intensity factors, or they may be anti-symmetric with respect to the crack plane, in which case

---

† An even simpler solution was presented by Kuo (1993), who used Fourier transforms and the Wiener-Hopf technique.

4.13 THREE-DIMENSIONAL CRACK PROBLEMS

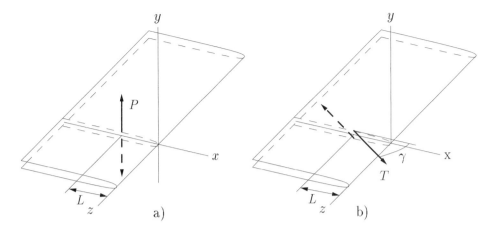

Fig. 4.13.1 Semi-infinite crack, subjected to a) concentrated normal crack face forces, and b) concentrated tangential forces.

they produce modes II and III stress intensity factors, but not mode I. The forces may be normal to the crack faces or tangential, and in the latter case, they may be decomposed into components normal and parallel to the crack edge. In all, six distinct fundamental problems are possible. Meade and Keer (1984) investigated all of them. Here, only the cases of opposed normal tractions,

$$\sigma_y^0 = -P\delta(x+L)\delta(z) \tag{4.13.11}$$

and opposed shear tractions,

$$\tau_{xy}^0 = -T\sin\gamma\,\delta(x+L)\delta(z), \quad \tau_{yz}^0 = -T\cos\gamma\,\delta(x+L)\delta(z) \tag{4.13.12}$$

will be considered; see Fig. 4.13.1. The problems consist of finding stresses and displacements, and, in particular, the stress intensity factors along the crack edge.

In addition to conditions of crack face tractions, there are conditions at infinity and at the crack edge. A concentrated force is balanced by stresses that decay in proportion to $1/R^2$ as $R \to \infty$, where R is the distance from the point of force application. Thus, the function Y cannot decay slower than in proportion to $1/R$, and the same applies to the derivatives of functions F, ψ_x and ψ_z. At the crack edge, the displacements must be bounded, and the strongest singularity for stresses must be the inverse square root of the distance r from the crack edge.

Opposed concentrated normal crack face forces
Under symmetric loading, the displacement components u and w are symmetric and v anti-symmetric with respect to y. Thus, F is symmetric and Y, ψ_x, ψ_z anti-symmetric with respect to y. This implies that

$$\frac{\partial F}{\partial y} = Y = 0, \quad \psi_x = \psi_z = 0 \text{ for } y = 0, x > 0 \tag{4.13.13}$$

Crack face tractions are given by (4.13.11). Consider the half-plane $y \geq 0$. The boundary conditions on $y = 0$ are

$$v = 0, \quad \tau_{xy} = \tau_{yz} = 0 \text{ for } x > 0 \tag{4.13.14}$$

$$\sigma_y = -P\delta(x+L)\delta(z) \text{ for } x < 0 \tag{4.13.15}$$

$$\tau_{xy} = 0, \quad \tau_{yz} = 0 \text{ for all } x \tag{4.13.16}$$

Insertion of (4.13.4)-(4.13.10) into these conditions gives, for $y = 0$,

$$\frac{\partial F}{\partial y} - (3 - 4\nu)Y + 2\left(\frac{\partial \psi_x}{\partial z} - \frac{\partial \psi_z}{\partial x}\right) = 0 \text{ for } x > 0 \tag{4.13.17}$$

$$\frac{\partial V}{\partial y} = -P\delta(x+L)\delta(z) \text{ for } x < 0 \tag{4.13.18}$$

$$\frac{\partial U}{\partial x} + \frac{\partial^2 \psi_z}{\partial y^2} = 0, \quad \frac{\partial U}{\partial z} - \frac{\partial^2 \psi_x}{\partial y^2} = 0 \text{ for all } x \tag{4.13.19}$$

where

$$U = \frac{\partial F}{\partial y} - (1 - 2\nu)Y + \frac{\partial \psi_x}{\partial z} - \frac{\partial \psi_z}{\partial x} \tag{4.13.20}$$

$$V = \frac{\partial F}{\partial y} - 2(1 - \nu)Y + 2\frac{\partial \psi_x}{\partial z} - 2\frac{\partial \psi_z}{\partial x} \tag{4.13.21}$$

For this case, the solution procedure does not need to involve all four potentials. Meade and Keer (1984) established a general relation between the potentials by assuming that U vanishes for all x, y, z. Then $\partial^2 \psi_x / \partial y^2$ and $\partial^2 \psi_z / \partial y^2$ vanish on $y = 0$, according to (4.13.19). They satisfy the Laplace equation, and, considering also the conditions at infinity, it follows that they vanish everywhere and that integrations result in $\psi_x = \psi_z = 0$ (additional constants have no significance for potentials). Thus, only two potentials remain, F and Y, and, moreover, the identity $U = 0$ implies that

$$\frac{\partial F}{\partial y} - (1 - 2\nu)Y = 0 \tag{4.13.22}$$

Then, the problem can be formulated in terms of Y, from (4.13.17)-(4.13.18), as

$$\Delta Y = 0 \tag{4.13.23}$$

$$Y = 0 \text{ for } y = 0, x > 0 \tag{4.13.24}$$

$$\frac{\partial Y}{\partial y} = -P\delta(x+L)\delta(z) \text{ for } y = 0, x > 0 \tag{4.13.25}$$

where the second condition alternatively could have been taken from (4.13.13).

In cylindrical coordinates, the Laplace equation reads:

$$\frac{\partial^2 Y}{\partial r^2} + \frac{1}{r} \cdot \frac{\partial Y}{\partial r} + \frac{1}{r^2} \cdot \frac{\partial^2 Y}{\partial \varphi^2} + \frac{\partial^2 Y}{\partial z^2} = 0 \tag{4.13.26}$$

Variable separation through the ansatz $Y = R(r)\Phi(\varphi)Z(z)$ gives

$$\frac{d^2 Z}{dz^2} = -\alpha^2 Z, \quad \frac{d^2 \Phi}{d\varphi^2} = \beta^2 \Phi, \quad \frac{d^2 R}{dr^2} + \frac{1}{r} \cdot \frac{dR}{dr} - \left(\alpha^2 - \frac{\beta^2}{r^2}\right)R = 0 \tag{4.13.27}$$

where, to satisfy the conditions at infinity, the constants α^2 and β^2 must be real

4.13 THREE-DIMENSIONAL CRACK PROBLEMS

and positive. Then, considering the symmetry conditions $Y(r, \varphi, -z) = Y(r, \varphi, z)$ and $Y = 0$ for $\varphi = 0$, the solution may be written as

$$Y = \int_0^\infty \int_0^\infty A(\alpha, \beta) \cos(\alpha z) \sinh(\beta\varphi) K_{i\beta}(\alpha r) \mathrm{d}\beta \, \mathrm{d}\alpha \qquad (4.13.28)$$

where $K_{i\beta}(\cdot)$ is the modified Bessel function of second order with imaginary argument (also called the Macdonald function). Now, from (4.13.25),

$$\frac{1}{x} \int_0^\infty \cos(\alpha z) \int_0^\infty A(\alpha, \beta) \beta \cosh(\pi\beta) K_{i\beta}(\alpha|x|) \mathrm{d}\beta \, \mathrm{d}\alpha = -P\delta(z)\delta(|x| - L) \quad (4.13.29)$$

Fourier inversion gives

$$\int_0^\infty A(\alpha, \beta) \beta \cosh(\pi\beta) K_{i\beta}(\alpha|x|) \mathrm{d}\beta$$

$$= -\frac{2P}{\pi} \int_0^\infty \cos(\alpha z) \delta(z) \mathrm{d}z \cdot x \delta(|x| - L) = \frac{PL}{\pi} \delta(|x| - L) \qquad (4.13.30)$$

Note that $\int_0^\infty \delta(z) \mathrm{d}z = 1/2 \int_{-\infty}^\infty \delta(z) \mathrm{d}z = 1/2$.

Now, the inversion by Lebedev (1949) may be used. It states that,

$$\text{if} \quad h(s) = \frac{2}{\pi^2} \int_0^\infty H(\beta) \beta \sinh(\pi\beta) K_{i\beta}(s) \mathrm{d}\beta \qquad (4.13.31)$$

$$\text{then} \quad H(\beta) = \int_0^\infty \frac{1}{s} K_{i\beta}(s) h(s) \mathrm{d}s \quad \text{or vice versa} \qquad (4.13.32)$$

$H(\beta)$ is known as the Kontorovich-Lebedev transform of $h(s)$; see, e.g., Sneddon (1972). The inversion yields

$$A(\alpha, \beta) = \frac{2P}{\pi^3} \cdot \frac{\sinh(\pi\beta)}{\cosh(\pi\beta)} K_{i\beta}(\alpha L) \qquad (4.13.33)$$

so that

$$Y = \frac{2P}{\pi^3} \int_0^\infty \int_0^\infty \cos(\alpha z) \frac{\sinh(\beta\varphi) \sinh(\pi\beta)}{\cosh(\pi\beta)} K_{i\beta}(\alpha r) K_{i\beta}(\alpha L) \mathrm{d}\beta \, \mathrm{d}\alpha \qquad (4.13.34)$$

Now (see e.g. Sneddon 1972),

$$K_{i\beta}(\alpha r) K_{i\beta}(\alpha L) = \int_0^\infty K_0\left(\alpha\sqrt{r^2 + 2Lr \cosh\xi + L^2}\right) \cos(\beta\xi) \mathrm{d}\xi \qquad (4.13.35)$$

Insertion gives

$$Y = \frac{2P}{\pi^3} \int_0^\infty \Bigg[\int_0^\infty \frac{\sinh(\beta\varphi) \sinh(\pi\beta)}{\cosh(\pi\beta)} \cos(\beta\xi) \mathrm{d}\beta$$

$$\times \int_0^\infty \cos(\alpha z) K_0\left(\alpha\sqrt{r^2 + 2Lr \cosh\xi + L^2}\right) \mathrm{d}\alpha \Bigg] \mathrm{d}\xi \qquad (4.13.36)$$

Both inner integrals may be found in Gradshteyn and Ryzhik (1980), the first one after writing $\sinh(\beta\varphi) \sinh(\pi\beta) = 1/2\{\cosh[\beta(\pi + \varphi)] - \cosh[\beta(\pi - \varphi)]\}$. The result is

$$Y = -\frac{P}{\pi^2} \int_0^\infty \frac{\sin(\varphi/2) \cosh(\xi/2) \, \mathrm{d}\xi}{(\cosh\xi - \cos\varphi)\sqrt{r^2 + 2Lr + L^2 + z^2 + 2Lr \cosh\xi}} \qquad (4.13.37)$$

By making the substitution $\sinh(\xi/2) = u$, the integral is readily calculated, and the solution is found to be

$$Y = -\frac{P}{\pi^2}\phi, \quad \phi = \frac{1}{\rho}\operatorname{atan}\frac{\sqrt{2L(r-x)}}{\rho} \tag{4.13.38}$$

where $\rho = \sqrt{(x+L)^2 + y^2 + z^2}$.

The potential F may now be obtained from (4.13.22). The displacements and selected stresses are found from the relations

$$2\mu u = \frac{\partial F}{\partial x} + y\frac{\partial Y}{\partial x}, \quad 2\mu v = y\frac{\partial Y}{\partial y} - 2(1-\nu)Y, \quad 2\mu w = \frac{\partial F}{\partial z} + y\frac{\partial Y}{\partial z} \tag{4.13.39}$$

$$\sigma_y = y\frac{\partial^2 Y}{\partial y^2} - \frac{\partial Y}{\partial y}, \quad \tau_{xy} = y\frac{\partial^2 Y}{\partial x \partial y}, \quad \tau_{yz} = y\frac{\partial^2 Y}{\partial y \partial z} \tag{4.13.40}$$

Obviously, Y becomes proportional to $r^{1/2}$ as $r \to 0$. Thus, the potential F becomes proportional to $r^{3/2}$ as $r \to 0$. It then follows that all displacements are bounded as $r \to 0$, as required. The stress intensity factors are

$$K_I = \lim_{x \to +0}\left[\sqrt{2\pi x}(\sigma_y)_{y=0}\right] = \frac{P\sqrt{2L}}{\sqrt{\pi^3(z^2 + L^2)}} \tag{4.13.41}$$

$$K_{II} = K_{III} = 0 \tag{4.13.42}$$

Opposed concentrated tangential crack face forces

Under anti-symmetric loading, the displacement components u and w are anti-symmetric and v symmetric with respect to y. Thus, F is anti-symmetric and Y, ψ_x, ψ_z symmetric with respect to y. Hence,

$$F = 0, \quad \frac{\partial Y}{\partial y} = 0, \quad \frac{\partial \psi_x}{\partial y} = \frac{\partial \psi_z}{\partial y} = 0 \text{ for } y = 0, x > 0 \tag{4.13.43}$$

Crack face tractions are given by (4.13.12) and shown in Fig. 4.13.1. Consider the half-plane $y \geq 0$. The boundary conditions on $y = +0$ are

$$u = w = 0 \text{ for } x > 0 \tag{4.13.44}$$

$$\sigma_y = 0 \text{ for all } x \tag{4.13.45}$$

$$\tau_{xy} = -T\cos\gamma\,\delta(x+L)\delta(z), \quad \tau_{yz} = -T\sin\gamma\,\delta(x+L)\delta(z) \text{ for } x < 0 \tag{4.13.46}$$

The conditions for displacements are automatically satisfied by (4.13.43). Insertion of (4.13.4)-(4.13.10) into the conditions for stresses gives, for $y = 0$,

$$\frac{\partial V}{\partial y} = 0 \text{ for all } x \tag{4.13.47}$$

$$\frac{\partial U}{\partial x} + \frac{\partial^2 \psi_z}{\partial y^2} = -T\cos\gamma\,\delta(x+L)\delta(z) \text{ for } x < 0 \tag{4.13.48}$$

$$\frac{\partial U}{\partial z} - \frac{\partial^2 \psi_x}{\partial y^2} = -T\sin\gamma\,\delta(x+L)\delta(z) \text{ for } x < 0 \tag{4.13.49}$$

where U and V are given by (4.13.20)-(4.13.21).

Because $\partial V/\partial y$ satisfies the Laplace equation and decays as $1/R^2$ as $R \to \infty$, condition (4.13.47) implies that V reduces to a constant, which may be set to zero:

$$V = \frac{\partial F}{\partial y} - 2(1-\nu)Y + 2\frac{\partial \psi_x}{\partial z} - 2\frac{\partial \psi_z}{\partial x} = 0 \quad (4.13.50)$$

This is a result of the problem degeneration caused by the absence of normal loads, and it should not be confused with the intentional reduction of potentials in the previous problem (normal, but not tangential loading) by putting $U = 0$. Here, it is not possible to put $U = 0$ for all x, y, z, but all conditions can be satisfied by requiring that U vanishes on the crack face:

$$U = \frac{\partial F}{\partial y} - (1-2\nu)Y + \frac{\partial \psi_x}{\partial z} - \frac{\partial \psi_z}{\partial x} = 0 \text{ for } y = 0, x < 0 \quad (4.13.51)$$

Note that then also $\partial U/\partial x$ and $\partial U/\partial z$ vanish on the crack face.

The condition (4.13.51) can be complemented with a condition for U on $y = 0$, $x > 0$. Thus, consider

$$\frac{\partial U}{\partial y} = \frac{\partial}{\partial y}\left[\frac{\partial F}{\partial y} - (1-2\nu)Y + \frac{\partial \psi_x}{\partial z} - \frac{\partial \psi_z}{\partial x}\right] \quad (4.13.52)$$

Use of (4.13.43) and the relation $\partial^2 F/\partial y^2 = -\partial^2 F/\partial x^2 - \partial^2 F/\partial z^2 = 0$ shows that this expression vanishes for $y = 0$, $x > 0$.

It is now possible to extract three mixed boundary value problems from (4.13.48)-(4.13.49), (4.13.51) and the condition for $\partial U/\partial y$, namely

Problem A: $\quad \Delta\left(\dfrac{\partial \psi_z}{\partial y}\right) = 0 \quad (4.13.53)$

$\dfrac{\partial \psi_z}{\partial y} = 0$ for $y - 0, x > 0 \quad (4.13.54)$

$\dfrac{\partial}{\partial y}\left(\dfrac{\partial \psi_z}{\partial y}\right) = -T\cos\gamma\,\delta(x+L)\delta(z)$ for $y = 0, x < 0 \quad (4.13.55)$

Problem B: $\quad \Delta\left(\dfrac{\partial \psi_x}{\partial y}\right) = 0 \quad (4.13.56)$

$\dfrac{\partial \psi_x}{\partial y} = 0$ for $y = 0, x > 0 \quad (4.13.57)$

$\dfrac{\partial}{\partial y}\left(\dfrac{\partial \psi_x}{\partial y}\right) = T\sin\gamma\,\delta(x+L)\delta(z)$ for $y = 0, x < 0 \quad (4.13.58)$

Problem C: $\quad \Delta U = 0 \quad (4.13.59)$

$\dfrac{\partial U}{\partial y} = 0$ for $y = 0, x > 0 \quad (4.13.60)$

$U = 0$ for $y = 0, x < 0 \quad (4.13.61)$

These problems, which also should include the appropriate behaviour at infinity, are essentially the same as the one given by (4.13.23)-(4.13.25). However, if solutions are obtained in the same way, it turns out that, contrary to intuition, the stress intensity factor K_{II} vanishes for all z if $\gamma = \pi/2$ and K_{III} vanishes if $\gamma = 0$. This gives the hint that the solution is not complete, and that singular terms need to be added (Kassir and Sih 1973). Such terms do not violate the physical conditions of the problem, if their singular parts cancel each other in the displacement representations. This is possible to achieve, because four potentials are used to represent three displacement components. With inclusion of singular terms, the solutions of Problems A, B and C can be written as

$$\frac{\partial \psi_z}{\partial y} = \frac{T \cos \gamma}{\pi^2} \phi(r, \varphi, z) + f_1(r, \varphi, z) \quad (4.13.62)$$

$$\frac{\partial \psi_x}{\partial y} = -\frac{T \sin \gamma}{\pi^2} \phi(r, \varphi, z) \quad (4.13.63)$$

$$U = f_2(r, \varphi, z) \quad (4.13.64)$$

where ϕ is given by (4.13.38), and

$$f_1 = \frac{\sin(\varphi/2)}{\sqrt{r}} \Re[g_1(\zeta)], \quad f_2 = \frac{\cos(\varphi/2)}{\sqrt{r}} \Re[g_2(\zeta)] \quad (4.13.65)$$

where $\zeta = r + L + iz$ (the inclusion of L for later convenience), and $g_i(\zeta)$ are analytic functions, so that $\Re[g_i(\zeta)]$ are harmonic functions. A singular solution cannot be added to $\partial \psi_x/\partial y$, because this would cause w to be unbounded at $r = 0$. This follows from (4.13.7), noting that potentials differentiated with respect to z remain bounded.

Function f_1 vanishes for $y = 0$, $x > 0$ ($\varphi = 0$) and its derivative $\partial f_1/\partial y = 0$ for $y = 0$, $x < 0$, because $(\partial f_1/\partial y)_{y=0} = x^{-1}(\partial f_1/\partial \varphi)_{y=0}$. Insertion shows that f_1 satisfies the Laplace equation. Thus, (4.13.62) satisfies Problem A, and, similarly, (4.13.64) satisfies Problem C.

For the determination of $\Re[g_1(\zeta)]$ and $\Re[g_2(\zeta)]$, two relations are needed. The first one is obtained from the requirement that u and v remain bounded as $r \to 0$. Use of the relation (4.13.50) to eliminate Y gives, for $y = 0$, $x < 0$:

$$2\mu \frac{\partial u}{\partial x} = \frac{\partial}{\partial y}\left[2\frac{\partial \psi_z}{\partial x} - \frac{\partial F}{\partial y} + \mathcal{R}\right] \quad (4.13.66)$$

$$2\mu v = \frac{1 - 2\nu}{2(1 - \nu)}\left[2\frac{\partial \psi_z}{\partial x} - \frac{\partial F}{\partial y} + \mathcal{R}\right] \quad (4.13.67)$$

where $\mathcal{R}$ represents terms that are bounded at $r = 0$. In the derivation, the equality $\partial^2 F/\partial x^2 = -\partial^2 F/\partial y^2 + \mathcal{R}$ has been used. After integration of (4.13.62), the part of ψ_z contributing to singular $\partial \psi_z/\partial x$ is found to be equal to $2\sqrt{r}\Re[g_1(\zeta)]$ for $y = 0$, $x < 0$. Then, the singular part of $\partial F/\partial y$ is obtained from (4.13.64) after elimination of Y. Insertion shows that vanishing of the singular part of the expressions within [ ] requires that $\Re[g_2(\zeta)] = \Re[g_1(\zeta)]$.

The second relation needed for determination of $\Re[g_1(\zeta)]$ and $\Re[g_2(\zeta)]$ is $F = 0$ for $y = 0$, $x > 0$ (4.13.43). Use of (4.13.50) for elimination of Y in equation (4.13.64) gives

$$\frac{\partial F}{\partial y} + 2\nu\left(\frac{\partial \psi_x}{\partial z} - \frac{\partial \psi_z}{\partial x}\right) = 2(1 - \nu)\frac{\cos(\varphi/2)}{\sqrt{r}} \Re[g(\zeta)] \quad (4.13.68)$$

Then, for $y = 0$, $x > 0$, noting that $\int_{-\infty}^{0}(\partial F/\partial y)dy = 0$, insertion of (4.13.62)-(4.13.63)

4.13 THREE-DIMENSIONAL CRACK PROBLEMS

yields

$$-\frac{2\nu T \sin\gamma}{\pi^2}\int_\infty^0\int_\infty^y \frac{\partial\phi}{\partial z}dy'dy - \frac{2\nu T\cos\gamma}{\pi^2}\int_\infty^0\int_\infty^y \frac{\partial\phi}{\partial x}dy'dy$$

$$-2\nu\int_\infty^0\int_\infty^y \frac{\partial}{\partial x}\left\{\frac{\sin(\varphi/2)}{\sqrt{r}}\Re[g(\zeta)]\right\}dy'dy$$

$$= 2(1-\nu)\int_\infty^0 \frac{\cos(\varphi/2)}{\sqrt{r}}\Re[g(\zeta)]dy \quad (4.13.69)$$

where y' is used to replace y in the integrands of inner integrals.

For calculation of the first double integral, it is useful to note that differentiation leads to

$$\frac{\partial\phi}{\partial\zeta} = \frac{z}{y}\cdot\frac{\partial\phi}{\partial y} - \sqrt{\frac{L}{2}}\cdot\frac{z}{r\sqrt{r-x}|\zeta|^2} \quad (4.13.70)$$

Then,

$$\int_\infty^0\int_\infty^y \frac{\partial\phi}{\partial z}dy'dy = \left[y\int_\infty^y \frac{\partial\phi}{\partial z}dy'\right]_\infty^0 - \int_\infty^0 y\frac{\partial\phi}{\partial z}dy + \sqrt{\frac{L}{2}}z\int_\infty^0 \frac{y\,dy}{r\sqrt{r-x}|\zeta|^2}$$

$$= -\sqrt{\frac{L}{2}}z\int_x^\infty \frac{dr}{\sqrt{r-x}[(r+L)^2+z^2]} \quad (4.13.71)$$

The last integral is conveniently determined by residue calculus. The result is

$$\int_\infty^0\int_\infty^y \frac{\partial\phi}{\partial z}dy'dy = \sqrt{\frac{L}{2}}\cdot\pi\cdot\Im\frac{1}{\zeta_0^{1/2}} \quad (4.13.72)$$

where $\zeta_0 = (\zeta)_{y=0} = x + L + iz$.

In essentially the same way, the second double integral in (4.13.69) is determined as

$$\int_\infty^0\int_\infty^y \frac{\partial\phi}{\partial x}dy'dy = -\sqrt{\frac{L}{2}}\cdot\pi\cdot\Re\frac{1}{\zeta_0^{1/2}} \quad (4.13.73)$$

The third double integral is determined by first making a partial integration leading to the expression

$$-\int_\infty^0 y\frac{\partial}{\partial x}\left\{\frac{\sin(\varphi/2)}{\sqrt{r}}\Re[g(\zeta)]\right\}dy \quad (4.13.74)$$

Carrying out the differentiation with respct to x and then making a second partial integration, noting that $\sin\varphi\, g'(\zeta) = \partial[g(\zeta)]/\partial y$, yields

$$\int_\infty^0\int_\infty^y \frac{\partial}{\partial x}\left\{\frac{\sin(\varphi/2)}{\sqrt{r}}\Re[g(\zeta)]\right\}dy'dy = \frac{1}{2}\Re\int_\infty^0 \frac{\cos(\varphi/2)}{\sqrt{r}}g(\zeta)dy \quad (4.13.75)$$

By changing the integration variable from y to ζ, noting that $dy = r\,d\zeta/\sqrt{r^2-x^2}$ and $\sqrt{2}\cos(\varphi/2) = \sqrt{1+x/r}$, it is found that the last integral equals

$$\Re\left[\int_\infty^0 \frac{g(\zeta)\cos(\varphi/2)}{\sqrt{r}}dy\right] = \frac{1}{\sqrt{2}}\Re\left[\int_\infty^{\zeta_0} \frac{g(\zeta)d\zeta}{\sqrt{\zeta-\zeta_0}}\right] \quad (4.13.76)$$

where $\zeta_0 = x + L + iz$. Note that $\sqrt{\zeta - \zeta_0} = \sqrt{r-x}$.

Insertion of the expressions found for the integrals in (4.13.69) gives

$$\Re \int_{\infty}^{\zeta_0} \frac{g(\zeta)}{\sqrt{\zeta - \zeta_0}} d\zeta = \frac{T\sqrt{L}}{\pi} \cdot \frac{2\nu}{2-\nu} \left(\cos\gamma \, \Re \frac{1}{\zeta_0^{1/2}} - \sin\gamma \, \Im \frac{1}{\zeta_0^{1/2}} \right)$$

$$= \frac{T\sqrt{L}}{\pi} \cdot \frac{2\nu}{2-\nu} \cdot \Re \frac{e^{i\gamma}}{\zeta_0^{1/2}} \quad (4.13.77)$$

Thus, $g(\zeta)$ can be found from the Abelian integral equation

$$\int_{\infty}^{\zeta_0} \frac{g(\zeta)}{\sqrt{\zeta - \zeta_0}} d\zeta = \frac{T\sqrt{L}}{\pi} \cdot \frac{2\nu}{2-\nu} \cdot \frac{e^{i\gamma}}{\zeta_0^{1/2}} \quad (4.13.78)$$

The solution (see e.g. Kanwal 1971) is

$$g(\zeta) = -\frac{T\sqrt{L}}{\pi^2} \cdot \frac{2\nu}{2-\nu} \cdot \frac{e^{i\gamma}}{\zeta} \quad (4.13.79)$$

All four potentials can now be found, ψ_x and ψ_z from (4.13.62)-(4.13.63), and then F and Y from (4.13.50) and (4.13.64). Stresses and displacements may thus, in principle, be determined for an arbitrary point, although several cumbersome integrations are needed. However, the non-zero stress intensity factors K_{II} and K_{III} can be directly calculated by insertion of expressions previously found. Thus, it is found from (4.13.9) and (4.13.10) that, for $y = 0$, $x > 0$,

$$\tau_{xy} = \frac{1}{2(1-\nu)} \frac{\partial}{\partial x}\left[\frac{\partial F}{\partial y} + 2\nu\left(\frac{\partial \psi_x}{\partial z} - \frac{\partial \psi_z}{\partial x}\right)\right] + \frac{\partial^2 \psi_z}{\partial y^2} \quad (4.13.80)$$

$$\tau_{yz} = \frac{1}{2(1-\nu)} \frac{\partial}{\partial z}\left[\frac{\partial F}{\partial y} + 2\nu\left(\frac{\partial \psi_x}{\partial z} - \frac{\partial \psi_z}{\partial x}\right)\right] - \frac{\partial^2 \psi_x}{\partial y^2} \quad (4.13.81)$$

where the expressions within [ ] are found in (4.13.68), and

$$\frac{\partial^2 \psi_z}{\partial y^2} = \frac{T\cos\gamma}{\pi^2}\left(\frac{\partial \phi}{\partial y}\right)_{y\to 0} + \frac{\partial f_1}{\partial y}$$

$$= \frac{T\cos\gamma}{\pi^2} \cdot \frac{\sqrt{L}}{\sqrt{x}(L^2+z^2)} + \frac{\partial}{\partial y}\left\{\frac{\sin(\varphi/2)\Re[g(\zeta)]}{\sqrt{r}}\right\} \quad (4.13.82)$$

$$\frac{\partial^2 \psi_x}{\partial y^2} = -\frac{T\sin\gamma}{\pi^2}\left(\frac{\partial \phi}{\partial y}\right)_{y\to 0} = -\frac{T\sin\gamma}{\pi^2} \cdot \frac{\sqrt{L}}{\sqrt{x}(L^2+z^2)} \quad (4.13.83)$$

After straightforward calculations it is found that

$$(\tau_{xy})_{y=0} = \Re[g'(\zeta_0)]\sqrt{x} + \frac{T\sqrt{L}\cos\gamma}{\pi^2(L^2+z^2)} \quad (4.13.84)$$

$$(\tau_{yz})_{y=0} = -\Im[g'(\zeta_0)]\sqrt{x} + \frac{T\sqrt{L}\sin\gamma}{\pi^2(L^2+z^2)} \quad (4.13.85)$$

From these expressions the stress intensity factors are found to be

$$K_{II} = \sqrt{\frac{2}{\pi^3}} \cdot \frac{T\sqrt{L}}{L^2 + z^2} \left\{ \cos\gamma \left[1 + \frac{2\nu}{2-\nu} \cdot \frac{L^2 - z^2}{L^2 + z^2}\right] \right.$$
$$\left. + \sin\gamma \cdot \frac{2\nu}{2-\nu} \cdot \frac{2Lz}{L^2 + z^2} \right\} \quad (4.13.86)$$

$$K_{III} = \sqrt{\frac{2}{\pi^3}} \cdot \frac{T\sqrt{L}}{L^2 + z^2} \left\{ \cos\gamma \cdot \frac{2\nu}{2-\nu} \cdot \frac{2Lz}{L^2 + z^2} \right.$$
$$\left. + \sin\gamma \left[1 - \frac{2\nu}{2-\nu} \cdot \frac{L^2 - z^2}{L^2 + z^2}\right] \right\} \quad (4.13.87)$$

Note that K_{II} and K_{III} may be positive or negative, depending on γ and z. The interpretation of different signs follows from the sign convention for K_{II}, page 80, and K_{III}, page 75.

Crack with slightly curved front

In two-dimensional analysis of crack propagation, a simple geometry of the crack front is assumed, straight or circular. Deviations from these geometries require three-dimensional analysis. Of special interest is the almost straight crack front, appearing, for instance, on a microscale when a crack is retarded at particles, dislocations or other obstacles (Rice 1989, Gao and Rice 1989) or when the question arises whether the straight configuration is stable or not (Rice 1985b, Gao and Rice 1986). Here, the variation of the mode I stress intensity factor along a non-uniformly advancing crack front will be considered, assuming, in particular, an originally straight front advancing to a slightly curved or wavy position. The analysis essentially follows Rice (1985b).

Consider first a planar crack with smoothly curved crack front in a body that is symmetric with respect to the crack plane, $y = 0$, and subjected to mode I loading; see Fig. 4.13.2a. Let the position along the crack front be described by an arc length s along the contour C in the xz plane. The stress intensity factor $K_I = K(s)$ is assumed to be known. Add now, temporarily, two opposed normal crack forces, P, to the crack faces. The forces P are assumed to be uniformly distributed over a circular area with radius ϵ, much smaller than the distance to the crack front and with centre at $(x_P, \pm 0, z_P)$. P may be positive or negative, but the combined load system must keep the crack open everywhere. The crack front is now assumed to advance an infinitesimally small distance $\delta a(s)$. This causes changes of the stress-strain energy U in the body and of the potential Π_0 of the forces in the original load system, so that

$$\delta(U + \Pi_0) = P\delta[\Delta v(x_P, z_P)] - \int_C \mathcal{G}(s)\delta a(s)\,ds \quad (4.13.88)$$

where Δv is the average crack opening over the area $\pi\epsilon^2$ and $\mathcal{G}$ is the energy flux into the crack edge, which, according to (3.5.14), is related to the plane strain stress intensity factor $K(s)$ by

$$\mathcal{G}(s) = \frac{1-\nu}{2\mu}[K(s)]^2 \quad (4.13.89)$$

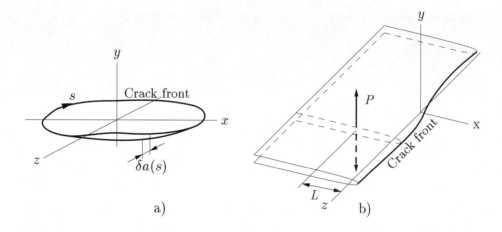

Fig. 4.13.2 Planar crack subjected to mode I loading. a) General contour shape. b) Crack with almost straight front.

Separate for a moment magnitude and shape of the crack advance by writing $\delta a(s) = \delta A \cdot g(s)$. Then,

$$\delta(P\Delta v - U - \Pi_0) = \Delta v \cdot \delta P + \int_C \mathcal{G}(s)g(s)\,ds \cdot \delta A \qquad (4.13.90)$$

For given $g(s)$, the argument on the left side is a function of P and A. Thus, the right side is a perfect differential, so that

$$\Delta v = \frac{\partial}{\partial P}(P\Delta v - U - \Pi_0), \quad \int_C \mathcal{G}(s)g(s)\,ds = \frac{\partial}{\partial A}(P\Delta v - U - \Pi_0) \qquad (4.13.91)$$

and, hence,

$$\frac{\partial(\Delta v)}{\partial A} = \frac{\partial}{\partial P}\left[\int_C \mathcal{G}(s)g(s)\,ds\right] = \int_C \frac{\partial \mathcal{G}}{\partial P}g(s)\,ds \qquad (4.13.92)$$

Multiplication by δA yields

$$\delta[\Delta v(x_P, z_P)] = \int_C \frac{\partial \mathcal{G}}{\partial P}\delta a(s)\,ds = \frac{1-\nu}{\mu}\int_C K(s)\frac{\partial K(s)}{\partial P}\delta a(s)\,ds \qquad (4.13.93)$$

Here, $K(s)$ is the sum of the stress intensity factor along the advanced crack front for the original load system, and the stress intensity factor that would be calculated for P only, and which will be written as $Pk(s; x_P, z_P)$. For $P = 0$ the relation reads

$$\delta[\Delta v(x_P, z_P)] = \frac{1-\nu}{\mu}\int_C K(s)k(s; x_P, z_P)\delta a(s)\,ds \qquad (4.13.94)$$

It is now possible to let $\epsilon \to 0$, i.e., $\Delta v(x_P, z_P)$ may now be interpreted as the crack opening displacement at the point (x_P, z_P) rather than an average displacement. Then, the relation gives the change of the crack opening displacement, due to a crack advance $\delta a(s)$, at any point of the crack faces under the action of the original load system. The force P has now served its purpose, and, to emphasize this, the point (x_P, z_P) on the crack faces will be denoted (x, z).

Semi-infinite crack
Consider the semi-infinite crack $x < 0$, $y = 0$; see Fig. 4.13.2b. Thus, the contour C before crack advance is the line $x = y = 0$. Equation (4.13.94) then reads, using $z' = s$ as integration variable:

$$\delta[\Delta v(x,z)] = \frac{1-\nu}{\mu} \int_{-\infty}^{+\infty} K(z') k(z'; x, z) \delta a(z') \, dz' \quad (4.13.95)$$

where, according to (4.13.41),

$$k(z'; x, z) = \sqrt{\frac{-2x}{\pi^3}} \cdot \frac{1}{x^2 + (z' - z)^2} \quad (4.13.96)$$

Assume that $\delta a(z)$ does not vary appreciably along portions of the z axis that are much smaller than some characteristic length λ, as would, for instance, be the case if $\delta a(z) = A[1 + \cos(2\pi z/\lambda)]$, $A \ll \lambda$. Consider a point $(x, 0, z)$, such that $\delta a(z) \ll -x \ll \lambda$, irrespective of z, and that both $-x$ and $\delta a(z)$ are infinitesimal. A suitable example is $-x = \sqrt{\lambda \delta a(z)}$. Use of (3.3.79) then shows that

$$\Delta v(x,z) \approx \frac{4(1-\nu)}{\mu} \sqrt{\frac{\delta a - x}{2\pi}} K(z) \quad (4.13.97)$$

The change of $\Delta v(x, z)$ during crack advance depends both on the advance $\delta a(z)$, and on the change $\delta K(z)$, due to the deviation of the crack front from a straight line. Then, from (4.13.97) and (4.13.95), because $\delta a \ll -x$,

$$\frac{\mu}{4(1-\nu)} \delta \Delta v \approx \sqrt{\frac{-x}{2\pi}} \delta K(z) - \sqrt{\frac{-x}{2\pi}} \cdot \frac{\delta a}{2x} K^0(z)$$

$$\approx \frac{1}{2\pi} \sqrt{\frac{-x}{2\pi}} \int_{-\infty}^{+\infty} \frac{K^0(z') \delta a(z')}{x^2 + (z' - z)^2} dz' \quad (4.13.98)$$

Write now

$$K(z) = \int_{-\infty}^{+\infty} K(z') \delta_D(z' - z) dz' \quad (4.13.99)$$

where $\delta_D(\cdot)$ is the Dirac delta function (subscript D used to avoid confusion in the present context). Recall now that $-x$ is infinitesimal and note that

$$\delta_D(z' - z) = \frac{1}{\pi} \lim_{x \to -0} \frac{-x}{x^2 + (z' - z)^2} \quad (4.13.100)$$

Insertion into (4.13.98) and some cleaning up leads to

$$\delta K(z) \approx \frac{1}{2\pi} \int_{-\infty}^{+\infty} \frac{K(z')[\delta a(z') - \delta a(z)]}{x^2 + (z' - z)^2} dz' \quad (4.13.101)$$

This expression becomes exact in the limit $x \to -0$ (and, of course, $\delta K(z)$ should not depend on x). Then the integral must be intepreted in the sense of the Cauchy principal value. Thus,

$$\delta K(z) = \frac{1}{2\pi} \fint_{-\infty}^{+\infty} \frac{K(z')[\delta a(z') - \delta a(z)]}{(z' - z)^2} dz' \quad (4.13.102)$$

Note that this expression is valid even if $\delta a(z)$ is measured from a reference line $x = a_0$ rather than from $x = 0$, and, in fact, the association of the reference line with an originally straight crack is not necessary. The expression is valid for any crack front with infinitesimal deviations from a straight line. Following Rice (1985b), the crack front position in relation to a chosen reference line may be written as a crack length measure,

$$a(z) = a_0 + \delta a(z) \qquad (4.13.103)$$

Then $[\delta a(z') - \delta a(z)]$ may be substituted by $[a(z') - a(z)]$ in (4.13.102). It is observed that the case $a(z') = a(z)$, i.e., a translation of a straight crack front, leads to $\delta K(z) = 0$. Thus, following notations suggested by Rice (1985b),

$$\delta K(z) = K(z) - K^0[z; a(z)] \qquad (4.13.104)$$

where $K(z)$ is the stress intensity factor at position z on the "wavy" crack front and $K^0[z; a(z)]$ is the stress intensity factor at position z on an assumed crack with straight front and with crack length measure $a(z)$. As a result, the stress intensity factor of the wavy crack front may be written as

$$K(z) = K^0[z; a(z)] + \frac{1}{2\pi} \fint_{-\infty}^{+\infty} \frac{K^0[z'; a(z)][a(z') - a(z)]}{(z' - z)^2} dz' \qquad (4.13.105)$$

where $a(z') - a(z)$ should be infinitesimal, but the relation holds approximately to the first order in $\max |a(z') - a(z)|$.

Configurational stability of a straight crack front

Consider a crack $x < a(z) = a_0 + A\cos(2\pi z/\lambda)$, $y = 0$, in an infinite body. Assume a load system causing mode I crack opening and such that the stress intensity factor for $A = 0$ is uniform, equal to $K^0(a_0)$. Equation (4.13.105) then leads to

$$K(z) = K^0(a_0) + \frac{dK^0}{da_0} \cdot A\cos(2\pi z/\lambda)$$
$$+ \frac{AK^0(a_0)}{2\pi} \fint_{-\infty}^{+\infty} \frac{\cos(2\pi z'/\lambda) - \cos(2\pi z/\lambda)}{(z' - z)^2} dz' \qquad (4.13.106)$$

After partial integration, the integral reduces to (cf., Gradshteyn and Ryzhik 1980):

$$-\frac{2\pi}{\lambda} \fint_{-\infty}^{+\infty} \frac{\sin(2\pi z'/\lambda)}{z' - z} dz' = -\frac{2\pi^2}{\lambda} \cos(2\pi z/\lambda) \qquad (4.13.107)$$

and, thus,

$$K(z) = K^0(a_0) + A\left[\frac{dK^0(a_0)}{da_0} - \frac{\pi}{\lambda} K^0(a_0)\right] \cos(2\pi z/\lambda) \qquad (4.13.108)$$

The result has some implications for the configurational stability of a straight crack front. It is reasonable to assume that, as for instance in fatigue crack growth, the growth rate increases with increasing stress intensity factor. Then, the result indicates that a slightly wavy crack front would tend towards straightness during crack growth, if

$$\frac{dK^0(a_0)}{da_0} - \frac{\pi}{\lambda} K^0(a_0) < 0 \qquad (4.13.109)$$

because then advanced parts of the front, $\cos(2\pi z/\lambda) > 0$, would grow slower than retarded parts, $\cos(2\pi z/\lambda) < 0$. On the other hand, configurational instability would occur if there would be a crack advance (say) over a portion of the order of λ along the crack front and the left side of (4.13.109) were positive.

Cracks with curved fronts are frequently observed during laboratory testing, but are generally associated with the variation from plane strain to plane stress dominance, as for edge cracks in plates. The present analysis may then, at best, be appropriate for a curved crack edge in the plane strain region, so that λ may be taken to be of the order of the plate thickness. However, analysis of some different load situations shows that it is difficult, though not impossible, to find conditions under which configurational instability of an advancing crack front would occur. The conclusion is therefore that deviations from crack front straightness can rarely be attributed to configurational instability.

Similar conclusions as for mode I cracks have been drawn for planar crack growth with curved front, propagating in modes II and III (Gao and Rice 1986).

Crack in a plate with consideration of transverse shear

Consider a crack $-a < x < 0$, $y = 0$, in a plate subjected to mode I loading. The crack length a is assumed to be much larger than the plate thickness $2h$. Approximative plane stress condition prevail in the plate, except in the neighbourhood of the crack edge(s). It is obvious that $\epsilon_z < 0$ along the crack edge, because tensile traction would be needed on the plate surfaces to achieve plane strain conditions, $\epsilon_z = 0$. If the extension of the dissipative region from the crack edge is much smaller than $2h$, then ϵ_z is of the order of elastic strains, and thus $|\epsilon_z| \ll 1$.

An attempt to use the plane stress approximation even for the neighbourhood of $x = 0$ gives, for $x > 0$, $y = 0$,

$$\sigma_x \approx \sigma_y(x) \approx \frac{K}{2\pi x}, \quad \epsilon_z = -\frac{\nu}{2(1+\nu)}(\sigma_x + \sigma_y) \approx -\frac{\nu}{1+\nu}\frac{K}{\sqrt{2\pi x}} \qquad (4.13.110)$$

i.e., the physically impossible result that $\epsilon_z < -1$ in a certain crack edge vicinity is obtained, except for $\nu = 0$. This implies that the conditions near the crack edge are much more similar to those of plane strain than to those of plane stress.

As a first approximation, it may be assumed that ϵ_z is uniformly distributed over each cross-section. This assumption makes it possible to formulate approximate equations of motion that involve only the two in-plane displacements u and v; see Appendix A9. For static cases such equations can be used after neglect of inertia terms. This was done by Yang and Freund (1985), who used a formulation by Kane and Mindlin (1956). The problem could then be solved approximately by matching an inner solution for the region near the crack edge to an outer solution for the plane stress region. The results showed that the plane stress solution is reasonably accurate outside a radius equal to about half the plate thickness. This result appears to carry over also to the dynamic case, according to estimates by Broberg (1995b). However, it should be noted that the approximate equations used do not permit satisfaction of traction free plate surfaces, and this shortcoming introduces an error that might be significant near the crack edge.

Bentham (1977, 1980) discovered the curious effect that the stress singularity along the edge of a crack in a plate suddenly changes from an inverse square root singularity to another singularity when the plate surface is reached. For Poisson's ratio 0.3, Bentham calculated the singularity exponent to about -0.4523. Other investigations have arrived at similar conclusions (Bažant and Estenssoro 1979, 1980 and Burton et al. 1984). Gudmundson and Östlund (1990) analysed the stress behaviour at the intersection between the front of a dynamically moving crack and the free surface for different crack velocities and different angles between the crack front and the free surface. Angles other than normal could, for instance, go with a curved front of a crack propagating in a plate. Angles larger than $90°$ were considered, i.e. angles for which the intersection is trailing the crack edge in the interior. Gudmundson and Östlund found that the singularity exponent equalled -0.5 for an angle of about $101°$, almost independently of the crack velocity, whereas it was about -0.548 for the angle $90°$ and vanishining velocity. For angles between $90°$ and about $101°$, it decreased with increasing velocity (i.e., the singularity became stronger), and for larger angles than about $101°$, it increased with the velocity.

4.14 Cracks in anisotropic linearly elastic media

Introduction

Anisotropy denotes different properties in different directions. A sphere does not in general deform to another sphere under hydrostatic loading. Most crystalline materials are anisotropic on the scale of individual grains, but polycrystalline materials often show elastic isotropy on a macroscale, due to varying orientation of crystal directions between the grains. Materials like stone and wood, as well as several composite materials, laminated or with non-random fiber directions, are elastically anisotropic on a macroscale. Further examples are rolled, particularly cold-rolled, metal plates. In general the elastic anisotropy is accompanied by a strength anisotropy, well-known to, for instance, stone- and woodcutters. This kind of anisotropy has obvious implications for crack propagation properties.

Instead of Hooke's law with its two material constants, the constitutive equations for linearly elastic anisotropic materials may contain up to 21 material constants. The figure 21 appears for general anisotropy, and the constitutive equations are then called the *generalized Hooke's law*. In many materials, the degree of anisotropy is lower, implying fewer material constants. Some such *degenerate* materials possess one or more *symmetry planes*. A cube with two opposed sides parallel to a symmetry plane and subjected to uniformly distributed normal loads deforms so that these sides remain normal to the other four sides. One example is wood, that is an *orthotropic* (also called rhombic or orthorhombic) material, i.e. the anisotropy may be described by the properties in three orthogonal directions. The number of elastic constants is then 9 and may be viewed as one modulus of elasticity and two Poisson's ratios for loading in each principal direction. An orthotropic material possesses three orthogonal symmetry planes. It is a special case of a *monoclinic* material, which possesses one symmetry plane and has 13 elastic constants. A special case of an orthotropic material is a *transversely isotropic* (also called hexagonal) material, for which the properties in all directions parallel to one principal symmetry plane are equal. It has 5 elastic constants, which can be viewed as one modulus of elasticity and one Poisson's ratio for loading normal to the principal symmetry plane and one modulus of elasticity and two Poisson's ratios for loading in any direction parallel to the principal symmetry plane. There are also several other classes of degenerate anisotropic materials; see e.g. Ting (1996).

Plane stress and plane strain cases, which play such an important role in the theory of cracks in isotropic materials, do not have exact correspondence for general anisotropic materials. They do not arise simply as a matter of the loading composition, but require certain degenerate and suitably oriented anisotropic materials. However, a case of *generalized plane strain* can be defined for general anisotropy, implying that all three displacement components are constant along the transverse direction. The foundation for studies of this case was laid by Lekhnitskii (1950) and, apparently independently, by Eshelby, Read and Shockley (1953) and Stroh (1958).

General relations

A general linear relation between the 9 stress components σ_{ij} and the 9 deformation components ϵ_{ij} would contain 81 constants C_{ijkl}:

$$\boxed{\sigma_{ij} = C_{ijkl}\epsilon_{kl}} \qquad i,j,k,l = 1,2,3 \qquad (4.14.1)$$

but the equilibrium relations $\sigma_{ji} = \sigma_{ij}$, the symmetry identities $\epsilon_{ji} = \epsilon_{ij}$, and the fact that the way (the loading order) in which a certain elastic state has been arrived at from an earlier state does not influence the accompanying stress-strain energy change, implies that

$$C_{ijkl} = C_{jikl} = C_{klij} = C_{ijlk} \qquad (4.14.2)$$

This is 60 different relations, i.e. the number of independent material constants, the *elastic stiffnesses*, C_{ijkl}, is 21 in a general anisotropic, linearly elastic material. Relations (4.14.1) and (4.14.2) constitute the generalized Hooke's law.

The matrix C_{ijkl}, which is a fourth rank tensor, is positive definite:

$$C_{ijkl}f_{ij}f_{kl} > 0 \text{ for all } f_{ji} = f_{ij} \neq 0 \qquad (4.14.3)$$

A different form of the constitutive relation is obtained after writing

$$\epsilon_{kl} = (u_{k,l} + u_{l,k})/2 \qquad (4.14.4)$$

The equations (4.14.1) and (4.14.2) then give

$$\sigma_{ij} = \frac{1}{2}C_{ijkl}u_{k,l} + \frac{1}{2}C_{ijkl}u_{l,k} = \frac{1}{2}C_{ijkl}u_{k,l} + \frac{1}{2}C_{ijlk}u_{l,k} \qquad (4.14.5)$$

and mere exchange of dummy indices k and l in the last term leads to

$$\sigma_{ij} = C_{ijkl}u_{k,l} \qquad (4.14.6)$$

Solution for the strains yields

$$\boxed{\epsilon_{ij} = S_{ijkl}\sigma_{kl}} \qquad (4.14.7)$$

where S_{ijkl} are the 21 *elastic compliances*, which possess the same symmetries as the elastic stiffnesses C_{ijkl}. Obviously the matrix S_{ijkl} is the inverse of the matrix C_{ijkl}, and *vice versa*. The elastic compliances are easier to determine experimentally than the elastic stiffnesses, because of the ease to control load compositions compared to displacement compositions.

It is sometimes convenient to use *contracted notations* (Nye 1957), implying that the constitutive equations are written in the form

$$\sigma_\alpha = c_{\alpha\beta}\epsilon_\beta \text{ or } \quad \epsilon_\alpha = s_{\alpha\beta}\sigma_\beta \tag{4.14.8}$$

Thus,

$$\sigma_1 = \sigma_{11} = \sigma_x, \ \sigma_2 = \sigma_{22} = \sigma_y, \ \sigma_3 = \sigma_{33} = \sigma_z,$$
$$\sigma_4 = \sigma_{23} = \tau_{yz}, \ \sigma_5 = \sigma_{31} = \tau_{zx}, \ \sigma_6 = \sigma_{12} = \tau_{xy}$$

and

$$\epsilon_1 = \epsilon_{11} = \epsilon_x, \ \epsilon_2 = \epsilon_{22} = \epsilon_y, \ \epsilon_3 = \epsilon_{33} = \epsilon_z,$$
$$\epsilon_4 = 2\epsilon_{23} = \gamma_{yz}, \ \epsilon_5 = 2\epsilon_{31} = \gamma_{zx}, \ \epsilon_6 = 2\epsilon_{12} = \gamma_{xy}$$
$$c_{\beta\alpha} = c_{\alpha\beta}, \qquad s_{\beta\alpha} = s_{\alpha\beta}$$

where $\alpha, \beta = 1, 2, \ldots 6$. The definitions of σ_α and ϵ_β ensure that $c_{\alpha\beta}$ is symmetric. The constants $c_{\alpha\beta}$ are related to C_{ijkl} through the replacement rules for $ij \leftrightarrow \alpha$ and $kl \leftrightarrow \beta$,

$$11 \leftrightarrow 1, \ 22 \leftrightarrow 2, \ 33 \leftrightarrow 3, \ 23 \leftrightarrow 4, \ 31 \leftrightarrow 5, \ 12 \leftrightarrow 6 \tag{4.14.9}$$

i.e.

$$\alpha = \begin{cases} i & \text{if } i = j \\ 9 - i - j & \text{if } i \neq j \end{cases} \quad \beta = \begin{cases} k & \text{if } k = l \\ 9 - k - l & \text{if } k \neq l \end{cases} \tag{4.14.10}$$

so that, for instance, $C_{1233} = c_{63}$, which, due to the symmetry, equals c_{36}. The constants $s_{\alpha\beta}$ are related to S_{ijkl} through the same replacement rules, after S_{ijkl} has been substituted by

$$\begin{aligned} & S_{ijkl} && \text{if } i = j \text{ and } k = l \\ & 2S_{ijkl} && \text{if } i = j \text{ and } k \neq l \text{ or if } i \neq j \text{ and } k = l \\ & 4S_{ijkl} && \text{if } i \neq j \text{ and } k \neq l \end{aligned} \tag{4.14.11}$$

Thus, for instance, $s_{52} = 2S_{1322} = s_{25}$.

All diagonal elements of the matrix $c_{\alpha\beta}$ and all its principal minors (determinants made up of elements symmetrically situated with respect to the diagonal of the matrix) are non-zero and positive. This can be shown to follow from the fact that $c_{\alpha\beta}$ is symmetric and positive definite, the latter property a consequence of the fact that the stress-strain energy

$$W = \frac{1}{2}\sigma_{ij}\epsilon_{ij} = \frac{1}{2}C_{ijkl}\epsilon_{ij}\epsilon_{kl} = \frac{1}{2}S_{ijkl}\sigma_{ij}\sigma_{kl} = \frac{1}{2}c_{\alpha\beta}\epsilon_\alpha\epsilon_\beta = \frac{1}{2}s_{\alpha\beta}\sigma_\alpha\sigma_\beta \tag{4.14.12}$$

is positive if not all strain or stress components vanish.

Let $x_3 = 0$ be the symmetry plane for a monoclinic material. Then,

$$c_{14} = c_{15} = c_{24} = c_{25} = c_{34} = c_{35} = c_{46} = c_{56} = 0 \tag{4.14.13}$$

4.14 CRACKS IN ANISOTROPIC LINEARLY ELASTIC MEDIA

For an orthotropic material with $x_1 = 0$, $x_2 = 0$ and $x_3 = 0$ as symmetry planes, the same relations hold, and in addition

$$c_{16} = c_{26} = c_{36} = c_{45} = 0 \qquad (4.14.14)$$

For a transversely isotropic material with the symmetry planes $x_3 = 0$ and any plane containing the x_3 axis, the same matrix elements vanish as for the orthotropic material, and in addition

$$c_{11} = c_{22}, \quad c_{13} = c_{23}, \quad c_{44} = c_{55} \quad c_{66} = (c_{11} - c_{12})/2 \qquad (4.14.15)$$

For an isotropic material,

$$C_{ijkl} = \frac{2\nu\mu}{1 - 2\nu}\delta_{ij}\delta_{kl} + \mu(\delta_{ik}\delta_{jl} + \delta_{il}\delta_{jk}) \qquad (4.14.16)$$

$$S_{ijkl} = -\frac{\nu}{2(1+\nu)\mu}\delta_{ij}\delta_{kl} + \frac{1}{4\mu}(\delta_{ik}\delta_{jl} + \delta_{il}\delta_{jk}) \qquad (4.14.17)$$

where ν is Poisson's ratio, μ the modulus of rigidity and δ_{ij} the Kronecker delta, equal to unity if $i = j$, vanishing otherwise. In contracted notation the non-vanishing matrix elements for an isotropic material are

$$c_{11} = c_{22} = c_{33} = \frac{2(1-\nu)\mu}{1-2\nu}, \quad c_{44} = c_{55} = c_{66} = \mu$$

$$c_{12} = c_{21} = c_{13} = c_{31} = c_{23} = c_{32} = \frac{2\nu\mu}{1-2\nu} \qquad (4.14.18)$$

$$s_{11} = s_{22} = s_{33} = \frac{1}{2(1+\nu)\mu}, \quad s_{44} = s_{55} = s_{66} = \frac{1}{\mu}$$

$$s_{12} = s_{21} = s_{13} = s_{31} = s_{23} = s_{32} = -\frac{\nu}{2(1+\nu)\mu} \qquad (4.14.19)$$

Note that all elastic constants are related to the orientation of the coordinate system with respect to the body dimensions. Thus, for instance, c_{14} is not in general zero for a monoclinic material if the coordinate system is oriented so that $x_1 = 0$, instead of $x_3 = 0$, is placed along the symmetry plane. However, it is not always suitable to orientate the coordinate system with respect to material symmetry planes. Most often the orientation is made with respect to body boundaries in order to obtain simple expressions for the boundary conditions. It is therefore sometimes necessary to transform the elastic stiffness or compliance matrix to fit the coordinate system chosen. Because C_{ijkl} and S_{ijkl} are tensors, tensor transformation rules are applicable. Thus, for a rotation of the coordinate axes x_1 and x_2 through an angle φ clockwise around the positive x_3 axis the transformation is achieved by

$$C^*_{ijkl} = \Omega_{ip}\Omega_{jq}\Omega_{kr}\Omega_{ls}C_{pqrs} \qquad (4.14.20)$$

where * marks the transformed elements and

$$\Omega = \begin{bmatrix} \cos\varphi & \sin\varphi & 0 \\ -\sin\varphi & \cos\varphi & 0 \\ 0 & 0 & 1 \end{bmatrix} \qquad (4.14.21)$$

Certain macroscopically heterogeneous materials, such as laminated or fibre-reinforced composites may be considered as anisotropic homogeneous materials on a sufficiently large scale. They are then represented by their effective elastic constants, which may be calculated for a representative volume of the composed material by relating average stresses over the volume to average strains; see e.g. Sun et al. (1968), Sun and Li (1988) and Sun and Wu (1996).

Plane cases

Plane cases, in particular the simple case of anti-plane strain and cases of in-plane stress and strain play an important part in the analysis of cracks in isotropic materials. These cases do not have exact counterparts for general anisotropic materials, but they do for certain degenerate materials, even for the monoclinic materials, which possess only one symmetry plane. For such materials, anti-plane and in-plane strains can be decoupled, i.e. it is possible to decompose a general state of strain into anti-plane and in-plane strain states.

General relations for anti-plane strain
Anti-plane strain is defined by

$$u_1 = u_2 = 0, \qquad u_{3,3} = 0 \qquad (4.14.22)$$

Insertion into the equilibrium equations

$$\sigma_{ij,j} = C_{ijkl} u_{k,lj} = 0 \qquad (4.14.23)$$

gives

$$C_{i131} u_{3,11} + C_{i231} u_{3,12} + C_{i132} u_{3,21} + C_{i232} u_{3,22} = 0, \quad i = 1, 2, 3 \qquad (4.14.24)$$

In contracted notation, these three equations read:

$$\begin{aligned} c_{15} u_{3,11} + (c_{56} + c_{14}) u_{3,12} + c_{46} u_{3,22} &= 0 \\ c_{56} u_{3,11} + (c_{25} + c_{46}) u_{3,12} + c_{24} u_{3,22} &= 0 \\ c_{55} u_{3,11} + 2 c_{45} u_{3,12} + c_{44} u_{3,22} &= 0 \end{aligned} \qquad (4.14.25)$$

Thus, three equations are found for the single unknown displacement u_3. These cannot all be satisfied for general anisotropy, because they are linearly independent if all coefficients are non-zero and independent. However, the first two equations are satisfied for a material with

$$c_{14} = c_{15} = c_{24} = c_{25} = c_{46} = c_{56} = 0 \qquad (4.14.26)$$

These relations are fulfilled for a monoclinic material oriented so that its symmetry plane is $x_3 = 0$, but express, in fact, a somewhat more general material, cf. (4.14.13).

Note that the coefficients c_{44} and c_{55} in the third of equations (4.14.25) are non-zero. This equation may be written in the form

$$c_{55} \frac{\partial^2 w}{\partial x^2} + 2 c_{45} \frac{\partial^2 w}{\partial x \partial y} + c_{44} \frac{\partial^2 w}{\partial y^2} = 0 \qquad (4.14.27)$$

to compare with (3.3.4), $\Delta w = 0$, for isotropic materials. For orthotropic and transversely isotropic materials, $c_{45} = 0$, so that the mixed derivative disappears from the equation, and for isotropic materials $c_{44} = c_{55} = \mu$.

4.14 CRACKS IN ANISOTROPIC LINEARLY ELASTIC MEDIA

For anti-plane strain the constitutive equations reduce to

$$\tau_{xz} = c_{55}\gamma_{xz} + c_{45}\gamma_{yz}, \quad \tau_{yz} = c_{45}\gamma_{xz} + c_{44}\gamma_{yz}, \quad (4.14.28)$$

$$\sigma_z = c_{35}\gamma_{xz} + c_{34}\gamma_{yz} \quad (4.14.29)$$

Equation (4.14.27) possesses the solution $w = w(x+py)$, if p is a root to the equation

$$c_{44}p^2 + 2c_{45}p + c_{55} = 0 \quad (4.14.30)$$

Note that the roots

$$p = p_{1,2} = -\frac{c_{45} \mp i\sqrt{c_{44}c_{55} - c_{45}^2}}{c_{44}} = \alpha \pm i\beta \quad (4.14.31)$$

cannot be real, because $c_{44}c_{55} - c_{45}^2$ is a principal minor of $c_{\alpha\beta}$ and thus non-zero and positive. Then, the most general solution to the equation for w, that satisfies the physical condition that w must be real, may be written as

$$w = \frac{1}{\mu_a}\Im[f(\zeta)] \quad (4.14.32)$$

where $\mu_a = \sqrt{c_{44}c_{55} - c_{45}^2}$ and $\zeta = x + p_1 y = x + \alpha y + i\beta y$. This form is chosen for agreement with the isotropic expression (4.3.5), for which ζ defaults to $z = x + iy$ and μ_a to μ, the modulus of rigidity.

Because $\gamma_{xz} = \partial w/\partial x$ and $\gamma_{yz} = \partial w/\partial y$, equation (4.14.28) gives

$$\tau_{xz} = \frac{1}{\mu_a}\Im[(c_{55} + c_{45}p_1)f'(\zeta)] = \Im[-ip_1 f'(\zeta)] \quad (4.14.33)$$

$$\tau_{yz} = \frac{1}{\mu_a}\Im[(c_{45} + c_{44}p_1)f'(\zeta)] = \Re[f'(\zeta)] \quad (4.14.34)$$

which default to (4.3.6) for an isotropic material. Here, the equalities

$$-\frac{c_{55} + c_{45}p_1}{p_1} = c_{44}p_1 + c_{45} = i\sqrt{c_{44}c_{55} - c_{45}^2} = i\mu_a \quad (4.14.35)$$

were used.

Put $\zeta = \xi + i\eta$. Then, $f(\zeta)$ is analytic in the regions of the ζ $(\xi, i\eta)$ plane where it is differentiable. Thus, an anti-plane strain problem consists of finding an analytic function $f(\zeta)$ that satisfies given boundary conditions.

Note that the representation (4.14.32) for w coincides with the representation (4.3.5) for an isotropic material, if ζ is substituted by z and μ_a by μ. Likewise, the representation (4.14.34) for τ_{yz} coincides with the corresponding representation (4.3.6) for an isotropic material, if ζ is substituted by z. *This implies that the solution of an anti-plane strain problem, which may be formulated in terms of w and τ_{yz}, only, can be taken from the solution for the corresponding isotropic problem, after substituting z by ζ and μ by μ_a.* The expressions for w and τ_{yz} are found directly from the corresponding expressions for the isotropic case, whereas τ_{xz} is found by using the solution for $f'(\zeta)$ together with (4.14.34). Finally, from (4.14.29), $\sigma_z = c_{35}\partial w/\partial x + c_{34}\partial w/\partial y$.

General relations for plane strain
Plane strain is defined by

$$u_{1,3} = u_{2,3} = 0, \qquad u_3 = 0 \tag{4.14.36}$$

Insertion into the equilibrium equations leads to three equations similar to those obtained for anti-plane strain, (4.14.25), but there are now two unknown displacements, u_1 and u_2. Two equations contain diagonal, and thus non-zero, elements of the stiffness matrix. The third equation may be satisfied by putting its coefficients equal to zero. This leads to the same relations as needed for anti-plane strain, (4.14.26). Thus, plane strain can also be realized for a monoclinic material.

For plane strain, the constitutive equations may be written in a convenient form by decomposition into two parts, one for in-plane stresses and strains and one for the transverse normal stress σ_{33}. Thus, using contracted notations, the relation $\epsilon_3 = s_{3\beta}\sigma_\beta = 0$ leads to $s_{33}\sigma_3 = -(s_{3\beta}\sigma_\beta)_{\beta=1,2,6}$, so that

$$\epsilon_\alpha = \left(s_{\alpha\beta} - \frac{s_{\alpha 3}s_{3\beta}}{s_{33}}\right)\sigma_\beta, \quad \alpha, \beta = 1, 2, 6 \tag{4.14.37}$$

$$\sigma_3 = -\frac{s_{3\beta}\sigma_\beta}{s_{33}} = c_{3\beta}\epsilon_\beta \quad \alpha, \beta = 1, 2, 6 \tag{4.14.38}$$

By introducing the reduced elastic stiffnesses

$$a_{\alpha\beta} = s_{\alpha\beta} - s_{\alpha 3}s_{3\beta}/s_{33} \text{ for } \alpha, \beta = 1, 2, 6 \tag{4.14.39}$$

the first equation may be written as

$$\epsilon_\alpha = a_{\alpha\beta}\sigma_\beta \tag{4.14.40}$$

The symmetry properties of $s_{\alpha\beta}$ obviously carry over to $a_{\alpha\beta}$, which is also positive definite.

The plane strain constitutive equations (4.14.40) and (4.14.38) may be written as

$$\epsilon_x = a_{11}\sigma_x + a_{12}\sigma_y + a_{16}\tau_{xy}, \quad \epsilon_y = a_{12}\sigma_x + a_{22}\sigma_y + a_{26}\tau_{xy} \tag{4.14.41}$$

$$\gamma_{xy} = a_{16}\sigma_x + a_{26}\sigma_y + a_{66}\tau_{xy} \tag{4.14.42}$$

for the in-plane strains, and

$$\sigma_z = c_{13}\epsilon_x + c_{23}\epsilon_y + c_{36}\gamma_{xy} \tag{4.14.43}$$

for the transverse normal stress.

Introduce the Airy stress function Φ, so that

$$\sigma_x = \frac{\partial^2 \Phi}{\partial y^2}, \quad \sigma_y = \frac{\partial^2 \Phi}{\partial x^2}, \quad \tau_{xy} = -\frac{\partial^2 \Phi}{\partial x \partial y} \tag{4.14.44}$$

Insertion into the constitutive equations for in-plane strains and use of the compatibility equation

$$\frac{\partial^2 \epsilon_y}{\partial x^2} + \frac{\partial^2 \epsilon_x}{\partial y^2} = \frac{\partial^2 \gamma_{xy}}{\partial x \partial y} \tag{4.14.45}$$

gives

$$a_{22}\frac{\partial^4 \Phi}{\partial x^4} - 2a_{26}\frac{\partial^4 \Phi}{\partial x^3 \partial y} + (2a_{12} + a_{66})\frac{\partial^4 \Phi}{\partial x^2 \partial y^2} - 2a_{16}\frac{\partial^4 \Phi}{\partial x \partial y^3} + a_{11}\frac{\partial^4 \Phi}{\partial y^4} = 0 \tag{4.14.46}$$

A general solution of this equation is

$$\Phi = \sum_{i=1}^{4} \Phi_i(x + p_i y) \tag{4.14.47}$$

where $\Phi_i(\cdot)$ are arbitrary functions, and p_i are the roots of

$$a_{11}p^4 - 2a_{16}p^3 + (2a_{12} + a_{66})p^2 - 2a_{26}p + a_{22} = 0 \tag{4.14.48}$$

Using the property that $a_{\alpha\beta}$ is positive definite, Lekhnitskii (1950) showed that this equation cannot possess real roots, and therefore the roots come in complex conjugate pairs,

$$p_{1,2} = \alpha_{1,2} + i\beta_{1,2}, \quad p_{3,4} = \overline{p_{1,2}} = \alpha_{1,2} - i\beta_{1,2} \tag{4.14.49}$$

where $\beta_{1,2}$ are non-zero and taken to be positive. Then, because the stresses must be real, it follows from (4.14.44) and (4.14.47) that Airy's stress function may be taken to be real and equal to

$$\Phi = \Re[\Phi_1(\zeta_1) + \Phi_2(\zeta_2)] \tag{4.14.50}$$

where $\zeta_1 = x + p_1 y$, $\zeta_2 = x + p_2 y$. By analogy with the anti-plane case, put $\zeta_1 = \xi_1 + i\eta_1$ and $\zeta_2 = \xi_2 + i\eta_2$. For convenience, make the substitutions

$$\phi(\zeta_1) = \Phi_1'(\zeta_1), \quad \psi(\zeta_2) = \Phi_2'(\zeta_2) \tag{4.14.51}$$

so that

$$\sigma_x = \Re[p_1^2 \phi'(\zeta_1) + p_2^2 \psi'(\zeta_2)], \quad \sigma_y = \Re[\phi'(\zeta_1) + \psi'(\zeta_2)] \tag{4.14.52}$$
$$\tau_{xy} = -\Re[p_1 \phi'(\zeta_1) + p_2 \psi'(\zeta_2)] \tag{4.14.53}$$

and, by integration of the expressions for $\epsilon_x = \partial u/\partial x$ and $\epsilon_y = \partial v/\partial y$,

$$u = \Re[(a_{11}p_1^2 + a_{12} - a_{16}p_1)\phi(\zeta_1) + (a_{11}p_2^2 + a_{12} - a_{16}p_2)\psi(\zeta_2)] \tag{4.14.54}$$
$$v = \Re\left[(a_{12}p_1^2 + a_{22} - a_{26}p_1)\frac{\phi(\zeta_1)}{p_1} + (a_{12}p_2^2 + a_{22} - a_{26}p_2)\frac{\psi(\zeta_2)}{p_2}\right] \tag{4.14.55}$$

The representations of stresses and displacements in terms of $\phi(\zeta_1)$ and $\psi(\zeta_2)$ allow similar treatments of crack problems as representations in complex potentials in the isotropic case. This analogy is particularly obvious for problems involving cracks situated in the plane $y = 0$, because this enables use of two new functions, $g(z)$ and $h(z)$, analytical in the same plane, $z = x + iy$, and such that $g(x) = \phi(x)$ and $h(x) = \psi(x)$. After $g(z)$ and $h(z)$ have been determined from the mathematical problem, the solution of the physical problem is obtained after putting $\phi(\zeta_1) = g(\zeta_1)$ and $\psi(\zeta_2) = h(\zeta_2)$. This will not be pursued here – the more general Lekhnitskii-Stroh formalism will be used – but the method may be convenient for certain degenerate materials, particularly orthotropic materials. Reference may be made to a review paper by Sih and Liebowitz (1968).

Plane stress
Plane stress conditions for a thin plate with plate surfaces parallel to $x_3 = 0$ require that the stresses σ_{3i} can be neglected. This requirement cannot be satisfied for any material defined by (4.14.26), as for anti-plane strain or plane strain, but a monoclinic material, oriented so that its symmetry plane is $x_3 = 0$, is needed; see (4.14.13). Thus, also c_{34} and c_{35} must be zero. Then, putting $\sigma_{3i}=C_{3ikl}u_{k,l} = 0$, the following equations in contracted notation are obtained:

$$i = 1: \quad c_{55}\epsilon_{13} + 2c_{45}\epsilon_{23} = 0$$
$$i = 2: \quad c_{45}\epsilon_{13} + 2c_{44}\epsilon_{23} = 0 \quad (4.14.56)$$
$$i = 3: \quad c_{13}\epsilon_{11} + 2c_{36}\epsilon_{23} + c_{23}\epsilon_{22} + c_{33}\epsilon_{33} = 0$$

From the first two, it follows that $\epsilon_{13} = \epsilon_{23} = 0$, because the determinant $c_{44}c_{55} - c_{45}^2$ is a principal minor of $c_{\alpha\beta}$ and thus non-zero. Note, however, that the vanishing of ϵ_{13} and ϵ_{23} is approximative, a result of the plane stress approximation that $\sigma_{3i} = 0$. From the third equation, it follows that

$$\epsilon_{33} = -\frac{1}{c_{33}}(c_{13}\epsilon_{11} + 2c_{36}\epsilon_{12} + c_{23}\epsilon_{22}) \quad (4.14.57)$$

By analogy with the plane strain case, the plane stress constitutive equations can be written in a convenient form by decomposition into two parts, one for in-plane stresses and strains and one for the transverse normal strain ϵ_{33}, by writing, in contracted notation,

$$\sigma_\alpha = c_{\alpha\beta}\epsilon_\beta + c_{\alpha 3}\epsilon_3, \quad \alpha, \beta = 1, 2, 6 \quad (4.14.58)$$

and then, noting that (4.14.57) can be written as $\epsilon_3 = -(c_{3\beta}\epsilon_\beta/c_{33})_{\beta=1,2,6}$:

$$\sigma_\alpha = (c_{\alpha\beta} - \frac{c_{\alpha 3}c_{3\beta}}{c_{33}})\epsilon_\beta = b_{\alpha\beta}\epsilon_\beta, \quad \alpha, \beta = 1, 2, 6 \quad (4.14.59)$$

The symmetry properties of $c_{\alpha\beta}$ obviously carry over to $b_{\alpha\beta}$, which also is positive definite. Thus, the constitutive equations (4.14.59) for plane stress agree with those for plane strain, (4.14.40), if the elements $a_{\alpha\beta}$ are substituted by the elements in the inverse matrix $[b_{\alpha\beta}]^{-1}$. This implies that the solution of a plane strain problem for a monoclinic material (or special cases of such a material) provides the solution of the corresponding plane stress problem and *vice versa*, in essentially the same way as for isotropic materials. Note that solutions are first worked out for the in-plane components, whereupon σ_{33} is obtained by (4.14.38) for the plane strain problem and ϵ_{33} by (4.14.57) for the plane stress problem.

Generalized plane strain
Finally, it should be noted that it is possible to impose conditions of so-called *generalized plane strain* for general anisotropic materials. In this case, the displacements depend only on x_1 and x_2, say, so that

$$u_1 = u_1(x_1, x_2), \quad u_2 = u_2(x_1, x_2), \quad u_3 = u_3(x_1, x_2) \quad (4.14.60)$$

As a consequence, $\epsilon_{33} = 0$, whereas ϵ_{13} and ϵ_{23} in general do not vanish. Furthermore, the stress components σ_{ij} depend on x_1 and x_2, only, but in general $\sigma_{3i} \neq 0$. Note that the converse is not true: the assumption that the stresses depend on x_1 and x_2,

4.14 CRACKS IN ANISOTROPIC LINEARLY ELASTIC MEDIA

only, does not imply that the displacements also do so, only that the strains do. This assumption therefore leads to more general stress-strain states than generalized plane strain.

Generalized plane strain is often used in analysis of plane anisotropic problems, due to its suitability for the Lekhnitskii-Stroh formalism, to be discussed later. It contains, of course, as special cases, anti-plane strain, plane strain and essentially also plane stress.

Basic relations for cracks in anisotropic media

Most relations established in Chapter 3 assume anti-plane strain or plane strain or stress. Some also assume the constitutive equations for isotropic materials, but those who do not are also valid for those anisotropic materials that can support plane stress or strain. The loading geometry must then be appropriately chosen with respect to the material directions, for instance so that in-plane loading under plane stress occurs in a material symmetry plane.

The possibility of considering the stress-strain state in the vicinity of a crack edge as a superposition of the three symmetry modes, I, II and III, does not hold for general linearly elastic anisotropy, because the symmetries must include both stresses and displacements. On the other hand, it holds for appropriately oriented orthotropic materials.

The path-independent integrals discussed in section 3.2 retain their path-independence for appropriately oriented orthotropic materials. The same is the case for the weight functions discussed in Section 3.5.

Cracks under anti-plane strain

Consider a crack with length $2a$ in a large plate, subjected to remote loading $\tau_{yz} = \tau_{yz}^\infty$. Thus, the boundary conditions are

$$\tau_{yz} = 0 \text{ for } |x| > a, \ y = 0, \quad \tau_{yz} \to \tau_{yz}^\infty \text{ as } \sqrt{x^2 + y^2} \to \infty \qquad (4.14.61)$$

Recall now the result on page 211 that the solution of a problem for an anisotropic body can be obtained from the corresponding problem for an isotropic body, if it can be formulated in terms of τ_{yz} (and w though this is not relevant here). Thus, cf. (4.6.21)-(4.6.22),

$$f'(\zeta) = \tau_{yz}^\infty \frac{\zeta}{(\zeta^2 - a^2)^{1/2}}, \quad \tau_{yz} = \tau_{yz}^\infty \Re \frac{\zeta}{(\zeta^2 - a^2)^{1/2}} \qquad (4.14.62)$$

$$\tau_{xz} = \tau_{yz}^\infty \Im \frac{-ip_1\zeta}{(\zeta^2 - a^2)^{1/2}}, \quad w = \frac{\tau_{yz}^\infty}{\mu_a} \Im(\zeta^2 - a^2)^{1/2} \qquad (4.14.63)$$

where the branch is chosen so that $(\zeta^2 - a^2)^{1/2} = \sqrt{\xi^2 - a^2}$ for $\zeta = \xi > a$.

Consider the vicinity of the right crack edge, $z = a$. Put $x = a + r\cos\varphi$, $y = r\sin\varphi$, implying $\zeta = a + r(\cos\varphi + p_1 \sin\varphi)$ and giving, for $r/a \to 0$,

$$\tau_{yz} = \tau_{yz}^\infty \sqrt{\frac{a}{2r}} \Re \frac{1}{(\cos\varphi + p_1 \sin\varphi)^{1/2}} \quad (4.14.64)$$

$$\tau_{xz} = \tau_{yz}^\infty \sqrt{\frac{a}{2r}} \Re \frac{p_1}{(\cos\varphi + p_1 \sin\varphi)^{1/2}} \quad (4.14.65)$$

$$w = \frac{\tau_{yz}^\infty}{\mu_a} \sqrt{2ar} \Re(\cos\varphi + p_1 \sin\varphi)^{1/2} \quad (4.14.66)$$

$$\sigma_z = c_{35} \frac{\partial w}{\partial x} + c_{34} \frac{\partial w}{\partial y} \quad (4.14.67)$$

Note that the solution does not possess mode III symmetry, except if $\Re(p_1) = 0$, implying $c_{45} = 0$, which is satisfied for suitably oriented orthotropic materials and specializations thereof. In general, the stress τ_{xz} does not vanish for $\varphi = 0$, and τ_{yz} is not a local maximum at $\varphi = 0$ for given r. The maximum at given r of the stress $\tau_{\varphi z} = \tau_{yz} \cos\varphi - \tau_{xz} \sin\varphi$ occurs in general at some non-zero angle φ. The direction in which the crack would propagate initially may to some extent depend upon this stress, but generally more on the material directions given by the strength anisotropy which usually accompanies elastic anisotropy. In this context it should be observed that cracks also may appear as a result of the stress σ_z, which, in general, is inverse square-root singular.

If the crack propagates straightforward, the energy flux into the crack edge may be calculated from the stresses and displacements on the symmetry plane in the crack edge vicinity, i.e. knowledge of offside stresses or displacements is not needed. Therefore, expression (3.5.16) will be modified, simply by substituting μ for μ_a, to

$$\mathcal{G} = \frac{K_{III}^2}{2\mu_a} \quad (4.14.68)$$

for the anisotropic anti-plane strain case. Note that straightforward crack growth could occur, in spite of a lack of stress symmetry, if the crack plane is a weak plane.

Cracks under generalized plane strain

The Lekhnitskii-Stroh formalism

It turns out that the analysis of generalized plane strain can be performed with essentially the same formulations and solution methods as for plane strain. A major difficulty is, of course, the handling of a hugh number of material constants. There is therefore a need for a compact formulation of problems, and a compact solution procedure. Such formulations and procedures were embedded in the formalisms by Lekhnitskii (1950), Eshelby, Read and Shockley (1953) and Stroh (1958). The Lekhnitskii formalism is based on the use of elastic compliances, whereas the other two use elastic stiffnesses. The work by Eshelby, Read and Shockley laid the foundation for the Stroh formalism, which will be introduced here, without expressing any opinion about its sometimes claimed superiority over the Lekhnitskii formalism.

The Stroh formalism is based on the generalized Hooke's law, the equations of

4.14 CRACKS IN ANISOTROPIC LINEARLY ELASTIC MEDIA

equilibrium, and the condition for generalized plane strain, in the following forms:

$$\sigma_{ij} = C_{ijkl}\frac{\partial u_k}{\partial x_l}, \qquad (i,j,k,l = 1,2,3) \tag{4.14.69}$$

$$\frac{\partial \sigma_{ij}}{\partial x_j} = C_{ijkl}\frac{\partial^2 u_k}{\partial x_j \partial x_l} = 0 \tag{4.14.70}$$

$$\frac{\partial u_k}{\partial x_3} = 0 \tag{4.14.71}$$

A general solution is found by inserting the ansatz

$$u_k = a_k f(x_1 + px_2) = a_k f(\zeta) \tag{4.14.72}$$

into the equations of equilibrium. It is assumed that $\overline{f}(\zeta) = f(\zeta)$, which is the case in most applications. Because

$$\frac{\partial u_k}{\partial x_j} = [f'(\zeta)\delta_{1j} + pf'(\zeta)\delta_{2j}]a_k \tag{4.14.73}$$

$$\frac{\partial^2 u_k}{\partial x_j \partial x_l} = f''(\zeta)[\delta_{1j}\delta_{1l} + p(\delta_{1j}\delta_{2l} + \delta_{2j}\delta_{1l}) + p^2\delta_{2j}\delta_{2l}]a_k \tag{4.14.74}$$

this gives

$$C_{ijkl}\frac{\partial^2 u_k}{\partial x_j \partial x_l} = f''(\zeta)[C_{i1k1} + p(C_{i1k2} + C_{i2k1}) + p^2 C_{i2k2}]a_k = 0 \tag{4.14.75}$$

with non-trivial solutions a_k if

$$\det[C_{i1k1} + p(C_{i1k2} + C_{i2k1}) + p^2 C_{i2k2}] = 0 \tag{4.14.76}$$

Recall that $i, k = 1, 2, 3$, so this is a sixth degree equation for p. Assume that it has six distinct roots – possible equal roots are considered as limiting cases of originally distinct roots. Using the fact that the stress-strain energy is positive, Lekhnitskii (1950) showed that this type of equation cannot have real roots, so the roots come in three complex conjugate pairs, those with positive imaginary parts denoted p_α, and those with negative imaginary parts denoted $\overline{p_\alpha}$. The corresponding values of a_k will be denoted $A_{k\alpha}$ and $\overline{A_{k\alpha}}$, which are elements of matrices but not of tensors: consequently summation over α will always be indicated explicitly, whereas the summation convention will be used for the tensor indices i, j, k, l. Note that each $a_k = A_{k\alpha}$ only can be determined to within a constant factor, but it is assumed here that such factors have been fixed by suitable normalization. The 15 quantities constituting $A_{k\alpha}$ and $\overline{A_{k\alpha}}$ (the 3 normalization factors not counted) may be considered as an alternative to the 15 stiffnesses $(C_{ijkl})_{j,l\neq 3}$ for characterization of the material.

The general solution for u_k is now found from (4.14.72) by summation over α. Because u_k is real, this can be written as

$$u_k = \Re \sum_{\alpha=1}^{3} A_{k\alpha} f_\alpha(\zeta_\alpha) q_\alpha \tag{4.14.77}$$

where $\zeta_\alpha = \xi_\alpha + i\eta_\alpha = x_1 + p_\alpha x_2$ and q_α a (generally complex) constant, that allows $A_{k\alpha}$ and $f_\alpha(\zeta_\alpha)$ to be suitably normalized. The relation (4.14.77) can be considered as a representation for the displacements in terms of the unknown functions $f_\alpha(\zeta_\alpha)$ and the unknown constants q_α. A similar representation for the stresses is now sought.

From (4.14.77) it is found that

$$\sigma_{ij} = C_{ijkl}\frac{\partial u_k}{\partial x_l} = \Re\sum_{\alpha=1}^{3} C_{ijkl}A_{k\alpha}f'_\alpha(\zeta_\alpha)\frac{\partial \zeta_\alpha}{\partial x_l}q_\alpha \qquad (4.14.78)$$

and, considering that $\partial \zeta_\alpha/\partial x_l = \partial(x_1 + p_\alpha x_2)/\partial x_l = \delta_{1l} + p_\alpha \delta_{2l}$, this gives

$$\sigma_{ij} = \Re\sum_{\alpha=1}^{3} (C_{ijk1} + p_\alpha C_{ijk2})A_{k\alpha}f'_\alpha(\zeta_\alpha)q_\alpha \qquad (4.14.79)$$

From the equilibrium equations (4.14.70), it follows that the stresses may be represented in terms of a potential, a vector ϕ_i, such that

$$\sigma_{i1} = -\frac{\partial \phi_i}{\partial x_2}, \quad \sigma_{i2} = \frac{\partial \phi_i}{\partial x_1} \qquad (4.14.80)$$

whereas the remaining stress is

$$\sigma_{33} = C_{33kl}u_{k,l} \qquad (4.14.81)$$

and is found after u_k has been determined. It was remarked by Stroh (1958) that the use of Airy's stress function, instead of the three potential functions ϕ_i, would lead to less symmetrical expressions in the solution procedure.

Now, from (4.14.79), after putting $j = 2$, it is found by integration of the expression $\sigma_{i2} = \partial \phi_i/\partial x_1$ that

$$\phi_i = \Re\sum_{\alpha=1}^{3} B_{i\alpha}f_\alpha(\zeta_\alpha)q_\alpha \qquad (4.14.82)$$

which also satisfies the expression $\sigma_{i1} = -\partial \phi_i/\partial x_2$, so there should be no additional integration function of x_2. Here†,

$$B_{i\alpha} = (C_{i2k1} + p_\alpha C_{i2k2})A_{k\alpha} = -\frac{1}{p_\alpha}(C_{i1k1} + p_\alpha C_{i1k2}) \qquad (4.14.83)$$

where the last equation follows from (4.14.75). The expressions $\sigma_{12} = \partial \phi_1/\partial x_1 = -\partial \phi_2/\partial x_2$, give the relation

$$B_{1\alpha} + p_\alpha B_{2\alpha} = 0 \qquad (4.14.84)$$

The basic equations, (4.14.77) and (4.14.82) are written in component form. In the more compact matrix form they read

$$\boxed{\boldsymbol{u} = \Re(\boldsymbol{AFq})} \qquad (4.14.85)$$

$$\boxed{\boldsymbol{\phi} = \Re(\boldsymbol{BFq})} \qquad (4.14.86)$$

---

† Notations vary somewhat in the literature. The elements $B_{i\alpha}$ were denoted $L_{i\alpha}$ in Stroh (1958), but in recent literature they are often considered as elements of a matrix denoted $\boldsymbol{B}$.

4.14 CRACKS IN ANISOTROPIC LINEARLY ELASTIC MEDIA

where A and B are the matrices with elements $A_{i\alpha}$ and $B_{i\alpha}$, respectively, F is the diagonal matrix† $\mathbf{diag}[f_\alpha(\zeta_\alpha)]$ and q is the vector with elements q_α.

The solution of a problem formulated by using the representations (4.14.77) and (4.14.82) consists of finding $f_\alpha(\zeta_\alpha)$ and q_α so that the boundary conditions are satisfied. In regions where the stresses are non-singular, the displacements, and thereby also $f_\alpha(\zeta_\alpha)$ are differentiable, and consequently $f_\alpha(\zeta_\alpha)$ is analytic.

Solution procedures may vary. Very often, solutions can be found by adapting procedures used for corresponding isotropic cases, e.g., Hwu (1991). Stroh (1958) applied the obvious ansatz $f_\alpha(\zeta_\alpha) = \ln(z_\alpha)$ when treating a dislocation problem, and a Fourier transform, leading to dual integral equations, when treating a crack problem. Barnett and Asaro (1972) considered a crack as an array of continuous dislocations; this leads, as in the isotropic case, to a singular integral equation. Yiantai and Xinghua (1996) used conformal mapping for a problem on a periodic array of collinear cracks.

A crack subjected to remote loading

Consider a crack $|x_1| < a$, $x_2 = 0$ in an infinite body, subjected to remote loads $\sigma_{2i} = \sigma_{ij}^\infty \delta_{2j}$, $i = 1, 2, 3$. No slip has taken place in the region outside the crack. Superpose a constant stress state so that the remote loads vanish and the crack faces become subjected to the uniform traction

$$\sigma_{2i} = -\sigma_{ij}^\infty \delta_{2j} \text{ for } |x_1| < a, x_2 = 0 \tag{4.14.87}$$

Stress continuity outside the crack implies that the functions $f_\alpha(\zeta_\alpha)$ are analytic in the respective plane cut by the straight line $|\xi_\alpha| \leq a$, $\eta_\alpha = 0$. Then, $f_\alpha(\zeta_\alpha)$ must be singular at $\zeta_\alpha = \pm a$, and if the singularity exponent is s at one crack edge, it must be $(n - s)$, $n =$ integer, at the other to ensure analyticity outside the branch cut. Furthermore, the condition of finite stress-strain energy requires that $s > 0$ and $n - s > 0$, and thus $n > 0$. Hence, considering that the stresses must vanish at infinity, the only choice, apart from an irrelevant constant factor, is $n = 1$, leading to

$$f_\alpha(\zeta_\alpha) = \zeta_\alpha - (\zeta_\alpha - a)^s (\zeta_\alpha + a)^{1-s} \tag{4.14.88}$$

where the branch is chosen so that $f_\alpha(\zeta_\alpha)$ is real and positive for $\zeta_\alpha = x_1 > a$. Then,

$$f'_\alpha(\zeta_\alpha) = se^{\mp i\pi s}\left(\frac{x_1 + a}{a - x_1}\right)^s + (1-s)e^{\pm i\pi s}\left(\frac{a - x_1}{x_1 + a}\right)^s$$

$$= \cos(\pi s)g_1(x_1) \pm i\sin(\pi s)g_2(x_1) \text{ for } |x_1| < a, x_2 = \pm 0 \tag{4.14.89}$$

Use of this expression, together with the boundary conditions (4.14.87), the expression (4.14.80) for σ_{2i} and the representation (4.14.82) for ϕ_i results in

$$\Re \sum_{\alpha=1}^{3} B_{i\alpha}[1 + \cos(\pi s)g_1(x_1)]q_\alpha \pm \sin(\pi s)g_2(x_1)\Re \sum_{\alpha=1}^{3} iB_{i\alpha}q_\alpha = -\sigma_{ij}^\infty \delta_{2j} \tag{4.14.90}$$

Because this relation holds for all $|x_1| < a$, it follows that $\cos \pi s = 0$, i.e. $s = 1/2$ and,

---

† With $\mathbf{diag}[a_1, a_2, a_3]$ is denoted a 3×3 matrix whose diagonal elements are a_1, a_2, a_3, the other elements being zero. To save space, when all three diagonal elements can be expressed by using a subscript $\alpha = 1, 2, 3$, a shorter notation, such as $\mathbf{diag}[a_\alpha]$, will be used.

hence,

$$\Re \sum_{\alpha=1}^{3} B_{i\alpha} q_\alpha = -\sigma_{ij}^\infty \delta_{2j}, \quad \Re \sum_{\alpha=1}^{3} i B_{i\alpha} q_\alpha = -\Im \sum_{\alpha=1}^{3} B_{i\alpha} q_\alpha = 0 \quad (4.14.91)$$

Thus,

$$\sum_{\alpha=1}^{3} B_{i\alpha} q_\alpha = -\sigma_{ij}^\infty \delta_{2j} \quad (4.14.92)$$

which in matrix form reads

$$\boldsymbol{Bq} = \boldsymbol{t_0} \quad (4.14.93)$$

where $\boldsymbol{t_0}$ is the traction vector on the lower crack face, with components $-\sigma_{ij}^\infty$. Thus, the remaining unknown quantity is found to be

$$\boldsymbol{q} = \boldsymbol{B}^{-1} \boldsymbol{t_0} \quad (4.14.94)$$

where $\boldsymbol{B}^{-1}$ is the inverse of $\boldsymbol{B}$.

The displacements and the stress potential are

$$\boldsymbol{u} = \Re(\boldsymbol{A}\boldsymbol{F_0}\boldsymbol{B}^{-1})\boldsymbol{t_0} \quad (4.14.95)$$
$$\boldsymbol{\phi} = \Re(\boldsymbol{B}\boldsymbol{F_0}\boldsymbol{B}^{-1})\boldsymbol{t_0} \quad (4.14.96)$$

where $\boldsymbol{F_0} = \mathbf{diag}[f_\alpha(\zeta_\alpha)]$.

Recall now that a stress state $-\sigma_{ij}^\infty \delta_{2j}$ was superposed. Resuperposition is effected by adding a stress state, given by

$$\frac{\partial \boldsymbol{\phi}}{\partial x_1} = \Re(\boldsymbol{B}\boldsymbol{D}\boldsymbol{F_\infty}\boldsymbol{B}^{-1})\boldsymbol{t_\infty} = \boldsymbol{t_\infty} \quad (4.14.97)$$

where $\boldsymbol{t_\infty}$ is the stress vector on a plane with outward normal in the x_2 direction and with components σ_{2i}^∞, $i = 1, 2, 3$. The matrix $\boldsymbol{D}\boldsymbol{F_\infty}$ clearly equals the identity matrix $\boldsymbol{I}$ so that $\boldsymbol{F_\infty}$ equals $\mathbf{diag}[\zeta_\alpha]$ – an integration constant would only imply a rigid-body translation. Thus, finally, because $\boldsymbol{t_0} = -\boldsymbol{t_\infty}$:

$$\boxed{\boldsymbol{u} = \Re(\boldsymbol{A}\boldsymbol{F}\boldsymbol{B}^{-1})\boldsymbol{t_\infty}} \quad (4.14.98)$$

$$\boxed{\boldsymbol{\phi} = \Re(\boldsymbol{B}\boldsymbol{F}\boldsymbol{B}^{-1})\boldsymbol{t_\infty}} \quad (4.14.99)$$

where $\boldsymbol{F} = \boldsymbol{F_\infty} - \boldsymbol{F_0} = \mathbf{diag}[(\zeta_\alpha^2 - a^2)^{1/2}]$. These expressions provide the full solution of the problem. The stresses are found from (4.14.80) and (4.14.81).

For calculation of the stresses σ_{2i} and σ_{1i}, the derivatives

$$\frac{\partial \boldsymbol{F}}{\partial x_1} = \boldsymbol{D}\boldsymbol{F}, \quad \frac{\partial \boldsymbol{F}}{\partial x_2} = \boldsymbol{D}\boldsymbol{F}\boldsymbol{P} \quad (4.14.100)$$

are needed. Here, $\boldsymbol{D}$ is the diagonal matrix operator $\mathbf{diag}[\partial/\partial \zeta_\alpha]$ and $\boldsymbol{P}$ is the matrix $\mathbf{diag}[p_\alpha]$.

Stresses and displacements on the crack plane

Consider now the stresses and displacements on $x_2 = 0$. There

$$\boldsymbol{F} = f(x_1)\boldsymbol{I}, \quad \boldsymbol{DF} = f'(x_1)\boldsymbol{I}, \quad \boldsymbol{DFP} = f'(x_1)\boldsymbol{P} \tag{4.14.101}$$

where $\boldsymbol{I}$ is the identity matrix $\text{diag}[1,1,1]$, and where, for $|x_1| < a$, with upper sign for upper, lower sign for lower crack face,

$$f(x_1) = \pm i\sqrt{a^2 - x_1^2}, \quad f'(x_1) = \mp\frac{ix_1}{\sqrt{a^2 - x_1^2}} \tag{4.14.102}$$

whereas for $x_1 > a$ (upper sign) and $x_1 < -a$ (lower sign),

$$f(x_1) = \pm\sqrt{x_1^2 - a^2}, \quad f'(x_1) = \frac{|x_1|}{\sqrt{x_1^2 - a^2}} \tag{4.14.103}$$

The displacements on the crack faces are now found to be

$$\boldsymbol{u} = \pm\sqrt{a^2 - x_1^2}\,\Re(i\boldsymbol{AB}^{-1})\boldsymbol{t}_\infty \tag{4.14.104}$$

where the upper sign refers to the upper and the lower sign to the lower crack face. The displacements on the crack plane outside the crack are

$$\boldsymbol{u} = \pm\sqrt{x_1^2 - a^2}\,\Re(\boldsymbol{AB}^{-1})\boldsymbol{t}_\infty \tag{4.14.105}$$

where the upper sign refers to $x_1 > a$, lower sign to $x_1 < -a$.

The matrix $\boldsymbol{AB}^{-1}$ can be decomposed into real and imaginary parts by using the *Barnett-Lothe tensors* (Barnett and Lothe 1973)

$$\boldsymbol{L} = 2i\boldsymbol{BB}^T \text{ and } \boldsymbol{S} = i(2\boldsymbol{AB}^T - \boldsymbol{I}) \tag{4.14.106}$$

where the superscript T indicates matrix transpose. These matrices, together with the third of the Barnett-Lothe tensors, $\boldsymbol{H} = 2i\boldsymbol{AA}^T$, can be shown to be real tensors (Barnett and Lothe 1973). Thus, the decomposition is

$$\boldsymbol{AB}^{-1} = \frac{1}{2}(\boldsymbol{I} - i\boldsymbol{S})(-2i\boldsymbol{L}^{-1}) = -\boldsymbol{SL}^{-1} - i\boldsymbol{L}^{-1} \tag{4.14.107}$$

so that the displacements on the crack plane are

$$\boldsymbol{u} = \begin{cases} \pm\sqrt{a^2 - x_1^2}\,\boldsymbol{L}^{-1}\boldsymbol{t}_\infty & \text{for } |x_1| < a \\ \mp\sqrt{x_1^2 - a^2}\,\boldsymbol{SL}^{-1}\boldsymbol{t}_\infty & \text{for } |x_1| > a \end{cases} \tag{4.14.108}$$

where $\pm$ refers to the upper and lower crack faces, respectively, and $\mp$ refers to $x > a$ and $x < -a$, respectively. It can be shown (see e.g. Ting 1996), that $\boldsymbol{L}^{-1}$ is symmetric and $\boldsymbol{SL}^{-1}$ is skew-symmetric.

Consider now the special case in which the only non-vanishing remote stress is σ_{22}^∞, corresponding to a mode I case for an isotropic material. Then, the normal displacement $u_2 = 0$ on the crack plane outside the crack, because $\boldsymbol{SL}^{-1}$ is skew-symmetric. However, in general, the tangential displacements will be non-zero, and so they will also be on the crack faces, which are displaced to the shape of an ellipse for each of the three displacement components. This implies, for instance, that the crack shape in the $x_1 x_2$ plane is an ellipse whose axes do not lie in the coordinate directions.

Note that the special case studied did not result in mode I displacements for a general anisotropic material. Similar studies for remote stresses corresponding to modes II or III lead to analogous conclusions: the resulting displacements do not in general satisfy the symmetry conditions for modes II or III.

The stresses σ_{2i} on the crack plane outside the crack are found from

$$\frac{\partial \phi}{\partial x_1} = \Re(BDFB^{-1})t_\infty = \frac{|x_1|}{\sqrt{x_1^2 - a^2}} t_\infty \qquad (4.14.109)$$

which is the same result as for the isotropic case. In the vicinity of the crack edge $x_1 = a$, the gradient $\partial \phi / \partial x_1 = \sqrt{a/(2r)} t_\infty$, where $r = x_1 - a$. This can be put equal to $k/\sqrt{2\pi r}$, where k is a vector with components K_{II}, K_I and K_{III} (in this order). This is done by analogy with the isotropic case, even though the symmetry relations defining modes I, II and III do not have full correspondence in general anisotropic cases. Then, t_∞ may be substituted by $k/\sqrt{\pi a}$ in the expressions for u and ϕ.

The energy flux into the crack edge

The energy flux into the crack edge, assuming crack growth to proceed straightforwards, is found by using (3.5.20), which is valid for all three modes. Because the derivation only uses stresses and displacements on the crack plane, it is valid also for anisotropic materials, and may be written as

$$\mathcal{G} = \lim_{r \to 0} \left\{ \frac{\pi r}{2} \cdot \frac{\partial \phi}{\partial r} \cdot \frac{\partial [u_+(r) - u_-(r)]}{\partial r} \right\} \qquad (4.14.110)$$

where the subscripts plus and minus refer to upper and lower crack faces, respectively, and r is the distance to the crack edge. Thus, $\mathcal{G}$ is found by inserting $\partial \phi / \partial x_1$ and $\partial u / \partial x_1$ from (4.14.109) and (4.14.108). The result is

$$\boxed{\mathcal{G} = \tfrac{1}{2} k^T L^{-1} k} \qquad (4.14.111)$$

Stresses in the crack edge vicinity

By using (4.14.80) and (4.14.99) and putting $t_\infty = k/\sqrt{\pi a}$, the expressions

$$\frac{\partial \phi}{\partial x_1} = \Re\left(B \frac{\partial F}{\partial x_1} B^{-1}\right) \frac{k}{\sqrt{\pi a}}, \quad -\frac{\partial \phi}{\partial x_2} = -\Re\left(B \frac{\partial F}{\partial x_2} B^{-1}\right) \frac{k}{\sqrt{\pi a}} \qquad (4.14.112)$$

are found to give the stresses σ_{2i} and σ_{1i}, respectively. Now,

$$F = \text{diag}[(\zeta_\alpha^2 - a^2)^{1/2}] \qquad (4.14.113)$$

which in the neigbourhood of the crack edge $x_1 = a$, $x_2 = 0$, may be written as

$$F \approx \sqrt{2a}\, \text{diag}[(\zeta_\alpha - a)^{1/2}] \qquad (4.14.114)$$

Then,

$$\frac{\partial F}{\partial x_1} \approx \sqrt{\frac{a}{2}}\, \text{diag}\left[\frac{1}{(\zeta_\alpha - a)^{1/2}}\right], \quad \frac{\partial F}{\partial x_2} \approx \sqrt{\frac{a}{2}}\, \text{diag}\left[\frac{p_\alpha}{(\zeta_\alpha - a)^{1/2}}\right] \qquad (4.14.115)$$

Introduce a polar coordinate system (r, φ) with origin at the crack edge. Then,

$$\zeta_\alpha - a = x_1 - a + p_\alpha x_2 = r(\cos\varphi + p_\alpha \sin\varphi) = r C_\alpha(\varphi) \qquad (4.14.116)$$

and, in the limit $r/a \to 0$,

$$\frac{\partial \boldsymbol{F}}{\partial x_1} = \sqrt{\frac{a}{2r}}\,\mathrm{diag}\Big\{\frac{1}{[C_\alpha(\varphi)]^{1/2}}\Big\}, \quad \frac{\partial \boldsymbol{F}}{\partial x_2} = \sqrt{\frac{a}{2r}}\,\mathrm{diag}\Big\{\frac{p_\alpha}{[C_\alpha(\varphi)]^{1/2}}\Big\} \quad (4.14.117)$$

The stresses σ_{2i} and σ_{1i} in the crack edge vicinity are now found to be

$$\frac{\partial \boldsymbol{\phi}}{\partial x_1} = \Re\Big[\boldsymbol{B}\,\mathrm{diag}\Big\{\frac{1}{[C_\alpha(\varphi)]^{1/2}}\Big\}\boldsymbol{B}^{-1}\Big]\frac{\boldsymbol{k}}{\sqrt{2\pi r}} \quad (4.14.118)$$

$$-\frac{\partial \boldsymbol{\phi}}{\partial x_2} = -\Re\Big[\boldsymbol{B}\,\mathrm{diag}\Big\{\frac{p_\alpha}{[C_\alpha(\varphi)]^{1/2}}\Big\}\boldsymbol{B}^{-1}\Big]\frac{\boldsymbol{k}}{\sqrt{2\pi r}} \quad (4.14.119)$$

To obtain the expression for the remaining stress, $\sigma_{33} = C_{33kl}u_{k,l}$, determine first

$$\frac{\partial \boldsymbol{u}}{\partial x_1} = \Re\Big[\boldsymbol{A}\,\mathrm{diag}\Big\{\frac{1}{[C_\alpha(\varphi)]^{1/2}}\Big\}\boldsymbol{B}^{-1}\Big]\frac{\boldsymbol{k}}{\sqrt{2\pi r}} \quad (4.14.120)$$

$$\frac{\partial \boldsymbol{u}}{\partial x_2} = \Re\Big[\boldsymbol{A}\,\mathrm{diag}\Big\{\frac{p_\alpha}{[C_\alpha(\varphi)]^{1/2}}\Big\}\boldsymbol{B}^{-1}\Big]\frac{\boldsymbol{k}}{\sqrt{2\pi r}} \quad (4.14.121)$$

Then, introduce two vectors, $\boldsymbol{C_1}$ with components C_{3311}, C_{3321}, C_{3331}, and $\boldsymbol{C_2}$ with components C_{3312}, C_{3322}, C_{3332}. This gives

$$\sigma_{33} = \Re\Big[\boldsymbol{C_1}\boldsymbol{A}\,\mathrm{diag}\Big\{\frac{1}{[C_\alpha(\varphi)]^{1/2}}\Big\}\boldsymbol{B}^{-1}\Big]\frac{\boldsymbol{k}}{\sqrt{2\pi r}}$$
$$+ \Re\Big[\boldsymbol{C_2}\boldsymbol{A}\,\mathrm{diag}\Big\{\frac{p_\alpha}{[C_\alpha(\varphi)]^{1/2}}\Big\}\boldsymbol{B}^{-1}\Big]\frac{\boldsymbol{k}}{\sqrt{2\pi r}} \quad (4.14.122)$$

It is found that the largest hoop stress, $\sigma_\varphi(r,\varphi)$ for constant r, is not in general obtained at $\varphi = 0$, even if $K_{II} = K_{III} = 0$. The direction of crack growth is therefore not obvious, and it might depend more strongly on anisotropy as regards strength properties than on the stress distribution at the crack edge. Note also, as for the antiplane strain case, that cracks due to σ_{33} may appear in planes normal to the crack edge, if the fracture toughness is comparatively small in these planes.

Crack face loading
The previous crack problem involved the subproblem of uniform crack face loading. Here, the more general case of non-uniform crack face loading will be considered, for convenience again involving a subproblem: two opposed concentrated forces $\boldsymbol{Q}$ and $-\boldsymbol{Q}$ acting on the lower and upper crack faces, respectively, at $x = r_0$.

The solution procedure is rather similar to the procedure leading to the solution (4.14.95)-(4.14.96), the difference essentially consisting of a different choice of function $\boldsymbol{F_0}$, which will be denoted by $\boldsymbol{F}$ here. An ansatz for the functions $f_\alpha(\zeta_\alpha)$ will be made by analogy with the corresponding isotropic mode I problem, for which the complex potential $p'(z)$ is given by (4.4.37), (4.4.40) and (4.4.42) after putting $\sigma_y^0(x) = P\delta(x-x_0)$, $\sigma_x^\infty = \sigma_y^\infty = 0$. $b = -a$ and $c = a$. This results in

$$p'(z) = \frac{Pa(a^2 - x_0^2)^{1/2}}{\pi(z^2 - a^2)^{1/2}(x_0 - z)} \quad (4.14.123)$$

which integrates to

$$p(z) = \frac{Pa}{\pi} \operatorname{asin} \frac{x_0 z - a^2}{a(z - x_0)} \tag{4.14.124}$$

Therefore, the choice

$$f_\alpha(\zeta_\alpha) = \frac{1}{\pi} \operatorname{asin} \frac{x_0 \zeta_\alpha - a^2}{a(\zeta_\alpha - x_0)} \tag{4.14.125}$$

is made, with a branch cut along the portion $|x_1| \leq a$ of the x_1 axis. Note that the absolute value of the argument of the arcsine function is larger than unity for $|x_1| < a$, which implies that the function is complex in this interval, whereas it is real for $|x_1| > a$. In fact, with $s = (x_0 x_1 - a^2)/[a(x_1 - x_0)]$,

$$\operatorname{asin}(s) = \operatorname{sgn}(s)\frac{\pi}{2} \pm i \cdot \operatorname{acosh}(s) \text{ for } |s| > 1 \tag{4.14.126}$$

where the upper sign refers to $x_2 = +0$ (upper crack face), the lower sign refers to $x_2 = -0$ and $\operatorname{sgn}(\cdot)$ equals -1 for negative and $+1$ for positive argument. Note that $\operatorname{sgn}(s) = -\operatorname{sgn}(x_1 - x_0)$.

Consider now $x_2 = 0$ and note that the argument of the arcsine function changes from $-\infty$ to $+\infty$ with increasing x_1 at $x_1 = x_0$, which implies a step increase π of the function itself. Thus, the derivative of the arcsine function contains a delta function $\pi\delta(x_1 - x_0)$. This could also have been obtained by putting $z = x$, $x \neq x_0$ in (4.4.37): the delta function appears as a result of the necessary indentation at $x = x_0$. Hence, because $\sigma_{2i} = Q_i \delta(x_1 - x_0)$ on $|x_1| < a$, $x_2 = +0$, use of (4.14.82) results in

$$\sigma_{2i} = \frac{\partial \phi_i}{\partial x_1} = \delta(x - x_0) \Re \sum_{\alpha=1}^{3} B_{i\alpha} q_\alpha$$

$$+ \frac{\sqrt{a^2 - x_0^2}}{\pi(x_0 - x_1)\sqrt{a^2 - x_1^2}} \Re \sum_{\alpha=1}^{3} i B_{i\alpha} q_\alpha = Q_i \delta(x_1 - x_0) \tag{4.14.127}$$

Because this relation holds for all $|x_1| < a$, it follows that

$$\sum_{\alpha=1}^{3} B_{i\alpha} q_\alpha = Q_i \tag{4.14.128}$$

or, in matrix notation,

$$\boldsymbol{B}\boldsymbol{q} = \boldsymbol{Q} \tag{4.14.129}$$

from which it follows that $\boldsymbol{q} = \boldsymbol{B}^{-1}\boldsymbol{Q}$, and, consequently, the solution is given by

$$\boldsymbol{u} = \Re(\boldsymbol{AFB}^{-1})\boldsymbol{Q} \tag{4.14.130}$$
$$\boldsymbol{\phi} = \Re(\boldsymbol{BFB}^{-1})\boldsymbol{Q} \tag{4.14.131}$$

where $\boldsymbol{F} = \operatorname{\mathbf{diag}}[f_\alpha(\zeta_\alpha)]$ is given by (4.14.125).

The solution is the response to two opposed crack face forces $\pm \boldsymbol{Q}$ acting at $x_1 = x_0$. For two forces $\pm \boldsymbol{t_0}(x_0)\mathrm{d}x_0$, the solution is obtained by substituting $\boldsymbol{Q}$ by $\boldsymbol{t_0}(x_0)\mathrm{d}x_0$.

Integration over x_0 then gives the response to a traction vector $\boldsymbol{t_0}$ on the lower and $-\boldsymbol{t_0}$ on the upper crack face:

$$\boxed{\boldsymbol{u} = \int_{-a}^{a} \Re(\boldsymbol{AFB}^{-1})\boldsymbol{t_0}(x_0)\mathrm{d}x_0} \qquad (4.14.132)$$

$$\boxed{\boldsymbol{\phi} = \int_{-a}^{a} \Re(\boldsymbol{BFB}^{-1})\boldsymbol{t_0}(x_0)\mathrm{d}x_0} \qquad (4.14.133)$$

Consider now displacements and stresses on $x_2 = 0$. Recognizing, from (4.14.126), that

$$\mathrm{asin}\frac{x_0 x_1 - a^2}{a(x_1 - x_0)} \qquad (4.14.134)$$

is real for $|x_1| > a$ and complex for $|x_1| < a$, and that $\Re(i\boldsymbol{AB}^{-1}) = \boldsymbol{L}^{-1}$, $\Re(\boldsymbol{AB}^{-1}) = -\boldsymbol{SL}^{-1}$ and $\Re(\boldsymbol{BB}^{-1}) = \boldsymbol{I}$, the following expressions, given by Stroh (1958), are obtained:

$$\boldsymbol{u} = \frac{1}{2}\int_{-a}^{a} \mathrm{sgn}(x_1 - x_0)\boldsymbol{SL}^{-1}\boldsymbol{t_0}(x_0)\mathrm{d}x_0$$

$$\pm \frac{1}{\pi}\int_{-a}^{a} \mathrm{acosh}\frac{x_0 x_1 - a^2}{a(x_1 - x_0)}\boldsymbol{L}^{-1}\boldsymbol{t_0}(x_0)\mathrm{d}x_0 \text{ for } |x_1| < a \qquad (4.14.135)$$

$$\boldsymbol{u} = -\frac{1}{\pi}\int_{-a}^{a} \mathrm{asin}\frac{x_0 x_1 - a^2}{a(x_1 - x_0)}\boldsymbol{SL}^{-1}\boldsymbol{t_0}(x_0)\mathrm{d}x_0 \text{ for } |x_1| \geq a \qquad (4.14.136)$$

$$\boldsymbol{\phi} = \frac{1}{\pi}\int_{-a}^{a} \mathrm{asin}\frac{x_0 x_1 - a^2}{a(x_1 - x_0)}\boldsymbol{t_0}(x_0)\mathrm{d}x_0 \text{ for } |x_1| > a \qquad (4.14.137)$$

where the upper sign refers to the upper and the lower sign to the lower crack face. Note that $\boldsymbol{t_0}(x_0)$ must be such that the crack opening displacement is non-negative everywhere: a sufficient condition is that $\boldsymbol{t_0}(x_0)\hat{\boldsymbol{x}}_2 \leq 0$ for all x_0.

As for the uniformly distributed crack face load, it is obvious that $u_2 = 0$ on the crack plane outside the crack if the only non-zero component of $\boldsymbol{t_0}(x_0)$ is $\boldsymbol{t_0}(x_0)\hat{\boldsymbol{x}}_2$, because $\boldsymbol{SL}^{-1}$ is skew-symmetric. Furthermore, the stresses σ_{2i} on $x_2 = 0$ are the same as in the isotropic case, because the expression (4.14.137) does not contain any material parameter.

The Barnett-Lothe tensors

Recall the assumption that p_1, p_2 and p_3 were distinct, implying that $\boldsymbol{A}$ and $\boldsymbol{B}$ can be determined. For some degenerate materials, including isotropic materials, they are not. Then, the definitions (4.14.106) of the Barnett-Lothe tensors $\boldsymbol{S}$, $\boldsymbol{L}$ and $\boldsymbol{H}$ may not be usable. However, they may be defined without use of $\boldsymbol{A}$ and $\boldsymbol{B}$. Thus, Barnett and Lothe (1973) showed that they can be computed as

$$\boldsymbol{S} = \frac{1}{\pi}\int_0^{\pi}\boldsymbol{N_1}(\varphi)\mathrm{d}\varphi, \quad \boldsymbol{H} = \frac{1}{\pi}\int_0^{\pi}\boldsymbol{N_2}(\varphi)\mathrm{d}\varphi, \quad \boldsymbol{L} = -\frac{1}{\pi}\int_0^{\pi}\boldsymbol{N_3}(\varphi)\mathrm{d}\varphi \qquad (4.14.138)$$

where

$$N_1(\varphi) = -T^{-1}(\varphi)R^T(\varphi), \quad N_2(\varphi) = T^{-1}(\varphi),$$
$$N_3(\varphi) = R(\varphi)T^{-1}(\varphi)R^T(\varphi) - Q(\varphi) \tag{4.14.139}$$
$$Q(\varphi) = Q(0)\cos^2\varphi + [R(0) + R^T(0)]\cos\varphi\sin\varphi + T(0)\sin^2\varphi \tag{4.14.140}$$
$$R(\varphi) = R(0)\cos^2\varphi + [T(0) - Q(0)]\cos\varphi\sin\varphi + R^T(0)\sin^2\varphi \tag{4.14.141}$$
$$T(\varphi) = T(0)\cos^2\varphi - [R(0) + R^T(0)]\cos\varphi\sin\varphi + Q(0)\sin^2\varphi \tag{4.14.142}$$

$Q(0)$ is the matrix with elements C_{i1j1}, $R(0)$ is the matrix with elements C_{i1j2}, and $T(0)$ is the matrix with elements C_{i2j2}, $i,j = 1, 2, 3$. Note that $R^T(0)$ is the matrix with elements $R_{ij}^T = C_{j1i2} = C_{i2j1}$. Obviously, the Barnett-Lothe tensors may be calculated directly from the elastic stiffnesses without prior calculation of p_1, p_2, p_3. (The 3×3 matrix $Q(0)$ should not be confused with the previously used vector Q.)

For monoclinic materials with the symmetry plane at $x_3 = 0$, the Barnett-Lothe tensors can be determined explicitly and conveniently without use of A or B (Ting 1992). Thus, with the positive quantities

$$a = \Re(p_1 + p_2), \quad b = \Im(p_1 + p_2), \quad c = \Re(p_1 p_2), \quad d = \Im(p_1 p_2),$$
$$e = ad - bc, \quad g = \frac{a_{12}}{a_{11}} - c, \quad h = be - d^2 \tag{4.14.143}$$

where $a_{11} > 0$ and a_{12} are given by (4.14.39), the tensors can be written as

$$S = \frac{g}{h}\begin{bmatrix} d & -b & 0 \\ e & -d & 0 \\ 0 & 0 & 0 \end{bmatrix}, \quad L = \frac{1}{a_{11}h}\begin{bmatrix} e & -d & 0 \\ -d & b & 0 \\ 0 & 0 & \mu a_{11}h \end{bmatrix}$$

$$H = \frac{a_{11}(h - g^2)}{h}\begin{bmatrix} b & d & 0 \\ d & e & 0 \\ 0 & 0 & \frac{h}{\mu a_{11}(h - g^2)} \end{bmatrix} \tag{4.14.144}$$

Two particularly useful matrices are

$$L^{-1} = a_{11}\begin{bmatrix} b & d & 0 \\ d & e & 0 \\ 0 & 0 & \frac{1}{\mu a_{11}} \end{bmatrix}, \quad SL^{-1} = ga_{11}\begin{bmatrix} 0 & -1 & 0 \\ 1 & 0 & 0 \\ 0 & 0 & 0 \end{bmatrix} \tag{4.14.145}$$

For an isotropic material, $p_1 = p_2 = p_3 = i$, giving $a = d = 0$, $c = -1$, $b = e = 2$ and $h = 4$. Furthermore, $a_{11} = (1-\nu)/(2\mu)$ and $a_{12} = -\nu/(2\mu)$, giving $g = (1-2\nu)/(1-\nu)$.

Interface cracks

Increasing use of laminated composites as well as of surface coatings has spawned much interest in interface cracks between anisotropic materials; see e.g. Willis (1971), Bassani and Qu (1989), Qu and Bassani (1989), Suo (1990), Ni and Nemat-Nasser (1991), Wu (1991), Erdogan and Wu (1993), Suo et al. (1992), Xu et al. (1993) and Ting (1996).

Generalized plane strain in the x_3 direction is assumed and a bimaterial body with

4.14 CRACKS IN ANISOTROPIC LINEARLY ELASTIC MEDIA

the interface at $x_2 = 0$ is considered. Consider the neighbourhood of the edge $x_1 = 0$ of a crack at $x_1 \leq 0$, $x_2 = 0$ and with traction free faces. Such a study may be performed by assuming the stresses to vanish at infinity. Subscripts or superscripts plus and minus will refer to the material on the upper and lower half, respectively. The treatment will follow, essentially, Suo (1990).

The stresses σ_{2i} are continuous across $x_2 = 0$, and traction free crack faces imply that the stresses $\sigma_{2i} = 0$ on $x_1 < 0$, $x_2 = 0$. These conditions can be formulated by using the stress potential ϕ:

$$\phi_+ = \phi_- \text{ for all } x_1, x_2 = 0 \qquad (4.14.146)$$
$$\phi_+ = 0 \text{ for } x_1 < 0, x_2 = 0 \qquad (4.14.147)$$

The displacements are continuous across $x_2 = 0$ on $x_1 > 0$. Thus,

$$\boldsymbol{u}_+ - \boldsymbol{u}_- = 0 \text{ for } x_1 > 0, x_2 = 0 \qquad (4.14.148)$$

The expressions for the displacements and the stress potential, (4.14.85)-(4.14.86), contain a constant vector $\boldsymbol{q}$ which will not be needed here. Therefore, the vector $\boldsymbol{Fq}$ is replaced by a vector $\boldsymbol{f}$ with elements $f_1(\zeta_1)$, $f_2(\zeta_2)$ and $f_3(\zeta_3)$, so that $\boldsymbol{q}$ is absorbed in this vector. Thus,

$$\boldsymbol{u} = \Re(\boldsymbol{A}_\pm \boldsymbol{f}_\pm), \qquad \boldsymbol{\phi}_\pm = \Re(\boldsymbol{B}_\pm \boldsymbol{f}_\pm) \qquad (4.14.149)$$

Then, the condition of stress continuity across $x_2 = 0$, (4.14.146), gives

$$\boldsymbol{B}_+\boldsymbol{f}_+(x_1) + \overline{\boldsymbol{B}}_+\overline{\boldsymbol{f}}_+(x_1) = \boldsymbol{B}_-\boldsymbol{f}_-(x_1) + \overline{\boldsymbol{B}}_-\overline{\boldsymbol{f}}_-(x_1) \text{ for all } x_1 \qquad (4.14.150)$$

Introduce now a vector $\boldsymbol{g} = \boldsymbol{g}(z)$, $z = x_1 + ix_2$, with elements $f_1(z)$, $f_2(z)$ and $f_3(z)$. Thus, $\boldsymbol{g} = \boldsymbol{f}$ for $x_2 = 0$. The function $f_\alpha^+(\zeta_\alpha)$ is analytic in the upper ζ_α half-plane, and consquently $\boldsymbol{g}_+(z)$ is analytic in the upper z half-plane. Similarly, it follows that $\boldsymbol{g}_-(z)$ is analytic in the lower z half-plane.

Now, substitute $\boldsymbol{f}$ by $\boldsymbol{g}$ in (4.14.150) and make a rearrangement to

$$\boldsymbol{B}_+\boldsymbol{g}_+(x_1) - \overline{\boldsymbol{B}}_-\overline{\boldsymbol{g}}_-(x_1) = \boldsymbol{B}_-\boldsymbol{g}_-(x_1) - \overline{\boldsymbol{B}}_+\overline{\boldsymbol{g}}_+(x_1) \text{ for all } x_1 \qquad (4.14.151)$$

The left member may be considered as the value when z approaches $x_2 = 0$ from above of a function that is analytic in $x_2 \geq 0$. Note that $\overline{\boldsymbol{g}}_-(z)$ is analytic in the upper half-plane as a consequence of the fact that $\boldsymbol{g}_-(z)$ is analytic in the lower half-plane, cf. Appendix A2. Similarly, the right member expresses the limiting value of a function, that is analytic in the lower half-plane, when z approaches $x_2 = 0$ from below. Thus, by analytic continuation, the two members equal the value at $x_2 = 0$ of a function that is analytic in the whole plane. By Liouville's theorem, Appendix A4, such a function is a polynomial, which here can be set to zero, because the stresses vanish at infinity† and an additional constant to the stress potential ϕ is optional. Thus,

$$\boldsymbol{B}_+\boldsymbol{g}_+(z) = \overline{\boldsymbol{B}}_-\overline{\boldsymbol{g}}_-(z), \qquad \boldsymbol{B}_-\boldsymbol{g}_-(z) = \overline{\boldsymbol{B}}_+\overline{\boldsymbol{g}}_+(z) \qquad (4.14.152)$$

where the first equality is valid for $\Im(z) \geq 0$ and the second for $\Im(z) \leq 0$. Use of this equation in the condition of traction free crack faces leads to

$$\boldsymbol{B}_+\boldsymbol{g}_+(x_1) + \boldsymbol{B}_-\boldsymbol{g}_-(x_1) = 0 \text{ for } x_1 < 0, x_2 = 0 \qquad (4.14.153)$$

---
† Note that $\boldsymbol{g}_\pm(z)$ equals $f_\alpha^\pm(\zeta_\alpha)$ not only at $x_2 = 0$ but also at infinity.

The condition of displacement continuity across $x_2 = 0$, $x_1 > 0$, (4.14.148), gives

$$\boldsymbol{A}_+\boldsymbol{g}_+(x_1) + \overline{\boldsymbol{A}}_+\overline{\boldsymbol{g}}_+(x_1) = \boldsymbol{A}_-\boldsymbol{g}_-(x_1) + \overline{\boldsymbol{A}}_-\overline{\boldsymbol{g}}_-(x_1) \text{ for } x_1 > 0 \quad (4.14.154)$$

Now, $\overline{\boldsymbol{g}}_+(x_1)$ can be eliminated by using (4.14.152). In order to do so, the identity matrix $I = \overline{\boldsymbol{B}}_+^{-1}\overline{\boldsymbol{B}}_+$ is inserted into the second term, giving

$$\boldsymbol{A}_+\boldsymbol{g}_+(x_1) = \overline{\boldsymbol{A}}_+\overline{\boldsymbol{B}}_+^{-1}\overline{\boldsymbol{B}}_+\overline{\boldsymbol{g}}_+(x_1) = \overline{\boldsymbol{A}}_+\overline{\boldsymbol{B}}_+^{-1}\boldsymbol{B}_-\boldsymbol{g}_-(x_1) \text{ for } x_1 > 0 \quad (4.14.155)$$

Similarly, $\overline{\boldsymbol{g}}_-(x_1)$ is eliminated, and, after inserting $\boldsymbol{B}_+^{-1}\boldsymbol{B}_+$ in the first term and $\boldsymbol{B}_-^{-1}\boldsymbol{B}_-$ in the third term, the resulting equation is

$$\boldsymbol{M}\boldsymbol{B}_+\boldsymbol{g}_+(x_1) = \overline{\boldsymbol{M}}\boldsymbol{B}_-\boldsymbol{g}_-(x_1) \text{ for } x_1 > 0 \quad (4.14.156)$$

where $\boldsymbol{M}$ is the mixed matrix

$$\boldsymbol{M} = \frac{i}{2}(\boldsymbol{A}_+\boldsymbol{B}_+^{-1} - \overline{\boldsymbol{A}}_-\overline{\boldsymbol{B}}_-^{-1}) \quad (4.14.157)$$

which also may be expressed in terms of the Barnett-Lothe tensors as

$$\boldsymbol{M} = \frac{1}{2}[\boldsymbol{L}_+^{-1} + \boldsymbol{L}_-^{-1} - i(\boldsymbol{S}_+\boldsymbol{L}_+^{-1} - \boldsymbol{S}_-\boldsymbol{L}_-^{-1})] \quad (4.14.158)$$

Non-oscillatory solutions

Consider first the special case when $\boldsymbol{M}$ is real, which implies $\boldsymbol{S}_+\boldsymbol{L}_+^{-1} = \boldsymbol{S}_-\boldsymbol{L}_-^{-1}$ and that $\boldsymbol{M} = (\boldsymbol{L}_+^{-1} + \boldsymbol{L}_-^{-1})/2$. This occurs not only for identical materials, but also for certain material combinations, at least obtainable in theory. Then, equations (4.14.153) and (4.14.156), expressing the conditions of traction free crack faces and displacement continuity across the interface outside the crack, can be written as

$$\boldsymbol{h}_+(x_1) + \boldsymbol{h}_-(x_1) = 0 \text{ for } x_1 < 0 \quad (4.14.159)$$
$$\boldsymbol{h}_+(x_1) - \boldsymbol{h}_-(x_1) = 0 \text{ for } x_1 > 0 \quad (4.14.160)$$

where $\boldsymbol{h}_+(z) = \boldsymbol{B}_+\boldsymbol{g}_+(z)$ and $\boldsymbol{h}_-(z) = \boldsymbol{B}_-\boldsymbol{g}_-(z)$. The solution of this Hilbert problem may be written in the form of an analytic function $\boldsymbol{h}(z)$ that is analytic in the plane cut along $x_1 < 0$, $x_2 = 0$ and equals $\boldsymbol{h}_+(z)$ in the upper and $\boldsymbol{h}_-(z)$ in the lower half-plane:

$$\boldsymbol{h}(z) = z^{1/2}\boldsymbol{q} \quad (4.14.161)$$

where $\boldsymbol{q}$ is a constant vector and the branch is chosen so that $z^{1/2} = \sqrt{x_1}$ for $x_1 > 0$, $x_2 = 0$.

For $x_2 = 0$, the stress potential is

$$\phi_\pm(x_1) = \Re[\boldsymbol{B}_\pm\boldsymbol{g}_\pm(x_1)] = \Re[\boldsymbol{h}_\pm(x_1)] \quad (4.14.162)$$

and thus,

$$\sigma_{2i} = \begin{cases} \dfrac{1}{2\sqrt{x_1}}\Re(q_i) = \dfrac{k_i}{\sqrt{2\pi x_1}} & \text{for } x_1 > 0 \\ = \pm\dfrac{1}{2\sqrt{|x_1|}}\Im(q_i) = 0 & \text{for } x_1 < 0, \, x_2 = \pm 0 \end{cases} \quad (4.14.163)$$

4.14 CRACKS IN ANISOTROPIC LINEARLY ELASTIC MEDIA

where k_i obviously is related to the stress intensity factors by $k_1 = K_{II}$, $k_2 = K_I$ and $k_3 = K_{III}$. Thus,

$$q_i = \sqrt{\frac{2}{\pi}} k_i \qquad (4.14.164)$$

In order to find the solution even for points outside $x_2 = 0$, use is made of (4.14.161), written as

$$\boldsymbol{B}_\pm \boldsymbol{g}_\pm = z^{1/2} \boldsymbol{q} = \sqrt{\frac{2}{\pi}} z^{1/2} \boldsymbol{k} \qquad (4.14.165)$$

where $\boldsymbol{k}$ is the vector with elements k_i. This implies

$$\boldsymbol{g}_\pm = \sqrt{\frac{2}{\pi}} z^{1/2} \boldsymbol{B}_\pm^{-1} \boldsymbol{k} \qquad (4.14.166)$$

so that

$$\boldsymbol{f}_\pm = \sqrt{\frac{2}{\pi}} \boldsymbol{F}_\pm \boldsymbol{B}_\pm^{-1} \boldsymbol{k} \qquad (4.14.167)$$

where $\boldsymbol{F} = \mathbf{diag}[\zeta_\alpha^{1/2}]$ with branch cuts and branches defined as for $\boldsymbol{h}$.

The solutions may now be written in the form

$$\boxed{\boldsymbol{\phi}_\pm = \sqrt{\frac{2}{\pi}} \Re(\boldsymbol{B}_\pm \boldsymbol{F}_\pm \boldsymbol{B}_\pm^{-1}) \boldsymbol{k}} \qquad (4.14.168)$$

$$\boxed{\boldsymbol{u}_\pm = \sqrt{\frac{2}{\pi}} \Re(\boldsymbol{A}_\pm \boldsymbol{F}_\pm \boldsymbol{B}_\pm^{-1}) \boldsymbol{k}} \qquad (4.14.169)$$

and then σ_{33} is obtained from (4.14.81). For $x_2 = 0$, the solutions default to

$$\boldsymbol{\phi}_\pm = \sqrt{\frac{2x_1}{\pi}} \boldsymbol{k} \quad \text{for } x_1 > 0 \qquad (4.14.170)$$

$$\boldsymbol{u}_\pm = \pm \sqrt{\frac{2|x_1|}{\pi}} \Re(i \boldsymbol{A}_\pm \boldsymbol{B}_\pm) \boldsymbol{k} = \pm \sqrt{\frac{2|x_1|}{\pi}} \boldsymbol{L}_\pm^{-1} \boldsymbol{k} \quad \text{for } x_1 < 0 \qquad (4.14.171)$$

Use of these expressions and (4.14.110) gives the energy flux into the crack edge,

$$\mathcal{G} = \frac{1}{4} \boldsymbol{k}^T (\boldsymbol{L}_+^{-1} + \boldsymbol{L}_-^{-1}) \boldsymbol{k} = \frac{1}{2} \boldsymbol{k}^T \boldsymbol{M} \boldsymbol{k} \qquad (4.14.172)$$

Recall the restriction to real $\boldsymbol{M}$. This implied non-oscillatory solutions for all traction and displacement components in the crack plane. Note that such solutions also occur for interface cracks under anti-plane strain in the isotropic case and for in-plane strain for certain combinations of isotropic materials, cf. page 181.

At this point, it may be appropriate to make a remark about the very close resemblance between expressions for displacements and for stress potentials; see e.g. (4.14.98)-(4.14.99), (4.14.132)-(4.14.133) and (4.14.168)-(4.14.169). Such similarilities in a wide range of elastic problems prompted Nemat-Nasser and Ni (1995) and Ni and Nemat-Nasser (1996) to make a systematic examination, which led to the formulation of a *general duality principle*. Using this principle, a number of elegant solutions of elastic problems was obtained by Nemat-Nasser (1995).

Oscillatory solutions

Assume now that M is complex, $\overline{M} \neq M$. Let $h_+(z)$ be equal to $B_+ g_+(z)$ as earlier, but put $h_-(z) = M^{-1}\overline{M} B_- g_-(z)$, so that equation (4.14.160), expressing displacement continuity across the interface outside the crack, remains unchanged,

$$h_+(x_1) - h_-(x_1) = 0 \text{ for } x_1 > 0 \qquad (4.14.173)$$

whereas (4.14.159), expressing traction free crack faces, changes to

$$h_+(x_1) + \overline{M}^{-1} M h_-(x_1) = 0 \text{ for } x_1 < 0 \qquad (4.14.174)$$

By analogy with the isotropic case, the solution of this Hilbert problem is written as

$$h(z) = z^{1/2+i\epsilon} q \qquad (4.14.175)$$

It will turn out that ϵ is real, but this is not assumed *a priori*. A branch cut is made along $x_1 < 0$, $x_2 = 0$, and the branch is chosen so that $h(z) = \sqrt{x_1} \exp[i\epsilon \ln(x_1)] q$ for $x_1 > 0$, $x_2 = 0$. Then,

$$h_+(x_1) = i\sqrt{|x_1|} e^{i\epsilon \ln |x_1| - \pi\epsilon} q \text{ for } x_1 < 0, \; x_2 = +0 \qquad (4.14.176)$$
$$h_-(x_1) = -i\sqrt{|x_1|} e^{i\epsilon \ln |x_1| + \pi\epsilon} q \text{ for } x_1 < 0, \; x_2 = -0 \qquad (4.14.177)$$

Insertion into (4.14.174) gives

$$e^{-\pi\epsilon} q - e^{\pi\epsilon} \overline{M}^{-1} M q = 0 \qquad (4.14.178)$$

This is an eigenvalue problem, which will be written as

$$(\overline{M} - e^{2\pi\epsilon} M) q = 0 \qquad (4.14.179)$$

From the definition of M, it follows that it is Hermitian, i.e., $\overline{M} = M^T$. Then, also $\overline{M}$ is Hermitian. Both M and $\overline{M}$ can be shown to be positive definite, cf. Ting (1996). Then, the eigenvalues $\exp(2\pi\epsilon)$ are positive (see *e.g.*, Wylie and Barrett 1985), and, consequently, ϵ is real. Note now that if $\exp(2\pi\epsilon) = \exp(2\pi\epsilon_*)$ is one eigenvalue and q_* the corresponding eigenvector, then $\exp(2\pi\epsilon) = \exp(-2\pi\epsilon_*)$ is another eigenvalue with the corresponding eigenvector $\overline{q}_*$. This follows from (4.14.179) after complex conjugation. But there are only three eigenvalues, and therefore the third eigenvalue must be unity, corresponding to $\epsilon = 0$, and consequently the corresponding eigenvector q_3 is real. Without loss of generality, ϵ_* can be assumed to be positive, and, to avoid too many symbols, it will be denoted simply by ϵ. The corresponding eigenvector is $q_* = q_1 + iq_2$, where q_1 and q_2 are real. Thus, the solutions to (4.14.179) are the eigenvalues $\exp(2\pi\epsilon)$, $\exp(-2\pi\epsilon)$ and 1, with the corresponding eigenvectors q_*, $\overline{q}_*$ and q_3. Note that these eigenvalues and eigenvectors can be determined without prior calculation of p_1, p_2 and p_3, because M can be expressed in terms of the Barnett-Lothe tensors.

Complex vectors, such as the eigenvectors q_* and $\overline{q}_*$ are also called bivectors (Gibbs 1881). They allow compact and elegant representation and handling of diverse physical phenomena (Boulanger and Hayes 1993).

The solution of the Hilbert problem may now be written as

$$h(z) = a_1 z^{1/2+i\epsilon} q_* + a_2 z^{1/2-i\epsilon} \overline{q}_* + a_3 z^{1/2} q_3 \qquad (4.14.180)$$

where a_1, a_2 and a_3 are constants.

4.14 CRACKS IN ANISOTROPIC LINEARLY ELASTIC MEDIA

For $x_2 = 0$, the stress potential is

$$\phi_+ = \phi_- = \Re[\boldsymbol{B}_+\boldsymbol{g}_+(x_1)] = \Re[\boldsymbol{h}_+(x_1)]$$
$$= \begin{cases} \sqrt{x_1}\Re[a_1 x_1^{i\epsilon}\boldsymbol{q}_* + a_2 x_1^{-i\epsilon}\overline{\boldsymbol{q}}_* + a_3\boldsymbol{q}_3] & \text{for } x_1 > 0 \\ \sqrt{|x_1|}\Re[ia_1 x_1^{i\epsilon} e^{-\pi\epsilon}\boldsymbol{q}_* + ia_2 x_1^{-i\epsilon} e^{\pi\epsilon}\overline{\boldsymbol{q}}_* + ia_3\boldsymbol{q}_3] & \text{for } x_1 < 0 \end{cases} \quad (4.14.181)$$

Now, traction free crack faces imply that $\phi(x_1) = 0$ for $x_1 < 0$, and therefore the expression inside [ ] in the last equation must be imaginary. This leads to $a_2 = \overline{a_1}\exp(-2\pi\epsilon)$ and that a_3 is real, because $\boldsymbol{q}_3$ is real.

The displacement on $x_2 = +0$ is

$$\boldsymbol{u}_+ = \Re[\boldsymbol{A}_+\boldsymbol{g}_+(x_1)] = \Re[\boldsymbol{A}_+\boldsymbol{B}_+^{-1}\boldsymbol{B}_+\boldsymbol{g}_+(x_1)] = \Re[\boldsymbol{A}_+\boldsymbol{B}_+^{-1}\boldsymbol{h}_+(x_1)] \quad (4.14.182)$$

and, similarly, on $x_2 = -0$,

$$\boldsymbol{u}_- = \Re[\boldsymbol{A}_-\boldsymbol{g}_-(x_1)] = \Re[\boldsymbol{A}_-\boldsymbol{B}_-^{-1}\overline{\boldsymbol{M}}^{-1}\boldsymbol{M}\boldsymbol{h}_-(x_1)]$$
$$= \begin{cases} -\Re[\boldsymbol{A}_-\boldsymbol{B}_-^{-1}\boldsymbol{h}_+(x_1)] & \text{for } x_1 < 0 \\ \Re[\boldsymbol{A}_-\boldsymbol{B}_-^{-1}\overline{\boldsymbol{M}}^{-1}\boldsymbol{M}\boldsymbol{h}_+(x_1)] & \text{for } x_1 > 0 \end{cases} \quad (4.14.183)$$

The traction vector on the plane $x_2 = 0$ is

$$\boldsymbol{t}(x_1) = \frac{\partial \phi}{\partial x_1} = \frac{1}{2\sqrt{x_1}}\Re[a_1(1+2i\epsilon)x_1^{i\epsilon}\boldsymbol{q}_* + \overline{a}_1(1-2i\epsilon)x_1^{-i\epsilon}\overline{\boldsymbol{q}}_* + a_3\boldsymbol{q}_3] \quad (4.14.184)$$

for $x_1 > 0$. With the stress intensity factors

$$K = K_1 + iK_2 = \sqrt{\frac{\pi}{2}}a_1(1+2i\epsilon)e^{-\pi\epsilon}\cosh(\pi\epsilon)r_0^{i\epsilon}, \quad K_3 = \sqrt{\frac{\pi}{2}}a_3 \quad (4.14.185)$$

this reads

$$\boldsymbol{t}(x_1) = \frac{1}{\sqrt{2\pi x_1}}\left[K\left(\frac{x_1}{r_0}\right)^{i\epsilon}\boldsymbol{q}_* + \overline{K}\left(\frac{x_1}{r_0}\right)^{-i\epsilon}\overline{\boldsymbol{q}}_* + K_3\boldsymbol{q}_3\right] \quad (4.14.186)$$

The complex stress intensity factor K and the arbitrary positive length parameter r_0 are chosen by analogy with the complex stress intensity factor in the isotropic case, (4.12.21).

Now, because

$$\boldsymbol{g}_+(x_1) = \boldsymbol{B}_+^{-1}\boldsymbol{h}_+(x_1) \quad (4.14.187)$$

it follows that

$$\boldsymbol{f}_+(\zeta_\alpha) = \boldsymbol{g}_+(\zeta_\alpha) = \sqrt{\frac{2}{\pi}}\left\{\frac{e^{\pi\epsilon}K}{(1+2i\epsilon)\cosh(\pi\epsilon)}\zeta_\alpha^{1/2}\left(\frac{\zeta_\alpha}{r_0}\right)^{i\epsilon}\boldsymbol{B}_+^{-1}\boldsymbol{q}_* \right.$$
$$\left. + \frac{e^{-\pi\epsilon}\overline{K}}{(1-2i\epsilon)\cosh(\pi\epsilon)}\zeta_\alpha^{1/2}\left(\frac{\zeta_\alpha}{r_0}\right)^{-i\epsilon}\boldsymbol{B}_+^{-1}\overline{\boldsymbol{q}}_* + K_3\zeta_\alpha^{1/2}\boldsymbol{B}_+^{-1}\boldsymbol{q}_3\right\} \quad (4.14.188)$$

and, hence,

$$\boldsymbol{u}_+ = \frac{e^{\pi\epsilon}}{\cosh(\pi\epsilon)}\sqrt{\frac{2}{\pi}}\Re\left\{\frac{K}{1+2i\epsilon}\boldsymbol{A}_+\,\mathrm{diag}\left[\zeta_\alpha^{1/2}\left(\frac{\zeta_\alpha}{r_0}\right)^{i\epsilon}\right]\boldsymbol{B}_+^{-1}\boldsymbol{q}_*\right\}$$

$$+ \frac{e^{-\pi\epsilon}}{\cosh(\pi\epsilon)}\sqrt{\frac{2}{\pi}}\Re\left\{\frac{\overline{K}}{1-2i\epsilon}\boldsymbol{A}_+\,\mathrm{diag}\left[\zeta_\alpha^{1/2}\left(\frac{\zeta_\alpha}{r_0}\right)^{-i\epsilon}\right]\boldsymbol{B}_+^{-1}\overline{\boldsymbol{q}}_*\right\}$$

$$+ \sqrt{\frac{2}{\pi}}\Re\{K_3\boldsymbol{A}_+\,\mathrm{diag}[\zeta_\alpha^{1/2}]\boldsymbol{B}_+^{-1}\boldsymbol{q}_3\} \quad (4.14.189)$$

The expression for $\boldsymbol{u}_-$ is obtained from $\boldsymbol{u}_+$ after the obvious changes of $\boldsymbol{A}_+$ to $\boldsymbol{A}_-$ and $\boldsymbol{B}_+^{-1}$ to $\boldsymbol{B}_-^{-1}$, but it must also be observed that $\boldsymbol{h}_-(z) = \boldsymbol{M}^{-1}\overline{\boldsymbol{M}}\boldsymbol{B}_-\boldsymbol{g}_-(z)$. This implies that $\boldsymbol{q}_*$ has to be replaced by $\boldsymbol{M}^{-1}\overline{\boldsymbol{M}}\boldsymbol{q}_*$, which then, because of (4.14.179), can be substituted by $e^{-2\pi\epsilon}\boldsymbol{q}_*$. After a similar change for $\overline{\boldsymbol{q}}_*$ (for $\boldsymbol{q}_3$ there is no net change), the result is that $e^{\pm\pi\epsilon}$ in the expression for $\boldsymbol{u}_+$ shall be substituted by $e^{\mp\pi\epsilon}$.

The expression for the potential $\boldsymbol{\phi}_\pm$ is obtained by changing $\boldsymbol{A}_\pm$ to $\boldsymbol{B}_\pm$ in the expressions for $\boldsymbol{u}_\pm$.

The displacement jump across the crack

By specializing to $x_1 < 0$, $x_2 = \pm 0$, the displacement jump across the crack is found to be

$$\boldsymbol{u}_+ - \boldsymbol{u}_- = \sqrt{\frac{2|x_1|}{\pi}}(\boldsymbol{M} + \overline{\boldsymbol{M}})$$

$$\times \left[\frac{K(|x_1|/r_0)^{i\epsilon}}{(1+2i\epsilon)\cosh(\pi\epsilon)}\boldsymbol{q}_* + \frac{\overline{K}(|x_1|/r_0)^{-i\epsilon}}{(1-2i\epsilon)\cosh(\pi\epsilon)}\overline{\boldsymbol{q}}_* + K_3\boldsymbol{q}_3\right] \quad (4.14.190)$$

For calculation of the energy flux into the crack edge, the gradient of the displacement jump across the crack is needed:

$$\frac{\mathrm{d}(\boldsymbol{u}_+ - \boldsymbol{u}_-)}{\mathrm{d}x_1} = \frac{\boldsymbol{M} + \overline{\boldsymbol{M}}}{\sqrt{2\pi|x_1|}}\left[\frac{K(|x_1|/r_0)^{i\epsilon}}{\cosh(\pi\epsilon)}\boldsymbol{q}_* + \frac{\overline{K}(|x_1|/r_0)^{-i\epsilon}}{\cosh(\pi\epsilon)}\overline{\boldsymbol{q}}_* + K_3\boldsymbol{q}_3\right] \quad (4.14.191)$$

Note that the traction and displacement vectors are real and that their components are not given along the coordinate directions, but rather along the directions of the real vectors $\boldsymbol{q}_1$, $\boldsymbol{q}_2$ and $\boldsymbol{q}_3$, which in general are not orthogonal to each other. The complex constant K and the real constant K_3 do not possess a physical significance similar to the corresponding constants in the isotropic case, unless the vectors $\boldsymbol{q}_1$, $\boldsymbol{q}_2$ and $\boldsymbol{q}_3$ are suitably normalized. This is assumed to be done.

Definition of the stress intensity factors

Because no physical length parameter is involved, the result holds for a sufficiently close vicinity of an edge of any interface crack, i.e. not necessarily a semi-infinite crack. This fact suggests that a general definition of K and K_3 be obtained. To this end, it is first observed that $\boldsymbol{q}_*$ and $\boldsymbol{q}_3$ satisfy the generalized orthogonality conditions (e.g. Wylie and Barrett 1985)

$$\overline{\boldsymbol{q}}_*^T\boldsymbol{M}\overline{\boldsymbol{q}}_* = \overline{\boldsymbol{q}}_*^T\boldsymbol{M}\boldsymbol{q}_3 = \boldsymbol{q}_3^T\boldsymbol{M}\boldsymbol{q}_* = \boldsymbol{q}_3^T\boldsymbol{M}\overline{\boldsymbol{q}}_* = 0 \quad (4.14.192)$$

Now, premultiply both sides of (4.14.186) by $\bar{\boldsymbol{q}}_*^T \boldsymbol{M}$. This gives

$$\bar{\boldsymbol{q}}_*^T \boldsymbol{M} \boldsymbol{t}(x_1) = \frac{1}{\sqrt{2\pi x_1}} K \left(\frac{x_1}{r_0}\right)^{i\epsilon} \bar{\boldsymbol{q}}_*^T \boldsymbol{M} \boldsymbol{q}_* \qquad (4.14.193)$$

and, consequently, K can be defined through the relation

$$K = K_1 + iK_2 = \lim_{r \to 0} \left[\sqrt{2\pi r} \left(\frac{r}{r_0}\right)^{-i\epsilon} \frac{\bar{\boldsymbol{q}}_*^T \boldsymbol{M} \boldsymbol{t}(r)}{\bar{\boldsymbol{q}}_*^T \boldsymbol{M} \boldsymbol{q}_*} \right] \qquad (4.14.194)$$

where r is the distance ahead of the crack edge. Note that for an edge to the left of a crack, the exponent $i\epsilon$ should be changed to $-i\epsilon$, because ϵ changes sign if the assembly is turned upside down.

A similar operation as for K yields the definition

$$K_3 = \lim_{r \to 0} \left[\sqrt{2\pi r} \frac{\boldsymbol{q}_3^T \boldsymbol{M} \boldsymbol{t}(r)}{\boldsymbol{q}_3^T \boldsymbol{M} \boldsymbol{q}_3} \right] \qquad (4.14.195)$$

Compare these definitions with the definitions for stress intensity factors in the isotropic and homogeneous case, for which

$$\boldsymbol{t}(r) = \frac{1}{2\pi r}(\sigma_{21}\hat{\boldsymbol{x}}_1 + \sigma_{22}\hat{\boldsymbol{x}}_2 + \sigma_{23}\hat{\boldsymbol{x}}_3) \qquad (4.14.196)$$

on the body below a cut along the interface close to the crack edge. Thus, for instance, using a similar format to that in (4.14.194)-(4.14.195),

$$K_I = \lim_{r \to 0} [\sqrt{2\pi r} \hat{\boldsymbol{x}}_2^T \boldsymbol{t}(r)] = \lim_{r \to 0} \left[\sqrt{2\pi r} \frac{\hat{\boldsymbol{x}}_2^T \boldsymbol{M} \boldsymbol{t}(r)}{\hat{\boldsymbol{x}}_2^T \boldsymbol{M} \hat{\boldsymbol{x}}_2} \right] \qquad (4.14.197)$$

where $\boldsymbol{M}$ could be inserted, because it is a diagonal matrix, $\boldsymbol{M} = \boldsymbol{L}^{-1}$, in the isotropic and homogeneous case.

The energy flux into the crack edge
The energy flux into the crack edge can now be calculated, using the Irwin formula (3.5.19), which after formal modification reads

$$\mathcal{G} = -\lim_{\Delta a \to 0} \left\{ \frac{1}{2} \int_0^{\Delta a} \boldsymbol{t}^T(r) \frac{d[\boldsymbol{u}_+(\Delta a - r) - \boldsymbol{u}_-(\Delta a - r)]}{dr} dr \right\} \qquad (4.14.198)$$

where the argument r for $\boldsymbol{t}$ is the distance ahead of the crack edge, and the argument $\Delta a - r$ for $\boldsymbol{u}$ is the distance behind the crack edge. The integration involves three integrals, conveniently calculated by residue calculus,

$$\int_0^{\Delta a} \frac{dr}{\sqrt{r(\Delta a - r)}} = \pi \quad \text{and} \quad \int_0^{\Delta a} \frac{dr}{r^{1/2 \pm i\epsilon}(\Delta a - r)^{1/2 \mp i\epsilon}} = \frac{\pi}{\cosh(\pi\epsilon)} \qquad (4.14.199)$$

Due to the factor $(\boldsymbol{M} + \overline{\boldsymbol{M}})$, products such as $\boldsymbol{q}_*^T \boldsymbol{M} \bar{\boldsymbol{q}}_*$ appear in the calculation. The generalized orthogonal relations (4.14.197) then imply that the mixed terms, such as $\boldsymbol{q}_*^T \boldsymbol{M} \boldsymbol{q}_3$, etc. disappear. Note that terms like $\bar{\boldsymbol{q}}_*^T \boldsymbol{M} \boldsymbol{q}_*$ are real, because $\boldsymbol{M}$

is Hermitian: $(\bar{q}_*^T M) q_* = q_*^T (\bar{q}_*^T M)^T = q_*^T (M^T \bar{q}_*) = q_*^T \overline{M} \bar{q}_*$. Considering these facts, the result is found to be

$$\mathcal{G} = \frac{K \overline{K} \bar{q}_*^T (M + \overline{M}) q_*}{2 \cosh^2(\pi\epsilon)} + \frac{1}{4} K_3^2 q_3^T (M + \overline{M}) q_3 \qquad (4.14.200)$$

Interface crack subjected to remote loading

Consider a crack $|x_1| < a$ along the interface between two anisotropic materials, that extend sufficiently far from the crack to be considered as semi-infinite. The body is subjected to remote loading, $\sigma_{2i} = \sigma_{2i}^\infty$, $i = 1, 2, 3$. On planes with outward normals in the $\hat{x}_2$ direction, this load may be given as a vector t_∞ with components σ_{2i}^∞. Stresses and displacements are sought, particularly on the plane $x_2 = 0$. It is assumed that the mixed matrix M is not real, i.e., only oscillatory solutions are considered.

It turns out to be convenient to first superpose a stress state $\sigma_{2i} = -\sigma_{2i}^\infty$ everywhere and thus consider the problem of a crack subjected to the traction t_∞ on its lower face and $-t_\infty$ on its upper face, whereas the stresses vanish at infinity.

The treatment differs only in a few respects from that for a semi-infinite crack. Thus, from (4.14.149), (4.14.146) and (4.14.152),

$$u_\pm = \Re(A_\pm f_\pm), \quad \phi_\pm = \Re(B_\pm f_\pm), \quad \phi_+(x_1) = \phi_-(x_1) \text{ for all } x_1 \qquad (4.14.201)$$

$$B_+ g_+(z) = \overline{B}_- \bar{g}_-(z), \quad B_- g_-(z) = \overline{B}_+ \bar{g}_+(z) \qquad (4.14.202)$$

Further, as for the semi-infinite crack, $g(x_1) = f(x_1)$.

The condition that the traction vector on the upper crack face is $-t_\infty$ gives

$$\phi_+ = -x_1 t_\infty \text{ for } |x_1| < a, x_2 = +0 \qquad (4.14.203)$$

because $\sigma_{2i} = \partial \phi_i / \partial x_1$ and an additional constant only implies a rigid-body translation.

As for the semi-infinite crack, introduce

$$h_+(z) = B_+ g_+(z), \quad h_-(z) = M^{-1} \overline{M} B_- g_-(z) \qquad (4.14.204)$$

Then, by analogy with (4.14.173)-(4.14.174), the following Hilbert problem is obtained:

$$h_+(x_1) - h_-(x_1) = 0 \text{ for } |x_1| > a \qquad (4.14.205)$$

$$h_+(x_1) + \overline{M}^{-1} M h_-(x_1) = -2 x_1 t_\infty \text{ for } |x_1| < a \qquad (4.14.206)$$

Now, the structure of the singularities at edges of interface cracks between anisotropic media is known from the treatment of the semi-infinite crack, and therefore a suitable ansatz is

$$h(z) = [(z - a)^{1/2 + i\epsilon} (z + a)^{1/2 - i\epsilon} - z] q \qquad (4.14.207)$$

where a branch cut is made along $|x_1| < a$, $x_2 = 0$ and the branch is chosen so that $(z-a)^{1/2+i\epsilon}(z+a)^{1/2-i\epsilon} = \sqrt{x_1^2 - a^2} \exp\{i\epsilon \ln[(x_1 - a)/(x_1 + a)]\}$ for $x_1 > a$, $x_2 = 0$. Note that the different signs for the exponent $i\epsilon$ were chosen to ensure analyticity in the cut z-plane and that the term z ensures that stresses vanish at infinity. The analyticity at $|x_1| > a$, $x_2 = 0$, implies that (4.14.205) is satisfied. Now,

$$h_\pm(x_1) = \{\pm i \sqrt{a^2 - x_1^2} e^{i\epsilon \ln[(a-x_1)/(a+x_1)] \mp \pi\epsilon} - x_1\} q \qquad (4.14.208)$$

which, inserted into (4.14.206), gives

$$x_1(\boldsymbol{q} + \overline{\boldsymbol{M}}^{-1}\boldsymbol{M}\boldsymbol{q})$$
$$- i\sqrt{a^2 - x_1^2}\mathrm{e}^{i\epsilon\ln[(a-x_1)/(a+x_1)]}[\mathrm{e}^{-\pi\epsilon}\boldsymbol{q} - \mathrm{e}^{\pi\epsilon}\overline{\boldsymbol{M}}^{-1}\boldsymbol{M}\boldsymbol{q}] = 2x_1\boldsymbol{t}_\infty \quad (4.14.209)$$

A necessary condition for this relation to hold for all $|x_1| < a$ is that

$$\overline{\boldsymbol{M}}\boldsymbol{q} - \mathrm{e}^{2\pi\epsilon}\boldsymbol{M}\boldsymbol{q} = 0 \quad (4.14.210)$$

which is the same eigenvalue problem as (4.14.179). Thus, there are three real and positive eigenvalues, $\exp(\pi\epsilon)$, $\exp(-\pi\epsilon)$ and 1, with the corresponding eigenvectors $\boldsymbol{q}_*$, $\overline{\boldsymbol{q}}_*$ and $\boldsymbol{q}_3$. This leaves

$$x_1(1 + \mathrm{e}^{-2\pi\epsilon})\boldsymbol{q} = 2x_1\mathrm{e}^{-\pi\epsilon}\cosh(\pi\epsilon)\boldsymbol{q} = 2x_1\boldsymbol{t}_\infty \quad (4.14.211)$$

to be satisfied. But this requires that $\boldsymbol{t}_\infty$ is collinear with one eigenvector $\boldsymbol{q}$, which, of course, in general, is not the case. Therefore, the general solution of the Hilbert problem has to be used,

$$h(z) = a_1[(z-a)^{1/2+i\epsilon}(z+a)^{1/2-i\epsilon} - z]\boldsymbol{q}_*$$
$$+ a_2[(z-a)^{1/2-i\epsilon}(z+a)^{1/2+i\epsilon} - z]\overline{\boldsymbol{q}}_*$$
$$+ a_3[(z-a)^{1/2}(z+a)^{1/2} - z]\boldsymbol{q}_3 \quad (4.14.212)$$

where a_1, a_2 and a_3 are constants. Insertion of this expression into (4.14.206), recalling that terms containing $\sqrt{a^2 - x_1^2}$ disappear, and that the remaining terms should be of the form (4.14.211), leads to

$$a_1\mathrm{e}^{-\pi\epsilon}\cosh(\pi\epsilon)\boldsymbol{q}_* + a_2\mathrm{e}^{\pi\epsilon}\cosh(\pi\epsilon)\overline{\boldsymbol{q}}_* + a_3\boldsymbol{q}_3 = \boldsymbol{t}_\infty \quad (4.14.213)$$

because the fact that $\boldsymbol{M}$ and $\overline{\boldsymbol{M}}$ are Hermitian and positive definite enables any vector to be expressed as a linear combination of the eigenvectors $\boldsymbol{q}_*$, $\overline{\boldsymbol{q}}_*$ and $\boldsymbol{q}_3$; see e.g. Wylie and Barrett (1985). Premultiply this relation in turn by $\overline{\boldsymbol{q}}_*^T\boldsymbol{M}$, $\boldsymbol{q}_*^T\boldsymbol{M}$ and $\boldsymbol{q}_3^T\boldsymbol{M}$, observing the generalized orthogonality conditions (4.14.192). This gives

$$a_1 = \frac{\mathrm{e}^{\pi\epsilon}\overline{\boldsymbol{q}}_*^T\boldsymbol{M}\boldsymbol{t}_\infty}{\cosh(\pi\epsilon)\overline{\boldsymbol{q}}_*^T\boldsymbol{M}\boldsymbol{q}_*}, \quad a_2 = \frac{\mathrm{e}^{-\pi\epsilon}\boldsymbol{q}_*^T\boldsymbol{M}\boldsymbol{t}_\infty}{\cosh(\pi\epsilon)\boldsymbol{q}_*^T\boldsymbol{M}\overline{\boldsymbol{q}}_*}, \quad a_3 = \frac{\boldsymbol{q}_3^T\boldsymbol{M}\boldsymbol{t}_\infty}{\boldsymbol{q}_3^T\boldsymbol{M}\boldsymbol{q}_3} \quad (4.14.214)$$

Obviously, $a_2 = \mathrm{e}^{-2\pi\epsilon}\overline{a}_1$, because $\boldsymbol{M}$ can be replaced by $\overline{\boldsymbol{M}}$ in the generalized orthogonality conditions, which also implies that a_3 is real.

Resuperposition of the stress $\sigma_{2i} = \sigma_{2i}^\infty$ everywhere implies, as expected, removal of the term $-z$ in (4.14.212). In fact, if $\phi_\infty = x_1\Re(a_1\boldsymbol{q}_* + a_2\overline{\boldsymbol{q}}_* + a_3\boldsymbol{q}_3)$, then $\partial\phi_\infty/\partial x_1$ equals the left member of (4.14.213). Thus, after resuperposition,

$$\boldsymbol{u}_+ = \Re(\boldsymbol{A}_+\boldsymbol{f}_+) = \Re\{a_1\boldsymbol{A}_+\,\mathrm{diag}[(\zeta_\alpha - a)^{1/2+i\epsilon}(\zeta_\alpha + a)^{1/2-i\epsilon}]\boldsymbol{B}_+^{-1}\boldsymbol{q}_*$$
$$+ a_2\boldsymbol{A}_+\,\mathrm{diag}[(\zeta_\alpha - a)^{1/2-i\epsilon}(\zeta_\alpha + a)^{1/2+i\epsilon}]\boldsymbol{B}_+^{-1}\overline{\boldsymbol{q}}_*$$
$$+ a_3\boldsymbol{A}_+\,\mathrm{diag}[(\zeta_\alpha^2 - a^2)^{1/2}]\boldsymbol{B}_+^{-1}\boldsymbol{q}_3\} \quad (4.14.215)$$

and the expressions for $\boldsymbol{u}_-$ and $\phi_\pm$ are obtained from $\boldsymbol{u}_+$ as for the semi-infinite crack, cf. page 232.

In particular, for $x_2 = 0$, the displacement jump across the crack is found to be

$$\boldsymbol{u}_+ - \boldsymbol{u}_- = \sqrt{a^2 - x_1^2}\{(\boldsymbol{M} + \overline{\boldsymbol{M}})[a_1 e^{-\pi\epsilon} e^{i\epsilon \ln[(a-x_1)/(a+x_1)]}\boldsymbol{q}_*$$
$$+ a_2 e^{\pi\epsilon} e^{-i\epsilon \ln[(a-x_1)/(a+x_1)]}\overline{\boldsymbol{q}}_* + a_3\boldsymbol{q}_3]\} \quad (4.14.216)$$

and the stress potential along the interface outside the crack is

$$\phi = \pm\sqrt{x_1^2 - a^2}\Re\{a_1 e^{i\epsilon[(x_1-a)/(x_1+a)]}\boldsymbol{q}_* + a_2 e^{-i\epsilon[(x_1-a)/(x_1+a)]}\overline{\boldsymbol{q}}_* + a_3\boldsymbol{q}_3\} \quad (4.14.217)$$

where the upper sign refers to $x > a$, the lower to $x < -a$. Thus, the stress vector on the body below a cut along the interface is

$$\boldsymbol{t}(x_1) = \sigma_{21}\hat{\boldsymbol{x}}_1 + \sigma_{22}\hat{\boldsymbol{x}}_2 + \sigma_{23}\hat{\boldsymbol{x}}_3$$
$$= \frac{\partial\phi}{\partial x_1} = \pm\frac{1}{\sqrt{x_1^2 - a^2}}\Re\{a_1(x_1 + 2i\epsilon a)e^{i\epsilon[(x_1-a)/(x_1+a)]}\boldsymbol{q}_*$$
$$+ a_2(x_1 - 2i\epsilon a)e^{-i\epsilon[(x_1-a)/(x_1+a)]}\overline{\boldsymbol{q}}_* + a_3\boldsymbol{q}_3\} \quad (4.14.218)$$

Then, the stress intensity factors at the right crack edge are found to be

$$K = \frac{\overline{\boldsymbol{q}}_*^T \boldsymbol{M} \boldsymbol{t}_\infty}{\overline{\boldsymbol{q}}_*^T \boldsymbol{M} \boldsymbol{q}_*}\sqrt{\pi a}(1 + 2i\epsilon)\left(\frac{r_0}{2a}\right)^{i\epsilon}, \quad K_3 = \frac{\boldsymbol{q}_3^T \boldsymbol{M} \boldsymbol{t}_\infty}{\boldsymbol{q}_3^T \boldsymbol{M} \boldsymbol{q}_3}\sqrt{\pi a} \quad (4.14.219)$$

Finally, it is remarked that non-oscillatory solutions can appear even if $\boldsymbol{M}$ is not real, but $a_1 = a_2 = 0$, which implies that $\boldsymbol{t}_\infty$ is coaxial with $\boldsymbol{q}_3$, cf. (4.14.213).

4.15 Cracks in isotropic viscoelastic media

General definitions and relations

As its name implies, a viscoelastic material exhibits both viscous and elastic properties. The viscous component implies that the material does not respond reversibly to loading followed by unloading, in general not even after infinite waiting time. The elastic component brings about partial reversibility.

Formally, the mechanical behaviour of a linearly viscoelastic material is very similar to that of a linearly elastic material, for which Hooke's law can be written in the form

$$\sigma_{ij} = \mu(u_{j,i} + u_{i,j}) + \left(\kappa - \frac{2\mu}{3}\right)u_{k,k}\delta_{ji} \quad (4.15.1)$$

where μ is the modulus of rigidity, κ is the modulus of compression and δ_{ji} is Kronecker's delta, that equals unity if $i = j$ and zero if $i \neq j$. The viscoelastic case is obtained after time differentiation and replacement of μ and κ by functions of time: the material does not possess a perfect memory. Thus,

$$\sigma_{ij} = \int_{-\infty}^{\infty}\left\{\mu(t-s)[\dot{u}_{j,i}(s) + \dot{u}_{i,j}(s)] + \left[\kappa(t-s) - \frac{2}{3}\mu(t-s)\right]\dot{u}_{k,k}(s)\delta_{ji}\right\}ds \quad (4.15.2)$$

where a dot denotes differentiation with respect to time. Note that the stress state depends on the previous history of the displacement gradients, so that the contribution from a certain previous time interval is weighted by the memory dependence of the stiffnesses, expressed by $\mu(t)$ and $\kappa(t)$. The resulting convolution integral is therefore called a *hereditary integral*.

The functions $\mu(t)$ and $\kappa(t)$ are obviously zero for $t < 0$; otherwise the material would possess prophetic power. The upper limit of the integral may therefore be replaced by t. A simple example is

$$\begin{aligned} \mu(t) &= [\mu_\infty + (\mu_0 - \mu_\infty)e^{-t/t_0}]U(t) \\ \kappa(t) &= [\kappa_\infty + (\kappa_0 - \kappa_\infty)e^{-t/t_0}]U(t) \end{aligned} \quad (4.15.3)$$

where $U(t)$ is the unit step function, μ_0, μ_∞, κ_0 and κ_∞ are constants and t_0 is the relaxation time. The larger t_0 is, the better is the material's memory.

Energy considerations imply that $\mu(t)$ and $\kappa(t)$ are non-negative, and the resonable assumption that the material's memory is fading or at least not improving with time, implies that $d\mu/dt$ and $d\kappa/dt$ are non-positive. From this, it follows that $\mu(0) = \lim_{t\to+0}\mu(t)$ and $\kappa(0) = \lim_{t\to+0}\kappa(t)$ are positive and that $\mu(0) > \mu(\infty)$, $\kappa_0 > \kappa(\infty)$. A further restriction on $\mu(t)$ and $\kappa(t)$ seems to hold for real materials, namely that $d^2\mu/dt^2 \geq 0$ and $d^2\kappa/dt^2 \geq 0$. These properties are obvious if t is large enough, and has been argued for all $t > 0$ on the basis of a hypothesis concerning the rate by which the memory is fading (see Christensen 1982).

From (4.15.2), it follows that the short-time response will depend mainly on the *short-time moduli* $\mu(0)$ and $\kappa(0)$. If they are finite, then, as follows from (4.15.2), the initial response equals the one in an elastic solid with elastic constants $\mu = \mu(0)$ and $\kappa = \kappa(0)$. The short-time moduli are therefore also called the *glassy moduli*.

The general long-time behaviour of a viscoelastic material is best illustrated by assuming constant stresses. It follows from (4.15.2) that such stresses will eventually produce strain rates approaching finite constants or zero. In the former case, defined by

$$0 < \int_0^t \mu(\tau)d\tau < \infty \quad (4.15.4)$$

the material behaves as a viscous fluid. Such a material is by convention called a *viscoelastic fluid*. Note that it is not synonomous with a viscous fluid: its short-term behaviour is different. Note also that $\int_0^t \kappa(\tau)d\tau$ need not be finite. A divergent integral over κ implies that the rate of volume change approaches zero as $t \to \infty$, which is a requirement for cohesive materials.

If

$$\int_0^t \mu(\tau)d\tau = \infty \text{ and } \int_0^\infty \kappa(\tau)d\tau = \infty \quad (4.15.5)$$

then all strain rates approach zero as $t \to \infty$. By convention such a material is called a *viscoelastic solid*.

A sufficient condition for a material to be a viscoelastic solid is that the *long-time moduli* $\mu(\infty)$ and $\kappa(\infty)$ are non-zero. An example of a viscoelastic solid with zero long-time moduli is one for which $\mu(t)$ and $\kappa(t)$ become proportional to $t^{-\alpha}$, $0 < \alpha < 1$ as $t \to \infty$. All viscoelastic fluids have vanishing long-time moduli. The long-time moduli are also called *rubbery moduli*.

A viscoelastic material subjected to suddenly applied constant stress exhibits *creep*, i.e., increasing strain with time. If, instead, it is subjected to suddenly applied constant strain, it exhibits *relaxation*, i.e., the instantaneously produced stress decreases with increasing time.

There are several works covering the basic theory of viscoelastic materials, for instance the books by Pipkin (1972), Christensen (1982) and by Golden and Graham (1988). Early works on viscoelastic fracture were published by Williams (1965), Knauss (1970a, 1973, 1974), Wnuk and Knauss (1970), Knauss and Dietmann (1970) and Kostrov and Nikitin (1970). Other works, specializing in dynamic crack propagation will be referred to later in the appropriate context.

Viscoelastic models

The most widely used models for viscoelastic materials are the *standard linear solid*, (4.15.3), and the *power law model*, defined by

$$\mu(t) = \mu_p \cdot \left(\frac{t}{t_0}\right)^{-\alpha} U(t), \quad \kappa(t) = \kappa_p \cdot \left(\frac{t}{t_0}\right)^{-\alpha} U(t), \quad 0 < \alpha < 1 \quad (4.15.6)$$

The behaviour of the power law model for short times is hardly representative for real materials, and it is therefore best suited for cases where the time range of interest is not too short.

Generalization of the standard linear solid leads to *spectrum models*, either discrete models, containing several exponential terms, with different time constants like t_0 in 4.15.3, instead of one, or continuous models, in which the sum of exponential terms is replaced by an integral. The time constants need not be the same for $\mu(t)$ as for $\kappa(t)$. The generalizations increase the possibility of accurate simulation of real material behaviour, though at the expense of increased mathematical complexity. Also fractional derivatives and fractional integrals have been used to obtain accurate simulations with relatively few parameters. An overview is given by Rossikhin and Shitikova (1997).

The mathematical complexity can be reduced by chosing $\mu(t)$ and $\kappa(t)$ proportional to each other. The standard linear solid may be specialized to a *Maxwell material*:

$$\mu(t) = \mu_0 e^{-t/t_0} U(t), \quad \kappa(t) = \kappa_0 e^{-t/t_0} U(t) \quad (4.15.7)$$

or to a *Kelvin material*:

$$\mu(t) = \mu_\infty U(t) + \eta_m \delta(t), \quad \kappa(t) = \kappa_\infty U(t) + \eta_k \delta(t) \quad (4.15.8)$$

where $\delta(t)$ is Dirac's delta function, which is obtained by putting the constant in front of the exponential term equal to η/t_0, and then letting $t_0 \to 0$: this yields $\lim_{t_0 \to 0}[\exp(-t/t_0)/t_0] \to \delta(t)$.

Note that the Maxwell material is a viscoelastic fluid and that the Kelvin material, like the power law model, does not give an instantaneous reponse to a suddenly applied load.

The elastic-viscoelastic correspondence principle

Laplace transformation of Eq. (4.15.2) gives:

$$S_{ij} = M(p)(U_{j,i} + U_{i,j}) + \left(K(p) - \frac{2}{3}M(p)\right)U_{k,k}\delta_{ij} \quad (4.15.9)$$

where S_{ij}, U_i, $M(p)$ and $K(p)$ are the Laplace transforms of σ_{ij}, u_i, $\mu(t)$ and $\kappa(t)$, respectively. Note that this equation could have been obtained through Laplace transformation of Hooke's law, (3.2.51), followed by exchange of μ and κ by $M(p)$ and $K(p)$.

A boundary value problem is given by the constitutive equations, the equilibrium conditions and the boundary conditions, and, if the constitutive equations are given in terms of strains rather than stresses, also the compatibility equations. In the so-called first boundary value problems, tractions are prescribed on the boundary, and in the second boundary value problem, displacements are prescribed. In mixed boundary value problems, tractions and displacements are prescribed on different parts of the boundary, either separately or in appropriate combinations.

For viscoelastic mixed boundary value problems, it is assumed that the boundaries on which tractions are prescribed do not change with time and that the same applies to boundaries on which displacements are prescribed. If tractions and displacements can be prescribed without involvement of material parameters, then only the constitutive equations contain such parameters, so that Laplace transformations produce identical sets of equations between an elastic and the corresponding viscoelastic problem if M and K are identified with μ and κ, respectively. This implies that the continued solution procedure, apart from the final Laplace transform inversion, runs in parallel between the elastic and the viscoelastic problem. In particular, the solution of the viscoelastic problem may be found by inverting the Laplace transformed solution of the elastic problem, after substituting μ and κ by $M(p)$ and $K(p)$. This is the *elastic-viscoelastic correspondence principle*. This principle can be extended to certain classes of problems in which the boundaries are time dependent (Graham 1968, Graham and Sabin 1973). This extension is important for problems concerning moving cracks, but certain limitations apply. Thus, the crack front must not retreat and the crack plane displacement in the elastic problem may only depend on the material parameters through an amplitude factor.

The simplest case occurs if $\mu(t)$ and $\kappa(t)$ are proportional to each other. Then, also their Laplace transforms will be proportional to each other, which implies that the Laplace transforms of Poisson's ratio ν or of the ratio $k = c_S/c_P$, introduced on page 79, are constants.

Stationary cracks

Suddenly applied constant load

"Suddenly applied" refers to a comparison between the time of load application and any characteristic time of the viscoelastic processes: the former should be much shorter than the latter. On the other hand, the time of load application must not be so short that it stirs up waves in the body. Inertia effects can thus be neglected, as they have been throughout the current chapter.

The correspondence principle is immediately applicable for stationary cracks. Consider the problem of a traction free crack of length $2a$ in a large viscoelastic plate, subjected to suddenly applied constant remote mode I loading, $\sigma_y = \sigma_y^\infty$, $\sigma_x = \sigma_x^\infty$. Stresses and displacements in the corresponding elastic case are given by (4.6.3)-(4.6.7). One observes that the stresses do not contain any material parameters, and then the correspondence principle tells that the stresses in the viscoelastic plate are the same as in the elastic plate.

The elastic displacements depend on material parameters μ and k^2. In order to apply the correspondence principle, k^2 is written as

$$k^2 = \frac{3\kappa}{3\kappa + 4\mu} \qquad (4.15.10)$$

which relation is given in Appendix A1. The Laplace transform of the viscoelastic solution is obtained by substituting μ and κ in the elastic solution by the Laplace transforms $M(p)$ and $K(p)$ of $\mu(t)$ and $\kappa(t)$.

Consider, for instance, the displacement v, which for the elastic case, (4.6.7), is

$$v = \frac{\sigma_y^\infty}{2\mu} \Im\left[\frac{1}{1-k^2}(z^2-a^2)^{1/2} - \frac{iyz}{(z^2-a^2)^{1/2}}\right] + \frac{\sigma_y^\infty - \sigma_x^\infty}{4\mu} y \qquad (4.15.11)$$

where $z = x+iy$. To avoid complications, $\mu(t)$ and $\kappa(t)$ are assumed, here and in the following, to be proportional to each other, but this is not essential for the procedures. Then, the Laplace transform of v is

$$V = \frac{\sigma_y^\infty}{2M(p)}\left\{\Im\left[\frac{1}{1-k_v^2}(z^2-a^2)^{1/2} - \frac{iyz}{(z^2-a^2)^{1/2}}\right] + \frac{1-m}{4}y\right\} \qquad (4.15.12)$$

where $m = \sigma_x^\infty/\sigma_y^\infty$ and the constant

$$k_v^2 = \frac{3\kappa(t)}{3\kappa(t) + 4\mu(t)} \qquad (4.15.13)$$

Thus, the inversion consists essentially of inversion of $1/M(p)$, i.e., to find a compliance function

$$C(t) = \mathcal{L}_{pt}^{-1}\left(\frac{1}{M(p)}\right) \qquad (4.15.14)$$

This will be performed for a standard linear solid, (4.15.3), and for a power law model, (4.15.6).

1. Consider first the standard linear solid, (4.15.3). Then,

$$M(p) = \mu_0 \frac{p + \frac{\mu_\infty}{\mu_0} \cdot \frac{1}{t_0}}{p + \frac{1}{t_0}}, \qquad \frac{1}{M(p)} = \frac{1}{\mu_0} \cdot \frac{p + \frac{1}{t_0}}{p + \frac{\mu_\infty}{\mu_0} \cdot \frac{1}{t_0}} \qquad (4.15.15)$$

with the inversion

$$C(t) = \mathcal{L}_{pt}^{-1}\left(\frac{1}{M(p)}\right) = \left[\frac{1}{\mu_\infty} - \left(\frac{1}{\mu_\infty} - \frac{1}{\mu_0}\right)e^{-\mu_\infty t/(\mu_0 t_0)}\right]U(t) \qquad (4.15.16)$$

Recall that inertia was neglected. This implies that the whole plate responds instantaneously to changes. Before $t = 0$, the cracked plate is unloaded. At $t = +0$, the remote loads σ_x^∞ and σ_y^∞ are applied. The plate responds immediately with the static response of an elastic plate with constants $\mu = \mu_0$ and $k = k_v$. Thereupon, all displacements grow toward an enlargement by the factor μ_0/μ_∞ from their values at $t = +0$. The instantaneous response over the whole plate might appear unrealistic, but the infinite plate is only a convenient model of a large plate, in which the response delay due to inertia is much smaller than characteristic times of viscoelastic deformations, such as t_0.

For the special case $y = 0$, the upper crack face displacement is

$$v_+ = \frac{\sigma_y^\infty \sqrt{a^2 - x^2}}{2\mu(1 - k^2)} \quad \text{for } |x| < a \tag{4.15.17}$$

for an elastic material. Thus, with $T(t) = \mu_0 C(t)$,

$$v_+ = \frac{\sigma_y^\infty T(t) \sqrt{a^2 - x^2}}{2\mu_0 (1 - k_v^2)} \quad \text{for } |x| < a \tag{4.15.18}$$

for the standard linear solid. The dimensionless time function $T(t)$ expresses the response in comparison to that in a body of elastic material.

Obviously, the inversion is possible for all standard linear solids, even those without proportionality between $\mu(t)$ and $\kappa(t)$ and for which the time constant is not the same (equal to t_0) for $\mu(t)$ and $\kappa(t)$. The Laplace transforms involved will always consist of rational functions. Furthermore, a solution can be found also for problems involving stationary cracks other than a straight crack in a large plate, provided that the solution of the corresponding elastic problem is known.

2. Consider next the power-law model, (4.15.6). Then,

$$M(p) = \mu_p \Gamma(1 - \alpha) p^\alpha t_0^\alpha, \quad \frac{1}{M(p)} = \frac{p^{-\alpha} t_0^{-\alpha}}{\mu_p \Gamma(1 - \alpha)} \tag{4.15.19}$$

with the inversion

$$C(t) = \mathcal{L}_{pt}^{-1}\left(\frac{1}{M(p)}\right) = \frac{(t/t_0)^\alpha U(t)}{\mu_p \Gamma(1-\alpha)\Gamma(1+\alpha)} = \frac{\sin(\pi\alpha)(t/t_0)^\alpha}{\pi \mu_p \alpha} U(t) \tag{4.15.20}$$

In this case, the response, $v(x, y, t)$, increases in absolute value smoothly from zero toward infinity. However, the power law model is not intended for simulations of the behaviour during very short or very long times, only for an intermediate range. What this range is has to be judged for each material from the procedure of fitting the constants α, t_0, μ_p and κ_p to experimental data.

The crack face displacement is found to be

$$v_+ = \frac{\sigma_y^\infty T(t) \sqrt{a^2 - x^2}}{2\mu_p (1 - k_v^2)} \quad \text{for } |x| < a \tag{4.15.21}$$

where $T(t) = \mu_p C(t)$.

For the power law model, proportionality between $\mu(t)$ and $\kappa(t)$ (implying the same exponent α) is essential for convenient Laplace transform inversion. Then, a solution can be found also for problems involving stationary viscoelastic cracks other than a straight crack in a large plate, provided that the solution of the corresponding elastic problem is known.

Suddenly applied constant displacement

The viscoelastic counterpart to the elastic problem of an infinite strip with edges $y = \pm h$, containing a central semi-infinite crack, $x < 0$, $y = 0$, and subjected to constant transversal displacement $\pm\delta$ on its edges, is considered. The elastic problem is treated on pages 127ff. The solution of this mode III problem shows that the displacements do not contain any material parameters, but the stresses do. Thus, the correpondence principle tells that the displacements are the same as in the elastic case. The stress τ_{yz} on $y = 0$, $x > 0$, and the stress intensity factor are given by (4.5.28) and (4.5.30) in the elastic case. From this, they are found for the viscoelastic case simply by substituting μ by $\mu(t)$, giving

$$\tau_{yz} = \frac{\mu(t)\delta}{h\sqrt{1 - \exp\left(\frac{-\pi x}{h}\right)}} \text{ for } x > 0, \quad K_{III} = \frac{\sqrt{2}\mu(t)\delta}{\sqrt{h}} \quad (4.15.22)$$

which shows stress relaxation.

Time dependent crack face loading

Consider a crack, $b < x < c$, $y = 0$, in a large plate, subjected to symmetrical crack face loads $\sigma_y = \sigma_y^0(x,t)$, applied at $t = 0$. Inertia effects are neglected with the same motivation as for suddenly applied constant load and with the additional requirement that the load variations after application are sufficiently slow. The corresponding elastic solution, for crack face load $\sigma_y = \sigma_y^0(x)$, is given by (4.4.43) for the stress intensity factor and by (4.4.44) for the displacement gradient $\partial v_+(x)/\partial x$ of the upper crack face. The expression for the stress intensity factor does not contain material parameters, and thus the correspondence principle implies that it equals

$$K_I = -\frac{1}{\sqrt{\pi a}} \int_b^c \sigma_y^0(\xi,t)\sqrt{\frac{c-\xi}{\xi-b}}\,d\xi \quad (4.15.23)$$

for the viscoelastic plate, i.e., the same expression as for the elastic case, except that the crack face load varies with t. Here, $a = (c-b)/2$.

After Laplace transformation of the elastic solution, (4.4.44), and substitution of $\sigma_y^0(x)$ by the Laplace transform $S_y^0(\xi,p)$ of $\sigma_y^0(x,t)$, the correspondence principle gives the Laplace transform of $\partial v_+/\partial x$ for a viscoelastic material as

$$\frac{dV_+}{dx} = \frac{1}{2\pi(1-k_v^2)M(p)\sqrt{(x-b)(c-x)}}\oint_b^c \frac{S_y^0(\xi,p)\sqrt{(\xi-b)(c-\xi)}}{\xi-x}d\xi \quad (4.15.24)$$

Inversion is accomplished by using the inversion formula (A5.10) for a product of Laplace transforms, which gives

$$\mathcal{L}_{pt}^{-1}\frac{S_y^0(\xi,p)}{M(p)} = \mathcal{L}_{pt}^{-1}\left[\frac{1}{p}\cdot\frac{p}{M(p)}S_y^0(\xi,p)\right] = \int_0^t \dot{C}(t-\tau)\sigma_y^0(\xi,\tau)d\tau \quad (4.15.25)$$

where a dot denotes differentiation with respect to t.

Consider as an example $\sigma_y^0(x,t) = -p_0\exp(-t/\theta)U(t)$, where p_0 and θ are positive constants, and assume a standard linear solid, (4.15.3). Then, the right member of

(4.15.25) equals $-p_0 T(t,\theta)/\mu_0$, where

$$T(t,\theta) = \frac{\mu_0}{\mu_0 t_0 - \mu_\infty \theta}\left[(t_0-\theta)e^{-t/\theta} + (1-\frac{\mu_\infty}{\mu_0})\theta e^{-\mu_\infty t/(\mu_0 t_0)}\right]U(t) \quad (4.15.26)$$

Thus, from (4.15.24):

$$\frac{\partial v_+}{\partial x} = -\frac{p_0 T(t,\theta)[x-(b+c)/2]}{2(1-k_v^2)\mu_0 \sqrt{(x-b)(c-x)}} \quad (4.15.27)$$

The time function $T(t,\theta)$ equals unity for $t = +0$. If $\theta > \mu_0 t_0/(\mu_0 - \mu_\infty)$ it is initially increasing and thereafter decreasing, otherwise it is decreasing for all finite t. The special case $\theta \to \infty$ corresponds to a suddenly applied constant load.

Slowly propagating crack

Consider a propagating crack, $b(t) < x < c(t)$, $y = 0$, in a large plate, subjected to crack face loading, $\sigma_y = \sigma_y^0(x,t)$. The crack is assumed to propagate so slowly that inertia effects may be neglected. The functions $b(t)$ and $c(t)$ are non-increasing and non-decreasing, respectively. The problem consists of finding the gradient of the upper crack face displacement $\partial v_+/\partial x$.

The problem may be solved by superposition of elementary solutions. Thus, the crack face load is considered as a superposition of impulsive loads, $\sigma_y = I_y^0(x,\tau)\delta(t-\tau)$, where $I_y^0(x,\tau) = \sigma_y^0(x,\tau)d\tau$. The neglect of inertia effects in spite of the use of impulsive loads is justified by the same motivation as for suddenly applied loads. The elementary solutions are obtained from the response to a suddenly applied crack face load, $\sigma_y = q(x)U(t)$. This response is given by (4.15.24) in combination with (4.15.14) as

$$\frac{\partial v_+}{\partial x} = \frac{C(t)}{2\pi(1-k^2)\sqrt{(x-b)(c-x)}} \oint_b^c \frac{q(\xi)\sqrt{(\xi-b)(c-\xi)}}{\xi - x}d\xi \quad (4.15.28)$$

for $b < x < c$, whereas $v = 0$ for $x < b$, $x > c$. Then, the response to the impulsive crack face load

$$\sigma_y = I_y^0(x,\tau)\delta(t-\tau) = \frac{d}{dt}[I_y^0(x,\tau)U(t-\tau)] \quad (4.15.29)$$

is

$$\frac{\partial v_+}{\partial x} = \frac{\dot{C}(t-\tau)}{2\pi(1-k^2)\sqrt{(x-b)(c-x)}} \oint_b^c \frac{I_y^0(\xi,\tau)\sqrt{(\xi-b)(c-\xi)}}{\xi - x}d\xi \quad (4.15.30)$$

The solution of the problem is now obtained by choosing $I_y^0(\xi,\tau) = \sigma_y^0(\xi,\tau)d\tau$ and integrating:

$$\frac{\partial v_+}{\partial x} = \int_{t_*}^t \frac{\dot{C}(t-\tau)}{2\pi(1-k^2)\sqrt{[x-b(\tau)][c(\tau)-x]}}$$
$$\times \oint_{b(\tau)}^{c(\tau)} \frac{\sigma_y^0(\xi,\tau)\sqrt{[\xi-b(\tau)][c(\tau)-\xi]}}{\xi - x}d\xi\, d\tau \quad (4.15.31)$$

where, denoting the inverse functions of $b(t)$ and $c(t)$ by $b^{-1}(x)$ and $c^{-1}(x)$, respectively,

$$t_* = \begin{cases} b^{-1}(x) & \text{if } b(t) < x < b(0) \\ 0 & \text{if } b(0) < x < c(0) \\ c^{-1}(x) & \text{if } c(0) < x < c(t) \end{cases} \qquad (4.15.32)$$

This ensures that no values of x outside and all values of x inside the current crack position are considered.

Note that an impulsive load applied at $t = \tau$ produces a contribution to the normal displacement on $y = 0$ that is zero outside the crack position $b(\tau) < x < c(\tau)$ for all following times, i.e., for $t > \tau$. Note also that an expression for the upper crack face displacement $v_+(x,t)$ can be obtained by integration over x in (4.15.31).

The stress intensity factor can be taken directly from (4.15.23), because this expression does not contain any material parameter, and the fact that the crack is not stationary does not have any significance when inertia effects are neglected. Thus,

$$K_I = -\frac{1}{\sqrt{\pi a(t)}} \int_{b(t)}^{c(t)} \sigma_y^0(\xi, t) \sqrt{\frac{c(t) - \xi}{\xi - b(t)}} \, d\xi \qquad (4.15.33)$$

where $a(t) = [c(t) - b(t)]/2$.

Crack meeting an interface
The stress singularity for a crack, moving slowly toward an interface between two viscoelastic media, and the related problem of dislocation interaction with such an interface, has been considered by Atkinson and Bourne (1989, 1990). They found that the stress intensity factor, just after the crack has reached the interface, evolved in a quite complicated way. However, both for short and for long times, the stress intensity factor coincided with the one for a crack meeting an interface between two elastic materials, provided that the viscoelastic moduli were replaced by the corresponding short-time and long-time moduli, respectively. A shortened version of these results is given in an overview paper on theoretical aspects of fracture mechanics by Atkinson and Craster (1995).

Energy flow into a cohesive region

During crack growth under small scale yielding conditions in an elastic-plastic solid, energy is dissipated only in the process region and in the plastic region. In a viscoelastic material, energy is dissipated everywhere outside a growing crack. This dissipation is not essential for crack growth, except for the part that reaches the process region. This part is needed for material separation. It can, in principle, be estimated by using a Barenblatt model of a cohesive region.

The appropriate structure and physical data for a Barenblatt region in a viscoelastic material are poorly known, but it appears appropriate to make some simple assumptions. Thus, the extension r_p of the region is assumed to be independent of the crack velocity V. Recall that only crack velocities that are sufficiently small to justify neglect of inertia effects are considered in the present section. Further, the normal stress across the Barenblatt region is assumed to be dependent only on the distance $-X$ to the front edge of the region and thus not on the crack velocity. Thus, it may be given as a function $\sigma_y^0(X)$, that decreases with X and reaches zero at $X = -r_p$.

Assume that the crack length is much larger than r_p. Then, the energy flux into the cohesive region may be determined by considering a steadily propagating semi-infinite crack, $x < Vt$, $y = 0$, subjected to a stress intensity factor K_I. The Galilean† transformation, $X = x - Vt$, $Y = y$, is introduced. Then, because inertia effects are neglected, the governing equations, expressed in X and Y are the same as the governing equations expressed in x and y in the case of a stationary crack.

The front of the cohesive region reaches $x = 0$ at $t = 0$, and its rear end leaves $x = 0$ at $t = r_p/V$. Consider the material particles at $x = y = 0$, i.e., $X = -Vt$. They are subjected to the stress $\sigma_y^0(-Vt)$. During a time interval dt, this stress performs the work

$$\sigma_y^0(-Vt)\frac{\partial(v_+ - v_-)}{\partial t}dt = 2\sigma_y^0(-Vt)\frac{\partial v_+}{\partial t}dt = -2V\sigma_y^0(-Vt)\frac{\partial v_+}{\partial x}dt \quad (4.15.34)$$

where the last equality follows from the condition of steady state. Integration gives the energy flux into the cohesive region:

$$\mathcal{G} = -2V\int_0^{r_p/V}\sigma_y^0(-Vt)\frac{\partial v_+}{\partial x}dt \quad (4.15.35)$$

The displacement gradient $\partial v_+/\partial x$ in the elastic case may be taken from (4.4.62), which, after change of x to X and of the lower integration limit to $-r_p$, reads

$$\frac{\partial v_+}{\partial x} = -\frac{\sqrt{-X}}{2\pi(1-k^2)\mu}\oint_{-r_p}^0 \frac{\sigma_y^0(\xi)}{\sqrt{-\xi}(\xi - X)}d\xi \text{ for } X < 0 \quad (4.15.36)$$

with the condition for smooth closing telling that

$$K_I = \sqrt{\frac{2}{\pi}}\int_{-r_p}^0 \frac{\sigma_y^0(\xi)}{\sqrt{-\xi}}d\xi \quad (4.15.37)$$

Then, the correspondence principle gives the viscoelastic response

$$\frac{\partial v_+}{\partial x} = -\frac{C(t)\sqrt{-X}}{2\pi(1-k_v^2)}\oint_{-r_p}^0 \frac{\sigma_y^0(\xi)}{\sqrt{-\xi}(\xi - X)}d\xi \text{ for } X < 0 \quad (4.15.38)$$

whereupon the energy flux into the cohesive region is found by using (4.15.35) and putting $X = -Vt = s$. The result is

$$\mathcal{G} = \frac{1}{\pi(1-k_v^2)}\int_{-r_p}^0 C(-s/V)\sigma_y^0(s)\sqrt{-s}\oint_{-r_p}^0 \frac{\sigma_y^0(\xi)}{\sqrt{-\xi}(\xi - s)}d\xi\, ds \quad (4.15.39)$$

The dependence of the energy flux on V is apparent from the behaviour of $C(-s/V)$. It follows that $\mathcal{G}$ decreases with increasing V, because $C(t)$ is an increasing function. This is, of course, the expected behaviour in view of the assumptions made about the model of the cohesive region, i.e., the stress distribution remains the same, whereas the displacements increase with time.

If V is so high that r_p/V is very small compared to a characteristic time of the

---

† In the present context, 'Galilean' refers to Galileo Galilei, 1564-1642, Italian scientist and philosopher.

material, like t_0 for a standard linear solid, then $C(-s/V)$ can be substituted by $C(0)$ in (4.15.39). Then,

$$\mathcal{G} = \frac{C(0)}{\pi(1-k_v^2)} I \qquad (4.15.40)$$

where

$$I = \int_{-r_p}^{0} \oint_{-r_p}^{0} \frac{\sigma_y^0(s)\sigma_y^0(\xi)}{\xi - s} \cdot \frac{\sqrt{-s}}{\sqrt{-\xi}} d\xi \, ds \qquad (4.15.41)$$

Swapping integration variables ξ and s, followed by addition of the two expressions for I, gives

$$2I = \int_{-r_p}^{0} \int_{-r_p}^{0} \frac{\sigma_y^0(s)\sigma_y^0(\xi)}{\xi - s} \cdot \left[\frac{\sqrt{-s}}{\sqrt{-\xi}} - \frac{\sqrt{-\xi}}{\sqrt{-s}}\right] d\xi \, ds$$

$$= \left[\int_{-r_p}^{0} \frac{\sigma_y^0(\xi)}{\sqrt{-\xi}} d\xi\right]^2 = \frac{\pi}{2} K_I^2 \qquad (4.15.42)$$

so that

$$\mathcal{G} = \frac{C(0)}{4(1-k_v^2)} K_I^2 \qquad (4.15.43)$$

i.e. the same expression as in the elastic case, (3.5.14), if $C(0)$ is replaced by $1/\mu$ and k_v by k. This is expected, because $C(0)$ is the inverse of the glassy modulus.

If V is low enough, then $C(-s/V)$ can be substituted by $C(\infty)$, the inverse of the rubbery modulus. The energy flux into the cohesive region is then given by (4.15.40), after substituting $C(0)$ by $C(\infty)$.

If r_p/V is of the same order of magnitude as a characteristic time of the material, then, as indicated by the factor $C(-s/V)$ in the integrand of (4.15.39), the energy flux depends on both r_p and V.

The assumption that the normal stress $\sigma_y^0(X)$ across the Barenblatt region, and by implication also the length r_p, are independent of the crack velocity, is not obvious from a physical point of view. It leads to a solution for each velocity that is sufficiently low to justify neglect of inertia effects. It would appear reasonable to assume that the propagation velocity increases with increasing load, but the load is manifested through the velocity-independent stress intensity factor (4.15.37), so it remains the same for all crack velocities. It is simply a critical stress intensity factor for slow crack propagtion, but the crack velocity cannot be determined.

Assume now that the cohesive stress is dependent on the crack velocity, so that $\sigma_y = \sigma_y^0(X, V)$ for $-r_p(V) < X < 0$. The derivation leading to (4.15.39) will still be valid, after substituting $\sigma_y^0(\cdot)$ by $\sigma_y^0(\cdot, V)$ and r_p by $r_p(V)$. But now the stress intensity factor is

$$K_I = \sqrt{\frac{2}{\pi}} \int_{-r_p(V)}^{0} \frac{\sigma_y^0(\xi, V)}{\sqrt{-\xi}} d\xi \qquad (4.15.44)$$

and thus dependent on V. If it increases with V, then increased load implies increased crack velocity, and steady state propagation at any desired velocity can be obtained within the range that allows inertia effects to be neglected.

5
Elastoplastic Stresses and Strains

5.1 Introduction

The necessity of considering plastic deformations in analyses of cracks is obvious in cases of large scale yielding. However, even at small scale yielding, an understanding of crack behaviour often requires examination of the processes in the plastic region. This is the case, for instance, when stable crack growth is considered or when the role of the process region is investigated, particularly the implications of different embedments in the plastic region.

There exist very few analytical results regarding plastic deformations in cracked bodies. Such results concern mainly small scale yielding or a close neighbourhood of the crack edge. The latter are often too idealized to correspond to real cases, because they are generally valid only in a region which in reality would be occupied by the process region to a significant extent. The difficulties are less pronounced for mode III than for modes I and II. The classical solution of an elastic-plastic mode III crack problem was given by Hult and McClintock (1956). The problem concerned mode III, small scale yielding and perfect plasticity. The main difficulty with this and similar problems is to find the location of the elastic-plastic boundary. In this respect the problem is similar to so-called Stefan problems of discontinuous phase changes, for instance melting of a metal.

Analytical solutions also exist for mode III at large scale yielding, both for perfect plasticity (Rice 1966) and for strain hardening (Rice 1967). A solution for growing mode III cracks was given by Chitaley and McClintock (1971); results for the strain on the symmetry plane had been given earlier by Hult (1957) and by McClintock (1963).

For modes I and II, there exists a number of analytical solutions for the plastic field in the vicinity of a crack edge, including, as for mode III, solutions for slowly moving cracks. For a stationary mode I crack, solutions were given by Cherepanov (1967). Hutchinson (1968a), Rice (1967, 1968b) and Rice and Rosengren (1968). A solution for a stationary mode II crack was given by Hutchinson (1968b). An early, but still very useful overview of elastic-plastic cracks was presented by McClintock (1968b). The pioneering work on slowly moving mode I and II cracks was made by Slepyan (1974). Later work on slowly moving mode I cracks includes analyses by Gao (1980), Rice at al. (1980), Drugan et al. (1982), Drugan and Chen (1989) and Chen and Drugan (1991).

All available analytical elastoplastic solutions assume a point-size process region.

5.2 Basic relations

Plastic deformations occur in accordance with some flow rule,

$$\mathrm{d}\epsilon_{ij}^p = \mathrm{d}\lambda \frac{\partial g(\sigma_{ij})}{\partial \sigma_{ij}} \quad (5.2.1)$$

where $\mathrm{d}\lambda$ is a non-negative multiplier, $g(\sigma_{ij})$ is the *plastic potential* and $\mathrm{d}\epsilon_{ij}^p$ the increment of the *plastic strain*,

$$\epsilon_{ij}^p = \epsilon_{ij} - \epsilon_{ij}^e \quad (5.2.2)$$

where ϵ_{ij}^e is the elastic strain. The multiplier $\mathrm{d}\lambda = 0$ when the state of stress is inside the *yield surface*,

$$f = f(\sigma_{ij}) = 0 \quad (5.2.3)$$

and then the strain increments during loading are purely elastic. For plastic flow to take place, the *yield condition* (5.2.3) must be satisfied, i.e. the state of stress is represented by a point on the yield surface. This surface may be graphically represented in the 3-dimensional principal stress space $(\sigma_1, \sigma_2, \sigma_3)$. In general, both the plastic potential $g(\sigma_{ij})$ and the *yield function* $f(\sigma_{ij})$ also contain internal variables, controlling the strain hardening, and sometimes other variables, controlling the current state of the material, for instance the volume fraction of voids. The yield function always contains material specific parameters, such as the yield strength σ_Y.

The *normality rule*, which may be derived from a postulate by Drucker (1951) and appears to hold for most elastic-plastic materials, states that the plastic strain increment must be directed along the outward normal to the yield surface, assuming directions of $\mathrm{d}\epsilon_{ij}^p$ in a geometrical representation, for instance in 6-dimensional space, to coincide with those of σ_{ij}. In mathematical terms this implies that the strain increment is proportional to the gradient of the yield function, i.e.

$$\mathrm{d}\epsilon_{ij}^p = \mathrm{d}\lambda \frac{\partial f(\sigma_{ij})}{\partial \sigma_{ij}} \quad (5.2.4)$$

which is the *flow rule*. Obviously, assuming $\mathrm{d}\lambda$ to be the same in (5.2.4) as in (5.2.1), the functions g and f may be taken to coincide; (5.2.4) is then a so-called *associated flow rule*, i.e., the flow rule is associated with the yield condition.

The normality rule imposes some restrictions on $f(\sigma_{ij})$. First, it must be differentiable (even though some corners on the yield surface may be allowed). Second, the yield surface must be non-concave outwards. Third, $f < 0$ inside the yield surface. The normality rule also ensures that plastic deformation is dissipative:

$$\sigma_{ij} \mathrm{d}\epsilon_{ij}^p > 0 \text{ if } \mathrm{d}\epsilon_{ij}^p \neq 0 \quad (5.2.5)$$

For analytical work, a *small strain theory* is most common, as in previously treated elastic cases, even for strains arbitrarily close to a crack edge. This does not exclude working with large strains; it is a simplification based on neglect of the influence of strains when defining stresses, with the consequence that the equilibrium equations are only approximatively satisfied for finite strains. Moreover, simple yield functions of the type

$$f(\sigma_{ij}) = \sigma_{eff}(\sigma_{ij}) - \sigma_{ref}(\epsilon^p_{eff}) \qquad (5.2.6)$$

are often chosen, sometimes in the form $f = \sigma^2_{eff} - \sigma^2_{ref}$, if this is more convenient. σ_{eff} is the *effective stress*, σ_{ref} is a *reference stress* and ϵ^p_{eff} is the *effective plastic strain*.

A very popular yield function, capable of describing plastic flow in a material with voids, is the Gurson condition (Gurson 1977),

$$f(\sigma_{ij}, v_f) = \left[\frac{\sigma_{eff}(v_f)}{\sigma_{eff}(0)}\right]^2 + 2q_1 v_f \cosh\left(\frac{q_2 \sigma_{kk}}{2\sigma_{eff}(0)}\right) - q_1^2 v_f^2 - 1 \qquad (5.2.7)$$

where v_f is the void volume fraction, $\sigma_{eff}(v_f)$ is the macroscopic effective stress, i.e. the stress over a volume containing many voids, and $\sigma_{eff}(0)$ is the current effective stress in the matrix, acting as a reference stress. q_1 and q_2 are parameters introduced by Tvergaard (1981) to give a good fit with experimental results for the material.

The first term of the right member of (5.2.6) depends on the current stress state, but not necessarily on the material, and the second term depends on the material and on the current strain state, but not on the stress state, because there is no unique relation between ϵ^p_{eff} and σ_{ij}. The function $\sigma_{ref}(\epsilon^p_{eff})$ is usually, but not necessarily, taken from a tensile test, i.e. a uniaxial stress state, and then the argument ϵ^p_{eff} is the plastic strain in the direction of tension and σ_{ref} is the uniaxial stress. Then, $\sigma_{ref}(0) = \sigma_Y$, where σ_Y is the yield stress. The result from a test for a reference stress state may be used for extrapolation to general stress states by assuming that the effect of the stress tensor on the plastic deformation may be described by means of a scalar stress quantity, the effective stress. Similarly, the state of plastic deformation, it is assumed, may be described by means of one scalar quantity, the effective strain. The relation between effective stress and effective strain us then taken from the stress-strain relation recorded for the reference state.

Insertion of (5.2.6) into the flow rule (5.2.4) gives:

$$\boxed{d\epsilon^p_{ij} = d\lambda \frac{\partial \sigma_{eff}(\sigma_{ij})}{\partial \sigma_{ij}}} \qquad (5.2.8)$$

For a uniaxial stress state, $\sigma_{ij} = 0$ if $ij \neq 11$, the effective stress is defined as σ_{11} and the effective plastic strain increment, $d\epsilon^p_{eff}$, is defined as $d\epsilon^p_{11}$. This implies that

$$d\epsilon^p_{eff} = d\lambda \qquad (5.2.9)$$

But

$$d\epsilon^p_{ij} d\epsilon^p_{ij} = (d\lambda)^2 \frac{\partial \sigma_{eff}}{\partial \sigma_{ij}} \cdot \frac{\partial \sigma_{eff}}{\partial \sigma_{ij}} \qquad (5.2.10)$$

so that

$$d\epsilon^p_{eff} = \sqrt{\frac{d\epsilon^p_{ij} d\epsilon^p_{ij}}{(\partial \sigma_{eff}/\partial \sigma_{kl}) \cdot (\partial \sigma_{eff}/\partial \sigma_{kl})}} \qquad (5.2.11)$$

It is then logical to put

$$\epsilon^p_{eff} = \int d\epsilon^p_{eff} \qquad (5.2.12)$$

where the integration shall be extended over the whole plastic deformation history.

For an isotropic material, $\sigma_{eff}(\sigma_{ij})$ must be a function of the stress invariants, which may be taken to be:

$$J_1 = \sigma_{ii}, \quad J_2 = \frac{3}{2} s_{ij} s_{ij}, \quad J_3 = \det(s_{ij}) \qquad (5.2.13)$$

where s_{ij} is the *deviatoric stress*, i.e. the stress after subtraction of the hydrostatic stress:

$$s_{ij} = \sigma_{ij} - \frac{1}{3} \sigma_{kk} \delta_{ij} \qquad (5.2.14)$$

Because a superposed hydrostatic stress does not seem to cause plastic flow in common isotropic elastic-plastic materials, the effective stress is not influenced by J_1. The simplest case is the *Huber-von Mises* material (Huber 1904, von Mises 1913, 1928), for which the effective stress is a function of J_2, only, and is taken as

$$\sigma_{eff}(\sigma_{ij}) = \sqrt{J_2} = \sqrt{\frac{3}{2} s_{ij} s_{ij}} \qquad (5.2.15)$$

The factor $3/2$ is explained by the satisfaction of the special case of uniaxial tension, $\sigma_{11} = \sigma_{eff}, \sigma_{22} = \sigma_{33} = 0$, which gives $s_{11} = 2\sigma_{eff}/3, s_{22} = s_{33} = -\sigma_{eff}/3$. Because

$$\frac{\partial \sigma_{eff}}{\partial \sigma_{ij}} = \frac{3 s_{ij}}{2 \sigma_{eff}} \qquad (5.2.16)$$

the effective plastic strain increment for a Huber-von Mises material, according to (5.2.11), is

$$d\epsilon^p_{eff} = \sqrt{\frac{2}{3} d\epsilon^p_{ij} d\epsilon^p_{ij}} \qquad (5.2.17)$$

The flow rule (5.2.4), in association with the Huber-von Mises yield condition, takes the form:

$$d\epsilon^p_{ij} = \frac{3 d\epsilon^p_{eff}}{2 \sigma_{eff}} s_{ij} = \frac{3 d\epsilon^p_{eff}}{2 \sigma_{ref}(\epsilon^p_{eff})} s_{ij} \qquad (5.2.18)$$

This is the *Prandtl-Reuss flow rule*, sometimes also referred to as the J_2 *flow theory*.

For a *Tresca* material, the effective stress is usually expressed by means of the principal stresses σ_1, σ_2 and σ_3:

$$\sigma_{eff}(\sigma_{ij}) = \max\{|\sigma_1 - \sigma_3|, |\sigma_3 - \sigma_2|, |\sigma_2 - \sigma_1|\} \qquad (5.2.19)$$

Note that a superposed hydrostatic stress state, $\sigma_1 = \sigma_2 = \sigma_3$ does not change σ_{eff}. Assume that all principal stresses are unequal, and that $\sigma_1 > \sigma_2 > \sigma_3$. Then,

$$\frac{\partial \sigma_{eff}}{\partial \sigma_{ij}} = \begin{cases} 1 & \text{for } \sigma_{ij} = \sigma_1 \\ 0 & \text{for } \sigma_{ij} = \sigma_2 \\ -1 & \text{for } \sigma_{ij} = \sigma_3 \end{cases} \quad (5.2.20)$$

and thus, according to (5.2.11):

$$d\epsilon^p_{eff} = \sqrt{\frac{1}{2}d\epsilon^p_{ij}d\epsilon^p_{ij}} \quad (5.2.21)$$

The flow rule (5.2.4) becomes

$$\begin{cases} d\epsilon^p_1 = d\epsilon^p_{eff} \\ d\epsilon^p_2 = 0 \\ d\epsilon^p_3 = -d\epsilon^p_{eff} \end{cases} \quad (5.2.22)$$

where $d\epsilon^p_1$, $d\epsilon^p_2$ and $d\epsilon^p_3$ are the plastic strain increments in the directions of the principal stresses σ_1, σ_2 and σ_3, respectively.

It is observed from (5.2.18) for a Huber-von Mises material, and from (5.2.22) for a Tresca material, that there is no volume change, $d\epsilon^p_{ii}$, associated with plastic straining. This result agrees with the apparent fact that a hydrostatic pressure only causes elastic deformation. In view of (5.2.1), the same holds in general, if the plastic potential $g(\sigma_{ij})$ is independent of a superposed hydrostatic stress. Conversely, the condition of *plastic incompressibility* (neither volume decrease nor increase) restricts plastic potentials to be independent of a superposed hydrostatic stress. Use of this condition and the flow rule shows that the plastic strain increments are zero for a Tresca material if two principal stresses are equal, and for both a Tresca and a Huber-von Mises material if all three principal stresses are equal. Note that there are material models which do not require plastic incompressibility – one example is the Gurson model, (5.2.7).

The flow rule (5.2.8), together with expressions of $\sigma_{eff}(\sigma_{ij})$ of the type (5.2.15) or (5.2.19), is capable of handling *isotropic hardening* (the yield surface grows, but retains the same symmetry axis), but not *kinematic hardening* (the yield surface is translated in parallel with its original axis). Kinematic hardening, which must be considered when the Bauschinger effect is important, may be incorporated by introducing an internal variable, the *back stress* α_{ij}, which, for a Huber-von Mises material is to be subtracted from s_{ij}. The back stress has to be "updated" according to some *evolution law*, just as the effective plastic strain ϵ^p_{eff} is continually updated, according to the evolution law (5.2.12). In the analytical studies to follow, kinematic hardening is not considered.

At linear elasticity, the elastic strains obey Hooke's law, so that for an isotropic material

$$d\epsilon_{ij} = d\epsilon^e_{ij} + d\epsilon^p_{ij} = \frac{1}{2\mu}\left(d\sigma_{ij} - \frac{\nu}{1+\nu}d\sigma_{kk}\delta_{ij}\right) + d\epsilon^p_{ij} \quad (5.2.23)$$

In the present section, elastic-plastic crack problems are studied, assuming, for simplicity, that the process region is infinitesimally small, i.e., the whole body outside the crack edge may be treated as a continuum. This implies that the constitutive equations must be extrapolated to infinite strains, just as Hooke's law in the preceding chapter was assumed to be valid regardless of the strain magnitude. The extrapolation may or may not also involve infinite stresses, and although it may be chosen in different ways, one would generally choose the most convenient one. Suppose, for instance, that the result of a tensile test is

$$\sigma_{11} = \begin{cases} E\epsilon_{11} & \text{for } \epsilon_{11} \leq \sigma_Y/E \\ \sigma_Y & \text{for } \sigma_Y/E < \epsilon_{11} < \epsilon_M \end{cases} \tag{5.2.24}$$

where ϵ_M is the maximum plastic strain at which continuum behaviour has been found to prevail in tests. Then†,

$$\sigma_{ref}(\epsilon^p_{eff}) = \sigma_Y \text{ for } 0 < \epsilon^p_{eff} < \epsilon_M - \sigma_Y/E \tag{5.2.25}$$

Obviously, it is convenient to perform the desired extrapolation so that the validity of (5.2.25) is extended to $\epsilon^p_{eff} = \infty$. This model, perfect plasticity, plays a very important part. Also other analytical expressions, particularly piece-wise linear relations and power-law strain hardening, both of which include as special cases linear elasticity and perfect plasticity are very important. When elastic deformations play an insignificant role, simplifications may be obtained by combining perfect plasticity with rigidity in the elastic region or assuming power-law strain hardening for all strains.

The simplest piece-wise linear relation is

$$\sigma_{11} = \begin{cases} E\epsilon_{11} & \text{for } \epsilon_{11} \leq \sigma_Y/E \\ \sigma_Y + E_T(\epsilon_{11} - \sigma_Y/E) & \text{for } \sigma_Y/E < \epsilon_{11} \end{cases} \tag{5.2.26}$$

where E_T is the so-called *tangent modulus*. The reference stress is related to the effective strain by the relation:

$$\sigma_{ref}(\epsilon^p_{eff}) = \sigma_Y + E_T \epsilon^p_{eff} \text{ for } \epsilon^p_{eff} > 0 \tag{5.2.27}$$

For a power-law hardening material,

$$\sigma_{11} = \begin{cases} E\epsilon_{11} & \text{for } \epsilon_{11} \leq \sigma_Y/E \\ \sigma_Y \left(\dfrac{E\epsilon_{11}}{\sigma_Y}\right)^N & \text{for } \sigma_Y/E < \epsilon_{11} \end{cases} \tag{5.2.28}$$

where $0 \leq N < 1$. The relation between σ_{ref} and ϵ^p_{eff} may be given implicitly:

$$\epsilon^p_{eff} = \frac{\sigma_Y}{E}\left(\frac{\sigma_{ref}}{\sigma_Y}\right)^{1/N} - \frac{\sigma_{ref}}{E} \text{ for } \epsilon^p_{eff} > 0 \tag{5.2.29}$$

Another material model is the Ramberg-Osgood relation, particularly useful when elastic strains are small compared to plastic strains. For uniaxial stress it takes the form

$$\epsilon_{11} = \epsilon_0 \frac{\sigma_{11}}{\sigma_0} + \alpha\epsilon_0 \left(\frac{\sigma_{11}}{\sigma_0}\right)^n \tag{5.2.30}$$

where $n > 1$. The reference stress and strain, here denoted by σ_0 and ϵ_0, respectively, and α are material constants. To fit materials having a stress-strain curve that is essentially linear below the yield point and possesses a pronounced "knee" at the yield point, the constant α should be small compared to unity. The yield stress is then

---

† In accordance with experience, it is here assumed that the strain increments are purely plastic for strains larger than σ_Y/E.

about equal to σ_0. For materials with a more smoothly bending stress-strain curve, the best fit may be found for larger values of α, and the yield stress is usually significantly smaller than σ_0. Thus, in general, σ_0, and consequently also ϵ_0, do not possess any direct physical significance.

For materials with sufficiently small α, the plastic part of the strains may be identified with the last term in (5.2.30), whereas the elastic part corresponds to the linear term. Generalization to multiaxial stress, according to J_2 flow theory, then yields

$$\frac{\epsilon_{ij}^p}{\epsilon_0} = \frac{3}{2}\alpha\left(\frac{\sigma_{eff}}{\sigma_0}\right)^{n-1}\frac{s_{ij}}{\sigma_0} \tag{5.2.31}$$

Differentiation gives

$$\frac{d\epsilon_{ij}^p}{\epsilon_0} = \frac{3}{2}(n-1)\alpha\left(\frac{\sigma_{eff}}{\sigma_0}\right)^{n-2}\frac{s_{ij}}{\sigma_0}\cdot\frac{d\sigma_{eff}}{\sigma_0} + \frac{3}{2}\alpha\left(\frac{\sigma_{eff}}{\sigma_0}\right)^{n-1}\frac{ds_{ij}}{\sigma_0} \tag{5.2.32}$$

From the sequence

$$\sigma_{eff}^2 = \frac{3}{2}s_{ij}s_{ij} \Rightarrow 2\sigma_{eff}^2 d\sigma_{eff} = 3\sigma_{eff}s_{ij}ds_{ij} \Rightarrow s_{ij}d\sigma_{eff} = \sigma_{eff}ds_{ij} \tag{5.2.33}$$

it follows that

$$\frac{d\epsilon_{ij}}{\epsilon_0} = (1+\nu)\frac{d\sigma_{ij}}{\sigma_0} - \nu\frac{d\sigma_{kk}}{\sigma_0}\delta_{ij} + \frac{3}{2}n\alpha\left(\frac{\sigma_{eff}}{\sigma_0}\right)^{n-2}\frac{s_{ij}}{\sigma_0}\cdot\frac{d\sigma_{eff}}{\sigma_0} \tag{5.2.34}$$

where the expression for $d\epsilon_{ij}^e$ from Hooke's law is added to $d\epsilon_{ij}^p$.

For mode III and other cases where the stress state can be expressed in terms of shear stresses, only, it is usually convenient to relate effective stresses and strains to pure shear rather than to uniaxial stress. The yield condition is therefore written in the form:

$$\tau_{eff}(\tau_{ij}) - \tau_{ref}(\gamma_{eff}^p) = 0 \tag{5.2.35}$$

where $\tau_{ref}(\gamma_{eff}^p)$ may be taken from a uniaxial tensile test, $\sigma_{11} \neq 0$, as the relation between τ_{12} and γ_{12}, or directly from a torsion test. By analogy with (5.2.12), the effective plastic shear strain

$$\gamma_{eff}^p = \int d\gamma_{eff}^p \tag{5.2.36}$$

where the integration shall be extended over the whole plastic deformation history. If all stress components other than τ_{xz} and τ_{yz} vanish, $\tau_{eff} = \sqrt{\tau_{xz}^2 + \tau_{yz}^2}$ and the flow rule (5.2.8) may be written in the form:

$$\frac{d\gamma_{xz}^p}{\tau_{xz}} = \frac{d\gamma_{yz}^p}{\tau_{yz}} = \frac{d\gamma_{eff}^p}{\tau_{ref}(\gamma_{eff}^p)} \tag{5.2.37}$$

For a Huber-von Mises material,

$$\sigma_{eff} = \sqrt{\frac{3}{2}s_{ij}s_{ij}} = \sqrt{3}\tau_{eff} \tag{5.2.38}$$

which may be derived by, for instance, specializing to a case with $\sigma_{12} = \sigma_{21}$ as the only non-vanishing components of the stress tensor.

For a Tresca material,

$$\sigma_{eff} = 2\tau_{eff} \tag{5.2.39}$$

which follows from (5.2.19).

In mode III, the two materials obviously differ only as regards τ_{ref}; a perfectly plastic material, for instance, yields $\tau_{ref} = \sigma_Y/\sqrt{3}$ for a Huber-von Mises material and $\tau_{ref} = \sigma_Y/2$ for a Tresca material.

The flow rule (5.2.1), as well as its particular expressions (5.2.4), (5.2.8), (5.2.18) and (5.2.22), contains strain increments rather than strains. In general, therefore, different paths of loading produce different strains, even if they start and end at the same states of stress. Plastic deformation thus depends on the history of loading. However, in some applications, *proportional* (or *radial*) *loading* prevails during the loading history, i.e. all components of the stress tensor increase in the same proportion at each point, and then the simpler *total strain* (or *deformation*) *theory* may be used. A certain stress state can be reached along only one stress path (one possible history) as soon as this proportion is established, and thus the relation between stress and (total) strain is uniquely determined by this proportion in any given material. Thus, the total strain theory makes no difference between a non-linear elastic material and an elastic-plastic material. Nearly proportional loading appears to prevail in the plastic region at a crack edge during a proportional increase of the outer loads at all points of application before crack advance, and in some idealized cases exact proportional loading prevails.

If a total strain theory is not used, this may be emphasized by using the expression *incremental strain theory*.

5.3 Stationary mode III cracks

General considerations for perfect plasticity

Perfectly plastic materials obeying either the Huber-von Mises or the Tresca yield condition are assumed. For anti-plane strain, the yield condition may be written as

$$\tau_{xz}^2 + \tau_{yz}^2 = \tau_{ref}^2 \tag{5.3.1}$$

for both materials, even though the value of τ_{ref} is different, if it is determined via a tensile test. In the following, the notation τ_Y, the yield stress in shear, will be used rather than τ_{ref}.

Because of the assumption of perfect plasticity, (5.3.1) holds at every point of the plastic region. Thus, it is possible to use the representations

$$\tau_{xz} = \tau_Y \cos \psi \tag{5.3.2}$$
$$\tau_{yz} = \tau_Y \sin \psi \tag{5.3.3}$$

where $\psi = \psi(x, y)$. Note that the representations may be contracted to the equation

$$\boxed{\boldsymbol{\tau} = \tau_Y \cos \psi \cdot \hat{\boldsymbol{x}} + \tau_Y \sin \psi \cdot \hat{\boldsymbol{y}}} \tag{5.3.4}$$

i.e. τ_{xz} and τ_{yz} may be interpreted as components of the stress vector $\boldsymbol{\tau}$ and ψ as the angle between this vector and the positive x direction. $\hat{x}$ and $\hat{y}$ are the unit vectors in the positive x and y directions. Obviously,

$$|\tau| = \tau_Y \tag{5.3.5}$$

A remarkable property of the anti-plane stress field at perfect plasticity is the existence of *straight characteristic lines*. The stress vector along such a characteristic line is constant and directed perpendicularly to the line. This property, which will now be proved, follows from (5.3.2)-(5.3.3) and the condition of equilibrium,

$$\frac{\partial \tau_{xz}}{\partial x} + \frac{\partial \tau_{yz}}{\partial y} = 0 \tag{5.3.6}$$

Because τ_{xz} and τ_{yz}, according to (5.3.2)-(5.3.3), can be expressed as functions of $\psi = \psi(x,y)$, this equation may be written as

$$\frac{\partial \tau_{xz}}{\partial \psi} \cdot \frac{\partial \psi}{\partial x} + \frac{\partial \tau_{yz}}{\partial \psi} \cdot \frac{\partial \psi}{\partial y} = 0 \tag{5.3.7}$$

which, with further use of (5.3.2)-(5.3.3), takes the form

$$\sin \psi \frac{\partial \psi}{\partial x} = \cos \psi \frac{\partial \psi}{\partial y} \tag{5.3.8}$$

The stress vector τ is constant along curves of constant ψ. Along such a curve

$$d\psi = \frac{\partial \psi}{\partial x} dx + \frac{\partial \psi}{\partial y} dy = 0 \tag{5.3.9}$$

Comparison with (5.3.8) now gives the equation of the curve:

$$\frac{dy}{dx} = -\cot \psi = \tan(\psi - \pi/2) \tag{5.3.10}$$

Thus, the curve of constant ψ has constant slope, i.e. it is a straight line. The line makes the angle $\psi - \pi/2$ with the x axis. As the stress vector τ makes the angle ψ to the x axis, it is obviously perpendicular to the characteristic line. Note that two solutions of (5.3.10) exist if dy/dx is given.

So far no assumption has been made as regards the history of loading. Thus, *in every perfectly plastic region, subjected to static loading, there exist straight characteristic lines, along which the stress vector is constant and perpendicular to the line.*

The strains

$$\gamma_{xz} = \frac{\partial w}{\partial x}, \qquad \gamma_{yz} = \frac{\partial w}{\partial y} \tag{5.3.11}$$

are now decomposed into elastic and plastic parts:

$$\gamma_{xz} = \gamma_{xz}^e + \gamma_{xz}^p, \qquad \gamma_{yz} = \gamma_{yz}^e + \gamma_{yz}^p \tag{5.3.12}$$

Then, from (5.2.37):

$$\boxed{\frac{d\gamma_{xz}^p}{\tau_{xz}} = \frac{d\gamma_{yz}^p}{\tau_{yz}}} \tag{5.3.13}$$

This relation holds for each material point in the plastic region. Furthermore, Hooke's law gives

$$\gamma_{xz}^e = \frac{\tau_{xz}}{\mu}, \qquad \gamma_{yz}^e = \frac{\tau_{yz}}{\mu} \qquad (5.3.14)$$

and, if the direction of the characteristic line through the material point considered has been unchanged during the whole history of loading in the plastic region, then τ_{xz} and τ_{yz} have remained unchanged, so that

$$\frac{\mathrm{d}\gamma_{xz}^e}{\tau_{xz}} = \frac{\mathrm{d}\gamma_{yz}^e}{\tau_{yz}} \qquad (5.3.15)$$

and thus the relation

$$\boxed{\frac{\mathrm{d}\gamma_{xz}}{\tau_{xz}} = \frac{\mathrm{d}\gamma_{yz}}{\tau_{yz}}} \qquad (5.3.16)$$

holds along a characteristic line, *provided that the direction of this line, through any material point on the line, has been the same during the whole history of loading in the plastic region.*

Orientating, for a moment, the x axis along an arbitrarily chosen characteristic line implies that $\psi = \pi/2$ along the x axis. Thus, $\tau_{xz} = 0$, $\tau_{yz} = \tau_Y$ along the x axis. Equation (5.3.16) then gives $\gamma_{xz} = 0$. Note that no integration constant should be added, because the elastic strain $\gamma_{xz}^e = \tau_{xz}/\mu = 0$ and the strain increments have been zero during the whole history of loading. Now, (5.3.11) shows that w is constant along the x axis. But the x axis was oriented along an arbitrary characteristic line. Thus, *w is constant along each characteristic line.*

Now, considering two characteristic lines, infinitesimally close, it is obvious that

$$\begin{cases} |\mathrm{grad}\, w| = \dfrac{\text{constant}}{s} \\ \mathrm{grad}\, w \text{ is perpendicular to the characteristic line} \end{cases} \qquad (5.3.17)$$

where s is the distance to the intersection point of the two lines. This point can, of course, not be situated inside the plastic region considered. Because $\mathrm{grad}\, w = \gamma_{xz}\hat{x} + \gamma_{yz}\hat{y}$, where a hat denotes unit vector, equation (5.3.11) shows that

$$\gamma_{xz} = \frac{\text{constant}}{s}, \qquad \gamma_{yz} = \frac{\text{constant}}{s} \qquad (5.3.18)$$

along a characteristic line.

Again it should be noted that the derivation asumes that the direction of the characteristic line through any material point has remained fixed during the whole history of loading. This is, for instance, not true for the plastic region near a moving crack edge.

Because the elastic strains are constant along a characteristic line, cf. (5.3.14), it follows from (5.3.18) that the plastic strains vary monotonically along such lines. At each regular point of the boundary between a plastic region and a virginly elastic region, the plastic strains are zero, and thus it follows that a characteristic line cannot intersect such a boundary twice, except when one intersection is a singular point.

The plastic region at a crack edge will now be considered. The positive x axis is oriented along the crack direction and the origin is placed at the crack edge, see

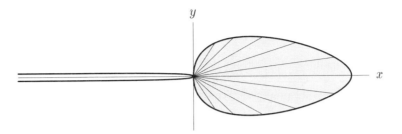

Fig. 5.3.1 Plastic region at a mode III crack edge, showing the centered fan characteristics. The crack is symbolically shown with an opening, although only sliding takes place.

Fig. 5.3.1. In view of the preceding discussion, no alternative seems to exist to the possibility that the characteristics are straight rays from the singular crack edge, i.e. forming a system of *centered fan characteristics*. This possibility leads to a solution for which the stress-strain field fulfills all requirements previously discussed. In particular, the direction of the characteristic line through any material point in the plastic region has been unchanged during the whole history of loading.

As an alternative to the Cartesian coordinate system, a cylindrical one, r, φ, z, with the same origin and $\varphi = 0$ along the positive x axis, is introduced. Then, the characteristic lines are lines of constant φ and the stresses are

$$\boxed{\begin{aligned} \tau_{rz} &= 0 \\ \tau_{\varphi z} &= \tau_Y \end{aligned}} \qquad (5.3.19)$$

in the plastic region. Then, noting that $\gamma_{\varphi z} = \gamma_{\varphi z}^e - \tau_{\varphi z}/\mu$ at the elastic-plastic boundary, (5.3.18) gives:

$$\boxed{\begin{aligned} \gamma_{rz} &= 0 \\ \gamma_{\varphi z} &= \frac{\tau_Y}{\mu} \cdot \frac{R(\varphi)}{r} \end{aligned}} \qquad (5.3.20)$$

where $R(\varphi)$ is the distance from the crack edge to the elastic-plastic boundary; see Fig. 5.3.2.

Equations (5.3.19)-(5.3.20) give the complete stress-strain field in a perfectly plastic region at a crack edge, provided that the location $R(\varphi)$ of its common boundary with the elastic region is known. The result is very simple. The stresses are regular, but there is a $1/r$ strain singularity. The strain energy per unit of volume becomes singular like $1/r$, just as in the elastic case. *No autonomy is found for the strains near the crack edge*: even in a very close vicinity the angular strain distribution depends on the location of the elastic-plastic boundary. The information of this location is carried to the crack edge vicinity by the characteristics: this is a consequence of the hyperbolicity of the governing equations.

An important observation is that *the plastic strains are smaller than the elastic ones in most of the plastic region*: in fact, for the perfectly plastic region, as evident from (5.3.20), they are smaller in 75% of the plastic region. Near the elastic-plastic boundary, they are very small in comparison, and consequently this boundary, which

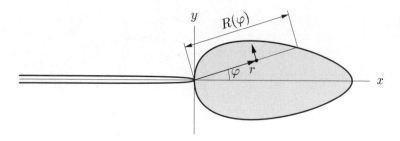

Fig. 5.3.2 The stress-strain field in the plastic region consists of the stresses $\tau_{\varphi z} = \tau_Y$ and the strains $\gamma_{\varphi z} = \tau_Y R(\varphi)/(\mu r)$, both symbolized by the small arrow perpendicular to the radius vector r.

appears deceptively smooth and sharp in analytical treatments, may be difficult to identify accurately in laboratory experiments or numerical calculations.

The stress-strain field (5.3.19)-(5.3.20) was found by using incremental strain theory. However, in this case – mode III, perfect plasticity and stationary crack – *the same result would have been obtained with a total strain theory*! This follows immediately from the fact that proportional loading prevails, according to (5.3.2)-(5.3.3) and (5.3.16) and the unchangeability of centered fan characteristics during loading. Because a total strain theory coincides formally with a non-linear elastic theory, *the J-integral is path independent even for paths inside the plastic region in the case studied.*

It should be remembered that the process region size is assumed to be infinitesimally small in the present chapter. If it is not, the equivalence with the total strain theory and the consequential path independence of the J-integral are only approximately prevalent.

A simple consideration shows that *the crack edge has become blunted*, due to plastic flow. Using (5.3.20) the finite crack edge slip is found:

$$\delta_0 = \int_{-\pi}^{+\pi} \frac{\partial w}{\partial \varphi} d\varphi = \int_{-\pi}^{+\pi} r\gamma_{\varphi z} d\varphi = \frac{\tau_Y}{\mu} \int_{-\pi}^{+\pi} R(\varphi) d\varphi \qquad (5.3.21)$$

where, of course, $R(\varphi)$ may be zero in some φ-intervals. Note that "crack edge slip" is defined by analogy with "crack edge opening" as the absolute value of the difference between the displacements on the two crack faces at the crack edge. The complete blunting is characteristic of perfectly plastic behaviour and stationary crack. However, equation (5.3.21) holds only in the idealized case of an infinitesimally small process region. *In a real case, the process region itself may contribute as much to the crack face displacements near the crack edge as the plastic flow in the continuum outside the process region.*

The hodograph transform

In the preceding, subsection the location $R(\varphi)$ of the elastic-plastic boundary was considered as known. However it is rarely known *a priori*. Problems involving unknown boundary locations are generally very difficult to handle, but in the mode III case, the difficulty may be circumvented by a re-formulation of the problem. To this, end a

5.3 STATIONARY MODE III CRACKS

hodograph transform is introduced, i.e. the roles of independent and dependent variables are switched, so that strains or stresses are considered as independent variables and the space coordinates as the dependent ones. This is done with the hope that a problem with known boundaries (in the strain or stress space) can be formulated.

The hodograph transform will be considered for a rather general case, which allows for non-linear elasticity: a single-valued relationship is assumed to hold between stresses and strains in the elastic region, i.e.

$$\tau = \tau(\gamma) \tag{5.3.22}$$

where τ is the magnitude of the stress vector $\boldsymbol{\tau}$ and γ the magnitude of the strain vector $\boldsymbol{\gamma}$. These vectors are assumed to be parallel to each other at each point considered.

Because both $\boldsymbol{\tau}$ and $\boldsymbol{\gamma}$ make the angle ψ with the x axis, the conditions of equilibrium, (5.3.6), which may be written

$$\operatorname{div} \boldsymbol{\tau} = 0 \tag{5.3.23}$$

are transformed to

$$\operatorname{div}_\tau \boldsymbol{r} = 0 \tag{5.3.24}$$

where $\boldsymbol{r}$ is the position vector and the index τ indicates that τ and ψ (or two other variables for which the stress vector may be specified, for instance τ_{xz} and τ_{yz}) are independent variables. That (5.3.23) actually implies (5.3.24) may appear obvious for physical or intuitive reasons; however, it may be shown in the following way:

$$\mathrm{d}\boldsymbol{\tau} = \boldsymbol{T}\,\mathrm{d}\boldsymbol{r}, \qquad \mathrm{d}\boldsymbol{r} = \boldsymbol{R}\,\mathrm{d}\boldsymbol{\tau} \tag{5.3.25}$$

where $\mathrm{d}\boldsymbol{r} = (\mathrm{d}x, \mathrm{d}y)$, $\mathrm{d}\boldsymbol{\tau} = (\mathrm{d}\tau_{xz}, \mathrm{d}\tau_{yz})$, and

$$\boldsymbol{T} = \begin{bmatrix} \dfrac{\partial \tau_{xz}}{\partial x} & \dfrac{\partial \tau_{xz}}{\partial y} \\ \dfrac{\partial \tau_{yz}}{\partial x} & \dfrac{\partial \tau_{yz}}{\partial y} \end{bmatrix}, \qquad \boldsymbol{R} = \begin{bmatrix} \dfrac{\partial x}{\partial \tau_{xz}} & \dfrac{\partial x}{\partial \tau_{yz}} \\ \dfrac{\partial y}{\partial \tau_{xz}} & \dfrac{\partial y}{\partial \tau_{yz}} \end{bmatrix} \tag{5.3.26}$$

These expressions show that $\operatorname{div} \boldsymbol{\tau} = \operatorname{tr} \boldsymbol{T}$ and $\operatorname{div}_\tau \boldsymbol{r} = \operatorname{tr} \boldsymbol{R}$. Now, $\operatorname{div} \boldsymbol{\tau} = 0$ so that

$$\operatorname{tr} \boldsymbol{T} = 0 \tag{5.3.27}$$

But from (5.3.25) it follows that

$$\mathrm{d}\boldsymbol{r} = \boldsymbol{T}^{-1}\mathrm{d}\boldsymbol{\tau} = \boldsymbol{R}\,\mathrm{d}\boldsymbol{\tau} \tag{5.3.28}$$

so that

$$\boldsymbol{R} = \boldsymbol{T}^{-1} = \frac{1}{\det \boldsymbol{T}} \begin{bmatrix} \dfrac{\partial \tau_{yz}}{\partial y} & -\dfrac{\partial \tau_{xz}}{\partial y} \\ -\dfrac{\partial \tau_{yz}}{\partial x} & \dfrac{\partial \tau_{xz}}{\partial x} \end{bmatrix} \tag{5.3.29}$$

Hence,

$$\operatorname{div}_\tau \boldsymbol{r} = \operatorname{tr} \boldsymbol{R} = \operatorname{tr} \boldsymbol{T}^{-1} = \frac{1}{\det \boldsymbol{T}} \operatorname{tr} \boldsymbol{T} = 0 \tag{5.3.30}$$

which completes the proof.

After noting that τ and ψ, or γ and ψ, may be considered as polar coordinates in the respective hodograph planes, equation (5.3.24) leads to

$$\operatorname{div}_\tau \boldsymbol{r} = \frac{\partial \boldsymbol{r}}{\partial \tau}\hat{\boldsymbol{\tau}} + \frac{1}{\tau}\cdot\frac{\partial \boldsymbol{r}}{\partial \psi}\hat{\boldsymbol{\psi}}$$

$$= \frac{\partial \boldsymbol{r}}{\partial \tau}\hat{\boldsymbol{\gamma}} + \frac{1}{\tau}\cdot\frac{\partial \boldsymbol{r}}{\partial \psi}\hat{\boldsymbol{\psi}} = \frac{1}{\tau'(\gamma)}\cdot\frac{\partial \boldsymbol{r}}{\partial \gamma}\cdot\hat{\boldsymbol{\gamma}} + \frac{\gamma}{\tau}\cdot\frac{1}{\gamma}\cdot\frac{\partial \boldsymbol{r}}{\partial \psi}\cdot\hat{\boldsymbol{\psi}}$$

$$= \frac{\gamma}{\tau}\left\{\operatorname{div}_\gamma \boldsymbol{r} + \left[\frac{\tau(\gamma)}{\gamma\tau'(\gamma)} - 1\right]\hat{\boldsymbol{\gamma}}\cdot\frac{\partial \boldsymbol{r}}{\partial \gamma}\right\} = 0 \quad (5.3.31)$$

where, as previously, a hat denotes a unit vector.

From (5.3.11) it follows that the compatibility equation

$$\operatorname{curl}\boldsymbol{\gamma} = 0 \tag{5.3.32}$$

is transformed to

$$\operatorname{curl}_\gamma \boldsymbol{r} = 0 \tag{5.3.33}$$

with essentially the same proof as for the divergence. Here, index γ indicates that $\boldsymbol{r}$ is considered as a function of γ and ψ (or, for instance, γ_{xz} and γ_{yz}).

Equation (5.3.33) suggests the existence of a scalar potential function F such that

$$\boldsymbol{r} = \operatorname{grad}_\gamma F \tag{5.3.34}$$

Insertion into (5.3.31) gives:

$$\operatorname{div}_\gamma\operatorname{grad}_\gamma F + \left[\frac{\tau(\gamma)}{\gamma\tau'(\gamma)} - 1\right]\hat{\boldsymbol{\gamma}}\cdot\frac{\partial \operatorname{grad}_\gamma F}{\partial \gamma} = 0 \tag{5.3.35}$$

or

$$\Delta_\gamma F + \left[\frac{\tau(\gamma)}{\gamma\tau'(\gamma)} - 1\right]\frac{\partial^2 F}{\partial \gamma^2} = \frac{\tau(\gamma)}{\gamma\tau'(\gamma)}\cdot\frac{\partial^2 F}{\partial \gamma^2} + \frac{1}{\gamma}\cdot\frac{\partial F}{\partial \gamma} + \frac{1}{\gamma^2}\cdot\frac{\partial^2 F}{\partial \psi^2} = 0 \tag{5.3.36}$$

where Δ_γ is the Laplace operator in the strain plane. This equation was derived by Rice (1967); see also Neuber (1961).

In a linearly elastic region

$$\gamma\tau'(\gamma) = \tau(\gamma) \tag{5.3.37}$$

and thus

$$\Delta_\gamma F = 0 \tag{5.3.38}$$

The Hult-McClintock problem

The solution of this problem by Hult and McClintock (1956) is the first solution to an elastic-plastic crack problem. It concerns mode III, linear elasticity, perfect plasticity and infinitesimally small scale yielding. In the present chapter, infinitesimally small scale yielding may be envisaged as the limiting case of infinitesimally small load on the cracked body, so that, even though the size of the plastic region is vanishingly small compared with the crack length, the elastic stresses and strains are limited to certain finite values, dependent on stress composition and material strength characteristics.

A solution of (5.3.38), which holds for the elastic region, satisfying the symmetry properties of mode III, is:
$$F_n = A_n \gamma^n \cos n\psi \tag{5.3.39}$$
Thus,
$$\boldsymbol{r} = \text{grad}_\gamma F = \sum n A_n \gamma^{n-1} (\cos n\psi \cdot \hat{\boldsymbol{\gamma}} - \sin n\psi \cdot \hat{\boldsymbol{\psi}}) \tag{5.3.40}$$
where the summation is taken over all n. On the crack faces, $\boldsymbol{r}$ is parallel to $\boldsymbol{\gamma}$. Thus,
$$\sin n\psi = 0 \quad \text{for } \psi = 0, \pi \tag{5.3.41}$$
which relations are satisfied only for integer values of n. Further, due to the anti-symmetry,
$$(\cos n\psi)_{\psi=0} = -(\cos n\psi)_{\psi=\pi} \tag{5.3.42}$$
and thus n is an odd integer. Hence,
$$\boldsymbol{r} = \sum_{1,3,5,\ldots} n\{[A_n \gamma^{n-1} - A_{-n}\gamma^{-n-1}] \cos n\psi \cdot \hat{\boldsymbol{\gamma}}$$
$$- [A_n\gamma^{n-1} + A_{-n}\gamma^{-n-1}] \sin n\psi \cdot \hat{\boldsymbol{\psi}}\} \tag{5.3.43}$$
On the elastic-plastic boundary, $\gamma = \tau_Y/\mu$ and $\boldsymbol{r}$ is perpendicular to $\boldsymbol{\gamma}$ according to equations (5.3.20). Thus,
$$\sum_{1,3,5,\ldots} n\left[A_n\left(\frac{\tau_Y}{\mu}\right)^{n-1} - A_{-n}\left(\frac{\tau_Y}{\mu}\right)^{-n-1}\right]\cos n\psi = 0 \tag{5.3.44}$$
i.e. $A_{-n} = (\tau_Y/\mu)^{2n} A_n$. Then, because $\gamma = \tau/\mu$ in the elastic region,
$$\boldsymbol{r} = \sum_{1,3,5,\ldots} n A_n \left(\frac{\tau_Y}{\mu}\right)^{n-1}\left\{\left[\left(\frac{\tau}{\tau_Y}\right)^{n-1} - \left(\frac{\tau}{\tau_Y}\right)^{-n-1}\right]\cos n\psi \cdot \hat{\boldsymbol{\tau}}\right.$$
$$\left. - \left[\left(\frac{\tau}{\tau_Y}\right)^{n-1} + \left(\frac{\tau}{\tau_Y}\right)^{-n-1}\right]\sin n\psi \cdot \hat{\boldsymbol{\psi}}\right\} \tag{5.3.45}$$
As $\tau \to 0$, $r \to \infty$. Because infinitesimally small scale yielding is considered, the elastic field far from the crack edge should show an inverse square root singularity:
$$\tau \to r^{-1/2} \quad \text{as } r \to \infty \tag{5.3.46}$$
and thus
$$r \to \tau^{-2} \quad \text{as } \tau \to 0 \tag{5.3.47}$$
Then, (5.3.45) gives
$$A_n = 0 \quad \text{for } n \neq 1 \tag{5.3.48}$$
and thus
$$\boldsymbol{r} = -R_0\left\{\left[\left(\frac{\tau_Y}{\tau}\right)^2 - 1\right]\cos\psi \cdot \hat{\boldsymbol{\tau}} + \left[\left(\frac{\tau_Y}{\tau}\right)^2 + 1\right]\sin\psi \cdot \hat{\boldsymbol{\psi}}\right\} \tag{5.3.49}$$
where $R_0 = A_1$ is a constant. The relation gives (implicitly) the stresses as functions of

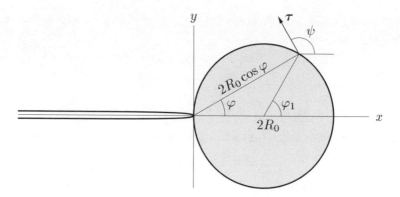

Fig. 5.3.3 The stress vector τ and its angle ψ from the positive x direction.

position in the elastic region. The elastic-plastic boundary is found by putting $\tau = \tau_Y$. Note that $\sin\psi = \cos\varphi$ and $\hat{\psi} = -\hat{r}$ on this boundary (see Fig. 5.3.3), so that

$$r = -2R_0 \sin\psi \cdot \hat{\psi} = 2R_0 \cos\varphi \cdot \hat{r} \tag{5.3.50}$$

Thus, *the plastic region is circular* with R_0 as radius. This fact suggests that the continued treatment – to give explicit expression for the elastic stress field – might be simplified if the coordinate origin is moved to the point $x = R_0$, $y = 0$. A new cylindrical coordinate system, r_1, φ_1, z with this origin is introduced.

Observing that

$$r_1 = r - R_0\hat{x} = r - R_0 \cos\psi \cdot \hat{r} + R_0 \sin\psi \cdot \hat{\psi} \tag{5.3.51}$$

and then using equation (5.3.49), results in the expression

$$r_1 = -R_0\left(\frac{\tau_Y}{\tau}\right)^2 (\cos\psi \cdot \hat{r} + \sin\psi \cdot \hat{\psi})$$
$$= -R_0\left(\frac{\tau_Y}{\tau}\right)^2 [(\cos^2\psi - \sin^2\psi)\hat{x} + 2\cos\psi\sin\psi \cdot \hat{y}] \tag{5.3.52}$$

because $\hat{r} = \cos\psi \cdot \hat{x} + \sin\psi \cdot \hat{y}$ and $\hat{\psi} = -\sin\psi \cdot \hat{x} + \cos\psi \cdot \hat{y}$. Thus,

$$r_1 \cos\varphi_1 = -R_0\left(\frac{\tau_Y}{\tau}\right)^2 \cos 2\psi \tag{5.3.53}$$

$$r_1 \sin\varphi_1 = -R_0\left(\frac{\tau_Y}{\tau}\right)^2 \sin 2\psi \tag{5.3.54}$$

These equations give

$$r_1 = R_0\left(\frac{\tau_Y}{\tau}\right)^2, \qquad \varphi_1 = 2\psi - \pi \tag{5.3.55}$$

and then, finally, because τ makes the angle ψ with the x axis, (see Fig. 5.3.4),

$$\tau_{r_1 z} = \tau_Y \sqrt{\frac{R_0}{r_1}} \sin\frac{\varphi_1}{2} \tag{5.3.56}$$

$$\tau_{\varphi_1 z} = \tau_Y \sqrt{\frac{R_0}{r_1}} \cos\frac{\varphi_1}{2} \tag{5.3.57}$$

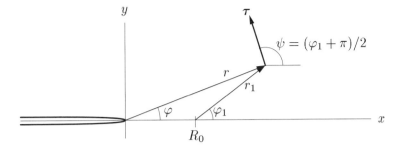

Fig. 5.3.4 Stress vector τ in the elastic region. At the elastic-plastic boundary, $r_1 = R_0$ and $\varphi_1 = 2\varphi$; cf. Fig. 5.3.3.

Except for different notations, these expressions are identical to those given by (3.3.35), although here they are valid only for $r_1 \geq R_0$. Comparison gives:

$$R_0 = \frac{K_{III}^2}{2\pi \tau_Y^2} = \frac{\mu J_{III}}{\pi \tau_Y^2} \qquad (5.3.58)$$

where J_{III} was taken from (3.3.40).

The strains in the *plastic region* are now found from (5.3.20):

$$\gamma_{rz} = 0, \qquad \gamma_{\varphi z} = \frac{K_{III}^2}{\pi \tau_Y \mu} \cdot \frac{\cos \varphi}{r} = \frac{2 J_{III} \cos \varphi}{\pi \tau_Y r} \qquad (5.3.59)$$

The crack edge slip may be obtained directly from (5.3.21). However, the more general concept of crack face slip at an arbitrary distance r behind the crack edge may be found by using the relation $\partial w / \partial \varphi_1 = \gamma_{\varphi_1 z} r = \tau_{\varphi_1 z} r / \mu$ and integrating along a circle with radius $r_1 \geq R_0$. This gives the crack face slip on $y = 0$, $x \leq 0$, where $r_1 = r + R_0$:

$$\delta_0 = \frac{4\tau_Y}{\mu} \sqrt{R_0 r_1} = \frac{4\tau_Y R_0}{\mu} \sqrt{\frac{r}{R_0} + 1} \qquad (5.3.60)$$

Thus, by putting $r = 0$, the crack edge slip is found:

$$\delta_0 = \frac{2}{\pi} \cdot \frac{K_{III}^2}{\tau_Y \mu} = \frac{4}{\pi} \cdot \frac{J_{III}}{\tau_Y} \qquad (5.3.61)$$

Large scale yielding at a mode III crack in a finite elastic-perfectly plastic body

The case of a finite body presents a more difficult problem than the one treated in the preceding subsection. It turns out that the series representation (5.3.45) is usually improper, due to lack of convergence. Another method – in fact the one used by Hult and McClintock (1956) – is to obtain a solution by representing F as the real or imaginary part of an analytic function of a complex strain variable. This possibility follows from the fact that F satisfies the Laplace equation (5.3.38).

As an example, a case studied by Rice (1966) will be considered. The body has finite width, b, between $x = -a$ and $x = b - a$, and contains an edge crack on $y = 0$,

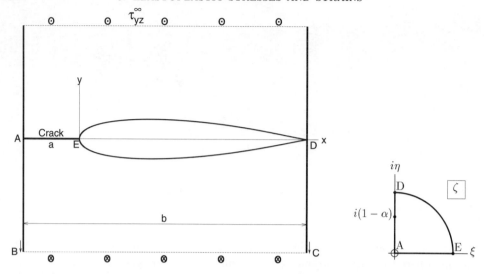

Fig. 5.3.5 Limit load plastic region from a mode III edge crack and the image of the half-plane $y \leq 0$ on the ζ-plane. Points A,B,C,D,E in the ζ-plane correspond to points with the same label in the physical plane.

$-a \leq x < 0$. The body is infinite in the y direction (thus not strictly "finite"). The remote stress is $\tau_{yz} = \tau_{yz}^{\infty}$.

In order to simplify the treatment, the *limit load* case is assumed, *i.e.* the plastic region traverses the net section $b - a$ of the body; see Fig. 5.3.5. (Rice, 1966, did not make this restriction, and he considered notches as well as cracks.) Thus,

$$\tau_{yz}^{\infty} = \frac{b-a}{b}\tau_Y \tag{5.3.62}$$

By introducing the coordinates

$$\xi = \left(\frac{\mu}{\tau_Y}\right)\gamma_{xz}, \quad \eta = \left(\frac{\mu}{\tau_Y}\right)\gamma_{yz}, \quad \zeta = \xi + i\eta \tag{5.3.63}$$

the image in the ζ plane of the elastic-plastic boundary, which is specified by the yield condition $|\tau| = |\tau_Y|$, i.e. $\mu|\gamma|/\tau_Y = 1$, is given as

$$|\zeta| = 1, \quad \eta = 0 \tag{5.3.64}$$

Only the lower half of the body is studied: its image on the ζ plane is contained in $\xi \geq 0$. Fig. 5.3.5 shows the boundaries of the image of the elastic region in the ζ plane with letters ABCDE indicating corresponding points in the xy plane.

Equation (5.3.34) gives:

$$x = \frac{\mu}{\tau_Y} \cdot \frac{\partial F}{\partial \xi} \tag{5.3.65}$$

$$y = \frac{\mu}{\tau_Y} \cdot \frac{\partial F}{\partial \eta} \tag{5.3.66}$$

Now, because $F(\xi, \eta)$ satisfies the Laplace equation, it will be considered as the real part of an analytic function $\tau_Y b f(\zeta)/\mu$:

$$F = \frac{\tau_Y b}{\mu} \Re[f(\zeta)] \qquad (5.3.67)$$

After putting

$$H = \frac{\tau_Y b}{\mu} \Im[f(\zeta)] \qquad (5.3.68)$$

and then using the Cauchy-Riemann equations,

$$\frac{\partial H}{\partial \xi} = -\frac{\partial F}{\partial \eta}, \qquad \frac{\partial H}{\partial \eta} = \frac{\partial F}{\partial \xi} \qquad (5.3.69)$$

the following relation is obtained:

$$x - iy = \frac{\mu}{\tau_Y}\left[\frac{\partial F}{\partial \xi} + i\frac{\partial H}{\partial \xi}\right] = bf'(\zeta) \qquad (5.3.70)$$

Differentiation of H gives

$$dH = \frac{\partial H}{\partial \xi}d\xi + \frac{\partial H}{\partial \eta}d\eta = -\frac{\partial F}{\partial \eta}d\xi + \frac{\partial F}{\partial \xi}d\eta = -y\,d\gamma_{xz} + x\,d\gamma_{yz} \qquad (5.3.71)$$

and then the variation of dH along the boundary ABCDE is found from the following considerations:

Along DE: $\gamma = \tau_Y/\mu$ and $\boldsymbol{\gamma}$ is perpendicular to $\boldsymbol{r}$. This implies that $d\boldsymbol{\gamma}$ is parallel to $\boldsymbol{r}$, i.e. $d\gamma_{xz}/d\gamma_{yz} = x/y$. Thus, $dH = 0$.
Along EA: $y = 0$ and $\gamma_{yz} = 0$. Thus, $dH = 0$.
Along AB: $x = -a$ and $\gamma_{xz} = 0$. Thus, $dH = -a\,d\gamma_{yz}$.
Along CD: $x = b - a$ and $\gamma_{xz} = 0$. Thus, $dH = (b-a)d\gamma_{yz}$.

Integration along the boundary, fixing the constants of integration so that H vanishes along DEA and also so that H will be continuous along the boundary, gives the following boundary condition:

$$\Im[f(\zeta)] = \begin{cases} 0 & \text{for } |\zeta| = 1, \xi \geq 0, \eta \geq 0 \\ 0 & \text{for } 0 \leq \xi < 1, \eta = 0 \\ -\alpha\eta & \text{for } \xi = 0, 0 < \eta \leq 1-\alpha \\ (1-\alpha)(\eta-1) & \text{for } \xi = 0, 1-\alpha < \eta < 1 \end{cases} \qquad (5.3.72)$$

where $\alpha = a/b$.

By means of the mapping function

$$\Omega = \zeta^2, \quad \Omega = \omega + i\lambda \qquad (5.3.73)$$

the region ABCDE is mapped onto the unit semi-circle $|\Omega| \leq 1$, $\lambda \geq 0$ in the Ω plane; see Fig. 5.3.6.

The function $f(\zeta)$ is transformed to the function $S(\Omega)$, i.e.

$$S(\Omega) = f(\zeta) \qquad (5.3.74)$$

Thus, the boundary conditions (5.3.72) are transformed to:

$$\Im[S(\Omega)] = 0 \quad \text{for } |\Omega| = 1, \lambda > 0 \qquad (5.3.75)$$
$$\Im[S(\Omega)] = g(\omega) \text{ for } |\Omega| < 1, \lambda = 0 \qquad (5.3.76)$$

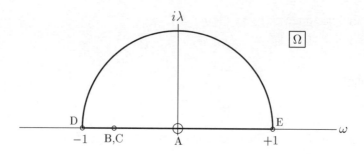

Fig. 5.3.6 The mapping from the ζ-plane on the Ω-plane.

where (because $\xi = 0$ gives $\lambda = 0$, $\omega = -\eta^2$)

$$g(\omega) = \begin{cases} (1-\alpha)[\sqrt{-\omega} - 1] & \text{for } -1 < \omega < -(1-\alpha)^2 \\ -\alpha\sqrt{-\omega} & \text{for } -(1-\alpha)^2 \leq \omega < 0 \\ 0 & \text{for } 0 \leq \omega < 1 \end{cases} \quad (5.3.77)$$

Note that $g(\omega)$ is a real function.

The problem now consists of finding the function $S(\Omega)$, regular in the region $|\Omega| < 1$, $\lambda > 0$ and with its imaginary part approaching the values given by (5.3.75)-(5.3.76) as Ω approaches the boundary of the region. The solution will be found by first considering the two boundary values (5.3.75) and (5.3.76) separately, and then combining the general solutions.

First, equation (5.3.75) is considered. Let $G(\Omega)$ be a function which is regular in the upper half-plane. Then, the imaginary part of

$$S(\Omega) = G(\Omega) + G(1/\Omega) \quad (5.3.78)$$

vanishes on $|\Omega| = 1$, $\lambda > 0$ if

$$\overline{G(\Omega)} = G(\overline{\Omega}), \quad \lambda > 0 \quad (5.3.79)$$

The proof is simple. As $1/\Omega = \overline{\Omega}$ for $|\Omega| = 1$, then

$$S(\Omega) = G(\Omega) + G(\overline{\Omega}) = G(\Omega) + \overline{G(\Omega)} \quad \text{for } |\Omega| = 1 \quad (5.3.80)$$

i.e. $\Im[S(\Omega)] = 0$ for $|\Omega| = 1$, $\lambda > 0$. Thus, a general solution, satisfying condition (5.3.75) may be written in the form (5.3.78).

In the second step, function $G(\Omega)$ is made to satisfy the boundary condition (5.3.76). To this end, the integral

$$G(\Omega) = \frac{1}{\pi} \int_{-1}^{+1} \frac{g(t)}{t - \Omega} dt, \quad \Im\Omega > 0 \quad (5.3.81)$$

is considered. When Ω approaches the real axis, an infinitesimally small indentation of the path of integration has to be made if $|\Omega| < 1$, see Fig. 5.3.7. Then, the whole integrand becomes real, except on the indentation, which therefore gives the only imaginary contribution when Ω approaches the real axis. This contribution consists of the residue

$$\pi i \cdot \frac{1}{\pi} g(\omega) = i g(\omega) \quad (5.3.82)$$

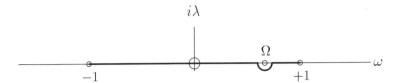

Fig. 5.3.7 Path of integration in the Ω plane.

Hence, $G(\Omega)$ satisfies the boundary condition (5.3.76). Use of the same argument shows that $G(1/\Omega)$ has no imaginary part when Ω approaches the real axis and $|\Omega| < 1$. Thus, the function $[G(\Omega) + G(1/\Omega)]$, which is analytic inside $|\Omega| < 1$, $\lambda > 0$, satisfies the boundary condition (5.3.76). But, because $G(\Omega)$, as given by (5.3.81), satisfies condition (5.3.79), it is obvious that the complete solution of the problem of finding $S(\Omega)$ is given by equations (5.3.78) and (5.3.81).

Insertion of (5.3.77) into (5.3.81) gives, after putting $t = -s^2$ and $\Omega = \zeta^2$:

$$G(\zeta^2) = \frac{2\alpha}{\pi}\int_0^{1-\alpha}\frac{s^2 ds}{s^2 + \zeta^2} - \frac{2(1-\alpha)}{\pi}\int_{1-\alpha}^1 \frac{(s^2-s)ds}{s^2+\zeta^2}$$

$$= \frac{1-\alpha}{\pi}\ln\frac{1+\zeta^2}{(1-\alpha)^2+\zeta^2}$$

$$+ \frac{2(1-\alpha)}{\pi}\zeta\,\text{atan}\frac{1}{\zeta} - \frac{2}{\pi}\zeta\,\text{atan}\frac{1-\alpha}{\zeta} \quad (5.3.83)$$

The position of the elastic-plastic boundary in space coordinates is found from (5.3.70):

$$R(\varphi)e^{-i\varphi} = bf'(e^{i\psi}) \quad (5.3.84)$$

where φ and ψ are the same coordinates as used in the preceding subsection. As $\psi = \varphi + \pi/2$ along the elastic-plastic boundary, this relation implies that

$$R(\varphi) = -ibe^{i\psi}f'(e^{i\psi}) = -ib\left[\zeta f'(\zeta)\right]_{|\zeta|=1}$$

$$= -2ib\left[\zeta^2 G'(\zeta^2) - \frac{1}{\zeta^2}G'\left(\frac{1}{\zeta^2}\right)\right]_{|\zeta|=1} = 2b\Im\left[\zeta\frac{dG}{d\zeta}\right]_{|\zeta|=1} \quad (5.3.85)$$

Then, from (5.3.83):

$$R(\varphi) = b\cos\varphi\left\{1 - \alpha - \frac{2}{\pi}\text{atan}\frac{2(1-\alpha)|\sin\varphi|}{\alpha(2-\alpha)}\right\}$$

$$+ \frac{2b}{\pi}|\sin\varphi|\left\{\frac{1}{2}\ln\frac{1+2(1-\alpha)\cos\varphi+(1-\alpha)^2}{1-2(1-\alpha)\cos\varphi+(1-\alpha)^2} + (1-\alpha)\ln\left|\tan\frac{\varphi}{2}\right|\right\} \quad (5.3.86)$$

This expression has been derived for $\varphi < 0$ (the lower half-plane was studied), but has been written so that it is symmetric with respect to φ, and consequently valid also for $\varphi > 0$.

Equation (5.3.86) is the limit load solution. Lower load solutions are found in the same manner, cf. Rice (1966), but the mapping function is more sophisticated and the

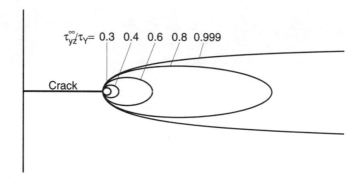

Fig. 5.3.8 Plastic region at a mode III edge crack in a semi-infinite plate for different remote loads, $\tau_{yz} = \tau_{yz}^\infty$, ranging from small scale to (almost) net section yielding.

treatment becomes more laborious. These cases were previously studied by Koskinen (1963), who used a numerical method and also, like Rice, included notched specimens. Some examples of the position of the elastic-plastic boundary for cracked specimens at different load levels, ranging from small scale yielding to net section yielding, are shown in Fig. 5.3.8. The strip width is assumed to be infinite, i.e., an edge crack in a semi-infinite plate is considered. Formulae given by Rice (1966) are used. It can be shown that the width of the plastic region when the limit load τ_{yz}^∞ approaches the yield stress in shear, τ_Y, approaches $4a/\pi$, where a is the length of the edge crack.

Loss of autonomy at large scale yielding

In elastic-perfectly plastic materials, the position of the elastic-plastic boundary determines completely the stress-strain field in the plastic region, according to equations (5.3.19)-(5.3.20). The field near the crack edge is of special interest. A study of Fig. 5.3.8 reveals that the strain field near the crack edge depends on the scale of yielding and on the body geometry. Autonomy of a pure mode III field near the crack edge prevails, if only cases of small scale yielding are considered, but it gets lost when larger scales of yielding are included.

Another way to illustrate how autonomy is restricted to small scale yielding for elastic-perfectly plastic materials, consists of comparing distances ahead of the crack where a specified strain appears at the same amount of crack edge slip but different scales of yielding. For an edge crack in a large plate, Rice (1966) has given the following expressions for the forwards extension R_0 of the plastic region, and for the crack edge slip δ_0:

$$\frac{R_0}{a} = \frac{2}{\pi} \cdot \frac{1+(\tau_{yz}/\tau_{yz}^\infty)^2}{1-(\tau_{yz}/\tau_{yz}^\infty)^2} \boldsymbol{E}\left(\frac{2\tau_{yz}/\tau_{yz}^\infty}{1+(\tau_{yz}/\tau_{yz}^\infty)^2}\right) - 1 = f(\tau_{yz}/\tau_{yz}^\infty) \qquad (5.3.87)$$

$$\frac{\mu\delta_0}{2\tau_Y a} = \frac{2}{\pi} \cdot [1+(\tau_{yz}/\tau_{yz}^\infty)^2]\boldsymbol{K}[(\tau_{yz}/\tau_{yz}^\infty)^2] - 1 = g(\tau_{yz}/\tau_{yz}^\infty) \qquad (5.3.88)$$

where a is the length of the crack and $\boldsymbol{K}$ and $\boldsymbol{E}$ are the complete elliptic integrals

5.3 STATIONARY MODE III CRACKS

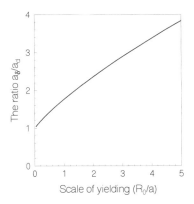
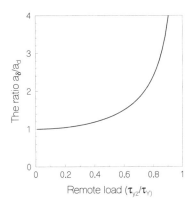

Fig. 5.3.9 Ratio between two critical crack lengths for onset of crack growth, one, a_δ, determined by the crack edge slip criterion, and the other, a_d, determined by the criterion of the attainment of a specified strain at a given distance ahead of the crack edge.

of the first and second kind. Using relation (5.3.20), it is now possible to calculate the distance d ahead of the crack where a specified strain, γ, occurs. In dimensionless form, the result is

$$\frac{2\gamma d}{\delta_0} = \frac{f(\tau_{yz}/\tau_{yz}^\infty)}{g(\tau_{yz}/\tau_{yz}^\infty)} \tag{5.3.89}$$

The left part of Fig. 5.3.9 shows the dimensionless distance $2\gamma d/\delta_0$ as a function of the scale of yielding, which is taken as R_0/a. It is shown as a function of τ_{yz}^∞/τ_Y in the right part of the figure. For $\tau_{yz}^\infty/\tau_Y = 0.5$, the distance $2\gamma d/\delta$ is about 1.3 and for $\tau_{yz}^\infty/\tau_Y = 0.75$, it is about 2.1.

Suppose now that the crack would start growing when the crack edge slip δ_0 overshoots a certain material-specific value, independently of the scale of yielding. Then, from (5.3.88)-(5.3.89), the largest crack length $a = a_{max}$ for which no crack growth occurs at a given remote load can be predicted:

$$a_{max} = a_\delta = \frac{\mu \delta_0}{2\tau_Y g(\tau_{yz}^\infty/\tau_Y)} \tag{5.3.90}$$

If, on the other hand, crack growth starts when the strain γ overshoots a certain material-specific value at given distance d ahead of the crack edge, then the prediction of a_{max} gives

$$a_{max} = a_d = \frac{\mu \gamma d}{\tau_Y f(\tau_{yz}^\infty/\tau_Y)} \tag{5.3.91}$$

The ratio between the two predicted crack lengths is

$$\frac{a_\delta}{a_d} = \frac{\delta_0}{2\gamma d} \cdot \frac{f(\tau_{yz}^\infty/\tau_Y)}{g(\tau_{yz}^\infty/\tau_Y)} \tag{5.3.92}$$

and, if the predictions are set to be equal at infinitesimally small scale of yielding, $\tau_{yz}^\infty/\tau_Y \to 0$, then the material-specific ratio $2\gamma d/\delta_0$ has to be taken as unity. This implies that the ratio a_δ/a_d is found from the ordinates in Fig. 5.3.9. Thus, for instance, a_{max} would be predicted to be about twice as long by the crack edge slip criterion as by the criterion of the attainment of a certain strain at a given distance ahead of the crack edge, if the remote stress is 75% of the yield stress, and about four times as long at 90% of the yield stress. The crack edge slip criterion is obviously less conservative compared to the criterion of attainment of a critical strain at a prescribed distance ahead of the crack edge.

Conditions under which proportional loading prevails in the plastic region

Proportional loading ensures that the plastic deformations agree with those predicted by a total strain theory. Consequently, this simpler theory may be used, and, more importantly, the J-integral is then path-independent, even for paths inside the plastic region. In all elastic-perfectly plastic mode III cases treated so far in this section, proportional loading has prevailed, provided that the boundary loading has been applied in a proportional manner. It was shown that $\psi = \pi/2 + \varphi$ for such cases, but other relations hold for proportional loading of materials with other constitutive relations.

Proportional loading at mode III implies that the angle ψ at a given material point inside the plastic region stays constant during loading, while the strain magnitude γ increases. The variables r, φ, $\hat{\psi}$ and $\hat{\gamma}$ are obviously all constant at a fixed material point, and a study of Fig. 5.3.10 shows that

$$\boldsymbol{r} \cdot \hat{\boldsymbol{\gamma}} = \frac{\partial F}{\partial \gamma} = r\cos(\psi - \varphi) \tag{5.3.93}$$

$$\boldsymbol{r} \cdot \hat{\boldsymbol{\psi}} = \frac{1}{\gamma} \cdot \frac{\partial F}{\partial \psi} = -r\sin(\psi - \varphi) \tag{5.3.94}$$

where the middle members follow from the relation $\boldsymbol{r} = \text{grad}_\gamma F$. Division gives

$$\frac{\gamma \partial F/\partial \gamma}{\partial F/\partial \psi} = -\cot(\psi - \varphi) \tag{5.3.95}$$

Because the governing equation, (5.3.36), for F, can be solved through variable separation, solutions of the equation may be written in the form $F = f(\gamma) \cdot s(\psi)$. Equation (5.3.95) then gives

$$\gamma \cdot f' \cdot s = -\cot(\psi - \varphi) \cdot f \cdot s' \tag{5.3.96}$$

Now, because a fixed material point is considered, φ does not change during loading, and thus

$$\frac{f'}{f} = \frac{\alpha}{\gamma} \tag{5.3.97}$$

where α does not depend on γ. Integration shows that $f \propto \gamma^\alpha$, and if $F = \gamma^\alpha \cdot s(\psi)$ is inserted into the governing equation (5.3.36), the relation

$$\frac{\tau(\gamma)}{\gamma \tau'(\gamma)} \cdot \alpha(\alpha - 1) + \alpha = -\frac{s''}{s} \tag{5.3.98}$$

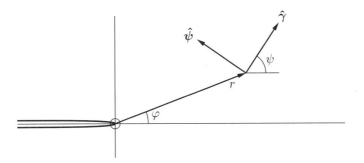

Fig. 5.3.10 The variables r, φ, $\hat{\gamma}$ and $\hat{\psi}$ in a plastic region.

is obtained, showing that

$$\frac{\tau(\gamma)}{\gamma \tau'(\gamma)} = constant = \frac{1}{N}, \text{ say} \qquad (5.3.99)$$

which integrates to the expression for *power-law strain hardening*:

$$\tau = \tau_Y \left(\frac{\gamma}{\gamma_Y}\right)^N \qquad (5.3.100)$$

Apart from trivial exceptions, this condition for proportional loading is necessary, but in general not sufficient. A problem also involves boundary conditions, and if these contain a finite length parameter, such as a crack length or a boundary coordinate of dimension length, the property that $F \propto \gamma^\alpha$ cannot be maintained. Otherwise, the boundary conditions are homogeneous, and then power-law strain hardening is a necessary and sufficient condition for proportional loading.

Thus, for materials whose strain hardening can be described by a power law, total and incremental strain theories lead to the same result for the plastic region at small scale yielding and for a sufficiently close vicinity of a crack edge at large scale yielding. One special case is $N = 0$, perfect plasticity, and another is $N = 1$, linear elasticity. Materials like structural steels may be approximately described by a power-law strain hardening with the exponent N around 0.05.

Mode III small scale yielding in strain hardening materials

Rice (1967) investigated anti-plane elastic-plastic strains at cracks and sharp notches in strain hardening materials, using a total strain theory. Then, equations (5.3.34) and (5.3.36) are applicable. For infinitesimally small scale yielding, he obtained the following equations for the strains in the plastic region at a crack edge, written in the present notations:

$$x = X(\gamma) - R(\gamma)\cos 2\psi, \quad y = -R(\gamma)\sin 2\psi \qquad (5.3.101)$$

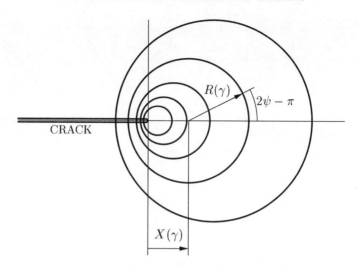

Fig. 5.3.11 Circles along which the strain magnitude is constant in the plastic region surrounding a crack edge under small scale yielding. The positions of the circle centres depend on the strain hardening characteristics.

where

$$R(\gamma) = \frac{J_{III}}{\pi \gamma \tau(\gamma)} \qquad (5.3.102)$$

$$X(\gamma) = \left[2\gamma\tau(\gamma) \int_\gamma^\infty \frac{du}{u^2 \tau(u)} - 1\right] \cdot R(\gamma) \qquad (5.3.103)$$

These relations may be considered as generalizations of the ones obtained for an elastic-perfectly plastic material; cf. (5.3.55) and (5.3.58). Plastic strains now appear also at the crack faces near the crack edge, if strain hardening is present, see Fig. 5.3.11. The strain magnitude, γ, is constant along circles with radii $R(\gamma)$ and centres at the distance $X(\gamma)$ ahead of the crack edge. Note that the radius of the plastic region, where $\gamma = \gamma_Y$, is independent of the strain hardening at given J_{III} and γ_Y.

The value of the J-integral, J_{III}, may be used to describe the state of loading. Because a total strain theory was used, it is path-independent. The solution is approximately correct, compared to the one which would have been obtained by the incremental strain theory, if the strain hardening of the material can be described approximately by a power law. Specialization to power-law strain hardening gives

$$R(\gamma) = \frac{J_{III}}{\pi \tau_Y \gamma_Y} \cdot \left(\frac{\gamma_Y}{\gamma}\right)^{1+N} \qquad (5.3.104)$$

$$X(\gamma) = \frac{1-N}{1+N} R(\gamma) \qquad (5.3.105)$$

so that

$$r\cos\varphi = R(\gamma)\left(\frac{1-N}{1+N} - \cos 2\psi\right) = \frac{2R(\gamma)}{1+N}(\sin^2\psi - N\cos^2\psi) \qquad (5.3.106)$$

$$r\sin\varphi = -R(\gamma)\sin 2\psi = -2R(\gamma)\sin\psi\cos\psi \qquad (5.3.107)$$

These equations may be used for the inversion from the hodograph plane to the real plane. They lead to the expressions:

$$\tan \psi = -\frac{1 + N + \sqrt{(1+N)^2 + 4N \tan^2 \varphi}}{2 \tan \varphi} \tag{5.3.108}$$

$$r = \frac{2}{1+N}\sqrt{\sin^2 \psi + N^2 \cos^2 \psi} \cdot R(\gamma) = \frac{2}{1+N}\sqrt{\frac{N^2 + \tan^2 \psi}{1 + \tan^2 \psi}} \cdot R(\gamma) \tag{5.3.109}$$

By the aid of (5.3.104), the explicit expression for γ is found to be:

$$\gamma = \gamma_Y \left[\frac{2\mu J_{III}}{(1+N)\pi \tau_Y^2 r}\right]^{1/(1+N)} \cdot [f(\varphi)]^{1/(1+N)}$$

$$= \gamma_Y \left[\frac{K_{III}^2}{(1+N)\pi \tau_Y^2 r}\right]^{1/(1+N)} \cdot [f(\varphi)]^{1/(1+N)} \tag{5.3.110}$$

where

$$f(\varphi) = \frac{1 - N + \sqrt{(1+N)^2 + 4N \tan^2 \varphi}}{2\sqrt{1 + \tan^2 \varphi}} \tag{5.3.111}$$

Then, for instance,

$$\gamma_{rz} = \gamma \cos(\psi - \varphi) \tag{5.3.112}$$

and

$$\gamma_{\varphi z} = \gamma \sin(\psi - \varphi) \tag{5.3.113}$$

may be expressed in terms of r and φ. The crack face slip, $2w$ in the crack edge vicinity, is obtained from (5.3.110) by putting $\varphi = \pi$ and using the relation $\partial w/\partial r = \gamma_{rz} = \gamma \cos(\psi - \varphi)$: the slip is found to be proportional to $r^{N/(1+N)}$, which shows how blunting develops as $N \to 0$.

Strains in the crack edge vicinity at large scale yielding in strain hardening materials

Rice (1967) showed that in cases of large scale yielding, a total strain theory and power-law strain hardening leads to exactly the same stress-strain field near a mode III crack edge as for small scale yielding, equations (5.3.112)-(5.3.110). The proof is based on expansion into eigenfunctions and identification of the dominant term, in a manner analogous to the one used in Section 3.3 for the elastic field near a mode III crack edge, although the present analysis is performed essentially in the hodograph plane.

For power-law strain hardening, equation (5.3.36) reduces to

$$\frac{1}{N} \cdot \frac{\partial^2 F}{\partial \gamma^2} + \frac{1}{\gamma} \cdot \frac{\partial F}{\partial \gamma} + \frac{1}{\gamma^2} \cdot \frac{\partial^2 F}{\partial \psi^2} = 0 \tag{5.3.114}$$

with solutions

$$F = -C\gamma^\beta \sin[\alpha(\pi/2 - \psi)], \qquad C = const \tag{5.3.115}$$

which satisfy the symmetry requirement that r is perpendicular to γ for $\varphi = 0$, i.e. $\psi = \pi/2$, which implies $\partial F/\partial \gamma = 0$, because

$$r = \frac{\partial F}{\partial \gamma}\hat{\gamma} + \frac{1}{\gamma} \cdot \frac{\partial F}{\partial \psi}\hat{\psi} \tag{5.3.116}$$

Substitution of F into equation (5.3.114) gives

$$\beta = \frac{1}{2}[1 - N \pm \sqrt{(1-N)^2 + 4N\alpha^2}] \tag{5.3.117}$$

and it is noted that the plus sign in front of the square root implies $\beta > 0$, the minus sign $\beta < 0$.

Relations between r and γ, and between φ and ψ are obtained through the equations

$$r \cdot \hat{\gamma} = -C\beta\gamma^{\beta-1}\sin[\alpha(\pi/2 - \psi)] = r\cos(\psi - \varphi) \tag{5.3.118}$$
$$r \cdot \hat{\psi} = C\alpha\gamma^{\beta-1}\cos[\alpha(\pi/2 - \psi)] = -r\sin(\psi - \varphi) \tag{5.3.119}$$

which give

$$r \propto \gamma^{\beta-1} \tag{5.3.120}$$

and

$$\tan(\pi/2 - \psi + \varphi) = \frac{\beta}{\alpha}\tan[\alpha(\pi/2 - \psi)] \tag{5.3.121}$$

Obviously, the sign of α is irrelevant; it is here assumed to be positive.

Equation (5.3.120) shows that $\gamma \propto r^{1/(\beta-1)}$ and, consequently, the stress-strain energy per unit volume is proportional to $\gamma\tau \propto \gamma^{1+N} \propto r^{(1+N)/(\beta-1)}$ in the crack edge vicinity. Then, because this energy must be bounded, permissible values of β must be situated in the intervals

$$\beta > 1 \text{ and } \beta < \frac{1-N}{2} \tag{5.3.122}$$

and it is obvious that the first interval implies regular and the second one singular strains.

For $0 < \varphi \ll 1$, the angle $\psi \approx \pi/2$. Thus,

$$\pi/2 - \psi + \varphi \approx \beta(\pi/2 - \psi) \implies \varphi \approx (\beta - 1)(\pi/2 - \psi) \tag{5.3.123}$$

and it can be concluded that ψ decreases with increasing φ if $\beta > 1$, but increases with φ if $\beta < 1$. Assume that ψ goes from $\pi/2$ to $\pi/2 + k\pi/2$ as φ goes from 0 to π. Then, because r is parallel with $\hat{\gamma}$, i.e. $r \cdot \hat{\psi} = 0$ for $\varphi = \pi$, the vanishing of the right member of (5.3.119) shows that

$$\cos(k\pi/2) = 0 \tag{5.3.124}$$

i.e. k is an odd integer, which obviously is negative if $\beta > 1$ and positive if $\beta < 1$.

The arguments in the two tangent functions in (5.3.121) must grow by the same amount when φ goes from 0 to π. Thus,

$$\alpha = \text{sgn}(\beta) \cdot \frac{k-2}{k} \tag{5.3.125}$$

and

$$1 < \beta: \quad k \leq -1, \quad \alpha = \frac{k-2}{k} \tag{5.3.126}$$

$$0 < \beta < 1: \quad k \geq 3, \quad \alpha = \frac{k-2}{k} \tag{5.3.127}$$

$$\beta < 0: \quad k = 1, \quad \alpha = 1 \tag{5.3.128}$$

It is easy to verify, from (5.3.117), that values according to (5.3.127) do not satisfy the requirement of bounded stress-strain energy, and that (5.3.126) leads to regular strains. Thus, the only singular strains are those for which $k = 1$, which implies an $r^{1/(1+N)}$ singularity in the strains and an $r^{N/(1+N)}$ singularity in the stresses. For $k = 1$, the equations (5.3.118)-(5.3.119) are reduced to

$$CN\gamma^{-(1+N)} \cos\psi = r\cos(\psi - \varphi) \tag{5.3.129}$$

$$C\gamma^{-(1+N)} \sin\psi = -r\sin(\psi - \varphi) \tag{5.3.130}$$

enabling the solution for r and ψ:

$$\tan\psi = -\frac{1 + N + \sqrt{(1+N)^2 + 4N\tan^2\varphi}}{2\tan\varphi} \tag{5.3.131}$$

$$r = C\sqrt{\sin^2\psi + N^2\cos^2\psi} \cdot \gamma^{-(1+N)} \tag{5.3.132}$$

These equations agree with (5.3.108)-(5.3.109), apart from the fact that r now contains an undeterminable constant factor. Hence, apart from such a factor, the strains in the crack edge vicinity are the same as in first two members of equation (5.3.110), with J_{III} interpreted as the J-integral for paths in the crack edge vicinity.

From expressions (5.3.101) and (5.3.105), the ratio between the extensions of the plastic region backwards and forwards from the crack edge is found to be $(R-X)/(R+X) = N$ at small scale yielding. This indicates that the crack edge vicinity in which the solution is valid becomes very small when perfect plasticity is approached. Now, the results in this subsection were obtained by (tacitly) assuming the extension of the plastic region to be infinite, in order to be able to use a homogeneous boundary condition on the crack faces. Therefore, the results may be expected to be accurate only in a crack edge vicinity which is much smaller than the backwards extension of the plastic region. A demonstration of this fact is provided by the perfectly plastic case, in which the backwards extension of the plastic region has shrunk to zero. Letting $N \to 0$ in the expression for $f(\varphi)$, equation (5.3.111), which, as shown, is valid also for the crack edge vicinity at large scale yielding, results in $f(\varphi) = \cos\varphi$, but, as shown by (5.3.20), the angular distribution of perfectly plastic strains is not unique at large scale of yielding in any vicinity of the crack edge.

For $N \ll 1$, equation (5.3.117) gives $\beta \approx -N\alpha^2 \ll 1$, for the dominating term in the crack edge vicinity. This results, approximately, in

$$r \approx C\alpha\gamma^{\beta-1} \cos[\alpha(\pi/2 - \psi)] \tag{5.3.133}$$

If the crack edge vicinity of interest is larger than the backwards extension of the plastic region, then a larger part of the plastic region boundary than a small part along the crack faces has to be used for boundary conditions. A full solution is obtained by matching solutions for elastic and plastic regions, as described on pages 276ff. The value of α in (5.3.133) may turn out to be very sensitive to boundary conditions at large scale of yielding.

The hodograph representation of the complete field in the plastic region is obtained by summation over all permissible terms of the form (5.3.115)-(5.3.116), obtained after the substitution $k = 2 - m$:

$$r = \sum_{1,3,5,\ldots} C_m \gamma^{\beta_m - 1} \left\{ \beta_m \sin\left[\frac{m}{m-2}\left(\frac{\pi}{2} - \psi\right)\right] \hat{\gamma} - \frac{m}{m-2} \cos\left[\frac{m}{m-2}\left(\frac{\pi}{2} - \psi\right)\right] \hat{\psi} \right\} \quad (5.3.134)$$

where

$$\beta_m = \frac{1}{2}\left[1 - N \pm \sqrt{(1-N)^2 + 4N\left(\frac{m}{m-2}\right)^2}\right] \quad (5.3.135)$$

with minus-sign chosen for $m = 1$ and plus-sign otherwise.

A note on the near edge autonomy for mode III elastic-plastic cracks

The fact that a unique solution, apart from an amplitude constant, could be found for the crack edge vicinity of a mode III crack in a power-law strain hardening material, implies that near edge autonomy prevails. Power-law strain hardening justifies the use of a total strain theory. For other kinds of hardening behaviour, autonomy is also to be expected, but analytical solutions, using total strain theory, are only approximative, even in the vicinity of the crack edge.

Autonomy is in general no longer present, when the strain hardening disappears, i.e. it is not present for an elastic-perfectly plastic material. This was discussed in connection with the general equations (5.3.19)-(5.3.20) for stresses and strains in the plastic region; the stresses are autonomous, but the strains are not. However, restriction to the class of small scale yielding situations, ensures autonomy near the crack edge, due to the fixed shape – circle, according to the Hult-McClintock solution – of the plastic region.

For materials with low strain hardening, the region where the asymptotic solution dominates is very small. This was shown by Gudmundson (1989) for linear and power law strain hardening materials and for a viscoplastic material with high fluidity. He investigated both modes I and III under small scale yielding and found that the stress-strain field in the main part of the plastic region is virtually the same as for an elastic-perfectly plastic material. Thus, considering the presence of a process region, the asymptotic solution for a strain hardening material, and the stronger singularity it predicts compared to an elastic-perfectly plastic material, may not have any significance at all for real materials.

Matching solutions for elastic and plastic regions

For a linearly elastic region, the position vector r is given by (5.3.43), i.e.

$$r = \sum_{1,3,5,\ldots} n\{[A_n\gamma^{n-1} - A_{-n}\gamma^{-n-1}]\cos n\psi \cdot \hat{\gamma}$$
$$- [A_n\gamma^{n-1} + A_{-n}\gamma^{-n-1}]\sin n\psi \cdot \hat{\psi}\} \quad (5.3.136)$$

Because this vector is continuous across the elastic-plastic boundary, the matching condition is given by equalizing the right members of (5.3.134) and (5.3.136) for $\gamma = \gamma_Y$. The procedure will be demonstrated for infinitesimally small scale yielding. Then, $r \propto \gamma^{-2}$ as $\gamma \to 0$, so that (5.3.136) may be written as

$$r = (A_1 - A_{-1}\gamma^{-2})\cos\psi \cdot \hat{\gamma} - (A_1 + A_{-1}\gamma^{-2})\sin\psi \cdot \hat{\psi}$$
$$+ \sum_{3,5,\ldots} n\gamma^{n-1}\{A_n\cos n\psi \cdot \hat{\gamma} - A_n\gamma^{-2}\sin n\psi \cdot \hat{\psi}\} \quad (5.3.137)$$

and by writing (5.3.134) in a similar way,

$$r = C_1\gamma^{-(1+N)}(N\cos\psi \cdot \hat{\gamma} + \sin\psi \cdot \hat{\psi})$$
$$+ \sum_{3,5,\ldots} C_m\gamma^{\beta_m-1}\left\{\beta_m\sin\left[\frac{m}{m-2}\left(\frac{\pi}{2}-\psi\right)\right]\hat{\gamma}\right.$$
$$\left. - \frac{m}{m-2}\cos\left[\frac{m}{m-2}\left(\frac{\pi}{2}-\psi\right)\right]\hat{\psi}\right\} \quad (5.3.138)$$

it is detected that matching can be accomplished only by choosing $A_n = 0$ for $n \geq 3$ and $C_m = 0$ for $m \geq 3$. Equalizing coefficients for $\cos\psi\cdot\hat{\gamma}$ and $\sin\psi\cdot\hat{\psi}$, and expressing coefficients A_1 and A_{-1} in terms of C_1, results in the expression

$$r = C_1\gamma_Y^{-(1+N)}(N\cos\psi \cdot \hat{\gamma} + \sin\psi \cdot \hat{\psi}) \quad (5.3.139)$$

and, because a study of (5.3.137) shows that

$$A_1 = -\frac{K_{III}^2}{2\pi\mu^2} \quad (5.3.140)$$

the position vector is

$$r = \frac{K_{III}^2}{(1+N)\pi\tau_Y^2}\cdot\left(\frac{\gamma_Y}{\gamma}\right)^{1+N}(N\cos\psi \cdot \hat{\gamma} + \sin\psi \cdot \hat{\psi}) \quad (5.3.141)$$

It is easy to verify that this expression may be inverted to the previously given result for infinitesimally small scale yielding, (5.3.110).

Fully plastic solutions

No full solutions exist for large scale yielding crack problems in elastic-plastic strain hardening materials. However, some insight about the behaviour at very large scale of yielding might be gained from solutions obtained under the assumption of a power-law stress-strain relation for $0 < \gamma < \infty$ (a "fully plastic" material). This assumption was used by Amazigo (1974, 1975) for mode III cracks. In the earlier paper, he studied a crack in an infinite body and in the later paper a centre-cracked strip. Using a hodograph transformation and the Wiener-Hopf technique, he succeeded in solving the problems analytically. The Wiener-Hopf equation was obtained after Mellin transformation. The results are not easily readable, containing rather complicated expressions, but they may be numerically evaluated to any desired accuracy. Of particular interest is the J-integral, which Amazigo evaluated. He found that for the infinite plate and $N < 0.1$, the numerical results could be approximated with good accuracy by the formula

$$J_{III} \approx \frac{1}{2\mu}\sqrt{\frac{\pi^3}{2N}}\left(\frac{\tau_{yz}^\infty}{\tau_0}\right)^2 \gamma_0 \tau_0 a \qquad (5.3.142)$$

which he assumed to be correct in the limit $N \to 0$. γ_0 is a reference shear strain and τ_0 a reference shear stress, such that $\tau/\tau_0 = (\gamma/\gamma_0)^N$. The formula indicates that, at large scale yielding of elastic-plastic materials with low strain hardening, the J-integral at a crack edge is very sensitive to the strain hardening exponent N, increasing rapidly with decreasing N. Such a sensitivity was also noted by Goldman and Hutchinson (1975) who treated the fully plastic problem of a centre-cracked strip under plane strain and mode I loading by means of a numerical method. The formula also indicates that the J-integral is approximately proportional to the applied load at very low strain hardening.

For power-law strain hardening, there is a complete equivalence between total and incremental strain theories, thus not only in the region near the crack edge. Thinking of the potential function F as the height of a surface in the γ_{xz}, γ_{yz} plane, this equivalence becomes obvious. Increasing the magnitude of the boundary strains by a factor c, implies radial expansion of the F-surface by the same factor, due to the homogeneous character of the left member of (5.3.114), the governing equation for F. From the relation $\boldsymbol{\tau} = \operatorname{grad}_\gamma F$, the right member of which may be interpreted as the slope of the F-surface, it can be seen that γ_{xz} and γ_{yz} at each point (x,y) become increased by the factor c, i.e. proportional loading prevails everywhere. This result is also valid for mode I, as shown by Goldman and Hutchinson (1975).

5.4 Stationary mode I cracks

Cracks in perfectly plastic materials. Slip line solutions

In the study of elastic-perfectly plastic mode III cracks, it was found that autonomy near the crack edge exists for the stresses, though not in general for the displacements. For mode I, the situation is less simple, because the remote normal stress, acting in parallel with the crack direction, is not fixed *a priori*. Rather, it may vary widely between different body and loading geometries, and it may be of the same order of magnitude as the yield strength. If this stress, usually called the T-stress (see page 80), is positive, it contributes a tendency towards increasing the mean stress and decreasing the effective stress in the plastic region, thereby constraining plastic flow. On the other hand, if the T-stress is negative, larger plastic strains result. Thus, the stress-strain state in the plastic region may vary considerably between different body and loading geometries for small scale yelding in one and the same material. Thus, autonomy is lost in the general case, but it ought to prevail if only cases with the same T-stress are compared.

Plane strain *slip line theory*, see e.g. Hill (1950), offers a particularly expedient way to construct perfectly-plastic stress fields. The material model is a rigid-perfectly plastic one, i.e. incompressibility prevails for both elastic and plastic strains. Either the Huber-von Mises or the Tresca yield condition may be assumed – they lead to identical solutions in terms of the flow stress in shear, τ_Y, which, however, equals $\sigma_Y/\sqrt{3}$ for a Huber-von Mises material and $\sigma_Y/2$ for a Tresca material. Two families of slip lines (characteristic lines, in the directions of maximum shear stress) exist, the α lines in the direction $-\pi/4$ and the β lines in the direction $+\pi/4$ from the direction of the largest principal stress.

The stresses may be written as

$$\sigma_x = \sigma_m - \tau_Y \sin 2\psi \qquad (5.4.1)$$
$$\sigma_y = \sigma_m + \tau_Y \sin 2\psi \qquad (5.4.2)$$
$$\tau_{xy} = \tau_Y \cos 2\psi \qquad (5.4.3)$$

where $\sigma_m = (\sigma_x + \sigma_y)/2$ is the mean in-plane stress and ψ is the angle between the α line and the x axis. For a Huber-von Mises material, the flow rule together with incompressibility implies that $\sigma_m = (\sigma_x + \sigma_y + \sigma_z)/3$, because $d\epsilon_z^p = 0$.

Consider an arbitrary point P and orient the coordinate system so that its origin is at P and the x axis is tangent to the α line through P. The equilibrium equation $\partial \sigma_x/\partial x + \partial \tau_{xy}/\partial y = 0$ gives

$$\frac{\partial \sigma_m}{\partial x} - 2\tau_Y \cos 2\psi \frac{\partial \psi}{\partial x} - 2\tau_Y \sin 2\psi \frac{\partial \psi}{\partial y} = 0 \qquad (5.4.4)$$

Thus, in the vicinity of the origin, the relation

$$\frac{\partial}{\partial x}(\sigma_m - 2\tau_Y \psi) = 0 \qquad (5.4.5)$$

holds approximately. In this vicinity, integration along the α line through P is approximately equivalent to integration with respect to x, which leads to the result that $\sigma_m - 2\tau_Y \psi$ is constant. But P is an arbitrary point, so this result holds along the whole α line. A similar result is obtained for β lines, and thus:

$$\sigma_m - 2\tau_Y \psi = \text{constant along an } \alpha \text{ line} \qquad (5.4.6)$$
$$\sigma_m + 2\tau_Y \psi = \text{constant along a } \beta \text{ line} \qquad (5.4.7)$$

Asymptotic stresses. Constant stress and centered fan sectors

One difficulty when using slip line theory is that elastic strains are not considered, so that good approximations are to be expected only in regions where the plastic strains are considerably larger than the elastic ones. Thus, the approximation is less good near a boundary towards an elastic region. When analysing the plastic stress field near a crack edge, this difficulty may be overcome by assuming that the plastic field extends all around the crack edge, i.e. also on the crack faces near the edge. Then, the plastic strains may be expected to be large sufficiently close to the crack edge. Using this assumption, Cherepanov (1967) and Rice (1968a,b) applied the Prandtl field for

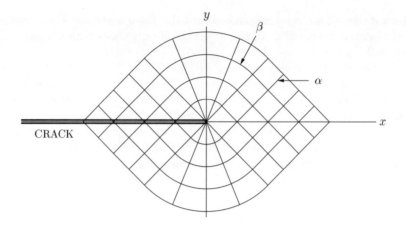

Fig. 5.4.1 The Prandtl slip line field, applied to the case of a mode I crack in an incompressible perfectly plastic material.

a crack edge vicinity. This field is known to give good results for the inverse problem, punching. It consists (on each side of the crack line) of two triangles with straight α and β lines and one fan with straight α lines, radiating from the crack edge; see Fig. 5.4.1.

At the crack surface,

$$\sigma_y = 0, \qquad \tau_{xy} = 0, \qquad \psi = 3\pi/4 \qquad (5.4.8)$$

because ψ is the angle between an α line and the x axis. Thus, according to (5.4.2),

$$\sigma_m - \tau_Y = 0 \qquad (5.4.9)$$

whereas (5.4.3) is satisfied identically. Then, as seen from (5.4.1),

$$\sigma_x = 2\tau_Y \qquad (5.4.10)$$

along the crack faces, and, actually, the stress state is uniform in the whole triangular region behind the crack edge.

By following a β line from the crack face to the triangular region ahead of the crack edge, the mean stress in this region is obtained by using (5.4.7):

$$\sigma_m + 2\tau_Y \cdot \frac{\pi}{4} = \tau_Y + 2\tau_Y \cdot \frac{3\pi}{4} \qquad (5.4.11)$$

i.e.

$$\sigma_m = (1 + \pi)\tau_Y \qquad (5.4.12)$$

Hence, in the triangular region ahead of the crack edge, the stresses are

$$\sigma_x = \pi\tau_Y, \qquad \sigma_y = (2 + \pi)\tau_Y \qquad (5.4.13)$$
$$\tau_{xy} = 0 \qquad (5.4.14)$$

Thus, σ_y equals about $2.97\sigma_Y$ in a Huber-von Mises material and about $2.57\sigma_Y$ in a Tresca material.

Expressed in cylindrical coordinates, the stresses in the centered fan region are

$$\sigma_r = \sigma_\varphi = (1 + 3\pi/2 - 2\varphi)\tau_Y \tag{5.4.15}$$

$$\tau_{r\varphi} = \tau_Y \tag{5.4.16}$$

It is then found that all stress components are continuous across the region boundaries $\varphi = \pi/4$ and $\varphi = 3\pi/4$. However, continuity of σ_r is not a necessary condition and should thus be considered as a feature of the Prandtl slip line field. In fact, Drugan and Miao (1992) showed that use of yield condition and associated flow rule for a type of porous material, results in radial normal stress discontinuities when the porosity exceeds about 3%. However, they later showed (Miao and Drugan 1993) that in such cases, insertion of elastic sectors could be done so as to remove stress discontinuities everywhere.

Note that two types of sectors may appear at a crack edge: *constant stress sectors* and *centered fan sectors*. In a centered fan sector where the α lines are radial at the angle $\psi = \varphi$, the general expressions for the stresses are

$$\sigma_r = \sigma_\varphi = \text{constant} - 2\tau_Y \varphi, \qquad \tau_{r\varphi} = \tau_Y \tag{5.4.17}$$

The same expressions hold for σ_r and σ_φ if the β lines are radial at the angle φ, but then $\tau_{r\varphi} = -\tau_Y$ because $\psi = \pi/2 - \varphi$.

The division into constant stress and centered fan sectors will be seen to be a general feature of asymptotic plastic stress fields for stationary, slowly moving (with neglection of inertia) and dynamically moving mode I, II or III cracks.

Asymptotic strains

Incremental and total strain theories give the same result under proportional loading. An obvious consequence of the general assumption of plastic incompressibility under proportional loading is that the plastic strain is zero in the direction of a principal shear stress. This follows from the fact that a "fibre" in such a direction is subjected to shear tractions, which do not cause fibre extension, and, in addition, to a hydrostatic pressure, which does not cause plastic strains. This property is used in slip line theory to determine deformation (or displacement rate) fields, cf. Hill (1950). Rice (1968b) pointed out that severe strain concentration can occur only in the fan, and that the vanishing of ϵ_r implies that u_r is independent of r, so that the displacements in the fan close to the crack edge may be represented by two functions, $f(\varphi)$ and $g(r)$:

$$u_r = f'(\varphi) \tag{5.4.18}$$

$$u_\varphi = -\int u_r \, d\varphi = -f(\varphi) + g(r) \tag{5.4.19}$$

The last relation follows from incompressibility, $\epsilon_r + \epsilon_\phi = 0$. Now, the shear strain in the fan near the crack edge is found from a general expression,

$$\gamma_{r\varphi} = \frac{1}{r} \cdot \frac{\partial u_r}{\partial \varphi} + \frac{\partial u_\varphi}{\partial r} - \frac{u_\varphi}{r} = g'(r) + \frac{1}{r} \cdot [f''(\varphi) + f(\varphi) - g(r)] \tag{5.4.20}$$

The uniform stress state in the triangular region ahead of the fan implies that all strains, even the shear strains, and the displacement gradients are bounded in this region. Because u_φ is continuous across the boundary to the centered fan, the expression (5.4.19) shows that both $g(r)$ and $g'(r)$ are bounded, and consequently equation (5.4.20) gives, for sufficiently small r,

$$\gamma_{r\varphi} = \frac{f''(\varphi) + f(\varphi)}{r} \quad \text{for} \quad \frac{\pi}{4} < \varphi < \frac{3\pi}{4} \tag{5.4.21}$$

Following Rice (1968b), though with insignificant changes in notations, this equation is written as

$$\boxed{\gamma_{r\varphi} = \gamma_Y \cdot \frac{R(\varphi)}{r} \quad \text{for} \quad \frac{\pi}{4} < \varphi < \frac{3\pi}{4}} \tag{5.4.22}$$

by analogy with the general expression for the mode III shear strain in the perfectly plastic region at a crack edge, if $\gamma_Y = \tau_Y/\mu$ is the yield strain in shear. Here, however, the interpretation of $R(\varphi)$ as the distance to the elastic-plastic boundary is rather uncertain, because the slip line field cannot be expected to accurately represent the actual field in peripheral parts of the plastic region. However, equation (5.4.22) shows that *the shear strains exhibit a $1/r$ singularity, just as in the perfectly plastic mode III case*, although a remarkable difference is that the strains, according to the presently used small strain theory, are found to be non-singular straight ahead of the crack, in the sector $-\pi/4 \leq \varphi \leq \pi/4$. This rather unexpected outcome will soon be discussed, but first another similarity with the mode III case, *complete crack edge blunting* will be demonstrated. To this end the displacement v in the y direction is considered:

$$v = u_r \sin\varphi + u_\varphi \cos\varphi \tag{5.4.23}$$

Thus,

$$\frac{\mathrm{d}v}{\mathrm{d}\varphi} = [f''(\varphi) + f(\varphi)] \sin\varphi = \gamma_Y R(\varphi) \sin\varphi \tag{5.4.24}$$

so that the crack edge opening displacement is

$$\delta_0 = 2\gamma_Y \int_{\pi/4}^{3\pi/4} R(\varphi) \sin\varphi \, \mathrm{d}\varphi \tag{5.4.25}$$

$R(\varphi)$ is certainly positive, because $\tau_{r\varphi}$ is positive in the fan, and therefore the crack edge opening displacement is certainly finite, just as in the perfectly plastic mode III case. There is also a difference in the displacement u in the positive x direction between the point $r = 0, \varphi = 0$ (which remains at the same position as the point $r = 0, \varphi = \pi/4$) and $r = 0, \varphi = 3\pi/4$, but it is substantially smaller than $\delta_0/2$, because it is obtained by replacing $\sin\varphi$ by $\cos\varphi$ in (5.4.25). The blunting profile is thus rounded, and it is obvious that the small strain theory applied here might not give a good picture of strains and displacements near the crack edge, a fact that prompted Rice (1968b) to modify the slip line field in order to investigate the role of large geometry changes in the crack edge vicinity.

Consideration of large geometry changes
The modification of the slip line fields, to account for large geometry changes, makes use of the logarithmic spiral,

$$r = r_0 e^{\varphi \cot\theta} \tag{5.4.26}$$

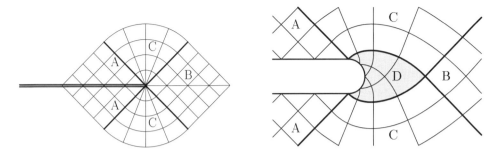

Fig. 5.4.2 Slip line field, modified to consider large geometry changes at the crack edge (right figure). Compared to the field for small geometry changes (left figure), a new, highly strained region (D), has appeared. The slip lines in this region consist of logarithmic spirals, if the blunting profile is semicircular.

which is defined by the property that radius vector from the origin cuts the spiral at the constant angle θ. Choosing $\theta = \pm \pi/4$ gives two families of spirals, cutting each other at right angle, and they may therefore be used as α and β lines to construct a slip line field (Hill 1950), provided that boundary conditions are satisfied. For a traction free blunted crack edge, these conditions are that the slip lines meet the boundary at the angle $\pm \pi/4$, which is satisfied by logarithmic spirals, if semicircular blunting, $r = \delta_0/2$, $-\pi/2 \leq \varphi \leq \pi/2$, is assumed. Even though this shape is less blunt than predicted by the small geometry change theory, it leads to an estimate of the size of the region of high stresses ahead of the crack edge.

Figure 5.4.2 shows the slip line field after modification.

The logarithmic spirals emanate from the semicircular boundary, and consequently the equation for such a spiral is

$$r = \frac{\delta_0}{2} e^{\pm(\varphi_0 - \varphi)}, \quad -\frac{\pi}{2} \leq \varphi \leq \frac{\pi}{2} \tag{5.4.27}$$

where φ_0 indicates the angular position at the semicircle from which the spiral emanates. This solution is valid in the region containing both α and β spiral lines, i.e. in the region D in Fig. 5.4.2. The spiral boundaries of this region are

$$r = \frac{\delta_0}{2} e^{\pi/2 - \varphi} \text{ for } 0 \leq \varphi \leq \frac{\pi}{2} \tag{5.4.28}$$

$$r = \frac{\delta_0}{2} e^{\pi/2 + \varphi} \text{ for } -\frac{\pi}{2} \leq \varphi \leq 0 \tag{5.4.29}$$

and, consequently, the length of region D is

$$(e^{\pi/2} - 1)\delta_0/2 \approx 1.9\delta_0 \tag{5.4.30}$$

This might be some overestimate, because the crack edge profile is probably more blunt than a semicircular shape, but it shows that the length dimensions of region D are of the same order as the crack edge opening displacement δ_0. Region A in Fig. 5.4.2 agrees completely with region A in Fig. 5.4.1 and the same applies to region B, even though it is moved forwards from the crack edge. The slip lines in region C, on the other hand, form now a *non-centered* fan with a spiralling continuation into region D to end at the blunted edge.

Coincidence between incremental and total perfectly plastic theory requires that the stress state in the plastic region remains constant during loading. This requirement was fulfilled for the slip line field when small geometry changes were assumed, but after the modification, it is apparently only approximately satisfied in region D and its vicinity.

It is easy to see that region D is much smaller than the whole plastic region: equation (5.4.25) shows that the ratio between the linear extension of D and the maximum radius $R_{max} = [R(\varphi)]_{max}$ of the plastic region equals about γ_Y, if $R(\varphi)$ is assumed to vary reasonably smoothly between $\varphi = \pi/4$ and $\varphi = 3\pi/4$. The shear strains in region C are therefore still given, approximately, by (5.4.22), except in the peripheral parts where slip line theory is less accurate, and near the region D. However, it is obvious that the strains increase monotonically along a slip line in regions C and D towards the blunted crack edge. The maximum shear strain at the boundary of region D is thus found to be

$$\boxed{\gamma \approx \gamma_Y \cdot \frac{R_{max}}{\delta_0} \approx 1} \qquad (5.4.31)$$

and inside region D, it increases towards the crack edge. The appearence of such high strains gives a clear signal that a small strain theory cannot give accurate results for the strains in region D. However, in spite of the very approximate character of the modified slip line field, some conclusions may be drawn from the investigation.

One conclusion is simply that *very large strains appear even in front of the crack edge* in contrast to what was found when large geometry changes were disregarded. This is consistent with observations of a "stretched zone" ahead of the crack edge before stable crack growth occurs. The region of large strains may be small, compared to the maximum radius of the plastic region, but its position suggests that it should rather be compared with relevant micro-structural dimensions, for instance the distance between particles from which voids can nucleate (Rice and Johnson 1970). When void growth occurs, the stress-strain field near the blunted edge changes dramatically and becomes dominated by the development of a process region, and later on by stable crack growth.

Because the tractions on the blunted edge are zero, the σ_φ stresses are considerably larger than the σ_r stresses in region D. The same then also applies to the strains: the ϵ_φ strains are considerably larger than the ϵ_r strains. In region B, the σ_y stress is larger than the σ_x stress: the small geometry change slip line field gives their ratio as

$$\frac{\sigma_y}{\sigma_x} = \frac{2+\pi}{\pi} \approx 1.6 \qquad (5.4.32)$$

This contrasts to the relation $\sigma_y = \sigma_x$ in front of a crack edge in a linearly elastic field dominated by a $1/\sqrt{r}$ singularity, and probably provides an explanation of the fact that straightforward mode crack growth direction at mode I loading is generally very stable. Consideration of large geometry changes would increase the ratio σ_y/σ_x in the vicinity of the blunted edge, because $\sigma_x \to 0$ as the edge is approached on the symmetry line.

Consideration of the T-stress
Both slip line fields discussed so far are based on the assumption that the plastic region completely surrounds the crack edge. This fact implies, unfortunately, that only a very special case was considered, namely a high and positive T-stress. This is obvious by studying the region A in Fig. 5.4.1. The stress in this region is $\sigma_x = 2\tau_Y$, indicating that the T-stress is about equal to the yield stress. Such high T-stresses are rare in engineering applications, where they are often negative (compressive) or positive but small, compared to the yield stress. High positive T-stresses are mostly associated with deeply cracked geometries, particularly when loaded in bending, such as some popular test specimens.

When the T-stress is negative or positive but small enough, the plastic region does not extend to the crack faces. Numerical investigations by Du and Hancock (1991) indicate that the crack faces remain elastic when the T-stress is about $0.446\sigma_Y$ (in a Huber-von Mises material) or smaller. They also claimed that the Prandtl field is still valid, except that it is truncated behind a straight slip line in the fan and that the stresses in the fan and the region A are reduced by a superposed compressive hydrostatic stress field. For $T = 0$, they found that the fan extends to $\varphi = 130°$ and the stresses are reduced from those in the full Prandtl field by about $0.07\sigma_Y$. For $T = -0.443\sigma_Y$, the corresponding figures are about $105°$ and $0.5\sigma_Y$, and for $T = -0.7\sigma_Y$ they are $90°$ and $1.2\sigma_Y$. This implies rather drastic reductions in the stresses in the plastic region near the crack edge in one and the same material when the T-stress is decreased from a positive value to a stress lower than about $-0.5\sigma_Y$. However, such low T-stresses are not likely to be reached during loading of the body before other drastic changes have occurred, for instance induced by formation and growth of voids or onset of stable crack growth.

Cracks in strain hardening materials

The HRR singularity
In 1968, two papers that have had a significant impact on the fracture mechanics community, appeared after each other in the same issue of the *Journal of the Mechanics and Physics of Solids*, addressing the problem of finding the stress-strain field in the vicinity of a mode I crack edge in a strain hardening material. One was authored by Rice and Rosengren (1968), who assumed power-law strain hardening, (5.2.28), and the other one by Hutchinson (1968a), who used the Ramberg-Osgood model, (5.2.31), and also considered linear strain hardening and plane stress. Note that a Ramberg-Osgood material and a power-law strain hardening material are equivalent when only asymptotic stress-strain fields are required. Then, the exponent n in the Ramberg-Osgood material corresponds to $1/N$ in the power-law material. This identification makes sense only for very large strains, because otherwise the two materials exhibit different strain hardening characteristics. Both materials yield solutions belonging to the small strain theory of plasticity, which implies stress and strain singularity at the crack edge.

Both Hutchinson (1968a) and Rice and Rosengren (1968) assumed that a total strain theory could be used, which appears to be a reasonable assumption for the edge vicinity of a stationary crack in a power-law type strain hardening material†. They used numerical methods to calculate the angular distribution of stresses and strains. Because the J-integral is path-independent under the assumptions made (total strain theory and no unloading), and consequently finite for a path surrounding the crack edge, the singular character of stresses and strains is immediately found by choosing a circular path with radius r:

$$J = \int_{-\pi}^{+\pi} [W \cos \varphi - n_j \sigma_{ij}(\partial u_i/\partial x_1)] r \, d\varphi \qquad (5.4.33)$$

The integrand is obviously independent of r in the crack edge vicinity, if the product of stresses and displacement gradients exhibits a $1/r$ singularity, at least in some sector from the crack edge. By using the relation between stresses and strains, the power-law model or the Ramberg-Osgood model for large strains, the singularities can be separated. For power-law strain hardening, it is an $N/(N+1)$ singularity for stresses and a $1/(N+1)$ singularity for strains. For the Ramberg-Osgood model, there is a $1/(n+1)$ singularity for stresses and an $n/(n+1)$ singularity for strains. With an ansatz according to this finding, stresses and strains in the crack edge vicinity may be obtained by using an Airy stress function and appropriate boundary conditions (traction free crack surfaces).

Assuming the Ramberg-Osgood material, the results for plane strain may be expressed in the form‡:

$$\sigma_{ij} = \sigma_0 \left(\frac{J_I}{\alpha \sigma_0 \epsilon_0 I_n r} \right)^{1/(n+1)} f_{ij}(\varphi; n) \qquad (5.4.34)$$

$$\epsilon_{ij} = \alpha \epsilon_0 \left(\frac{J_I}{\alpha \sigma_0 \epsilon_0 I_n r} \right)^{n/(n+1)} g_{ij}(\varphi; n) \qquad (5.4.35)$$

Here, according to common usage, the reference stress and strain are denoted by σ_0 and ϵ_0, respectively; see (5.2.31). The angular distributions of stresses and strains, here given by $f_{ij}(\varphi; n)$ and $g_{ij}(\varphi; n)$, were calculated for different values of n (or N) and φ by Hutchinson (1968a). They are normalized by setting $[f_{eff}(\varphi; n)]_{max} = 1$, where $f_{eff}(\varphi; n)$ is the angular variation of the Huber-von Mises effective stress. Then, the dimensionless integration constant I_n becomes uniquely defined. The results (5.4.34) and (5.4.35) are generally referred to as the *HRR singularity*. As later shown by Rice (1988a), the calculation of HRR type fields can be greatly simplified by using two general integrals of the governing equations for singular deformation fields.

Some further results belonging to the family of HRR fields, were presented by Shih (1974), who considered pure and combined mode I and II conditions. Pan and Shih examined such fields for orthotropic materials in plane strain (Pan and Shih 1986), and later, (Pan and Shih 1988), they gave solutions for plane stress plastic stress-strain fields for mode I in orthotropic materials. In these papers, accurate calculations of the integration constant I_n can be found. Plastic orthotropy was described by a quadratic yield function proposed by Hill (1948). Pan and Shih (1992) also determined combined mode I, II and III asymptotic plastic fields.

---

† Compare with the discussion on page 270 for mode III, and with the results for fully plastic solutions in mode I by Goldman and Hutchinson (1975), briefly described on page 278.
‡ Expressions for the power-law material are obtained by the substitutions $n \to 1/N$, $\alpha \epsilon_0 \to \epsilon_Y$, $\sigma_0 \to \sigma_Y$.

The HRR field is an asymptotic field that is found without assuming small scale yielding. It should therefore also be valid for large scale yielding. However, in the limit $N \to 0$ ($n \to \infty$), the asymptotic field is not unique between different cases of large scale yielding, because of a loss of autonomy in elastic-perfectly plastic materials, as discussed for mode III on pages 257 and 268. Thus, it depends on body and loading geometry. As $r \to 0$, the stresses become bounded when $N \to 0$ ($n \to \infty$), but any small amount of strain hardening leads to infinite stresses, and autonomy then prevails in a certain vicinity of the crack edge. However, it is obvious that this vicinity shrinks continuously toward zero when N appoaches zero ($n \to \infty$), cf. Gudmundson (1989). For low degrees of strain hardening, the region of autonomy may therefore be too small to encompass a process region.

In equations (5.4.34) and (5.4.35), the distance r to the crack edge appears in the context $r\sigma_0/J_I$, which is logical, because $r\sigma_0/J_I$ is the only non-dimensional distance that can be found for the plastic region in the crack edge vicinity, where elastic strains are negligible†. Distances deep inside the plastic region thus scale as J_I/σ_0. Another scale will later be introduced for the plastic region region as a whole in small scale yielding: it scales primarily as EJ_I/σ_0^2 or K_I^2/σ_0^2, but there is also some influence of Poisson's ratio (page 291).

Whereas the small strain theory of plasticity (see page 248) was used for establishing the HRR singularity, a large strain theory (or finite deformation theory) will provide more accurate results in a close vicinity of the crack edge. Application of a large strain theory results in a lower ratio between mean stress and maximum shear stress in the highly strained region, where σ_x is very small. Even σ_y is typically decreasing towards the blunted crack edge, as schematically shown in Fig. 5.4.3. Several finite element calculations, e.g. O'Dowd and Shih (1991), assuming finite strain plasticity, have shown such relations and how they depend on strain hardening. Note that σ_y is now bounded, and so also are the other stresses.

As in the discussion of slip line field solutions, the results are expected to hold during loading of the body only as long as the continuum approach may be used even for the crack edge vicinity. Thus, when micro-structural mechanisms start operating to form a process region (which occurs at a significantly lower load than at onset of stable crack growth and generally at a much smaller load than at onset of unstable crack growth), the solutions obtained are at best approximately valid at some distance outside the process region, if this is deeply embedded in the plastic region. The investigations by Gudmundson (1989), page 276, indicate that the region of dominance of the asymptotic solution is so small that it can only accommodate a real process region in cases of considerable strain hardening. For low strain hardening, he found that the stress-strain field in the main part of the plastic region is virtually the same as for elastic-perfectly plastic material.

Consideration of the T-stress. The Q-parameter
The assumptions made about the boundary conditions for HRR fields imply that the plastic region envelopes the crack edge. For $N \to 0$ ($n \to \infty$), perfect plasticity, the

---

† The reason why also α, σ_0 and ϵ_0 appear together with r in (5.4.34) and (5.4.35) is that they are constants in the particular constitutive equation used. Recall from the discussion about (5.2.31) that σ_0 and ϵ_0 do not in general have any obvious physical significance, even though σ_0 for some materials may be approximately equal to the yield stress.

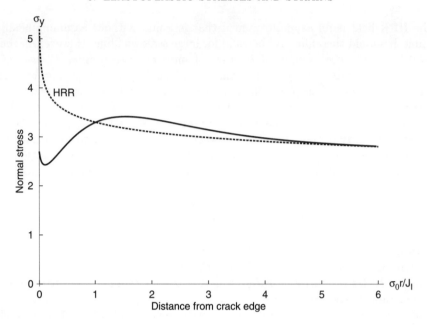

Fig. 5.4.3 Typical relation between the normal stress σ_y straight ahead of the crack and the distance from the crack edge for a strain hardening material (solid line). The (singular) asymptotic HRR field is also schematically shown for comparison (dashed line).

slip line solution for the full Prandtl field is closely approached. With reference to the discussion about the influence of the T-stress on the slip line solution, this fact indicates that the HRR singularity represents a field characterized by a positive, and apparently very large T-stress. This has been confirmed by a number of investigations (Bilby et al. 1986, Al-Ani and Hancock 1991, Betegón and Hancock 1991, Du and Hancock 1991, O'Dowd and Shih 1991, 1992, 1994). These investigations also show that the stresses ahead of the crack edge are reduced more and more the lower the T-stress is, and that the angular distributions of the stresses also change, as well as the stress composition at a given location, for instance straight ahead of the crack edge.

Now, the T-stress is defined for small scale yielding (page 80), whereas the HRR field is an asymptotic field that is not restricted to small scale yielding. It is therefore desirable to extend the notion of higher-order terms to large scale of yielding. This was done by Li and Wang (1986), who found that second-order terms were only slowly varying with r. Such results were also reported by Sharma and Arawas (1991) and by O'Dowd and Shih (1991). Corresponding results for mode III were obtained by Arawas and Blazo (1991).

The first-order term may be defined as the HRR field. This was also done originally by O'Dowd and Shih (1991, 1992), but they later found (O'Dowd and Shih 1994) that a more convenient definition would be the asymptotic stress-strain field in the plastic region for small scale yielding under the T-stress $T = 0$ (whereas the HRR field corresponds to a T-stress about equal to the yield stress). When investigating the difference between this field and fields for cases other than small scale yielding and $T = 0$, in the sector $|\varphi| < \pi/2$ of the annulus $J_I/\sigma_0 < r < 5J_I/\sigma_0$, they found that the difference field, $(\sigma_{ij})_{diff}$, approximates a hydrostatic field, i.e.,

$$(\sigma_{ij})_{diff} \approx Q\sigma_0 \delta_{ij} \qquad (5.4.36)$$

where Q is a constant for each material and loading situation. The reason for choosing the particular sector and annulus for comparisons was said to be that "this zone encompasses the microstructurally significant length scales for both brittle and ductile materials"† (O'Dowd and Shih 1994).

O'Dowd and Shih (1994) suggested a precise definition of Q:

$$Q = \frac{\sigma_\varphi - (\sigma_\varphi)_{SSY;T=0}}{\sigma_0} \quad \text{at } \varphi = 0, \ r = 2J_I/\sigma_0 \qquad (5.4.37)$$

where SSY denotes small scale yielding. This definition may be convenient for testing standards, etc., but it should be recalled that a difference term of the form $Q\sigma_0\delta_{ij}$ is only approximate. Note that J_I and Q together specify, approximately, the state in the crack edge vicinity during loading. For small scale yielding, this two-parameter field characterization may be substituted by the K-T characterization suggested by e.g. Betegón and Hancock (1991). A relation between Q and T, based on numerical calculations, using a large strain theory and incremental strain plasticity, is given by O'Dowd and Shih (1991) as

$$Q \approx a_1\left(\frac{T}{\sigma_0}\right) + a_2\left(\frac{T}{\sigma_0}\right)^2 + \left(\frac{T}{\sigma_0}\right)^3 \qquad (5.4.38)$$

with tabulated valus of a_1, a_2 and a_3 for different values of n. For example, $a_1 \approx 0.76$, $a_2 \approx -0.52$, $a_3 \approx 0.00$ for $n = 10$.

With consideration of their dependence on Q, stresses, strains and displacements in the crack edge vicinity may be written in the general form

$$\sigma_{ij} = \sigma_0 f_{ij}(r\sigma_0/J_I, Q, \varphi), \quad \epsilon_{ij} = \epsilon_0 g_{ij}(r\sigma_0/J_I, Q, \varphi) \qquad (5.4.39)$$

$$u_i = \frac{J_I}{\sigma_0} d_i(r\sigma_0/J_I, Q, \varphi) \qquad (5.4.40)$$

where $f_{ij}(\cdot)$ and $g_{ij}(\cdot)$ are tensor functions and $d_i(\cdot)$ is a vector function.

Plane stress fields
The stress-strain field near a crack edge under plane stress conditions is quite different from the stress-strain field under plane strain, but the main features prevail, i.e. a strain singularity approaching a $1/r$ singularity at low degrees of strain hardening, a stretched zone before onset of crack growth and very high strains ahead of the crack. Also, the field is dependent on the T-stress or a Q-parameter. One example of a stress-strain field under plane stress is given by Hutchinson (1968a). According to his results for strain hardening, the angular distribution of the effective stress at

---

† This may be the case for many important materials, but the height of the process region is generally given by the distance between dominating kernels of micro-separation and the cohesion-decohesion characteristics, which are not related to σ_0; see Section 1.5. Numerical calculations by Tvergaard and Hutchinson (1992) rather suggest that the process region length in a power-law hardening material, with $N = 0.1$ and with void growth as micro-separation mechanism, may be between $8J_I/\sigma_Y$ and $40J_I/\sigma_Y$, depending on the cohesive strength.

plane stress does not possess a clear minimum in the direction straight ahead of the crack ($\varphi = 0$), as it does under plane strain, but shows rather very little variation with the angle around $\varphi = 0$, with, perhaps, a very flat maximum instead of a minimum. This fact is worth remembering when studying the Leonov-Panasyuk-Dugdale model, which will be discussed later, and in which perfectly plastic flow is confined entirely to a thin strip, because the effective stress on both sides falls just short of what is needed for plastic flow. Whether the plastic region will be distributed over a rounded area or concentrated in a necking region appears to depend on the ultimate strain of the material (cf. page 42).

The crack opening displacement
The crack opening displacement before development of a process region has been determined in several numerical investigations, associated with the HRR singularity, e.g. McMeeking (1977) and Shih (1981). Later calculations, also considering different Q-values have been made by O'Dowd and Shih (1992). The crack opening displacement (COD) is generally defined as the opening where two orthogonal planes through the crack edge, forming 45° with the crack plane, meet the crack faces.

The results obtained by O'Dowd and Shih (1992) indicate that the plane strain crack opening displacement is smaller than J_I/σ_Y, and that it is strongly dependent on the strain hardening exponent n (or N), and less strongly dependent on Q, except for large negative values. Thus, for $n=10$ it equals about $0.5 J_I/\sigma_Y$ for Q larger than about -0.5 (corresponding to $T > -0.5$ for small scale yielding), raising to about $0.7 J_I/\sigma_Y$ for $Q = -1.3$ ($T \approx -\sigma_Y$). For $n = 5$ it is smaller, about $0.33 J_I/\sigma_Y$ for Q larger than about -0.5 (about $T > -0.5$), raising to about $0.4 J_I/\sigma_Y^2$ for $Q = -1$ ($T \approx -\sigma_Y$). For plane stress, Shih (1981) estimates the crack opening displacements associated with the HRR field to about $0.7 J_I/\sigma_Y^2$ for $n = 10$.

For low strain hardening, the loss of autonomy at large scale of yielding may imply considerable sensitivity of the crack opening displacement to body and loading geometry. Another factor that influences the crack opening displacement is, of course, the opening of a process region.

Mode I plastic region at small scale yielding

Equation (5.4.35) shows that the asymptotic total shear strains along a radius from the crack edge decrease approximately in proportion to $1/r$. Even for a strain hardening as high as $N = 0.2$ ($n = 5$), the decrease is as rapid as in proportion to $r^{-5/6}$. By analogy with the result (5.3.20) for mode III, the asymptotic r dependence in (5.4.35) may be assumed to hold in the whole plastic region. Because the strains are purely elastic along the periphery of the plastic region, an approximate $1/r$-dependence implies that the plastic strains in most parts of the plastic region are smaller than the elastic ones (cf. Fig. 2.8.3). This fact suggests that the shape of the plastic region could be approximately obtained from the purely elastic case as the region in which the effective stress is larger than the yield stress. For small scale yielding, these stresses are given by the asymptotic expressions (3.3.54)-(3.3.57) for $T = 0$. These expressions may be written in the form

$$\sigma_r = \frac{K_I}{\sqrt{2\pi r}} f_r(\varphi), \quad \sigma_\varphi = \frac{K_I}{\sqrt{2\pi r}} f_\varphi(\varphi), \quad \tau_{r\varphi} = \frac{K_I}{\sqrt{2\pi r}} f_{r\varphi}(\varphi) \tag{5.4.41}$$

$$\sigma_z = \frac{K_I}{\sqrt{2\pi r}} f_z(\varphi) = \nu(\sigma_r + \sigma_\varphi) \tag{5.4.42}$$

Consideration of the T-stress leads to

$$\sigma_r = \frac{K_I}{\sqrt{2\pi r}} f_r(\varphi) + T\cos^2\varphi, \quad \sigma_\varphi = \frac{K_I}{\sqrt{2\pi r}} f_\varphi(\varphi) + T\sin^2\varphi \tag{5.4.43}$$

$$\tau_{r\varphi} = \frac{K_I}{\sqrt{2\pi r}} f_{r\varphi}(\varphi) - T\cos\varphi\sin\varphi, \quad \sigma_z = \frac{K_I}{\sqrt{2\pi r}} f_z(\varphi) + \nu T \tag{5.4.44}$$

and thus, the Huber-von Mises effective stress at the plastic region boundary, $R_p = R_p(\varphi)$, is given by

$$2\sigma_{eff}^2 = [\kappa\sigma_Y(f_r - f_\varphi) + T(\cos^2\varphi - \sin^2\varphi)]^2$$
$$+ [\kappa\sigma_Y(f_\varphi - f_z) + T(\sin^2\varphi - \nu)]^2$$
$$+ [\kappa\sigma_Y(f_z - f_r) + T(\nu - \cos^2\varphi)]^2$$
$$+ 6[\kappa\sigma_Y f_{r\varphi} - T\cos\varphi\sin\varphi]^2 = 2\sigma_Y^2 \tag{5.4.45}$$

where $\kappa = K_I/(\sqrt{2\pi R_p}\sigma_Y)$. Solution of κ for different values of T/σ_y gives $R_p(\varphi)$ in the form

$$R_p(\varphi) = \frac{1}{\kappa^2} \cdot \frac{K_I^2}{2\pi\sigma_Y^2} \tag{5.4.46}$$

Note that Poisson's ratio ν only enters in the plane strain expression $\sigma_z = \nu(\sigma_r + \sigma_\varphi)$. Thus, plane stress results are obtained simply by choosing $\nu = 0$. Expression (5.4.46) shows that the linear size of the plastic region primarily scales as K_I^2/σ_Y^2, although there is also a dependence on Poisson's ratio.

Figure 5.4.4 shows the plastic region boundary, as approximately determined from (5.4.45) and (5.4.46), for different values of Poisson's ratio and the T-stress. Note that the result for Poisson's ratio $\nu = 0$ also is valid for plane stress. The difference between plastic region shapes for $\nu = 0.3$ and $\nu = 0.5$ (not shown in the figure) is only marginal, except for the extension of the plastic region in a narrow sector ahead of the crack.

The plastic regions shown in Fig. 5.4.4 are approximately in agreement with numerically determined extensions of the plastic region for different material models and parameters in the constitutive equations. Some examples are shown by Shih et al. (1993). One prominent feature is the shift backwards of the plastic region with increasing T-stress, observed already by Larsson and Carlsson (1973). Another one is the strong dependence on the Poisson's ratio in a sector ahead of the crack, i.e., where the process region develops. The straightforward extension is found to be

$$R_p(0) = \frac{8(1/2 - \nu)^2}{\pi[\sqrt{4 - 3T^2/\sigma_Y^2} - (1 - 2\nu)T/\sigma_Y]^2} \cdot \frac{K_I^2}{\sigma_Y^2} \tag{5.4.47}$$

which, for instance, implies that the forward extension of the plastic region for $T = 0$ is 2.25 times larger for $\nu = 1/4$ than for $\nu = 1/3$.

292   5. ELASTOPLASTIC STRESSES AND STRAINS

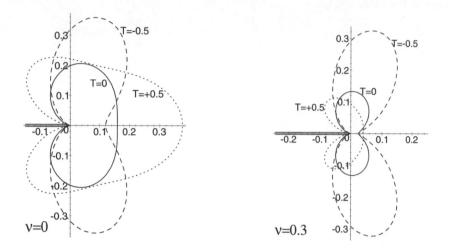

Fig. 5.4.4 The extension of the plane strain plastic region under small scale yielding conditions for a material with Poisson's ratio, ν, equal to 0 (left figure) and 0.3 (right figure). The distance from the crack edge is given in units of K_I^2/σ_Y^2. The T-stress equals $-0.5\sigma_Y$, 0 and $+0.5\sigma_Y$ as indicated in the figure, where it is given in units of σ_Y. The left figure, for $\nu = 0$, also applies to plane stress.

The mode I Leonov-Panasyuk-Dugdale model

Most often, the Leonov-Panasyuk-Dugdale model is called the Dugdale model after Dugdale (1960), appparently because the contributions by Leonov and Panasyuk (1959) and by Panasyuk (1960) were largely unknown in Western literature. The full solution, including crack opening displacement, was given by Panasyuk (1960) and by Vitvisky and Leonov (1961). The model was originally introduced for a finite mode I crack in a large body. It may alternatively be established for other geometrical configurations, for instance a row of coplanar cracks or a crack in a strip. Other models, including modes II and III, were considered by Bilby et al. (1963, 1964).

Although the Leonov-Panasyuk-Dugdale model may be considered as a special case of the Barenblatt model from the mathematical point of view, it is not a model of a decohesive region, but of a plastic region in a thin sheet. The plastic flow in this region is given a two-dimensional formulation by the assumption that it is confined to an infinitesimally thin strip straight ahead of the crack†. Plane stress and perfect plasticity, obeying the Tresca yield condition are assumed.

Analytical solutions are possible for certain body and loading geometries, without restriction to small scale yielding. Here, a crack, $|x| < a$, $y = 0$, in an infinite sheet, subjected to a remote load, $\sigma_y = \sigma_y^\infty$, will be studied. Because the sheet is thin, plane stress is assumed. This implies that the Tresca effective stress equals σ_Y whenever $0 \leq \sigma_x \leq \sigma_y$, and then $\sigma_y = \sigma_Y$ in a plastic region. Hence, the stress σ_y equals σ_Y on the two segments $a < |x| < (a+R_p)$, $y = 0$, where R_p is the length of the plastic region

---

† The model is therefore often referred to as the strip-yield model.

5.4 STATIONARY MODE I CRACKS

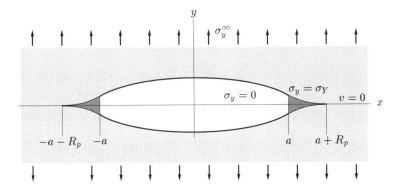

Fig. 5.4.5 Notations and boundary conditions for the Leonov-Panasyuk-Dugdale model.

(the strip). The shear stress, τ_{xy}, vanishes along the x axis, and the displacement v in the y direction vanishes for $|x| > (a + R_p)$, $y = 0$. The material outside the strip is *a priori* assumed to behave linearly elastically, see Fig. 5.4.5.

According to (4.6.4) and (4.6.11), the case $R_p = 0$ yields

$$\sigma_y(x,0) = \frac{\sigma_y^\infty |x|}{\sqrt{x^2 - a^2}} \quad \text{for } |x| > a \tag{5.4.48}$$

$$\delta_0(x) = \frac{\sigma_y^\infty}{(1-k^2)\mu} \sqrt{a^2 - x^2} \quad \text{for } |x| < a \tag{5.4.49}$$

where $k^2 = (1-\nu)/2$ and $\delta_0(x)$ is the crack opening displacement. Thus, for a crack of half-length c, subjected to a remote load $\sigma_y = q'(c)\mathrm{d}c$,

$$\mathrm{d}\sigma_y = \frac{|x| q'(c) \mathrm{d}c}{\sqrt{x^2 - c^2}} \quad \text{for } |x| > c, y = 0 \tag{5.4.50}$$

$$\mathrm{d}\delta_0 = \frac{q'(c)\mathrm{d}c}{(1-k^2)\mu} \sqrt{c^2 - x^2} \quad \text{for } |x| < c \tag{5.4.51}$$

Superposition of solutions for cracks with different half-lengths c ($a < c < a + R_p$), under different remote loads $q'(c)\mathrm{d}c$, gives the following stress, σ_y, and separation distance, $\delta_0(x)$, along $y = 0$:

$$\sigma_y(x,0) = \begin{cases} 0 & \text{for } |x| < a \\ \displaystyle\int_a^{|x|} \frac{|x| q'(c)\mathrm{d}c}{\sqrt{x^2-c^2}} & \text{for } a < |x| < a + R_p \\ \displaystyle\int_a^{a+R_p} \frac{|x| q'(c)\mathrm{d}c}{\sqrt{x^2-c^2}} & \text{for } a + R_p < |x| \end{cases} \tag{5.4.52}$$

$$\delta_0(x) = \begin{cases} \dfrac{1}{(1-k^2)\mu} \displaystyle\int_a^{a+R_p} q'(c)\sqrt{c^2-x^2}\,dc & \text{for } |x| < a \\[2mm] \dfrac{1}{(1-k^2)\mu} \displaystyle\int_{|x|}^{a+R_p} q'(c)\sqrt{c^2-x^2}\,dc & \text{for } a < |x| < a+R_p \\[2mm] 0 & \text{for } a+R_p < |x| \end{cases} \qquad (5.4.53)$$

The case arrived at may be described as a crack extending along the segment $|x| < (a+R_p)$ on the x axis, subject to crack face loading according to (5.4.52) on $a < |x| < (a+R_p)$, and to a remote load

$$\sigma_y = \int_a^{a+R_p} q'(c)\,dc \qquad (5.4.54)$$

Then, by choosing $q'(c)$ so that $\sigma_y(x,0) = \sigma_Y$ for $a < |x| < (a+R_p)$ and the remote load equal to σ_y^∞, the problem as originally posed is obtained. Thus,

$$\int_a^{|x|} \frac{|x|q'(c)\,dc}{\sqrt{x^2-c^2}} = \sigma_Y \qquad (5.4.55)$$

$$\int_a^{a+R_p} q'(c)\,dc = \sigma_y^\infty \qquad (5.4.56)$$

The first equation is an Abelian integral equation, the solution of which is, cf., for instance, Kanwal (1971),

$$q'(c) = \frac{2\sigma_Y}{\pi} \cdot \frac{d}{dc}\int_a^c \frac{du}{\sqrt{c^2-u^2}} = \frac{2\sigma_Y}{\pi} \cdot \frac{a}{c\sqrt{c^2-a^2}} \qquad (5.4.57)$$

and thus,

$$\sigma_y^\infty = \int_a^{a+R_p} q'(c)\,dc = \frac{2\sigma_Y}{\pi}\operatorname{acos}\frac{a}{a+R_p} \qquad (5.4.58)$$

giving the length of the plastic region,

$$\boxed{R_p = a\left(\sec\frac{\pi\sigma_y^\infty}{2\sigma_Y} - 1\right)} \qquad (5.4.59)$$

and, according to (5.4.53), the height increase of the plastic region,

$$\delta_0(x) = \frac{2\sigma_Y a}{\pi(1-k^2)\mu}\int_{|x|}^{a+R_p}\frac{\sqrt{c^2-x^2}\,dc}{c\sqrt{c^2-a^2}} \qquad (5.4.60)$$

Specialization to $|x|=a$ gives the crack edge opening displacement

$$\delta_0 = \delta_0(a) = \frac{2\sigma_Y a}{\pi(1-k^2)\mu}\int_a^{a+R_p}\frac{dc}{c} = \frac{2\sigma_Y a}{\pi(1-k^2)\mu}\ln(1+R_p/a) \qquad (5.4.61)$$

which equals

$$\boxed{\delta_0 = \frac{2\sigma_Y a}{\pi(1-k^2)\mu}\ln\left(\sec\frac{\pi\sigma_y^\infty}{2\sigma_Y}\right)} \qquad (5.4.62)$$

For $|x| \to (a + R_p - 0)$ the displacement (5.4.60) may be written as

$$\delta_0(x) \to \frac{2\sqrt{2}\sigma_Y a}{\pi(1-k^2)\mu} \cdot \frac{1}{\sqrt{R_p(a+R_p)(2a+R_p)}} \int_{|x|}^{a+R_p} \sqrt{c-|x|}\,dc$$

$$= \frac{4\sqrt{2}}{3\pi(1-k^2)\mu} \frac{a}{\sqrt{R_p(a+R_p)(2a+R_p)}}(a+R_p-|x|)^{3/2} \quad (5.4.63)$$

showing that the height increase of the plastic region approaches zero smoothly at $|x| = (a + R_p)$.

By choosing an integration path Γ very close to the crack edge, the value of the J-integral is immediately found to be

$$J = \int_\Gamma W\,dy = \int_0^{\delta_0(a)} \sigma_Y\,d\delta = \sigma_Y \delta_0 = \frac{2\sigma_Y^2 a}{\pi(1-k^2)\mu} \ln\left(\sec\frac{\pi\sigma_y^\infty}{2\sigma_Y}\right) \quad (5.4.64)$$

The results obtained may be specialized to small scale yielding. Alternatively, equations (4.4.61) and (4.4.62) may be used, by putting $\sigma_y^0(x) = \sigma_Y$ and $r_p = R_p$. The result, after moving the coordinate origin, $x = 0$, $y = 0$, to the front of the Leonov-Panasyuk-Dugdale region, is

$$R_p = \frac{\pi K_I^2}{8\sigma_Y^2}, \quad \frac{\partial \delta_0(x)}{\partial x} = -\frac{\sigma_Y}{\pi(1-k^2)\mu} \ln \frac{1+\sqrt{-x/R_p}}{1-\sqrt{-x/R_p}} \quad (5.4.65)$$

The only possible physical realization of the Leonov-Panasyuk-Dugdale model appears to be a necking region in a thin sheet. But continuing necking is associated with decreasing stress just outside the necking region, as in a tensile test. This is a geometry effect which is not considered in the small strain theory of two-dimensional (plane stress) plastic flow. Thus, a more realistic model would not lead to a constant stress $\sigma_y = \sigma_Y$ along an infinitesimally thin strip, as in the Leonov-Panasyuk-Dugdale model, but to a stress decreasing from $\sigma_y = \sigma_Y$ at the front end of the necking region to $\sigma_y = 0$ at the crack edge. Making distinction, as previously, between plastic regions and process regions, this would clearly be a process region, although governed by a characteristic length dimension which is related to the geometry (the sheet thickness) and not to the material (an intrinsic material dimension); cf. Section 1.3. A Barenblatt model (Barenblatt 1959a,b,c), which is a model of a process region, would therefore possess a greater potential of accurate modelling of a necking region than the Leonov-Panasyuk-Dugdale model, which, on the other hand, is more simple to handle.

The presupposition that $0 \leq \sigma_x \leq \sigma_y$ imposes restrictions on the remote stress $\sigma_x = \sigma_x^\infty$, namely that $-(\sigma_x^\infty - \sigma_y^\infty) \leq \sigma_x^\infty \leq \sigma_y^\infty$; see (4.6.12). Moreover, recall the *a priori* assumption that no plastic flow occurs outside the strip $a < |x| \leq (a + R_p)$ of the x axis. As pointed out by Drucker and Rice (1970), the Tresca effective stress is maximum on the strip, and the *a priori* assumption is thus justified. However, for a Huber-von Mises material a similar solution, forcing σ_y to be constant along a plastically deforming strip, would show offside yielding, and, consequently, such a solution would not be valid. The plane stress plastic region at the crack edge in a Huber-von Mises material is therefore not expected to be concentrated to a strip, but rather to extend over a somewhat rounded region; cf. Fig. 5.4.4. This considerable difference in plastic region shape, a strip or a broad region, is quite surprising, but, as mentioned in Section 2.8, experiments have shown that both types can appear (Bergkvist and Andersson 1972). However, as will be discussed shortly, the observed differences are hardly related to such subtle distinctions as the one between Tresca and Huber-von Mises materials, but rather to the distinction between plastic flow and decohesion.

To elucidate the question as to why a broad dissipative region appears in thin plates of some materials, whereas it is concentrated to a narrow necking region in others, the analogy with necking in a tensile test may be helpful. In a material with a small elongation, necking takes place in an environment with comparatively low strain. Then, because necking would be unstable under load control, the subsequent straining under grip control in a tensile test becomes confined to the narrow region of necking, whereas outside strains will remain small and actually even decrease during continued necking. Similarly, a necking region at the edge of a crack in such a material will be formed in an environment with small plastic strains and decreasing elastic stresses.

In a material with a large elongation in a tensile test, necking appears as localization in a highly strained environment. Large plastic strains are therefore expected in a broad plastic region near a crack edge. A process region inside this region may appear either as necking from the crack edge or as a material related process region, (i.e., governed by intrinsic material length parameters).

Experimental evidence supports this analogy with the tensile test. In Section 2.8, the results by Bergkvist and Andersson (1972) were mentioned: a rounded plastic region appeared in a thin sheet of celluloid with 25% ultimate strain, whereas a necking type region appeared in PVC with 3.5% ultimate strain.

5.5 Stationary mode II cracks

Cracks in perfectly plastic materials. Slip line solutions

Asymptotic mode II stresses and strains for perfect plasticity were found by Hutchinson (1968b), using slip line theory. The slip line field contains the following stress sectors, given in the cylindrical coordinate φ from the straightforward direction; see Fig. 5.5.1:

P1: centered fan sector for $0 \leq |\varphi| < \pi/8 + 1/4$
P2: constant stress sector for $\pi/8 + 1/4 \leq |\varphi| \leq 5\pi/8 + 1/4$
P3: centered fan sector for $5\pi/8 + 1/4 < |\varphi| < 3\pi/4$
P4: constant stress sector for $3\pi/4 \leq |\varphi| \leq \pi$

The asymptotic stresses are independent of the radius r from the crack edge:

P1: $\sigma_r = \sigma_\varphi = -\tau_Y$, $\quad \tau_{r\varphi} = \tau_Y$
P2: $\sigma_x = -\tau_Y[1 + \sin(\pi/4 + 1/2)]$, $\quad \sigma_y = -\tau_Y[1 - \sin(\pi/4 + 1/2)]$,
$$\tau_{xy} = \tau_Y \cos(\pi/4 + 1/2)$$
P3: $\sigma_r = \sigma_\varphi = -\tau_Y$, $\quad \tau_{r\varphi} = -\tau_Y$
P4: $\sigma_x = -2\tau_Y$, $\quad \sigma_y = 0$, $\quad \tau_{xy} = 0$

5.5 STATIONARY MODE II CRACKS

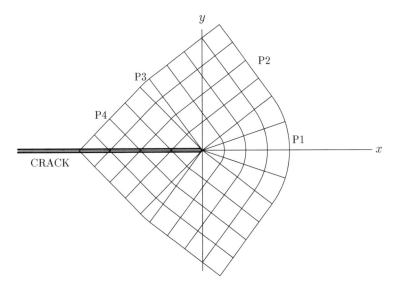

Fig. 5.5.1 Asymptotic sectors appearing at slip line solution for mode II.

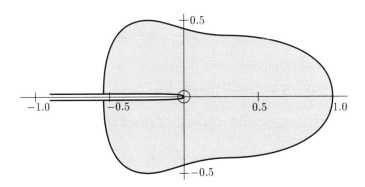

Fig. 5.5.2 Sketch of the small scale yielding plastic region at the edge of a mode II crack. The distance from the crack edge is given in units of K_{II}^2/σ_Y^2. Size and shape are somewhat dependent on the strain hardening and of the elastic properties.

The corresponding strain fields exhibit $1/r$ singularities in the centered fan sectors. Note that such a singularity appears straight ahead of the crack, in contrast to the mode I case.

For strain hardening, the asymptotic stress-strain field may be written in the same form as for the HRR field in mode I, equations (5.4.34)-(5.4.35); see Hutchinson (1968a). At small scale yielding, the shape of the mode II plastic region is very different from mode I (see Fig. 5.5.2), and its extension straightforwards is much larger for the same value of the stress intensity factor.

The mode II Leonov-Panasyuk-Dugdale model

The Leonov-Panasyuk-Dugdale model for mode II is obtained by only cosmetic and obvious changes from the mode I solution (5.4.59) and (5.4.62). These changes consist essentially of replacing σ_y^∞ by the remote load $\tau_{xy} = \tau_{xy}^\infty$, and σ_Y by the yield stress in shear, τ_Y. Thus,

$$R = a\left(\sec\frac{\pi\tau_{xy}^\infty}{2\tau_Y} - 1\right) \qquad (5.5.1)$$

$$\delta_0 = \frac{2\tau_Y a}{\pi(1-k^2)\mu}\ln\left(\sec\frac{\pi\tau_{xy}^\infty}{2\tau_Y}\right) \qquad (5.5.2)$$

As for mode I, the Tresca yield condition is assumed, but a difference is that the mode II result is valid for both plane stress and plane strain. The deformation consists of plastic slip in a strip, often appearing as shearbanding. This would also be the case if overall compressive stresses (σ_y and σ_x) are superposed. Such stress states do not, strictly, belong to mode II, but they are very often found together with mode II. In fact, superposed compressive stresses are generally even required for straightforward mode II growth (Nemat-Nasser and Horii 1982, Horii and Nemat-Nasser 1985,1986, Melin 1986, 1992, Nemat-Nasser 1995).

Concentration of non-elastic deformations to a narrow strip ahead of a mode II crack is probably most often a question of decohesion rather than of plastic flow. Such narrow strips of decohesion are observed in earthquake slip.

5.6 Slowly moving mode III cracks

General considerations for perfect plasticity and steady state

Steady state conditions are expected to be approached at small scale yielding toward the end of the stable phase, cf. Section 2.5. The basic theory for mode III cracks moving slowly under steady state conditions in an elastic-perfectly plastic material was given by Chitaley and McClintock (1971), who studied the neighbourhood of a crack edge, assuming infinitesimally small scale yielding. A systematic and comprehensive treatment, also including mode I, plane stress or strain, and anisotropy, has been given by Rice (1982). Questions about the exact nature of perfect plasticity were raised by Ståhle (1989, 1993), who showed that the asymptotic solution in the limit of vanishing hardening is different from the solution obtained by assuming zero strain hardening *a priori*.

As described in Section 2.8, two separate active plastic regions are, in general, created during crack growth, the primary plastic region (P) and the secondary plastic region (S), the latter characterized by reverse plastic flow. Behind the primary plastic region there is a wake (W), in which the active deformations are purely elastic, whereas the plastic deformations stay constant. These regions are shown in Fig. 5.6.1. It is convenient to use a moving coordinate system, Cartesian, x, y, z, or cylindrical, r, φ, z, with $x = 0$ or $r = 0$ along the crack edge and $x < 0$, $y = \pm 0$ or $\varphi = \pm\pi$ along the crack faces. All equations (5.3.1)-(5.3.16) hold even for slowly moving cracks, and, in

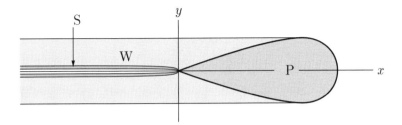

Fig. 5.6.1 Primary (P) and secondary (S) plastic regions and the wake (W) region.

particular, straight characteristics exist in active plastic regions, as in the stationary case. However, because steady state conditions now prevail, the history of a material point is displayed along the line parallel to the x axis from $x = \infty$ to the point considered, and thus the flow rule (5.3.13),

$$\frac{\mathrm{d}\gamma^p_{xz}}{\tau_{xz}} = \frac{\mathrm{d}\gamma^p_{yz}}{\tau_{yz}} \tag{5.6.1}$$

may be written in the form

$$\frac{1}{\tau_{xz}} \cdot \frac{\partial \gamma^p_{xz}}{\partial x} = \frac{1}{\tau_{yz}} \cdot \frac{\partial \gamma^p_{yz}}{\partial x} \tag{5.6.2}$$

Constant stress and centered fan stress sectors

At this point, it might be useful to consider some general features of active plastic regions. In cylindrical coordinates, the equilibrium condition reads:

$$\frac{\partial \tau_{rz}}{\partial r} + \frac{\tau_{rz}}{r} + \frac{1}{r} \cdot \frac{\partial \tau_{\varphi z}}{\partial \varphi} = 0 \tag{5.6.3}$$

When the crack edge is approached, $r \to 0$, the equilibrium condition defaults to

$$\tau_{rz} + \frac{\partial \tau_{\varphi z}}{\partial \varphi} = 0 \tag{5.6.4}$$

because stresses are bounded so that $r\partial \tau_{rz}/\partial r \to 0$ as $r \to 0$. Now, this coincides with the equilibrium condition that would be obtained by assuming *a priori* that τ_{rz} depends only on φ, and then $\tau_{\varphi z}$ can depend on r only through an additive function of only r. But, if it does, the yield condition, (5.3.1), is violated, and consequently it has to be concluded that, as the crack edge is approached from a plastic region, *the asymptotic stresses do not depend on r, only on φ*. As a consequence, active plastic regions appear as sectors (wedges with plane boundaries) near the crack edge, and so does then, of course, also the wake. Note that this does not necessarily hold true for material models with hardening at large strains.

Following Rice (1982), the yield condition (5.2.3) is written in the differential form

$$\frac{\partial f}{\partial \tau_{xz}} \cdot \frac{\partial \tau_{xz}}{\partial \varphi} + \frac{\partial f}{\partial \tau_{yz}} \cdot \frac{\partial \tau_{yz}}{\partial \varphi} = 0 \tag{5.6.5}$$

Then, use of the equilibrium condition, specialized to the crack edge vicinity, where the stresses do not depend on r,

$$\sin\varphi \frac{\partial \tau_{xz}}{\partial \varphi} = \cos\varphi \frac{\partial \tau_{yz}}{\partial \varphi} \qquad (5.6.6)$$

results in the relation

$$\left(\frac{\partial f}{\partial \tau_{xz}}\cos\varphi + \frac{\partial f}{\partial \tau_{yz}}\sin\varphi\right) \cdot \frac{\partial \tau_{xz}}{\partial \varphi} = 0 \qquad (5.6.7)$$

Thus,

$$\frac{\partial \tau_{xz}}{\partial \varphi} = 0 \quad \text{or} \quad d\gamma^p_{xz}\cos\varphi + d\gamma^p_{yz}\sin\varphi = 0 \qquad (5.6.8)$$

where use was made of the associated flow rule, (5.2.4). The first condition, together with (5.6.6), implies constant stresses, and the second condition, with use of (5.6.2), requires that

$$\frac{\tau_{yz}}{\tau_{xz}} = -\cot\varphi \qquad (5.6.9)$$

This shows a centered fan field. Active plastic sectors may thus be of two different types near the edge of a steadily moving mode III crack: *constant stress fields* or *centered fan stress fields*. This result may immediately be generalized to more remote parts of mode III active plastic regions: the existence of straight characteristics in general implies that a constant stress region extends to all boundaries of the plastic region in question, and that a centered fan region from the crack edge extends to all points in the plastic region in question, that may be reached by straight characteristics from the crack edge, without crossing a boundary of this plastic region. Thus, for instance, a centered fan field extends to all points of an outward convex plastic region from a crack edge.

Stresses and strains in the primary plastic region

It is obvious that the stress distribution in the plastic region in front of the crack edge, the primary plastic region, must be of the centered fan type, with characteristics radiating from the crack edge in the xy-plane, i.e.

$$\tau_{xz} = -\tau_Y \cdot \frac{y}{r}, \qquad \tau_{yz} = \tau_Y \cdot \frac{x}{r} \qquad (5.6.10)$$

as found from, for instance (5.6.9) and the yield condition, assuming that the remote stress τ_{yz} is positive. However, this result may not be valid for a possible rear part of the primary plastic region, which cannot be reached from the crack edge by a straight line inside the region. In cylindrical coordinates the result is

$$\boxed{\tau_{rz} = 0, \qquad \tau_{\varphi z} = \tau_Y} \qquad (5.6.11)$$

in agreement with (5.3.19). Because steady state prevails, the second of equations (5.6.8) may be written in the form

$$x\frac{\partial \gamma^p_{xz}}{\partial x} + y\frac{\partial \gamma^p_{yz}}{\partial x} = 0 \qquad (5.6.12)$$

5.6 SLOWLY MOVING MODE III CRACKS

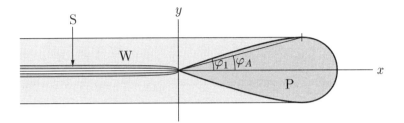

Fig. 5.6.2 The angles φ_1 and φ_A. The centered fan region is $0 \leq \varphi \leq \varphi_1$.

from which γ_{yz}^p may be eliminated by using the compatibility equation

$$\frac{\partial \gamma_{xz}}{\partial y} = \frac{\partial \gamma_{yz}}{\partial x} \tag{5.6.13}$$

which is decomposed into plastic and elastic parts,

$$\frac{\partial \gamma_{xz}^p}{\partial y} - \frac{\partial \gamma_{yz}^p}{\partial x} = -\frac{1}{\mu} \cdot \frac{\partial \tau_{xz}}{\partial y} + \frac{1}{\mu} \cdot \frac{\partial \tau_{yz}}{\partial x} = \frac{\tau_Y}{\mu r} \tag{5.6.14}$$

Thus, an equation for γ_{xz}^p is obtained,

$$x \cdot \frac{\partial \gamma_{xz}^p}{\partial x} + y \cdot \frac{\partial \gamma_{xz}^p}{\partial y} = \frac{\tau_Y y}{\mu r} \tag{5.6.15}$$

which in cylindrical coordinates reads

$$\frac{\partial \gamma_{xz}^p}{\partial r} = \frac{\tau_Y}{\mu} \cdot \frac{\sin \varphi}{r} \tag{5.6.16}$$

Integration gives

$$\boxed{\gamma_{xz}^p = \frac{\tau_Y}{\mu} \sin \varphi \cdot \ln \frac{r}{C(\varphi)} + \frac{\tau_Y}{\mu} \cdot g(\varphi)} \tag{5.6.17}$$

where $C(\varphi)$ and $g(\varphi)$ are constant along each characteristic line. These constants may be determined by using the condition that $\gamma_{xz}^p = 0$ at the boundary to the virgin elastic region and that $\partial \gamma_{xz}^p / \partial x = 0$ at the boundary to the wake. For $0 \leq \varphi \leq \varphi_A$, where φ_A is the angle shown in Fig. 5.6.2, the result is

$$C(\varphi) = R(\varphi), \qquad g(\varphi) = 0 \tag{5.6.18}$$

where $R(\varphi)$ is the radius vector to the elastic-plastic boundary. For $\varphi > \varphi_A$ the derivative of $g(\varphi)$ is found to be

$$g'(\varphi) = \cos \varphi + \sin \varphi \cdot \frac{R'}{R} \tag{5.6.19}$$

for the region which may be reached by straight lines from the crack edge without crossing the boundary to the wake. Here, the argument φ for $R(\varphi)$ is omitted.

Equations (5.6.2) and (5.6.9) lead to the relation

$$\frac{\partial \gamma_{yz}^p}{\partial x} = -\cot \varphi \cdot \frac{\partial \gamma_{xz}^p}{\partial x} \tag{5.6.20}$$

giving

$$\frac{\partial \gamma_{yz}^p}{\partial x} = \frac{\tau_Y}{\mu r} \cdot \cos^2 \varphi [\ln(r/R) - 1] - \frac{\tau_Y}{\mu r} \cos \varphi \sin \varphi \cdot \frac{R'}{R} \qquad (5.6.21)$$

for $0 \leq \varphi < \varphi_A$. Larger values of φ result, somewhat surprisingly, in a simpler expression,

$$\frac{\partial \gamma_{yz}^p}{\partial x} = \frac{\tau_Y}{\mu r} \cdot \cos^2 \varphi \cdot \ln \frac{r}{R} \qquad (5.6.22)$$

but the study of this region will not be pursued any further.

Integration of (5.6.21) leads to an expression which is not very useful as long as the shape of the forwards elastic-plastic boundary is not known, and therefore the integration will be performed only for $\varphi \ll 1$. After addition of the elastic strain, $\gamma_{yz}^e = \tau_Y/\mu \cos \varphi$, the result is

$$\boxed{\gamma_{yz} = \frac{\tau_Y}{\mu} \left\{ \frac{1}{2} \left[\ln \frac{R(0)}{r} \right]^2 + \ln \frac{R(0)}{r} + 1 \right. \\ \left. - \frac{\varphi^2}{8} \left[2 \ln \frac{R(0)}{r} + 5 - \frac{6R''(0)}{R(0)} - \left(1 - 2\frac{R''(0)}{R(0)}\right) \frac{r^2}{[R(0)]^2} \right] \right\}} \qquad (5.6.23)$$

This result shows a remarkable difference from the stationary case. *The strain singularity is now of logarithmic type, and such a singularity is much weaker than a $1/r$ or a $1/\sqrt{r}$ singularity.* The solution (5.6.23) for $\varphi = 0$ was obtained by McClintock (1963), albeit in a non-integrated form – it was given in integrated form by Rice (1968b).

McClintock and Irwin (1965) pointed out that the considerable weakening of the singularity from incipient to unstable crack growth implies a substantial increase of load during the stable phase. In their paper they show examples of numerically calculated relations between applied stress and crack growth, from incipient crack growth to instability, based on an earlier investigation by McClintock (1958). They used a simple crack growth criterion to demonstrate this: the achievement of a critical plastic strain at a fixed distance from the crack edge.

Rice (1968b) used the criterion to show that the extension of the plastic region ahead of a mode III crack would need to be, for instance, 18 times larger at steady state than at incipient crack growth if the strain required is $25\tau_Y/\mu$ and 167 times larger if it is $50\tau_Y/\mu$. This may be readily checked from (5.6.23). Note that for many ductile materials, $50\tau_Y/\mu$ is not an unrealistically large plastic shear strain. However, because a process region develops before crack growth, quantitative results obtained by a continuum approach become questionable for growing cracks, but, qualitatively, they serve the purpose of emphasizing that the changing strain distribution works in the direction of increasing the "resistance" to crack growth, the terminology used by Krafft et al. (1961), later widely used in the context of "resistance curves"; see Section 8.4.

It is also seen from (5.6.23) that the strain component γ_{yz} at a fixed distance, much smaller than $R(0)$, from the crack edge, is maximum at $\varphi = 0$, and, making use of (5.6.17), it is found that this also holds for the resulting shear strain, $\sqrt{\gamma_{xz}^2 + \gamma_{yz}^2}$.

Stresses and strains in the secondary plastic region

The secondary plastic regions (one at each side of the crack) touch the crack faces, where $\tau_{yz} = 0$, because the faces are traction free, and $\tau_{xz} = \pm \tau_Y$, because the yield condition has to be satisfied in an active plastic region. It is clearly of the constant stress type, and then

$$\tau_{xz} = \tau_Y, \qquad \tau_{yz} = 0 \tag{5.6.24}$$

in the whole region for $y \geq 0$. The flow rule implies that

$$\frac{\partial \gamma^p_{yz}}{\partial x} = 0 \tag{5.6.25}$$

showing that $\gamma^p_{yz} = \gamma^p_{yz}(y)$. Then, the compatibility equation (5.6.13), decomposed into plastic and elastic parts, reads:

$$\frac{\partial \gamma^p_{xz}}{\partial y} - \frac{\partial \gamma^p_{yz}}{\partial x} = -\frac{1}{\mu} \cdot \frac{\partial \tau_{xz}}{\partial y} + \frac{1}{\mu} \cdot \frac{\partial \tau_{yz}}{\partial x} = 0 \tag{5.6.26}$$

so that

$$\frac{\partial \gamma^p_{xz}}{\partial y} = 0 \tag{5.6.27}$$

showing that $\gamma^p_{xz} = \gamma^p_{xz}(x)$.

Note that the possibility $\tau_{xz} = -\tau_Y$ for $y \geq 0$ is not realistic, because it is obvious that the plastic flow must be reverse to that in the primary plastic region, where $\tau_{xz} < 0$. It may also be shown, by following the procedure in the next subsection that assumption of plastic flow in the same sense as in the primary plastic region leads to a physically unacceptable result.

Assembly of sectors close to the crack edge

The previously deduced fact that active perfectly-plastic regions appear as sectors (wedges with plane boundaries) in the vicinity of a crack edge, and that the stresses do not depend on r, only on φ, will now be used to find the sector angle for each one of the three regions, the primary plastic region (P), the wake (W) and the secondary plastic region (S). The primary plastic region is assumed to occupy an angle $2\varphi_1$, and the secondary plastic region an angle $\pi - \varphi_2$ on each side of the crack; see Fig. 5.6.3. Following Chitaley and McClintock (1971), the wake region will be studied, using a stress potential Φ, such that

$$\tau_{xz} = \frac{\partial \Phi}{\partial y}, \qquad \tau_{yz} = -\frac{\partial \Phi}{\partial x} \tag{5.6.28}$$

or, in cylindrical coordinates,

$$\tau_{rz} = \frac{1}{r} \cdot \frac{\partial \Phi}{\partial \varphi}, \qquad \tau_{\varphi z} = -\frac{\partial \Phi}{\partial r} \tag{5.6.29}$$

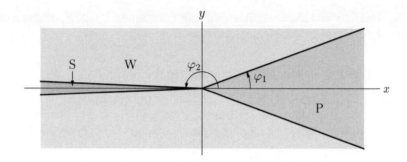

Fig. 5.6.3 The sectors of the primary plastic region (P), the secondary plastic region (S) and the wake (W) in an asymptotic vicinity of a slowly moving mode III crack edge. The sector angle for S is very small and overexaggerated in the figure.

These representations follow from the equilibrium condition. Now, because

$$\gamma_{xz} = \gamma^p_{xz} + \frac{\tau_{xz}}{\mu} = \gamma^p_{xz} + \frac{1}{\mu} \cdot \frac{\partial \Phi}{\partial y} \tag{5.6.30}$$

$$\gamma_{yz} = \gamma^p_{yz} + \frac{\tau_{yz}}{\mu} = \gamma^p_{yz} - \frac{1}{\mu} \cdot \frac{\partial \Phi}{\partial x} \tag{5.6.31}$$

the stresses may be eliminated:

$$\Delta \Phi = -\mu \left(\frac{\partial \gamma^p_{xz}}{\partial y} - \frac{\partial \gamma^p_{yz}}{\partial x} \right) = -\mu \frac{\partial \gamma^p_{xz}}{\partial y} \tag{5.6.32}$$

where use was made of the fact that $\partial \gamma^p_{yz}/\partial x = 0$ in the wake.

Consider now a small rectangle with the sides parallel with the Cartesian axes and with two diagonal corners in W infinitesimally close to the P-W boundary. The condition that $\partial \gamma^p_{xz}/\partial x$ is continuous across the P-W boundary and vanishes in W implies that the difference between the values of γ^p_{xz} at two corners with the same y coordinate is zero or vanishes with the size of the rectangle. Consequently, the difference between the values of γ^p_{xz} at two corners with the same x coordinate approaches the same value as the size of the rectangle vanishes. Thus, $\partial \gamma^p_{xz}/\partial y$ is continuous across the P-W boundary. But, according to (5.6.14), $\mu \partial \gamma^p_{xz}/\partial y = \tau_Y/r$, in P near the P-W boundary where $\partial \gamma^p_{yz}/\partial x = 0$, and therefore

$$\Delta \Phi = -\frac{\tau_Y \sin \varphi_1}{r \sin \varphi} \tag{5.6.33}$$

in W, because $\partial \gamma^p_{xz}/\partial y$ is a function of y, only, in W.

Note that all points in P, in the asymptotic crack edge vicinity considered, may be reached by straight lines from the edge, and consequently (5.6.11) holds. Because stresses are continuous across the P-W boundary, two boundary conditions for W at $\varphi = \varphi_1$ are

$$\frac{\partial \Phi}{\partial \varphi} = 0, \qquad \frac{\partial \Phi}{\partial r} = -\tau_Y \tag{5.6.34}$$

For all points in S, the stresses are $\tau_{rz} = \tau_Y \cos \varphi$ and $\tau_{\varphi z} = -\tau_Y \sin \varphi$, and therefore two boundary conditions for W at $\varphi = \varphi_2$ are obtained,

$$\frac{\partial \Phi}{\partial \varphi} = \tau_Y r \cos \varphi, \qquad \frac{\partial \Phi}{\partial r} = \tau_Y \sin \varphi \qquad (5.6.35)$$

Because the stresses in the asymptotic crack edge vicinity considered, depend only on φ, a solution of (5.6.33) may be written in the form

$$\Phi = -\tau_Y \sin \varphi_1 \cdot r \cdot f(\varphi) \qquad (5.6.36)$$

giving

$$f''(\varphi) + f(\varphi) = \frac{1}{\sin \varphi} \qquad (5.6.37)$$

with the solution for $\varphi > 0$,

$$f(\varphi) = A \cos \varphi + B \sin \varphi + \sin \varphi \cdot \ln(\sin \varphi) - \varphi \cos \varphi \qquad (5.6.38)$$

where A and B are constants.

Expressed in terms of $f(\varphi)$, the boundary conditions are

$$f(\varphi) = \frac{1}{\sin \varphi_1}, \quad f'(\varphi) = 0 \quad \text{for } \varphi = \varphi_1 \qquad (5.6.39)$$

$$f(\varphi) = -\frac{\sin \varphi_2}{\sin \varphi_1}, \quad f'(\varphi) = \frac{\cos \varphi_2}{\sin \varphi_1} \quad \text{for } \varphi = \pi - \varphi_2 \qquad (5.6.40)$$

These conditions give four equations for A, B, φ_1 and φ_2. Solutions for A and B are

$$A = \varphi_2 = \varphi_1 + \cot \varphi_1, \qquad B = 1 - \ln(\sin \varphi_1) \qquad (5.6.41)$$

whereupon φ_1 is obtained from the equation

$$\varphi_1 + \cot \varphi_1 - \mathrm{asin}\left\{ \sin \varphi_1 / \exp\left(\frac{1 + \sin \varphi_1}{\sin \varphi_1}\right) \right\} = 0 \qquad (5.6.42)$$

giving $\varphi_1 \approx 19.711°$ and then $\pi - \varphi_2 \approx 0.3666°$, as the only physically acceptable solution.

It is immediately noticed that the sector angle for S is very small, indeed, and the whole secondary plastic region would certainly be absorbed in the wake of the process region in most real materials. Also noticeable is that the sector angle for P appears to be rather small, even though it is obvious that it should be smaller than for the stationary case. In fact, numerical calculations seem constantly to arrive at a substantially larger angle, see e.g. Sorensen (1978), Dean and Hutchinson (1980) and Ponte Castañeda (1987b). So do also analytical investigations for materials with very low strain hardening (Ståhle 1989, 1993). Even though Ståhle found that the angle decreases with the strain hardening, the lowest value, obtained for vanishing strain hardening, is as high as $\varphi_1 = 32.8453°$. He determined $\pi - \varphi_2$ to be 0.2809°. Ståhle's results are particularly interesting, because they show clearly that the asymptotic solution for vanishing strain hardening does not approach the asymptotic solution by Chitaley and McClintock (1971) for perfect plasticity. Moreover, Ståhle (1993) shows that inertia effects cannot be neglected *a priori* even for vanishing strain hardening. His results for the noninertial case (Ståhle 1989) will be discussed in the next subsection, and his results for the inertial case (Ståhle 1993) will be discussed in Section 7.2.

The latter showed different possible solutions depending on how the crack velocity approaches zero at vanishing strain hardening.

Numerical results like those of Sorensen (1978) and Dean and Hutchinson (1980) also indicate that the active plastic region may not be outward convex everywhere, as assumed by Chitaley and McClintock (1971). Its trailing boundary may be outward concave, except far away from the symmetry plane. In such a case, those characteristics which intersect the plastic boundary twice (this is possible for a boundary to a wake region, although not for a boundary to a virgin elastic region) need not point toward the crack edge. In fact, the computations by Dean and Hutchinson (1980) indicate that they don't.

The important result, as regards steady state elastic-perfectly plastic mode III crack growth, concerns the logarithmic strain field near the crack edge, which obviously explains the occurrence of stable crack growth. The angle of sector P is, as it appears, less interesting.

The stresses in the wake are found from the relations

$$\tau_{xz} = \frac{\partial \Phi}{\partial y} = \frac{\partial \Phi}{\partial r} \sin \varphi + \frac{1}{r} \cdot \frac{\partial \Phi}{\partial \varphi} \cos \varphi$$
$$= -\tau_Y \sin \varphi_1 [f'(\varphi) \cos \varphi + f(\varphi) \sin \varphi] \quad (5.6.43)$$

$$\tau_{yz} = -\frac{\partial \Phi}{\partial x} = -\frac{\partial \Phi}{\partial r} \cos \varphi + \frac{1}{r} \cdot \frac{\partial \Phi}{\partial \varphi} \sin \varphi$$
$$= -\tau_Y \sin \varphi_1 [f'(\varphi) \sin \varphi - f(\varphi) \cos \varphi] \quad (5.6.44)$$

which result in the stress field

$$\tau_{xz} = -\tau_Y \sin \varphi_1 \left(1 + \ln \frac{\sin \varphi}{\sin \varphi_1}\right), \quad \tau_{yz} = \tau_Y \sin \varphi_1 \cdot (\varphi_2 - \varphi) \quad (5.6.45)$$

The asymptotic crack face sliding $\delta_0 = w(r, \pi) - w(r, -\pi)$ may be determined by integration of strains, along
1) a path starting at a point $(x_1, 0)$ on the symmetry plane ahead of the crack and following constant x to a point $(x_1, x_1 \tan \varphi_1)$ on the P-W boundary, then
2) proceeding towards the W-S boundary on constant y to a point $(x_1 \tan \varphi_1 / \tan \varphi_2, x_1 \tan \varphi_1)$, and thereupon
3) continuing along constant x to the upper crack face, so that it ends at the point $(x_1 \tan \varphi_1 / \tan \varphi_2, 0)$.
The integrand should be $\partial w / \partial y = \gamma_{yz}$ on the first and third legs, and $\partial w / \partial x = \gamma_{xz}$ on the second leg.

Now, because $\tau_{yz} = 0$ in S, and thus $\gamma_{yz}^e = 0$, it follows from (5.6.2) that $d\gamma_{yz}/\partial x = 0$ and, consequently γ_{yz} is a function of y, only, say $\gamma_{yz}(y) = \gamma_{yz}^p(y)$. But no plastic strain is produced in W, so this function must equal $\gamma_{yz}^p(y)$ on the P-W boundary. But then integration of $\partial w / \partial y = \gamma_{yz}(y)$ along the third leg gives the same result as integration of $\gamma_{yz}^p(y)$ along the P-W boundary from the point $(x_1, x_1 \tan \varphi_1)$ to the origin. Then, the first and third integrations may be replaced by an integration along the broken path, starting from $(x_1, 0)$ on the symmetry plane, proceeding to $(x_1, x_1 \tan \varphi_1)$ on the P-W boundary and then following the P-W boundary to the origin, $(0, 0)$. But this integration starts at a point where $w = 0$ and ends at a point where $w = 0$, i.e.

it gives no contribution. Note that the crack edge is not blunted as in the case of a stationary crack, because shear strain singularities are at most $(\ln r)^2$.

The asymptotic crack face sliding is obviously found by integration along the second leg, only. Because γ_{xz}^p is constant in W, and, according to (5.6.17), asymptotically (to within a constant term) equal to

$$\gamma_{xz}^p = \frac{\tau_Y}{\mu} \sin\varphi_1 \cdot \ln\frac{x_1}{R(0)} \qquad (5.6.46)$$

the integration gives the crack face sliding

$$\delta_0 = 2\frac{\tau_Y}{\mu}(\sin\varphi_1 - \cos\varphi_1 \tan\varphi_2)r\ln\frac{R(0)}{r} \approx 0.6866\frac{\tau_Y}{\mu}r\ln\frac{R(0)}{r} \qquad (5.6.47)$$

Because the stresses are bounded, and $\delta_0 \to 0$ as $r \to 0$, the energy flux into the crack edge region cannot be determined from the present treatment, which assumes an infinitesimally small process region. However, energy *would* flow to a finite size process region, because the sign of δ_0 is the same as that of τ_{yz} ahead of the crack.

Strain hardening

Strain hardening has been considered for a steadily moving mode III crack by e.g. Slepyan (1973), Amazigo and Hutchinson (1977), Dunayevsky and Achenbach (1982b), Ponte Castañeda (1987a,b) and Ståhle (1989, 1993).

Most investigations of asymptotic fields in a strain hardening material assume linear strain-hardening. They also assume or postulate that the different regions appear asymptotically as sectors, as in the elastic-perfectly-plastic case. The stresses then take the form

$$\tau_{rz} = \tau_Y \cdot f_{rz}(\varphi) \cdot (r/r_0)^s, \qquad \tau_{\varphi z} = \tau_Y \cdot f_{\varphi z}(\varphi) \cdot (r/r_0)^s \qquad (5.6.48)$$

where r_0 is a reference length. The strains may be written in the form

$$\gamma_{rz} = \frac{\tau_Y}{\mu} \cdot g_{rz}(\varphi) \cdot \frac{(r/r_0)^s}{s}, \qquad \gamma_{\varphi z} = \frac{\tau_Y}{\mu} \cdot g_{\varphi z}(\varphi) \cdot \frac{(r/r_0)^s}{s} \qquad (5.6.49)$$

where the division by s has been found convenient; cf. Ponte Castañeda (1987b). The exponent s is in the interval $-1/2 < s < 0$. The strain hardening is expressed by $\alpha = \mu_T/\mu$, where μ_T is the tangent modulus of rigidity in the plastic region. According to results by Dunayevsky and Achenbach (1982b), Ponte Castañeda (1987b) and Ståhle (1989, 1993), s approaches zero for vanishing strain hardening as

$$s \propto \sqrt{\alpha} \qquad (5.6.50)$$

Representative values of α are around $10^{-4} - 10^{-3}$ for many modern steels, although much larger values have usually been assumed in theoretical work. Assumption of perfectly-plastic behaviour appears to provide good results when strain hardening is low, except in a region close to the crack edge, where consideration of strain hardening is essential in analytical treatments. The size of this region decreases with the strain hardening, according to estimates by Gudmundson (1989), who used a perturbation method. In fact, the difference in asymptotic behaviour between materials with low strain hardening and perfectly-plastic materials is probably insignificant in real materials, because of the presence of a process region (which is considered as vanishingly small in asymptotic analyses).

Note that an asymptotic stress-strain field in a linear strain hardening material deviates very much from that in a perfectly-plastic material: the ratio between stresses and strains tends toward a finite limit when the singularity at the crack edge is approached at linear hardening, but toward zero at perfect plasticity, and actually also at power-law hardening. It cannot be taken for granted that the asymptotic field obtained for vanishing strain hardening should equal the asymptotic perfectly-plastic field, because its region of validity may disappear with the strain hardening. However, for low strain hardening the *angular distribution* of stresses and strains seems to approach the perfectly-plastic distribution, even though the *radial distribution* (the r-dependence) is quite different (Ponte Castañeda 1987b).

Ståhle (1989) found that the angular distribution of the asymptotic field for vanishing strain hardening agrees closely with the field in the primary plastic region at perfect plasticity (Chitaley and McClintock 1971). However, the primary plastic region, obtained in the limit of vanishing strain hardening, extends to a larger angle, 32.8453°, from the symmetry plane. Close to the boundary between the primary plastic region and its wake, the stress gradient $\partial \tau_{rz}/\partial x$ decreases rapidly with increasing angle from the symmetry plane, and in the limit of vanishing strain hardening this increase turns into a discontinuity (Ståhle 1989). This implies that τ_{rz}, which equals τ_Y in the primary plastic region, will abruptly begin to decrease linearly with x when the wake is reached from the primary plastic region. At perfect plasticity, this decrease proceeds smoothly, *i.e.*, the gradient $\partial \tau_{rz}/\partial x$ is continuous. Further insight into this phenomenon was given by Ståhle (1993), who incorporated inertia effects, and considered, in particular, vanishing crack velocity at vanishing strain hardening. This will be discussed in Section 7.2.

5.7 Slowly moving mode I and II cracks

General considerations

The asymptotic stress and deformation fields for moving mode I and II cracks in elastic-plastic media was first given by Slepyan (1974) for plane strain, steady growth, perfect plasticity and the Tresca yield condition. Slepyan showed that different sectors, active plastic sectors and wakes, appear even in mode I and II cases. Corresponding analyses for a Huber-von Mises material were given for mode I by Gao (1980), Rice et al. (1980), Rice (1982) and, in more detail, by Drugan et al. (1982) and by Drugan and Chen (1989). Rice (1982) also gave the general framework for analysis of the asymptotic fields at the edge of slowly moving cracks in elastic-perfectly plastic solids, including elastic anisotropy. In a later paper (Rice 1987), he presented solutions for opening mode asymptotic fields in elastic-plastic fcc and bcc crystals, with the crack moving in certain lattice directions. Ponte Castañeda (1986) gave a solution for mode II, plane stress in a perfectly plastic Huber-von Mises material. No analytical solution for mode I, plane stress has yet been given. Strain hardening cases have been treated analytically by Gao and Nemat-Nasser (1983) for modes I, II and III. Numerical investigations were performed by Ponte Castañeda, for mode I, plane strain (1987a), and for both plane stress and plane strain in modes I and II (1987b).

Even though plane strain and plane stress cases present considerably more difficulties than the anti-plane case, there are some close similarities. Thus, in the perfectly-plastic case, the asymptotic stresses do not depend on the distance r to the crack edge, only on the angle φ to the symmetry plane, and the asymptotic active plastic field contains centered fan and constant stress regions. However, for a Huber-von Mises material, an active plastic sector with non-singular strains also appears, and even though this sector turns out to be rather narrow, its presence complicates the analysis considerably (Rice 1982, Drugan et al. 1982). Here, consideration will be limited to modes I and II under plane strain, perfect plasticity and an isotropic Tresca material, i.e. the problems solved by Slepyan (1974). However, as pointed out by Rice (1982), for the special case of elastic as well as plastic incompressibility, that is Poisson's ratio $\nu = 1/2$, the solution for the Huber-von Mises material is identical with the one for the Tresca material.

For a Tresca material, the yield function at plane strain has to be given with three different expressions, depending on the combination of σ_x, σ_y, σ_z and τ_{xy}. One principal stress is σ_z, so that the other two are found from a second degree equation. Then, the yield function may be written as the largest of the functions

$$f_1 = \sqrt{\sigma_d^2 + \tau_{xy}^2} - \tau_Y, \qquad f_{2,3} = \frac{1}{2}\left|\frac{\sigma_x + \sigma_y}{2} - \sigma_z \pm \sqrt{\sigma_d^2 + \tau_{xy}^2}\right| - \tau_Y \qquad (5.7.1)$$

where $\sigma_d = \frac{1}{2}(\sigma_y - \sigma_x)$. The first yield function, valid when σ_z is neither the largest nor the least principal stress, corresponds to slip caused by shear stresses in planes normal to the xy plane; the other two also involve slip caused by shear stresses making 45° angle to this plane. The first slip system cannot produce plastic strains in the z direction. This is consistent with the fact that f_1 does not contain σ_z, so that the flow rule implies $d\epsilon_z^p = 0$, if f_1 is larger than f_2 and f_3. However, if f_2 or f_3 is larger than or equal to f_1, then non-zero plastic strain increments ϵ_z^p may be produced.

It will be assumed that f_1 is actually larger than or equal to f_2 or f_3 in the present problems, and this assumption will be controlled after a solution is found. Then, the flow rule (5.2.4) will be

$$\frac{4d\epsilon_x^p}{\sigma_x - \sigma_y} = \frac{4d\epsilon_y^p}{\sigma_y - \sigma_x} = \frac{4d\gamma_{xy}^p}{\tau_{xy}} \qquad (5.7.2)$$

As $r \to 0$, the equilibrium conditions show, just as in the anti-plane case, that the asymptotic stresses do not depend on r. The equations corresponding to (5.6.4) read

$$\frac{\partial \sigma_x}{\partial \varphi} = \cot\varphi \frac{\partial \tau_{xy}}{\partial \varphi}, \qquad \frac{\partial \sigma_y}{\partial \varphi} = \tan\varphi \frac{\partial \tau_{xy}}{\partial \varphi} \qquad (5.7.3)$$

Thus, because τ_{xy} is non-singular, due to the yield condition, all three in-plane stresses are non-singular in r, and then also the remaining stress, σ_z, must be non-singular. Moreover, if one of the stress derivatives in (5.7.3) is known, the other two are also known. Alternatively, they may all be expressed in the combination $\partial\sigma_x/\partial\varphi + \partial\sigma_y/\partial\varphi$:

$$\frac{\partial \sigma_x}{\partial \varphi} = \cos^2 \varphi \left(\frac{\partial \sigma_x}{\partial \varphi} + \frac{\partial \sigma_y}{\partial \varphi} \right) \tag{5.7.4}$$

$$\frac{\partial \sigma_y}{\partial \varphi} = \sin^2 \varphi \left(\frac{\partial \sigma_x}{\partial \varphi} + \frac{\partial \sigma_y}{\partial \varphi} \right) \tag{5.7.5}$$

$$\frac{\partial \tau_{xy}}{\partial \varphi} = \cos \varphi \sin \varphi \left(\frac{\partial \sigma_x}{\partial \varphi} + \frac{\partial \sigma_y}{\partial \varphi} \right) \tag{5.7.6}$$

A useful relation may be found between the change of stress at a material point with respect to a crack advance and the change of stress with respect to the angle φ. Writing $\sigma_{ij} = \sigma_{ij}(r, \varphi, a)$, where a is the length of the crack from a fix point to the crack edge, and differentiating with respect to a leads to

$$\frac{d\sigma_{ij}}{da} = \frac{\partial \sigma_{ij}}{\partial r} \cdot \frac{dr}{da} + \frac{\partial \sigma_{ij}}{\partial \varphi} \cdot \frac{d\varphi}{da} + \frac{\partial \sigma_{ij}}{\partial a} = -\frac{\partial \sigma_{ij}}{\partial r} \cos \varphi + \frac{1}{r} \cdot \frac{\partial \sigma_{ij}}{\partial \varphi} \sin \varphi + \frac{\partial \sigma_{ij}}{\partial a} \tag{5.7.7}$$

Thus, because σ_{ij} is bounded,

$$\boxed{r\frac{d\sigma_{ij}}{da} \to \frac{\partial \sigma_{ij}}{\partial \varphi} \sin \varphi \quad \text{as } r \to 0} \tag{5.7.8}$$

By using Hooke's law, corresponding expressions for the changes of elastic strains with a may be found:

$$r\frac{d\epsilon^e_{ij}}{da} = \frac{1}{2\mu} \left[\frac{\partial \sigma_{ij}}{\partial \varphi} - \frac{\nu}{1+\nu} \cdot \frac{\partial \sigma_{kk}}{\partial \varphi} \delta_{ij} \right] \sin \varphi \tag{5.7.9}$$

Thus, for example,

$$r\frac{d}{da}(\epsilon^e_x + \epsilon^e_y + \epsilon^e_z) = \frac{1-2\nu}{2(1+\nu)\mu} \cdot \frac{\partial}{\partial \varphi}(\sigma_x + \sigma_y + \sigma_z) \sin \varphi \tag{5.7.10}$$

Constant stress and centered fan sectors

A differential form of the yield condition, corresponding to (5.6.7), is

$$\left(\frac{\partial f_1}{\partial \sigma_x} \cot \varphi + \frac{\partial f_1}{\partial \sigma_y} \tan \varphi + \frac{\partial f_1}{\partial \tau_{xy}} \right) \frac{\partial \tau_{xy}}{\partial \varphi} = 0 \tag{5.7.11}$$

showing, together with (5.7.3), that, for an active plastic region, one possibility is *constant stress*. In combination with the flow rule (5.2.4), the other possibility gives

$$d\epsilon^p_x \cot \varphi + d\epsilon^p_y \tan \varphi + \frac{1}{2} d\gamma^p_{xy} = 0 \tag{5.7.12}$$

Use of the flow rule (5.7.2) then leads to a relation between the stresses,

$$\boxed{\sigma_x - \sigma_y + 2\tau_{xy} \tan 2\varphi = 0} \tag{5.7.13}$$

The same expression follows from the relations (5.4.1)-(5.4.3) in the introduction of slip line theory, if ψ, the direction of maximum principal shear stress, is identified with φ. Thus, two families of characteristic lines exist, one along the curves $\varphi = constant$, i.e. a *centered fan field*, and the other orthogonal to the first family. The absolute value of the shear stress $|\tau_{r\varphi}|$ is maximal, equal to the yield stress in shear, along the characteristic lines, and σ_r equals σ_φ on these lines. These relations hold in the whole asymptotic centered fan region, and comparison with (5.4.15) shows that

$$\tau_{r\varphi} = \pm \tau_Y, \qquad \sigma_r = \sigma_\varphi = \sigma_0 \mp 2\tau_Y \cdot \varphi \qquad (5.7.14)$$

where σ_0 is a constant stress. For simplicity, the upper halfplane, $y \geq 0$ will be studied. Mode I symmetry implies that the upper sign should be chosen, and so is the case also for mode II if the remote shear stress $\tau_{yz} > 0$. In the following this upper sign is assumed.

Apart from an additive constant stress, σ_0, the stresses and thereby also the elastic strains are now known for a centered fan region. In Cartesian coordinates, the in-plane stresses and their derivatives with respect to φ are

$$\sigma_x = \sigma_0 - \tau_Y(2\varphi + \sin 2\varphi), \qquad \frac{\partial \sigma_x}{\partial \varphi} = -4\tau_Y \cos^2 \varphi \qquad (5.7.15)$$

$$\sigma_y = \sigma_0 - \tau_Y(2\varphi - \sin 2\varphi), \qquad \frac{\partial \sigma_y}{\partial \varphi} = -4\tau_Y \sin^2 \varphi \qquad (5.7.16)$$

$$\tau_{xy} = \tau_Y \cos 2\varphi, \qquad \frac{\partial \tau_{xy}}{\partial \varphi} = -4\tau_Y \cos \varphi \sin \varphi \qquad (5.7.17)$$

It will be assumed here, to be controlled later, that f_1 is larger than f_2 and f_3 in centered fan sectors. Then, as discussed on page 309, $d\epsilon_z^p = 0$, which implies that σ_z must change inside a centered fan sector so that

$$\epsilon_z = \epsilon_0^p + \epsilon_z^e = \epsilon_0^p + \frac{1}{2(1+\nu)\mu}[\sigma_z - \nu(\sigma_x + \sigma_y)] = 0 \qquad (5.7.18)$$

where ϵ_0^p is the plastic strain ϵ_z^p at the front sector boundary on the y level studied. Thus,

$$\sigma_z = \nu(\sigma_x + \sigma_y) - 2(1+\nu)\mu\epsilon_0^p, \qquad \frac{\partial \sigma_z}{\partial \varphi} = \nu\left(\frac{\partial \sigma_x}{\partial \varphi} + \frac{\partial \sigma_y}{\partial \varphi}\right) \qquad (5.7.19)$$

The elastic strains then follow from Hooke's law.

The rate of deformation tensor

The analysis of slow crack growth requires great care. One difficulty concerns the displacement change at a material point when cylindrical coordinates are used. With Cartesian coordinates, this change follows from the expressions

$$\frac{du_1}{da} = -\frac{\partial u_1}{\partial x_1}, \qquad \frac{du_2}{da} = -\frac{\partial u_2}{\partial x_1} \qquad (5.7.20)$$

Now, for example,

$$u_r = u_1 \cos \varphi + u_2 \sin \varphi \qquad (5.7.21)$$

so that

$$u_r(r,\varphi) \to \left[u_1(x_1,x_2) - \frac{\partial u_1}{\partial x_1}da\right]\cos\varphi + \left[u_2(x_1,x_2) - \frac{\partial u_2}{\partial x_1}da\right]\sin\varphi$$

$$= u_r(r,\varphi) - \left(\frac{\partial u_1}{\partial x_1}\cos\varphi + \frac{\partial u_2}{\partial x_1}\sin\varphi\right)da \quad (5.7.22)$$

as $a \to a + da$. Note that the last term is different from

$$-\frac{\partial}{\partial x_1}(u_1\cos\varphi + u_2\sin\varphi)da = -\frac{\partial u_r}{\partial x_1}da \quad (5.7.23)$$

and thus $du_r/da \neq -\partial u_r/\partial x_1$. Similarly, care should be exercised when expressing strain changes involving cylindrical coordinates, and a good strategy is to introduce the tensor

$$\boxed{D_{ij} = \frac{1}{2}\left(\frac{\partial U_i}{\partial x_j} + \frac{\partial U_j}{\partial x_i}\right)} \quad (5.7.24)$$

where $U_i = du_i/da = -\partial u_i/\partial x_1$. This tensor will be called the *rate of deformation tensor*. Here, "rate", as, for instance, in the context of "energy release rate", refers to change with respect to crack advance. This deviates somewhat from the common usage of this concept as change of deformation with respect to time, then motivating the image of the *displacement rate* U_i as a velocity, even though, actually, no time is involved. However, if this image is preferred, one should include a factor $\dot{a} = da/dt$ in quite a few of the following expressions.

Indexing of D_{ij} and U_i is made in the same way as for ϵ_{ij} and u_i; for example D_r is the rate of deformation in the r direction, and

$$D_r = \frac{\partial U_r}{\partial r} = \frac{\partial (U_1\cos\varphi + U_2\sin\varphi)}{\partial r} \quad (5.7.25)$$

in complete analogy with the expression for the strain ϵ_r in terms of the displacement u_r. Note that D_r is not equal to

$$-\frac{\partial \epsilon_r}{\partial x_1} = -\frac{\partial}{\partial x_1}\cdot\frac{\partial u_r}{\partial r} = -\frac{\partial}{\partial x_1}\cdot\frac{\partial(u_1\cos\varphi + u_2\cos\varphi)}{\partial r}$$

For plane strain, the component $U_3 = 0$. A relation between the two other components of the displacement rate may be established by using the condition of plastic incompressibility:

$$\operatorname{div}\boldsymbol{U} = \frac{1}{r}\cdot\frac{\partial(rU_r)}{\partial r} + \frac{1}{r}\cdot\frac{\partial U_\varphi}{\partial \varphi} = \frac{d}{da}(\epsilon_x^e + \epsilon_y^e + \epsilon_z^e) \quad (5.7.26)$$

which by (5.7.10) results in

$$\frac{\partial(rU_r)}{\partial r} + \frac{\partial U_\varphi}{\partial \varphi} = \frac{1-2\nu}{2(1+\nu)\mu}\frac{\partial}{\partial \varphi}(\sigma_x + \sigma_y + \sigma_z)\sin\varphi \quad (5.7.27)$$

Asymptotic deformations in centered fan sectors

Apart from still undetermined constants, the stresses and therefore also the elastic strains are known in the centered fan by (5.7.15)-(5.7.19), but the plastic strains are still unknown. These may be found from the asymptotic displacement rate U_i. One of its components is already known ($U_z = 0$) and (5.7.27) provides a relation between the other two:

$$\boxed{\frac{\partial(rU_r)}{\partial r} + \frac{\partial U_\varphi}{\partial \varphi} = -\frac{2(1-2\nu)\tau_Y}{\mu} \sin\varphi} \qquad (5.7.28)$$

Another equation is found by using the flow rule and the yield condition. Consider the rate of deformation tensor. In an active plastic region it consists of two parts, one elastic and one plastic:

$$D_{ij} = D_{ij}^e + D_{ij}^p = D_{ij}^e + \frac{d\lambda}{da} \cdot \frac{\partial f_1}{\partial \sigma_{ij}} \qquad (5.7.29)$$

Here, the associated flow rule (5.2.4) was used. The elastic part is known: it equals $d\epsilon_{ij}^e/da$, given by (5.7.9). Now, it is possible to eliminate D_{ij}^p by using the differential form of the yield condition (5.2.3),

$$\frac{\partial f_1}{\partial \sigma_{ij}} \cdot \frac{\partial \sigma_{ij}}{\partial \varphi} = 0 \qquad (5.7.30)$$

Thus, multiplying both members of (5.7.29) by $\partial \sigma_{ij}/\partial \varphi$ results in

$$\frac{\partial \sigma_{ij}}{\partial \varphi} D_{ij} = \frac{\partial \sigma_{ij}}{\partial \varphi} D_{ij}^e \qquad (5.7.31)$$

Insertion of $D_{ij}^e = -\partial \epsilon_{ij}^e/\partial x_1$ from (5.7.9), and use of (5.7.15)-(5.7.17) and (5.7.19), gives

$$r\frac{\partial \sigma_{ij}}{\partial \varphi} D_{ij} = \frac{8(1-\nu)\tau_Y^2}{\mu} \sin\varphi \qquad (5.7.32)$$

The left member may be written as

$$r\left(\frac{\partial \sigma_x}{\partial \varphi} D_x + \frac{\partial \sigma_y}{\partial \varphi} D_y + \frac{\partial \tau_{xy}}{\partial \varphi} D_{xy}\right)$$

which after insertion of the expressions for the stress derivatives, (5.7.15)-(5.7.17), is found to be

$$-4\tau_Y r(D_x \cos^2\varphi + D_y \sin^2\varphi + D_{xy}\cos\varphi \sin\varphi) = -4\tau_Y r D_r = -4\tau_Y r \frac{\partial U_r}{\partial r} \qquad (5.7.33)$$

so that

$$r\frac{\partial U_r}{\partial r} = -\frac{2(1-\nu)\tau_Y}{\mu}\sin\varphi \qquad (5.7.34)$$

which is the second differential equation for the asymptotic displacement rates in the fan. This equation may be immediately integrated to yield

$$\boxed{U_r = \frac{2(1-\nu)\tau_Y}{\mu} \sin\varphi \cdot \ln\frac{R}{r} + \frac{\partial f(\varphi)}{\partial \varphi}} \qquad (5.7.35)$$

where the length parameter R and the function $f(\varphi)$ must be left undetermined in an asymptotic analysis. Equation (5.7.27) may then be reduced to

$$\frac{\partial U_\varphi}{\partial \varphi} = -\left[\frac{2(1-2\nu)\tau_Y}{\mu}\sin\varphi + \frac{2(1-\nu)\tau_Y}{\mu}\sin\varphi\cdot\left(\ln\frac{R}{r}-1\right) + \frac{\partial f(\varphi)}{\partial \varphi}\right] \quad (5.7.36)$$

which integrates to

$$\boxed{U_\varphi = \frac{2(1-\nu)\tau_Y}{\mu}\left[\frac{\nu}{1-\nu} - \ln\frac{R}{r}\right](\cos\varphi_0 - \cos\varphi) - f(\varphi) + g(r)} \quad (5.7.37)$$

where φ_0 is an angle used to express a constant of integration, and the function $g(r)$ should satisfy the condition $g(0) = 0$. Rice (1982) suggests the form $g(r) \propto r$; this will contribute a rigid rotation near the crack edge in the fan. Note, however, that the logarithmic term is asymptotically dominating.

The asymptotic field in elastically deforming sectors

For an elastically deforming sector, $D_{ij}^p = 0$, and then, because $D_z = 0$ for plane strain, the elastic part $D_z^e = 0$, so that Hooke's law, (5.7.9), leads to

$$\frac{\partial \sigma_z}{\partial \varphi} = \nu\left(\frac{\partial \sigma_x}{\partial \varphi} + \frac{\partial \sigma_y}{\partial \varphi}\right) \quad (5.7.38)$$

As a further consequence of the fact that $D_{ij}^p = 0$, the left member of (5.7.9) equals rD_{ij} for $r \to 0$. With the use of (5.7.4)-(5.7.6) and (5.7.38), this equation may then be written in the form

$$rD_x = \frac{1}{2\mu}(\cos^2\varphi - \nu)\left(\frac{\partial \sigma_x}{\partial \varphi} + \frac{\partial \sigma_y}{\partial \varphi}\right)\sin\varphi \quad (5.7.39)$$

$$rD_y = \frac{1}{2\mu}(\sin^2\varphi - \nu)\left(\frac{\partial \sigma_x}{\partial \varphi} + \frac{\partial \sigma_y}{\partial \varphi}\right)\sin\varphi \quad (5.7.40)$$

$$rD_{xy} = \frac{1}{\mu}(\cos\varphi\sin\varphi)\left(\frac{\partial \sigma_x}{\partial \varphi} + \frac{\partial \sigma_y}{\partial \varphi}\right)\sin\varphi \quad (5.7.41)$$

where D_{xy} is defined by analogy with γ_{xy} so that the tensor component is $D_{xy}/2$. Also the cylindrical component D_r will be needed:

$$D_r = D_x\cos^2\varphi + D_y\sin^2\varphi + D_{xy}\cos\varphi\sin\varphi = \frac{1-\nu}{2\mu}\left(\frac{\partial \sigma_x}{\partial \varphi} + \frac{\partial \sigma_y}{\partial \varphi}\right)\frac{\sin\varphi}{r} \quad (5.7.42)$$

In cylindrical coordinates, the in-plane compatibility equation is

$$\frac{\partial}{\partial r}\left(r^2\frac{\partial D_\varphi}{\partial r}\right) - r\frac{\partial D_r}{\partial r} + \frac{\partial^2 D_r}{\partial \varphi^2} = \frac{\partial}{\partial r}\left(r\frac{\partial D_{r\varphi}}{\partial \varphi}\right) \quad (5.7.43)$$

Because the stresses are bounded, the components of rD_{ij} are bounded according to (5.7.9) and thus they are functions of φ only. So, put

$$rD_r = F_r(\varphi), \qquad rD_\varphi = F_\varphi(\varphi), \qquad rD_{r\varphi} = F_{r\varphi}(\varphi) \quad (5.7.44)$$

Then, the compatibility equations show that $F_r(\varphi)$ is determined by the equation

$$F_r''(\varphi) + F_r(\varphi) = 0 \quad (5.7.45)$$

whereas $F_\varphi(\varphi)$ and $F_{r\varphi}(\varphi)$ are left undetermined. Solution gives

$$F_r(\varphi) = A_1 \cos \varphi + A_2 \sin \varphi \tag{5.7.46}$$

whereupon equation (5.7.42) leads to

$$\frac{\partial \sigma_x}{\partial \varphi} + \frac{\partial \sigma_y}{\partial \varphi} = \frac{2\mu}{1-\nu}(A_1 \cot \varphi + A_2) \tag{5.7.47}$$

where A_1 and A_2 are constants. Thereby, the rate of deformation tensor becomes determined through (5.7.39)-(5.7.41), and then transformation to cylindrical coordinates gives

$$rD_\varphi = -\frac{\nu}{1-\nu}F_r(\varphi), \qquad rD_{r\varphi} = 0 \tag{5.7.48}$$

The displacement rates are now found by integration. Use of (5.7.25) and the relation

$$D_\varphi = \frac{1}{r} \cdot \frac{\partial U_\varphi}{\partial \varphi} + \frac{U_r}{r} \tag{5.7.49}$$

leads to

$$U_r = F_r(\varphi) \int \frac{dr}{r} = -F_r(\varphi) \ln \frac{R}{r} + C(\varphi) \tag{5.7.50}$$

$$U_\varphi = \ln \frac{R}{r} \int F_r(\varphi) d\varphi - \int \left[\frac{\nu}{1-\nu}F_r(\varphi) + C(\varphi)\right] d\varphi + R(r) \tag{5.7.51}$$

where $C(\varphi)$ and $R(r)$ should satisfy the condition $D_{r\varphi} = 0$. This gives rise to one differential equation for $R(r)$, resulting in $R(r) \propto r$, which is a rigid rotation, and one differential equation for $C(\varphi)$, giving, apart from a rigid translation,

$$C(\varphi) = \frac{1-2\nu}{2(1-\nu)}\varphi(A_1 \sin \varphi - A_2 \cos \varphi) \tag{5.7.52}$$

Then, after returning to Cartesian coordinates, the displacement rates are found to be

$$U_1 = -\frac{1}{4(1-\nu)}\{A_1[4(1-\nu)\ln(R/r) + \cos 2\varphi] + A_2[2(1-2\nu)\varphi + \sin 2\varphi]\} \tag{5.7.53}$$

$$U_2 = \frac{1}{4(1-\nu)}\{A_1[2(1-2\nu)\varphi - \sin 2\varphi] - A_2[4(1-\nu)\ln(R/r) - \cos 2\varphi]\} \tag{5.7.54}$$

to within a rigid body motion.

With use of (5.7.15)-(5.7.17) and (5.7.38), all components of the stress derivatives may be found. Integrations give

$$\sigma_x = \frac{\mu}{2(1-\nu)}[4A_1 \ln(\sin \varphi) + A_1 \cos 2\varphi + A_2(2\varphi + \sin 2\varphi)] + C_{11} \tag{5.7.55}$$

$$\sigma_y = \frac{\mu}{2(1-\nu)}[-A_1 \cos 2\varphi + A_2(2\varphi - \sin 2\varphi)] + C_{22} \tag{5.7.56}$$

$$\tau_{xy} = \frac{\mu}{2(1-\nu)}[A_1(2\varphi + \sin 2\varphi) - A_2 \cos 2\varphi] + C_{12} \tag{5.7.57}$$

$$\sigma_z = \frac{\nu\mu}{2(1-\nu)}[4A_1 \ln(\sin \varphi) + 4A_2\varphi] + C_{33} \tag{5.7.58}$$

316   5. ELASTOPLASTIC STRESSES AND STRAINS

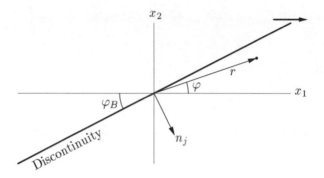

Fig. 5.7.1 Plane discontinuity surface, moving in the positive x_1 direction.

Matching conditions at sector boundaries

The sector boundaries are examples of moving surfaces of discontinuity in continua. The stress-displacement fields have different expressions on the two sides of the discontinuity surface, but all field quantities are not discontinuous at the surface. Some of the discontinuities concern gradients or higher derivatives of stresses and strains and may be considered as weak in comparison with those in stresses or strains, so-called strong discontinuities. General *jump conditions* exist, telling whether or not a certain field quantity is always continuous and, if it may be discontinuous, how its change – jump – across the discontinuity surface is related to other field quantities (Hadamard 1903, Hill 1962, Drugan and Rice 1984, Drugan and Shen 1987, Leighton et al. 1987, Shen and Drugan 1990, Varias and Shih 1994). A distinction may be made between *mechanical conditions* and *kinematic conditions*. For generality, inertia forces are here taken into account in establishing mechanical conditions, even though such forces are not involved in the present study of slowly moving cracks.

Quantities in front of a discontinuity surface will be labelled with plus, quantities behind with minus. Introduce the notation $[\![q]\!] = q^+ - q^-$ for the decrease of a quantity q at the passage of the surface of discontinuity (i.e., from plus to minus).

Consider the part $x_2 > 0$ of a plane discontinuity surface moving in the positive x_1 direction and oriented so that its forwards normal, n_j, makes the angle $\varphi_B - \pi/2$ with the positive x_1 direction. Plane stress or strain is assumed. Introduce a cylindrical coordinate system (r, φ, z), following the surface, with $\varphi = 0$ in the positive x_1 direction, so that the surface is at $\varphi = \varphi_B$; see Fig. 5.7.1.

Material integrity demands that

$$\boxed{[\![u_i]\!] = 0} \tag{5.7.59}$$

or, in cylindrical coordinates,

$$\boxed{[\![u_r]\!] = 0, \quad [\![u_\varphi]\!] = 0, \quad [\![u_z]\!] = 0} \tag{5.7.60}$$

These are kinematic conditions. Other kinematic conditions follow from the fact that continuity of u_i also implies continuity of $\partial u_i/\partial r$ and $\partial u_i/\partial z$, which, in combination

with the relation
$$\frac{\partial u_i}{\partial r} = \frac{\partial u_i}{\partial x_1}\cos\varphi + \frac{\partial u_i}{\partial x_2}\sin\varphi \tag{5.7.61}$$

leads to
$$\boxed{[\![\partial u_i/\partial x_2]\!] = -[\![\partial u_i/\partial x_1]\!]\cot\varphi_B = [\![U_i]\!]\cot\varphi_B, \qquad [\![\partial u_i/\partial x_3]\!] = 0} \tag{5.7.62}$$

where U_i is the displacement rate, defined in the same way as previously, so that
$$U_i = \frac{du_i}{da} = -\frac{\partial u_i}{\partial x_1} \tag{5.7.63}$$

for an advance da in positive x_1 direction of the discontinuity surface.

Then, by making use of the definition
$$\epsilon_{ij} = \frac{1}{2}\left(\frac{\partial u_i}{\partial x_j} + \frac{\partial u_j}{\partial x_i}\right) \tag{5.7.64}$$

the jump conditions for the strains are found to be
$$[\![\epsilon_{11}]\!] = -[\![U_1]\!], \qquad [\![\epsilon_{22}]\!] = [\![U_2]\!]\cot\varphi_B, \qquad [\![\epsilon_{33}]\!] = 0 \tag{5.7.65}$$
$$[\![\epsilon_{12}]\!] = \frac{1}{2}[[\![U_1]\!]\cot\varphi_B - [\![U_2]\!]], \qquad [\![\epsilon_{23}]\!] = 0, \qquad [\![\epsilon_{31}]\!] = 0 \tag{5.7.66}$$

Transformation to cylindrical coordinates gives
$$\boxed{[\![\epsilon_r]\!] = 0, \qquad [\![\epsilon_\varphi]\!] = \frac{[\![U_\varphi]\!]}{\sin\varphi_B}, \qquad [\![\gamma_{r\varphi}]\!] = \frac{[\![U_r]\!]}{\sin\varphi_B}} \tag{5.7.67}$$

Note that $[\![U_\varphi]\!]$ vanishes in plane strain if elastic as well as plastic incompressibility is assumed. Now, it is possible to interpret the condition for $[\![\epsilon_\varphi]\!]$ so that $[\![U_\varphi]\!]$ must be non-singular. This follows from the fact that all elastic strains are bounded, because the stresses are bounded. Thus, $[\![\epsilon_r^p]\!] = [\![\epsilon_r]\!] - [\![\epsilon_r^e]\!]$ is bounded and, because of plastic incompressibility and plane strain, equal to $-[\![\epsilon_\varphi^p]\!] - [\![\epsilon_z^p]\!] = -[\![\epsilon_\varphi^p]\!] + [\![\epsilon_z^e]\!]$, implying that $[\![\epsilon_\varphi]\!]$ is also bounded, which, by (5.7.67) proves that $[\![U_\varphi]\!]$ is bounded. Similarly, it is evident that $[\![U_r]\!]$ is bounded at the boundary between an active plastic sector followed by an elastically deforming sector, because cessation of plastic flow occurs when $d\epsilon_r^p = d\epsilon_\varphi^p = d\gamma_{r\varphi}^p = 0$. Likewise, at the boundary between an elastically deforming sector followed by a constant stress sector, $[\![U_r]\!]$ is bounded because $[\![\gamma_{r\varphi}^p]\!] = 0$ according to the flow rule, because $[\![\epsilon_r^p]\!] = 0$ and $[\![\epsilon_\varphi^p]\!] = 0$. However, at the boundary between a centered fan plastic sector and a constant stress sector, $[\![U_r]\!]$ is not necessarily bounded. Note that the flow rule does not relate $[\![\gamma_{r\varphi}^p]\!]$ to $[\![\epsilon_r^p]\!]$, because $\sigma_\varphi = \sigma_r$ on such a boundary.

Note that boundedness of, for instance $[\![U_\varphi]\!]$, does not exclude singularity of U_r on each side of the boundary.

Mechanical conditions will now be studied. Assume that the mass velocity is VU_i^- just behind and VU_i^+ just ahead of the discontinuity surface. Consider, at time t, a mass layer bounded by two planes parallel with the discontinuity surface, one behind and the other ahead, but sufficiently close to the surface that stresses and strains within the mass on each side of the discontinuity surface may be considered as constants in the thickness direction of the layer; see Fig. 5.7.2.

After a time dt, the rear boundary of the mass layer has swept over a volume

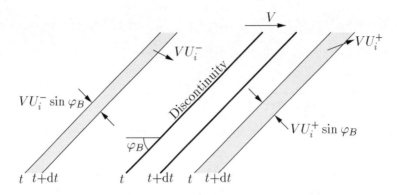

Fig. 5.7.2 A mass layer between two parallel planes, parallel to and on each side of a discontinuity surface which is moving with velocity V in the positive x direction. The mass velocity in the layer is VU_i^- behind and VU_i^+ ahead of the discontinuity surface. Two positions are shown, one at time t and one at time $t + \mathrm{d}t$.

$VU_1^- \mathrm{d}t \sin \varphi_B$ per unit of area along the boundary. Thus, neglecting the influence of dilatation on the mass density ϱ, the change of momentum per unit area consists of a loss $\varrho V U_1^- \mathrm{d}t \sin \varphi_B \cdot VU_i^-$ at the rear end of the mass, a gain $\varrho V U_1^+ \mathrm{d}t \sin \varphi_B \cdot VU_i^+$ at the front end and a gain $\varrho V \mathrm{d}t \sin \varphi_B (VU_i^- - VU_i^+)$ at the discontinuity surface. Now, because the force equals the change of momentum per unit of time, the force per unit area on the mass layer equals

$$(\sigma_{ij}^- - \sigma_{ij}^+) n_j = [-\varrho V U_1^- \mathrm{d}t \cdot VU_i^- + \varrho V U_1^+ \mathrm{d}t \cdot VU_i^+$$
$$+ \varrho V \mathrm{d}t (VU_i^- - VU_i^+)] \sin \varphi_B / \mathrm{d}t \quad (5.7.68)$$

which in more compact form reads

$$\boxed{[\![\sigma_{ij}]\!] n_j = -\varrho V^2 [\![U_i]\!] \sin \varphi_B + \varrho V^2 [\![U_1 U_i]\!] \sin \varphi_B} \quad (5.7.69)$$

The last term can usually be neglected, even though it might be asymptotically dominating if the discontinuity surface is a boundary between different sectors at the edge of a propagating crack. Then, U_i is usually singular and therefore $[\![U_1 U_i]\!]$ is generally more strongly singular than $[\![U_i]\!]$. However, the region of dominance of the last term is extremely small. For a logarithmic singularity, such as the one in (5.7.35), it is of the order of

$$e^{-\mu/\tau_Y} R \quad (5.7.70)$$

It is therefore neglected in the following.

In (5.7.69), $n_1 = \sin \varphi$, $n_2 = -\cos \varphi$ and $n_3 = 0$. The two equations for $i, j = 1, 2$ include $[\![\sigma_{11}]\!]$, $[\![\sigma_{22}]\!]$, $[\![\sigma_{12}]\!]$, $[\![U_1]\!]$ and $[\![U_2]\!]$. Transformation to cylindrical coordinates results in two equations that do not contain $[\![\sigma_r]\!]$, which allows separation of $[\![\sigma_\varphi]\!]$ and $[\![\tau_{r\varphi}]\!]$:

$$\boxed{[\![\sigma_\varphi]\!] = \varrho V^2 [\![U_\varphi]\!] \sin \varphi_B = \varrho V^2 [\![\epsilon_\varphi]\!] \sin^2 \varphi_B} \quad (5.7.71)$$

$$\boxed{[\![\tau_{r\varphi}]\!] = \varrho V^2 [\![U_r]\!] \sin \varphi_B = \varrho V^2 [\![\gamma_{r\varphi}]\!] \sin^2 \varphi_B} \qquad (5.7.72)$$

Here, use was also made of (5.7.67).

So far, the requirements of material integrity and conservation of mass have been used. Conservation of energy is also required, but leads to a more complicated picture. In general, there has to be a temperature jump across a discontinuity surface. This creates some problems which are usually avoided for small temperature gradients by using adiabatic or isothermal constitutive equations.

Heat is transferred across a layer with a high temperature gradient by conduction and radiation. The amount of heat conducted across a layer per unit of area and time is proportional to the temperature gradient in the layer. For a discontinuity surface, this would imply an infinite heat flux for any finite temperature jump, and thus the conclusion has to be that detailed discussions of energy relations must consider a thin layer rather than an ideal discontinuity. In fact, the effective thickness of the layer is determined by the temperature jump (and material constants such as the coefficient of heat conduction). As a consequence, the principles of maximum plastic work and of plastic incompressibility cannot be expected to hold in such a layer. Thus, it might not be possible to stretch the jump conditions much further without a detailed thermodynamic analysis. Any attempt to use continuous stress-strain relations, such as adiabatic or isothermal, assuming a thin layer instead of a discontinuity surface, would lead, even in the limit of vanishing thickness, to stress continuity. This is different from assuming *a priori* the existence of a discontinuity and applying, say, adiabatic stress-strain relations on each side. That such discontinuities exist, with both stress and strain jumps, has been demonstrated numerically by Varias and Shih (1994) for dynamic elastic-plastic crack propagation.

For vanishingly small velocity, V, equations (5.7.71) and (5.7.72) show continuity of σ_φ and $\tau_{r\varphi}$, except in a vanishingly small region near a strain singularity. (Note that, strictly, $V=0$ is not compatible with crack growth). Isothermal changes may be assumed (which simply implies use of "ordinary" elastic and plastic stress-strain relations).

For the case of vanishingly small V, it can be shown that the yield function $f(\sigma_{ij})$ is continuous, and then equal to zero, even across a boundary between one active plastic region and one elastically deforming region. Its continuity across the boundary between two active plastic regions (constant stress and constant strain) is obvious. For an elastically deforming region, continuity of the yield function f would imply that it is maximal on the boundary to an active plastic region and, consequently negative inside the elastically deforming region. In order to prove its continuity, assume the opposite, namely that σ_r is not continuous across such a boundary. (If it is, full stress continuity prevails and then also the yield function is continuous). For simplicity the flow rule in the form (5.7.2) is assumed. In cylindrical coordinates it reads

$$\frac{4 \mathrm{d}\epsilon_r^p}{\sigma_r - \sigma_\varphi} = -\frac{4 \mathrm{d}\epsilon_\varphi^p}{\sigma_r - \sigma_\varphi} = \frac{\mathrm{d}\gamma_{r\varphi}^p}{\tau_{r\varphi}} \qquad (5.7.73)$$

Continuity of $\tau_{r\varphi}$ implies that the changes of $\gamma_{r\varphi}$ are purely plastic across the boundary. Then, the flow rule tells that

$$\mathrm{d}\epsilon_r^p = -\mathrm{d}\epsilon_\varphi^p \neq 0 \qquad (5.7.74)$$

except, perhaps, if $\sigma_r = \sigma_\varphi$, which, however, would imply $[\![\sigma_r]\!] = [\![\sigma_\varphi]\!] = 0$ and thus contradict the assumption at the outset that σ_r is not continuous. But if $[\![\sigma_r]\!] \neq 0$, whereas $[\![\sigma_\varphi]\!] = [\![\tau_{r\varphi}]\!] = 0$, then the yield function $f(\sigma_r, \sigma_\varphi, \tau_{r\varphi})$, which equals zero at the plastic side of the discontinuity, decreases to a negative value at the elastic side, which implies a purely elastic stress-strain path through the discontinuity surface. But this contradicts the result that the changes of $\gamma_{r\varphi}$ are purely plastic, with the consequence (5.7.74). Thus, the yield function must be continuous across a boundary between an active plastic and an elastically deforming region, which also implies full stress continuity across such a boundary. This has been shown to be the case in general at vanishingly slow motion of discontinuity surfaces by Drugan and Rice (1984).

A quantity which experiences a jump may be imagined to follow a certain path within the discontinuity. This path is not arbitrary, even if the jump remains fixed. Thus, for instance, a stress following different stress-strain paths could produce different plastic strain jumps. In some cases the path is obvious. Thus, the strain ϵ_r must remain constant within the discontinuity, because this is a condition of material integrity. Likewise, for vanishing velocity V, σ_φ and $\tau_{r\varphi}$ must remain constant, because this is a condition of static equilibrium.

Assembly of sectors

Now, turn back to the study of slowly moving mode I and II cracks. Recall that the upper half $y \geq 0$, was considered. Thus, there are boundary conditions at $\varphi = 0$ and $\varphi = \pi$. The latter consist of traction free crack faces, whereas the former are given by the mode I or mode II symmetry relations. Between these angles there are section boundaries, where matching conditions apply. These matching conditions are given by the requirement of continuity of the stress components. In addition, the conditions of non-singular jumps of the deformation rates will be used (page 317).

With a view toward slowly moving mode III cracks, it may be assumed that a primary plastic region, P, is followed by a wake, W, and a secondary plastic region, S. For the secondary plastic region, a constant stress field may be assumed:

$$\sigma_y = 0, \qquad \tau_{xy} = 0, \qquad \sigma_x = 2\tau_Y \tag{5.7.75}$$

Here, the first two stresses follow from the boundary conditions at $\varphi = \pi$ and then the third in-plane stress follows from the yield condition $f_1 = 0$, if it is assumed that $\sigma_x > 0$. As will be obvious later, the sign chosen for σ_x is compatible with mode I loading and for mode II with a remote load $\tau_{xy} > 0$.

The validity of all assumptions made about different regions and about the yield condition must be controlled after a solution is found. The continued investigation has to be made separately for modes I and II.

Mode I. With a view toward the stationary mode I crack, it is assumed that the primary plastic region consists of one constant stress region, P1, in front of the crack, followed by a centered fan region, P2; see Fig. 5.7.3. Assuming that $\sigma_y > \sigma_x$ in P1, the in-plane stresses in the region are

$$\tau_{xy} = 0, \qquad \sigma_x = \sigma_x^0, \qquad \sigma_y = \sigma_x^0 + 2\tau_Y \tag{5.7.76}$$

where the first stress follows from the boundary condition at $\varphi = 0$ and the relation

5.7 SLOWLY MOVING MODE I AND II CRACKS   321

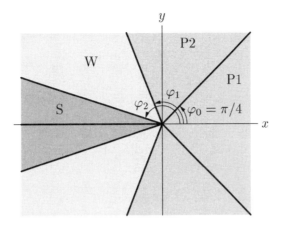

Fig. 5.7.3 The sectors of the primary plastic regions (P1 and P2), the wake (W) and the secondary plastic region (S) in an asymptotic vicinity of a slowly moving mode I crack edge. P1 is a constant stress sector, P2 is a centered fan sector, W is elastically deforming, and S is a constant stress sector.

between the other two, which contains the as yet undetermined constant stress σ_x^0, follows from the yield condition. The stress σ_z depends on whether $f_1 > f_2$ (or f_3) or $f_1 = f_2$ (or f_3). If $f_1 = f_2$ (or f_3), then σ_z is determined from the condition $f_2 = 0$, giving

$$\sigma_z = \sigma_x^0 \tag{5.7.77}$$

On the other hand, if $f_1 > f_2$ (or f_3), then σ_z cannot be determined from the yield condition, but from the plane strain condition that $\epsilon_z = 0$, which implies $\epsilon_z^e = 0$, because the flow rule shows that $d\epsilon_z^p = 0$ and no accumulated plastic strain prevails, because P1 is the front plastic region. Thus,

$$\epsilon_z^e = \frac{1}{2(1+\nu)\mu}[\sigma_z - \nu(\sigma_x + \sigma_y)] = 0 \tag{5.7.78}$$

so that

$$\sigma_z = 2\nu(\sigma_x^0 + \tau_Y) \tag{5.7.79}$$

Now, the case $f_1 > f_2$ (or f_3) implies that $f_1 = 0$, f_2 (or f_3) < 0, which, in turn, implies that $\sigma_z > (\sigma_x + \sigma_y)/2 - \tau_Y = \sigma_x^0$ in P1. Thus, it is applicable if $2\nu(\sigma_x^0 + \tau_Y) > \sigma_x^0$, and, consequently,

$$\sigma_z = \sigma_x^0 \text{ if } \nu \leq \nu^* = \frac{\sigma_x^0}{2(\sigma_x^0 + \tau_Y)} \tag{5.7.80}$$

$$\sigma_z = 2\nu(\sigma_x^0 + \tau_Y) \text{ if } \nu > \nu^* \tag{5.7.81}$$

Consider next the P1-P2 boundary. The boundary angle, φ_0, must equal $\pi/4$, because the coefficient of the $\ln(R/r)$ term in U_φ, see (5.7.37), must vanish at the boundary, because U_φ is less singular than $\ln(R/r)$ in the constant stress sector, and, as shown previously, $[\![U_\varphi]\!]$ must be bounded.

The in-plane stresses in P2 are given by (5.7.15)-(5.7.17). From (5.7.17) it follows that $\tau_{xy} = 0$ for $\varphi = \pi/4$, i.e., τ_{xy} is continuous across the P1-P2 boundary. Then, continuity of the other two in-plane stresses is found to give

$$\sigma_0 = \sigma_x^0 + \tau_Y(\pi/2 + 1) \tag{5.7.82}$$

where σ_0 is the additive constant stress in the centered fan region introduced in connection with (5.7.15)-(5.7.16).

Next, use will be made of the conditions that $[\![U_\varphi]\!]$ and $[\![U_r]\!]$, or, equivalently, $[\![U_1]\!]$ and $[\![U_2]\!]$, must be bounded at the P2-W boundary. Change to Cartesian components from (5.7.35) and (5.7.37) gives

$$U_1 = \frac{2(1-\nu)\tau_Y}{\mu} \cos\varphi_0 \sin\varphi \ln\frac{R}{r} + \cdots \tag{5.7.83}$$

$$U_2 = \frac{2(1-\nu)\tau_Y}{\mu}(1 - \cos\varphi_0 \cos\varphi) \ln\frac{R}{r} + \cdots \tag{5.7.84}$$

Comparison with (5.7.53) and (5.7.54) and equalization of the respective coefficients for $\ln(R/r)$ for $\varphi = \varphi_1$, leads to

$$A_1 = -\frac{2(1-\nu)\tau_Y}{\mu} \cdot \frac{\sin\varphi_1}{\sqrt{2}}, \qquad A_2 = -\frac{2(1-\nu)\tau_Y}{\mu} \cdot \frac{\sqrt{2} - \cos\varphi_1}{\sqrt{2}} \tag{5.7.85}$$

where $\varphi = \varphi_1$ is the P2-W boundary. Then, denoting the boundary between sectors W and S by $\varphi = \varphi_2$, in-plane stress continuity at the P2-W and W-S boundaries gives six equations from (5.7.15)-(5.7.17) for P2, (5.7.55)-(5.7.57) for W and (5.7.75) for S. These may be arranged so that the constants C_{11}, C_{22} and C_{12} are eliminated, i.e. as $(\sigma_{ij})_{P1-W} = 0 = (\sigma_{ij})_{W-S}$:

$$1 - \frac{\pi}{2} - \frac{\sigma_x^0}{\tau_Y} + \sin 2\varphi_1 + 2\varphi_1 + \frac{\sin\varphi_1}{\sqrt{2}}\left(4\ln\frac{\sin\varphi_2}{\sin\varphi_1} + \cos 2\varphi_2 - \cos 2\varphi_1\right)$$
$$+ \frac{\sqrt{2}-\cos\varphi_1}{\sqrt{2}}[2(\varphi_2 - \varphi_1) + \sin 2\varphi_2 - \sin 2\varphi_1] = 0 \tag{5.7.86}$$

$$-1 - \frac{\pi}{2} - \frac{\sigma_x^0}{\tau_Y} - \sin 2\varphi_1 + 2\varphi_1 + \frac{\sin\varphi_1}{\sqrt{2}}(\cos 2\varphi_1 - \cos 2\varphi_2)$$
$$+ \frac{\sqrt{2}-\cos\varphi_1}{\sqrt{2}}[2(\varphi_2 - \varphi_1) - \sin 2\varphi_2 + \sin 2\varphi_1] = 0 \tag{5.7.87}$$

$$-\cos 2\varphi_1 + \frac{\sin\varphi_1}{\sqrt{2}}[2(\varphi_2 - \varphi_1) + \sin 2\varphi_2 - \sin 2\varphi_1]$$
$$+ \frac{\sqrt{2}-\cos\varphi_1}{\sqrt{2}}(\cos 2\varphi_1 - \cos 2\varphi_2) = 0 \tag{5.7.88}$$

These equations were derived by Slepyan (1974). Elimination of σ_x^0 gives two equations for determination of φ_1 and φ_2. The Newton-Raphson method, starting with $\varphi_1 = \pi/2$, $\varphi_2 = 3\pi/4$, leads to $\varphi_1 \approx 112.08°$, $\varphi_2 \approx 162.10°$, and then it is found that $\sigma_x^0 \approx 3.10536\tau_Y$, which, by (5.7.80) determines ν^* to about 0.37821. (The high accuracy is motivated only by the possibility of using the solution for controlling the accuracy of numerical methods.)

All in-plane stresses may now be calculated from the expressions for the different kinds of regions and from the condition of stress continuity. Note that they are independent of Poisson's ratio. The highest stress turns out to be $\sigma_y \approx 5.10536\tau_Y$ in P1, which is only about 0.7% smaller than the maximum stress in the Prandtl field for a stationary crack, page 280. In fact, rather surprisingly, this close similarity with the Prandtl field stresses holds for all angles, even though the strains are very different. A conspicuous difference from the Prandtl field is the existence of an elastically deforming sector, the wake, W, about 40° wide.

For the stress σ_z, a distinction has to be made for $\nu \leq \nu^*$ and $\nu > \nu^*$. The respective values in P1 are given by (5.7.80)-(5.7.81). In P2, the stress σ_z is found by using (5.7.19). The strain ϵ_0^p appearing in this equation vanishes if $\nu > \nu^*$, because then f_2 and f_3 are negative; see page 321. Thus, $\epsilon_z^e = 0$ and then, by Hooke's law, $\sigma_z = \nu(\sigma_x + \sigma_y)$ in P2 if $\nu > \nu^*$. If $\nu \leq \nu^*$, the strain

$$\epsilon_0^p = -\epsilon_z^e = -\frac{1}{2(1+\nu)\mu}[\sigma_x^0 - 2\nu(\sigma_x^0 + \tau_Y)] \tag{5.7.89}$$

which is found by inserting the stresses on the P1-P2 boundary into Hooke's law. Thus, by (5.7.19),

$$\sigma_z = \nu(\sigma_x + \sigma_y) + (1-2\nu)\sigma_x^0 - 2\nu\tau_Y \tag{5.7.90}$$

in P2.

The distinction between the cases $\nu > \nu^*$ and $\nu \leq \nu^*$ carries over also to regions W and S. Note that the boundary conditions for $\varphi = \pm\pi$ imply that $\sigma_y = \tau_{xy} = 0$ in S, and that the yield condition then implies that $\sigma_x = 2\tau_Y$. No additional plastic strain ϵ_z^p is produced in these regions. Thus, because $\epsilon_z^p = \epsilon_z^e = 0$ in P1 and P2 for $\nu > \nu^*$, the stress $\sigma_z = 2\nu\tau_Y$ in S. For $\nu \leq \nu^*$, the stress σ_z follows from (5.7.90): it equals $(1-2\nu)\sigma_x^0$. For W, the stress σ_z is found by using expression (5.7.58) and the continuity conditions at the P2-W or W-S boundary.

It is now possible to control the assumptions about the branches f_1, f_2 and f_3 of the Tresca yield function. The result is affirmative: $f_2 = f_1 = 0$ for $\varphi \leq \pi/4$ if $\nu \leq \nu^*$, otherwise f_2 and f_3 are negative, whereas $f_1 = 0$ in active plastic regions.

The asymptotic deformation rates are found from (5.7.35) and (5.7.37) for P2 and by (5.7.53) and (5.7.54) for W. Note that A_1 and A_2 are now known. The signs of the plastic parts of these rates at the sector boundaries must relate to the signs of the stresses so that the plastic work is positive, and an investigation shows that so is the case. Other possibilities, for instance the choice $\sigma_x = -2\tau_Y$ in (5.7.75), might have led to a solution satisfying continuity conditions, but violating the requirement of positive plastic work.

A correct solution has been found, but is it unique? The answer appears to be that it is not. As shown by Drugan and Chen (1989) and Chen and Drugan (1991), using the Huber-von Mises yield condition and associated flow rule, the sector boundaries do not necessarily need to be straight, but they may be curved near the crack edge, with the consequence that a family of valid solutions can be obtained, even though such solutions are rather similar to those obtained by Drugan et al. (1982). It may also be noticed that the result $\sigma_x = 2\tau_Y$ near the crack faces implies a high T-stress, i.e. a high constraint to plastic flow in front of the crack. The analysis by Drugan and Chen (1989) and Chen and Drugan (1991), which is not limited to small scale yielding, makes it possible also to consider other degrees of constraint.

The rates of deformation follow from U_r and U_φ as deformations from u_r and u_φ. For the trailing constant strain sector, S, the singular part follows from the boundedness of the deformation rate jump at the W-S boundary. It turns out that the most singular rate of deformation is situated in P2 and is

$$D_{r\varphi} = \frac{2(1-\nu)\tau_Y}{\mu} \cdot \frac{1}{r} \ln \frac{R}{r} \cos \varphi_0 + \cdots \quad (5.7.91)$$

showing only the most singular term. Note that $D_{r\varphi}$ is defined in analogy with $\gamma_{r\varphi}$, so that the tensor component is $D_{r\varphi}/2$.

As Hooke's law shows, $D_{r\varphi}^e = 0$ in P2, and thus the most singular plastic strain is found from the relation $d\gamma_{r\varphi}^p/da = D_{r\varphi}$. An expression for the effective plastic shear strain may then be found by analogy with the effective plastic extensional strain, (5.2.11):

$$\frac{d\gamma_{eff}^p}{da} = \sqrt{2 D_{ij}^p D_{ij}^p} = \sqrt{2[(D_r^p)^2 + (D_\varphi^p)^2 + (D_z^p)^2] + (D_{r\varphi}^p)^2} \to D_{r\varphi} \quad (5.7.92)$$

Then, by integrating the expression

$$\frac{\partial \gamma_{eff}^p}{\partial x} = -D_{r\varphi} = -\frac{2(1-\nu)\tau_Y}{\mu} \cdot \frac{\cos \varphi_0}{r} \ln \frac{R}{r} \quad (5.7.93)$$

the asymptotic effective plastic shear strain is found. Because y is assumed to be constant during the integration, this may preferably be performed over φ rather than over x, after writing $r = y/\sin\varphi$. The result is

$$\gamma_{eff}^p = \frac{2(1-\nu)\tau_Y}{\mu} \left[1 + \frac{1}{\sqrt{2}} \cdot \ln\left(\frac{\tan \varphi/2}{\tan \pi/8}\right) \right] \ln \frac{R}{r} \quad (5.7.94)$$

where an integration constant was determined by the shear strain jump at the P1-P2 boundary, which follows from (5.7.67):

$$[\![\gamma_{r\varphi}]\!] = \frac{[\![U_\varphi]\!]}{\sin \varphi_0} = \frac{2(1-\nu)\tau_Y}{\mu} \ln \frac{R}{r} \quad (5.7.95)$$

Thus, as in the mode III case, *a logarithmic type singularity is obtained, i.e. a much weaker singularity than a $1/r$ or $1/\sqrt{r}$ singularity as found for stationary cracks*. The singularity is also weaker than in mode III. However, for mode I this singularity does not appear in front of the crack, but anyway large displacement changes are expected to occur ahead of the crack edge if the existence of a finite process region is considered.

In region P1, the plastic deformation increments are found from the flow rule:

$$d\epsilon_x^p = -d\epsilon_y^p, \quad d\gamma_{xy}^p = 0 \quad (5.7.96)$$

$$d\epsilon_z = \begin{cases} d\epsilon_x^p & \text{for } \nu < \nu^* \\ 0 & \text{for } \nu > \nu^* \end{cases} \quad (5.7.97)$$

Thus, because of plastic incompressibility, $d\epsilon_x^p = d\epsilon_y^p = d\epsilon_z^p = 0$ for $\nu \leq \nu^*$, i.e. P1 is then not an active plastic sector, but a virgin elastic sector, even though the yield condition is satisfied. With the benefit of hindsight, this could have been found by using (5.7.55)-(5.7.58) for elastically deforming sectors: conditions of boundedness and symmetry for σ_x give $A_1 = A_2 = 0$, and after determination of C_{ij} it is found that the general condition $f_1 \leq 0$ is not violated if $\nu \leq \nu^*$.

Because no plastic strain is produced in P1 if $\nu \leq \nu^*$ or in W, the plastic strain at the W-S boundary is produced in P2. At this boundary the singular part of the displacement rate equals the one at the P1-W boundary at the same y level. Addition of a constant non-singular part gives

$$U_2 = -A_2 \ln \frac{R}{r} + C = \frac{2(1-\nu)\tau_Y}{\mu} \cdot \frac{\sqrt{2}-\cos\varphi_1}{\sqrt{2}} \ln \frac{R}{r} + C \qquad (5.7.98)$$

By using the plane strain compatibility equation it may be shown that no additional singular strain is produced in S, and so the crack opening displacement δ_0 is found by integrating $U_2(r)$ and multiplying by 2, i.e.,

$$\delta_0 = \frac{4(1-\nu)\tau_Y}{\mu} \cdot \frac{\sqrt{2}-\cos\varphi_1}{\sqrt{2}} r \ln \frac{R_1}{r} \qquad (5.7.99)$$

where the constant C is absorbed by introducing a constant R_1, which, like R, cannot be determined by asymptotic analysis, but would increase with the size of the plastic region in small scale yielding.

Note that, by analogy with the crack face slip in mode III, the crack opening displacement in mode I increases smoothly from zero, i.e. in contrast to the case of a stationary crack, there is no complete blunting of the edge of a slowly moving crack in an elastic-plastic material. The same result follows for the case $\nu > \nu^*$. The energy flux into the crack edge region cannot be determined from the present analysis, because an infinitesimally small process region was assumed. However, energy *would* flow into a finite process region, because the signs of δ_0 and of σ_y straight ahead of the crack are both positive.

Miao and Drugan (1995) extended the analysis of asymptotic stresses and strains for a growing mode I crack by assuming a yield condition and associated flow rule for a type of porous material. The results were similar to those for the Huber-von Mises material discussed here, although sector angles were different. However, the hydrostatic stress was considerably reduced for porosity levels of a few per cent.

Mode II. For mode II, an attempt to use the same layout of sector types as for mode I does not work. Instead, the front sector, P, should be of the centered fan type, followed by an elastically deforming sector, the wake, W, and a constant stress secondary plastic sector, S; see Fig. 5.7.4. The yield condition is assumed to be expressed by f_1, i.e. $f_1 > f_2$ (or f_3), so that $\sigma_z = \nu(\sigma_x + \sigma_y)$ everywhere. The in-plane stresses in S are already determined by (5.7.75) and those in W by (5.7.55)-(5.7.57). The stresses in P are given by (5.7.15)-(5.7.17) and the mode II symmetry conditions. The latter imply $\sigma_0 = 0$, and thus

$$\sigma_x = -\tau_Y(2\varphi + \sin 2\varphi), \qquad \sigma_y = -\tau_Y(2\varphi - \sin 2\varphi) \qquad (5.7.100)$$

$$\tau_{xy} = \tau_Y \cos 2\varphi, \qquad \sigma_z = -4\nu\tau_Y\varphi \qquad (5.7.101)$$

Before using the conditions of stress continuities at the P-W boundary, $\varphi = \varphi_1$, and the W-S boundary, $\varphi = \varphi_2$, the constants A_1 and A_2 will be determined in the same

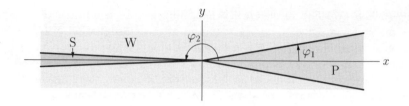

Fig. 5.7.4 The sectors of the primary plastic region (P), the wake (W) and the secondary plastic region (S) in an asymptotic vicinity of a slowly moving mode II crack edge. P is a centered fan sector, W is elastically deforming, and S is a constant stress sector. The sector angle for S is very small and overexaggerated in the figure.

manner as for mode I:

$$A_1 = -\frac{2(1-\nu)}{\mu} C_0 \sin\varphi_1 \tag{5.7.102}$$

$$A_2 = -\frac{2(1-\nu)}{\mu}(1 - C_0 \cos\varphi_1) \tag{5.7.103}$$

where the integration constant $\cos\varphi_0$ has been denoted by C_0.

The conditions of in-plane stress continuity now give three equations for the P-W boundary and three for the W-S boundary. Elimination of the constants C_{11}, C_{22} and C_{12}, appearing in (5.7.55)-(5.7.57), gives three equations containing the unknown angles φ_1 and φ_2, and the unknown constant C_0. Again, as for mode I, Newton-Raphson's method may be used, eventually after elimination of C_0, but it has to be damped and very good guesses about initial values have to be made. The result (Slepyan 1974) is $\varphi_1 \approx 9.566°$, $\pi - \varphi_2 \approx 0.1800°$ and $C_0 = 1.0136$. All stresses may now be determined. The in-plane stresses are independent of Poisson's ratio. The highest shear stress is $\tau_{r\varphi} = \tau_Y$ in P, and the highest normal stress is $\sigma_x \approx 1.543\tau_Y$ for $\varphi = -\pi/2$.

The angles occupied by the plastic sectors are very small, only about half of those at mode III. The secondary plastic region would probably be absorbed more or less completely in the wake of a finite process region.

From the deformation rates, expressions for strains may be found. The most singular strains appear in P. Straight ahead of the crack edge, the maximum shear strain is

$$\gamma_{xy} = \frac{\tau_Y}{\mu}\left[C_0(1-\nu)\left(\ln\frac{R}{x}\right)^2 + O\left(\ln\frac{R}{x}\right)\right] \tag{5.7.104}$$

Note the similarity with the corresponding mode III expression, (5.6.23). Recall that the strongest strain singularity in mode I is weaker and occurs offside the straightforward direction, see (5.7.94).

In the same way as for mode I, the assumptions made about sector layout and branches of the yield condition may now be controlled, and it may also be asserted that the condition of positive plastic work is not violated at the sector boundaries.

The crack face sliding $\delta_0 = u(r,\pi) - u(r,-\pi)$ is obtained in a similar way as for mode III. Thus, it may be determined from the asymptotic displacement rate U_1 in

W, which by (5.7.35) and (5.7.37), is found to be

$$U_1 = \frac{2(1-\nu)\tau_Y}{\mu} C_0 \sin\varphi_1 \ln\frac{R}{y} \qquad (5.7.105)$$

Integration along constant y in W and multiplication by 2 gives the asymptotic crack face sliding

$$\delta_0 = \frac{4(1-\nu)\tau_Y}{\mu} C_0(\sin\varphi_1 - \cos\varphi_1 \tan\varphi_2) r \ln\frac{R}{r} \approx \frac{0.686(1-\nu)\tau_Y}{\mu} r \ln\frac{R}{r} \qquad (5.7.106)$$

Again, as for mode I, it is seen that the energy flux into the crack edge region cannot be determined, because of the assumption of an infinitesimally small process region. However, the signs of δ_0 and of σ_x straight ahead of the crack are the same, so that energy *would* flow to a finite process region.

The significance of a finite process region

The analysis of slowly moving elastic-plastic cracks was made under the assumption of an infinitesimally small process region, leading to strain singularities. This assumption does not allow determination of the energy flux into the crack edge region, but there are also other shortcomings. In reality, the process region size is not always small compared to the extension of the plastic region. Under small scale yielding, it may penetrate most of the plastic region in some materials. If it does, then the actual stress and deformation fields in the process region vicinity, may be very different from the asymptotic fields obtained here. Conclusions drawn on the basis of the asymptotic solution about possible criteria for steady state elastic-plastic crack growth, could be very misleading in cases of poor embedment of the process region in the plastic region. Thus, the load increase during stable crack growth and the amount of stable crack growth may be very much overestimated, as discussed on page 587.

6
Elastodynamic Crack Mechanics

6.1 Introduction

In the present context, the concept of dynamics refers to processes in which inertia plays a part. This excludes, for instance, slowly moving cracks of the kind discussed in the two preceding chapters.

Dynamic processes in a cracked body fall into two basic categories: dynamic crack propagation and dynamic loading of bodies with stationary cracks. Often a mixture of these two kinds of processes appears, for instance when dynamic loading on a body containing a stationary crack causes dynamic crack propagation, or when dynamic crack propagation generates waves, which, after reflection at an outer boundary of the body, impinge on the crack, perhaps even after the crack, for some reason, has been arrested.

Unstable crack growth in a structure is often synonymous with a catastrophic event – fracture. Usually it accelerates to a high velocity, which may be several hundred m/s, and it may also proceed long distances, for instance all the way through a ship's hull or several kilometers along a pipeline or an earthquake fault. Even though design priority normally aims at measures to prevent unstable crack growth, the possibility of such growth cannot always be ruled out and might call for additional design precautions. These could consist of segmentation of the structure with crack arresting barriers between the segments. In some cases, crack arrest may simply follow as a result of decreasing ambient stresses, for instance when a crack is leaving a region dominated by local, short range residual or thermal stresses.

Certain processes, such as mining operations and stone cutting, require or otherwise involve dynamic crack propagation. At impact loading on bodies with preexisting cracks, stress waves are impinging on the cracks, eventually causing dynamic crack growth. Questions regarding crack paths and branching as well as suitable dynamic characteristics of loading devices are of interest in such connections.

In virtually all engineering applications, dynamic crack propagation in a homogeneous, isotropic medium proceeds, as in slow growth cases, under mode I conditions. Earthquake sliding motion constitutes a prominent example of dynamic crack propagation in modes II or III or in a mixed mode II/III. In these cases, frictional forces act on the mating surfaces. Of particular interest are, of course, circumstances that lead to onset and arrest of such events, but also the radiated waves, both near the source and far away, are of great significance.

Because one additional dimension, time, enters when going from a static to a corresponding dynamic case, analysis of the latter is, of course, more difficult. However, in several dynamic problems of fundamental significance, the number of independent variables does not exceed those in the static case. This occurs for *steady crack propagation* and for some cases of crack growth with constant edge velocities in infinite bodies. In the former case, one space coordinate, say x_1, and time, t, always appear in the context $x_1 - Vt$, where V is the crack edge velocity. Introduction of a moving coordinate system $X_1 = x_1 - Vt$, $X_i = x_i$, $i = 2, 3$, the so-called *Galilean transformation*, then reduces the number of independent variables to the same as for static deformations. In certain other cases, each space coordinate, x_i, $i = 1, 2$ in two-dimensional, $i = 1, 2, 3$ in three-dimensional cases, always appears in the context x_i/Vt. Thus, Vt acts as a scaling factor, and the term *self-similarity* is used to describe this type of dynamic processes.

Actually, virtually all known analytical solutions to elastodynamic crack problems may be based on steady state crack propagation or self-similarity. If the problem at hand is neither steady state nor self-similar itself, it can often be formulated in terms of superposition of solutions to such problems.

It is sometimes argued that continuing unstable crack growth inevitably leads to small scale yielding. The obvious reason would be the increasing crack length. However, the experimental results described in Section 2.7, showing that the energy dissipation in the process region increases considerably with crack edge velocity, imply a tendency towards increased scale of yielding. Furthermore, in some cases, not even the increase of crack length works in the direction of decreasing scale of yielding. This scale might instead be influenced by another length dimension such as the diameter of a pipe, implying that large scale yielding may prevail even for crack lengths of several kilometers.

The present chapter will deal mainly with the idealization of infinitesimally small scale yielding and with time- and rate-independent linear materials, i.e. with materials which obey Hooke's (generalized) law. However, also some dynamic crack processes in viscoelastic materials will be discussed, and in some cases the existence of a finite process region, will be taken into account, using the Barenblatt model (cf. Section 1.6) which was introduced by Barenblatt (1959a,b,c) and applied to dynamic cases by Barenblatt et al. (1962).

Basic equations for stress waves in linearly elastic solids are given in Appendix A9. In particular, it is shown that two fundamental types of waves appear in an infinite body, either *irrotational* waves, travelling with velocity c_P, or *equivoluminal* waves, travelling with velocity $c_S < c_P$. Expressions for c_P and c_S are given in Appendix A9. These two fundamental waves have counterparts in wave propagation in a thin plate, following the so-called plane stress approximation. One of these waves travels with velocity c_P^{pss} and the other with velocity c_S. Expression for c_P^{pss} is given in Appendix A9. The equations of motion are formally the same for the plane stress approximation as for plane longitudinal waves in an infinite medium if c_P is replaced by c_P^{pss}, whereas c_S is left unchanged.

The very close similarity between plane strain and plane stress results, together with the frequent appearance of these plane cases, suggests simultaneous treatment. To this end, the terminology will also be chosen so as to be applicable to both cases. Thus, waves of the fundamental type travelling with velocity c_P or, in the plane stress approximation, with velocity c_P^{pss}, will be called P waves, and waves of the fundamental type travelling with velocity c_S will be called S waves. This is the terminology used in seismology, where it is usually interpreted as Primary and Secondary waves, because the P waves, being faster, arrive earlier at an observation point than the S waves. Note that P waves are not irrotational in the plane stress approximation. In the literature, P waves are sometimes called dilatational waves, because they involve dilatation in addition to shear, or longitudinal, because they are longitudinal in some special cases. S waves are always equivoluminal. They are sometimes called shear waves, distortional waves, rotational waves or transverse waves.

For brevity, the expression "in-plane" is used in this work to denote not only perfect in-plane deformation, but also the predominantly in-plane wave motion in thin plates, the plane stress approximation, which is referred to simply as "plane stress".

Simultaneous treatment of plane strain and plane stress problems is achieved, in dynamic as well as in static cases, simply by using μ and k as elastic constants. Recall that k is defined as the ratio between the propagation velocities of S and P waves, and that

$$k^2 = \begin{cases} \dfrac{1-2\nu}{2(1-\nu)} & \text{for plane strain} \\ \dfrac{1-\nu}{2} & \text{for plane stress} \end{cases} \quad (6.1.1)$$

where ν is Poisson's ratio. The equations show that $c_S < c_P$. As shown in Appendix A9,

$$c_S = \sqrt{\frac{\mu}{\varrho}} \quad (6.1.2)$$

where ϱ is the density.

For most materials, c_S is about half as large as c_P and c_P^{pss} is about 5-15% smaller than c_P. Typical values for steel are $c_S = 3200$m/s, $c_P=5900$m/s, $c_P^{pss} = 5400$m/s, for granite $c_S = 2700$m/s, $c_P = 4400$m/s, $c_P^{pss} = 4300$m/s, and for PMMA $c_S = 1300$m/s, $c_P = 2600$m/s, $c_P^{pss} = 2250$m/s.

In the continuation, to make simultaneous treatment of plane strain and plane stress cases easier, the notation c_P will be used also in the plane stress approximation, i.e. instead of c_P^{pss}, except when a distinction is called for.

6.2 The Yoffe problem

Introduction

The first solution of a dynamic steady state crack problem was given by Yoffe (1951). She studied a crack of constant length, moving with constant velocity in an infinite solid. Thus, the crack has one leading edge, where the process is *tearing*, and one trailing edge, where the process is *healing*. The solution shows most of the features pertinent to dynamically moving cracks in general. It will be demonstrated here for mode II, whereas the original Yoffe solution concerned mode I. A travelling mode II crack with constant length comes closer to physical realizations, because healing in mode I does not appear to be a spontaneous physical process. For modes II and III, more or less well realized in seismic events, healing simply means cessation of

6.2 THE YOFFE PROBLEM

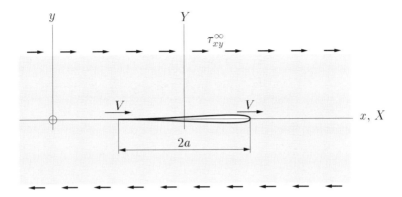

Fig. 6.2.1 The mode II Yoffe problem. The crack is shown in conventional symbolic manner, but in reality there is no crack opening in mode II. Thus, the displacement u, rather than the displacement v, is shown. Energy-neutral healing, which is equivalent to smooth closing at the trailing end, is assumed. Two coordinate systems are shown, one, x, y, fixed, and the other, X, Y, moving with the crack.

slip, although this process is not a unique one, because the energy exchange with the ambient stress-strain field may vary, depending on material and situation. The process of healing is different from that of tearing in that it gives away, rather than dissipates, energy. The amount of energy radiated is, however, probably much smaller than the energy dissipated at the leading edge, and, for simplicity, it will be assumed here to be vanishingly small, and then the healing process becomes uniquely specified.

Statement of the mode II Yoffe problem

Consider a crack (or a slipping region) $-a < x - Vt < a$, moving with constant velocity V in positive x direction along $y = 0$; see Fig. 6.2.1. Plane strain or plane stress is assumed. The remote load consists of a shear stress $\tau_{xy} = \tau_{xy}^\infty$. In addition, there might be a remote normal stress, $\sigma_y = \sigma_y^\infty$, but this stress will be assumed to be zero for a while. Then, there is no friction in the slipping region.

Because slip takes place between the leading and trailing edges, an amount of slip will be deposited along $y = 0$ for $x < Vt - a$. Thus, two adjacent particles situated on each side of the plane $y = 0$ and in contact before the arrival of the leading edge, will be displaced an amount 2Δ, say, from each other after the passage of the trailing edge. It is assumed that Δ is constant on $x < Vt - a, y = 0$.

It is further assumed that $V < c_S$. Thus, the crack velocity is *subsonic*. The significance of this assumption will be obvious later when *intersonic* crack velocities are considered, i.e. the crack velocity is between the propagation velocities c_S of S waves and c_P of P waves.

By considering the semi-infinite solid $y \geq 0$, a mixed boundary value problem results.

The boundary conditions for $y = 0$ are

$$\sigma_y = 0 \text{ for all } x \tag{6.2.1}$$
$$\tau_{xy} = 0 \text{ for } |x - Vt| < a \tag{6.2.2}$$
$$\frac{\partial u}{\partial x} = 0 \text{ for } |x - Vt| > a \tag{6.2.3}$$

Solution for a moving line load

Solve first a simple boundary value problem: a line shear force, $\tau_{xy} = T\delta(x - Vt)$ is moving with constant velocity V in the positive x direction on the surface $y = 0$ of the semi-infinite solid $y \geq 0$. Find the displacement u (or $\partial u/\partial x$ if this is simpler) on the surface.

The displacement potential functions $\phi(x - Vt, y)$ and $\psi(x - Vt, y)$ are used for the representations (see Appendix A1)

$$u = \frac{\partial \phi}{\partial x} + \frac{\partial \psi}{\partial y} \tag{6.2.4}$$

$$v = \frac{\partial \phi}{\partial y} - \frac{\partial \psi}{\partial x} \tag{6.2.5}$$

$$\sigma_x = \frac{\mu}{k^2}\left[\frac{\partial^2 \phi}{\partial x^2} + (1 - 2k^2)\frac{\partial^2 \phi}{\partial y^2} + 2k^2\frac{\partial^2 \psi}{\partial x \partial y}\right] \tag{6.2.6}$$

$$\sigma_y = \frac{\mu}{k^2}\left[(1 - 2k^2)\frac{\partial^2 \phi}{\partial x^2} + \frac{\partial^2 \phi}{\partial y^2} - 2k^2\frac{\partial^2 \psi}{\partial x \partial y}\right] \tag{6.2.7}$$

$$\tau_{xy} = \mu\left[2\frac{\partial^2 \phi}{\partial x \partial y} - \frac{\partial^2 \psi}{\partial x^2} + \frac{\partial^2 \psi}{\partial y^2}\right] \tag{6.2.8}$$

Now, the Galilean transformation $X = x - Vt$, $Y = y$, i.e. a moving coordinate system with origin at the load application, is introduced. The representations (6.2.4)-(6.2.8) need only the changes $x \to X$ and $y \to Y$, whereas the equations of motion (A9.18) will be

$$a_P^2 \frac{\partial^2 \phi}{\partial X^2} + \frac{\partial^2 \phi}{\partial Y^2} = 0 \tag{6.2.9}$$

$$a_S^2 \frac{\partial^2 \psi}{\partial X^2} + \frac{\partial^2 \psi}{\partial Y^2} = 0 \tag{6.2.10}$$

where

$$a_P = \sqrt{1 - \beta^2}, \quad \beta = V/c_P, \quad a_P > 0 \tag{6.2.11}$$
$$a_S = \sqrt{1 - \beta^2/k^2}, \quad a_S > 0 \tag{6.2.12}$$

Note that a_P and a_S are related.

Solutions of the equations of motion, satisfying the requirement of boundedness as $Y \to \infty$, are

$$\phi = \int_0^\infty A(\alpha) e^{-\alpha a_P Y} \sin \alpha X \, d\alpha \tag{6.2.13}$$

$$\psi = \int_0^\infty C(\alpha) e^{-\alpha a_S Y} \cos \alpha X \, d\alpha \tag{6.2.14}$$

With a slight modification, the boundary conditions on $Y = 0$ are

$$\tau_{xy} = \frac{T}{\pi} \cdot \frac{\delta}{\delta^2 + X^2}, \qquad \sigma_y = 0 \qquad (6.2.15)$$

The modification, changing the delta function (temporarily) to a function which equals the delta function when $\delta \to 0$, is made in order to avoid a non-convergent integral during the solution procedure. Using the representations for the stresses, the boundary conditions for $Y = 0$ read:

$$\frac{k^2}{\mu}\sigma_y = \int_0^\infty \{\alpha^2 A [a_P^2 - (1 - 2k^2)] + 2\alpha^2 k^2 C\} \sin \alpha X \, d\alpha = 0 \qquad (6.2.16)$$

$$\frac{1}{\mu}\tau_{xy} = \int_0^\infty [-2\alpha^2 A a_P + \alpha^2 C(1 + a_S^2)] \cos \alpha X \, d\alpha = \frac{T}{\pi\mu} \cdot \frac{\delta}{\delta^2 + X^2} \qquad (6.2.17)$$

Inversion, and some cleaning up, gives

$$(1 + a_S^2) A - 2 a_S C = 0$$
$$-2 a_P A + (1 + a_S^2) C = \frac{T}{\pi^2 \mu \alpha^2} \int_0^\infty \frac{\delta}{\delta^2 + X^2} \cos \alpha X \, dX = \frac{T}{\pi \mu \alpha^2} e^{-\delta \alpha} \qquad (6.2.18)$$

This system determines $A(\alpha)$ and $C(\alpha)$:

$$A = -\frac{T}{\pi\mu} \cdot \frac{2 a_S}{4 a_P a_S - (1 + a_S^2)^2} \cdot \frac{e^{-\delta\alpha}}{\alpha^2} \qquad (6.2.19)$$

$$C = -\frac{T}{\pi\mu} \cdot \frac{1 + a_S^2}{4 a_P a_S - (1 + a_S^2)^2} \cdot \frac{e^{-\delta\alpha}}{\alpha^2} \qquad (6.2.20)$$

Insertion into

$$\frac{\partial u_+}{\partial x} = \int_0^\infty [-\alpha^2 A + a_S \alpha^2 C] \sin \alpha X \, d\alpha, \qquad Y = 0 \qquad (6.2.21)$$

where index plus refers to the upper half-plane, gives

$$\frac{\partial u_+}{\partial x} = \frac{T}{\pi\mu} \cdot \frac{a_S(1 - a_S^2)}{R(a_P, a_S)} \int_0^\infty e^{-\delta\alpha} \sin \alpha X \, d\alpha, \qquad Y = 0 \qquad (6.2.22)$$

with $R(a_P, a_S) = 4 a_P a_S - (1 + a_S^2)^2$ is one form of the *Rayleigh function*, suitable for steady state cases. It is discussed in Appendix A9 in the form

$$R(\beta) = 4k^3 \sqrt{1 - \beta^2} \sqrt{k^2 - \beta^2} - (2k^2 - \beta^2)^2 = k^4 R(a_P, a_S) \qquad (6.2.23)$$

$R(\beta)$ approaches $2k^2(1 - k^2)\beta^2$ as $\beta \to 0$ and vanishes at the *Rayleigh velocity* c_R, which is somewhat lower than the S wave velocity (Rayleigh 1885). A useful formula is

$$1 - c_R/c_S \approx \frac{0.135}{3 - 4k^2} \qquad (6.2.24)$$

which determines c_R to within $\pm 0.5\%$.

The integral in (6.2.22) equals $X/(\delta^2 + X^2)$, so that returning to the originally posed load, $\tau_{xy} = T\delta(x - Vt)$, by letting $\delta \to 0$, leads to the response

$$\boxed{\frac{\partial u_+}{\partial x} = -\frac{T}{\pi\mu} \cdot \frac{Y_{II}(\beta)}{2(1 - k^2)} \cdot \frac{1}{X}, \qquad X \neq 0, \ Y = 0} \qquad (6.2.25)$$

where

$$Y_{II}(\beta) = 2(1-k^2)\frac{a_S(1-a_S^2)}{R(a_P, a_S)} = \frac{2k(1-k^2)\beta^2\sqrt{k^2-\beta^2}}{R(\beta)} \qquad (6.2.26)$$

is normalized so that $Y_{II}(0) = 1$.

The potential functions ϕ and ψ cannot be determined – the integrals in (6.2.13) and (6.2.14) are actually not convergent. An inspection shows, however, that these expressions were used only formally: operationally, the convergent second derivatives of ϕ and ψ were used.

Stresses, as well as mass velocities may now be determined even for points outside the X axis. With the aid of (6.2.4), (6.2.5), (6.2.13), (6.2.14), (6.2.19) and (6.2.20) the mass velocities $\partial u/\partial t = -V\partial u/\partial x$ and $\partial v/\partial t = -V\partial v/\partial x$ in the upper half-plane are found to be

$$\frac{1}{V}\cdot\frac{\partial u_+}{\partial t} = -\frac{2T}{\pi\mu}\cdot\frac{k^4 a_S}{R(a_P, a_S)}\cdot u_0(X) \qquad (6.2.27)$$

$$\frac{1}{V}\cdot\frac{\partial v_+}{\partial t} = -\frac{2T}{\pi\mu}\cdot\frac{k^4 a_S}{R(a_P, a_S)}\cdot v_0(X) \qquad (6.2.28)$$

where

$$u_0(X) = \frac{X}{X^2 + a_P^2 Y^2} - \frac{1+a_S^2}{2}\cdot\frac{X}{X^2 + a_S^2 Y^2} \qquad (6.2.29)$$

$$v_0(X) = \frac{a_P^2 Y}{X^2 + a_P^2 Y^2} - \frac{1+a_S^2}{2}\cdot\frac{Y}{X^2 + a_S^2 Y^2} \qquad (6.2.30)$$

Here, it may be remarked that an alternative to displacement potentials for solution of plane dynamic steady state problems is the use of the so-called Sneddon-Radok equations (Radok 1956, Sneddon 1958). These are, however, exclusively specialized to steady state problems.

Solution of the mode II Yoffe problem

The result obtained for the moving line load is very simple. The dynamic subsonic response, $\partial u_+/\partial x$, $Y = 0$, is obtained from the static response simply by multiplication by the factor $Y_{II}(\beta)$. The principle of superposition then immediately tells that the response to the shear load $\tau_{xy} = \tau_{xy}^0(X)$ on $Y = 0$ is obtained from the static response to the shear load $\tau_{xy} = \tau_{xy}^0(x)$ after multiplication by $Y_{II}(\beta)$.

In particular, if the shear stress distribution along $y = 0$ is $\tau_{xy}^0(x)$ for a static crack or array of cracks along $y = 0$, then it is $\tau_{xy}^0(X)$ on $Y = 0$ in the corresponding steady state dynamic subsonic case, and the dynamic displacements are obtained from the static ones after multiplication by $Y_{II}(\beta)$, and change of x to X. Thus, *for each such dynamic problem, the solution for $y = 0$ becomes immediately known once the corresponding static solution is known*. This was pointed out by Sih (1968), and the correspondence holds for all three modes.

Here, a crack of finite length is considered; cf. Fig. 6.2.1. The condition of energy-neutral healing implies smooth closing at the trailing edge, $\partial u/\partial x = 0$, so that, for

6.2 THE YOFFE PROBLEM

$y = 0$, the static solution, given by (4.4.95) and (4.4.96) is

$$\tau_{xy} = \tau_{xy}^{\infty}\sqrt{\frac{x+a}{x-a}} \quad \text{for } |x| > a \tag{6.2.31}$$

$$\frac{\partial u_+}{\partial x} = -\frac{\tau_{xy}^{\infty}}{2(1-k^2)\mu}\sqrt{\frac{a+x}{a-x}} \quad \text{for } |x| < a \tag{6.2.32}$$

Thus, the solution of the dynamic problem for shear stresses and tangential displacement gradient on $Y = 0$ is

$$\tau_{xy} = \tau_{xy}^{\infty}\sqrt{\frac{X+a}{X-a}} \quad \text{for } |X| > a \tag{6.2.33}$$

$$\frac{\partial u_+}{\partial x} = -\frac{\tau_{xy}^{\infty} Y_{II}(\beta)}{2(1-k^2)\mu}\sqrt{\frac{a+X}{a-X}} \quad \text{for } |X| < a \tag{6.2.34}$$

The slip deposited on each side of the crack trace behind the trailing region is

$$\Delta = \int_{+a}^{-a} \frac{\partial u_+}{\partial x} dX = \frac{\pi \tau_{xy}^{\infty} a Y_{II}(\beta)}{2(1-k^2)\mu} \tag{6.2.35}$$

Note that this result could have been obtained directly from the corresponding static result, (4.4.93), after multiplication by $Y_{II}(\beta)$.

Knowledge of $(\tau_{xy})_{Y=0} = \tau_{xy}^0(X)$ for the moving crack enables determination of the mass velocities at an arbitrary point in the upper half-plane from the solution for the line load, (6.2.27)-(6.2.28):

$$\frac{1}{V} \cdot \frac{\partial u_+}{\partial t} = -\frac{2}{\pi\mu} \cdot \frac{a_S}{R(a_P, a_S)} \int_{-\infty}^{\infty} \tau_{xy}^0(s) u_0(X-s) ds \tag{6.2.36}$$

$$\frac{1}{V} \cdot \frac{\partial v}{\partial t} = -\frac{2}{\pi\mu} \cdot \frac{a_S}{R(a_P, a_S)} \int_{-\infty}^{\infty} \tau_{xy}^0(s) v_0(X-s) ds \tag{6.2.37}$$

which, after evaluation of the integrals, may be written as

$$\frac{\mu}{V\tau_{xy}^{\infty}} \cdot \frac{\partial u_+}{\partial t} = -\frac{\sqrt{2}a_S}{R(a_P, a_S)} f(\xi, \eta) \tag{6.2.38}$$

$$\frac{\mu}{V\tau_{xy}^{\infty}} \cdot \frac{\partial v}{\partial t} = -\frac{1+a_S^2 - 2a_P a_S}{R(a_P, a_S)} - \frac{2\sqrt{2}a_S}{R(a_P, a_S)} g(\xi, \eta) \tag{6.2.39}$$

where

$$\xi = \frac{X}{a}, \quad \eta = \frac{Y}{a}$$

$$f(\xi,\eta) = a_P \eta D_1 - (\xi+1)S_1 - \frac{1+a_S^2}{2}[a_S \eta D_2 - (\xi+1)S_2]$$

$$g(\xi,\eta) = a_P[(\xi+1)D_1 + a_P \eta S_1] - \frac{1+a_S^2}{2a_S}[(\xi+1)D_2 + a_S \eta S_2]$$

$$D_1 = \frac{\xi}{|\xi|R_1}\sqrt{R_1 + \xi^2 - 1 - a_P^2 \eta^2}, \quad D_2 = (D_1)_{a_P \to a_S}$$

$$S_1 = \frac{1}{R_1}\sqrt{R_1 - \xi^2 + 1 + a_P^2 \eta^2}, \quad S_2 = (S_1)_{a_P \to a_S}$$

$$R_1 = \sqrt{(\xi^2 + a_P^2 \eta^2 - 1)^2 + 4a_P^2 \eta^2}$$

The crack edge vicinity in modes I and II

Asymptotic stresses and displacements on the symmetry plane
For mode II, specialization of (6.2.33)-(6.2.34) to $Y=0$ and the leading crack edge vicinity, $|X-a|=r \ll a$, leads to

$$\boxed{\tau_{xy} = \frac{K_{II}}{\sqrt{2\pi r}}, \quad X > a} \qquad (6.2.40)$$

$$\boxed{\frac{\partial u_+}{\partial x} = -\frac{K_{II}Y_{II}(\beta)}{2(1-k^2)\mu\sqrt{2\pi r}}, \quad X < a} \qquad (6.2.41)$$

where $K_{II} = 2\tau_{xy}^\infty \sqrt{\pi a}$ is the mode II stress intensity factor. The energy flux is found from the expression for the slowly moving crack, (3.5.15), by multiplication with $Y_{II}(\beta)$ or, alternatively, in the same way as used for a slow crack, cf. Section 3.5. The result is

$$\boxed{\mathcal{G} = \frac{K_{II}^2 Y_{II}(\beta)}{4(1-k^2)\mu}} \qquad (6.2.42)$$

Thus, the stress-displacement field in the crack edge vicinity and the crack edge velocity specify the energy flux uniquely, as found originally by Broberg (1964, 1967) by studying the dissipation in a Barenblatt type process region. Another approach to determine this energy flux consists of integration along a path around the crack edge, assuming a vanishingly small process region; this approach was used first by Craggs (1963) (though with incorrect result due to a missing term), and later by Erdogan (1968), Atkinson and Eshelby (1968) and by Kostrov and Nikitin (1970). However, this approach is less general, because the assumption of a vanishingly small process region gives unrealistic results in certain cases, for instance for intersonic crack propagation, cf. page 353, where it predicts zero energy flow to the process region. On the other hand, Kostrov and Nikitin (1970), who included all three modes in their expression, also considered influx of heat energy.

Consider a certain crack edge vicinity, much smaller than the crack length. If the crack is moving with constant velocity, then approximate steady state conditions prevail in this vicinity, cf. page 68. This is also the case if the crack speed is changing sufficiently slowly. Examples are the vicinity of the edge of a crack expanding in both directions or propagating from a plate edge, but then the approximate steady state prevails only within some limited time interval. If the dissipative region is very small compared to the steady state region, then infinitesimally small scale yielding may be assumed in order to find the asymptotic stress field, approximately realized in some annular region outside the crack edge or at least controlling the state in the dissipative region, cf. Section 3.3. Note that the steady state region, and the time interval during which it prevails, may be chosen arbitrarily small in the idealized case of infinitesimally small scale yielding.

The steady state assumption for the vicinity of the edge of a crack with traction free faces leads to a boundary value problem for this vicinity, which leaves the conditions at the peripheral boundary of the steady state region unspecified:

$$a_P^2 \frac{\partial^2 \phi}{\partial X^2} + \frac{\partial^2 \phi}{\partial Y^2} = 0 \tag{6.2.43}$$

$$a_S^2 \frac{\partial^2 \psi}{\partial X^2} + \frac{\partial^2 \psi}{\partial Y^2} = 0 \tag{6.2.44}$$

$$\tau_{xy} = 0 \text{ for } X < 0, Y = 0, \quad \sigma_y = 0 \text{ for all } X, Y = 0 \tag{6.2.45}$$

$$u = 0 \text{ for } X > 0, Y = 0 \tag{6.2.46}$$

But a problem concerning the vicinity of the leading edge of a Yoffe crack would be posed in exactly the same way. Thus, the solution for the dominating (inverse square-root singular) term will be the same as for the Yoffe crack, although an amplitude factor, which may be written in the form of a stress intensity factor, will be left undetermined. However, in contrast to the Yoffe crack, the difficulties associated with the physical realization of healing at a mode I trailing edge are no longer present. There is also another difference, which makes the crack edge vicinity approach more general, although the result for the dominating term happens to be the same as for the leading edge of a Yoffe crack. Stress waves are radiated out from a Yoffe crack, whereas a steady state problem formulated for the crack edge vicinity, only, also allows for incoming stress waves, which appear, for instance, during crack propagation in a strip, where waves are reflected at the strip boundaries.

For elastodynamic, as for elastostatic cases, once the solution of a mode II problem is known, the solution of the corresponding mode I problem is found after simple and obvious modifications. Thus, the mode I equivalent to relations (6.2.40)-(6.2.42) may be written as

$$\boxed{\sigma_y = \frac{K_I}{\sqrt{2\pi r}}} \tag{6.2.47}$$

$$\boxed{\frac{\partial v_+}{\partial x} = -\frac{K_I Y_I(\beta)}{2(1-k^2)\mu\sqrt{2\pi r}}} \tag{6.2.48}$$

$$\boxed{\mathcal{G} = \frac{K_I^2 Y_I(\beta)}{4(1-k^2)\mu}} \tag{6.2.49}$$

where $Y_I(\beta)$ is found to be

$$Y_I(\beta) = 2(1-k^2)\frac{a_P(1-a_S^2)}{4a_P a_S - (1+a_S^2)^2} \tag{6.2.50}$$

Note that the functions $Y_I(\beta)$ and $Y_{II}(\beta)$, which appropriately could be called the *Yoffe functions*, may alternatively be written in the form

$$Y_I(\beta) = \frac{2k^2(1-k^2)\beta^2\sqrt{1-\beta^2}}{R(\beta)} \tag{6.2.51}$$

$$Y_{II}(\beta) = \frac{2k(1-k^2)\beta^2\sqrt{k^2-\beta^2}}{R(\beta)} \tag{6.2.52}$$

Both functions are normalized so that $Y_I(0) = Y_{II}(0) = 1$.

Because $Y_I(\beta)$ and $Y_{II}(\beta)$ become infinite for the Rayleigh wave velocity, the expressions (6.2.49) and (6.2.42) show that the stress intensity factor cannot approach a non-zero value when this velocity is approached. Thus, *the Rayleigh wave velocity is an upper limit of modes I and II crack edge velocities in the subsonic region*. This was originally shown by Barenblatt and Cherepanov (1960) and Broberg (1960) for mode I and by Craggs (1960) for modes I and II. Note that, theoretically, the crack can reach velocities arbitrarily close to the Rayleigh wave velocity, but not propagate with this velocity. See further discussion in Section 9.1.

Note that dynamic stress intensity factors are found from the expressions

$$\boxed{K_I = \lim_{r \to 0}(\sqrt{2\pi r}\,\sigma_y)_{y=0}, \quad K_{II} = \lim_{r \to 0}(\sqrt{2\pi r}\,\tau_{xy})_{y=0}} \tag{6.2.53}$$

$$\boxed{K_I = \frac{(1-k^2)\mu}{Y_I(\beta)} \lim_{r \to 0}\left(\sqrt{\frac{2\pi}{r}} v_+\right)_{y=0}, \quad K_{II} = \frac{(1-k^2)\mu}{Y_{II}(\beta)} \lim_{r \to 0}\left(\sqrt{\frac{2\pi}{r}} u_+\right)_{y=0}} \tag{6.2.54}$$

where r is forwards distance from the crack edge. These expressions are generally valid for constant or continuously changing crack velocity.

Asymptotic stress fields around the crack edge

The stress and displacement fields are found in essentially the same way as the mass velocity field for the mode II Yoffe crack. Thus, if, for instance, the shear stress component $\tau_{xy} = \tau_{xy}^{(C)}(X,Y)$ is sought, then the corresponding stress component $\tau_{xy} = \tau_{xy}^{(T)}(X,Y)$ for the case of the moving line load, is first to be calculated. It is found by use of (6.2.8), (6.2.13), (6.2.14), (6.2.19) and (6.2.20). The result is

$$\tau_{xy}^{(T)}(X,Y) = \frac{T}{\pi R(a_P, a_S)} \int_0^\infty [4a_P a_S e^{-\alpha a_P Y} - (1+a_S^2)^2 e^{-\alpha a_S Y}] \cos \alpha X \, d\alpha$$

$$= \frac{T}{\pi R(a_P, a_S)}\left[\frac{4a_P a_S a_P Y}{X^2 + a_P^2 Y^2} - \frac{(1+a_S^2)^2 a_S Y}{X^2 + a_S^2 Y^2}\right] \tag{6.2.55}$$

6.2 THE YOFFE PROBLEM

This is the response to the half-plane load $\tau_{xy} = T\delta(X)$. Now, specialization to the crack edge vicinity may be achieved by considering the semi-infinite propagating crack $X > 0$, $Y = 0$. For this crack,

$$\tau_{xy}^{(C)}(X,0) = \begin{cases} 0 & \text{for } X < 0 \\ \dfrac{K_{II}}{\sqrt{2\pi X}} & \text{for } X > 0 \end{cases} \tag{6.2.56}$$

This may be written as

$$\tau_{xy}^{(C)}(X,0) = K_{II} \int_0^\infty \frac{\delta(X-s)}{\sqrt{2\pi s}} \, ds \tag{6.2.57}$$

Thus,

$$\tau_{xy}^{(C)}(X,Y) = K_{II} \int_0^\infty \frac{\tau_{xy}^{(T)}(X-s,Y)\,ds}{T\sqrt{2\pi s}} \tag{6.2.58}$$

Insertion of $\tau_{xy}^{(T)}$ gives integrals of the form

$$\int_0^\infty \frac{ds}{[(X-s)^2 + a_P^2 Y^2]\sqrt{s}} = \frac{\pi}{a_P Y}\sqrt{\frac{\sqrt{X^2 + a_P^2 Y^2} + X}{X^2 + a_P^2 Y^2}} \tag{6.2.59}$$

as conveniently determined by residue calculus (or use of a program for symbolic computation). After changing to a cylindrical coordinate system (r, φ, z), with $r = 0$ along the crack edge and $\varphi = \pm\pi$ on the crack faces, the asymptotic shear stress τ_{xy} at the crack edge is found to be

$$\tau_{xy} = \frac{K_{II}}{\sqrt{2\pi r}R(a_P, a_S)}\left[4a_P a_S \cdot \frac{\cos(\omega_P/2)}{T_P} - (1 + a_S^2)^2 \cdot \frac{\cos(\omega_S/2)}{T_S}\right] \tag{6.2.60}$$

where

$$\tan \omega_i = a_i \tan \varphi, \quad T_i = \sqrt{\cos^2 \varphi + a_i^2 \sin^2 \varphi} \tag{6.2.61}$$

Insertion of the expressions for a_P and a_S gives

$$\tau_{xy} = \frac{K_{II}}{\sqrt{2\pi r}R(\beta)}[4k^3\sqrt{1-\beta^2}\sqrt{k^2-\beta^2}F_2(\beta,\varphi) - (2k^2 - \beta^2)^2 F_2(\gamma,\varphi)] \tag{6.2.62}$$

where

$$F_2(\beta, \varphi) = \sqrt{\frac{\sqrt{1-\beta^2 \sin^2 \varphi} + \cos\varphi}{2(1-\beta^2 \sin^2 \varphi)}} \tag{6.2.63}$$

and $\gamma = \beta/k = V/c_S$, i.e. γ is the dimensionless velocity normalized with respect to the S wave velocity, whereas β is normalized with respect to the P wave velocity. These notations will be used throughout the continuation. Note that $F_2(\beta, \varphi)$ is normalized so that $F_2(\beta, 0) = 1$ and that

$$F_2(\beta, \varphi) \to \cos\frac{\varphi}{2}\left(1 + \frac{\beta^2}{2}\sin\frac{\varphi}{2}\sin\frac{3\varphi}{2}\right) \text{ as } \beta \to 0 \tag{6.2.64}$$

In the same way, the remaining mode II asymptotic Cartesian stress components are found to be

$$\sigma_x = -\frac{2k\sqrt{k^2-\beta^2}K_{II}}{\sqrt{2\pi r}R(\beta)}\{[2k^2+(1-2k^2)\beta^2]F_1(\beta,\varphi)$$
$$-(2k^2-\beta^2)F_1(\gamma,\varphi)\} \qquad (6.2.65)$$

$$\sigma_y = \frac{2k(2k^2-\beta^2)\sqrt{k^2-\beta^2}K_{II}}{\sqrt{2\pi r}R(\beta)}[F_1(\beta,\varphi)-F_1(\gamma,\varphi)] \qquad (6.2.66)$$

$$\sigma_z = \begin{cases} \dfrac{1-2k^2}{2(1-k^2)}(\sigma_x+\sigma_y) = \nu(\sigma_x+\sigma_y) & \text{for plane strain} \\ 0 & \text{for plane stress} \end{cases}$$

where

$$F_1(\beta,\varphi) = \text{sgn}(\varphi)\sqrt{\frac{\sqrt{1-\beta^2\sin^2\varphi}-\cos\varphi}{2(1-\beta^2\sin^2\varphi)}} \qquad (6.2.67)$$

$$\text{sgn}(\varphi) = \begin{cases} +1 & \text{if } \varphi > 0 \\ -1 & \text{if } \varphi < 0 \end{cases} \qquad (6.2.68)$$

Note that

$$F_1(\beta,\varphi) \to \sin\frac{\varphi}{2}\left(1-\frac{\beta^2}{2}\cos\frac{\varphi}{2}\cos\frac{3\varphi}{2}\right) \quad \text{as } \beta \to 0 \qquad (6.2.69)$$

With the, as usual modest, modifications when the solution of a mode I problem is sought from the solution for the corresponding mode II problem, the asymptotic stresses at a mode I crack edge are found to be

$$\sigma_x = \frac{K_I}{\sqrt{2\pi r}R(\beta)}\{(2k^2-\beta^2)[2k^2+(1-2k^2)\beta^2]F_2(\beta,\varphi)$$
$$-4k^3\sqrt{1-\beta^2}\sqrt{k^2-\beta^2}F_2(\gamma,\varphi)\} \qquad (6.2.70)$$

$$\sigma_y = \frac{K_I}{\sqrt{2\pi r}R(\beta)}\{[4k^3\sqrt{1-\beta^2}\sqrt{k^2-\beta^2}]F_2(\gamma,\varphi)$$
$$-(2k^2-\beta^2)^2F_2(\beta,\varphi)\} \qquad (6.2.71)$$

$$\tau_{xy} = \frac{2k^2(2k^2-\beta^2)\sqrt{1-\beta^2}K_I}{\sqrt{2\pi r}R(\beta)}[F_1(\beta,\varphi)-F_1(\gamma,\varphi)] \qquad (6.2.72)$$

$$\sigma_z = \begin{cases} \dfrac{1-2k^2}{2(1-k^2)}(\sigma_x+\sigma_y) = \nu(\sigma_x+\sigma_y) & \text{for plane strain} \\ 0 & \text{for plane stress} \end{cases}$$

These components of the mode I asymptotic stress field were determined by Yoffe (1951) and subsequently confirmed by Broberg (1960), Craggs (1960) and Baker (1962) for different crack problems.

The asymptotic cylindrical coordinate stress components are found from the Cartesian components by the usual transformations,

$$\sigma_r = \sigma_x \cos^2 \varphi + \sigma_y \sin^2 \varphi + 2\tau_{xy} \cos \varphi \sin \varphi \qquad (6.2.73)$$
$$\sigma_\varphi = \sigma_x \sin^2 \varphi + \sigma_y \cos^2 \varphi - 2\tau_{xy} \cos \varphi \sin \varphi \qquad (6.2.74)$$
$$\tau_{r\varphi} = (\sigma_y - \sigma_x) \cos \varphi \sin \varphi + \tau_{xy}(\cos^2 \varphi - \sin^2 \varphi) \qquad (6.2.75)$$

Thus, for instance, for mode I,

$$\begin{aligned}\sigma_\varphi = \frac{K_I}{\sqrt{2\pi r} R(\beta)} \{&4k^3 \sqrt{k^2 - \beta^2} \sqrt{1 - \beta^2} F_2(\gamma, \varphi)(\cos^2 \varphi - \sin^2 \varphi) \\
&-4k^2(2k^2 - \beta^2)\sqrt{1 - \beta^2}[F_1(\beta, \varphi) - F_1(\gamma, \varphi)] \cos \varphi \sin \varphi \\
&+(2k^2 - \beta^2)[2k^2 + (1 - 2k^2)\beta^2] F_2(\beta, \varphi) \sin^2 \varphi \\
&-(2k^2 - \beta^2)^2 F_2(\beta, \varphi) \cos^2 \varphi\} \end{aligned} \qquad (6.2.76)$$

This component, the hoop stress, is of particular interest. As found already by Yoffe (1951), the hoop stress develops maxima offside the symmetry plane at high crack velocities. For lower velocities – though not at all low in comparison with most crack speeds obtainable in experiments – the maximum occurs in the straightforward direction, but is actually very flat. This may be seen by specializing to $\varphi^2 \ll 1$:

$$\sigma_\varphi \approx \frac{K_I}{\sqrt{2\pi r}} \left[1 - \frac{s(\beta)}{8R(\beta)} \varphi^2 \right] \qquad (6.2.77)$$

where

$$s(\beta) = 4k(9k^2 + \beta^2)\sqrt{1 - \beta^2}\sqrt{k^2 - \beta^2} \\ - (2k^2 - \beta^2)[18k^2 - (1 + 6k^2)\beta^2 + 3\beta^4] \qquad (6.2.78)$$

$s(\beta)$ is positive for crack edge velocities less than about $0.6c_S$ at $k^2 = 0.375$ (corresponding to Poisson's ratio $\nu = 0.2$ for plane strain, 0.25 for plane stress) and less than about $0.7c_S$ when $k^2 = 1/6$ (corresponding to $\nu = 0.4$ for plane strain, unattainable in plane stress). For $k^2 = 1/3$ ($\nu = 1/4$ for plane strain, $1/3$ for plane stress), which is used in Fig. 6.2.2, $s(\beta)$ reaches zero for approximately $0.628c_S$. The figure shows the angular distribution of σ_φ for a few different crack edge velocities.

Crack branching in mode I
In accord with the intuitive hypothesis that mode I crack growth occurs in the direction of *maximum asymptotic hoop stress*, offside maxima would indicate branching. This was also suggested by Yoffe (1951). Note, however, that the limiting crack speed $0.628c_S$ in Fig. 6.2.2, which equals about $0.68c_R$, is a very high velocity, indeed, very seldom obtained in experiments. Branching virtually always occurs at crack speeds below the Yoffe velocity, as the limiting speed may be referred to for short. There are obviously also factors involved other than the character of asymptotic stress fields, as will be discussed in Section 9.2. One essential ingredience seems to be attempted branching, which is known to occur repeatedly, already for comparatively low velocities, before successful branching occurs. Such attempted branching might perhaps to some extent be associated with the fact that the hoop stress maximum for all crack speeds below the Yoffe velocity, is too flat to give a clear preference for straightforward crack growth, but the situation is more complex.

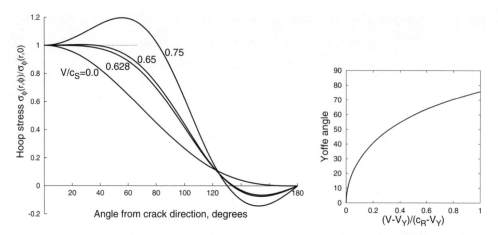

Fig. 6.2.2 Left figure: Angular distribution of the mode I hoop stress σ_φ at constant radial distance from the crack edge for different velocities. $k^2 = 1/3$, corresponding to Poisson's ratio 1/4 in plane strain and 1/3 in plane stress. The hoop stress is maximum in the straightforward direction if the crack speed, V, is less than $V_Y \approx 0.628 c_S$. When the crack speed is increased beyond this value, the maximum occurs offside the symmetry line as shown in the right figure, where the angle for the maximum is called the Yoffe angle, for short. This angle equals about 26° for $V/c_S = 0.65$ and about 56° for $V/c_S = 0.75$, and the corresponding maxima overshoot the hoop stress in the straightforward direction by about 0.5% and 20%, respectively.

The rationale behind the intuitive hypothesis that relates mode I crack growth direction to maximum hoop stress, seems to be the idealized notion of continuous crack edge motion, not the real stepwise motion consisting of opening of micro-separations and subsequent coalescences. In reality, the crack path is lined up by the larger micro-separations, and this point of view indicates that the *maximum asymptotic principal stress*, at given distance from the crack edge, could provide a more plausible explanation of attempted branching. However, a closer inspection casts some doubt also on this possibility.

During mode I crack growth, the maximum asymptotic principal stress at a given distance from the crack edge, actually occurs offside the symmetry plane for all velocities. However, somewhat surprisingly, it is most pronounced for zero velocity, at which it occurs at 60° to the straightforward direction and is about 30% larger than the largest principal stress in the straightforward direction. For $k^2 = 1/3$, used in Fig. 6.2.2, it occurs at about 90° for $V = 0.65 c_S$ and is only about 5% larger than at $\varphi = 0$. For velocities above about $0.77 c_S$, the largest principal stress is found at $\varphi = 0$. Therefore, it seems unlikely that branching would be initiated in the direction of maximum principal stress.

A noteworthy feature of mode I crack propagation, with expected implications on micro-separation morphology, is that *the ratio σ_x/σ_y in the straightforward direction* increases with the crack speed from its static value, unity. For the case shown in Fig. 6.2.2, it is about 1.06 for $0.25 c_S$, 1.12 for $0.53 c_S$ and 1.70 for $0.65 c_S$. It seems likely that a ratio above unity could cause opening of micro-separations, elongated in planes approximately normal to the symmetry plane, and these, in turn, could cause attempted branching. This tendency would then be more pronounced for high than for low crack velocities, and it seems to explain the increasing fracture surface roughness with crack speed, cf. Sections 1.7 and 9.2.

Even though the influence of the crack speed on the mode I asymptotic stress field is considerable when the Rayleigh speed is approached, it is quite modest for normally obtained crack speeds, around 50% of the Rayleigh speed or lower. With one exception, the deviations from the static field do not show any features with clearly recognizable implications on crack propagation properties. *The exception is the ratio σ_x/σ_y in the straightforward direction, which therefore appears to be the strongest candidate for causing branching, in comparison with the two other candidates discussed, the maximum hoop stress and the maximum principal stress.* Crack branching will be further discussed in Section 9.2.

Kink formation in mode II
The influence of the crack speed on the asymptotic stress field is even less pronounced for mode II than for mode I. The tendency towards formation of a kink, propagating in mode I, is very strong and requires very high compressive stresses to be suppressed, cf. page 37. Kink formation is obviously a result of high tensile stresses on one side of the crack path. Near the edge of a mode II crack with positive K_{II} (see page 80 for sign convention), the maximum principal stress occurs at $-180°$ for all velocities, i.e. at a location which cannot influence the crack propagation. For the angle $-45°$, the largest principal stress is about 22% larger than in the straightforward direction for the crack speed $0.4c_S$, and for the velocities $0.6c_S$ and $0.8c_S$ it is about 32% and 80% larger, respectively. The hoop stress for $-45°$ is substantially smaller; its maximum occurs at about $-74°$ for $0.4c_S$ and about $-93°$ for $0.8c_S$. The figures indicate that the kinking tendency increases with the crack velocity, i.e., if the ambient compressive stress is sufficiently strong to suppress kinking for low crack velocities, it might not be so for higher velocities.

Higher order terms and crack propagation along a curved path
In a similar way as for stationary cracks, non-singular stresses, particularly T-stresses, generally appear together with the singular stresses in the vicinity of mode I crack edges. In contrast to the singular stresses, they are dependent on body and loading geometry. Such stresses have been studied for particular dynamic cases by Rosakis *et al.* (1991), Freund and Rosakis (1992), Liu, Lambros and Rosakis (1993) and by Liu, Rosakis and Freund (1993).

Although not strictly belonging to the shearing modes, constant nonsingular stresses play a significant part in some cases of modes II and III crack propagation, especially when this takes place in a three-dimensional compressive stress environment, as for earthquake slip, where the compressive stresses are caused by the weight of the overburden. In particular, the normal stress at the crack faces (the slip interface) is important, because it controls the friction properties.

It was generally accepted in the 1960s that the singular asymptotic stresses and displacements for each mode (I, II or III) are uniquely determined as functions of the crack edge velocity, apart from an amplitude factor, just as they are in the static case. Thus, they would be independent of body and loading geometry. It was also generally accepted that the same asymptotic stresses would prevail even for smoothly varying crack velocities, i.e., the singular field would be dependent only on the current (instantaneous) velocity. Thus, if one accepts that this is the case if the velocity varies "slowly", it must be so even for arbitrarily fast but continuous variations, because it is always possibly to consider such a small region near the crack edge that the crack velocity stays constant within any prescibed accuracy, when the crack traverses this region. In analytical terms, a term of the character $-\mathrm{d}V/\mathrm{d}t \cdot \partial \phi/\partial X$ would be added to (6.2.9), and such a term is one order of magnitude smaller, counted in powers of r, than the other terms in a crack edge vicinity. However, some doubt seems to have arisen in the 1970s about the uniqueness for each instantaneous crack velocity of the singular terms, even if this velocity is (continuously) varying. Detailed investigations were performed by Freund and Clifton (1974), Nilsson (1974a) and Achenbach and Bažant (1975), but they confirmed the uniqueness. In later work, it is explicitly shown how the angular distributions of *nonsingular* terms are not unique. They depend not only on the instantaneous crack velocity, but also on the crack edge acceleration (Rosakis et al. 1991, Freund and Rosakis 1992).

The asymptotic field, including higher order terms, for mixed mode II-III crack propagation along a curved path, was considered by Liu and Rosakis (1994). They gave explicit expressions for the stresses up to the third order term. The results showed the same radial dependence, $r^{-1/2}$, r^0 and $r^{1/2}$, as for a non-curved crack. Singular terms, expressed in the stress intensity factors K_I and K_{II} for locally mode I and mode II, respectively, are the same as for cracks propagating along a straight path, but the $r^{1/2}$ terms show different angular distributions. Moreover, as expected, the $r^{1/2}$ terms are dependent not only on the instantaneous crack edge velocity, $V(t)$, but also on the crack edge accelerations $\dot{V}(t)$ and $\kappa(t) V^2(t)$, where $\kappa(t)$ is the crack path curvature at the crack edge. Another result, perhaps less expected, is that the coefficients for the $r^{1/2}$ terms include both K_I and K_{II}, even for mode I type (symmetric) displacements and for mode II type (anti-symmetric) displacements. Therefore, as Liu and Rosakis remark, the mode II part of the $r^{1/2}$ term will in general survive if the crack is following a curved path, even if crack propagation occurs under the condition $K_{II} = 0$. This might cause some difficulty for interpretation of experimental observations.

Other analytical investigations of non-planar crack propagation have been made by Willis and Movchan (1997), but they concern three-dimensional perturbations (crack direction varying along the crack front) and will be discussed in the context of three-dimensional crack propagation, page 500.

Crack face loads. The Craggs problem

As previously formulated, the correspondence between static and dynamic solutions for cracks propagating steadily along a symmetry plane in an infinite medium, is not restricted to traction free crack faces. One example will be given, the Craggs problem (Craggs 1960). In this problem, the crack is semi-infinite, extending from $X = -\infty$ to $X = 0$ and driven by moving normal and tangential loads on the crack faces. Only normal loads, $\sigma_y = \sigma_y^0(X)$ are considered here – the treatment for tangential loads follows after trivial and obvious changes.

From the static solution, (4.4.51)-(4.4.52), the normal stress and displacement gradient are obtained as

$$\sigma_y(X) = \frac{1}{\pi\sqrt{X}} \int_{-\infty}^{0} \frac{\sqrt{-\xi}\sigma_y^0(\xi)}{\xi - X} d\xi \text{ for } X > 0 \tag{6.2.79}$$

$$\frac{\partial v_+}{\partial x} = -\frac{Y_I(\beta)}{2\pi(1-k^2)\mu\sqrt{-X}} \oint_{-\infty}^{0} \frac{\sqrt{-\xi}\sigma_y^0(\xi)}{\xi - X} d\xi \text{ for } X < 0 \tag{6.2.80}$$

where the integral in the second expression is taken in the sense of the Cauchy principal value. The stress intensity factor is

$$K_I = -\sqrt{\frac{2}{\pi}} \int_{-\infty}^{0} \frac{\sigma_y^0(\xi)}{\sqrt{-\xi}} d\xi \tag{6.2.81}$$

and the energy supply from the moving load, per unit crack growth and per unit width along the crack edge, is

$$2\int_{-\infty}^{0} \sigma_y^0(X) \frac{\partial v_+}{\partial X} dX = -\frac{Y_I(\beta)}{(1-k^2)\pi\mu} \int_{-\infty}^{0}\int_{-\infty}^{0} \frac{\sqrt{-\xi}\sigma_y^0(X)\sigma_y^0(\xi)}{\sqrt{-X}(\xi - X)} d\xi\, dX \tag{6.2.82}$$

The double integral may be evaluated by adding the double integral obtained after swopping ξ and X and then dividing by two. The resulting energy supply is then found to be

$$\frac{Y_I(\beta)}{2(1-k^2)\pi\mu}\left[\int_{-\infty}^{0} \frac{\sigma_y^0(\xi)}{\sqrt{-\xi}} d\xi\right]^2 = \frac{Y_I(\beta)K_I^2}{4(1-k^2)\mu} \tag{6.2.83}$$

Note that this is the same result as calculated for the energy flux to an infinitesimally small process region, expression (6.2.49). This is logical, because all energy supplied goes to dissipation at the crack edge. The other possible energy sink, infinity, does not receive energy, because the remote stresses and strains decay faster than in proportion to the inverse distance from the crack edge, cf. Appendix A2. Craggs (1960) calculated the energy flux for a specific case, constant load along a portion of the crack faces from the crack edge, but he did not recognize the result (6.2.83), which relates the energy flux to quantities in the crack edge vicinity, only.

6.3 Friction and cohesion in mode II crack propagation

Sub-Rayleigh crack velocities

Mode II crack propagation – slip propagation – was studied in the previous section under the idealized assumption of friction free crack surfaces and infinitesimally small scale yielding. In reality both friction and finite dimensions of the dissipative region are significant. Thus, for instance, if it is assumed that slip occurs when the shear stress reaches some (usually velocity dependent) finite value, then the previously obtained solutions are contradictory, because they contain unbounded shear stresses in the non-slipping region ahead of the crack.

By considering the linear Coulomb friction model and the Barenblatt process region model, the great advantage of linearity can be retained, and these idealizations will therefore be adopted; see Fig. 6.3.1.

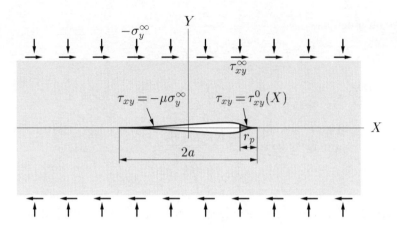

Fig. 6.3.1 The mode II Yoffe problem with friction and cohesion. The crack is shown in conventional symbolic manner, but in reality there is no crack opening in mode II. Thus, the displacement u, rather than the displacement v, is shown. Energy-neutral healing, which is equivalent to smooth closing at the trailing end, is assumed. By assuming Coulomb friction and a Barenblatt process region, linearity is preserved.

If the normal stress on the crack faces is $\sigma_y^\infty < 0$, then the friction stress is

$$\tau_{xy} = \tau_f = -\mu_d \sigma_y^\infty \tag{6.3.1}$$

where μ_d is the dynamic friction coefficient. But then, if a homogeneous stress state, consisting of the shear stress $\tau_{xy} = -\tau_f$, is superposed everywhere, the boundary conditions default to those of the frictionless problem. Because the equations remain unchanged, the solution of the friction problem is readily obtained after this superposition, from the solution of the frictionless problem, followed by the consequential resuperposition. Thus, for example, consideration of Coulomb friction in the mode II Yoffe problem (but not yet a finite cohesive region), implies that the solutions (6.2.33)-(6.2.35) are changed to

$$\tau_{xy} = (\tau_{xy}^\infty - \tau_f)\sqrt{\frac{X+a}{X-a}} + \tau_f \quad \text{for } |X| > a \tag{6.3.2}$$

$$\frac{\partial u_+}{\partial x} = -\frac{(\tau_{xy}^\infty - \tau_f)Y_{II}(\beta)}{2(1-k^2)\mu}\sqrt{\frac{a+X}{a-X}} \quad \text{for } |X| < a \tag{6.3.3}$$

$$\Delta = \frac{\pi(\tau_{xy}^\infty - \tau_f)aY_{II}(\beta)}{2(1-k^2)\mu} \tag{6.3.4}$$

From the first equation, it is noted that

$$K_{II} = 2(\tau_{xy}^\infty - \tau_f)\sqrt{\pi a} \tag{6.3.5}$$

Incorporation of a Barenblatt process region model was made for a slowly moving mode II crack in Section 4.7. The results obtained will be used here for the dynamic sub-Rayleigh case, with incorporation also of Coulomb friction. The crack length (in-

6.3 FRICTION AND COHESION IN MODE II CRACK PROPAGATION

cluding the process region) is $2a$ and the shear stress on $Y = 0$ is

$$\tau_{xy} = \begin{cases} \tau_f & \text{for } -a < X < a - r_p \\ \tau_{xy}^0(X) & \text{for } a - r_p < X < a \end{cases} \quad (6.3.6)$$

where r_p is the length of the process region. $\tau_{xy}^0(X)$ is subject to the smooth closing condition, cf. (4.7.21),

$$\frac{1}{\pi} \int_{a-r_p}^{a} \frac{[\tau_{xy}^0(\xi) - \tau_f]\mathrm{d}\xi}{\sqrt{a^2 - \xi^2}} = \tau_{xy}^\infty - \tau_f \quad (6.3.7)$$

For the dynamic sub-Rayleigh case, use of the results (4.7.22) and (4.7.27), which assume energy-neutral healing (so that there is no process region at the trailing edge), gives

$$\tau_{xy} = \frac{X\sqrt{X^2 - a^2}}{\pi|X|} \int_{-a}^{a} \frac{\tau_{xy}(\xi)\mathrm{d}\xi}{\sqrt{a^2 - \xi^2}(X - \xi)}$$

$$= \frac{X\sqrt{X^2 - a^2}}{\pi|X|} \int_{a-r_p}^{a} \frac{[\tau_{xy}^0(\xi) - \tau_f]\mathrm{d}\xi}{\sqrt{a^2 - \xi^2}(X - \xi)} + \tau_f \quad \text{for } |X| > a \quad (6.3.8)$$

$$\frac{\partial u_+}{\partial x} = -\frac{\sqrt{a^2 - X^2}Y_{II}(\beta)}{2\pi(1-k^2)\mu} \oint_{a-r_p}^{a} \frac{[\tau_{xy}^0(\xi) - \tau_f]\mathrm{d}\xi}{\sqrt{a^2 - \xi^2}(\xi - X)} \quad \text{for } |X| < a \quad (6.3.9)$$

for $Y = 0$, and, if $r_p \ll a$, the slip deposited on each side behind the crack is, cf. (4.7.33),

$$\Delta \approx \frac{T_{II} Y_{II}(\beta)}{2(1-k^2)\mu} \sqrt{\frac{a}{2}} \quad (6.3.10)$$

which becomes exact in the limit $r_p/a \to 0$ and then agrees with (6.3.4), because T_{II}, the Barenblatt cohesion modulus in mode II, cf. (3.6.18), is

$$T_{II} = \sqrt{\frac{\pi}{2}} K_{II} = \pi(\tau_{xy}^\infty - \tau_f)\sqrt{2a} \quad (6.3.11)$$

The value of T_{II}, and thus also of K_{II}, during dynamic crack propagation, is dependent on β and probably also of the crack propagation history; cf. Section 9.1.

Equation (6.3.11) shows that the length of the sliding region (the crack length) is

$$2a = \frac{T_{II}^2}{\pi^2(\tau_{xy}^\infty - \tau_f)^2} \quad (6.3.12)$$

and that the deposited slip on each side is

$$\Delta = \frac{T_{II}^2 Y_{II}(\beta)}{4\pi(1-k^2)\mu(\tau_{xy}^\infty - \tau_f)} \quad (6.3.13)$$

These expressions are dependent on the cohesion modulus T_{II}, the overstress $\tau_{xy}^\infty - \tau_f$ and the velocity. During slow crack propagation, the crack length is uniquely specified, if T_{II} and $\tau_{xy}^\infty - \tau_f$ are known, but in the dynamic case, even if there would be a unique and known relation between T_{II} and the velocity, the equations presented so far do not enable calculation of the velocity. *The crack length and the deposited slip cannot be* determined in the dynamic case from steady state relations alone: the history through which the steady state is arrived at must also be known.

The energy dissipation at the leading edge is calculated in the same way as for slow crack growth, Section 3.5, but the total energy dissipation also involves friction energy along the whole sliding region. Thus, the energy dissipation per unit of area along the interface, after a pulse of unidirectional sliding has passed by, is

$$\frac{dW}{dS} \approx 2\tau_f \Delta + \frac{T_{II}^2 Y_{II}(\beta)}{2\pi(1-k^2)\mu} \tag{6.3.14}$$

Note that the expression does not contain r_p. It becomes exact in the limit $r_p/a \to 0$. As will be shown later, the corresponding expression for energy dissipation under intersonic crack speed depends on r_p/a and vanishes in the limit $r_p/a \to 0$.

A remarkably simple result is obtained by determining the ratio w_{pr} between the energy dissipation in the process region (i.e. at the leading edge) and the total energy dissipation:

$$\boxed{w_{pr} = \frac{\tau_{xy}^\infty - \tau_f}{\tau_f}} \tag{6.3.15}$$

If slip propagation takes place along a weak interface rather than through a homogeneous material, it might be inadmissible to neglect the modulus of healing, H_{II}. It turns out that the changes needed by retaining a non-zero value of H_{II} are quite modest; see Broberg (1978). Thus, the substitution $T_{II} \to (T_{II} - H_{II})$ has to be made in equations (6.3.10) and (6.3.12), and the substitution $T_{II}^2 \to (T_{II}^2 - H_{II}^2)$ in (6.3.13) and (6.3.14), whereas (6.3.15) holds without change†. Equation (6.3.11) has to be changed to

$$T_{II} = \sqrt{\frac{\pi}{2}} K_{II} \tag{6.3.16}$$

$$T_{II} - H_{II} = \pi(\tau_{xy}^\infty - \tau_f)\sqrt{2a} \tag{6.3.17}$$

Intersonic crack velocities

Stresses and displacements on the symmetry plane

For intersonic crack velocities, $c_S < V < c_P$, the notation

$$b_S = \sqrt{\beta^2/k^2 - 1} \tag{6.3.18}$$

will be used, whereas $a_P = \sqrt{1-\beta^2}$ as in the subsonic case.

Consider a crack edge vicinity, in which steady state may be assumed to prevail. It turns out to be necessary to include a finite size process region, because, in contrast to the sub-Rayleigh case, there will be no energy flux to a point size process region. In order to retain linearity, a Barenblatt region is assumed, with length r_p and tangential

---

† There are errors in the counterparts of (6.3.14)-(6.3.15) in Broberg (1978).

6.3 FRICTION AND COHESION IN MODE II CRACK PROPAGATION

stress $\tau_{xy} = \tau_{xy}^0(X)$). Thus, the boundary conditions at $Y = 0$ may be taken to be

$$\tau_{xy} = \begin{cases} 0 & \text{for } X < -r_p \\ \tau_{xy}^0(X) & \text{for } -r_p \leq X \leq 0 \end{cases} \quad (6.3.19)$$

$$\frac{\partial u}{\partial x} = 0 \text{ for } X > 0 \quad (6.3.20)$$

$$\sigma_y = 0 \text{ for all } X \quad (6.3.21)$$

The last equation implies friction free crack face sliding. It is assumed that the linear size of the steady state region is much larger than r_p in all directions, but at the same time much smaller than linear in-plane dimensions of the body. The coordinates within this steady state region can then be stretched to infinity without appreciably influencing the stress-strain field in the steady state region. As a consequence, the remote stresses are considered as vanishingly small.

Consider the lower half-plane, $Y < 0$. After the coordinate transformations

$$X = \xi, \quad a_P Y = \eta \quad (6.3.22)$$

the equations of motion (6.2.9)-(6.2.10) become

$$\frac{\partial^2 \phi}{\partial \xi^2} + \frac{\partial^2 \phi}{\partial \eta^2} = 0 \quad (6.3.23)$$

$$\alpha^2 \frac{\partial^2 \psi}{\partial \xi^2} - \frac{\partial^2 \psi}{\partial \eta^2} = 0 \quad (6.3.24)$$

where $\alpha = b_S/a_P$. The first equation is satisfied by $\phi = \Re F(\zeta)$, where $F(\zeta)$ is analytic in the half-plane $\zeta = \xi + i\eta$, $\eta < 0$. The second equation is satisfied by $\psi = f(\xi - \alpha\eta)$ and $\psi = f(\xi + \alpha\eta)$, where $f(\cdot)$ is one not yet determined function. Because $\eta < 0$, the second expression for f does not satisfy the problem: it corresponds to a wave travelling towards rather than away from the crack. Obviously, $f(\xi - \alpha\eta) = 0$ for $\xi > \alpha\eta$, $\eta < 0$, because S waves travel slower than the crack edge, so that their front is $\eta = \xi/\alpha$, $\eta < 0$. This corresponds to the angle $\varphi = -\varphi_f$, where φ is the angular coordinate in a cylindrical coordinate system (r, φ, z) with $\varphi = 0$ along the positive X axis, and φ_f is given by

$$\sin \varphi_f = k/\beta, \quad \frac{\pi}{2} < \varphi_f < \pi \quad (6.3.25)$$

Insertion of the expressions for ϕ and ψ into the representations (see Appendix A1)

$$u_- = \frac{\partial \phi}{\partial X} + \frac{\partial \psi}{\partial Y} \quad (6.3.26)$$

$$\sigma_x = \frac{\mu}{k^2} \left[\frac{\partial^2 \phi}{\partial X^2} + (1 - 2k^2) \frac{\partial^2 \phi}{\partial Y^2} + 2k^2 \frac{\partial^2 \psi}{\partial X \partial Y} \right] \quad (6.3.27)$$

$$\sigma_y = \frac{\mu}{k^2} \left[(1 - 2k^2) \frac{\partial^2 \phi}{\partial X^2} + \frac{\partial^2 \phi}{\partial Y^2} - 2k^2 \frac{\partial^2 \psi}{\partial X \partial Y} \right] \quad (6.3.28)$$

$$\tau_{xy} = \mu \left[2 \frac{\partial^2 \phi}{\partial X \partial Y} - \frac{\partial^2 \psi}{\partial X^2} + \frac{\partial^2 \psi}{\partial Y^2} \right] \quad (6.3.29)$$

gives the following relations (note that $\phi = [F(\zeta) + \overline{F(\zeta)}]/2$, $\partial \phi/\partial X = [F'(\zeta) + \overline{F'(\zeta)}]/2$

and that $\partial \phi / \partial Y = ia_P[F'(\zeta) - \overline{F'(\zeta)}]/2$:

$$\frac{\partial u_-}{\partial x} = \frac{1}{2}[F''(\zeta) + \overline{F''(\zeta)}] - b_S f''(\xi - \alpha \eta) \tag{6.3.30}$$

$$\frac{2k^2}{\mu}\sigma_x = [1 - (1 - 2k^2)a_P^2][F''(\zeta) + \overline{F''(\zeta)}] - 4k^2 b_S f''(\xi - \alpha \eta) \tag{6.3.31}$$

$$\frac{2}{\mu}\sigma_y = (b_S^2 - 1)[F''(\zeta) + \overline{F''(\zeta)}] + 4b_S f''(\xi - \alpha \eta) \tag{6.3.32}$$

$$\frac{1}{\mu}\tau_{xy} = ia_P[F''(\zeta) - \overline{F''(\zeta)}] + (b_S^2 - 1)f''(\xi - \alpha \eta) \tag{6.3.33}$$

Insertion of the boundary conditions into these expressions gives for $\eta = -0$,

$$\frac{1}{2}[F''(\xi) + \overline{F''(\xi)}] - b_S f''(\xi) = 0 \text{ for } \xi > 0 \tag{6.3.34}$$

$$(b_S^2 - 1)[F''(\xi) + \overline{F''(\xi)}] + 4b_S f''(\xi) = 0 \text{ for all } \xi \tag{6.3.35}$$

$$ia_P[F''(\xi) - \overline{F''(\xi)}] + (b_S^2 - 1)f''(\xi) = \frac{\tau_{xy}^0(\xi)}{\mu} \text{ for } \xi < 0 \tag{6.3.36}$$

Note that $f''(\xi) = 0$ in the first equation. According to the second equation

$$f''(\xi) = -\frac{b_S^2 - 1}{4b_S}[F''(\xi) + \overline{F''(\xi)}] \text{ for all } \xi \tag{6.3.37}$$

which, inserted into the other two equations, yields

$$\overline{F''(\xi)} + \frac{(b_S^2 - 1)^2 - 4ia_P b_S}{(b_S^2 - 1)^2 + 4ia_P b_S}F''(\xi) = -\frac{4b_S}{(b_S^2 - 1)^2 + 4ia_P b_S} \cdot \frac{\tau_{xy}^0(\xi)}{\mu} \text{ for } \xi < 0 \tag{6.3.38}$$

$$F''(\xi) + \overline{F''(\xi)} = 0 \text{ for } \xi > 0 \tag{6.3.39}$$

These equations may be used to formulate a Hilbert problem. To this end, the function $F(\zeta)$, will be considered as a sectionally analytic function. Its part for $\Im \zeta < 0$, already defined, will be denoted by $F_-(\zeta)$ for clarity, and the part for $\Im \zeta \geq 0$ will be defined as

$$F_+(\zeta) = -\overline{F_-(\zeta)}, \quad \Im \zeta > 0 \tag{6.3.40}$$

Then,

$$\overline{F_-(\xi)} = \overline{F_-}(\xi) = -F_+(\xi) \tag{6.3.41}$$

Inserting these expressions into (6.3.38)-(6.3.39) and noting that the coefficient for the second term in (6.3.38) may be written in the form $\exp(-2\pi i g)$, where

$$g = \frac{1}{\pi}\operatorname{atan}\frac{4a_P b_S}{(b_S^2 - 1)^2}, \quad 0 < g \leq 1/2 \tag{6.3.42}$$

leads to the Hilbert problem

$$F_+''(\xi) - e^{-2\pi i g}F_-''(\xi) = e^{-\pi i g}\sin \pi g \cdot \frac{\tau_{xy}^0(\xi)}{\mu a_P} \text{ for } \xi < 0 \tag{6.3.43}$$

$$F_+''(\xi) - F_-''(\xi) = 0 \text{ for } \xi > 0 \tag{6.3.44}$$

As usual in treatment of Hilbert problems, the two equations are now written in the common form

$$G_+(\xi)F''_+(\xi) - G_-(\xi)F''_-(\xi) = G_+(\xi)T(\xi) \tag{6.3.45}$$

where the function $G(\zeta)$ is sectionally analytic and such that the ratio $G_-(\xi)/G_+(\xi)$ equals $\exp(-2\pi i g)$ for $\xi < 0$ and $+1$ for $\xi > 0$, and

$$T(\xi) = \begin{cases} e^{-\pi i g} \sin \pi g \cdot \dfrac{\tau^0_{xy}(\xi)}{\mu a_P} & \text{for } \xi < 0 \\ 0 & \text{for } \xi > 0 \end{cases} \tag{6.3.46}$$

The solution of the Hilbert problem is then found from the expression

$$F''(\zeta) = \frac{1}{2\pi i G(\zeta)} \int_{-r_p}^0 \frac{G_+(w)T(w)}{w - \zeta} dw + \frac{P(\zeta)}{G(\zeta)} \tag{6.3.47}$$

where $P(\zeta)$ is a polynomial.

Because no singularity shall appear, it is convenient to choose

$$G(\zeta) = \zeta^{g-1} \tag{6.3.48}$$

with a branch cut along the negative real axis and the branch chosen so that $G(\xi)$ is positive for $\xi > 0$. Then, because $\tau_{xy} \to 0$ as $\xi \to \infty$, the polynomial $P(\zeta) = 0$, and the solution of the Hilbert problem is

$$F''(\zeta) = \frac{i \sin \pi g \, \zeta^{1-g}}{2\pi \mu a_P} \int_{-r_p}^0 \frac{\tau^0_{xy}(w)}{|w|^{1-g}(w-\zeta)} dw \tag{6.3.49}$$

after replacing the integration variable ξ by w.

The shear stress τ_{xy} on $X > 0$, $Y = 0$ and the displacement gradient $\partial u/\partial x$ on $X < 0$, $Y = 0$, will first be determined. Because $F''_+(\xi) = F''_-(\xi) = F''(\xi)$ for $\xi > 0$, $Y = 0$, the shear stress is

$$\tau_{xy} = 2ia_P\mu F''(\xi) = \frac{\sin \pi g}{\pi} X^{1-g} \int_0^{r_p} \frac{\tau^0_{xy}(-w)}{w^{1-g}(w+X)} dw \quad \text{for } X > 0 \tag{6.3.50}$$

and, because $f''(\xi) = 0$ for $\xi < 0$, the displacement gradient for $X < 0$ is

$$\frac{\partial u_-}{\partial x} = -\frac{1}{4}(b_S^2 + 1)[F''_+(\xi) - F''_-(\xi)]$$

$$- \frac{Y_{II}(\beta)}{2(1-k^2)\mu}\left[\sin \pi g \cdot |X|^{1-g} \fint_0^{r_p} \frac{\tau^0_{xy}(-w)}{w^{1-g}(w+X)} dw + \pi \cos \pi g \, \tau^0_{xy}(X)\right] \tag{6.3.51}$$

where the integral is taken in the sense of the Cauchy principal value, and

$$Y_{II}(\beta) = \frac{(1-k^2)\beta^2 \sin \pi g}{2k^2\sqrt{1-\beta^2}} \quad \text{for } k < \beta < 1 \tag{6.3.52}$$

Note that the subsonic expression for Y_{II} is given by (6.2.26).

Angular stress distribution

Recall that the process region is assumed to be small compared to the extension of the steady state region. For $r_p \ll |X|$ the expressions for the shear stress and the tangential displacement gradient on $Y = -0$ may be simplified to

$$\tau_{xy} = \frac{A}{X^g} \quad \text{for } X > 0 \tag{6.3.53}$$

$$\frac{\partial u_-}{\partial x} = \frac{A Y_{II}(\beta)}{2(1-k^2)\mu |X|^g} \quad \text{for } X < 0 \tag{6.3.54}$$

where

$$A = \frac{\sin \pi g}{\pi} \int_0^{r_p} \frac{\tau_{xy}^0(-w)}{w^{1-g}} dw \tag{6.3.55}$$

These are the expressions, which would have been obtained if a point size process region had been assumed, and then the amplitude factor A had been left undetermined, like the stress intensity factor in the corresponding subsonic treatment. Note that a singular factor $|X|^{-g}$ appears both for stresses and displacement gradients (thus also for strains). Because $g < 1/2$, except in a special case, the singularity is too weak to allow energy to flow to the crack edge. The exception, see (6.3.42), is for $b_S = 1$, i.e. $\beta = \sqrt{2}k$, and then $g = 1/2$, as in the subsonic case, so that energy flow to a point size process region is possible for this velocity, as for sub-Rayleigh velocities.

The angular stress distribution for $r_p \ll r$ will now be determined. Equation (6.3.49) simplifies to

$$F''(\zeta) = -\frac{iA}{2\mu a_P \zeta^g} \tag{6.3.56}$$

and by writing $\zeta = \rho e^{i\omega}$, where ρ and ω are polar coordinates in the $\xi\eta$-plane, giving

$$\rho = \sqrt{X^2 + a_P^2 Y^2} = r\sqrt{\cos^2 \varphi + a_P^2 \sin^2 \varphi} \tag{6.3.57}$$

$$\tan \omega = \tan(a_P \tan \varphi) \tag{6.3.58}$$

the following expressions are found:

$$F''(\zeta) + \overline{F''(\zeta)} = -\frac{A \sin g\omega}{\mu a_P \rho^g} \tag{6.3.59}$$

$$F''(\zeta) - \overline{F''(\zeta)} = -i \frac{A \cos g\omega}{\mu a_P \rho^g} \tag{6.3.60}$$

$$f''(\xi - a\eta) = \begin{cases} 0 & \text{for } |\varphi| < \varphi_f \\ \dfrac{b_S^2 - 1}{4 a_P b_S} \cdot \dfrac{A \sin \pi g}{\mu |\xi - a\eta|^g} & \text{for } |\varphi| > \varphi_f \end{cases} \tag{6.3.61}$$

Here, the last equation was obtained by using (6.3.37).

Use of the expressions (6.3.27)-(6.3.29) for the Cartesian stress components, and of the relations for transformation to polar stress components, (6.2.73)-(6.2.75), gives

$$\left.\begin{array}{c}\sigma_r \\ \sigma_\varphi\end{array}\right\} = -\frac{A}{r^g} \cdot \frac{(1-k^2)(1+b_S^2)}{2a_P} \cdot \frac{\sin g\omega}{T_P}$$
$$\mp \frac{A}{r^g} \left\{ \frac{\cos 2\varphi}{2a_P} \left[(1+a_P^2) \cdot \frac{\sin g\omega}{T_P} + (b_S^2 - 1) \cdot \frac{\sin \pi g}{T_S} \cdot U(|\varphi| - \varphi_f) \right] \right.$$
$$\left. - \sin 2\varphi \left[\frac{\cos g\omega}{T_P} + \frac{(b_S^2 - 1)^2}{4a_P b_S} \cdot \frac{\sin \pi g}{T_S} \cdot U(|\varphi| - \varphi_f) \right] \right\} \quad (6.3.62)$$

$$\tau_{r\varphi} = \frac{A}{r^g} \left\{ \frac{\sin 2\varphi}{2a_P} \left[(1+a_P^2) \cdot \frac{\sin g\omega}{T_P} + (b_S^2 - 1) \cdot \frac{\sin \pi g}{T_S} \cdot U(|\varphi| - \varphi_f) \right] \right.$$
$$\left. + \cos 2\varphi \left[\frac{\cos g\omega}{T_P} + \frac{(b_S^2 - 1)^2}{4a_P b_S} \cdot \frac{\sin \pi g}{T_S} \cdot U(|\varphi| - \varphi_f) \right] \right\} \quad (6.3.63)$$

where

$$T_P = (\cos^2 \varphi + a_P^2 \sin^2 \varphi)^{g/2} \quad (6.3.64)$$
$$T_S = (|\cos \varphi| - b_S |\sin \varphi|)^g \quad (6.3.65)$$

and $U(\cdot)$ is the unit step function. The expressions are written in a form valid for $-\pi < \varphi < +\pi$. Note that $T_S = 0$ for $|\varphi| = \varphi_f + 0$.

The contribution from S waves consists of the terms containing the unit step function. Note that these terms disappear for the curious velocity $V = \sqrt{2}c_S$ ($b_S = 1$). The same phenomenon appears for dynamically propagating edge dislocations, as noted by Eshelby (1949).

The S wave contribution to σ_r and σ_φ is continuous across the S wave front, in spite of the singularity of $1/T_S$. This is shown by specializing to $\varphi = \varphi_f$. The S wave contribution to $\tau_{r\varphi}$ is continuous for $g = 1/2$ ($b_S = 1$), but discontinuous for $g < 1/2$. In both cases the energy per unit volume changes continuously during the passage of the S wave front, so there is no energy dissipation as at the passage of a shock front. Moreover, the front is sharp only in the approximation $r_p \ll r$.

Intersonic crack propagation was first analysed by Burridge et al. (1979) and by Freund (1979).

Energy flux into the process region

It was shown in Sections 3.5 and 6.2 that the energy flux to the process region is to a first approximation, independent of the size of this region for cracks moving slowly or dynamically under sub-Rayleigh crack speeds. This is not so for intersonic crack speeds. The energy flux is directly obtained from (6.3.51):

$$\mathcal{G} = 2 \int_{-r_p}^{0} \tau_{xy}^0(X) \cdot \frac{\partial u_-}{\partial x} dX$$
$$- \frac{Y_{II}(\beta) \sin \pi g}{\pi(1-k^2)\mu} \int_{-r_p}^{0} \tau_{xy}^0(X) \Big[|X|^{1-g} \oint_0^{r_p} \frac{\tau_{xy}^0(-w)}{w^{1-g}(w+X)} dw$$
$$+ \pi \cot \pi g \, \tau_{xy}^0(X) \Big] dX \quad (6.3.66)$$

This expression will be written in a form more suitable for interpretations and numerical calculations. To this end, the stress distribution in the Barenblatt region is written as

$$\tau_{xy}^0(X) = \tau_D \cdot D(X/r_p) \tag{6.3.67}$$

where τ_D is the decohesive strength and $D(X/d)$ a shape function, which equals $+1$ for $X/d = 0$, is zero for $X/d < -1$ and $X/d > 0$, and is increasing in the interval $-1 < X/d < 0$. One simple example is $D(X/d) = 1 + X/d$. The term $\pi \cot \pi g$ in (6.3.66) is now substituted by an expression involving the integral

$$\int_0^\infty \frac{dw}{w^{1-g}(w+X)} = -\pi \cot \pi g \, |X|^{g-1} \text{ if } X < 0 \tag{6.3.68}$$

Then, after some rearrangement and redefinition of the integration variable w, it follows that

$$\mathcal{G} = \tau_D^2 r_p \cdot \frac{Y_{II}(\beta) \sin \pi g}{\pi(1-k^2)\mu} w_D(g) \tag{6.3.69}$$

where

$$w_D(g) = \int_0^1 D(-v)v^{1-g} \left[\int_0^1 \frac{D(-v) - D(-w)}{w^{1-g}(w-v)} dw \right.$$
$$\left. + D(-v) \int_1^\infty \frac{dw}{w^{1-g}(w-v)} \right] dv \tag{6.3.70}$$

Note that both integrands of the inner integrals are positive, so that $w_D(g) > 0$.

Because only the crack edge vicinity is considered, there is no length parameter involved other than r_p. For discussion of the result, it is helpful to introduce a length L, the distance ahead of the crack edge where a given stress $(\tau_{xy})_{y=0} = \tau_{xy}^L$ prevails. It is assumed that τ_{xy}^L is chosen small enough to yield $L \gg r_p$. Then, equations (6.3.53) and (6.3.55) give

$$A = (\tau_{xy})_{y=0} X^g = \frac{\sin \pi g}{\pi} \tau_D r_p^g \int_0^1 \frac{D(-w)}{w^{1-g}} dw = \tau_{xy}^L \cdot L^g \tag{6.3.71}$$

so that, after elimination of τ_D,

$$\mathcal{G} = \frac{\pi(\tau_{xy}^L)^2 L}{2(1-k^2)\mu} \cdot Y_{II}(\beta) \cdot \Gamma_D(g) \cdot \left(\frac{r_p}{L}\right)^{1-2g} \tag{6.3.72}$$

where

$$\Gamma_D(g) = \frac{2w_D(g)}{\sin \pi g \left[\int_0^1 \frac{D(-w)}{w^{1-g}} dw \right]^2} \tag{6.3.73}$$

Note that $\mathcal{G}$ is composed of four factors, the first one dependent on the outer field through τ_{xy}^L, the second on the crack speed, the third on the shape of the process region and the velocity-dependent constant g, and the fourth on the size of the process region. It is immediately seen that the assumption of a point size process region, $r_p = 0$, results in zero energy flux into the crack edge, if $g < 1/2$, but a non-zero flux, if $g = 1/2$, i.e.

6.3 FRICTION AND COHESION IN MODE II CRACK PROPAGATION

for $\beta = \sqrt{2}k$. On the other hand, for most velocities in the intersonic region, g is not much different from $1/2$, and then the factor $(r_p/L)^{1-2g}$ is not very small compared to unity, until r_p/L becomes extremely small. Thus, for instance, using the values $k^2 = 1/3$ and $r_p/L = 10^{-6}$, for the velocity $1.3c_S$ it is about 0.66, for $1.2c_S$ it is about 0.24, and for $1.1c_S$ it is about 0.026.

Note further, from (6.3.53), that the ordinary stress intensity factor appears in the case $g = 1/2$. Then, because $K_{II} = \sqrt{2\pi} A = \tau_{xy}^L \sqrt{2\pi L}$ and, because it can be shown that $\Gamma_D(g) \to 1$ as $g \to 1/2$ by using the procedure leading to (4.15.42), the energy flux becomes

$$\mathcal{G} \to \frac{K_{II}^2 Y_{II}(k\sqrt{2})}{4(1-k^2)\mu} \tag{6.3.74}$$

which is the same expression as for the subsonic case!

An attempt to calculate the energy flux to an intersonic mode I crack results in a negative value, which, of course, implies that mode I intersonic crack propagation is physically impossible. The change of sign is due to a factor $-1/\sqrt{\beta^2 - k^2}$ in $Y_I(\beta)$ instead of the factor $1/\sqrt{1-\beta^2}$ in $Y_{II}(\beta)$. It could also be mentioned that a formal calculation determines the exponent g for mode I to be $(1 - g_{II})$, where g_{II} is the value of the exponent for mode II under the same crack speed.

The intersonic mode II Yoffe problem

With only modest changes, the procedure used for the edge vicinity of a semi-infinite crack can be applied to a crack of finite length, $2a$, propagating under steady state conditions in an infinite body under the remote stress τ_{xy}^∞, i.e. the Yoffe problem adapted to mode II and intersonic velocities. The results for tangential stress and displacement gradient on $Y = 0$ are found to be (Broberg 1989a)

$$\tau_{xy} = -\text{sgn}(X) \cdot \frac{\sin \pi g}{\pi} \cdot |a-X|^{1-g}|a+X|^g$$
$$\times \int_{-a}^{a} \frac{\tau_{xy}^0(\xi) d\xi}{(a-\xi)^{1-g}(a+\xi)^g(\xi-X)} \quad \text{for } |X| > a \tag{6.3.75}$$

$$\frac{\partial u_\pm}{\partial x} = \mp \cdot \frac{Y_1(\beta)}{\pi\mu} \Big[\sin \pi g \cdot (a-X)^{1-g}(a+X)^g$$
$$\times \oint_{-a}^{a} \frac{\tau_{xy}^0(\xi) d\xi}{(a-\xi)^{1-g}(a+\xi)^g(\xi-X)} - \pi \cos \pi g \cdot \tau_{xy}^0(X)\Big] \quad \text{for } |X| < a$$
$$\tag{6.3.76}$$

These expressions are subject to the condition

$$\frac{\sin \pi g}{\pi} \int_{-a}^{a} \frac{\tau_{xy}^0(\xi) d\xi}{(a-\xi)^{1-g}(a+\xi)^g} = \tau_{xy}^\infty \tag{6.3.77}$$

The stress τ_{xy}^0 on the crack faces may be rather arbitrary, thus, for instance, allowing for a healing process associated with energy release and for varying friction stress along the slipping region. For energy-neutral healing and a constant friction stress, τ_f, the shear stress on the crack faces may be written as

$$\tau_{xy} = \begin{cases} \tau_f & \text{for } -a < X < a - r_p \\ \tau_{xy}^0(X) & \text{for } a - r_p < X < a \end{cases} \qquad (6.3.78)$$

so that condition (6.3.77) reads:

$$\frac{\sin \pi g}{\pi} \int_{a-r_p}^{a} \frac{[\tau_0(\xi) - \tau_f] d\xi}{(a-\xi)^{1-g}(a+\xi)^g} = \tau_{xy}^\infty - \tau_f \qquad (6.3.79)$$

Assuming the process region to be small, so that $r_p/a \ll 1$, gives for $Y = 0$, approximately,

$$\tau_{xy} = (\tau_{xy}^\infty - \tau_f) \left(\frac{X+a}{X-a}\right)^g \quad \text{for } X < -a, \ X - r_p \gg a \qquad (6.3.80)$$

$$\frac{\partial u_-}{\partial x} = \frac{(\tau_{xy}^\infty - \tau_f) Y_{II}(\beta)}{2(1-k^2)\mu} \left(\frac{a+X}{a-X}\right)^g \quad \text{for } r_p \ll a - X < 2a \qquad (6.3.81)$$

and the amount of slip deposited on each side is

$$\Delta = \int_{-a}^{a} \frac{\partial u_-}{\partial x} dX = \frac{\pi(\tau_{xy}^\infty - \tau_f)a}{(1-k^2)\mu} \cdot \frac{g Y_{II}(\beta)}{\sin \pi g} \qquad (6.3.82)$$

6.4 Steady state mode III crack propagation

The first solution of a mode III steady state problem was given by McClintock and Sukhatme (1960), who considered a semi-infinite crack under crack face loading, *i.e.*, the same problem as considered by Craggs (1960) for the in-plane case. Their results are analogous to those of Craggs, and they showed that the upper limit for mode III crack velocities, as expected, is the S wave velocity.

As for modes I and II, the response to mode III steady state crack propagation along $y = 0$ with velocity V can be directly obtained from the corresponding static response. It is assumed that the slip deposited behind any trailing crack edge is constant. By analogy with modes I and II, the equation of motion (A9.24) may be written as

$$a_S^2 \frac{\partial^2 w}{\partial X^2} + \frac{\partial^2 w}{\partial Y^2} = 0 \qquad (6.4.1)$$

where, as previously, $a_S = \sqrt{1-\gamma^2}$, $\gamma = V/c_S$. The substitutions $X = \xi$, $a_S Y = \eta$, give

$$\frac{\partial^2 W}{\partial \xi^2} + \frac{\partial^2 W}{\partial \eta^2} = 0 \qquad (6.4.2)$$

where $W = W(\xi, \eta) = w(X, Y)$. Boundary conditions on $Y = 0$ consist of prescribed shear stress, $\tau_{yz} = \tau_{yz}^0(X)$, on some part or parts of the X axis, say the portion L, and zero displacement gradient, $\partial w/\partial x = 0$, on the remaining part or parts. Thus,

6.4 STEADY STATE MODE III CRACK PROPAGATION

because $\tau_{yz} = \mu \partial w/\partial y = \mu a_S \partial W/\partial \eta$, the problem can be stated by the equations

$$\frac{\partial^2(a_S W)}{\partial \xi^2} + \frac{\partial^2(a_S W)}{\partial \eta^2} = 0 \tag{6.4.3}$$

$$\frac{\partial(a_S W)}{\partial \eta} = \frac{\tau_{yz}^0(\xi)}{\mu} \text{ for } \xi \in L, \eta = 0 \tag{6.4.4}$$

$$\frac{\partial(a_S W)}{\partial \xi} = 0 \text{ for } \xi \notin L, \eta = 0 \tag{6.4.5}$$

By letting $a_S \to 1$ the static case is obtained. Obviously then, these equations agree exactly with those for the static case, if $a_S W$ is replaced by w, ξ by x and η by y. This implies that the solution of any dynamic steady state problem, involving mode III cracks moving on the symmetry plane, and allowing for constant slip deposition behind each trailing edge, is found from the corresponding static solution simply by the replacements

$$x \to \xi X, \quad y \to \eta = a_S Y, \quad w(x,y) \to a_S W(\xi, \eta) = a_S w(X, Y),$$
$$\tau_{yz} \to \tau_{yz}, \quad \tau_{xz} = \mu[\partial(a_S W)/\partial \xi]/a_S \to a_S \tau_{xz}$$

or, if the static solution is written in cylindrical coordinates, r, φ, z,

$$r \to \rho, \quad \varphi \to \omega, \quad w(r, \varphi) \to a_S w(\rho, \omega)$$

where ρ, ω are polar coordinates in the $\xi\eta$-plane:

$$\rho = \sqrt{X^2 + a_S^2 Y^2} = r\sqrt{1 - \gamma^2 \sin^2 \varphi} \tag{6.4.6}$$

$$\tan \omega = \tan \frac{a_S Y}{X} = \tan(a_S \tan \varphi) \tag{6.4.7}$$

Note that this is a stronger result than for modes I and II, where the correspondence between static and dynamic steady state results is limited to the symmetry plane. Thus, for instance, the dynamic steady state asymptotic stress field at a mode III crack edge is found from the static results (3.3.36), after making the appropriate substitutions, as

$$\tau_{xz} = -\frac{K_{III}}{a_S\sqrt{2\pi\rho}} \sin \frac{\omega}{2} = -\frac{K_{III} Y_{III}(\gamma)}{\sqrt{2\pi r}} F_1(\gamma, \varphi) \tag{6.4.8}$$

$$\tau_{yz} = \frac{K_{III}}{\sqrt{2\pi\rho}} \cos \frac{\omega}{2} = \frac{K_{III}}{\sqrt{2\pi r}} F_2(\gamma, \varphi) \tag{6.4.9}$$

where $F_1(\cdot, \cdot)$ and $F_2(\cdot, \cdot)$ are given by (6.2.67) and (6.2.63), and, by analogy with mode I and II cases, a velocity-dependent function

$$Y_{III}(\gamma) = \frac{1}{a_S} = \frac{1}{\sqrt{1-\gamma^2}} \tag{6.4.10}$$

is introduced. Note that $Y_{III}(\gamma)$ is a universal function, whereas the corresponding functions for modes I and II, Y_I and Y_{II} depend on the material through the ratio k between S and P wave velocities.

The asymptotic displacement field is found from (3.3.37) as

$$w = \frac{K_{III}}{a_S \mu}\sqrt{\frac{2\rho}{\pi}}\sin\frac{\omega}{2} = \frac{K_{III}Y_{III}(\gamma)}{\mu}\sqrt{\frac{r}{\pi}}\sqrt{\sqrt{1-\gamma^2\sin^2\varphi}-\cos\varphi} \qquad (6.4.11)$$

Dynamic stress intensity factors are found from the expressions

$$\boxed{K_{III} = \lim_{r\to 0}(\sqrt{2\pi r}\tau_{yz})_{y=0}} \qquad (6.4.12)$$

$$\boxed{K_{III} = \frac{\mu}{Y_{III}(\gamma)}\lim_{r\to 0}\left(\sqrt{\frac{\pi}{2r}}w_+\right)_{y=0}} \qquad (6.4.13)$$

where r is forwards distance from the crack edge. These expressions are generally valid for constant or continuously changing crack velocity.

By analogy with mode I and II cases, the energy flux to an infinitesimally small process region is found from the static expression (3.5.16) by multiplication with $Y_{III}(\gamma)$:

$$\boxed{\mathcal{G} = \frac{K_{III}^2 Y_{III}(\gamma)}{2\mu}} \qquad (6.4.14)$$

In a similar way as for the asymptotic fields the solution for $Y = 0$ to the moving mode III Yoffe crack, with energy-neutral healing, is obtained from the static solutions (4.3.45) and (4.3.47), as

$$\tau_{yz} = \tau_{yz}^\infty\sqrt{\frac{X+a}{X-a}} \quad \text{for } |X| > a \qquad (6.4.15)$$

$$w_+(x) = \frac{\tau_{yz}^\infty Y_{III}(\gamma)}{\mu}[\sqrt{a^2-X^2}+a\cdot\mathrm{acos}(X/a)] \quad \text{for } |X| < a \qquad (6.4.16)$$

By a slight modification, a static/dynamic correspondence is obtained also in the general case of non-uniformly deposited slip, i.e. when $\partial w_+/\partial x$ equals a prescribed function $w_0'(X)$ on $Y = 0$. Such a non-uniform slip distribution might be a result of slip depositions during previous events.

For future reference, a remarkable property of the asymptotic field will be demonstrated. Consider a crack edge vicinity, in which steady state prevails during a certain time interval, and in which the inverse square-root singular term is dominating so that other terms may be neglected. Then, the displacement at a radius $c_S t$ from the crack edge position at time $t = 0$ may be written in the form

$$w = \frac{K_{III}(\gamma)}{\mu\sqrt{1-\gamma}}\sqrt{\frac{2c_S t}{\pi}}\sin\frac{\varphi}{2} \qquad (6.4.17)$$

where $K_{III}(\gamma)$ is the stress intensity factor for the moving crack (with velocity $V = \gamma c_S$). Note that the expression agrees exactly with the displacement at radius $r = c_S t$ for a stationary crack with stress intensity factor

$$K_{III}(0) = \frac{K_{III}(\gamma)}{\sqrt{1-\gamma}} \qquad (6.4.18)$$

6.4 STEADY STATE MODE III CRACK PROPAGATION

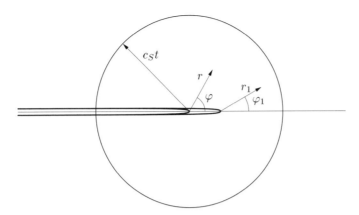

Fig. 6.4.1 Cylindrical coordinates r_1 and φ_1 with origin at the crack edge position at time t, in relation to r and φ with origin at the crack edge position at time zero. The circle shows the position of the front of the cylindrical wave radiating out from the crack edge at time zero.

and that it may be interpreted as the displacement radiated out with velocity c_S from the crack edge.

This result follows, of course, from the displacement field at time t on the circular cylinder with radius $c_S t$ and with its axis located where the crack edge was at time $t = 0$. Two cylindrical coordinate systems are introduced, one, r, φ, z, with axis along the crack edge position at $t = 0$, and the other, r_1, φ_1, z, with axis along the crack edge position at time t; see Fig. 6.4.1. Then, from geometrical relations,

$$r_1 \cos \varphi_1 = r \cos \varphi - \gamma c_S t \qquad (6.4.19)$$
$$r_1 \sin \varphi_1 = r \sin \varphi \qquad (6.4.20)$$

so that r_1 and φ_1 may be expressed in terms of r, φ and γ. Introduce now temporarily $\alpha = c_S t/r$, so that $\alpha = 1$ corresponds to the position of the cylindrical wave that radiates out from the crack edge at time zero. Insertion into (6.4.11) gives (6.4.17)-(6.4.18) when $\alpha = 1$. In fact, noticing that

$$\frac{r_1^2}{r^2} = (1 - \alpha\gamma \cos \varphi)^2 + \alpha^2 \gamma^2 \sin^2 \varphi \qquad (6.4.21)$$

$$\cos \varphi_1 = \frac{r}{r_1}(\cos \varphi - \alpha\gamma) \qquad (6.4.22)$$

(6.4.11), after replacing r by r_1 and φ by φ_1, gives

$$w = \frac{K_{III} Y_{III}(\gamma)}{\mu} \sqrt{\frac{r}{\pi}} F(\alpha) \qquad (6.4.23)$$

where

$$F(\alpha) = \sqrt{\sqrt{(1-\alpha\gamma \cos \varphi)^2 - (1-\alpha^2)\gamma^2 \sin^2 \varphi} + \alpha\gamma - \cos \varphi} \qquad (6.4.24)$$

Furthermore,
$$\tau_{rz}(r,\varphi,t) = \mu\frac{\partial w}{\partial r} = K_{III}Y_{III}\cdot\frac{F(\alpha)}{2\sqrt{\pi r}} - K_{III}Y_{III}\cdot\frac{c_S t}{r\sqrt{\pi r}}\cdot\frac{\partial F(\alpha)}{\partial \alpha} \quad (6.4.25)$$

For $\alpha = 1$, the last term coincides with $-\rho c_S \partial w/\partial t$, because $\rho c_S = \mu/c_S$. Thus, for $r = c_S t$,

$$(\tau_{rz})_{r=c_S t} = (\tau_{rz}^0)_{r=c_S t} - \rho c_S \frac{\partial w}{\partial t} \quad (6.4.26)$$

where

$$(\tau_{rz}^0)_{r=c_S t} = \frac{K_{III}(\gamma)}{\sqrt{2(1-\gamma)\pi c_S t}}\sin\frac{\varphi}{2} \quad (6.4.27)$$

coincides with the expression for the stress τ_{rz} near a stationary crack edge with stress intensity factor $K_{III}(0)$ as given by (6.4.18). Relations (6.4.17) and (6.4.26) show, as will be discussed on page 462, that a static stress field is radiated out with the S wave velocity from a crack that is suddenly arrested.

6.5 Steady state crack propagation in anisotropic materials

Steady state crack propagation under generalized plane strain in anisotropic materials may be treated by using the Lekhnitskii-Stroh formalism, in essentially the same way as for correponding static cases. Here, the stress and displacement fields are sought in the vicinity of a crack edge, travelling with constant velocity V in the positive x_1 direction with the crack edge at $x_1 = Vt$. In reality, such a propagation requires in general that $x_2 = 0$ is a weak plane. The study will be performed by assuming that the crack faces are traction free and that the stresses vanish at infinity. This problem was solved by Wu (1989).

By analogy with the corresponding static case, page 216ff., it is found that

$$\sigma_{ij} = C_{ijkl}\frac{\partial u_k}{\partial x_l}, \quad (i,j,k,l = 1,2,3) \quad (6.5.1)$$

$$\frac{\partial \sigma_{ij}}{\partial x_j} = C_{ijkl}\frac{\partial^2 u_k}{\partial x_j \partial x_l} = \varrho\frac{\partial^2 u_i}{\partial t^2} \quad (6.5.2)$$

$$\frac{\partial u_k}{\partial x_3} = 0 \quad (6.5.3)$$

Still by analogy, the obvious ansatz is

$$u_k = a_k f(x_1 - Vt + px_2) = a_k f(X_1 + pX_2) = a_k f(\mathcal{Z}) \quad (6.5.4)$$

where $X_1 = x_1 - Vt$, $X_2 = x_2$ (the Galilean transformation), and the symbol $\mathcal{Z} = X_1 + pX_2$, corresponding to ζ in the static case, is introduced to avoid confusion, because the corresponding Greek capital letter is indistinguishable from latin letter $Z = X_1 + iX_2$. Insertion into (6.5.2) gives

$$f''(\mathcal{Z})[C_{i1k1} - \varrho V^2 \delta_{ik} + p(C_{i1k2} + C_{i2k1}) + p^2 C_{i2k2}]a_k = 0 \quad (6.5.5)$$

with nontrivial solutions a_k if

$$\det[C_{i1k1} - \varrho V^2 \delta_{ik} + p(C_{i1k2} + C_{i2k1}) + p^2 C_{i2k2}] = 0 \qquad (6.5.6)$$

This is a sixth degree equation for p, but in contrast to the static case, p will now depend not only on the material and its orientation, but also on the crack velocity V. Moreover, the static result that no root p is real does not hold for arbitrary V, but only for V smaller than some critical value V_c. Here, only $V < V_c$ is considered. Then, the roots to (6.5.6) come in conjugated pairs, p_α and $\overline{p_\alpha}$, with p_α denoting the three roots with positive imaginary part. They are assumed to be distinct as in the static case – possible equal roots are considered as limiting cases of distinct roots. Furthermore, the corresponding values of a_k will be denoted by $A_{k\alpha}$ and $\overline{A_{k\alpha}}$. Note that $A_{k\alpha}$ is a function of p_α and thus velocity dependent.

In the continuation summation over α is always indicated explicitly.

By analogy with the static case, the displacement u_k is written as

$$u_k = \Re \sum_{\alpha=1}^{3} A_{k\alpha} f_\alpha(Z_\alpha) q_\alpha \qquad (6.5.7)$$

where the introduction of q_α allows $A_{k\alpha}$ to be suitably normalized. Then,

$$\sigma_{ij} = C_{ijkl} \frac{\partial u_k}{\partial X_l} = \Re \sum_{\alpha=1}^{3} C_{ijkl} A_{k\alpha} f'_\alpha(Z_\alpha) \frac{\partial Z_\alpha}{\partial X_l} q_\alpha$$

$$= \Re \sum_{\alpha=1}^{3} (C_{ijk1} + p_\alpha C_{ijk2}) A_{k\alpha} f'_\alpha(Z_\alpha) q_\alpha \qquad (6.5.8)$$

and, in particular,

$$\sigma_{1i} = \Re \sum_{\alpha=1}^{3} (C_{i1k1} + p_\alpha C_{i1k2}) A_{k\alpha} f'_\alpha(Z_\alpha) q_\alpha \qquad (6.5.9)$$

$$\sigma_{2i} = \Re \sum_{\alpha=1}^{3} (C_{i2k1} + p_\alpha C_{i2k2}) A_{k\alpha} f'_\alpha(Z_\alpha) q_\alpha \qquad (6.5.10)$$

Introduce

$$B_{i\alpha} = (C_{i2k1} + p_\alpha C_{i2k2}) A_{k\alpha} = \frac{1}{p_\alpha}[\varrho V^2 A_{i\alpha} - (C_{i1k1} + p_\alpha C_{i1k2}) A_{k\alpha}] \qquad (6.5.11)$$

where the last equality follows from (6.5.5). Then,

$$\sigma_{1i} = \Re \sum_{\alpha=1}^{3} (\varrho V^2 A_{i\alpha} - p_\alpha B_{i\alpha}) f'_\alpha(Z_\alpha) q_\alpha, \quad \sigma_{2i} = \Re \sum_{\alpha=1}^{3} B_{i\alpha} f'_\alpha(Z_\alpha) q_\alpha \qquad (6.5.12)$$

Traction free crack faces and vanishing stresses at infinity imply that

$$\sigma_{2i} = \Re \sum_{\alpha=1}^{3} B_{i\alpha} f'_\alpha(Z_\alpha) q_\alpha = 0 \text{ for } X_1 < 0, X_2 = 0 \qquad (6.5.13)$$

$$f'_\alpha(Z_\alpha) \to 0 \text{ as } |Z_\alpha| \to \infty \qquad (6.5.14)$$

The only unknowns are $f_\alpha(\mathcal{Z}_\alpha)$ and q_α. The obvious ansatz for the former is

$$f'_\alpha(\mathcal{Z}_\alpha) = \mathcal{Z}^s \tag{6.5.15}$$

where s is a constant. A branch cut is made along $X_1 < 0$, $X_2 = 0$ and the branch is chosen so that $\mathcal{Z}^s = \exp[s\ln(X_1)]$ for $X_1 > 0$, $X_2 = 0$. Insertion into the condition for traction free crack faces gives

$$\sigma_{2i} = \Re \sum_{\alpha=1}^{3} B_{i\alpha}|X_1|^s e^{\pm i\pi s} q_\alpha = 0 \text{ for } X_1 < 0, X_2 = \pm 0 \tag{6.5.16}$$

This holds for all $X_1 < 0$ only if $B_{i\alpha}q_\alpha \exp(\pm i\pi s) + \overline{B_{i\alpha}q_\alpha}\exp(\mp i\pi s) = 0$, where the upper sign refers to the upper crack face, the lower sign to the lower crack face. Thus,

$$\overline{B_{i\alpha}q_\alpha} = -B_{i\alpha}q_\alpha e^{2i\pi s} = -B_{i\alpha}q_\alpha e^{-2i\pi s} \tag{6.5.17}$$

which is satisfied only for $s = 0, \pm 1/2, \pm 1, \ldots$. The dominant singularity, which does not violate the condition of bounded stress-strain energy, is $s = -1/2$. Note that this implies that $B_{i\alpha}q_\alpha$ is real.

Now, for $X_2 = 0$,

$$\sigma_{2i} = \frac{1}{\sqrt{X_1}} B_{i\alpha}q_\alpha = \frac{k_i}{\sqrt{2\pi X_1}} \text{ for } X_1 > 0 \tag{6.5.18}$$

where the real constants k_i are introduced by analogy with the stress intensity factors in the isotropic case, i.e. $k_1 = K_{II}$, $k_2 = K_I$, $k_3 = K_{III}$. However, the subscripts I, II and III do not imply modes with symmetry properties as in the isotropic case.

Obviously, $\sum_{\alpha=1}^{3} B_{i\alpha}q_\alpha = k_i/\sqrt{2\pi}$, so that $q_\alpha = B_{\alpha i}^{-1}k_i/\sqrt{2\pi}$, where $B_{\alpha i}^{-1}$ is the inverse of $B_{i\alpha}$, i.e. $\sum_{\alpha=1}^{3} B_{i\alpha}B_{\alpha j}^{-1} = \delta_{ij}$.

The stresses and the displacements are now found to be

$$\sigma_{2i} = \frac{1}{\sqrt{2\pi}}\Re \sum_{\alpha=1}^{3} B_{i\alpha}B_{\alpha j}^{-1}k_j \mathcal{Z}_\alpha^{-1/2} \tag{6.5.19}$$

$$\sigma_{1i} = \frac{1}{\sqrt{2\pi}}\Re \sum_{\alpha=1}^{3} [(\varrho V^2 A_{i\alpha} - p_\alpha B_{i\alpha})B_{\alpha j}^{-1}k_j \mathcal{Z}_\alpha^{-1/2}] \tag{6.5.20}$$

$$u_i = \sqrt{\frac{2}{\pi}}\Re \sum_{\alpha=1}^{3} A_{i\alpha}B_{\alpha j}^{-1}k_j \mathcal{Z}_\alpha^{1/2} \tag{6.5.21}$$

and

$$\sigma_{33} = C_{33kl}\frac{\partial u_k}{\partial X_l} \tag{6.5.22}$$

The traction vector on the lower half of the body equals

$$\mathbf{t} = \frac{1}{\sqrt{2\pi X_1}}\mathbf{k} \text{ for } X_1 > 0, X_2 = 0 \tag{6.5.23}$$

and the displacement jump across the crack is

$$\mathbf{u}_+ - \mathbf{u}_- = 2\sqrt{\frac{2|X_1|}{\pi}}\Re[i\mathbf{A}\mathbf{B}^{-1}]\mathbf{k} \text{ for } X_1 < 0, X_2 = 0 \tag{6.5.24}$$

Here, $\mathbf{k}$ is the vector with elements k_i, and $\mathbf{A}$ and $\mathbf{B}$ are the matrices with elements A_{ij} and B_{ij}, respectively.

The energy flux into the crack edge is found by using (3.5.20), which was shown by Broberg (1964) to hold also in the dynamic case. The treatment follows closely the one for static cracks in anisotropic media, which gave the result (4.14.111). Thus,

$$\mathcal{G} = \frac{1}{2}\mathbf{k}^T \mathbf{L}^{-1} \mathbf{k} \tag{6.5.25}$$

where $\mathbf{L}^{-1} = -\Im(\mathbf{AB}^{-1})$, by analogy with the static case, although $\mathbf{L}^{-1}$ is now dependent on V and in general not a tensor. Crack propagation requires that $\mathcal{G}$ is positive, which it is, irrespective of $\mathbf{k}$, if $\mathbf{L}^{-1}$ is positive definite. This requirement limits the validity of the analysis to an interval $0 < V < c_*$, where c_* is the velocity for which $\mathcal{G}$ vanishes. Compare the isotropic mode I case, in which the analysis may be formally valid for all crack velocities smaller than the S wave velocity, but physically unacceptable for velocities above the Rayleigh wave velocity. Note, however, that the analysis is only valid in the interval $0 < V < V_c$ if $V_c \leq c_*$.

As a simple illustrative example, a transversely isotropic solid was chosen by Wu (1989). With the transverse isotropy in planes parallel with the x_3 plane, the non-vanishing elastic stiffnesses are

$$C_{1111} = C_{2222}, \quad C_{1212} = (C_{1111} - C_{1122})/2,$$
$$C_{1313} = C_{2323}, \quad C_{1133} = C_{2233}, \quad C_{3333} \tag{6.5.26}$$

The last three are excluded from (6.5.6), which thus may be written in terms of C_{1111}, C_{1212} and C_{1313}, only. All three solutions for p^2 turn out to be real and negative for sufficiently small V. With $p_\alpha = i\beta_\alpha$, $\beta_\alpha > 0$, they† are given by

$$\beta_1 = \sqrt{1 - \frac{V^2}{c_1^2}}, \quad \beta_2 = \sqrt{1 - \frac{V^2}{c_2^2}}, \quad \beta_3 = \sqrt{1 - \frac{V^2}{c_3^2}} \tag{6.5.27}$$

where

$$c_1 = \sqrt{\frac{C_{1111}}{\varrho}}, \quad c_2 = \sqrt{\frac{C_{1212}}{\varrho}}, \quad c_3 = \sqrt{\frac{C_{1313}}{\varrho}} \tag{6.5.28}$$

These velocities are the three propagation velocities for plane waves with propagation directions parallel with $x_3 = 0$, the first one (c_1) is the velocity of a longitudinal wave and the two other are velocities of transverse waves, one (c_2) with displacements parallel to $x_3 = 0$ and the other (c_3) with displacements perpendicular to $x_3 = 0$. This may be found by putting the propagation direction $g_i = 1$ in (A9.72) to give c_i, $i = 1, 2, 3$.

The matrices $\mathbf{A}$, $\mathbf{B}$, $\mathbf{B}^{-1}$ and $\mathbf{L}^{-1}$ are found to be

$$\mathbf{A} = \begin{bmatrix} 1 & -i\beta_2 & 0 \\ i\beta_1 & 1 & 0 \\ 0 & 0 & 1 \end{bmatrix}, \quad \mathbf{B} = C_{1212}\begin{bmatrix} 2i\beta_1 & 1+\beta_2^2 & 0 \\ -(1+\beta_2^2) & 2i\beta_2 & 0 \\ 0 & 0 & i\frac{C_{1313}}{C_{1212}}\beta_3 \end{bmatrix} \tag{6.5.29}$$

---

† The notation β_α, which is commonly used for the imaginary part of p_α, should not be confused with the dimensionless velocity β, frequently appearing in the present chapter.

$$\boldsymbol{B}^{-1} = -\frac{1}{C_{1212}} \begin{bmatrix} \dfrac{2i\beta_2}{R(\beta_1,\beta_2)} & -\dfrac{1+\beta_2^2}{R(\beta_1,\beta_2)} & 0 \\ \dfrac{1+\beta_2^2}{R(\beta_1,\beta_2)} & \dfrac{2i\beta_1}{R(\beta_1,\beta_2)} & 0 \\ 0 & 0 & \dfrac{iC_{1212}}{C_{1313}\beta_3} \end{bmatrix} \qquad (6.5.30)$$

$$\boldsymbol{L}^{-1} = \begin{bmatrix} \dfrac{\beta_2(1-\beta_2^2)}{C_{1212}R(\beta_1,\beta_2)} & 0 & 0 \\ 0 & \dfrac{\beta_1(1-\beta_2^2)}{C_{1212}R(\beta_1,\beta_2)} & 0 \\ 0 & 0 & \dfrac{1}{C_{1313}\beta_3} \end{bmatrix} \qquad (6.5.31)$$

where $R(\beta_1,\beta_2) = 4\beta_1\beta_2 - (1+\beta_2^2)^2$ corresponds to the Rayleigh function $R(a_P, a_S)$ in the isotropic case, cf. (6.2.23).

Stresses, displacements and energy flux may now be obtained from (6.5.19)-(6.5.21) and (6.5.25). Thus, for instance, with polar coordinates (r, φ), such that $X_1 + p_\alpha X_2 = r(\cos\varphi + p_\alpha \sin\varphi) = rC_\alpha(\varphi)$, it is found that

$$\sigma_{22} = \frac{2k_1\beta_2(1+\beta_2^2)}{\sqrt{2\pi r}}\Im\left\{\frac{1}{[C_2(\varphi)]^{1/2}} - \frac{1}{[C_1(\varphi)]^{1/2}}\right\}$$
$$+ \frac{k_2}{\sqrt{2\pi r}}\Re\left\{\frac{4\beta_1\beta_2}{[C_2(\varphi)]^{1/2}} - \frac{(1+\beta_2^2)^2}{[C_1(\varphi)]^{1/2}}\right\} \qquad (6.5.32)$$

$$\sigma_{13} = \frac{k_3}{\sqrt{2\pi r}\beta_3}\Im\frac{1}{[C_3(\varphi)]^{1/2}}, \qquad \sigma_{23} = \frac{k_3}{\sqrt{2\pi r}}\Re\frac{1}{[C_3(\varphi)]^{1/2}} \qquad (6.5.33)$$

$$u_2 = \sqrt{\frac{2}{\pi}}\cdot\frac{1}{C_{1213}R(\beta_1,\beta_2)}\Big[k_1\Re\{2\beta_1\beta_2[C_1(\varphi)]^{1/2} - (1+\beta_2^2)[C_2(\varphi)]^{1/2}\}$$
$$- k_2\beta_1\Im\{(1+\beta_2^2)[C_1(\varphi)]^{1/2} - 2[C_2(\varphi)]^{1/2}\}\Big] \qquad (6.5.34)$$

$$\mathcal{G} = \frac{(k_1^2\beta_2 + k_2^2\beta_1)(1-\beta_2^2)}{2C_{1212}R(\beta_1,\beta_2)} + \frac{k_3^2}{2C_{1313}\beta_3} \qquad (6.5.35)$$

A positive energy flux is obtained for all possible combinations of k_1, k_2 and k_3 if β_1, β_2, β_3 and $R(\beta_1,\beta_2)$ are all positive. This is equivalent to stating that $\boldsymbol{L}^{-1}$ must be positive definite. In terms of crack velocities, this implies that V must be smaller than the smallest of c_1, c_2, c_3 and c_R, where c_R is the velocity of surface waves propagating along the crack faces in the x_2 direction. This velocity is obtained from the equation $R(\beta_1,\beta_2) = 0$. Compare the isotropic case, in which $c_2 = c_3 = c_S < c_P = c_1$ and where c_R, the Rayleigh wave velocity, is smaller than c_S.

6.6 Steady state crack propagation in viscoelastic materials

The analysis of dynamic crack propagation in viscoelastic materials cannot, in general, avail of the convenient elastic-viscoelastic correspondence principle, which was used in Section 4.15; see page 239. Consequently, only few problems have been solved, first by Willis (1967) and later by Atkinson and Coleman (1977), Atkinson and Popelar (1979), Popelar and Atkinson (1980) and Walton (1982, 1985, 1987); see also Atkinson (1979). Early results for crack propagation and fracture in viscoelastic materials were obtained by Williams (1963) and Knauss (1974). Energy relations were studied by Willis (1967), Kostrov and Nikitin (1970), Knauss (1973), Schapery (1975), Atkinson and Popelar (1979), Walton (1987) and Golden and Graham (1990). Crack propagation along the interface between two Maxwell materials was considered by Sills and Benveniste (1981). With few exceptions, available solutions for viscoelastic dynamic crack propagation concern mode III. Modes I and II present considerable difficulties. Thus, for instance, more than one Rayleigh (surface) wave may be possible in viscoelastic materials (Currie et al. 1977).

Formulation of a Wiener-Hopf equation for mode III crack propagation

Here, steady state mode III crack propagation will be considered, using the problem of a semi-infinite crack, running with constant velocity V in an infinite viscoelastic body in the plane $y = 0$, with the crack edge along $x = Vt$. The crack is driven by crack face loads $\tau_{yz} = \tau_{yz}^0(x - Vt)$, moving with the crack edge. The boundary conditions are

$$\tau_{yz} = \tau_{yz}^0(x - Vt) \text{ for } x < Vt, \qquad w = 0 \text{ for } x > Vt \qquad (6.6.1)$$

where w is the displacement in the z direction. It is also assumed that stresses vanish at infinity. This is the problem that was solved by Willis (1967) for the standard linear solid and by Walton (1982) for a general viscoelastic material.

After specialization to anti-plane strain, the constitutive equation (4.15.2) reads

$$\tau_{xz} = \int_{-\infty}^{+\infty} \mu(t-\tau) U(t-\tau) \frac{\partial \dot{w}(x,y,\tau)}{\partial x} \mathrm{d}\tau \qquad (6.6.2)$$

$$\tau_{yz} = \int_{-\infty}^{+\infty} \mu(t-\tau) U(t-\tau) \frac{\partial \dot{w}(x,y,\tau)}{\partial y} \mathrm{d}\tau \qquad (6.6.3)$$

where a dot denotes differentiation with respect to τ, and the unit step function serves as a reminder that the stiffness function $\mu(t)$ vanishes for negative argument.

Insertion into the equation of motion,

$$\frac{\partial \tau_{xz}}{\partial x} + \frac{\partial \tau_{yz}}{\partial y} = \varrho \frac{\partial^2 w}{\partial t^2} \qquad (6.6.4)$$

where ϱ is the density, gives

$$\int_{-\infty}^{+\infty} \mu(t-\tau) U(t-\tau) \left[\frac{\partial^2 \dot{w}(x,y,\tau)}{\partial x^2} + \frac{\partial^2 \dot{w}(x,y,\tau)}{\partial y^2} \right] \mathrm{d}\tau = \varrho \frac{\partial^2 w(x,y,t)}{\partial t^2} \qquad (6.6.5)$$

Introduce the Galilean transformation,

$$X = x - Vt, \quad y = Y, \quad \xi = x - V\tau, \quad \eta = y \tag{6.6.6}$$

Then, the equation of motion becomes

$$-\int_{-\infty}^{+\infty} \mu\left(\frac{X-\xi}{-V}\right) U\left(\frac{X-\xi}{-V}\right) \frac{\partial}{\partial \xi}\left[\frac{\partial^2 w(\dot{\xi},\eta)}{\partial \xi^2} + \frac{\partial^2 w(\dot{\xi},\eta)}{\partial \eta^2}\right] d\xi = \varrho V^2 \frac{\partial^2 w(X,Y)}{\partial X^2} \tag{6.6.7}$$

where the same function symbol w is used as previously, but now with two arguments. Laplace transformation with respect to X gives

$$M(-qV)\left(q^2 W + \frac{\mathrm{d}^2 W}{\mathrm{d}Y^2}\right) = \varrho V^2 q^2 W, \quad \Re(q) = 0 \tag{6.6.8}$$

where it is assumed that the strip of convergence contains the imaginary axis, and

$$W = W(q, Y) = q \int_{-\infty}^{+\infty} e^{-qX} w(X, Y) \mathrm{d}X \tag{6.6.9}$$

$$M(p) = p \int_0^\infty e^{-pt} \mu(t) \mathrm{d}t, \quad \Re(p) \geq 0 \tag{6.6.10}$$

Thus, $M(-qV)$ is regular for $\Re(q) \leq 0$, and is therefore, for clarity, written as $M_-(-qV)$.

Equation (6.6.8) is now written in the form

$$\frac{\mathrm{d}^2 W}{\mathrm{d}Y^2} = \left[\gamma^2 \frac{\mu(0)}{M_-(-qV)} - 1\right] q^2 W \tag{6.6.11}$$

where

$$\gamma = \frac{V}{c_S^0}, \quad c_S^0 = \sqrt{\frac{\mu(0)}{\varrho}} \tag{6.6.12}$$

c_S^0 is the highest anti-plane wave propagation in the solid, and γ is the dimensionless crack velocity, normalized with respect to c_S^0, so that $\gamma^2 < 1$. As will be evident later, there exists another propagation velocity in anti-plane strain, $c_S^\infty = \sqrt{\mu(\infty)/\varrho} < c_S^0$. Thus, the highest wave propagation velocity is determined by the glassy modulus and the lowest by the rubbery modulus. Consequently, there exists one subsonic velocity region for crack propagation, $V < c_S^\infty$, and one intersonic region, $c_S^\infty < V < c_S^0$.

The solution of (6.6.11) is

$$W = A(q)e^{-\lambda Y} + B(q)e^{+\lambda Y}, \quad \lambda = \left[\gamma^2 \frac{\mu(0)}{M_-(-qV)} - 1\right]^{1/2} q, \quad \Re(q) = 0 \tag{6.6.13}$$

Consider now the upper half of the body, $Y \geq 0$, and assume that, after suitable branch cuts in the q plane, a branch can be found so that $\lambda = \lambda(q)$ has non-negative real part for $\Re(q) = 0$. Then, $B(q) = 0$, which implies that

$$W(q, Y) = W(q, 0)e^{-\lambda Y}, \quad \Re(q) = 0 \tag{6.6.14}$$

A Wiener-Hopf technique will be used. Proceed therefore as if $\tau_{yz}(X,0)$ is known and equal to $\tau_{yz}^+(X)$ for $X > 0$ and $w(X,0)$ is known and equal to $w_-(X)$ for $X < 0$.

6.6 STEADY STATE CRACK PROPAGATION IN VISCOELASTIC MATERIALS

The Laplace transform of $\tau_{yz}(X,Y)$ is found from (6.6.3) by using the procedure leading to (6.6.8). It is

$$T_{yz}(q,Y) = M_-(-qV)\frac{\mathrm{d}W}{\mathrm{d}Y}, \quad \Re(q) = 0 \tag{6.6.15}$$

The boundary condition (6.6.1) then gives

$$M_-(-qV)\frac{\mathrm{d}W}{\mathrm{d}Y} = T_{yz}^-(q) + T_{yz}^+(q), \quad W(q,0) = W_-(q), \quad \Re(q) = 0 \tag{6.6.16}$$

where

$$T_{yz}^-(q) = q\int_{-\infty}^0 e^{-qX}\tau_{yz}^0(X)\mathrm{d}X \quad \Re(q) \leq 0 \tag{6.6.17}$$

$$T_{yz}^+(q) = q\int_0^\infty e^{-qX}\tau_{yz}^+(X)\mathrm{d}X \quad \Re(q) \geq 0 \tag{6.6.18}$$

$$W_-(q) = q\int_{-\infty}^0 e^{-qX}w_-(X)\mathrm{d}X \quad \Re(q) \leq 0 \tag{6.6.19}$$

Use of expression (6.6.14) leads to the Wiener-Hopf equation

$$-\lambda(q)M_-(-qV)W_-(q) = T_{yz}^-(q) + T_{yz}^+(q), \quad \Re(q) = 0 \tag{6.6.20}$$

Solution for the standard linear solid

Specialize now to the standard linear solid, see page 238,

$$\mu(t) = \mu_\infty + (\mu_0 - \mu_\infty)e^{-t/t_0} \tag{6.6.21}$$

$$M(p) = \mu_0\frac{p+a}{p+b}, \quad a = \frac{\mu_\infty}{\mu_0 t_0}, \quad b = \frac{1}{t_0}, \quad \Re(q) = 0 \tag{6.6.22}$$

Note that $a < b$, because $\mu_\infty < \mu_0$. Insertion into (6.6.13) gives

$$\lambda^2 = -(1-\gamma^2)\frac{q^2(q-g)}{q-h}, \quad \lambda = \pm i\sqrt{1-\gamma^2}\frac{q(q-g)^{1/2}}{(q-h)^{1/2}}, \quad \Re(q) = 0 \tag{6.6.23}$$

where $g = (a - \gamma^2 b)/[(1-\gamma^2)V] = (a/V)(1-\gamma^2 b/a))/(1-\gamma^2) < a/V = h$. Note that $g > 0$ if $a - \gamma^2 b > 0$, which implies $\gamma^2 < a/b = \mu_\infty/\mu_0$, i.e., $V < \sqrt{\mu_\infty/\varrho} = c_S^\infty$. Thus, $g > 0$ in the subsonic region, $V < c_S^\infty$, and $g < 0$ in the intersonic region, $c_S^\infty < V < c_S^0$, whereas $h > 0$ and $h > g$ for all V.

In the following, a function of type $(q - c)^{1/2}$, where c is a positive constant, will be defined by a branch cut from $q = c$ to $q = \infty$, with the branch chosen so that $(q-c)^{1/2} = i\sqrt{c-q}$ for $q = \Re(q) < c$. It is then analytic in the left half-plane and will therefore, for clarity, be denoted $(q-c)_-^{1/2}$. If $c < 0$, the function will be defined by a branch cut from $q = c$ to $q = -\infty$, with the branch chosen so that $(q-c)^{1/2} = \sqrt{q-c}$ for $q = \Re(q) > c$. It is analytic in the right half-plane and will be denoted $(q-c)_+^{1/2}$.

Because the sign of g differs between the subsonic and intersonic cases, the way to ensure that $\lambda(q)$ is analytic with real non-negative part for $\Re(q) = 0$ is also different. The intersonic case, for which $g < 0$, is quite simple in this respect: it is sufficient to choose the negative sign in (6.6.23). Note that the factor $-iq$ alone does not possess the required property: it equals $|q|$ for $\Im(q) > 0$, but $-|q|$ for $\Im(q) < 0$. The factor

Fig. 6.6.1 Branch cuts in the q plane for the subsonic (left) and intersonic (right) cases.

$(q-g)_+^{1/2}/(q-h)_-^{1/2}$ switches the sign of the real part of $\lambda(q)$ along the imaginary axis at $q = 0$. In the subsonic case, $g > 0$, and then the factor $(q-g)_-^{1/2}/(q-h)_-^{1/2}$ does not possess this property, which, however, may be introduced through multiplication of the right member of (6.6.11) by the factor $(q+\epsilon)/(q-\epsilon)$, where ϵ is a temporarily used positive quantity, smaller than g. This implies multiplication by a factor $(q+\epsilon)_+^{1/2}/(q-\epsilon)_-^{1/2}$ in the expression for $\lambda(q)$, which then will be non-negative along the imaginary axis by choosing the negative sign in (6.6.23). After a solution has been obtained, the original problem is restored by letting $\epsilon \to 0$.

The branch cuts are shown in Fig. 6.6.1 for the subsonic and intersonic cases. These cases will be treated separately.

1. The subsonic case, $V < c_S^\infty$. Here, $g > 0$ and

$$\lambda = \lambda(q) = -i\sqrt{1-\gamma^2}\frac{q(q+\epsilon)_+^{1/2}(q-g)_-^{1/2}}{(q-\epsilon)_-^{1/2}(q-h)_-^{1/2}}, \quad \Re(q) = 0 \qquad (6.6.24)$$

Solution of (6.6.20) requires the factorization $\lambda(q) = \lambda_+(q)\lambda_-(q)$, where

$$\lambda_+(q) = (q+\epsilon)_+^{1/2}, \quad \lambda_-(q) = -i\sqrt{1-\gamma^2}\frac{q(q-g)_-^{1/2}}{(q-\epsilon)_-^{1/2}(q-h)_-^{1/2}} \qquad (6.6.25)$$

Then, from (6.6.20),

$$-\lambda_-(q)M_-(-qV)W_-(q) = \frac{T_{yz}^+(q)}{\lambda_+(q)} + \frac{T_{yz}^-(q)}{\lambda_+(q)}, \quad \Re(q) = 0 \qquad (6.6.26)$$

where the last term has to be decomposed into one part, $A_-(q)$, analytic in the left half-plane, and one part, $A_+(q)$, analytic in the right half-plane. This is achieved by using Cauchy's integral formula, see Appendix A4. The contour of integration is chosen as shown in Fig. 6.6.2. After expanding the path to infinity, the result is found to be

$$A_-(q) = \frac{T_{yz}^-(q)}{(q+\epsilon)_+^{1/2}} + \frac{1}{\pi}\int_{-\infty}^{-\epsilon}\frac{T_{yz}^-(s)}{\sqrt{-s-\epsilon}(s-q)}ds$$

$$= \frac{1}{\pi}\int_{-\infty}^{-\epsilon}\frac{T_{yz}^-(s) - T_{yz}^-(q)}{\sqrt{-s-\epsilon}(s-q)}ds, \quad \Re(q) \leq 0 \qquad (6.6.27)$$

$$A_+(q) = -\frac{1}{\pi}\int_{-\infty}^{-\epsilon}\frac{T_{yz}^-(s)}{\sqrt{-s-\epsilon}(s-q)}ds, \quad \Re(q) \geq 0 \qquad (6.6.28)$$

6.6 STEADY STATE CRACK PROPAGATION IN VISCOELASTIC MATERIALS

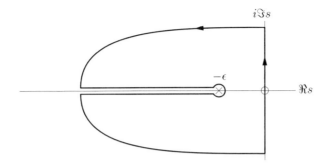

Fig. 6.6.2 Contour of integration.

and then the Wiener-Hopf equation can be written in the form

$$-\lambda_-(q)M_-(-qV)W_-(q) - A_-(q) = \frac{T_{yz}^+(q)}{\lambda_+(q)} + A_+(q), \quad \Re(q) = 0 \quad (6.6.29)$$

so that the left member is analytic for $\Re(q) \leq 0$ and the right member is analytic for $\Re(q) \geq 0$. This implies that the two members together define a function that is analytic in the whole plane, and thus, by Liouville's theorem (Appendix A4) equal to a polynomial of finite degree, $K_0 + K_1 q + \cdots + K_n q^n$. This gives

$$W_-(q) = -\frac{K_0 + K_1 q + \cdots + K_n q^n + A_-(q)}{\lambda_-(q)M_-(-qV)}, \quad \Re(q) \leq 0 \quad (6.6.30)$$

$$T_{yz}^+(q) = [K_0 + K_1 q + \cdots + K_n q^n - A_+(q)]\lambda_+(q), \quad \Re(q) \geq 0 \quad (6.6.31)$$

The coefficients K_i are to be determined from the conditions that $w(X,0) \to 0$ as $X \to 0$ and that $\tau_{yz}(X,0) \to 0$ as $X \to \infty$. According to Abel theorems, Appendix A5, this corresponds to $W_-(-\infty) = 0$ and $T_{yz}^+(0) = 0$, which gives

$$W_-(q) \to -\frac{K_n q^n}{\sqrt{1-\gamma^2}\mu(0)(-q)^{1/2}} \to 0 \text{ as } q \to -\infty \Longrightarrow n = 0 \quad (6.6.32)$$

$$T_{yz}^+(0) = [K_0 - A_+(0)]\sqrt{\epsilon} = 0 \Longrightarrow K_0 = A_+(0) \quad (6.6.33)$$

Thus,

$$W_-(q) = -\frac{A_-(q) + A_+(0)}{\lambda_-(q)M_-(-qV)}, \quad \Re(q) \leq 0 \quad (6.6.34)$$

$$T_{yz}^+(q) = -[A_+(q) - A_+(0)]\lambda_+(q), \quad \Re(q) \geq 0 \quad (6.6.35)$$

Insertion of the expressions for $A_+(q) - A_+(0)$ and $\lambda_+(q)$ into the last equation gives, for $\Re(q) \geq 0$,

$$T_{yz}^+(q) = \frac{q(q+\epsilon)_+^{1/2}}{\pi} \int_{-\infty}^{-\epsilon} \frac{T_{yz}^-(s)}{s\sqrt{-s-\epsilon}(s-q)} ds$$

$$= \frac{q(q+\epsilon)_+^{1/2}}{\pi} \int_{-\infty}^{-\epsilon} \int_{-\infty}^{0} e^{-su} \tau_{yz}^0(u) du \cdot \frac{ds}{\sqrt{-s-\epsilon}(s-q)}$$

$$= -\frac{q(q+\epsilon)_+^{1/2}}{\pi} \int_{0}^{\infty} e^{-\epsilon u} \tau_{yz}^0(-u) \int_{0}^{\infty} \frac{e^{-us} ds}{\sqrt{s}(s+q+\epsilon)} du \quad (6.6.36)$$

By using (A5.23), the inversion of $T_{yz}^+(q)$ can be written as

$$\tau_{yz}(X, 0) = -\frac{e^{-\epsilon X}}{\pi} \int_{0}^{\infty} \frac{e^{-\epsilon u} \tau_{yz}^0(-u)}{u} \mathcal{L}_{qX}^{-1}\{\mathcal{L}_{us}[g(s/q)]\} du, \quad X > 0 \quad (6.6.37)$$

where $g(s/q) = q(q/s)^{1/2}/(s+q)$. $\mathcal{L}_{qX}^{-1}$ and $\mathcal{L}_{us}$ are inverse and direct Laplace transforms, respectively, with the first letter in the subscript indicating the transform variable and the second letter the argument of the original function.

Note now that $\mathcal{L}_{us}[g(s/q)]$ is a function of uq, say $G(uq)$, and its inverse, $\mathcal{L}_{qX}^{-1}[G(qu)]$, is a function of X/u, which obviously equals $g(X/u)$. Thus, finally, for $X > 0$:

$$\tau_{yz}(X, 0) = -\frac{e^{-\epsilon X}}{\pi\sqrt{X}} \int_{0}^{\infty} \frac{e^{-\epsilon u}\sqrt{u} \tau_{yz}^0(-u)}{u+X} du$$

$$\to -\frac{1}{\pi\sqrt{X}} \int_{0}^{\infty} \frac{\sqrt{u} \tau_{yz}^0(-u)}{u+X} du \text{ as } \epsilon \to 0 \quad (6.6.38)$$

But this result does not contain any material parameters. Consequently, it is the same as for the corresponding elastic problem. The stress intensity factor is

$$K_{III} = \lim_{X \to 0} \sqrt{2\pi X} \tau_{yz}(X, 0) = -\sqrt{\frac{2}{\pi}} \int_{0}^{\infty} \frac{e^{-\epsilon u} \tau_{yz}^0(-u)}{\sqrt{u}} du$$

$$\to -\sqrt{\frac{2}{\pi}} \int_{0}^{\infty} \frac{\tau_{yz}^0(-u)}{\sqrt{u}} du \text{ as } \epsilon \to 0 \quad (6.6.39)$$

An expression for $w(X, 0)$, $X < 0$, may be found in the same way as for $\tau_{yz}(X, 0)$, but it is more complicated. The Laplace transform of $w(X, 0)$ is

$$W_-(q) = \frac{q}{\pi \lambda_-(q) M_-(-qV)} \int_{0}^{\infty} \frac{1}{\sqrt{s}(s+q+\epsilon)}$$

$$\times \int_{0}^{\infty} [e^{-(s+\epsilon)u} - e^{qu}] \tau_{yz}^0(-u) du \, ds, \quad \Re(q) \leq 0 \quad (6.6.40)$$

For the crack edge vicinity, which correponds to $q \to -\infty$,

$$W_-(q) \to \frac{1}{\pi\sqrt{1-\gamma^2}\mu(0)(-q)^{1/2}} \int_{0}^{\infty} \frac{1}{\sqrt{s}} \int_{0}^{\infty} e^{-(s+\epsilon)u} \tau_{yz}^0(-u) du \, ds$$

$$= \frac{1}{\sqrt{\pi}\sqrt{1-\gamma^2}\mu(0)(-q)^{1/2}} \int_{0}^{\infty} \frac{e^{-\epsilon u} \tau_{yz}^0(-u)}{\sqrt{u}} du \quad (6.6.41)$$

6.6 STEADY STATE CRACK PROPAGATION IN VISCOELASTIC MATERIALS

with the inversion

$$w(X, +0) \to -\frac{2\sqrt{-X}}{\pi\sqrt{1-\gamma^2}\mu(0)} \int_0^\infty \frac{e^{-\epsilon u}\tau_{yz}^0(-u)}{\sqrt{u}} du = \frac{K_{III}}{\sqrt{1-\gamma^2}\mu(0)}\sqrt{\frac{-2X}{\pi}} \quad (6.6.42)$$

as $X \to -0$. For $\epsilon \to 0$ this agrees with the corresponding expression for an elastic solid, if $\gamma = V/c_S^0$ is identified with V/c_S in the elastic case.

Note that the non-dependence of $\tau_{yz}(X,0)$ for $X > 0$ on material parameters and the fact that $w(X,0) = 0$ for $X > 0$ implies that the solution for anti-plane strain crack propagation along an interface between two dissimilar viscoelastic media is directly obtained by considering the two semi-infinite bodies separately, cf, page 185. The crack velocity has to be subsonic with respect to both media.

2. The intersonic case, $c_S^\infty < V < c_S^0$. Here, $g < 0$ and

$$\lambda(q) = \lambda_+(q)\lambda_-(q) = -i\sqrt{1-\gamma^2}\frac{q(q-g)_+^{1/2}}{(q-h)_-^{1/2}}, \quad \Re(q) = 0 \quad (6.6.43)$$

$$\lambda_+(q) = (q-g)_+^{1/2}, \quad \lambda_-(q) = -i\sqrt{1-\gamma^2}\cdot\frac{q}{(q-h)_-^{1/2}} \quad (6.6.44)$$

The treatment for the subsonic case can be followed with almost formal adjustments. The factor $(q+\epsilon)_+^{1/2}$ in the intersonic case corresponds to the factor $(q-g)_+^{1/2}$ here. Thus, the stress ahead of the crack, the stress intensity factor and the crack face displacement are found from the middle members of (6.6.38), (6.6.39) and (6.6.42), respectively, by substituting ϵ for $-g$.

At this point, it may be noted that the elastic-viscoelastic principle does not in general cover inertial problems. This follows, for instance, from (6.6.40), after the change $\epsilon \to -g$, because the quantity $\exp(gu)$ cannot be recovered from the elastic solution.

Solution for models other than the standard linear solid

The properties of the standard linear solid, that were instrumental in obtaining the solution of the crack problem, are

a) No zeros of $M_-(-qV)$ are present in the left half-plane – if there would be such zeros, then they would contribute singularities of $\lambda(q)$ in the left half-plane, see (6.6.13).

b) One, and only one, branch point, $q = g$, for $\lambda_+(q)$ is present in the left half-plane in the intersonic case. The absence of such a branch point in the subsonic case may be compensated by introduction of the factor $(q+\epsilon)_+^{1/2}/(q-\epsilon)_-^{1/2}$, where ϵ eventually is shrunk to zero.

Investigate first whether $M(p)$ possesses zeros for $\Re(p) > 0$. From the expression

$$\mu(t) = \mu(0) + \int_0^t \mu'(\tau)d\tau = \mu(0) + \int_0^\infty \mu'(\tau)d\tau \quad (6.6.45)$$

it follows that

$$M(p) = \mu(0) + \int_0^\infty \mu'(\tau)e^{-p\tau}\mathrm{d}\tau \tag{6.6.46}$$

A possible zero of $M(p)$ for $\Re(p) \geq 0$ is found from

$$\Re[M(p)] = \mu(0) + \int_0^\infty \mu'(\tau)e^{-\alpha\tau}\cos(\beta\tau)\mathrm{d}\tau = 0 \tag{6.6.47}$$

where $\alpha = \Re(p) \geq 0$ and $\beta = \Im(p)$. But $\mu'(\tau) \leq 0$ and $\exp(-\alpha\tau)\cos(\beta\tau) \leq 1$ if $\alpha \geq 0$. Hence,

$$\Re[M(p)] \geq \mu(0) + \int_0^\infty \mu'(\tau)\mathrm{d}\tau = \mu(\infty) \tag{6.6.48}$$

and thus $M(p)$ has no zeros for $\Re(p) \geq 0$ if $\mu(\infty) \geq 0$. Consequently, $M_-(-qV)$ does not possess zeros in $\Re(q) \leq 0$.

Investigate next possible branch points for $\lambda(q)$ in $\Re(q) \leq 0$. Write first

$$\lambda(q) = \frac{[\gamma^2\mu(0) - M_-(-qV)]^{1/2}q}{[M_-(-qV)]^{1/2}} \tag{6.6.49}$$

Clearly λ^2 does not possess any singularities in $\Re(q) \leq 0$. Thus, the only possible such singularities are introduced by the square root operation, and they are found from the equation

$$\gamma^2\mu(0) - M_-(-qV) = 0, \quad \Re(q) \leq 0 \tag{6.6.50}$$

which, after the substitution $qV = -p$ is equivalent to

$$\Re[M(p)] = \gamma^2\mu(0), \quad \Im[M(p)] = 0, \quad \Re(p) \geq 0 \tag{6.6.51}$$

Assume now that $\beta > 0$. Then, cf. (6.6.47),

$$\Re[M(p)] = \mu(0) + \sum_{n=0}^\infty \int_{2n\pi/\beta}^{2(n+1)\pi/\beta} \mu'(\tau)e^{-\alpha\tau}\cos(\beta\tau)\mathrm{d}\tau \tag{6.6.52}$$

$$\Im[M(p)] = \sum_{n=0}^\infty \int_{2n\pi/\beta}^{2(n+1)\pi/\beta} \mu'(\tau)e^{-\alpha\tau}\sin(\beta\tau)\mathrm{d}\tau \tag{6.6.53}$$

For $n = 0$, the integral in the expression for $\Im[M(p)]$ may be written as

$$\int_0^{\pi/\beta} [\mu'(\tau) - \mu'(\tau - \pi/\beta)e^{-\alpha\pi/\beta}]\sin(\beta\tau)\mathrm{d}\tau \tag{6.6.54}$$

which obviously is non-positive, because $\mu'(\tau)$ is non-increasing, see page 237. Obviously, the same holds for $n = 1, 2, 3, \ldots$, which implies that $\Im[M(p)] < 0$ for $\beta > 0$, because $\mu'(t)$ is not identically zero for a viscoelastic material. Thus, $\Im[M(p)] = 0$ only if $\beta = \Im(p) = 0$. This implies that (6.6.51) cannot be satisfied unless $\Im(p) = 0$, and thus possible roots of (6.6.51) are real. The same conclusion is obtained after the initial assumption that $\beta < 0$.

It will now be shown that the root $q = g$ of (6.6.51) is real and negative if and only

6.6 STEADY STATE CRACK PROPAGATION IN VISCOELASTIC MATERIALS 373

if $\mu(\infty) < \Re[M_-(-qV)] \leq \mu(0)$, which is the intersonic region $c_S^\infty < \gamma V < c_S$. To this end, the integral for $n = 0$ in (6.6.52) is written as

$$I_0 = \int_0^{\pi/(2\beta)} [\mu'(\tau) - \mu'(\pi/\beta - \tau)e^{-\alpha\pi/(2\beta)} - \mu'(\tau + \pi/\beta)e^{-\alpha\pi/\beta}$$
$$+ \mu'(2\pi/\beta - \tau)e^{-3\alpha/(\beta\tau)}]e^{-\alpha\tau}\cos(\beta\tau)d\tau \quad (6.6.55)$$

Now, because $\mu'(\tau) \leq 0$ and $\mu''(\tau) \geq 0$, cf. page 237, it follows that $\mu'(\pi/\beta - \tau)$ and $\mu'(\tau + \pi/\beta)$ are greater than or equal to $\mu'_m(\tau)$, where

$$\mu'_m(\tau) = \frac{1}{2}[\mu'(\tau) + \mu'(2\pi/\beta - \tau))] \quad (6.6.56)$$

Thus,

$$I_0 < \int_0^{\pi/(2\beta)} \left[\mu'(\tau) - \mu'_m(\tau)\left(e^{-\alpha\pi/(2\beta)} - e^{-\alpha\pi/\beta}\right)\right.$$
$$\left. + \mu'(2\pi/\beta - \tau)e^{-3\alpha/(2\beta)}\right]e^{-\alpha\tau}\cos(\beta\tau)d\tau$$
$$= \int_0^{\pi/(2\beta)} \{\mu'_m(\tau)(1 - e^{-\alpha\pi/(2\beta)})(1 - e^{-\alpha\pi/\beta})$$
$$+ [\mu'(\tau) - \mu'(2\pi/\beta - \tau)](1 - e^{-3\alpha/(2\beta)})\}e^{-\alpha\tau}\cos(\beta\tau)d\tau \leq 0 \quad (6.6.57)$$

Obviously, the corresponding integral I_n for $n = 1, 2, 3, \cdots$ is also non-positive. Thus, it follows from (6.6.48) and (6.6.52), together with the fact that $\mu'(t)$ is not identically zero, that $\mu(\infty) \leq \Re[M(p)] < \mu(0)$ if $\Re(p) \geq 0$. On the other hand, if $\Re(p) < 0$, $\Im(p) = 0$, equation (6.6.47) shows that $\Re[M(p)] < \mu(\infty)$. These results imply that the root $q = g$ of (6.6.50) is real and negative if and only if $\mu(\infty) < \Re[M_-(-qV)] < \mu(0)$, which is the intersonic region $c_S^\infty < \gamma V < c_S$. Thus, it has been shown that properties a) and b) are also present for viscoelastic material models other than the standard linear solid. Consequently, solutions for steady state crack propagation in viscoelastic media may be obtained in the same way as here used for the standard linear solid.

Energy flux into the crack edge and energy dissipation

In contrast to crack propagation in elastic bodies, energy dissipation takes place not only as energy flux into the crack edge, but also as energy dissipation in the part of the body reached by stress waves. A face loaded crack for the standard linear solid may serve as an illustration of this fact. It will be shown that the work performed by the crack face load exceeds the energy flow into the crack edge from the stress-strain field. Recall that in the corresponding elastic case, all work done by the crack face load is eventually dissipated at the crack edge, because of the condition of steady state and the fact that infinity cannot be an energy sink, cf. page 345.

Because an inverse square root singularity appears for stresses and strains at the crack edge, the energy flux into the edge can be calculated in exactly the same way as for the elastic case. Thus, from (6.6.39), (6.6.42) and (6.4.14) the energy flux is found to be

$$\mathcal{G} = \frac{K_{III}^2}{2\sqrt{1-\gamma^2}\mu_0} \quad (6.6.58)$$

Assume, for simplicity, that the crack face load is

$$\tau_{yz}^0(X, 0) = -\tau_0 e^{X/L}, \quad X < 0 \qquad (6.6.59)$$

Then, for intersonic crack propagation,

$$K_{III} = \sqrt{\frac{2}{\pi}} \tau_0 \int_0^\infty \frac{e^{gu} e^{-u/L}}{\sqrt{u}} du = \frac{\sqrt{2} \tau_0 \sqrt{L}}{\sqrt{1 - gL}} \qquad (6.6.60)$$

$$\mathcal{G} = \frac{\tau_0^2 L}{\sqrt{1-\gamma^2} \mu_0 (1 - gL)} = \frac{\tau_0^2 L \sqrt{1-\gamma^2}}{\mu_0 [1 - \gamma^2 + (\gamma^2 - \mu_\infty/\mu_0) \delta]} \qquad (6.6.61)$$

where $\delta = L/(V t_0)$. The elastic result is recovered by letting $t_0 \to \infty$. For subsonic crack propagation, g shall be put equal to $-\epsilon \to 0$. This, likewise, leads to the elastic result, which also is obvious from (6.6.39) and (6.6.42).

The work performed by the load per unit of crack advance and edge length is

$$\mathcal{A} = 2 \int_{-\infty}^0 \tau_{yz}^0 \frac{\partial w}{\partial X} dX = -2\tau_0 \int_{-\infty}^0 e^{X/L} \frac{\partial w}{\partial X} dX \qquad (6.6.62)$$

Now, the Laplace transform of the integrand is

$$\mathcal{L}_{qX}\left(e^{X/L} \frac{\partial w}{\partial X}\right) = \frac{q}{q - 1/L} \cdot (q - 1/L) W_-(q - 1/L) \qquad (6.6.63)$$

and, hence, using Abel theorems, Appendix A5,

$$\mathcal{A} = 2\tau_0 \lim_{q \to 0} W_-(q - 1/L) \qquad (6.6.64)$$

After insertion of the expression for τ_{yz}^0 in (6.6.40) and integrating first over u and then over s, the result is

$$W_-(q) = \frac{\tau_0 \sqrt{L} q}{\sqrt{1 - gL} \lambda_-(q) M_-(-qV)(q - 1/L)} \qquad (6.6.65)$$

for the intersonic case. The expression for the subsonic case is obtained by putting $g = 0$.

For the subsonic case,

$$\mathcal{A} = \frac{\tau_0^2 L (1 + \delta)}{\mu_0 \sqrt{1 - \gamma^2} \sqrt{1 + \mu_\infty \delta/\mu_0}}, \qquad \frac{\mathcal{G}}{\mathcal{A}} = \frac{\sqrt{1 + \mu_\infty \delta/\mu_0}}{1 + \delta} \qquad (6.6.66)$$

and for the intersonic case,

$$\mathcal{A} = \frac{\tau_0^2 L (1 + \delta)}{\mu_0 \sqrt{1 - \gamma^2 + (\gamma^2 - \mu_\infty/\mu_0)\delta} \sqrt{1 + \mu_\infty \delta/\mu_0}} \qquad (6.6.67)$$

$$\frac{\mathcal{G}}{\mathcal{A}} = \sqrt{\frac{1 - \gamma^2}{1 - \gamma^2 + (\gamma^2 - \mu_\infty/\mu_0)/\delta}} \cdot \frac{\sqrt{1 + \mu_\infty \delta/\mu_0}}{1 + \delta} \qquad (6.6.68)$$

It is obvious that $\mathcal{G}/\mathcal{A} < 1$ for both the subsonic and the intersonic case, unless $t_0 \to \infty$ ($\delta \to 0$), which is the elastic case, giving $\mathcal{G}/\mathcal{A} = 1$. This proves that energy is dissipated both in the crack edge and its surroundings.

Note that all work performed by external forces is dissipated during steady state. For

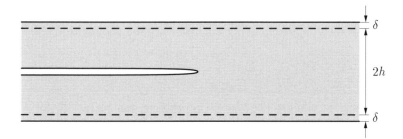

Fig. 6.7.1 Crack propagating in a long strip. The edges are subjected to fixed grip, moved apart the distance 2δ from the unloaded position.

another case, for instance an expanding crack, some part also goes to kinetic energy. However, for a viscoelastic material, it is obvious that energy may also be dissipated in a body interior, not only at the crack edge as for an elastic material.

6.7 Crack propagation in a strip

A very simple example of crack propagation is provided by the long strip with edges subjected to fixed grip and with a crack moving with constant velocity along the symmetry plane. For analysis, the strip is assumed to be infinitely long and the crack semi-infinite, see Fig. 6.7.1. Mode I conditions and plane stress or plane strain are assumed.

The stress-strain energy per unit of strip length and strip thickness far ahead of the crack is denoted by W_∞. Far behind the crack edge, this energy is vanishingly small. Because there cannot be any energy flux through the strip edges, the stress-strain energy released goes to the dissipative region at the crack edge, and because the steady state prevails, the energy flux is constant:

$$\mathcal{G} = W_\infty \tag{6.7.1}$$

Note that this result is valid irrespective of the constitutive equations of the continuum and the scale of yielding.

If the extension of the dissipative region is much smaller than the strip width and the material otherwise is linearly elastic and isotropic, then, because

$$\mathcal{G} = \frac{K_I^2 Y_I(\beta)}{4(1-k^2)\mu} \tag{6.7.2}$$

the stress intensity factor is found to be

$$K_I = \sqrt{\frac{4(1-k^2)\mu W_\infty}{Y_I(\beta)}} \tag{6.7.3}$$

Now,

$$W_\infty = 4(1-k^2)\mu \frac{\delta^2}{h} \tag{6.7.4}$$

where δ is the displacement imposed on each crack edge (see Fig. 6.7.1) and h is the half-width of the strip. Thus,

$$K_I = \frac{4(1-k^2)\mu\delta}{\sqrt{h}Y_I(\beta)} \tag{6.7.5}$$

This simple way to analyse the long strip problem was suggested (for mode II) by Rivlin and Thomas (1952). Due to the equally simple result, the long strip configuration is very well suited for experimental determination of energy dissipation during crack propagation under different speeds. It has been used for this purpose by, for instance, Paxson and Lucas (1973) and by Nilsson (1974b). The method of analysis gives only the asymptotic crack edge vicinity solution. The full solution – stresses and displacements at any point in the strip – presents a more difficult problem; see Nilsson (1972, 1973b). The corresponding mode III problem, however, is readily solved (Field and Baker 1962).

Using moving coordinates $\xi = X$ and $\eta = a_S Y$, the governing equation for mode III may be written as (6.4.3). Consider the upper half of the strip, $0 \leq \eta \leq a_S h$. The displacement of the upper edge of the strip is $w = \delta$. Then, by analogy with the problem formulation in connection with (6.4.3), the boundary conditions are written as

$$\frac{\partial(a_S W)}{\partial \eta} = 0 \text{ for } \xi < 0,\ \eta = 0 \tag{6.7.6}$$

$$a_S W = 0 \text{ for } \xi > 0,\ \eta = 0 \tag{6.7.7}$$

$$a_S W = a_S \delta \text{ for } \eta = a_S h \tag{6.7.8}$$

But these equations are the same as (4.5.10)-(4.5.11), if $a_S W$ is identified with w, ξ with x, η with y, $a_S h$ with h and $a_S \delta$ with δ. The stresses in the strip are therefore immediately found from the static solution, (4.5.27), as

$$\tau_{yz} + i a_S \tau_{xz} = \frac{\mu \delta}{h\left[1 - \exp\left(\frac{-\pi(X + i a_S Y)}{a_S h}\right)\right]^{1/2}} \tag{6.7.9}$$

Note that, as in Section 6.4, the stress τ_{xz} from the static solution was replaced by $a_S \tau_{xz}$. The stress on the symmetry plane is

$$\tau_{yz} = \frac{\mu \delta}{h\sqrt{1 - \exp\left(\frac{-\pi X}{a_S h}\right)}} \tag{6.7.10}$$

and the stress intensity factor is

$$K_{III} = \frac{\sqrt{2}\mu\delta}{\sqrt{h}Y_{III}(\gamma)} \tag{6.7.11}$$

a result, which, of course, could have been obtained in the same simple way as demonstrated for mode I.

6.8 Crack propagation along an interface

Introduction

The theory for dynamic interface cracks was first given by Gol'dshtein (1966, 1967) and later by, for instance, Willis (1971), Atkinson (1977), Wu (1991) and Yang et al. (1991). Earlier, a somewhat related study, motion of an edge dislocation along an interface, was presented by Weertman (1963). Dynamic crack extension of cracks along curvilinear interfaces under the action of both crack face loads and thermal strains has been considered by Herrmann and Noe (1995). Interface cracks, propagating at a velocity above the S wave velocity for one of the media, have been studied by Yu and Yang (1995) and by Huang et al. (1996, 1998) and observed experimentally by Lambros and Rosakis (1995). Analytical solutions have been obtained for steady state problems, only. No such solutions appear to be possible for self-similar cases.

The subsonic anti-plane case

Consider a horizontal semi-infinite crack propagating under anti-plane deformation with velocity V to the right – positive x direction – in the interface $y = 0$ between two linearly elastic dissimilar media with S wave velocities c_S^- and c_S^+. Here, as in the following, superscript (or subscript) minus refers to the lower, superscript (or subscript) plus to the upper medium. It is assumed that $V < \min(c_S^-, c_S^+)$, so that the crack velocity is subsonic with respect to both media. With the Galilean transformation $x - Vt = X$, $y = Y$, the wave equations are, cf. (6.4.1),

$$(a_S^-)^2 \frac{\partial^2 w_-}{\partial X^2} + \frac{\partial^2 w_-}{\partial Y^2} = 0 \tag{6.8.1}$$

$$(a_S^+)^2 \frac{\partial^2 w_+}{\partial X^2} + \frac{\partial^2 w_+}{\partial Y^2} = 0 \tag{6.8.2}$$

where $a_S^- = \sqrt{1 - \gamma_-^2}$ and $a_S^+ = \sqrt{1 - \gamma_+^2}$, with $\gamma_- = V/c_S^-$ and $\gamma_+ = V/c_S^+$.

Assuming the crack faces to be traction free, the boundary conditions on $Y = 0$ are

$$\tau_{yz}^- = \tau_{yz}^+ \text{ for all } X \tag{6.8.3}$$

$$\tau_{yz}^- = \tau_{yz}^+ = 0 \text{ for } X < 0 \tag{6.8.4}$$

$$w_- = w_+ \text{ for } X > 0 \tag{6.8.5}$$

Note that $\tau_{yz}^\pm = \mu_\pm \partial w_\pm / \partial y$.

Introduce now $\zeta_- = \xi_- + i\eta_-$, $\zeta_+ = \xi_+ + i\eta_+$, where

$$\xi_- = \xi_+ = \xi = X, \quad \eta_- = a_S^- Y, \quad \eta_+ = a_S^+ Y \tag{6.8.6}$$

Then, the equations of motion may be written as Laplace equations:

$$\frac{\partial^2 w_-}{\partial \xi_-^2} + \frac{\partial^2 w_-}{\partial \eta_-^2} = 0 \tag{6.8.7}$$

$$\frac{\partial^2 w_+}{\partial \xi_+^2} + \frac{\partial^2 w_+}{\partial \eta_+^2} = 0 \tag{6.8.8}$$

The solutions are taken as the real parts of harmonic functions:

$$w_- = \Re F(\zeta_-) = \frac{1}{2}[F(\zeta_-) + \overline{F(\zeta_-)}] \tag{6.8.9}$$

$$w_+ = \Re G(\zeta_+) = \frac{1}{2}[G(\zeta_+) + \overline{G(\zeta_+)}] \tag{6.8.10}$$

Then,

$$\frac{\partial w_-}{\partial y} = \frac{ia_S^-}{2}[F'(\zeta_-) - \overline{F'(\zeta_-)}] \tag{6.8.11}$$

$$\frac{\partial w_+}{\partial y} = \frac{ia_S^+}{2}[G'(\zeta_+) - \overline{G'(\zeta_+)}] \tag{6.8.12}$$

Functions F and G are defined in the lower ζ_- half-plane and the upper ζ_+ half-plane, respectively, including the real axis in each plane. The boundary conditions (6.8.3)-(6.8.5) give

$$\mu_- a_S^-[F'(\xi) - \overline{F'(\xi)}] = \mu_+ a_S^+[G'(\xi) - \overline{G'(\xi)}] \text{ for all } \xi \tag{6.8.13}$$

$$F'(\xi) - \overline{F'(\xi)} = G'(\xi) - \overline{G'(\xi)} \text{ for } \xi < 0 \tag{6.8.14}$$

$$F(\xi) + \overline{F(\xi)} = G(\xi) + \overline{G(\xi)} \text{ for } \xi > 0 \tag{6.8.15}$$

Because $F(\zeta_-)$ is defined for $\Im\zeta \leq 0$, it will be called $F_-(\zeta_-)$ for clarity, but subscript minus for ζ_- will be dropped temporarily. Define now $F_+(\zeta) = \overline{F_-(\zeta)}$ for $\Im\zeta \geq 0$, and consider $F(\zeta)$ as a sectionally analytic function. Similarly, consider $G(\zeta)$ as a sectionally analytic function, with $G_-(\zeta) = \overline{G_+(\zeta)}$ for $\Im\zeta < 0$. Then, after integrations, the boundary conditions may be written in the form

$$\mu_- a_S^- F_+(\xi) + \mu_+ a_S^+ G_+(\xi) = \mu_- a_S^- F_-(\xi) + \mu_+ a_S^+ G_-(\xi) \text{ for all } \xi \tag{6.8.16}$$

$$F_+(\xi) + G_+(\xi) = F_-(\xi) + G_-(\xi) \text{ for } \xi < 0 \tag{6.8.17}$$

$$F_+(\xi) - G_+(\xi) = -[F_-(\xi) - G_-(\xi)] \text{ for } \xi > 0 \tag{6.8.18}$$

Note that integration constants correspond to rigid-body motion, only.

The first equation shows that the function

$$\mu_- a_S^- F(\zeta) + \mu_+ a_S^+ G(\zeta) \tag{6.8.19}$$

is analytic in the whole plane and thus equal to a polynomial. This polynomial can be set to zero, because linear terms correspond to a rigid-body motion, whereas second degree terms or higher violate the condition of finite stresses at infinity. Thus,

$$G(\zeta) = -\frac{\mu_- a_S^-}{\mu_+ a_S^+} F(\zeta) \tag{6.8.20}$$

Substitution into (6.8.17)-(6.8.18) gives

$$F_+(\xi) - F_-(\xi) = 0 \text{ for } \xi < 0 \tag{6.8.21}$$

$$F_+(\xi) + F_-(\xi) = 0 \text{ for } \xi > 0 \tag{6.8.22}$$

which constitutes a Hilbert problem for $F(\zeta)$, the solution of which may be written in the form (suitable for the continuation)

$$F(\zeta) = \frac{iC}{\mu_- a_S^-} \zeta^{1/2} \tag{6.8.23}$$

6.8 CRACK PROPAGATION ALONG AN INTERFACE

where C is a constant, and a branch cut is made along the positive real axis. Then,

$$G(\zeta) = -\frac{iC}{\mu_+ a_S^+}\zeta^{1/2} \tag{6.8.24}$$

and the displacements are

$$w_- = \frac{1}{\mu_- a_S^-}\Re(iC\zeta_-^{1/2}) \tag{6.8.25}$$

$$w_+ = -\frac{1}{\mu_+ a_S^+}\Re(iC\zeta_+^{1/2}) \tag{6.8.26}$$

Obviously, C is real – otherwise the condition $w_- = w_+$ on the interface in front of the crack cannot be satisfied. Then, $w_- = w_+ = 0$ for $X > 0$, $Y = 0$, and the full asymptotic expression for the displacement in the upper half-plane is

$$w_+(X,Y) = \frac{C}{\mu_+ a_S^+}\Im(X + ia_S^+ Y)^{1/2} \tag{6.8.27}$$

With $X = r\cos\varphi$, $Y = r\sin\varphi$, the result may be written as

$$w_+(r,\varphi) = \frac{C\sqrt{r}}{\sqrt{2}\mu_+ a_S^+}\sqrt{\sqrt{1 - \gamma_+^2 \sin^2\varphi} - \cos\varphi} \tag{6.8.28}$$

$$w_-(r,\varphi) = -\frac{C\sqrt{r}}{\sqrt{2}\mu_- a_S^-}\sqrt{\sqrt{1 - \gamma_-^2 \sin^2\varphi} - \cos\varphi} \tag{6.8.29}$$

In each half-plane, these expressions are identical to those for the asymptotic mode III displacement for a crack in a homogeneous medium, (6.4.11). Thus, as for the static case, the solution could have been obtained simply by joining the halves from two asymptotic homogeneous mode III regions with the same stress intensity factor. However, this possibility is restricted to steady state crack propagation, because otherwise the shear stress variation along the symmetry plane outside the crack depends in general on the crack velocity, see, for instance, (6.9.146).

The subsonic in-plane case

Basic relations

The asymptotic field may be obtained by analogy with the anti-plane case. The crack velocity V is assumed to be smaller than the smaller S wave velocity. The crack edge vicinity will be studied, so that the body can be taken as infinite, with vanishing remote stress, and the crack semi-infinite. After the Galilean transformation, the equations of motion read, see (6.2.43)-(6.2.44):

$$(a_P^\pm)^2\frac{\partial^2 \phi_\pm}{\partial X^2} + \frac{\partial^2 \phi_\pm}{\partial Y^2} = 0 \tag{6.8.30}$$

$$(a_S^\pm)^2\frac{\partial^2 \psi_\pm}{\partial X^2} + \frac{\partial^2 \psi_\pm}{\partial Y^2} = 0 \tag{6.8.31}$$

where $a_P^\pm = \sqrt{1 - \beta_\pm^2}$, $a_S^\pm = \sqrt{1 - \beta_\pm^2/k_\pm^2}$, subscripts and superscripts plus refer to the upper and subscripts and superscripts minus to the lower half-plane. Other notations are the same as previously used, so that, for instance, $\beta_\pm = V/c_P^\pm$, where $c_P^\pm$ are the P wave velocities in the media. By use of the complex variables

$$\zeta_\pm^{P,S} = \xi + i\eta_\pm^{P,S}, \quad \xi = X, \quad \eta_\pm^{P,S} = a_{P,S}^\pm Y \quad (6.8.32)$$

the equations of motion may be written in the Laplace form:

$$\frac{\partial^2 \phi_\pm}{\partial \xi^2} + \frac{\partial^2 \phi_\pm}{\partial (\eta_\pm^P)^2} = 0 \quad (6.8.33)$$

$$\frac{\partial^2 \psi_\pm}{\partial \xi^2} + \frac{\partial^2 \psi_\pm}{\partial (\eta_\pm^S)^2} = 0 \quad (6.8.34)$$

The solutions may be taken as the real parts of analytic functions:

$$\phi_+ = \Re F_+(\zeta_+^P) = \frac{1}{2}[F(\zeta_+^P) + \overline{F_+(\zeta_+^P)}] \quad (6.8.35)$$

$$\phi_- = \Re G_-(\zeta_-^P) = \frac{1}{2}[G(\zeta_-^P) + \overline{G_-(\zeta_-^P)}] \quad (6.8.36)$$

$$\psi_+ = \Re S_+(\zeta_+^S) = \frac{1}{2}[S(\zeta_+^S) + \overline{S_+(\zeta_+^S)}] \quad (6.8.37)$$

$$\psi_- = \Re T_-(\zeta_-^S) = \frac{1}{2}[T(\zeta_-^S) + \overline{T_-(\zeta_-^S)}] \quad (6.8.38)$$

where F_+ and S_+ are defined in the upper half-plane and G_- and T_- in the lower half-plane.

Boundary conditions
The boundary conditions are:

$$\tau_{xy}^+ = \tau_{xy}^-, \quad \sigma_y^+ = \sigma_y^- \text{ for all } \xi \quad (6.8.39)$$
$$\tau_{xy}^+ = 0, \quad \sigma_y^+ = 0 \text{ for } \xi < 0 \quad (6.8.40)$$
$$u_+ = u_-, \quad v_+ = v_- \text{ for } \xi > 0 \quad (6.8.41)$$

Derivatives of $\phi_\pm$ and $\psi_\pm$ will be needed. For ϕ_+ differentiations with respect to y give

$$\frac{\partial \phi_+}{\partial Y} = \frac{i a_P^+}{2}[F'_+(\zeta_+^P) - \overline{F'_+(\zeta_+^P)}] \quad (6.8.42)$$

$$\frac{\partial^2 \phi_+}{\partial Y^2} = -\frac{(a_P^+)^2}{2}[F''_+(\zeta_+^P) + \overline{F''_+(\zeta_+^P)}] \quad (6.8.43)$$

and the corresponding derivatives for the other three functions are analogous. Subscripts and superscripts for ζ are now dropped until needed after the functions F_+, G_-, S_+ and T_- have been determined.

Introduce the sectionally analytic functions $F(\zeta)$ and $S(\zeta)$, equal to $F_+(\zeta)$ and $S_+(\zeta)$ in the upper half-plane and defined for the lower half-plane as $F_-(\zeta)$ and $S_-(\zeta)$ by the relations

$$F_-(\zeta) = \overline{F_+}(\zeta) \quad (\zeta = \zeta_+^P), \quad S_-(\zeta) = -\overline{S_+}(\zeta) \quad (\zeta = \zeta_+^S) \quad (6.8.44)$$

6.8 CRACK PROPAGATION ALONG AN INTERFACE

Similarly, the sectionally analytic functions G and T are introduced, with

$$G_+(\zeta) = \overline{G_-(\bar\zeta)} \quad (\zeta = \zeta_-^P), \qquad T_+(\zeta) = -\overline{T_-(\bar\zeta)} \quad (\zeta = \zeta_-^S) \qquad (6.8.45)$$

The choice of signs in the definitions will be explained later.

Stresses and displacements are now expressed in terms of the potential functions. Thus, from Appendix A1 and (6.8.44), with $P_\pm = 1 + (a_S^\pm)^2$,

$$\tau_{xy}^+ = \mu_+ \left(2\frac{\partial^2 \phi_+}{\partial X \partial Y} + \frac{\partial^2 \psi_+}{\partial Y^2} - \frac{\partial^2 \psi_+}{\partial X^2} \right)$$

$$= \frac{\mu_+}{2} \{ 2ia_P^+ [F''_+(\xi) - F''_-(\xi)] - P_+[S''_+(\xi) - S''_-(\xi)] \} \qquad (6.8.46)$$

$$\sigma_y^+ = -\frac{\mu_+}{2} \{ P_+[F''_+(\xi) + F''_-(\xi)] + 2ia_S^+[S''_+(\xi) + S''_-(\xi)] \} \qquad (6.8.47)$$

$$u_+ = \frac{1}{2} \{ F'_+(\xi) + F'_-(\xi) + ia_S^+[S'_+(\xi) + S'_-(\xi)] \} \qquad (6.8.48)$$

$$v_+ = \frac{1}{2} \{ ia_P^+[F'_+(\xi) - F'_-(\xi)] - S'_+(\xi) + S'_-(\xi) \} \qquad (6.8.49)$$

for $Y = 0$. Use of the expression for τ_{xy}^+ and the analogous one for τ_{xy}^- in the first boundary condition (6.8.39) gives, for all ξ,

$$\frac{\mu_+}{2} \{ 2ia_P^+ [F''_+(\xi) - F''_-(\xi)] - P_+[S''_+(\xi) - S''_-(\xi)] \}$$

$$= \frac{\mu_-}{2} \{ 2ia_P^- [G''_-(\xi) - G''_+(\xi)] - P_-[T''_-(\xi) - T''_+(\xi)] \} \qquad (6.8.50)$$

This may be rearranged as

$$\mu_+ \{ 2ia_P^+ F''_+(\xi) - P_+ S''_+(\xi) \} + \mu_- \{ 2ia_P^- G''_+(\xi) - P_- T''_+(\xi) \}$$
$$= \mu_+ \{ 2ia_P^+ F''_-(\xi) - P_+ S''_-(\xi) \} + \mu_- \{ 2ia_P^- G''_-(\xi) - P_- T''_-(\xi) \} \qquad (6.8.51)$$

The form of this equation explains the choice of signs in the definitions of the sectionally analytic functions. Because it is valid for all ξ, it implies that the function

$$\Theta_1(\zeta) = 2i\mu_+ a_P^+ F''(\zeta) - \mu_+ P_+ S''(\zeta) + 2i\mu_- a_P^- G''(\zeta) - \mu_- P_- T''(\zeta) \qquad (6.8.52)$$

is analytic in the whole plane and thus equal to a polynomial of finite degree, which, because stresses, and thereby the functions $F''(\zeta)$, $G''(\zeta)$, $S''(\zeta)$ and $T''(\zeta)$ vanish at infinity, reduces to zero. Hence, one relation, $\Theta_1(\zeta) = 0$, between F, G, S and T is obtained.

Similarly, the second boundary condition (6.8.39) gives

$$\mu_+ \{ P_+ F''_+(\xi) + 2ia_S^+ S''_+(\xi) \} - \mu_- \{ P_-[G''_+(\xi) + 2ia_S^- T''_+(\xi)] \}$$
$$= -\mu_+ \{ P_+ F''_-(\xi) + 2ia_S^+ S''_-(\xi) \} + \mu_- \{ P_-[G''_-(\xi) + 2ia_S^- T''_-(\xi)] \} \qquad (6.8.53)$$

for all ξ. This shows the analyticity in the whole plane of the function $\theta(\zeta)\Theta_2(\zeta)$, where $\theta(\zeta)$ is sectionally analytic and such that $\theta_+(\zeta) = 1$, $\theta_-(\zeta) = -1$, and

$$\Theta_2(\zeta) = \mu_+ \{ P_+ F''(\zeta) + 2ia_S^+ S''(\zeta) \} - \mu_- \{ P_-[G''(\zeta) + 2ia_S^- T''(\zeta)] \} \qquad (6.8.54)$$

Moreover, because stresses vanish at infinity, $\theta(\zeta)\Theta_2(\zeta) = 0$. Hence, $\Theta_2(\zeta) = 0$, which is a second relation between F'', G'', S'' and T''. This enables $G''(\zeta)$ and $T''(\zeta)$ to be expressed in terms of $F''(\zeta)$ and $S''(\zeta)$.

The remaining boundary conditions, (6.8.40)-(6.8.41), give two relations between $F''(\xi)$, $G''(\xi)$, $S''(\xi)$ and $T''(\xi)$ for $\xi < 0$ and two for $\xi > 0$. Because $G''(\xi)$ and $T''(\xi)$ can be expressed in terms of $F''(\xi)$ and $S''(\xi)$, they can be eliminated. The resulting equations are

$$\sigma_y^+ = 0 \implies F_+''(\xi) + iA_N S_+''(\xi) = -[F_-''(\xi) + iA_N S_-''(\xi)] \text{ for } \xi < 0 \quad (6.8.55)$$
$$\tau_{xy}^+ = 0 \implies F_+''(\xi) + iB_N S_+''(\xi) = F_-''(\xi) + iB_N S_-''(\xi) \text{ for } \xi < 0 \quad (6.8.56)$$
$$u_+ = u_- \implies F_+''(\xi) + iA_P S_+''(\xi) = -[F_-''(\xi) + iA_P S_-''(\xi)] \text{ for } \xi > 0 \quad (6.8.57)$$
$$v_+ = v_- \implies F_+''(\xi) + iB_P S_+''(\xi) = F_-''(\xi) + iB_P S_-''(\xi) \text{ for } \xi > 0 \quad (6.8.58)$$

where

$$A_N = \frac{2a_S^+}{P_+}, \quad A_P = \frac{\mu_- a_S^+ R_- + \mu_+(2a_S^+ C_- + a_S^- P_+ Q_-)}{\mu_- R_- + \mu_+(P_+ C_- + 2a_P^+ a_S^- Q_-)} \quad (6.8.59)$$

$$B_N = \frac{P_+}{2a_P^+}, \quad B_P = \frac{\mu_- R_- + \mu_+(P_+ C_- + 2a_P^- a_S^+ Q_-)}{\mu_- a_P^+ R_- + \mu_+(2a_P^+ C_- + a_P^- P_+ Q_-)} \quad (6.8.60)$$

$$R_- = 4a_P^- a_S^- - P_-^2, \quad C_- = P_-^2 - 2a_P^- a_S^-, \quad Q_- = 1 - (a_S^-)^2 \quad (6.8.61)$$

Obvious changes of subscripts and superscripts from minus to plus define the quantities R_+, C_+ and Q_+, needed later. Note that R_- is the Rayleigh function in the form $R(a_P^-, a_S^-)$, introduced in connection with (6.2.23). R_- vanishes at the Rayleigh wave velocity and approaches $2(1-k_-^2)V^2/(c_S^-)^2$ as $V \to 0$.

Formulation and solution of Hilbert problems

Equations (6.8.55) and (6.8.56) show that the sectionally analytic functions $L(\zeta)$ and $M(\zeta)$, defined by the relations

$$L_+(\zeta) = F_+''(\zeta) + iA_N S_+''(\zeta), \quad L_-(\zeta) = -[F_-''(\zeta) + iA_N S_-''(\zeta)] \quad (6.8.62)$$
$$M_+(\zeta) = F_+''(\zeta) + iB_N S_+''(\zeta), \quad M_-(\zeta) = F_-''(\zeta) + iB_N S_-''(\zeta) \quad (6.8.63)$$

are analytic in the plane cut along the positive real axis, because

$$L_+(\xi) = L_-(\xi) \text{ for } \xi < 0, \quad M_+(\xi) = M_-(\xi) \text{ for } \xi < 0 \quad (6.8.64)$$

Express now $F_\pm''(\zeta)$ and $S_\pm''(\zeta)$ in terms of $L_\pm(\zeta)$ and $M_\pm(\zeta)$:

$$F_\pm(\zeta) = \frac{A_N M_\pm(\zeta) \mp B_N L_\pm(\zeta)}{A_N - B_N}, \quad S_\pm(\zeta) = i\frac{M_\pm(\zeta) \mp L_\pm(\zeta)}{A_N - B_N} \quad (6.8.65)$$

Insertion into equations (6.8.57) and (6.8.58) then gives

$$(A_P - B_N)[L_+(\xi) - L_-(\xi)] + (A_N - A_P)[M_+(\xi) + M_-(\xi)] = 0 \quad (6.8.66)$$
$$(B_P - B_N)[L_+(\xi) + L_-(\xi)] + (A_N - B_P)[M_+(\xi) - M_-(\xi)] = 0 \quad (6.8.67)$$

for $\xi > 0$. The coefficients may be written in the form

$$A_P - B_N = \frac{N_0}{2a_P^+ D_1}, \quad A_N - A_P = \frac{N_2}{P_+ D_1} \quad (6.8.68)$$

$$B_P - B_N = \frac{N_1}{2a_P^+ D_2}, \quad A_N - B_P = \frac{N_0}{P_+ D_2} \quad (6.8.69)$$

where

$$N_0 = \mu_+ C_- R_+ - \mu_- C_+ R_-, \quad N_2 = \mu_- a_S^+ Q_+ R_- + \mu_+ a_S^- Q_- R_+$$
$$N_1 = \mu_- a_P^+ Q_+ R_- + \mu_+ a_P^- Q_- R_+, \quad D_1 = \mu_- R_- + \mu_+(P_+ C_- + 2 a_P^+ a_S^- Q_-)$$
$$D_2 = \mu_- a_P^+ R_- + \mu_+(2 a_P^+ C_- + a_P^- P_+ Q_-)$$

Note that N_1 and N_2 are positive if R_+ and R_- are positive, which implies sub-Rayleigh velocity with respect to both media. Note further that $N_0 = 0$ if the two media are identical.

Recall now that the aim is to obtain the asymptotic field, i.e. to extract the strongest singularity. Thus, the functions F'', G'', S'' and T'' and thereby also the functions L and M will differ only by amplitude factors. It is evident that such functions L and M can be found after writing (6.8.66) and (6.8.67) in the form

$$L_+(\xi) - \lambda L_-(\xi) = 0 \text{ for } \xi > 0, \quad M_+(\xi) - \lambda M_-(\xi) = 0 \text{ for } \xi > 0 \qquad (6.8.70)$$

which is possible if λ is properly chosen. Insertion into (6.8.66) and (6.8.67) gives

$$\frac{N_0}{2 a_P^+}(1 - \lambda)L_-(\xi) - \frac{N_2}{P_+}(1 + \lambda)M_-(\xi) = 0 \qquad (6.8.71)$$

$$\frac{N_1}{2 a_P^+}(1 + \lambda)L_-(\xi) - \frac{N_0}{P_+}(1 - \lambda)M_-(\xi) = 0 \qquad (6.8.72)$$

Vanishing determinant leads to the two possible values

$$\lambda = \begin{cases} \lambda_1 = -\dfrac{(N_1 N_2)^{1/2} - N_0}{(N_1 N_2)^{1/2} + N_0} & \Longrightarrow \dfrac{L_-(\xi)}{M_-(\xi)} = \dfrac{L(\zeta)}{M(\zeta)} = +\dfrac{2 a_P^+}{P_+}\left(\dfrac{N_2}{N_1}\right)^{1/2} \\ \lambda_2 = \dfrac{1}{\lambda_1} & \Longrightarrow \dfrac{L_-(\xi)}{M_-(\xi)} = \dfrac{L(\zeta)}{M(\zeta)} = -\dfrac{2 a_P^+}{P_+}\left(\dfrac{N_2}{N_1}\right)^{1/2} \end{cases} \qquad (6.8.73)$$

If the crack velocity is sub-Rayleigh with respect to both media, then N_1 and N_2 are positive and $\sqrt{N_1 N_2} > |N_0|$, implying that λ is real and negative. For velocities between the lower Rayleigh wave velocity and the lower S wave velocity, several situations are possible, depending on the signs of N_1 and N_2, on whether the higher Rayleigh wave velocity is smaller or larger than the lower S wave velocity, and also on the magnitudes of the lower P wave velocity in relation to the higher Rayleigh and S wave velocities. Adding to this complex picture are the implications of Stoneley waves (Stoneley 1924, Achenbach 1973), i.e. surface waves along the interface like Rayleigh waves along a free surface. Their velocity is found from $\min(|\lambda_1|, |\lambda_2|) = 0$, $N_1 < 0$, $N_2 < 0$. Discussions about these situations were given by Gol'dshtein (1966, 1967). Here, it is assumed that the crack velocity is sub-Rayleigh with respect to both media.

Equations (6.8.64) and (6.8.70) constitute two Hilbert problems of the same form as equations (4.12.10) and (4.12.12), particularly after introduction of the dynamic Dunders' parameter

$$\beta_d = \frac{1 + \lambda_1}{1 - \lambda_1} = \frac{N_0}{\sqrt{N_1 N_2}} \qquad (6.8.74)$$

Thus, by analogy with (4.12.19), the solutions of the Hilbert problems for $\lambda = \lambda_1$ are

$$L_1(\zeta) = \kappa P_1 \zeta^{-1/2 - i\epsilon}, \quad M_1(\zeta) = \frac{P_+}{2 a_P^+} \cdot \frac{P_1}{\kappa} \zeta^{-1/2 - i\epsilon} \qquad (6.8.75)$$

where $\kappa = \sqrt[4]{N_2/N_1}$ is introduced for symmetry reasons, P_1 is a constant, and

$$\epsilon = \frac{1}{2\pi}\ln\frac{1+\beta_d}{1-\beta_d} = -\frac{1}{2\pi}\ln|\lambda_1| \qquad (6.8.76)$$

The branch is chosen so that $\zeta^{-1/2-i\epsilon} = |\xi|^{-1/2-i\epsilon} = [\cos(\epsilon \ln|\xi|) - i\sin(\epsilon \ln|\xi|)]/\sqrt{\xi}$ for $\eta = 0$, $\xi < 0$. For simplicity, the introduction of a length parameter x_0 to make the logarithm argument dimensionless is postponed for a while.

Similarly, the solutions for $\lambda = \lambda_2$ are found to be

$$L_2(\zeta) = \kappa P_2 \zeta^{-1/2+i\epsilon}, \quad M_2(\zeta) = -\frac{P_+}{2a_P^+}\cdot\frac{P_2}{\kappa}\zeta^{-1/2+i\epsilon} \qquad (6.8.77)$$

where P_2 is a constant. The branch is chosen so that $\zeta^{-1/2+i\epsilon} = |\xi|^{-1/2+i\epsilon}$ for $\eta = 0$, $\xi < 0$.

Note that $\zeta_\pm^{-1/2-i\epsilon} = \pm i e^{\mp\pi\epsilon}\xi^{-1/2-i\epsilon}$, and $\zeta_\pm^{-1/2+i\epsilon} = \pm i e^{\pm\pi\epsilon}\xi^{-1/2+i\epsilon}$ for $\eta = 0$, $\xi > 0$.

The interface stresses and the crack separation gradient

Use of (6.8.47) and (6.8.65) gives the normal stress on the symmetry plane ahead of the crack as

$$\sigma_y = -\frac{\mu_+}{2(A_N-B_N)}\{P_+[A_N(M_+(\xi)+M_-(\xi))-B_N(L_+(\xi)-L_-(\xi))]$$
$$- 2a_S^+[M_+(\xi)+M_-(\xi)-L_+(\xi)+L_-(\xi)]\} = -\frac{\mu_+ P_+}{2}(L_+(\xi)-L_-(\xi)) \qquad (6.8.78)$$

Now, $L(\zeta) = L_1(\zeta) + L_2(\zeta)$, giving

$$\sigma_y = -i\frac{\mu_+ P_+}{\sqrt{\xi}}\kappa\cosh\pi\epsilon\,[P_1\xi^{-i\epsilon}+P_2\xi^{i\epsilon}] \qquad (6.8.79)$$

The fact that σ_y is real requires $P_1 = -\overline{P_2}$, and then the result, after the substitution $2\mu_+ P_+ \cosh\pi\epsilon P_2 = i\overline{K}x_0^{-i\epsilon}/\sqrt{2\pi}$, becomes

$$\sigma_y = \frac{\kappa\Re[K\,(X/x_0)^{i\epsilon}]}{\sqrt{2\pi X}} \qquad (6.8.80)$$

Similarly, the shear stress on the symmetry plane ahead of the crack is found to be

$$\tau_{xy} = i\mu_+ a_P^+(M_+ - M_-) = \frac{\mu_+ P_+}{\kappa\sqrt{\xi}}\cosh\pi\epsilon\,[P_1\xi^{-i\epsilon} - P_2\xi^{i\epsilon}] \qquad (6.8.81)$$

leading to

$$\tau_{xy} = \frac{\Im[K\,(X/x_0)^{i\epsilon}]}{\kappa\sqrt{2\pi X}} \qquad (6.8.82)$$

Finally, the crack separation gradient

$$\frac{d\delta_c(X)}{dX} = u'_+(X) - u'_-(X) + i[v'_+(X) - v'_-(X)]$$
$$= 2\mu_+ a_P^+\left[\frac{a_S^- Q_-}{\mu_- R_-} + \frac{a_S^+ Q_+}{\mu_+ R_+}\right]M - \mu_+ P_+\left[\frac{a_P^- Q_-}{\mu_- R_-} + \frac{a_P^+ Q_+}{\mu_+ R_+}\right]L \qquad (6.8.83)$$

Now,
$$L = \frac{2i\kappa}{\sqrt{|\xi|}}\Im[P_2|\xi|^{i\epsilon}], \quad M = -\frac{P_+}{a_P^+}\cdot\frac{1}{\kappa\sqrt{|\xi|}}\Re[P_2|\xi|^{i\epsilon}] \tag{6.8.84}$$

and thus,
$$\frac{\mathrm{d}\delta_c(X)}{\mathrm{d}X} = \frac{1}{\kappa}\cdot\left[\frac{a_S^- Q_-}{\mu_- R_-} + \frac{a_S^+ Q_+}{\mu_+ R_+}\right]\frac{\Im[\overline{K}\,(|X|/x_0)^{-i\epsilon}]}{\cosh\pi\epsilon\sqrt{2\pi|X|}}$$
$$- i\kappa\left[\frac{a_P^- Q_-}{\mu_- R_-} + \frac{a_P^+ Q_+}{\mu_+ R_+}\right]\frac{\Re[\overline{K}\,(|X|/x_0)^{-i\epsilon}]}{\cosh\pi\epsilon\sqrt{2\pi|X|}} \tag{6.8.85}$$

The derivation of (6.8.83) may preferably be made for, say, $\partial u_+/\partial X$ and $\partial v_+/\partial X$ separately. These contributions are clearly recognized from the composed result, and they may be written in the form
$$\frac{\partial u_+}{\partial X} = -\frac{Y_{II}^+}{2(1-k_+^2)}\cdot\frac{\Im[\overline{K}\,(|X|/x_0)^{i\epsilon}]}{\kappa\cosh\pi\epsilon\sqrt{2\pi|X|}} \tag{6.8.86}$$
$$\frac{\partial v_+}{\partial X} = -\frac{Y_I^+}{2(1-k_+^2)}\cdot\frac{\kappa\Re[\overline{K}\,(|X|/x_0)^{i\epsilon}]}{\cosh\pi\epsilon\sqrt{2\pi|X|}} \tag{6.8.87}$$

where
$$Y_{II}^+ = \frac{2(1-k_+^2)a_S^+ Q_+}{R_+}, \quad Y_I^+ = \frac{2(1-k_+^2)a_P^+ Q_+}{R_+} \tag{6.8.88}$$

The functions Y_I and Y_{II} are, of course, the Yoffe functions, defined by (6.2.26) and (6.2.50); see also (6.2.51) and (6.2.52). Note the similarity with the homogeneous case. The similarity with the expression for the stationary interface case, (4.12.25), is obvious after introduction of
$$C_m^{I,II} = \frac{Y_{I,II}^-}{(1-k_-^2)} + \frac{Y_{I,II}^+}{(1-k_+^2)} \tag{6.8.89}$$

which gives
$$\frac{\mathrm{d}\delta_c(X)}{\mathrm{d}X} = -\frac{i}{2\cosh\pi\epsilon}\left\{\frac{\kappa C_m^I \Re[\overline{K}\,(|X|/x_0)^{-i\epsilon}]}{\sqrt{2\pi|X|}} + \frac{iC_m^{II}\Im[\overline{K}\,(|X|/x_0)^{-i\epsilon}]}{\kappa\sqrt{2\pi|X|}}\right\} \tag{6.8.90}$$

Note that κ, $Y_I^\pm$ and $Y_{II}^\pm$ equal unity in the stationary case.

The energy flux
The energy flux into the crack edge is determined in the same way as for the stationary interface crack. The result is
$$\mathcal{G} = \frac{\kappa^2 C_m^I + C_m^{II}/\kappa^2}{16\cosh^2\pi\epsilon}K\overline{K} \tag{6.8.91}$$

In the derivation it is helpful to note that $\kappa^4 = C_m^{II}/C_m^I$. Note also that the result, as expected, is independent of x_0.

Discussion

The complex quantity $\mathcal{K}$ does not play the same strong role as the complex stress intensity factor K for the stationary interface crack. However, when the crack velocity approaches zero, the two coincide.

The energy flux changes sign when the velocity exceeds the lower Rayleigh wave velocity, because both C_m^I and C_m^{II} change sign. Thus, the lower Rayleigh wave velocity is an upper limit for subsonic interface cracks.

The stress distributions near the crack edge are found by using the expressions for L and M, (6.8.75) and (6.8.77) and their connections with F'' and S''. The arguments for the latter functions, related to X and Y, are taken from (6.8.32). Thus, for instance, for τ_{xy}^+, expression (6.8.46) is used (with ζ_+^P in F'' and ζ_+^S in S'' instead of ξ), whereupon F'' and S'' are expressed with the aid of (6.8.75) and (6.8.77).

Only a few steady state interface problems are of physical interest. One problem with reasonable physical significance would be propagation of frictionless slip along a portion of an interface with energy-neutral healing at the trailing end. The condition of zero normal stress on the crack faces would then be changed to the condition of zero crack opening. This would prevent oscillating singularities. However, complications arise, because a strong presence of remote shear stresses implies opening of the crack towards its trailing end. An infinite compressive stress would be needed for contact along the whole crack.

A curiosity is that oscillating singularites may be absent even if the two materials are not identical. This obviously requires $\epsilon = 0$, which can occur for certain material combinations at the velocity for which

$$\frac{\mu_- R_-}{C_-} = \frac{\mu_+ R_+}{C_+} \tag{6.8.92}$$

The intersonic in-plane case

Formulation of Hilbert problems

The same notations as for the in-plane subsonic case will be adopted, except when otherwise defined. It is assumed that

$$\max(c_S^+, c_S^-) < V < \min(c_P^+, c_P^-) \tag{6.8.93}$$

Define the following constants:

$$b_S^\pm = \sqrt{\beta_\pm^2/k_\pm^2 - 1}, \quad P_\pm = -[(b_S^\pm)^2 - 1], \quad Q_\pm = 1 + (b_S^\pm)^2$$
$$R_\pm = 4ia_P^\pm b_S^\pm - P_\pm^2, \quad C_\pm = P_\pm - 2ia_P^\pm b_S^\pm \tag{6.8.94}$$

Note that these notations are consistent with those for the subsonic case if $a_S^\pm$ is identified with $ib_S^\pm$. The analysis follows essentially the same lines as for the subsonic case. Equations (6.8.31) will be changed to

$$(b_S^\pm)^2 \frac{\partial^2 \psi_\pm}{\partial X^2} - \frac{\partial^2 \psi_\pm}{\partial Y^2} = 0 \tag{6.8.95}$$

with the solutions

$$\psi_+ = s(\xi + \eta_+^S), \quad \psi_- = t(\xi - \eta_-^S) \tag{6.8.96}$$

where $\xi = X$ and $\eta_\pm^S = b_S^\pm Y$. Equations (6.8.46)-(6.8.49) will then be changed to

$$\tau_{xy}^+ = \mu_+\left(2\frac{\partial^2\phi_+}{\partial X \partial Y} + \frac{\partial^2\psi_+}{\partial Y^2} - \frac{\partial^2\psi_+}{\partial X^2}\right)$$
$$= \mu_+\{ia_P^+[F_+''(\xi) - F_-''(\xi)] - P_+ s''(\xi + \eta_+^S)\} \quad (6.8.97)$$

$$\sigma_y^+ = -\frac{\mu_+}{2}\{P_+[F_+''(\xi) + F_-''(\xi)] + 4b_S^+ s''(\xi)\} \quad (6.8.98)$$

$$u_+ = \frac{1}{2}\{F_+'(\xi) + F_-'(\xi) + 2b_S^+ s'(\xi)\} \quad (6.8.99)$$

$$v_+ = \frac{1}{2}\{ia_P^+[F_+'(\xi) - F_-'(\xi)] - 2s'(\xi)\} \quad (6.8.100)$$

for $Y = 0$. Similar expressions, using $G_\pm(\cdot)$ and $t(\cdot)$ are obtained for the lower half-plane.

The conditions of traction free crack faces may preferably be formulated as

$$P_+\sigma_y^+ - 2b_S^+\tau_{xy}^+ = 0, \quad P_-\sigma_y^- + 2b_S^-\tau_{xy}^- = 0 \quad (6.8.101)$$

for $\xi = 0$. Then, the functions $s''(\xi)$ and $t''(\xi)$ will disappear, and the resulting equations will be

$$\overline{R}_+ F_+''(\xi) + R_+ F_-''(\xi) = 0 \text{ for } \xi < 0 \quad (6.8.102)$$
$$\overline{R}_- G_+''(\xi) + R_- G_-''(\xi) = 0 \text{ for } \xi < 0 \quad (6.8.103)$$

The continuity conditions for the interface stresses and displacements read:

$$\mu_+ P_+[F_+''(\xi) + F_-''(\xi)] + 4\mu_+ b_S^+ s''(\xi)$$
$$= \mu_- P_-[G_+''(\xi) + G_-''(\xi)] - 4\mu_- b_S^- t''(\xi) \quad (6.8.104)$$

$$i\mu_+ a_P^+[F_+''(\xi) - F_-''(\xi)] - \mu_+ P^+ s''(\xi)$$
$$= \mu_- a_P^-[G_+''(\xi) - G_-''(\xi)] - \mu_- P_- t''(\xi) \quad (6.8.105)$$

$$F_+''(\xi) + F_-''(\xi) = G_+''(\xi) + G_-''(\xi)] \text{ for } \xi > 0 \quad (6.8.106)$$
$$a_P^+[F_+''(\xi) - F_-''(\xi)] = -a_P^-[G_+''(\xi) - G_-''(\xi)] \text{ for } \xi > 0 \quad (6.8.107)$$

Elimination of $s''(\xi)$ and $t''(\xi)$ gives

$$A_P F_+''(\xi) + \overline{A}_P F_-''(\xi) = B_N G_+''(\xi) + \overline{B}_N G_-''(\xi) \text{ for } \xi > 0 \quad (6.8.108)$$
$$B_P F_+''(\xi) + \overline{B}_P F_-''(\xi) = A_N G_+''(\xi) + \overline{A}_N G_-''(\xi) \text{ for } \xi > 0 \quad (6.8.109)$$

where

$$A_P = \mu_+[b_S^+ Q_+ P_- + b_S^- Q_- P_+ + 2ia_P^+ b_S^-(b_S^+ Q_- + b_S^- Q_+)] \quad (6.8.110)$$
$$A_N = \mu_-[b_S^- Q_- P_+ + b_S^+ Q_+ P_- + 2ia_P^- b_S^+(b_S^- Q_+ + b_S^+ Q_-)] \quad (6.8.111)$$
$$B_P = \mu_+(b_S^+ + b_S^-)\overline{R}_+ + 2\mu_-(b_S^+ P_- + b_S^- P_+)(1 + ia_P^+ b_S^+) \quad (6.8.112)$$
$$B_N = \mu_-(b_S^- + b_S^+)\overline{R}_- + 2\mu_+(b_S^- P_+ + b_S^+ P_-)(1 + ia_P^- b_S^-) \quad (6.8.113)$$

Equations (6.8.102) and (6.8.103) show that the sectionally analytic functions $L(\zeta)$ and $M(\zeta)$, defined by the relations

$$L_+(\zeta) = \overline{R}_+ F_+''(\zeta), \quad L_-(\zeta) = -R_+ F_-''(\zeta) \quad (6.8.114)$$
$$M_+(\zeta) = \overline{R}_- G_+''(\zeta), \quad M_-(\zeta) = -R_- G_-''(\zeta) \quad (6.8.115)$$

are analytic in the plane cut along the positive real axis, because

$$L_+(\xi) - L_-(\xi) = 0, \quad M_+(\xi) - M_-(\xi) = 0 \text{ for } \xi < 0 \qquad (6.8.116)$$

Now, as for the subsonic case, a constant λ is sought such that (6.8.108)-(6.8.109) can be written in the form

$$L_+(\xi) - \lambda L_-(\xi) = 0, \quad M_+(\xi) - \lambda M_-(\xi) = 0 \text{ for } \xi > 0 \qquad (6.8.117)$$

thereby, together with (6.8.116), forming two Hilbert problems. Insertion of $F''_\pm(\xi)$ and $G''_\pm(\xi)$ expressed in $L_\pm(\xi)$ and $M_\pm(\xi)$ into (6.8.108)-(6.8.109), together with use of (6.8.117), gives

$$\left(\frac{A_P}{\overline{R}_+}\lambda - \frac{\overline{A}_P}{R_+}\right)L_-(\xi) = \left(\frac{B_N}{\overline{R}_-}\lambda - \frac{\overline{B}_N}{R_-}\right)M_-(\xi) \qquad (6.8.118)$$

$$\left(\frac{B_P}{\overline{R}_+}\lambda - \frac{\overline{B}_P}{R_+}\right)L_-(\xi) = \left(\frac{A_N}{\overline{R}_-}\lambda - \frac{\overline{A}_N}{R_-}\right)M_-(\xi) \qquad (6.8.119)$$

Vanishing determinant gives

$$\lambda^2 - \frac{UR_+\overline{R}_- + \overline{U}\,\overline{R}_+R_-}{VR_+R_-}\lambda + \frac{\overline{V}\,\overline{R}_+\overline{R}_-}{VR_+R_-} = 0 \qquad (6.8.120)$$

where

$$U = A_P\overline{A}_N - B_P\overline{B}_N, \quad V = A_PA_N - B_PB_N \qquad (6.8.121)$$

The equation for λ may be written in the form

$$\lambda^2 - 2\frac{B}{A}\lambda + \frac{\overline{A}}{A} = 0 \qquad (6.8.122)$$

where $A = VR_+R_-$ and $B = \Re(UR_+\overline{R}_-)$. The solutions are

$$\lambda_{1,2} = \frac{B \pm \sqrt{B^2 - A\overline{A}}}{A} \quad \text{if } |A| < |B|. \quad \text{Note that } \lambda_2 = \frac{1}{\overline{\lambda}_1} \qquad (6.8.123)$$

$$\lambda_{1,2} = \frac{B \pm i\sqrt{A\overline{A} - B^2}}{A} \quad \text{if } |A| \geq |B|. \quad \text{Note that } |\lambda_1| = |\lambda_2| = 1 \qquad (6.8.124)$$

The roots λ_1 and λ_2 are complex, except when $A = \overline{A} = B$. Each root implies a relation between L_- and M_-:

$$M_-(\xi) = \frac{A_P\lambda/\overline{R}_+ - \overline{A}_P/R_+}{B_N\lambda/\overline{R}_- - \overline{B}_N/R_-}L_-(\xi) \qquad (6.8.125)$$

Because asymptotic solutions are sought, the ratio $M_-(\xi)/L_-(\xi)$ is also the ratio between the functions $M(\zeta)$ and $L(\zeta)$, which differ only by their amplitude factors. If $L_1(\zeta)$ and $M_1(\zeta)$ belong to $\lambda = \lambda_1$, then inspection of (6.8.125) shows that

$$M_1(\zeta) = \kappa L_1(\zeta), \text{ where } \kappa = \frac{A_P\lambda_1/\overline{R}_+ - \overline{A}_P/R_+}{B_N\lambda_1/\overline{R}_- - \overline{B}_N/R_-} \qquad (6.8.126)$$

If $L_2(\zeta)$ and $M_2(\zeta)$ belong to $\lambda = \lambda_2$ and $\lambda_2 = 1/\overline{\lambda}_1$, then

$$M_2(\zeta) = \overline{\kappa}L_2(\zeta) \qquad (6.8.127)$$

Solution of the Hilbert problems

Distinction is made between the two cases $\lambda_2 = 1/\overline{\lambda_1}$ and $|\lambda_{1,2}| = 1$. In the first case, analogy with the subsonic case shows that $L_{1,2}(\zeta)$ and $M_{1,2}(\zeta)$ are found to be proportional to $\zeta^{-g_{1,2}}$, where

$$g_1 = \frac{1}{2\pi i}\ln(\lambda_1) = \frac{1}{2\pi}\arg(\lambda_1) + i\epsilon = g, \quad g_2 = \overline{g} \qquad (6.8.128)$$

$$\epsilon = -\frac{1}{2\pi}\ln|\lambda_1|, \quad 0 \le \arg(\cdot) < 2\pi \qquad (6.8.129)$$

if $\lambda_2 = 1/\overline{\lambda_1}$. The branch is chosen so that $\zeta^{-g} = |\xi|^{-g}$ and $\zeta^{-\overline{g}} = |\xi|^{-\overline{g}}$ for $\eta = 0$, $\xi < 0$. Then, $\zeta_\pm^{-g} = \xi^{-g}e^{\pm \pi i g}$ and $\zeta_\pm^{-\overline{g}} = \xi^{-\overline{g}}e^{\pm \pi i \overline{g}}$ for $\eta = 0$, $\xi > 0$. Note that $0 \le \Re(g) < 1$, so that the strongest possible singularity is obtained which does not violate the condition of bounded stress-strain energy.

The solutions for $L(\zeta) = L_1(\zeta) + L_2(\zeta)$ and $M(\zeta) = M_1(\zeta) + M_2(\zeta)$ are now written as

$$L_1(\zeta) = P_1\zeta^{-g}, \quad M_1(\zeta) = \kappa P_1\zeta^{-g} \qquad (6.8.130)$$
$$L_2(\zeta) = P_2\zeta^{-\overline{g}}, \quad M_2(\zeta) = \overline{\kappa} P_2\zeta^{-\overline{g}} \qquad (6.8.131)$$

Then, the combination

$$P_+\sigma_y^+ - 2b_S^+\tau_{xy}^+ = \frac{\mu_+}{2}[L_+(\xi) - L_-(\xi)] = \frac{i\mu_+}{2}[P_1\xi^{-g}(e^{\pi i g} - e^{-\pi i g})$$
$$+ P_2\xi^{-\overline{g}}(e^{\pi i \overline{g}} - e^{-\pi i \overline{g}})] \quad \text{for } \eta = 0, \xi > 0 \qquad (6.8.132)$$

The condition that this expression is real requires that $P_2 = \overline{P}_1$.

In the second case, $|\lambda_{1,2}| = 1$, the singularities are real:

$$g_{1,2} = \frac{1}{2\pi}\arg(\lambda_{1,2}) \qquad (6.8.133)$$

and, because the asymptotic field is asked for, only the strongest singularity, given by $g = \max(g_1, g_2)$ is of interest. The branch is chosen so that $\zeta^{-g} = |\xi|^{-g}$ for $\eta = 0$, $\xi = 0$. Then, $\zeta_\pm^{-g} = \xi^{-g}e^{\pm i\pi g}$ for $\eta = 0$, $\xi = 0$.

Assume that indices are chosen so that $\arg(\lambda_1) > \arg(\lambda_2)$ when $|\lambda_{1,2}| = 1$, the case for which g is real. This gives $g = g_1$. Then, $L(\zeta) = L_1(\zeta)$, $M(\zeta) = M_1(\zeta)$, because only the strongest singularity is to be represented in the asymptotic solutions. These solutions are now written as

$$L(\zeta) = P_1\zeta^{-g}, \quad M(\zeta) = \kappa P_1\zeta^{-g} \qquad (6.8.134)$$

Then, the combination

$$P_+\sigma_y^+ - 2b_S^+\tau_{xy}^+ = \frac{\mu_+}{2}[L_+(\xi) - L_-(\xi)] = P_1\mu_+\xi^{-g}\sin\pi g \quad \text{for } \eta = 0, \xi > 0 \qquad (6.8.135)$$

The condition that this expression is real requires that P_1 is real. Note that κ is real in this case, perhaps most easily seen if λ_1 is written as $e^{2i\alpha}$, where α is real, in the expression (6.8.126) for κ.

Recall that the presence of an imaginary component to the stress singularity implies an oscillatory region near the crack edge. Obviously, such oscillations do not occur if $\epsilon = 0$, which implies $|\lambda_{1,2}| = 1$, which in turn happens if $|A| = |VR_+R_-| \ge |B| = |\Re(UR_+\overline{R}_-)|$.

Stresses, displacements and energy flux

Expressions for stresses and displacements may now be found in the same way as for the subsonic case, in particular those on the symmetry plane. It then remains to show that the crack propagation is energetically possible. In the homogeneous intersonic case, it is possible for mode II, but not for mode I. This result was obtained by assuming a Barenblatt region at the crack edge. The same technique is, in principle, also possible for interface cracks. It can be expected to show that the energy flow to a Barenblatt region is positive when there is a strong presence of shear stresses on the symmetry plane ahead of the crack, at a distance much larger than the length of the oscillatory region, but much smaller than the crack length or other similar characteristic length dimensions.

Actually, determination of the sign of the energy flux may be done more conveniently without introduction of a Barenblatt region. To demonstrate this, note first from expression (4.12.29) that the energy flux is positive if $\Im[s(X) \cdot d\delta_c(X)/dX]$ is negative for X within the Barenblatt region. Make the reasonable, not very critical, assumption that the shape of the Barenblatt region, represented by $D(\cdot)$ in (6.3.67), may be taken to be the same for shear and opening components. Then, the sign of $\Im[s(X) \cdot d\delta_c(X)/dX]$ should not be affected if X in $s(X)$ is taken from a position a small distance ΔX_0 ahead of the Barenblatt region and X in $d\delta_c(X)/dX$ is taken from a position the same distance behind the Barenblatt region. Furthermore, this sign should be the same if the Barenblatt region shrinks to zero, then being the sign of $\Im\{[s(X)]_{X=\Delta X_0} \cdot [d\delta_c(X)/dX]_{X=-\Delta X_0}\}$. The sign of this quantity is then in general the same as the sign of the constant

$$\Im\left\{[s(X) \cdot X^g]_{X \to +0} \cdot \left[\frac{d\delta_c(X)}{dX} \cdot |X|^{\bar{g}}\right]_{X \to -0}\right\} \quad (6.8.136)$$

and the conclusion would be that a negative sign of this constant indicates positive energy flow to a finite Barenblatt region, although the expression for the constant is based on the simplifying assumption of a point-sized process region.

In contrast to the intersonic case in homogeneous materials, S waves appear in front of the crack. They are generated at the interface by P waves – their presence is required to satisfy all four continuity conditions for the interface ahead of the crack.

The subsonic/supersonic anti-plane case

Huang *et al.* (1996) demonstrated that an interface crack can propagate under antiplane deformation with velocities above the sound (S wave) velocity for one of the two media. There is no oscillatory stress at the crack edge as for subsonic in-plane deformation, but the stress singularity is weaker than an inverse square root singularity. They found a new form of propagating singularity connected with crack motion: a singularity along an entire plane moving with the crack edge.

Consider a horizontal crack propagating under anti-plane deformation with velocity V to the right – positive x direction – in the interface $y = 0$ between two linearly elastic dissimilar media with S wave velocities c_S^- and c_S^+. Here, as in the following, the minus index refers to the lower, the plus index to the upper medium. It is assumed that $c_S^+ < V < c_S^-$, so that the crack velocity is subsonic with respect to the lower medium, supersonic with respect to the upper medium. The crack edge vicinity will

6.8 CRACK PROPAGATION ALONG AN INTERFACE

be studied, so that the body may be taken as infinite, with vanishing remote stress, and the crack semi-infinite. With the Galilean transformation $x - Vt = X$, $y = Y$, the wave equations are, cf. (6.4.1),

$$(a_S^-)^2 \frac{\partial^2 w_-}{\partial X^2} + \frac{\partial^2 w_-}{\partial Y^2} = 0 \tag{6.8.137}$$

$$(b_S^+)^2 \frac{\partial^2 w_+}{\partial X^2} - \frac{\partial^2 w_+}{\partial Y^2} = 0 \tag{6.8.138}$$

where $a_S^- = \sqrt{1 - \gamma_-^2}$ and $b_S^+ = \sqrt{\gamma_+^2 - 1}$, with $\gamma_- = V/c_S^-$ and $\gamma_+ = V/c_S^+$.
Assuming the crack faces to be traction free, the boundary conditions on $Y = 0$ are

$$\tau_{yz}^- = \tau_{yz}^+ \text{ for all } X \tag{6.8.139}$$

$$\tau_{yz}^- = 0 \text{ for } X < 0 \tag{6.8.140}$$

$$w_- = w_+ \text{ for } X > 0 \tag{6.8.141}$$

Note that $\tau_{yz}^{\pm} = \mu_{\pm} \partial w_{\pm}/\partial y$.
The solution of equation (6.8.138) may be written as

$$w_+ = h(X + b_S^+ Y) \tag{6.8.142}$$

A function $h(X - b_S^+ Y)$, also satisfying equation (6.8.138), is discarded because it implies an incoming wave to the crack.

With the substitutions $\xi = X$ and $\eta = a_S^- Y$, equation (6.8.137) becomes the Laplace equation

$$\frac{\partial^2 w_-}{\partial \xi^2} + \frac{\partial^2 w_-}{\partial \eta^2} = 0 \tag{6.8.143}$$

Because the real part of an analytic function satisfies the Laplace equation, the solution can be written as

$$w_- = \Re F_-(\zeta) = \frac{1}{2}[F_-(\zeta) + \overline{F_-(\zeta)}] \tag{6.8.144}$$

where $\zeta = \xi + i\eta$ and $F_-(\zeta)$ is analytic for $\eta \leq 0$, with the possible exception of isolated points on $\eta = 0$. Note that

$$\frac{\partial w_-}{\partial y} = \frac{ia_S^-}{2}[F_-'(\zeta) - \overline{F_-'(\zeta)}] \tag{6.8.145}$$

Define now the function $F_+(\zeta) = \overline{F_-(\zeta)}$, which is analytic for $\eta \geq 0$, with the possible exception for isolated points on $\eta = 0$, cf. A2.22. Then, the boundary conditions (6.8.139)-(6.8.141) can be expressed as

$$\mu_+ b_S^+ h'(\xi) = -i\frac{\mu_- a_S^-}{2}[F_+'(\xi) - F_-'(\xi)] \text{ for all } \xi \tag{6.8.146}$$

$$F_+'(\xi) - F_-'(\xi) = 0 \text{ for } \xi < 0 \tag{6.8.147}$$

$$h(\xi) = \frac{1}{2}[F_+(\xi) + F_-(\xi)] \text{ for } \xi > 0 \tag{6.8.148}$$

Elimination of $h(\xi)$ gives

$$\mu_+ b_S^+ [F_+'(\xi) + F_-'(\xi)] = -i\mu_- a_S^- [F_+'(\xi) - F_-'(\xi)] \text{ for } \xi > 0 \tag{6.8.149}$$

resulting in the Hilbert problem

$$F'_+(\xi) - F'_-(\xi) = 0 \text{ for } \xi < 0 \tag{6.8.150}$$

$$F'_+(\xi) + \frac{\mu_+ b_S^+ - i\mu_- a_S^-}{\mu_+ b_S^+ + i\mu_- a_S^-} F'_-(\xi) = 0 \text{ for } \xi > 0 \tag{6.8.151}$$

Write the last equation as

$$F'_+(\xi) - e^{2\pi i g} F'_-(\xi) = 0 \text{ for } \xi > 0 \tag{6.8.152}$$

where

$$g = \frac{n}{2} - \frac{1}{\pi}\operatorname{atan}\frac{\mu_- a_S^-}{\mu_+ b_S^+} \tag{6.8.153}$$

and n is an odd number. Introduce now the sectionally analytic function $G(\zeta) = \zeta^{g-1}$ defined through a branch cut along the positive real axis with the branch chosen so that $G_+(\xi)$ is real and positive for $\xi > 0$. Then,

$$\frac{G_-(\xi)}{G_+(\xi)} = \begin{cases} 1 & \text{for } \xi < 0 \\ e^{2\pi i g} & \text{for } \xi > 0 \end{cases} \tag{6.8.154}$$

so that both equations of the Hilbert problem can be written in the common form

$$G_+(\xi)F'_+(\xi) - G_-(\xi)F'_-(\xi) = 0 \text{ for all } \xi \tag{6.8.155}$$

with the solution

$$G(\zeta)F(\zeta) = P(\zeta) \tag{6.8.156}$$

where $P(\zeta)$ is a finite polynomial, which reduces to a constant, P_0, because the remote stresses are zero. Hence,

$$F(\zeta) = P_0 \zeta^{1-g} \tag{6.8.157}$$

and then the function $h(\xi)$ is found from (6.8.146):

$$h(\xi) = -i\frac{\mu_- a_S^-}{2\mu_+ b_S^+}[F_+(\xi) - F_-(\xi)]$$

$$= \begin{cases} 0 & \text{for } \xi < 0 \\ -i\dfrac{\mu_- a_S^-}{2\mu_+ b_S^+}\xi^{1-g} P_0(1 - e^{-2\pi i g}) & \text{for } \xi > 0 \end{cases} \tag{6.8.158}$$

where an integration constant would only imply a rigid body translation. For convenient inspection of the results, the substitution

$$P_0 = \frac{e^{\pi i g}\tau_0 a^g}{(1-g)\mu_- a_S^- \sin \pi g} \tag{6.8.159}$$

is made, where two interrelated constants have been introduced, τ_0 with dimension stress and a with dimension length. Then the displacement and the stresses in the

6.8 CRACK PROPAGATION ALONG AN INTERFACE

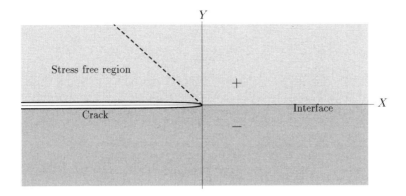

Fig. 6.8.1 The vicinity of the edge of a dynamically propagating intersonic interface crack under anti-plane deformation. The figure shows a section in a vertical plane normal to the crack edge, illustrating the plane of stress singularity and the stress free wedge behind the plane of stress singularity (dashed). The crack is symbolically shown with an opening, but in reality only sliding takes place in mode III. The crack velocity is supersonic with respect to the upper medium and subsonic with respect to the lower medium.

upper half-plane can be written in the form

$$w_+(X,Y) = h(X + b_S^+ Y)$$
$$= \begin{cases} 0 & \text{for } X + b_S^+ Y < 0 \\ \dfrac{\tau_0 a}{(1-g)\mu_+ b_S^+} \left(\dfrac{X + b_S^+ Y}{a}\right)^{1-g} & \text{for } X + b_S^+ Y > 0 \end{cases} \quad (6.8.160)$$

$$\tau_{yz}^+(X,Y) = \begin{cases} 0 & \text{for } X + b_S^+ Y < 0 \\ \tau_0 \cdot \left(\dfrac{a}{X + b_S^+ Y}\right)^g & \text{for } X + b_S^+ Y > 0 \end{cases} \quad (6.8.161)$$

$$\tau_{xz}^+(X,Y) = \begin{cases} 0 & \text{for } X + b_S^+ Y < 0 \\ \dfrac{\tau_0}{b_S^+} \cdot \left(\dfrac{a}{X + b_S^+ Y}\right)^g & \text{for } X + b_S^+ Y > 0 \end{cases} \quad (6.8.162)$$

These results fix n to be 1, so that $0 < g < 1/2$, giving the strongest admissible singularity at he crack edge. Obviously, both constants τ_0 and a may be taken as real.

An interesting result is that the wedge $0 \leq b_S^+ Y < -X$ is stress free and that there is a *stress singularity along the whole plane* $X + b_S^+ Y = 0$, not just at the crack edge; see Fig. 6.8.1. Thus, consider a plane parallel to and ahead of the front plane of the wedge. The shear stress on this plane is

$$\tau_{yz} \cos^2 \varphi - \tau_{xz} \sin^2 \varphi = -\tau_0 \cdot \dfrac{[1+(b_S^+)^2]^{(1-g)/2}}{b_S^+} \cdot \left(\dfrac{a}{\Delta s}\right)^g \quad (6.8.163)$$

where Δs is the distance between the two planes and φ is their angular direction from the positive X axis.

Consider now the lower half-plane. Introduce for a moment polar coordinates (ϱ, ω) in the ξ, η plane. Then, recalling that a branch cut is made along $\omega = 0$, so that

$\pi < \omega < 2\pi$,

$$F_-(\zeta) = P_0 \varrho^{1-g} e^{i\omega(1-g)} = \frac{\tau_0 e^{\pi i g}}{(1-g)\mu_- a_S^-\sin\pi g} \cdot \left(\frac{\varrho}{a}\right)^{1-g} e^{i\omega(1-g)} \quad (6.8.164)$$

By using (6.8.144), the displacement in the lower half-plane is found to be

$$w_-(\varrho,\omega) = \frac{\tau_0 a}{(1-g)\mu_- a_S^-\sin\pi g} \cdot \left(\frac{\varrho}{a}\right)^{1-g} \cos[(\pi-\omega)g+\omega] \quad (6.8.165)$$

The crack sliding displacement is

$$w_-(\varrho,\pi) = -\frac{\tau_0 a}{(1-g)\mu_- a_S^-\sin\pi g} \cdot \left(\frac{\varrho}{a}\right)^{1-g}$$

$$= -\frac{\tau_0 a}{1-g} \cdot \left(\frac{|X|}{a}\right)^{1-g} \sqrt{\frac{1}{(\mu_- a_S^-)^2} + \frac{1}{(\mu_+ b_S^+)^2}} \quad (6.8.166)$$

Now, because $\partial w_-/\partial X > 0$ on the crack face if $\tau_{yz} > 0$ on the interface, positive energy would flow to a process region, such as a Barenblatt region, cf. discussion on page 390. Thus, intersonic crack propagation along an interface crack under anti-plane deformation is, in principle, possible.

The polar coordinates (ϱ,ω), $\pi \leq \omega \leq 2\pi$, are related to the polar coordinates r,φ, $\pi \leq \varphi \leq 2\pi$, in the X,Y plane by the equations

$$\varrho = r\sqrt{\cos^2\varphi + (a_S^-\sin\varphi)^2} \quad (6.8.167)$$

$$\omega = \operatorname{atan}(a_S^- \tan\varphi) \quad (6.8.168)$$

If so desired, the strains in the lower half-plane may be found from the displacement by differentiations, whereupon the stresses are found from Hooke's law. Note that the stress singularity is weaker than the subsonic inverse square root singularity. The singularity exponent, g, increases from 0 to 1/2 when the crack velocity increases from c_S^+ to c_S^-. Unlike τ_{yz}, the shear stress τ_{xz} is discontinuous for $Y=0$, so that $\tau_{xz}^+(X,0) = (a_S^-/b_S^+)\tau_{xz}^-(X,0)$ for $X>0$.

Figure 6.8.2 shows displacements and slips at the interface as well as the moving singularity plane. Data used for the figure are: $\mu_+/\mu_- = 0.357$, $V = 0.75 c_S^- = 1.5 c_S^+$, giving the stress singularity exponent $g \approx 0.173$.

The subsonic/intersonic in-plane case

The subsonic-intersonic interface crack problem has been treated by Yu and Yang (1995) and by Huang et al. (1996). Yu and Yang used an integral equation approach, whereas Huang et al. used differential equations. Velocities between the two S wave velocities were considered, if the higher S wave velocity was lower than the lower P wave velocity, but otherwise velocities in the range between the lower S and P wave velocities. These intervals may include the Rayleigh wave velocity for the medium with higher S wave velocity. It was found that the stress singularity can be either oscillatory as in the subsonic case or non-oscillatory, but then with a weaker singularity than the inverse square root. Oscillatory singularities appear in some middle part of the velocity region studied. A pronounced discontinuity of the behaviour appeared

6.8 CRACK PROPAGATION ALONG AN INTERFACE 395

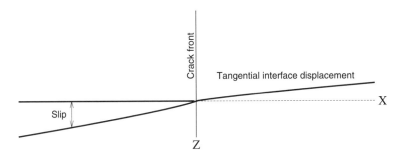

Fig. 6.8.2 The vicinity of the edge of a dynamically propagating intersonic interface crack under anti-plane deformation. The figure shows the interface, as if looked upon from above, illustrating how an originally straight line (dashed) on the interface in the crack direction is deformed. Behind the crack front, slip occurs along the interface, so that such a line splits into two parts. One part, the lower branch in the figure, is attached to the lower medium (in which the crack velocity is subsonic), and the other part, which is straight, is attached to the upper medium (in which the crack velocity is supersonic). The weak singularity is demonstrated by the deviation from the parabolic crack sliding displacement in subsonic cases. The interface vicinity of the upper medium is stress free behind the crack front.

for the Rayleigh wave velocity, if this is involved. As for the anti-plane case, a stress singularity is connected with a plane through the crack edge, moving away from the crack with the S wave velocity in the medium with the lower S wave velocity. If the crack edge singularity is oscillatory, this singularity is also oscillatory.

The mixed case is, indeed, very complicated, and requires skilful administration of the interpretation procedure. The theoretical results appeared to agree well with experimental results found by Liu, Lambros and Rosakis (1993) and by Lambros and Rosakis (1995). For a PMMA/steel combination, they found crack velocities exceeding the S wave velocity in the PMMA. Lambros and Rosakis could also observe a finite length contact region, obviously associated with decohesion, i.e. a process region in the interface adhesive and the roughened bimaterial surfaces. The existence of such a region was later incorporated in an analytical model by Huang et al. (1998), thus enabling determination of the energy flux.

Here, some features of the analysis (without a cohesive region model) will be discussed, following essentially the same lines as for the subsonic and intersonic in-plane cases and using the same notations as for corresponding quantities in these cases. In particular:

$$c_S^+ < V < c_S^-, \quad \beta_\pm = \frac{V}{c_P^\pm} = \frac{k_\pm V}{c_S^\pm} \qquad (6.8.169)$$

$$a_P^\pm = \sqrt{1 - \beta_\pm^2}, \quad a_S^- = \sqrt{1 - \frac{\beta_-^2}{k_-^2}}, \quad b_S^+ = \sqrt{\frac{\beta_+^2}{k_+^2} - 1} \qquad (6.8.170)$$

Expressions for stresses and displacements may now be taken from the subsonic case for the lower half-plane and from the intersonic case for the upper half-plane.

The boundary conditions are:

$$\sigma_y^+ = \sigma_y^- \Longrightarrow \mu_+ P_+[F_+'''(\xi) + F_-'''(\xi)] + 4\mu_+ b_S^+ s''(\xi)$$
$$= \mu_- P_-[G_+'''(\xi) + G_-'''(\xi)] + 2i\mu_- a_S^-[T_+'''(\xi) + T_-'''(\xi)] \text{ for all } \xi \quad (6.8.171)$$

$$\tau_{xy}^+ = \tau_{xy}^- \Longrightarrow 2i\mu_+ a_P^-[F_+'''(\xi) - F_-'''(\xi)] - 2\mu_+ P^+ s''(\xi)$$
$$= \mu_- P_-[G_+'''(\xi) - G_-'''(\xi)] + 2i\mu_- a_S^-[T_+'''(\xi) - T_-'''(\xi)] \text{ for all } \xi \quad (6.8.172)$$

$$\sigma_y^+ = 0 \Longrightarrow P_-[G_+'''(\xi) + G_-'''(\xi)] + 2ia_S^-[T_+'''(\xi) + T_-'''(\xi)] = 0 \text{ for } \xi < 0 \quad (6.8.173)$$

$$\tau_{xy}^+ = 0 \Longrightarrow 2ia_P^-[G_+'''(\xi) + G_-'''(\xi)] - P_-[T_+'''(\xi) + T_-'''(\xi)] = 0 \text{ for } \xi < 0 \quad (6.8.174)$$

$$u_+' = u_-' \Longrightarrow F_+'''(\xi) + F_-'''(\xi) + 2b_S^+ s''(\xi) - G_+'''(\xi) - G_-'''(\xi)$$
$$+ ia_S^-[T_+'''(\xi) + T_-'''(\xi)] \text{ for } \xi > 0 \quad (6.8.175)$$

$$v_+' = v_-' \Longrightarrow ia_P^+[F_+'''(\xi) - F_-'''(\xi)] - 2s''(\xi)$$
$$+ ia_P^-\mu_- P_-[G_+'''(\xi) - G_-'''(\xi)] - a_S^-[T_+'''(\xi) - T_-'''(\xi)] \text{ for all } \xi \quad (6.8.176)$$

Function s'' can be eliminated from the two first equations, yielding $\Theta_+(\xi) = \Theta_-(\xi)$, where

$$\Theta_+(\zeta) = \mu_+ \overline{R}_+ F_+'''(\zeta) + \mu_-(P_+ P_- - 4ia_P^- b_S^+) F_+'''(\zeta) + 2\mu_- \Omega T_+'''(\zeta)$$
$$\Theta_-(\zeta) = -\mu_+ R_+ F_-'''(\zeta) - \mu_-(P_+ P_- + 4ia_P^- b_S^+) G_-'''(\zeta) + 2\mu_- \overline{\Omega} T_-'''(\zeta) \quad (6.8.177)$$

and $\Omega = 2\mu_-(b_S^+ P_- + ia_S^- P_+)$. Obviously, $\Theta_+(\zeta)$ and $\Theta_-(\zeta)$ are analytic in their respective half-plane, and because they are equal on the real axis, they define together a function which is analytic in the whole plane. Because stresses vanish at infinity, this function is zero. Hence, $T_\pm'''(\zeta)$ can be expressed in terms of $F_\pm'''(\zeta)$ and $G_\pm'''(\zeta)$. Insertion of $T_\pm'''(\xi)$ into, for instance, (6.8.171), makes it possible to express $s''(\xi)$ in $F_\pm'''(\xi)$ and $G_\pm'''(\xi)$. Thus, the remaining four equations can be written in terms of $F_\pm'''(\xi)$ and $G_\pm'''(\xi)$, only. The result is

$$\overline{R}_+ F_+'''(\xi) + R_+ F_-'''(\xi) = 0 \text{ for } \xi < 0 \quad (6.8.178)$$

$$2i\mu_+ a_S^- P_- \overline{R}_+ F_+'''(\xi) + \overline{\Omega} R_- G_+'''(\xi) = 2i\mu_+ a_S^- P_- R_+ F_-'''(\xi) - \Omega R_- G_-'''(\xi)$$
$$\text{for } \xi < 0 \quad (6.8.179)$$

$$A_P F_+'''(\xi) + \overline{A}_P F_-'''(\xi) = B_N G_+'''(\xi) + \overline{B}_N G_-'''(\xi) \text{ for } \xi > 0 \quad (6.8.180)$$

$$B_P F_+'''(\xi) + \overline{B}_P F_-'''(\xi) = A_N G_+'''(\xi) + \overline{A}_N G_-'''(\xi) \text{ for } \xi > 0 \quad (6.8.181)$$

where

$$A_P = \mu_+ \overline{\Omega}[Q_+ \Omega + 2i(\mu_+ - \mu_-) a_s^- \overline{R}_+] \quad (6.8.182)$$

$$A_N = -\mu_- P_+ \overline{\Omega}[\mu_- R_- + \mu_+ P_+ C_- - 2i\mu_+ a_P^- b_S^+ Q_-] \quad (6.8.183)$$

$$B_P = \mu_+ \overline{\Omega}[(\mu_+ P_+ - \mu_- P_-) \overline{R}_+ - ia_P^+ \Omega] \quad (6.8.184)$$

$$B_N = 2\mu_- \overline{\Omega}[\mu_- b_S^+ R_- - 2\mu_+ b_S^+ C_- + i\mu_+ a_S^- P_+ Q_-] \quad (6.8.185)$$

Introduce now the functions

$$L_+(\zeta) = \overline{R}_+ F_+'''(\zeta) \text{ for } \Im(\zeta) \geq 0 \quad (6.8.186)$$

$$L_-(\zeta) = -R_+ F_-'''(\zeta) \text{ for } \Im(\zeta) \leq 0 \quad (6.8.187)$$

$$M_+(\zeta) = 2i\mu_+ a_S^- P_- \overline{R}_+ F_+'''(\zeta) + \overline{\Omega} R_- G_+'''(\zeta) \text{ for } \Im(\zeta) \geq 0 \quad (6.8.188)$$

$$M_-(\zeta) = 2i\mu_+ a_S^- P_- R_+ F_-'''(\zeta) - \Omega R_- G_-'''(\zeta) \text{ for } \Im(\zeta) \leq 0 \quad (6.8.189)$$

Then, from (6.8.178)-(6.8.179),

$$L_+(\xi) - L_-(\xi) = 0, \quad M_+(\xi) - M_-(\xi) = 0 \text{ for } \xi < 0 \qquad (6.8.190)$$

Now, as for the subsonic and intersonic cases, a constant λ is sought so that (6.8.180)-(6.8.181) can be written in the form

$$L_+(\xi) - \lambda L_-(\xi) = 0, \quad M_+(\xi) - \lambda M_-(\xi) = 0 \qquad (6.8.191)$$

thereby, together with (6.8.190) forming two Hilbert problems. Insertion of $F''_\pm(\xi)$ and $G''_\pm(\xi)$ expressed in $L_\pm(\xi)$ and $M_\pm(\xi)$ into (6.8.180)-(6.8.181), together with use of (6.8.191), gives

$$(U\lambda - \overline{U})L_-(\xi) - \left(\frac{B_N}{\overline{\Omega}R_-}\lambda - \frac{\overline{B}_N}{\Omega R_-}\right)M_-(\xi) = 0 \text{ for } \xi > 0 \qquad (6.8.192)$$

$$(V\lambda - \overline{V})L_-(\xi) - \left(\frac{A_N}{\overline{\Omega}R_-}\lambda - \frac{\overline{A}_N}{\Omega R_-}\right)M_-(\xi) = 0 \text{ for } \xi > 0 \qquad (6.8.193)$$

where

$$U = \frac{A_P}{R_+} + \frac{2i\mu_+ a_S^- P_- B_N}{\overline{\Omega}R_-}, \quad V = \frac{B_P}{R_+} + \frac{2i\mu_+ a_S^- P_- A_N}{\overline{\Omega}R_-} \qquad (6.8.194)$$

Vanishing determinant gives

$$\lambda^2 - \frac{(A_N\overline{U} - B_N\overline{V})\Omega + (\overline{A}_N U - \overline{B}_N V)\overline{\Omega}}{(A_N U - B_N V)\Omega}\lambda + \frac{(\overline{A}_N U - \overline{B}_N V)\overline{\Omega}}{(A_N U - B_N V)\Omega} = 0 \qquad (6.8.195)$$

Note that this equation is of the same form as the corresponding equation, (6.8.122), for the intersonic case. Consequently, its roots, $\lambda_{1,2}$, are either connected so that $\lambda_2 = 1/\overline{\lambda_1}$ or such that $|\lambda_{1,2}| = 1$. In the former case, the singularity exponent is complex, implying an oscillating singularity, in the latter case it is real so that the singularity is non-oscillating.

The continuing analysis – determination of stresses and displacements and the investigation of the sign of the energy flux to a finite process region – follows the same lines as discussed for the intersonic case. For a complete analysis with incorporation of a cohesive region model, reference is made to Huang et al. (1998).

Huang et al. (1996) made numerical calculations for a few cases and found that oscillating singularities occur in a middle region of the subsonic/intersonic velocity interval, whereas non-oscillating singularities occurred in the remaining parts. The real part of the singularity exponent was claimed to be larger than $-1/2$ in all cases studied, although it appears that it could be smaller in some cases. Thus, if the Rayleigh velocity for the medium with higher S wave velocity is within the crack velocity interval studied, then there is a discontinuity – a step of $1/2$ – of the singularity exponent. An exponent between $-1/2$ and -1 would lead to infinite energy flux to a point-sized process region, but a finite region, such as the Barenblatt model, would receive a finite energy flux.

6.9 Crack expanding with constant velocity

Self-similar dynamic problems

Introduction
Previously treated dynamic crack problems have all been characterized by steady state propagation, allowing use of the Galilean transformation $x - Vt = X$. Here, *self-similar* problems will be considered. These are two- or three-dimensional problems for which dynamic processes start at time $t = 0$, and in which the geometric configuration (including the crack) and the dynamic stress field, including boundary tractions, remain unchanged, apart from a spatial scaling, i.e. a scaling of the spatial coordinates, with a scaling factor increasing in proportion to time. The same applies to the dynamic strain field. Consequently, the dynamic displacement field is subjected to both spatial scaling and to temporal scaling, i.e. scaling of the displacement amplitudes in addition to scaling of the spatial coordinates, with the same scaling factor. Only linearly elastic self-similar problems will be considered.

A typical example of self-similar problems is the expansion of a mode I crack at constant velocity in an infinite medium. The crack length at $t = 0$ is assumed to be zero. In the case of symmetric expansion, the crack length (or diameter in a three-dimensional case) at $t > 0$ is $2a = 2Vt$, where V is the velocity of each crack edge. The problem might not be realistic, because it assumes constant velocity crack growth starting from a point, but it is analytically solvable and can therefore be used for control of numerical methods intended for more realistic cases of dynamic crack propagation. It could also reflect the growth of an expanding crack after an initial period of acceleration. Experiments have shown that accelerating cracks seem to approach a constant terminal velocity; cf. Section 9.1).

The plane strain problem was solved in Broberg (1960) with an integral equation approach. It was later also solved by Craggs (1963), who took immediate advantage of the self-similarity by reducing the number of independent variables from three (for instance r, φ, t) to two, r/t and φ. The substitutions†

$$\text{sech}^{-1}\left(\frac{r}{c_P t}\right) = -\rho_p \tag{6.9.1}$$

$$\text{sech}^{-1}\left(\frac{r}{c_S t}\right) = -\rho_s \tag{6.9.2}$$

transform the governing equations (A9.18) to Laplace equations,

$$\frac{\partial^2 \phi}{\partial \rho_p^2} + \frac{\partial^2 \phi}{\partial \varphi^2} = 0 \tag{6.9.3}$$

$$\frac{\partial^2 \psi}{\partial \rho_s^2} + \frac{\partial^2 \psi}{\partial \varphi^2} = 0 \tag{6.9.4}$$

Due to the symmetry, ϕ need be determined only in the semi-circle $0 \leq \varphi \leq \pi$, $r/c_P t \leq 1$. Because ϕ satisfies the Laplace equation it can be considered as the real part of an analytic function, whereupon mapping of the semicircle onto a half-plane can be performed. After carrying out the same operations for ψ, the problem is readily

---

† These substitutions are a form of the Chaplygin transformations.

solved, as shown by Craggs (1963). The method used by Craggs was earlier described by Ward (1955) and, in a slightly different form, for a crack problem, by Maue (1954). A similar but somewhat simpler method, avoiding conformal mapping, was devised by Smirnov and Sobolev (1932) for self-similar problems in general. A description of the method is also given in Smirnov (1964). Another similar and very convenient method was used by Cherepanov and Afanasev (1974). General methods have also been discussed by Willis (1973). A rather direct method, leading to a Hilbert problem, was used by Atkinson (1965), who solved the problem of a mode I crack, propagating with constant velocity along a symmetry plane of an orthotropic material.

Double Laplace transforms for self-similar crack problems
Here, a different approach to self-similar problems will be taken, using double Laplace transforms. It is based on a general feature of self-similarity: homogeneous expressions, which prevail, not only in the physical region, but also in the Laplace transform region.

Suppose that one-sided Laplace transforms are made with respect to $\tau = c_P t$ (note that τ has dimension length, but may nevertheless be considered as a time variable) with transform variable p, and two-sided transforms are made with respect to x with transform variable q. It can then be shown that the double Laplace transform of a homogeneous function of degree zero, say $f(\tau/x)$, is a homogeneous function of degree zero, say $F(q/p)$, and that $F(\cdot)$ is odd if $f(\cdot)$ is even, and *vice versa*. Further, it can be shown that the double Laplace transform of a homogeneous function of degree n, say $\tau^n f(\tau/x)$ (or $x^n g(\tau/x)$), can be written in the form $p^{-n}F(q/p)$ (or $q^{-n}G(q/p)$). All proofs are similar. For instance, the double Laplace transform of a function $f(\tau/|x|)$ is

$$\mathcal{L}_{qx}\mathcal{L}_{p\tau}f(\tau/|x|) = q\int_{-\infty}^{+\infty} e^{-qx} p \int_0^{\infty} e^{-pt} f(\tau/|x|)\mathrm{d}\tau\,\mathrm{d}x$$

$$= q\int_{-\infty}^0 e^{-qx} f_1(-px)\mathrm{d}x + q\int_0^{\infty} e^{-qx} f_1(px)\mathrm{d}x \quad (6.9.5)$$

where $f_1(p|x|) = \mathcal{L}_{p\tau}f(\tau/|x|)$. It is convenient to assume that p is real and positive; cf. Appendix A5. The last integral in (6.9.5) can obviously be written in the form $p^{-1}h(q/p)$, so that

$$\mathcal{L}_{qx}\mathcal{L}_{p\tau}f(\tau/|x|) = q/p[h(-q/p) + h(q/p)] = F(q/p) \quad (6.9.6)$$

where $F(\cdot)$ is an odd function.

A simple inversion procedure is possible for double Laplace transforms appearing in self-similar problems. For S waves, radiating from $y = 0$ into the half-space $y \geq 0$, such a transform will have the following form:

$$\mathcal{L}_{qx}\mathcal{L}_{p\tau}f^{(S)}(x,y,\tau) = q^n F^{(S)}(q/p) e^{-(p^2/k^2 - q^2)^{1/2}y} \quad (6.9.7)$$

where n is an integer. This will later be shown for a particular case. Alternatively, an expression with p^n rather than q^n may be used. Note that multiplication by q or p implies, simply, differentiation with respect to x or τ, respectively. It is therefore sufficient to discuss the inversion for one specific value of n, and it is convenient to

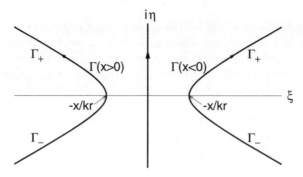

Fig. 6.9.1 The path of integration can be deformed to the left hyperbola branch, if $x > 0$, or to the right hyperbola branch, if $x < 0$. Each branch consists of one part, Γ_+, for $\eta > 0$ and one part, Γ_-, for $\eta < 0$.

choose $n = 1$. Then, a formal inversion with respect to q gives

$$\mathcal{L}_{p\tau} f^{(S)}(x,y,\tau) = \frac{1}{2\pi i} \int_{-i\infty}^{i\infty} e^{qx - (p^2/k^2 - q^2)^{1/2} y} F^{(S)}(q/p) dq$$

$$= \frac{1}{2\pi i} \int_{-i\infty}^{i\infty} p e^{p[\zeta x - (1/k^2 - \zeta^2)^{1/2} y]} F^{(S)}(\zeta) d\zeta \quad (6.9.8)$$

where the substitution $q = p\zeta$, with $\zeta = \xi + i\eta$, was made. It is assumed that the strip of convergence for the Laplace transform includes the imaginary axis. Branch cuts are made along $|\xi| > 1/k$, $\eta = 0$, and the branch is chosen so that $(1/k^2 - \zeta^2)^{1/2} = \sqrt{1/k^2 - \xi^2}$ for $|\xi| < 1/k$, $\eta = 0$. This implies that $(1/k^2 - \zeta^2)^{1/2} \to \mp i\zeta$ as $|\zeta| \to \infty$, where the minus sign refers to $\Im\zeta > 0$, the plus sign to $\Im\zeta < 0$.

Assume that the path of integration can be deformed to the path Γ, the hyperbola branch

$$\frac{\xi^2}{x^2} - \frac{\eta^2}{y^2} = \frac{1}{k^2 r^2}, \quad r^2 = x^2 + y^2, \quad \frac{\xi}{x} < 0 \quad (6.9.9)$$

i.e. the left branch if $x > 0$ and the right branch if $x < 0$; see Fig. 6.9.1. Then, $s = \zeta x - (1/k^2 - \zeta^2)^{1/2} y$ is real on Γ, and decreasing from $-r/k$ to $-\infty$ as ζ goes from $-x/(kr)$ to infinity along either the branch Γ_+ (the part of Γ for $\eta > 0$) or the branch Γ_- (the part of Γ for $\eta < 0$). Subscript plus will be used to denote quantities on Γ_+, subscript minus quantities on Γ_-. Note that

$$\zeta = \frac{xs}{r^2} \pm \frac{iy}{kr^2} \sqrt{k^2 s^2 - r^2} = \zeta_\pm^{(S)}(s) \text{ on } \Gamma_\pm \quad (6.9.10)$$

Thus,

$$\mathcal{L}_{p\tau} f^{(S)}(x,y,\tau) = \frac{1}{2\pi i} \left[\int_{\Gamma_+} p e^{ps} F^{(S)}(\zeta) d\zeta + \int_{\Gamma_-} p e^{ps} F^{(S)}(\zeta) d\zeta \right]$$

$$= \frac{1}{2\pi i} \left[-\int_{-\infty}^{-r/k} p e^{ps} F_+^{(S)}(\zeta) \frac{d\zeta_+^{(S)}}{ds} ds + \int_{-\infty}^{-r/k} p e^{ps} F_-^{(S)}(\zeta) \frac{d\zeta_-^{(S)}}{ds} ds \right] \quad (6.9.11)$$

where ζ is chosen according to (6.9.10). $F_+^{(S)}(\zeta)$ is the value of $F^{(S)}(\zeta)$ in the upper half-plane, and $F_-^{(S)}(\zeta)$ is its value in the lower half-plane. The derivative

$$\frac{d\zeta_\pm^{(S)}}{ds} = \frac{x}{r^2} \pm \frac{iky s}{r^2\sqrt{k^2 s^2 - r^2}} = g_\pm^{(S)}(s) \qquad (6.9.12)$$

where the plus sign refers to Γ_+ and the minus sign to Γ_-. Now, inversion with respect to p involves only $p\exp(ps)$, the inversion of which is $\delta(\tau + s)$, and hence, because the delta function sifts out the value for $s = -\tau$ of the integrand, the full Laplace transform inversion is

$$f^{(S)}(x, y, \tau) = -\frac{U(\tau - r/k)}{2\pi i}\{g_+^{(S)}(-\tau)[F_+^{(S)}(\zeta)]_{\zeta=\zeta_+^{(S)}(-\tau)} \\ - g_-^{(S)}(-\tau)[F_-^{(S)}(\zeta)]_{\zeta=\zeta_-^{(S)}(-\tau)}\} \qquad (6.9.13)$$

This method of using self-similarity properties for inverting a Laplace transform is essentially due to Cagniard (1939). As $y \to 0$, the inversion formula takes the form

$$f^{(S)}(x, 0, \tau) = -\frac{U(\tau - |x|/k)}{2\pi i x}[F_+^{(S)}(\xi) - F_-^{(S)}(\xi)]_{\xi=-\tau/x} \qquad (6.9.14)$$

and the hyperbola has degenerated to the upper $(+)$ and lower $(-)$ sides of the part $\xi < -1/k$ of the real axis for $x > 0$ and of the part $\xi > 1/k$ for $x < 0$. The unit step function is optional for $y = 0$: omitting it is equivalent to extending the path Γ to the origin, $\xi = \eta = 0$.

Note that the inversion formulae may alternatively be written in the form

$$f^{(S)}(x, y, \tau) = -\frac{U(\tau - r/k)}{\pi}\Im\{g_+^{(S)}(-\tau)[F_+^{(S)}(\zeta)]_{\zeta=\zeta_+^{(S)}(-\tau)}\} \qquad (6.9.15)$$

$$f^{(S)}(x, 0, \tau) = -\frac{U(\tau - |x|/k)}{\pi x}\Im\{[F_+^{(S)}(\xi)]_{\xi=-\tau/x}\} \qquad (6.9.16)$$

because $g_-^{(S)}(-\tau)$ and $\zeta_-^{(S)}(-\tau)$ are complex conjugates of $g_+^{(S)}(-\tau)$ and $\zeta_+^{(S)}(-\tau)$, respectively.

For P waves (superscript P) the inversions are given by the same formulae, but k should be replaced by 1 everywhere, i.e. in the unit step function, in the equation for the hyperbola and in the definitions of $\zeta_\pm^{(P)}(s)$ and $g_\pm^{(P)}(s)$. Omitting the unit step function leads to the same expression for the inversions of P and S waves when $y = 0$ and is therefore convenient when the two wave types appear together in one single Laplace transform. Then, the inversion formula reads

$$\boxed{f(x, 0, \tau) = -\frac{1}{2\pi i x}[F_+(\xi) - F_-(\xi)]_{\xi=-\tau/x}} \qquad (6.9.17)$$

where $qF(q/p)$ is the Laplace transform of $f(x, 0, \tau)$, and subscripts plus and minus refer to the upper and lower sides, respectively, of the negative real ξ-axis if $x > 0$, and of the positive real ξ-axis if $x < 0$.

Note that the branch of the hyperbola used for the inversion of the Laplace transform appears for self-similar problems in general, not only for crack problems. An example regarding waves in a half-space is given in Broberg (1959); see also de Hoop (1961).

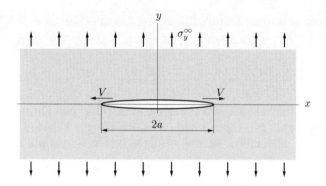

Fig. 6.9.2 Mode I crack expanding at constant velocity.

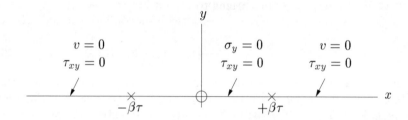

Fig. 6.9.3 Boundary conditions.

Symmetric mode I crack expansion

Figure 6.9.2 shows the problem: a crack is expanding with constant velocity $V = \beta c_P < c_S$ in both directions in an infinite medium, subjected to remote loading $\sigma_y = \sigma_y^\infty$. Plane strain or plane stress is assumed.

Due to the symmetry, it is sufficient to consider the upper half, $y \geq 0$, see Fig. 6.9.3. Obviously a mixed boundary value problem is posed: τ_{xy} vanishes on the whole plane $y = 0$, $\sigma_y = 0$ for $|x| < Vt = \beta\tau$ and $v = 0$ for $|x| > \beta\tau$. Primarily unknown functions are σ_y for $|x| > \beta\tau$, $y = 0$ and v for $|x| < \beta\tau$, $y = 0$.

Represent now displacements and stresses with two potentials ϕ and ψ, so that, according to Appendix A1,

$$u = \frac{\partial \phi}{\partial x} + \frac{\partial \psi}{\partial y}, \quad v = \frac{\partial \phi}{\partial y} - \frac{\partial \psi}{\partial x} \tag{6.9.18}$$

$$\sigma_y = \frac{\mu}{k^2}\left[(1 - 2k^2)\frac{\partial^2 \phi}{\partial x^2} + \frac{\partial^2 \phi}{\partial y^2} - 2k^2\frac{\partial^2 \psi}{\partial x \partial y}\right] \tag{6.9.19}$$

$$\tau_{xy} = \mu\left[2\frac{\partial^2 \phi}{\partial x \partial y} + \frac{\partial^2 \psi}{\partial y^2} - \frac{\partial^2 \psi}{\partial x^2}\right] \tag{6.9.20}$$

6.9 CRACK EXPANDING WITH CONSTANT VELOCITY

Expressed in terms of ϕ and ψ, the equations of motion read

$$\Delta\phi = \frac{\partial^2 \phi}{\partial \tau^2} \tag{6.9.21}$$

$$\Delta\psi = \frac{1}{k^2} \cdot \frac{\partial^2 \psi}{\partial \tau^2} \tag{6.9.22}$$

The following Laplace transforms are introduced:

$$\left.\begin{array}{ll} F = q \int_{-\infty}^{\infty} e^{-qx} p \int_0^{\infty} e^{-p\tau} \phi \, d\tau \, dx \\ S = & \psi \\ S_y = & \sigma_y \\ T_{xy} = & \tau_{xy} \\ V_+ = & v_+ \end{array}\right\} \tag{6.9.23}$$

where subscript plus in the last transform serves as a reminder that $y \geq 0$. The equations of motion are then transformed to

$$q^2 F + \frac{d^2 F}{dy^2} = p^2 F \tag{6.9.24}$$

$$q^2 S + \frac{d^2 S}{dy^2} = \frac{p^2}{k^2} S \tag{6.9.25}$$

with the solutions

$$F = A \exp[-(p^2 - q^2)^{1/2} y] \tag{6.9.26}$$
$$S = C \exp[-(p^2/k^2 - q^2)^{1/2} y] \tag{6.9.27}$$

after removal of terms that violate the radiation condition. Such terms are recognized because they are unbounded: the Laplace transform with respect to τ of a wave function $f(\tau + y)$ for a wave travelling in the negative y direction contains the factor $\exp(py)$, and increases without bound when $y \to \infty$. It is convenient to consider p as real and positive, cf. Appendix A5. The exponents in (6.9.26) and (6.9.27) are defined by branch cuts in the q plane from $q = \pm p$ and $q = \pm p/k$, respectively, to infinity along the real axis, with the branches chosen so that the exponents are negative on the imaginary axis. The exponents are then non-positive in the entire q plane.

Equations (6.9.18)-(6.9.20) give

$$V_+ = \frac{dF}{dy} - qS \tag{6.9.28}$$

$$S_y = \frac{\mu}{k^2} \left[(1 - 2k^2) q^2 F + \frac{d^2 F}{dy^2} - 2k^2 q \frac{dS}{dy} \right] \tag{6.9.29}$$

$$T_{xy} = \mu \left[2q \frac{dF}{dy} + \frac{d^2 S}{dy^2} - q^2 S \right] \tag{6.9.30}$$

Here, it is advantageous to "pretend" that one is dealing with a simple rather than a mixed boundary value problem. To this end, σ_y is written as a function of x and τ for $y = 0$ and – without specifying – it is assumed that this function is known. Actually, one property of this function is known: $(\sigma_y)_{y=0}$ must have the form $\sigma(\tau/|x|)$. This follows from symmetry and self-similarity. Thus, as shown in the preceding subsection, its double Laplace transform can be written as

$$\mathcal{L}_{qx}\mathcal{L}_{p\tau}(\sigma_y)_{y=0} = \sigma_y^\infty \cdot q/p\, H(q/p) \tag{6.9.31}$$

where $H(\cdot)$ is an even function. Insertion into (6.9.29) and (6.9.30) gives for $y = 0$:

$$[(1 - 2k^2)q^2 + p^2 - q^2]A + 2k^2q(p^2/k^2 - q^2)^{1/2}C = \frac{k^2\sigma_y^\infty q}{\mu p} H(q/p) \tag{6.9.32}$$

$$2q(p^2 - q^2)^{1/2}A - (p^2/k^2 - 2q^2)C = 0 \tag{6.9.33}$$

giving

$$A = -\frac{k^4 \sigma_y^\infty p(1/k^2 - 2q^2/p^2)}{\mu q^3 R(p/q)} H(q/p) \tag{6.9.34}$$

$$C = -\frac{2k^4 \sigma_y^\infty (1 - q^2/p^2)^{1/2}}{\mu q^2 R(p/q)} H(q/p) \tag{6.9.35}$$

where $R(\cdot)$ is the Rayleigh function. Thus,

$$\mathcal{L}_{qx}\mathcal{L}_{p\tau}(v_+)_{y=0} = -(p^2 - q^2)^{1/2}A - qC = \frac{k^2 \sigma_y^\infty}{\mu p} \cdot \frac{(1 - \zeta^2)^{1/2} H(\zeta)}{\zeta^3 R(1/\zeta)} \tag{6.9.36}$$

where $\zeta = q/p$.

It turns out to be simpler to consider $(\partial\sigma_y/\partial\tau)_{y=0}$ and $(\partial^2 v_+/\partial\tau^2)_{y=0}$ rather than $(\sigma_y)_{y=0}$ and $(v)_{y=0}$. The Laplace transform of these quantities is obtained simply by multiplication by p and p^2, respectively, in (6.9.31) and (6.9.36). Inversions are then obtained according to (6.9.17), so that

$$\left(\frac{\partial\sigma_y}{\partial\tau}\right)_{y=0} = -\frac{\sigma_y^\infty}{2\pi i x}[H_+(\xi) - H_-(\xi)]_{\xi=-\tau/x} \tag{6.9.37}$$

$$\left(\frac{\partial^2 v_+}{\partial\tau^2}\right)_{y=0} = -\frac{k^2\sigma_y^\infty}{2\pi i \mu x}[G_+(\xi) - G_-(\xi)]_{\xi=-\tau/x} \tag{6.9.38}$$

where

$$G(\zeta) = \frac{(1 - \zeta^2)^{1/2}}{\zeta^4 R(1/\zeta)} H(\zeta) \tag{6.9.39}$$

Note that $\zeta = 0$ is not a pole in the factor preceding $H(\zeta)$, but simple poles are found at $\zeta = \pm 1/k_R = \pm c_P/c_R$, and there are branch-points at $\zeta = \pm 1/k$ and $\zeta = 1$. Branch cuts are made according to Fig. 6.9.4, where

$$a(\xi) = 4k^3\xi^2\sqrt{\xi^2 - 1}\sqrt{k^2\xi^2 - 1} - (2k^2\xi^2 - 1)^2 \tag{6.9.40}$$

and, consequently

$$b(\xi) = -4k^3\xi^2\sqrt{1 - \xi^2}\sqrt{1 - k^2\xi^2} - (1 - 2k^2\xi^2)^2 \tag{6.9.41}$$

Now, the boundary conditions

$$\left(\frac{\partial\sigma_y}{\partial\tau}\right)_{y=0} = 0 \text{ for } \tau/|x| > 1/\beta \tag{6.9.42}$$

$$\left(\frac{\partial^2 v_+}{\partial\tau^2}\right)_{y=0} = 0 \text{ for } \tau/|x| < 1/\beta \tag{6.9.43}$$

6.9 CRACK EXPANDING WITH CONSTANT VELOCITY

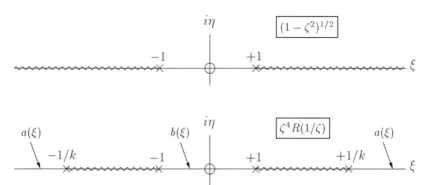

Fig. 6.9.4 Branch cuts for function definitions.

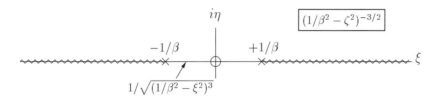

Fig. 6.9.5 Branch cuts for function definition.

lead to the equations

$$H_+(\xi) - H_-(\xi) = 0 \text{ for } \beta^{-2} < \xi^2 \quad (6.9.44)$$
$$G_+(\xi) - G_-(\xi) = 0 \text{ for } \xi^2 < \beta^{-2} \quad (6.9.45)$$

The first condition, (6.9.44), implies that

$$G_+(\xi) + G_-(\xi) = 0 \text{ for } \beta^{-2} < \xi^2 \quad (6.9.46)$$

because the factor in front of $H(\zeta)$ in (6.9.39) possesses the property

$$\left[\frac{(1-\xi^2)^{1/2}}{\xi^4 R(1/\xi)}\right]_+ = -\left[\frac{(1-\xi^2)^{1/2}}{\xi^4 R(1/\xi)}\right]_- \quad (6.9.47)$$

for $k^{-2} < \beta^{-2} < \xi^2$, if branch cuts are introduced as shown in Fig. 6.9.4.

Equations (6.9.45) and (6.9.46) formulate a Hilbert problem. With regard to the condition of bounded stress-strain energy the solution is

$$G(\zeta) = P(\zeta)(\beta^{-2} - \zeta^2)^{-3/2} \quad (6.9.48)$$

where, because $G(\zeta)$ is even, $P(\zeta)$ is an even polynomial, which, due to the boundedness of the mass velocity at $x = 0$ reduces to a constant, say $P(\zeta) = D_1$. The factor $(\beta^{-2} - \zeta^2)^{-3/2}$ is defined by the branch cuts in Fig. 6.9.5. It is now possible to confirm that $H(\zeta)$ has singularities only on the real axis, and the path deformation to the hyperbola, discussed on page 400, is therefore legitimate.

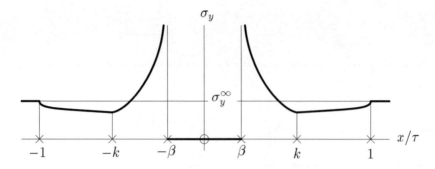

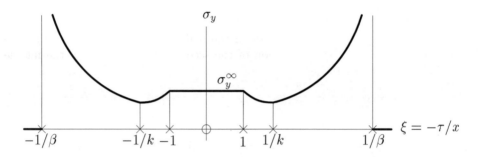

Fig. 6.9.6 The normal stress on $y = 0$.

From (6.9.38) and (6.9.48) the normal mass acceleration on $y = +0$ is obtained:

$$\frac{\partial^2 v_+}{\partial \tau^2} = -\frac{k^2 \sigma_y^\infty D_1}{2\pi i \mu x} \{[(\beta^{-2} - \xi^2)^{-3/2}]_+ - [(\beta^{-2} - \xi^2)^{-3/2}]_-\}_{\xi = -\tau/x} \qquad (6.9.49)$$

A glance on Fig. 6.9.5 reveals that this acceleration is zero for $\tau/|x| < 1/\beta$, i.e. $|x| > Vt = a$, where a is the half-length of the crack, and that

$$\frac{\partial^2 v_+}{\partial \tau^2} = -\frac{k^2 \sigma_y^\infty D_1}{\pi \mu |x|} \cdot \frac{1}{\sqrt{(\xi^2 - \beta^{-2})^3}} = -\frac{k^2 \sigma_y^\infty D_1}{\pi \mu} \cdot \frac{\beta^3 x^2}{\sqrt{(\beta^2 \tau^2 - x^2)^3}} \qquad (6.9.50)$$

for $\xi = \tau/|x| > 1/\beta$. Integration gives

$$\frac{\partial v_+}{\partial \tau} = \frac{k^2 \sigma_y^\infty D_1}{\pi \mu} \cdot \frac{\beta^3 \tau}{\sqrt{\beta^2 \tau^2 - x^2}} \qquad (6.9.51)$$

and a further integration results in

$$v_+ = \frac{k^2 \sigma_y^\infty D_1 \beta}{\pi \mu} \sqrt{a^2 - x^2} \quad \text{for} \quad -a < x < a \qquad (6.9.52)$$

showing that the crack opening is elliptic, as in the static case.

The stress $(\sigma_y)_{y=0}$ is obtained by integration of (6.9.37) from $|x| = \tau + 0$, where $(\sigma_y)_{y=0} = \sigma_y^\infty$, see Fig. 6.9.6. This corresponds to $\xi = 1 - 0$, if $x < 0$ is chosen. Then,

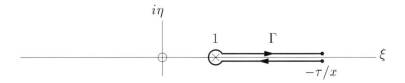

Fig. 6.9.7 Integration path.

since $\partial/\partial\tau = -(\partial/\partial\xi)/x$, the result is

$$(\sigma_y)_{y=0} = \sigma_y^\infty + \frac{\sigma_y^\infty}{2\pi i}\int_{1-0}^{-\tau/x}[H_+(\xi) - H_-(\xi)]d\xi = \sigma_y^\infty + \frac{\sigma_y^\infty}{2\pi i}\int_\Gamma H(\zeta)d\zeta \quad (6.9.53)$$

where Γ is the path along the lower side of the real axis from $\xi = -\tau/x = \tau/|x|$ to $\xi = 1+0$, corresponding to integration of the second term inside [ ] in the middle member of (6.9.53), then rounding $\zeta = 1$ and returning to $\xi = -\tau/x$ along the upper side of the real axis, which corresponds to integration of the first term; see Fig. 6.9.7.

The solution for $y = 0$ is now complete, except for an amplitude factor, because D_1 is not determined.

Determination of the constant D_1
The constant D_1 is to be determined by using the condition that $\sigma_y = 0$ for $\xi > 1/\beta$, see Fig. 6.9.6. An expression for this stress is obtained by extending the path Γ in (6.9.53) to $\xi = \infty$, and using (6.9.39):

$$(\sigma_y)_{y=0, |x|<\beta\tau} = \sigma_y^\infty + \frac{\sigma_y^\infty D_1}{2\pi i}\int_\Gamma \frac{\zeta^4 R(1/\zeta)}{(1-\zeta^2)^{1/2}(1/\beta^2 - \zeta^2)^{3/2}}d\zeta = 0 \quad (6.9.54)$$

where Γ now is the path along the lower side of the real axis from $\xi = \infty$ to $\xi = 1+0$, then rounding $\zeta = 1$ and returning to $\xi = \infty$ along the upper side of the real axis. All singular points on $|\xi| > 0$ are situated on the portion $\xi \geq 1$ of the real axis and are avoided by infinitesimally small indentations. The path of integration can thus be deformed to the imaginary axis, i.e. from $\zeta = i\eta = -i\infty$ to $\zeta = i\eta = i\infty$, so that

$$\sigma_y^\infty = -\frac{\sigma_y^\infty D_1}{\pi}\int_0^\infty \frac{\eta^4 R(-i/\eta)}{(1+\eta^2)^{1/2}(1/\beta^2 + \eta^2)^{3/2}}d\eta \quad (6.9.55)$$

After insertion of the expression for $\eta^4 R(-i/\eta)$, see Fig. 6.9.4, the integral can be written as

$$4k^4 \int_0^\infty \frac{\eta^2\sqrt{(\eta^2+1)(\eta^2+1/k^2)} - [\eta^2 + 1/(2k^2)]^2}{\sqrt{(\eta^2+1)(\eta^2+1/\beta^2)^3}}d\eta \quad (6.9.56)$$

The integral is elliptic and can be brought to standard form for complete elliptic integrals of the first and second kind. Then, the constant D_1 is found to be

$$D_1 = \frac{\pi(1-\beta^2)}{\beta g_1(\beta)} \quad (6.9.57)$$

with

$$g_1(\beta) = [(1-4k^2)\beta^2 + 4k^4]K(\sqrt{1-\beta^2})$$
$$- \beta^{-2}[\beta^4 - 4k^2(1+k^2)\beta^2 + 8k^4]E(\sqrt{1-\beta^2})$$
$$- 4k^2(1-\beta^2)K(\sqrt{1-\beta^2/k^2})$$
$$+ 8k^4\beta^{-2}(1-\beta^2)E(\sqrt{1-\beta^2/k^2}) \quad (6.9.58)$$

where K and E are the complete elliptic integrals of the first and second kind. For $\beta = 0$, the function $g_1(\beta) = 2k^2(1-k^2)$.

Stresses and displacements

From a purely mathematical point of view, the solution is valid for $V < c_S$. Physical considerations, on the other hand, require $V < c_R$, because the Rayleigh wave velocity is the upper limit for mode I crack speed, as previously shown.

After insertion of D_1 in (6.9.52), the normal displacement on $y = +0$, $|x| < a$, is found to be

$$v_+ = \frac{k^2 \sigma_y^\infty (1-\beta^2)}{\mu g_1(\beta)} \sqrt{a^2 - x^2} \text{ for } -a < x < a \quad (6.9.59)$$

The excentricity of the ellipse increases with V, for the same a and σ_y^∞. At the Rayleigh wave velocity†, the crack opening is between about 57.5% and 58.6% of the crack opening for a static crack in the interval $0.2 < k^2 < 0.5$, with a maximum at $k^2 \approx 0.35$.

The tangential displacement, u, and the stress σ_x on the crack faces may be determined by using the simple expression

$$\sigma_x + \sigma_y = \frac{2(1-k^2)\mu}{k^2}\Delta\phi = \frac{2(1-k^2)\mu}{k^2}\frac{\partial^2\phi}{\partial\tau^2} \quad (6.9.60)$$

Thus,

$$\mathcal{L}_{qx}\mathcal{L}_{p\tau}(\sigma_x+\sigma_y)_{y=0} = \frac{2(1-k^2)\mu}{k^2}p^2 A \quad (6.9.61)$$

where A is given by (6.9.34). Insertion gives

$$\mathcal{L}_{qx}\mathcal{L}_{p\tau}(\sigma_x+\sigma_y)_{y=0} = -\frac{4k^2(1-k^2)\sigma_y^\infty\zeta[1/(2k^2)-\zeta^2]}{(1-\zeta)^{1/2}(1/\beta^2-\zeta^2)^{3/2}} \quad (6.9.62)$$

where $\zeta = q/p$. From this relation there follows, as (6.9.53) follows from (6.9.31), the inversion

$$(\sigma_x+\sigma_y)_{y=0} = \sigma_x^\infty + \sigma_y^\infty - \frac{4k^2(1-k^2)\sigma_y^\infty D_1}{2\pi i}$$
$$\times \int_\Gamma \frac{1/(2k^2)-\zeta^2}{(1-\zeta^2)^{1/2}(1/\beta^2-\zeta^2)^{3/2}}d\zeta \quad (6.9.63)$$

where Γ is the same path as in (6.9.53). By extending Γ beyond $\xi = 1/\beta$, the expression gives $\sigma_x+\sigma_y = \sigma_x$ on the crack faces, and because the integrand is analytic for $\xi > 1/\beta$,

---

† Recall that, theoretically, this velocity can be approached, but not reached.

$\eta = 0$, the stress σ_x is constant on the crack faces. By choosing the extension of Γ to $\xi = \infty$ and then deforming the path to the imaginary axis, this stress is found to be

$$\sigma_x = \sigma_x^\infty + \sigma_y^\infty - \frac{4k^2(1-k^2)\sigma_y^\infty D_1}{\pi} \int_0^\infty \frac{\eta^2 + 1/(2k^2)}{\sqrt{(\eta^2+1)(\eta^2+1/\beta^2)}} d\eta$$

$$= \sigma_x^\infty + \sigma_y^\infty - \frac{2(1-k^2)\sigma_y^\infty}{g_1(\beta)}$$

$$\times \left[(1-2k^2)\beta^2 \boldsymbol{K}(\sqrt{1-\beta^2}) + (2k^2-\beta^2)\boldsymbol{E}(\sqrt{1-\beta^2})\right] \quad (6.9.64)$$

As $\beta \to 0$, $\sigma_x \to \sigma_x^\infty - \sigma_y^\infty$ at the crack faces, as also previously shown for the static case. The ratio $(\sigma_x - \sigma_x^\infty)/\sigma_y^\infty$ increases from -1 at $\beta = 0$ to about -0.642 at the Rayleigh wave velocity if $k^2 = 1/3$.

The displacement u at the crack faces follows from Hooke's law for plane strain or plane stress, considering that $\sigma_y = 0$:

$$\epsilon_x = \frac{\sigma_x}{4(1-k^2)\mu} \quad (6.9.65)$$

Thus,

$$u = \frac{\sigma_x x}{4(1-k^2)\mu} \quad (6.9.66)$$

After specialization to the vicinity of the crack edge $x = a$ and insertion of D_1 in (6.9.53), which for $a < |x| \leq \tau$ can be written in the form

$$\left(\frac{\partial \sigma_y}{\partial \tau}\right)_{y=0} = \frac{\sigma_y^\infty D_1}{\pi |x|} \left\{ \frac{\xi^4 \Re[R(1/\xi)]}{\sqrt{(\xi^2-1)(1/\beta^2-\xi^2)^3}} \right\}_{\xi=\tau/|x|} \quad (6.9.67)$$

the normal stress on $y = 0$ for $0 < x - a \ll a$ is found to be

$$\sigma_y \approx \sigma_y^\infty \frac{\sqrt{1-\beta^2} R(\beta)}{\beta^2 g_1(\beta)} \sqrt{\frac{a}{2(x-a)}} \quad (6.9.68)$$

so that the stress intensity factor, which alternatively can be obtained from (6.9.59), is

$$K_I = \frac{\sigma_y^\infty \sqrt{1-\beta^2} R(\beta)}{\beta^2 g_1(\beta)} \sqrt{\pi a} \quad (6.9.69)$$

The stress intensity factor, normalized with respect to the stress intensity factor at the same crack length but vanishingly low crack speed is shown in Fig. 6.9.8.

By integration of (6.9.67), the normal stress in the symmetry plane outside the crack is found from the expressions

$$\frac{(\sigma_y)_{y=0} - \sigma_y^\infty}{\sigma_y^\infty} = -\frac{1}{\beta^2 g_1(\beta)} \Big\{ \beta^2 [4k^4 + (1-4k^2)\beta^2] F(\kappa_1, q_1)$$

$$- [8k^4 - 4k^2(1+k^2)\beta^2 + \beta^4] E(\kappa_1, q_1)$$

$$+ [8k^4 - 4k^2(1+k^2)\beta^2 + \beta^4 - 4k^4\beta^2(1-\beta^2)\xi^2] \frac{\sqrt{1-1/\xi^2}}{\sqrt{1-\beta^2\xi^2}} \Big\}$$

$$\text{for } 1 < \xi < 1/k \quad (6.9.70)$$

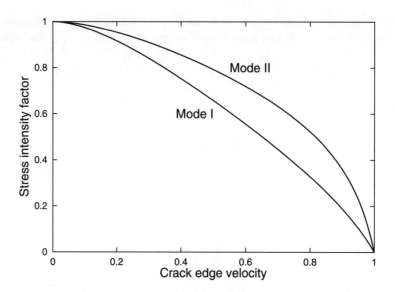

Fig. 6.9.8 Stress intensity factor as function of the crack edge velocity for symmetrically expanding mode I and mode II cracks. The stress intensity factor is normalized with respect to the stress intensity factor for the same crack length but vanishingly low crack speed, and the crack edge velocity is normalized with respect to the Rayleigh wave velocity. The material constant $k^2 = 1/3$, corresponding to Poisson's ratio $\nu = 1/4$ for plane strain and $\nu = 1/3$ for plane stress. The dependence on k is only marginal for mode I: at the velocity $0.5c_R$ the stress intensity factor is a maximum for about $k^2 = 1/3$, and it is only about 0.4% lower for $k^2 = 1/4$ ($\nu = 1/3$ in plane strain, $\nu = 1/2$ in plane strss) and about 2% lower for $k^2 = 1/2$ ($\nu = 0$). For mode II, the corresponding figures are about 3% higher and 8% lower than for $k^2 = 1/3$.

$$\frac{(\sigma_y)_{y=0} - \sigma_y^\infty}{\sigma_y^\infty} = -\frac{1}{\beta^2 g_1(\beta)} \left\{ \beta^2[4k^4 + (1-4k^2)\beta^2]F(\kappa_1, q_1) \right.$$
$$- [8k^4 - 4k^2(1+k^2)\beta^2 + \beta^4]E(\kappa_1, q_1)$$
$$\left. + [8k^4 - 4k^2(1+k^2)\beta^2 + \beta^4 - 4k^4\beta^2(1-\beta^2)\xi^2]\frac{\sqrt{1-1/\xi^2}}{\sqrt{1-\beta^2\xi^2}} \right\}$$
$$+ \frac{4k^2(1-\beta^2)}{\beta^2 g_1(\beta)} \left\{ \beta^2 F(\kappa_2, q_2) - 2k^2 E(\kappa_2, q_2) + k^2(2-\beta^2\xi^2)\frac{\sqrt{1-1/(k^2\xi^2)}}{\sqrt{1-\beta^2\xi^2}} \right\}$$
$$\text{for } 1/k < \xi < 1/\beta \quad (6.9.71)$$

where $\xi = \tau/|x|$ and

$$\sin \kappa_1 = \frac{\sqrt{1-1/\xi^2}}{\sqrt{1-\beta^2}}, \qquad \sin \kappa_2 = \frac{\sqrt{k^2 - 1/\xi^2}}{\sqrt{k^2 - \beta^2}}$$

$$q_1 = \sqrt{1-\beta^2} \qquad \text{and} \qquad q_2 = \sqrt{1-\beta^2/k^2}$$

and $F(\cdot, \cdot)$ and $E(\cdot, \cdot)$ are the (incomplete) elliptic integrals of the first and second kind, respectively.

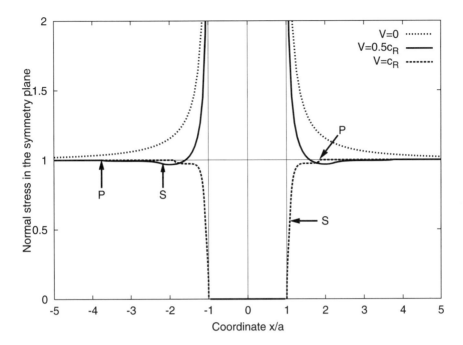

Fig. 6.9.9 The normalized normal stress $(\sigma_y)_{y=0}/\sigma_y^\infty$ in the symmetry plane of a mode I crack expanding with constant velocity $V = 0.5c_R$, i.e. half the Rayleigh wave velocity. For comparison the same stress is shown for $V = 0$ and $V = c_R$. Labels P and S indicate the arrivals of P and S waves, respectively. The stress for $V = 0.5c_R$ is negative after the arrival of the P wave and remains negative until some time after the arrival of the S wave. The material constant $k^2 = 1/3$.

The normalized stress $(\sigma_y)_{y=0}/\sigma_y^\infty$ is shown in Fig. 6.9.9 for the crack velocity $V = 0.5c_R$. For comparison, the same stress is also shown for $V = 0$ and $V = c_R$. Note that $(\sigma_y)_{y=0}$ is initially negative after the arrival of stress waves. Note also that there is no singularity for $V = c_R$. This indicates that the stress peak in the crack edge vicinity is very narrow for crack velocities just below the Rayleigh wave velocity.

Stresses and displacements at points outside the symmetry plane are found by using (6.9.15). As an example, σ_y is found by integration of $\partial \sigma_y/\partial \tau$ with respect to τ, after inversion of pS_y has been made. Insertion of (6.9.26) and (6.9.27) into (6.9.29) gives

$$pS_y = -q\frac{k^6\sigma_y^\infty(1/k^2 - 2\zeta^2)^2 H(\zeta)}{\mu\zeta^4 R(1/\zeta)}\exp[-p(1-\zeta^2)^{1/2}]$$
$$-q\frac{4k^6\sigma_y^\infty(1-\zeta^2)^{1/2}(1/k^2-\zeta^2)^{1/2}H(\zeta)}{\mu\zeta^2 R(1/\zeta)}\exp[-p(1/k^2-\zeta^2)^{1/2}] \quad (6.9.72)$$

where the first term in the right member represents a P wave and the second term an S wave: they are clearly distinguishable from the arguments of the exponential functions. Then, the inversion is found from (6.9.15) as

$$\frac{\partial \sigma_y}{\partial \tau} = -\frac{\sigma_y^\infty D_1}{\mu} \Im\left\{U(\tau-1)\left[g_+^{(P)}(-\tau)\frac{k^6(1/k^2 - 2\zeta^2)^2}{(1-\zeta^2)^{1/2}(1/\beta^2 - \zeta^2)^{3/2}}\right]_{\zeta=\zeta_+^{(P)}(-\tau)} \right.$$
$$\left. + U(\tau - 1/k)\left[g_+^{(S)}(-\tau)\frac{4k^6\zeta^2(1/k^2 - \zeta^2)^{1/2}}{(1/\beta^2 - \zeta^2)^{3/2}}\right]_{\zeta=\zeta_+^{(S)}(-\tau)}\right\} \quad (6.9.73)$$

where use was made of (6.9.39).

It is possible to extend the analysis to include a finite length process region by using the Barenblatt model. This can be done with a superposition technique, as in Section 3.5, assuming the length of the process region to be small compared with the crack length (Broberg 1967) or, in addition, with specialization to the Leonov-Panasyuk-Dugdale model (Broberg 1964). This model was also used by Atkinson (1967), without restriction to small size. Self-similarity, of course, implies that the length of the process region increases in proportion to the crack length, which might not be in accord with physical requirements, except, perhaps, at very high crack speeds; see pages 628ff.

The solution for the expanding mode I crack was extended to orthotropic materials, with the crack propagating along a symmetry plane, by Atkinson (1965). The result turned out to be very similar to the isotropic solution. Thus, for instance, the crack shape is elliptic as in the isotropic case.

Energy relations
Use of (6.2.49) gives the energy flux to each process region,

$$\mathcal{G}(\beta) = \frac{K_I^2 Y_I(\beta)}{4(1-k^2)\mu} = \frac{\pi(\sigma_y^\infty)^2 a \cdot k^2 \sqrt{(1-\beta^2)^3} R(\beta)}{2\mu\beta^2 [g_1(\beta)]^2} \quad (6.9.74)$$

This can also be written in the form

$$\mathcal{G}(\beta) = \mathcal{G}(0) \cdot w_1(\beta) \quad (6.9.75)$$

where $\mathcal{G}(0) = \pi(\sigma_y^\infty)^2 a/[4(1-k^2)\mu]$ is the energy flux at vanishingly low crack speed, and thus the factor

$$w_1(\beta) = \frac{2k^2(1-k^2)\sqrt{(1-\beta^2)^3} R(\beta)}{\beta^2 [g_1(\beta)]^2} \quad (6.9.76)$$

expresses the energy flux into the crack edge in relation to the energy flux at vanishingly low crack edge velocity. The function $w_1(\beta)$ is shown in Fig. 6.9.10.

The total energy balance for an expanding crack involves mechanical work of outer forces, accumulated stress-strain energy, kinetic energy and energy dissipation at the crack edges. Consider an expanding crack $|x| < a$, $y = 0$, driven by crack face loads $\sigma_y = -\sigma_y^\infty$, so that superposition of the stress $\sigma_y = \sigma_y^\infty$ everywhere would lead to an expanding crack with traction free faces in an infinite body subjected to a remote load $\sigma_y = \sigma_y^\infty$. The work performed by the crack face loads during extension of the crack from $|x| = a - da$ to $|x| = a$ is formally the same as in the corresponding static case, (4.6.31):

$$dU^* = 2\int_{-a}^{a} \sigma_y^\infty \frac{\partial v(x,a)}{\partial a} dx\, da \quad (6.9.77)$$

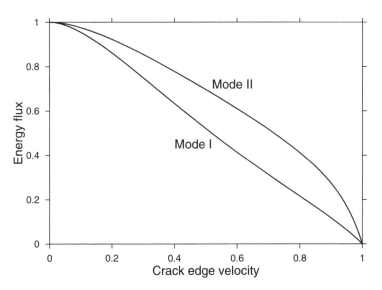

Fig. 6.9.10 Energy flow per unit crack growth and length along the crack edge to one process region of symmetrically expanding mode I and II cracks. This energy flux is normalized with respect to the energy flux for vanishingly low speed, and the crack wedge velocity is normalized with respect to the Rayleigh wave velocity. The material constant $k^2 = 1/3$, corresponding to Poisson's ratio $\nu = 1/4$ for plane strain, $\nu = 1/3$ for plane stress. The dependence on k is only marginal for mode I: at the velocity $0.5c_R$, the energy flux is a maximum for about $k^2 = 1/3$, and it is only about 1% lower for $k^2 = 1/4$ ($\nu = 1/3$ in plane strain, $\nu = 1/2$ in plane strss) and about 3% lower for $k^2 = 1/2$ ($\nu = 0$). For mode II, the corresponding figures are about 5% higher and 12% lower than for $k^2 = 1/3$.

where the asterisk serves as a reminder that crack face loading is studied. The crack face displacement $v(x,a)$ depends on time and crack edge velocity, because $a = Vt$. Recall now that the crack opening is elliptic as in the static case. Thus, it is possible to write, cf. (6.2.54),

$$v(x,a) = \frac{Y_I(\beta) K_I(a,\beta)}{K_I(a,0)} v_0(x,a) \qquad (6.9.78)$$

where $v_0(x,a)$ is half the crack opening for a stationary crack, subjected to crack face loads $\sigma_y = -\sigma_y^\infty$, given by (4.6.11) as

$$v_0(x,a) = \frac{\sigma_y^\infty}{2(1-k^2)\mu} \sqrt{a^2 - x^2} \qquad (6.9.79)$$

Thus, because $K_I(a,\beta)/K_I(a,0)$ is independent of a,

$$\frac{dU^*}{da} = \frac{Y_I(\beta) K_I(a,\beta)}{K_I(a,0)} \cdot \frac{dU_0^*}{da} \qquad (6.9.80)$$

where dU_0^*/da is the work performed in the corresponding static case.

The energy dissipated at the crack edges per unit crack growth and unit length along the edges is, cf. (6.2.49),

$$\frac{dD^*}{da} = \mathcal{G} = \frac{Y_I(\beta)K_I^2(a,\beta)}{K_I^2(a,0)} \cdot \frac{dD_0^*}{da} \tag{6.9.81}$$

where dD_0^*/da is the energy dissipation in the corresponding static case. According to (4.6.33),

$$\frac{dD_0^*}{da} = \frac{dW_0^*}{da} = \frac{1}{2} \cdot \frac{dU_0^*}{da} \tag{6.9.82}$$

where W_0^* is the increase of stress strain energy. From (6.9.80) and (6.9.81), the increase of energy dissipation at the two crack edges is found to be

$$\frac{dD^*}{da} = \frac{Y_I(\beta)K_I^2(a,\beta)}{2K_I^2(a,0)} \cdot \frac{dU_0^*}{da} = \frac{K_I(a,\beta)}{2K_I(a,0)} \cdot \frac{dU^*}{da} \tag{6.9.83}$$

The energy balance tells that the work supplied equals the increase of stress-strain energy plus energy dissipation and kinetic energy. Thus,

$$dU^* = dW^* + dD^* + dT^* \tag{6.9.84}$$

where T^* is the kinetic energy. However, in spite of the simple relation (6.9.83) between dD^* and dU^*, there does not seem to be a simple way to determine the distribution between stress-strain energy increase and kinetic energy, and integrations using the stress-strain distribution in the whole body appear to be extremely time consuming. On the other hand, the general principles of energy relations together with dimensional analysis has led to certain interesting results in connection with crack expansion (Mott 1948).

Symmetric and nonsymmetric mode II crack expansion

Only marginal adjustments are needed to obtain results for a symmetrically expanding mode II crack from those for mode I. The problem is posed in the same way, except that the remote stress is $\tau_{xy} = \tau_{xy}^\infty$ and the boundary conditions on $y = 0$ are $\sigma_y = 0$ for all x, $\tau_{xy} = 0$ for $|x| < a$ and $u = 0$ for $|x| > a$. In essentially the same way as discussed in Section 6.3, dynamic Coulomb friction, $\tau_{xy} = \tau_f$, can be added on the crack faces, and cohesive regions at the crack edges can be taken into account, cf. Broberg (1978), but for simplicity this is not done here.

The changes needed consist essentially of exchanging $(1-\zeta^2)^{1/2}$ in (6.9.36) and (6.9.54) by $(1/k^2 - \zeta^2)^{1/2}$, and of exchanging $(1+\eta^2)^{1/2}$ by $(1/k^2 + \eta^2)^{1/2}$ in (6.9.55) and (6.9.56). Results for $y = 0$ are

$$u_+ = \frac{\tau_{xy}^\infty(k^2 - \beta^2)}{\mu g_2(\beta)}\sqrt{a^2 - x^2} \text{ for } -a < x < a \tag{6.9.85}$$

$$\left(\frac{\partial \tau_{xy}}{\partial \tau}\right)_{y=0} = -\frac{\tau_{xy}^\infty D_2}{\pi|x|}\Re\left\{\frac{\xi^4 R(1/\xi)}{(\xi^2 - 1/k^2)^{1/2}(1/\beta^2 - \xi^2)^{3/2}}\right\}_{\xi = -\tau/x} \tag{6.9.86}$$

and the similarity with (6.9.59) and (6.9.67) is obvious. The function

$$g_2(\beta) = \beta^{-2}[8k^2(k^2-\beta^2)]\boldsymbol{E}(\sqrt{1-\beta^2}) - 4k^2(k^2-\beta^2)\boldsymbol{K}(\sqrt{1-\beta^2})$$
$$- \beta^{-2}[8k^4 - 8k^2\beta^2 + \beta^4]\boldsymbol{E}(\sqrt{1-\beta^2/k^2})$$
$$+ (4k^2 - 3\beta^2)\boldsymbol{K}(\sqrt{1-\beta^2/k^2}) \quad (6.9.87)$$

and the constant

$$D_2 = \frac{\pi(k^2-\beta^2)}{k^2\beta g_2(\beta)} \quad (6.9.88)$$

are found from the equation

$$\tau_{xy}^\infty = -\frac{\tau_{xy}^\infty D_2}{\pi}\int_0^\infty \frac{\eta^4 R(-i/\eta)}{(1/k^2+\eta^2)^{1/2}(1/\beta^2+\eta^2)^{3/2}}d\eta \quad (6.9.89)$$

The function $g_2(\beta) \to 2k^2(1-k^2)$ as $\beta \to 0$.
The stress intensity factor is found to be

$$K_{II} = \frac{\tau_{xy}^\infty \sqrt{k^2-\beta^2}R(\beta)}{k\beta^2 g_2(\beta)}\sqrt{\pi a} \quad (6.9.90)$$

and the energy flux to each process region is

$$\mathcal{G}(\beta) = \mathcal{G}(0) \cdot w_2(\beta) \quad (6.9.91)$$

where $\mathcal{G}(0) = \pi(\tau_{xy}^\infty)^2 a/[4(1-k^2)\mu]$ is the energy flux to the process region at vanishingly low crack speed. Thus, the factor

$$w_2(\beta) = \frac{2(1-k^2)\sqrt{(k^2-\beta^2)^3}R(\beta)}{k\beta^2[g_2(\beta)]^2} \quad (6.9.92)$$

expresses the energy flux into the crack edge in relation to the energy flux at vanishingly low crack speed. The function $w_2(\beta)$ is shown in Fig. 6.9.10. Note that the enery flux into the process region is considerable larger for mode II than for mode I, particularly for higher velocities. Therefore, a mode II crack can be expected to accelerate faster than a mode I crack; this might be of interest in a seismological context.

Integration of (6.9.86) gives, after some calculations, expressions for the shear stress in the symmetry plane outside the crack:

$$\frac{(\tau_{xy})_{y=0} - \tau_{xy}^\infty}{\tau_{xy}^\infty} = \frac{4k^2(k^2-\beta^2)}{\beta^2 g_2(\beta)}\left\{\beta^2 F(\kappa_1,q_1) - 2E(\kappa_1,q_1) + (2-\beta^2\xi^2)\frac{\sqrt{1-1/\xi^2}}{\sqrt{1-\beta^2\xi^2}}\right\}$$
$$\text{for } 1 < \xi < 1/k \quad (6.9.93)$$

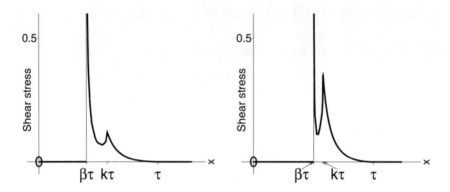

Fig. 6.9.11 Shear stress in front of an extensional mode II crack, given as the normalized overstress $[(\tau_{xy})_{y=0} - \tau_{xy}^\infty]/\tau_{xy}^\infty$. x is the distance from the midpoint of the crack. The material constant $k^2 = 1/3$, and the two crack speeds are about 75% and 94% of the Rayleigh wave speed, which is about $0.5308 c_P$. The S wave speed is about $0.5774 c_P$.

$$\frac{(\tau_{xy})_{y=0} - \tau_{xy}^\infty}{\tau_{xy}^\infty} = \frac{4k^2(k^2-\beta^2)}{\beta^2 g_2(\beta)}\left\{\beta^2 F(\kappa_1, q_1) - 2E(\kappa_1, q_1) + (2-\beta^2\xi^2)\frac{\sqrt{1-1/\xi^2}}{\sqrt{1-\beta^2\xi^2}}\right\}$$

$$- \frac{1}{\beta^2 g_2(\beta)}\left\{(4k^2 - 3\beta^2)\beta^2 F(\kappa_2, q_2) - [8k^2(k^2-\beta^2) + \beta^4]E(\kappa_2, q_2)\right.$$

$$\left. + [8k^2(k^2-\beta^2) + \beta^4 - 4k^2\beta^2(k^2-\beta^2)\xi^2]\frac{\sqrt{1-1/(k^2\xi^2)}}{\sqrt{1-\beta^2\xi^2}}\right\}$$

$$\text{for } 1/k < \xi < 1/\beta \quad (6.9.94)$$

where $\xi = \tau/|x|$ and

$$\sin \kappa_1 = \frac{\sqrt{1-1/\xi^2}}{\sqrt{1-\beta^2}}, \quad \sin \kappa_2 = \frac{\sqrt{k^2-1/\xi^2}}{\sqrt{k^2-\beta^2}}$$

$$q_1 = \sqrt{1-\beta^2} \quad \text{and} \quad q_2 = \sqrt{1-\beta^2/k^2}$$

and $F(\cdot, \cdot)$ and $E(\cdot, \cdot)$ are the (incomplete) elliptic integrals of the first and second kind, respectively. These expressions are reproduced in order to demonstrate a presumably important feature of an expanding mode II crack, first noticed by Burridge (1973). Numerical calculations reveal that the shear stress on $y = 0$ drops to a minimum some distance ahead of the crack, at about the arrival of Rayleigh waves, whereupon it increases towards a peak at the arrival of S waves. Two examples are shown in Fig. 6.9.11.

The peak becomes more pronounced the higher the crack speed is, and increases indefinitely when the crack speed approaches the Rayleigh wave speed. It also becomes more "fat", especially in relation to the stress concentration at the crack edge, which, of course, is inverse square-root singular only in mathematical models. Initiation of slip could then, perhaps, be triggered by this peak, and this might be a mechanism by which the "forbidden" velocity region between the Rayleigh speed and the S wave speed could be surpassed. Numerical calculations, first by Andrews (1976, 1985) and subsequently by Johnson (1990, 1992a), indicate, indeed, that this mechanism might be possible.

Analysis of nonsymmetric crack expansion needs only small adjustments of the procedure for symmetric crack expansion, although the final numerical work may be more demanding. The adjustments are essentially consequences of the fact that the region of validity of the boundary condition for $\partial \tau_{xy}/\partial \tau$ is changed to $\xi < -\beta_L^{-1}$, $\xi > \beta_L^{-1}$, where β_L is the velocity to the left of the left crack edge and β_R the velocity to the right of the right crack edge, and, further, that the region of validity of the boundary condition for $\partial^2 u_+/\partial \tau^2$ is changed to $-\beta_L^{-1} < \xi < \beta_R^{-1}$. Then, the function $G(\zeta)$ will be changed to $(\beta_L^{-1} - \zeta)^{3/2}(\beta_R^{-1} + \zeta)^{3/2}$, so that the tangential mass acceleration of the upper crack face becomes

$$\frac{\partial^2 u_+}{\partial \tau^2} = -\frac{k^2 \tau_{xy}^\infty D_2'}{\pi \mu} \cdot \frac{(\beta_L \beta_R)^{3/2} x^2}{(\beta_L \tau + x)^{3/2}(\beta_R \tau - x)^{3/2}} \text{ for } -\beta_L \tau < x < \beta_R \tau \qquad (6.9.95)$$

where D_2' corresponds to constant D_2 in the symmetric case. Assume first that the crack edges move in different directions, i.e. both β_L and β_R are positive. Then, two integrations give

$$u_+ = \frac{4k^2 \tau_{xy}^\infty D_2'(\beta_L \beta_R)^{3/2}}{\pi \mu (\beta_L + \beta_R)^2} \sqrt{(\beta_L \tau + x)(\beta_R \tau - x)} \text{ for } -\beta_L \tau < x < \beta_R \tau \qquad (6.9.96)$$

Actually, symmetry reasons do not imply that the polynomial $P(\zeta)$ in (6.9.48) must be even in this case, but, nevertheless it reduces to a constant by the condition that $(u_+)_{x=-\beta_L \tau} = (u_+)_{x=\beta_R \tau} = 0$. The constant D_2' is determined in essentially the same way as the constant D_2, although there is no advantage of deforming the branch to the imaginary axis. Rather, integration can be performed along the negative real axis, from $\xi = -1$ to $\xi = -1/\beta_R + \epsilon$, and then along a semi-circular indentation with radius ϵ above $\zeta = -1/\beta_R$, where ϵ eventually shall go to zero. The integrand is found from the right member of (6.9.53), after substituting the mode I factor $(1 - \zeta^2)^{1/2}$ by the mode II factor $(1/k^2 - \zeta^2)^{1/2}$ and the factor $(1/\beta^2 - \zeta^2)^{3/2}$ by $(1/\beta_L - \zeta)^{3/2}(1/\beta_R + \zeta)^{3/2}$.

It is interesting to note that the crack shape is elliptic, even though the two stress intensity factors are different. For the left crack edge the stress intensity factor is found from the expression for u_+ by using (6.2.41):

$$K_{II}^{(L)} = \frac{4\tau_{xy}^\infty D_2'}{\sqrt{\pi}} \cdot \frac{R(\beta_L)}{(\beta_L + \beta_R)^2 \sqrt{k^2 - \beta_L^2}} \cdot \left(\frac{\beta_R}{\beta_L}\right)^{3/2} \sqrt{a} \qquad (6.9.97)$$

where a is the half-length of the crack. Note, however, that D_2' depends on β_L and β_R.

The ratio between the stress intensity factors at the left and right crack edges is

$$\frac{K_{II}^{(L)}}{K_{II}^{(R)}} = \frac{\beta_R^2 R(\beta_L)}{\beta_L^2 R(\beta_R)} \sqrt{\frac{k^2 - \beta_R^2}{k^2 - \beta_L^2}} \qquad (6.9.98)$$

which shows that the slower crack edge gets the higher stress intensity factor, a fact which indicates a tendency towards symmetric expansion in a homogeneous medium.

In particular, when the left crack edge is stationary,

$$\frac{K_{II}^{(L)}}{K_{II}^{(R)}} = \frac{2k(1-k^2)\beta_R^2\sqrt{k^2-\beta_R^2}}{R(\beta_R)} \qquad (6.9.99)$$

This ratio increases from unity to infinity when the velocity of the right crack edge increases from 0 to the Rayleigh wave velocity. The increase is rather slow: for $k^2 = 1/3$ (Poisson's ratio = 1/4 for plane strain, 1/3 for plane stress) the ratio is about 1.11 at $0.5c_R$, about 1.29 at $0.7c_R$ and about 3.3 at $0.95c_R$.

If $\beta_L < 0$, i.e. if both crack edges move to the right, slip may be deposited behind the trailing edge. In such a case the polynomial $P(\zeta)$ in (6.9.48) does not reduce to a constant, but to a linear function, so that

$$\frac{\partial^2 u_+}{\partial \tau^2} = -\frac{k^2 \tau_{xy}^\infty D_2'}{\pi \mu} \cdot \frac{(\beta_L \beta_R)^{3/2}(x^2 + A\tau x)}{(\beta_L \tau + x)^{3/2}(\beta_R \tau - x)^{3/2}} \quad \text{for } -\beta_L\tau < x < \beta_R\tau \qquad (6.9.100)$$

where A is a constant. If it is assumed that the trailing edge is energy-neutral, so that its stress intensity factor is zero, then $A = \beta_L$, and integration of $\partial^2 u_+/\partial \tau^2$ gives

$$\frac{\partial u_+}{\partial \tau} = \frac{2k^2 \tau_{xy}^\infty D_2' |\beta_L \beta_R|^{3/2}}{\pi\mu(\beta_L + \beta_R)} \sqrt{\frac{\beta_L\tau + x}{\beta_R\tau - x}} \quad \text{for } -\beta_L\tau < x < \beta_R\tau \qquad (6.9.101)$$

The slip deposited on each side behind the trailing edge is

$$\Delta = \int_{-\beta_L}^{\beta_R} \frac{\partial u_+}{\partial x} dx = \frac{2k^2 \tau_{xy}^\infty D_2'|\beta_L\beta_R|^{3/2}}{\mu(\beta_L + \beta_R)} \tau \qquad (6.9.102)$$

The constant D_2' is found in the same way as constant D_2, leading to the equality for vanishing ϵ,

$$\frac{4k^4 \tau_{xy}^\infty D_2'}{\pi}\left\{\int_1^{1/\beta_R - \epsilon} \frac{\xi^2\sqrt{\xi^2-1}(1+A\xi)}{\sqrt{(1/\beta_L - \xi)^3(1/\beta_R + \xi)^3}}d\xi\right.$$
$$-\int_{1/k}^{1/\beta_R - \epsilon} \frac{[\xi^2 - 1/(2k^2)]^2(1+A\xi)}{\sqrt{\xi^2 - 1/k^2}\sqrt{(1/\beta_L - \xi)^3(1/\beta_R + \xi)^3}}d\xi$$
$$\left.-\frac{R(\beta_R)(A+\beta_R)\sqrt{|\beta_L\beta_R|^3}}{4k^3\beta_R^4\sqrt{k^2 - \beta_R^2}\sqrt{(\beta_L + \beta_R)^3}} \cdot \frac{2}{\sqrt{\epsilon}}\right\} = -\tau_{xy}^\infty \qquad (6.9.103)$$

where $A = \beta_L < 0$. The last term within $\{\ \}$ is due to integration along the indentation above $\zeta = -1/\beta_R$. By putting $A = 0, \beta_L > 0$, D_2' is determined for crack growth in both directions, and the symmetric case is then obtained, of course, by putting $\beta_L = \beta_R$, leading to a somewhat different way of calculating D_2 than previously described.

If numerical integration is chosen, it is suitable to add terms proportional to $(1/\beta_R - \xi)^{-3/2}$ to the integrands in (6.9.103), such that the integration interval can be extended to $1/\beta_R$. The (elementary) integrals of these terms are then subtracted, and ϵ can be set to zero throughout.

Intersonic mode II crack expansion

In the preceding subsection, symmetric subsonic mode II crack growth was analysed by making simple modifications in the corresponding mode I case. For symmetric intersonic mode II crack expansion (Broberg 1994, 1995c), i.e., $k < \beta < 1$, the same procedure can be used initially, leading to the equations, cf. (6.9.44)-(6.9.45),

$$H_+(\xi) - H_-(\xi) = 0 \text{ for } 1/\beta^2 < \xi^2 \qquad (6.9.104)$$
$$G_+(\xi) - G_-(\xi) = 0 \text{ for } \xi^2 < 1/\beta^2 \qquad (6.9.105)$$

where

$$H(\zeta) = \frac{1}{\tau_{xy}^\infty \zeta} \mathcal{L}_{p\tau} \mathcal{L}_{qx}(\tau_{xy})_{y=0} \qquad (6.9.106)$$

$$G(\zeta) = \frac{(1/k^2 - \zeta^2)^{1/2}}{\zeta^4 R(1/\zeta)} H(\zeta) \qquad (6.9.107)$$

Use of the last equation implies that (6.9.105) can be written as

$$H_+(\xi) - c(\xi) H_-(\xi) = 0 \text{ for } \xi^2 < 1/\beta^2 \qquad (6.9.108)$$

where

$$c(\xi) = \frac{[\xi^4 R(1/\xi)]_+}{[\xi^4 R(1/\xi)]_-} \qquad (6.9.109)$$

Actually, by defining $c(\xi) = 1$ for $\xi^2 > 1/\beta^2$, the equality (6.9.108) holds for all ξ. Then,

$$c(\xi) = \begin{cases} 1 & \text{for} \quad \xi^2 < 1 \\ e^{2\pi i g(\xi)} & \text{for} \quad -1/\beta < \xi < -1 \\ e^{-2\pi i g(\xi)} & \text{for} \quad 1 < \xi < 1/\beta \\ 1 & \text{for} \quad 1/\beta^2 < \xi^2 \end{cases} \qquad (6.9.110)$$

where

$$g(\xi) = \frac{1}{\pi} \operatorname{atan} \frac{4k^3 \sqrt{1 - 1/\xi^2} \sqrt{1/\xi^2 - k^2}}{(1/\xi^2 - 2k^2)^2} \qquad (6.9.111)$$

Note that $0 < g(\xi) \leq 1/2$ for $1 < |\xi| < 1/k$ and that $g(\xi) \to 0$ as $\xi \to \pm 1$. Thus, $c(\xi)$ is continuous in the interval $\xi^2 < 1/\beta^2$. Note further that (6.9.108) is an equation for $H(\zeta)$, whereas the corresponding equation in the subsonic case is an equation for $G(\zeta)$. This difference is, of course, related to the position of branch points of the factor preceding $H(\zeta)$ in (6.9.107).

Equation (6.9.108), which constitutes a Hilbert problem, is now written in logarithmic form,

$$[\ln H(\xi)]_+ - [\ln H(\xi)]_- = \ln c(\xi) \qquad (6.9.112)$$

where $\ln c(\xi)$ is continuous along the real axis and zero along the portion $|\xi| > 1/\beta$. Then, according to the Plemelj formulae, one particular solution is

$$\ln H_0(\zeta) = \int_{-1/\beta}^{-1} \frac{g(w)}{w - \zeta} dw - \int_{1}^{1/\beta} \frac{g(w)}{w - \zeta} dw = -2 \int_{1}^{1/\beta} \frac{w g(w)}{w^2 - \zeta^2} dw \qquad (6.9.113)$$

which, for brevity is written as

$$H_0(\zeta) = e^{-I(\zeta)} \tag{6.9.114}$$

where

$$I(\zeta) = 2\int_1^{1/\beta} \frac{wg(w)}{w^2 - \zeta^2} dw \tag{6.9.115}$$

For $1 < \xi < 1/\beta$, $\eta = 0$, this reads

$$I(\xi) = 2\oint_1^{1/\beta} \frac{wg(w)}{w^2 - \xi^2} dw \pm \pi i g(\xi) \tag{6.9.116}$$

where C on the integral sign expresses the Cauchy principal value, and the upper sign is valid if ξ approaches the real axis from the upper half-plane, the lower sign if ξ approaches the real axis from the lower half-plane. For $-1/\beta < \xi < -1$, the same result, but with $\mp$ instead of $\pm$, is obtained, and for $\xi^2 < 1$ the last term shall be omitted.

In order to exhibit its singular character at $\xi = 1/\beta$, the function $I(\xi)$ is written as

$$I(\xi) = I_0(\xi) + \frac{g(1/\beta)}{\beta\xi} \ln\left|\frac{(1/\beta - \xi)(\xi + 1)}{(1/\beta + \xi)(\xi - 1)}\right| \pm \pi i g(\xi) \tag{6.9.117}$$

for $1 < \xi < 1/\beta$. Here,

$$I_0(\xi) = 2\oint_1^{1/\beta} \frac{wg(w) - 1/\beta g(1/\beta)}{w^2 - \xi^2} dw \tag{6.9.118}$$

Equations (6.9.114), (6.9.117) and (6.9.118) now give

$$[H_0(\xi)]_+ - [H_0(\xi)]_- \rightarrow \left[\frac{2(1-\beta)}{\beta(1+\beta)}\right]^g \cdot \frac{e^{-I_0(1/\beta)}[e^{-\pi i g} - e^{\pi i g}]}{(1/\beta - \xi)^g}$$

$$= -2i\left[\frac{2(1-\beta)}{\beta(1+\beta)}\right]^g \cdot \frac{e^{-I_0(1/\beta)}\sin(\pi g)}{(1/\beta - \xi)^g} \quad \text{as } \xi \rightarrow 1/\beta - 0 \tag{6.9.119}$$

which shows a singularity factor $(1/\beta - \xi)^{1/2}$ for $[H_0(\xi)]_+ - [H_0(\xi)]_-$. Here, the simplified notation $g(1/\beta) = g$ is used, i.e. when the argument of the function $g(\cdot)$ is omitted, $g(1/\beta)$ is implied. This is the same function as given in (6.3.42). Now, because $(\partial \tau_{xy}/\partial \tau)_{y=0}$ is proportional to $[H(\xi)]_+ - [H(\xi)]_-$, and a singular factor $(1/\beta - \xi)^{-\alpha}$, where $0 < \alpha < 1$ can be allowed for $(\tau_{xy})_{y=0}$, the appropriate solution is

$$H(\zeta) = \frac{D_i}{1/\beta^2 - \zeta^2} \cdot H_0(\zeta) \tag{6.9.120}$$

where D_i (with subscript i for "intersonic") is a polynomial, which, to ensure finite mass velocity at $x = 0$, is of zero order, i.e. a constant. Note that the factor $D_i/(1/\beta^2 - \zeta^2)$ is regular in the region of validity of both (6.9.104) and (6.9.105).

The constant D_i is determined in essentially the same way as the constant D_2', although reduction to standard form elliptic integrals cannot be made: direct numerical integrations are needed.

6.9 CRACK EXPANDING WITH CONSTANT VELOCITY

Discussion of the solution
After determination of D_i, the functions $H(\cdot)$ and $G(\cdot)$ are known and all quantities of interest can be found, particularly $(\tau_{xy})_{y=0}$ and $(u_+)_{y=0}$. The expressions are complicated. Details are given by Broberg (1994). The crack shape, which is very simple for subsonic crack expansion – elliptic for crack propagation in both directions – is here quite complicated, with a logarithmic velocity singularity travelling at Rayleigh speed. The expression for the acceleration is

$$\left(\frac{\partial^2 u_+}{\partial \tau^2}\right)_{y=0} = \begin{cases} \dfrac{\tau_{xy}^\infty}{4\pi\mu|x|} \cdot \dfrac{\sin^2 \pi g}{k^2 \xi^2 \sqrt{\xi^2-1}} H(\xi) & \text{for } 1/\beta < \xi = \tau/|x| < 1/k \\[2ex] -\dfrac{k^2 \tau_{xy}^\infty}{\pi\mu|x|} \cdot \dfrac{\sqrt{\xi^2 - 1/k^2}}{\xi^4 R(1/\xi)} H(\xi) & \text{for } \xi = \tau/|x| > 1/k \end{cases} \quad (6.9.121)$$

For the crack edge vicinity, $x \approx \beta\tau = a$, the shear stress on $y = 0$,

$$\tau_{xy} \to \frac{\tau_{xy}^\infty}{N} \cdot \frac{\beta \sin \pi g \cdot e^{-I_0(1/\beta)}}{2g} \cdot \left(\frac{1-\beta}{1+\beta}\right)^g \cdot \left(\frac{2a}{x-a}\right)^g \text{ as } x \to a+0 \quad (6.9.122)$$

where $N = -\pi/D_i$ is approximately 1.182 for $k^2 = 1/3$, corresponding to Poisson's ratio $= 1/4$ at plane strain.

The displacement gradient on the upper crack face,

$$\frac{\partial u_+}{\partial x} \to -\frac{Y_{II}(\beta)}{2(1-k^2)\mu} \cdot \frac{\tau_{xy}^\infty}{N} \cdot \frac{\beta \sin \pi g \cdot e^{-I_0(1/\beta)}}{2g} \cdot \left(\frac{1-\beta}{1+\beta}\right)^g \cdot \left(\frac{2a}{a-x}\right)^g$$
$$\text{as } x \to a-0 \quad (6.9.123)$$

where $Y_{II}(\beta)$ is given by (6.3.52).

Energy flux into the process region
Assume now that a Barenblatt type process region exists at each crack edge and that its length r_p is much smaller than the current crack length a. Then, the energy flux can be obtained from equation (6.3.72). Note that a factor $\tau_{xy}^L L^g = A = x^g \tau_{xy}$ appears in this equation, corresponding to $\lim_{x\to a+0}[(x-a)^g(\tau_{xy})_{y=0}]$ in the present case. Thus,

$$\mathcal{G} = \frac{\pi(\tau_{xy}^\infty)^2 a}{\mu} \cdot B_i(\beta) \cdot \Gamma_D(g) \cdot \left(\frac{r_p}{a}\right)^{1-2g} \quad (6.9.124)$$

where

$$B_i(\beta) = \frac{\beta^4 e^{-2I_0(1/\beta)} \sin^2 \pi g}{2^{2(2-g)} k^2 N^2 g^2 \sqrt{1-\beta^2}} \cdot \left(\frac{1-\beta}{1+\beta}\right)^{2g} \quad (6.9.125)$$

and $\Gamma_D(g)$ is the same function as in (6.3.72). Recall that subscript i denotes "intersonic".

Numerical calculations, assuming that the function in (6.3.67), $D(-\xi) = 1-\xi$, show that the energy flux is a maximum for a crack velocity somewhat lower than $\sqrt{2}c_S$, and it vanishes when the velocity approaches c_S from above and c_P from below. This is shown in Fig. 6.9.12 by the function $w_i(\beta) = \mathcal{G}/\mathcal{G}(0)$, where $\mathcal{G}(0)$ is the energy flux at vanishingly low velocity.

Recall that $c_R \le V \le c_S$ is a "forbidden" region. If this region can be bypassed, for

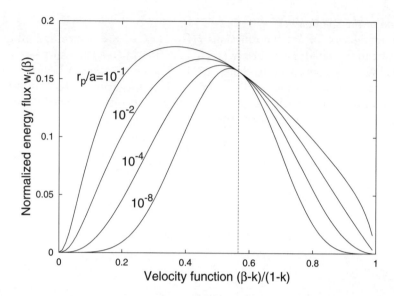

Fig. 6.9.12 The dependence on intersonic velocities of the energy flux to a Barenblatt process region for a symmetrically expanding mode II crack. The energy flux is calculated for different values of r_p/a, where r_p is the length of the Barenblatt region and a is the half-length of the crack. The energy flux is normalized with respect to the flux at vanishingly low crack velocity. The material constant $k^2 = 1/3$, which corresponds to Poisson's ratio equal to $1/4$ in plane strain and $1/3$ in plane stress. The dashed line indicates the crack velocity $2^{1/2}c_S$ for which the stress and strain fields are square root singular at the crack edge, if the process region is assumed to be point-sized. All curves are calculated under the assumption that $r_p/a \ll 1$.

instance by the mechanism described in connection with Fig. 6.9.11, then sustained intersonic crack growth might result, perhaps with acceleration to a velocity beyond the one for maximum energy flux; cf. Section 9.3. However, nothing is known about the velocity dependence of the energy requirement in the intersonic region.

Symmetric mode III crack expansion

For anti-plane deformation, the equation of motion, (A9.24), is

$$\Delta w = \frac{1}{c_S^2} \cdot \frac{\partial^2 w}{\partial t^2} \qquad (6.9.126)$$

so that a double Laplace transformation yields

$$q^2 W + \frac{d^2 W}{dy^2} = p^2 W \qquad (6.9.127)$$

where $W = \mathcal{L}_{p\tau}\mathcal{L}_{qx}w$, with $\tau = c_S t$. The solution for the upper half-plane is

$$W_+ = (W_+)_{y=0} \exp[-(p^2 - q^2)^{1/2} y] \qquad (6.9.128)$$

6.9 CRACK EXPANDING WITH CONSTANT VELOCITY

Further, because $\tau_{yz} = \mu \partial w/\partial y$, the double Laplace transform

$$\mathcal{L}_{p\tau}\mathcal{L}_{qx}\tau_{yz} = \mu\frac{dW_+}{dy} = -\mu(W_+)_{y=0}(p^2-q^2)^{1/2}\exp(-(p^2-q^2)^{1/2}y) \quad (6.9.129)$$

The problem of a mode III crack, expanding with constant velocity $V = \gamma c_S$ along the y-axis, symmetrically from $y = 0$, is solved in close analogy with the corresponding mode I problem, and only the major steps will be discussed. The friction at the sliding crack faces is neglected, but linear Coulomb friction, as well as a finite Barenblatt region, may be incorporated in the same way as for a mode II crack. The remote load is $\tau_{yz} = \tau_{yz}^\infty$. The shear stress $(\tau_{yz})_{y=0}$ and the mass velocity $(\partial w_+/\partial \tau)$ are homogeneous functions of $|x|$ and τ, and thus their double Laplace transforms can be written as

$$\mathcal{L}_{p\tau}\mathcal{L}_{qx}(\tau_{yz})_{y=0} = \tau_{yz}^\infty \cdot \frac{q}{p}H\left(\frac{q}{p}\right) \quad (6.9.130)$$

$$\mathcal{L}_{p\tau}\mathcal{L}_{qx}\left(\frac{\partial w_+}{\partial \tau}\right)_{y=0} = p(W_+)_{y=0} = \frac{\tau_{yz}^\infty q}{\mu p}G\left(\frac{q}{p}\right) \quad (6.9.131)$$

where $H(\cdot)$ and $G(\cdot)$ are even functions. Then, use of (6.9.129) gives a relation between $H(\cdot)$ and $G(\cdot)$:

$$\frac{H(\zeta)}{(1-\zeta^2)^{1/2}} = -G(\zeta) \quad (6.9.132)$$

where $\zeta = \xi + i\eta = q/p$ and branch cuts are made on the portions $|\xi| > 1$ of the real axis, with the branch chosen so that $(1-\zeta^2)^{1/2} = \sqrt{1-\xi^2}$ for $|\xi| < 1$.

The inversion formula (6.9.17) gives

$$\frac{\partial \tau_{yz}}{\partial \tau} = -\frac{\tau_{yz}^\infty}{2\pi i x}[H_+(\xi) - H_-(\xi)]_{\xi=-\tau/x} \quad (6.9.133)$$

$$\frac{\partial^2 w}{\partial \tau^2} = -\frac{\tau_{yz}^\infty}{2\pi\mu i x}[G_+(\xi) - G_-(\xi)]_{\xi=-\tau/x} \quad (6.9.134)$$

so that the boundary conditions, $\partial \tau_{yz}/\partial \tau = 0$ for $|x| < \gamma\tau$ and $\partial^2 w_+/\partial \tau^2 = 0$ for $|x| < \gamma\tau$, lead to

$$H_+(\xi) - H_-(\xi) = 0 \text{ for } \gamma^{-2} < \xi^2 \quad (6.9.135)$$
$$G_+(\xi) - G_-(\xi) = 0 \text{ for } \xi^2 < \gamma^{-2} \quad (6.9.136)$$

where the first equation, by using (6.9.132), can be replaced by

$$G_+(\xi) + G_-(\xi) = 0 \quad (6.9.137)$$

so that a Hilbert problem for $G(\zeta)$ is obtained. The solution, with regard to symmetry and boundedness conditions, is

$$G(\zeta) = \frac{D_3}{(1/\gamma^2 - \zeta^2)^{3/2}} \quad (6.9.138)$$

where D_3 is a constant, and branch cuts are made on the portions $|\xi| > 1/\gamma$ of the real axis, with the branch chosen so that $(1/\gamma^2 - \zeta^2)^{3/2} = \sqrt{(1/\gamma^2 - \xi^2)^3}$ for $|\xi| < 1/\gamma$. Thus, from (6.9.17),

$$\left(\frac{\partial^2 w_+}{\partial \tau^2}\right)_{y=0} = -\frac{\tau_{yz}^\infty D_3}{2\pi i \mu x}[(1/\gamma^2 - \xi^2)_+^{-3/2} - (1/\gamma^2 - \xi^2)_-^{-3/2}]_{\xi=-\tau/x} \quad (6.9.139)$$

which, after two integrations yields

$$w_+ = \frac{\tau_{yz}^\infty D_3 \gamma}{\pi \mu}\sqrt{a^2 - x^2} \text{ for } |x| < a \quad (6.9.140)$$

where $a = \gamma\tau$ is the half-length of the crack. For the shear stress, (6.9.17) gives

$$\left(\frac{\partial \tau_{yz}}{\partial \tau}\right)_{y=0} = \frac{\tau_{yz}^\infty D_3}{2\pi i x}\left[\frac{(1-\xi^2)_+^{1/2}}{(1/\gamma^2 - \xi^2)_+^{3/2}} - \frac{(1-\xi^2)_-^{1/2}}{(1/\gamma^2 - \xi^2)_-^{3/2}}\right]_{\xi=-\tau/x} \quad (6.9.141)$$

By analogy with (6.9.54), the constant D_3 is determined from the condition

$$(\tau_{yz})_{y=0,|x|<a} = \tau_{yz}^\infty - \frac{\tau_{yz}^\infty D_3}{2\pi i}\int_\Gamma \frac{(1-\zeta^2)^{1/2}}{(1/\gamma^2 - \zeta^2)^{3/2}}d\zeta = 0 \quad (6.9.142)$$

where Γ is the path along the lower side of the real axis from $\xi = \infty$ to $\xi = 1+0$, corresponding to integration of the second term inside $[\,]$ in (6.9.141), then rounding $\zeta = 1$ and returning to $\xi = \infty$ along the upper side of the real axis, corresponding to integration of the first term. This path can be deformed to the imaginary axis, which leads to

$$\frac{D_3}{\pi}\int_0^\infty \sqrt{\frac{1+\eta^2}{(1/\gamma^2 + \eta^2)^3}}d\eta = 1 \quad (6.9.143)$$

giving

$$D_3 = \frac{\pi}{\gamma \boldsymbol{E}(\sqrt{1-\gamma^2})} \quad (6.9.144)$$

where $\boldsymbol{E}(\cdot)$ is the complete elliptic integral of the second kind. Thus,

$$w_+ = \frac{\tau_{yz}^\infty}{\mu \boldsymbol{E}(\sqrt{1-\gamma^2})}\sqrt{a^2 - x^2} \text{ for } |x| < a \quad (6.9.145)$$

For $a < |x| < \tau$, $y = 0$, integration of (6.9.141) and insertion of D_3 yields

$$\tau_{yz} = \frac{\tau_{yz}^\infty}{\boldsymbol{E}(\sqrt{1-\gamma^2})}\left[\frac{|x|}{\tau}\sqrt{\frac{\tau^2-x^2}{x^2-a^2}} + E(\sqrt{1-\gamma^2}) - E(\kappa, \sqrt{1-\gamma^2})\right] \quad (6.9.146)$$

where $\kappa = \mathrm{asin}\sqrt{(1-x^2/\tau^2)/(1-\gamma^2)}$, and $E(\cdot,\cdot)$ is the (incomplete) elliptic integral of the second kind.

The stress intensity factor is

$$K_{III} = \tau_{yz}^\infty \sqrt{\pi a} \cdot \frac{\sqrt{1-\gamma^2}}{\boldsymbol{E}(\sqrt{1-\gamma^2})} \quad (6.9.147)$$

It is shown graphically in Fig. 6.9.13.

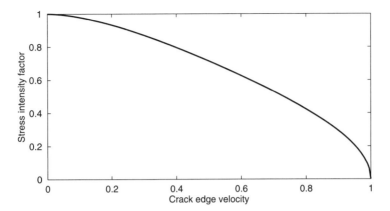

Fig. 6.9.13 Stress intensity factor for a symmetrically expanding mode III crack, normalized with respect to the stress intensity factor for vanishingly low speed. The crack edge velocity is normalized with respect to the S wave velocity.

The energy flux to each process region is

$$\mathcal{G} = \frac{K_{III}^2 Y_{III}}{2\mu} = \mathcal{G}(0) \cdot w_3(\gamma) \tag{6.9.148}$$

where $\mathcal{G}(0) = \pi(\tau_{yz}^\infty)^2 a/(2\mu)$ is the energy flux for vanishingly low crack speed, and thus the factor

$$w_3(\gamma) = \frac{\sqrt{1-\gamma^2}}{[E(\sqrt{1-\gamma^2})]^2} \tag{6.9.149}$$

expresses the energy flux into the crack edge in relation to the energy flux at vanishingly low crack speed. The function $w_3(\beta)$ is shown in Fig. 6.9.14. Comparison with Fig. 6.9.10 shows that the energy flux is about the same as for mode I (considering that S and Rayleigh wave velocities are not very different), but clearly smaller than for mode II.

Self-similar expansion of circular and elliptic cracks

The rotationally symmetric counterpart to the self-similar in-plane problem of a symmetrically expanding crack under remote tensile loading – also described as the expansion of a penny or disk shaped crack – was first solved by Kostrov (1964a), see also Craggs (1966), Atkinson (1968) and Bergkvist (1977). The result, using cylindrical coordinates r, φ, z, is very simple as regards the crack shape during subsonic growth: it is elliptic in a cut containing the z-axis:

$$w_+ = \frac{K_I Y_I(\beta)}{2(1-k^2)\mu} \cdot \sqrt{\frac{a^2 - r^2}{\pi a}} \tag{6.9.150}$$

where $a = Vt = \beta c_S t$. Note that only the plane strain value of k is relevant in this case. In contrast to the in-plane case, the stress intensity factor can be expressed in

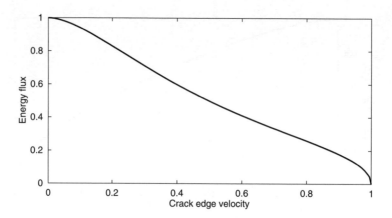

Fig. 6.9.14 Energy flow per unit of crack growth and length along the crack edge to one process region of a symmetrically expanding mode III crack. This energy flux is normalized with respect to the energy flux at vanishingly low speed and the crack edge velocity is normalized with respect to the S wave velocity.

terms of elementary factions. Thus, following Bergkvist (1977), though with different notations,

$$K_I = \frac{\sqrt{1-\beta^2}R(\beta)}{\beta^2 g_r(\beta)}\sigma_y^\infty\sqrt{\pi a} \qquad (6.9.151)$$

where σ_y^∞ is the remote stress, and

$$g_r(\beta) = [4k^3(3-2k)(1-\beta^2) - (\beta^2 - 2k^2)^2]/\beta$$
$$+ \frac{(\beta^2 - 2k^2)[6k^2 + (1-8k^2)\beta^2]}{\beta^2\sqrt{1-\beta^2}}\left(\frac{\pi}{2} - \operatorname{atan}\frac{\beta}{\sqrt{1-\beta^2}}\right)$$
$$+ \frac{4k^3(3k^2 - 2\beta^2)(1-\beta^2)}{\beta^2\sqrt{k^2-\beta^2}}\left(\frac{\pi}{2} - \operatorname{atan}\frac{\beta}{\sqrt{k^2-\beta^2}}\right) \quad (6.9.152)$$

The function $g_r(\beta) = \pi k^2(1-k^2)$ for $\beta = 0$. The stress intensity factor is shown as a function of the velocity in Fig. 6.9.15.

Knowledge of K_I enables immediate determination of the energy flux from the general expression

$$\mathcal{G} = \frac{K_I^2 Y_I(\beta)}{4(1-k^2)\mu} \qquad (6.9.153)$$

given previously as equation (6.2.49). By analogy with previous cases, a normalized energy flux w_r is introduced:

$$\mathcal{G} = \mathcal{G}(0) \cdot w_r(\beta) \qquad (6.9.154)$$

The function w_r is shown in Fig. 6.9.16. Comparison with Fig. 6.9.10 shows that the energy flux is considerably larger than for a mode I crack expanding in a plane, which indicates a faster acceleration than for plane expansion. The explanation seems to be

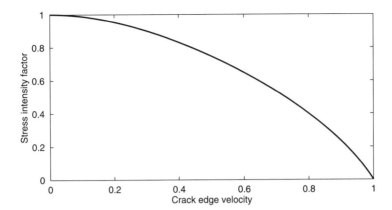

Fig. 6.9.15 Stress intensity factor for a symmetrically expanding mode I penny shaped crack, normalized with respect to the stress intensity factor for vanishingly low speed. The crack edge velocity is normalized with respect to the Rayleigh velocity. The material constant $k^2 = 1/3$, corresponding to Poisson's ratio $\nu = 1/4$.

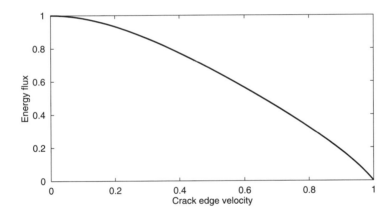

Fig. 6.9.16 Energy flow per unit of crack growth and length along the crack edge to the process region of a symmetrically expanding penny shaped crack. This energy flux is normalized with respect to the energy flux for vanishingly low speed and the crack edge velocity is normalized with respect to the Rayleigh speed. The material constant $k^2 = 1/3$ corresponds to Poisson's ratio $\nu = 1/4$.

that the circular geometry contributes to the convergence of energy flow to the process region.

Growth of penny shaped cracks under remote shear loading is of particular interest for the early development of earthquake slip. Kostrov (1964b) studied self-similar expansion of a circular crack. Although this problem can be solved in essentially the same way as the corresponding problem for tensile loading, some specific assumptions of a physical nature have to be made. Only two points along the crack edge, diametrically opposed, experience pure mode II growth, and two other points, on the perpendicular diameter, experience pure mode III growth. All other points are subjected to mixed mode propagation. The stress-strain fields near the points of pure mode II growth are different from those near the points of pure mode III growth in all respects during circular expansion, for instance as regards the energy flux into the process region. In addition, the material dependent energy requirement cannot be expected to be the same for the tearing mechanisms in modes II and III, even though these obviously are more closely related to each other than to the mode I mechanism.

Whatever the criterion for mixed mode II/III crack growth should be for the material and expansion velocity studied, it is obvious that self-similar expansion of an elliptic rather than a circular crack, would be a more adaptable model for shear loading. Such a model, even including anisotropy, motion of the crack centre and a phase change in an expanding ellipsoidal region, was considered by Burridge and Willis (1969).

6.10 Face loaded mode III cracks

A general remark

Direct dynamical crack face loading in mode III would require application of tangential forces with opposite direction on the two crack faces. This situation is hardly encountered in practical situations, and it appears to be extremely difficult to realize in experiments. However, in problems regarding stress wave interaction with traction free cracks, crack face loading is often used as part in a superposition scheme (see Section 6.13), and analysis of face loaded mode III cracks is therefore motivated.

A self-similar problem with wide applicability

First, a basic problem will be treated: a semi-infinite mode III crack, moving from an unloaded stationary position, $x < 0$, $y = 0$, with constant velocity $V = \gamma c_S$ along the positive x axis, under the action of a crack face load

$$\tau_{yz} = \tau_{yz}^0 U(x - \lambda\tau) \text{ for } x < \gamma\tau \qquad (6.10.1)$$

where $U(\cdot)$ is the unit step function, $\tau = c_S t$ and $\lambda < \gamma$. Fig. 6.10.1 shows the lower half of the body and the stress $(\tau_{yz})_{y=0}$.

The initial procedure is virtually identical with the one used for a self-similar expanding mode III crack, cf. (6.9.126)-(6.9.132). Thus,

$$\left(\frac{\partial \tau_{yz}}{\partial \tau}\right)_{y=0} = -\frac{\tau_{yz}^0}{2\pi i x}[H_+(\xi) - H_-(\xi)]_{\xi=-\tau/x} \qquad (6.10.2)$$

$$\left(\frac{\partial^2 w_+}{\partial \tau^2}\right)_{y=0} = -\frac{\tau_{yz}^0}{2\pi \mu i x}[G_+(\xi) - G_-(\xi)]_{\xi=-\tau/x} \qquad (6.10.3)$$

$$H(\zeta) = -(1 - \zeta^2)^{1/2} G(\zeta) \qquad (6.10.4)$$

6.10 FACE LOADED MODE III CRACKS

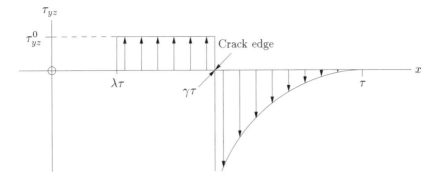

Fig. 6.10.1 The stress τ_{yz} for $y = 0$. The crack extends from $x = -\infty$ to $x = Vt = \gamma\tau$. The displacement $w = 0$ for $\gamma\tau < x < \infty$.

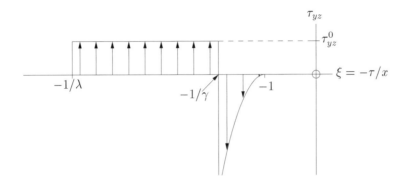

Fig. 6.10.2 Basically the same as the preceding figure, but with $\xi = -\tau/x$ instead of x as coordinate. The displacement $w = 0$ for $-1/\gamma < \xi < 0$.

The boundary conditions are (cf. Fig. 6.10.2)

$$\left(\frac{\partial \tau_{yz}}{\partial \tau}\right)_{y=0} = -\frac{\tau_{yz}^0}{x}\delta(\xi + 1/\lambda) \text{ for } \xi < -1/\gamma \text{ and } \xi > 0 \quad (6.10.5)$$

$$\left(\frac{\partial^2 w_+}{\partial \tau^2}\right)_{y=0} = 0 \text{ for } -1/\gamma < \xi < 0 \quad (6.10.6)$$

where $\delta(\cdot)$ is the Dirac delta function. Insertion of (6.10.2)-(6.10.3) gives

$$H_+(\xi) - H_-(\xi) = 2\pi i \delta(\xi + 1/\lambda) \text{ for } \xi < -1/\gamma \text{ and } \xi > 0 \quad (6.10.7)$$

$$G_+(\xi) - G_-(\xi) = 0 \text{ for } -1/\gamma < \xi < 0 \quad (6.10.8)$$

By using (6.10.4), the first of these equations can be replaced by

$$G_+(\xi) + G_-(\xi) = -\frac{2\pi\lambda}{\sqrt{1-\lambda^2}}\delta(\xi + 1/\lambda) \text{ for } \xi < -1/\gamma \text{ and } \xi > 0 \quad (6.10.9)$$

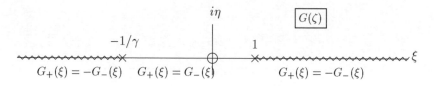

Fig. 6.10.3 Branch cuts for $G(\zeta)$.

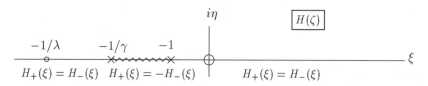

Fig. 6.10.4 Branch cut for $H(\zeta)$.

which forms a Hilbert problem for $G(\zeta)$ together with (6.10.8). The solution is

$$G(\zeta) = -\frac{\sqrt{(\gamma-\lambda)^3}}{\lambda\sqrt{1-\lambda\gamma}\sqrt{\gamma}} \cdot \frac{1}{(\zeta+1/\gamma)^{3/2}(1-\zeta)^{1/2}} \left(\frac{1}{\zeta+1/\lambda} + P_0\right) \quad (6.10.10)$$

where the singularities are the strongest ones allowed, corresponding to an inverse square-root singularity for stresses and strains at the crack edge and to bounded mass velocity at $x = -\tau$. P_0 is a polynomial in ζ, which reduces to a constant because the mass acceleration at $x = 0$ is zero (constant velocity). Branch cuts are made along the real axis from $\xi = -\infty$ to $\xi = -1/\gamma$ and from $\xi = 1$ to $\xi = \infty$, and the branch is chosen so that $(\zeta+1/\gamma)^{3/2} = 1/\sqrt{\gamma^3}$ and $(1-\zeta)^{1/2} = 1$ for $\zeta = 0$; see Fig. 6.10.3. Then, from (6.10.4) it follows that

$$H(\zeta) = \frac{\sqrt{(\gamma-\lambda)^3}}{\lambda\sqrt{1-\lambda\gamma}\sqrt{\gamma}} \cdot \frac{(1+\zeta)^{1/2}}{(\zeta+1/\gamma)^{3/2}} \left(\frac{1}{\zeta+1/\lambda} + P_0\right) \quad (6.10.11)$$

with branch as shown in Fig. 6.10.4 and the branch chosen so that $(\zeta+1/\gamma)^{3/2} = 1/\sqrt{\gamma^3}$ and $(1+\zeta)^{1/2} = 1$ for $\zeta = 0$.

The constant P_0 is determined from the condition that $(\tau_{yz})_{y=0}$ equals τ_{yz}^0 for $-1/\lambda < \xi < -1/\gamma$ and vanishes for $x > \tau$, i.e., for $\xi < 1$. Thus, integration of (6.10.2) leads to

$$\tau_{yz} = \tau_{yz}^0 + \frac{\tau_{yz}^0}{2\pi i} \int_{-1/\gamma-0}^{-1+0} [H_+(\xi) - H_-(\xi)]d\xi = \tau_{yz}^0 + \frac{\tau_{yz}^0}{2\pi i} \int_\Gamma H(\zeta)d\zeta = 0 \quad (6.10.12)$$

where Γ is a path around the portion $-1/\gamma \le \xi \le -1$ of the real axis. Because $H(\zeta) \propto P_0/\zeta$ as $|\zeta| \to \infty$, and $\zeta = -1/\lambda$ is a pole, residue calculus gives $P_0 = 0$.

It is now possible to calculate quantities of interest. The stress rate on $y = 0$ is found to be

$$\frac{\partial \tau_{yz}}{\partial \tau} = -\frac{\tau_{yz}^0\sqrt{(\gamma-\lambda)^3}}{\pi\sqrt{1-\lambda}} \cdot \frac{x\sqrt{\tau-x}}{\sqrt{(x-\gamma\tau)^3(x-\lambda\tau)}} \quad (6.10.13)$$

for $\gamma\tau < x < \tau$, and the mass acceleration on the upper crack face is

$$\frac{\partial^2 w_+}{\partial \tau^2} = \frac{\tau_{yz}^0 \sqrt{(\gamma-\lambda)^3}}{\pi\mu\sqrt{1-\lambda}} \cdot \frac{x^2}{\sqrt{(\gamma\tau-x)^3}\sqrt{\tau+x(x-\lambda\tau)}} \qquad (6.10.14)$$

for $-\tau < x < \gamma\tau$.

The crack edge vicinity
Specialization to the crack edge vicinity, $x \to \gamma\tau$, gives

$$\left(\frac{\partial \tau_{yz}}{\partial \tau}\right)_{y=0} \to -\frac{\tau_{yz}^0 \sqrt{(\gamma-\lambda)\tau}\sqrt{1-\gamma}}{\pi\sqrt{1-\lambda}} \cdot \frac{1}{\sqrt{(x-\gamma\tau)^3}} \qquad (6.10.15)$$

$$\left(\frac{\partial^2 w_+}{\partial \tau^2}\right)_{y=0} \to \frac{\tau_{yz}^0 \gamma^2 \sqrt{(\gamma-\lambda)\tau}}{\pi\mu\sqrt{1-\lambda}\sqrt{1+\gamma}} \cdot \frac{1}{\sqrt{(\gamma\tau-x)^3}} \qquad (6.10.16)$$

which integrate to

$$(\tau_{yz})_{y=0} \to -\frac{2\tau_{yz}^0 \sqrt{(\gamma-\lambda)\tau}\sqrt{1-\gamma}}{\pi\sqrt{1-\lambda}} \cdot \frac{1}{\sqrt{x-\gamma\tau}} \qquad (6.10.17)$$

$$(w_+)_{y=0} \to -\frac{4\tau_{yz}^0 \sqrt{(\gamma-\lambda)\tau}}{\pi\mu\sqrt{1-\lambda}\sqrt{1+\gamma}}\sqrt{\gamma\tau-x} \qquad (6.10.18)$$

Thus, the stress intensity factor is

$$K_{III} = \sqrt{2\pi}\lim_{x\to\gamma\tau}[\tau_{yz}\sqrt{x-\gamma\tau}] = -\frac{2\sqrt{2(1-\gamma)}\tau_{yz}^0\sqrt{(\gamma-\lambda)\tau}}{\sqrt{\pi}\sqrt{1-\lambda}} \qquad (6.10.19)$$

The minus sign depends on the convention for sign of mode III stress intensity factors, cf. definition (3.3.33), and is typical for crack face loading.

The wave pattern near the crack edge is shown schematically in Fig. 6.10.5. Note that cylindrical waves emanate from the crack edges, their front travelling with the S wave velocity. Plane waves, also with front velocity c_S, appear outside these cylindrical waves when the crack face load moves away from the crack edge with supersonic velocity ($\lambda < -1$).

Special cases
Four specializations are:

Case 1. The load extends from $x = 0$, i.e. $\lambda = 0$:

$$G(\zeta) = -\frac{1}{(\zeta+1/\gamma)^{3/2}(1-\zeta)^{1/2}} \qquad (6.10.20)$$

$$H(\zeta) = \frac{(1+\zeta)^{1/2}}{(\zeta+1/\gamma)^{3/2}} \qquad (6.10.21)$$

$$K_{III} = -\frac{2\sqrt{2}\tau_{yz}^0\sqrt{\gamma(1-\gamma)\tau}}{\sqrt{\pi}} \qquad (6.10.22)$$

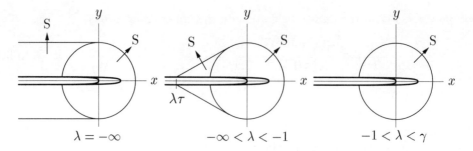

Fig. 6.10.5 Wave fronts travelling with the velocity of S waves away from the crack edge and the crack faces at mode III crack face loading. Cylindrical waves originate at the crack edges; the axis of their front coincides with the original crack edge position. Plane S waves originate from the applied loads.

Case 2. The load extends in the negative x-direction, and the crack edge remains at $x = 0$, i.e. $\lambda < 0$, $\gamma = 0$:

$$K_{III} = -\frac{2\sqrt{2}\tau_{yz}^0 \sqrt{|\lambda|\tau}}{\sqrt{\pi}\sqrt{1+|\lambda|}} \tag{6.10.23}$$

Case 3. The load appears suddenly everywhere on the crack faces, i.e. $\lambda \to -\infty$:

$$K_{III} \to -\frac{2\sqrt{2(1-\gamma)}\tau_{yz}^0 \sqrt{\tau}}{\sqrt{\pi}} \tag{6.10.24}$$

If the crack is stationary, $\gamma = 0$, the displacement of the upper crack face, found by integration twice of (6.10.14), is

$$w_+ = \begin{cases} -\dfrac{\tau_{yz}^0 \tau}{\mu}\left\{1 - \dfrac{2}{\pi}\left[\operatorname{atan}\sqrt{\dfrac{\tau+x}{|x|}} - \sqrt{\dfrac{|x|}{\tau}\left(1+\dfrac{x}{\tau}\right)}\right]\right\} & \text{for } -\tau \leq x < 0 \\ -\dfrac{\tau_{yz}^0 \tau}{\mu} & \text{for } x < -\tau \end{cases} \tag{6.10.25}$$

In the other extreme, propagation velocity approaching the S wave velocity, $\gamma \to 1$, the displacement is found to be

$$w_+ = \begin{cases} -\dfrac{\tau_{yz}^0 \tau}{\mu}\left\{1 - \dfrac{1}{\pi}\left[\operatorname{asin}\dfrac{x}{\tau} - \sqrt{1 - \dfrac{x^2}{\tau^2}} + \dfrac{\pi}{2}\right]\right\} & \text{for } -\tau \leq x < \tau \\ -\dfrac{\tau_{yz}^0 \tau}{\mu} & \text{for } x < -\tau \end{cases} \tag{6.10.26}$$

The expressions show clearly the contribution from the cylindrical wave emanating from the crack edge and travelling with the S wave velocity, see also Fig. 6.10.6.

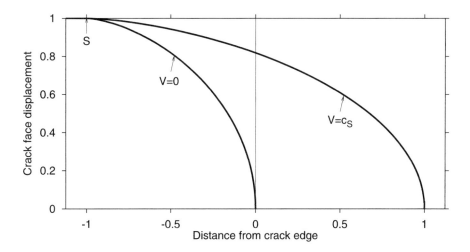

Fig. 6.10.6 Displacement of the upper crack face after a sudden load application, $\tau_{yz} = \tau_{yz}^0$, on the crack faces, both when the crack is stationary (marked $V = 0$) and when the crack speed is approaching the S wave velocity (marked $V = c_S$). The distance from the crack edge and the crack face displacement are given in units of $|\tau_{yz}^0|c_S t/\mu$ and $x/c_S t$, respectively. The arrow marked S indicates the arrival of the cylindrical S wave from the crack edge. Note that, although the stress intensity factor vanishes when the crack speed approaches the S wave velocity, smooth closing is not approached.

Case 4. The velocity λ is only infinitesimally smaller than the velocity γ. This implies a steady state when τ reaches infinity in such a way that the extension of the load is finite, $\gamma\tau - \lambda\tau = L$:

$$K_{III} \to -\frac{2\sqrt{2}\tau_{yz}^0 \sqrt{L}}{\sqrt{\pi}} \quad (6.10.27)$$

In the first two cases, the static solution for a crack face load $\tau_{yz} = \tau_{yz}^0$, extending a distance L from the crack edge, is obtained by setting γ and λ to zero, after L has been identified with $\gamma\tau$ in the first case and with $|\lambda|\tau$ in the second case. This static solution is obtained directly from the third case, in agreement with a general property of steady state problems of this kind, see page 334. The static solution for mode III crack face loading was previously given by (4.3.39).

It will be shown later (Section 6.13) that cases 2 and 3 provide the essential solution of problems concerning plane S wave interaction with a stationary crack.

For case 1, the mass acceleration is

$$\left(\frac{\partial^2 w_+}{\partial \tau^2}\right)_{y=0} = \frac{\tau_{yz}^0}{\pi\mu} \cdot \frac{x}{\sqrt{(\gamma\tau - x)^3}\sqrt{\tau + x}} \quad \text{for} \quad -\tau < x < \gamma\tau \quad (6.10.28)$$

which integrates to

$$\left(\frac{\partial w_+}{\partial \tau}\right)_{y=0} = -\frac{2\gamma\sqrt{\gamma}\tau_{yz}^0}{\pi(1+\gamma)\mu}\sqrt{\frac{\tau+x}{\gamma\tau-x}} \tag{6.10.29}$$

$$(w_+)_{y=0} = -\frac{\tau_{yz}^0}{\pi\mu}\left[\frac{2\sqrt{\gamma}}{1+\gamma} + x\ln\frac{2\sqrt{\gamma}(\tau+x)(\gamma\tau-x) - x + 2\gamma\tau + \gamma x}{(1+\gamma|x|)}\right] \tag{6.10.30}$$

Superposition of elementary solutions

The dynamic crack problems treated so far have been either steady state problems or self-similar problems. As shown in particular by Freund (1972a,b, 1973, 1974a,b, 1990), superposition techniques are powerful tools for obtaining solutions to more general kinds of dynamic crack problems. The starting point is then finding solutions to steady state or self-similar problems. Such solutions may be considered as elementary solutions, which then can be superposed in suitable ways to obtain solutions to new problems, not necessarily steady state or self-similar. These solutions can, in turn, be used as elementary solutions for further superposition, and so on.

Moving pair of opposed concentrated crack face forces

Consider again, as in the beginning of this section, a semi-infinite mode III crack, moving from an unloaded stationary position, $x < 0$, $y = 0$, with constant velocity $V = \gamma c_S$ along the positive x axis, under the action of a crack face load

$$\tau_{yz} = \tau_{yz}^0 U(x - \lambda\tau) \text{ for } x < \gamma\tau \tag{6.10.31}$$

where $U(\cdot)$ is the unit step function, $\tau = c_S t$ and $\lambda < \gamma$. The rear end of the load is at $x = \lambda\tau$ and the crack edge is at $x = \gamma\tau$. Thus, the length of the loading region at time $\tau = \tau_0$ equals L if $\tau_0 = L/(\gamma - \lambda)$. If, in addition, another step load, with magnitude $-\tau_{yz}^0$ and trailing edge velocity λc_S, appears behind the crack edge at $x = x_0 = \gamma\tau_0$ and $\tau = \tau_0$, then the result is a rectangular load with constant extension L, after time $\tau = \tau_0$, moving with velocity λc_S; see Fig. 6.10.7.

During the time $0 < \tau < \tau_0$, the process is self-similar, as studied in the preceding section, but thereafter it is not. The solution for the stress rate $(\partial \tau_{yz}/\partial \tau)_{y=0}$ for $\gamma\tau < x < \tau$, $\tau > \tau_0$ is found by superposition of the elementary solution (6.10.13) for the two loads:

$$\left(\frac{\partial\tau_{yz}}{\partial\tau}\right)_{y=0} = -\frac{\sqrt{(\gamma-\lambda)^3}\tau_{yz}^0}{\pi\sqrt{1-\lambda}}\left\{\frac{x\sqrt{\tau-x}}{\sqrt{(x-\gamma\tau)^3}(x-\lambda\tau)}\right.$$
$$\left.-\frac{(x-x_0)\sqrt{\tau-\tau_0-(x-x_0)}U(x-x_0)}{\sqrt{(x-x_0-\gamma(\tau-\tau_0))^3}[x-x_0-\lambda(\tau-\tau_0)]}\right\} \tag{6.10.32}$$

After time $\tau = \tau_0$ the total shear force on each crack face remains constant, equal to $T_{yz}^0 = \tau_{yz}^0 L$ per unit of width. Then, for $\tau \gg L/(\gamma-\lambda)$, the stress rate for $\gamma\tau < x < \tau$, $\tau > \tau_0$, is

$$\left(\frac{\partial\tau_{yz}}{\partial\tau}\right)_{y=0} \approx -\frac{\sqrt{\gamma-\lambda}T_{yz}^0}{\pi\sqrt{1-\lambda}}\cdot\frac{(1-2\lambda-\gamma)x^2 + \lambda(3\gamma+1)\tau x - 2\lambda\gamma\tau^2}{2\sqrt{(x-\gamma\tau)^3}\sqrt{\tau-x}(x-\lambda\tau)^2} \tag{6.10.33}$$

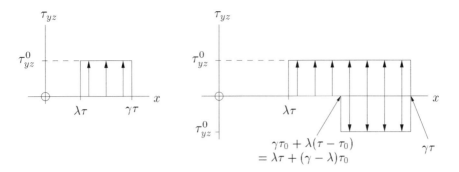

Fig. 6.10.7 Step function crack face load (left figure) and superposition of two staggered step function loads of different signs, leading to a moving crack face load with constant length and amplitude.

and this expression becomes exact in the limit $L \to 0$, corresponding to a pair of opposed concentrated shear forces, moving with velocity λc_S on the crack faces. This is the mode III version of what Freund (1972a) called the *fundamental solution*, which is a particularly useful and versatile elementary solution in creation of solutions to other problems through superposition. The stress intensity factor is

$$K_{III} = -\frac{\sqrt{2}T_{yz}^0\sqrt{1-\gamma}}{\sqrt{\pi}\sqrt{1-\lambda}\sqrt{(\gamma-\lambda)\tau}} \qquad (6.10.34)$$

Because stress intensity factors are superposable, this result could have been obtained in a simpler way by superposition directly from (6.10.19) if the stress intensity factor, only, had been required. Thus,

$$K_{III} = [K(\tau) - K(\tau - \tau_0)]_{\tau_0/\tau \to 0} = \tau_0 \frac{\mathrm{d}K}{\mathrm{d}\tau} = \frac{T_{yz}^0}{(\gamma-\lambda)\tau_{yz}^0}\cdot\frac{\mathrm{d}K}{\mathrm{d}\tau} \qquad (6.10.35)$$

where K is the stress intensity factor given by (6.10.19).

General time-independent crack face loading

A more elaborate superposition scheme than one involving only two elementary solutions, as in the previous subsection, may be used to obtain rather general types of crack face loading. Here, time-independent loading at each point of the crack faces created after the start of crack motion, will be considered. One such case was studied in the previous section, a step function load with its rear end stationary, $\lambda = 0$, and appearing behind the moving crack. Solutions were obtained for mass acceleration and stress rate on $y = 0$. By integrations, mass velocities, displacements and stresses can be obtained. Let one such solution for the step function load $\tau_{yz}^0 U(x)$, $x < \gamma\tau$, be $\tau_{yz}^0 s_0(x,\tau)$. Then, the solution for an elementary step function load,

$$\mathrm{d}h(x_0)U(x-x_0) \text{ for } x < \gamma\tau \qquad (6.10.36)$$

is

$$\mathrm{d}h(x_0)s_0(x-x_0,\tau-x_0/\gamma) \qquad (6.10.37)$$

and thus the solution for a general time-independent crack face load, $\tau_{yz}(x) = h(x)$ for $0 < x < \gamma\tau$ is

$$\int_0^{\gamma\tau} s_0(x - x_0, \tau - x_0/\gamma) h'(x_0) \mathrm{d}x_0 \qquad (6.10.38)$$

Alternatively, the superposition elements may be considered as functions of their starting time τ_0 rather than their starting position x_0, which implies that the solution for a general time-independent crack face load is written as

$$\gamma \int_0^{\tau} s_0(x - \gamma\tau_0, \tau - \tau_0) h'(\gamma\tau_0) \mathrm{d}\tau_0 \qquad (6.10.39)$$

This superposition technique is sometimes referred to as the Duhamel integral. The result becomes particularly simple, if s_0 is the stress intensity factor, which then can be written as a function of τ, only. Use of (6.10.22) gives

$$\begin{aligned} K_{III} &= -\frac{2\sqrt{2}\gamma\sqrt{\gamma(1-\gamma)}}{\sqrt{\pi}} \int_0^{\tau} \sqrt{\tau - \tau_0}\, h'(\gamma\tau_0)\mathrm{d}\tau_0 \\ &= -\frac{2\sqrt{2}\sqrt{\gamma(1-\gamma)}}{\sqrt{\pi}} \left\{ [\sqrt{\tau - \tau_0}\, h(\gamma\tau_0)]_0^{\tau} + \int_0^{\tau} \frac{h(\gamma\tau_0)}{2\sqrt{\tau - \tau_0}} \mathrm{d}\tau_0 \right\} \\ &= -\frac{\sqrt{2}\sqrt{1-\gamma}}{\sqrt{\pi}} \int_0^{a} \frac{h(u)}{\sqrt{a-u}} \mathrm{d}u \qquad (6.10.40) \end{aligned}$$

where $\gamma\tau$ was identified with the extension a of the load and the substitution $\tau_0 = u/\gamma$ was made. The result is very interesting. The crack velocity appears only in the factor $\sqrt{1-\gamma}$, so that the result can be written as

$$K_{III} = K_{III}(a, \gamma) = k_{III}(\gamma) K_{III}^{stat}(a), \qquad k_{III}(\gamma) = \sqrt{1-\gamma} \qquad (6.10.41)$$

where $K_{III}^{stat}(a)$ obviously, as is also verified by (4.3.39), is the stress intensity factor for a stationary crack, subjected to the same crack face load as the running crack at the same edge position. Thus, the resulting stress intensity factor equals the corresponding static stress intensity factor multiplied by the universal function $k_{III}(\gamma)$.

Crack motion from an equilibrium state

So far, the semi-infinite crack has been assumed to start running from an unloaded state. However, the results obtained can be generalized to crack motion starting from an equilibrium state of a finite size body, and the loads are not necessarily restricted to crack face loads as previously in this section, but may also contain volume forces and loads or prescribed displacements on outer boundaries.

Before the crack starts running, there is a shear stress $\tau_{yz} = h_a(x)$, say, in the symmetry plane ahead of the crack edge, $x = 0$; cf. Fig. 6.10.8a. This shear stress turns out to provide sufficient information of the equilibrium state for determination of certain crack propagation properties, before the arrival of waves reflected from an outer boundary or another crack edge.

The problem at hand consists of finding the stress intensity factor after the crack suddenly starts running, at time $t = 0$, with constant velocity $V = \gamma c_S$. Thus, the crack faces are traction free on $0 \leq x < \gamma\tau$, where $\tau = c_S t$; cf. Fig 6.10.8b. This

6.10 FACE LOADED MODE III CRACKS

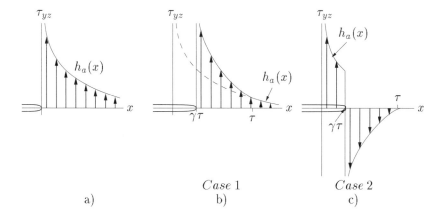

Fig. 6.10.8 a) Stationary crack in a body, subjected to loads that generate the stress $\tau_{yz} = h_a(x)$ on the plane ahead of the crack. b) Case 1: Crack moving from its stationary position in b) with velocity $\gamma c_S \tau$. c) Case 2: A crack moving with constant velocity γc_S in an originally stress free body, under the action of the crack face load $\tau_{yz} = h_a(x)$, appearing behind the moving crack edge.

case will be referred to as *Case* 1. Consider now the superposition of another case, *Case* 2, for the same cracked body, but initially unloaded, and then, at $t = 0$ starting to move with constant velocity V under the action of the load $h_a(x)$, $0 < x < \gamma \tau$, appearing behind the crack edge on the crack faces; cf. Fig. 6.10.8c. The result of the superposition is obviously again the equilibrium crack (Fig. 6.10.8a), with the crack edge still at $x = 0$, although one may imagine, if so desired, that the crack edge is at $x = \gamma \tau$ (thus running with velocity V), but the crack faces are prevented from sliding against each other between $x = 0$ and $x = \gamma \tau$ by sufficiently strong friction. The stress intensity factor at this (imagined) crack edge is obviously zero, which implies that the superposition yields

$$(K_{III})_{Case1} + (K_{III})_{Case2} = 0 \tag{6.10.42}$$

Now, $(K_{III})_{Case2}$ can be obtained by using (6.10.40):

$$(K_{III})_{Case2} = -\frac{\sqrt{2}\sqrt{1-\gamma}}{\sqrt{\pi}} \int_0^a \frac{h_a(u)}{\sqrt{a-u}} du \tag{6.10.43}$$

where $a = \gamma \tau$. Thus, the stress intensity asked for (*Case* 1) is

$$K_{III}(a, \gamma) = \frac{\sqrt{2}\sqrt{1-\gamma}}{\sqrt{\pi}} \int_0^a \frac{h_a(u) du}{\sqrt{a-u}} \tag{6.10.44}$$

By writing

$$K_{III}^{ref}(a) = \sqrt{\frac{2}{\pi}} \int_0^a \frac{h_a(u) du}{\sqrt{a-u}} \tag{6.10.45}$$

the relation (6.10.44) can be brought to the same form as (6.10.41),

$$K_{III}(a, \gamma) = k_{III}(\gamma) K_{III}^{ref}(a) \tag{6.10.46}$$

The function $K_{III}^{ref}(a)$ is obviously independent both of time and of crack edge velocity and is therefore a fixed reference stress intensity factor to be associated with the cracked body under equilibrium loads. Note that the results are subject to the condition that no reflection has arrived from the outer boundary or another crack edge, and therefore they cannot in general be extrapolated to arbitrarily low velocities. If, however, the body is infinite and the crack semi-infinite, no such reflection can arrive in finite time, which also implies that the expression (6.10.44) for $K_{III}(a, \gamma)$ is valid even if γ is infinitesimally small. Then, $K_{III}^{ref}(a) = K_{III}^{stat}(a)$, the stress intensity factor for a stationary crack with its edge at $x = a$ and subjected to the action of the loads $h_b(x)$ on $0 < x < a$.

The solution (6.10.44) may be considered as an elementary solution, which may take part in superposition together with solutions to other problems concerning a crack starting to move with velocity V from the crack edge position $x = 0$. Thus, in combination with (6.10.40) the stress intensity factor is obtained for the case when an equilibrium crack starts moving with constant velocity under the action of both the equilibrium loads and a crack face loading $\tau_{yz} = h_b(x)$, appearing behind the crack edge during its motion from $x = 0$ to $x = \gamma\tau$:

$$K_{III}(a, \gamma) = -\frac{\sqrt{2}\sqrt{1-\gamma}}{\sqrt{\pi}} \int_0^a \frac{h_b(u) - h_a(u)}{\sqrt{a-u}} du \qquad (6.10.47)$$

where $a = \gamma\tau$. The subscripts of h can be interpreted so that a indicates "ahead" and b "behind", as related to the load location with respect to the crack edge. Note that the action of the equilibrium loads can be represented by a crack face load, appearing behind the moving crack edge and equalling the *negated* stress on the symmetry plane ahead of the crack edge before crack motion. By writing

$$K_{III}^{ref}(a) = -\frac{\sqrt{2}}{\sqrt{\pi}} \int_0^a \frac{h_b(u) - h_a(u)}{\sqrt{a-u}} du \qquad (6.10.48)$$

the relation (6.10.47) can be written in the same form as (6.10.41)

$$K_{III}(a, \gamma) = k_{III}(\gamma) K_{III}^{ref}(a) \qquad (6.10.49)$$

Again, the function $K_{III}^{ref}(a)$, here given by (6.10.48), is independent both of time and of crack edge velocity. It is thus a fixed reference stress intensity factor to be associated with the cracked body under equilibrium loads. If the body is infinite and the crack semi-infinite, then $K_{III}^{ref}(a) = K_{III}^{stat}(a)$, the stress intensity factor for a stationary crack with its edge at $x = a$ and subjected to the loads $h_b(x)$, appearing behind the crack edge on $0 < x < a$, and, in addition, to the action of equilibrium loads which give rise to the stress distribution $h_a(x)$ on the symmetry plane for $0 < x < a$. In the general case of a finite body, there is no such obvious connection of the definition (6.10.48) with an existing static stress intensity factor for the body.

Time-dependent crack face loads

Various time-dependent cases may be obtained by superposition techniques. One example is the response to a pair of opposed concentrated forces, $T_{yz}^0 = T_{yz}'t$ ($T_{yz}' = $ constant), moving with velocity λc_S behind a crack edge, which in turn is moving

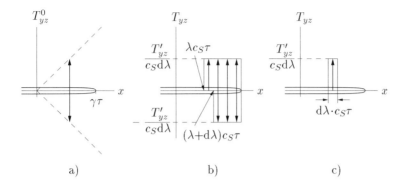

Fig. 6.10.9 a) Two opposed moving concentrated crack face forces, increasing linearly with time. Figures b) and c) show how one of the concentrated forces may be obtained by a superposition procedure.

with velocity $V = \gamma c_S > \lambda c_S$. The response is found by superposition of two crack face loads of the kind (6.10.1). One load should then have the amplitude $T'_{yz}/(c_S d\lambda)$, and move with velocity λc_S, and the other load should have the amplitude $-T'_{yz}/(c_S d\lambda)$ and move with velocity $(\lambda + d\lambda)c_S$; cf. Fig. 6.10.9. It is then obvious that the total force on each crack face is $T^0_{yz} = T'_{yz} t$ and that it becomes concentrated and moving with velocity λc_S when $d\lambda \to 0$.

The stress intensity factor is found from (6.10.19), after substituting τ^0_{yz} by $-T'_{yz}/c_S$ and differentiating with respect to λ:

$$K_{III} = -\frac{\sqrt{2}T'_{yz}}{\sqrt{\pi c_S}} \cdot \frac{\sqrt{(1-\gamma)^3 \tau}}{\sqrt{\gamma - \lambda}\sqrt{(1-\lambda)^3}} \qquad (6.10.50)$$

This result is valid for a finite body as long as no reflections have arrived from an outer boundary or another crack edge.

6.11 Face loaded in-plane mode cracks

The analytical treatment of crack face loaded mode I cracks is very similar to the one for mode III in Section 6.10. Therefore, the discussion will follow the same course as for mode III, avoiding unnecessary repetitions. Furthermore, because the formal difference between modes I and II is very small, a separate treatment for mode II is not considered as essential. However, the adjustments needed for obtaining mode II solutions from mode I cases will be pointed out.

In-plane mode crack face loading, as for mode III, is often part of a superposition scheme in problems regarding stress wave interaction with traction free cracks. The unphysical result of crack face interpenetration can then be tolerated for individual parts of such a scheme, if it disappears after superposition. Very often, the composed problem concerns traction free crack faces, but it may nevertheless be advantageous to consider crack face loading in individual parts. One example is shown by Fig. 6.13.2.

6. ELASTODYNAMIC CRACK MECHANICS

The special case of interpenetration at mode I loading will be discussed in the next subsection.

Face loaded mode I cracks

A general remark
Unlike tangential crack face loading, direct dynamic loading of crack faces by normal forces occurs in several practical situations, such as hydraulic cracking (pressure applied to liquid-filled cracks) and explosive loading (by detonating explosives or electrically induced bursting of copper strips, etc.). However, analytical solutions exploiting mode I symmetry to only consider half the body, often lead to interpenetration of the crack faces already before waves have reached the crack edge. This phenomenon is related to a particular feature of wave motion along a plane surface (Lamb 1904): downwards directed impact results in initially upwards directed surface displacement at points outside the impacted area, particularly pronounced just before the arrival of the Rayleigh wave. To investigate whether or not interpenetration has occurred in an analytical solution is often difficult and may require numerical inspection. However, in practical cases of direct crack face load application, the crack faces are initially somewhat separated. This situation may be analysed by using an idealized model of crack faces in initial contact, and interpenetration as a result of such analysis may then be tolerated, if it is smaller than the original real separation.

A self-similar problem with wide applicability
Consider a semi-infinite crack moving from an unloaded stationary position, $x < 0$, $y = 0$, with constant velocity $V = \beta c_P$ along the positive x axis, under the action of a crack face load

$$\sigma_y = \sigma_y^0 U(x - \lambda \tau) \text{ for } x < \beta \tau \qquad (6.11.1)$$

where $U(\cdot)$ is the unit step function, $\tau = c_P t$, $\lambda < \beta$ and the load $\sigma_y^0 < 0$. This corresponds to equation (6.10.1) for mode III.

The initial procedure is virtually identical with the one used for a self-similar expanding mode I crack, cf. (6.9.18)-(6.9.39). Thus,

$$\left(\frac{\partial \sigma_y}{\partial \tau}\right)_{y=0} = -\frac{\sigma_y^0}{2\pi i x}[H_+(\xi) - H_-(\xi)]_{\xi=-\tau/x} \qquad (6.11.2)$$

$$\left(\frac{\partial^2 v_+}{\partial \tau^2}\right)_{y=0} = -\frac{k^2 \sigma_y^0}{2\pi \mu i x}[G_+(\xi) - G_-(\xi)]_{\xi=-\tau/x} \qquad (6.11.3)$$

$$G(\zeta) = \frac{(1-\zeta^2)^{1/2}}{\zeta^4 R(1/\zeta)} H(\zeta) = C(\zeta)H(\zeta) \qquad (6.11.4)$$

See Fig. 6.9.4 for the definition of branch cuts and branches for $C(\zeta)$.

The boundary conditions are

$$\left(\frac{\partial \sigma_y}{\partial \tau}\right)_{y=0} = -\frac{\sigma_y^0}{x}\delta(\xi + 1/\lambda) \text{ for } \xi < -1/\beta \text{ and } \xi > 0 \qquad (6.11.5)$$

$$\left(\frac{\partial^2 v_+}{\partial \tau^2}\right)_{y=0} = 0 \text{ for } -1/\beta < \xi < 0 \qquad (6.11.6)$$

where $\delta(\cdot)$ is the Dirac delta function. Insertion of (6.11.2)-(6.11.3) gives

$$H_+(\xi) - H_-(\xi) = 2\pi i\delta(\xi + 1/\lambda) \text{ for } \xi < -1/\beta \text{ and } \xi > 0 \tag{6.11.7}$$

$$G_+(\xi) - G_-(\xi) = 0 \text{ for } -1/\beta < \xi < 0 \tag{6.11.8}$$

By using (6.11.4) the second of these equations can be substituted by

$$C_+(\xi)H_+(\xi) - C_-(\xi)H_-(\xi) = 0 \text{ for } -1/\beta < \xi < 0 \tag{6.11.9}$$

Define now

$$c(\xi) = \begin{cases} \dfrac{C_-(\xi)}{C_+(\xi)} & \text{for } -1/\beta < \xi < 0 \\ 1 & \text{for } \xi < -1/\beta \text{ and } \xi > 0 \end{cases} \tag{6.11.10}$$

Then,

$$H_+(\xi) - c(\xi)H_-(\xi) = 2\pi i\delta(\xi + 1/\lambda) \text{ for all } \xi \tag{6.11.11}$$

because $\delta(\xi + 1/\lambda) = 0$ for $-1/\beta < \xi < 0$. Note that

$$\frac{C_-(\xi)}{C_+(\xi)} = \begin{cases} -1 & \text{for } -1/\beta < \xi < -1/k \\ -e^{2\pi i g(\xi)} & \text{for } -1/k < \xi < -1 \\ +1 & \text{for } -1 < \xi < 0 \end{cases} \tag{6.11.12}$$

where, cf. (6.9.111),

$$g(\xi) = \frac{1}{\pi}\text{atan}\frac{4k^3\xi^2\sqrt{\xi^2-1}\sqrt{1-k^2\xi^2}}{(1-2k^2\xi^2)^2} \tag{6.11.13}$$

In order to solve equation (6.11.11), the function

$$S(\zeta) = \exp\left\{\int_{-1/k}^{-1}\frac{g(u)du}{u-\zeta}\right\} = \exp\left\{-\int_1^{1/k}\frac{g(u)du}{u+\zeta}\right\} \tag{6.11.14}$$

is introduced. Then,

$$S_+(\xi) = S_-(\xi) \text{ for } \xi < -1/k \text{ and } \xi > -1 \tag{6.11.15}$$

$$S_\pm(\xi) = \exp\left\{-\oint_1^{1/k}\frac{g(u)du}{u+\xi} \pm \pi i g(\xi)\right\} \text{ for } -1/k < \xi < -1 \tag{6.11.16}$$

where C on the integral sign expresses the Cauchy principal value. Note that $S(\zeta)$ has branch points at $\zeta = -1/k$ and $\zeta = -1$ and is defined through a branch cut between these points, outside which cut the function is analytic; see Fig. 6.11.1. The branch is chosen so that $S(\pm\infty) = 1$.

Note that $S_+(\xi)/S_-(\xi) = +\exp 2\pi i g(\xi)$ for $-1/k < \xi < -1$. Thus, it is possible to write

$$c(\xi) = \frac{Q_-(\xi)}{Q_+(\xi)} \tag{6.11.17}$$

where the function

$$Q(\zeta) = \frac{(\zeta + 1/\beta)^{3/2}(1+\zeta)^{1/2}}{S(\zeta)} \tag{6.11.18}$$

is analytic outside a branch cut from $\zeta = -1/\beta$ to $\zeta = -1$ along the real axis. The

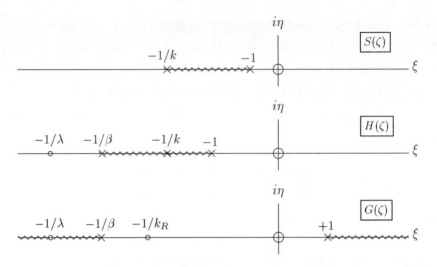

Fig. 6.11.1 Branch cuts for function definitions.

exponent 3/2 is chosen so that $\partial\sigma_y/\partial\tau \propto (\xi + 1/\beta)^{-3/2}$ as $\xi \to -1/\beta$, which will be found to ensure bounded stress-strain energy. Equation (6.11.11) can then be written in the form

$$Q_+(\xi)H_+(\xi) - Q_-(\xi)H_-(\xi) = 2\pi i Q_+(\xi)\delta(\xi + 1/\lambda) \text{ for all } \xi \quad (6.11.19)$$

The solution of this equation is found from the Plemelj formulae, with incorporation of a finite polynomial to consider solutions to the homogeneous part. Thus,

$$Q(\zeta)H(\zeta) = \int_{-\infty}^{\infty} \frac{Q_+(u)\delta(u + 1/\lambda)}{u - \zeta} du + \text{polynomial} \quad (6.11.20)$$

which leads to

$$H(\zeta) = \frac{F_1(\lambda,\beta)S(\zeta)}{\beta\sqrt{\beta}(\zeta + 1/\beta)^{3/2}(1 + \zeta)^{1/2}}\left[\frac{1}{\zeta + 1/\lambda} + P(\zeta)\right] \quad (6.11.21)$$

$$G(\zeta) = \frac{F_1(\lambda,\beta)(1 - \zeta)^{1/2}S(\zeta)}{\beta\sqrt{\beta}(\zeta + 1/\beta)^{3/2}\zeta^4 R(1/\zeta)}\left[\frac{1}{\zeta + 1/\lambda} + P(\zeta)\right] \quad (6.11.22)$$

where

$$F_1(\lambda,\beta) = \frac{\sqrt{1-\lambda}\sqrt{(\beta-\lambda)^3}}{\lambda^2 S(-1/\lambda)} \quad (6.11.23)$$

and the polynomial $P(\zeta)$ is of first order or less, because the mass acceleration is zero at $x = 0$. Branches and branch cuts for the irrational functions are defined as for the same functions in the mode III case and in equation (6.9.39). Note, however, that $S(\zeta)/R(1/\zeta)$ does not have branch points at $\zeta = -1/k$ and $\zeta = -1$; this follows for instance from the fact that $v_+ = 0$, and consequently $G_+(\xi) = G_-(\xi)$, for $-1/\beta < \xi < 1$.

A more generally applicable way of finding $Q(\zeta)$ involves solving the homogeneous part of equation (6.11.11) by the method used for equation (6.9.108).

Put $P(\zeta) = P_0 + P_1\zeta$, where P_0 and P_1 are constants. The constant P_1 is found in the same way as the constant P_0 for the mode III case, here by integrating $H(\zeta)$ along a path around the portion $-1/\beta < \xi < -1$ of the real axis: residue calculus shows that $P_1 = 0$. The present constant P_0, however, does not affect the integral: it has to be determined from equation (6.11.8), which is not satisfied if $G(\zeta)$ has a pole at $\zeta = -1/k_R$, where $k_R = c_R/c_P$. As can be seen from (6.11.22), this implies that

$$P_0 = \frac{1}{1/k_R - 1/\lambda} = -\frac{k_R\lambda}{k_R - \lambda} \qquad (6.11.24)$$

Note that a pole of $G(\zeta)$ within the interval specified in equation (6.11.8) would imply a delta function in the right member, instead of zero, just as the pole $\zeta = -1/\lambda$ of $H(\zeta)$ implies a delta function in (6.11.7).

It is now possible to calculate interesting quantities. The stress rate on $y = 0$ is found to be

$$\frac{\partial \sigma_y}{\partial \tau} = -\frac{\sigma_y^0}{\pi} \cdot \frac{\lambda F_1(\lambda, \beta)[S_+(-\tau/x) + S_-(-\tau/x)]x}{2\sqrt{\tau - x}\sqrt{(x - \beta\tau)^3}} \left(\frac{x}{x - \lambda\tau} - \frac{k_R}{k_R - \lambda} \right) \qquad (6.11.25)$$

for $\beta\tau < x \leq \tau$. Note that $S_+(-\tau/x) = S_-(-\tau/x)$ for $\beta\tau < x < k\tau$. The mass acceleration on the upper crack face is

$$\frac{\partial^2 v_+}{\partial \tau^2} = \frac{k^2\sigma_y^0}{\pi\mu} \cdot \frac{\lambda F_1(\lambda, \beta)S(-\tau/x)x^4\sqrt{\tau + x}}{\tau^4 R(x/\tau)\sqrt{(\beta\tau - x)^3}} \left(\frac{x}{x - \lambda\tau} - \frac{k_R}{k_R - \lambda} \right)$$

$$\text{for } -k\tau \leq x < \beta\tau \qquad (6.11.26)$$

$$\frac{\partial^2 v_+}{\partial \tau^2} = -\frac{k^2\sigma_y^0}{\pi\mu} \cdot \frac{\lambda F_1(\lambda, \beta)S(-\tau/x)x^4\sqrt{\tau + x}(x^2 - 2k^2\tau^2)^2}{\tau^8|R(x/\tau)|^2\sqrt{(\beta\tau - x)^3}} \left(\frac{x}{x - \lambda\tau} - \frac{k_R}{k_R - \lambda} \right)$$

$$\text{for } -\tau \leq x < -k\tau \qquad (6.11.27)$$

The wave pattern near the crack edge is schematically shown in Fig. 6.11.2. Cylindrical P and S waves emanate from the crack edges. Plane S waves are generated by the cylindrical P wave due to the rotational motion behind its front at the crack faces. Plane P and S waves appear outside the cylindrical P wave as a direct action of the crack face load if this moves away with supersonic velocity ($\lambda < -1$), except for suddenly applied crack face load everywhere ($\lambda = -\infty$), when only P waves are generated, because then there is no rotational motion.

The crack edge vicinity

Specialization to the crack edge vicinity, $x \to \beta\tau$, gives

$$\left(\frac{\partial \sigma_y}{\partial \tau}\right)_{y=0} \to -\frac{\sigma_y^0}{\pi} \cdot \frac{F_1(\lambda, \beta)S(-1/\beta)\beta\sqrt{\tau}}{\sqrt{1-\beta}\sqrt{(x-\beta\tau)^3}} \cdot m(\lambda, \beta) \qquad (6.11.28)$$

$$\left(\frac{\partial^2 v_+}{\partial \tau^2}\right)_{y=0} \to \frac{k^2\sigma_y^0}{\pi\mu} \cdot \frac{F_1(\lambda, \beta)S(-1/\beta)\beta^4\sqrt{1+\beta}\sqrt{\tau}}{R(\beta)\sqrt{(\beta\tau - x)^3}} \cdot m(\lambda, \beta) \qquad (6.11.29)$$

where

$$m(\lambda, \beta) = \frac{\lambda^2(k_R - \beta)}{(\beta - \lambda)(k_R - \lambda)} \qquad (6.11.30)$$

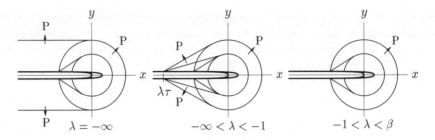

Fig. 6.11.2 Wave fronts with the velocity of P and S waves, travelling away from the crack edge and the crack faces at mode I crack face loading. Cylindrical P and S waves originate at the crack edges; the axes of their fronts coincide with the original crack edge position. Plane P and S waves originate from the applied loads and plane S waves are generated by the cylindrical P wave. P wave fronts are marked P; all other fronts are S wave fronts. For $-1 < \lambda < -k$ there is also a plane wave front from $x = \lambda \tau$, tangent to the cylindrical S wave front.

Integrations give

$$\sigma_y \to -\frac{2\sigma_y^0}{\pi} \cdot \frac{F_1(\lambda,\beta)S(-1/\beta)\sqrt{\tau}}{\sqrt{1-\beta}\sqrt{x-\beta\tau}} \cdot m(\lambda,\beta) \tag{6.11.31}$$

$$(v_+)_{y=0} \to -\frac{4\sigma_y^0 \beta^2 F_1(\lambda,\beta) S(-1/\beta)\sqrt{1+\beta}\sqrt{\tau} m(\lambda,\beta)\sqrt{\beta\tau - x}}{\pi\mu R(\beta)} \tag{6.11.32}$$

The stress intensity factor is

$$K_I = -\sqrt{2\pi} \lim_{x \to \beta\tau}[\sigma_y\sqrt{x-\beta\tau}]$$

$$= -\frac{2\sqrt{2}\sigma_y^0}{\sqrt{\pi}} \cdot \frac{\sqrt{1-\lambda}\sqrt{(\beta-\lambda)\tau}}{S(-1/\lambda)} \cdot \frac{S(-1/\beta)(k_R - \beta)}{(k_R - \lambda)\sqrt{1-\beta}} \tag{6.11.33}$$

The minus sign is due to the fact that $\sigma_y^0 < 0$.

Interpenetration

The mass velocity $(\partial v_+/\partial \tau)_{y=0}$ equals zero for $x < -\tau$ if $\lambda > -1$ and $-k^2\sigma_y^0/\mu$ for $\lambda\tau < x < -\tau$, because the mass velocity $\partial v/\partial t$ at a sudden load application is found by dividing the load by the acoustic impedance, ϱc_P. The displacement near $x = -\tau$ can now be determined by simplifying (6.11.27) and integrating twice. The result is

$$(v_+)_{y=0} \to -\frac{4k^2\sigma_y^0 F_1(\lambda,\beta)S(1)}{15\pi(1-2k^2)^2\mu\sqrt{(1+\beta)^3}} \cdot \left[\frac{\lambda}{1+\lambda} - \frac{\lambda}{1-\lambda/(1-\lambda/k_R)}\right]\sqrt{\frac{(\tau+x)^5}{\tau}}$$
$$-\frac{k^2\sigma_y^0\tau\, U(-1-\lambda)}{\mu} \quad \text{as } x \to -\tau + 0 \tag{6.11.34}$$

If $\lambda > -1$ this displacement is negative, showing interpenetration, and, strictly, the result can only be used as part of certain superposition schemes. If $\lambda < -1$, both terms in (6.11.34) become positive. In combination with the result that $(\partial^2 v_+/\partial \tau^2)_{y=0}$ is negative for $x > -k_R\tau$ and positive for $-\tau < x < -k_R\tau$, this fact ensures that there is no interpenetration for $\lambda < -1$.

Special cases
Four specializations are:

Case 1. The load extends from $x = 0$, i.e. $\lambda = 0$:

$$K_I = -\frac{2\sqrt{2}\sigma_y^0\sqrt{\tau}}{\sqrt{\pi}} \cdot \frac{\sqrt{\beta}S(-1/\beta)(k_R - \beta)}{k_R\sqrt{1-\beta}} \qquad (6.11.35)$$

Case 2. The load extends in negative x-direction and the crack edge remains at $x = 0$, i.e. $\lambda < 0$, $\beta = 0$:

$$K_I = -\frac{2\sqrt{2}\sigma_y^0\sqrt{\tau}}{\sqrt{\pi}} \cdot \frac{\sqrt{1+|\lambda|}\sqrt{|\lambda|}}{S(1/|\lambda|)} \cdot \frac{k_R}{(k_R + |\lambda|)} \qquad (6.11.36)$$

Case 3. The load appears suddenly everywhere on the crack faces, i.e. $\lambda \to -\infty$:

$$K_I = -\frac{2\sqrt{2}\sigma_y^0\sqrt{\tau}}{\sqrt{\pi}S(0)} \cdot \frac{S(-1/\beta)(k_R - \beta)}{\sqrt{1-\beta}} \qquad (6.11.37)$$

The constant $S(0)$ can be evaluated with residue calculus. After a change of integration variable, the expression (6.11.14) reads, for $\zeta = 0$:

$$S(0) = \exp\left\{-\frac{1}{2\pi}\int_{k^2}^1 \mathrm{atan}\frac{4k^3\sqrt{1-s}\sqrt{s-k^2}}{(s-2k^2)^2} \cdot \frac{ds}{s}\right\} \qquad (6.11.38)$$

Here, it is convenient to introduce the function

$$\phi(s) = \frac{1}{2\pi i}\ln\frac{a(s)}{b(s)} \qquad (6.11.39)$$

where

$$a(s) = (s - 2k^2)^2 - 4k^3(1-s)^{1/2}(k^2-s)^{1/2} \qquad (6.11.40)$$
$$b(s) = (s - 2k^2)^2 + 4k^3(1-s)^{1/2}(k^2-s)^{1/2} \qquad (6.11.41)$$

$a(s)$ and $b(s)$ are related to the Rayleigh function. Thus, (see pages 694ff.), $a(s)$ possesses the zeros 0 and k_R^2, and $b(s)$ has two complex conjugate zeros, if $k < k_0 \approx 0.5760\ldots$, and otherwise two real zeros larger than unity. Note that $a(s)b(s)=s(s-k_R^2)(s-v_1^2)(s-v_2^2)$, where v_1^2 and v_2^2 are the zeros of $b(s)$, and that $\phi(s)$ has branch points at the zeros of $a(s)$ and $b(s)$ as well as at k^2 and 1. $\phi(s)$ is defined by a branch cut between $s = 0$ and $s = v_2^2$, with the branch chosen so that $\phi(s) = \ln|a(s)/b(s)|/(2\pi i)$ for $s < 0$. When moving from left to right along the real axis, $\phi_+(s) - \phi_-(s)$ jumps by the amount -1 at $s = 0$ and $s = k_R^2$, and then by the amount $+1$ at $s = v_1^2$ and $s = v_2^2$, assuming the zeros of $b(s)$ to be real (if not, the slight modification needed will lead to the same result). Now, using the fact that

$$\phi_\pm = \pm\frac{1}{\pi}\mathrm{atan}\frac{4k^3\sqrt{1-s}\sqrt{s-k^2}}{(s-2k^2)^2} \mp 1 = \mp 2\ln[S(0)] \mp 1 \qquad (6.11.42)$$

for $k^2 < s < 1$, a closed path of integration of $\phi(s)/s$, embracing the real axis between $s = 0$ and $s = v_2^2$, gives $4\ln S(0)$ from the fact that the integral vanishes. To avoid possible confusion from the coincidence of branch point and pole at $s = 0$, the somewhat more general integral of $\phi(s)/(s-\epsilon)$, where ϵ is real and positive and afterwards chosen as zero, is used to give

$$4\ln S(0) = \oint_0^{v_2^2} \frac{(-1)\cdot ds}{s-\epsilon} - \ln\frac{a(\epsilon)}{b(\epsilon)} + \int_{k_R^2}^{v_1^2}\frac{(-1)\cdot ds}{s-\epsilon} \qquad (6.11.43)$$

$$\to \ln\frac{k_R^2 b(0)}{2k^2(1-k^2)v_1^2 v_2^2} \quad \text{as } \epsilon \to 0$$

and, because $v_1^2 v_2^2 = \lim_{s\to 0}[a(s)b(s)/(-sk_R^2)] = 2k^2(1-k^2)b(0)/k_R^2$,

$$S(0) = \frac{k_R}{k\sqrt{2(1-k^2)}} \qquad (6.11.44)$$

Insertion of the expression for $S(0)$ into (6.11.37) gives

$$K_I(\tau,\beta) = -\frac{4k\sqrt{1-k^2}\sigma_y^0\sqrt{\tau}}{\sqrt{\pi}} \cdot \frac{S(-1/\beta)(k_R-\beta)}{k_R\sqrt{1-\beta}} \qquad (6.11.45)$$

which can be written in the form

$$K_I(\tau,\beta) = K_I^{statn}(\tau)k_I(\beta) \qquad (6.11.46)$$

where $K_I^{statn}(\tau)$ is the stress intensity factor for a stationary crack, and the material dependent factor

$$k_I(\beta) = \frac{S(-1/\beta)(k_R-\beta)}{k_R\sqrt{1-\beta}} \qquad (6.11.47)$$

expresses the velocity dependence in the same way as the previously encountered universal factor $k_{III}(\gamma)$. Thus, $k_I(0) = 1$.

If the crack is stationary, $\beta = 0$, the displacement of the upper crack face, found by numerical integration twice of (6.11.26)-(6.11.27) after putting $\lambda = -\infty$, is shown in Fig. 6.11.3. Note the overshoot of the displacement around the position where Rayleigh waves arrive from the crack edge. Such an overshoot is not found at shear loading of the crack faces, cf. Figs 6.10.6 and 6.11.5. The significance of this overshoot will be discussed in connection with stress wave incidence on a crack of finite length, page 487.

Case 4 The velocity λ is only infinitesimally smaller than the velocity β. This implies a steady state when τ reaches infinity in such a way that the extension of the load is finite, $\beta\tau - \lambda\tau = L$:

$$K_I \to -\frac{2\sqrt{2}\sigma_y^0\sqrt{L}}{\sqrt{\pi}} \qquad (6.11.48)$$

In the first two cases, the static solution for a crack face load $\sigma_y = \sigma_y^0$, extending a distance L from the crack edge is obtained by setting β and λ to zero, after L has been identified with $\beta\tau$ in the first case and with $|\lambda|\tau$ in the second case. This static

6.11 FACE LOADED IN-PLANE MODE CRACKS

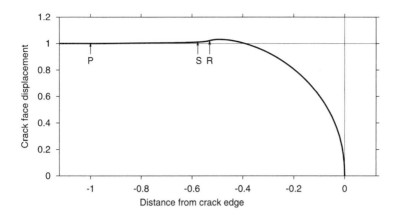

Fig. 6.11.3 Displacement of the upper crack face after suddenly applied load $\sigma_y = \sigma_y^0$ on the faces of a stationary crack. The material constant $k^2 = 1/3$, which corresponds to Poisson's ratio $1/4$ in plane strain and $1/3$ in plane stress. The distance from the crack edge and the crack face displacement are given in units of $|k^2\sigma_y^0|c_P t/\mu$ and $x/c_P t$, respectively. The arrows marked P and S indicate the arrival of the cylindrical P and S waves from the crack edge, whereas the arrow marked R indicates the arrival of Rayleigh waves from the crack edge. Note in particular the overshoot, which reaches a maximum somewhat later than the arrival of Rayleigh waves (the curve possesses an inflexion point at this arrival). This overshoot prevails for all values of Poisson's ratio.

solution is obtained directly from the third case, in agreement with a general property of steady state problems of this kind, see page 334. The static solution for mode I crack face loading was previously given by (4.4.53).

It will be shown later that Case 3 provides the essential solution of the problem of normal incidence on the crack by a P wave and that Case 2 provides part of the solution of the problem of oblique incidence of an in-plane stress wave on a stationary crack.

Case 3 was solved by Baker (1962), who presented full details of the stress field. The same case, but for a stationary crack, $\beta = 0$, had been solved previously by Maue (1954), Fillipov (1956) and Ang (1958). Maue, Ang and Baker actually treated the problem of a plate under tension, in which the crack suddenly appeared. Maue had previously shown (Maue 1953) that solutions to such problems, after trivial superposition, also provided solutions of problems concerning stress wave incidence on cracks: he studied oblique incidence of harmonic pressure and shear waves on a semi-infinite crack. Maue (1953, 1954), Ang (1958) and Baker (1962) used Wiener-Hopf techniques, Maue for solution of an integral equation, Ang after solution of dual integral equations and Baker by establishing a Wiener-Hopf equation directly from Laplace-Fourier transforms. The Wiener-Hopf method has been adopted in further extensions towards non-constant crack velocity, with or without delay time for the start of crack propagation and with oblique stress wave incidence, by Freund (1973, 1974a, 1990). However, in the present work, Hilbert problem formulation is preferred.

The function $S(\zeta)$, defined by (6.11.14), plays an important part in problems concerning in-plane crack face loading. Its structural elements show a close association

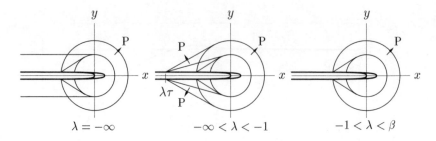

Fig. 6.11.4 Wave fronts travelling with the velocities of P and S waves away from the crack edge and the crack faces at mode II crack face loading. Cylindrical P and S waves originate at the crack edges; the axes of their fronts coincide with the original crack edge position. Plane P and S waves originate from the applied loads and plane S waves are generated by the cylindrical P wave. P wave fronts are marked P; all other fronts are S wave fronts. For $-1 < \lambda < -k$ there is also a plane wave front from $x = \lambda \tau$, tangent to the cylindrical S wave front.

with the Rayleigh function. It was first established by Maue (1953), albeit in a slightly different form.

Face loaded mode II cracks

In the mode II case, the crack face load is expressed as for mode I after changing σ_y^0 to τ_{xy}^0, which does not need to be restricted to a negative value. The only other change consists of the replacement of $(1 - \zeta)^{1/2}$ by $(1/k^2 - \zeta)^{1/2}$ in (6.11.4) and following consequential changes. The latter consists of a change of sign in (6.11.12) for $c(\xi)$ in the interval $-1/k < \xi < -1$, leading to the change of $(1 + \zeta)^{1/2}$ to $(1/k + \zeta)^{1/2}$ in the expression for $Q(\zeta)$ in (6.11.18), and the changes $(1 \pm \zeta)^{1/2} \to (1/k \pm \zeta)^{1/2}$ in equations (6.11.21)-(6.11.22), $\sqrt{1-\lambda} \to \sqrt{k-\lambda}/\sqrt{k}$ in equation (6.11.23), $\sqrt{\tau \pm x} \to \sqrt{\tau \pm x/k}$ in equations (6.11.25)-(6.11.26), and, finally, $\sqrt{1 \pm \beta} \to \sqrt{1 \pm \beta/k}$ in equations (6.11.28)-(6.11.37). Then, the general expression for the mode II stress intensity factor, corresponding to (6.11.33) for mode I, becomes

$$K_{II} = -\frac{2\sqrt{2}\tau_{xy}^0}{\sqrt{\pi}} \cdot \frac{\sqrt{k-\lambda}\sqrt{(\beta-\lambda)\tau}}{S(-1/\lambda)} \cdot \frac{S(-1/\beta)(k_R - \beta)}{(k_R - \lambda)\sqrt{k - \beta}} \quad (6.11.49)$$

and the stress intensity factor in Case 3 is found to be

$$K_{II} = K_{II}^{statn}(\tau)k_{II}(\beta) = -\frac{4\sqrt{k(1-k^2)}\tau_{xy}^0\sqrt{\tau}}{\sqrt{\pi}} \cdot \frac{\sqrt{k}S(-1/\beta)(k_R - \beta)}{k_R\sqrt{k - \beta}} \quad (6.11.50)$$

where $K_{II}^{statn}(\tau)$ is the stress intensity factor for a stationary crack and the material dependent factor

$$k_{II}(\beta) = \frac{\sqrt{k}S(-1/\beta)(k_R - \beta)}{k_R\sqrt{k - \beta}} \quad (6.11.51)$$

expresses the velocity dependence. Note that $k_{II}(0) = 1$.

6.11 FACE LOADED IN-PLANE MODE CRACKS

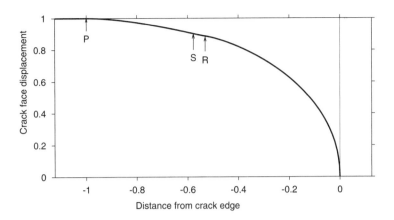

Fig. 6.11.5 Displacement of the upper crack face after suddenly applied load $\tau_{xy} = \tau_{xy}^0$ on the faces of a stationary crack. The material constant $k^2 = 1/3$, which corresponds to Poisson's ratio 1/4 in plane strain and 1/3 in plane stress. The distance from the crack edge and the crack face displacement are given in units of $|\tau_{xy}^0|c_St/\mu$ and x/c_Pt, respectively. The arrows marked P and S indicate the arrival of the cylindrical P and S waves from the crack edge, whereas the arrow marked R indicates the arrival of Rayleigh waves from the crack edge.

The wave pattern near the crack edge is schematically shown in Fig. 6.11.4. Cylindrical P and S waves emanate from the crack edges. Plane S waves are generated by the cylindrical P wave due to the rotational motion behind its front at the crack faces. Plane P and S waves appear outside the cylindrical P wave as a direct action of the crack face load if this moves away with supersonic velocity ($\lambda < -1$), except for suddenly applied crack face load everywhere ($\lambda = -\infty$), when only S waves are generated, because such a load does not create any dilatation.

If the crack is stationary, $\beta = 0$, the displacement of the upper crack face at a suddenly applied crack face load is found by numerical integration by analogy with the corresponding mode I case, page 446. It is shown in Fig. 6.11.5.

Moving pair of opposed concentrated crack face forces

The stress intensity factors for mode I and II cracks, subjected to a pair of moving opposed concentrated crack face forces, are found in direct analogy with the mode III case, see (6.10.35). Let the mode I and II forces be S_y^0 and T_{xy}^0, respectively, moving from the original position of the crack edge with velocity λc_P, whereas the crack moves with velocity $\beta c_P > \lambda c_P$. Then, the analogy gives

$$K_I = \frac{S_y^0}{(\beta - \lambda)\sigma_y^0} \cdot \frac{\partial K_1}{\partial \tau} \tag{6.11.52}$$

$$K_{II} = \frac{T_{xy}^0}{(\beta - \lambda)\tau_{xy}^0} \cdot \frac{\partial K_2}{\partial \tau} \tag{6.11.53}$$

where K_1 and K_2 are the stress intensity factors given by (6.11.33) and (6.11.49), respectively. This results in

$$K_I = -\frac{\sqrt{2}S_y^0}{\sqrt{\pi}} \cdot \frac{\sqrt{1-\lambda}S(-1/\beta)(k_R-\beta)}{S(-1/\lambda)(k_R-\lambda)\sqrt{1-\beta}\sqrt{(\beta-\lambda)\tau}} \quad (6.11.54)$$

$$K_{II} = -\frac{\sqrt{2}T_{xy}^0}{\sqrt{\pi}} \cdot \frac{\sqrt{k-\lambda}S(-1/\beta)(k_R-\beta)}{S(-1/\lambda)(k_R-\lambda)\sqrt{k-\beta}\sqrt{(\beta-\lambda)\tau}} \quad (6.11.55)$$

These two expressions belong to the mode I and II *fundamental solutions*, so named by Freund (1972a, 1990), because they are very useful and versatile as elementary solutions for creating solutions of other problems through superposition.

General time-independent in-plane mode crack face loading

For a general normal time-independent crack face load, $\sigma_y = \sigma_y(x) = h(x)$ for $0 < x < \beta\tau$, the mode I stress intensity factor is found in the same way as the corresponding mode III stress intensity factor, given by (6.10.40) and (6.10.41). Thus, after using (6.11.35), the same kind of remarkably simple relation as (6.10.41) is found:

$$K_I = K_I(a,\beta) = k_I(\beta)K_I(a,0) \quad (6.11.56)$$

where $\beta\tau$ is identified with the extension a of the load, $k_I(\beta)$ is the material dependent velocity factor, given by (6.11.47), and $K_I(a,0)$ is the stress intensity factor for a stationary crack subjected to the same crack face load as the running crack at the same edge position. The expression for this factor is the same as for mode III in (6.10.40),

$$K_I(a,0) = -\sqrt{\frac{2}{\pi}}\int_0^a \frac{h(u)}{\sqrt{u-a}}du \quad (6.11.57)$$

and, although it is obvious that it is the static stress intensity factor, because $k_I(0) = 1$, this fact can be verified by comparison with (4.4.45).

The corresponding mode II stress intensity factors are found from (6.11.56) and (6.11.57) by replacing subscript I by II. The factor $k_{II}(\beta)$ is given by (6.11.51).

In-plane mode crack motion from an equilibrium state

For crack motion starting from an equilibrium state of a finite size body, with loads that are not necessarily restricted to the crack faces, as previously in this section, but may also contain volume forces and loads or prescribed displacements on outer boundaries, results for modes I and II are found in the same way as for mode III, page 438. Thus, for a crack edge starting to move from an initial position $x = 0$ to a current position $x = a$ with constant velocity $V = \beta c_S$ the stress intensity factor is

$$K_I(a,\beta) = k_I(\beta)K_I^{ref}(a) \quad (6.11.58)$$

where $k_I(\beta)$ is given by (6.11.47) and

$$K_I^{ref}(a) = -\frac{\sqrt{2}}{\sqrt{\pi}}\int_0^a \frac{h_b(u)-h_a(u)}{\sqrt{a-u}}du \quad (6.11.59)$$

Here, $h_a(x)$ is the normal stress on the symmetry plane ahead of the crack edge before crack motion and $h_b(x)$ is the normal stress (if any) appearing on the crack faces behind the crack edge. The result is valid until reflections from the outer boundary or another crack edge arrive. Note that $K_I^{ref}(a)$ is a reference stress intensity factor that is given by (6.11.59), which, as in the mode III case, only equals the stress intensity factor for a corresponding stationary crack if the solid is infinite and the crack semi-infinite.

The corresponding mode II stress intensity factors are found by replacing subscript I by II. The factor $k_{II}(\beta)$ is given by (6.11.51).

Self-similar time-dependent crack face loading

Cases of time-dependent crack face loads were discussed for mode III on pages 438ff. This discussion carries over to modes I and II with only obvious changes in notations. Thus, for a pair of opposed crack face forces, $S'_y t$ for mode I, $T'_{xy} t$ for mode II, appearing at $t = 0$, $x = 0$, and travelling with velocity λc_P behind a crack edge, which begins moving from $x = 0$ with velocity $V = \beta c_P > \lambda c_P$, the stress intensity factors are

$$K_I = \frac{2\sqrt{2} S'_y}{\sqrt{\pi c_P}} \cdot \frac{\partial A_I(\lambda, \beta, \tau)}{\partial \lambda} k_I(\beta) \qquad (6.11.60)$$

$$K_{II} = \frac{2\sqrt{2} T'_{xy}}{\sqrt{\pi c_P}} \cdot \frac{\partial A_{II}(\lambda, \beta, \tau)}{\partial \lambda} k_{II}(\beta) \qquad (6.11.61)$$

where

$$A_I = A_{II} = \frac{k_R \sqrt{1-\lambda} \sqrt{(\beta - \lambda)\tau}}{(k_R - \lambda) S(-1/\lambda)} \qquad (6.11.62)$$

and $k_I(\beta)$, $k_{II}(\beta)$ are given by (6.11.47) and (6.11.51).

General time-dependent crack face loading

General types of time-dependent crack face loading were considered by Freund and Rice (1974), Kostrov (1975) and Burridge (1976). Here, the stress intensity factor caused by a pair of suddenly applied opposed and normal crack face forces, $PU(t)$ per unit of length, acting at a distance a from the edge of a stationary crack, will be considered. This problem was solved by Freund (1974b, 1990), using a superposition technique involving moving dislocations, by Freund and Rice (1974), using dynamic weight functions, and by Kuo and Chen (1992), using the Wiener-Hopf technique. The analysis by Kuo and Chen is followed here. The solution can be used together with superposition techniques to generate rather general types of time-dependent loading. This problem differs from previously treated elastodynamic problems insofar as it cannot be constructed from self-similar or steady state problems.

Statement of the problem
Consider a semi-infinite stationary crack $x < 0$, $y = 0$, in an infinite body. Plane strain or plane stress prevails. Due to mode I symmetry, it is sufficient to study the upper

half, $y \geq 0$. The boundary conditions for $y = 0$ are

$$\tau_{xy} = 0 \text{ for all } x \tag{6.11.63}$$
$$\sigma_y = -P\delta(x+a)U(t) \text{ for } x < 0 \tag{6.11.64}$$
$$v = 0 \text{ for } x > 0 \tag{6.11.65}$$

The stress intensity factor is sought.

Introduce the double Laplace transforms $T_{xy} = \mathcal{L}_{qx}\mathcal{L}_{p\tau}(\tau_{xy})$, $S_y = \mathcal{L}_{qx}\mathcal{L}_{p\tau}(\sigma_y)$ and $V = \mathcal{L}_{qx}\mathcal{L}_{p\tau}(v)$, where $\tau = c_P t$. Then, according to (6.9.18)-(6.9.30),

$$(T_{xy})_{y=0} = -\mu[2q(p^2-q^2)^{1/2}A - (p^2/k^2 - 2q^2)C] \tag{6.11.66}$$
$$(S_y)_{y=0} = \frac{\mu}{k^2}[(p^2 - 2k^2q^2)A + 2k^2q(p^2/k^2 - q^2)^{1/2}C] \tag{6.11.67}$$
$$(V)_{y=0} = -(p^2 - q^2)^{1/2}A - qC \tag{6.11.68}$$

The Wiener-Hopf method will be applied. Therefore it is assumed that $V = V_- + V_+$, where

$$V_- = q\int_{-\infty}^{0} e^{-qx} p \int_{0}^{+\infty} e^{-p\tau}(v)_{y=0} d\tau \, dx \tag{6.11.69}$$

$$V_+ = q\int_{0}^{+\infty} e^{-qx} p \int_{0}^{+\infty} e^{-p\tau}(v)_{y=0} d\tau \, dx \tag{6.11.70}$$

Similarly, $S_y = S_y^- + S_y^+$. The boundary conditions (6.11.63)-(6.11.65) then give

$$2q(p^2-q^2)^{1/2}A - (p^2/k^2 - 2q^2)C = 0 \tag{6.11.71}$$
$$(p^2 - 2k^2q^2)A + 2k^2q(p^2/k^2 - q^2)^{1/2}C = \frac{k^2}{\mu}(S_y^+ - Pqe^{qa}) \tag{6.11.72}$$
$$-(p^2-q^2)^{1/2}A - qC = V_- + V_+ = V_- \tag{6.11.73}$$

Solution of the problem

Elimination of A and C results in the Wiener-Hopf equation (cf. Section 4.9)

$$\frac{\mu\zeta^4 R(1/\zeta)}{k^2(1-\zeta^2)^{1/2}} V_- = \frac{S_y^+}{p} - P\zeta e^{p\zeta a} \tag{6.11.74}$$

where $\zeta = \xi + i\eta = q/p$ and $R(\cdot)$ is the Rayleigh function, (A9.57), so that

$$\zeta^4 R(1/\zeta) = 4k^4[\zeta^2(1-\zeta^2)^{1/2}(1/k^2 - \zeta^2)^{1/2} - (\zeta^2 - 1/2k^2)^2] \tag{6.11.75}$$

$R(1/\zeta)$ is defined through branch cuts between $\zeta = -1/k$ and $\zeta = -1$, and between $\zeta = 1$ and $\zeta = 1/k$, with the branch chosen so that $R(\zeta) \to 2k^2(1-k^2)/\zeta^2$ as $\zeta \to \infty$. Consider the Laplace transform variable p as real and positive (see Appendix A5), so that ζ only differs from the Laplace transform variable q by a scale factor.

The first step toward the solution of the Wiener-Hopf equation for V_- and S_y^+ is to factorize $R(1/\zeta)$ and $(1-\zeta^2)^{1/2}$. The latter function can be factored by inspection to $(1-\zeta)^{1/2}(1+\zeta)^{1/2}$, where the two factors are regular overlapping half-planes, the first one in $\Re\zeta \leq 1$ and the second one in $\Re\zeta \geq -1$. Factorization of $R(1/\zeta)$ is achieved by

6.11 FACE LOADED IN-PLANE MODE CRACKS

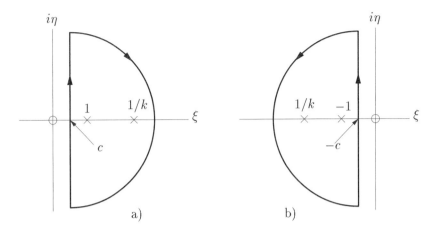

Fig. 6.11.6 Integration paths for a) $\ln S_-(\zeta)$ and b) $\ln S_+(\zeta)$.

using the general method described on pages 679ff. Thus, partition is sought of the function

$$\ln S(\zeta) = \ln \frac{\zeta^4 R(1/\zeta)}{2k^2(1-k^2)(\zeta^2 - 1/k_R^2)} \qquad (6.11.76)$$

which possesses the desired property of approaching zero sufficiently fast as $\zeta \to \pm i\infty$. Note also that the argument of the logarithm does not contain any zeros – the two zeros, $\zeta = \pm 1/k_R$, of $R(1/\zeta)$ are counterbalanced by the factor $(\zeta^2 - 1/k_R^2)$. From (A4.51), it is found that, for $\Re\zeta < 0$,

$$\ln S_-(\zeta) = \frac{1}{2\pi i} \int_{c-i\infty}^{c+i\infty} \ln S(s) \frac{ds}{s-\zeta} \qquad (6.11.77)$$

where $0 < c < 1$. The integration path is now completed to a closed path by an infinite semi-circle around the right half-plane (Fig. 6.11.6a). This will not result in any contribution to the integral. The only singularities of the integrand inside this contour are the branch points $s = 1$ and $s = 1/k$. Thus, the path of integration may be shrunk to enclose the branch cut between these points. Let $S_U(\zeta)$ and $S_L(\zeta)$ be the values of $S(\zeta)$ on the upper and lower sides of the branch cut, respectively. Then,

$$\ln S_-(\zeta) = \frac{1}{2\pi i} \int_1^{1/k} \{\ln[S_U(s)] - \ln[S_L(s)]\} \frac{ds}{s-\zeta}$$

$$= -\frac{1}{\pi} \int_1^{1/k} \operatorname{atan} \frac{s^2\sqrt{s^2-1}\sqrt{1/k^2-s^2}}{(s^2-1/2k^2)^2} \cdot \frac{ds}{s-\zeta} \qquad (6.11.78)$$

where the last member is found from the observation that

$$\left.\begin{array}{l} S_U(s) \\ S_L(s) \end{array}\right\} = |S(s)| \exp\left[\mp i \operatorname{atan} \frac{s^2\sqrt{s^2-1}\sqrt{1/k^2-s^2}}{(s^2-1/2k^2)^2}\right] \qquad (6.11.79)$$

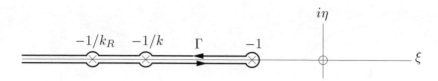

Fig. 6.11.7 Deformed integration path for $A_+(\zeta)$.

In the same way, using the integration path shown in Fig. 6.11.6b, it is found that

$$\ln S_+(\zeta) = -\frac{1}{\pi} \int_1^{1/k} \operatorname{atan} \frac{s^2\sqrt{s^2-1}\sqrt{1/k^2-s^2}}{(s^2-1/2k^2)^2} \cdot \frac{ds}{s+\zeta} \qquad (6.11.80)$$

so that

$$S_\pm(\zeta) = \exp\left[-\frac{1}{\pi} \int_1^{1/k} \operatorname{atan} \frac{s^2\sqrt{s^2-1}\sqrt{1/k^2-s^2}}{(s^2-1/2k^2)^2} \cdot \frac{ds}{s\pm\zeta}\right] \qquad (6.11.81)$$

where $S_-(\zeta)$ is analytic for $\zeta < 1$ and $S_+(\zeta)$ is analytic for $\zeta > -1$. Thus, the required factorization of $R(1/\zeta)$ is

$$R(1/\zeta) = 2k^2(1-k^2)(\zeta^2 - 1/k_R^2)S_-(\zeta)S_+(\zeta)/\zeta^4 \qquad (6.11.82)$$

and the Wiener-Hopf equation (6.11.74) now reads

$$\frac{2\mu(1-k^2)(\zeta^2-1/k_R^2)S_-(\zeta)}{(1-\zeta)^{1/2}}V_- = \frac{(1+\zeta)^{1/2}S_y^+}{pS_+(\zeta)} - \frac{P\zeta(1+\zeta^{1/2})e^{p\zeta a}}{S_+(\zeta)} \qquad (6.11.83)$$

Because $\exp(p\zeta a)$ becomes singular as $\zeta \to +\infty$, it remains to partition the last member of the Wiener-Hopf equation to obtain the desired form: the left member regular in the left half-plane and the right member regular in the right half-plane. The function to be partitioned is chosen as

$$A(\zeta) = \frac{(1+\zeta)^{1/2}e^{p\zeta a}}{(\zeta+1/k_R)S_+(\zeta)} \qquad (6.11.84)$$

so that $A(\zeta) \to 0$ sufficiently fast as $\zeta \to \pm\infty$. The partition $A(\zeta) = A_-(\zeta) + A_+(\zeta)$ is sought. Then, from (A4.50)-(A4.51),

$$A_-(\zeta) = \frac{1}{2\pi i} \int_{c-i\infty}^{c+i\infty} \frac{(1+s)^{1/2}e^{pas}}{(s+1/k_R)S_+(s)} \cdot \frac{ds}{s-\zeta} \quad \text{for } \Re\zeta < 0 \qquad (6.11.85)$$

$$A_+(\zeta) = -\frac{1}{2\pi i} \int_{-c-i\infty}^{-c+i\infty} \frac{(1+s)^{1/2}e^{pas}}{(s+1/k_R)S_+(s)} \cdot \frac{ds}{s-\zeta} \quad \text{for } \Re\zeta > 0 \qquad (6.11.86)$$

Now, fold the path of integration for $A_+(\zeta)$ to a path Γ, shown in Fig. 6.11.7. This results in

$$A_+(\zeta) = -\frac{1}{2\pi i} \int_\Gamma \frac{(1+s)^{1/2}e^{pas}}{(s+1/k_R)S_+(s)} \cdot \frac{ds}{s-\zeta} \qquad (6.11.87)$$

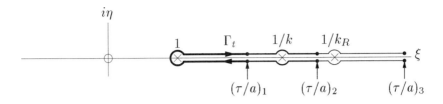

Fig. 6.11.8 Truncated integration path for K_I, shown for three different values of τ/a.

The desired form of the Wiener-Hopf equation is now found to be

$$\frac{2\mu(1-k^2)(\zeta-1/k_R)S_-(\zeta)}{(1-\zeta)^{1/2}}V_- + \zeta A_-(\zeta) = \frac{(1+\zeta)^{1/2}S_y^+}{p(\zeta+1/k_R)S_+(\zeta)} - P\zeta A_+(\zeta) \quad (6.11.88)$$

Both members are regular in the strip $-1 < \Re\zeta < 1$. Together they define a function that is analytic in the whole plane. From Abel theorems (see Appendix A5) it is found that $S_y^+ \to const \cdot \zeta^{1/2}$ as $\zeta \to +\infty$, $V_- \to 0$ as $\zeta \to -\infty$ and both S_y^+ and V_- vanish for $\zeta = 0$. This implies that both members are bounded as $|\zeta| \to \infty$ and equal zero for $\zeta = 0$, so that, by Liouville's theorem, they vanish in the entire plane. Thus,

$$\begin{aligned}S_y^+ &= \frac{P\zeta p(\zeta+1/k_R)S_+(\zeta)A_+(\zeta)}{(1+\zeta)^{1/2}} \\ &= -\frac{1}{2\pi i} \cdot \frac{P\zeta p(\zeta+1/k_R)S_+(\zeta)}{(1+\zeta)^{1/2}} \int_\Gamma \frac{(1+s)^{1/2}e^{pas}}{(s+1/k_R)S_+(s)} \cdot \frac{ds}{s-\zeta} \quad (6.11.89)\end{aligned}$$

The stress $(\sigma_y)_{y=0}$ for $0 < x \ll a$ is found from

$$(S_y^+)_{\zeta \to \infty} = \frac{Pq^{1/2}}{2\pi i}\int_\Gamma \frac{(1+s)^{1/2}p^{1/2}e^{pas}}{(s+1/k_R)S_+(s)}ds \quad (6.11.90)$$

where the relation $\zeta = q/p$ was used. Inversion is now possible, both with respect to q and, by using (A5.14), with respect to p:

$$\mathcal{L}_{qx}^{-1}(q^{1/2}) = \frac{1}{\sqrt{\pi x}}, \quad \mathcal{L}_{p\tau}^{-1}(p^{1/2}e^{pas}) = \frac{U(\tau+as)}{\sqrt{\pi}(\tau+as)^{1/2}} \quad (6.11.91)$$

where $U(\cdot)$ is the unit step function. Thus, for $x \to +0$,

$$(\sigma_y)_{y=0} = \frac{P}{2\pi^2 i\sqrt{ax}}\int_\Gamma \frac{U(s+\tau/a)(1+s)^{1/2}}{(s+1/k_R)(s+\tau/a)^{1/2}S_+(s)}ds \quad (6.11.92)$$

and the stress intensity factor K_I is found by multiplication with $\sqrt{2\pi x}$. For convenience the integration variable change $s \to -s$ is made, so that

$$K_I = -\frac{P}{\pi i\sqrt{2\pi a}}\int_{\Gamma_t} \frac{(1-s)^{1/2}}{(s-1/k_R)(\tau/a-s)^{1/2}S_-(s)}ds \quad (6.11.93)$$

where Γ_t is the truncated path shown in Fig. 6.11.8. Note that $S_+(-s) = S_-(s)$. The truncation at $s = \tau/a$ justifies omission of the factor $U(\tau/a - s)$ in the integrand.

Features of the stress intensity factor
The integrand in (6.11.93) has a pole at $s = 1/k_R$ and branch points at $s = 1$, $s = 1/k$ and $s = \tau/a$. It is defined by a branch cut between $s = 1$ and $s = \tau/a$, if $\tau/a > 1/k$ and by a branch cut between $s = 1$ and $s = 1/k$ if $1 < \tau/a < 1/k$. The branch is chosen so that the integrand approaches $1/s$ as $s \to \infty$.

For $\tau/a > 1/k_R$, the integration path can be closed at $s = \tau/a + 0$ and then expanded to infinity, giving

$$K_I = \frac{\sqrt{2}P}{\sqrt{\pi a}} \text{ for } \tau/a > 1/k_R, \text{ i.e., } c_R t > a \qquad (6.11.94)$$

which equals the static stress intensity factor for two opposed crack face loads P per unit length at the distance a from a crack edge, as can be seen from (4.4.61) after putting $\sigma_y(\xi) = -P\delta(\xi + a)$. It is, of course, the expected long-time response, which thus becomes fully established as soon as the Rayleigh wave arrives.

For $1/k < \tau/a < 1/k_R$, the integration path can again be closed at $s = \tau/a + 0$, but now the pole at $s = 1/k_R$ contributes a residue, so that

$$K_I = \frac{\sqrt{2}P}{\sqrt{\pi a}}\left[1 - \frac{\sqrt{1-k_R}}{\sqrt{1-k_R\tau/a}\,S_-(1/k_R)}\right] \text{ for } 1/k < \tau/a < 1/k_R \qquad (6.11.95)$$

i.e. in the time interval between the arrivals of S waves and Rayleigh waves. In the preceding time interval, $1 < \tau/a < 1/k$, i.e., between the arrivals of P waves and S waves, the path cannot be closed at the truncation point. The stress intensity factor can be evaluated from

$$K_I = -\frac{\sqrt{2}P}{\sqrt{\pi a}} \cdot \frac{1}{\pi} \int_1^{\tau/a} \frac{\sqrt{1-s}\,\Re[1/S_-(s)]}{(1/k_R - s)\sqrt{\tau/a - s}}\,ds \qquad (6.11.96)$$

where use was made of the fact that $[S_-(s)]_L = \overline{[S_-(s)]}_U$, where the bar denotes complex conjugation and subscripts U and L denote upper and lower side of the real axis, respectively.

An inspection of the results reveals that the stress intensity factor is negative until the Rayleigh wave has arrived, and it approaches $-\infty$ at its arrival. This result indicates that the crack faces interpenetrate each other, and, in fact, this follows from the phenomenon that a point on the surface of a semi-infinite body initially moves upwards as a result of a downwards impact at another location on the surface. Thus, the solution obtained is not physically acceptable, except as a part solution in certain superposition schemes. See further the discussion on page 440.

6.12 Non-constant crack velocity and crack arrest

Several different kinds of crack propagation at non-constant speed can be identified. A few clarifying distinctions may be appropriate.

The simplest – though not simple – analytical treatments concern cases when waves are only radiated outwards from the moving crack edge, just as for the cases of constant velocity, analysed in the preceding section. One example is non-constant velocity motion of a crack edge, starting – more or less abruptly – from an equilibrium state, or

6.12 NON-CONSTANT CRACK VELOCITY AND CRACK ARREST

from a motion with vanishingly small velocity, and proceeding until a reflected wave arrives from another crack edge or from a body boundary. Common to all such cases seems to be a one-to-one association between the dynamic field near the crack edge and a certain reference static field. Thus, if the crack suddenly stops, the stress intensity factor immediately takes on the value associated with this field. The strongest result appears for mode III: the entire static field radiates out from the arrested crack edge with the S wave velocity. This was shown by Eshelby (1969), and was implied by an earlier solution by Kostrov (1966), the first solution of a problem involving dynamic crack growth at non-constant velocity. For modes I and II, the result is less strong, but the normal (tangential stress) on the symmetry plane takes on a certain reference static distribution within the distance reached by S waves from the arrested crack edge, and the crack face opening (sliding) takes on the corresponding static distribution within the distance reached by Rayleigh waves. This was shown by Freund (1972b). By means of superposition techniques, these arrest properties for modes I, II and III can be used to construct solutions for crack propagation at non-constant velocity.

The next simplest case concerns crack propagation at non-constant crack velocity, starting from a steady state motion, during which the crack edge is receiving incoming waves. One example is a semi-infinite crack, moving with constant velocity in an infinite strip. Waves are then continuously reflected at the strip boundaries and reach the crack edge. If a crack motion suddenly changes from steady state to some other type of motion, then the stress intensity factor will to some extent continue to depend on the character of the steady state motion (Nilsson 1977a), although its value immediately after the change is dependent only on the velocity change. This was shown by Nilsson for modes III (1977c) and I (1977d).

More complex cases concern crack propagation accelerating from an equilibrium state and then receiving incoming waves from outer boundaries. These cases seem to be too complex to allow meaningful analytical treatment.

Sudden arrest of a mode III crack subjected to crack face loading

Crack arrest after motion starting from an unloaded state

The arrest of a mode III crack, propagating under the action of a crack face loading that appears behind its moving edge, will be analysed. Consider a stationary crack on $y = 0$ with its right edge at $x = 0$. No loads are acting on the body as long as the crack is stationary. Suppose now that the crack starts moving with constant velocity $V = \gamma c_S$ at time $t = 0$ under the action of the crack face loading

$$\tau_{yz} = h(x) \text{ for } 0 < x < \gamma \tau \quad (6.12.1)$$

where $\tau = c_S t$ (see Fig. 6.12.1). When considering events occurring before reflections from an outer boundary or another crack edge arrive, the body may be considered as infinite and the crack as semi-infinite.

By using (6.10.3) and (6.10.39), the mass acceleration on the upper crack face is found to be

$$\left(\frac{\partial^2 w_+}{\partial \tau^2}\right)_{y=0} = -\frac{\gamma}{2\pi \mu i} \int_0^\tau \frac{1}{x - \gamma \tau_0} [G_+(\xi_0) - G_-(\xi_0)] h'(\gamma \tau_0) d\tau_0 \quad (6.12.2)$$

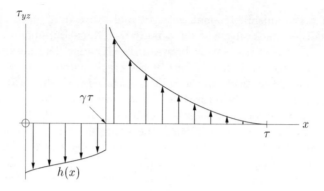

Fig. 6.12.1 Crack, starting to move with constant velocity under the action of a crack face load, $h(x)$, appearing behind the moving crack edge.

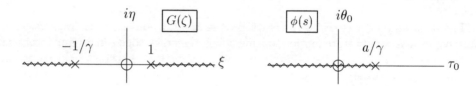

Fig. 6.12.2 Branch cuts for $G(\zeta)$ and $\phi(s)$.

where

$$\xi_0 = -\frac{\tau - \tau_0}{x - \gamma\tau_0}, \quad \text{which implies} \quad \tau_0 = \frac{\tau + x\xi_0}{1 + \gamma\xi_0} \quad (6.12.3)$$

and $G(\zeta)$ is found from (6.10.20). $G(\zeta)$ is analytic in the plane cut along the real axis from $-\infty$ to $-1/\gamma$ and from 1 to $+\infty$; see the left part of Fig. 6.12.2.

It is now assumed that the crack edge suddenly stops at $x = a$, i.e. after time $\tau = a/\gamma$. This problem may be considered as the special case when $h(x)$ in (6.12.2) is chosen so for $x > a$ that $(w_+)_{y=0} = 0$ for $x > a$; see Fig. 6.12.3. Then, with the notations

$$h(x) = h_b(x) \text{ for } 0 < x < a, \quad h(x) = h_a(x) \text{ for } a < x < \gamma\tau \quad (6.12.4)$$

the expression (6.12.2) leads to the following equation for determination of $h_a(x)$:

$$\int_0^{a/\gamma} \frac{1}{x - \gamma\tau_0}[G_+(\xi_0) - G_-(\xi_0)]h_b'(\gamma\tau_0)d\tau_0$$
$$+ \int_{a/\gamma}^{\tau} \frac{1}{x - \gamma\tau_0}[G_+(\xi_0) - G_-(\xi_0)]h_a'(\gamma\tau_0)d\tau_0 = 0 \quad (6.12.5)$$

for $a < x < \gamma\tau$. It is possible to reformulate this integral equation as a Hilbert problem. To this end, consider a complex plane $s = \tau_0 + i\theta_0$, and introduce a function $\phi(s)$, analytic in the s-plane cut along the real axis from $\tau_0 = -\infty$ to $\tau_0 = a/\gamma$; see Fig.

6.12 NON-CONSTANT CRACK VELOCITY AND CRACK ARREST   459

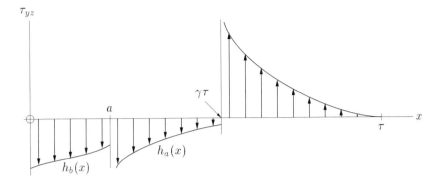

Fig. 6.12.3 Crack arrest at $x = a$ may be envisaged as continued crack growth under the crack face load $h_a(x)$ for $x > a$, so chosen that the crack sliding displacement is zero for $x > a$.

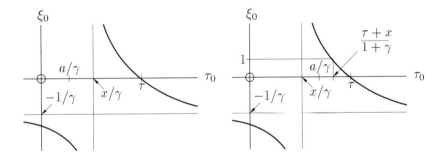

Fig. 6.12.4 Sketch of ξ_0 as function of τ_0 for $a < x < \gamma\tau$ (left figure) and for $a - (\tau - a/\gamma) < x < a$ (right figure).

6.12.2. Then, with suitable choice of $\phi(s)$, equation (6.12.5) can be written in the form

$$\int_{-\infty}^{\tau} \frac{1}{x - \gamma\tau_0}[G_+(\xi_0)\phi'_+(\tau_0) - G_-(\xi_0)\phi'_-(\tau_0)]d\tau_0$$

$$- \int_{a/\gamma}^{\tau} \frac{1}{x - \gamma\tau_0}[G_+(\xi_0) - G_-(\xi_0)][\phi'(\tau_0) - h'_a(\gamma\tau_0)]d\tau_0 = 0 \text{ for } a < x < \gamma\tau \quad (6.12.6)$$

This suitable choice of $\phi(s)$ implies that

$$G_+(\xi_0)\phi'_+(\tau_0) - G_-(\xi_0)\phi'_-(\tau_0) = \begin{cases} [G_+(\xi_0) - G_-(\xi_0)]h'_b(\gamma\tau_0) \\ \qquad\qquad\qquad\qquad \text{for } 0 < \tau_0 < a/\gamma \\ 0 \qquad\qquad\qquad \text{for } \tau_0 < 0 \end{cases} \quad (6.12.7)$$

A sketch of ξ_0 as function of τ_0, assuming $a < x < \gamma\tau$, shows that $\xi_0 < -1/\gamma$ for $\tau_0 < x/\gamma$, and thus also for $\tau_0 < a/\gamma$, if $a < x, \gamma\tau$; see the left part of Fig. 6.12.4. Thus, as shown in Fig. 6.12.2, $G_+(\xi_0) = -G_-(\xi_0)$ for $\tau_0 < a/\gamma$, and then the condition

(6.12.7) requires that

$$\phi_+(\tau_0) + \phi_-(\tau_0) = \begin{cases} 2h_b(\gamma\tau_0)/\gamma & \text{for } 0 < \tau_0 < a/\gamma \\ 0 & \text{for } \tau_0 < 0 \end{cases} \quad (6.12.8)$$

For simplicity, no integration constant was included. Checking the final result turns out to justify this omission.

The first integral in equation (6.12.6) can be written in the form

$$\int_\Gamma N(s)\,ds \quad (6.12.9)$$

where Γ encloses the portion $\tau_0 < \tau$ of the real axis and

$$N(s) = \frac{1}{x - \gamma s} G\left(\frac{s - \tau}{x - \gamma s}\right) \phi'(s) \quad (6.12.10)$$

Now, because $a < x < \gamma\tau$ is considered, and

$$\phi_+(\tau_0) = \phi_-(\tau_0) = \phi(\tau_0) \text{ for } \tau_0 > a/\gamma \quad (6.12.11)$$
$$G_+(\xi_0)\phi_+(\tau_0) - G_-(\xi_0)\phi_-(\tau_0) = 0 \text{ for } \tau < 0 \quad (6.12.12)$$

it is found that the function $N(s)$ is analytic in the s-plane cut along some portion of the real axis between $\tau_0 = 0$ and $\tau_0 = \tau$, depending on the value of x. Thus, the path Γ can be shrunk to enclose this portion of the real axis. Consequently, the integral (6.12.9) and thereby also the first integral in equation (6.12.6) equals zero, if $sN(s) \to 0$ as $|s| \to \infty$. This condition requires that

$$\phi(s) \to \text{const} \cdot |s|^{-\alpha} \text{ as } |s| \to \infty \quad (6.12.13)$$

where $\alpha > 1/2$. The remaining integral in equation (6.12.6) must then vanish for all x in the interval $a < x < \gamma\tau$, implying that

$$h_a(\gamma\tau_0) = \gamma\phi(\tau_0) \text{ for } a < \gamma\tau_0 < \gamma\tau \quad (6.12.14)$$

Thus, the problem consists of finding $\phi(s)$, which is done by solving the Hilbert problem posed by equations (6.12.8) and (6.12.11). Because an inverse square-root singularity must be allowed, and $s^{1/2}\phi(s) \to 0$ as $|s| \to \infty$, the solution is

$$\phi(s) = \frac{1}{2\pi i(s - a/\gamma)^{1/2}} \int_0^{a/\gamma} \frac{i\sqrt{a/\gamma - u} \cdot 2h_b(\gamma u)}{\gamma(u - s)}\,du \quad (6.12.15)$$

and, hence,

$$h_a(x) = \frac{1}{\pi\sqrt{x - a}} \int_0^a \frac{\sqrt{a - u}\, h_b(u)}{u - x}\,du \text{ for } a < x < \gamma\tau \quad (6.12.16)$$

where the integration variable u was redefined. But this expression is independent of τ, and $h_a(x)$ is thus a static stress on $a < x < \gamma\tau$, $y = 0$. It is also independent of γ, which implies that it equals the stress in a reference static field characteristic of the body under equilibrium with the crack edge at $x = a$ and the crack face loading $h_b(x)$ on $0 < x < a$. In the special case of an infinite body and a semi-infinite crack, it equals, as can also be directly verified, the stress on the portion $a < x < \gamma\tau$ of the symmetry plane ahead of the edge of a stationary crack, subjected to the crack face load $h_b(x)$.

6.12 NON-CONSTANT CRACK VELOCITY AND CRACK ARREST

The stress intensity factor of the arrested crack is

$$K_{III}^{arrest}(a) = K_{III}^{ref}(a) = \lim_{x \to a}[\sqrt{2\pi(x-a)}h_a(x)] = -\sqrt{\frac{2}{\pi}}\int_0^a \frac{h_b(u)}{\sqrt{a-u}}du \quad (6.12.17)$$

The problem is now solved, but it is also interesting to investigate how far ahead of the crack the static stress prevails (so far it has been shown that it prevails from the arrested crack edge to the position where the crack edge should have been if it had not been arrested), and what the mass acceleration is behind the arrested crack edge.

Crack face displacement behind the arrested crack
The mass acceleration on the crack faces is immediately found to be zero for $x_S^- = a - (\tau - a/\gamma) < x < a$, where x_S^- is the x position, behind which no information has arrived about the arrest of the crack at $x = a$. It is found that

$$\left(\frac{\partial^2 w_+}{\partial \tau^2}\right)_{y=0} = -\frac{\gamma}{2\pi\mu i}\int_\Gamma N(s)ds \quad (6.12.18)$$

holds for the whole interval $a - (\tau - a/\gamma) < x < \gamma\tau$. This is because the condition $G_+(\xi_0) = -G_-(\xi_0)$ for $\tau_0 < a/\gamma$, required for obtaining (6.12.8), is valid not only for the interval previously considered, $a < x < \gamma\tau$, extending only ahead of the arrested crack, but, in fact, for the larger interval $a - (\tau - a/\gamma) < x < \gamma\tau$, extending also some distance behind the edge of the arrested crack. This may be understood by studying the sketch of ξ_0 as a function of τ_0, assuming $a - (\tau - a/\gamma) < x < a$; see the right part of Fig. 6.12.4. Then, it is seen that $\xi_0 < -1/\gamma$ for $\tau_0 < x/\gamma$ and $\xi_0 > 1$ for $x/\gamma < \tau_0 < (\tau+x)/(1+\gamma)$. But $(\tau+x)/(1+\gamma)$ is larger than a/γ for $x > a-(\tau-a/\gamma)$. Thus, if $\tau_0 < a/\gamma$, then either $\xi_0 < -1/\gamma$ or $\xi_0 > 1$. In both cases $G_+(\xi_0) = -G_-(\xi_0)$; see Fig. 6.12.2.

The fact that the mass acceleration is zero may be interpreted so that the crack face position becomes frozen after the arrival of the stress wave from the arrested crack edge. Thus, no further motion occurs after this wave arrival, which implies a static condition: *the crack face displacement on $x_S^- = a - (\tau - a/\gamma) < x < a$ is a static displacement, which is radiating out in the negative x direction with the S wave velocity.*

Shear stress ahead of the arrested crack
The stress rate on the symmetry plane for $\gamma\tau < x < \tau$ is found from the expression

$$\left(\frac{\partial \tau_{yz}}{\partial \tau}\right)_{y=0} = -\frac{\gamma}{2\pi i}\int_0^\tau \frac{1}{x - \gamma\tau_0}[H_+(\xi_0) - H_-(\xi_0)]h'(\gamma\tau_0)d\tau_0 \quad (6.12.19)$$

by analogy with (6.12.2). $H(\zeta)$ is given by (6.10.21) with branch cut shown in Fig. 6.12.5. From a study of the sketch of ξ_0 (see Fig. 6.12.6) as function of τ_0 for $\gamma\tau < x < a + (\tau - a\gamma)$, it is seen that $-1/\gamma < \xi_0 < -1$ for $\tau_0 < (\tau - x)/(1 - \gamma)$ and thus also for $\tau_0 < a/\gamma$. Thus, $H_+(\xi_0) = -H_-(\xi_0)$ for $\tau_0 < a/\gamma$, if $\gamma\tau < x < a + (\tau - a/\gamma) = x_S^+$, where x_S^+ is the x position, ahead of which no information has arrived about the arrest

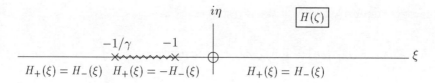

Fig. 6.12.5 Branch cut for $H(\zeta)$.

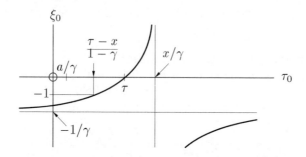

Fig. 6.12.6 Sketch of ξ_0 as function of τ_0 for $\gamma\tau < x < a + (\tau - a/\gamma)$.

of the crack at $x = a$. Then, equation (6.12.8) leads to

$$\left(\frac{\partial \tau_{yz}}{\partial \tau}\right)_{y=0} = -\frac{\gamma}{2\pi i}\int_{-\infty}^{\tau}\frac{1}{x - \gamma\tau_0}[H_+(\xi_0)\phi'_+(\tau_0) - H_-(\xi_0)\phi'_-(\tau_0)]d\tau_0$$

$$+ \frac{\gamma}{2\pi i}\int_{a/\gamma}^{\tau}\frac{1}{x - \gamma\tau_0}[H_+(\xi_0) - H_-(\xi_0)][\phi'(\tau_0) - h'_a(\gamma\tau_0)]d\tau_0 \quad (6.12.20)$$

where, through (6.12.14), the last integral vanishes. Thus, it is possible to write

$$\left(\frac{\partial \tau_{yz}}{\partial \tau}\right)_{y=0} = -\frac{\gamma}{2\pi i}\int_\Gamma M(s)ds \quad (6.12.21)$$

where Γ is a path enclosing the portion $0 < \tau_0 < \tau$ of the real axis, and

$$M(s) = \frac{1}{x - \gamma s}H\left(\frac{s - \tau}{x - \gamma s}\right)\phi'(s) \quad (6.12.22)$$

But the integral vanishes, because $M(s)$ is analytic outside Γ and decays faster than $\mathrm{const}\cdot|s|^{-1}$ as $|s| \to \infty$. Thus, $(\partial \tau_{yz}/\partial \tau)_{y=0} = 0$ for $\gamma\tau < x < a + (\tau - a/\gamma)$, which shows that *a static stress radiates out on the symmetry plane in the positive x direction from the arrested crack edge with the S wave velocity*. It is sufficient to know this feature for the continuation, but it can be shown that *an entire static stress-strain field radiates out from the arrested crack edge with the S wave velocity*, cf. Eshelby (1969). Actually, this follows from an earlier discussion of a property of the asymptotic dynamic mode III field, pages 358ff.

Crack arrest after motion starting from an equilibrium state
In the previous analysis, it was assumed that the body was unloaded before crack growth and that loads $\tau_{yz} = h_b(x)$ were supplied to the crack faces behind the mov-

ing crack edge. If the body had been initially in a loaded equilibrium state, then there would have been a stress $h_{a,before}(x)$ on the symmetry plane ahead of the crack edge, before the crack motion started. Then, it is obvious that the negated stress $-h_{a,before}(x)$ should be added to $h_b(x)$, as shown by (6.10.47). With this modification, the results obtained previously will still hold, except that the stress $h_{a,before}(x)$, which also prevailed for $t < 0$ on $x > a$, should be added to $h_a(x)$ in (6.12.16), so that the stress on the symmetry plane ahead of the arrested crack is

$$h_a(x) = h_{a,before}(x) + \frac{1}{\pi\sqrt{x-a}} \int_0^a \frac{\sqrt{a-u}[h_b(u) - h_{a,before}(u)]}{u-x} du \quad (6.12.23)$$

Note that the expression (6.12.17) for the stress intensity factor of the arrested crack will be modified to

$$K_{III}^{arrest} = K_{III}^{ref}(a) = \lim_{x \to a} \left[\sqrt{2\pi(x-a)} h_a(x)\right]$$

$$= -\sqrt{\frac{2}{\pi}} \int_0^a \frac{h_b(u) - h_{a,before}(u)}{\sqrt{a-u}} du \quad (6.12.24)$$

Sudden arrest of in-plane mode cracks

The analysis of sudden crack arrest in modes I and II follows closely the analysis for mode III in the initial stages, from equation (6.12.1) to equation (6.12.17). The crack is supposed to start moving at $\tau = 0$ with velocity $V = \beta c_P$ and being arrested at $x = a$. Necessary changes consist essentially of changing the dimensionless crack velocity from γ to β, the crack face load from $\tau_{yz} = h(x)$ for $0 < x < \gamma\tau$ to $\sigma_y = h(x)$, $0 < x < \beta\tau$, for mode I, and to $\tau_{xy} = h(x)$, $0 < x < \beta\tau$ for mode II and the crack face displacement from w_+ to v_+ for mode I and to u_+ for mode II. Note also that the functions $H(\zeta)$ and $G(\zeta)$ are different – for mode I they are given by (6.11.21) and (6.11.22) after putting $P(\zeta) = P_0 = -k_R\lambda/(k_R - \lambda)$ and letting $\lambda \to 0$, and for mode II after the modifications from mode I described on page 448. The stress intensity factor for the arrested crack is found to be

$$K_{I,II}^{arrest} = K_{I,II}^{ref}(a) = -\sqrt{\frac{2}{\pi}} \int_0^a \frac{h_b(u)}{\sqrt{a-u}} du \quad (6.12.25)$$

where $h_b(x) = h(x)$ for $0 < x < a$. Note that this result is independent of τ and β – it is a reference static stress intensity factor, which equals the static stress intensity factor for a stationary crack in an infinite medium, subjected to the same crack face load, $h_b(x)$.

The continuing analysis, investigating the radiation along the plane straight ahead of the crack of the static normal (mode I) or tangential (mode II) stresses, as well as the radiation along the crack faces of the static normal (mode I) or tangential (mode II) displacements, differs somewhat from the mode III analysis and produces a less strong result.

It was shown for mode III that the static crack face displacement radiates out from the arrested crack edge with the velocity of S waves, i.e. the mode III propagation velocity. The reason, given on page 461, was found to be that the validity of the condition $G_+(\xi_0) = -G_-(\xi_0)$ for $\tau_0 < a/\gamma$ could be extended towards the x value corresponding to the branch point $\zeta = 1$ of $G(\zeta)$, i.e. to $x = a - (\tau - a/\gamma)$. For modes I and II, the validity of the corresponding condition $G_+(\xi_0) = -G_-(\xi_0)$ for $\tau_0 < a/\beta$, can also be extended towards lower x values than $x = a$. However, $G(\zeta)$ possesses a pole at $\zeta = 1/k_R$, which implies a pole of $N(s)$. Thus, the integral corresponding to (6.12.18),

$$\int_\Gamma N(s)\,ds \qquad (6.12.26)$$

where Γ embraces the portion $0 < \tau_0 < \tau$ of the real axis, only vanishes if $a - k_R(\tau - a/\beta) \leq x < a$, in which case the pole is situated inside Γ and $N_-(\xi_0) = -N_+(\xi_0)$ at that point. Otherwise, a residue appears. This implies that *the static normal displacement (mode I) or the static tangential displacement (mode II) radiates out from the crack edge backwards along the crack faces with the Rayleigh wave velocity* rather than with the S wave velocity as for mode III.

The stress rate at the symmetry plane ahead of the crack is found in close analogy with equations (6.12.19)-(6.12.22). The essential difference is that $H(\zeta)$ now possesses a branch point at $\zeta = -1/k$. This implies that the relation $[H_+(\xi_0) - H_-(\xi_0) = 0$ for $\tau_0 < (k\tau - x)/(k - \beta)$ and $(k\tau - x)/(k - \beta) > a/\beta]$ if $x < a + k(\tau - a/\beta)$, but this relation does not hold if $a + k(\tau - a/\beta) < x < a + (\tau - a/\beta)$. Thus, *the static normal stress (mode I) or the static tangential stress (mode II) radiates out ahead of the crack edge along the symmetry plane with the S wave velocity.*

The result (6.12.25) may be generalized to include equilibrium loads existing before crack growth, manifested by a normal stress $\sigma_y = h_{a,before}$ (mode I) or a tangential stress $\tau_{xy} = h_{a,before}$ that prevailed on the symmetry plane. Then, the same result as for mode III, (6.12.24), is obtained, so that (6.12.25) is modified by the expressions

$$K_{I,II}^{ref} = K_{I,II}^{ref}(a) = \lim_{x \to a}\left[\sqrt{2\pi(x-a)}h_a(x)\right]$$
$$= -\sqrt{\frac{2}{\pi}}\int_0^a \frac{h_b(u) - h_{a,before}(u)}{\sqrt{a-u}}\,du \qquad (6.12.27)$$

Although this is a less strong result than for mode III, because the entire stress field is not radiated out from the crack edge with the highest possible sound velocity (here c_P), the properties of crack arrest at modes I and II are remarkable. As will be seen in the next subsection, they are also quite sufficient for constructing certain solutions for crack propagation at non-constant velocities. These remarkable results were discovered by Freund and published in the second of a series of four papers (Freund 1972a,b, 1973, 1974a) of fundamental importance in the whole field of crack elastodynamics.

Crack growth at non-constant velocity

The results in the two preceding subsections make it possible to find an expression for the stress intensity factor at a crack edge running with non-constant velocity. To this end, it is illustrative first to consider mode III crack advance at constant velocity $V_1 = \gamma_1 c_S$ from an equilibrium state with the crack edge at $x = 0$, and a stress $\tau_{yz} = h_0(x)$ on the symmetry plane ahead of the crack edge. For simplicity, no loads are assumed to appear on the crack faces, even though the action of such loads could

6.12 NON-CONSTANT CRACK VELOCITY AND CRACK ARREST

be considered in a rather straightforward manner. When the crack edge is at $x = a_1$, its stress intensity factor is

$$K_{III}(a_1, \gamma_1) = k_{III}(\gamma_1) K_{III}^{ref}(a_1) \tag{6.12.28}$$

where $k_{III}(\cdot)$ is given by (6.10.41), $K_{III}^{ref}(\cdot)$ is given by (6.10.48) and $h_b(x) - h_a(x) = -h_0(x)$. Thus,

$$K_{III}^{ref}(a_1) = \sqrt{\frac{2}{\pi}} \int_0^{a_1} \frac{h_0(u)}{\sqrt{a_1 - u}} du \tag{6.12.29}$$

Suppose that the crack suddenly stops at $x = a_1$. Then, the stress intensity factor changes to $K_{III}^{ref}(a_1)$, and a static stress radiates out in the positive x direction from $x = a_1$ with velocity c_S. According to (6.12.23), this stress is

$$h_1(x) = h_0(x) - \frac{1}{\pi \sqrt{x - a_1}} \int_0^{a_1} \frac{\sqrt{a_1 - u} h_0(u)}{u - x} du \tag{6.12.30}$$

Assume further that the crack edge starts running again after some rest, now with velocity $V_2 = \gamma_2 c_S$. The situation is then almost identical with the first crack motion: a crack edge starts moving from a static equilibrium environment. Even if the duration of crack edge rest is arbitrarily small, the presence of dynamically changing stresses and strains at positions ahead of the crack edge not yet reached by the static stress is irrelevant for this second crack advance. This advance can be assumed to stop again at some crack edge position $x = a_2$. Then, the stress intensity factor is

$$K_{III}^{ref}(a_2) = \sqrt{\frac{2}{\pi}} \int_{a_1}^{a_2} \frac{h_1(u)}{\sqrt{a_2 - u}} du$$

$$= \sqrt{\frac{2}{\pi}} \int_{a_1}^{a_2} \frac{1}{\sqrt{a_2 - u}} \left[h_0(x) - \frac{1}{\pi \sqrt{u - a_1}} \int_0^{a_1} \frac{\sqrt{a_1 - \xi} h_0(\xi)}{\xi - u} d\xi \right] du \tag{6.12.31}$$

After change of the order of integration, the inner integral can be calculated by residue calculus:

$$\int_{a_1}^{a_2} \frac{du}{\sqrt{(a_2 - u)(u - a_1)}(u - \xi)} = \frac{\pi}{\sqrt{(a_2 - \xi)(a_1 - \xi)}} \tag{6.12.32}$$

Thus, the stress intensity factor is

$$K_{III}^{ref}(a_2) = \sqrt{\frac{2}{\pi}} \int_0^{a_2} \frac{h_0(u)}{\sqrt{a_2 - u}} du \tag{6.12.33}$$

i.e. differing from (6.12.29) only in the replacement of a_1 by a_2. Thus, the process may be repeated indefinitely and the durations of the crack edge rests may be made arbitrarily small as well as the periods of crack advance. In this way, the case of a crack edge moving with continuously changing crack edge velocity, even interspersed with velocity jumps, may be approached: the current stress intensity factor is simply given by the fundamental formula

$$\boxed{K_{III}(a, \gamma) = k_{III}(\gamma) K_{III}^{ref}(a)} \tag{6.12.34}$$

where $\gamma = \dot{a}/c_S$ is the instantaneous velocity, $k_{III}(\gamma)$ is given by (6.10.41) and

$$K_{III}^{ref}(a) = \sqrt{\frac{2}{\pi}} \int_0^a \frac{h_0(u)}{\sqrt{a-u}} du \qquad (6.12.35)$$

where $h_0(x)$ is the stress on the symmetry plane ahead of the stationary crack assumed to exist before the start of crack motion. Note, however, that a complete analytical solution is not obtained until $h_0(x)$ is found. This could, of course, be very difficult or even impossible by analytical methods, especially if the body is finite. If the body is infinite and the crack semi-infinite, then $K_{III}^{ref}(a) = K_{III}^{stat}(a)$, the static stress intensity factor for a stationary crack with the edge at $x = a$.

The fundamental formula (6.12.34) shows that the stress intensity factor, at any time during the crack propagation, is dependent only on the instantaneous velocity and on the original stress distribution ahead of the stationary equilibrium crack before crack growth. It is thus independent on the history of crack propagation.

The whole discussion for mode III can be carried over to modes I and II with only appropriate changes of notation, except that intersonic crack velocities (only possible at mode II) cannot be considered. The less strong result for modes I and II is irrelevant in this context, because what matters is only that the speed of radiation of static normal and tangential stresses on the symmetry plane ahead of an arrested crack edge is greater than the maximum subsonic crack edge velocity, the Rayleigh wave speed. Thus, the fundamental formula for non-constant crack propagation at modes I and II, before waves are arriving from an outer boundary or another crack edge, is

$$K_{I,II}(a, \beta) = k_{I,II}(\beta) K_{I,II}^{ref}(a) \qquad (6.12.36)$$

where $\beta = \dot{a}/c_P$ is the instantaneous velocity, $k_I(\beta)$ and $k_{II}(\beta)$ are given by (6.11.47) and (6.11.51), respectively, and

$$K_{I,II}^{ref}(a) = \sqrt{\frac{2}{\pi}} \int_0^a \frac{h_0(u)}{\sqrt{a-u}} du \qquad (6.12.37)$$

where $h_0(x)$ is the normal (mode I) or tangential (mode II) stress on the symmetry plane ahead of the stationary crack assumed to exist before the start of crack motion. Note, as for mode III, that a complete analytical solution is not obtained until $h_0(x)$ is found, which could be very difficult or impossible by analytical methods, especially if the body is finite. If the body is infinite and the crack semi-infinite, then $K_{I,II}^{ref}(a) = K_{I,II}^{stat}(a)$, the static stress intensity factor for a stationary crack with the edge at $x = a$.

The result (6.12.34) was derived in a more direct way by Kostrov (1966). The procedure followed here is due to Eshelby (1969) for mode III and to Freund (1972b) for modes I and II. Kostrov (1975) remarks that it involves two limit procedures, one to obtain the stress intensity factor and one to obtain a smooth velocity variation from stepwise changes. He claims that the necessary change of the order in which these limits are taken is not in general a permissible operation. However, he shows, again with a direct method, that the procedure leads to correct result also for modes I and II.

Sudden change of crack speed

If a mode III crack edge suddenly changes speed, say from $V_1 = \gamma_1 c_S$ to $V_2 = \gamma_2 c_S$, then, as shown by the fundamental formula (6.12.34), its stress intensity factor changes from $K_{III}(\gamma_1)$ to

$$K_{III}(\gamma_2) = \frac{k_{III}(\gamma_2)}{k_{III}(\gamma_1)} K_{III}(\gamma_1) \qquad (6.12.38)$$

The interesting feature of this result is that it does not contain any reference to a static stress state, and it is therefore valid for any crack propagation, not only propagation starting from a stationary crack under time-independent equilibrium loads, until reflections from an outer boundary or another crack edge arrive. Thus, for instance, any sudden arrest of a crack will lead to an immediate change of the stress intensity factor from its value $K_{III}(\gamma)$ just before the arrest to

$$K_{III}^{arrest} = \frac{K_{III}(\gamma)}{k_{III}(\gamma)} = \frac{K_{III}(\gamma)}{\sqrt{1-\gamma}} \qquad (6.12.39)$$

which always implies a step increase. If no waves are coming in to the crack edge region, this arrest stress intensity factor will stay constant, but otherwise it will vary with time after the arrest.

Similarly, if a mode I or II crack edge suddenly changes speed from $V_1 = \beta_1 c_P$ to $V_2 = \beta_2 c_P$, then the stress intensity factor changes from $K_{I,II}(\beta_1)$ to

$$K_{I,II}(\beta_2) = \frac{k_{I,II}(\beta_2)}{k_{I,II}(\beta_1)} K_{I,II}(\beta_1) \qquad (6.12.40)$$

and, as for mode III, the result is valid for any crack propagation. Any sudden arrest will lead to an immediate change of the stress intensity factor from its value $K_{I,II}(\beta)$ just before the arrest to

$$K_{I,II}^{arrest} = \frac{K_{I,II}(\beta)}{k_{I,II}(\beta)} \qquad (6.12.41)$$

which always implies a step increase. If no waves are coming in to the crack edge region, this arrest stress intensity factor will stay constant, but otherwise it will vary with time after the arrest.

Arrest from steady state crack propagation

As an example, consider arrest from steady state motion with velocity $V = \gamma c_S$ of a semi-infinite mode III crack in a strip with width $2h$, the edges of which are displaced a distance $\Delta w = 2\delta$ from each other (w is displacement in the strip thickness direction). By using (6.7.11) the stress intensity factor can be written as

$$K_{III}(\gamma) = \frac{K_{III}^{stat}}{\sqrt{Y_{III}(\gamma)}} \qquad (6.12.42)$$

where $Y_{III}(\gamma) = 1/\sqrt{1-\gamma^2}$ and K_{III}^{stat}, the stress intensity factor for a stationary crack, equals $K_{III}(0) = \sqrt{2/h}\,\mu\delta$. Then, it follows from (6.12.39) that

$$K_{III}^{arrest} = \frac{K_{III}^{stat}}{k_{III}(\gamma)\sqrt{Y_{III}(\gamma)}} = \left(\frac{1+\gamma}{1-\gamma}\right)^{1/4} K_{III}^{stat} \qquad (6.12.43)$$

showing that the ratio $K_{III}^{arrest}/K_{III}^{stat}$ increases indefinitely with increasing crack speed. It equals about 1.543 at $\gamma = 0.7$ and about 2.088 at $\gamma = 0.9$.

For mode I, making use of (6.7.5) and (6.12.41), the arrest stress intensity factor for a steady state semi-infinite crack, moving with velocity $V = \beta c_P$ in a strip with width $2h$, loaded through a displacement $\Delta v = 2\delta$ between the crack edges (v is the displacement in the width direction), is found to be

$$K_I^{arrest} = \frac{K_I^{stat}}{k_I(\beta)\sqrt{Y_I(\beta)}} \qquad (6.12.44)$$

where, according to (6.7.5), $K_I^{stat} = 4(1-k^2)\mu\delta/\sqrt{h}$. The functions $Y_I(\beta)$ and $k_I(\beta)$ are given by (6.2.51) and (6.11.47), respectively. Numerical calculations, for $k^2 = 1/3$, show that the ratio K_I^{arrest}/K_I^{stat} increases from unity at $\beta = 0$ to about 1.887 at $\beta = 0.7 k_R$ and about 3.217 at $\beta = 0.9 k_R$.

For mode II, the result is completely analoguous, and numerical calculations, for $k^2 = 1/3$, show that the ratio $K_{II}^{arrest}/K_{II}^{stat}$ increases from unity at $\beta = 0$ to about 1.565 at $\beta = 0.7 k_R$ and about 2.313 at $\beta = 0.9 k_R$.

Because waves are reflected continuously from the strip edges to the crack edge region both before and after crack arrest, the stress intensity factor does not stay constant after arrest. As shown by Nilsson (1977a) for mode III and (1977b) for mode I, it decreases initially, whereupon, as his numerical calculations indicate, it performs damped oscillations around its static value. The method used by Nilsson was introduced by Freund (1974b); it consists of solving the arrest problem by negating the crack face displacement, which would have appeared ahead of the point of arrest, if the constant velocity crack edge motion had continued uninterrupted. Such a negation is based on superposition of the solution of the problem of a suddenly appearing dislocation, moving with constant velocity, thus a dynamic correspondence to the use of continuous arrays of dislocations described for static problems in Section 4.6. This problem was solved by Freund (1974b) for in-plane loading of an infinite body. The solutions by Nilsson (1977a,b) involved transient dislocation motion in a strip, treated with a method used in a previous study (Nilsson 1973c). Nilsson also treated the problem of steady propagation, followed by non-steady propagation for mode III (1977c) and mode I (1977d), making use of results for time-dependent loading on the crack faces, given by Kostrov (1975) and Burridge (1976).

Arrest of an expanding crack

Another example of crack arrest in the presence of incoming waves to the crack edge region, concerns an expanding mode III crack, for which the motion of one edge is dependent on the motion of the other one. Use is made of the expression (6.9.147),

$$K_{III}(\gamma) = K_{III}^{stat} \frac{\sqrt{1-\gamma^2}}{E(\sqrt{1-\gamma^2})} \qquad (6.12.45)$$

where $E(\cdot)$ is the complete elliptic integral of the second kind, $K_{III}^{stat} = \tau_{yz}^\infty \sqrt{\pi a}$ and τ_{yz}^∞ is the remote stress and a is the current half-length of the crack. Thus, from (6.12.39),

$$K_{III}^{arrest} = \frac{K_{III}^{stat}\sqrt{1-\gamma^2}}{k_{III}(\gamma)E(\sqrt{1-\gamma^2})} = \frac{K_{III}^{stat}\sqrt{1+\gamma}}{E(\sqrt{1-\gamma^2})} \qquad (6.12.46)$$

Calculations show that the ratio $K_{III}^{arrest}/K_{III}^{stat}$ increases from unity at $\gamma = 0$ to a maximum of about 1.043 at $\gamma \approx 0.222$, whereupon it decreases towards $2\sqrt{2}/\pi \approx 0.900$ as $\gamma \to 1$.

For mode I, use is made of (6.9.69), which can be written in the form

$$K_I = K_I^{stat}\frac{\sqrt{1-\beta^2}R(\beta)}{\beta^2 g_1(\beta)} \qquad (6.12.47)$$

where $K_I^{stat} = \sigma_y^\infty \sqrt{\pi a}$, σ_y^∞ is the remote stress, and a is the current half-length of the crack. Thus, from (6.12.41),

$$K_I^{arrest} = K_I^{stat}\frac{\sqrt{1-\beta^2}R(\beta)}{k_I(\beta)\beta^2 g_1(\beta)} \qquad (6.12.48)$$

This result was obtained by Achenbach and Tolikas (1977); see also Freund (1976). Numerical calculations show that the ratio K_I^{arrest}/K_I^{stat} increases from unity at $\beta = 0$ towards a somewhat higher value at $\beta = k_R$; for $k^2 = 1/3$ this value is about 1.19.

Similarly, for mode II, making use of (6.9.90),

$$K_{II} = K_{II}^{stat}\frac{\sqrt{k^2-\beta^2}R(\beta)}{k\beta^2 g_2(\beta)} \qquad (6.12.49)$$

where $K_{II}^{stat} = \tau_{xy}^\infty\sqrt{\pi a}$, τ_{xy}^∞ is the remote stress and a is the current half-length of the crack. The arrest stress intensity factor is

$$K_{II}^{arrest} = K_{II}^{stat}\frac{\sqrt{k^2-\beta^2}R(\beta)}{k_{II}(\beta)k\beta^2 g_2(\beta)} \qquad (6.12.50)$$

Numerical calculations show that the ratio $K_{II}^{arrest}/K_{II}^{stat}$ increases from unity at $\beta = 0$ towards a somewhat higher value at $\beta = k_R$; for $k^2 = 1/3$ this value is about 1.32.

6.13 Stress wave loading of cracks

General considerations

Stress waves are capable of creating stress intensity factors which may be larger than those produced by static stresses of the same magnitudes. However, a general feature is that the stress intensity factor does not even experience a step increase for an incoming step pulse, but it is gradually increasing from zero. Typically, the stress intensity factor produced by the impact of a step pulse of constant magnitude, overshoots the corresponding static stress intensity factor after a while, whereupon oscillations occur around this static value.

In the present section, plane cases are considered, with either anti-plane or in-plane

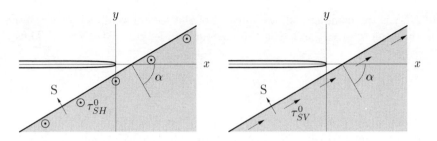

Fig. 6.13.1 SH wave (left figure), and an SV wave (right figure) incident on a semi-infinite crack, at the angle α, such that $\alpha = \pi/2$ implies normal incidence.

deformations. The front of an incoming plane wave is assumed to hit all points of a crack edge simultaneously, and its propagation direction can be described by its angle of incidence. If the incident wave is an S wave, then it is suitable to make a decomposition into two components, one with mass velocities parallel with the crack edge and the other with mass velocities in a plane normal to the crack edge. It is customary to envisage the crack plane as horizontal and therefore consider the first wave component as *horizontally polarized* and the second one as *vertically polarized*, abbreviated as SH and SV waves, respectively; see Fig. 6.13.1. Note that the displacement is normal to the stress wave propagation direction in both cases, because plane S waves are transversal.

Stress wave interaction with stationary cracks was first considered by Friedman (1949), Maue (1953, 1954), Fillipov (1956) and Ang (1958).

Step function SH wave impinging on a crack

Stationary crack

The problem of an SH wave incident on a crack was solved by Achenbach (1970a), who also considered crack propagation after the incidence. Later, Brock and Achenbach (1973) extended the solution to SH wave incidence on an interface crack. For the homogeneous case, consider the neighbourhood of a crack edge towards which a step function SH wave is incident. The analysis will first be performed for a semi-infinite crack, $x < 0$, $y = 0$, in an infinite body: this implies that solutions obtained are valid until waves reflected from boundaries or another crack edge arrive. Oblique incidence is assumed, with the angle $\pi - \alpha$, $0 \leq \alpha \leq \pi/2$, between the propagation direction of the stress wave front and the prospective straightforward propagation direction of the crack; see Fig. 6.13.1, left part. Thus, normal incidence (wave front parallel with the crack plane) occurs for $\alpha = \pi/2$. If the shear stress behind the front of the incoming stress wave is τ_{SH}^0, then its component in a plane parallel with the crack faces is $\tau_{yz} = \tau_{SH}^0 \sin \alpha$.

Figure 6.13.2 shows the situation after the wave front has reached the crack edge, which is assumed to occur at $t = 0$. The crack faces are assumed to slide without friction against each other. A wave is reflected from the lower crack face and another wave is propagated from the crack edge in all directions. All wave front velocities are, of course, equal to c_S.

6.13 STRESS WAVE LOADING OF CRACKS

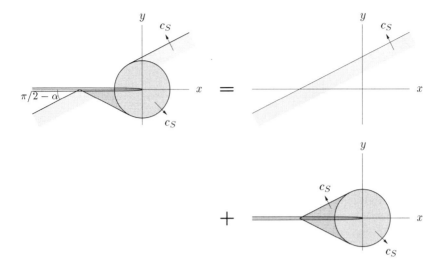

Fig. 6.13.2 Stress waves impinging on a stationary crack as shown to the left. To the right is shown how this situation can be envisaged as a superposition of a stress wave in a crack-free body and a crack face loading. The crack faces in the left figure are traction free. In the lower right figure, the crack face tractions equal the negated tangential stresses on $y = 0$, $x < 0$ in the stress wave shown in the upper right figure.

Because the crack faces are traction free, $\tau_{yz} = 0$ on $y = \pm 0$, $x < 0$. Thus, the situation can be envisaged as the superposition of 1) a stress wave in a crack-free body, i.e. the incoming stress wave propagated as if no crack were present, and 2) a crack face loading

$$\tau_{yz} = -\tau_{SH}^0 \sin \alpha\, U(\tau + \cos \alpha) \text{ for } x < 0 \tag{6.13.1}$$

where $\tau = c_S t$ and $U(\cdot)$ is the unit step function. The stress intensity factor is obviously found from the crack face loading problem. But this problem has already been solved: the solution is given by (6.10.23) after putting $\lambda = -1/\cos \alpha$ and $\tau_{yz}^0 = -\tau_{SH}^0 \sin \alpha$, so that

$$K_{III} = \frac{2\tau_{SH}^0 \sin \alpha \sqrt{2\tau}}{\sqrt{\pi}\sqrt{1 + \cos \alpha}} = \frac{2\tau_{SH}^0 \sin \alpha \sqrt{2 c_S t}}{\sqrt{\pi}\sqrt{1 + \cos \alpha}} \tag{6.13.2}$$

Note that K_{III} increases in proportion to $\sqrt{t}$. The increase continues until reflected waves arrive.

By means of superposition techniques (the Duhamel integral) the response to stress wave profiles other than a step followed by constant stress can be found.

Crack moving after impact

Suppose now that the crack starts moving with constant velocity, $V = \gamma c_S$, when the impact occurs. The stress intensity factor for this case is given from (6.10.19) by putting $\lambda = -1/\cos\alpha$ and $\tau_{yz}^0 = -\tau_{SH}^0 \sin\alpha$:

$$K_{III}(\tau,\gamma) = \frac{2\tau_{SH}^0 \sqrt{1-\gamma}\sin\alpha\sqrt{2(\gamma\cos\alpha+1)\tau}}{\sqrt{\pi}\sqrt{1+\cos\alpha}}$$

$$= \frac{2\tau_{SH}^0 \sqrt{1-V/c_S}\sin\alpha\sqrt{2(V\cos\alpha+c_S)t}}{\sqrt{\pi}\sqrt{1+\cos\alpha}} \quad (6.13.3)$$

In the special case of normal incidence, $\alpha = \pi/2$, the result can be written in the form similar to (6.10.41),

$$K_{III}(\tau,\gamma) = k_{III}(\gamma) K_{III}(\tau,0) \quad (6.13.4)$$

where $K_{III}(\tau,0)$ is given by (6.13.2), and is obviously independent of the instantaneous crack edge position and of the fact that the crack edge has been moving. It is, for instance, equal to the stress intensity factor $K_{III}^{statn}(\tau)$ of a stationary crack at time τ after the incidence of the stress pulse. Thus,

$$\boxed{K_{III}(\tau,\gamma) = k_{III}(\gamma) K_{III}^{statn}(\tau)} \quad (6.13.5)$$

The mass acceleration on the upper crack surface at normal incidence, found from (6.10.14) by putting $\tau_{yz}^0 = -\tau_{SH}^0$ and $\lambda = -\infty$, is

$$\left(\frac{\partial^2 w_+}{\partial \tau^2}\right)_{y=0} = -\frac{\tau_{SH}^0}{\pi\mu} \cdot \frac{x^2}{\sqrt{(\gamma\tau-x)^3}\sqrt{\tau+x}\,\tau} \quad (6.13.6)$$

for $-\tau < x < \gamma\tau$. In front of the crack, for $x > \gamma\tau$, $y = 0$, the displacement is the one carried by the stress wave, i.e. $(w)_{y=0} = -\tau_{SH}^0/\mu$, and behind the crack, for $x < -\tau$, $y = 0$, the displacement $(w_+)_{y=0} = 0$ (the upper crack face on this portion has not yet been reached by the stress wave), whereas the displacement $(w_-)_{y=0} = -2\tau_{SH}^0\tau/\mu$, the factor 2 appearing as a result of the reflection.

The stress rate on the symmetry plane ahead of the crack is found from (6.10.13) by putting $\tau_{yz}^0 = -\tau_{SH}^0$ and $\lambda = -\infty$. It is

$$\left(\frac{\partial \tau_{yz}}{\partial \tau}\right)_{y=0} = \frac{\tau_{SH}^0}{\pi} \cdot \frac{x\sqrt{\tau-x}}{\sqrt{(x-\gamma\tau)^3}\,\tau} \quad (6.13.7)$$

for $\gamma\tau < x < \tau$. For $x > \tau$, $(\tau_{yz})_{y=0} = \tau_{SH}^0$, the stress carried by the stress pulse. For $\gamma = 0$, equation (6.13.7) can be written as

$$\left(\frac{\partial \tau_{yz}}{\partial \tau}\right)_{y=0} = \tau_{SH}^0 \cdot \frac{\partial [h(x/\tau)]}{\partial \tau} = -\tau_{SH}^0 \cdot \frac{x}{\tau^2} h'\left(\frac{x}{\tau}\right) \quad (6.13.8)$$

where

$$h'\left(\frac{x}{\tau}\right) = -\frac{1}{\pi} \cdot \frac{\sqrt{1-x/\tau}}{\sqrt{x^3/\tau^3}} \quad (6.13.9)$$

Step function SV wave impinging on a crack

Consider the neighbourhood of a crack edge towards which a step function SV wave is incident, such that all points of the crack edge are hit by the wave front simultaneously.

6.13 STRESS WAVE LOADING OF CRACKS

The analysis is rather similar as for an SH wave, with the important difference that the mass velocities are in-plane instead of anti-plane and that consequently two stress intensity factors appear, one in mode I, the other in mode II. First, a semi-infinite crack, $x < 0$, $y = 0$, in an infinite body will be considered: this implies that solutions obtained are valid until waves reflected from boundaries or another crack edge arrive. Oblique incidence is assumed, with the angle $\pi - \alpha$, $0 \leq \alpha \leq \pi/2$, between the propagation direction of the stress wave front and the prospective straightforward propagation direction of the crack; see Fig. 6.13.1, right part. Thus, normal incidence (wave front parallel with the crack plane) occurs for $\alpha = \pi/2$.

If the shear stress behind the front of the incoming stress wave is τ_{SV}^0, then the stress components in a plane parallel with the crack faces are $\sigma_y = \tau_{SV}^0 \sin 2\alpha$ and $\tau_{xy} = -\tau_{SV}^0 \cos 2\alpha$. It is assumed that the crack edge starts propagating with velocity $V = \beta\tau$, where $\tau = c_P t$, immediately after it is reached by the stress wave. By analogy with (6.13.1) for incident SH waves, with generalization to a moving crack, the stress intensity factors are found by assuming the crack face loads

$$\sigma_y = -\tau_{SV}^0 \sin 2\alpha\, U(c_S t + \cos \alpha) \text{ for } x < \beta\tau \tag{6.13.10}$$

$$\tau_{yz} = \tau_{SV}^0 \cos 2\alpha\, U(c_S t + \cos \alpha) \text{ for } x < \beta\tau \tag{6.13.11}$$

where the unit step function also can be written as $U(\tau + \cos\alpha/k)$. The stress intensity factors are found from the corresponding crack face loading problems. Thus, for the normal load a mode I stress intensity factor is found from (6.11.33), and for the tangential load a mode II stress intensity factor is found from (6.11.49), in both cases after putting $\lambda = -k/\cos\alpha$. In the first case, σ_y^0 equals $-\tau_{SV}^0 \sin 2\alpha$, and in the second case, τ_{yz}^0 equals $\tau_{SV}^0 \cos 2\alpha$. The result is

$$K_I = \frac{2\sqrt{2}\tau_{SV}^0 \sin 2\alpha}{\sqrt{\pi}} \cdot \frac{\sqrt{k + \cos\alpha}\sqrt{(k + \beta\cos\alpha)c_P t}}{S(\cos\alpha/k)}$$
$$\times \frac{S(-1/\beta)(k_R - \beta)}{(k + k_R \cos\alpha)\sqrt{1 - \beta}} \tag{6.13.12}$$

$$K_{II} = -\frac{2\sqrt{2}\tau_{SV}^0 \cos 2\alpha}{\sqrt{\pi}} \cdot \frac{\sqrt{k + \cos\alpha}\sqrt{(k + \beta\cos\alpha)c_P t}}{S(\cos\alpha/k)}$$
$$\times \frac{\sqrt{k}S(-1/\beta)(k_R - \beta)}{(k + k_R \cos\alpha)\sqrt{k - \beta}} \tag{6.13.13}$$

Note, however, that $K_I = 0$ if τ_{SV}^0 is negative, in which case the expression for K_{II} is valid only in the absence of crack face friction.

Step function P wave impinging on a crack

If the normal stress behind the front of the incoming stress wave is σ_P^0, then the stress components in a plane parallel with the crack faces are $\sigma_y = (1 - 2k^2 \cos^2\alpha)\sigma_P^0$ and $\tau_{xy} = -2k^2 \sigma_P^0 \cos\alpha \sin\alpha$. It is assumed that the crack edge starts propagating with velocity $V = \beta c_P$ immediately after it is reached by the stress wave; see Fig. 6.13.3. By analogy with (6.13.10) and 6.13.11 for incident SV waves, the stress intensity factors are found by assuming the crack face loads

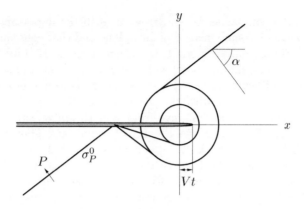

Fig. 6.13.3 P wave incident on a stationary semi-infinite crack, at the angle α, such that $\alpha = \pi/2$ implies normal incidence. The fronts of the cylindrical and plane P and S waves, resulting from the interaction between the incoming P wave and the crack are also shown. The crack starts moving with velocity $V = \beta c_P$ immediately after it is reached by the P wave.

$$\sigma_y = -\sigma_P^0(1 - 2k^2 \cos^2 \alpha)U(\tau + \cos \alpha) \text{ for } x < \beta\tau \quad (6.13.14)$$

$$\tau_{yz} = 2k^2 \sigma_P^0 \cos \alpha \sin \alpha U(\tau + \cos \alpha) \text{ for } x < \beta\tau \quad (6.13.15)$$

where $\tau = c_P t$. In the same way as for an incident SV wave, this leads to the stress intensity factors

$$K_I = \frac{2\sqrt{2}\sigma_P^0(1 - 2k^2 \cos^2 \alpha)}{\sqrt{\pi}} \cdot \frac{\sqrt{1 + \cos \alpha}\sqrt{(1 + \beta \cos \alpha)c_P t}}{S(\cos \alpha)}$$

$$\times \frac{S(-1/\beta)(k_R - \beta)}{(1 + k_R \cos \alpha)\sqrt{1 - \beta}} \quad (6.13.16)$$

$$K_{II} = -\frac{4\sqrt{2}k^2 \sigma_P^0 \cos \alpha \sin \alpha}{\sqrt{\pi}} \cdot \frac{\sqrt{1 + \cos \alpha}\sqrt{(1 + \beta \cos \alpha)c_P t}}{S(\cos \alpha)}$$

$$\times \frac{\sqrt{k}S(-1/\beta)(k_R - \beta)}{(1 + k_R \cos \alpha)\sqrt{k - \beta}} \quad (6.13.17)$$

Note that $K_I = 0$ if σ_P^0 is negative, in which case the expression for K_{II} is valid only if the crack face friction is zero. Note also that, for positive σ_P^0, the stress intensity factor $K_I > 0$, even if the angle of incidence is zero. This is, of course, explained by the crack opening caused by lateral contraction behind the stress wave.

Crack growth after a delay time

Antiplane strain

Because the stress intensity factor increases gradually from zero, no crack growth is likely to be initiated immediately after the stress wave has reached the crack edge. If a condition for crack growth is that the stress intensity factor should reach a certain

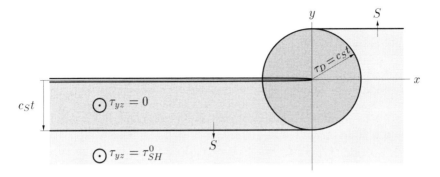

Fig. 6.13.4 Normal incidence of an SH wave on a stationary semi-infinite crack, which starts propagating after a delay time $\tau = \tau_D$. The figure shows the situation at that time.

critical value, independent of its rate (or more generally, its history), then the delay time is found simply by setting K_{III} equal to this critical value in (6.13.2). In any case, it is obvious that a more realistic situation than the one just studied, would involve a delay time before the crack starts moving. This case will now be analysed for normal incidence.

Assume that an SH wave with stress τ_{SH}^0 behind its front reaches the crack at $t = 0$ ($\tau = c_S t = 0$) and that the crack starts moving with constant velocity $V = \gamma c_S$ at time $t = t_D = \tau_D/c_S$. The difference between this case and the one leading to (6.13.6) is thus the delay before the crack starts moving. This implies that when the crack starts moving, a stress, varying with time, is already distributed on the plane of symmetry ahead of the crack; see Fig. 6.13.4. Following the same superposition scheme as in the previous case, this stress can be considered as the sum of the stress $\tau_{yz} = \tau_{SH}^0$, imposed on the symmetry plane ahead of the crack at the passage of the stress wave front, and the stress radiated ahead from the crack edge, as a result of the step load $\tau_{yz} = -\tau_{SH}^0$ on the crack faces. With use of (6.13.8) and (6.13.9), after integration, this stress is found to be

$$\tau_{yz} = \tau_{SH}^0 - \frac{2\tau_{SH}^0}{\pi}\left[\sqrt{\frac{\tau}{x} - 1} - \operatorname{atan}\sqrt{\frac{\tau}{x} - 1}\right] = \tau_{SH}^0[1 + h(x/\tau)] \quad (6.13.18)$$

where $x/\tau > 0$, $x < \tau$, and a function $h(x/\tau)$ is introduced, defined by relation (6.13.18)

The problem at hand may be solved by means of superposition of elementary solutions already obtained. Thus, essentially following Freund (1973), who treated the corresponding mode I case, the problem can be looked upon as the superposition of

Problem A: a step pulse, with stress $\tau_{yz} = \tau_{SH}^0$, moving in a crack-free body.
Problem B: crack face loading $\tau_{yz} = -\tau_{SH}^0$ on a stationary crack, starting at $\tau = 0$.
Problem C: crack propagation with velocity γ, starting at $\tau = \tau_D$ and driven by the load $-\tau_{SH}^0[1 + h(x/\tau)]$, appearing on the crack faces behind the crack edge.

Problem A is trivial, and Problem B has already been solved – the crack face acceleration and the stress rate on the symmetry plane ahead of the crack are the special cases of (6.13.6) and (6.13.7) for $\gamma = 0$:

$$\left(\frac{\partial^2 w_+}{\partial \tau^2}\right)_{y=0} = \frac{\tau_{SH}^0}{\pi\mu} \cdot \frac{\sqrt{-x}}{\sqrt{\tau + x\tau}} \quad \text{for } -\tau < x < 0 \tag{6.13.19}$$

$$\frac{\partial \tau_{yz}}{\partial \tau} = \frac{\tau_{SH}^0}{\pi} \cdot \frac{\sqrt{\tau - x}}{\tau\sqrt{x}} \quad \text{for } 0 < x < \tau \tag{6.13.20}$$

Problem C is more difficult. It is not a self-similar problem, but it can be reduced to a superposition of an infinity of such problems, each component involving a step load $\tau_{SH}^0 h'(\lambda)d\lambda$, moving with dimensionless velocity λ. At time τ, one such component extends from $x = \lambda\tau$ and forwards to the moving crack edge at $x = \gamma(\tau - \tau_D)$. It appears as a crack face load first at $x = x_0$, $\tau = \tau_0$, where

$$x_0 = \lambda\tau_0 = \gamma(\tau_0 - \tau_D) \tag{6.13.21}$$

i.e.,

$$\tau_0 = \frac{\gamma\tau_D}{\gamma - \lambda}, \quad x_0 = \frac{\lambda\gamma\tau_D}{\gamma - \lambda} \tag{6.13.22}$$

The response to this moving crack face load is found directly from (6.10.14) after the substitutions $\tau_{yz}^0 \to -\tau_{SH}^0 h'(\lambda)d\lambda$, $x \to x - x_0$, and $\tau \to \tau - \tau_0$. This gives

$$d\left(\frac{\partial^2 w_+}{\partial \tau^2}\right)_{y=0} = \frac{\tau_{SH}^0}{\pi^2\mu\sqrt{[(\gamma(\tau - \tau_D) - x]^3}}$$

$$\times \frac{[\gamma x - \lambda(x + \gamma\tau_D)]^2 q\,d\lambda}{\lambda\sqrt{\lambda}\sqrt{\gamma(\tau + x - \tau_D) - \lambda(\tau + x + \gamma\tau_D)}(x - \lambda\tau)} \tag{6.13.23}$$

Because expression (6.10.14) is valid for $-\tau < x < \gamma\tau$, the present expression is valid for $-(\tau - \tau_0) < x - x_0 < \gamma(\tau - \tau_0)$, which implies

$$\lambda < \gamma\frac{\tau + x - \tau_D}{\tau + x + \gamma\tau_D} = \lambda_0 < \gamma \tag{6.13.24}$$

and thus λ takes on values between 0 and λ_0. Note that the last square-root in the denominator of (6.13.23) vanishes for $\lambda = \lambda_0$.

Now, a difficulty presents itself in an attempt to integrate (6.13.23) from $\lambda = 0$ to $\lambda = \lambda_0$, because the integral would not be convergent. This difficulty can be traced to the impossibility of integrating $h'(\lambda) = \sqrt{1-\lambda}/(\pi\lambda\sqrt{\lambda})$ over the same interval, and then the remedy is found to consist of using the expression

$$h'(\lambda) = \frac{1}{2} \cdot [L_+(\lambda) - L_-(\lambda)], \quad \lambda > 0 \tag{6.13.25}$$

where, temporarily considering λ to be a complex variable,

$$L(\lambda) = \frac{(1-\lambda)^{1/2}}{\pi\lambda^{3/2}} \tag{6.13.26}$$

which is defined by a branch cut along the real axis from $\lambda = 0$ to $\lambda = 1$, and the branch chosen so that $L_+(\lambda)$ is real and positive for $0 < \Re\lambda < 1$, $\Im\lambda = 0$. Then,

$$\frac{1}{2}\int_\Gamma L(\lambda)d\lambda = 1 + h(\lambda_0) \tag{6.13.27}$$

where Γ is a path running below the real axis from $\lambda = \lambda_0$ towards $\lambda = 0$, rounding this point clockwise, and then following the upper part of the real axis to $\lambda = \lambda_0$. This result can, for instance, be obtained by assuming that Γ is circular near $\lambda = 0$, with radius $\epsilon \ll 1$ and otherwise infinitesimally close to the real axis.

The mass acceleration on the upper crack face for $-(\tau - \tau_D) < x < \gamma(\tau - \tau_D)$ is now given as

$$\left(\frac{\partial^2 w_+}{\partial \tau^2}\right)_{y=0} = \frac{\tau_{SH}^0}{2\pi^2 \mu \sqrt{[(\gamma(\tau - \tau_D) - x]^3}}$$
$$\times \int_\Gamma \frac{[\gamma x - \lambda(x + \gamma \tau_D)]^2 \, d\lambda}{\lambda^{3/2}[\gamma(\tau + x - \tau_D) - \lambda(\tau + x + \gamma \tau_D)]^{1/2}(x - \lambda \tau)} \quad (6.13.28)$$

The integral may be calculated by residue calculus. Note that Γ now can be closed at $\lambda = \lambda_0 + 0$, because the integrand is analytic to the right of the branch point $\lambda = \lambda_0$. If $x < 0$, then the pole at $\lambda = x/\tau$ contributes to the integral, otherwise the only contribution is the one obtained by expanding Γ to infinity. The result for $0 < x < \gamma(\tau - \tau_D)$ is

$$\left(\frac{\partial^2 w_+}{\partial \tau^2}\right)_{y=0} = \frac{\tau_{SH}^0}{\pi \mu \sqrt{[\gamma(\tau - \tau_D) - x]^3}} \cdot \frac{(x + \gamma \tau_D)^2}{\sqrt{\tau + x + \gamma \tau_D \, \tau}} \quad (6.13.29)$$

Recall that this refers to Problem C. For $-(\tau - \tau_D) < x < 0$, the contribution from the pole at $\lambda = x/\tau$ has to be added, giving

$$\left(\frac{\partial^2 w_+}{\partial \tau^2}\right)_{y=0} = \frac{\tau_{SH}^0}{\pi \mu \sqrt{[\gamma(\tau - \tau_D) - x]^3}} \cdot \frac{(x + \gamma \tau_D)^2}{\sqrt{\tau + x + \gamma \tau_D \, \tau}} - \frac{\tau_{SH}^0}{\pi \mu} \cdot \frac{\sqrt{-x}}{\sqrt{\tau + x \, \tau}} \quad (6.13.30)$$

The last term turns out to cancel the response (6.13.19) to Problem B, which, however, still remains for $-\tau < x < -(\tau - \tau_D)$. Consequently, the mass acceleration for the composed problem (A+B+C) is given by (6.13.19) for $-\tau < x < -(\tau - \tau_D)$ and by (6.13.30) for $-(\tau - \tau_D) < x < \gamma(\tau - \tau_D)$.

Note now that the delay time τ_D appears only in the context $x + \gamma \tau_D$ in (6.13.29). This suggests a variable substitution so that the spatial coordinate is changed fom x to r, the distance to the crack edge in front of the point considered. Thus, for $\tau > \tau_D$,

$$x = \gamma(\tau - \tau_D) - r \quad (6.13.31)$$

giving, for $0 < r < (1 + \gamma)(\tau - \tau_D)$,

$$\left(\frac{\partial^2 w_+}{\partial \tau^2}\right)_{y=0} = \frac{\tau_{SH}^0}{\pi \mu r \sqrt{r}} \cdot \frac{(\gamma \tau - r)^2}{\sqrt{(1 + \gamma)\tau - r \, \tau}} \quad (6.13.32)$$

This is a remarkable result, discovered for mode I by Freund (1973). It does not contain the delay time τ_D, which implies that *the crack profile, up to a distance $r = (1+\gamma)(\tau - \tau_D)$ behind the crack edge, is independent of the delay time!* Consequently, this crack profile is identical with the one found in a simpler way for $\tau_D = 0$, relation (6.13.6).

In particular, of course, *the stress intensity factor becomes independent of the delay time, as soon as the crack starts moving*, i.e. it is then dependent only on the time τ

after the incidence of the stress pulse on the crack. Thus, for $\tau > \tau_D$,

$$K_{III}(\tau, \gamma) = \frac{2\sqrt{2(1-\gamma)}\tau_{SH}^0\sqrt{\tau}}{\sqrt{\pi}} \qquad (6.13.33)$$

which also can be written in the same form as (6.13.5),

$$\boxed{K_{III}(\tau, \gamma) = k_{III}(\gamma) K_{III}^{statn}(\tau)} \qquad (6.13.34)$$

Thus, this formula is now generalized to allow a delay time before crack motion.

The stress rate in front of the crack in Problem C is determined in essentially the same way as the crack face acceleration. Thus, with the substitutions $\tau_{SH}^0 \to -\tau_{SH}^0 h'(\lambda)$, $x \to x - x_0$, and $\tau \to \tau - \tau_0$, equation (6.10.13) leads to

$$d\left(\frac{\partial \tau_{yz}}{\partial \tau}\right) = \frac{\tau_{SH}^0[\gamma x - \lambda(x + \gamma \tau_D)]\sqrt{\gamma(\tau - x - \tau_D) - \lambda(\tau - x - \gamma \tau_D)}}{\pi^2 \lambda \sqrt{\lambda} \sqrt{[x - \gamma(\tau - \tau_D)]^3}(x - \lambda \tau)} d\lambda \qquad (6.13.35)$$

for $y = 0$. This is valid for $\gamma(\tau - \tau_0) < x < \tau - \tau_0$, which implies

$$\lambda < \gamma \frac{\tau - x - \tau_D}{\tau - x - \gamma \tau_D} = \lambda_1 \qquad (6.13.36)$$

Note that the square-root in the nominator of (6.13.35) vanishes for $\lambda = \lambda_1$.

Integration as in (6.13.28) leads to

$$\left(\frac{\partial \tau_{yz}}{\partial \tau}\right)_{y=0} = \frac{\tau_{SH}^0}{\pi} \cdot \frac{(x + \gamma \tau_D)\sqrt{\tau - x - \gamma \tau_D}}{\sqrt{[x - \gamma(\tau - \tau_D)]^3} \tau} - \frac{\tau_{SH}^0}{\pi} \cdot \frac{\sqrt{\tau - x}}{\tau \sqrt{x}} \qquad (6.13.37)$$

for $\gamma(\tau - \tau_D) < x < \tau - \tau_D$. Recall that this refers to Problem C, so that the stress rate ahead of the crack for Problem (A+B+C) is given by the first term, only, because the second term cancels the stress rate (6.13.20) of Problem (A+B). Note that this first term contains x and τ_D only in the context $x + \gamma \tau_D$, as for the mass acceleration, so that the stress ahead of the crack on the symmetry plane up to the distance $(1-\gamma)(\tau - \tau_D)$ from the crack edge is independent of the delay time. This is confirmed by the fact that the substitution $(x + \gamma \tau_D) \to x$ brings (6.13.37) to coincide with (6.13.7). For $\tau - \tau_D < x < \tau$, the stress rate is given by (6.13.20).

In-plane modes

Crack propagation with constant velocity following the normal incidence of a stress wave after a delay time may be analysed in the same way for the in-plane modes as for mode III (Achenbach and Nuismer 1971). Here, a simplified approach will be taken, direct determination of the stress intensity factor without first calculating the crack profile. Later, the determination of crack face acceleration and stress rate straight ahead of the crack will be briefly discussed.

Assume first that the incoming wave is a P wave, reaching the crack at time $t = 0$. By analogy with mode III, equation (6.13.9), the expression for the stress rate (6.11.25), after putting $\beta = 0$, $\lambda = -\infty$ and $\sigma_y^0 = \sigma_P^0$, is written in the form

$$\left(\frac{\partial \sigma_y}{\partial \tau}\right)_{y=0} = -\sigma_P^0 \cdot \frac{x}{\tau^2} h'\left(\frac{x}{\tau}\right) \qquad (6.13.38)$$

6.13 STRESS WAVE LOADING OF CRACKS

where

$$h'\left(\frac{x}{\tau}\right) = -\frac{1}{\pi} \cdot \frac{S(-\tau/x)(k_R - x/\tau)}{S(0)\sqrt{x^3/\tau^3}\sqrt{1-x/\tau}} \qquad (6.13.39)$$

Then, the same superposition scheme as for mode III is applied, but here for the stress intensity factor rather than for the mass acceleration. It is obvious that this factor is determined from the problem corresponding to Problem C for mode III, page 475, i.e. crack propagation with velocity β, starting at time $\tau = \tau_D$, where $\tau = c_P t$, and driven by the load $-\sigma_P^0[1+h(x/\tau)]$, appearing on the crack faces behind the crack edge. This problem can be reduced to superposition of an infinity of self-similar problems, each component involving a step load $\sigma_P^0 h'(\lambda)d\lambda$, moving with dimensionless velocity λ. At time τ, one such component extends from $x = \lambda\tau$ and forwards to the moving crack edge at $x = \beta(\tau - \tau_D)$. It appears as a crack face load first at $x = \lambda\tau_0$, $\tau = \tau_0$, where

$$\tau_0 = \frac{\beta\tau_D}{\beta - \lambda} \qquad (6.13.40)$$

The contribution to the stress intensity factor from this moving crack face load is found directly from (6.11.33) following the substitutions $\sigma_P^0 \to -\sigma_P^0 h'(\lambda)d\lambda$, and $\tau \to \tau - \tau_0$. This gives

$$dK_I = \frac{2\sqrt{2}\sigma_P^0}{\pi\sqrt{\pi}} \cdot \frac{S(-1/\beta)(k_R - \beta)}{S(0)\sqrt{1-\beta}} \cdot \frac{\sqrt{\beta(\tau - \tau_D) - \lambda\tau}}{\lambda\sqrt{\lambda}} d\lambda \qquad (6.13.41)$$

The integration has to be performed over all values of λ, from $\lambda = 0$ for the first appearing infinitesimal component to $\lambda = \beta(\tau - \tau_D) = \lambda_0$ for the component, appearing at time τ. The same integration technique is used as for mode III, leading to

$$K_I = \frac{2\sqrt{2}\sigma_P^0}{2\pi\sqrt{\pi}} \cdot \frac{S(-1/\beta)(k_R - \beta)}{S(0)\sqrt{1-\beta}} \int_\Gamma \frac{[\beta(\tau-\tau_D) - \lambda\tau]^{1/2}}{\lambda^{3/2}} d\lambda \qquad (6.13.42)$$

where Γ is a path running below the real axis from $\lambda = \lambda_0$ towards $\lambda = 0$, rounding this point clockwise, and then following the upper part of the real axis to $\lambda = \lambda_0$. As in the mode III case, the path can be closed around $\lambda = \lambda_0$, because the integrand is analytic to the right of this point. Then, residue calculus gives the stress intensity factor after the crack starts moving:

$$K_I = \frac{2\sqrt{2}\sigma_P^0}{\sqrt{\pi}} \cdot \frac{S(-1/\beta)(k_R - \beta)\sqrt{\tau}}{S(0)\sqrt{1-\beta}}$$
$$= \frac{2\sqrt{2}\sigma_P^0}{\sqrt{\pi}} \cdot \frac{k\sqrt{2(1-k^2)}S(-1/\beta)(k_R - \beta)\sqrt{c_P t}}{k_R\sqrt{1-\beta}} \qquad (6.13.43)$$

As for mode III, the result is independent of the delay time τ_D and obviously equals (6.13.16) for $\alpha = \pi/2$. Note that the result can be written in the form

$$K_I(\tau, \beta) = k_I(\beta) K_I(\tau, 0) \qquad (6.13.44)$$

where $k_I(\beta)$ is given by (6.11.47). Because $K_I(\tau, 0)$ does not contain τ_D, it equals the stress intensity factor $K_I^{statn}(\tau)$ for a stationary crack at time τ after it was hit by the incident stress pulse. Thus,

$$K_I(\tau,\beta) = k_I(\beta) K_I^{statn}(\tau) \tag{6.13.45}$$

which is the same form as (6.13.5) for mode III.

Similarly, for normal incidence of an SV wave, carrying the shear stress τ_{SV}^0, the stress intensity factor is found by using the expression (6.11.49) for the stress intensity factor, whereas function $h'(x/\tau)$ is found from the mode I case by making the changes described on page 6.11. Because the result is again independent of time, the stress intensity factor after the crack starts moving may be taken from (6.13.13) by putting $\alpha = \pi/2$:

$$K_{II} = \frac{2\sqrt{2}\tau_{SV}^0}{\sqrt{\pi}} \cdot \frac{\sqrt{k}S(-1/\beta)(k_R - \beta)\sqrt{\tau}}{S(0)\sqrt{k - \beta}}$$

$$= \frac{2\sqrt{2}\tau_{SV}^0}{\sqrt{\pi}} \cdot \frac{k\sqrt{2k(1-k^2)}S(-1/\beta)(k_R - \beta)\sqrt{c_P t}}{k_R\sqrt{k - \beta}} \tag{6.13.46}$$

Note that the result may be written in the form

$$K_{II}(\tau,\beta) = k_{II}(\beta) K_{II}^{statn}(\tau) \tag{6.13.47}$$

where $k_{II}(\beta)$ is given by (6.11.51).

The result obtained for the stress intensity factor, either in mode I or mode II, can be used to determine the stress rate radiated out ahead of the crack edge after the crack starts moving, but the radiation velocity is left undetermined. The same applies to the crack profile (given through the crack face acceleration $\partial^2 v/\partial t^2$ in mode I, $\partial^2 u/\partial t^2$ in mode II). In mode III, both radiation velocities are the S wave velocity. Here, it can be shown that the stress rate radiates out forwards from the crack edge with the S wave velocity and the crack profile with the Rayleigh wave velocity. This result is obtained by calculating the stress rate and the crack profile by analogy with the corresponding calculations for mode III, Problems B and C, page 475. For an incident P wave carrying the stress σ_P^0, equation (6.11.25) is used together with (6.13.38) and (6.13.40) to give

$$\frac{\partial \sigma_y}{\partial \tau} = \frac{\sigma_P^0}{\pi^2 \sqrt{[x - \beta(\tau - \tau_D)]^3}} \int_\Gamma \frac{S(-\chi)[\beta x - \lambda(x + \beta\tau_D)]}{\lambda^{5/2}[\beta(\tau - x - \tau_D) - \lambda(\tau - x - \beta\tau_D)]^{1/2}}$$

$$\times \left\{ \frac{[\beta x - \lambda(x + \beta\tau_D)](k_R - \lambda)}{x - \lambda\tau} - k_R(\beta - \lambda) \right\} d\lambda \tag{6.13.48}$$

where Γ is the same path as in (6.13.42) and

$$\chi = \frac{(\beta - \lambda)\tau - \beta\tau_D}{(\beta - \lambda)x - \lambda\beta\tau_D} \tag{6.13.49}$$

The branch point given by $\chi = 1/k$ is situated between the two other branch points, $\lambda = 0$ and

$$\lambda = \beta \frac{\tau - x - \tau_D}{\tau - x - \beta\tau_D} \tag{6.13.50}$$

if $x < k(\tau - \tau_D)$, and therefore the path can be closed to the right of the latter branch point. Residue calculus then shows that the result contains τ_D only in the context $x + \beta\tau_D$. The response to stress pulse incidence on a stationary crack, corresponding to Problem B, page 475, shall then be added. The result shows that the stress rate as a function of the distance $r = x - \beta(\tau - \tau_D)$ ahead of the moving crack edge does not contain τ_D for $0 < r < (k-\beta)(\tau - \tau_D)$. This implies that it radiates out forwards from the moving crack edge with the S wave velocity.

Analogous calculations for the crack face acceleration $\partial^2 v/\partial t^2$ show that a crack profile, independent of τ_D, radiates out backwards from the moving crack edge with the Rayleigh wave velocity. This difference in result depends on a simple pole in the integrand which corresponds to the one in (6.13.48) for the stress rate. The pole is

$$\lambda = \beta \frac{k_R(\tau - \tau_D) + x}{k_R\tau + \beta\tau_D + x} \tag{6.13.51}$$

which, if $x < -k_R(\tau - \tau_D)$, falls outside the branch cut for the integrand, which extends from $\lambda = 0$ to

$$\lambda = \beta \frac{\tau - \tau_D + x}{\tau + \beta\tau_D + x} \tag{6.13.52}$$

In this case the integrand is more complicated than for the stress rate, with more branch points and poles requiring attention.

The principal results carry over to the case of an incident SV wave, i.e. the stress rate and the crack profile are independent of the delay time within a distance from the moving crack edge covered by S waves and Rayleigh waves, respectively.

Crack arrest after stress pulse loading

Freund (1973) devised a very simple way to study arrest of the moving crack. Consider first the normal incidence at $\tau = 0$ of an SH step pulse, carrying the stress $\tau_{yz} = \tau_{SH}^0$, on a semi-infinite crack, $x < 0$, $y = 0$, which then immediately starts moving with constant velocity $V = \gamma c_S$. Compare this situation with the one of Problem C, page 475, with the change that the edge of the crack before $\tau = 0$ is at a position $x = a_0 = \gamma\tau_D$. This implies that the two cracks edges are running synchronously after time $\tau = \tau_D$, i.e., the instantaneous position of both crack edges is the same, $x = \gamma\tau$; see Fig. 6.13.5. Also, according to the previous result (6.13.32), the crack profiles are identical for $a_0 < x < \gamma\tau$. Hence, if the stress-displacement field of the latter situation is subtracted from the former, the crack face displacement is the same for the upper and the lower crack face for $a_0 < x < \gamma\tau$, which implies that the composed situation is the one of a crack, $x < 0$, $y = 0$, set in motion at time $\tau = 0$ with velocity V by an incident stress pulse and arrested at $x = a_0$.

After arrest, the acceleration of the upper crack face on $-(1-\gamma)\tau < x < a_0$ is given by subtraction of the right member of (6.13.30) from the right member of (6.13.6), after replacing τ_D by a_0/γ and x to $x - a_0$ in the latter equation:

$$\left(\frac{\partial^2 w_+}{\partial \tau^2}\right)_{y=0} = \frac{\tau_{SH}^0}{\pi\mu} \cdot \frac{\sqrt{a_0 - x}}{\sqrt{\tau + x - a_0\tau}} \tag{6.13.53}$$

The fact that arrest could be obtained by subtracting the stress-displacement field of the modified Problem C leads to a simple way to determine the stress radiated out

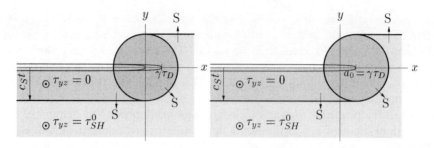

Fig. 6.13.5 Left figure: The initial crack edge position is at $x = 0$. The crack starts moving at $\tau = 0$ with velocity γc_S. The situation at $\tau = \tau_D$ is shown. Right figure: The initial crack edge position is at $x = a_0 = \gamma \tau_D$. The crack starts moving at $\tau = \tau_D$ with velocity γc_S. The situation at $\tau = \tau_D$ is shown. The crack is symbolically shown with an opening, although, in reality, only sliding takes place.

on the symmetry plane ahead of the arrested crack. Thus, this subtraction operation implies that, for $a_0 < x < \gamma \tau$, this stress equals the negated crack face load of the modified Problem C. It is thus found from differentiation of (6.13.18) with respect to τ. Equivalently, it may be taken from (6.13.8)-(6.13.9), after substituting x by $x - a_0 = x - \gamma \tau_D$. Thus, it is

$$\left(\frac{\partial \tau_{yz}}{\partial \tau}\right)_{y=0} = \frac{\tau_{SH}^0}{\pi} \cdot \frac{\sqrt{\tau - x + a_0}}{\tau \sqrt{x - a_0}} \tag{6.13.54}$$

for $a_0 < x < \gamma \tau$. For $\gamma \tau < x < \tau - (1 - \gamma)\tau_D$, it is given by subtraction of (6.13.37) from (6.13.7), after changing x to $x - a_0 = x - \gamma \tau_D$ in the former equation (also in its region of validity). This results in the same expression as in (6.13.54), which thus gives the stress rate on the plane of symmetry for $a_0 < x < \tau - (1 - \gamma)\tau_D$, i.e. $a_0 < x < \tau - \tau_D + a_0$. This stress rate is thus radiated out from the edge of the arrested crack with the S wave velocity, and the stress can be obtained by integration.

The stress intensity factor of the arrested crack is found from (6.13.54) to be

$$K_{III}^{arrest}(\tau) = \lim_{x \to a_0}\left[\sqrt{2\pi(x - a_0)} \int_0^\tau \left(\frac{\partial \tau_{yz}}{\partial \tau}\right)_{y=0} d\tau\right] = \frac{2\sqrt{2}\tau_{SH}^0 \sqrt{\tau}}{\sqrt{\pi}} \tag{6.13.55}$$

Note that this result is independent of where the crack was arrested and also of the crack velocity before arrest. It is only dependent on the time τ after the incidence of the stress pulse. It equals, for instance, the stress intensity factor $K_{III}^{statn}(\tau)$ of a stationary crack, hit by the stress pulse at $\tau = 0$. Thus, the result, which is valid before stress waves arrive from an outer boundary or another crack edge, can also be written in the form

$$K_{III}^{arrest}(\tau) = K_{III}^{statn}(\tau) \tag{6.13.56}$$

For the in-plane modes, the situation is analogous. For an incident P wave, the stress rate ahead of the crack after arrest is found from (6.13.38), after changing x to $x - a_0$:

$$\left(\frac{\partial \sigma_y}{\partial \tau}\right)_{y=0} = \frac{\sigma_P^0}{\pi} \cdot \frac{S(-\tau/(x - a_0))(k_R \tau - x + a_0)}{S(0)\tau \sqrt{x - a_0}\sqrt{\tau - x + a_0}} \tag{6.13.57}$$

Thus, the stress intensity factor of the arrested crack is

$$K_I^{arrest}(\tau) = \lim_{x \to a_0}\left[\sqrt{2\pi(x-a_0)}\int_0^\tau \left(\frac{\partial \sigma_y}{\partial \tau}\right)_{y=0} d\tau\right]$$

$$= \frac{2\sqrt{2}\sigma_P^0 k_R \sqrt{\tau}}{\sqrt{\pi}S(0)} = \frac{4}{\sqrt{\pi}}\sigma_P^0 k\sqrt{(1-k^2)c_P t} \quad (6.13.58)$$

where $S(0)$ is given by (6.11.44). Again, as in the mode III case, the result is independent of where the crack was arrested and also of the velocity before arrest. It is only dependent on the time elapsed after the incidence of the stress pulse. Thus, the result can be written in the form

$$K_I^{arrest}(\tau) = K_I^{statn}(\tau) \quad (6.13.59)$$

where $K_I^{statn}(\tau)$ is the stress intensity factor for a stationary crack, hit by a stress pulse at $\tau = 0$, cf. (6.13.16).

For normal incidence at time $t = 0$ of an SV wave, carrying the shear stress τ_{SV}^0, the result is, cf. (6.13.13)

$$K_{II}^{arrest}(\tau) = K_{II}^{statn}(\tau) = \frac{4}{\sqrt{\pi}}\tau_{SV}^0 k\sqrt{(1-k^2)c_P t} \quad (6.13.60)$$

The results are valid until stress waves arrive from an outer boundary or another crack edge.

Non-constant crack velocity after stress pulse loading

Mode III

In Section 6.12 it was shown that a static stress field radiates out from an arrested mode III crack that had started to propagate from an equilibrium state under the action of time-independent crack face loading, appearing behind the moving crack edge; see page 462. As a contrast, the stress field radiated from an arrested mode III crack, that had started to propagate after being hit by a stress pulse, is not a static field; cf. (6.13.53) and (6.13.54). However, it possesses features which make it possible to derive an expression for non-constant crack propagation as a result of stress pulse loading in virtually the same way as in Section 6.12. For mode III, these features might be more clearly expressed by (6.13.53) and (6.13.54) if the coordinate origin is temporarily moved to the edge of the arrested crack. For (6.13.54) this leads to

$$\left(\frac{\partial \tau_{yz}}{\partial \tau}\right)_{y=0} = \frac{\tau_{SH}^0}{\pi}\cdot\frac{\sqrt{\tau-x}}{\tau\sqrt{x}}, \qquad \tau > x \quad (6.13.61)$$

This stress rate is radiated out in the positive x direction with the S-wave velocity, which implies that any trace of the previous history disappears with this velocity from the crack edge vicinity. The same applies to the crack face acceleration. An observer of the events on the symmetry plane, appearing at time τ in this vicinity, would not be able to find out where the crack edge was before the incidence of the stress wave (except that it must have been somewhere between $x = -\tau$ and $x = 0$ or even at $x = 0$), what the crack speed was just before arrest (except that it must have been smaller than c_S, possibly even 0), or how long time had elapsed since crack arrest (except that it must have been less than τ).

Assume now that the crack immediately after arrest continues to propagate with the dimensionless velocity γ_1. The crack velocity before arrest might have been different, but the environment which is known by the crack edge after arrest is the same as if this velocity also had been γ_1. Thus, assuming this had been the case, it is obvious that the continued crack growth can be considered as having started at $\tau = 0$ and proceeded with velocity γ_1 all the time. Suppose now that the crack is arrested at some edge position $x > 0$. This is the same situation as previously considered, but now it can alternatively be considered as the result of one crack propagation with speed γ_0, say, from $\tau = 0$, followed by arrest at $x = 0$ and then immediately continuing with velocity γ_1 until it is again arrested. This process can be repeated indefinitely, and in this way even continuous changes of the crack speed, interspersed with discontinuous changes, can be handled. Because the stress intensity factor at any moment is independent of the previous history, except for the time elapsed since the incidence of the stress pulse, it is governed by the same expression as for constant velocity propagation without, (6.13.5), or with, (6.13.34), delay time,

$$K_{III}(\tau,\gamma) = k_{III}(\gamma) K_{III}^{statn}(\tau) \qquad (6.13.62)$$

where $K_{III}^{statn}(\tau)$, the stress intensity factor for a stationary crack at time τ after it was hit by the stress pulse, is given by (6.13.55). This result obviously contains as special cases previous results, crack propagtion at constant velocity immediately after it was hit by the stress pulse or after a delay time and crack arrest after such a propagation.

Equation (6.13.62) enables determination of the stress intensity factor history if the instantaneous velocity, $V = \gamma c_S = da/dt$, is known as a function of the distance of travel, a, or time, $t = \tau/c_S$. Alternatively, it may be used together with an energy based equation for the motion of the crack edge, cf. Section 9.1 to find out how the crack speed evolves rather than prescribing it. Note that (6.13.62) is valid until the arrival of stress waves, reflected from an outer boundary or another crack edge.

The case of crack propagation with non-constant velocity after oblique incidence of a stress pulse can be treated in the same manner, by using (6.10.14) and (6.10.13) as starting points, after substituting τ_{SH} by its component $\tau_{SH} \sin\alpha$ and λ by $-1/\cos\alpha$ as in the derivation of (6.13.3). A more direct, though somewhat intuitive way, is to rely upon the feature found for normal incidence, that the stress intensity factor during motion at non-constant velocity is not dependent on the history of crack growth. Comparison between the stress intensity factor given by (6.10.19) and (6.13.3), for oblique incidence on a stationary crack, followed by constant crack velocity without delay, and the stress intensity factor for normal incidence ($\alpha = \pi/2$) shows that the former is obtained from the latter after multiplication by

(1) the universal dynamic factor $k_{III}(\gamma)$,
(2) the stress projection factor $\sin\alpha$,
(3) the attenuation factor $1/\sqrt{1-\lambda} = \sqrt{\cos\alpha}/\sqrt{1+\cos\alpha}$, and
(4) the load extension factor $\sqrt{(\gamma-\lambda)\tau} = \sqrt{\tau/\cos\alpha + \gamma\tau}$

If constant velocity propagation after a delay time, τ_D, is considered, then the load extension changes to $\tau/\cos\alpha + \gamma(\tau - \tau_D)$, and, for the general case of non-constant velocity, it changes to $\tau/\cos\alpha + a(\tau)$, where $a(\tau)$ is the distance of crack edge travel.

Thus, it is expected that the stress intensity factor at oblique incidence followed by crack growth at non-constant velocity is

$$K_{III}(\tau,\gamma) = k_{III}(\gamma)\frac{2\sqrt{2}\tau_{SH}^0 \sin\alpha \sqrt{\tau + a(\tau)\cos\alpha}}{\sqrt{1+\cos\alpha}} \quad (6.13.63)$$

The argument of the square-root in the nominator has a simple interpretation. It is the distance from the crack edge to the stress wave front. Thus, compared to the case of normal incidence and the same crack edge travel, a longer portion of the stress wave has been acting on the crack edge after it was struck by the pulse, and this fact is reflected in the change of argument from τ to the larger $\tau + a(\tau)\cos\alpha$. The significance of the distance $\tau + a(\tau)\cos\alpha$ from the crack edge to the stress wave front might be more obvious, if it is considered that the displacement $|w|$ increases in proportion to this distance in an undisturbed step pulse. Note also that the previous interpretation of the time dependence of stress intensity factors at normal stress pulse incidence could as well have been expressed by a distance dependence.

As a further consequence of the significance of the distance from the crack edge to the stress wave front, it might be safer to use this distance as independent variable, rather than time, when establishing a Duhamel integral to consider other stress wave profiles than a step function. Note that the stress intensity factor at crack arrest is

$$K_{III}^{arrest}(\tau) = \frac{2\sqrt{2}\tau_{SH}^0 \sin\alpha \sqrt{\tau + a(\tau)\cos\alpha}}{\sqrt{1+\cos\alpha}} \quad (6.13.64)$$

even at constant velocity, $a(\tau) = \gamma\tau$, i.e. to obtain $K_{III}^{arrest}(\tau)$ from the expression for $K_{III}(\tau,\gamma)$, the velocity γ should be put equal to zero only in $k_{III}(\gamma)$. Here, the fundamental relation (6.13.62) would not be appropriate, but the general relation

$$\boxed{K_{III}(\tau,\gamma) = k_{III}(\gamma)K_{III}^{arrest}(\tau)} \quad (6.13.65)$$

would hold.

In-plane modes

For non-constant crack propagation after normal incidence of the stress pulse, the discussion for mode III (incoming SH wave) carries over to modes I (incoming P wave) and II (incoming SV wave) with only minor and obvious modifications. Thus, the stress intensity factor is given by the same kind of fundamental relation as for mode III,

$$\boxed{K_{I,II}(\tau,\beta) = k_{I,II}(\beta)K_{I,II}^{statn}(\tau)} \quad (6.13.66)$$

where $k_I(\beta)$ and $k_{II}(\beta)$ are given by (6.11.47) and (6.11.51), respectively. $K_{I,II}^{statn}(\tau) = K_{I,II}^{arrest}(\tau)$ are found from (6.13.58) and (6.13.60).

For oblique incidence, the same discussion applies as for an incoming S wave. Thus, for an incoming SV wave, the change

$$\sqrt{(k+\beta\cos\alpha)c_P t} \rightarrow \sqrt{c_S t + a(t)\cos\alpha} \quad (6.13.67)$$

where $a(t)$ is the amount of crack advance at time t, should be made in equations

(6.13.12) and (6.13.13). For an incoming P wave, the change

$$\sqrt{(1+\beta\cos\alpha)c_P t} \to \sqrt{c_P t + a(t)\cos\alpha} \qquad (6.13.68)$$

should be made in equations (6.13.16) and (6.13.17).

Stress wave incidence on a crack of finite length

The previously found expressions for stress wave interaction with a crack are valid until reflected waves arrive from an outer boundary or another crack edge. However, in the case of a finite crack, it is possible in principle to follow successive reflections of waves from one crack edge to the other with methods proposed by Flitman (1963) and Kostrov (1964b), but the complexity of the problem increases enormously with each reflection. An extensive and very readable account of the basic theory for the transient response of cracks of different geometries, for instance penny-shaped cracks, is given by Chen and Sih (1977). Experimental results are reported by e.g. Shockey and Curran (1973), Kalthoff (1983) and Shockey et al. (1983).

Assume normal incidence of an SH wave at time $t = 0$ on a crack $-a < x < a$, $y = 0$. The same notations are used as in the previous subsection. Immediately after the incidence, the lower crack face moves with velocity $\partial w_-/\partial\tau = -2\tau_{SH}^0/\mu$ (the factor 2 is due to wave reflection), whereas the upper crack face is not affected: $w_+ = 0$. It is convenient to consider the crack face (sliding) separation $w_{sep} = w_+ - w_-$, which thus equals $2\tau_{SH}^0\tau/\mu$ immediately after the incidence. Thereupon, cylindrical waves are emerging from each crack edge, causing a crack face motion in the opposite direction, thus acting to diminish w_{sep}, cf. equation (6.10.25) and Fig. 6.10.6.

After having reached the opposite crack edges at time $\tau = 2a$, the cylindrical waves will obviously cause a slower growth of the stress intensity factor than given by the expression (6.13.2) for $\alpha = 0$,

$$K_{III} = \frac{2\sqrt{2}\tau_{SH}^0\sqrt{\tau}}{\sqrt{\pi}} \qquad (6.13.69)$$

for a semi-infinte crack. Thus, for $2a < \tau \le 4a$, the stress intensity factor is smaller than given by (6.13.69), at least before the possible arrival of waves reflected from an outer boundary. After time $\tau = 4a$, new cylindrical waves, caused by reflections from the crack edges, will be reflected from the opposite crack edges, and such reflections will continue repeatedly. Due to their spatial divergence, each successive reflected cylindrical wave ought to have a smaller influence on the stress intensity factor than its predecessor, and oscillations with decreasing amplitude about the static value are expected to occur until waves arrive from an outer boundary. The static value is

$$K_{III}^{stat} = \tau_{SH}^0\sqrt{\pi a} \qquad (6.13.70)$$

for a remote stress $\tau_{yz} = \tau_{SH}^0$. The oscillating behaviour is in agreement with theoretical investigations as well as with numerical calculations and experiments, see e.g. Sih (1968), Ravera and Sih (1969), Achenbach (1970b), Sih and Embley (1972), Sih et al. (1972), Chen and Sih (1977), Shockey et al. (1986) and Freund (1990).

In relation to the static value (6.13.70), the stress wave induced stress intensity factor is

$$\frac{K_{III}}{K_{III}^{stat}} = \frac{2\sqrt{2}}{\pi}\sqrt{\frac{\tau}{a}} \text{ for } 0 < \tau \le 2a \qquad (6.13.71)$$

$$\frac{K_{III}}{K_{III}^{stat}} < \frac{2\sqrt{2}}{\pi}\sqrt{\frac{\tau}{a}} \text{ for } 2a < \tau \le 4a \qquad (6.13.72)$$

This ratio equals $4/\pi \approx 1.273$ for $\tau = 2a$, whereas it is smaller than $4\sqrt{2}/\pi \approx 1.8$ for $\tau = 4a$. Thus, a dynamic overshoot of at least a factor 1.273 occurs, and, due to the presumed oscillating character, it might be assumed that the overshoot never exceeds a factor of about 1.8.

In-plane mode cases are similar to the mode III case, although a somewhat unexpected difference appears: the original cylindrical waves from the crack edges are not consistently opposing the separation motion of the crack faces. For a P wave with normal incidence the initial action of such cylindrical waves is actually to *increase* the crack face separation; see Fig. 6.11.3. Numerical calculations show that this increase persists even when half the cylindrical wave has passed by, irrespective of the value of k. For $k^2 = 1/3$, corresponding to Poisson's ratio $1/4$ at plane strain and $1/3$ at plane stress, the rear end of this increase travels with a velocity of about $0.4c_P$. Assume for a moment that the increase is vanishingly small. This is equivalent to assuming that more than the first half of the cylindrical wave is absent – for $k^2 = 1/3$, approximately the first 60% is absent. Consequently, it could not influence the stress intensity factor at the crack edge opposite to its origin before some time after $\tau = 4a$ – for $k^2 = 1/3$ not before the time $\tau \approx 2a/0.4 = 5a$. Thus, this factor would be the same as for a semi-infinite crack, found from (6.11.45) after putting $\sigma_y^0 = -\sigma_P^0$, $\beta = 0$ and, for $k^2 = 1/3$, $t \approx 5a/c_P$:

$$K_I \approx \frac{4\sqrt{5}k\sqrt{1-k^2}\sigma_P^0\sqrt{a}}{\sqrt{\pi}} \qquad (6.13.73)$$

Now, because the increase is not vanishingly small (although it is not very significant) there will actually be some increase of the maximum stress intensity factor beyond this value, which is about 1.34 times the static value $\sigma_P^0\sqrt{\pi a}$ for a remote load $\sigma_y = \sigma_P^0$.

For an incident SV wave, the original cylindrical waves from the crack edges oppose the crack face separation consistently, as for incident SH waves, see Fig. 6.11.5. This implies that the stress intensity factor before time $\tau = 4a$ reaches a maximum value between the values for the edge of a semi-infinite crack at times $\tau = 2a$ and $\tau = 4a$. From (6.11.50), after putting $\tau_{xy}^0 = -\tau_{SH}^0$, $\beta = 0$ and $t = 2a/c_P$ or $t = 4a/c_P$, this gives the ratio between $(K_{II})_{max}$ and $K_{II}^{stat} = \tau_{xy}^0\sqrt{\pi a}$:

$$\frac{4\sqrt{2}\sqrt{k(1-k^2)}}{\pi} < \frac{(K_{II})_{max}}{K_{II}^{stat}} < \frac{8\sqrt{k(1-k^2)}}{\pi} \qquad (6.13.74)$$

For $k^2 = 1/3$ the ratio is between 0.85 and 1.20, indicating modest oscillations towards the static value. For low values of k there might not be any dynamic overshoot at all, so that the stress intensity factor would approach its static value from below.

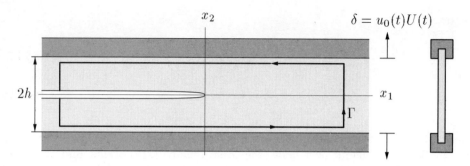

Fig. 6.13.6 The path Γ in the strip.

Stress wave incidence on a crack in a strip

Consider an infinite strip, $|x_2| \leq h$, with a stationary semi-infinite crack, $x_1 < 0$, $x_2 = 0$. The strip edges, $|x_2| = h$, are clamped in rigid frames which can be moved in the x_2 direction, only. Assume that they are moved so that the strip height is being extended symmetrically by the amount

$$2\delta = 2u_0(t)U(t) \qquad (6.13.75)$$

where $U(\cdot)$ is the unit step function. This implies that stress waves propagate symmetrically from each strip edge towards the crack and hit the crack at time $t = h/c_P$. Thereupon, repeated wave reflections will occur at the strip edges and at the crack. The problem, which consists of finding the stress intensity factor as a function of time, is ideally suited for the Nilsson integral (3.2.56), and it was chosen by Nilsson (1973a) in his demonstration of the integral. Modified for use of $\tau = c_P t$ as time variable rather than t, the integral reads

$$J_N = \int_\Gamma \left\{ \left[W^*(p) + \frac{\mu}{2k^2} p^2 U_i U_i \right] n_1 - n_j S_{ji} \frac{\partial U_i}{\partial x_1} \right\} ds \qquad (6.13.76)$$

where $i = 1, 2$, $j = 1, 2$, s is the arc length along the path Γ, $S_{ij} = \mathcal{L}_{p\tau}(\sigma_{ij})$, $U_i = \mathcal{L}_{p\tau}(u_i)$ and $W^*(p)$ is a function of $U_{i,j}$ such that

$$\mathcal{L}_{p\tau}^{-1} \frac{\partial W^*(p)}{\partial U_{i,j}} = \sigma_{ji} \qquad (6.13.77)$$

Here, $\mathcal{L}_{p\tau}$, as in Appendix A5, indicates Laplace transformation.

The path Γ is now chosen according to Fig. 6.13.6. It crosses the strip far enough from the crack edge not to be reached by P waves from this edge during the time of interest, say at $|x_1| = x_\infty$. Then, $n_1 ds = |dx_2| = 0$ and $\partial u_i / \partial x_1 = 0$ along the strip edges, so that the only contribution to the Nilsson integral derives from the parts at $|x_1| = x_\infty$. For these parts the horizontal displacement $u_1 = 0$, the displacement gradient $\partial u_2 / \partial x_1 = 0$ and the stress $\sigma_{22} = \mu/k^2 \partial u_2 / \partial x_2$. Thus, from (6.13.77),

$$\frac{dW^*(p)}{dU_{2,2}} = S_{22} = \frac{\mu}{k^2} U_{2,2} \qquad (6.13.78)$$

so that
$$W^*(p) = \frac{\mu}{2k^2}U_{2,2}^2 \qquad (6.13.79)$$

Then,
$$J_N = \frac{\mu}{2k^2}\int_\Gamma [U_{2,2}^2 + p^2 U_2^2]\mathrm{d}x_2 \qquad (6.13.80)$$

where the path Γ is reduced to the two parts at $|x_1| = x_\infty$. Thus, U_2 has to be determined for $x_1 = \pm x_\infty$. In both cases, the wave equation is one-dimensional and takes the simple form

$$\frac{\partial^2 u_2}{\partial x_2^2} = \frac{\partial^2 u_2}{\partial \tau^2} \qquad (6.13.81)$$

which, after Laplace transformation, reads
$$\frac{\mathrm{d}^2 U_2}{\mathrm{d}x_2^2} = p^2 U_2 \qquad (6.13.82)$$

Consider the upper half, $x_2 \geq 0$. The Laplace transformed boundary conditions are

$$\begin{array}{llll} x_1 = -x_\infty: & U_2 = U_0(p) \text{ for } x_2 = h & S_{22} = 0 \Longrightarrow U_{2,2} = 0 & \text{for } x_2 = 0 \\ x_1 = +x_\infty: & U_2 = U_0(p) \text{ for } x_2 = h & U_2 = 0 & \text{for } x_2 = 0 \end{array} \qquad (6.13.83)$$

where $U_0(p) = \mathcal{L}_{p\tau}[u_0(\tau/c_P)]$. Then, the solutions of the wave equation, satisfying the boundary conditions, are

$$U_2 = U_0(p) \cdot \frac{\cosh(px_2)}{\cosh(ph)} \quad \text{for } x_1 = -x_\infty \qquad (6.13.84)$$

$$U_2 = U_0(p) \cdot \frac{\sinh(px_2)}{\sinh(ph)} \quad \text{for } x_1 = +x_\infty \qquad (6.13.85)$$

Insertion into the Nilsson integral (6.13.80) gives

$$J_N = \frac{4\mu p^2 [U_0(p)]^2}{k^2 \sinh^2(2ph)}\int_0^h [\cosh^2(px_2) + \sinh^2(px_2)]\mathrm{d}x_2 = \frac{2\mu p [U_0(p)]^2}{\sinh(2ph)} \qquad (6.13.86)$$

Hence, from (3.5.28), the Laplace transform of the stress intensity factor is

$$\mathcal{L}_{p\tau}(K_I) = \frac{4\sqrt{1-k^2}\,\mu}{k}\cdot\frac{p^{1/2}U_0(p)}{[2\sinh(2ph)]^{1/2}} \qquad (6.13.87)$$

Now,
$$[2\sinh(2ph)]^{-1/2} = e^{-ph}(1-e^{-4ph})^{-1/2} = e^{-ph}\sum_{n=0}^{\infty}(-1)^n\binom{-1/2}{n}e^{-4nph} \qquad (6.13.88)$$

and
$$\mathcal{L}_{p\tau}(p^{1/2}) = \frac{1}{\sqrt{\pi\tau}} \qquad (6.13.89)$$

The theorem of Laplace transform of convolution integrals, (A5.10), gives

$$\mathcal{L}_{p\tau}^{-1}\left[\frac{1}{p}\cdot p^{1/2}\cdot pU_0(p)\right] = \int_0^\tau \frac{\mathrm{d}u_0(\tau'/c_P)/\mathrm{d}\tau'}{\sqrt{\pi(\tau-\tau')}}\mathrm{d}\tau'\cdot U(\tau) \qquad (6.13.90)$$

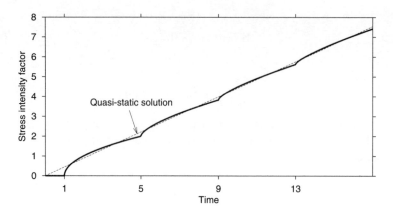

Fig. 6.13.7 The stress intensity factor at the edge of a semi-infinite crack, symmetrically situated in a strip, whose edges are moved apart with constant velocity, V_0. The strip height is $2h$. The stress intensity factor is given in units of $\pi^{1/2} k c_P / [8(1-k^2)^{1/2} h^{1/2} \mu V_0]$ and the time in units of h/c_P. The straight line indicates the development of the stress intensity factor in a quasi-static treatment, i.e. with neglect of stress wave propagation across the strip.

and a multiplication of a Laplace transform by $\exp(-p\tau_n)$ implies a time shift $\tau \to \tau_n$, cf. (A5.14). Thus,

$$K_I = \frac{4\sqrt{1-k^2}\,\mu}{\sqrt{\pi}\,k} \sum_{n=0}^{\infty} (-1)^n \binom{-1/2}{n}$$
$$\times \int_0^{\tau-\tau_n} \frac{du_0(\tau'/c_P)/d\tau'}{\sqrt{\pi(\tau-\tau_n-\tau')}} d\tau' \cdot U(\tau-\tau_n) \quad (6.13.91)$$

where $\tau_n = (4n+1)h$. This solution is obviously valid even for a strip of finite length until disturbances arrive to the crack edge from the vertical strip boundaries.

Specialization to $u_0(t) = V_0 t\, U(t)$, where V_0 is a constant velocity, gives

$$K_I = \frac{8\sqrt{1-k^2}\sqrt{h}V_0\mu}{\sqrt{\pi}\,kc_P}$$
$$\times \sum_{n=0}^{\infty} (-1)^n \binom{-1/2}{n} \sqrt{\tau/h - 1 - 4n}\, U(\tau/h - 1 - 4n) \quad (6.13.92)$$

This relation is shown in Fig. 6.13.7 together with the quasi-static solution

$$K_I = \frac{2\sqrt{(1-k^2)h}\,\mu V_0}{kc_P} \cdot \frac{c_P t}{h} \quad (6.13.93)$$

Note how well the stress intensity factor is approximated by the quasi-static solution after a few wave reflections. This is the case also for reasonably smooth relations $u_0(t)$ other than the linear one. Note also that the time between two reflections is very short compared to the response time of most loading devices – for example, $h = 200$mm and $c_P = 4000$m/s give $4h/c_P = 200\mu$s.

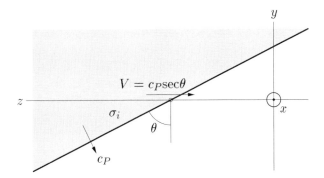

Fig. 6.14.1 Oblique incidence of a P wave on a crack edge (the z axis). The crack faces are situated on the xz-plane for $x < 0$. Positive x direction points toward the reader.

It might appear strange that the response contains time intervals of length $4h/c_P$ rather than $2h/c_P$. The explanation seems to be that reflected stress wave fronts cause both a crack face velocity decrease at times $3h/c_P$, $7h/c_P$, ..., which tends to decrease K_I, and also an increase of the stress ahead of the crack, which tends to increase K_I. Apparently, these two actions cancel each other.

Before time $t = h/c_P$, the stress carried by the stress wave is $\sigma_P^0 = \mu V_0/(k^2 c_P)$. Then, writing the stress intensity factor for the time interval $h/c_P < t < 5h/c_P$ as

$$K_I = \frac{8}{\sqrt{\pi}} k \sqrt{1-k^2}\, \sigma_P^0 \sqrt{c_P(t - h/c_P)} \qquad (6.13.94)$$

shows agreement with (6.13.58), apart from a time shift and a factor 2, due to incidence of *two* symmetrical stress waves in the present case. The agreement is obvious for the time interval $h/c_P < t < 3h/c_P$, i.e. before the existence of strip edges is known at the crack edge.

6.14 Three-dimensional elastodynamic crack problems

Oblique P wave incidence on a crack edge

Statement of the problem

Consider a stationary semi-infinite crack $x < 0$, $y = 0$, in an infinite body. A plane P wave, with step front and principal stress σ_P^0 on planes parallel with the wave front (the other principal stresses are then equal to $[\nu/(1-2\nu)]\sigma_P^0$), is obliquely incident on the crack plane. The wave front is parallel with the x axis and the propagation direction forms the angle $\pi - \theta$ with the positive z axis; see Fig. 6.14.1. Thus, normal incidence (wave front parallel with the crack plane) occurs for $\theta = \pi/2$. The stress intensity factors along the crack edge are sought. They are obviously dependent on t and z. This problem was solved for time-harmonic plane waves by Achenbach and Gautesen (1977) and for a wave pulse by Freund (1990) on the basis of a previous investigation of oblique reflection of Rayleigh waves from a crack edge (Freund 1971).

The wave front cuts the xz plane along the line $z = -Vt$, where $V = c_P/\cos\theta =$

$c_P \sec\theta$ and time t is set to zero when the wave reaches the x axis. Because $V > c_P$, and thus supersonic, head waves are created from the upper crack face into the body, viz., P waves with Mach number $\sec\theta$ and S waves with Mach number $\sec\theta/k$, where k is related to Poisson's ratio ν by $k^2 = (1-2\nu)/[2(1-\nu)]$. Two different shapes of head waves appear: plane head waves emanating from the line where the wave front meets the upper crack face and conical head waves, emanating from the point where the wave front meets the crack edge.

In the absence of the crack, the normal and tangential stresses on the xz plane would be

$$\sigma_y = \sigma_y^0 = \sigma_P^0 \{\sin^2\theta + [\nu/(1-\nu)]\cos^2\theta\} \qquad (6.14.1)$$

$$\tau_{yz} = \tau_{yz}^0 = \sigma_P^0[(1-2\nu)/(1-\nu)]\cos\theta\sin\theta \qquad (6.14.2)$$

The same kind of superposition technique as shown by Fig. 6.13.2 is now used, so that the problem is considered as the superposition of three parts:

Problem A: a plane P wave, with the largest principal stress σ_P^0, moving in a crack-free body.

Problem B: a normal traction $\sigma_y = -\sigma_y^0$ extending over the part $z \geq -Vt$ of the crack faces.

Problem C: a tangential traction $\tau_{yz} = -\tau_{yz}^0$ extending over the part $z \geq -Vt$ of the crack faces.

Problem A is trivial. Problems B and C are similar. The stress intensity factors follow from their solution, K_I from Problem B and K_{III} from Problem C. Note that the front of the tractions in Problems B and C is perpendicular to the crack edge.

Because problems B and C are steady state problems, the Galilean transformation $X = x$, $Y = y$, $Z = z + Vt$ is introduced. The problems may then also be considered as self-similar with respect to the origin, $X = Y = Z = 0$, because no length parameter is involved. In both problems, the half-plane $Z < 0$ is stress free. Fig. 6.14.2 shows the intersection of different wave fronts with the XZ plane.

The equations of motion, (A9.16)-(A9.17), take the form

$$\frac{\partial^2\phi}{\partial X^2} + \frac{\partial^2\phi}{\partial Y^2} - (\sec^2\theta - 1)\frac{\partial^2\phi}{\partial Z^2} = 0 \qquad (6.14.3)$$

$$\frac{\partial^2\psi}{\partial X^2} + \frac{\partial^2\psi}{\partial Y^2} - (\sec^2\theta/k^2 - 1)\frac{\partial^2\psi}{\partial Z^2} = 0 \qquad (6.14.4)$$

$$\frac{\partial\psi_x}{\partial X} + \frac{\partial\psi_y}{\partial Y} + \frac{\partial\psi_z}{\partial Z} = 0 \qquad (6.14.5)$$

where the last equation expresses that div $\psi = 0$; see (A9.14).

Consider the upper half-plane $Y \geq 0$. The boundary conditions for $Y = 0$ are

Problem B:
All X: $\tau_{xy} = 0$
All X: $\tau_{yz} = 0$
$X < 0$: $\sigma_y = -\sigma_y^0 U(Z)$
$X > 0$: $v_+ = 0$

Problem C:
All X: $\tau_{xy} = 0$
All X: $\sigma_y = 0$ \qquad (6.14.6)
$X < 0$: $\tau_{yz} = -\tau_{yz}^0 U(Z)$
$X > 0$: $w_+ = 0$

where subscript plus on v_+ and w_+ indicates displacements on the upper half-plane, $Y \geq 0$.

6.14 THREE-DIMENSIONAL ELASTODYNAMIC CRACK PROBLEMS

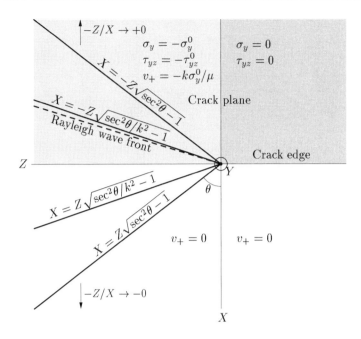

Fig. 6.14.2 Section through $Y = 0$, showing the lower crack face, $X < 0$ (shadowed), and the intersections of the two conical head waves. Positive Y direction points toward the reader. The Rayleigh wave on the crack face is shown as a dashed line. The only load in Problem B is the normal traction $\sigma_y = -\sigma_y^0$ on the crack faces to the left of $Z=0$, and in Problem C it is the shear traction $\tau_{yz} = -\tau_{yz}^0$.

Solution of Problems B and C

Introduce the double Laplace transform

$$F = \mathcal{L}_{qX}\mathcal{L}_{sZ}\phi = q\int_{-\infty}^{+\infty} e^{-qX} s \int_0^\infty e^{-sZ} \phi(X, Y, Z) \mathrm{d}Z\, \mathrm{d}X \tag{6.14.7}$$

and, similarly, the transforms

$$\begin{aligned}
\mathcal{L}_{qX}\mathcal{L}_{sZ}\psi_x &= Q_1 & \mathcal{L}_{qX}\mathcal{L}_{sZ}\tau_{xy} &= T_{xy} \\
\mathcal{L}_{qX}\mathcal{L}_{sZ}\psi_y &= Q_2 & \mathcal{L}_{qX}\mathcal{L}_{sZ}\tau_{yz} &= T_{yz} \\
\mathcal{L}_{qX}\mathcal{L}_{sZ}\psi_z &= Q_3 & \mathcal{L}_{qX}\mathcal{L}_{sZ}v_+ &= V_+ \\
\mathcal{L}_{qX}\mathcal{L}_{sZ}\sigma_y &= S_y & \mathcal{L}_{qX}\mathcal{L}_{sZ}w_+ &= W_+
\end{aligned} \tag{6.14.8}$$

Note that one-sided Laplace transforms could be used for transformations with respect to Z, because all stresses and displacements vanish for $Z < 0$. This is important, because it makes it possible to consider the Laplace transform variable s as real and positive; cf. Appendix A5.

After Laplace transformation, equations (6.14.3)-(6.14.4) read:

$$\frac{\mathrm{d}^2 F}{\mathrm{d}Y^2} = [(\sec^2\theta - 1)s^2 - q^2]F \tag{6.14.9}$$

$$\frac{\mathrm{d}^2 Q_i}{\mathrm{d}Y^2} = [(\sec^2\theta/k^2 - 1)s^2 - q^2]Q_i, \quad i = 1, 2, 3 \tag{6.14.10}$$

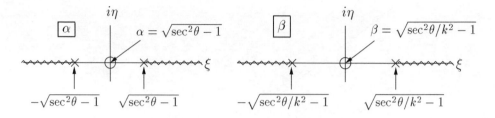

Fig. 6.14.3 Branch cuts for α and β.

Consider s real and positive, and introduce the complex coordinate $\zeta = \xi + i\eta = q/s$ together with the dimensionless quantities

$$\alpha = (\sec^2\theta - 1 - \zeta^2)^{1/2}, \quad \beta = (\sec^2\theta/k^2 - 1 - \zeta^2)^{1/2} \quad (6.14.11)$$

with α and β defined through branch cuts along the real axis in the q plane from their branch points to infinity, and with their branches chosen so that α and β are positive on the imaginary axis; see Fig. 6.14.3. They are then non-negative for all ζ. The acceptable solutions for $Y \geq 0$ are

$$F = Ae^{-\alpha sY}, \quad Q_i = C_i e^{-\beta sY}, \quad i = 1, 2, 3 \quad (6.14.12)$$

where A and C_i are constants, though dependent on ζ and s.

After Laplace transformation of (6.14.5), insertion of (6.14.12) gives

$$\zeta C_1 - \beta C_2 + C_3 = 0 \quad (6.14.13)$$

Next, the boundary conditions (6.14.6) will be used. Expressions for stresses and displacements in terms of ϕ and ψ are found in Appendix A1. After Laplace transformation and insertion of (6.14.12), the condition $\tau_{xy} = 0$, which is common to Problems B and C, reads:

$$2\alpha\zeta A - \zeta C_1 - \beta C_2 - (\beta^2 - \zeta^2)C_3 = 0 \quad (6.14.14)$$

The treatment is now temporarily focused on Problem B. The boundary condition $\tau_{yz} = 0$ leads to:

$$2\alpha A + (\beta^2 - 1)C_1 + \beta\zeta C_2 + \zeta C_3 = 0 \quad (6.14.15)$$

The two remaining boundary conditions for Problem B are mixed on the X axis. It is, however, advantageous to "pretend" that a simple boundary value problem is at hand. To this end, a formal expression, valid for all X, will be given for $(S_y)_{Y=0}$. With reference to the procedure outlined on pages 398ff., the self-similarity suggests the convenient form

$$(S_y)_{Y=0} = \sigma_y^0 \zeta H(\zeta) \quad (6.14.16)$$

where function $H(\zeta)$ is still unknown. This expression leads to the relation

$$(\beta^2 - \zeta^2 - 1)A - 2\beta C_1 + 2\beta\zeta C_3 = \frac{\sigma_y^0}{\mu s^2} \cdot \zeta H(\zeta) \quad (6.14.17)$$

which, together with (6.14.13)-(6.14.15), makes it possible to determine the constants

6.14 THREE-DIMENSIONAL ELASTODYNAMIC CRACK PROBLEMS

A, C_1, C_2, C_3. It is found that $C_2 = 0$, showing that $\psi_y = 0$, a fact which could, perhaps, have been anticipated because of the mode I symmetry.

The two remaining boundary conditions for Problem B will now be used. An expression for $(V_+)_{Y=0}$ is needed. Use of Appendix A1 leads to

$$(V_+)_{Y=0} = -\alpha s A + s C_1 - \zeta s C_3 = \frac{\sec^2 \theta \, \sigma_y^0}{k^2 \mu s} \cdot \frac{\alpha(\zeta) \cdot \zeta H(\zeta)}{R(\zeta, \theta)} \quad (6.14.18)$$

where

$$R(\zeta, \theta) = 4(\sec^2 \theta - 1 - \zeta^2)^{1/2}(\sec^2 \theta / k^2 - 1 - \zeta^2)^{1/2}$$
$$+ (\sec^2 \theta / k^2 - 2 - 2\zeta^2)^2 = 4\alpha\beta(\zeta^2 + 1) + (\beta^2 - \zeta^2 - 1)^2 \quad (6.14.19)$$

Write now, in order to obtain formal similarity with (6.11.4),

$$(V_+)_{Y=0} = -\frac{k^2 \sigma_y^0}{\mu s} \cdot \frac{\zeta}{\sec \theta} \cdot G(\zeta), \quad G(\zeta) = \frac{\sec^3 \theta \, \alpha(\zeta) H(\zeta)}{k^4 R(\zeta, \theta)} \quad (6.14.20)$$

It may be noted that $R(\zeta, \theta)$ is related to the Rayleigh function $R(\cdot)$. In fact, it is possible to write

$$R(\zeta, \theta) = -\frac{\sec^4 \theta}{k^4} \chi^4 R(1/\chi), \quad \chi^2 = \frac{\zeta^2 + 1}{\sec^2 \theta} \quad (6.14.21)$$

so that

$$G(\zeta) = \frac{(1 - \chi^2)^{1/2} H(\zeta)}{\chi^4 R(1/\chi)} \quad (6.14.22)$$

which shows the analogy with (6.11.4).

Inversions are obtained according to (6.9.17), so that

$$\left(\frac{\partial \sigma_y}{\partial Z}\right)_{Y=0} = -\frac{\sigma_y^0}{2\pi i X}[H_+(\xi) - H_-(\xi)]_{\xi=-Z/X} \quad (6.14.23)$$

$$\left(\frac{\partial^2 v_+}{\partial Z^2}\right)_{Y=0} = -\frac{k^2 \sigma_y^0}{2\pi \mu \sec \theta \, iX}[G_+(\xi) - G_-(\xi)]_{\xi=-Z/X} \quad (6.14.24)$$

The last two of the boundary conditions (6.14.6) for Problem B may be written as

$$\left(\frac{\partial \sigma_y}{\partial Z}\right)_{Y=0} = -\sigma_y^0 \delta(Z) = -\sigma_y^0 \delta(\xi|X|) = \frac{\sigma_y^0}{X}\delta(\xi) \text{ for } \xi \geq 0 \quad (6.14.25)$$

$$\left(\frac{\partial^2 v_+}{\partial Z^2}\right)_{Y=0} = 0 \text{ for } \xi < 0 \quad (6.14.26)$$

This gives

$$H_+(\xi) - H_-(\xi) = -2\pi i \delta(\xi) \text{ for } \xi \geq 0 \quad (6.14.27)$$
$$G_+(\xi) - G_-(\xi) = 0 \text{ for } \xi < 0 \quad (6.14.28)$$

Note the surprising analogy with (6.11.5)-(6.11.6). Thus, a three-dimensional problem gives rise to equations that are very similar to those for a two-dimensional problem! In fact, the only difference, after specializing to a stationary crack and a crack face load extending to $x = -\infty$ in (6.11.5), is that the functions $G(\zeta)$ are not identical, but their structure is the same. The continued treatment is therefore closely similar to that for two-dimensional crack face loading, pages 440ff., and only a few steps will be shown.

The last equation can be written in the form

$$C_+(\xi)H_+(\xi) - C_-(\xi)H_-(\xi) = 0 \text{ for } \xi < 0, \quad C(\zeta) = \frac{\alpha(\zeta)}{R(\zeta,\theta)} \tag{6.14.29}$$

Note that

$$\frac{C_-(\xi)}{C_+(\xi)} = \begin{cases} -1 & \text{for } \xi < -\sqrt{\sec^2\theta/k^2 - 1} \\ -e^{2\pi i g(\xi)} & \text{for } -\sqrt{\sec^2\theta/k^2 - 1} < \xi < -\sqrt{\sec^2\theta - 1} \\ +1 & \text{for } -\sqrt{\sec^2\theta - 1} < \xi < 0 \end{cases} \tag{6.14.30}$$

where

$$g(\xi) = \frac{1}{\pi}\operatorname{atan}\frac{4|\alpha|\beta(\xi^2+1)}{(\sec^2\theta/k^2 - 2 - 2\xi^2)^2} \tag{6.14.31}$$

Then, defining

$$c(\xi) = \begin{cases} C_-(\xi)/C_+(\xi) & \text{for } \xi < 0 \\ 1 & \text{for } \xi \geq 0 \end{cases} \tag{6.14.32}$$

it is possible to write

$$H_+(\xi) - c(\xi)H_-(\xi) = 2\pi i\delta(\xi) \text{ for all } \xi \tag{6.14.33}$$

Then, following the procedure outlined on pages 440ff., it is found that

$$c(\xi) = \frac{Q_-(\xi)}{Q_+(\xi)} \tag{6.14.34}$$

where

$$Q(\zeta) = \frac{\left(\sqrt{\sec^2\theta - 1} + \zeta\right)^{1/2}}{S(\zeta,\theta)} \tag{6.14.35}$$

$$S(\zeta,\theta) = \exp\left[-\int_{\sqrt{\sec^2\theta - 1}}^{\sqrt{\sec^2\theta/k^2 - 1}} \frac{g(u)\,du}{u + \zeta}\right] \tag{6.14.36}$$

The function $S(\zeta,\theta)$, which possesses branch points at $\zeta = -\sqrt{\sec^2\theta/k^2 - 1}$ and at $\zeta = -\sqrt{\sec^2\theta/k^2 - 1}$, is defined through a branch cut between these points, outside which the function is analytic. The branch is chosen so that $S(\pm\infty,\theta) = 1$. Then, it follows that

$$\frac{S_+(\xi,\theta)}{S_-(\xi,\theta)} = \begin{cases} 1 & \text{for } \xi < -\sqrt{\sec^2\theta/k^2 - 1} \\ e^{2\pi i g(\xi)} & \text{for } -\sqrt{\sec^2\theta/k^2 - 1} < \xi < -\sqrt{\sec^2\theta - 1} \\ 1 & \text{for } -\sqrt{\sec^2\theta - 1} < \xi \end{cases} \tag{6.14.37}$$

which, together with (6.14.35), confirms relation (6.14.34).

Equation (6.14.33) can be written in the form

$$Q_+(\xi)H_+(\xi) - Q_-(\xi)H_-(\xi) = -2\pi i Q_+(0)\delta(\xi) \tag{6.14.38}$$

6.14 THREE-DIMENSIONAL ELASTODYNAMIC CRACK PROBLEMS

The solution is found by using the Plemelj formulae, (A4.45)-(A4.46), and adding a polynomial to consider solutions of the homogeneous part. Thus,

$$Q(\zeta)H(\zeta) = -Q_+(0)\int_{-\infty}^{\infty}\frac{\delta(u)du}{u-\zeta} + polynomial \qquad (6.14.39)$$

which leads to

$$H(\zeta) = \frac{\sqrt[4]{\sec^2\theta - 1}S(\zeta,\theta)}{S(0,\theta)(\sqrt{\sec^2\theta - 1} + \zeta)^{1/2}}\left[\frac{1}{\zeta} + P_0\right] \qquad (6.14.40)$$

$$G(\zeta) = -\frac{\sec^3\theta\sqrt[4]{\sec^2\theta - 1}(\sqrt{\sec^2\theta - 1} - \zeta)^{1/2}S(\zeta,\theta)}{k^4 S(0,\theta)R(\zeta,\theta)}\left[\frac{1}{\zeta} + P_0\right] \qquad (6.14.41)$$

because the added polynomial, as for two-dimensional crack face loading, turns out to be a constant, P_0, which is found to be $k_R/\sqrt{\sec^2\theta - k_R^2}$ from the condition that $G(\zeta)$ must not have a pole for $\xi > 0$.

The stress $(\sigma_y)_{Y=0}$ and the displacement $(v_+)_{y=0}$ are now found (as derivatives) from (6.14.23)-(6.14.24). Thus, the stress derivative

$$\left(\frac{\partial \sigma_y}{\partial Z}\right)_{Y=0} = \frac{\sigma_y^0}{2\pi i X} \cdot \frac{\sqrt[4]{\sec^2\theta - 1}(\sqrt{\sec^2\theta - k_R^2} + k_R\zeta)}{\sqrt{\sec^2\theta - k_R^2}\,S(0,\theta)\,\zeta}$$

$$\times \left[\frac{S_+(\zeta,\theta)}{(\sqrt{\sec^2\theta - 1} + \zeta)_+^{1/2}} - \frac{S_-(\zeta,\theta)}{(\sqrt{\sec^2\theta - 1} + \zeta)_-^{1/2}}\right] \qquad (6.14.42)$$

In the crack edge vicinity, $\xi \to -\infty$, the expression reduces to

$$\left(\frac{\partial \sigma_y}{\partial Z}\right)_{Y=0} \to \frac{\sigma_y^0}{\pi} \cdot \frac{k_R\sqrt[4]{\sec^2\theta - 1}}{\sqrt{\sec^2\theta - k_R^2}\,S(0,\theta)} \cdot \frac{1}{\sqrt{ZX}} \qquad (6.14.43)$$

Integration gives

$$(\sigma_y)_{Y=0} \to \frac{2\sigma_y^0}{\pi} \cdot \frac{k_R\sqrt[4]{\sec^2\theta - 1}}{\sqrt{\sec^2\theta - k_R^2}\,S(0,\theta)} \cdot \frac{\sqrt{Z}}{\sqrt{X}} \quad \text{as } X \to +0,\ Z > 0 \qquad (6.14.44)$$

from which the stress intensity factor is found to be

$$K_I = \frac{2\sqrt{2}\sigma_y^0 k_R\sqrt[4]{\sec^2\theta - 1}\sqrt{Z}}{\sqrt{\pi}\sqrt{\sec^2\theta - k_R^2}\,S(0,\theta)} = \frac{2\sqrt{2}\sigma_y^0 k_R\sqrt[4]{\sec^2\theta - 1}\sqrt{z + \sec\theta\,c_P t}}{\sqrt{\pi}\sqrt{\sec^2\theta - k_R^2}\,S(0,\theta)} \qquad (6.14.45)$$

It is possible to determine $S(0,\theta)$ in terms of primary quantities. To this end the integration variable change to $s = (u^2 + 1)/\sec^2\theta$ is introduced in (6.14.36), after putting $\zeta = 0$, giving

$$S(0,\theta) = \exp\left[\frac{\sec^2\theta}{2\pi}\int_{k^2}^{1}\operatorname{atan}\frac{4k^3\sqrt{1-s}\sqrt{s-k^2}}{(s-2k^2)^2} \cdot \frac{ds}{s(s-\sec^2\theta)}\right] \qquad (6.14.46)$$

Note the similarity with the expression for $S(0)$ in the two-dimensional (2D) case, (6.11.38). This similarity, which will now be exploited, motivated the choice of the integration variable s, trusting that there will be no confusion with the Laplace transform variable s, previously used in the present section. Thus,

$$4\ln S(0,\theta) = -\sec^2\theta \int_\Gamma \frac{\phi(s)}{(s-\epsilon)(s-\sec^2\theta)}ds \qquad (6.14.47)$$

where $\phi(s)$ is given by (6.11.39) and the path Γ encircles all singular points, which are the same as in the 2D case, except for the addition of a pole at $s = \sec^2\theta$. The integration proceeds in the same manner as for the 2D case. Using the same notations as for this case, it results in

$$4\ln S(0,\theta) = \ln\frac{k_R^4|v_1^2 - \sec^2\theta|\cdot|v_2^2 - \sec^2\theta|a(\sec^2\theta)}{(1-k^2)^2\sec^2\theta(\sec^2\theta - k_R^2)b(\sec^2\theta)} \qquad (6.14.48)$$

and then, noting that $a(s)b(s) = s(s - k_R^2)(s - v_1^2)(s - v_2^2)$ and that $a(\sec^2\theta) = -R(\sec\theta)$, where $R(\cdot)$ is the Rayleigh function (A9.57), the final result is

$$S(0,\theta) = \frac{k_R\sqrt{|R(\sec\theta)|}}{k\sqrt{2(1-k^2)}\sec\theta\sqrt{\sec^2\theta - k_R^2}} \qquad (6.14.49)$$

Insertion of the expression for $S(0,\theta)$ into (6.14.45), gives

$$K_I = \frac{4\sigma_y^0 k\sqrt{1-k^2}\sec\theta\sqrt[4]{\sec^2\theta - 1}\sqrt{z + \sec\theta\cdot c_P t}}{\sqrt{\pi|R(\sec\theta)|}} \qquad (6.14.50)$$

For normal impact, $\theta = \pi/2$ ($\sec\theta \to \infty$), this expression coincides with (6.11.45) for $\beta \to 0$, as expected (the sign difference depends on different sign of the crack face load). Expressing k in terms of Poisson's ratio, ν, leads to

$$K_I = \frac{2\sigma_P^0[(1-\nu)\tan^2\theta + \nu]\sqrt{\sin\theta}\sqrt{c_P t + \cos\theta\cdot z}}{(1-\nu)^2\sqrt{\pi|R(1/\cos\theta)|}} \qquad (6.14.51)$$

which result was given be Freund (1990), although in different form.

The result shows that the stress intensity factor disappears as $\theta \to 0$, i.e. at grazing incidence. This goes against intuition, because the incident wave carries a lateral stress, suggesting opening of the crack. The phenomenon is closely related to the one of grazing incidence of a plane P wave on the free surface of a half-plane: assumption of steady state implies that the reflected wave annihilates the incident wave, leaving the body stress and displacement free. A non-steady state excitation of the incident wave (for instance by imposing a suddenly applied uniformly distributed body force on a plane parallell with the intended wave front) would prevent such annihilation at finite distances from the plane of excitation.

Problem C is rather similar to Problem B. After writing

$$(T_{yz})_{Y=0} = \tau_{yz}^0 \zeta H(\zeta) \qquad (6.14.52)$$

equations (6.14.15) and (6.14.17) are changed to

$$(\beta^2 - \zeta^2 - 1)A - 2\beta C_1 + 2\beta\zeta C_3 = 0 \qquad (6.14.53)$$

$$-2\alpha A + (1-\beta^2)C_1 - \beta\zeta C_2 - \zeta C_3 = \frac{\tau_{yz}^0}{\mu s^2}\cdot \zeta H(\zeta) \qquad (6.14.54)$$

These equations, together with (6.14.12) and (6.14.14) determine A, C_1, C_2, C_3 (here all four are non-zero), whereupon it follows that

$$(W_+)_{Y=0} = sA + qC_2 + \beta C_3 = \frac{\tau_{yz}^0}{\mu s} \cdot \frac{T(\zeta,\theta) \cdot \zeta H(\zeta)}{R(\zeta,\theta)} \quad (6.14.55)$$

$$T(\zeta,\theta) = 4\alpha\beta\zeta^2 + 2\beta^2 + (\beta^2 - \zeta^2 - 1)(\beta^2 - \zeta^2) \quad (6.14.56)$$

Following essentially the same procedure as for Problem B, the resulting mode III stress intensity factor, given by Freund (1990), is found to be

$$K_{III} = \frac{2\sqrt{2}\tau_{yz}^0 \sqrt[4]{1 - k^2 \cos^2\theta}\sqrt{c_S t + k\cos\theta \cdot z}}{\cos^2\theta\sqrt{\pi|R(1/\cos\theta)|}}$$

$$= \frac{4\sqrt{2}\sigma_P^0 k \tan\theta \sqrt[4]{1 - k^2 \cos^2\theta}\sqrt{c_S t + k\cos\theta \cdot z}}{\sqrt{\pi|R(1/\cos\theta)|}} \quad (6.14.57)$$

For normal incidence, $\theta = \pi/2$, the first of these expressions agrees with (6.10.24) for $\gamma = 0$, as expected.

Three-dimensional face loading on semi-infinite cracks

Although the problem of oblique wave incidence on a crack edge is three-dimensional in the sense that stresses and displacements depend on all three coordinates, x, y, z, the solution procedure turned out to be essentially the same as for related two-dimensional problems. The z dependence could be suppressed by the Galilean transformation. In more general cases of crack face loading, the z dependence may be suppressed by Laplace transformation, but the three-dimensional character re-appears during the inversion. The simplest (though not simple) cases concern problems in which no length parameter is involved. Such cases were discussed by Freund (1987, 1990), who also solved the particular problem of opposed suddenly applied uniform normal line loads along the negative x-axis on the faces of a semi-infinite crack $x < 0, y = 0$. After Laplace transformations with respect to t (time), x and z, Freund formulated and solved a Wiener-Hopf equation. The solution shows certain similarities with the two-dimensional problem of opposed normal forces suddenly applied on the crack faces at a distance from the crack edge, pages 451ff. Thus, only mode I deformation results, and after the P wave arrival at a point on the crack edge, the stress intensity factor for that point becomes initially negative and reaches negative infinity at the arrival of the Rayleigh wave, whereupon it increases to reach positive values. The implications of such results were discussed on page 440.

A related problem, concentrated and opposed normal forces, suddenly appearing at the edge of a moving semi-infinite crack and then remaining stationary while the crack edge moves away, was solved by Champion (1988). A stationary crack subjected to concentrated and opposed normal forces, suddenly appearing at the crack edge and moving on the crack faces away from the edge was considered by Ramirez (1987). Champion and Ramirez used the technique devised by Freund (1987) and their solutions show the same general features.

A more advanced problem, concentrated and opposed shear forces, suddenly appearing on the faces of a stationary semi-infinite crack at a distance from the crack edge, was solved by Li and Liu (1995). They considered shear force directions both parallel and normal to the crack edge, and they used Laplace transforms combined with the Wiener-Hopf technique. No advantage can be taken of similarity, because a fixed length is involved. An additional difficulty appears because the resulting deformation modes, II and III, are coupled. This leads to matrix Wiener-Hopf formulation, which in general is much more difficult to handle than scalar formulation, cf. pages 170ff. However, the particular matrix encountered could be transformed "after some manipulation" into a diagonal matrix, whereupon standard Wiener-Hopf technique could be used. The results show rather complex stress intensity factors, exhibiting initial oscillations until eventually their static values are approached. The oscillations involve sign changes, but, in contrast to mode I, these are physically acceptable for modes II and III. However, if friction is present, obvious physical complications occur, rendering the validity of the solutions questionable.

A problem involving a crack $|x| < a$, $y = 0$, situated in an infinite linearly elastic solid and subjected to face tractions of the form $\sigma_y = g(x,z)f(t)$, where $g(x,z)$ is symmetric with respect to x, was analysed by Li and Liu (1994). They used a Laplace transform technique to establish a Fredholm equation in the Laplace transform variable, and obtained solutions for the stress intensity factors after numerical inversion of Laplace transforms.

Arbitrary time-dependent loading was considered by Willis and Movchan (1995, 1997) and by Movchan and Willis (1995) for a semi-infinite crack, moving at constant speed. They constructed dynamic weight functions for opening (Willis and Movchan 1995) and shear (Movchan and Willis 1995) modes. They used Fourier transforms together with the Wiener-Hopf technique. The coupling between modes II and III lead to matrix rather than scalar formulation of the Wiener-Hopf equations, but, as in the case considered by Li and Liu (1995), a reasonably friendly matrix was encountered, and an explicit solution could be obtained.

Willis and Movchan (1995) and Movchan and Willis (1995) also applied an in-plane first-order perturbation of the wave front. Thus, the crack occupies the region $x - Vt < \epsilon\phi(t,z)$, $y = 0$, where V is the unperturbed velocity and ϵ is sufficiently small to permit first-order perturbation theory when $\phi(t,z)$ is given. They were able to give explicit results for changes of the stress intensity factors. These changes are proportional to ϵ. Mode I changes are decoupled from modes II and III, whereas changes in modes II and III are coupled. The same applies to the unperturbed stress intensity factors. By letting $V \to 0$, Movchan and Willis (1995) recovered the results of Gao and Rice (1986); see page 201.

The results by Willis and Movchan (1995) were used by Ramanathan and Fischer (1997) to consider crack front waves, i.e. waves propagating along a crack front. Such waves will be discussed later (page 505).

Willis and Movchan (1997) considered out-of-plane perturbations. Thus, the crack occupies the region $x - Vt < 0$, $y = \epsilon\psi(x,z)$, where ϵ is sufficiently small to permit first-order perturbation theory when $\psi(x,z)$ is given. Note that ψ is a function of x and z, but not of t, so that the crack faces become permanently "wrinkled". Examples of such wrinkles are the Wallner lines appearing on fracture surfaces of glass (Wallner 1939). Willis and Movchan were able to calculate the first-order perturbation of all three stress intensity factors. Because linearity prevails, these results can be added to the results for in-plane crack front perturbation, to give a complete solution for the stress intensity factors at a general perturbation of the crack front.

Finally, it should be noted that physical reasons may demand that propagation of a semi-infinite crack with straight front proceeds under constant stress intensity factor along the front, i.e., three-dimensional crack face loading might not be compatible with a moving straight front. Three-dimensional load perturbations of two-dimensional loading, resulting in perturbations of crack front straightness may, however, be allowed. The following subsection will address such problems and the problem of dynamic planar crack expansion with curved front.

General three-dimensional loading on planar cracks

Several elastodynamic crack problems of practical significance involve crack propagation in non-homogeneous media, for instance such that contain particles, like most metal alloys, or such that contain larger inclusions or obstacles to wave propagation, like seismic fault regions. The combination of three-dimensionality and inhomogeneity offers little hope of arriving at meaningful analytical solutions, but even numerical treatments might appear to be forbiddingly laborious and costly. However, as shown by Rice and coworkers (Perrin et al. 1995, Geubelle and Rice 1995, Cochard and Rice 1997, and Morrisey and Rice 1998), the amount of work can be considerably reduced for planar crack propagation, by working with spectral representations of relations between tractions and displacement discontinuities. These relations can be handled to great length for each Fourier term independently of the others, an ideal setting for massively parallel computers. As in most linear crack problems, partition into different modes is advantageous. For the three-dimensional case, where the crack front in general is not straight, modes II and III are coupled, so that partition is made into two parts, one opening mode (mode I) and one sliding mode (mixed modes II and III). The relations between tractions and displacement discontinuities will be analysed first for the opening mode and then for the sliding mode.

The opening mode: the half-space problem for one Fourier term
The crack plane is assumed to be $y = 0$, and there is one or more mode I cracks on this plane. Consider one term in a double Fourier series with wave numbers m for the x dependence and n for the z dependence, so that the crack opening $\delta_y(x, z, \tau)$ takes the form $D_y(\tau; m, n)e^{imx+inz}$, where $\tau = c_P t$, and the traction needed to maintain this opening, $\sigma_y^0(x, z, \tau)$, can be written as $s_y^0(\tau; m, n)e^{imx+inz}$. All other stresses and displacements can be written in a similar way. No dynamic loads are acting before $t = 0$. There is no need to worry about material interpenetration until the final Fourier synthesis is accomplished. Due to the symmetry, it suffices to consider the upper half-space, $y \geq 0$. The relation between $s_y^0(\tau; m, n)$ and $D_y(\tau; m, n)$ is sought.

The three-dimensional character of the problem implies that three potentials, ϕ, ψ_x and ψ_z, are needed, obeying the equations of motion (A9.16)-(A9.17),

$$\Delta \phi = \frac{\partial^2 \phi}{\partial \tau^2} \tag{6.14.58}$$

$$\Delta \psi_x = \frac{1}{k^2} \cdot \frac{\partial^2 \psi_x}{\partial \tau^2}, \quad \Delta \psi_z = \frac{1}{k^2} \cdot \frac{\partial^2 \psi_z}{\partial \tau^2} \tag{6.14.59}$$

With obvious notations, the following Laplace transforms are introduced:

$$\mathcal{L}_{p\tau} v_+(x,y,z,\tau) = V_+(y;p,m,n)e^{imx+inz} \tag{6.14.60}$$

$$\mathcal{L}_{p\tau} \sigma_y(x,y,z,\tau) = S_y(y;p,m,n)e^{imx+inz} \tag{6.14.61}$$

$$\mathcal{L}_{p\tau} \tau_{xy}(x,y,z,\tau) = T_{xy}(y;p,m,n)e^{imx+inz} \tag{6.14.62}$$

$$\mathcal{L}_{p\tau} \tau_{yz}(x,y,z,\tau) = T_{yz}(y;p,m,n)e^{imx+inz} \tag{6.14.63}$$

$$\mathcal{L}_{p\tau} \phi(x,y,z,\tau) = \Phi(y;p,m,n)e^{imx+inz} \tag{6.14.64}$$

$$\mathcal{L}_{p\tau} \psi_x(x,y,z,\tau) = \Psi_x(y;p,m,n)e^{imx+inz} \tag{6.14.65}$$

$$\mathcal{L}_{p\tau} \psi_z(x,y,z,\tau) = \Psi_z(y;p,m,n)e^{imx+inz} \tag{6.14.66}$$

After Laplace transformation, the equations of motion yield the acceptable solutions

$$\Phi = \Phi_0(p,m,n)e^{-(p^2+q^2)^{1/2}y} \tag{6.14.67}$$

$$\Psi_x = \Psi_x^0(p,m,n)e^{-(p^2/k^2+q^2)^{1/2}y} \tag{6.14.68}$$

$$\Psi_z = \Psi_z^0(p,m,n)e^{-(p^2/k^2+q^2)^{1/2}y} \tag{6.14.69}$$

where $q = \sqrt{m^2+n^2}$.

For convenience, introduce by analogy with (6.2.11)-(6.2.12), the abbreviations $A_P = (p^2+q^2)^{1/2}$ and $A_S = (p^2/k^2+q^2)^{1/2}$. Let subscript or superscript 0 denote quantities on $y=0$, so that $V_+^0(p,m,n) = V_+(0;p,m,n)$, etc. Use of Appendix A1 leads to the following expressions for $y=0$:

$$V_+^0(p,m,n) = -A_P\Phi_0 + in\Psi_x^0 - im\Psi_z^0 \tag{6.14.70}$$

$$S_y^0(p,m,n) = \mu[(A_S^2+q^2)\Phi_0 - 2inA_S\Psi_x^0 + 2imA_S\Psi_z^0] \tag{6.14.71}$$

$$T_{xy}^0(p,m,n) = \mu[-2imA_P\Phi_0 - mn\Psi_x^0 + (A_S^2+m^2)\Psi_z^0] \tag{6.14.72}$$

$$T_{yz}^0(p,m,n) = \mu[-2inA_P\Phi_0 - (A_S^2+n^2)\Psi_x^0 + mn\Psi_z^0] \tag{6.14.73}$$

Due to mode I symmetry, $T_{xy}^0 = T_{yz}^0 = 0$. This condition gives

$$\Psi_x^0 = -\frac{2inA_P}{A_S^2+q^2}\Phi_0, \quad \Psi_z^0 = \frac{2imA_P}{A_S^2+q^2}\Phi_0 \tag{6.14.74}$$

Insertion into (6.14.70) and (6.14.71) leads to the relation

$$\frac{1}{k^2\mu}S_y^0 = \frac{1}{p} \cdot S(p,q)V_+^0(p,m,n) \tag{6.14.75}$$

where

$$S(p,q) = [4q^2A_PA_S - (A_S+q^2)^2]/(pA_P) \tag{6.14.76}$$

It may be noted that the expression within [] equals $q^4R(ip/q)/k^4$, where $R(\cdot)$ is the Rayleigh function, (A9.57).

According to (A5.10) the inversion of (6.14.75) may be written as

$$s_y^0(\tau;m,n) = \frac{k^2\mu}{2}\int_0^\tau [\mathcal{L}_{p\tau}^{-1}S(p,q)]_{\tau\to\tau'} D_y(\tau-\tau';m,n)d\tau' \tag{6.14.77}$$

where $\mathcal{L}_{p\tau}^{-1}S(p,q)$ is found after first writing

$$S(p,q) = \frac{4q^2p}{k^2A_S} + \frac{4q^4}{pA_S} - \frac{p^3}{k^4A_P} - \frac{4q^2p}{k^2A_P} - \frac{4q^4}{pA_P} \tag{6.14.78}$$

6.14 THREE-DIMENSIONAL ELASTODYNAMIC CRACK PROBLEMS

and then using the general relations (A5.17), (A5.21) and (A5.3) applied to the inversion formula

$$\mathcal{L}_{p\tau}^{-1} \frac{p}{(p^2+1)^{1/2}} = J_0(\tau)U(\tau) \tag{6.14.79}$$

Here, $U(\tau)$ is the unit step function and $J_0(\tau)$ the Bessel function of zeroth order. This inversion may be found in common Laplace transform tables. Then, using the recursion formula $J_0''(\tau) = -J_1'(\tau) = J_1(\tau)/\tau - J_0(\tau)$ (e.g. Gradshteyn and Ryzhik 1980), it follows that

$$\mathcal{L}_{p\tau}^{-1} \frac{p^3}{(p^2+1)^{1/2}} = J_0''(\tau)U(\tau) + \delta'(\tau) = \left[\frac{J_1(\tau)}{\tau} - J_0(\tau)\right]U(\tau) + \delta'(\tau) \tag{6.14.80}$$

$$\mathcal{L}_{p\tau}^{-1} \frac{1}{p(p^2+1)^{1/2}} = \int_0^\tau \int_0^s J_0(u)du\,ds\,U(\tau) = \int_0^\tau \int_0^s \frac{J_1(u)}{u} du\,ds\,U(\tau)$$
$$+ \int_0^\tau \int_0^s J_1'(u)du\,ds\,U(\tau) = \tau \int_0^\tau \frac{J_1(s)}{s} ds\,U(\tau) \tag{6.14.81}$$

where $\delta(\cdot)$ is the Dirac delta function. A partial integration with respect to s was performed on the middle double integral.

The result of the inversion of $S(p,q)$ is

$$\mathcal{L}_{p\tau}^{-1} S(p,q) = \frac{q^2}{k}\left[\frac{J_1(q\tau)}{q\tau} + 4kq\tau \int_{kq\tau}^{q\tau} \frac{J_1(s)}{s} ds\right.$$
$$\left. + \left(\frac{4}{k} - \frac{1}{k^3}\right)J_0(q\tau) - 4J_0(kq\tau)\right]U(\tau) - \frac{q^2}{k^4}\delta'(q\tau)$$
$$= -\frac{q^2}{k}C_I(kq\tau)U(\tau) - \frac{1}{k^4}\delta'(\tau) \tag{6.14.82}$$

where the notation $C_I(\cdot)$ follows Geubelle and Rice (1995).

Finally, the inversion (6.14.77) gives

$$s_y^0(\tau;m,n) = -\frac{\mu}{2k^2}\dot{D}_y(\tau;m,n) - \frac{\mu}{2}kq^2\int_0^\tau C_I(kq\tau')D_y(\tau-\tau';m,n)d\tau' \tag{6.14.83}$$

where a dot denotes differentiation with respect to τ. This is the relation sought between traction and displacement on the crack plane. Note that $C_I(kq\tau)$ is a mode I convolution kernel.

The opening mode: Fourier synthesis and numerical scheme

Consider now a rectangular part $-L/2 < x < L/2$, $-H/2 < z < H/2$, of the crack plane, large enough around the region of central interest that no disturbances will arrive from the outside during the time of interest. Make the notation change $m \to 2\pi m/L$, $n \to 2\pi n/H$, implying that the new quantities m and n are integers. Fourier synthesis gives

$$\delta_y(x,z,\tau) = \sum_{m=-M/2}^{M/2} \sum_{n=-N/2}^{N/2} D_y(\tau;m,n)e^{2\pi i(mx/L+nz/H)} \quad (6.14.84)$$

$$f_y(x,z,\tau) = \sum_{m=-M/2}^{M/2} \sum_{n=-N/2}^{N/2} F_y(\tau;m,n)e^{2\pi i(mx/L+nz/H)} \quad (6.14.85)$$

where $M = L/(2\pi)$, $N = H/(2\pi)$, the function $f_y(x,z,\tau)$ is dependent on the previous history of crack opening, and

$$F_y(\tau;m,n) = -\frac{\mu}{2}kq^2 \int_0^\tau C_I(kq\tau')D_y(\tau-\tau';m,n)\mathrm{d}\tau' \quad (6.14.86)$$

M and N, which must be even, should be chosen as powers of two in order to take advantage of the FFT (Fast Fourier Transform) algorithm.

In the derivation of the relation between tractions and crack openings, it was tacitly assumed that only crack face loading prevails. In order also to allow offside loading such as from stress waves (obeying mode I symmetry), a component $\sigma_y^{00}(x,z,\tau)$ has to be added, being equal to the normal stress that would prevail on the crack plane if crack opening were prevented. This is the same kind of superposition as previously used for stress wave loading and illustrated in Fig. 6.13.2. The added term may also include a static load prevailing before $t = 0$. The most common situation for mode I consists of traction free crack faces, except near the crack front, where a cohesive model like the Barenblatt model should be assumed, or at an obstacle (assumed to be flat and situated in the crack plane), where some other cohesive model may be appropriate. The relation between tractions and crack opening now reads

$$\sigma_y^0(x,z,\tau) = \sigma_y^{00}(x,z,\tau) - \frac{\mu}{2k^2}\dot\delta_y(x,z,\tau) + f_y(x,z,\tau) \quad (6.14.87)$$

and (6.14.83) changes to

$$s_y^0(\tau;m,n) = s_y^{00}(\tau;m,n) - \frac{\mu}{2k^2}\dot D_y(\tau;m,n)$$
$$- \frac{\mu}{2}kq^2 \int_0^\tau C_I(kq\tau')D_y(\tau-\tau';m,n)\mathrm{d}\tau' \quad (6.14.88)$$

where $s_y^{00}(\tau;m,n)$ is the Fourier coefficient with wave numbers m and n for $\sigma_y^{00}(x,z,\tau)$.

The numerical procedure is described by Geubelle and Rice (1995). Note that the load term $\sigma_y^{00}(x,z,\tau)$ is considered to be known. It involves time-stepping with steps $\Delta\tau$ in the following sequence for sample points x, z:
(1) Calculate $\delta_y(x,z,\tau) = \delta_y(x,z,\tau-\Delta\tau) + \Delta\tau\dot\delta_y(x,z,\tau-\Delta\tau)$.
(2) Calculate the cohesive stress from the cohesive model, $\sigma_y^0(x,z,\tau) = g(x,z;\delta_y)$, for points where cohesion prevailed at the previous step. For other points $\sigma_y^0(x,z,\tau) = 0$ (assuming that no loads are acting on open crack faces).
(3) Use FFT to determine the Fourier coefficients $D_y(\tau;m,n)$ from (6.14.84).
(4) Calculate $F_y(\tau;m,n)$ from (6.14.86).
(5) Use inverse FFT to find f_y from (6.14.85).
(6) Find $\dot\delta_y(x,z,\tau)$ from (6.14.87). Three possibilities exist:

a if complete decohesion has previously occurred, then $\dot{\delta}_y(x,z,\tau)$ is obtained from (6.14.87) after putting $\sigma_y^0(x,z,\tau) = 0$ (assuming that no loads are acting on open crack faces).

b if cohesion prevailed in the previous step, but the stress $\sigma_y^0(x,z,\tau) < \sigma_y^{00}(x,z,\tau)$ $+ f_y(x,z,\tau)$, then cohesion has ceased, so that $\sigma_y^0(x,z,\tau)$ should be set to zero, whereupon $\dot{\delta}_y(x,z,\tau)$ is obtained from (6.14.87).

c if cohesion prevailed in the previous step, and the stress $\sigma_y^0(x,z,\tau) > \sigma_y^{00}(x,z,\tau)$ $+ f_y(x,z,\tau)$, then cohesion still prevails, so that $\dot{\delta}_y(x,z,\tau) = 0$.

(7) Repeat the sequence.

Modifications are needed if the dependence of the cohesive relation on the crack velocity is taken into account. In the simplest case, this can be achieved by incorporating the opening velocity $\dot{\delta}_y$ into the cohesive relation, so that function $g = g(x,z;\delta_y,\dot{\delta}_y)$. The dependence on $\dot{\delta}_y$, which has to be found empirically, is, in general, such that g increases with $\dot{\delta}_y$, for some materials very strongly, but it is questionable whether a unique relation exists, cf. Section 9.1.

Some results for the opening mode
The method has been used by Geubelle and Rice (1995) and by Morrissey and Rice (1998) to produce some stunning results. Geubelle and Rice studied the passage of a mode I crack front through two asperities (tough obstacles) at the side of each other. The result shows how the front is first retarded at each asperity, but, after rupture of the asperities, the retarded parts of the crack front accelerate and actually jump ahead of the rest of the front.

Morrisey and Rice (1998) demonstrated how a disturbance, caused by an asperity, is propagating along a moving mode I crack front. This phenomenon shows one remarkable feature: the disturbance does not seem to decay during its motion (in two opposite directions) along the front – Morrisey and Rice call it a "persistent" wave. The velocity of the wave from the point of origin (the asperity) on the crack face was found to be slightly smaller than the Rayleigh wave velocity, c_R, in agreement with a result obtained in a different way by Ramanathan and Fischer (1997). These authors used the general results by Movchan and Willis (1995) on crack speed perturbation along the crack front. They were able to estimate the velocity V_f of the disturbance from the source to be such that $\sqrt{V_f^2 - V^2}/\sqrt{c_R^2 - V^2}$, where V is the unperturbed crack front velocity, increases from about 0.94 to 1.00 as V increases from 0 to c_R. Note that $\pm\sqrt{V_f^2 - V^2}$ is the disturbance velocity in the crack front direction.

Two conclusions may be drawn, at least tentatively, from the fact that disturbances proceed along the crack front as persistent waves. One is that each inhomogeneity (which in reality is not completely flat) contributes a trace on the crack face, even far away, so that the crack face will be rather covered with such traces after the passage through several asperities has been negotiated. Another is that the line pattern often seen on the crack faces of fractured glass plates, the Wallner lines (Wallner 1939), may be traces of persistent waves originating from inhomogeneities (Morrisey and Rice 1998).

The sliding mode

The analysis of the sliding mode parallels the one for the opening mode. The crack slips $\delta_x(x,z,\tau)$ and $\delta_z(x,z,\tau)$ take the form $D_x(\tau;m,n)e^{imx+inz}$ and $D_z(\tau;m,n)e^{imx+inz}$, respectively, and the tractions required to maintain these slips can be written as $\tau_{xy}^0(x,z,\tau) = t_{xy}^0(\tau;m,n)e^{imx+inz}$ and $\tau_{yz}^0(x,z,\tau) = t_{yz}^0(\tau;m,n)e^{imx+inz}$. Expressions are sought for $t_{xy}^0(\tau;m,n)$ and $t_{yz}^0(\tau;m,n)$ as functions of $D_x(\tau;m,n)$ and $D_z(\tau;m,n)$.

Let
$$\mathcal{L}_{p\tau} u_+(x,y,z,\tau) = U_+(y;p,m,n)e^{imx+inz} \tag{6.14.89}$$
$$\mathcal{L}_{p\tau} w_+(x,y,z,\tau) = W_+(y;p,m,n)e^{imx+inz} \tag{6.14.90}$$

Use of Appendix A1 then gives
$$U_+^0 = im\Phi^0 - A_S \Psi_z^0, \quad W_+^0 = in\Phi_0 + A_S \Psi_x^0 \tag{6.14.91}$$

These expressions replace expression (6.14.70) for the opening mode, whereas expressions (6.14.71)-(6.14.73) remain unchanged. For the sliding mode, $S_y^0 = 0$. Thus, from (6.14.71),
$$\Psi_x^0 = \frac{C\Phi_0 + 2ik^2 mA_S \Psi_z^0}{2ik^2 n A_S} \tag{6.14.92}$$

which leads to
$$W_+^0 = \frac{A_S^2 + m^2 - n^2}{2in}\Phi_0 + \frac{m}{n}A_S\Psi_z^0 \tag{6.14.93}$$

Now, Φ_0 and Ψ_z^0 can be expressed in U_+^0 and W_+^0:
$$\Phi_0 = \frac{2ik^2}{p^2}(mU_+^0 + nW_+^0) \tag{6.14.94}$$
$$\Psi_z^0 = -\frac{k^2}{p^2 A_S}[(A_S^2 + m^2 - n^2)U_+^0 + 2mnW_+^0] \tag{6.14.95}$$

Insertion of (6.14.92) and (6.14.94)-(6.14.95) into the expressions (6.14.72)-(6.14.73) gives
$$\frac{1}{k^2\mu}T_{xy}^0 = \frac{1}{p}T_{11}U_+^0 + \frac{1}{p}T_{12}W_+^0 = \frac{P-n^2Q}{p^2 A_S}U_+^0 + \frac{mnQ}{p^2 A_S}W_+^0 \tag{6.14.96}$$
$$\frac{1}{k^2\mu}T_{yz}^0 = \frac{1}{p}T_{21}U_+^0 + \frac{1}{p}T_{22}W_+^0 = \frac{mnQ}{p^2 A_S}U_+^0 + \frac{P-m^2Q}{p^2 A_S}W_+^0 \tag{6.14.97}$$

where $P = P(p,q)$ and $Q = Q(p,q)$ are given by the expressions
$$P = 4q^2 A_P A_S - (A_S^2 + q^2)^2, \quad Q = 4A_P A_S - 3A_S^2 - q^2 \tag{6.14.98}$$

It may be noted that $P = q^4 R(ip/q)/k^4$, where $R(\cdot)$ is the Rayleigh function, (A9.57). By analogy with (6.14.77), the inversion can be written as
$$t_{xy}^0(\tau;m,n) = \frac{k^2\mu}{2}\int_0^\tau [\mathcal{L}_{p\tau}^{-1} T_{11}(p,m,n)]_{\tau\to\tau'} D_x(\tau - \tau';m,n)d\tau'$$
$$+ \frac{k^2\mu}{2}\int_0^\tau [\mathcal{L}_{p\tau}^{-1} T_{12}(p,m,n)]_{\tau\to\tau'} D_z(\tau - \tau';m,n)d\tau' \tag{6.14.99}$$

$$t_{yz}^0(\tau; m, n) \frac{k^2 \mu}{2} \int_0^\tau [\mathcal{L}_{p\tau}^{-1} T_{21}(p, m, n)]_{\tau \to \tau'} D_x(\tau - \tau'; m, n) \mathrm{d}\tau'$$

$$+ \frac{k^2 \mu}{2} \int_0^\tau [\mathcal{L}_{p\tau}^{-1} T_{22}(p, m, n)]_{\tau \to \tau'} D_z(\tau - \tau'; m, n) \mathrm{d}\tau' \quad (6.14.100)$$

The same procedure as for the opening mode, leads to the inversions

$$\mathcal{L}_{p\tau}^{-1}\left(\frac{P}{pA_S}\right) = -\frac{q^2}{k} C_{II}(kq\tau) U(\tau) - \frac{1}{k^3} \delta'(\tau) \quad (6.14.101)$$

$$\mathcal{L}_{p\tau}^{-1}\left(\frac{Q}{pA_S}\right) = -\frac{1}{k}[C_{II}(kq\tau) U(\tau) - C_{III}(kq\tau)] U(\tau) \quad (6.14.102)$$

where

$$C_{II}(kq\tau) = 3 J_0(kq\tau) + \frac{J_1(kq\tau)}{kq\tau} - 4k J_0(q\tau) - 4kq\tau \int_{kq\tau}^{q\tau} \frac{J_1(s)}{s} \mathrm{d}s \quad (6.14.103)$$

$$C_{III}(kq\tau) = \frac{J_1(kq\tau)}{kq\tau} \quad (6.14.104)$$

The notations $C_{II}(\cdot)$ and $C_{III}(\cdot)$ follow Geubelle and Rice (1995). By letting $n = 0$, $W_+^0 = 0$, in (6.14.96) and (6.14.97), it is found that $T_{xy}^0 = PU_+^0/(p^2 A_S)$, $T_{yz}^0 = 0$, which shows that C_{II} is associated with mode II. Then, it can be concluded that C_{III} is associated with mode III. This may also be found directly from mode III propagation in the x direction, $m = 0$: $\tau_{yz} = \partial w / \partial y$ leads to $T_{yz}^0 = A_S W_+^0 = 1/p \cdot p A_S W_+^0$ and the inversion of pA_S equals $kq^2 C_{III}(kq\tau) + \delta'(\tau)$.

The inversions of T_{ij} can now be written as

$$\mathcal{L}_{p\tau}^{-1}(T_{11}) = -\frac{1}{k}[m^2 C_{II}(kq\tau) + n^2 C_{III}(kq\tau)] U(\tau) - \frac{1}{k^3} \delta'(\tau) \quad (6.14.105)$$

$$\mathcal{L}_{p\tau}^{-1}(T_{12}) = \mathcal{L}_{p\tau}^{-1}(T_{21}) = -\frac{mn}{k}[C_{II}(kq\tau) - C_{III}(kq\tau)] U(\tau) \quad (6.14.106)$$

$$\mathcal{L}_{p\tau}^{-1}(T_{22}) = -\frac{1}{k}[n^2 C_{II}(kq\tau) + m^2 C_{III}(kq\tau)] U(\tau) - \frac{1}{k^3} \delta'(\tau) \quad (6.14.107)$$

Insertion into (6.14.99)-(6.14.100) gives

$$\begin{bmatrix} t_{xy}^0 \\ t_{yz}^0 \end{bmatrix} = -\frac{\mu}{2k}\begin{bmatrix} \dot{D}_x \\ \dot{D}_z \end{bmatrix} - \frac{k\mu}{2}\begin{bmatrix} m^2 & mn \\ mn & n^2 \end{bmatrix}\begin{bmatrix} I_{IIx} \\ I_{IIz} \end{bmatrix} - \frac{k\mu}{2}\begin{bmatrix} n^2 & -mn \\ -mn & m^2 \end{bmatrix}\begin{bmatrix} I_{IIIx} \\ I_{IIIz} \end{bmatrix} \quad (6.14.108)$$

where a dot denotes differentiation with respect to τ, and

$$I_{IIx} = \int_0^\tau C_{II}(kq\tau') D_x(\tau - \tau') \mathrm{d}\tau', \text{ etc.} \quad (6.14.109)$$

This is the sought relation between tractions and displacements on the crack plane. Note that $C_{II}(kq\tau)$ and $C_{III}(kq\tau)$ are convolution kernels associated with modes II and III, respectively.

The Fourier synthesis and the numerical scheme follow closely what was described for the opening mode.

Elimination of disturbances from ghost events
The use of a Fourier series leads to a double periodic solution with periods L and H, representing a spatial array of events, rather than one single event, on the crack plane. This necessitates L and H to be chosen large enough, compared to the linear dimesions of the region of interest, to avoid disturbances from the unwanted neighbouring events during the time of interest. To overcome this drawback of the method, a technique was devised by Cochard and Rice (1997), whereby the two periods of the Fourier series may be chosen only twice as large as the sides of the rectangular region of interest, with complete elimination of disturbances from ghost events. However, the convolution kernel has to be truncated in space, which leads to a far more complicated expression than for $C_I(\cdot)$, $C_{II}(\cdot)$ and $C_{III}(\cdot)$.

7
Elastoplastic Crack Dynamics

7.1 Introduction

Compared to elastic crack dynamics, only a few analytical results are available for the elastic-plastic counterpart. The first solutions were given by Slepyan (1976), who presented asymptotic expressions for stresses and strains at the edge of steadily moving mode I and mode III cracks in an elastic-perfectly plastic body. Later, such expressions were also given for mode II (Lo 1982). Stresses and strains in the symmetry plane ahead of a steadily moving mode III crack, from the crack edge to the elastic-plastic boundary, were determined by Freund and Douglas (1982). They also resolved a previous paradox: when the crack velocity approaches zero, the asymptotic solutions for dynamically propagating elastic-plastic cracks (Slepyan 1976) do not default to the asymptotic solutions obtained by neglecting inertia effects *a priori* for slowly moving cracks (McClintock 1963, Chitaley and McClintock 1971). The explanation, also given by Dunayevsky and Achenbach (1982a), is that the asymptotic solution in the dynamic case is dominating in a region whose size shrinks to zero with vanishing crack velocity.

The justification of combining perfect plasticity and inertia was questioned by Ståhle (1993), who considered linear strain hardening. He found that, by studying a mode III crack edge vicinity, an asymptotic solution cannot be obtained, if the crack velocity exceeds a certain value, decreasing with the strain hardening. For vanishing strain hardening, a solution can only be obtained for an infinitesimally small velocity. Other discussions of asymptotic dynamic solutions for modes I, II and III in strain hardening materials have been made by Gao and Nemat-Nasser (1983) and by Östlund and Gudmundson (1988).

Full-field solutions, obtained by numerical methods (Lam and Freund 1985, Deng and Rosakis 1991, 1992, and Varias and Shih 1994), indicate that there are different sectors at the crack edge. For mode I, one sector, remarkably, is sufficiently narrow to be indistinguishable from a line discontinuity, a plastic wave front at 90° angle from the crack plane, immediately followed by unloading (Varias and Shih 1994). The discontinuity appears to occur in the shear strain γ_{xy}.

Available asymptotic solutions for the in-plane cases seem to be dependent on certain assumptions, and other solutions might be possible. A discussion of mode I solutions is given for plane strain by Leighton *et al.* (1987) and for plane stress by Deng and Rosakis (1991). Only mode I and II plane strain and mode III will be considered here.

7.2 Mode III crack propagation

Governing equations

Consider steady state propagation with velocity V of a mode III crack edge in a perfectly plastic material. Following Slepyan (1976) and Freund and Douglas (1982), the analysis will be performed for a Huber-von Mises or a Tresca material. The notations follow those used for slowly moving mode III cracks, Section 5.6. It is possible to write the yield condition for both materials in the common form

$$\tau_{xz} = \tau_Y \cos\psi, \qquad \tau_{yz} = \tau_Y \sin\psi \qquad (7.2.1)$$

although τ_Y takes on different values if it is determined from the yield stress σ_Y.

The equation of motion is

$$\frac{\partial \tau_{xz}}{\partial x} + \frac{\partial \tau_{yz}}{\partial y} = \varrho \frac{\partial^2 w}{\partial t^2} \qquad (7.2.2)$$

which, with the Galilean transformation $X = x - Vt$, $Y = y$, implying that $\partial/\partial t$ can be replaced by $-V\partial/\partial X$, becomes

$$\tau_Y\left(-\sin\psi \cdot \frac{\partial \psi}{\partial X} + \cos\psi \cdot \frac{\partial \psi}{\partial Y}\right) = \varrho V^2 \frac{\partial \gamma_{xz}}{\partial X} \qquad (7.2.3)$$

where (7.2.1) is also used and $\partial w/\partial x$ is replaced by γ_{xz}.

Then, noting that $\varrho V^2/\mu = V^2/c_S^2 = \gamma^2$ and introducing the substitutions

$$\gamma_{xz} = \frac{\tau_Y}{\mu} g, \qquad \psi = \frac{\pi}{2} + \vartheta \qquad (7.2.4)$$

the equation of motion for an active plastic region reads

$$\cos\vartheta \frac{\partial \vartheta}{\partial X} + \sin\vartheta \frac{\partial \vartheta}{\partial Y} + \gamma^2 \frac{\partial g}{\partial X} = 0 \qquad (7.2.5)$$

Next, the constitutive equations will be used:

$$\frac{d\gamma_{xz}}{dt} = \frac{d\lambda}{dt} \cdot \frac{\partial f}{\partial \tau_{xz}} + \frac{1}{\mu} \cdot \frac{d\tau_{xz}}{dt} \qquad (7.2.6)$$

$$\frac{d\gamma_{yz}}{dt} = \frac{d\lambda}{dt} \cdot \frac{\partial f}{\partial \tau_{yz}} + \frac{1}{\mu} \cdot \frac{d\tau_{yz}}{dt} \qquad (7.2.7)$$

Here, the plastic deformation rate follows from the associated flow rule (5.2.4), and f is the yield function, which may be written as $\sqrt{\tau_{xz}^2 + \tau_{yz}^2} - \tau_Y$. Then,

$$\frac{\partial f}{\partial \tau_{xz}} = \cos\psi, \qquad \frac{\partial f}{\partial \tau_{yz}} = \sin\psi \qquad (7.2.8)$$

Furthermore, $d\gamma_{xz}/dt = -V\partial\gamma_{xz}/\partial X$, etc. Thus, also eliminating τ_{xz} and τ_{yz} in favour of ψ, the constitutive equations become

$$\frac{\partial \gamma_{xz}}{\partial X} = -\frac{1}{V} \cdot \frac{d\lambda}{dt} \cos\psi - \frac{\tau_Y}{\mu} \cdot \sin\psi \cdot \frac{\partial \psi}{\partial X} \qquad (7.2.9)$$

$$\frac{\partial \gamma_{yz}}{\partial X} = -\frac{1}{V} \cdot \frac{d\lambda}{dt} \sin\psi + \frac{\tau_Y}{\mu} \cdot \cos\psi \cdot \frac{\partial \psi}{\partial X} \qquad (7.2.10)$$

7.2 MODE III CRACK PROPAGATION

The shear strain γ_{yz} can now be eliminated by using the compatibility equation

$$\frac{\partial \gamma_{yz}}{\partial X} = \frac{\partial^2 w}{\partial X \partial Y} = \frac{\partial \gamma_{xz}}{\partial Y} \qquad (7.2.11)$$

From the resulting two equations, $d\lambda/dt$ can be eliminated and also expressed in a convenient form:

$$\frac{\partial \vartheta}{\partial X} + \cos\vartheta \frac{\partial g}{\partial X} + \sin\vartheta \frac{\partial g}{\partial Y} = 0 \qquad (7.2.12)$$

$$\frac{d\lambda}{dt} = \frac{\tau_Y V}{\mu}\left(\sin\vartheta \frac{\partial g}{\partial X} - \cos\vartheta \frac{\partial g}{\partial Y}\right) \qquad (7.2.13)$$

The last relation will only be used to control that $d\lambda/dt$ is non-negative.

Equations (7.2.5) and (7.2.12) form a system from which the stresses and strains in active plastic regions can, in principle, be found, assuming the boundary conditions to be known. First, the asymptotic solution for the crack edge vicinity (Slepyan 1976) will be discussed.

Asymptotic solution

Centered fan and constant stress fields

Asymptotic relations for the crack edge vicinity are sought. Simplifications can be made by using the facts that the angle ψ and the shear strain γ_{xz} are bounded, the latter because of (7.2.3) and because it vanishes exactly on the positive X axis. Then, transformation to cylindrical coordinates, keeping the same function notations despite the argument change from (X, Y) to (r, φ), yields the following asymptotic relations,

$$\frac{\partial Q}{\partial X} = \frac{\partial Q}{\partial r}\cos\varphi - \frac{\partial Q}{\partial \varphi} \cdot \frac{\sin\varphi}{r} \to -\frac{\partial Q}{\partial \varphi} \cdot \frac{\sin\varphi}{r} \qquad (7.2.14)$$

$$\frac{\partial Q}{\partial Y} = \frac{\partial Q}{\partial r}\sin\varphi + \frac{\partial Q}{\partial \varphi} \cdot \frac{\cos\varphi}{r} \to \frac{\partial Q}{\partial \varphi} \cdot \frac{\cos\varphi}{r} \qquad (7.2.15)$$

where Q equals ψ, γ_{xz}, ϑ or g. With these relations, equations (7.2.5) and (7.2.12) take the form

$$\sin(\vartheta - \varphi)\frac{\partial \vartheta}{\partial \varphi} = \gamma^2 \sin\varphi \cdot \frac{\partial g}{\partial \varphi} \qquad (7.2.16)$$

$$\sin\varphi \cdot \frac{\partial \vartheta}{\partial \varphi} = \sin(\vartheta - \varphi) \cdot \frac{\partial g}{\partial \varphi} \qquad (7.2.17)$$

whereas equation (7.2.13) becomes

$$\frac{d\lambda}{dt} = -\frac{\tau_Y V}{\mu r} \cdot \sin\varphi \cdot \cot(\vartheta - \varphi) \cdot \frac{\partial \vartheta}{\partial \varphi} \qquad (7.2.18)$$

Elimination of $\partial g/\partial \varphi$ between (7.2.16) and (7.2.17) gives

$$[\sin^2(\vartheta - \varphi) - \gamma^2 \sin^2\varphi]\frac{\partial \vartheta}{\partial \varphi} = 0 \qquad (7.2.19)$$

Thus, either

$$\vartheta = \varphi \pm \operatorname{asin}(\gamma \sin\varphi) \qquad (7.2.20)$$

or
$$\frac{\partial \vartheta}{\partial \varphi} = 0 \Longrightarrow \vartheta = \text{constant} \tag{7.2.21}$$

Note that ϑ cannot be a function of r, because τ_{xz} is anti-symmetric with respect to φ.

The first solution shows a centered fan stress field and the second solution a constant stress field. The first solution satisfies the condition $\vartheta = \psi - \pi/2 = 0$ for $\varphi = 0$, but not the condition $|\vartheta| = |\psi - \pi/2| = \pi/2$ for $\varphi = \pm\pi$. Because the opposite holds for the second solution, the situation is indeed very similar to the one for a slowly moving mode III crack (Section 5.4), with one frontal centered fan region and one constant stress region at each crack face. There are, however, as will be evident later, some significant differences in other respects.

The stresses and strains in the centered fan region – the primary plastic region – can be determined after the sign ambiguity for ϑ is resolved. This is done by using (7.2.18):

$$\frac{d\lambda}{dt} = \mp \frac{\tau_Y V}{\mu \gamma r}\left[\sqrt{1 - \gamma^2 \sin^2 \varphi} \pm \gamma \cos \varphi\right] \tag{7.2.22}$$

The expression within the brackets [ ] is positive, and thus the lower sign shall be used. Hence,

$$\vartheta = \varphi - \operatorname{asin}(\gamma \sin \varphi), \quad \psi = \frac{\pi}{2} + \varphi - \operatorname{asin}(\gamma \sin \varphi) \tag{7.2.23}$$

The asymptotic stresses and strains in the primary plastic region are now found from ψ and g. The stresses are

$$\tau_{xz} = \tau_Y \cos \psi = -\tau_Y \sin \varphi \left[\sqrt{1 - \gamma^2 \sin^2 \varphi} - \gamma \cos \varphi\right] \tag{7.2.24}$$

$$\tau_{yz} = \tau_Y \sin \psi = \tau_Y \left[\cos \varphi \sqrt{1 - \gamma^2 \sin^2 \varphi} + \gamma \sin^2 \varphi\right] \tag{7.2.25}$$

Insertion of the centered fan expression for ϑ into (7.2.17) gives

$$\frac{\partial g}{\partial \varphi} = -\frac{1}{\gamma} \frac{\partial \vartheta}{\partial \varphi} \tag{7.2.26}$$

so that use of (7.2.4) and integration yields

$$\gamma_{xz} = \frac{\tau_Y g}{\mu} = -\frac{\tau_Y \vartheta}{\mu \gamma} = -\frac{\tau_Y}{\mu \gamma}[\varphi - \operatorname{asin}(\gamma \sin \varphi)] \tag{7.2.27}$$

The integration constant (or function of r) must be zero, because $\gamma_{xz} = 0$ for $\varphi = 0$. Differentiation with respect to φ gives

$$\frac{\partial \gamma_{xz}}{\partial \varphi} = \frac{\tau_Y}{\mu}\left[\frac{\cos \varphi}{\sqrt{1 - \gamma^2 \sin^2 \varphi}} - \frac{1}{\gamma}\right] \tag{7.2.28}$$

whereupon use of (7.2.11) and (7.2.15) leads to

$$\frac{\partial \gamma_{yz}}{\partial X} = \frac{\tau_Y \cos \varphi}{\mu r}\left[\frac{\cos \varphi}{\sqrt{1 - \gamma^2 \sin^2 \varphi}} - \frac{1}{\gamma}\right] \tag{7.2.29}$$

The strain γ_{yz} can now be found by integration. For the symmetry plane ahead of the crack, $\varphi = 0$, the result is

$$\gamma_{yz} = \frac{\tau_Y(1-\gamma)}{\mu\gamma} \cdot \ln\frac{R}{r} \qquad (7.2.30)$$

where R is a length parameter, undetermined at asymptotic analysis, but asymptotic analysis requires that $r \ll R$. The result shows a weaker singularity than for the slowly moving crack, an $\ln(R/r)$ rather than an $[\ln(R/r)]^2$ singularity; see (5.6.23). It cannot be extrapolated to $\gamma = 0$ while R remains finite. It would otherwise be anticipated that the result should default to the one obtained for the slowly moving crack (5.6.23), as $\gamma \to 0$. This paradox will be discussed when a full solution for the strain on the symmetry plane ahead of the crack is obtained.

Integration of (7.2.29) with respect both to X and Y gives the asymptotic displacement in the primary plastic region. The result, a lengthy expression given by Slepyan (1976), reduces to

$$w \to \frac{\tau_Y(1-\gamma)}{\mu\gamma} \cdot r\sin\varphi \cdot \ln\frac{R}{r} \qquad (7.2.31)$$

as $r \to 0$, indicating that, as for the slowly moving crack, the crack edge is not blunted.

The second solution of (7.2.19), $\vartheta = constant$, corresponds to a constant stress region at each crack face, if the constant is chosen to be $+\pi/2$ or $-\pi/2$. For the upper crack face, the plus sign implies $\psi = 0$, i.e., $\tau_{xz} = \tau_Y$ and the minus sign implies $\psi = \pi$, i.e., $\tau_{xz} = -\tau_Y$.

Assembly of sectors

As shown by Slepyan (1976), the solution $\tau_{xz} = -\tau_Y$, $\tau_{yz} = 0$, in the constant stress sectors allows a full asymptotic solution consisting of one frontal centered fan sector followed by a constant stress sector, thus with no unloading (elastically deforming) sector between the two. The angle, φ^*, between the two sectors is found from (7.2.25) as the angle for which $\tau_{yz} = 0$ in the centered fan sector, which gives stress continuity. Thus, for the upper half-plane,

$$\varphi^* = \operatorname{atan}\left(-\frac{1}{\gamma}\right) \qquad (7.2.32)$$

showing that $\pi/2 < \varphi^* < 3\pi/4$. However, this solution corresponds to a rather peculiar loading. To understand this, consider the development of γ_{xz} for a material point at constant $y > 0$ as the crack edge sweeps by. In the passage of the point through the centered fan region, this strain decreases from a vanishingly small value if $Y \to +0$ to a finite negative value, dependent on the velocity γ, but not on Y. The absolute value of its plastic part is larger than the absolute value of its elastic part, which is also negative. Then, because $\tau_{xz} < 0$ in the constant stress region, γ_{xz}^p cannot increase ($|\gamma_{xz}^p|$ cannot decrease) when the point travels further in the negative X direction. Thus, a non-increasing finite permanent strain $\gamma_{xz}^p < 0$ will prevail along the upper crack face, and consequently the crack sliding displacement increases at least linearly towards $X = -\infty$. Such a displacement cannot be supported by a remote stress field of the usual boundary layer loading type, $\tau_{xz} = K_{III}\sin(\varphi/2)/\sqrt{2\pi r}$,

$\tau_{yz} = K_{III}\cos(\varphi/2)/\sqrt{2\pi r}$. It could possibly be obtained for crack propagation in a strip whose edges are subjected to moving loads.

Solution for the symmetry plane ahead of the crack

The hodograph transform

Freund and Douglas (1982) showed that the strain γ_{yz} can be determined for the symmetry plane from the crack edge to the elastic-plastic boundary ahead of the crack. Here, their treatment is essentially followed, with some differences in notations and sign conventions, due to previous commitments in the present work.

For convenience, the original equations, (7.2.5) and (7.2.12), which form a hyperbolic system, are repeated here:

$$\cos\vartheta \frac{\partial \vartheta}{\partial X} + \sin\vartheta \frac{\partial \vartheta}{\partial Y} + \gamma^2 \frac{\partial g}{\partial X} = 0 \tag{7.2.33}$$

$$\frac{\partial \vartheta}{\partial X} + \cos\vartheta \frac{\partial g}{\partial X} + \sin\vartheta \frac{\partial g}{\partial Y} = 0 \tag{7.2.34}$$

Because the location of the elastic-plastic boundary in the (X,Y) plane is not known, a hodograph transform is suitable. Thus, X and Y are regarded as functions of ϑ and g:

$$X = X(\vartheta, g), \qquad Y = Y(\vartheta, g) \tag{7.2.35}$$

The transformation of equations (7.2.33) and (7.2.34) can preferably be made in two steps: first the original equation system is transformed into an intermediate system of equations along characteristic lines and then the hodograph transform is conveniently obtained.

Multiply (7.2.34) by, say, Γ and add (7.2.33). Write the result in the form

$$\frac{\partial \vartheta}{\partial X} + \frac{\sin\vartheta}{\cos\vartheta + \Gamma} \cdot \frac{\partial \vartheta}{\partial Y} + \frac{\Gamma\cos\vartheta + \gamma^2}{\cos\vartheta + \Gamma}\left(\frac{\partial g}{\partial X} + \frac{\Gamma\sin\vartheta}{\Gamma\cos\vartheta + \gamma^2} \cdot \frac{\partial g}{\partial Y}\right) = 0 \tag{7.2.36}$$

This is equivalent to

$$\frac{\partial \vartheta}{\partial X} + \frac{\partial \vartheta}{\partial Y} \cdot \frac{dY}{dX} + q\left(\frac{\partial g}{\partial X} + \frac{\partial g}{\partial Y} \cdot \frac{dY}{dX}\right) = 0 \tag{7.2.37}$$

i.e.,

$$d\vartheta + q\,dg = 0 \tag{7.2.38}$$

if

$$\frac{dY}{dX} = \frac{\sin\vartheta}{\cos\vartheta + \Gamma} = \frac{\Gamma\sin\vartheta}{\Gamma\cos\vartheta + \gamma^2}, \qquad q = \frac{\Gamma\cos\vartheta + \gamma^2}{\cos\vartheta + \Gamma} \tag{7.2.39}$$

From equations (7.2.39) it follows that

$$\Gamma = \pm\gamma, \qquad q = \pm\gamma \tag{7.2.40}$$

Thus, both Γ and q turn out to be constants and the original equation system has been transformed to the system

$$d\vartheta + q\,dg = 0, \qquad dX + p\,dY = 0 \tag{7.2.41}$$

where
$$q = \pm\gamma, \qquad p = -\frac{\cos\vartheta \pm \gamma}{\sin\vartheta} \qquad (7.2.42)$$

Note the symmetry between equations (7.2.41) – they give no indication about which variable set is independent and which is dependent. They may be interpreted so that the quantities $\int(\mathrm{d}\vartheta + q\,\mathrm{d}g)$ are constant along the characteristic lines $\mathrm{d}X + p\,\mathrm{d}Y = 0$ in the (X,Y) plane (recall that p and q are bounded), or alternatively so that the quantities $\int(\mathrm{d}X + p\,\mathrm{d}Y)$ are constant along the characteristic lines $\mathrm{d}\vartheta + q\,\mathrm{d}g = 0$ in the (ϑ, g) plane, the hodograph plane. The latter interpretation leads to

$$\mathrm{d}X = \frac{\partial X}{\partial\vartheta}\mathrm{d}\vartheta + \frac{\partial X}{\partial g}\mathrm{d}g = \left(\frac{\partial X}{\partial\vartheta} - \frac{1}{q}\cdot\frac{\partial X}{\partial g}\right)\mathrm{d}\vartheta$$
$$= -p\,\mathrm{d}Y = -p\left(\frac{\partial Y}{\partial\vartheta}\mathrm{d}\vartheta + \frac{\partial Y}{\partial g}\mathrm{d}g\right) = \left(-p\frac{\partial Y}{\partial\vartheta} + \frac{p}{q}\cdot\frac{\partial Y}{\partial g}\right)\mathrm{d}\vartheta \quad (7.2.43)$$

i.e.,
$$q\frac{\partial X}{\partial\vartheta} - \frac{\partial X}{\partial g} + pq\frac{\partial Y}{\partial\vartheta} - p\frac{\partial Y}{\partial g} = 0 \qquad (7.2.44)$$

These are two equations, one for the upper, one for the lower sign in the expressions for p and q. Addition and subtraction leads to the sought hodograph transform,

$$\sin\vartheta\frac{\partial X}{\partial g} + \gamma^2\frac{\partial Y}{\partial\vartheta} - \cos\vartheta\frac{\partial Y}{\partial g} = 0 \qquad (7.2.45)$$

$$\sin\vartheta\frac{\partial X}{\partial\vartheta} - \cos\vartheta\frac{\partial Y}{\partial\vartheta} + \frac{\partial Y}{\partial g} = 0 \qquad (7.2.46)$$

Specialization to the symmetry plane vicinity
As shown by Freund and Douglas (1982), it is convenient to use polar coordinates in the hodograph plane,

$$\varrho = \sqrt{\vartheta^2 + g^2}, \qquad \omega = \mathrm{atan}\frac{g}{\vartheta} \qquad (7.2.47)$$

so that
$$\vartheta = \varrho\cos\omega, \qquad g = \varrho\sin\omega \qquad (7.2.48)$$

With X and Y now understood to be functions of ϱ and ω, the equations become

$$\sin\vartheta\frac{\partial X}{\partial\varrho}\sin\omega + \sin\vartheta\frac{\partial X}{\partial\omega}\cdot\frac{\cos\omega}{\varrho} + \gamma^2\frac{\partial Y}{\partial\varrho}\cos\omega$$
$$-\gamma^2\frac{\partial Y}{\partial\omega}\cdot\frac{\sin\omega}{\varrho} - \cos\vartheta\frac{\partial Y}{\partial\varrho}\sin\omega - \cos\vartheta\frac{\partial Y}{\partial\omega}\cdot\frac{\cos\omega}{\varrho} = 0 \quad (7.2.49)$$

$$\sin\vartheta\frac{\partial X}{\partial\varrho}\cos\omega - \sin\vartheta\frac{\partial X}{\partial\omega}\cdot\frac{\sin\omega}{\varrho} - \cos\vartheta\frac{\partial Y}{\partial\varrho}\cos\omega$$
$$+ \cos\vartheta\frac{\partial Y}{\partial\omega}\cdot\frac{\sin\omega}{\varrho} + \frac{\partial Y}{\partial\varrho}\sin\omega + \frac{\partial Y}{\partial\omega}\cdot\frac{\cos\omega}{\varrho} = 0 \quad (7.2.50)$$

The transformation to polar coordinates is particularly useful because $\tau_{xz} = 0$ and

516   7. ELASTOPLASTIC CRACK DYNAMICS

$\gamma_{xz} = 0$, and therefore $\vartheta = 0$, $g = 0$, on the symmetry plane ahead of the crack. Hence, $\varrho = 0$ on $0 < X \leq R_p$, $Y = 0$, where R_p is the straightforward extension of the plastic region. This fact allows crucial simplifications for the symmetry plane vicinity. First, it is obvious that $\sin \vartheta = \sin(\varrho \cos \omega)$ can be put equal to $\varrho \cos \omega$ and $\cos \vartheta$ can be put equal to 1. Second, for $0 < X \leq R_p$, $|Y| \ll X$,

$$X(\varrho, \omega) = X(0, \omega) + \text{ terms proportional to } \varrho \text{ or smaller} \qquad (7.2.51)$$

which implies that

$$\left| \varrho \frac{\partial X}{\partial \varrho} \right| \ll \left| \frac{\partial X}{\partial \omega} \right| \qquad (7.2.52)$$

in the immediate vicinity of the symmetry plane, so that terms containing $\partial X / \partial \varrho$ can be neglected. With these simplifications and a straightforward elimination procedure, the following equations are obtained:

$$\frac{\partial Y}{\partial \varrho}[1 - (1 + \gamma^2) \cos \omega \sin \omega] - \frac{1}{\varrho} \cdot \frac{\partial Y}{\partial \omega}[\cos^2 \omega - \gamma^2 \sin^2 \omega] = 0 \qquad (7.2.53)$$

$$\frac{1 - \gamma^2}{\cos \omega (\cos^2 \omega - \gamma^2 \sin^2 \omega)} \cdot \frac{\partial Y}{\partial \varrho} - \frac{\partial X}{\partial \omega} = 0 \qquad (7.2.54)$$

Solution of the equations

The function $X(0, \omega)$ is sought for $0 < X \leq R_p$. To this end, equation (7.2.53) is integrated by separation of variables. Put

$$Y = R(\varrho) \Omega(\omega) \qquad (7.2.55)$$

and normalize $R(\varrho)$ so that $R(1) = 1$. Then,

$$\frac{\varrho}{R} \cdot \frac{dR}{d\varrho} = \frac{\cos^2 \omega - \gamma^2 \sin^2 \omega}{1 - (1 + \gamma^2) \cos \omega \sin \omega} \cdot \frac{1}{\Omega} \cdot \frac{d\Omega}{d\omega} = \alpha \qquad (7.2.56)$$

where α is constant, because the left member is not a function of ω and the middle member is not a function of ϱ. Integration of $R(\varrho)$ gives

$$R(\varrho) = \varrho^\alpha \qquad (7.2.57)$$

Insertion into (7.2.54) shows that $\partial X / \partial \omega$, and hence also X, is proportional to $\varrho^{\alpha-1}$. But, $\varrho = 0$ must yield $0 < X \leq R_p$ and thus $\alpha = 1$.

Using $\alpha = 1$ in (7.2.56) gives

$$d(\ln \Omega) = \frac{1 - (1 + \gamma^2) \cos \omega \sin \omega}{\cos^2 \omega - \gamma^2 \sin^2 \omega} d\omega \qquad (7.2.58)$$

which integrates to

$$\Omega = C \frac{(\cos \omega + \gamma \sin \omega)^{(1+\gamma)/(2\gamma)}}{(\cos \omega - \gamma \sin \omega)^{(1-\gamma)/(2\gamma)}} \qquad (7.2.59)$$

where C is a constant. Insertion of $Y = \varrho \Omega(\omega)$ into (7.2.54) gives

$$\frac{\partial X}{\partial \omega} = C(1 - \gamma^2) \frac{(\cos \omega + \gamma \sin \omega)^{(1-\gamma)/(2\gamma)}}{\cos \omega (\cos \omega - \gamma \sin \omega)^{(1+\gamma)/(2\gamma)}} \qquad (7.2.60)$$

7.2 MODE III CRACK PROPAGATION

By integrating with respect to ω and then introducing the substitution

$$u = \frac{1 + \gamma \tan \omega}{1 - \gamma \tan \omega} \tag{7.2.61}$$

the integral can be written in the form

$$X(0,\omega) = -\frac{C(1-\gamma^2)}{\gamma} \int \frac{u^{(1-\gamma)/(2\gamma)}}{1+u} du = -\frac{C(1-\gamma^2)}{\gamma} I(u) \tag{7.2.62}$$

Use of boundary conditions
The expression for the function $X = X(0,\omega)$ contains the as yet undetermined constant C and an indefinite integral. In order to proceed, the conditions at the ends of the interval $0 < X(0,\omega) \leq R_p$ are needed. Thus, the values $\omega = \omega_0$, as the crack edge is approached from the symmetry plane ahead of the edge, and $\omega = \omega_1$, at the elastic-plastic boundary on the symmetry plane, have to be determined. In the hodograph plane, these boundary conditions for the unknown function $X(0,\omega)$ read:

$$X(0,\omega) = 0 \text{ for } \omega = \omega_0, \quad X(0,\omega) = R_p \text{ for } \omega = \omega_1 \tag{7.2.63}$$

The value ω_0 can be found from the asymptotic relations (7.2.24) and (7.2.27) for vanishing φ. They give

$$\tau_{xz} \approx -\tau_Y(1-\gamma)\varphi, \quad \gamma_{xz} \approx -\frac{\tau_Y(1-\gamma)}{\mu\gamma}\varphi \tag{7.2.64}$$

Then, from $\tau_{xz} = \tau_Y \cos\psi = -\tau_Y \sin\vartheta$, $\gamma_{xz} = \tau_Y g/\mu$, it follows that, as $\varphi \to 0$,

$$\omega = \operatorname{atan}\frac{g}{\vartheta} \to -\operatorname{atan}\frac{1}{\gamma} = \omega_0 \tag{7.2.65}$$

Thus, $-\pi/2 < \omega_0 < -\pi/4$.

The value ω_1 is to be determined from the conditions at the elastic-plastic boundary near the symmetry plane. There the stress τ_{xz} is found from its value when approaching the plastic region from the inside, and γ_{xz} from its value when approaching from the outside:

$$\tau_{xz} = \tau_Y \cos\psi = -\tau_Y \sin\vartheta, \quad \gamma_{xz} = \frac{\tau_{xz}}{\mu} = \frac{\tau_Y g}{\mu} \tag{7.2.66}$$

Because $\tau_{xz} = 0$ for $Y = 0$, both ϑ and g approach zero, but their ratio approaches -1. Thus,

$$\omega_1 = \operatorname{atan}\frac{g}{\vartheta} = -\frac{\pi}{4} \tag{7.2.67}$$

Use of (7.2.61) shows that $u = u_0$ corresponds to $\omega = \omega_0$ and $u = u_1$ to $\omega = \omega_1$, where

$$u_0 = 0, \quad u_1 = \frac{1-\gamma}{1+\gamma} \tag{7.2.68}$$

The first of the boundary conditions (7.2.63) implies that the lower limit of the integral in (7.2.62) is zero. The second boundary condition determines the constant C from

$$X(0,\omega_1) = -\frac{C(1-\gamma^2)}{\gamma} \int_0^{u_1} \frac{u^{(1-\gamma)/(2\gamma)}}{1+u} du = R_p \tag{7.2.69}$$

so that
$$C = -\frac{\gamma R_p}{(1-\gamma^2)I(u_1)} \qquad (7.2.70)$$

Thus,
$$\frac{X(0,\omega)}{R_p} = \frac{I(u)}{I(u_1)}, \quad u = \frac{1+\gamma\tan\omega}{1-\gamma\tan\omega} \qquad (7.2.71)$$

This gives X as a function of ω for $Y = 0$. However, the relation of primary interest is between γ_{yz} and X for $0 < X \leq R_p$, $Y = 0$. Thus, ω has to be expressed in terms of γ_{yz}. To this end, the compatibility equation (7.2.11) is first used:
$$\frac{\partial \gamma_{yz}}{\partial X} = \frac{\partial \gamma_{xz}}{\partial Y} \qquad (7.2.72)$$

For $X > 0$ and $Y \to 0$, antisymmetry implies that $\gamma_{xz} \to f(X)Y$. Thus, recalling that $\gamma_{xz} = \tau_Y g/\mu$ and using (7.2.48) and (7.2.55) results in
$$\frac{\partial \gamma_{yz}}{\partial X} = \frac{\tau_Y g}{\mu Y} = \frac{\tau_Y \sin\omega}{\mu \Omega(\omega)} \qquad (7.2.73)$$

so that
$$\gamma_{yz}(X,0) = \frac{\tau_Y}{\mu} \int \frac{\sin\omega}{\Omega(\omega)} \cdot \frac{\partial X(0,\omega)}{\partial \omega} d\omega$$
$$= -\frac{\tau_Y(1-\gamma^2)}{2\mu\gamma^2} \ln\left(\frac{1-\gamma^2\tan^2\omega}{1-\gamma^2}\right) + A \qquad (7.2.74)$$

where A is a constant. Now, $\gamma_{yz}(X,0) = \tau_Y/\mu$ for $X = R_p$, corresponding to $\omega = -\pi/4$. Thus, $A = \tau_Y/\mu$. Note that ω is a function of X/R_p given implicitly by (7.2.71), and the function $\gamma_{yz}(X,0)$ for $0 < X \leq R_p$ can be calculated numerically. Results are shown in Fig. 7.2.1.

The result shows a remarkably strong influence of the crack velocity. The shear strain near the crack edge is much smaller for a high than for a low velocity, assuming the same forwards extension, R_p, of the plastic region. One consequence appears to be that high velocity crack propagation requires a much larger R_p, and thereby a much higher remote load, than low velocity crack propagation. This is discussed in some detail by Freund (1990), using some continuum-mechanical criteria for crack propagation. However, for real cases, several factors, not considered in the present treatment, have to be kept in mind.

One factor of importance is the existence of a process region, which, at small scale yielding, generally extends to more than a tenth of R_p. If the size and other properties of this region were roughly independent of the crack speed, then there would be a tendency towards levelling the differences in load requirements between high and low velocities. However, the process region size appears to increase with the velocity, in some materials very strongly, cf. Section 9.1, and this works in the opposite direction.

Another factor of importance is the size of the plastic region, R_p. This size depends on the size and other properties of the process region, but also on time dependent effects on plastic flow, which can be very strong at high crack velocities. These effects include delayed yield (onset of plastic flow requires a higher stress), which acts to

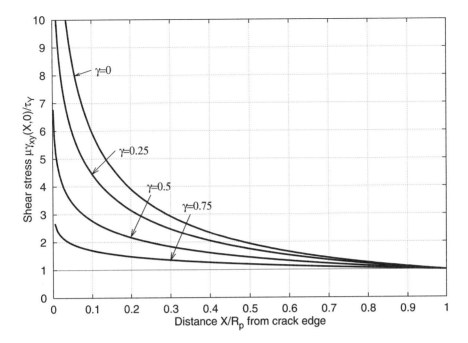

Fig. 7.2.1 The shear strain in the plastic region in the symmetry plane ahead of a mode III crack propagating steadily with velocity $V = \gamma/c_S$ in an elastic-perfectly plastic medium. R_p is the extension of the plastic region straight ahead of the crack.

diminish R_p, and viscoplasticity, which implies resistance to high speed crack growth. In all, the situation is very complex. Experiments on mode I crack propagation (e.g. Paxson and Lucas 1973, Ravi-Chandar 1982, Rosakis et al. 1984, Zehnder and Rosakis 1990) show pronounced increase with the crack velocity of the required energy flux into the dissipative region (see further discussion in Section 9.1).

The paradox resolved

It was pointed out that the asymptotic solution could not be brought to agreement with the solution for the slowly moving (inertia-free) crack by letting the crack velocity vanish. This limit can, however, be obtained from the full solution (7.2.74), if $\tan \omega$ is expressed in X/R_p. This is done by using (7.2.71).

The asymptotic solution will first be extracted from the full solution for arbitrary γ. As the crack edge is approached, $\omega \to \omega_0$ and thus $\gamma \tan \omega \to -1$. Hence, from (7.2.71),

$$\frac{X}{R_p} \approx \frac{I(u)}{I(u_1)}, \qquad u \approx \frac{1 + \gamma \tan \omega}{2} \ll 1 \tag{7.2.75}$$

From this, it follows that

$$I(u) \approx \frac{2\gamma}{1+\gamma} \cdot \left(\frac{1 + \gamma \tan \omega}{2}\right)^{(1+\gamma)/(2\gamma)} \tag{7.2.76}$$

and then, from (7.2.75),

$$\frac{1+\gamma\tan\omega}{2} \approx \left[\frac{1+\gamma}{2\gamma}I(u_1)\frac{X}{R_p}\right]^{2\gamma/(1+\gamma)} \tag{7.2.77}$$

Note that this approximation assumes that

$$\left[\frac{1+\gamma}{2\gamma}I(u_1)\frac{X}{R_p}\right]^{2\gamma/(1+\gamma)} \ll 1 \tag{7.2.78}$$

because the left member of (7.2.77) is much smaller than unity. Insertion of $\gamma\tan\omega$ into (7.2.74) gives

$$\gamma_{yz} \to \frac{\tau_Y(1-\gamma)}{\mu\gamma}\ln\frac{R}{X} \tag{7.2.79}$$

where

$$R = \frac{\gamma R_p}{2^{(1-\gamma)/(2\gamma)}(1+\gamma)I(u_1)} \tag{7.2.80}$$

This is the asymptotic result (7.2.30) by Slepyan (1976), but it is obtained under the condition (7.2.78), which specifies the region in which the result is valid. This region shrinks with γ and disappears when $\gamma \to 0$. The factor $(1+\gamma)I(u_1)/(2\gamma) \approx 0.2$ for velocities up to about one third of the S wave velocity and decreases at higher velocities. Numerical estimates show that the asymptotic result is only dominating in a very small region, extending towards less than a tenth of R_p for velocities below about one quarter of the S wave velocity. In a real case, such a region would be occupied by the process region, and it is doubtful whether the asymptotic solution has any practical significance at all, except at very high velocities, above half the S wave velocity.

It has now been shown that the region of validity of the asymptotic solution (7.2.30) by Slepyan (1976) disappears when $\gamma \to 0$. It remains to show that the solution (7.2.74) by Freund and Douglas (1982) outside this region of validity tends towards the solution (5.6.23) by McClintock (1963) when $\gamma \to 0$. To this end, the solution (7.2.74) will be specialized to vanishing crack velocity, $\gamma \to 0$ for arbitrary X. Then, $u_1 \to 1 - 2\gamma$, and the integral $I(u_1)$ is dominated by values close to 1 of the integration variable u. Consequently,

$$I(u_1) \to \frac{1}{2}\int_0^{1-2\gamma} u^{(1-\gamma)/(2\gamma)}du \to \frac{\gamma}{1+\gamma}(1-2\gamma)^{1/(2\gamma)} \to \gamma e^{-1}$$
$$\text{as } \gamma \to 0 \quad (7.2.81)$$

From (7.2.71), it is found that

$$\frac{X}{R_p} \to \frac{I(1+2\gamma\tan\omega)}{I(u_1)} \tag{7.2.82}$$

and, in a similar way as for $I(u_1)$, it is found that

$$I(1+2\gamma\tan\omega) \to \gamma e^{-\tan\omega} \text{ as } \gamma \to 0 \tag{7.2.83}$$

The function $\tan \omega$ can now be expressed in X/R_p and then insertion into (7.2.74) gives the shear strain on $0 < X \leq R_p$, $Y = 0$ by the expression

$$\gamma_{yz}(X,0) \to \frac{\tau_Y}{\mu}\left[\frac{1}{2}\left(\ln \frac{R_p}{X}\right)^2 + \ln \frac{R_p}{X} + 1\right] \text{ as } \gamma \to 0 \qquad (7.2.84)$$

which, apart from a slight difference in notation, equals (5.6.23). The paradox is now resolved. This was originally done by Freund and Douglas (1982) and also, with a different method, by Dunayevsky and Achenbach (1982a).

A technique of obtaining a solution on the symmetry plane ahead of the crack for dynamic elastic-plastic problems was introduced by Achenbach and Li (1985) and also used by Guo et al. (1988) and by Zhu and Li (1991).

On the validity of asymptotic analyses. Ståhle's solution

Ståhle (1993) considered dynamic mode III crack propagation in a strain hardening material obeying the Huber-von Mises yield condition and its associated flow rule. Linear strain hardening was assumed, expressed by the constant $\alpha = \mu_T/\mu$, where μ is the modulus of rigidity in the elastic region and μ_T is the tangent modulus in the strain hardening region. Thus, the propagation velocity of elastic waves is $c_S = \sqrt{\mu/\varrho}$ and the propagation velocity of plastic waves is $\sqrt{\alpha}c_S = \sqrt{\mu_T/\varrho}$.

By asymptotic analysis, Ståhle (1993) found a frontal primary plastic region in the angular interval $0 \leq |\varphi| \leq \varphi_1$ and a secondary plastic region in $\varphi_2 \leq |\varphi| \leq \pi$, where φ_1 and φ_2 both increased with the crack velocity V. Vanishingly small strain hardening, i.e., infinitesimally small α was considered. The stresses were found to be singular as r^s, $s < 0$, where $s \propto \sqrt{\alpha}$. For each crack velocity, V, the ratio s^2/α and the angles φ_1 and φ_2 were determined. Ståhle found that the highest velocity for which a solution can be obtained is $V = V_{max} \approx 1.241\sqrt{\mu_T/\varrho} = 1.241\sqrt{\alpha}c_S$, thus somewhat higher than the propagation velocity for plastic waves. Note that this velocity is infinitesimally small, because infinitesimally small α is considered.

Ståhle's solution shows that s increases from about $-0.8122\sqrt{\alpha}$ at $V = 0$ to about $-0.697\sqrt{\alpha}$ for $V = V_{max}$, that φ_1 increases from about $33.8453°$ for $V = 0$ to about $37.7632°$ for $V = V_{max}$, and that φ_2 increases from about $180° - 0.2809° = 179.7191°$ for $V = 0$ to about $180° - 0.214° = 179.786°$ for $V = V_{max}$. In an earlier paper (Ståhle 1989), the same figures for $V = 0$ were obtained by a non-inertial analysis, cf. page 305.

As in the non-inertial analysis for $V = 0$, Ståhle (1989), a discontinuity of the stress gradient $\partial \tau_{rz}/\partial x$ is found at the boundary of the primary plastic region for $0 < V < V_{max}$ as the hardening approaches zero (Ståhle 1993).

Note that different solutions are obtained, depending on how V approaches zero in relation to α, i.e. on the choice of $V/\sqrt{\alpha}$. The reason why a solution cannot be obtained for velocities above V_{max}, which is of the same order as the propagation velocity of plastic waves, seems to be connected with the assumption of an infinite primary plastic region. A finite size would allow elastic waves to interact with the plastic region not only from behind, but also on the frontal parts of the elastic-plastic boundary.

A comparison may be made with mode I crack propagation in an elastic layer embedded in an infinite elastic body. Even if the elastic constants are such that the

P wave velocity in the layer is lower than the Rayleigh wave velocity in the infinite body, crack propagation exceeding the P wave velocity in the layer is still possible (Broberg 1973b). The stresses near the crack edge are finite, and the crack edge is completely blunted. Note that an asymptotic analysis, obtained by stretching the region close to the crack edge, and fully inside the layer, to infinity, would not yield a solution for velocities higher than the Rayleigh wave velocity in the layer. Obviously, elastic wave interaction with the layer ahead of the crack is supplying the energy needed for crack propagation. An asymptotic analysis must take this interaction into account.

7.3 Asymptotic solutions for in-plane crack propagation

General relations

Constitutive equations
Consider a mode I or mode II crack, propagating steadily under plane strain conditions in the positive x direction in the xz plane with velocity $V = \gamma c_S$, where c_S is the S wave velocity. The Galilean transformation $X = x - Vt$, $y = Y$, $z = Z$, is introduced. Perfect plasticity is assumed, obeying the Tresca yield condition. The yield function is to be taken as the largest of the functions f_1, f_2 and f_3 in Section 5.7. However, it will be assumed here that f_1 is larger than the other two. The validity of this assumption will be investigated later. Recall that

$$f_1 = \sqrt{\sigma_d^2 + \tau_{xy}^2} - \tau_Y, \qquad \sigma_d = \frac{1}{2}(\sigma_y - \sigma_x) \qquad (7.3.1)$$

For an active plastic region, $f_1 = 0$, admitting the representation

$$\sigma_d = \tau_Y \cos\Omega, \qquad \tau_{xy} = \tau_Y \sin\Omega \qquad (7.3.2)$$

Note that both σ_d and τ_{xy} are bounded.

The flow rule gives

$$\frac{d\epsilon_x^p}{dt} = \frac{d\lambda}{dt} \cdot \frac{\partial f_1}{\partial \sigma_x} = -\frac{d\lambda}{dt} \cdot \frac{\sigma_d}{2\tau_Y} = -\frac{1}{2}\frac{d\lambda}{dt} \cos\Omega \qquad (7.3.3)$$

and analogous expressions for other strain rates. After use of the Galilean transformation, these expressions take the forms

$$\frac{\partial \epsilon_x^p}{\partial X} = +\frac{1}{2V} \cdot \cos\Omega \cdot \frac{d\lambda}{dt} \qquad (7.3.4)$$

$$\frac{\partial \epsilon_y^p}{\partial X} = -\frac{1}{2V} \cdot \cos\Omega \cdot \frac{d\lambda}{dt} \qquad (7.3.5)$$

$$\frac{\partial \gamma_{xy}^p}{\partial X} = -\frac{1}{V} \cdot \sin\Omega \cdot \frac{d\lambda}{dt} \qquad (7.3.6)$$

$$\epsilon_z^p = 0 \qquad (7.3.7)$$

From the last equation and plastic incompressibility, it follows that

$$\epsilon_x + \epsilon_y + \epsilon_z = \epsilon_x + \epsilon_y = \epsilon_x^e + \epsilon_y^e \qquad (7.3.8)$$

7.3 ASYMPTOTIC SOLUTIONS FOR IN-PLANE CRACK PROPAGATION

so that, according to Hooke's law,

$$\sigma_z = \nu(\sigma_x + \sigma_y) \tag{7.3.9}$$

$$\frac{\partial u}{\partial X} + \frac{\partial v}{\partial Y} = \frac{3(1-2\nu)}{2(1+\nu)\mu}\sigma_m \tag{7.3.10}$$

where

$$\sigma_m = \frac{1}{3}(\sigma_x + \sigma_y + \sigma_z) = \frac{1+\nu}{3}(\sigma_x + \sigma_y) \tag{7.3.11}$$

Further use of Hooke's law gives

$$\frac{\partial \epsilon_x}{\partial X} = +\frac{1}{2V}\cdot\cos\Omega\cdot\frac{d\lambda}{dt} + \frac{1}{2\mu}\cdot\frac{\partial}{\partial X}\left(\sigma_x - \frac{3\nu}{1+\nu}\sigma_m\right) \tag{7.3.12}$$

$$\frac{\partial \epsilon_y}{\partial X} = -\frac{1}{2V}\cdot\cos\Omega\cdot\frac{d\lambda}{dt} + \frac{1}{2\mu}\cdot\frac{\partial}{\partial X}\left(\sigma_y - \frac{3\nu}{1+\nu}\sigma_m\right) \tag{7.3.13}$$

$$\frac{\partial \gamma_{xy}}{\partial X} = -\frac{1}{V}\cdot\sin\Omega\cdot\frac{d\lambda}{dt} + \frac{1}{\mu}\cdot\frac{\partial \tau_{xy}}{\partial X} \tag{7.3.14}$$

from which it follows that

$$\frac{\partial}{\partial X}\left(\frac{\partial u}{\partial X} - \frac{\partial v}{\partial Y}\right) = +\frac{\cos\Omega}{V}\cdot\frac{d\lambda}{dt} + \frac{\tau_Y}{\mu}\cdot\sin\Omega\cdot\frac{\partial\Omega}{\partial X} \tag{7.3.15}$$

$$\frac{\partial}{\partial X}\left(\frac{\partial u}{\partial Y} + \frac{\partial v}{\partial X}\right) = -\frac{\sin\Omega}{V}\cdot\frac{d\lambda}{dt} + \frac{\tau_Y}{\mu}\cdot\cos\Omega\cdot\frac{\partial\Omega}{\partial X} \tag{7.3.16}$$

From these equations, $d\lambda/dt$ can be eliminated and also expressed in a convenient form:

$$\frac{\tau_Y}{\mu}\cdot\frac{\partial\Omega}{\partial X} = \sin\Omega\cdot\frac{\partial}{\partial X}\left(\frac{\partial u}{\partial X} - \frac{\partial v}{\partial Y}\right) + \cos\Omega\cdot\frac{\partial}{\partial X}\left(\frac{\partial u}{\partial Y} + \frac{\partial v}{\partial X}\right) \tag{7.3.17}$$

$$\frac{d\lambda}{dt} = V\left[\cos\Omega\cdot\frac{\partial}{\partial X}\left(\frac{\partial u}{\partial X} - \frac{\partial v}{\partial Y}\right) - \sin\Omega\cdot\frac{\partial}{\partial X}\left(\frac{\partial u}{\partial Y} + \frac{\partial v}{\partial X}\right)\right] \tag{7.3.18}$$

Equations of motion

The Galilean transformation implies that $\partial^2/\partial t^2 = V^2\partial^2/\partial X^2$, so that the equations of motion read:

$$\frac{\partial\sigma_x}{\partial X} + \frac{\partial\tau_{xy}}{\partial Y} = \varrho V^2\frac{\partial^2 u}{\partial X^2} \tag{7.3.19}$$

$$\frac{\partial\sigma_y}{\partial Y} + \frac{\partial\tau_{xy}}{\partial X} = \varrho V^2\frac{\partial^2 v}{\partial X^2} \tag{7.3.20}$$

In order to write these equations in terms of Ω, u and v, the normal stresses are first expressed in σ_d and σ_m:

$$\sigma_x = \frac{\sigma_x + \sigma_y}{2} - \sigma_d = \frac{3\sigma_m}{2(1+\nu)} - \sigma_d, \quad \sigma_y = \frac{3\sigma_m}{2(1+\nu)} + \sigma_d \tag{7.3.21}$$

Use of (7.3.2) and (7.3.10) then gives the desired form:

$$\sin\Omega\frac{\partial\Omega}{\partial X} + \cos\Omega\frac{\partial\Omega}{\partial Y} = \frac{\mu}{\tau_Y}\left[\gamma^2\frac{\partial^2 u}{\partial X^2} - \frac{1}{1-2\nu}\cdot\frac{\partial}{\partial X}\left(\frac{\partial u}{\partial X} + \frac{\partial v}{\partial Y}\right)\right] \tag{7.3.22}$$

$$\cos\Omega\frac{\partial\Omega}{\partial X} - \sin\Omega\frac{\partial\Omega}{\partial Y} = \frac{\mu}{\tau_Y}\left[\gamma^2\frac{\partial^2 v}{\partial X^2} - \frac{1}{1-2\nu}\cdot\frac{\partial}{\partial Y}\left(\frac{\partial u}{\partial X} + \frac{\partial v}{\partial Y}\right)\right] \tag{7.3.23}$$

where $\gamma^2 = \varrho V^2/\mu = V^2/c_S^2$.

Boundedness of the asymptotic stresses

It will be shown that all stress components are bounded. Because σ_d is bounded and $|\sigma_z|$ is never larger than both $|\sigma_x|$ and $|\sigma_y|$, it follows from (7.3.11) that σ_x, σ_y and σ_z are bounded if σ_m is bounded.

Assume tentatively that σ_m is not bounded. Then, asymptotically, σ_d and τ_{xy} can be neglected, so that $\sigma_x \approx \sigma_y \approx 3\sigma_m/[2(1+\nu)]$. Then, from the equations of motion (7.3.19) and (7.3.20), it follows that

$$\frac{3}{2(1+\nu)}\Delta\sigma_m = \varrho V^2 \frac{\partial^2}{\partial X^2}\left(\frac{\partial u}{\partial X} + \frac{\partial v}{\partial Y}\right) = \frac{3(1-2\nu)}{2(1+\nu)} \cdot \gamma^2 \frac{\partial^2 \sigma_m}{\partial X^2} \quad (7.3.24)$$

which is simplified to

$$a_*^2 \frac{\partial^2 \sigma_m}{\partial X^2} + \frac{\partial^2 \sigma_m}{\partial Y^2} = 0 \quad (7.3.25)$$

where $a_*^2 = 1 - (1-2\nu)\gamma^2$. By making the substitutions

$$\frac{X}{a_*} = \xi, \qquad Y = \eta \quad (7.3.26)$$

a Laplace equation is obtained:

$$\frac{\partial^2 \sigma_m}{\partial \xi^2} + \frac{\partial^2 \sigma_m}{\partial \eta^2} = 0 \quad (7.3.27)$$

It is convenient to temporarily move to polar coordinates r_ζ, φ_ζ in the $\xi\eta$ plane. The unbounded solution compatible with bounded stress-strain energy can then be written as

$$\sigma_m = \frac{A}{\sqrt{r_\zeta}} f_1(\varphi_\zeta) \quad (7.3.28)$$

where $f_1(\varphi_\zeta) = \cos(\varphi_\zeta/2)$ for mode I, $f_1(\varphi_\zeta) = \sin(\varphi_\zeta/2)$ for mode II. Putting $\varphi_\zeta = 0$ shows that $A > 0$ for mode I.

For mode I, equation (7.3.20) gives

$$\frac{2(1+\nu)\varrho V^2}{3} \cdot \frac{\partial^2 v}{\partial X^2} = \frac{\partial \sigma_m}{\partial Y} = -\frac{AY\cos(\varphi_\zeta/2)}{2r_\zeta^2\sqrt{r_\zeta}} - \frac{AX\sin(\varphi_\zeta/2)}{a_* r_\zeta^2 \sqrt{r_\zeta}} \quad (7.3.29)$$

Thus, for the upper crack face, $\varphi_\zeta = \pi$ ($X = -a_* r_\zeta$), the following inequality holds:

$$\frac{2(1+\nu)\varrho V^2}{3} \cdot \frac{\partial^2 v_+}{\partial X^2} = \frac{A}{2r_\zeta\sqrt{r_\zeta}} > 0 \quad (7.3.30)$$

Then, by integration,

$$\frac{2(1+\nu)\varrho V^2}{3} \cdot \frac{\partial v_+}{\partial X} = \frac{a_* A}{\sqrt{r_\zeta}} > 0 \quad (7.3.31)$$

This inequality implies interpenetration of the crack faces, so that A must vanish and the assumption of an unbounded solution was false.

7.3 ASYMPTOTIC SOLUTIONS FOR IN-PLANE CRACK PROPAGATION

For mode II,

$$\sigma_x + \sigma_y = \frac{3\sigma_m}{1+\nu} = \frac{3A}{(1+\nu)\sqrt{r_\zeta}} \sin \frac{\varphi_\zeta}{2} \qquad (7.3.32)$$

For $\varphi_\zeta = \pi$, the normal stress $\sigma_y = 0$ and then

$$\sigma_d = -\frac{\sigma_x}{2} = -\frac{3A}{2(1+\nu)\sqrt{r_\zeta}} \qquad (7.3.33)$$

But σ_d is bounded, and thus $A = 0$. Again, the assumption of an unbounded solution was false.

It has now been shown that σ_m is bounded both for mode I and for mode II, implying that all stress components are bounded. It then follows that the dilatation, $\partial u/\partial X + \partial v/\partial Y$, is bounded, due to (7.3.10). It also follows, from the equations of motion (7.3.19) and (7.3.20), that the second derivatives of the displacements are singular as $\tau_Y/(\varrho V^2 r)$, at most. Here, r is a polar coordinate, together with φ, in the XY plane. Integration of these second derivatives leads to the following general form:

$$\frac{\mu}{\tau_Y} \cdot \frac{\partial u}{\partial X} = B \ln \frac{R_0}{r} + g(\varphi), \qquad \frac{\mu}{\tau_Y} \cdot \frac{\partial v}{\partial X} = C \ln \frac{R_0}{r} + h(\varphi) \qquad (7.3.34)$$

where $g(\varphi)$ and $h(\varphi)$ are bounded. The boundedness of the dilatation implies that

$$\frac{\mu}{\tau_Y} \cdot \frac{\partial v}{\partial Y} = -B \ln \frac{R_0}{r} + m(\varphi) \qquad (7.3.35)$$

where $m(\varphi)$ is bounded. Here, B, C and R_0 are constants which cannot all be determined in an asymptotic analysis, whereas the functions $g(\varphi)$, $h(\varphi)$ and $m(\varphi)$ are, in principle, determinable.

It is further assumed that the remaining displacement gradient can be written in the same form as the other ones. Thus,

$$\frac{\mu}{\tau_Y} \cdot \frac{\partial u}{\partial Y} = H \ln \frac{R_0}{r} + n(\varphi) \qquad (7.3.36)$$

where $n(\varphi)$ is bounded.

Asymptotic solution for mode I

Governing equations

Mode I symmetry implies that $\Omega = 0$ on $X > 0$, $Y = 0$, anticipating that $\sigma_d > 0$. It also implies that $v = 0$ for $\varphi = 0$, so that constant $C = 0$ in (7.3.34), and $h(0) = 0$. Furthermore, for $|\varphi| \ll 1$, $\partial v/\partial Y = \epsilon_y > 0$, implying that constant B in equation (7.3.35) is non-positive. Differentiation of the displacement gradients in

(7.3.34)-(7.3.35), only retaining terms proportional to r, gives

$$\frac{\mu r}{\tau_Y} \cdot \frac{\partial^2 u}{\partial X^2} = -B \cdot \frac{\partial r}{\partial X} + rg'(\varphi) \cdot \frac{\partial \varphi}{\partial X} = -B \cos \varphi - g'(\varphi) \sin \varphi \qquad (7.3.37)$$

$$\frac{\mu r}{\tau_Y} \cdot \frac{\partial^2 u}{\partial X \partial Y} = -B \sin \varphi + g'(\varphi) \cos \varphi = -H \cos \varphi - n'(\varphi) \sin(\varphi) \qquad (7.3.38)$$

$$\frac{\mu r}{\tau_Y} \cdot \frac{\partial^2 v}{\partial X^2} = -h'(\varphi) \sin \varphi \qquad (7.3.39)$$

$$\frac{\mu r}{\tau_Y} \cdot \frac{\partial^2 v}{\partial X \partial Y} = B \cos \varphi - m'(\varphi) \sin \varphi = h'(\varphi) \cos \varphi \qquad (7.3.40)$$

$$\frac{\mu r}{\tau_Y} \cdot \frac{\partial^2 v}{\partial Y^2} = B \sin \varphi + m'(\varphi) \cos \varphi = \frac{B - h'(\varphi) \cos^2 \varphi}{\sin \varphi} \qquad (7.3.41)$$

The last equation implies that $B = h'(0)$, because $\partial^2 v / \partial Y^2$ is bounded.

Because Ω is bounded, the following relations hold asymptotically, cf. (7.2.14)-(7.2.15):

$$r \frac{\partial \Omega}{\partial X} = -\frac{\partial \Omega}{\partial \varphi} \sin \varphi, \qquad r \frac{\partial \Omega}{\partial Y} = \frac{\partial \Omega}{\partial \varphi} \cos \varphi \qquad (7.3.42)$$

Insertion of the asymptotic expressions (7.3.37)-(7.3.42) into (7.3.17), (7.3.22), (7.3.23) and (7.3.18) gives

$$g'(\varphi) \cos(\Omega + \varphi) - h'(\varphi) \sin(\Omega + \varphi) + \Omega'(\varphi) \sin \varphi = B \sin(\Omega + \varphi) \qquad (7.3.43)$$

$$g'(\varphi)[1 - (1 - 2\nu)\gamma^2] \sin \varphi - h'(\varphi) \cos \varphi - (1 - 2\nu)\Omega'(\varphi) \cos(\Omega + \varphi)$$
$$= -B[1 - (1 - 2\nu)\gamma^2] \cos \varphi \qquad (7.3.44)$$

$$g'(\varphi) \cos \varphi - h'(\varphi) \left[\frac{\cos^2 \varphi}{\sin \varphi} - (1 - 2\nu)\gamma^2 \sin \varphi\right] - (1 - 2\nu)\Omega'(\varphi) \sin(\Omega + \varphi)$$
$$= -B \frac{\cos^2 \varphi}{\sin \varphi} \qquad (7.3.45)$$

$$\frac{d\lambda}{dt} = -\frac{\tau_Y V}{\mu r} \{g'(\varphi) \sin(\Omega + \varphi) + [h'(\varphi) + B] \cos(\Omega + \varphi)\} \qquad (7.3.46)$$

Also, from (7.3.10), (7.3.34), (7.3.35) and (7.3.41):

$$\frac{d\sigma_m}{d\varphi} = \frac{2(1 + \nu)\tau_Y}{3(1 - 2\nu)} \{g'(\varphi) + [B - h'(\varphi)] \cot(\varphi)\} \qquad (7.3.47)$$

Constant stress and centered fan sectors

Following Slepyan (1976), the case $B = 0$ is considered, implying through (7.3.46) that $d\lambda/dt = 0$ for $\varphi = 0$, because $h'(0) = B$ and $\Omega(0) = 0$. (The notations used by Slepyan are not followed: in particular the sign of B is different, whereas g and h equal

7.3 ASYMPTOTIC SOLUTIONS FOR IN-PLANE CRACK PROPAGATION

Slepyan's p and q, respectively.) Use of (7.3.44)-(7.3.45) then gives

$$g'(\varphi) = -\frac{\Omega'(\varphi)}{D(\varphi)}[\cos(\Omega + 2\varphi)\cot\varphi - (1-2\nu)\gamma^2\cos(\Omega+\varphi)\sin\varphi] \qquad (7.3.48)$$

$$h'(\varphi) = -\frac{\Omega'(\varphi)}{D(\varphi)}[\cos(\Omega + 2\varphi) + (1-2\nu)\gamma^2\sin(\Omega+\varphi)\sin\varphi] \qquad (7.3.49)$$

$$D(\varphi) = \gamma^2[1 - (1-2\nu)\gamma^2\sin^2\varphi] \qquad (7.3.50)$$

Insertion into (7.3.43), (7.3.46) and (7.3.47) leads to an equation for Ω and expressions for $d\lambda/dt$ and $\sigma'_m(\varphi)$:

$$\Omega'(\varphi)\{\cos^2(\Omega+2\varphi) - \gamma^2[2 - 2\nu - (1-2\nu)\gamma^2\sin^2\varphi]\sin^2\varphi\} = 0 \qquad (7.3.51)$$

$$\frac{d\lambda}{dt} = \frac{\tau_Y V}{\mu r D(\varphi)} \cdot \frac{\Omega'(\varphi)}{\sin\varphi} \cdot \cos(\Omega+2\varphi)\sin(\Omega+2\varphi) \qquad (7.3.52)$$

$$\frac{d\sigma_m}{d\varphi} = \frac{2(1+\nu)\gamma^2\tau_Y}{3D(\varphi)} \cdot \Omega'(\varphi)\sin(\Omega+2\varphi) \qquad (7.3.53)$$

The solutions of the equation for Ω are

$$\Omega'(\varphi) = 0 \qquad (7.3.54)$$

$$\cos(\Omega+2\varphi) = \pm\gamma\sin\varphi\sqrt{2 - 2\nu - (1-2\nu)\gamma^2\sin^2\varphi} \qquad (7.3.55)$$

The first solution corresponds to constant stress sectors, because it implies $\sigma'_d(\varphi) = \tau'_{xy}(\varphi) = \sigma'_m(\varphi) = 0$. It also implies that $d\lambda/dt = 0$, i.e., no plastic deformations are taking place in constant stress sectors. The second solution corresponds to centered fan sectors, because stresses are only dependent on φ. Differentiation of (7.3.55) shows, after some investigation, that $\Omega'(\varphi)$ is always negative, in fact smaller than $-(2 - \sqrt{2-2\nu})$, in a centered fan.

Assembly of sectors. The asymptotic stresses
Consider the upper half of the crack edge vicinity. Solution (7.3.55) does not exist for $\varphi = 0$, and therefore the sector around $\varphi = 0$ must be of the constant stress type. Because $\Omega = 0$ for $\varphi = 0$, $\sigma_d = \tau_Y$, the stresses are found from (7.3.21) to be

$$\sigma_x = \frac{3\sigma_0}{2(1+\nu)} - \tau_Y, \quad \sigma_y = \frac{3\sigma_0}{2(1+\nu)} + \tau_Y, \quad \sigma_z = \frac{3\nu\sigma_0}{1+\nu}, \quad \tau_{xy} = 0 \qquad (7.3.56)$$

Here, σ_0 is a constant stress, equal to the mean stress σ_m in the sector. Because $\sigma_y > 0$ for $\varphi = 0$, this sector does not satisfy the boundary conditions at $\varphi = \pi$, so it is only valid within some interval $\varphi \leq \varphi_* < \pi$. Thus, a centered fan sector will follow and extend from $\varphi = \varphi_*$ to, say, $\varphi = \varphi_{**}$.

The condition $\tau_{xy} = 0$ at $\varphi = \pi$ cannot be satisfied by the centered fan solution, because it would imply $\cos\Omega = 0$ through (7.3.55). Thus, $\varphi_{**} < \pi$, so there will be a constant stress solution in a sector $\varphi_{**} < \varphi \leq \pi$; cf. Fig. 7.3.1. There, the Cartesian stresses are the same as at $\varphi = \pi$:

$$\sigma_x = 2\tau_Y, \quad \sigma_y = 0, \quad \sigma_z = 2\nu\tau_Y \quad \tau_{xy} = 0 \qquad (7.3.57)$$

From this, it follows that $\sigma_m(\varphi) = 2(1+\nu)\tau_Y/3$. Furthermore, $\Omega = -\pi$ in the sector,

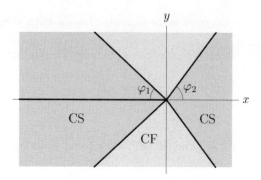

Fig. 7.3.1 Asymptotic constant stress and centered fan sectors at the edge of a dynamically propagating mode I crack. CS denotes constant stress sector and CF denotes centered fan sector.

where the minus sign is due to the fact that $\Omega(\varphi)$ is zero in the frontal constant stress sector and decreases with increasing φ in the centered fan sector.

Stress continuity across the sector boundaries is now assumed. This is equivalent to continuity of $\Omega(\varphi)$ and $\sigma_m(\varphi)$. Thus,

$$\Omega(\varphi_*) = 0, \qquad \sigma_m(\varphi_*) = \sigma_0 \tag{7.3.58}$$

$$\Omega(\varphi_{**}) = -\pi, \qquad \sigma_m(\varphi_{**}) = \frac{2(1+\nu)\tau_Y}{3} \tag{7.3.59}$$

The first continuity condition for $\Omega(\varphi)$ gives

$$\cos 2\varphi_* = \pm \gamma \sin \varphi_* \sqrt{2 - 2\nu - (1-2\nu)\gamma^2 \sin^2 \varphi_*} \tag{7.3.60}$$

There are four candidate solutions,

$$\varphi_* = \varphi_{1,2}, \qquad \varphi_* = \pi - \varphi_{1,2} \tag{7.3.61}$$

where $0 < \varphi_1 < \pi/4$ and $\pi/4 < \varphi_2 < \pi/2$. The solution φ_1 is obtained for the upper sign in (7.3.60), the solution φ_2 for the lower sign. The solutions $\varphi_* = \varphi_1$ and $\varphi_* = \pi - \varphi_2$ must be discarded, because they imply negative $d\lambda/dt$ for $\varphi > \varphi_*$ if $\varphi - \varphi_*$ is sufficiently small. It is therefore assumed that the centered fan sector starts at $\varphi = \varphi_* = \varphi_2$. Inspection of the sign of $d\lambda/dt$ in this sector can be done for $\varphi = \varphi_2 + 0$. Equation (7.3.52) gives for $\varphi = \varphi_2$:

$$\frac{d\lambda}{dt} = \frac{2\tau_Y V}{\mu r D(\varphi)} \Omega'(\varphi_2) \cos 2\varphi_2 \cos \varphi_2 \tag{7.3.62}$$

Because $\cos 2\varphi_2 < 0$, the sign of $d\lambda/dt$ is opposite that of $\Omega'(\varphi_2)$. Now, $\Omega'(\varphi)$ was found to be negative in a centered fan sector, so $d\lambda/dt$ is positive, implying that the centered fan solution is permissible.

The second continuity condition for $\Omega(\varphi)$ gives an equation for φ_{**} which is identical to equation (7.3.60) for φ_*, using the upper sign. Thus, the solution is $\varphi_{**} = \pi - \varphi_1$. It can be shown from (7.3.55), upper sign, that $d\lambda/dt \geq 0$ throughout the centered fan sector $\varphi_2 \leq \varphi \leq \pi - \varphi_1$, because $2\varphi_2 < \Omega + 2\varphi < \pi - \varphi_1$ in the sector.

7.3 ASYMPTOTIC SOLUTIONS FOR IN-PLANE CRACK PROPAGATION

As $\gamma \to 0$, the angle $\varphi_2 \to \pi/4$ and the angle $\pi - \varphi_1 \to 3\pi/4$. Both angles increase with the crack velocity, so that the angular range of the centered fan sector only decreases marginally when the crack velocity is increased to about half the S wave velocity.

In order to use the continuity condition (7.3.59) for $\sigma_m(\varphi)$, an expression for the mean stress in the centered fan is needed: this is found by integration of (7.3.53):

$$\sigma_m(\varphi) = \sigma_0 + \frac{2(1+\nu)\gamma^2 \tau_Y}{3} \int_{\varphi_2}^{\varphi} \frac{\Omega'(\varphi)}{D(\varphi)} \sin(\Omega + 2\varphi) d\varphi$$

$$\text{for } \varphi_2 \leq \varphi \leq \pi - \varphi_1 \quad (7.3.63)$$

The functions $\Omega'(\varphi)$ and $\sin(\Omega + 2\varphi)$ are found from (7.3.55), upper sign. It can be shown that $\sin(\Omega + 2\varphi) > 0$ in the centered fan sector. Insertion into (7.3.59) gives, for $\varphi = \varphi_{**} = \pi - \varphi_1$,

$$\frac{2(1+\nu)\tau_Y}{3} = \sigma_0 + \frac{2(1+\nu)\tau_Y}{3} \int_{\varphi_2}^{\pi - \varphi_1} \frac{\gamma^2 \Omega'(\varphi)}{D(\varphi)} \sin(\Omega + 2\varphi) d\varphi \quad (7.3.64)$$

This equation determines σ_0, and then all asymptotic stresses can be found. For the centered fan sector, they can be written in the form

$$\sigma_x = \frac{3\sigma_m}{2(1+\nu)} - \tau_Y \cos\Omega, \quad \sigma_y = \frac{3\sigma_m}{2(1+\nu)} + \tau_Y \cos\Omega, \quad (7.3.65)$$

$$\sigma_z = \frac{3\nu\sigma_m}{1+\nu}, \quad \tau_{xy} = \tau_Y \sin\Omega \quad (7.3.66)$$

The angle Ω is found from (7.3.55):

$$\Omega = -2\varphi + \operatorname{asin}\sqrt{1 - \gamma^2[2 - 2\nu - (1 - 2\nu)\gamma^2 \sin^2\varphi]} \quad (7.3.67)$$

As $\gamma \to 0$, it follows from (7.3.64) $\sigma_0 \to 2(1+\pi)(1+\nu)\tau_Y/3$, so that $\sigma_y \approx 4.7\tau_Y$, $\sigma_x \approx 2.7\tau_Y$ in front of the crack for small crack velocities and Poisson's ratio about 1/3. According to calculations by Slepyan (1976), these values decrease somewhat with increasing crack velocity. The stress σ_y is virtually constant, about $4.6\tau_Y - 4.7\tau_Y \approx 2.3\sigma_Y$ for Poisson's ratio around 1/3 and crack velocities smaller than about half the S wave velocity, which is well inside the range encountered in practical situations. This is only 10% lower than what was obtained for the Prandtl field, page 280, and for the slowly moving mode I crack after *a priori* neglect of inertia effects, page 322.

Note now that $d\lambda/dt = 0$ in the frontal asymptotic constant stress sector. This implies that deformations in this sector are purely elastic. For common external types of mode I loading they would also be elastic everywhere in front of this asymptotic region, so that the deformations in the frontal constant stress sector can be regarded as virginly elastic.

Now, the assumption that f_1 is the appropriate yield function can be investigated. For the frontal constant stress sector, this is found to require that

$$f_2 = \frac{3(1-2\nu)\sigma_0}{4(1+\nu)} - \frac{\tau_Y}{2} < 0 \quad (7.3.68)$$

Because numerical calculations show that σ_0 decreases with increasing crack velocity, this condition is satisfied for all velocities if it is satisfied for $\gamma = 0$. Then, it can be written as a sufficient condition for Poisson's ratio:

$$\nu > \nu_* = \frac{\pi - 1}{2(\pi + 1)} \approx 0.2585 \qquad (7.3.69)$$

This condition for the frontal sector can be immediately extended to the other sectors, because $\sigma_m < \sigma_0$ in these sectors, as follows from the fact that the right hand side of (7.3.53) is negative for the centered fan sector.

The asymptotic strains

Because the deformations are elastic in the frontal constant stress sector, they follow from Hooke's law:

$$\epsilon_x = \frac{3(1-2\nu)\sigma_0}{4(1+\nu)\mu} - \frac{\tau_Y}{2\mu}, \quad \epsilon_y = \frac{3(1-2\nu)\sigma_0}{4(1+\nu)\mu} + \frac{\tau_Y}{2\mu}, \quad \epsilon_z = \gamma_{xy} = 0 \qquad (7.3.70)$$

It is now assumed that the constant $H = 0$ in equation (7.3.36). Then, the displacement gradient $\partial u/\partial Y$ is a function of φ, only, and it is found from (7.3.38):

$$\frac{\partial}{\partial X}\left(\frac{\partial u}{\partial Y}\right) = \frac{d}{d\varphi}\left(\frac{\partial u}{\partial Y}\right) \cdot \left(-\frac{\sin\varphi}{r}\right) = \frac{\tau_Y}{\mu r} g'(\varphi) \cos\varphi \qquad (7.3.71)$$

so that

$$\frac{\partial u}{\partial Y} = -\frac{\tau_Y}{\mu}\int_{\varphi_2}^{\varphi} g'(\varphi)\cot\varphi \, d\varphi + \text{constant} \qquad (7.3.72)$$

Expressions for the strains follow from this equation and from equations (7.3.34)-(7.3.35):

$$\epsilon_x(\varphi) = \frac{\tau_Y}{\mu}\int_{\varphi_2}^{\varphi} g'(\varphi) d\varphi + \epsilon_x(0) \qquad (7.3.73)$$

$$\epsilon_y(\varphi) = \frac{\tau_Y}{\mu} m(\varphi) = -\frac{\tau_Y}{\mu}\int_{\varphi_2}^{\varphi} h'(\varphi)\cot\varphi \, d\varphi + \epsilon_y(0) \qquad (7.3.74)$$

$$\gamma_{xy}(\varphi) = \frac{\tau_Y}{\mu}\int_{\varphi_2}^{\varphi} [h'(\varphi) - g'(\varphi)\cot\varphi] d\varphi \qquad (7.3.75)$$

Thus, the strains can be calculated by using the expressions for $g'(\varphi)$ and $h'(\varphi)$ in (7.3.48)-(7.3.49). Note that the expressions give the strains everywhere, because $g'(\varphi)\cot\varphi = h'(\varphi)\cot\varphi = 0$ in constant stress sectors.

From (7.3.34) it follows that

$$\frac{\partial v_+}{\partial X} - \frac{\tau_Y}{\mu} h(\pi - \varphi_1) \qquad (7.3.76)$$

because $h'(\varphi) = 0$ in constant stress sectors. Thus, the crack face opening occurs with a finite angle at the crack edge.

For vanishingly small crack velocities, the strains behind the constant stress sector turn out to increase in proportion to $1/\gamma$ as $\gamma \to 0$. This is analogous with the corresponding mode III case, expression (7.2.30), except for the fact that the strains are here non-singular in r. Therefore, the same conclusion has to be made as for mode III, pages 519ff., namely that the asymptotic relations found are valid in a region

7.3 ASYMPTOTIC SOLUTIONS FOR IN-PLANE CRACK PROPAGATION

which shrinks towards zero with vanishing crack velocity. This resolves the paradox that the present asymptotic solution does not approach the one for a slowly moving mode I crack, obtained under the *a priori* assumption of no inertia effects.

Note that the solution obtained possesses full stress and strain continuity at the boundaries between constant stress and centered fan sectors. Sophisticated jump conditions, as discussed in Section 5.7, might have appeared if a strain had shown a dependence on r in some sector.

Asymptotic solution for mode II

Governing equations
Mode II symmetry implies that $\Omega = \pi/2$ on $X > 0$, $Y = 0$ for a remote load $\tau_{xy} = \tau_{xy}^\infty > 0$. It also implies that $u = 0$ for $\varphi = 0$, so that the constant $B = 0$ in equations (7.3.34)-(7.3.35), and $g(0) = 0$. Moreover, because $\Omega'(\varphi)$ is non-positive, as previously found, it can be concluded that $\Omega(\varphi)$ decreases from $\pi/2$ to 0 as φ increases from 0 to π, giving $\sigma_x = -2\tau_Y$, $\sigma_y = \tau_{xy} = 0$ on the upper crack face, as required. Furthermore, for $|\varphi| \ll 1$, $\partial u/\partial Y + \partial v/\partial X = \gamma_{xy} > 0$, implying through (7.3.34) and (7.3.36) that $C + H \geq 0$. Differentiation of the displacement gradients in equations (7.3.34)-(7.3.35), only retaining terms proportional to r, gives

$$\frac{\mu r}{\tau_Y} \cdot \frac{\partial^2 u}{\partial X^2} = -g'(\varphi) \sin \varphi \qquad (7.3.77)$$

$$\frac{\mu r}{\tau_Y} \cdot \frac{\partial^2 u}{\partial X \partial Y} = g'(\varphi) \cos \varphi = -H \cos \varphi - n'(\varphi) \sin \varphi \qquad (7.3.78)$$

$$\frac{\mu r}{\tau_Y} \cdot \frac{\partial^2 v}{\partial X^2} = -C \cos \varphi - h'(\varphi) \sin \varphi \qquad (7.3.79)$$

$$\frac{\mu r}{\tau_Y} \cdot \frac{\partial^2 v}{\partial X \partial Y} = -m'(\varphi) \sin \varphi = -C \sin \varphi + h'(\varphi) \cos \varphi \qquad (7.3.80)$$

$$\frac{\mu r}{\tau_Y} \cdot \frac{\partial^2 v}{\partial Y^2} = m'(\varphi) \cos \varphi = C \cos \varphi - \frac{h'(\varphi) \cos^2 \varphi}{\sin \varphi} \qquad (7.3.81)$$

The last equation and the boundedness of $\partial^2 v/\partial Y^2$ for $\varphi = 0$, $r \neq 0$, implies that $C = \lim_{\varphi \to 0}[h'(\varphi) \cot \varphi]$, so that $h'(\varphi) \to C\varphi$ as $\varphi \to 0$.

The left hand sides of equations (7.3.43), (7.3.44) and (7.3.45) remain unchanged, but the right hand sides change to $C \cos(\Omega + \varphi)$, $-C \sin \varphi$ and $-C \cos \varphi$, respectively. In addition, the following relations, corresponding to (7.3.46) and (7.3.47), are obtained:

$$\frac{d\lambda}{dt} = -\frac{\tau_Y V}{\mu r}\{[g'(\varphi) - C]\sin(\Omega + \varphi) + h'(\varphi)\cos(\Omega + \varphi)\} \qquad (7.3.82)$$

$$\frac{d\sigma_m}{d\varphi} = \frac{2(1+\nu)\tau_Y}{3(1-2\nu)}[g'(\varphi) + C - h'(\varphi)\cot(\varphi)] \qquad (7.3.83)$$

It is now observed that the shear strain γ_{xy} cannot be expected to increase with X on $Y = 0$, $X > 0$. Thus, $\partial\gamma_{xy}/\partial X \leq 0$ for $\varphi = 0$, which implies $g'(0) - C \leq 0$. It will later be shown that $g'(0) = -H$, so that this condition is equivalent to the previously found relation $C + H \geq 0$.

The assumption made for mode I that $\partial u/\partial Y$ is independent of r turns out to lead

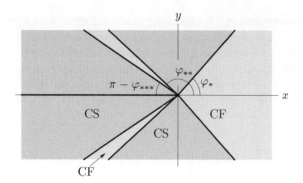

Fig. 7.3.2 Asymptotic constant stress (CS) and centered fan (CF) sectors at the edge of a dynamically propagating mode II crack.

to an irregularity for mode II. Thus, $H \neq 0$, and equation (7.3.38) gives the relation

$$n'(\varphi) = -[H + g'(\varphi)]\cot\varphi \qquad (7.3.84)$$

Following Lo (1982), the case $C = 0$ is studied. Then, $h'(0) = 0$, $g'(0) \leq 0$ and $d\lambda/dt = -\tau_Y V g'(0)/(\mu r) \geq 0$ for $\varphi = 0$. Equations (7.3.48)-(7.3.55) remain unchanged, and so do the conclusions about the interpretations of (7.3.54) and (7.3.55) as constant stress and centered fan sectors, respectively.

Assembly of sectors
The assumption that $\partial u/\partial Y$ may be singular in r does not violate the condition of stress continuity. This can be seen from (5.7.69), which does not include a jump in $\partial u/\partial Y$.

A difference from the mode I case is that the solution $\Omega = \pi/2$ for $\varphi = 0$ of equation (7.3.55) is now permissible. This makes a centered fan sector around $\varphi = 0$ possible, and it is also assumed that this is the case. By trial and error it is found that a constant stress frontal sector does not work. Likewise, the possibility of a centered fan sector extending to $\varphi = \pi$ can be ruled out and also the possibility of a frontal centered fan sector followed by a rear constant stress sector. However, the next possibility will be shown to satisfy the requirements of stress continuity and of positive $d\lambda/dt$ in active plastic regions. In the upper half-plane, it consists of

P1: a centered fan sector, from $\varphi = 0$ to $\varphi = \varphi_*$ (say)
P2: a constant stress sector, from $\varphi = \varphi_*$ to $\varphi = \varphi_{**}$ (say)
P3: a centered fan sector, from $\varphi = \varphi_{**}$ to $\varphi = \varphi_{***}$ (say)
P4: a constant stress sector, from $\varphi = \varphi_{***}$ to $\varphi = \pi$

This is shown in Fig. 7.3.2.

Stress continuity can be expressed through continuity of $\Omega(\varphi)$ and $\sigma_m(\varphi)$. Boundary conditions for the upper half of the crack edge vicinity are

$$\Omega(0) = \frac{\pi}{2}, \qquad \sigma_m(0) = 0 \qquad (7.3.85)$$

$$\Omega(\pi) = 0, \qquad \sigma_m(\pi) = -\frac{2(1+\nu)}{3}\tau_Y \qquad (7.3.86)$$

Continuity conditions give

$$\Omega(\varphi_{***}) = 0, \quad \sigma_m(\varphi_{***}) = -\frac{2(1+\nu)}{3}\tau_Y \quad (7.3.87)$$

$$\Omega(\varphi_{**}) = \Omega(\varphi_{*}), \quad \sigma_m(\varphi_{**}) = \sigma_m(\varphi_{*}) \quad (7.3.88)$$

where use has been made of the fact that $\sigma_m = 3(\sigma_x + \sigma_y)/[2(1+\nu)]$ and thus is constant in constant stress regions (which, for instance, is not true about σ_r).

The continuity conditions involve $\Omega(\varphi)$ and $\sigma_m(\varphi)$, which are found from (7.3.55) and (7.3.53):

$$\Omega(\varphi) = -2\varphi - \mathrm{acos}\left[\gamma \sin\varphi\sqrt{2 - 2\nu - (1-2\nu)\gamma^2\sin^2\varphi}\right] + n\pi \quad (7.3.89)$$

$$\sigma_m(\varphi) = \frac{2(1+\nu)\gamma^2\tau_Y}{3}\int\frac{\Omega'(\varphi)}{D(\varphi)}\sin(\Omega + 2\varphi)\mathrm{d}\varphi + constant \quad (7.3.90)$$

where n in the first equation should be chosen so that the flow condition $\mathrm{d}\lambda/\mathrm{d}t > 0$ is not violated and so that $0 \leq \Omega \leq \pi/2$. The function $\mathrm{acos}(\cdot)$ is here assumed to run from 0 to $\pi/2$. The first equation is valid for centered fan sectors, only.

The first continuity condition for $\Omega(\varphi)$ gives an equation for determination of φ_{***}: it is identical to (7.3.60). One solution is $\varphi_{***} = \pi - \varphi_1$. For this solution, equation (7.3.52) gives

$$\frac{\mathrm{d}\lambda}{\mathrm{d}t} = -\frac{2\tau_Y V}{\mu r D(\varphi)}\Omega'(\pi - \varphi_1)\cos 2\varphi_1 \cos\varphi_1 \quad (7.3.91)$$

Because $\cos 2\varphi_1 > 0$, the sign of $\mathrm{d}\lambda/\mathrm{d}t$ is opposite that of Ω', which was found to be negative. Thus, $\mathrm{d}\lambda/\mathrm{d}t$ is positive, which implies that $\varphi_{***} = \pi - \varphi_1$ is an acceptable solution. The only other possibility associated with positive $\mathrm{d}\lambda/\mathrm{d}t$, $\varphi_{***} = \varphi_2$, does not lead to a permissible sector assembly. Thus, $\varphi_{***} = \pi - \varphi_1$ and the upper sign of (7.3.55) shall be used for P3. This implies that $n = 2$ in (7.3.89). As regards sector P1, the appropriate sign in (7.3.55) is found by considering $0 < \varphi \ll 1$. Then, $\sin(\Omega + 2\varphi) \approx 1$ and thus positive, which implies that the lower sign should be used. It then follows that $n = 1$ in (7.3.89).

Insertion of the expressions for $\Omega(\varphi)$ and $\sigma_m(\varphi)$ into (7.3.88) gives two equations for φ_* and φ_{**}, and then $\Omega(\varphi)$ and $\sigma_m(\varphi)$ are known for all φ, so that all asymptotic stresses become known everywhere. It remains, of course, to ensure that f_1, as assumed, is the appropriate yield function, a condition which, as for mode I, might impose restrictions on Poisson's ratio. The procedure will now be demonstrated for the particular case of elastic incompressibility, $\nu = 1/2$, the case studied by Lo (1982).

For $\nu = 1/2$, equation (7.3.60), upper sign, for determination of φ_1, will read:

$$\cos 2\varphi = \gamma \sin\varphi \quad (7.3.92)$$

The acceptable solution is

$$\varphi_1 = \mathrm{asin}\frac{\sqrt{\gamma^2 + 8} - \gamma}{4} \quad (7.3.93)$$

Note that $\pi/6 < \varphi_1 < \pi/4$, so that $3\pi/4 < \varphi_{***} < 5\pi/6$.

For sector P3, equation (7.3.89) will be simplified to

$$\text{P3:}\quad \Omega(\varphi) = 2\pi - 2\varphi - \mathrm{acos}(\gamma\sin\varphi) \quad (7.3.94)$$

The integrand in (7.3.90) can be expressed as a function of φ, by differentiating (7.3.55), which now is simplified to

$$\text{P3:} \quad \cos(\Omega + 2\varphi) = \gamma \sin \varphi \tag{7.3.95}$$

This gives the integrand, apart from a factor $1/\gamma^2$, as:

$$\text{P3:} \quad \Omega'(\varphi) \sin(\Omega + 2\varphi) = -2\sin(\Omega + 2\varphi) - \gamma \cos \varphi$$
$$= 2\sqrt{1 - \gamma^2 \sin^2 \varphi} - \gamma \cos \varphi \tag{7.3.96}$$

where the last member follows from the fact that $\cos 2\varphi_{***}$ and $\sin 2\varphi_{***}$ are both positive, so that $\sin(\Omega + 2\varphi)$ is negative and therefore equal to $-\sqrt{1 - \gamma^2 \sin^2 \varphi}$.

After performing the integration, equation (7.3.90) gives

$$\text{P3:} \quad \sigma_m(\varphi) = \tau_Y[2E(\varphi, \gamma) - 2E(\pi - \varphi_1, \gamma) - \gamma \sin \varphi + \gamma \sin \varphi_1 - 1] \tag{7.3.97}$$

because $\sigma_m(\pi - \varphi_1) = -\tau_Y$. Here, $E(\cdot, \cdot)$ is the elliptic integral of the second kind. Note that $E(\pi - \varphi_1, \gamma) = 2\boldsymbol{E}(\gamma) - E(\varphi_1, \gamma)$, where $\boldsymbol{E}(\cdot)$ is the complete elliptic integral of the second kind. The integration constant was determined from condition (7.3.87).

The expression for $\Omega(\varphi)$ in sector P3 is given by (7.3.89), upper sign, and $n = 2$. Thus,

$$\text{P3:} \quad \Omega(\varphi) = 2\pi - 2\varphi - \text{acos}(\gamma \sin \varphi) \tag{7.3.98}$$

For sector P1, equation (7.3.89) reduces to

$$\text{P1:} \quad \Omega(\varphi) = \pi - 2\varphi - \text{acos}(\gamma \sin \varphi) \tag{7.3.99}$$

and the integrand in (7.3.90) can be expressed as a function of φ, by differentiating (7.3.55), which now is simplified to

$$\text{P1:} \quad \cos(\Omega + 2\varphi) = -\gamma \sin \varphi \tag{7.3.100}$$

This gives the integrand, apart from a factor $1/\gamma^2$, as:

$$\text{P1:} \quad \Omega'(\varphi) \sin(\Omega + 2\varphi) = -2\sin(\Omega + 2\varphi) + \gamma \cos \varphi$$
$$= -2\sqrt{1 - \gamma^2 \sin^2 \varphi} + \gamma \cos \varphi \tag{7.3.101}$$

After performing the integration, (7.3.90) becomes

$$\text{P1:} \quad \sigma_m(\varphi) = -\tau_Y[2E(\varphi, \gamma) - \gamma \sin \varphi] \tag{7.3.102}$$

because $\sigma_m(0) = 0$.

Now, expressions have been obtained for $\Omega(\varphi)$ and $\sigma_m(\varphi)$ in both P1 and P3, and then the continuity conditions (7.3.88) give two equations for determination of φ_* and φ_{**}. Then, $\Omega(\varphi)$ and $\sigma_m(\varphi)$ become determined everywhere, and thereby also all stress components. These have been calculated by Lo (1982) for different crack velocities.

There is no need to ensure that f_1 is the appropriate yield condition, because $f_2 = f_3 = f_1$ for $\nu = 1/2$, which also implies that Huber-von Mises and Tresca yield conditions coincide. Investigation of the signs of $\cos(\Omega + 2\varphi)$ and $\sin(\Omega + 2\varphi)$ shows, together with (7.3.52), that $d\lambda/dt$ is positive in each centered fan sector.

7.3 ASYMPTOTIC SOLUTIONS FOR IN-PLANE CRACK PROPAGATION

The asymptotic strains

In spite of the fact that $\partial u/\partial Y$ was allowed to be singular in r, see (7.3.36), there will be no jump in $\partial u/\partial Y$, because this would imply a jump also in $\partial u/\partial X$ by condition (5.7.62). Thus, the constant H in (7.3.36) is the same in all sectors.

The strains are found in the same way as for mode I. They can be written as

$$\epsilon_x(\varphi) = \frac{T_Y}{\mu} \int_0^\varphi g'(\varphi) d\varphi \qquad (7.3.103)$$

$$\epsilon_y(\varphi) = \frac{T_Y}{\mu} m(\varphi) = -\frac{T_Y}{\mu} \int_0^\varphi h'(\varphi) \cot\varphi \, d\varphi \qquad (7.3.104)$$

$$\gamma_{xy}(\varphi) = \frac{T_Y}{\mu} [h(\varphi) + H \ln \frac{R_0}{r} + n(\varphi)]$$

$$= \frac{T_Y}{\mu} \int_0^\varphi \{h'(\varphi) - [g'(\varphi) + H]\cot\varphi\} d\varphi + \frac{T_Y}{\mu} \cdot H \ln \frac{R_0}{r} \qquad (7.3.105)$$

Convergence of the last integral demands that

$$H = -g'(0) = \frac{2-\gamma}{\gamma} \qquad (7.3.106)$$

Note that $g'(\varphi)\cot(\varphi) = h'(\varphi)\cot(\varphi) = 0$ in constant stress sectors. As in the mode I case, $d\lambda/dt = 0$ in these sectors, so no plastic strain increments are produced, even though the yield condition is fulfilled.

The result shows, in contrast to the mode I solution, that singular strains appear: γ_{xy} exhibits a logarithmic singularity in r.

For vanishingly small crack velocities, the strains turn out to increase in proportion to $1/\gamma$ as $\gamma \to 0$. This is analogous with the corresponding mode III case, see expression (7.2.30). Therefore, the same conclusion has to be made as for mode III, pages 519ff., namely, that the asymptotic relations found are valid in a region which shrinks towards zero when crack velocity zero is approached. This resolves the paradox that the present asymptotic solution does not approach the one for a slowly moving mode II crack, obtained under the *a priori* assumption of no inertia effects.

Following Lo (1982), the particular case of elastic as well as plastic incompressibility, $\nu = 1/2$, will be considered. Then, differentiation of (7.3.99) gives

$$\text{P1: } \Omega'(\varphi) = -2 + \frac{\gamma \cos\varphi}{\sqrt{1 - \gamma^2 \sin^2\varphi}} \qquad (7.3.107)$$

Use of this expression and relations (7.3.99)-(7.3.100) together with (7.3.48)-(7.3.49) gives

$$\text{P1: } g'(\varphi) = -\frac{2}{\gamma}\cos\varphi + \frac{\cos^2\varphi}{\sqrt{1 - \gamma^2 \sin^2\varphi}}, \quad h'(\varphi) = -g'(\varphi)\tan\varphi \qquad (7.3.108)$$

The strains in sector P1 are then found after insertion of these expressions and integrations:

$$\epsilon_x = -\epsilon_y = \frac{T_Y}{\gamma^2 \mu}[E(\varphi,\gamma) - (1-\gamma^2)F(\varphi,\gamma) - 2\gamma \sin\varphi] \qquad (7.3.109)$$

$$\gamma_{xy} = \frac{T_Y}{\mu}\left[\ln\left(1 + \sqrt{1-\gamma^2 \sin^2\varphi}\right) - \frac{2}{\gamma}\ln(1+\cos\varphi) + \frac{2-\gamma}{\gamma}\ln\frac{2R_0}{r}\right] \qquad (7.3.110)$$

Here, $F(\cdot, \cdot)$ and $E(\cdot, \cdot)$ are the elliptic integrals of the first of second kinds, respectively.

The Cartesian strain components are constant in the constant stress sectors, and the strains in sector P3 are found in the same way as for the strains in P1, although the expressions for $g'(\varphi)$ and $h'(\varphi)$ are slightly different.

7.4 Temperature changes at moving crack edges

General considerations

In general, the greater part of the irreversible energy associated with loading and unloading is eventually dissipated as heat. The remaining part is trapped in the material, due to transformation of the material structure and to constraints caused by dislocation pile-ups, etc. It has been estimated that the fraction β dissipated as heat may be as high as 0.85-0.95 for metals (Taylor and Quinney 1934, Bever et al. 1973, Rosakis et al. 1992), but it is dependent on the magnitude and rate of deformation (Bever et al. 1973, Mason et al. 1994). Thus, Mason et al. found that β may be as low as about 0.5 for aluminium and steel at low strains and for titanium at both high and low strains. For simplicity, it will be assumed here that β is a material constant, i.e. independent ot the magnitude and the rate of deformation, as well as of the temperature.

The large amount of heat generation per unit of volume at the edge of a running crack may give rise to a substantial temperature increase, in some cases several hundred degrees Celsius, as shown both by careful numerical studies (e.g. Li et al 1996) and by a number of skilfully performed experiments by Fuller et al. (1975), Zehnder and Rosakis (1991), Rosakis et al. (1992), Mason and Rosakis (1993), Mason et al. (1994), Kallivayalil and Zehnder (1994) and Kallivayalil et al. (1996). However, if the crack is moving slowly, the temperature rise will be small, because heat is conducted away from the crack edge. Temperature measurements are generally performed for mode I, but the highest temperatures are expected to occur in modes II and III, because of the large deformations accompanying shear banding, particularly under a high confining pressure, which for impact generated shear bands may be produced by the impact load itself. Measurements on impact produced shear bands (mode II) by Zhou et al. (1996b) and Rosakis et al. (1997) show temperatures rises of over 1400°C for a C-300 (a high strength maraging) steel. Very high temperatures may also be reached in impact processes in glass; cf. Weichert and Schönert (1978).

Also elastic dilatations or compressions contribute to temperature changes. Near a running crack, these are usually negative (temperature drop), due to thermoelastic cooling. This effect is smaller than the temperature rise due to heat generated by irreversible deformation, but, as shown by Rittel (1998a,b), it is generally not negligible.

The maximum temperature rise at the edge of a running crack occurs in the process region. In materials for which the cell model is appropriate, the total energy supply, $\mathcal{A}$, to a central cell is given by the area under the cohesion-decohesion curve, Fig. 1.3.2. Thus, the heat generated per unit of mass is $\beta \mathcal{A}/(\varrho d^3)$, where ϱ and d^3 are density and cell volume before loading. Then, the adiabatic temperature rise, i.e. the temperature rise occurring if no heat is conducted away from the cell, is

$$T_A = \frac{\beta \mathcal{A}}{\varrho c_p d^3} \tag{7.4.1}$$

where c_p is the specific heat at constant pressure, which is only weakly dependent on temperature and matrix volume in temperature intervals of interest.

Assume, as an example, that the maximum cell height increase $\delta_{max} = 0.3d$ and that the cohesive strength $\sigma_D = 4.7\sigma_Y$. Assume further that A/d^3 can be estimated as $(2/3)\sigma_D\delta_{max}/d$ from the cohesion-decohesion curve, and that $\beta = 0.85$. The adiabatic temperature rise is then

$$T_A = \frac{0.8\sigma_Y}{\varrho c_p} \qquad (7.4.2)$$

which equals about 390 °C for 4340 steel ($\sigma_Y = 1700\text{MPa}$, $\varrho = 7800\text{kg/m}^3$, $c_p = 450\text{Nm}/(\text{kg}°\text{C})$), and about 480 °C for titanium ($\sigma_Y = 1400\text{MPa}$, $\varrho = 4700\text{kg/m}^3$, $c_p = 500\text{Nm}/(\text{kg}°\text{C})$). A parallel estimate for a ceramic like glass, using $\sigma_D = E/10$, where E is the modulus of elasticity, leads to $T_A = 560\,°\text{C}$ with the values $E = 70\text{GPa}$, $\varrho = 2500\text{kg/m}^3$ and $c_p = 850\text{Nm}/(\text{kg}°\text{C})$.

Now, heat is not generated uniformly within the cell. It is concentrated in bands of localized plastic flow, occurring as slip band concentrations, void-sheets, etc. Therefore, the maximum adiabatic temperature rise may be much higher than predicted by (7.4.1): it will be

$$\frac{\beta A}{s\varrho c_p d^3} \qquad (7.4.3)$$

where s is the volume fraction of the bands of localized plastic flow, assuming the heat generation per volume to be the same in all bands. This fraction may well be as low as 0.1 or lower, cf. Faleskog and Shih (1997).

Will maximum temperatures reach the neighbourhood of the adiabatic temperatures for cell averages, (7.4.1), or for bands of localized plastic flow, (7.4.3)? The answer, of course, depends on how rapidly heat is generated in a cell or in a band of localized plastic flow and on how rapidly it is conducted away. A simple estimate may be obtained by studying the temperature decrease in a material layer, $-h < y < h$, subjected to instantaneously supplied and uniformly distributed heat, Q_V, per unit of volume, at time $t = 0$. Such a layer might represent a process region consisting of one layer of cells, or a band of localized plastic flow.

The solution may be obtained by a superposition technique, after solving the problem of heat conduction following the instantaneous heat generation Q_V per unit volume in the upper half-plane $y > 0$. The governing equation for heat conduction is

$$\Delta T - \frac{1}{a^2} \cdot \frac{\partial T}{\partial t} = -\frac{1}{\lambda}\frac{\partial Q_V}{\partial t} \qquad (7.4.4)$$

where T is the temperature, counted from the (constant) ambient temperature at $t < 0$, λ is the conductivity, $a^2 = \lambda/(\varrho c_p)$ is the diffusivity and $Q_V = Q_V(x,y,z,t)$ is heat per unit volume. For the one-dimensional case, the derivation is obtained by considering a unit area of an element between x and $x + dx$ and a unit of time: the heat influx is $\lambda(\partial^2 T/\partial x^2)dx$, the increase of heat is $\varrho c_p \partial T/\partial t\, dx$ and the heat generation within the element is $\partial Q_V/\partial t\, dx$. The three-dimensional heat equation is then obtained by obvious generalization.

For the half-plane problem, the heat conduction equation takes the form
$$\frac{\partial^2 T}{\partial y^2} - \frac{1}{a^2}\frac{\partial T}{\partial t} = -\frac{Q_V}{\lambda}\delta(t)U(y) \tag{7.4.5}$$
where Q_V is constant. Laplace transformations give
$$\mathcal{L}_{pt}\mathcal{L}_{ry}T = \frac{a^2 Q_V}{\lambda} \cdot \frac{p}{p - a^2 r^2} \tag{7.4.6}$$
where p and r are Laplace transform variables with respect to t and y, respectively. Inversion with respect to p yields, for $t > 0$,
$$\mathcal{L}_{ry}T = \frac{Q_V}{\varrho c_p} e^{a^2 r^2 t} \tag{7.4.7}$$
By using Laplace transform tables, the inversion is found to be
$$T(y) = \frac{Q_V}{2\varrho c_p}\{1 + \mathrm{erf}[y/(2a\sqrt{t})]\} \tag{7.4.8}$$
where $\mathrm{erf}(u) = (2/\sqrt{\pi})\int_0^u \exp(-s^2)ds$.

The superposition $T = T(y+h/2) - T(y-h/2)$ provides the solution of the problem of heat conduction in and from a layer $-h/2 < y < h/2$, subjected to an instantaneous heat supply Q_V per unit volume:
$$T = \frac{Q_V}{2\varrho c_p}\{\mathrm{erf}[(y+h/2)/(2a\sqrt{t})] - \mathrm{erf}[(y-h/2)/(2a\sqrt{t})]\} \tag{7.4.9}$$
In the mid-plane of the layer,
$$T = \frac{Q_V}{\varrho c_p}\mathrm{erf}[h/(4a\sqrt{t})] = T_A\,\mathrm{erf}[h/(4a\sqrt{t})] \tag{7.4.10}$$
where T_A is the temperature at $t = +0$.

The temperature in the mid-plane of the layer has dropped to $0.9T_A$ after the time $t_{0.9} \approx 0.05h^2/a^2$, to $0.5T_A$ after the time $t_{0.5} \approx 0.3h^2/a^2$ and to $0.1T_A$ after the time $t_{0.1} \approx 8h^2/a^2$. Two examples will be given. In the first, the layer thickness is 100 µm, which might be representative for cell dimensions in a 4340 steel. Then, using the value $a^2 = 10^{-5}\mathrm{m}^2/\mathrm{s}$, the time $t_{0.9} \approx 50\,\mu\mathrm{s}$, $t_{0.5} \approx 0.3\,\mathrm{ms}$ and $t_{0.1} \approx 8\,\mathrm{ms}$. For titanium the correponding times are about twice as large. The figures indicate that the maximum temperature in the layer will be close to the adiabatic temperature if the heat is supplied to the layer in less than about 50 µs. In this time, a crack travelling with 10 m/s has propagated 0.5 mm, which might be the order of the length of the process region.

Next, consider a layer thickness of 10 µm, which might be typical for bands of localized plastic flow in a 4340 steel. The values for $t_{0.9}$, etc., are now 100 times smaller, but still large enough that the maximum temperatures will be close to the adiabatic temperature if heat is supplied to the layer in less than about $1\mu\mathrm{s}$, which might correspond to the passage of a process region at a crack travelling with 500 m/s. The conclusion is that the cell average maximum temperatures come close to the adiabatically calculated temperatures for cracks moving faster than a few metres per second in metallic materials, whereas a few hundred metres per second might be needed for adiabatic conditions in bands of localized plastic flow. The relative smallness of h^2/a^2 in ceramic materials may increase these velocities considerably. In the next subsection an analysis of the heat conduction from the edge of a propagating crack will be considered, using essentially the approach by Rice and Levy (1969).

Heat conduction and temperature distribution near a moving crack edge

General relations

A propagating crack edge generates a heat source, moving through the material. Mode I symmetry is assumed. From a thermal point of view, it is then irrelevant that the source is generated by a propagating crack, if the heat transfer to the medium (usually air) between the crack faces is negligible. This is assumed to be the case and, similarly, heat transfer from other body boundaries is considered to be negligible. Furthermore, the heat generation is assumed to be uniform along lines parallel to the crack edge. Thus, the thermal problem can be reduced to the problem of finding the temperature field accompanying a cylindrical source, delivering the heat rate $\dot{Q}(x - Vt, y)$, where V is the crack velocity, which is assumed to be constant. The Galilean transformation $X = x - Vt$, $y = Y$, is introduced.

The solution is obtained by superposition, using the fundamental solution for a moving line heat source, $\dot{Q}_L \delta(X)\delta(Y)$, where $\dot{Q}_L$ is the heat rate per unit length and $\delta(\cdot)$ is the Dirac delta function. For this case, the heat conduction equation (7.4.4) reads

$$\frac{\partial^2 T}{\partial X^2} + \frac{\partial^2 T}{\partial Y^2} + \frac{V}{a^2} \cdot \frac{\partial T}{\partial X} = -\frac{\dot{Q}_L}{\lambda} \delta(X)\delta(Y) \tag{7.4.11}$$

Laplace transformations give

$$\mathcal{L}_{qX}\mathcal{L}_{rY} T = -\frac{\dot{Q}_L}{\lambda} \cdot \frac{qr}{q^2 + (V/a^2)q + r^2} \tag{7.4.12}$$

where q and r are Laplace transform variables with respect to X and Y, respectively. For inversion with respect to q, it is convenient to write

$$\frac{qr}{q^2 + (V/a^2)q + r^2} = r \frac{q}{q + 1/d_V} \cdot \frac{q + 1/d_V}{(q + 1/d_V)^2 + r^2 - 1/d_V^2} \tag{7.4.13}$$

where $d_V = 2a^2/V$. Use of (A5.23) then gives

$$T = -\frac{\dot{Q}_L}{\lambda} e^{-X/d_V} \mathcal{L}_{qX}^{-1} \mathcal{L}_{rY}^{-1} \frac{qr}{q^2 + r^2 - 1/d_V^2} \tag{7.4.14}$$

The inversion can be found in Laplace transform tables, for instance in van der Pol and Bremmer (1959, equation XVI,62), where it is given as $-i/4 \cdot H_0^{(1)}(i\sqrt{X^2 + Y^2}/d_V)$, which equals $-1/(2\pi)K_0(\sqrt{X^2 + Y^2}/d_V)$. Here, $H_0^{(1)}(\cdot)$ is a Hankel function and $K_0(\cdot)$ is a modified Hankel function. Hence, the solution of the fundamental problem is

$$T = \frac{\dot{Q}_L}{2\pi\lambda} e^{-X/d_V} K_0(\sqrt{X^2 + Y^2}/d_V) \tag{7.4.15}$$

Now, by superposition, the solution of the original problem is found to be

$$T = \frac{1}{2\pi\lambda} \int_{-\infty}^{+\infty} \int_{-\infty}^{+\infty} \dot{Q}(X - X_0, Y - Y_0) e^{-X_0/d_V}$$
$$\times K_0(\sqrt{X_0^2 + Y_0^2}/d_V)\,\mathrm{d}Y_0\,\mathrm{d}X_0 \quad (7.4.16)$$

The function $\dot{Q}(X,Y)$ has to be determined from the irreversible mechanical work, multiplied by the factor β, and from the negative component (cooling) due to thermoelastic dilatation. The latter part is usually neglected, but experimental research by Rittel (1998a,b) indicates that it may be quite substantial. In experiments concerning temperatures in front of a crack, either a stationary crack subjected to stress wave loading or a crack running at high speed, a cooling phase precedes the heating. This was noticed previously by Fuller et al. (1975) and, as pointed out by Rittel (1998a,b), it can also be seen in recordings by Weichert and Schönert (1978).

Thermoelastic cooling

The magnitude of cooling may be estimated from the expression for the heat flux, due to the elastic dilatation rate $\dot{\epsilon}_{kk}^e$. It equals approximately (see e.g. Boley and Weiner 1960)

$$-3\kappa\alpha T_0 \dot{\epsilon}_{kk}^e \quad (7.4.17)$$

where κ is the modulus of compression, T_0 is the original absolute temperature (it is assumed that relative temperature changes $|T|/T_0$ are small compared to unity) and α is the linear thermal expansion coefficient. Then, assuming that $\dot{\epsilon}_{kk} \geq 0$ in the elastic region in front of the crack, the minimum possible temperature will be

$$T = -3\frac{\kappa\alpha T_0 \epsilon_{kk}}{\varrho c_p} \quad (7.4.18)$$

if the dilatation increase from zero is so rapid that appreciable amounts of heat are not conducted away. Here, ϵ_{kk} is to be determined for the location just in front of the plastic region. The stresses in the plastic region straight ahead of the crack do not vary much between stationary, slowly moving or fast running cracks (cf. pages 280, 322 and 529). Roughly, for plane strain, the in-plane stresses equal $\sigma_y \approx 2.5\sigma_Y$ and $\sigma_x \approx 1.5\sigma_Y$, somewhat depending on Poisson's ratio and on whether the Tresca or the Huber-von Mises yield condition is used.

Now, use of Hooke's law gives the maximum possible total dilatation as

$$\epsilon_{kk} = \frac{(1+\nu)(\sigma_x + \sigma_y)}{3\kappa} \quad (7.4.19)$$

and the minimum possible temperature equals

$$T = -\frac{(1+\nu)\alpha T_0(\sigma_x + \sigma_y)}{\varrho c_p} \quad (7.4.20)$$

Insertion of the value $4\sigma_Y$ for $\sigma_x + \sigma_y$ and putting $4(1+\nu) = 5.3$ yields†

$$T \approx -\frac{5.3\alpha T_0 \sigma_Y}{\varrho c_p} \tag{7.4.21}$$

Rittel (1998b) gives some figures for PMMA and for steel C-300. He remarks that ϱc_p varies little between different materials, and may be taken to about $10^6 \text{ N/m}^2/°\text{C}$. The yield stress σ_Y may be estimated as 200 Mpa for PMMA and 2000 MPa for the steel, reflecting yield stress increase at rapid loading. The linear thermal expansion coefficient may be taken as $10^{-4}/°\text{C}$ for PMMA and $10^{-5}/°\text{C}$ for steel C-300. Insertion into (7.4.21), using the value $T_0 = 300°\text{K}$, yields the temperature drop

$$T \approx -32°\text{C} \tag{7.4.22}$$

for both PMMA and steel C-300. Rittel (1998a) reports temperature drops between a few degrees and about 80°C in experiments with PMMA.

Heat generation and conduction

Temperature changes due to thermoelastic cooling, (7.4.21), result effectively in a lowered ambient temperature for crack propagation. The continued analysis may therefore proceed without any consideration of thermoelastic cooling other than lowered ambient temperature. Suitable models of the process region and its surrounding plastic region have to be used, and in general the temperature field has to be calculated numerically. However, some analytical advances are possible, by assuming that $\dot{Q}(X,Y)$ can be written in the form $\dot{Q}_0 f(\xi)g(\eta)$, where $\xi = X/d_V$, $\eta = Y/d_V$ and $f(\cdot)$ and $g(\cdot)$ are dimensionless. Then, expressing $g(\eta)$, which is assumed to be even, by a Fourier integral,

$$g(\eta) = \int_0^\infty G(c)\cos(c\eta)\mathrm{d}c, \quad G(c) = \frac{2}{\pi}\int_0^\infty g(\eta)\cos(c\eta)\mathrm{d}\eta \tag{7.4.23}$$

the temperature field becomes

$$T = \frac{T_0}{2\pi}\int_{-\infty}^{+\infty} f(\xi-\xi_0)e^{-\xi_0}\int_{-\infty}^{+\infty} K_0(\sqrt{\xi_0^2+\eta_0^2})\int_0^\infty G(c)\cos[c(\eta-\eta_0)]\mathrm{d}c\,\mathrm{d}\eta_0\,\mathrm{d}\xi_0$$

$$= \frac{T_0}{\pi}\int_{-\infty}^{+\infty} f(\xi-\xi_0)e^{-\xi_0}\int_0^\infty G(c)\cos(c\eta)$$

$$\times \int_0^{+\infty} K_0(\sqrt{\xi_0^2+\eta_0^2})\cos(c\eta_0)\mathrm{d}\eta_0\,\mathrm{d}c\,\mathrm{d}\xi_0 \tag{7.4.24}$$

where

$$T_0 = \dot{Q}_0 d_V^2/\lambda = 4\dot{Q}_0 a^2/(\varrho c_p V^2) \tag{7.4.25}$$

The inner integral in (7.4.24) can be found, for instance, in Gradshteyn and Ryzhik (1980), leading to

$$T = \frac{T_0}{2}\int_0^\infty \frac{G(c)\cos(c\eta)}{\sqrt{1+c^2}}\int_{-\infty}^{+\infty} e^{-\xi_0-\sqrt{1+c^2}|\xi_0|}f(\xi-\xi_0)\mathrm{d}\xi_0\,\mathrm{d}c \tag{7.4.26}$$

---

† This approximation (for values of ν other than 0.325) is made because, as mentioned, there are other dependences on Poisson's ratio, so that keeping the factor $1+\nu$ might give a misleading indication of the temperature dependence on ν.

Consider, as an example, a process region consisting of one row of cells, i.e., forming a rectangular region with height h_p and length r_p. The heat generation outside this region is neglected. Guided by the previous estimate of the velocity for which the adiabatic temperature is approached (e.g. about 10 m/s for 4340 steel), it is assumed that the crack velocity is not higher than that it is possible to use elastostatic relations. An estimate of $\dot{Q}$ may be obtained by using a Barenblatt region, $0 < X < r_p$, with parabolic stress distribution, $\sigma_y^0(X) = \sigma_D s(2-s)$, where $s = X/r_p$. With $2v_+$ denoting the crack opening displacement, the energy dissipation per unit time and area in this region is

$$2\sigma_y^0 \cdot \frac{\partial v_+}{\partial t} = -2V\sigma_y^0 \cdot \frac{\partial v_+}{\partial x} = \frac{V\sigma_D^2}{\pi(1-k^2)\mu} s(2-s)$$

$$\times \left[s(2-s) \ln \frac{1+\sqrt{1-s}}{1-\sqrt{1-s}} + 2(4/3 - s)\sqrt{1-s} \right] \quad (7.4.27)$$

This may be obtained from (4.4.62). A plot shows that the function in (7.4.27) following $V\sigma_D^2/[\pi(1-k^2)\mu]$ equals $2\sin(\pi s)$, approximately, which gives the heat generation per unit time and volume as

$$\dot{Q} = \dot{Q}_0 f(\xi) g(\eta), \qquad \dot{Q}_0 = \frac{\beta V \sigma_D^2}{\pi(1-k^2)\mu h_p} \quad (7.4.28)$$

where

$$f(\xi) = 2\sin(\pi \xi d_V/r_p) \text{ for } 0 < \xi < r_p/d_v \quad (7.4.29)$$
$$g(\eta) = 1 \text{ for } -h_p/(2d_V) < \eta < h_p/(2d_V) \quad (7.4.30)$$

assuming the heat generation to be uniform across the layer thickness.

Integration of $2\partial v_+/\partial x$ gives the crack opening displacement, which is assumed to equal $0.3h_p$. The integration can be performed exactly and leads to a determination of h_p/r_p from the expression

$$20\sigma_D r_p 9\pi(1-k^2)\mu = 0.3 h_p \quad (7.4.31)$$

For simplicity, only the point $X = Y = 0$ will be considered; in this vicinity the maximum temperature is expected, except for very slowly moving cracks. For very fast running cracks, it will approach the adiabatic temperature,

$$T_A = \frac{1}{\varrho c_p V} \int_0^{r_p} \dot{Q} dX = \frac{4\dot{Q}_0 r_p}{\pi \varrho c_p V} = \frac{4\beta \sigma_D^2 r_p}{\pi^2(1-k^2)\varrho c_p \mu h_p} \quad (7.4.32)$$

Use of (7.4.25) and insertion into 7.4.26 gives, after the change $\xi_0 \to -\xi_0$,

$$T = \frac{\pi a^2}{V r_p} T_A \int_0^\infty \frac{G(c)}{\sqrt{1+c^2}} \int_0^{r_p/d_V} e^{-(\sqrt{1+c^2}-1)\xi_0} \sin(\pi \xi_0 d_V/r_p) d\xi_0 \, dc \quad (7.4.33)$$

where the inner integral is elementary. Evaluation and use of (7.4.23) leads to

$$T = \frac{\pi d_V^2 T_A}{r_p^2} \int_0^\infty \frac{\sin[ch_p/(2d_V)][1 + e^{-(\sqrt{1+c^2}-1)r_p/d_V}]}{c\sqrt{1+c^2}[(\sqrt{1+c^2}-1)^2 + \pi^2 d_V^2/r_p^2]} dc \quad (7.4.34)$$

Consider now cracks running fast enough that $\pi^2 d_V^2/r_p^2 = 4\pi^2 a^4/(V r_p)^2 \ll 1$. Then,

7.4 TEMPERATURE CHANGES AT MOVING CRACK EDGES

by changing the integration variable to $u = c\sqrt{r_p/(2d_V)}$, it is seen that the integral is dominated by values of $u \ll 1$, so that an estimate is given by

$$T \approx \frac{T_A}{\pi} \int_0^\infty \frac{\sin[uh_p\sqrt{V}/(2a\sqrt{r_p})](1 + e^{-u^2})}{u} du$$

$$= \frac{T_A}{2}\left[1 + \mathrm{erf}\left(\frac{h_p\sqrt{V}}{4a\sqrt{r_p}}\right)\right]$$

$$\approx T_A\left[1 - \frac{2a\sqrt{r_p}}{h_p\sqrt{\pi V}} e^{-h_p^2 V/(16a^2 r_p)}\right] \quad (7.4.35)$$

Obviously, the temperature deviates appreciably (more than about 10 per cent) from the adiabatic temperature for crack velocities $V < V_0$, $V_0 \approx 16a^2 r_p/h_p^2$. For 4340 steel, assuming $h_p = 100\,\mu\mathrm{m}$, $\sigma_D = 4.7\sigma_Y$, $\sigma_Y = 1700\,\mathrm{MPa}$, $\mu = 77\,\mathrm{GPa}$, $\nu = 0.3$ ($k^2 \approx 0.29$), $a^2 = 10^{-5}\mathrm{m}^2/\mathrm{s}$, $\beta = 0.85$, it is found that V_0 is approximately $10\,\mathrm{m/s}$, in agreement with the estimate following (7.4.10), and $T_A \approx 400°$, in approximate agreement with a previous estimate†. Note, however, that the adiabatic temperature rise in the present case occurs in an environment which has been exposed to thermoelastic cooling.

In the derivation of the results, no attempt was made to consider possible increased heat generation due to viscoplastic flow. Note also that elastostatic relations were used, because the stress-strain fields are not very sensitive to the crack velocity, except for very high velocities. The maximum temperature obtained near the edge of a running crack may therefore be expected to change very little over a wide range of velocities. This appears to be corroborated by experimental (e.g. Zehnder and Rosakis 1991) and numerical studies (Li et al. 1996).

---

† The slight difference is due to differences in the cohesive-decohesive curve, which was not explicitly given in the present case, but may be extracted from the properties assumed for the Barenblatt region.

8
Physical and Engineering Aspects of Fracture

8.1 Introduction

The preceding chapters have mainly dealt with descriptions of different phases in the fracture process and with crack mechanics. In the present chapter, the development of fracture will be discussed, with respect both to significant physical processes, such as void formation and growth, and to different ways of predicting the occurrence of fracture. In engineering contexts, fracture is usually an unwanted phenomenon, even though it may be desired for some purposes, for instance in mining. Regardless of whether it is wanted or not, it is often desirable to make some kind of an assessment about the likelihood of fracture. It is also desirable to find the cause of fracture that has occurred, in order to prevent further occurrences of a similar kind or, perhaps, to determine legal responsibilities in accidents involving fracturing. A common question is whether bad material or bad design is to blame. Another question may concern the load, for instance if a fractured structure had been exposed to more severe loading than it was intended for. Other questions may concern when, rather than if, unstable crack growth will occur, such as the onset of an earthquake or a landslide, even though terms like slip may be more appropriate than fracture in such contexts.

As in many other physical disciplines, fracture mechanics is based on the trilogy theory, experiment and practice. These three components are involved in perpetual interaction. Theories are tested against experiments and practical experiences, experiments are designed with respect to theoretical questions and practical needs, and the practical reality requires theoretical methods and experimental data. Advances in one of these components inspire revisions of the other two.

Certain practical design matters and certain testing methods are subject to rigorous regulations when the possibility of fracture is a concern. Most countries have developed codes or normative rules for different applications, for instance building codes and pressure vessel codes. Such codes also contain rules for maintenance and inspection of structures. Inspections include routines for crack detection and rules for accepting or correcting the faults found. There are also standards about testing procedures, etc.

One might argue that codes and standards often reflect yesterday's state of the art and that recent research has made some codes or standards obsolete. They are, however, extremely important for design and maintenance of engineering structures, even if some of them might be based more on rules of thumb than on scientific work, and they give a reasonably well-founded feeling of reliance both for designers and users. One should also realize that changes of codes and standards are not solely motivated by new research results. Reasons for changes may arise because of changing production methods or structural scales. One example is provided by ship-building, where the transition from riveted to welded structures and the huge size increase of large vessels made it necessary to make far-reaching code changes. Other reasons for code changes or introduction of new codes may be the advent of new materials, environmental concerns, etc.

The present work is only occasionally referring to codes and standards, since it is directed towards research in crack and fracture mechanics. On the other hand, it has a clear connection with codes and standards, because these will eventually be rewritten with a view toward research results.

The present chapter will deal with some aspects on void growth and other processes in the crack edge vicinity that were not covered in Chapter 1, because the framework presented in later chapters, for instance the theory of plastic flow, given in in Chapter 5, is needed. Thereupon follows a discussion about the development and the prediction of fracture, including statistical aspects. There is some discussion about crack directions, including cracks caused by compressive loads. Finally, some methods for tests in fracture mechanics and for prediction of fracture in engineering structures will be briefly discussed.

8.2 Processes in the crack edge vicinity

Void growth

The role of void formation and growth in fracture of ductile materials was recognized very early by Tipper (1949). The first analysis of hole growth appears to have been made by McClintock (1966, 1968a). The following analysis of the growth of a cylindrical hole is based on these papers.

Consider a circular cylindrical hole in an infinite rigid-perfectly plastic solid, satisfying the Huber-von Mises yield condition and the associated flow rule. A cylindrical coordinate system r, φ, z is introduced, with $r = 0$ along the axis of the hole; see Fig. 8.2.1. The body is subjected to a remote radial stress $\sigma_r = \sigma_r^\infty > 0$ and to a uniform strain rate $\dot{\epsilon}_z = \dot{\epsilon}_z^\infty > 0$ (as a result of a remote imposed strain rate $\dot{\epsilon}_z^\infty$). If σ_r^∞ exceeds a certain value, the hole will expand. Let a be the current hole radius and σ_Y the yield strength. From dimensional considerations it follows that the relative expansion rate of the hole radius, $\dot{a}/a$, must be of the form $\dot{\epsilon}_z^\infty f(\sigma_r^\infty/\sigma_Y)$. This relation will now be determined.

The strain rates are connected with the only non-zero displacement, u_r, through the relations

$$\dot{\epsilon}_r = \frac{\partial \dot{u}_r}{\partial r}, \quad \dot{\epsilon}_\varphi = \frac{\dot{u}_r}{r}, \quad \dot{\epsilon}_z = \dot{\epsilon}_z^\infty \tag{8.2.1}$$

Plastic incompressibility then implies that

$$\frac{\partial \dot{u}_r}{\partial r} + \frac{\dot{u}_r}{r} + \dot{\epsilon}_z^\infty = 0 \implies \frac{1}{r}\left(\frac{\partial (r\dot{u}_r)}{\partial r}\right) + \dot{\epsilon}_z^\infty = 0 \tag{8.2.2}$$

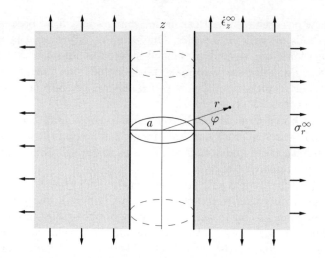

Fig. 8.2.1 Cylindrical hole in an infinite body, subjected to a remote radial stress $\sigma_r = \sigma_r^\infty$ and to a uniform longitudinal strain rate $\dot{\epsilon}_z = \dot{\epsilon}_z^\infty$.

which integrates to

$$\dot{u}_r = \frac{a\dot{a}}{r} - \frac{\dot{\epsilon}_z^\infty a}{2}\left(\frac{r}{a} - \frac{a}{r}\right) \tag{8.2.3}$$

Thus,

$$\dot{\epsilon}_r = -\dot{\epsilon}_z^\infty\left[\left(\frac{\dot{a}}{\dot{\epsilon}_z^\infty a} + \frac{1}{2}\right)\frac{a^2}{r^2} + \frac{1}{2}\right] \tag{8.2.4}$$

$$\dot{\epsilon}_\varphi = +\dot{\epsilon}_z^\infty\left[\left(\frac{\dot{a}}{\dot{\epsilon}_z^\infty a} + \frac{1}{2}\right)\frac{a^2}{r^2} - \frac{1}{2}\right] \tag{8.2.5}$$

The flow rule (5.2.18) gives

$$\dot{\epsilon}_r = \frac{3}{2}(\sigma_r - \sigma_m)\frac{\dot{\epsilon}_{eff}}{\sigma_Y}, \quad \dot{\epsilon}_\varphi = \frac{3}{2}(\sigma_\varphi - \sigma_m)\frac{\dot{\epsilon}_{eff}}{\sigma_Y} \tag{8.2.6}$$

where σ_m is the mean stress, $(\sigma_r + \sigma_\varphi + \sigma_z)/3$, and, cf. (5.2.17),

$$\dot{\epsilon}_{eff} = \sqrt{\frac{2}{3}[\dot{\epsilon}_r^2 + \dot{\epsilon}_\varphi^2 + (\dot{\epsilon}_z^\infty)^2]} = \dot{\epsilon}_z^\infty\sqrt{A^2 a^4/r^4 + 1} \tag{8.2.7}$$

where

$$A = \frac{2}{\sqrt{3}}\left(\frac{\dot{a}}{\dot{\epsilon}_z^\infty a} + \frac{1}{2}\right) \tag{8.2.8}$$

Use of the flow rules (8.2.6) and the equilibrium equation

$$\frac{\partial \sigma_r}{\partial r} + \frac{\sigma_r - \sigma_\varphi}{r} = 0 \tag{8.2.9}$$

leads to

$$\frac{\partial \sigma_r}{\partial r} = -\frac{2\sigma_y}{3} \cdot \frac{\dot{\epsilon}_r - \dot{\epsilon}_\varphi}{r\dot{\epsilon}_{eff}} = \frac{2\sigma_Y}{3} \cdot \frac{Aa^2}{r\sqrt{A^2 a^4 + r^4}} \qquad (8.2.10)$$

which integrates to

$$\sigma_r = \frac{2\sigma_Y A a^2}{\sqrt{3}} \int_a^r \frac{dr}{r^3 \sqrt{A^2 a^4 / r^4 + 1}} \qquad (8.2.11)$$

Make the variable substitution $s = Aa^2/r^2$, implying that $\mathrm{d}r/r^3 = -\mathrm{d}s/(2Aa^2)$. Then,

$$\sigma_r = \frac{\sigma_Y}{\sqrt{3}} \int_{Aa^2/r^2}^{A} \frac{\mathrm{d}s}{\sqrt{1+s^2}} = \frac{\sigma_Y}{\sqrt{3}} \left(\operatorname{asinh} A - \operatorname{asinh} \frac{Aa^2}{r^2} \right) \qquad (8.2.12)$$

Letting $r \to \infty$ leads to the sought expression for hole radius expansion:

$$\boxed{\frac{\dot{a}}{a} = \frac{\dot{\epsilon}_z^\infty}{2} \left(\sqrt{3} \sinh \frac{\sqrt{3}\sigma_r^\infty}{\sigma_Y} - 1 \right)} \qquad (8.2.13)$$

This expression was established (in a slightly different form) by McClintock (1966). It shows that the expansion rate becomes very high for stress levels that are encountered in a crack edge vicinity, because $\sinh(\cdot)$ increases exponentially for arguments substantially larger than unity. As remarked by Rice (1968b), by putting $\sigma_r^\infty = (1+\pi)\sigma_Y/\sqrt{3}$, the mean stress in front of a crack edge according to slip line theory, (5.4.12), the expansion rate becomes as high as $\dot{a}/a \approx 26.7\dot{\epsilon}_z^\infty$.

All stresses and strain rates can now be determined. The strain rates $\dot{\epsilon}_r$ and $\dot{\epsilon}_\varphi$ follow from (8.2.4)-(8.2.5), and the stress σ_φ is found by using (8.2.9). The stress σ_z is found from the yield condition, noting that the flow rule for $\dot{\epsilon}_z$ implies that $\sigma_z > \sigma_m$. As $r \to \infty$, $\sigma_\varphi \to \sigma_r^\infty$ and $\sigma_z \to \sigma_r^\infty + \sigma_Y$. From the expression for $\dot{\epsilon}_\varphi = \dot{u}_r/r$ it follows that hole growth requires that $\sigma_r/\sigma_Y > \sqrt{1/3}\operatorname{asinh}(1/\sqrt{3}) \approx 0.3171$, and that the radial velocity is positive everywhere if $\sigma_r/\sigma_Y > \sqrt{1/3}\operatorname{asinh}(2/\sqrt{3}) \approx 0.5696$. Finally, it may be noted that $\dot{a}/a > \dot{\epsilon}_z^\infty$ if $\sigma_r/\sigma_Y > \sqrt{1/3}\operatorname{asinh}(\sqrt{3}) \approx 0.7603$.

McClintock (1968a) showed that the Tresca yield condition leads to similar expressions for the hole expansion rate, i.e. an approximately exponential increase with σ_r/σ_Y. Note that the triaxiality (which can be defined as the ratio between the mean stress and the maximum difference of the principal stresses) increases with σ_r/σ_Y.

The approximately exponential dependence was also found by Rice and Tracey (1969) for the relative increase of the average radius of a spherical hole in a body, subjected to the same composition of remote strain rates as a tensile test specimen, but with a superposed hydrostatic stress. As could be expected, considering the lateral contraction, the hole will be extended more in the direction of maximum strain rate than laterally, although this difference is less pronounced at higher loads (mean stresses), because of the higher stress triaxiality. Rice and Tracey (1969) used the Rayleigh-Ritz method, but the number of terms afforded was not sufficient for an accurate estimate. Increase of the number of terms (Huang 1989, Needleman et al. 1992) leads to

$$\frac{\dot{a}}{a} \approx 0.427 \exp\left(\frac{3\sigma_m^\infty}{2\sigma_Y} \right) \qquad (8.2.14)$$

where a is an average hole radius and σ_m the remote mean stress.

Because Rice and Tracey (1969) assumed an ambient strain field not unlike the one in a tensile test piece, it is not surprising that they found a more rapid hole growth in the axial than in the lateral direction. An attempt to simulate the ambient strain composition for a hole in front of a crack edge was made by Andersson (1977). He considered axial straining of a circular cylinder with, originally, the same diameter as height and containing a central spherical hole. The cylinder surface is prevented from moving radially in order to simulate the approximately plane strain condition in front of a crack edge. The material is assumed to be rigid-perfectly plastic. He used a finite element method and found that the hole grows faster laterally than axially. Thus, the hole becomes flattened, in agreement with findings from fractographic studies (e.g., Broek 1971, 1972). Andersson (1977) obtained a relation between the increase of cylinder height and applied axial stress, i.e. a decohesion relation. Due to the assumption of rigid-perfectly plastic material, there is no increasing (cohesion) part of this relation.

As previously described (page 18), Xia and Shih (1995b) considered the growth of originally spherical voids in cubic cells, subjected to various macroscopic strains. Faleskog and Shih (1997) considered growth of one central cylindrical void in a representative cell, also containing a number of smaller, also cylindrical, voids.

The appropriateness of axisymmetric models was questioned by Kuna and Sun (1996). They considered representative cells containing spherical voids in different geometrical arrays, cubic primitive, body centered cubic and hexagonal together with an axisymmetric model. The cells were subjected to loads with varying triaxiality. For all cases, it was found that the three-dimensional models sustained a higher maximum load and exhibited a slower load decrease on the decohesive side than the axisymmetric model, whereas the differences on the cohesive side were small, except near the maximum load.

Growth of small voids. Strain gradient and non-local plasticity

Even the largest pre-existing voids in ductile materials may be very small, about 1μm or smaller. Considering that the average distance between dislocations may be of about the same order of magnitude, it is obvious that a continuum theory of plasticity may be inappropriate. Also, whereas more or less rounded large voids may exist as a result of macroscopic plastic flow, voids of micrometer size cannot be expected to be rounded, but rather to exhibit corners, perhaps atomistically sharp. Some voids may be located at a local interface, such as between a second phase particle and the matrix material.

For very small voids, growth by plastic deformation might not be possible by use of pre-existing dislocations, but by emission of dislocations from corners. Alternatively, cleavage may take place rather than dislocation motion. The competition between cleavage and plastic flow will be briefly discussed on pages 556ff.

Even for voids with a size that is about one order of magnitude larger than the average distance between dislocations, conventional plasticity theory might be insufficient. One reason is that the strain gradients near a void will be very high, due to the small scale over which large strain differences appear in the void vicinity. In conventional plasticity, strain gradients do not enter the constitutive equations. Thus, conventional plasticity predicts that two bodies with the same shape, but differing in scale, and subjected to the same boundary tractions at corresponding points, will exhibit the same, appropriately scaled, deformations. However, there seems to be experimental support for exceptions to that behaviour for small geometries (Fleck et al. 1994), even though some uncertainty exists about whether material properties are the same near a body surface as in its interior, or if they are independent of specimen size. The apparent influence of the strain gradient on plastic flow properties has given rise to theories of *strain gradient plasticity* (Aifantis 1984, Mühlhaus and Aifantis 1991, de Borst and Mühlhaus 1992, Mühlhaus 1994, Fleck and Hutchinson 1993, 1997, Fleck et al. 1994, Xia and Hutchinson 1996, Wei and Hutchinson 1997).

In gradient plasticity, the constitutive relations depend both on plastic strains and strain gradients by the incorporation of a length parameter, l. Gradient effects are important when the scale of considerable strain variations is of the order of l, but they are negligible when the scale is much larger. Then, the strain gradient plasticity coincides with conventional plasticity.

The material in the vicinity of a small void will be severely plastically distorted during void growth, and this implies an accumulation of dislocations to maintain this distorsion, so-called *geometrically necessary dislocations*. There are also dislocations present that are not geometrically necessary, but randomly distributed, so-called *statistically stored dislocations*. Such dislocations dominate in number when the strain gradients are sufficiently small, for instance in bent beams of millimetre thickness or larger, but on a scale of about $10\mu m$, the contribution by geometrically necessary dislocations may be considerable. The length parameter l may be chosen as the length scale for which both contributions are of the same order of magnitude, which may lead to $l \approx 5\mu m$ for some materials; cf. Fleck et al. (1994).

Hardening of ductile materials appears to increase with increasing dislocation density. This implies that the plastic flow stress depends both on the strain and the strain gradient, which might explain the basic mechanism for strain gradient influence (Fleck and Hutchinson 1993, 1997).

Consideration of small scale plasticity is also taken in *non-local plasticity theory* (e.g., Eringen 1981, 1983, Bažant et al. 1984, Bažant and Lin 1988, Strömberg and Ristinmaa 1996). In non-local theories, the dependent variables at one point depend not only on the independent variables at the point itself, as in local theories, but also on the independent variables at other points. The dependence decreases with increasing distance to the point, which implies that a length parameter can be defined for incorporation in the constitutive equations. Mathematically, non-local plasticity theory is rather similar to strain gradient plasticity (Mühlhaus 1994, Strömberg and Ristinmaa 1996).

Strain gradient plasticity and non-local plasticity appear to be useful not only in the context of the growth of small voids, but also in many other contexts where large strain gradients appear, for instance in shear bands (Zbib and Aifantis 1988, Mühlhaus 1994, Strömberg and Ristinmaa 1996).

Cavitation in metal foils sandwiched between ceramics
Under certain circumstances, a void may grow spontaneously without bounds (other than those set by boundaries, etc.), so-called cavitation. This occurs, in particular, under highly constrained plastic flow, usually appearing as the result of a very large mean stress (Ashby et al. 1989, Hill 1950, Huang et al. 1991, Tvergaard and Hutchinson 1993). Such conditions are usually not met in the process region at the edge of a crack in homogeneous bulk material, but they may occur in front of a crack in a thin

metal foil, sandwiched between ceramics or other comparatively rigid materials (e.g. Reimanis et al. 1991). This situation was analysed by Varias et al. (1991, 1992) and later by He et al. (1996). Varias et al. (1991) showed that the mean stress in the metal may reach more than 5 times the yield stress when the assembly is loaded normal to the foil, i.e. in mode I. They also showed that the maximum mean stresses are obtained at a distance from the crack edge of several times the foil thickness, so that cavitation would take place too far ahead to be involved in immediate coalescence with the crack†. The experiments by Reimanis et al. (1991) indicated rather that the crack was opened and propagated as an interface cleavage crack until it merged with voids, nucleated at tiny defects and grown by cavitation‡.

Varias et al. (1992) analysed mixed mode loading of a metal-ceramic assembly and found a considerable loss of constraint when a strong mode II component is present. High stresses could then occur in the ceramic, possibly large enough to lead to cleavage.

He et al. (1996) analysed interface crack formation and growth for the metal-ceramic assembly. They found an interesting size dependence of the process: a penny-shaped crack with a diameter about equal to half the foil thickness for a very high mean stress is subjected to significantly higher stress intensity at its edge than either smaller or larger cracks.

The basic mechanism for cavitation in solids was analysed by Hill (1950). It will be demonstrated here for the case of an elastic-perfectly plastic solid. It is assumed that a spherical hole has been opened, due to some local stress or strain concentraion at an inhomogeneity such as a particle, but otherwise the solid is assumed to be homogeneous. For simplicity it is assumed to be incompressible both in elastic and plastic regions.

Consider a hole with radius a_0, which will be expanded to a current radius a under the action of a remote radial stress, $\sigma_r = \sigma_r^\infty$. A mass point, originally situated at radius r_0, will move to a current radius r. The tangential strain, ϵ_φ, at radius r is found from the condition of incompressibility. Because large strains are expected, it is convenient to use the concept of logarithmic strain, so that

$$\epsilon_\varphi = \ln \frac{r}{r_0} \tag{8.2.15}$$

Incompressibility implies that

$$r^3 - r_0^3 = a^3 - a_0^3 \tag{8.2.16}$$

leading to

$$\epsilon_\varphi = -\frac{1}{3} \ln\left(1 - \frac{a^3 - a_0^3}{r^3}\right) \tag{8.2.17}$$

It also follows from incompressibility that the radial strain

$$\epsilon_r = -2\epsilon_\varphi \tag{8.2.18}$$

A further consequence of incompressibility is that the deformation at any point is independent of any hydrostatic component at the point. Then, by subtracting the

---

† Investigations (e.g. Aoki et al. 1984) show that voids nucleate and grow at distances less than the opening due to blunting from the edge of a crack in bulk material.
‡ This will be discussed on page 563.

hydrostatic stress σ_φ, the radial stress becomes $\sigma_r - \sigma_\varphi$ and the tangential stresses vanish, so that there is a uniaxial stress state. Thus, the simple uniaxial stress-strain relation for an elastic-perfectly plastic solid can be used:

$$\sigma_r - \sigma_\varphi = \begin{cases} \dfrac{\sigma_Y}{\epsilon_Y}\epsilon_r & \text{for } 0 \le \epsilon_r \le \epsilon_Y \\ -\sigma_Y & \text{for } \epsilon_r > \epsilon_Y \end{cases} \qquad (8.2.19)$$

The equilibrium conditions read:

$$\frac{d\sigma_r}{dr} + \frac{2(\sigma_r - \sigma_\varphi)}{r} = 0 \qquad (8.2.20)$$

and then it follows that

$$-\frac{r}{2}\cdot\frac{d\sigma_r}{dr} = \begin{cases} -\sigma_Y & \text{for } a \le r < r_Y \\ \dfrac{2\sigma_Y}{3\epsilon_Y}\ln\left(1 - \dfrac{a^3 - a_0^3}{r^3}\right) & \text{for } r \ge r_Y \end{cases} \qquad (8.2.21)$$

where r_Y is the radius for which the two expressions become equal.
At $r = a$, $\sigma_r = 0$, so that integration gives

$$\sigma_r = 2\sigma_Y \int_a^{r_Y} \frac{dr}{r} - \frac{4\sigma_Y}{3\epsilon_Y}\int_{r_Y/a}^{r/a} \ln\left(1 - \frac{1 - (a_0/a)^3}{s^3}\right)\frac{ds}{s} \qquad (8.2.22)$$

The remote stress, $\sigma_r^\infty = (\sigma_r^\infty)_{cav}$, at cavitation, $a_0/a \to 0$, can now be determined:

$$\left(\frac{\sigma_r^\infty}{\sigma_Y}\right)_{cav} = 2\ln s_Y - \frac{4}{3\epsilon_Y}\int_{s_Y}^\infty \ln\left(1 - \frac{1}{s^3}\right)\frac{ds}{s} \qquad (8.2.23)$$

Here, $s_Y = r_Y/a$ is determined from the relation

$$\frac{2}{3\epsilon_Y}\ln\left(1 - \frac{1}{s_Y^3}\right) = -1 \qquad (8.2.24)$$

which follows from (8.2.21).
Make now the substitution

$$\frac{2}{3}\ln\left(1 - \frac{1}{s^3}\right) = -\epsilon_Y\xi \qquad (8.2.25)$$

leading to the expression for the cavitation stress,

$$\left(\frac{\sigma_r^\infty}{\sigma_Y}\right)_{cav} = 2\ln s_Y + \epsilon_Y\int_0^1 \frac{\xi\,d\xi}{e^{3\epsilon_Y\xi/2} - 1}$$

$$= \frac{2}{3}\ln\frac{1}{1 - e^{-3\epsilon_Y/2}} + \epsilon_Y\int_0^1 \frac{\xi\,d\xi}{e^{3\epsilon_Y\xi/2} - 1} \approx \frac{2}{3}\left(\ln\frac{2}{3\epsilon_Y} + 1\right) \qquad (8.2.26)$$

where the reasonable assumption that $\epsilon_Y \ll 1$ was used for the approximation.

The cavitation stress depends only on ϵ_Y. It is about $5.00\sigma_Y$ for $\epsilon_Y = 0.001$ and about $3.47\sigma_Y$ for $\epsilon_Y = 0.01$. Note that these values are considerably higher than expected ahead of a crack in bulk material: slip line theory gives a mean stress of $(1+\pi)\sigma_Y/\sqrt{3} \approx 2.39\sigma_Y$. Consideration of strain hardening leads to even higher values, whereas a smaller Poisson's ratio than $1/2$ leads to somewhat lower values; cf. Hill (1950) and Huang et al. 1991.

Tvergaard and Hutchinson (1993) analysed the consequences of deviations from spherical symmetry for the cavitation stress, by assuming a void with originally ellipsoidal shape, subjected to axisymmetric loading. For the cases considered, they found very small dependence on the initial void shape.

Internal local instability in a continuum model

Coalescence between micro-separations is closely related to local instability or bifurcation. It is usually a spontaneous, fast process. Although micro-separations play an important part for internal local instabilities, such instablilities may also occur in homogeneous continua under certain circumstances. A simple discussion will illustrate this.

Suppose that the forces $\boldsymbol{P}_k$ on the boundary facets k of a material volume form an equilibrium system and that the whole volume is at rest. It is assumed that rate and inertia effects are negligible. In order to investigate whether the equilibrium state is stable or not, infinitesimal boundary displacement deviations, $\mathrm{d}\boldsymbol{u}_k$, from this equilibrium state are assumed. In general, the new state cannot be an equilibrium state if the original force system, $\boldsymbol{P}_k$, remains unchanged. Suppose that the additional forces, $\mathrm{d}\boldsymbol{P}_k$, needed to preserve equilibrium, perform negative work. Then, an excess of energy, compared to the energy in this new equilibrium state, would have been given to the material unit, if the boundary forces instead had remained unchanged. Part of this energy would go to kinetic energy and the system moves further away from the original equilibrium state. This state is thus unstable under load control.

The work performed by the additional forces is $\frac{1}{2}\mathrm{d}\boldsymbol{P}_k\,\mathrm{d}\boldsymbol{u}_k$, where $\mathrm{d}\boldsymbol{u}_k$ are the displacements at the forces $\mathrm{d}\boldsymbol{P}_k$. Summation over index k is assumed. Now, $\frac{1}{2}\mathrm{d}\boldsymbol{P}_k\,\mathrm{d}\boldsymbol{u}_k$ depends on the choice of the displacement configuration $\mathrm{d}\boldsymbol{u}_k$. One special choice makes $\frac{1}{2}\mathrm{d}\boldsymbol{P}_k\mathrm{d}\boldsymbol{u}_k$ minimum, provided that the displacement magnitudes are normalized in some way. If this minimum is negative, then the equilibrium is unstable, and if it is positive, the equilibrium is stable. The way of normalizing influences the magnitude and the position of the minimum, but obviously not its sign. It can therefore be disregarded in a criterion for stability or instability. The criterion for stability under load control conditions can thus be written as

$$\mathrm{d}\boldsymbol{P}_k\mathrm{d}\boldsymbol{u}_k > 0 \tag{8.2.27}$$

for all possible choices of $\mathrm{d}\boldsymbol{u}_k$. Similarly, the criterion for instability under load control can be written as

$$\mathrm{d}\boldsymbol{P}_k\mathrm{d}\boldsymbol{u}_k < 0 \tag{8.2.28}$$

for some choice of $\mathrm{d}\boldsymbol{u}_k$.

It may be noted that even if $\mathrm{d}\boldsymbol{P}_k\mathrm{d}\boldsymbol{u}_k$ is negative, the forces $\boldsymbol{P}_k$ perform positive work together with the displacements $\mathrm{d}\boldsymbol{u}_k$, the work associated with incipient unstable motion under load control.

At this point, specialization to uniaxial extension along a symmetry axis of an orthotropic material, may be illustrative. Consider a cylindrical volume between two planes perpendicular to the axis of extension, subjected to a displacement disturbance with magnitude $\mathrm{d}u$ in the direction of the axis. Inequality (8.2.27) may then be written as

$$\mathrm{d}P\,\mathrm{d}u > 0 \tag{8.2.29}$$

where $\mathrm{d}P$ is the magnitude of the force required for equilibrium. This is obviously equivalent to the condition

$$\frac{\mathrm{d}P}{\mathrm{d}u} > 0 \tag{8.2.30}$$

which is recognized as the condition that the slope of the force-elongation curve should be positive, as for the ascending part of a cohesion-decohesion curve. The corresponding instability condition, with $<$ sign rather than $>$ sign, is characteristic of the descending part of a cohesion-decohesion curve, which marks instability under load control.

The stability criterion (8.2.27) is formulated for an arbitrary volume of the material. For a connection with continuum models, it is suitable to choose an infinitesimally small volume $\mathrm{d}V = \mathrm{d}x_1\mathrm{d}x_2\mathrm{d}x_3$, with plane boundary surfaces parallel with the x_1x_2-, x_2x_3- and x_3x_1-planes in a Cartesian coordinate system. This volume is supposed to contain material in a loaded state. Thus, originally, i.e. before loading, the same material occupied a volume with some other shape.

Note that an infinitesimally small volume in a homogeneous and homogeneously loaded continuum is automatically subject to load control, even if the volume boundaries would be subjected to grip control. This may be understood by a comparison with a tensile test specimen loaded under grip control (slowly increasing elongation). Incipient necking occurs at some cross-section just after maximum force is reached. Rapid instability (fracture) will then occur if the specimen is much longer than its diameter, whereas instability in a short test piece only occurs after a substantial decrease of the force.

The stress σ_{11} is defined as the current normal force on the surface $\mathrm{d}x_2\mathrm{d}x_3$ divided by $\mathrm{d}x_2\mathrm{d}x_3$, and in the same manner the other components of the stress tensor σ_{ij} are defined. The displacement vector is called u_i. Then, the condition of stability, inequality (8.2.27), reads:

$$\mathrm{d}[\sigma_{11}(x_1+\mathrm{d}x_1)]\mathrm{d}x_2\mathrm{d}x_3 \cdot \mathrm{d}[u_1(x_1+\mathrm{d}x_1)] - \mathrm{d}[\sigma_{11}(x_1)]\mathrm{d}x_2\mathrm{d}x_3 \cdot \mathrm{d}[u_1(x_1)]$$
$$+ \mathrm{d}[\sigma_{12}(x_1+\mathrm{d}x_1)]\mathrm{d}x_2\mathrm{d}x_3 \cdot \mathrm{d}[u_2(x_1+\mathrm{d}x_1)] - \mathrm{d}[\sigma_{12}(x_1)]\mathrm{d}x_2\mathrm{d}x_3 \cdot \mathrm{d}[u_2(x_1)] + \cdots$$
$$= \mathrm{d}\Big[\sigma_{11}(x_1) + \frac{\partial \sigma_{11}}{\partial x_1}\mathrm{d}x_1\Big]\mathrm{d}x_2\mathrm{d}x_3 \cdot \mathrm{d}\Big[u_1(x_1) + \frac{\partial u_1}{\partial x_1}\mathrm{d}x_1\Big] - \cdots$$
$$= \mathrm{d}\Big(\frac{\partial \sigma_{11}}{\partial x_1}\Big)\mathrm{d}u_1\mathrm{d}V + \mathrm{d}\sigma_{11}\mathrm{d}\Big(\frac{\partial u_1}{\partial x_1}\Big)\mathrm{d}V + \mathrm{d}\Big(\frac{\partial \sigma_{12}}{\partial x_1}\Big)\mathrm{d}u_2\mathrm{d}V + \mathrm{d}\sigma_{12}\mathrm{d}\Big(\frac{\partial u_2}{\partial x_1}\Big)\mathrm{d}V + \cdots$$
$$= \mathrm{d}\sigma_{ij,i}\mathrm{d}u_j\mathrm{d}V + \mathrm{d}\sigma_{ij}\mathrm{d}u_{i,j}\mathrm{d}V = \mathrm{d}\sigma_{ij}\mathrm{d}u_{i,j}\mathrm{d}V > 0 \tag{8.2.31}$$

Here, use was made of the equilibrium condition

$$\sigma_{ij,i} = 0 \tag{8.2.32}$$

which is valid with the definition of σ_{ij} chosen. Thus, the stability criterion for an infinitesimally small volume in a continuum can be written as

$$\boxed{\mathrm{d}\sigma_{ij}\mathrm{d}u_{i,j} > 0} \tag{8.2.33}$$

for all possible choices of $\mathrm{d}u_{i,j}$.

Now, because $\mathrm{d}\sigma_{ij} = \mathrm{d}\sigma_{ji}$,

$$\mathrm{d}\sigma_{ij}\mathrm{d}u_{i,j} = \mathrm{d}\sigma_{ji}\mathrm{d}u_{i,j} = \mathrm{d}\sigma_{ij}\mathrm{d}[(u_{i,j}+u_{j,i})/2] \tag{8.2.34}$$

where the expression $(u_{i,j} + u_{j,i})/2$ is recognized as the strain tensor in small strain theory. Thus, for small strains, the stability criterion takes the form

$$d\sigma_{ij} d\epsilon_{ij} > 0 \qquad (8.2.35)$$

for all possible choices of $d\epsilon_{ij}$, as postulated by Drucker (1951, 1959).

In a region that is sufficiently small to be considered as homogeneously deformed, the stress increment $d\sigma_{ij}$ is related to the displacement gradient increments $du_{i,j}$ by a constitutive equation for increments:

$$d\sigma_{ij} = C_{ijkl} du_{k,l} \qquad (8.2.36)$$

The incremental moduli C_{ijkl} are generally not independent of the direction of $du_{i,j}$, and they can therefore be considered as constants only in distinct sectors of the $d\sigma_{ij}$-space. The uniaxial tensile case during plastic deformation is a simple example: unloading implies an abrupt change of the modulus $d\sigma_{11}/du_{1,1}$.

Inequality (8.2.33), combined with equation (8.2.36), gives

$$C_{ijkl} du_{k,l} du_{i,j} > 0 \qquad (8.2.37)$$

It is convenient to write the displacement gradients as vectors in 9-dimensional space and the modulus as a second order matrix. This may be done by introducing, for instance, $p = i + 3j - 3$ and $q = k + 3l - 3$, and letting

$$du_{i,j} \to dU_p, \quad du_{k,l} \to dU_q, \quad C_{ijkl} \to A_{pq} \qquad (8.2.38)$$

Then, inequality (8.2.37) can be written in the well-known quadratic form

$$dU_p A_{pq} dU_q > 0 \qquad (8.2.39)$$

for all possible dU_p. Thus, the stability criterion reduces to the statement that the matrix A_{pq} must be positive definite.

The transition from stability to instability will occur when the incremental modulus A_{pq} becomes positive semidefinite, i.e. when the equality

$$dU_p A_{pq} dU_q = 0 \qquad (8.2.40)$$

is satisfied for some direction of the vector dU_p. The eigenvalues of A_{pq} follow from the equation

$$(A_{pq} - \lambda \delta_{pq}) dU_q = 0 \qquad (8.2.41)$$

where λ is an eigenvalue and δ_{pq} is Kronecker's delta, which equals unity when its indices are equal, otherwise it is zero. Because it is known that at least one eigenvalue of a semidefinite matrix is zero and that the non-zero ones are positive,

$$dU_p A_{pq} dU_q = \lambda \delta_{pq} dU_p dU_q = \lambda dU_p dU_p \geq 0 \qquad (8.2.42)$$

The equal sign is valid if $\lambda = 0$ and then the equation coincides with equation (8.2.40). For a non-vanishing solution of equation (8.2.41) when $\lambda = 0$, it is required that

$$\boxed{\det(A_{pq}) = 0} \qquad (8.2.43)$$

where "*det*" denotes the determinant. This is the condition for transition from stability to instability. The mode of deformation during this transition is given by dU_p, which is found, apart from an arbitrary, unessential amplitude factor, as the eigenvector corresponding to eigenvalue zero. From dU_p the displacement gradients $du_{i,j}$ are uniquely found.

Special cases of instability in a continuum model
In two special cases, the displacement gradients $du_{i,j}$ express pure instability modes, i.e. either opening mode instability or shear mode instability. Rice (1980) considered the shear mode instability, assuming that $du_{i,j}$ can be written as $g_i n_j$ where n_j is the unit normal vector to the slip plane. After changing the inequality sign to an equality sign, inequality (8.2.37) gives

$$g_i n_j C_{ijkl} n_l g_k = 0 \qquad (8.2.44)$$

and, following the same procedure as for equation (8.2.40), the condition for transition to instability discussed by Rice (1980) is found†:

$$\boxed{\det(n_j C_{ijkl} n_l) = 0} \qquad (8.2.45)$$

Although the incremental moduli C_{ijkl}, alternatively written A_{pq}, were considered as constants, albeit different in different sectors of the incremental stress space, they are dependent on the current stress-strain state. Whether opening or shear mode instability, or possibly a mixed mode instability, will occur, would therefore depend on the current state. It seems logical to assume that a high ratio between hydrostatic stress (defined as positive at tensile stresses) and maximum absolute shear stress would favour opening mode instability. The method outlined in connection with equation (8.2.43) will provide the answer in specific cases.

Insufficiency of a continuum model for predicting plastic flow localization
Note that conditions (8.2.43) and (8.2.45) can only be satisfied for certain constitutive equations. As discussed by Rudnicki and Rice (1975), Rice (1977), Asaro (1979) and Needleman and Rice (1978), they cannot, for instance, be satisfied in plane strain for elastic-plastic solids with substantial hardening and a smooth yield surface. These authors also demonstrated a significant destabilization effect of yield surface vertices‡. Equation (8.2.43) gives a sufficient condition for transition to instability, which occurs when the right hand sign becomes negative. For a perfect continuum, it would also be a necessary condition. In reality, however, instability occurs earlier because of inevitable local variations. Imagine, for instance, an aggregate of several almost, but not exactly identical, cubic cells with central spherical holes. Instability would then be triggered by one of the first coalescences, and this would occur earlier than anticipated from a study of the force-displacement (P-δ) relation for the whole aggregate, and it may well occur even if (8.2.43) is not satisfied. Given the considerable variations between individual real cells, it is obvious that localization of plastic flow may seldom be correctly anticipated solely from constitutive equations for a continuum with or without yield surface vertices. Some discussion about different kinds of contribution to plastic flow localization is given by Tvergaard (1987).

---

† This condition was established already by Hadamard (1903) and later also by Thomas (1961), Hill (1962) and Rice (1977).
‡ Such vertices appear, for instance, in single-crystals.

Localization of plastic flow in shear

Shear mode instability is generally referred to as shear banding, cf. page 11. Mathematically, it is associated with a loss of ellipticity of the governing equations (Hill 1962, Rice 1977). Compare also pages 552ff. Shear banding is accompanied by strain softening, i.e., the yield limit decreases locally. Such behaviour is difficult to describe by means of conventional plasticity, but strain gradient or non-local plasticity may be used (Zbib and Aifantis 1988, Mühlhaus 1994, Strömberg and Ristinmaa 1996).

A particularly clear demonstration of the role of inhomogeneities in localization into shear bands or opening mode cracks is given by coalescence of micro-voids into void sheets, often through the linking of small holes between large voids. This was shown by, for instance, Rogers (1960), Hancock and Mckenzie (1976) and Hancock and Brown (1983). The latter note that localization is intimately connected with void growth, rather than with yield surface vertices. The mechanism was analysed by means of finite element methods by, e.g., Tvergaard (1982a), Needleman *et al.* (1992) and by Faleskog and Shih (1997).

Although shear bands are caused by shear stresses, they only follow principal shear stress directions approximately, because the resistance to shear causes deviations from these directions. In a single-crystal, they are generally not aligned with slip bands, because these bands follow crystallographic directions (Chang and Asaro 1980). Shear bands may even be non-planar. Shear banding will be further discussed in Section 9.2.

For thin sheets, localized necking from a crack edge, often not proceeding straightforward from the edge, may appear, as discussed on pages 13ff. The theory for necking of uniaxially loaded sheets (without cracks) was given by Hill (1950). An essential element in the theory is the recognition of lines of zero extension in the plane of the sheet. For biaxial loading, such that both the length and the width of the sheet are extended, no such line exists. However, experience tells that localized necking nevertheless may occur in such cases. This apparent contradiction was investigated by Stören and Rice (1975), who found that yield surface vertices could make localized necking possible, even under biaxial loading. They assumed that such vertices might be formed as a result of plastic flow. However, they remark that necking might alternatively develop as a result of small initial inhomogeneities.

Competition between decohesion and plastic flow

The embedment of the process region in the plastic region varies between different materials and different crack edge environments, for instance different T-stresses. In extreme cases, the dissipative region at the crack edge consists only of a process region or only of plastic flow. Thus, if the decohesive strength, σ_D, is sufficiently low, decohesion occurs before the yield condition is reached, and no plastic flow takes place. Similarly, if σ_D is sufficiently high, the maximum normal plastic flow stress may not reach σ_D, and then no cells in the crack edge neighbourhood reach the decohesive state, i.e., no process region is created (cf. Section 1.4).

Plastic flow without decohesion
Obviously, σ_D/σ_Y is an important parameter for the competition between decohesion and plastic flow, but also the stress state is important. For mode I and small scale yield-

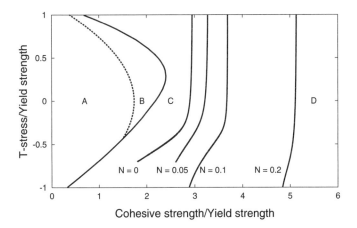

Fig. 8.2.2 Competition between decohesion and macroscopic plastic flow and its dependence on the T-stress. Decohesion, but no plastic flow, takes place in the region labelled A, some plastic flow occurs during the early stages of process region development in region B, both decohesion and plastic flow take place in region C, and plastic flow, but no decohesion, occurs in region D. The location of the boundary between C and D is dependent on the strain hardening: four examples are shown for different strain hardening exponents N, based on data reported by Du and Hancock (1991) and by O'Dowd and Shih (1991, 1994).

ing, this implies that the T-stress plays a part. The Prandtl field slip line solution, (5.4.10)-(5.4.13), indicates that the maximum stress in a non-hardening Huber-von Mises material equals about $3.0\sigma_Y$ (and about $2.6\sigma_Y$ in a non-hardening Tresca material), and thus no process region develops if $\sigma_D/\sigma_Y > 3.0$. Recall that the plastic Prandtl field envelopes the crack edge, which indicates a high T-stress: the plastic field envelops the crack edge for T-stresses higher than about $0.446\sigma_Y$, as determined by Du and Hancock (1991) for a non-hardening Huber-von Mises material. However, it follows from their analysis, that the upper limit, $\sigma_D/\sigma_Y = S_{Dmax}$, for which a process region develops, stays virtually constant for all positive T-stresses: it only decreases to about 2.9 for $T = 0$. For negative T-stresses, S_{Dmax} decreases with decreasing T, to about 2.5 for $T/\sigma_Y = -0.44$ and to about 1.8 for $T/\sigma_Y = -0.7$. These values lead to the approximate relation between S_{Dmax} and σ_D/σ_Y for non-hardening materials shown by the line marked $N = 0$ in Fig. 8.2.2.

Strain hardening implies higher overall flow stresses, resulting in higher values of S_{Dmax}, up to 4–5 for $T = 0$ and power law strain hardening strain hardening with exponent N about 0.1–0.2 (Tvergaard and Hutchinson 1992). Numerical calculations by O'Dowd and Shih (1991, 1994), using a large strain plasticity formulation, gave results for the hoop stress straight ahead of the crack as sketched in Fig. 5.4.3. Thus, the hoop stress exhibits a maximum at some distance ahead of the crack edge, and if the cohesive strength is higher than this maximum stress, no process region develops. From curves presented by O'Dowd and Shih (1991), this maximum stress can be found for $N = 0.1$ and $N = 0.2$† and for several values of the T-stress between $T = -1$ and

---

† Actually, O'Dowd and Shih used a strain hardening exponent n, which for large strains may be identified with $1/N$.

$T = +1$. These values allow drawing of the curves in Fig. 8.2.2 that show the minimum cohesive strength for which no process region develops, as functions of T for the strain hardening exponents $N = 0.1$ and $N = 0.2$. Similarly, from curves in O'Dowd and Shih (1994), results for $N = 0.05$ can be found for $T = 0$. The curve for $N = 0.05$ in Fig. 8.2.2 is, however, drawn by interpolating from the curves for $N = 0$, $N = 0.1$ and $N = 0.2$.

Now, what happens if $\sigma_D/\sigma_Y > S_{Dmax}$, corresponding to region D in Fig. 8.2.2? Obviously, plastic flow will cause blunting of the crack edge, increasing with increasing outer load, until plastic collapse occurs. This is a common experience with certain plastic foils: plastic flow, rather than crack growth, occurs. Even if a crack is cut, stretching of the foil produces an almost elliptical hole with the major axis eventually becoming parallel to the stretching direction.

Decohesion without plastic flow

The other extreme case, decohesion without plastic flow, may be investigated by assuming a Barenblatt region and using stress representations in complex potentials, which follow from (4.4.23)-(4.4.24) as

$$\sigma_x = 2\Re p'(z) - 2y\Im p''(z) - 4\Re s'(z) \qquad (8.2.46)$$

$$\sigma_y = 2\Re p'(z) + 2y\Im p''(z), \qquad \tau_{xy} = -2y\Re p''(z) \qquad (8.2.47)$$

It is assumed that the length r_p of the Barenblatt region is much smaller than the crack length and other significant in-plane dimensions. The vicinity of the crack edge is considered, and therefore the modified boundary layer approach, pages 119ff., may be used. Plane strain is assumed in the crack edge vicinity.

Consider a crack $x < -r_p$, $y = 0$, with a Barenblatt process region $-r_p < x < 0$, along which the decohesive stress is $\sigma_y = \sigma_y^0(x)$. The modified boundary layer solution is given by equations (4.4.63) and (4.4.64) as

$$p'(z) = -\frac{z^{1/2}}{2\pi} \int_{-r_p}^{0} \frac{\sigma_y^0(\xi)}{\sqrt{-\xi}(\xi - z)} d\xi \qquad (8.2.48)$$

$$s'(z) = -\frac{T}{4} \qquad (8.2.49)$$

For a fully developed process region, the decohesive stress distribution may be assumed to be parabolic – the exact shape is not expected to be critical for conclusions about the possibility of plastic flow. Thus,

$$\sigma_y^0(x) = \sigma_D (1 - x^2/r_p^2), \qquad -r_p < x \leq r_p \qquad (8.2.50)$$

leading to

$$p'(z) = \frac{\sigma_D}{\pi} \left[\left(1 - \frac{z^2}{r_p^2}\right) \operatorname{atan}\left(\frac{r_p}{z}\right)^{1/2} + \left(\frac{z}{r_p} - \frac{1}{3}\right)\left(\frac{z}{r_p}\right)^{1/2} \right] \qquad (8.2.51)$$

$$p''(z) = -\frac{2\sigma_D}{\pi r_p} \left[\frac{z}{r_p} \operatorname{atan}\left(\frac{r_p}{z}\right)^{1/2} - \left(\frac{z}{r_p} - \frac{1}{3}\right)\left(\frac{r_p}{z}\right)^{1/2} \right] \qquad (8.2.52)$$

By using (8.2.46) and (8.2.47), all in-plane stresses can be determined, whereupon the transverse stress follows as $\sigma_z = \nu(\sigma_x + \sigma_y)$, where ν is Poisson's ratio. The primary

8.2 PROCESSES IN THE CRACK EDGE VICINITY

interest is focused on the effective stress, which for a Huber-von Mises material is found from the expression

$$\sigma_{eff}^2 = (1 - \nu + \nu^2)(\sigma_x + \sigma_y)^2 - 3\sigma_x\sigma_y + 3\tau_{xy}^2 \qquad (8.2.53)$$

Numerical calculation of σ_{eff} may be made by a program that can handle complex functions, for instance Maple. For $T = 0$ and $\nu = 1/3$, the maximum effective stress turns out to occur at about $x = -0.4r_p$, $y = \pm 0.5r_p$. It equals σ_Y for $\sigma_D \approx 2.18\sigma_Y$, which implies that plastic flow takes place if $\sigma_D/\sigma_Y > 2.18$. The seemingly complementary conclusion, that no plastic flow takes place anywhere if $\sigma_D/\sigma_Y < 2.18$, turns out to be incorrect, as will be discussed later.

By repeating the calculations for different values of T in the region $-\sigma_Y < T < +\sigma_Y$, the results illustrated by the solid curve in Fig. 8.2.2 are obtained. The curve was obtained for $\nu = 1/3$, but the dependence on Poisson's ratio was found to be marginal in the interval $1/4 < \nu < 1/2$.

The case $T = 0$ was studied by Sjöberg and Ståhle (1992), assuming the Leonov-Panasyuk-Dugdale model. They found that the maximum effective stress occurred at the crack edge, i.e. at the rear end of the Leonov-Panasyuk-Dugdale region. This result suggests an investigation of the maximum effective stress not only for the load when the process region is fully developed, but also for the previous history, from the state when a process region first appears. With the mathematical Barenblatt model used also for this state, the stress across such a small process region is approximately constant, equal to σ_D. Thus, $\sigma_y^0(x) = \sigma_D$ for $-r_p^{(\epsilon)} < x < 0$, where superscript (ϵ) serves as a reminder that the process region length is much smaller than for a fully developed process region in the material. Then, from (8.2.48) it follows that

$$p'(z) = \frac{\sigma_D}{\pi} \operatorname{atan}\left(\frac{r_p^{(\epsilon)}}{z}\right)^{1/2} \qquad (8.2.54)$$

$$p''(z) = -\frac{\sigma_D}{\pi r_p^{(\epsilon)}} \cdot \frac{1}{1 + z/r_p^{(\epsilon)}} \left(\frac{r_p^{(\epsilon)}}{z}\right)^{1/2} \qquad (8.2.55)$$

The stresses are continuous across the process region, except at its rear point, $x = -r_p^{(\epsilon)}$, $y = 0$, where they are singular. The effective stress near this point depends on its angular position. Put $z = -r_p^{(\epsilon)} + re^{i\varphi}$. Then, for $|z + r_p^\epsilon| \ll r_p^{(\epsilon)}$,

$$p'(z) \approx \frac{\sigma_D}{2\pi i} \ln \frac{4r_p^{(\epsilon)} e^{i\varphi}}{z + r_p^{(\epsilon)}} = -\frac{i\sigma_D}{2\pi} \ln \frac{4r_p^{(\epsilon)}}{r} + \frac{\sigma_D}{2\pi}(\pi - \varphi) \qquad (8.2.56)$$

The in-plane stresses can now be determined from (8.2.46)-(8.2.47), and then the effective stress in the limit $r \to 0$ is found from the expression

$$\sigma_{eff}^2 = \frac{\sigma_Y^2}{\pi^2} \Big\{ [(1-2\nu)^2(\pi-\varphi)^2 + 3\sin^2\varphi] \frac{\sigma_D^2}{\sigma_Y^2}$$
$$+ [(1-2\nu)^2(\pi-\varphi) - 3\cos\varphi\sin\varphi] \frac{\sigma_D T}{\sigma_Y^2} + (1-\nu+\nu^2)\frac{T^2}{\sigma_Y^2} \Big\} \qquad (8.2.57)$$

For $T = 0, \nu = 1/3$, the maximum effective stress is found for $\varphi \approx 86.5°$. Thus, it is larger than along the process region ($\varphi = 0$). It equals σ_Y for $\sigma_D \approx 1.73\sigma_Y$. Note that this is lower than the value $2.18\sigma_Y$ obtained for the fully developed process region. Therefore, some plastic flow will occur in the region $1.73 < \sigma_D/\sigma_Y < 2.18$ during the early stages of process region development. The plastic region will come in the form of two lobes around $\varphi = 86.5°$.

By repeating the calculations for different values of T in the region $-\sigma_Y < T < +\sigma_Y$, the results illustrated by the dashed curve in Fig. 8.2.2 are obtained. The curve is obtained for $\nu = 1/3$, but the dependence on Poisson's ratio was found to be marginal in the interval $1/4 < \nu < 1/2$. The curve is not drawn below about $T = -0.4\sigma_Y$, because there the maximum effective stress is larger for the fully developed process region.

The embedment of the process region in the plastic region
The solid line to the left of the region labelled C in Fig. 8.2.2 may be taken to represent the lower limit, $\sigma_D/\sigma_Y = S_{Dmin}$, for which a plastic region develops, even though some plastic flow also occurs in the region labelled B in the figure. The fairly large extension of this region indicates that the limit is not very well defined in a real case. The plastic flow in B may be confined to some blunting, occurring at early stages of process region development. For combinations of σ_D/σ_Y and T-stress in region C in Fig. 8.2.2, both a process and a plastic region exists. It is reasonable to assume that the process region will be more deeply embedded in the plastic region for such combinations close to the C-D boundary than for those close to the B-C boundary.

The embedment of the process region in the plastic region may be expressed by the ratio r_p/R_p between the forward extensions of the process region and the plastic region. Close to the B-C boundary, this value may be put equal to unity. For an estimate of the ratio r_p/R_p close to the C-D boundary, it is observed that the process region must contain at least one cell in the straightforward direction. For a void growth mechanism, calculations indicate that a cell collapses under a strain of the order of 0.1. Then, taking the stress state in the plastic region equal to what is found from slip line theory, and using Hooke's law for a point just ahead of the plastic region, leads to a maximum strain of the order of σ_Y/μ. A typical value may be $\sigma_Y/\mu = 0.01$, i.e. about 10 times smaller strain at the front of the plastic region than for the process region. Then, assuming that the strains decay approximately in proportion to $1/r$ in the plastic region with the distance r from the rear end of the process region, gives $r_p/R_p \approx 0.1$ in the neighbourhood of the C-D boundary.

The direction of maximum effective stress at the early stages of process region development increases from about 70° for $T = -0.4\sigma_Y$ to about 130° for $T = \sigma_Y$, thus a considerable shift backwards with increasing T. The same tendency is found for the plastic regions appearing near the fully developed process region. It has also been observed in various numerical simulations of the plastic region at stationary cracks, e.g., O'Dowd et al. (1994), Tvergaard and Hutchinson (1994), and Xia and Shih (1995a). Somewhat less pronounced, the same phenomenon occurs for slowly moving cracks (O'Dowd et al. 1994).

Slowly moving cracks
So far a stationary crack has been assumed in the discussion of the competition between decohesion and plastic flow. The relations obtained are therefore pertinent for onset of crack growth. However, none or very little stable crack growth is expected

Fig. 8.2.3 Development of the crack profile during the stable phase.

to take place when there is no plastic region and when the process region consists of essentially only one layer of cells†. This may be understood by realizing that the energy flux into a Barenblatt region during crack growth will not be impeded in the absence of a plastic region; cf. (3.5.14). The Barenblatt region is simply translated forwards with the crack edge.

The situation when there is no process region has been described previously (page 558) as increasing and progressive blunting without crack growth. When both a plastic and a process region are present, stable crack growth will normally take place. A certain amount of energy is needed for the processes in the process region. This amount may not stay constant during the stable phase, but, for simplicity in the discussion, it may be assumed that it does, at least approximately, and that approximate process region autonomy prevails. Then, the stress-strain state immediately outside the process region must remain approximately the same during the whole stable phase. This may be perceived as a necessary condition for crack growth.

Now, the magnitudes of the strain gradients in the plastic region decrease considerably from onset of crack growth to steady state crack propagation. For a deeply embedded process region in a non-hardening material, this is evidenced by the change from an inverse to a logarithmic dependence of the strains on the distance to the crack edge; cf. Section 5.7. This gradient decrease implies that the outer load has to be increased considerably to maintain the stress-strain environment near the process region that is required for crack growth. As a result, an increasing amount of energy is dissipated in the plastic region per unit crack growth, whereas the amount of energy that reaches the process region stays approximately constant. For a deeply embedded process region, this required, approximately constant, amount may only be a tiny fraction of the totally dissipated energy. Thus, numerical results by Tvergaard and Hutchinson (1992) indicate a fraction of less than 5 per cent for a small scale yielding steady state case with deep embedment (power-law strain hardening with $N = 0.1$, $\sigma_D/\sigma_Y = 3.75$, $\sigma_Y = 0.0078/\mu$, $T = 0$). On the other hand, for poor embedment, the energy dissipation in the process region, in relation to the total energy dissipation, may be rather high at steady state growth, even in apparently ductile materials: Ståhle (1985) reports the fraction 80 per cent for a small scale yielding steady state case with linear strain hardening, $E_T/E = 0.001 - 0.01$, and cohesive strength $\sigma_D = 2.5\sigma_Y$.

In the cases associated with the region labelled C in Fig. 8.2.2, i.e. with both a process region and a plastic region, blunting occurs only at the onset of crack growth, so that the crack profile develops in a manner shown schematically in Fig. 8.2.3. If the process region is poorly embedded in the plastic region, the plastic flow may be essentially limited to blunting during the early stages of process region development,

---

† A process region consisting of a cluster of micro-cracks will be discussed on pages 568ff.

Fig. 8.2.4 Competition between cleavage and dislocation emission. Originally horizontal atom rows are symbolized by thin lines and the crack is symbolized by a thick line.

perhaps in a way as described for region B in Fig. 8.2.2. This may be interpreted as a change from ductile to brittle crack growth.

Competition between cleavage and plastic flow at a crack edge

Cleavage or dislocation emission?

In the preceding discussion of the competition between decohesion and plastic flow, it was tacitly assumed that both phenomena operate on a macro- or mesoscale (between macro- and microscale), as opposed to a nanoscale (the atomic scale). This does not exclude cleavage, but if cleavage occurs in a polycrystalline material, it takes place in facets, usually of about grain size, and most of the energy dissipation in the process region is spent in severing the bridges between different facets. Usually, this severing occurs by plastic slip. Here, the possibility of unobstructed cleavage, such as occurring in single-crystals, will be considered, and compared with the alternative possibility of plastic flow near the crack edge.

The first analysis of the competition between cleavage and plastic flow near the edge of a crack in a crystal seems to have been given by Rice and Thomson (1974). Previously, it was assumed that cleavage occurred when the theoretical tensile strength was sufficiently large compared to the theoretical shear strength. This macroscopic point of view was questioned by Rice and Thomson, who brought the focus to the nanoscopic scale, and argued that cleavage is favoured when the energy required to separate atoms (cleavage) is lower than the energy associated with dislocation emission from the crack edge (plastic flow). Very schematically, these two mechanisms are illustrated in Fig. 8.2.4. It should be emphasized that plastic flow on the nanoscale requires dislocations to be emitted from the crack edge, because there would only occasionally be any pre-existing dislocations available at nanoscopic distances from the crack edge.

The scales involved are indeed very different. Cleavage and dislocation emission take place on the atomic scale (of the order of 10^{-10}m), whereas the distance between pre-existing dislocations may amount to micrometers (10^{-6}m). Encounters between the crack edge and pre-existing dislocations would thus be extremely rare events. On the other hand, dislocations near a crack edge may be moved under the influence of high stresses in that neighbourhood.

Rice and Thomson (1974) found that spontaneous dislocation emission occurs for face-centered cubic metals, such as gold, silver, copper and aluminium, but is hampered in body-centered cubic metals (such as iron) and diamond-cubics (such as silicon and germanium). Later work, using more refined models of dislocation nucleation, confirms these results. Thus, among eight face-centered cubic metals investigated, Rice (1992) found that, most probably, only iridium, and possibly also nickel, would be incapable of cleaving under mode I conditions. Among ten body-centered metals and three diamond-cubics, cleavage would be the preferred mechanism, except in lithium and possibly also sodium (both body-centered metals). A relatively small amount of shear mode loading was found to have a significant impact, so that the preferred mechanism in "borderline" materials like nickel and sodium would be dislocation emission rather than cleavage.

In Rice (1992), the influence of thermal fluctuations in the crystal lattice was not investigated, but was later considered by Rice and Beltz (1994). They found that the energy required for spontaneous emission of dislocations from a crack edge is significantly decreased at room temperature, if thermal fluctuations are taken into account.

Note that the cell model for crack propagation in pure cleavage would use very small cells, each being a structure cell of the lattice (i.e. the smallest repetitive unit, in extreme cases containing only one atom). The process region would be a one cell thick layer of cells. If plastic flow by dislocation emission occurs, rather than cleavage, there will not be a process region in the sense earlier defined, i.e. a number of cells having reached the decohesive state.

The void growth to cleavage transition
It is often experienced that stable growth in ferritic steels by void growth and coalescence may be suddenly interrupted by unstable growth under cleavage. This phenomenon, "ductile-brittle" transition, may have serious practical impications, and it will be discussed on pages 600ff. in the context of statistical fracture theories.

Cleavage cracking in the presence of plastic flow

Gold has been recognized as a very ductile metal, both from millennia of practical experience of cold-forming the metal and from more recent studies of its nanoscopic properties. It was therefore a surprise when Reimanis et al. (1991) found that a sharp crack could grow along a gold-sapphire† interface, obviously without dislocation emission from the crack edge. They studied a very thin (a few hundredths of a millimetre) gold layer, sandwiched between two sapphire blocks. Interface debonding without plastic deformation was not the only outcome of the experiment, but the crack extended by a succession of void formation (cavitation), as described on page 550, and cleavage type interface debonding. Completed fracture required much larger energy than the work of adhesion, but significantly smaller than for ductile interface fracture. The energy dissipation increased with the layer thickness.

Similar experiments by Elssner et al. (1994) produced similar results, with clean

---

† Sapphire is single-crystal aluminium oxide, Al_2O_3. It is known as a precious stone, of course, usually coloured by metallic impurities, but it can also be produced artificially.

crack growth along the interface between single-crystals of niobium and sapphire. Thus, the fracture process consisted of cleavage type interface debonding, although plastic deformation occurred in the niobium. In one experiment, they determined the macroscopic energy flux to the dissipative region to $2100\,\mathrm{J/m^2}$ and the nanoscopic energy flux to the cleavage process to about $1\,\mathrm{J/m^2}$†.

A look at Fig. 8.2.2 suggests that no cleavage or other types of decohesion would occur if σ_D/σ_Y is larger than about 3.0 for a perfectly plastic material and considerably larger for strain hardening. However, even for quite large strain hardening, decohesion would hardly occur for values of σ_D/σ_Y larger than about 5 (Tvergaard and Hutchinson 1992). Now, for cleavage, σ_D is the atomic cohesive strength, about $E/10$, where E is the modulus of elasticity. Thus, σ_D/σ_Y is at least equal to 10, i.e. far exceeding values for which cleavage would be possible according to macroscopic continuum theories. This fact might appear to to contradict the results by Reimanis et al. (1991) and Elssner et al. (1994). However, consideration of the different scales involved will lead to a more nuanced view.

The scale required for a continuum theory of cleavage (for instance applied on a Barenblatt region in an elastic surrounding), and the scale required for a continuum theory of plastic flow, are vastly different. The continuum scale for cleavage is the nanoscale (about 10^{-9}m to cover 5-10 atoms) and the scale for plastic flow is a microscale of about 10^{-5}m (to cover 5-10 pre-existing dislocations). This great difference makes predictions of simultaneous occurrence of cleavage and plastic flow based on continuum theories very subtle. Note that the Rice-Thomson theory (page 562) did not consider pre-existing dislocations, but rather dislocations emitted from the crack edge, and therefore it is based on nanoscale considerations for both cleavage and plastic flow.

A small flaw may be considered to be situated in a plastic region, as defined on a continuum scale, but in an elastic region on the nanoscale. The ambient mean stress may be as high as 5 times the yield stress or more, as estimated for a metal foil between ceramic blocks (page 550). Therefore, in the highly stressed region near an edge of the flaw, conditions may be more favourable for cleavage than for dislocation emission.

On the basis of such scale arguments, Suo et al. (1993) assumed that no dislocations were emitted from the crack edge and they neglected the rare events when the crack edge occasionally encounters a pre-existing dislocation. In the same spirit, it may be assumed that the crack edge is situated inside an elastic cell, enveloping the edge and being free from dislocations. Three regions may then be recognized: the cleavage process region on the nanoscale, the elastic cell on the microscale and the background plastic region on the macroscale. In this model, concepts like stress and strain, and thereby a continuum theory, may be used for the elastic cell, except near the process region and near the background region, and for the background region, except near the elastic cell.

By introducing the Barenblatt model for the process region at the edge of a very small flaw, continuum theory can be used to show that the maximum stress is of the order of 10-100 times larger near the process region than near the dissipative background region, depending on the length of the process region (which would be a few

---

† The main aim with their study was to investigate the influence of impurities on the energy requirements. By introducing small amounts of silver to the interface, they found a 40 per cent reduction in the nanoscopic energy flux and a five-fold reduction in the macroscopic energy flux.

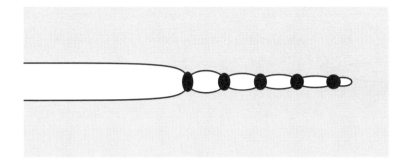

Fig. 8.2.5 Crack bridged by ductile particles in a ceramic matrix.

atomic distances) and on the constant stresses (which add to the inverse square root stresses at larger distances). Considering that the stresses near the process region may be of the order of $E/10$, the stresses in a substantial part of the background region will still be sufficiently high to move dislocations. Thus, the crack cannot move forwards (except perhaps a few atomic steps initially) without causing dissipation in the background region. Because of the vast difference in scale between the process region and the dissipative background region (the plastic region), the energy dissipation per unit crack growth is much larger in the plastic region than in the process region.

The model established by Suo *et al.* (1993) has been further developed by Beltz *et al.* (1996), and by He *et al.* (1996). In the latter work, which was previously mentioned in the context of cavitation (page 550), numerical simulations were performed and the whole process of debonding was discussed. This process does not occur smoothly, but exhibits jumps and debond nucleations ahead of the main crack.

Toughening mechanisms

Crack bridging

In certain materials, the opening of a crack may be opposed by physical bridges between the crack faces. One example is a ceramic containing ductile metal particles (a particulate-reinforced ceramic). Along some distance behind the front of an advancing crack, the crack faces are pinned together by particles, whereas the ceramic matrix is broken; see Fig. 8.2.5. The ductile particles become increasingly stretched with the increasing crack opening, i.e. with the distance to the crack front, until they rupture one after the other.

The bridge region does not exhibit the features of a process region as defined in Section 1.4, because the forces across the particles are increasing in the backwards direction, except, perhaps, near the trailing end. Rather, a case of crack face loading may be considered to prevail, with the crack edge at the front end of the bridge region and the process region situated in the ceramic matrix; see Fig. 8.2.5. This process region may be very small, perhaps of interatomic dimensions, and generally considerably smaller than the distances between the bridging particles.

Toughening of particulate-reinforced ceramics has been analysed by, for instance, Evans and McMeeking (1986), Nemat-Nasser and Hori (1987a), Rose (1987), Budian-

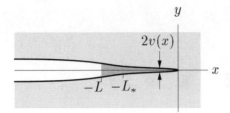
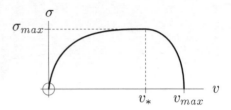

Fig. 8.2.6 Model of continuous bridging and point size process region, $x = y = 0$. The position $x = -L_*$, $y = 0$ corresponds to maximum bridging stress, σ_{max}.

Fig. 8.2.7 Relation between stress and height increase for the bridge region.

sky et al. (1988) and Erdogan and Joseph (1989). An analytical approach, allowing detailed calculation of the development of the fracture toughness during loading, has been presented by Rubinstein and Wang Peng (1998). The analysis here is essentially based on some main points in the paper by Budiansky et al. (1988), with the simplification that the linear size of the particles is assumed to be much smaller than the spacing between them.

A suitable model may assume that the bridging is continuous (often referred to as a continuous distribution of springs) and that the process region is infinitesimally small; see Fig. 8.2.6. The bridge region is then characterized by a relation between the stress, σ, normal to the crack faces, and the height increase, which is equal to the crack opening, $2v$ (Fig. 8.2.7). It is assumed to be situated along the portion $-L < x < 0$ of the x axis. Small scale bridging is assumed, i.e. $L \ll a$, where a is the crack length. Thus, a semi-infinite crack, $x < 0$, $y = 0$, may be considered. The remote load is given through the boundary layer approach, page 119, so that the stress $\sigma_y = K_I/\sqrt{2\pi x}$ for $x \gg L$, $y = 0$, where K_I is the stress intensity factor. Because the process region is situated in an essentially elastic environment, there is also an inner stress intensity factor, $K_{I(pr)}$, pertinent to the crack edge vicinity†. Obviously, $K_{I(pr)}$ is smaller than K_I.

It is reasonable to assume that the crack edge will start moving when a critical value $K_{I(pr)} = K^0_{I(pr)}$ has been reached, and that $K_{I(pr)}$ equals another critical value, $K^{SS}_{I(pr)}$, when local steady state conditions are reached. These critical values may be found from a test on particle free material, and they may be rather close to each other. Local steady state conditions around the whole bridge region may be approached for some critical value, K_{ISS}‡ of K_I. This critical value, which may be found from a test on particulate-reinforced material, using specimen dimensions much larger than L, is called the *fracture toughness*, a concept which will be discussed further on pages 585ff. Note that when local steady state conditions are approached in the dissipative region

---

† The subscript $I(pr)$ indicates mode I and the process region.
‡ This is traditionally denoted K_{Ic}, but the subscript ISS (SS for steady state) is preferred here to avoid confusion with the notation J_{Ic}, which for some reason or another has been introduced to denote a critical value of J for onset of crack growth rather than steady state. However, because of the well known definition of K_{Ic}, this notation will be used alongside K_{ISS} when this is judged to be helpful.

Fig. 8.2.8 *J*-integral path near the crack edge.

under small scale yielding, unstable crack growth (fracture) occurs, except in cases like wedging, etc.

Alternatively, critical values of the energy flux, $\mathcal{G}$, rather than of the stress intensity factor, may be used to characterize the crack resistance of the material. Note that the relation $\mathcal{G}^0_{I(pr)}/\mathcal{G}_{ISS} \approx (K^0_{I(pr)})^2/K^2_{ISS}$, which follows from (3.5.14), holds only approximately, because the elastic constants are not the same for material without particles as for material with particles.

The relation between σ and v is now written in the form

$$v = \begin{cases} f(\sigma) & \text{for } 0 < v < v_* \\ g(\sigma) & \text{for } v_* < v < v_{max} \end{cases} \quad (8.2.58)$$

where v_* is the half height increase of the bridge region for maximum stress, $\sigma = \sigma_{max}$; cf. Fig. 8.2.7. Now, according to (3.5.14), the *J*-integral for a remote path equals $K_I^2/[4(1-k_{cm}^2)\mu_{cm}]$, where subscript cm indicates the composite ceramic-metal material. Then, by choosing the path shown in Fig. 8.2.8, one contribution is found from the path encircling the crack edge, and one contribution is found from the paths on each side of the bridge region. Thus,

$$\frac{K_I^2}{4(1-k_{cm}^2)\mu_{cm}} = \frac{K_{I(pr)}^2}{4(1-k_c^2)\mu_c} + 2\int_{-L}^{0} \sigma \frac{\partial v}{\partial x} dx \quad (8.2.59)$$

where subscript c denotes the ceramic material†. With $x = -L_*$ denoting the position of maximum stress along the bridge region, the integral may be written as

$$\int_{-L}^{-L_*} \sigma \frac{\partial v}{\partial x} dx + \int_{-L_*}^{0} \sigma \frac{\partial v}{\partial x} dx = \int_{0}^{\sigma_{max}} \sigma[f'(\sigma) - g'(\sigma)] d\sigma = \int_{0}^{\sigma_{max}} [g(\sigma) - f(\sigma)] d\sigma \quad (8.2.60)$$

where partial integration was used. The result equals the area under the σ-v curve in Fig. 8.2.7 and may be written as $\alpha \sigma_{max} v_{max}$. For a linear relation, $\alpha = 1/2$, and for a rigid-plastic relation, $\alpha = 1$. Budiansky et al. (1988) make some estimates about the σ-v relation for different material combinations and compositions. It appears that α is generally closer to unity than to $1/2$. Other estimates may be made on the basis of detailed numerical analyses of the deformation of a bridging particle, as performed by Mataga (1989) and by Tvergaard (1995). Tvergaard also incorporated partial debonding and cavitation from a pre-existing microvoid inside the particle.

---

† Note that k_c should refer to plane strain and k_{cm} to plane stress for a crack in a plate, if plane stress prevails far from the crack edge and plane strain near the crack edge. The *J*-integral for a remote path will nevertheless be equal to the *J*-integral for a path close to the crack edge; see (3.2.12).

Putting $K_{ISS} = \lambda K^{SS}_{I(pr)}$, gives the expression for the *toughening ratio*

$$\lambda = \sqrt{\frac{(1-k_{cm}^2)\mu_{cm}}{(1-k_c^2)\mu_c} + \frac{8(1-k_{cm}^2)\alpha\mu_{cm}\sigma_{max}v_{max}}{[K^{SS}_{I(pr)}]^2}} \qquad (8.2.61)$$

Note that this expression is valid for an advancing crack at the onset of unstable crack growth.

An expression has now been found for the toughening ratio, but it might also be of some interest to know the length L of the bridge region when steady state conditions are approached. Here, equation (4.4.59) may be used, if either plane strain or plane stress prevails everywhere, and if it is asumed that $k_c = k_{cm} = k$ and $\mu_c = \mu_{cm} = \mu$, in agreement with the assumption of low particle density. With the present notations, this equation reads, after introducing the substitutions $s = -x$, $u = -\xi$, and using the relations $v = f(\sigma)$ and $K_{ISS} = \lambda K^{SS}_{I(pr)}$:

$$f[\sigma(s)] = \frac{\lambda K^{SS}_{I(pr)}\sqrt{s}}{\sqrt{2\pi}(1-k^2)\mu} + \frac{1}{2\pi(1-k^2)\mu}\oint_0^L \sigma(u)\ln\frac{\sqrt{u}+\sqrt{s}}{|\sqrt{u}-\sqrt{s}|}du \qquad (8.2.62)$$

Unknown quantities are L and the stress distribution, $\sigma(s)$, along the bridge region. Note that λ is known by (8.2.61). The integral equation (8.2.62) was investigated and solved numerically for some special cases by Budiansky et al. (1988). For a rigid-plastic bridging material, the integral can be evaluated exactly, giving

$$v = \frac{\lambda K^{SS}_{I(pr)}\sqrt{s}}{\sqrt{2\pi}(1-k^2)\mu} + \frac{\sigma_{max}}{2\pi(1-k^2)\mu}\left[(L-s)\ln\frac{\sqrt{L}+\sqrt{s}}{\sqrt{L}-\sqrt{s}} + 2\sqrt{Ls}\right] \qquad (8.2.63)$$

which for $s = L$, $v = v_{max}$, yields

$$L = \frac{\pi(\lambda-1)^2}{8}\cdot\frac{[K^{SS}_{I(pr)}]^2}{\sigma_{max}^2} \qquad (8.2.64)$$

A rough estimate of L may be obtained by using equation (4.4.60), which in present notations reads

$$K^{SS}_{I(pr)} = K_{ISS} - \sqrt{\frac{2}{\pi}}\int_0^L \frac{\sigma(s)}{\sqrt{s}}ds \qquad (8.2.65)$$

where use was made of the fact that $\sigma(x)\sqrt{2\pi x} \to K_{I(pr)}$ as $x \to +0$. Now,

$$\int_0^L \frac{\sigma(s)}{\sqrt{s}}ds = 2\theta\sigma_{max}\sqrt{L} \qquad (8.2.66)$$

where $0 < \theta < 1$. This leads to

$$L = \frac{\pi(\lambda-1)^2}{8\,\theta^2}\cdot\frac{[K^{SS}_{I(pr)}]^2}{\sigma_{max}^2} \qquad (8.2.67)$$

For a rigid-plastic σ-v relation, $\theta = 1$, and for other relations an estimate of θ may be based on a suitably averaged stress. For ductile particles, a reasonable estimate may be obtained by putting $\theta = \alpha$.

Micro-crack shielding

Certain polycrystalline and multiphase ceramics are highly heterogeneous on a microscale, with mismatch of elastic anisotropy and moduli as well as of thermal expansion properties. This causes a propensity for formation of micro-cracks during loading (and sometimes to some extent even before loading) as a result of residual stresses introduced during fabrication (or natural creation) of the material. Another important source of residual stresses consists of material transformations, by which a microstructural element, such as a particle, of one phase constituent transforms so as to occupy a larger volume, for instance martensite transformation (Evans and Cannon 1986).

Micro-cracking may be profuse in highly stressed regions such as the neighbourhood of a macro-crack edge. The intuitive reflection that the micro-cracks would have a negative effect on the resistance to crack growth has proved to be wrong in many cases, in which, on the contrary, a positive effect has been experienced (e.g. Evans and Faber 1980). This effect has been called shielding, i.e., the cluster of micro-cracks around a crack edge shield the crack from the outer stress-strain field so that the local stress intensity factor at the crack edge (corresponding to $K_{I(pr)}$ in the analysis of particulate-reinforced ceramics) becomes smaller than the outer stress stress intensity factor (corresponding to K_I). A necessary condition for shielding appears to be that the micro-cracks are arrested soon after formation by boundaries between microstructural constituents, for instance grain boundaries, so that a fairly stable cloud of micro-cracks is created.

Phase transformations occurring in the highly stressed region near a crack edge lead to so-called transformation toughening. Thus, for instance, high stresses may induce tetragonal-to-monoclinic phase transformation in zirconia particles in a brittle ceramic. This creates a transformation region around the crack edge. The particle volume may increase a few per cent as a result of the phase transformation (McMeeking and Evans 1982, Evans and Cannon 1986). This implies lowering of the local stress intensity factor. In addition, the transformed particles may constitute direct obstacles to crack propagation by crack front deflection and trapping (Evans and Cannon 1986, Møller and Karihaloo 1995). The toughening effects imply that the outer stress intensity factor may be an order of magnitude larger than the local stress intensity factor before fracture occurs. They also lead to a stable phase of crack growth, which makes the ceramic appear less brittle. Detailed numerical simulations of crack growth in transformable ceramics have been made by e.g. Stam and van der Giessen (1996a,b).

Continuum approach to micro-crack clusters

Budiansky and O'Connell (1975), Hoenig (1979), Evans and Faber (1980), Horii and Nemat-Nasser (1983), Hutchinson (1987) and Gudmundson (1990), among others, assumed that the micro-crack cloud can be treated as a continuum, obviously with smaller stiffness than intact material. This, together with some relief of residual stresses, could then explain the toughening effect in about the same way as the reduction of the strain singularity by plastic flow during stable crack growth, cf. page 324. An account for the detailed mechanisms suggested for such actions (see e.g. Hutchinson 1987) will not be pursued here, but a simple approach for estimating the reduction of the compression modulus due to micro-cracking will be given. Thus, following Hutchinson (1987), the micro-cracks are assumed to be penny-shaped and sufficiently far away from each other that interactions may be neglected. It is further assumed that the micro-cracks open in an environment with a hydrostatic stress σ^∞. From (4.8.24) the volume of an opened micro-crack is found to be

$$\Delta V = 2\int_0^a u_z^+ \cdot 2\pi r \mathrm{d}r = 8(1-\nu)\sigma^\infty \int_0^a r\sqrt{a^2 - r^2}\mathrm{d}r = \frac{8}{3}(1-\nu)\frac{\sigma^\infty}{\mu}a^3 \quad (8.2.68)$$

where a is the radius of the micro-crack. The average dilatation in a volume V containing n randomly oriented micro-cracks is then

$$\Delta\epsilon_{kk} = \frac{n\Delta V}{V} = \frac{8}{3}(1-\nu)\frac{\sigma^\infty}{\mu}\varrho_m \quad (8.2.69)$$

where $\varrho_m = na^3/V$ may be taken as a measure of the micro-crack density.

The modulus of compression for the uncracked solid is

$$\kappa = \frac{\sigma^\infty}{\epsilon_{kk}} \quad (8.2.70)$$

where ϵ_{kk} is the dilatation caused by a hydrostatic stress σ^∞. The presence of micro-cracks has increased the dilatation to $\epsilon_{kk} + \Delta\epsilon_{kk}$. This implies an effective compression modulus κ_{red}, given by

$$\frac{1}{\kappa_{red}} = \frac{\epsilon_{kk} + \Delta\epsilon_{kk}}{\sigma^\infty} = \frac{1}{\kappa} + \frac{8(1-\nu)\varrho_m}{3\mu} \quad (8.2.71)$$

i.e., by the introduction of micro-cracks, the compression modulus has been reduced to

$$\kappa_{red} = \frac{\kappa}{1 + 8(1-\nu)\kappa\varrho_m/(3\mu)} = \frac{\kappa}{1 + 16(1-\nu)\varrho_m/[9(1-2\nu)]} \quad (8.2.72)$$

where the relation $\kappa = 2(1+\nu)\mu/[3(1-2\nu)]$ (Appendix A1) was used. This and other relations were given by Budiansky and O'Connell (1975), who also included micro-crack interactions in their model. For a sufficiently dilute cluster of micro-cracks, their results give the modulus of rigidity,

$$\mu_{red} = \frac{\mu}{1 + 32(1-\nu)(5-\nu)\varrho_m/[45(2-\nu)]} \quad (8.2.73)$$

For more dense clusters, their results are more complicated. Hutchinson (1987) remarks that the estimates (8.2.72) and (8.2.73) appear to be reasonably accurate for ϱ_m less than about 0.2. $\varrho_m = 0.2$ implies $\kappa_{red} \approx 0.5\kappa$ and $\mu_{red} \approx 0.8\mu$ for $\nu \approx 1/3$. Note that $\kappa_{red}/\kappa < \mu_{red}/\mu$ for all values of $\nu > 0$.

Crack closure and load induced anisotropy may also be taken into account in a continuum model (Horii and Nemat-Nasser 1983, Gudmundson 1990). Crack closure, of course, implies stiffer material. Horii and Nemat-Nasser (1983) pointed out that crack closure in combination with frictional sliding implies sensitivity to the loading path. Also anisotropic nucleation of micro-cracks may be included in the models. This was done by Gudmundson (1990), who investigated, in particular, micro-cracking under uniaxial, biaxial and triaxial tension. Anisotropic micro-cracking was found to be most important at intermediate load levels: at high levels a saturation is reached, and the micro-cracking then becomes isotropic.

Discrete approach to micro-crack clusters
Several authors have treated micro-crack clusters as discrete arrays rather than as a continuum, e.g., Rubinstein (1986), Hori and Nemat-Nasser (1987), Hutchinson (1987), Kachanov (1986, 1987), Kachanov and Laures (1989), Shum and Hutchinson (1990), and Laures and Kachanov (1991). Some of these investigations show that micro-cracks quite clearly may have a shielding effect. Shum and Hutchinson (1990) considered certain special two-dimensional micro-crack configurations. The introduction of four micro-cracks, two on each side of the main crack, almost parallel to each other and inclined with respect to the plane of the main crack, was found to lower the stress intensity factor at the main crack to one half. Thus, the toughening ratio could be as high as $\lambda = K_{ISS}/K_{I(inner)}^{SS} = 2$†. However, this micro-crack configuration is rather peculiar, and other configurations yielded substantially lower toughening ratios. Even negative effects ($\lambda < 1$) may occur: an obvious example is a micro-crack that is coplanar with the main crack.

Laures and Kachanov (1991) considered the influence of three-dimensional configurations of randomly oriented penny-shaped micro-cracks on stress shielding. They were not able to detect any statistically convincing toughening effect. They found that only the micro-cracks closest to the edge of the main crack had a significant influence on its stress intensity factor, and they argued that this speaks against a continuum model. On the other hand, Laures and Kachanov (1991) showed that the stress intensity factor‡ varies along the crack front, which may cause the front to advance along portions with low stress intensity factor, until it is trapped by a local micro-crack array that causes a sufficiently high increase of this factor. There are also other possible mechanisms for crack trapping. Note that crack growth occurs through coalescence of micro-cracks with the main crack. Now, such coalescences may not be as straightforward as intuitively imagined. Rather it appears that head on collision between cracks does not occur: the cracks appear to avoid each other, and bypass each other somewhat until coalescence occurs between one crack edge and the flank of the other crack (Melin 1983; see page 606). This phenomenon obviously requires much higher energy than a head on collision. The significance for the growth of micro-crack clusters has been considered by Fleck (1991). The coalescence mechanism is also described by Yokobori *et al.* (1971), who studied interaction between staggered cracks.

A cell model interpretation to formation of micro-crack clusters
Soon after its formation, a micro-crack is generally arrested by some obstacle, for instance a grain boundary, and it will therefore be confined to a microstructural volume, for instance a grain. Such volumes may be taken to identify the cells to be used in the cell model. Knowledge of the sites of micro-crack nucleation and of the possible arresting boundaries, makes it possible to identify the cells even before the appearance of micro-cracks.

Because micro-crack formation does not require large amounts of plastic strain,

---

† Because the extension of the process region is not obvious, the notation $K_{I(inner)}^{SS}$, rather than $K_{I(pr)}^{SS}$, is used for the stress intensity factor at the edge of the main crack when onset of unstable crack growth occurs.
‡ Actually, there are three such factors, because all three modes are present in a three-dimensional model. However, for an applied stress intensity factor K_I, the factors $K_{II(inner)}$ and $K_{III(inner)}$ will be rather small.

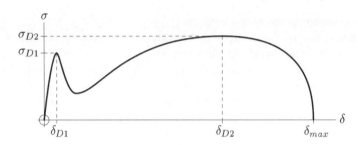

Fig. 8.2.9 Cohesion-decohesion curve for a cubic cell with one micro-crack, subjected to uniaxial straining.

whereas the process of breaking through the arresting boundaries generally does, the cell response to loading under grip control may be considered as two consecutive cohesion-decohesion relations. This may be illustrated by uniaxial straining of a representative cubic cell, cf. page 18. The result may look as shown by Fig. 8.2.9. Thus, there may be two local maxima, one corresponding to onset of contained decohesion inside the cell and the other to onset of complete decohesion of the cell as a whole. Now, the first maximum, σ_{D2} in the figure, may be lower or higher than the second maximum, σ_{D1}. The possibility that it is lower is increased if high residual stresses are present: in extreme cases such stresses may cause micro-cracking even before loading.

If the first maximum is higher than the second one, the process region height will be about the same as the cell height, because unloading of cells above and below cells in a central layer will occur as soon as the maximum load has been reached in this layer†. On the other hand, if the first maximum is sufficiently lower than the second one, the process region height will not be limited to the cell height, but may rather extend over several cells to create a profuse micro-crack cloud. This situation is somewhat similar to the one described for the process region at high crack speed (page 22), and it may have the similar consequence that the significance of the intrinsic length parameter, the cell height, becomes suppressed. Thus, the size of the cloud of microcracks may not be a material property. If a cloud contains sufficiently many microcracks, macroscopic stresses and strains may be defined for the cloud. A redistribution of these stresses and strains is expected to occur during crack advance, somewhat similar to the redistribution of strains in the plastic region of an advancing crack, page 324. This might lead to less concentrated macroscopic strains, which in turn would lead to a size increase of the cloud and thereby to some stable crack growth. Quite obviously, the energy dissipation in the process region will be considerably larger for a cloud than for a one cell thick layer of micro-cracks.

Fractal models for cracks

A number of attempts have been made to describe fracture surfaces by means of their fractal dimension. Different methods have been used. In one method, the fracture profile in a vertical section (for horizontal main direction of the fracture surface) is

---

† For this process region, the Barenblatt model might be appropriate, but the normal stress along this region will possess a local minimum.

recorded by a profilometer, and a Richardson plot† is made. A straight line often fits the results quite well, and the slope of this line gives the fractal dimension.

In the so-called slit-island technique, sections are made parallel to the general direction of the crack either on the fractured surfaces themselves or on a mould of these surfaces. Such sections contain islands from the cut surface peaks, and their number and size will increase with the depth of sectioning. The total perimeter of the islands is measured on images with the same resolution (corresponding to using a measuring rod of fixed length), and the result is plotted against the logarithm of the total area of the islands. Even in this case, a straight line can be fitted to give the fractal dimension.

For a section containing a pattern such as a two-dimensional array of micro-cracks, the so-called box-counting method may be used. The area considered is divided into identical squares, and the number of squares which contain some part of the pattern, is counted. This is repeated for squares of different sizes, and the result is plotted against the side length of the squares in lorgarithmic scale. In cases when a straight line fits the data reasonably well, a fractal dimension of the pattern can be obtained.

The fractal dimension may constitute a convenient way to characterize the appearence of fracture surfaces. It seems to be reasonably constant for the same material, environmental conditions, composition and time dependence of the load and crack speed.

Different aspects of fractal models in connection with fracture are covered in, for instance, Barenblatt (1993), Barenblatt and Botvina (1983), Borodich (1997), Mandelbrot et al. (1984), Mecholsky et al. (1989), Okubo and Aki (1987) and Underwood and Banerji (1986). An interesting discussion of the modelling of fracture from different perspectives, including fractal theory, is given by Mishnaevsky Jr (1997).

Fracto-emission

In his experiments on splitting of mica (see page 582), Obreimoff (1930) noticed light emission from the newly created surfaces. He referred to this phenomenon as triboluminescence, a term used for emission of light from rubbing or scratching certain substances. Such emission is also experienced when crushing rocks, etc. Light emission from fracture surfaces is only part of a broad spectrum of emission of photons, phonons and charged particles, collectively termed *fracto-emission* (Dickinson 1990). Fracto-emission occurs in many different materials, ceramics, polymers and metals, albeit with wide variations in energy and other properties.

Fracto-emission is generally most intense during fracture. In some materials it may also occur as a precursor of fracture, and it is then probably caused by micro-structural phenomena, such as opening and coalescence of micro-separations. Fracto-emission has therefore been suggested as a tool for prediction of imminent fracture or slip, for instance in the form of *acoustic emission*. Some fracto-emission may also take place after a fracture event – the duration is then highly material specific and may last as long as a few seconds.

Dickenson (1990) suggests sequences of events leading to fracto-emission. One key element appears to be charge separation accompanying material separation. Even a seemingly extremely small charge imbalance, such as one excess charge in a million

---

† A Richardson plot is based on measurements of the length of a curve by using rods of different lengths, L, and the result is plotted against $\ln L/L_0$, where L_0 is a constant length, for instance 1 mm.

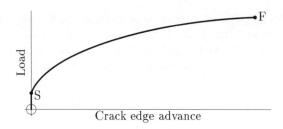

Fig. 8.3.1 Load versus crack edge advance. Forwards displacement of the crack edge, due to blunting, is not included. Onset of stable crack growth occurs at point S, unstable crack growth (fracture) takes place at point F.

surface sites, may cause very strong electric fields. At surface asperities, such fields may lead to electric discharges, even to the effect of ionizing atmospheric nitrogen. Thus, the high mechanical activity in the process region may give rise to strong electrical and chemical activity as well.

8.3 Prediction of onset of crack growth

General considerations. The concept of autonomy before crack growth

In the present section, monotone loading of a cracked body is assumed. For most materials and body and load geometries, crack growth begins with a stable phase. There are some conceptual difficulties in defining onset of stable crack growth (Section 2.4). However, these are of less importance for the following discussion, in which it is therefore assumed that onset of crack growth occurs suddenly and simultaneously along the entire crack edge.

In many practical situations, the onset of crack growth takes place at a load that is considerably lower than the fracture load, and it is then not a great concern. One typical example is illustrated by the relation between load and displacement under the load; see Fig. 8.3.1. However, such difference between the fracture load and the load at incipient crack growth is not always present, and, moreover, the onset of stable crack growth is important in connection with the so-called resistance to crack growth, a concept that is essentially explained by Fig. 8.3.1 and will be discussed in Section 8.4.

Any reasonably simple criterion for onset of unstable crack growth has to be based on a condition of autonomy, and use a suitable quantity to specify the state of the region of autonomy, which must, at least, cover the process region. Autonomy refers to a comparison between different situations, as regards body and loading geometry (called body-load cases). These may be vastly different: one case may be a large oil container, loaded by the pressure from the oil, and another case may be a console, supporting weights.

If an observer, who is able to observe what happens during loading in a region A near the crack edge, but not necessarily regions farther away, will notice the same processes in that region during loading in different body-load cases, then these cases belong to the same autonomy class, and the autonomy refers to the region A. The smaller A is, the wider the autonomy class.

Obviously, an autonomy class consists of situations in which the different bodies 1) are manufactured or created of the same material, 2) contain a crack oriented so that the same mode mix (usually either mode I, II or III) prevails in the region of autonomy. In addition, it may be necessary to specify temperature, rate of loading, etc., if these are not tacitly understood.

Several criteria for onset of crack growth have been suggested. Some of these involve physical quantities in the crack edge vicinity, such as the crack opening displacement, COD, defined on page 289, the Crack Opening Angle, COA† (Andersson 1973a), the *final stretch* (Wnuk 1971, 1973, 1974), which is essentially the stretching of the process region after it has reached the decohesive state‡, and the attainment of a critical strain at a specified distance ahead of the crack (McClintock 1958). Other criteria use quantities that are not tied to features in the crack edge vicinity, such as the J-integral (Broberg 1971) and, for small scale yielding, the stress intensity factor (Irwin 1957).

Small scale yielding

Mode III

Under pure mode III conditions, small scale yielding implies that the dissipative region near the crack edge (the process region, the plastic region and the small regions of unloading on each side of the process region) is controlled by the stress intensity factor, cf. pages 71ff. This implies that autonomy prevails for the dissipative region, i.e., this region develops in a manner that is characteristic of the material at given environmental conditions, such as temperature, irrespective of body and loading geometry. Hence, onset of stable crack growth occurs at a certain critical value, K_{III}^0, of the stress intensity factor.

Note that the autonomy implies that the attainment of a critical value of any quantity that changes with the state of the dissipative region, may be used as a criterion for onset of stable crack growth. The stress intensity factor is only one among several possible choices. Other examples are the shear strain at a given distance ahead of the crack edge and the crack edge sliding displacement just behind the crack edge. The J-integral for a path outside the dissipative region is an obvious alternative, because it is uniquely determinable from the stress intensity factor through the relation (3.3.40). Moreover, the J-integral alternative is more general, because it may be used even for materials which are non-linear in the elastic region. The critical stress intensity factor and the critical J-integral are, of course, easier to measure than candidates connected with the vicinity of the crack edge, because they can be determined from the outer load at onset of crack growth. Their relation to this load may be obtained numerically or experimentally by, for instance, using the expressions in Appendix A10.

The situation becomes more complicated when compressive stresses are acting in addition to pure anti-plane loading. This is a very common situation for mode III crack propagation, because otherwise crack growth tends to take place under mode I, by abrupt change of crack plane direction (Knauss 1970b, Broberg 1983b). The

---

† COD and COA are often written CTOD and CTOA, where T indicates the tip of a crack in a two-dimensional projection.
‡ This criterion is particularly useful for viscoelastic and viscoplastic materials.

dissipative region will now be controlled jointly by the stress intensity factor and the compressive stresses. It appears reasonable to assume that decohesion in the process region will be increasingly impeded with increasing compressive stresses. Also the plastic region will be influenced, particularly if the compressive stress state is not hydrostatic. Autonomy still exists, but only between cases with the same compressive stress state in the crack edge vicinity (whereas body and loading geometry may vary). Ideally, the critical stress intensity factor should be determined as a function of the compressive stress state. Assuming that this stress state is symmetric, it will have one component normal to the crack plane, one component parallel with the crack direction and one component parallel with the crack edge. The critical stress intensity factor will then depend on these three stresses, but, judging from experience about onset of crack growth in mode I, the dependence on the compressive stress state might be rather small, so that the main concern about this state may be reduced to its possibility to prevent conversion to mode I.

Mode II

As for mode III, autonomy of the dissipative region at small scale region is limited to cases with the same compressive stress. Furthermore, like the case for pure mode III, the stress intensity factor in pure mode II controls the state of the dissipative region at sufficiently small scale yielding. Thus, onset of crack growth is governed by a critical stress intensity factor, K_{II}^0.

For a crack to propagate in mode II, a sufficiently high compressive stress normal to the crack plane is generally required. The critical stress intensity factor for onset of crack growth in mode II should thus be determined as a function of the compressive stress state, which, as for mode III, usually has three components. On the other hand, judging from experience about onset of crack growth in mode I, the dependence on the compressive stress state might be rather small, so that the main concern about this state may be reduced to its possibility of preventing conversion to mode I. Mode II (as mode III) propagation is usually associated with geophysical phenomena, and therefore the special case of plane stress is much less common than for mode I.

Mode I, plane strain

Pure mode I differs from pure modes II and III in the presence of a T-stress. The two special cases of plane strain and plane stress are of particular interest, although they may appear side by side along the same crack edge, as in a plate with plane strain in central parts and plane stress near the plate surfaces (see Fig. 2.8.4). In the present discussion, it is assumed, for simplicity, that plane strain prevails in the entire crack edge vicinity. Plane stress, as approached in thin plates, will be discussed later (page 595).

The normal stresses acting on the plastic region straight ahead of the crack are predominantly directed normal to the crack plane. A larger T-stress would therefore lead to higher triaxiality than a smaller one, and thus impose a higher constraint to plastic flow. This would lead to a smaller size of the plastic region; cf. Fig. 5.4.4.

The T-stress also influences the micro-separation mechanism. High T-stresses would favour micro-crack formation, whereas low T-stresses would favour void growth. The void growth mechanism is also influenced by the T-stress. This is, for instance, evident from the difference between results obtained by Rice and Tracey (1969) and Andersson

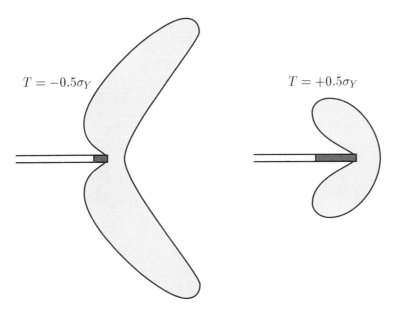

Fig. 8.3.2 The process region (heavily shadowed) and the plastic region (lightly shadowed) at the onset of stable crack growth under small scale yielding in a material with void growth in the process region, for two different T-stresses. The figure is schematic, although it reflects broadly results from numerical calculations by Xia and Shih (1995a).

(1977), due to different ambient stress states, page 548. The influence of the T-stress on the cohesion-decohesion curve is indicated by Fig. 1.3.4.

The T-stress appears to have a significance influence on the division of energy dissipation between the process region and the plastic region. This is illustrated for a void growth mechanism in Fig. 8.3.2, which is schematic, although based on results from numerical calculations for small scale yielding, particularly those reported by Xia and Shih (1995a), who used the cell model and considered both a very small and a considerable crack advance. Note that the process region is much larger and the plastic region much smaller for a large positive than for a large negative T-stress. It is therefore surprising that, as shown by several experimental and numerical investigations, the stress intensity factor, K_I^0, at which onset of stable crack growth occurs, is almost independent of the T-stress for a void growth mechanism. On the other hand, for a micro-crack mechanism, the larger process region for large negative T-stresses may imply larger critical stress intensity factor for positive than for negative T-stresses.

During loading under small scale yielding conditions, the T-stress is proportional to the stress intensity factor in the common situation of proportional increase of all loads on the structure. It is then usually considerably smaller at the onset of crack growth than at the onset of fracture, cf. Fig. 8.3.1. This fact contributes to the apparent independence of K_I^0 on the T-stress, as found from experiments with different specimen geometries.

Large scale yielding

Modes II and III
For large scale yielding, there is no autonomy of the dissipative region. There might, however, be satisfactory autonomy of the process region, at least between cases with approximately the same compressive stresses. If this is the case, then the onset of crack growth will occur when the process region has reached a certain critical state. Any quantity capable of characterizing the state of the process region may then be used for a criterion of onset of crack growth. The most convenient quantity seems to be the J-integral, which appears to be reasonably path-independent in the plastic region before crack growth, so that the criterion may be given as $J_{II} = J_{II}^0$ for mode II and $J_{III} = J_{III}^0$ for mode III. These critical values will be dependent on the compressive stress state. However, as for small scale yielding, they may be only weakly dependent on the compressive stress state, provided that this is sufficient to prevent conversion to mode I.

Now, there are cases without satisfactory autonomy for the process region at large scale of yielding. The autonomy becomes less reliable the larger the scale of yielding and the lower the strain hardening. This was shown (pages 268ff.) for a mode III crack in an elastic-perfectly plastic material. Different criteria then predict different instants for the onset of crack growth. It does not seem to be possible to establish general criteria for onset of crack growth in such cases. Each case has to be investigated individually, either by numerical simulations, using, for instance, the cell model for the crack edge vicinity, or by experiments.

Mode I, plane strain
As for modes II and III, there is no satisfactory autonomy for the dissipative region in mode I under large scale yielding (cf. page 286), but it might prevail for the process region. If so, it will prevail between cases with the same kind of constraint to plastic flow in front of the crack edge. This formulation has to be used, rather than a reference to a T-stress, because this concept is limited to linearized cases of small scale yielding. For large scale of yielding, the constraint may be measured by the triaxiality parameter Q (pages 288ff.) Now, consider a series of tests, involving different cases of large scale yielding (different body and loading geometry) in the same material, and suppose that the values of the J-integral, J_I^0, and of the Q-parameter, Q_0, are determined at the onset of crack growth. Suppose that, apart from some scatter, it would be possible to represent the outcome of the test series by a curve, showing a relation between J_I^0 and Q_0, as schematically shown in Fig. 8.3.3.

Figure 8.3.3 also shows the J_I-Q relation obtained during loading of a cracked body of the same material as in the test series. The relation may have been determined numerically, for instance by finite element calculations. Onset of crack growth would then be expected to occur for the J_I-Q values where the loading path crosses the J_I^0-Q_0 curve†. Under what conditions is this likely to occur?

There are two conditions. First, the J-integral must be sufficiently path-independent for paths shrinking from the boundary of the plastic region to the neighbourhood of

---

† This is the procedure suggested by O'Dowd and Shih (1994).

8.3 PREDICTION OF ONSET OF CRACK GROWTH

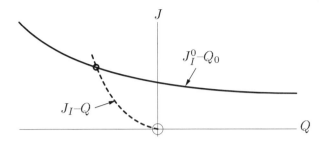

Fig. 8.3.3 Relation between J_I^0 and Q_0 at onset of crack growth, obtained from a series of tests (solid line), and the J_I-Q relation during loading in a real or simulated test with the same material (dashed line).

the process region†. Second, autonomy must prevail for the process region, so that the same value pair of J and Q implies the same state of the process region. The first condition may be reasonably well satisfied before crack growth. The second condition requires that the process region is well embedded in the plastic region, that the strain hardening is sufficiently high and that the scale of yielding is not too large. The two last mentioned requirements are coupled: the larger the strain hardening, the larger the scale of yielding that may be allowed. Now, onset of crack growth usually occurs well ahead of onset of unstable crack growth (cf. Fig. 8.3.1), and therefore the scale of yielding is seldom very large at the onset of crack growth. In fact, judging from experimental determinations of the J-integral at the onset of crack growth, even the dependence on the Q-parameter appears to be quite modest.

Figure 8.3.3 shows schematically a feature detected by numerical simulations: the curve approaches a horizontal line for positive Q, i.e. high constraint of the plastic flow. In this region, the value of Q is not needed for prediction of the onset of crack growth: the critical value J_I^0 is sufficient‡. Still the two conditions, path-independence of the J-integral and autonomy of the process region are required. These two conditions are often loosely lumped together under the term "J-dominance".

The process in the process region may be different in the same material for different constraints. It may be micro-crack formation and growth for high constraints (large Q), and void formation and growth for low constraints. The change may appear as a "knee" in the $J_I^0 - Q_0$ curve.

---

† It is assumed that the Q-parameter sufficiently accurately reflects the hydrostatic stress ahead of the crack.
‡ This is the so-called J-integral criterion (Broberg 1971). It is now reasonably well documented experimentally. Early attempts by Begley and Landes (1972), which received much attention, were made under the erroneous assumption that crack growth did not occur until maximum load was reached. The reason why they still obtained a good agreement with the J-integral criterion might have been due to numerical errors in the evaluation of the tests. For thin sections, early experimental confirmations of the J-integral criterion were obtained by Hickerson (1977).

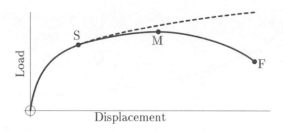

Fig. 8.4.1 Example of the relation between load and displacement at the point of load application for a specimen tested in a very stiff machine. Onset of stable crack growth occurs at point S and fracture occurs at point F. In a very soft machine, fracture would occur just after the maximum point, M, has been reached. The dashed line shows the curve if no crack growth occurs (this may be arranged in certain materials by cutter-produced blunting).

8.4 Prediction of fracture

Global *versus* local criteria

The onset of crack growth is governed by a local criterion, namely that a certain critical state of the process region is reached. On the other hand, onset of unstable crack growth, usually synonymous with fracture, occurs as a result of a global instability (Broberg 1973a). This fact may be demonstrated by loading a test specimen in a very stiff loading machine and recording the load versus the displacement under the load. The curve may look like the one shown in Fig. 8.4.1, i.e., it possesses a maximum, and fracture occurs at a lower load. It is then obvious that unstable crack growth in an identical specimen, tested in a very soft machine, will take place just after the maximum load is reached, but up to that instant, the load-displacement curve will be identical to the one from the test in the stiff machine. As a consequence, the state of the process region develops in exactly the same way in the two tests up to maximum load.

During unstable crack growth, the stiff machine cannot supply the energy needed for dissipation in the crack edge region, and therefore all this energy is supplied by energy release from the stress-strain field in the specimen. The soft machine, on the other hand, is capable of supplying all energy needed, beyond what is released from the stress-strain field, after the maximum load is reached.

During the stable phase, a certain amount of energy has to be supplied to the process region per unit crack growth. This is clearly a local condition, which is thus valid simultaneously with the global criterion for onset of unstable crack growth. A local criterion obviously governs crack growth, but it cannot decide whether global instability will occur or not. The amount of energy needed by the process region for crack growth may change during the whole stable phase, until unstable crack growth occurs or steady state conditions are approached†. Thereupon, some other local criterion, involving inertia effects, may take over during the unstable crack propagation.

As for the onset of crack growth, a reasonably simple criterion for onset of unstable

---

† Steady state conditions are only approached for small scale yielding, except for certain cases of crack face loading; this will be shown later, page 613.

crack growth has to be based on a condition of autonomy, and use a suitable quantity to specify the state of the region of autonomy, which must, at least, cover the process region. Several criteria have been suggested. The simplest ones relate to small scale yielding, in cases when it can be assumed that the whole dissipative region at the crack edge develops in an autonomous manner during crack growth. This implies that an observer, who is able to observe the whole dissipative region during loading, but not necessarily regions outside, will notice the same development of processes (increase of stresses and strains in the plastic region, formation and growth of micro-separations in the process region) between different cases as regards body and loading geometry. For large scale yielding, at least the process region has to develop in the same manner between cases belonging to the same autonomy class.

Small scale yielding

A convention for the validity of LEFM

Small scale yielding is a very useful concept, which has been used to describe that the linear extension of the plastic region is small compared to significant dimensions of the body, particularly the crack length. This is not a precise definition, and, moreover, the sense in which the plastic region shall be small is not stated. In previous chapters the unambiguous concept of infinitesimal small scale yielding has been extensively used, and "small scale yielding" has been used for cases which in some respect are assumed to show similar behaviour as those of infinitesimal small scale yielding.

Often, the term small scale yielding is associated with *Linear Elastic Fracture Mechanics*, LEFM (cf. page 615), which implies that linearly elastic behaviour can be assumed to hold arbitrarily close to the crack edge without seriously affecting the outcome of an analysis, aimed at prediction of the onset of unstable crack growth. On the basis of several tests on different specimen types and sizes, Brown and Srawley (1967) concluded that a reasonable upper limit for the validity of LEFM for mode I is given by the requirement that

$$a \geq 2.5 \frac{K_I^2}{\sigma_Y^2} \qquad (8.4.1)$$

where a is the half-length of a crack in a remotely loaded large plate, or an equivalent length, for instance the length of an edge crack. Brown and Srawley claimed that determinations of the *fracture toughness*, K_{ISS} (K_{Ic}), from tests on different specimen types gave the same result with reasonable accuracy, if (8.4.1) holds. This convention is very convenient, and it is codified in standards in several countries. For cracks in plates, it also requires that the largest ligament from the crack edge to an outer in-plane boundary must be larger than the right hand side of (8.4.1), and so also must the plate thickness be for plane strain LEFM in mode I. Relation 8.4.1 is often referred to as the ASTM (American Society for Testing and Materials) convention or limit for LEFM.

For a large plate with a central crack, $-a < x < a$, $y = 0$, and subjected to remote loading, $\sigma = \sigma_y^\infty$, $K_I = \sigma_y \sqrt{\pi a}$, as given by (4.6.1), and therefore the LEFM limit implies $\sigma_y^\infty \leq \sigma_Y/\sqrt{2.5\pi} \approx 0.36\sigma_Y$.

The ASTM limit does not include the T-stress. It is now well known that the linear size of the plastic region varies considerably with the T-stress; see Fig. 5.4.4.

8. PHYSICAL AND ENGINEERING ASPECTS OF FRACTURE

The ASTM limit may therefore be too conservative in some cases and perhaps too generous in others.

Different criteria. The Griffith-Orowan-Irwin approach

For modes II and III, autonomy may require the same compressive stress state between different cases, and for mode I cases, in which all loads increase in proportion to each other, it may require the same T-stress in relation to K_I, i.e. the same T/K_I (which has dimension $1/\sqrt{length}$). Then, criteria that are not tied to features in the crack edge vicinity may be used, in particular a critical value of the stress intensity factor, $K_I = K_{ISS}$ (K_{Ic}), $K_{II} = K_{IISS}$ (K_{IIc}) or $K_{III} = K_{IIISS}$ (K_{IIIc}). These criteria were established by Irwin (1957), an achievement that may be considered as the starting point of modern fracture mechanics. They are generally functions of the T-stress (for mode I) or of the ambient compressive stress (for modes II and III), but for small scale yielding the dependence is weak, except that the compressive stress in modes II and III must be sufficiently high to prevent conversion to mode I.

Historically, the fracture criterion for small scale of yielding (actually for brittle materials like glass, for which, using modern terminology, it may be assumed that only a process region and no plastic region exists) was established by Griffith (1920)†, using an energy argument.

Griffith used the Inglis (1913) solution of the canonical crack problem, pages 132ff. According to (4.6.29), the change of potential energy per unit plate thickness by opening a crack of length a in a large plate subjected to the remote mode I load $\sigma_y = \sigma_y^\infty$, is

$$\Delta \Pi = -\frac{\pi (\sigma_y^\infty)^2 a^2}{4(1-k^2)\mu} \tag{8.4.2}$$

i.e. a decrease.

The energy dissipation during crack growth consists of what is needed to create new surface: it is the *surface energy*, γ, per unit of area. Thus, an energy supply, $4\gamma \mathrm{d}a$, is needed for a crack extension $\mathrm{d}a$ at each end. Static equilibrium is assumed before this presumed extension takes place. Then, the change of total energy must be zero, i.e.,

$$\mathrm{d}\Pi + 4\gamma \mathrm{d}a + kinetic\ energy + heat = 0 \tag{8.4.3}$$

where the kinetic energy and the heat production are associated with the crack extension. Thus, crack extension is not possible, unless

$$\left[-\frac{\pi(\sigma_y^\infty)^2 a}{2(1-k^2)\mu} + 4\gamma\right]\mathrm{d}a < 0 \tag{8.4.4}$$

Neglecting kinetic energy and heat production, leads to the Griffith critical stress for unstable crack growth from the equilibrium state:

$$\sigma_y^\infty = \sigma_{crit} = 2\sqrt{\frac{2(1-k^2)\mu\gamma}{\pi a}} \tag{8.4.5}$$

Note that kinetic energy and heat production are expressions of the same phenomenon

---

† Griffith's paper appeared in the *Philosophical Transactions of the Royal Society (London)*, and is found in volume **A221**, which covers the period 1920-1921. This has caused some confusion about the proper year of publication. The exact date is October 21, 1920.

on the atomic scale. Neglecting them may be compensated by using a somewhat higher value for the surface energy.

Because $d^2\Pi/da^2 < 0$, the equilibrium state is unstable, i.e., unstable crack growth will result. Changing the sign of da implies crack retraction (healing) rather than crack extension. In this case, neglecting kinetic energy and heat production may be compensated by using a somewhat lower value for the surface energy. Now, healing would be very improbable for mode I, because of physical obstacles, induced by the condition of the crack faces. It is not impossible, however, as was shown in experiments by Obreimoff (1930) on the cleavage of mica. Obreimoff used a wedge with slightly rounded front end to split the mica and applied beam theory to determine the energy flux into the crack edge. The load required for crack growth was generally reduced after healing, but in some experiments performed in vacuum, it was completely restored after healing. For modes II and III, some healing is possible: it is simply a cessation of sliding, after which the more or less rough mating faces engage in a grip, especially in the presence of an ambient compressive load. In certain materials, for instance geological materials, some bonding of the mating faces may evolve with time.

Experimental evidence shows that one feature of the Griffith theory holds well, namely that the fracture load, σ_{crit}, is inversely proportional to $\sqrt{a}$ for sufficiently large values of a. However, quantitatively, σ_{crit} is usually grossly underestimated. This fact prompted E. Orowan at a symposium at Massachusetts Institute of Technology, 1950, to suggest that also plastic energy dissipation, Γ_{pl} per unit area, should be taken into account, thus replacing 2γ by $2\gamma + \Gamma_{pl}$ (Orowan 1952)†. Usually, γ may be neglected at the side of Γ_{pl}. Exceptions are very brittle single-crystals, in which cleavage occurs; see e.g. Obreimoff (1930).

In fact, the surface energy might not have been what determined the fracture strength in the experiments on glass that Griffith carried out to test his theory (Griffith 1920), although he could confirm the inverse square root dependence of the fracture load on the crack length. However, he also determined γ from the outcome of these experiments and compared with independent measurements of the surface energy in glass heated to different temperatures over 1000 K and extrapolated to room temperature. The agreement was reasonable, but it has afterwards been found that the energy flux into the crack edge in glass is considerably reduced in air, due to action of water vapor. It was this considerably reduced energy flux that Griffith identified with 2γ and interpreted γ as the surface energy of glass.

The Griffith (1920) approach was used in a new formulation by Irwin (1957), who introduced the concept of energy release rate (in the present work called energy flux into the crack edge), $\mathcal{G}$, and its critical value, $\mathcal{G}_c$. Thus, unstable crack growth occurs when $\mathcal{G} = \mathcal{G}_c$, which, for the case considered by Griffith, would read:

$$\frac{K_I^2}{4(1-k^2)\mu} = \frac{(\sigma_y^\infty)^2 \pi a}{4(1-k^2)\mu} = \mathcal{G}_c \qquad (8.4.6)$$

with obviously the same result as (8.4.5) if $\mathcal{G}_c = 2\gamma$. The expression for $\mathcal{G}$ is given by (3.5.14).

Now, the Irwin (1957) approach is more general than the Griffith (1920) theory, because it may be applied in all cases of steady state crack propagation under small scale

---

† A similar approach is present in an earlier work, Orowan (1945).

584   8. PHYSICAL AND ENGINEERING ASPECTS OF FRACTURE

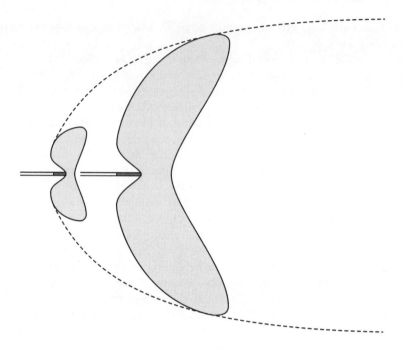

Fig. 8.4.2 The development of the process region (heavily shadowed) and the plastic region (lightly shadowed) during stable crack growth in small scale yielding. The dashed lines show the envelope of the plastic region during crack growth.

yielding, whereas the Griffith theory only deals with materials for which, with modern terminology, there is only a process region and no plastic region. This implies that there is no stable crack growth in the Griffith theory, and thus, there is no distinction between steady state growth and onset of crack growth. Interesting discussions of the Griffith-Irwin approach have been given by Sanders (1960) and Rice (1978).

K- and J-resistance curves

Consider stable crack growth in mode I†. For simplicity, it is assumed for a moment that the T-stress is the same, for instance $T = 0$, between the cases discussed. Approximate autonomy will then apply for the dissipative region. During loading, the stress intensity factor, K_I, increases. In fact, K_I may be used as a measure of loading. When K_I reaches K_I^0, stable crack growth occurs.

The development of the process region and the plastic region during stable crack growth may appear as shown schematically in Fig. 8.4.2, but great variations occur, depending on strain hardening, cohesive strength, micro-separation mechanism, etc. Note that both the process region and the plastic region increase in size during crack growth. Both Ståhle (1985) and Xia and Shih (1995a) found that the length of the process region may increase to more than twice its length at the onset of crack growth‡.

---
† The discussion will, in principle, also be valid for modes II and III.
‡ By contrast, Tvergaard and Hutchinson (1992) remark that the length of the process region "varies little once crack growth has been initiated".

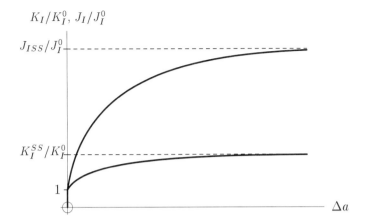

Fig. 8.4.3 K- and J-resistance curves. Only real crack propagation is shown as crack advance, but the crack edge displacement due to blunting is not.

This size increase as well as the size increase of the plastic region, is most pronounced during the early stages of stable crack growth (Ståhle 1985), and thereupon both regions tend gradually toward a maximum size as steady state is approached.

Let Δa denote the amount of crack growth. If this amount is measured in relation to a point on the crack faces far from the crack edge, it includes a part that is not really crack growth, but an effect of blunting; see Fig. 8.2.3. This part is not included here, only the real crack growth. The relation between K_I and Δa usually has the appearance shown in Fig. 8.4.3, which also shows the relation between J_I and Δa, where J_I is the J-integral for a path that surrounds the plastic region. These two curves are called K-resistance and J-resistance curves, respectively. Note that the relation between J_I and K_I is unique for small scale yielding; it follows from (4.6.13)-(4.6.15).

A typical feature is that the curves tend toward a horizontal line. This implies that local steady state is approached; cf. Fig. 8.4.2. The value of K_I at fracture under small scale yielding, K_{ISS} (K_{Ic}), gives one measure of the fracture toughness. Another measure is J_{ISS}, the value of the J-integral at onset of fracture in small scale yielding. Thus,

$$J_{ISS} = \frac{(1-\nu)K_{ISS}^2}{2\mu} \qquad (8.4.7)$$

if the dissipative region is subjected to plane strain.

Consider now small scale yielding cases under different T-stresses. Recall that autonomy between such cases apparently is lost already at the onset of stable crack growth, but that experience nevertheless shows that the criterion $K_I = K_I^0$ is in general applicable. The same happy result does not carry over to the onset of unstable crack growth.

The process region and the plastic region at the onset of unstable crack growth are shown for two different T-stresses in Fig. 8.3.2. Note that the process region size increases, whereas the plastic region size decreases, with increasing T. The figure shows

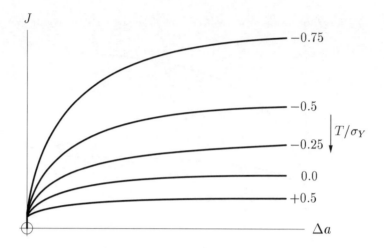

Fig. 8.4.4 J-resistance curves for different T-stresses. The figure is based upon numerical calculations by Tvergaard and Hutchinson (1994) and by Xia and Shih (1995a). Large variations in the T-stress influence may occur between different materials.

that the embedment of the process region may be very deep or very shallow in the same material, depending on the T-stress. The same micro-separation mechanism – void growth – is assumed for both T-stresses, but it is quite possible that for some materials, void growth may be the preferred mechanism for low T-stresses and micro-cracking (cleavage and rupture of bridges) for high T-stresses.

J-resistance curves for different T-stresses are schematically shown in Fig. 8.4.4, based on results by Tvergaard and Hutchinson (1994) and Xia and Shih (1995a). Such curves are, of course, dependent on material properties, but the main features, approximate independence of T for the onset of crack growth and a larger increase of J per unit crack growth for lower than for higher T-stresses, appears to be typical for a void growth mechanism. However, some results seem to indicate a different behaviour. Thus, Varias and Shih (1993) show that the steady state stress intensity factor is clearly higher for positive values of T than for $T = 0$, so that both positive and negative T-stresses seem to shield the moving crack edge. However, they consider propagation of a sharp crack, without explicit incorporation of a process region. This might be more typical for a micro-crack than for a void growth mechanism.

Interpretation of resistance curves in terms of energy flux
It may be argued that the energy flux to the process region equals the J-integral just after onset of stable crack growth, by comparing the real case with a hypothetical case, in which the material outside the process region is non-linearly elastic, with the same stress-strain relations as for the real material during loading (i.e. before unloading). The very first movement of the process region is then governed by the same conditions along its boundary in both cases. Then, because the energy flux into the process region equals J in the hypothetical case, it also equals J in the real case (Broberg 1975). This may be interpreted so that the total energy dissipation per unit crack growth is spent mainly in the process region just after the onset of crack growth.

The argument given is only approximate, because the J-integral is not exactly path-indpendent (there is, for instance, a small region of unloading on each side of the process region), the process region is not vanishingly small compared to the plastic region, and some finite movement of the process region is needed before it reaches approximate steady state conditions. However, the numerical calculations by Tvergaard and Hutchinson (1992), show surprisingly good agreement with an initial energy energy flux equal to J. On the other hand, the results by Xia and Shih (1995a) point toward a somewhat smaller value with a pronounced dependence on the strain hardening.

The apparent fact that most of the energy dissipation goes to the process region just after onset of crack growth, whereupon an increasing fraction takes place in the plastic region, may be interpreted as a gradual screening of the energy flow to the process region (Broberg 1974, 1975). Only a comparatively small fraction may reach the process region when a local steady state is approached, as shown by Andersson (1973b), who used finite element methods with boundary layer loading and a node relaxation technique. Andersson also showed that the fraction that reaches the process region decreases with increasing strain hardening. He kept the process region size constant (it was, in fact, one element size), but it is obvious that a smaller size would have resulted in a smaller fraction and *vice versa*. Numerous following calculations (e.g. Ståhle 1985, Tvergard and Hutchinson 1992, and Xia and Shih 1995a) have confirmed Andersson's results.

An estimate of the screening of the energy flow to the process region may be found by considering the strain gradient decrease that takes place in the plastic region during stable crack growth. As found from the mode I expressions (5.4.22) and (5.7.94) for elastic-perfectly plastic materials, the strain distribution changes from an approximately R_p/r dependence to an approximately $\ln(R_p/r)$ dependence, where R_p is a measure of the linear size of the plastic region in small scale yielding.

For $r = r_p$, where r_p is the length of the process region, the autonomy requires that the strains shall be approximately the same for steady state crack growth as for incipient crack growth†. Let R_p^0 and R_p^{SS} be the size of the plastic region for incipient and for steady state crack growth, respectively. Then it follows that

$$\frac{R_p^0}{r_p} \approx \ln \frac{R_p^{SS}}{r_p} \tag{8.4.8}$$

and, assuming the energy flux to the plastic region to be proportional to its size during steady state, gives

$$\frac{\mathcal{G}_{SS}}{\mathcal{G}_0} \approx \frac{\exp(R_p^0/r_p)}{e \cdot R_p^0/r_p} \tag{8.4.9}$$

where subscripts and superscripts 0 and ss denote incipient growth and steady state, respectively, and the denominator is chosen to give $\mathcal{G}_{SS}/\mathcal{G}_0 = 1$ for $R_p^0 = r_p$. The relation is, of course, very approximate, but it gives an idea about the large span of screening action that may occur between different perfectly plastic materials. Thus, for instance, for $R_p^0/r_p = 4$, about 20 per cent of the energy flux may be expected to reach the process region, whereas less than 1 per cent may do so for $R_p^0/r_p = 8$. A larger fraction is to be expected for strain hardening materials.

---

† It is assumed for simplicity that r_p stays constant during the stable phase.

It is obvious that the load increase and the amount of crack growth during the stable phase will be comparatively smaller the smaller the ratio R_p^0/r_p is. Serious overestimates may result from overestimates of the embedment of the process region in the plastic region.

The screening of the energy flux to the process region has been used in the formulation of a theory on the "essential work of fracture" (Cotterell 1977, Cotterell and Reddel 1977, Mai and Cotterell 1986, Levita 1996, Wu and Mai 1996). This theory is particularly interesting for large scale of yielding in polymers.

Large scale yielding

J-resistance curves

The basic requirement for a reasonably simple theory of fracture, autonomy of, at least, the process region, is more complicated at large than at small scale yielding. Autonomy implies that an observer, who is able to observe and follow the moving process region during loading, but not regions outside, will notice the same development of processes (formation and growth of micro-separations in the process region and its motion forwards) between cases that differ in body and loading geometry. Thus, the process region develops in the same manner between cases belonging to the same autonomy class.

Now, the situation for the observer of a large scale yielding event is not the same as for the observer following the development of the plastic region in small scale yielding. Even after following hundreds of cases within the same autonomy class, the observer would not be able to predict from previous experience when the catastrophic event of unstable crack growth will take place, because this is governed by a global criterion for large scale yielding and would consequently require observation of events far away from the process region. In some cases, the stable crack growth may end soon after its onset, and in other cases it may continue much further.

It was once believed that autonomy classes could be found, spanning both small and large (though not too large) scale of yielding. This could be the case if the extent of stable crack growth were significantly smaller than the forward extent of the plastic region in small scale yielding. Approximate path independence of the J-integral for paths outside the process region could then be assumed, so that the J-integral would be able to identify the current state in the process region neighbourhood, in particular the current extent of stable crack growth. Thus, a J-resistance curve could be established, and it would be the same as for small scale yielding up to the point of incipient unstable crack growth (fracture). However, the extension of the autonomy beyond small scale of yielding is questionable for many materials; cf. pages 592ff.

The mathematical expression for the J-resistance curve, for small scale yielding in mode I may be written in the form

$$\frac{J_{IR}(a-a_0)}{J_{ISS}} = F\left(\frac{\sigma_Y^2(a-a_0)}{EJ_{ISS}}\right) \qquad (8.4.10)$$

where $J_{ISS} = (1-\nu)K_{ISS}^2/(2\mu)$ in plane strain, a is the current and a_0 the original crack length, and $F(\cdot)$ is monotonically increasing toward unity. This relation is specific for the material (and temperature, etc.).

Assume that (8.4.10) holds even for some amount of large scale yielding. For each

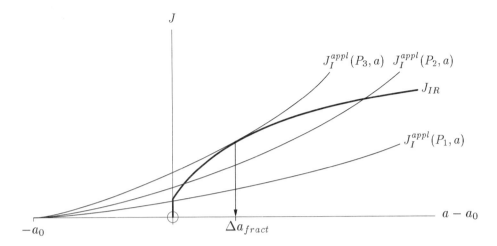

Fig. 8.4.5 J-resistance curve (J_{IR}) and the relation between the J-integral, $J_I^{appl}(P, a)$ and crack length, a, for different values, $P_1 < P_2 < P_3$, of the load parameter P.

specific geometry of the cracked body, the applied J_I, here denoted by J_I^{appl}, depends on the load and on the current crack length. For both load and grip control (see page 16), J_I^{appl} is an increasing function of a, but the functional form is different. For load control, it is convenient to write J_I^{appl} as a function of a load parameter† P and the crack length a, i.e., $J_I^{appl} = J_I^{appl}(P, a)$.

From (8.4.10), the amount of stable crack growth under a given load is found as the lowest root $a = a_1$ of the equation

$$J_I^{appl}(P, a) = J_{IR}(a - a_0) \tag{8.4.11}$$

provided that roots exist; cf. Fig. 8.4.5. Increasing load implies increasing stable crack growth, as long as $\partial J_I^{appl}(P, a)/\partial a < dJ_{IR}/da$ for $a - a_0 = \Delta a_1$. Unstable crack growth occurs for $J_I = J_I^{fract}$, $a - a_0 = \Delta a_1 = \Delta a_{fract}$, when, in addition to (8.4.11), the equation

$$\frac{\partial J_I^{appl}(P, a)}{\partial a} = \frac{dJ_{IR}}{da} \tag{8.4.12}$$

is satified. This corresponds to the point where the curve $J_I = J_I^{appl}(P, a)$ is tangent to the curve $J_I = J_{IR}(a - a_0)$.

Note that (8.4.11) constitutes a local condition for stable crack growth, and (8.4.12) constitutes a global condition for unstable crack growth. There are several cases for which the global condition cannot be satisfied. One example is shown in Fig. 2.5.2, which represents a configuration that allows stable but not unstable crack growth.

A study of Fig. 8.4.5 reveals that the root $a - a_0 = \Delta a_{fract}$ increases with the original crack length a_0, i.e., the amount of stable crack growth and the value J_I^{fract} of the J-integral, for which unstable crack growth occurs, both increase with decreasing

---

† If the structure is subjected to mutually independent loads, more than one load parameter is needed.

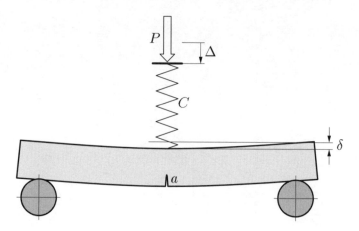

Fig. 8.4.6 Load transmitted to a cracked structural part via a device, here symbolized by a spring.

scale of yielding. The relative amount of stable crack growth, $(a-a_0)/a_0$, on the other hand, decreases with decreasing scale of yielding. For small scale yielding, a_0 is much larger than $a - a_0$, and then J_I^{fract} approaches J_{ISS}.

For grip control, the same relations hold, provided that $J_I^{appl}(P,a)$ is replaced by $J_I^{appl}(\delta,a)$, where δ is a load displacement parameter. Note that $J_I^{appl}(\delta,a)$ decreases with increasing a, if δ is kept constant. This implies curves that only intersect the J-resistance curve once for all values of δ, except in cases when the resistance curve is decreasing after a maximum is reached. If δ increases, the point of intersection moves along the resistance curve, and, eventually, fracture may occur, if the resistance curve possesses a maximum, but otherwise, plastic collapse may occur, rather than fracture.

Now, load control and grip control are only the two extreme conditions under which loads may be applied. A general case may be represented by load application through a spring, as shown in Fig. 8.4.6, with the remote end of the spring kept under grip control. Grip control would be realized by choosing a vanishing spring compliance, and load control by choosing a very large compliance. The spring may, for instance, represent a hydraulic testing machine or simply structural parts outside a cracked part.

Let δ be the displacement under the spring and Δ the displacement under the load P. The same relations as for load control and grip control apply, provided that J_I^{appl} is now replaced by $J_I^{appl}(\Delta, a)$. However, $J_I^{appl}(\Delta, a)$ must be found from consideration both of the J-integral in the cracked part and of the properties of the spring. Assume, for simplicity, that the spring is linearly elastic, so that

$$\Delta = \delta(P, a) + CP \qquad (8.4.13)$$

where C is the spring compliance. The J-integral may be written both as a function of Δ and a and as a function of P and a, i.e.,

$$J_I^{appl}(\Delta, a) = G(P, a) \qquad (8.4.14)$$

8.4 PREDICTION OF FRACTURE

Thus,

$$\frac{\partial J_I^{appl}(\Delta, a)}{\partial a} = \frac{\partial G(P, a)}{\partial a} + \frac{\partial G(P, a)}{\partial P} \cdot \frac{\partial P}{\partial a} \tag{8.4.15}$$

Now, from (8.4.13)

$$\frac{\partial \Delta}{\partial a} = \frac{\partial \delta(P, a)}{\partial a} + \frac{\partial \delta(P, a)}{\partial P} \cdot \frac{\partial P}{\partial a} + C \frac{\partial P}{\partial a} \tag{8.4.16}$$

But, grip control at the remote end of the spring implies that unstable crack growth occurs under the condition $\partial \Delta / \partial a = 0$. Thus,

$$\frac{\partial J_I^{appl}(\Delta, a)}{\partial a} = \frac{\partial G(P, a)}{\partial a} - \frac{\partial G(P, a)}{\partial P} \cdot \frac{\partial \delta}{\partial a} \left[C + \frac{\partial \delta}{\partial P} \right]^{-1} \tag{8.4.17}$$

Because $J_I^{appl}(\Delta, a) = G(P, a)$, this relation may alternatively be written in the notations used in, for instance, physical chemistry, i.e.,

$$\left(\frac{\partial J_I^{appl}}{\partial a} \right)_\Delta = \left(\frac{\partial J_I^{appl}}{\partial a} \right)_P - \left(\frac{\partial J_I^{appl}}{\partial P} \right)_a \cdot \left(\frac{\partial \delta}{\partial a} \right)_P \left[C + \left(\frac{\partial \delta}{\partial P} \right)_a \right]^{-1} \tag{8.4.18}$$

An alternative expression is found by letting G be a function of δ and a, rather than of P and a:

$$\left(\frac{\partial J_I^{appl}}{\partial a} \right)_\Delta = \left(\frac{\partial J_I^{appl}}{\partial a} \right)_\delta - C \left(\frac{\partial J_I^{appl}}{\partial \delta} \right)_a \cdot \left(\frac{\partial P}{\partial a} \right)_\delta \left[1 + C \left(\frac{\partial P}{\partial \delta} \right)_a \right]^{-1} \tag{8.4.19}$$

Onset of unstable crack growth is found at the point $J_I = J_I^{fract}$, $a - a_0 = \Delta a_{fract}$, where

$$J_I^{appl}(\Delta, a) = \frac{J_{IR}(a - a_0)}{J_{ISS}} \tag{8.4.20}$$

$$\frac{\partial J_I^{appl}(\Delta, a)}{\partial a} = \frac{dJ_{IR}}{da} \tag{8.4.21}$$

This is a generalization of equations (8.4.11)-(8.4.12), and the corresponding graphical representation is basically the same as shown by Fig. 8.4.5. Figure 8.4.7 shows the difference in the J-integral at fracture between loading via a very soft device ($C \to \infty$, load control) and a very stiff device ($C \to 0$, grip control).

Note that the J-integral depends on the history during crack growth. To make this more clear, it may be written as $J_{flow} = J(P, a; \Gamma; a_0)$, where Γ is the integration path, subscript $flow$ indicates the flow or incremental strain theory of plasticity, and subscript I is dropped for simplicity. In general,

$$J_{def} = J(P, a; \Gamma; a) \neq J_{flow} \tag{8.4.22}$$

where subscript def indicates deformation or total strain plasticity.

For $\Gamma = \Gamma_0$, the contour along the outer boundary, Rice et al. (1980) showed that

$$\frac{dJ_{def}}{da} + \frac{J_{def}}{b} = \frac{dJ_{flow}}{da} \tag{8.4.23}$$

for a rigid-perfectly plastic bend specimen.

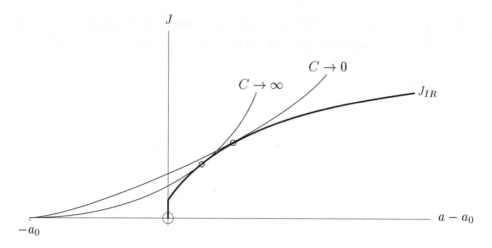

Fig. 8.4.7 Onset of unstable crack growth for a cracked body subjected to loading via either a very soft ($C \to \infty$) or a very stiff ($C \to 0$) loading device. Instability points are shown by the small circles.

The range of validity of J-resistance curves

The basic requirement for a simple theory of fracture, process region autonomy during stable crack growth, is confined to materials in which the total extent of stable crack growth, Δa_{fract}, is substantially smaller than the forward extension of the plastic region at small scale yielding. However, as may be inferred from a study of Fig. 8.4.2, this probably occurs only for materials with shallow embedment of the process region in small scale yielding, and then it is doubtful whether reasonable autonomy may exist, because the embedment changes from shallow to deep under increasing scale of yielding. Thus, unfortunately, it has to be concluded that reasonable autonomy may not extend very far beyond small scale of yielding, even if only geometries with the same plastic flow constraint ahead of the crack edge are concerned.

Coupled to the requirement of small crack growth compared to the forward extension of the plastic region, comes the poor path-independence of the J-integral for paths reasonably close to the process region, due to changes of the plastic strains and their principal directions during crack advance.

Two kinds of large scale yielding may be distinguished. One is pertinent to most practical cases in engineering structures or to shallowly cracked specimens for materials testing, where the crack is considerably smaller than the smallest ligament from the crack edge to an opposing outer boundary. The plastic region is then well contained in a region near the crack, rather far from opposing boundaries, even if the linear dimensions of this region is considerably larger than the crack length. Another kind of large scale yielding is pertinent to deeply cracked specimens for materials testing. In such cases, the plastic flow increases rapidlly with the load toward net section plasticity after it has reached a substantial portion of the unbroken ligament.

For a well contained plastic region, it is possible to calculate a J-integral that is at least reasonably path-independent outside a region that extends a few times Δa_{fract} from the process region, even if the scale of yielding is fairly large. For net section yielding, the J-integral appears to indicate a plastic flow effect rather than the state of the process region. To illustrate this, a cracked rigid-perfectly plastic bend specimen, loaded to net section plasticity, will be discussed.

Let b be the ligament straightforward from the crack edge. Thus, $b = W - a$, where W is the width of the bend specimen in the crack direction. Rice et al. (1980) determined the crack opening displacement during crack growth at the distance $r \ll b$ behind the crack edge to be

$$\delta_0 = 0.51 \frac{\mathrm{d}J/\mathrm{d}a}{\sigma_Y} r \qquad (8.4.24)$$

Autonomy in the crack edge vicinity prevails if $\delta/r = constant$. Integration gives

$$J = J_{a=a_0} + C(a - a_0), \quad \text{for } a > a_0 \qquad (8.4.25)$$

Thus, the J-integral increases linearly with $a - a_0$, i.e., the J-resistance curve is linear for $a > a_0$, rather than being constant as it should be if J reflected the state in the crack edge vicinity. This linear increase is thus a purely plastic effect that is not related to stable crack growth characteristics, which are governed by a local criterion.

The result was obtained for crack growth after net section plasticity has been obtained in a rigid-perfectly plastic material, but it may be expected to hold approximately for elastic-plastic materials with low strain hardening when net section plasticity is approached and thereafter. It appears to explain why experimentally determined J-resistance curves so often tend toward a linear increase, rather than approaching a horizontal maximum level, $J = J_{SS}$ (Broberg 1982). This view is confirmed by numerical calculations; see e.g. Fig. 19 in Xia and Shih (1995a) or Fig. 10 in Xia et al. (1995). Thus, by choosing a sufficiently large test specimen to obtain well contained yielding, a J-resistance curve approaching a horizontal level, is obtained, whereas the J-resistance curve obtained by a smaller test specimen may show approximately the same initial relation, but then deviate toward a linear increase; see Fig. 8.4.8. If only tests using the smaller specimens are available, it would be difficult to estimate where the J-resistance curve obtained begins to deviate appreciably from a curve obtained by a large specimen, i.e. where the curve is an expression for plastic flow, rather than crack growth.

The practice of using J_{def}, rather than J_{flow}, as a measure of the state in the crack edge vicinity, appears to be questionable when net section plasticity is approached. If $\mathrm{d}J_{flow}/\mathrm{d}a$ tends toward a constant value, then, from (8.4.23), for the rigid-perfectly plastic bend specimen,

$$\frac{\mathrm{d}J_{def}}{\mathrm{d}a} + \frac{J_{def}}{b} = -b \frac{\mathrm{d}(J_{def}/b)}{\mathrm{d}b} = \frac{\mathrm{d}J_{flow}}{\mathrm{d}a} \qquad (8.4.26)$$

which integrates to

$$J_{def} = J_0 \frac{b}{b_0} + \frac{\mathrm{d}J_{flow}}{\mathrm{d}a} b \ln \frac{b_0}{b} \qquad (8.4.27)$$

where J_0 is the J-integral and b_0 is the ligament at onset of crack growth. This relation shows that J_{def} decreases from J_0 toward zero during crack growth, if $b_0 < J_0/(\mathrm{d}J_{flow}/\mathrm{d}a)$; otherwise it increases initially toward a maximum, whereupon it de-

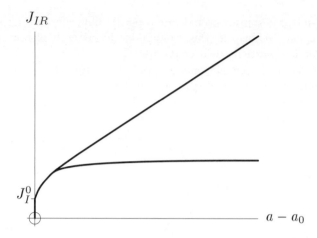

Fig. 8.4.8 J-resistance curves obtained for contained yielding (lower curve) and for net section plasticity, eventually occurring after some crack growth (upper curve).

creases. The corresponding J-resistance curves would thus be different for different original crack lengths.

The conclusion of the discussion of the possibility of extending the J-resistance curve approach to large scale yielding has to be that the theoretical basis for this is very weak, due to lacking autonomy of the region near the crack edge and to poor path-independence of the J-integral. However, both autonomy and path-independence will be lost gradually rather than abruptly when the scale of yielding increases, and it might be possible in certain cases to obtain reasonably reliable predictions of the onset of unstable crack growth for crack lengths that are considerably smaller than those specified by the ASTM convention for LEFM (8.4.1), if the plastic region is well contained. On the other hand, when net section plasticity is approached, the J-integral will depend more on plastic flow than on the state near the crack edge, and the J-resistance curve approach may lead to large errors.

Three-dimensional crack growth

Most analytical and numerical work on crack growth, from onset of stable crack growth to fracture has been devoted to two-dimensional cases, because of the complexity involved in three-dimensional cases. However, numerical simulations are possible for handling geometries which are either clearly three-dimensional, for instance semi-elliptical edge cracks, or geometries which traditionally have been considered to be two-dimensional by neglecting, for instance, crack tunnelling and the influence of plate thickness for crack growth in plates. Examples are given by, e.g., Narasimhan *et al.* (1992), Ruggieri *et al.* (1996) and Shih *et al.* (1997).

Dependence on plate thickness

It is not always possible to find a critical stress intensity factor, like K_{ISS} (K_{Ic}), or a critical J-integral, like J_{ISS}, or a J-resistance curve, that may be supposed to

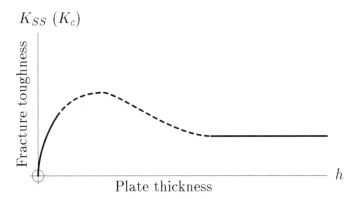

Fig. 8.4.9 Fracture toughness dependence on plate thickness. The solid curves represent the fracture toughness under complete slip penetration (left part) and under pure plane strain conditions (right part). The dashed part shows the intermediate fracture toughness.

be independent on the lateral dimensions of the body. If they are, they are usually referred to as the plane strain fracture toughness, etc. However, for cracks in sufficiently thin plates, it is well known that the plate thickness may have a very pronounced effect on the stress intensity factor or the J-integral at fracture, and on the J-resistance curve. This may be demonstrated by a simple dimensional argument (Broberg 1971).

Consider a crack in a plate with thickness h. Assume that necking develops ahead of the crack edge, if the plate is sufficiently thin and that small scale yielding prevails. This necking region is a geometry related process region (cf. pages 13ff.): its linear dimensions scale as h. Thus, the energy flux into the crack edge region is proportional to h. Hence, the critical value of the J-integral when steady crack growth is approached, is proportional to h, and, consequently, the fracture toughness, K_{ISS}, is proportional to $\sqrt{h}$.

If the plate is sufficiently thin, there will be complete slip penetration, as shown by the left part of Fig. 1.2.7. With increasing plate thickness, crack growth will occur under plane strain in the central part, whereas shear lips develop near the plate surfaces. For sufficiently thick plates these shear lips may contribute very little to the fracture toughness, which then may be considered as the plane strain fracture toughness. Generally, the small contribution of the shear lips results in a somewhat higher fracture toughness than for the case of pure plane strain fracture. Thus, the effect of the plate thickness on the fracture toughness, K_{ISS}, may be of the character sketched in Fig. 8.4.9.

It is obvious that tests for determination of the fracture toughness in thin plates, must be performed for the actual plate thickness in the application. This may not be necessary for the plane strain fracture toughness, although very often this is needed, because material properties may be dependent on the plate thickness and even vary across the thickness.

8.5 Weibull's statistical theory of the strength of materials

Introduction

The strength properties of a material may be expressed in many different ways. In some contexts, it may be appropriate to consider the yield stress, in others the ultimate stress, the fracture toughness, etc. Irrespective, however, of what strength measure is used, the strength data for a given material (and environment, etc.) are subject to scatter. The scatter may be extremely large – a twofold difference in strength between different ceramic components, with identical geometry and loading conditions, is not unusual (e.g. Lawn 1993). This implies that material testing must involve a large number of specimens, so that the reliability of the material in service can be estimated.

The scatter may be particularly large for brittle materials, like ceramics, but also for ductile materials large scatter may be encountered, and generally much more for the fracture toughness than for the yield strength. Note also that materials of the same description may be very different in such respects, even if fabricated in the same factory, but in different batches.

Different methods are used to cope with the scatter of strength data. A classical method is to apply a safety factor on some catalogue value of the ultimate stress. This method is somewhat arbitrary, and does not really deal with reliability†, and probabilistic methods appear to be preferable. The first statistical theory for the strength of materials was given by Weibull (1939a,b, 1951). Originally based on the concept of the weakest link, this is the most widely used statistical theory for fracture. Noticing that existing standard distribution functions are not sufficiently versatile to deal with several different data collections, he also introduced the statistical distribution function that now bears his name: the Weibull distribution.

A thorough discussion on the statistical approach to fracture of brittle materials and its theoretical background is given by Freudenthal (1968). He emphasizes that the weakest link concept by necessity requires an asymptotic to extreme values in statistical distribution and argues that this in turn requires physically relevant probability models. The alternative, to obtain a reasonably reliable asymptotic distribution by extrapolation from tests, would require several thousand test results. He shows, however, that the Weibull distribution is in agreement with theoretical requirements.

Weibull's theory

It appears that Weibull (1939a,b, 1951) got inspiration from experiments by Griffith (1920) on glass fibres. By breaking these fibres, which were originally 6 inches long, "in tension several times until pieces about 0.5 in. long remained", test specimens were obtained. Griffith noted that successively higher tensile force was needed for each generation fractures, which obviously was dependent on non-uniform distribution and

---

† One may agree with Orowan (1945), who states that: "The mathematical theory of such safety factors has been given by the Oxford mathematician, C.L. Dodgson, [Lewis Caroll], in his famous treatise on 'The Hunting of the Snark':
'Taking Three as the subject to reason about—   The result we proceed to divide, as you see,
A convenient number to state—   By Nine Hundred and Ninety and Two,
We add Seven and Ten, and then multiply out   Then subtract Seventeen, and the answer must be
By One Thousand diminished by Eight.   Exactly and perfectly true.' "

8.5 WEIBULL'S STATISTICAL THEORY OF THE STRENGTH OF MATERIALS

orientation of flaws (micro-cracks) at the specimen surfaces. He used fibres of different diameters and could notice a more than tenfold increase of the fracture strength of the small pieces when the diameter increased from 0.0013 inch to 0.042 inch. The stress concentration at the flaws would depend on flaw size, orientation and local ambient stress. This was an example of the weakest link hypothesis: onset of fracture could be related to the single flaw with the most severe stress concentration.

Interestingly, similar experiments were performed by Leonardo da Vinci (1452-1519). Leonardo used iron wires[†], first in a series of tests with recording of the weight a long wire could carry, then "a wire of one half the previous length is tested and the additional weight it can carry is recorded; then a wire of one-fourth length is tested and so forth, noting each time the ultimate strength and the location of the fracture" (see, e.g., Irwin and Wells 1965, Timoshenko 1983). The term "additional" indicates that Leonardo observed higher strength, at least on the average, for shorter wires. Whether he reused broken wires or not does not seem to be obvious. The difference between using fresh wires cut to desired length and reusing broken wires is that low-cycle fatigue will have an influence when reusing broken wires, and the strength of the test pieces would not with certainty increase for each new generation. For a brittle material, like glass, as used by Griffith, low-cycle fatigue would be absent (or negligible).

The weakest link hypothesis appears to hold well for most ceramics and other materials which fail in a brittle manner. A statistical theory, based on this hypothesis, is simplified if the defect density is sufficiently low that interactions between defects may be neglected. In certain so-called brittle materials, such as coarse-grained ceramics, the weakest link hypothesis and the assumption of a low defect density may not be applicable, because failure does not occur until formation and coalescence of micro-cracks has taken place.

In the introduction of his first paper on the subject, Weibull (1939a) considered the strength of a number of rods in series, loaded in tension, a simple example of weakest link behaviour. If the probability of fracture in one rod is $P(\sigma)$, where σ is the axial stress in the rod, then the probability that the rod will survive the stress σ is $1 - P(\sigma)$. For two rods in series, the probability of survival is $[1 - P_1(\sigma)][1 - P_2(\sigma)]$, where P_1 and P_2 are the two probabilities for fracture of each rod tested separately. Consequently, the probability of failure is

$$P(\sigma) = 1 - [1 - P_1(\sigma)][1 - P_2(\sigma)] \tag{8.5.1}$$

For n identical rods in series, each rod having the fracture probability $P_0(\sigma)$, the probability of fracture is

$$P(\sigma) = 1 - [1 - P_0(\sigma)]^n \tag{8.5.2}$$

Now, there would not be any difference in the probability of fracture between n rods of length L_0 in series and one rod of length $L = nL_0$. The probability of fracture for a rod with length L may thus be written as

$$P(\sigma) = 1 - [1 - P_0(\sigma)]^{L/L_0} \tag{8.5.3}$$

where $P_0(\sigma)$ is the fracture probability of a rod with the reference length L_0.

Consider now a volume V as composed of a large number of smaller volumes, each

---

[†] The distinction between iron and steel in the Renaissance is not obvious: after reduction of the ore, some excess carbon from the fuel could enter the melt, producing carbides (Tylecote 1976, 1987).

equal to a reference volume, V_0, but not necessarily connected in series. Assume that the fracture probability function $P_0(\sigma)$ is the same for all of the smaller volumes. Assume further that the appropriate stress measure, for instance the maximum principal stress, is uniformly distributed in the volume V. Then, with the weakest link concept in mind, the length ratio L/L_0 in (8.5.3) may be replaced by V/V_0, the ratio between the volume considered and a reference volume V_0. Thus,

$$P(\sigma) = 1 - [1 - P_0(\sigma)]^{V/V_0} = 1 - \exp[-f(\sigma)V/V_0] \qquad (8.5.4)$$

where

$$f(\sigma) = -\ln[1 - P_0(\sigma)] \qquad (8.5.5)$$

Note that the probability of fracture is dependent not only on the material and the acting stress, but also on the material volume.

Now, the appropriate stress measure is not in general uniformly distributed. A generalization of (8.5.4), assuming that the volume may be considered as a continuum from a macroscopic point of view, leads to

$$P(\sigma) = 1 - \exp\left[-(1/V_0)\int_V f(\sigma)\mathrm{d}V\right] \qquad (8.5.6)$$

where $\mathrm{d}V$ is a volume element.

Weibull (1939a) suggested the specific choice

$$f(\sigma) = \begin{cases} \left(\dfrac{\sigma - \sigma_{th}}{\sigma_s}\right)^m & \text{for } \sigma \geq \sigma_{th} \\ 0 & \text{for } \sigma < \sigma_{th} \end{cases} \qquad (8.5.7)$$

where m is the Weibull modulus, σ_{th} is a threshold stress and σ_s is a scaling stress.

From (8.5.4) or (8.5.6), it follows that σ_s depends on V_0, but that $\sigma_s^m V_0$ is a material constant. Thus, V_0 may be arbitrarily chosen. When the choice it made, the Weibull distribution function (8.5.6) contains three parameters, specific for the material, m, σ_{th} and σ_s.

The three parameters in the Weibull distribution function are to be determined from experiments. If tests can be performed so that σ is uniformly distributed over the volume, then it follows from the relation

$$\ln\ln\frac{1}{1 - P(\sigma)} = m\ln\left(\frac{\sigma - \sigma_{th}}{\sigma_s}\right) + m\ln\frac{V}{V_0} \qquad (8.5.8)$$

that a plot in a log-log log diagram will produce a straight line. Weibull (1939a) describes a procedure in which the tentative choice $\sigma_{th} = 0$ is first made, whereupon $P(\sigma)$ is plotted according to the experimental results. In general, a curved, rather than a straight line is obtained, and another choice of σ_{th} has to be made. The curvature is very sensitive to variations of σ_{th}, but after a few attempts a straight line provides a sufficiently accurate fit to the experimental data.

Such plots were made on special diagram paper, "Weibull paper", but nowadays there are computer programs available for determination of the three Weibull parameters and for producing plots if needed. In some cases, of course, the distribution does not give rise to a straight line. This is, for instance, the case when two different fracture mechanisms are operating, one for a lower range of stresses and one for higher stresses. Each of these mechanisms may give rise to a straight line.

Now, as Weibull remarked in his original paper (Webull 1939a), the function $f(\sigma)$ may vary not only with σ, but also with the geometrical location in the volume considered. Thus, it is very common for brittle materials like glass that the flaws are concentrated on body surfaces. In some cases, it may be assumed that all flaws are located at the surfaces and none in the interior. Then, (8.5.6) still holds, after replacing dV and V_0 by the area element dS and the reference area S_0, respectively. Weibull also suggested the more general formula

$$P(\sigma) = 1 - \exp\left[-(1/V_0)\int_V f_i(\sigma)dV\right] - \exp\left[-(1/V_0)\int_S f_e(\sigma)dS\right] \quad (8.5.9)$$

where $f_i(\sigma)$ and $f_e(\sigma)$ are related to the body interior and surface, respectively.

As an illustration of the size dependence of the strength of structures in brittle materials, some model experiments on glass plates (unpublished results) may be mentioned†. Several glass plates, 18 mm thick with 2 m² area (on each side) were to be fitted as bottom of a container, filled with water to the height of 0.40 m. In order to estimate the probabilty of failure, 40 model experiments were performed on 3 mm thick glass plates of approximately the same quality as the large plates. Thus, the area scale of the models was 1:36. The model plates failed at water levels between 1.60 m and 4.10 m. After determination of the Weibull parameters, estimates could be made for the full scale, assuming that all fractures were initiated at the surface. The Weibull plot is shown in Fig. 8.5.1 together with the cumulated probability of failure for the model scale, the full scale and, as an extrapolation, the infinite scale. The number of experiments is too small for a meaningful determination of the threshold water level, h_{th}, below which no failure would occur. However, the right part of Fig. 8.5.1 shows how the sensitivity to this determination of the prediction of failure probability increases with the scale.

The result indicated a statistical mean of 1.65 m, which came very close to the result of the only full scale experiment. It was also found that the probability of failure at the work load (0.4 m water pressure) was much smaller than 0.01, even under the conservative assumption of vanishing threshold value for failure. However, such predictions are very uncertain unless the number of experiments is very large.

Prediction of ductile-brittle transition, using Weibull statistics

It is well known that stable crack growth in ferritic steels may initially occur by void growth and coalescence (so-called "ductile" crack growth), but thereupon being suddenly interupted by unstable cleavage crack growth (so-called "brittle" crack growth). This may occur at a considerably lower load than expected if ductile crack growth had continued until fracture. In experimental work, it is found that the ductile-brittle transition is accompanied by large scatter, indicating a statistical nature of the phenomenon. It is influenced by stress triaxiality, loading rate and temperature, and appears to be triggered by cleavage of small particles, usually carbides. It may have serious practical implications.

Note that it is the phenomenon of ductile-brittle transition, that occurs after some stable crack growth during loading under constant temperature, that is discussed here.

---

† The experiments were performed and analysed by H. Andersson and T. Cervin in 1966 at the Department of Solid Mechanics, Lund Institute of Technology, Sweden.

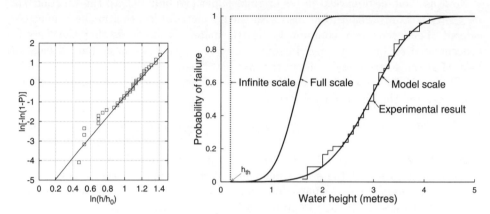

Fig. 8.5.1 Weibull plot of the probability of failure of glass plates (left figure, where h is the water height and $h_0 = 1\,\text{m}$) and the cumulated probability of fracture for the model and the full scale, determined by using the Weibull parameters (right figure). The stepped curve shows the outcome of the model experiments, for which the cumulated probability of failure is calculated as $P = (3n - 1)/(3N + 1)$, where N is the total number of tests and n is the number of an individual test in a list arranged after ascending h. The dotted curve, determined by the threshold level $h = h_{th}$ for failure, is the extrapolation to infinite scale.

A change of fracture mechanism from ductile to brittle is also known to occur in certain materials under constant load by lowering the temperature. The term ductile-brittle transition is used for both phenomena, and they are, of course related.

Ferritic steels generally contain two or more different populations of particles, such as relatively large and soft sulphides (typically a few μm in size), and smaller and harder carbides (typically less than a μm). It was described on page 17 that voids generally open first at the larger particles, but thereafter also at the smaller ones, which may contribute significantly to the coalescence mechanism. Void nucleation is often considered to be stress-controlled for larger particles and slip-induced (strain-controlled) for smaller particles.

Now, it may happen that a carbide particle suddenly fractures by cleavage under the action of a high stress. This sudden cleavage may then be propagated dynamically into the surrounding matrix. It was suggested by Ritchie et al. (1973) that this occurs if the tensile stress exceeds a critical value at a certain critical distance ahead of the crack edge. Curry and Knott (1979) added a statistical element to this view, looking for the probability of finding an intact carbide particle in this vicinity. This was followed by other similar considerations (e.g. Pineau 1981, Wallin et al. 1984, Wallin 1993, Ruggieri and Dodds 1996).

Xia and Shih (1996) performed numerical simulations of the ductile-brittle transition in a ferritic steel for different cracked specimens. They used the cell model, with each cell containing one large particle. The matrix material was described by the Gurson model (Gurson 1977; see page 249). Void nucleation was assumed to occur in accordance with suggestions by Chu and Needleman (1980) for stress-controlled and strain-controlled mechanisms.

8.5 WEIBULL'S STATISTICAL THEORY OF THE STRENGTH OF MATERIALS

The volume V around the crack edge, where initiation of cleavage fracture can be anticipated, is divided into a large number of smaller volumes V_i with the reference volume V_0. V_0 should be large enough so that each one of the smaller volumes would contain a statistically reasonably representative set of particles. In particular, it would be possible to determine certain probabilities for the smaller volumes. Xia and Shih (1996) consider the following probabilities:

P^i_{clnv}: the cleavage fracture probability, taking no account of void nucleation,
P^i_{clea}: the cleavage fracture probability, taking account of void nucleation,
P^i_{void}: the void nucleation probability.

Here, superscript i denotes the ith volume, V_i. Note that void nucleation may occur by debonding or by cracking of a particle. Thus, cracking of a particle does not necessarily imply propagation of cleavage fracture.

Now, cleavage fracture and void nucleation are mutually excluding processes. Thus, the probability of cleavage is

$$P^i_{clea} = P^i_{clnv}[1 - P^i_{void}] \qquad (8.5.10)$$

and the probability that no cleavage fracture (i.e. either void nucleation or survival) is initiated in volume V_i is $1 - P_{clea}$. For the whole volume V, this probability is

$$1 - P_f = \prod_{i=1}^{n}(1 - P^i_{clea}) = \prod_{i=1}^{n}[1 - (1 - P^i_{void})P_{clnv}] \qquad (8.5.11)$$

Thus,

$$\ln(1 - P_f) = \sum_{i=1}^{n} \ln[1 - (1 - P^i_{void})P_{clnv}] \approx \sum_{i=1}^{n}[-(1 - P^i_{void})P_{clnv}] \qquad (8.5.12)$$

because $(1 - P^i_{void})P_{clnv} \ll 1$, and, hence,

$$P_f \approx 1 - \prod_{i=1}^{n} \exp[1 - (1 - P^i_{void})P_{clnv}] \qquad (8.5.13)$$

The probability of void nucleation P^i_{void} is determined from the criteria by Chu and Needleman (1980) in terms of the appropriate stress and strain measures in each volume V_i. The appropriate measure for the probability of initiation of cleavage fracture from a particle in volume V_i was suggested by Xia and Shih (1996) to be the average maximum principal stress in the volume. For the hypothetical case of no void nucleation, they used the general Weibull form (8.5.7), so that

$$P^i_{clnv} = \left(\frac{\sigma^i_1 - \sigma_{th}}{\sigma_s}\right)^m \quad \sigma^i_1 \geq \sigma_{th} \qquad (8.5.14)$$

where σ^i_1 is the average maximum principal stress in volume V_i. Thus,

$$P_f \approx 1 - \prod_{i=1}^{n} \exp\left[1 - (1 - P^i_{void})\left(\frac{\sigma^i_1 - \sigma_{th}}{\sigma_s}\right)^m\right] \qquad (8.5.15)$$

Here, Xia and Shih (1996) define the Weibull stress as

$$\sigma_W = \left[\sum_{i=1}^{n}(1 - P_{void}^i)(\sigma_1^i - \sigma_{th})^m\right]^{1/m} \quad (8.5.16)$$

and the non-dimensional Weibull stress as $\overline{\sigma}_W = \sigma_W/\sigma_s$†.

The name "Weibull stress" becomes apparent after writing

$$P_f \approx 1 - \exp\left(\frac{\sigma_W}{\sigma_s}\right) = 1 - \exp(\overline{\sigma}_W) \quad (8.5.17)$$

Note that the Weibull stress is a cumulative quantity, taking into account the sampling volume V, as well as the load intensity in the neighbourhood of the main crack. The sampling volume is the volume around the crack edge where $\sigma_1 > \sigma_{th}$.

During loading of the body, the Weibull stress increases in general, but may reach a maximum for deeply cracked geometries. The Weibull stress may be considered as a material dependent parameter, from which the probability of cleavage fracture can be determined from (8.5.17).

Xia and Shih (1996) considered a few test specimen geometries for a certain steel (A533B). Their calculations are only exploratory, because existing experimental data do not allow determination of the statistical parameters m, σ_{th} and σ_s. However, by using estimated values, they obtained results from which several interesting conclusions may be drawn. Of particular interest is the dependence of the sampled volume on specimen size, loading and constraint to plastic flow, because the probability of cleavage fracture increases with the sampled volume.

The sampled volume increases, of course, with the specimen size. It also increases with the stress triaxiality (or the constraint to plastic flow). Furthermore, it increases in general during crack growth, but may reach a maximum for deeply cracked geometries. The increase may be related to both the increase of the load during stable crack growth and to the increase of stress triaxiality (constraint to plastic flow) during growth, when the crack edge traverses the stretched region at the blunted crack edge.

Xia and Shih (1996) also investigated rate effects on the ductile-brittle transition. In agreement with the general implications of viscoplasticity, they found that the resistance to ductile fracture increases with the loading rate. However, the probability of transition to cleavage fracture also increases. For the same loading rate, but different rate sensitivity of the material, Xia and Shih found that increasing rate sensitivity leads to decreasing probability of ductile fracture, but increasing probability of cleavage fracture.

The temperature dependence on the ductile-brittle transition during crack growth was not addressed directly in the investigation by Xia and Shih (1996), but it would be manifested in the temperature dependence of the statistical parameters m, σ_{th} and σ_s and of the void nucleation parameters. It is, of course, well known that there is a relatively narrow temperature region above which ductile fracture is almost completely dominant, and below which cleavage fracture is almost completely dominant. This

---

† Because it would be impractical to use elements of the same size in their finite element calculations, Xia and Shih (1996) allowed the volumes V_i (actually areas in their two-dimensional model) to vary. This implies modification of (8.5.15) by multiplying the exponent by V_i/V_0, and the consequential modification of the Weibull stress

transition temperature region depends on the size of the cracked body and on the constraint to plastic flow.

8.6 Crack directions

Introduction

Crack kinking and the directional stability of a crack in a large plate were considered on pages 171ff. Here, some additional aspects on crack paths will be discussed.

The macroscopic direction in which a crack is growing in an isotropic material may be influenced by various factors, some of which are related to features at some distance from the crack edge, such as severe material deformation that may attract the crack, for instance necking from the crack edge (pages 13ff.), micro-separations in a long process region or anisotropy introduced by flow in the plastic region. In other cases, it appears that the local conditions at the crack edge govern the direction of continued crack growth. The simplest condition for cracks in isotropic materials is that of symmetry, which, for instance, implies that a crack would propagate in the opening mode, unless this mode is suppressed by sufficiently high compressive stresses. It appears to be a general experience that the opening mode (mode I) does not readily mix with the shearing modes (modes II and III). Modes II and III, on the other hand, appear to mix well. Thus, for instance, a vertical planar crack of approximately circular shape, in the Earth's crust, subjected to horizontal shear stresses in its plane, will expand horizontally under mode II, vertically under mode III and in intermediate directions under a mixture of modes II and III.

The notions of mixed mode loading and mixed mode crack propagation (such as straightforward growth under mixed mode II/III) should not be confused. A crack may, for instance, propagate under mode I in an anisotropic solid, by following a weak plane, even if it is subjected to mixed mode loading. A stationary crack subjected to mixed mode loading may start propagating by kinking under local mode I conditions.

The apparent preference for mode I appears to be most pronounced for small scale yielding. Under massive plastic flow, a crack may propagate in a shearing mode, as, for instance, evidenced by shear lips in tensile tests.

Microscopically, the crack direction is seldom well defined, except in cases like cleavage of single-crystals. Propagation in shearing modes may be preferred, even if macroscopically the crack propagates in the opening mode. Whether microscopic crack growth occurs in a shearing mode or in the opening mode may also depend on the temperature; cf., Maccagno and Knott (1991), Bhattacharjee and Knott (1993, 1994) and Knott (1994).

The rate of loading often influences the direction of crack growth. Shear banding followed by crack propagation in a shearing mode may occur only if the rate of loading is sufficiently high, whereas otherwise an opening mode crack may develop.

Several criteria have been suggested for the growth direction of a crack subjected to mixed mode loading. For small scale yielding, the crack growth direction would depend on the micro-separation mechanism in the process region and, to some extent, on the stress and strain distribution in the plastic region. Note that the micro-separation mechanisms are fundamentally different in opening and shearing, so that general criteria lumping together such mechanisms in terms of energy may not reflect the physical processes properly.

Assume that high-cycle fatigue is used to produce a sharp crack in a large plate, which thereupon is subjected to monotone pure mode II loading until crack growth occurs. Small scale yielding is assumed. Before crack growth, the process region and the plastic region develop in a way typical for mode II. This implies that tensile macroscopic stresses develop on one side of the process region, which may lead to opening (as opposed to shearing) and growth of micro-separations. When the crack advances, it may bend towards the tensile side, and this tendency may continue with increasing deflection of the crack from the original straightforward direction, until it reaches local mode I symmetry. The crack may then have travelled a distance of about the plastic region size. Thereupon it is expected to continue under mode I symmetry, which may be expressed by either of the two equivalent criteria, suggested by Erdogan and Sih (1963):

$$K_I(\theta) = maximum \quad \text{or} \quad K_{II}(\theta) = 0 \qquad (8.6.1)$$

Here, θ is the current angle of crack growth direction. The second criterion may be preferable, because it is generally simpler to use, and in numerical schemes, a zero passage is more easily determined than the position of a maximum.

The criteria (8.6.1) are equivalent for a crack propagating under smoothly changing direction, i.e., if the crack propagates in a continuous but non-uniform stress field so that $K_I(\theta)$ is maximal, then $K_{II}(\theta) = 0$. However, for a stationary crack subjected to mixed mode loading, so that $K_{II} \neq 0$, a somewhat peculiar result is obtained. To avoid confusion, let K_I and K_{II} denote the stress intensity factors for the stationary crack and k_I and k_{II} be the stress intensity factors for an infinitesimally small kink. These stress intensity factors are interrelated through (4.10.3), which is repeated here for convenience:

$$\left. \begin{array}{l} k_I = R_{11}(\theta)K_I + R_{12}(\theta)K_{II} \\ k_{II} = R_{21}(\theta)K_I + R_{22}(\theta)K_{II} \end{array} \right\} \qquad (8.6.2)$$

The angle θ is taken as the angle for incipient crack growth from the straightforward direction, i.e. the kink angle. Now, it follows from (4.10.2) that the asymptotic shear stress $\tau_{r\varphi}$ at the edge of the mother crack before kinking, vanishes for

$$\frac{\sin\theta}{1 - 3\cos\theta} = \frac{K_{II}}{K_I} \qquad (8.6.3)$$

and it may be controlled from (4.10.1) that this implies maximum hoop stress, σ_φ. However, assuming an infinitesimally small kink and calculating the angle θ that maximizes k_I, by using the relations (8.6.2), gives as a result a different angle than found from (8.6.3). Moreover, it does not give $k_{II} = 0$.

The fact that the value of θ predicted from the asymptotic field at the edge of the mother crack before kinking is different from that after kinking may not be surprising, because (8.6.1) is only valid for a crack propagating with smoothly changing direction. What is asked for is, of course, the situation after kinking, and therefore the different results imply that the correct kink angle cannot be found from the maximum hoop stress criterion. But why does the angle that maximizes k_I differ from the one for which $k_{II} = 0$? The answer may be given by assuming, for a moment, that the kink is

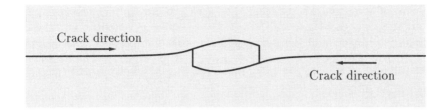

Fig. 8.6.1 Two coplanar cracks moving towards each other seem to avoid coalescence, which, however, eventually occurs by cross-fractures. The symmetry is quite typical in experiments, and, because the cracks in experiments obtain a high velocity, both cross-fractures become completed, so that the small piece in the middle becomes cut out.

formed by growth along a smooth curve, rather than by an abrupt change of direction. This could be achieved by removing the mode II component of the remote load and then gradually restoring it during a finite but arbitrarily small initial phase of crack growth. The criteria of maximal k_I and vanishing k_{II} would then coincide, as long as the radius of curvature remains finite, but, in the limit of vanishing radius of curvature, they cannot be expected to do so.

Fortunately, the differences between the two criteria is not great. For pure mode II loading, $K_I = 0$, $K_{II} > 0$, the maximal value of k_I occurs for $\theta \approx -76.58°$, and $k_{II} = 0$ for $\theta \approx -77.33°$†. In finite element calculations, Melin (1989) used the criterion of maximal k_I for the first step of crack growth and the criterion $k_{II} = 0$ for the following steps. The change in angle between the first and the second step was insignificant.

The simple symmetry criterion $K_{II} = 0$ for the direction of crack growth in small scale yielding appears to agree very well with experimental results (e.g. Erdogan and Sih 1963, Melin 1989, Richard 1982, Williams and Ewing 1972, 1974). For large scale yielding, on the other hand, it may not be possible to find general criteria, because of the large variation of factors that may influence the crack growth direction. Methods that are not dependent on criteria for crack growth, for instance numerical simulations based on the cell model, may be used.

Bergkvist and Guex (1979) point out that a small deviation from the small scale yielding condition $K_{II} = 0$ results in a deviation from the undisturbed crack growth direction by an angle $\Delta\theta = -2K_{II}/K_I$. This follows from equation (8.6.3).

Crack directions under small scale yielding

Coplanar cracks growing towards each other
Observations often show that cracks and micro-cracks growing in the same plane appear to avoid each other rather than coalesce edge to edge. Experiments by Melin (1983) show clearly a peculiar pattern, which is sketched in Fig. 8.6.1. This phenomenon was also analysed theoretically by Melin, assuming slow crack growth.

Melin (1983) used a method devised by Gol'dstein and Salganik (1974), who modelled the cracks as continuous arrays of dislocations. She actually considered a periodic

---

† As a comparison, the maximum hoop stress criterion, (8.6.3) gives $\cos\theta = 1/3$, i.e. $\theta \approx -70.53°$.

array of almost straight cracks, rather than just two cracks. The cracks are situated at $|x - nd| < a$, $n = 0, \pm 1 \pm 2, \ldots$, $y = y(x) = y(-x) = y(x + 2nd)$, i.e., the horizontal crack lengths are $2a$ and the distance between the midpoints of two neighbouring cracks is $2d$. It is assumed that $|y(x)| \ll d - a$ and $|y'(x)| \ll 1$. The load consists of remote normal stresses $\sigma_x = \sigma_x^\infty$, $\sigma_y = \sigma_y^\infty$ and $\tau_{xy} = \tau_{xy}^\infty$. Melin found that the stress intensity factors at $x = a$ are

$$K_I(a) = \tau_{xy}^\infty \sqrt{d\cosec\frac{\pi a}{d}} \sin\frac{\pi a}{2d} \left[-3y'(a) + 2\frac{\sigma_y^\infty}{\tau_{xy}^\infty} - \frac{\pi}{d} y(a) \cosec\frac{\pi a}{d}\right] \quad (8.6.4)$$

$$K_{II}(a) = \sigma_y^\infty \sqrt{d\cosec\frac{\pi a}{d}} \sin\frac{\pi a}{2d}$$
$$\times \left[y'(a) + 2\frac{\tau_{xy}^\infty}{\sigma_y^\infty} - \frac{\pi}{d} y(a) \cosec\frac{\pi a}{d} + \frac{1}{d}\left(1 - \frac{\sigma_x^\infty}{\sigma_y^\infty}\right) \int_{-a}^{a} \frac{y'(\xi)}{\omega(\xi)} \cos\frac{\pi\xi}{2d} \, d\xi\right] \quad (8.6.5)$$

where $\omega(x) = \sqrt{\sin^2[\pi a/(2d)] - \sin^2[\pi x/(2d)]}$. It may be noted that for straight cracks, $y(x) = 0$, and mode I loading, $\tau_{xy}^\infty = 0$, the stress intensity factor $K_I(a)$ coincides with (4.6.58), found for a periodic array of straight cracks, subjected to mode I loading.

The condition $K_{II}(a) = 0$ gives the integral equation

$$y'(a) = \frac{\pi y(a)}{d} \cosec\frac{\pi a}{d} - \frac{2}{d}\left(1 - \frac{\sigma_x^\infty}{\sigma_y^\infty}\right) \int_0^a \frac{y'(\xi)}{\omega(\xi)} \cos\frac{\pi\xi}{2d} \, d\xi - 2\frac{\tau_{xy}^\infty}{\sigma_y^\infty} \quad (8.6.6)$$

Assume now that $\tau_{xy}^\infty = 0$ and consider the directional stability of originally straight cracks. To this end, assume an infinitesimal disturbance near the crack edges, such that $y(a) = y_0$ for $a = a_0$, where a_0 is the half-length of the original crack. Instability of the straight path would occur if $[y(a)/a]/[y_0/a_0]$ exceeds any predetermined value as a/d approaches unity. Note that $y(a)/a$ expresses the straightness of the crack.

For $\sigma_x^\infty/\sigma_y^\infty = 1$, $y'(a) = (\pi/d)y(a)\cosec(\pi a/d)$, so that

$$y(a) = y_0 \tan\frac{\pi a}{2d} / \tan\frac{\pi a_0}{2d} \quad (8.6.7)$$

which implies that $[y(a)/a]/[y_0/a_0]$ exceeds any predetermined value as a/d approaches unity, i.e., the straight path is unstable. It is immediately obvious that this also is the case for $\sigma_x^\infty/\sigma_y^\infty > 1$. A more intricate analysis shows that the straight path is also unstable for $\sigma_x^\infty/\sigma_y^\infty < 1$ (Melin 1983, Broberg 1987b), i.e., the straight path is unstable and coalescences between the crack edges will not take place for any value of $\sigma_x^\infty/\sigma_y^\infty$. Note that the analysis cannot be used to determine the full crack path shown in Fig. 8.6.1, because the presupposition $|y(x)| \ll d - a$ becomes violated when a approaches d.

Intuition would suggest that coalescence of cracks moving towards each other would occur by mutual attraction of the crack edges. This may be true before their edges are very close to each other, because of increasing stress intensity factors; see Fig. 4.6.3. The cracks only avoid head on coalescence, and therefore bypass each other so that eventual coalescence occurs by cross-cracking or plastic tearing. This process obviously requires more energy than head on coalescence and seems to be one factor contributing to micro-crack shielding near a crack edge in certain materials; cf. page 571.

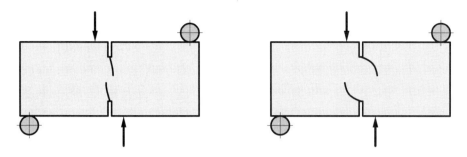

Fig. 8.6.2 Principal appearance of crack paths obtained by Bažant and Pfeiffer (1986) for concrete and mortar (left figure), and by Melin (1989) for PMMA (right figure).

Comparison between crack paths in two different materials
Laboratory experiments on symmetrically notched beam specimens, subjected to antisymmetric loading, have been performed by Bažant and Pfeiffer (1986) for concrete and mortar, and by Melin (1989) for PMMA. The results were completely different; see Fig. 8.6.2. Melin also performed finite element calculations, and showed that the crack paths in PMMA were in good agreement with the condition $K_{II} = 0$ for crack growth.

In the present context, the main difference between concrete and mortar, on one hand, and PMMA on the other, is that the extension of the process region is very small compared to the crack length in PMMA, but not in the other two materials. In both concrete and mortar, micro-cracks open up in a narrow band extending from the crack edge, most probably as a result of high tensile stresses. The crack will then grow by successive coalescences with micro-cracks in the band. Finite element calculations by Melin (1989) showed that the largest principal stresses at different fixed radial distances from the crack edge follow a curve which comes quite close to the crack path obtained by Bažant and Pfeiffer (1986). The comparison with the PMMA case shows that the criterion of maximum hoop stress is not reliable for large scale yielding.

Crack paths in a layer on top of a substrate
In an interesting paper on crack paths in a layer on top of a substrate, Gunnars *et al.* (1997) found that a crack in the layer may grow all the way to the interface or be deflected before it reaches the interface. If it reaches the interface, the crack may continue in the substrate or cause debonding. If the layer serves as a coating to protect the substrate, exposure of the interface is obviously not wanted, and it is important to know whether the crack may reach the interface or not.

Gunnars *et al.* (1997) performed experiments on a polycarbonate layer bonded to a steel plate. The thickness of the steel plate was either chosen to be thick, so that the substrate could be considered as rigid in comparison with the layer, or very thin, so that it could be considered to be fully flexible. Intermediate thicknesses were also used. The general result was that the crack reaches the interface for very thin substrates, but is deflected without reaching the interface for thick substrates.

In addition to the experiments, Gunnars *et al.* performed numerical calculation by means of a boundary element method, using the crack growth condition $K_{II} = 0$. For

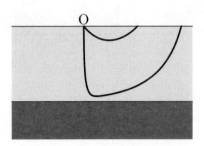 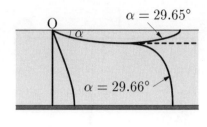

Fig. 8.6.3 Possible paths for cracks originating from the top of a layer on a rigid substrate (left figure), and possible paths originating from the top of a layer on a fully flexible substrate (right figure). Crack origin is marked with O.

an infinite layer on a rigid substrate, they found that a crack, originating from the top of the layer, is always deflected before it reaches the interface, and it returns to the top of the layer; see Fig. 8.6.3. Thus, the straight path, normal to the interface, is unstable.

For an infinite layer on a fully flexible substrate (but inextendable in the interface direction), it was found that a crack, originating from the top of the layer, reaches the interface, except if its original direction forms an angle with the layer surface that is smaller than about 29.66°; see Fig. 8.6.3. It appears that in theory, a crack originating at a certain angle between 29.65° and 29.66° may eventually reach a path parallel with the layer surface, but that such a path would be unstable, i.e., any small disturbance would cause either deflection down to the substrate or up to the top of the layer.

Several investigations have been made about crack propagation along or across interfaces, some of which have been referred to in Section 4.12. An extensive overview of cracking in layered media is given by Hutchinson and Suo (1991).

Crack growth in compression

Growth of a single crack under compression

It is generally assumed that cracks do not grow under compressive stresses, but experience in, for instance, crushing of rocks show that they can. Simple laboratory experiments on glass or photoelastic plates (Brace and Bombolakis 1963, Hoek and Bienawski 1965) demonstrate some growth of suitably oriented cracks that are subjected to uniaxial compressive loading. In these experiments, crack growth occurs from the edges of a straight pre-existing crack as shown in Fig. 8.6.4, by sliding of the original crack faces against each other. Obviously, kinks, propagating in mode I, develop. The kinks grow along a curved path, approaching the direction of compression, whereupon further growth is retarded and comes to a stop. The crack length (between the edges) may then have been roughly doubled. By comparison with compressive loading of a straight crack oriented in the direction of compression, this retardation appears logical. Note, however, that unstable growth may be caused by a comparatively small amount of lateral tension. Note also the different situation for an inclined crack in a

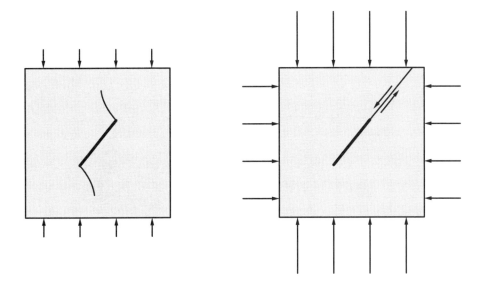

Fig. 8.6.4 Development of kinks from the edges of a pre-existing cracks, in a plate subjected to uniaxial compression (left figure). Such kinks can be suppressed by a high confining pressure, so that unstable mode II crack growth occurs (right figure).

plate subjected to tensile loading; cf. Fig. 4.10.1. In that case, kinks propagate under mode I and approach a direction normal to the direction of remote tension.

Similar experiments in PMMA have been performed with the addition of a confining pressure (in the plane of the plate) in order to suppress mode I growth (Broberg 1987b). The crack was therefore oriented so that mode II growth would occur in the straightforward direction. This direction depends on the friction coefficient for slip between the crack faces. A surprisingly high superposed pressure is needed before crack growth by mode I kinking could be suppressed so that mode II crack growth takes over (see Fig. 8.6.4). When this happens, the crack growth is unstable. Analysis, aided by numerical results from Melin (1986), showed that initiation of mode II growth requires a stress intensity factor, K_{II}, that is about 2.4 times the critical stress intensity factor, K_{ISS} (K_{Ic}), for straightforward mode I crack growth. Mode II crack growth, of course, occurs under shear stresses. Clear evidence of permanent deformation was visible near the crack surfaces. In fact, the plate was found to be somewhat thickened near the crack path (Broberg 1987b).

A note on the critical mode II stress intensity factor
The critical stress intensity factor $K_{II} \approx 2.4 K_{ISS}$ obtained in the experiment just described, is the fracture toughness in mode II for PMMA, i.e., it equals K_{IISS} (K_{IIc}), if it is accepted that small scale yielding prevails. This should not be confused with the mode II stress intensity factor that exists just before onset of unstable crack growth in the situation shown by Fig. 4.10.1, if the crack is oriented so that $K_I = 0$, $K_{II} > 0$, before crack growth. What is measured in that case would rather be the value of K_I for the kink just after it has been formed. As an example, consider the idealized case

of an infinitesimal kink from a crack edge subjected to pure mode II conditions. From (8.6.2) the kink angle is found to be $\theta \approx -77°$, by using the condition $k_{II} = 0$, and

$$k_I = R_{12}(\theta)K_{II} \approx 1.23K_{II} \qquad (8.6.8)$$

Thus, the mode I fracture toughness K_{ISS} would be about $1.23K_{II}$, where K_{II} is the mode II stress intensity factor just before unstable crack growth.

Now, of course, there is in general a stable phase, during which the crack edge is moving from $\theta = 0$ towards the direction $\theta \approx -77°$. Because of the adjustment of growth direction during this phase, the onset of unstable crack growth may take place under a somewhat higher value of K_I than for a crack subjected to pure mode I loading.

Growth and coalescences of micro-crack clusters under compression
In several materials with a highly heterogeneous structure, for instance rocks, microscopically small cracks may nucleate under axial compression with or without lateral confinement. These micro-cracks may eventually be linked together, causing macroscopic failure.

Depending on the magnitude of the confining pressure, the different phenomena of *axial splitting*, *faulting* and *brittle-ductile transition* may occur (Nemat-Nasser and Horii 1982, Horii and Nemat-Nasser 1985, 1986†, Nemat-Nasser 1995). The following description is based on these works.

Figure 8.6.5 shows a random array of micro-cracks in a large body subjected to uniaxial compression. For some micro-cracks with suitable orientation and size, kinks develop as previously described for a single crack. Increasing axial compression will involve more micro-cracks in the kinking procedure and also create more micro-cracks. However, large scale coalescences would be expected only for cases with a very dense array of micro-cracks. Recall that the kink growth stops after the length of the micro-cracks has been roughly doubled and that the kinks then are pointing in the compression direction, approximately.

1. Axial splitting. Assume now that a lateral tension is superposed. Already a small amount would be sufficient to cause continued mode I growth of some of the longer kinked micro-cracks. This growth may be retarded after a while, because of the highly inhomogeneous stress field and the presence of obstacles in the form of unsuitably oriented cracks, etc., but a few cracks may grow toward coalescence with similarly oriented cracks, thus creating an effectively longer crack, etc. In this way complete axial splitting of the body may result. In fact, this may happen even if no lateral tension is applied, because the inhomogeneous stress field may supply local mode I loading on some of the crack edges, particularly on those near a lateral boundary. Model experiments by Horii and Nemat-Nasser (1986) show the sequence of events very clearly.

2. Faulting. Assume now that a lateral compression is superposed, which may be due to axial compression under lateral confinement. During increasing compression, kinks are opened, first at the longer micro-cracks and then also at smaller ones. The lateral compression, however, prevents the continued crack growth, that is responsible

---

† This is an excellent overview paper, which also contains interesting and spectacular experimental results.

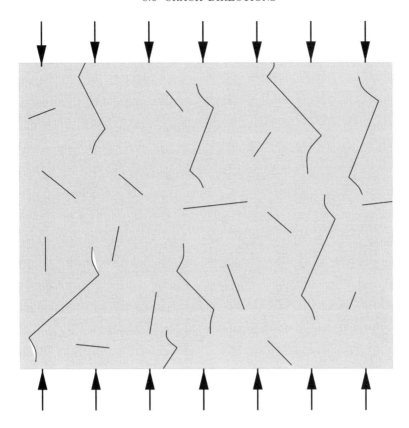

Fig. 8.6.5 Sketch of a cluster of originally straight micro-cracks, subjected to axial compression. Note the kinks on some cracks of suitable orientation and size.

for axial splitting by the action of weak lateral tensile forces, global or local. Here, another mechanism is operating. In the pattern of kinked micro-cracks, some arrays may be found, like the one shown in Fig. 8.6.6. This implies large shear stresses on the row of ligaments between the crack edges. At the crack edges, there will be both mode I and mode II stress intensity factors. The mode I shear stress intensity factors were calculated for a suitable analytical model by Horii and Nemat-Nasser (1985, 1986)[†]. As soon as this stress intensity factor reaches the critical stress intensity factor, crack growth toward coalescence will occur, creating an effectively longer crack, with increasing shear stresses on the remaining ligaments in the array, etc. In this way a fault will result, i.e. slipping in a slanting direction with respect to the axial compression with grinding of the matching surfaces during the procedure.

3. Transition brittle-ductile. The transition brittle-ductile occurs under a confining pressure that is sufficiently high to suppress both axial splitting and faulting. The micro-mechanisms associated with inelastic deformation may be different in different materials known to be brittle, and it is also temperature dependent. Assume, however, that the compressive load has created a dense array of micro-cracks. It was

---

[†] Similar calculations for dynamic cases were performed by Nemat-Nasser and Deng (1994).

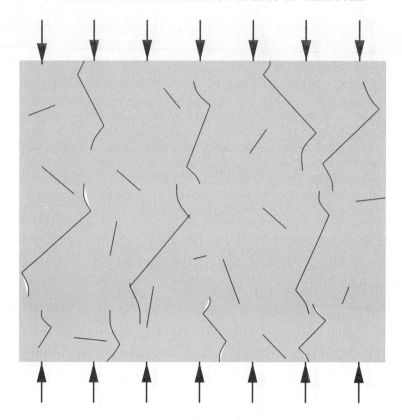

Fig. 8.6.6 Array of cracks, showing paths containing ligaments subjected to large stresses. It may be imagined that such paths will occur frequently in a large piece of material, leading to faulting when ligaments rupture, even though this possibility is overexaggerated in a two-dimensional view.

described in relation to Fig. 8.6.4 that mode I kink formation may be suppressed by a high lateral compression, so that mode II crack growth, with clear signs of plastic flow, takes over. The growth of a single shear crack in a highly hetereogeneous body with a high density of micro-cracks may be arrested at some obstacle in the neighbourhood, but similar mechanisms may occur for a large number of the micro-cracks in the body. This leads to transition from brittle to ductile behaviour; see Horii and Nemat-Nasser (1986) and, for dynamic cases, Deng and Nemat-Nasser (1994).

Tensile cracking under compression followed by unloading
A remarkable phenomenon is tensile cracking in a body of elastic-plastic material, that has never been exposed to tensile loading, but to very high compression deformation, followed by unloading (Nemat-Nasser and Chang 1990, Nemat-Nasser 1997). The cracks grow in a plane normal to the compression direction. The mechanism seems to begin with collapse of voids during the compression phase. A collapsed void will then be flattened to form a crack. High stresses arise in the plane of the crack, but

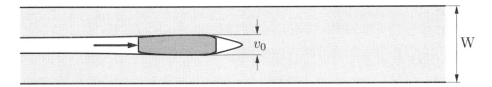

Fig. 8.6.7 Wedging of a plate.

they would be limited to the plastic flow stress, and therefore they would be roughly uniformly distributed along the crack plane.

During unloading, the material reverts to elastic deformations. This may be viewed as superposition of an elastic stress-strain state, with mode I tensile stresses, on the stress-strain state in the body just before load reversal. But such a stress-strain state implies high tensile stress concentration in the crack plane near the edge. Because the stresses during loading were limited to the flow stress, superposition leads to net tensile stresses near the crack edge, and the crack formed from the void will grow. If the load is removed sufficiently fast, there may also be some dynamic overshoot of tensile stresses.

The phenomenon is highly rate-dependent. Void collapse requires increasing compressive load at increasing loading rate. This implies larger tensile stresses during unloading. Analysis of the void collapse process has been made by Nemat-Nasser and Hori (1987b).

Directional stability during wedging

Assume that a wedge is forced through a large plate; see Fig. 8.6.7. There would be an open space between the crack edge and the wedge front. After some wedge advance, it would be expected that steady state conditions are approached, but this is often not the case. Rather the crack edge advances in a staccato manner, in rapid steps, obviously as sequences of unstable crack growth. This happens even if the wedge motion is very smooth. The reason might be that the energy requirement per unit advance of the process region decreases with increasing velocity, as is known to occur for some materials in the low velocity region; cf. Section 9.1. Another possibility is that the fracture toughness varies along the prospective crack path.

In wedging experiments in isotropic plates, a tendency has been found for the crack to veer out of a straightforward path. This tendency decreases with increasing width, W, of the plate across the crack direction. Finite element calculations by Melin (1991), assuming steady state conditions, indicate that a condition of directional stability in small scale yielding can be expressed as

$$v_0 < v_{0c} \approx 1.69 \frac{K_{ISS}\sqrt{W}}{E} \qquad (8.6.9)$$

where v_0 is the width of the front end of the wedge. Thus, as intuitively anticipated, the smaller the plate width, the higher the risk for directional instability.

In order to prevent cracks from veering out of a straight path, guiding grooves on both sides of the plate may be used. However, even if the crack follows the grooves, the crack faces might not be plane, but bulge considerably toward one side. This phenomenon may be related to the experience that the direction of the plane of a crack subjected to mode I loading, sometimes changes from being normal to the plate surfaces to being slanted at an angle of about 45°. Such a change may come abruptly and indicates a mode III, as well as a mode I component, near the crack edge. Laboratory experiments show that this change of crack plane direction does not lead to decreased energy requirement for crack growth: on the contrary, the slanted crack requires more energy.

8.7 Applications to engineering

Introduction

In Section 8.4, the basic tools used in fracture mechanics for prediction of fracture were discussed. Only a few remarks of the practical handling of these tools will be made here. A detailed and insightful description, with due attention to standards and codes, is given by Anderson (1995).

Conventional fracture mechanics seems, in my opinion, to be too much focused on the behaviour of test pieces subjected to monotone loading. A suggestion will therefore be given at the end of the present section about another approach, the J-N method, concentrating on structures and specimens subjected to several load applications, although not as many as in high-cycle fatigue.

Linear elastic fracture mechanics

The Griffith-Orowan-Irwin approach to fracture (Griffith 1920, Orowan 1952, Irwin 1957; see pages 581ff.), often referred to as LEFM, Linear Elastic Fracture Mechanics, is, of course, limited to small scale yielding. LEFM is simple to apply: cracked test pieces are subjected to monotone loading until fracture occurs, and the critical stress intensity factor K_{ISS} (K_{Ic}) is determined from the fracture load. This stress intensity factor is then used for design, often after estimation of the size of the largest (or most severe) pre-existing crack in the structure that is known or might have escaped detection. Naturally, appropriate consideration of scatter, both in tests and service (applications), has to be taken.

LEFM appears to offer a reasonably reliable method for fracture prediction, not only for monotone loading, but also for several load applications, generally up to 1000–10000, before corrections have to be made for fatigue†.

Normally, LEFM and the concept of stress intensity factor should be used only when the dissipative region is sufficiently small to allow description of test piece and structure behaviour by linear relations. Thus, K-resistance curves would normally not be needed for design purposes, only their maximum values, which would come close to K_{ISS}. When non-linear relations are needed, K-resistance curves are replaced by J-resistance curves, and then the full curve might be needed. However, it is common practice that, in some applications, K-resistance curves are used in a similar way

---

† LEFM is also extensively used for prediction of fatigue failure, but is then based on tests with a great number of load applications.

as J-resistance curves, after some extensions into the non-linear region. This, and other questions about the handling of fracture mechanics for engineering purposes under observations of appropriate standards and codes, are thoroughly described by Anderson (1995).

The J-resistance curve method

J-resistance curves were described on pages 587ff. A J-resistance curve is obtained by loading a cracked test specimen in a stiff machine. With a soft machine, the full curve may not be obtained, because of premature fracture.

Recall that the J-resistance curve is not a material property: it depends on the geometry of the cracked body (i.e. on the constraint to plastic flow) and on the scale of yielding, particularly on whether contained plastic flow prevails or net section plasticity is approached (pages 592ff.). The procedure to obtain J-resistance curves is subject to detailed prescriptions in standards. The crack advance, Δa, may be inferred by the unloading compliance technique rather than by direct measurements; cf. Fig. 2.5.4. Then, the advance due to blunting, before onset of crack growth, is included. An estimate of this advance is extracted from the $J - \Delta a$ curve so obtained by regulated procedures; see e.g. Andersson (1995).

The application of J-resistance curves for design is, as for LEFM, based on estimates of the largest (or most severe) flaw in the structure that is known or might have escaped detection, with appropriate consideration of scatter, both in test and service. The failure load is found as described on pages 587ff.

Whereas LEFM may be used by assuming monotone loading, even for structures subjected to a fairly large number of load applications, this practice may be very hazardous for non-linear fracture mechanics. Common experience tells that this is the case: to sever a steel wire, for instance, it may be sufficient to bend it back and forth a few times, whereas one load application at a much higher level may not work.

The J-N method

Complications in conventional fracture mechanics for large scale yielding
The J-resistance curve method was originally a logical extension of LEFM, using the J-integral rather than the stress intensity factor as a measure of the state near the crack edge. For LEFM, the J-integral may be used as well as the stress intensity factor, because the two are uniquely related. However, as previously discussed, with increasing scale of yielding several complications appear, which do not show up, or are of minor importance, in LEFM. These complications are:

1. A critical curve (the J-resistance curve) has to be considered, rather than a critical value (K_{ISS} or J_{SS}).
2. The autonomy of the process region becomes gradually more and more poor.
3. The path independence of the J-integral becomes gradually more and more poor and disappears when net section plasticity is approached.
4. The influence of the constraint to plastic flow becomes more pronounced.
5. Testing methods become more complicated.

6. The application of fracture mechanics to design of structures becomes more complicated.
7. Results from tests performed for monotone loading are only appropriate for a decreasing number of load applications.

Note that items 1, 2, 4 and 7 are independent of the method used for fracture prediction, and is thus present in, for instance, methods based on the crack opening displacement (COD or CTOD). So are probably also items 5 and 6. Obviously, the larger the scale of yielding, the more fragile the basis for conventional fracture mechanics, and when net section plasticity is approached, it disappears.

General principles of fracture mechanics

The safest method for predicting the probability of fracture in a structure would, of course, be to test a number of copies of the structure under the same conditions as intended for the structure in service. No fracture mechanics theory and method would then be needed. This method, full scale testing, is often chosen when possible for practical and economic reasons, but in most cases these circumstances are not at hand.

The next safest method seems to be testing of a scaled model of the structure. One example, testing of model glass plates, was discussed on page 599. The disadvantage with this method is that tests may be expensive, because of the cost of manufacturing the models and of the need to make separate models for each structure.

The next step would be to make tests on specimens that do not need to be scaled models of actual engineering structures, but would resemble such structures in important respects as regards crack geometry and loading conditions. Because of the differences between the specimens and an actual structure, a fracture mechanics method is needed. Obviously, it would be advantageous if the same method could be used for evaluation of the test results as for design of the structure with respect to the risk of fracture. This philosophy is schematically illustrated in Fig. 8.7.1; cf. Broberg (1982).

Now, how can standardized tests resemble a plethora of cracked engineering structures reasonably closely? There are certain conditions that are more common than others:

1. By far, the most common fracture mode in engineering structures is mode I.
2. The cracks are generally small compared to pertinent structural dimensions.
3. The structures are generally subjected to several load applications during their life-time.

Note that condition 2 implies that deeply cracked specimens should be avoided for the most common situations. It is interesting to note that ASTM, on the contrary, recommends deeply cracked specimens, because otherwise the unloading compliance method would not be sufficiently sensitive for measuring crack advance.

Condition 3 is often overlooked. Tests performed under monotone loading are used in numerous contexts in cases of relatively few load applications, less than a few thousand, say, i.e. cases that are not recognized as requiring the methods of high-cycle fatigue. A container for oil may be one example. During its life-time, it may be

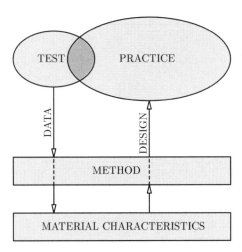

Fig. 8.7.1 Schematic illustration of how laboratory tests should resemble the actual service conditions as closely as feasible, and that the same method should preferably be used for test evaluation as for structure design.

emptied and filled several hundred times. In fact, it is very difficult to find examples of structures that are subject to monotone loading, i.e. one major load application, only.

Now, as previously mentioned, within the realm of small scale yielding, LEFM may be used with good results even for several thousand load applications. Non-linear fracture mechanics, on the other hand, may lead to considerable underestimate of the fracture resistance of a structure, because of low-cycle fatigue. Tests should therefore in general be performed for a number of load applications.

The J-N method for non-linear fracture mechanics. Testing procedure
The fragile theoretical basis for conventional non-linear fracture mechanics suggests that more direct methods could serve as suitable replacements in certain cases. One method, the J-N method, developed in previous work (Broberg 1982, 1983a, 1992, 1995a; see also Lundström and Tryding 1991 and Tryding 1991), will be discussed here. It addresses the fact that multiple load applications have to be considered, and that test conditions should come reasonably close to the conditions that apply to the loaded engineering structure, which will be referred to as the application.

It appears that most cracks in engineering structures are small compared to pertinent structural dimensions, such as the ligament from the crack edge to an opposing boundary. Thus, they are characterized by a shallowly cracked geometry and consequently also, in general, by a small constraint to plastic flow, corresponding to a usually negative T-stress. For simplicity in the discussion, the rare cases of deeply cracked structures in service will not be considered. Thus, also the test specimen should be shallowly cracked.

Some compromise has to be made. The length of the most severe crack in a large structure may be only a few per cent of pertinent structure lengths, or less, but such relations in a test specimen would imply unrealistically large specimens. A suitable test specimen may be a three point bend specimen (3PB) with width W, thickness

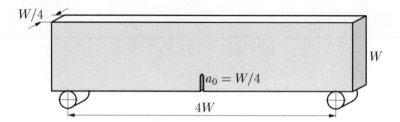

Fig. 8.7.2 Test specimen dimensions.

$W/4$ and original crack length $a_0 = W/4$; see Fig. 8.7.2. The specimen length may be $4W$. The specimen thickness should not differ so much from corresponding dimensions of the engineering structures under consideration that plane stress dominates crack growth in the test and plane strain in the application.

For the test procedure, it is suggested that the load is cycled between a maximum, P_{max}, and a minimum, P_{min}. P_{max} should not be so large that net section plastic flow is approached during the first load application. For general purposes, P_{min} may be chosen as zero, or, in order to avoid play in the test set-up, non-zero but very small compared to P_{max}. The test is completed when fracture occurs, and the total number of cycles, N, is recorded. By repeating the test with different values of P_{max}, a relation between P_{max} and N is obtained. This is the only result needed for evaluation of the test, except that, as will be discussed later, the total amount of crack growth should be controlled, so that it does not exceed about $0.25W$ (from $0.25W$ to $0.5W$). This control is easily done post-mortem.

The subscript max will now be dropped, for simplicity. Assuming that the J-integral $J(P, a_0)$ for the first load application, is reasonably path-independent, then P may be replaced by $J(P, a_0)$ as a measure of the load amplitude. Thus, the P-N relation is replaced by a $J(P, a_0)$-N relation. An example is shown in Fig. 8.7.3. Note that the dependence of $J(P, a_0)$ on N is considerable, which illustrates the hazard of using test results for monotone loading on structures that are subjected even to rather few load applications.

Fig. 8.7.3 also shows the relative amount $(a_f - a_0)/W$ of crack growth just before fracture. This is clearly visible on post-mortem specimens. Note that the average amount of crack growth per load cycle is very small compared to a_0. Considering also that the amount of crack growth per load application ought to increase during the test, because of the increasing crack-length, this implies that the amount of crack growth after the first load application is very small, and certainly much smaller than for monotone loading to fracture. Hence, the path-independence of the J-integral during the first load application is much improved compared to the J-integral during tests under monotone loading.

Procedure for prediction of fracture
Consider a structure containing a small crack, i.e. a crack with length much smaller than pertinent length dimensions of the structure. Let $J(P, a)$ be the J-integral for an original crack with length a. For dimensional reasons, this J-integral is approximately

8.7 APPLICATIONS TO ENGINEERING

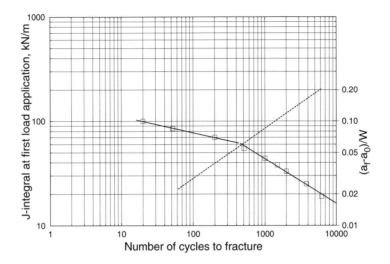

Fig. 8.7.3 Results from J-N tests on steel SS2132 (yield stress 395 MPa, ultimate stress 560 MPa). Original crack length is $a_0 = 20$ mm and the specimen hight $W = 80$ mm. The dashed line shows the total amount a_f of crack growth at fracture in dimensionless form, $(a_f - a_0)/W$. For a valid test, a_f should not exceed $0.5W$. Data for the J-N curve are obtained from experiments reported by Lundström and Tryding (1991), and data for the crack growth were obtained by Tryding (1991).

proportional to a. Thus, it may be assumed that

$$J(P,a) = J(P,a_0)\frac{a}{a_0} \tag{8.7.1}$$

where a_0 is a reference crack length.

Consider now crack growth from the original length a_0 to a length a, during a number of load applications, all with magnitude P. The amount of crack growth per cycle may be assumed to depend only on $J(P,a)$. Thus, the amount of crack growth per cycle may be written as

$$da/dN = f(a/a_0) \tag{8.7.2}$$

If $J(P,a_0) = J_0$ is the same as for another combination of P and a_0, then $f(a/a_0)$ is also the same. Thus, relation (8.7.2) holds irrespective of the original crack length a_0 as long as J_0 is the same and (8.7.1) holds. Integration yields

$$N = \int_{a_0}^{a_f} \frac{1}{f(a/a_0)} da = a_0 \int_1^{a_f/a_0} \frac{1}{f(a/a_0)} d(a/a_0) \tag{8.7.3}$$

where a_f is the crack length at onset of unstable crack growth (fracture). a_f is obtained from the relation

$$f(a/a_0) \to \infty \tag{8.7.4}$$

i.e., a_f/a_0 is independent of a_0. Consequently, N is proportional to a_0. Hence, if a test with original crack length a_0^T (T=Test) resulted in fracture after N^T cycles with the

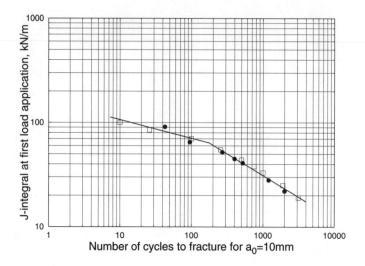

Fig. 8.7.4 Results from two series J-N tests on steel SS2132 (yield stress 395 MPa, ultimate stress 560 MPa). In one series (open symbols), the same as shown in Fig. 8.7.3, the original crack length is $a_0 = 20$ mm and the specimen height is $W = 80$ mm. In the other series (filled symbols), the original crack length is $a_0 = 9$ mm and the specimen height is $W = 36$ mm. The J-N data are adjusted to the common reference crack length $a_0 = 10$ mm by multiplying the N-values in the first series by $10/20$ and the N-values in the second series by $10/9$.

amplitude $J(P, a_0) = J_0$, then a structure in service, containing a crack with length a_0^S (S=Service) will fracture after N^S cycles with the amplitude J_0, where

$$N^S = \frac{a_0^S}{a_0^T} N^T \qquad (8.7.5)$$

Formula (8.7.5) makes it possible to use test results obtained for one particular original crack length for prediction of fracture in a structure that contains a crack of different length. It may also be used for adjustment of test results in a series of tests to one reference crack length: the original crack length generally varies somewhat between individual tests and becomes accurately known only after post-mortem inspection. An example of test results referred to a common reference crack length (10 mm), but obtained from test specimens of different sizes (W=36 mm and W=80 mm), is shown in Fig. 8.7.4. The results for the larger specimen are the same as shown in Fig. 8.7.3, but the N-values shown in Fig. 8.7.4 are only half of those in Fig. 8.7.3.

In a log-log diagram, as in Fig. 8.7.3, the J-N curve can generally be drawn as a broken line (Lundström and Tryding 1991). It appears that the straight line part for high loads corresponds to plane stress dominance, whereas the other part corresponds to plane strain dominance. This would imply that scaling within the part belonging to the plane stress region between different plate thicknesses is questionable, cf. Fig. 8.4.9.

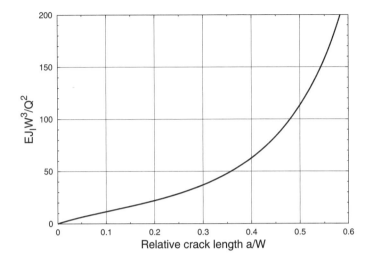

Fig. 8.7.5 The elastic J-integral as a function of the crack length for the 3PB specimen shown in Fig. 8.7.2, loaded by a central force Q. The curve is drawn with the aid of expression (A10.16).

The maximum crack growth allowed

The relations derived would hold approximately for any structure containing a crack that is sufficiently small for (8.7.1) to hold. This includes test specimens. However, the compromise made for the dimensions of the test specimen suggested implies that (8.7.1) only holds reasonably well for rather small amounts of crack growth; cf. Fig. 8.7.5. The maximum amount of crack growth before fracture should therefore be appropriately limited. The following considerations have lead to the recommendation that this limit should be $0.25a_0$, i.e. a crack growth from $0.25W$ to $0.5W$.

Most of the number of cycles to fracture are spent during the first half of crack growth. To illustrate this fact, consider the case for which fracture occurs when the total amount of crack growth is $0.25a_0$ (from $a_0 = 0.25W$ to $a = 0.5W$). For this amount of crack growth, N would be a large number (cf. Fig. 8.7.3). The average amount of crack growth per cycle is $0.25a_0/N$. When $a = 0.5W$ is reached, the amount of crack growth during the next cycle becomes unbounded. This indicates that the crack growth per cycle increases rapidly during the second half of the crack growth. The majority of the load cycles will then be spent in the region $0.25W < a < 0.35W$ (say). As shown by calculations, $J(P,a)$ is a reasonably linear function of a/a_0 in this region.

In an exploratory approach, the somewhat arbitrary but not unreasonable assumption is made that

$$\frac{\mathrm{d}a}{\mathrm{d}N} = C \frac{1 - a_0/a_f}{1 - a/a_f} \cdot \frac{a}{a_0} \tag{8.7.6}$$

where $a_f = 2a_0 = 0.5W$. Following integration, it is found that

$$\frac{N(a = 0.35W)}{N(a = 0.5W)} \approx 0.71 \tag{8.7.7}$$

i.e., only about 30 per cent of the number of load cycles is spent during crack growth from $0.35W$ to $0.5W$. The substantial deviations from linearity in this region will thus affect only a comparatively small number of load cycles.

The fact that most load cycles are spent before the crack has grown to $a = 0.35W$ also implies that net section plasticity that might be approached after this growth would not seriously affect the number of cycles to fracture. It could, however, lead to plastic collapse rather than fracture.

The limitation of maximum permissible crack growth for a valid test also sets a limit for the lowest $J(P, a_0)$ that can be used in the test. For the particular case shown in Fig. 8.7.3, the lowest permissible $J(P, a_0)$ is approximately $5\,\text{kN/m}$.

In general, because of the compromise made about test specimen geometry, the test results are obtained for a somewhat more deeply cracked body than found in the majority of engineering structures. There would thus be a larger deviation from the proportionality (8.7.1) for the test specimen than for the engineering structure. This deviation is not entirely unwelcome, because it leads to some conservatism in the prediction of fracture in the engineering structure. The same also applies to the constraint to plastic flow. This is very modest, as may be inferred from the biaxiality parameter B (3.3.76). This equals about -0.19 for $a/W = 0.25$ and becomes positive for a/W larger than about 0.34, i.e., B is negative, but $|B|$ is comparatively small in the region where the majority of the load cycles is spent. For more shallow cracks, as may be common in engineering structures, B is lower – it is about -0.39 for $a/W = 0.1$. This difference between test and application is also contributing somewhat to conservative predictions for engineering structures, because the resistance to fracture seems to increase with decreasing constraint; cf. Fig. 8.3.3.

It should be remarked, however, that results in the literature are somewhat ambiguous as regards the influence of the constraint on the resistance to crack growth. Not all results point toward decrease of the resistance with increasing constraint. For monotone loading, Varias and Shih (1991) found that K_{ISS} (K_{Ic}) possesses a minimum very close to $T = 0$; thus it increases with increasing T when $T > 0$. For fatigue loading, Miller (1977) found that the crack growth rate increases with decreasing T-stress (see also Howard 1981).

Discussion of the J-N method

As Fig. 8.7.3 indicates, there is a large difference in the scale of yielding between tests involving only a few load cycles and those involving many. However, even for cases with very few load cycles, the scale of yielding is substantially smaller than for corresponding tests under monotone loading, except for the last few cycles. This may be understood by comparison with a J-resistance curve (e.g. Fig. 8.4.5). During the first load application in a J-N test, the J-integral follows the J-resistance curve until $J = J(P, a_0)$ is reached, and this occurs for an amount of crack growth that is much smaller than the one for which $J = J_I^{fract}$ is reached under monotone loading. Consequently, $J(P, a)$ is, in general, substantially smaller than J_I^{fract}. Thus, both the J-integral and the amount of crack growth after the first load application are smaller than for monotone testing, which implies that the path independence of $J(P, a_0)$, the J-integral of interest, is much better. Also, the autonomy is more reliable than for a corresponding test under monotone loading, except during the end of the test, i.e. during a period that has very small influence on the final result.

It is a great advantage if the same method can be used for evaluation of the test results as for prediction of fracture in the application, cf. Fig. 8.7.1. One problem consists of the determination of the J-integral. It is known, for instance from circular enquiries, that finite element calculations of the J-integral for large scale of yielding may differ widely. This fact, together with the complications connected with such calculations for engineering structures and the risk of making mistakes, suggest that a simple estimate might be preferable to time consuming calculations. The stress intensity factor, K_I, and the scale of yielding, expressed by the ratio P/P_L, where P_L is the limit load, may be taken as the basis for such an estimate. Both are relatively easy to obtain. For the test specimen, the limit load is approximately (Green and Hundy 1956)

$$P_L \approx 1.261\sigma_u B(W-a)^2/L \tag{8.7.8}$$

where σ_u is the ultimate stress†. The expression may also be used to estimate the maximum load to be used in the test for avoiding net section plasticity: it should not exceed $P = \sigma_Y P_L/\sigma_u$.

Lundström and Tryding (1991) suggested the following expression as one possible way to estimate the J-integral:

$$J(P,a_0) = J_e(P,a_0)\left[1 + \ln\frac{1}{1-(P/P_L)^4}\right] \tag{8.7.9}$$

where J_e is the J-integral found by assuming linearity. Even though expressions of this kind are very simplified, they have the merit of being common to test and application, thus avoiding errors that may creep in by using different methods of estimating $J(P,a_0)$.

In the rare cases when the application actually is subjected to monotone loading, a J-N test could still be used, conservatively, by basing the prediction of fracture on the result for the smallest number of load cycles that could be obtained in the test without reaching net section plasticity during the first load applications. Recall that J-resistance curve methods would give a non-conservative result if net section plasticity is approached; cf. pages 592ff.

Note that the J-N method, as described, deals with cracks that grow to less than double their original length. This contrasts to the case of high-cycle fatigue, in which the crack grows to several times its original length.

Finally, it is remarked that a J-N test is very simple. No crack length measurements are needed, and the test may take place in the same machine and with the same set-up as for the fabrication of the original crack by high-cycle fatigue.

---

† By choosing the ultimate stress, rather than the yield stress, some conservatism is introduced.

9
Dynamic Processes in Fracture Mechanics

9.1 Dynamic crack propagation

Introduction

After onset of unstable crack growth, the crack generally accelerates to a high velocity, sometimes several hundred m/s or even over a thousand m/s. From a practical point of view, the main interest is generally focused on whether fracture will occur or not, rather than on what happens during the dynamic crack propagation. However, there are several exceptions. Analysis of dynamic crack growth is a vital part in designing methods for crack arrest, it is of importance in earthquake source physics, in certain mining operations, in evaluation of damage after vehicle collisions and other accidents, etc. An interesting discussion of the potential of methods in dynamic fracture mechanics dealing with various practical applications, is given by Kanninen and O'Donoghue (1995).

Historically, much attention has been focused on the maximum attainable crack velocity. Theoretical analysis indicates that it may approach the Rayleigh wave velocity for mode I (Barenblatt and Cherepanov 1960, Broberg 1960, Craggs 1960); see page 338. However, experimental results point stubbornly to a substantially lower velocity, at most up to about 70 per cent of the Rayleigh wave velocity; see e.g. Kerkhof (1970, Tabelle A4). It has long been known that cracks tend to accelerate to a maximum velocity (e.g. Schardin 1950, 1959), but this maximum velocity might not be a material property; cf. Ravi-Chandar (1982), Ravi-Chandar and Knauss (1984c), Takahashi and Arakawa (1987).

Laboratory determination of crack velocities are much more difficult for modes II and III than for mode I, particularly because of the difficulties of producing cracks running in either of these modes. Some recordings from earthquake events (Archuleta 1982, Scholtz 1990) appear to show evidence of intersonic crack propagation in mode II. Recall that analytical investigations indicate that such propagation is possible, because the energy flux into the crack edge region would be positive for intersonic velocities; cf. page 353.

The energy flux into the process region, required for crack propagation, has been found to increase with the crack velocity, at least in the higher velocity region. In some materials, the increase may be very dramatic: Paxson and Lucas (1973) report experimental data for PMMA† that point towards almost 50 times higher energy

---

† Many experimental investigations have been made on PMMA. It should be noted that PMMA is not a well defined material. Different batches may differ considerably in strength properties, and even

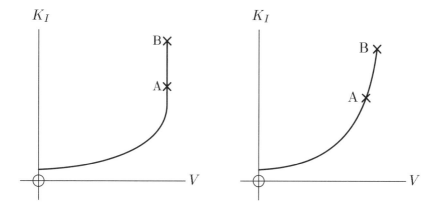

Fig. 9.1.1 Two examples of the development of the stress intensity factor as a function of the crack velocity during dynamic crack growth. In the left part, a constant maximum velocity is reached, but the stress intensity factor continues to increase. Eventually branching (labelled B) may occur. In the right part, a constant velocity is not reached before branching. Attempted branching (labelled A) begins well before successful branching. This might not happen before a constant terminal velocity is reached (left figure).

requirement at very high than at very low crack speeds. However, experiments in small scale yielding by Kalthoff et al. (1977), Kobayashi and Dally (1977), Ravi-Chandar (1982), Ravi-Chandar and Knauss (1984c), Takahashi and Arakawa (1987), Arakawa and Takahashi (1991a), Fineberg et al. (1991, 1992), Sharon et al. (1995, 1996), Takahashi and Kido (1995) and Sharon and Fineberg (1996), among others, demonstrate that there is no unique relationship between the energy flux into the crack edge region and the crack edge velocity. Some of these investigators, e.g. Kobayashi and Dally (1977), Ravi-Chandar (1982) and Ravi-Chandar and Knauss (1984c), find that the stress intensity factor increases during crack propagation under constant velocity. The opposite effect, decreasing stress intensity factor under constant velocity, is present in results reported by Kalthoff et al. (1977).

Figure 9.1.1 shows schematically two common ways in which the stress intensity factor develops during dynamic crack growth. Such results were obtained by, for instance, Kobayashi and Dally (1977). The possible occurrences of branching and attempted branching, shown in the figure, will be discussed in Section 9.2.

The lack of a unique relation between the energy flux into the crack edge and the crack velocity was particularly clearly demonstrated by Ravi-Chandar (1982) and Ravi-Chandar and Knauss (1984c). The experiments were performed on a geometry which could be envisaged as infinite, because the crack was subjected to crack face loading, and the interesting results were obtained before disturbances arrived from the outer boundaries. The crack accelerated to a constant maximum velocity, different for different load magnitudes. During each such constant velocity phase, an increasing stress intensity factor was recorded.

---

atmospheric humidity plays a part. Professor J. Kalthoff, Bochum, has informed me that he uses the comparatively inexpensive PMMA only for preliminary experiments, and more consistent materials, such as Araldite, for final investigations.

The increasing energy requirement with crack speed is accompanied by increasing roughness of the crack surfaces. Andrews (1959) distinguishes between the "mirror", "mist" and "hackle" zones in materials like glass, after the appearance of the roughness of fracture surfaces†. He also describes what he calls "conic lines" (hyperbolic, parabolic or elliptical) on the crack surfaces. Further investigations in this field have been made by e.g. Kerkhof (1970), Ravi-Chandar (1982, 1998), Ravi-Chandar and Knauss (1984b), Arakawa and Takahashi (1991a), Fineberg et al. (1991, 1992), Sharon et al. (1995, 1996), Sharon and Fineberg (1996) and Ravi-Chandar and Yang (1997). For a given material, the fracture surface roughness appears to depend both on the stress intensity factor and on the crack velocity. Note that the stress intensity factor and the crack velocity are not in general independent, so it might be difficult to separate their influences on the surface roughness.

Ravi-Chandar (1982), using Homalite 100, showed that both surface roughness and stress intensity factor may increase during a period of constant velocity,‡ and that this may occur for different values of the constant velocity in the same material. Ravi-Chandar and Knauss (1984b) determined the depth of the fracture surface markings by using a light section microscope and found a considerable increase along the crack path, which corresponded well with the increase in the stress intensity factor. They interpreted the depth of these markings as the depth of the process region.

By plotting the surface roughness against the stress intensity factor, Arakawa and Takahashi (1991a) showed that the surface roughness may be higher during deceleration than during acceleration for the same stress intensity factor but different crack velocity.§ However, physical reasons strongly suggest that the main correlation is between surface roughness and some measure of the local loading intensity, such as the stress intensity factor or the energy flux (note that these two quantities are connected by a velocity dependent factor).

The increase of the surface roughness during crack propagation under constant velocity provides a strong and independent support for the lack of a unique relation between stress intensity factor and crack velocity in the high velocity region. The experimental results could otherwise be questioned on the basis of certain experimental shortcomings, although only to a relatively insignificant extent (Freund 1983, Dally et al. 1985, Ma and Freund 1986, Liu et al. 1993, Aoki et al. 1995).

In some materials, for instance PMMA, part of the surface roughness consists of what may be interpreted as successive unsuccessful branching events. At higher velocities, a periodic pattern develops during crack propagation in plates; see e.g. Green and Pratt (1974), Fineberg et al. (1991), Washabaugh and Knauss (1993), Shioya and Zhou (1995), Sharon and Fineberg 1996, Ravi-Chandar and Yang 1997. In PMMA, the pattern is associated with crazing. The period, expressed in terms of length along the crack path, increases with the velocity. In this context, it is interesting to note that experiments with high resolution (e.g. Carlsson 1963, Sharon and Fineberg 1996, Ravi-Chandar and Yang 1997) show that crack velocities are irregular on a micro-scale.

---

† Ravi-Chandar (1998) remarks that the distinctions of mirror, mist and hackle regions is a result of visual impressions: in reality the surface roughness increases smoothly.
‡ Interestingly, Kerkhof (1973, p. 26) reports increasing surface roughness, from "fine" to "rough hackles" during the constant velocity phase for cracks in glass plates.
§ Actually, Arakawa and Takahashi (1991a) plotted the surface roughness against a monotone function of the stress intensity factor.

This may be attributed to microscopic processes in the process region. An interesting modelling of such behaviour, based on the Leonov-Panasyuk-Dugdale model, is given by Neimitz (1995).

Similar results as for glass and brittle plastics have been obtained for other materials, for instance structural steels; see, e.g., Carlsson (1963), Rosakis et al. (1984) and Zehnder and Rosakis (1990). Thus, for instance, a four-fold increase of the stress intensity factor from slow crack growth to a crack velocity of about 1000 m/s (about 30 per cent of the Rayleigh wave velocity) was observed for AISI 4340 steel (Zehnder and Rosakis 1990).

The experimental results for mode I dynamic crack propagation in small scale yielding, may be summarized as follows†:

1. A crack subjected to a constant remote load accelerates to a constant velocity. This velocity is not only dependent on the material but also on the experimental conditions‡. Thus, it appears to be history dependent.
2. Under given experimental conditions, the stress intensity factor and the energy flux into the process region increase with the crack edge velocity. For a crack subjected to remote loading, they may also increase during a period of constant velocity.
3. Under given experimental conditions, the surface roughness increases with the crack velocity, although not uniquely: there is, for instance, some difference in the surface roughness at a given velocity, depending on whether the crack is accelerating or decelerating.
4. After a certain velocity is reached, successive attempts at branching occur, until successful branching takes place at a higher velocity.
5. For some materials, like PMMA, the fracture surfaces exhibit a periodic ripple pattern, with a period that increases with the crack velocity above a certain rather well defined crack velocity.

There are other causes for the lack of a unique relation for a given material between stress intensity factor and crack velocity than the crack propagation history. Thus, there is, as for slow crack growth, a dependence on the T-stress, which manifests itself in experiments as a specimen dependence; see e.g. Kobayashi and Mall (1978). Dahlberg et al. (1980) and Kalthoff et al. (1980). In spite of such differences of K-V (or $\mathcal{G}$-V) relations, obtained under different experimental conditions, the main features are essentially common to these different cases. Thus, the lack of uniqueness will be tacitly understood rather than emphasized in the continuation, and references to a K-V or a $\mathcal{G}$-V relation should be understood as references to individual events rather than to specific materials.

It is probable that the main features of dynamic crack propagation in mode I, such as increasing stress intensity factor and surface roughness with increasing crack velocity, carry over to modes II and III.

---

† This summary is mainly based on experiments with glass and so-called brittle plastics, like PMMA, Homalite and Araldite, but it appears to be essentially pertinent, at least to some extent, also to other types of materials; cf., e.g., Rosakis et al. (1984), Rosakis and Zehnder (1985) and Zehnder and Rosakis (1990).
‡ Here, the term "experimental conditions" includes the experimental setup, the geometry of the cracked specimen and the load magnitude.

The crack edge equation of motion

Early attempts to analyse dynamic crack propagation, by using continuum mechanics, were based on what later has been called "the crack tip equation of motion". The simplest form of such an equation assumes that the energy dissipation Γ, per unit crack advance and length along the crack edge, is a material specific constant, independent of the velocity. For simplicity, this quantity will be called "the specific energy dissipation" in the continuation. A differential equation for crack edge motion is obtained by equalizing the specific energy dissipation with the energy flux $\mathcal{G}(a, \dot{a})$, where a is a crack length parameter, measuring the crack edge advance, and $\dot{a}$ is the crack edge velocity:

$$\mathcal{G}(a, \dot{a}) = \Gamma \qquad (9.1.1)$$

This is the natural extension of the Griffith approach, which may be written as $\mathcal{G} = 2\gamma$, where γ is the surface energy. As an example, consider a symmetrically expanding mode I crack, with the original length $2a_0$ and subjected to remote loading, $\sigma_y = \sigma_y^\infty$. Assume that the relation (6.9.75), which, with some change in notation, reads

$$\mathcal{G}(a, \dot{a}) = \mathcal{G}(a, 0) \cdot w_1(\dot{a}) \qquad (9.1.2)$$

holds approximately. Here, $w_1(\dot{a})$ is a known function, given by (6.9.76), and

$$\mathcal{G}(a, 0) = \frac{\pi(\sigma_\infty)^2 a}{4(1-k^2)\mu} = Ca \qquad (9.1.3)$$

Thus, a crack edge equation of motion is obtained:

$$Caw_1(\dot{a}) = \Gamma \qquad (9.1.4)$$

Integration of this differential equation, from $a = a_0$ to a current value of a, will determine the crack edge motion.

Even if Γ would be independent of the crack velocity, the crack edge equation of motion (9.1.1) is only approximative, because the energy flux $\mathcal{G}(a, \dot{a})$ is assumed to be independent of higher order time derivatives of a. Note that a crack edge equation of motion for the initial phase, until a P wave arrives from the other crack edge, could have been established by obtaining $\mathcal{G}(a, \dot{a})$ from (6.2.49) with K_I taken from (6.12.36). However, the crack velocity during this initial phase would in most cases be very small compared to the P wave velocity, so that only a very small amount of crack growth would be governed by such an equation. Thus, apart from the initial motion, equation (9.1.2) is expected to hold with somewhat better approximation until a P wave from the original crack position meets the crack after reflection from an outer boundary. On the other hand, as shown by Rose (1976b), the repeated disturbances from one crack edge on the propagation of the other has a very small effect on the motion of the crack edges.

Now, $w_1(\dot{a})$ is decreasing with increasing $\dot{a}$, reaching zero for the Rayleigh wave velocity (see Fig. 6.9.10). Thus, by writing (9.1.4) as $w_1(\dot{a}) = \Gamma/(Ca)$, it is immediately seen that an increase of a will lead to a decrease of $w_1(\dot{a})$ and thus to an increase of the velocity $\dot{a}$, and the Rayleigh wave velocity will be asymptotically approached as $a \to \infty$.

A further development would be to recognize that Γ is a function of the velocity. This would lead to an equation of the form

$$\frac{w_1(\dot{a})}{\Gamma(\dot{a})} = \frac{1}{Ca} \tag{9.1.5}$$

The experience that $\Gamma(\dot{a})$ is an increasing function (except perhaps for low velocities), again implies that the crack will reach the Rayleigh wave velocity asymptotically, provided that $\Gamma(\dot{a})$ is finite for all sub-Rayleigh velocities. However, it cannot be ruled out that $\Gamma(\dot{a})$ might increase without bound when some critical sub-Rayleigh velocity is reached, and in such a case, the crack will reach this velocity asymptotically; cf. Fig. 9.1.1. In fact, the experiments by Kobayashi and Dally (1977) indicate such an increase for an epoxy resin. Paxson and Lucas (1973) found a very steep increase of $\Gamma(\dot{a})$ when the maximum velocity in their experiments was approached. Bergkvist (1973) used their results together with (9.1.5) to determine the acceleration of a mode I crack in a large plate and obtained very good agreement with experimental data reported by Dulaney and Brace (1960). However, the crack edge equation of motion cannot itself explain the increase of $\Gamma(\dot{a})$ with the crack edge velocity. The same applies, of course, also to the more general assumption that the specific energy dissipation depends on the history as well as on the instantaneous velocity, by writing $\Gamma = \Gamma(a, \dot{a})$ in the crack edge equation of motion (Broberg 1967).

One suggestion for explaining why the highest crack velocities recorded in a homogeneous body are substantially lower than the Rayleigh wave velocity, has been that what matters could be some kind of a local Rayleigh wave velocity in the highly strained region near the crack edge (Broberg 1964), rather than the Rayleigh wave velocity in the undisturbed material. However, a later investigation (Broberg 1973b) showed that this explanation is clearly insufficient. So, also, are explanations based on the assumptions of oscillations of the crack propagation direction (Gao 1993, Slepyan 1993). In addition, these kinds of explanation are not compatible with the substantial increase of surface roughness that may take place during constant velocity crack propagation.

The results obtained from studies of a crack edge equation of motion indicate that continuum mechanics is not capable of reproducing and explaining the experimental results summarized in points 1-5 on page 628. In particular, they cannot explain why Γ is not a unique function of the crack velocity in the high velocity region and why the maximum velocity obtained under remote loading is not a material constant.

Finally, it is noted that there is no inertia term in a crack edge equation of motion such as (9.1.1). This is a consequence of the neglect of higher order time derivatives in the expression for $\mathcal{G}$. Moreover, it is also a consequence of the tacit assumption that the process region is point-sized and thus does not carry any mass.

A model for dynamic crack propagation

The cell model
In order to explain the very strong increase of the specific energy dissipation Γ with the crack velocity and the accompanying strong increase of surface roughness that take place in certain materials, a cell model of the material was introduced (Broberg

Fig. 9.1.2 Loss of localization in the process region at high crack velocities. The cells belonging to (or having belonged to) the process region are hatched.

1979a). This model has been described previously in the context of slow crack growth (Section 1.3).

During crack growth in the low velocity region, the process region will consist of only one central layer of cells, as discussed in Sections 1.4-1.5 for slow crack growth. Recall that the reason is that after a cell has reached the decohesive state, the macroscopic stress acting across this cell decreases, and thus it will also decrease across the adjacent cells above and below (the crack plane is considered to be horizontal), with the consequence that the cells outside the central layer do not reach the decohesive state.

Now, for a high crack velocity, the information of a macroscopic stress decrease in the central layer of cells may not arrive in time for preventing some offside cells reaching the decohesive state. The process region will then be higher than one layer of cells, and the higher the crack velocity, the higher the process region. Eventually, the process region will encompass so many cells that the intrinsic micro-structural length parameter, determining the cell size (usually the distance between dominant micro-separation kernels), can no longer determine the height of the process region. Thus, a loss of localization of the process region occurs; see Fig. 9.1.2.

Now, when the intrinsic length parameter of the material effectively disappears, the process region size must be determined in some other way. Obviously, for small scale yielding, the specific energy dissipation has to be supplied by the energy flux from the outer, elastic, field, and this flux is screened during its passage through the plastic region outside the process region (cf. page 586). This implies that the total energy dissipation is regulated by the dissipation in the process region, i.e., an increase of the size of the process region is accompanied by a related increase of the size of the plastic region.

After the effective loss of the intrinsic length parameter, there does not seem to be any other determinant of the size of the process region, than the interaction between the outer stress-strain field and the process region. This implies, for instance, that the evolution of the process region is history dependent, i.e., it would not be dependent solely on the material and the instantaneous crack velocity. This point of view is compatible with the experience that the specific energy dissipation Γ may increase during crack propagation under constant velocity.

In experiments so designed that a crack accelerates to a maximum velocity and then decelerates, Arakawa and Takahashi (1991a) showed that the stress intensity factor and the crack face roughness continued to increase for some time after the velocity had reached its maximum. Similarly, results reported by Kalthoff et al. (1977) show a delay effect: after a period of constant stress intensity factor and constant velocity, the crack velocity continues to stay constant for a while, although the stress intensity factor decreases. These results constitute further indications that the evolution of the process region is history dependent.

The similarity argument

Whereas the process region size, and thereby the energy dissipation Γ, may increase during crack growth with constant speed under remote loading, this is not possible for constant velocity crack propagation in a strip, whose edges are subjected to fixed grip loading; cf. Section 6.7. In this case, the energy into the dissipative region is by necessity a constant, uniquely determined by the displacement between the strip edges. Experiments (e.g. Paxson and Lucas 1973, Nilsson 1974b, Shioya and Zhou 1995) have shown that the crack can propagate under constant average velocity (reflections from the strip edges may cause some small ocillations). Even though it cannot be ruled out that this velocity may depend on the history of acceleration, it is assumed that such a dependence is insignificant in the following discussion of crack propagation in a strip.

Following Broberg (1979a), consider two different crack edge velocities, V_1 (case 1) and V_2 (case 2), obtained for the same material in the same strip configuration. Mode I loading and very small scale yielding is assumed. Both velocities are assumed to be high enough that the significance of the intrinsic material length parameter (the cell height) has disappeared. On the other hand, the velocities are not assumed to be higher than those encountered normally in experiments on high velocity crack propagation, i.e. less than about half the Rayleigh wave velocity. In this velocity range, there is only a moderate variation of the angular distribution of the elastic stress and strain terms with inverse square root radial decrease near the crack edge; see page 343.

To a first approximation, it is now assumed that there is no difference in the angular elastic stress-strain distribution at all between cases 1 and 2, and thus the elastic stress-strain fields in the crack edge vicinity differ only by a spatial scale factor. Then, in the absence of an intrinsic material length parameter, this similarity also prevails for the macroscopic stresses and strains in the dissipative region, as schematically shown in Fig. 9.1.3. Thus, the same (macroscopic) stresses and strains appear at a point $(V_2 x/V_1, V_2 y/V_1)$ in case 2 and at the point (x, y) in case 1.

Now, the (macroscopic) stress and strain histories for a point at coordinate y in case

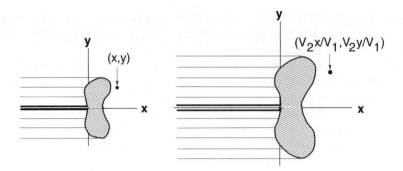

Fig. 9.1.3 Active dissipative regions (hatched) for two crack edges, propagating with different velocities. The wakes are shown by horizontal lines.

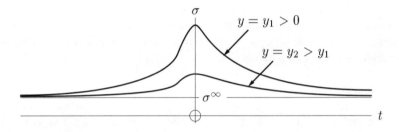

Fig. 9.1.4 Stress history for two particles, one at the distance y_1 and the other at the distance y_2 from the symmetry plane. Time $t = 0$ is arbitrary. σ^∞ is the remote stress.

1 and a point at coordinate $V_2 y/V_1$ in case 2 are the same. They might look as shown in Fig. 9.1.4 for some stress component σ and two different values of y. Note that the similarity property is independent on material properties, such as, for instance, viscoplasticity.

The minimum energy requirement
Obviously, the specific energy dissipation increases in proportion to the crack velocity in the velocity range considered, according to the simple similarity model used. Recall that an essential feature of this model is that there is no intrinsic length parameter. Extrapolation of the model predictions to lower velocities would lead to the physically unacceptable result of vanishing specific energy dissipation for vanishing crack velocity. However, there exists *a minimum energy requirement*: at least one layer of cells is needed for crack growth, and a minimum amount of energy dissipation consists of the dissipation in this layer together with dissipation in neighbouring cells that do not reach the decohesive state, even if micro-separations are opened. Incorporation of this fact in the model as a second approximation, would lead to a relation between specific energy dissipation and crack velocity as shown in Fig. 9.1.5.

The figure also shows estimated behaviour in the transition between very low velocities, for which the process region size is determined by an intrinsic material length parameter, and higher velocities, for which the process region size is dependent on

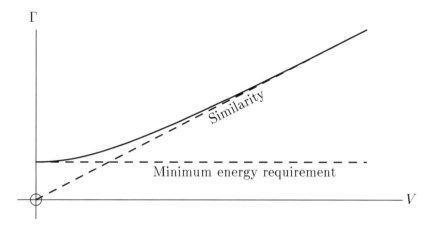

Fig. 9.1.5 Relation between specific energy dissipation, Γ, and crack edge velocity, V, resulting from a simple model, that only considers similarity for high velocities and the minimum energy requirement for low velocities. The figure also shows the estimated behaviour in the transition between these two velocity regions (dashed line).

interaction with the ambient stress-strain field. In addition, estimates are made for the influence of deviations from similarity in the high velocity region, including a possible shielding effect of peripheral micro-separations on central ones, and an estimated effect of the changing morphology of micro-separations (due to the increase of σ_x/σ_y with velocity; see page 343).

Shielding effects and rate sensitivity
The possible shielding effect of peripheral micro-separations, which was called the "barrier" effect in Broberg (1979a), might explain the history dependence on the specific energy dissipation Γ. More and more micro-separations may develop in the peripheral parts of the process region, without appreciable change of the state in the central parts. It appears that such an increase of the size of the process region, which may occur under constant crack velocity, might be responsible for the very steep increase of the stress intensity factor with velocity that has often been observed in the high velocity region (e.g. Kobayashi and Dally 1977 and Ravi-Chandar 1982).

Assume now that a significant amount of the specific energy dissipation takes place in the plastic region. The assumption of similarity, embracing the whole crack edge vicinity, still works as a first approximation for some intermediate velocity range. The deviations from similarity now also contain the increasing shielding effect of the plastic region with crack velocity; see Fig. 7.2.1 and connected comments†.

The effect of the rate sensitivity of the material is to some extent neutralized in the similarity region of the model, because a particle in case 1 at $y = y_0$ will experience the same time history as a particle in case 2 at $y = V_2 y_0/V_1$. For lower velocities, however, there may be a substantial influence, due to the phenomenon of delayed

---
† The figure is obtained for mode III, but the principal results are assumed to carry over to modes I and II.

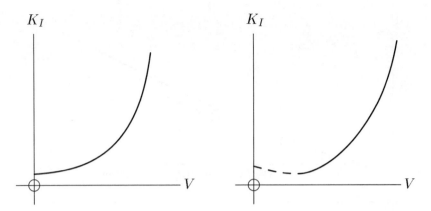

Fig. 9.1.6 Principal relation between specific energy dissipation, Γ, and crack edge velocity, V, shown for initially increasing plastic dissipation (left figure), and initially decreasing plastic dissipation (right figure). The dashed part in the right figure might not have a physical interpretation, and the solid part may have a physical interpretation only for decelerating cracks

yielding (Clark and Wood 1949) and of viscoplasticity. Note that the strain rates may be very high, even for low crack velocities. Consider, for instance, a plastic region size of 1 mm, a maximum strain of 10 per cent and a crack velocity of 10 m/s. Then the strain rate is higher than $10^3 \, \text{s}^{-1}$, which is high enough to have a considerable impact on the energy dissipation. The energy dissipation in the plastic region may be initially increasing or decreasing with increasing crack velocity, depending on the material. The implication on the Γ-V curve is shown in Fig. 9.1.6.

Note that the curve shown in the right part of Fig. 9.1.6 implies instability during acceleration from a stationary state, because the crack edge would very rapidly bypass the minimum of the $\mathcal{G}$-V curve. Similarly, during deceleration (assuming that the same curve is still relevant), crack arrest would occur when this minimum is reached.

Simulations, using the cell model

Numerical simulations of certain experiments by Ravi-Chandar (1982) and Ravi-Chandar and Knauss (1984c) were performed by Johnson (1992b,c, 1993), using a cell model in combination with finite elements. The cells, all cubic, were distributed along the prospected crack path in 12 layers on each side. Outside the cells, larger finite elements were used.

A simple relation was established for the mechanical behaviour of the cells. In terms of loads and displacements, this behaviour was illustrated by the cohesion-decohesion curve for uniaxial macroscopic strain that follows from this relation (Johnson 1992b). The Johnson model is thus a genuine cell model of the material and not a continuum damage model. Note that it is two-dimensional for obvious reasons: a three-dimensional model would not be feasible because the numerical computations would be too time consuming and expensive.

The computer resources did not allow modelling of cells on the appropriate scale for the micro-structure of the material used in the experiments (Homalite 100), and the limitation of the number of layers to 2×12 did influence the results to some extent. Nevertheless, the simulations were capable of reproducing qualitatively all of the experimentally found phenomena, labelled 1-5 on page 628. This is a strong indication that the cell model gives an appropriate description of material behaviour in the context of dynamic crack propagation.

The numerical simulations by Johnson (1992b, 1993) were to some extent not only qualitatively, but also quantitatively, in agreement with experimental results. Thus, for instance, the constant maximum velocities obtained, in a series of simulations with different load magnitudes, but otherwise the same model parameters, were found to vary from $0.34c_R$ to $0.44c_R$, where c_R is the Rayleigh wave velocity, in good agreement with the experiments by Ravi-Chandar (1982). Note, however, that the choice of cell model parameters is more or less arbitrary (although they may be determined from the desired shape and magnitude of the cohesion-decohesion relation), and some choices will give a better agreement with experimental data than others.

The maximum velocity obtained in the simulations was in most cases constant with a high degree of accuracy (Johnson 1993), apart from the fact that it oscillates on a micro-scale, due to the successive cell collapses. As expected, and in agreement with the experimental results, the crack acceleration and the constant maximum crack velocity increased in general with the load magnitude. However, in the series of simulations mentioned in the previous paragraph, the somewhat puzzling feature of a slight decrease of the constant maximum velocity (from $0.44c_R$ to $0.41c_R$) for an increase of the load magnitude above a certain level, was observed (Johnson 1992b). This might perhaps be explained by the screening effect: the highest constant maximum velocity is reached after a longer crack edge travel, which implies more time for evolution of the process region.

In one simulation, Johnson (1993) constrained the process region to one layer of cells. Then, also the specific energy dissipation is constrained to an almost constant value. The result was that the crack accelerated during the whole simulation and reached a final value $0.59c_R$ when the computations had to be interrupted because of limitations of computational resources. However, there is little doubt that the acceleration would have continued toward the neighbourhood of the Rayleigh wave velocity, if it had been possible to extend the computations further. The result should be compared with a constant maximum velocity $0.44c_R$ obtained with 12 layers of cells on each side and otherwise the same conditions (Johnson 1992b).

The simulations with the process region constrained to one layer of cells has an experimental counterpart. Washabaugh and Knauss (1994) merged two PMMA plates together to form a large plate connected with a very thin layer with considerably reduced strength. Thereby a new length parameter, the layer thickness, was effectively introduced. As a result, they succeeded in obtaining crack edge velocities up to about 90 per cent of the Rayleigh wave velocity. Such high velocities were also obtained by Field (1971) for crack propagation in anisotropic single-crystals along certain crystallographic directions.

The results obtained on attempted and successful branching by Johnson (1992b,c, 1993) will be discussed in Section 9.2.

Other simulations of dynamic crack propagation

For materials with a regular atomic lattice, numerical simulations of dynamic crack propagation may be made by considering the mechanical interactions between the atoms in a way known from molecular dynamics, with typical use of parallel computing. The body is divided into cells and each processor models a cell and its nearest-neighbour interactions. Dynamic crack propagation in two-dimensional atomic lattices were considered by Abraham et al. (1994, 1997), whereas Nakano et al. (1995) studied rapid crack propagation in a film of amorphous silicon nitride (Si_3N_4). The smallness of the scale implies that the simulations can cover only very small distances of crack growth, typically of the order of ten nanometres, even if the total number of atoms involved is of the order of a million (Abraham et al. used two million atoms in some simulations).

Molecular dynamics models are capable of reproducing some of the features listed on page 628, such as increasing surface roughness with increasing velocity and acceleration to a constant velocity. Abraham et al. (1994, 1997) also obtained dislocation emission from the crack edge, both in the directions ±30° and ±90°. However, as pointed out by Ravi-Chandar (1998), some other results, such as asymmetric crack propagation in a zigzag pattern, are somewhat paradoxical, because clean cleavage would be expected for a regular lattice (as also found experimentally for single-crystals).

A different approach was used by Xu and Needleman (1994, 1995, 1996) and Needleman (1997). They used a two-dimensional plane strain finite element formulation with originally triangular elements cut out along lines in the directions 0°, 45°, 90° and 135° from a symmetry plane of the body. In a few cases, other angles were chosen. Decohesion was only possible between the elements. Consequently, "ordinary" constitutive equations were assigned to the elements and cohesive-decohesive relations for the boundaries between the elements. These relations included both opening and in-plane sliding, and such displacement discontinuities were assumed to occur not only on the decohesive side, i.e. after the cohesive strength is reached, but also on the cohesive side at any tangential and positive normal load from a virgin state.

Numerical simulations in (Xu and Needleman 1994) were performed for suddenly applied velocities on the top and bottom boundaries of rectangular blocks, in most cases symmetrically. Elastic constants were taken to be representative for PMMA, but the mode I cohesive strength was chosen as $E/10=324$ MPa. This high value may be more typical for cleavage in a single-crystal than for decohesion in a polymer. The specimen height was about 2 mm in most simulations. The applied boundary velocities were generally 10 m/s, which implies a stress wave magnitude of about 25 MPa, which doubles when the two waves meet on the symmetry plane.

The simulations showed very rapid crack acceleration, generally reaching velocities above half the Rayleigh wave velocity in less than 1 μs after the crack was hit by the stress waves. The amount of crack edge travel covered by the simulations was generally shorter than 1 mm. Successful crack branching was observed for high crack velocities.

In some simulations, the decohesive directions were chosen as 0° and ±15°, with the result that the numerical computations led to a zigzag crack propagation, even under full mesh and loading symmetry. The same phenomenon occurred for the cohesive directions 0° and ±30°, but not for 0° and ±45° or for 0° and ±60°.

The maximum crack velocities obtained were much higher than encountered in experiments. In some simulations, the crack accelerated to velocities above the Rayleigh wave velocity. This was described as "a combination of crack-like propagation and lift-off". Lift-off, i.e. decohesion on the symmetry plane ahead of the crack, would obviously occur for a normal stress higher than the cohesive strength.

Super-Rayleigh velocities were also found in cases when cohesive boundaries were restricted to the symmetry plane. The model is then similar to the Barenblatt model, but with the significant difference that even the ascending side of the cohesive-decohesive relation is modelled, whereas the Barenblatt model includes decohesion properties only. The front of a Barenblatt model at a crack edge cannot propagate with the Rayleigh velocity or faster†. The super-Rayleigh speeds obtained by Xu and Needleman (1994) seem to be a result of the model feature that an additional compliance is introduced in the vicinity of the symmetry plane by the cohesive relation. Thus, there is no clearly defined front of a process region as in the Barenblatt model. The material ahead of a Barenblatt process region is a homogeneous continuum rather than an extended cohesive region along the symmetry plane. This difference from the Barenblatt model appears to explain why super-Rayleigh velocities are obtained by Xu and Needleman (1994) for the case when the cohesive boundaries are restricted to the symmetry plane, and the same principal difference would then explain such results also for cases without such restriction.

In some simulations, Xu and Needleman (1995, 1996) considered dynamic crack propagation along an interface between PMMA and aluminium, using their finite element model with interspersed potential surfaces of decohesion for PMMA. They found that debonding and crack deviation into the PMMA occurrred repeatedly. They also found that the crack velocity could overshoot the Rayleigh wave velocity in the PMMA, in agreement with theoretical results (see pages 379ff.) by Gol'dshtein (1966, 1967) and experimental results (see page 395) by Liu, Lambros and Rosakis (1993) and by Lambros and Rosakis (1995).

Needleman (1997) used the model by Xu and Needleman (1994, 1995, 1996) for simulations of dynamic crack growth in ductile materials, specifically a high strength steel. Stress wave loading was assumed, and they considered elastic-viscoplastic material behaviour. As in previous work, the simulations only covered very small amounts of crack growth, typically less than 1 mm.

Dynamic crack propagation in viscoplastic materials

Constitutive equations for viscoplasticity
Viscoplasticity plays a significant part in dynamic crack propagation in ductile materials, because of the high strains encountered in the crack edge vicinity. It may be described as the action by an overstress in relation to the stress under which static equilibrium prevails in the plastic region. In its simplest form, this overstress is related to a constant yield stress, σ_0. For uniaxial loading, Perzyna (1963, 1966) assumed that this overstress causes a plastic strain increase with the rate

$$\dot{\epsilon}_x^p = \frac{1}{\eta}\left[U\left(\frac{\sigma_x - \sigma_0}{\sigma_0}\right)\right]^n \qquad (9.1.6)$$

---

† An expanding crack with a Barenblatt region in an elastic material may be modelled by superpositions of expanding cracks with point-size process regions, each of these limited to sub-Rayleigh velocities; see e.g. Broberg (1964, 1967).

where η is a viscosity parameter, n is a hardening exponent ($n \geq 1$), and $U(\cdot)$ is the unit step function.

For a Huber-von Mises material (page 250) with linear elastic response, generalization to multiaxial stress yields

$$\dot{\epsilon}_{ij} = \frac{1}{2\mu}\left[\dot{\sigma}_{ij} - \frac{\nu}{1+\nu}\dot{\sigma}_{kk}\delta_{ij}\right] + \frac{1}{\eta}\left[U\left(\frac{\sqrt{J_2} - \sigma_0}{\sigma_0}\right)\right]^n \frac{s_{ij}}{\sqrt{J_2}} \quad (9.1.7)$$

where ν is Poisson's ratio, δ_{ij} is the Kronecker delta, $s_{ij} = \sigma_{ij} - \frac{1}{3}\sigma_{kk}\delta_{ij}$ is the stress deviator and $J_2 = 3s_{ij}s_{ij}/2$.

In order to obtain better agreement with experimental results in specific cases, the Perzyna relation may be sophisticated by choosing dependences of the overstress other than a power law, and strain hardening may be taken into consideration, in the simplest case by replacing σ_0 by a flow stress as a function of the effective plastic strain. A more elaborate form, which includes consideration of anisotropic hardening, was presented by Gilat (1985). It was shown to give good agreement with experimental results. Clifton (1990) emphasizes the significance of the rate sensitivity of strain hardening, particularly for very high rates of strain. This sensitivity may be modelled by using an internal variable formulation, in which the strain hardening rate depends on the deformation rate.

A different form of viscoplastic relation was suggested by Bodner and Partom (1975). They assumed that the plastic strain rate can be written as a function of the plastic work, W_p, in a strain hardening material. For pure shear and isotropic strain hardening, the shear strain rate is written as

$$\dot{\gamma}^p = 2D_0 \exp\left[-\frac{1}{2}\cdot\left(\frac{Z^2}{3\tau^2}\right)\right]^n \quad (9.1.8)$$

where D_0 is a constant strain rate, τ is the shear stress, n is a dimensionless strain rate sensitivity parameter, and Z is a state variable expressing strain hardening, given by

$$Z = Z_1 - (Z_1 - Z_0)e^{-mW_p} \quad (9.1.9)$$

where Z_0 is the initial value of Z, i.e., the value for $W_p = 0$, Z_1 is the saturation value of Z as $W_p \to \infty$, and m is a parameter that expresses the sensitivity of Z to the plastic work.

For both the Perzyna law and the Bodner-Partom law, the parameters, particularly the exponent n, are dependent on the temperature.

The asymptotic field at a running crack edge

The asymptotic field at the edge of a crack, running in a viscoplastic material, was first calculated by Lo (1983). The main features of his anaysis will be followed here.

Consider, for simplicity, a crack propagating in anti-plane strain and introduce cylindrical coordinates (r, φ, z) with the origin at the crack edge, i.e., the coordinate system travels with the crack. The Perzyna law (9.1.7) gives

$$\dot\gamma_{rz} = \frac{\dot\tau_{rz}}{\mu} + \frac{1}{\eta}\left[U\left(\sqrt{\tau_{rz}^2 + \tau_{\varphi z}^2}/\tau_Y - 1\right)\right]^n \cdot \frac{\tau_{rz}}{\sqrt{\tau_{rz}^2 + \tau_{\varphi z}^2}} \qquad (9.1.10)$$

$$\dot\gamma_{\varphi z} = \frac{\dot\tau_{\varphi z}}{\mu} + \frac{1}{\eta}\left[U\left(\sqrt{\tau_{rz}^2 + \tau_{\varphi z}^2}/\tau_Y - 1\right)\right]^n \cdot \frac{\tau_{\varphi z}}{\sqrt{\tau_{rz}^2 + \tau_{\varphi z}^2}} \qquad (9.1.11)$$

where τ_Y is the yield stress in shear.

In the vicinity of the crack edge, steady state conditions are approached, and the Galilean transformation is applicable. Thus, $\partial/\partial t = -V\partial/\partial X$, where V is the crack velocity and $X = x - Vt$. But, $\partial/\partial X = \cos\varphi\, \partial/\partial r - (\sin\varphi/r)\cdot\partial/\partial\varphi$. Hence, assuming the stresses to be unbounded as $r \to 0$, the Perzyna equations for $r \to 0$ may be written in the form

$$-\cos\varphi\frac{\partial\gamma_{rz}}{\partial r} + \frac{\sin\varphi}{r}\cdot\frac{\partial\gamma_{rz}}{\partial\varphi}$$
$$= -\frac{\cos\varphi}{\mu}\cdot\frac{\partial\tau_{rz}}{\partial r} + \frac{\sin\varphi}{\mu r}\cdot\frac{\partial\tau_{rz}}{\partial\varphi} + \frac{1}{\eta V \tau_Y^n}(\tau_{rz}^2 + \tau_{\varphi z}^2)^{(n-1)/2}\tau_{rz} \qquad (9.1.12)$$

$$-\cos\varphi\frac{\partial\gamma_{\varphi z}}{\partial r} + \frac{\sin\varphi}{r}\cdot\frac{\partial\gamma_{\varphi z}}{\partial\varphi}$$
$$= -\frac{\cos\varphi}{\mu}\cdot\frac{\partial\tau_{\varphi z}}{\partial r} + \frac{\sin\varphi}{\mu r}\cdot\frac{\partial\tau_{\varphi z}}{\partial\varphi} + \frac{1}{\eta V \tau_Y^n}(\tau_{rz}^2 + \tau_{\varphi z}^2)^{(n-1)/2}\tau_{\varphi z} \qquad (9.1.13)$$

In addition to the constitutive equations, there is a compatibility equation and an equation of motion. These may be written as

$$\gamma_{rz} = \frac{\partial w}{\partial r}, \quad \gamma_{\varphi z} = \frac{1}{r}\cdot\frac{\partial w}{\partial\varphi}, \quad \frac{\partial\tau_{rz}}{\partial r} + \frac{1}{r}\cdot\frac{\partial\tau_{\varphi z}}{\partial\varphi} = \varrho\frac{\partial^2 w}{\partial t^2} = \varrho V^2\frac{\partial^2 w}{\partial X^2} \qquad (9.1.14)$$

where the last member may be expressed in terms of r and φ as

$$\varrho V^2\left[\frac{\partial^2 w}{\partial r^2}\cos^2\varphi - \frac{2}{r}\cdot\frac{\partial^2 w}{\partial r\partial\varphi}\cos\varphi\sin\varphi + \frac{1}{r^2}\cdot\frac{\partial^2 w}{\partial\varphi^2}\sin^2\varphi\right.$$
$$\left. + \frac{1}{r}\cdot\frac{\partial w}{\partial r}\sin^2\varphi + \frac{2}{r^2}\cdot\frac{\partial w}{\partial\varphi}\cos\varphi\sin\varphi\right]$$

Assume that

$$\tau_{rz} \propto r^\kappa, \quad \tau_{\varphi z} \propto r^\kappa, \quad w \propto r^{\lambda+1} \qquad \text{as } r \to 0 \qquad (9.1.15)$$

Then, $\gamma_{rz} \propto r^\lambda$ and $\gamma_{\varphi z} \propto r^\lambda$. From equations (9.1.14), it follows that the singularities of stresses and strains are the same, i.e., $\kappa = \lambda$. The Perzyna equations (9.1.12) and (9.1.13) then show that the singularity of the asymptotic viscoplastic strain, which is the last term in each of the equations, is either weaker or of the same order as that of the elastic strain. Thus, the following possibilities exist:

1. The singularity of the viscoelastic strain is weaker than that of the elastic strain. Then, $r^{n\kappa} \ll r^{\kappa-1}$ as $r \to 0$. Elastic behaviour then dominates near the crack edge. Hence, $\kappa = \lambda = -1/2$ and $n\kappa > \kappa - 1$, which implies $n < 3$.

2. The singularity of the viscoplastic strain is of the same order as the elastic strain. Then, $n\kappa = \kappa - 1$, which implies $\kappa = \lambda = -1/(n-1)$. For $n > 3$ this implies a weaker singularity than the inverse square root, whereas for $n = 3$, the singularity of the viscoplastic strain is the same – inverse square root – as the singularity of the elastic strain. This was shown by Brickstad (1983) and Lo (1983), and, for the related problem of creep, by Hart† (1980), Hui and Riedel (1981); see also Riedel (1987).

For $n < 3$, the asymptotic stress and strain fields are the same as for the elastic case. For $n \geq 3$, they can be deduced from (9.1.12), (9.1.13) and (9.1.14). It is then convenient to write the displacement and the stresses in the form

$$w = \left(\frac{\eta V \tau_Y^n}{r\mu^n}\right)^{1/(n-1)} rF(\varphi) \qquad (9.1.16)$$

$$\tau_{rz} = \left(\frac{\eta V \tau_Y^n}{r\mu^n}\right)^{1/(n-1)} \mu G_r(\varphi), \qquad \tau_{\varphi z} = \left(\frac{\eta V \tau_Y^n}{r\mu^n}\right)^{1/(n-1)} \mu G_\varphi(\varphi) \qquad (9.1.17)$$

The equations (9.1.12), (9.1.13) and (9.1.14) then become ordinary differential equations for $F(\varphi)$, $G_r(\varphi)$ and $G_\varphi(\varphi)$, and they will only contain $\sqrt{\varrho V^2/\mu} = V/c_S$ and n as parameters. Numerical solutions for different values of the Mach number V/c_S and the parameter n were given by Lo (1983), who also considered mode I in plane strain and performed numerical calculations for Poisson's ratio $1/2$. Similar calculations for mode I and Poisson's ratio $1/4$ were performed by Gross et al. (1997).

A remarkable property of the asymptotic field is that its magnitude does not depend on the outer load, except that there is a minimum outer load below which the crack will not grow. The action of the load is thus to control the crack speed rather than to control stress and strain magnitudes. This property was noted for creep crack growth by Hui and Riedel (1981).

Lo (1983) remarked that "the region of dominance of the asymptotic fields may turn out to be so small compared to the prevailing microstructural size scale that the asymptotic results are rendered meaningless". For this reason he also discussed stress and strain fields away from the crack edge.

Energy relations and numerical simulations
Studies of viscoplastic crack growth are mainly performed by numerical simulations. Pioneering work was made by Brickstad (1983). He considered the case $n < 3$, and used the fact, connected to an elastically deforming crack edge vicinity, that a non-zero energy flux would reach a vanishingly small process region, so that an energy criterion for crack growth could be established in about the same way as for the elastic case. Several later contributions have used the same assumption, which results in a specific and material-related energy flux to the crack edge in the elastic enclave, whereas a larger energy flux arrives at the dissipative region. Brickstad (1983) was able to obtain results in excellent agreement with experimental results for steel (Dahlberg et al. 1980). Interesting discussions of certain consequences of such an approach were made by Freund and Hutchinson (1985) and Mataga et al. (1987).

As pointed out by Lo (1983), Broberg (1989b) and Östlund (1990, 1991), the attractive assumption of an enclave of elastically deforming material inside the region

---

† Hart used a different constitutive formulation than the Perzyna law.

of viscoplastically deforming material, may not reflect real material behaviour even if the exponent n would be smaller than 3. The reason is mainly that the extension of the elastically deforming region would be so small that it would even be smaller than the process region, contrary to the tacit assumption that it is substantially larger. The work by Östlund (1990) is particularly clear in this respect, with numerical estimates of the size of the elastically deforming region. It may also be remarked that the stresses in real materials reach a maximum at some distance ahead of the crack edge (cf. Fig. 5.4.3), and that experimental results on material behaviour at large rates of strain fail to reveal purely elastic behaviour for the largest strain rates.

Dynamic crack propagation in modes II and III. Shear banding

Very few experimental results exist for modes II and III, because of the difficulties in realizing these modes in the laboratory. They occur, for instance, in earthquake sliding motion, where kinking into mode I is impeded by a high ambient pressure.

The process region in shear mode earthquakes may be very long compared to its thickness (Rice 1980). It consists of localization to a shear band, which may be preceded by grain distortion and rotations. Shear banding was briefly discussed on page 556. It occurs in several contexts other than earthquake sliding motion or other forms of modes II or III crack propagation, and it appears to be invariably very thin compared to its length when fully developed. It may be identified with the process region at the edge of a crack propagating in mode II, mode III or mixed mode II/III. Material separation is completed behind the process region, but frictional sliding between the crack faces causes additional energy dissipation.

The phenomenon of shear banding has received much attention, among other reasons because of its importance in high speed machining and forming, and its relation to vehicle structural behaviour in collisions and to projectile penetration of armor. Shear banding may be a precursor to crack propagation in modes II and III. A shear band is usually initiated at some inhomogeneity from which it propagates; cf. Marchand and Duffy (1988). It may be arrested before a crack has been formed, i.e., before complete decohesion has occurred.

Zener and Hollomon (1944) suggest that strain softening (a stress-strain curve with negative slope in the plastic region) implies an intrinsic instability so that plastic deformation cannot be homogeneous. They argue that "for a region which, by chance, suffers more deformation than the surrounding region and becomes weakened thereby, will continue to deform while the surrounding region undergoes no further strain". Local strain softening may occur as a result of high temperatures caused by high strains and high strain rates (cf. Section 7.4). This leads to very rapid localization as soon as an embryonic shear band is formed. Overall strain softening may occur as a result of sufficiently low temperatures, as pointed out by Zener and Hollomon (1944), who also demonstrated an equivalence in mechanical behaviour of materials between low temperatures and high rates of strain†.

The temperature may be very high in shear bands. Marchand and Duffy (1988) estimated the temperature rise at above 1100°C for a low alloy structural steel, and

---

† A (perhaps oversimplified) explanation of this equivalence suggests that certain mechanical processes on the nanoscale, such as dislocation motion and atomic migration, may not have time to develop if the strain rate is sufficiently high or the temperature is sufficiently low.

Zhou et al. (1996b) measured a temperature rise of about 1400°C for a maraging steel. The high temperatures, which in many materials cause strain softening, are obtained during high speed operations, when there is little time for conducting heat away from the shear band region. Such shear bands are therefore often called *adiabatic shear bands*.

In theory, strain localization is possible for a continuum (cf. pages 552ff.), but in reality it appears to be initiated at some inhomogeneity and it is, of course, prevented from developing into an infinitesimally thin layer by particles or other material constituents, that introduce intrinsic lengths to the material, and by diffusion, mainly viscoplasticity and heat conduction and convection. Some numerical simulations consider finite deformations, viscoplasticity, heat conduction, thermal softening, strain hardening, strain-rate hardening and, more or less explicitly, void nucleation and deformation (e.g. Zhou et al. 1992, Tvergaard and Needleman 1993, Needleman and Tvergaard 1995, Zhou et al. 1996a; cf. also Needleman 1989). A review of the modelling of dynamic shear strain localization in inelastic solids is presented by Predeleanu (1992).

Experimental investigations of dynamic shear banding in metals or polymers are generally based on a method proposed by Kalthoff (1988, 1990). A cylindrical projectile is fired towards the edge of a plate, containing one prefabricated crack; see Fig. 9.1.7. In the original suggestion by Kalthoff, the plate contained two cracks to achieve symmetry. However, very little advantage may be obtained from the symmetry: analysis is not easier for the symmetric than for the asymmetric configuration. The advantages of obtaining a longer load duration without disturbances from another crack has later motivated an asymmetric arrangement. An analytical treatment of an idealized model for the initial wave-crack interactions has been performed by Lee and Freund (1990).

It is obvious that the region in front of the crack will be subjected to mode II loading. However, the main reason for the success of the arrangement suggested by Kalthoff is that the region in front of the crack will also be subjected to compressive mode I loading. Otherwise, kinking and mode I crack propagation cannot be prevented. Note that the higher the projectile impact velocity, the more intense both the mode II and the compressive mode I loading will be.

The asymmetric impact may be envisaged as the sum of one symmetric and one antisymmetric impact. Thus, there is one compressive wave component, propagating from the impact site, and, assuming that the stress-strain relations are reasonably symmetric with respect to tension-compression, there is one stress wave component that causes a pure mode II stress intensity factor after it has reached the crack edge. This stress intensity factor increases initially approximately in proportion to the square root of time after the stress wave arrival; see Section 6.13. Note that the stresses are biaxially compressive in the compressive stress wave component.

For a sufficiently high impact velocity, the mode II stress intensity factor reaches the critical value needed for shear band formation. At the same time, mode I crack growth through kinking is prevented by the compression carried by the stress wave. Thus, a shear band propagates approximately straight forward from the pre-existing crack. Near the front of the shear band, both high shear stresses and shear strains develop, but the shear stresses are quickly relaxed, due to decohesion. For earthquake sliding motion, this is known as a stress drop, which takes place in the process region

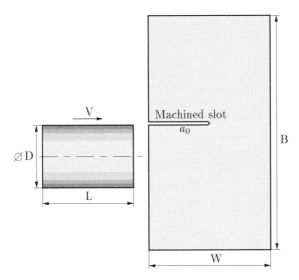

Fig. 9.1.7 The usual experimental setup after a suggestion by Kalthoff (1990). A cylindrical projectile is fired towards a cracked plate.

(the shear band) and reaches a level that is governed by the friction properties in the sliding region. It has also been referred to as a stress collapse (Wright and Walker 1987).

The shear band front may move with a considerable velocity. Zhou *et al.* (1996b) report propagation velocities of up to 1200 m/s in a maraging steel. This is about $0.4c_R$. However, this velocity is considerably smaller than the P wave velocity, which for a plate is approximately 5300 m/s. Thus, if $a_0 = W/2$ (see Fig. 9.1.7), as in their experiments, the P wave that is reflected at the remote plate boundary will meet the shear band before this has travelled more than about 25 per cent of the ligament from the pre-existing crack, depending on the delay time for shearband initiation. The S wave, which travels with a velocity of about 3100 m/s will meet the shearband later.

The reflected waves come with normal and shear stresses of different signs from those in the incoming waves. Near the boundary of reflection, incoming and reflected waves cancel each other, but further away the amplitude of the reflected way will dominate. The reversal of shear direction, or the decrease in shear magnitude, may cause arrest of the shear band, whereas the reversal from compression to tension may cause mode I crack propagation via kinking from the vicinity of the shear band edge. For sufficiently high impact velocities, the shear band may run through the whole specimen, ultimately by shearing off the decreasing ligament between the shear band edge and the plate boundary, under wave propagations back and forth along the ligament.

The sequence and nature of events in the Kalthoff experiments depend on the specimen configuration as well as on the thermal and mechanical properties of the material and the impact velocity. Arrest without kinking has been observed in polycarbonate (Ravi-Chandar 1995, Rittel *et al.* 1997) and in a titanium alloy, Ti-6Al-4V (Zhou *et al.* 1996b). Arrest with subsequent mode I kink formation, or, at very high impact velocities, shear band propagation all the way through the specimen was found in C-300 maraging steel (Zhou et al. 1996b).

In the experiments by Rittel et al. (1997), the pre-existing crack was systematically sharpened by fatigue, whereas other experiments seem to have been generally performed with a machined slot. There is some difference: for a machined slot the compression wave component creates a field with inverse square root stress components describable by a negative mode I stress intensity factor (Lee and Freund 1990), whereas the mode I stress intensity factor for the crack is zero. Thus, a mixed mode loading is obtained for the machined slot, whereas pure mode II loading in a compressive environment is obtained for the crack. If a mode I kink is formed rather than a shear band at the edge of a crack, the kink angle would be approximately $-77°$, according to (8.6.2), but, if the mode I crack emerges from the root of a machined slot, and the estimate $K_I/K_{II} \approx -0.40$ (Lee and Freund 1990) is used, then the angle would be about $-86°$. The minus sign indicates kinking away from the impacted side†. In this context, it is interesting to note that Lu et al. (1995) report kink angles around $-90°$.

A disturbance reaches the original crack edge position at the time $\sqrt{a_0^2 + D^2}/c_P$, where a_0 is the original crack length, D is the projectile diameter and c_P the P wave velocity in a plate. For $a_0 = D = 50$mm, this time equals about $8\mu s$ for steel and $16\mu s$ for PMMA. This might be shorter than the time for kink initiation, and the real kink angle magnitude may then be smaller than what is calculated without consideration of the disturbance. Kalthoff (1990) reports an angle of about $-70°$ for a high strength maraging steel and Ravi-Chandar (1995) reports a kink angle magnitude of about $-66°$ for polycarbonate.

If the delay time for kink formation is long enough, reflected waves from the far boundary will change the stress composition near the crack edge considerably, because such waves carry reversals of the stresses in the incoming waves. The kink angle magnitude may then be comparatively small. Rittel et al. (1997) observed angles whose magnitude did not exceed $40°$. Zhou et al. (1996b), who found kinking after some shear band propagation, report angles of about $30°$ away from the impacted side. The kink direction in that case would depend upon the direction of the maximum principal stress in the vicinity of the front of the arrested shear band.

The phenomenon of change from mode II to mode I, which, for certain materials also is a transition from ductile to brittle behaviour, is sometimes referred to as a failure mode transition (Ravi-Chandar 1995, Rittel et al. 1997). Kalthoff (1990) noticed that a phase transition occurred in the shear band in a chromium-molybdenum steel, appearing as a white edging band with strongly increased hardness. For many practical situations in which shear banding may result, for instance high speed machining, it is, of course important to know whether such transitions from mode II to mode I and from ductile to brittle behaviour might appear. A very interesting discussion on failure mode transitions in polymers, based on several different experimental results, is given by Ravi-Chandar et al. (1998).

Zhou et al. (1996b) show scanning electron micrographs of the shear band surfaces in C-300 maraging steel and in a titanium alloy, Ti-6Al-4V. Both surfaces show very

---

† Lee and Freund (1990), who use the maximum hoop stress criterion, find the value $63°$, but it appears that they picked the minimum rather than the maximum: the maximum occurs at about $-78°$.

clear evidence of sheared and coalesced voids. This might imply that the height of the shear band is about equal to the distance between the kernels (usually particles) of void formation. It also suggests that a cell model might be suitable for investigation of the processes inside shear bands. Another way of modelling dynamic propagation of shera bands with due attention to their finite thickness and the existence of a process region has been suggested by Mercier and Molinari (1998).

9.2 Branching

Introduction

The phenomenon of branching is not well understood. The principal difficulties concern why, how and when branching nucleates. The branching geometry may be relatively simple and well understood in cases of multiple branching of the kind shown in Fig. 2.7.4. This geometry, which is probably more common in metals than in ceramics and brittle polymers, can be simulated reasonably well for small scale yielding by assuming that branching occurs when the energy flux into the crack edge reaches a critical value, and that the branching geometry is the one for which the mode II stress intensity factor vanishes; cf. (8.6.1). This local mode II symmetry close to the crack edge has been convincingly demonstrated for dynamically running cracks in brittle materials by Rossmanith (1983), and it was assumed to hold by e.g. Kalthoff (1973) and Pärletun (1979). On the other hand, the branching morphology may vary considerably between different materials and between different loading situations in the same material. The number of branches may range from two to three, four or more, and branches may appear in rapid succession after the initial branching; see e.g. Schardin (1959) and Arakawa and Takahashi (1991b).

A hypothesis

When and why does branching occur? A possible hypothesis would be that branching of an accelerating crack can take place when the energy flux into the crack edge is sufficient for propagating two branches. Johnson and Holloway (1966) remark that the existence of small sub-surface cracks in the mist zone in glass "suggests that the mechanism for gross branching of the fracture front is available well in advance of its appearance", but is postponed until "the rate of release of strain energy by the expanding fracture is sufficient to create *four* new surfaces". This hypothesis would imply that the energy flux into each branch edge should be approximately $\mathcal{G}/2$ just after branching, if the flux into the edge of the mother crack is $\mathcal{G}$ just before branching. The approximation involved appears to be very good, judging from a static analysis: Isida and Noguchi (1992) analysed two infinitesimally small symmetric branches at the angles $\pm 27.3°$ from the mother crack, i.e., the angles for which $K_{II} = 0$ at the branch edges; cf. Fig. 4.10.4. They found that the stress intensity factors for the branches equal about $0.75 K_{Im}$, where K_{Im} is the stress intensity factor for the mother crack. This is only slightly higher than $1/\sqrt{2}$, which corresponds to half the energy flux into the mother crack just before branching.

The branch velocities just after branching would be lower than the velocity of the

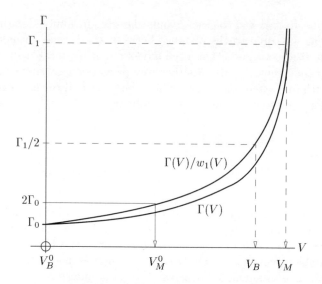

Fig. 9.2.1 Specific energy dissipation, $\Gamma(V)$, and the function $\Gamma(V)/w_1(V)$ as functions of the crack edge velocity V. The figure also shows a graphical construction (solid lines) for obtaining the crack velocity just before branching, according to the hypothesis that branching occurs as soon as the energy flux can support two branches. In addition, the same construction is used for obtaining the branch velocities if branching occurs at a high velocity (dashed lines). V_M and V_M^0 are velocities of the mother crack just before branching, and V_B and V_B^0 are the branch velocities just after branching.

mother crack just before branching. If branching would occur according to the hypothesis, then the branches would move with an infinitesimally small velocity, because this would be the least energy consuming option, provided that the specific energy dissipation $\Gamma(\dot{a})$ is monotone increasing and unique in the respect of being the same for the branches as for the mother crack.

As an example, consider a symmetrically expanding mode I crack subjected to mode I loading. Sudden arrest of the crack leads to an instantaneous change of the stress intensity factor to the value given by (6.12.48). From this equation, it follows that the ratio between the stress intensity factor immediately after crack arrest and the stress intensity factor for a stationary crack of the same length varies very little around about 1.06 for crack velocities around $0.2c_R$-$0.6c_R$. The stress intensity factor of the arrested crack will oscillate towards the value for the stationary crack. If these oscillations are neglected, an estimate of the crack velocity, for which branching would occur, according to the hypothesis, is found from (9.1.2), after replacing $\mathcal{G}(a,\dot{a})$ by $\Gamma(\dot{a})$ and $\mathcal{G}(a,0)$ by $2\Gamma(0)$. This gives

$$\frac{\Gamma(\dot{a})}{w_1(\dot{a})} = 2\Gamma(0) \tag{9.2.1}$$

This relation is shown schematically in Fig. 9.2.1. The velocities found in this way appear to be about $0.2c_R$-$0.3c_R$ for brittle plastics like PMMA and Homalite 100 and about $0.3c_R$ for steel 4340. These estimates are based on some $\Gamma(\dot{a})$ relations reported in the literature, but it may be noted that the scatter is in general considerable, so that accurate estimates cannot be made. Recall that two approximations were made in arriving at the estimates. One was based on the assumption that infinitesimally small slowly moving branches receive together the same energy flux as the mother crack, rather than slightly more, according to Isida and Noguchi (1992). The other was based on the assumption that the stress intensity factor of an arrested expanding crack equals the static stress intensity factor of a stationary crack with the same length, rather than a somewhat higher value. Thus, both approximations imply that the estimates made of the crack velocities predicted by the hypothesis are slightly too high.

It is obvious that the hypothesis leads to results at variance with experimental evidence. The velocity of branching is usually substantially higher than the predictions, and, more importantly, the initial branch velocity is not at all very low. On the contrary, observations have shown that the branched crack edges, immediately after branching, move almost as fast as the mother crack just before branching (Schardin 1959, high speed photography). In fact, although the observations by Schardin were very accurate, no slowing down could be detected in some cases. The hypothesis that branching occurs as soon as the energy available can feed two crack edges can therefore not be supported.

The astonishingly small reduction in crack velocities before and after branching, may be understood by the same graphic construction as in Fig. 9.2.1 for the limiting case, but carried out for a high velocity. This yields the result shown by the dashed line†. Obviously, the velocity difference between the branches and the mother crack may vanish, if the Γ-V relation approaches a vertical asymptote, as has been observed for some materials (e.g. Kobayashi and Dally 1977, Ravi-Chandar 1982).

Branching symmetry

A frequent experimental observation is that branching may be very symmetric, not only as regards the pattern produced after completed crack growth, but also, as high speed photography reveals (Schardin 1959), during each instant of crack growth. This fact appeared for a while to be amazing, because one of two branches would inevitably be somewhat shorter than the other and consequently, by being "in the shadow" of the longer branch, it would receieve a smaller energy flux. As a result, it would propagate less fast than the longer branch, leading to still larger difference in length between the two cracks, etc.

In fact, the intuitive appreciation that the shorter of two branches will be retarded in comparison with the other branch, is confirmed by calculations. Pärletun (1979), using the data for $\Gamma(\dot{a})$ reported by Paxson and Lucas (1973), found that the shorter branch will be rapidly arrested if a branching attempt occurs in the low velocity region, even if the difference in length between the two branches is as small as 1 per cent. However, for sufficiently high crack edge velocities, a branching attempt may be successful, even if the difference in length is as high as 10 per cent. This is schematically shown in Fig. 9.2.2.

Two facts contribute to sustained branch growth if branching occurs at a sufficiently

---

† Essentially the same discussion was made by Rose (1976a) in an excellent review article on fast crack growth and fracture.

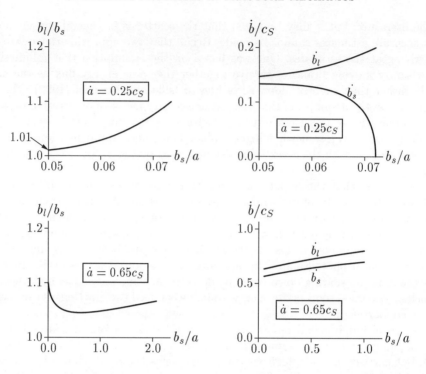

Fig. 9.2.2 The left figures show the length ratio b_l/b_s between two branches at one edge of a symmetrically expanding crack as function of the ratio b_s/a, where a is the half-length of the mother crack. The initial value of b_l/b_s is 1.01 (upper left figure) and 1.1 (lower left figure), respectively. The right figures show the velocities $\dot{b}_l$ and $\dot{b}_s$ of the branch edges as functions of b_s/a. The velocity of the mother crack is $0.2c_S$ in the upper figures and $0.65c_S$ in the lower figures.

high velocity. One is that the communication between the two branches, which is carried by stress waves, is impeded at high velocities, so that the "shadow" effect on the shorter branch becomes less pronounced. The other one is the rapid increase of the energy flux with the velocity in the high velocity region, which implies only insignificant differences between the velocities of the two branch edges, even if the difference between the energy fluxes is quite substantial.

Attempted branching

Attempted branching at lower velocities has often been observed. Anthony et al. (1970) and Congleton (1973) reported what they called "sub-branching" well before successful branching. They also showed a successive legthening of those precursory branches during the growth of the mother crack. For a tool steel, they measured branch lengths of about 0.1 mm half-way to successive branching and about 1 mm just prior to successive branching. Such precursory branching attempts have also been reported for other materials in several later investigations, for instance by Kobayashi and Dally (1977), Ravi-

Chandar (1982), Ravi-Chandar and Knauss (1984b,c), Sharon et al. (1995), Sharon and Fineberg (1996) and Ravi-Chandar and Yang (1997).

It has long been understood that successful branching develops from micro-structural events causing out-of-plane crack-like openings from the mother crack; see e.g. Carlsson (1962, 1963), Clark and Irwin (1966), Johnson and Holloway (1966), Congleton and Petch (1967) and Rose (1976a). It was discussed on page 343 that the most likely candidate for the cause of such events in brittle materials appears to be the ratio of σ_x/σ_y, which increases with velocity. As also discussed, the hypotheses that branching would occur as a result of the maximum hoop stress or the maximum principal stress are, for different reasons, less promising. Some support for the role of the ratio σ_x/σ_y is given by the observation of Carlsson (1962, 1963) that branching in steel plates often occurs by the action of a reflected dilatational stress wave with a strong σ_x component.

Branching synchronicity

A problem with the notion of branching as a result of micro-structural events is that individual such events have a very small spatial reach across the plate, compared to plate thicknesses for which branching is observed. How are these events coordinated and synchronized across the whole plate in the thickness direction? Observations on glass plates show that branching appears to occur along a line almost perpendicular to the plate surfaces, and the resulting edge of the part between the branches is generally sharp and well defined. Some light has been shed on this problem by an important discovery by Sharon and Fineberg (1996), working with PMMA. They studied the evolution of patterns on the fracture surfaces. For velocities below about $0.36c_R$, no particular patterns were observed, but for higher velocities, pattern islands develop. With increasing crack velocity, the density of such islands increases, and they are also organized into bands, roughly perpendicular to the crack direction. At the highest velocities recorded, about $0.63c_R$ ($c_R \approx 945\,\mathrm{m/s}$), such bands extend across the whole plate in the thickness direction, almost perpendicularly to the plate surfaces. For a plate thickness of 3 mm the distance between the bands is about 1 mm. The bands seem to consist of clusters of micro-cracks which develop in the process region (Ravi-Chandar and Yang 1997).

Sharon and Fineberg (1996) described the development of banded patterns as a transition from a three-dimensional to a two-dimensional state. In particular, it appears that this may occur also for the development of the small branches that have been referred to as attempted branching, and which Sharon and Fineberg call "micro-branches". If these are more or less aligned along narrow oblong regions across the plate in the thickness direction, then the necessary conditions for successful branching to take place virtually simultaneously across the plate thickness might be at hand.

The spacing between the micro-branches is of the order of a few tenths of a millimetre at higher velocities (Sharon and Fineberg 1996). Micro-branches appear at velocities of about $0.36c_R$ and their lengths were found to increase approximately linearly with the velocity, from a few μm to over 0.4 mm at $0.6c_R$. The branching angle for the micro-branches varied between about 20° and 40°. Note that these angles are more typical for micro-structural events than for continuum behaviour, so that the deviations from the maximum angle 27.3° found by Isida and Noguchi (1992) are quite natural.

It was suggested by Sharon et al. (1996) that the micro-branches are the main cause of the increase with velocity of the energy dissipation during crack propagation in PMMA. However, other investigations show that there is a considerable increase of energy dissipation per unit crack growth already before micro-branches are observed. Thus, the energy flux seems to increase by a factor of about 4 or more from incipient crack growth to $0.36c_R$ (e.g. Paxson and Lucas 1973, Takahashi and Arakawa 1987). On the average, i.e. apart from fluctuations, the energy flux appears to increase smoothly with the crack velocity, which indicates a gradual increase in the number of micro-cracks during crack growth, rather than a sudden appearance of new types of micro-structural events. Ravi-Chandar (1998) remarks that the recognition of the three distinct regions, "mirror", "mist" and "hackle", is simply based on visual observation, and it does not correspond to a similarly clear distinction based on surface roughness measurements. He finds that the micro-branches consist of clusters of micro-cracks. It appears likely that coalescences occur in such clusters.

Macroscopic criteria for the onset of branching

Even though the mechanism of branching is clearly related to micro-structural events and thus not primarily describable by continuum mechanics, the micro-structural events themselves might be related to certain continuum characteristics for each given material and loading conditions. Older theories connect branching with changes in the hoop stress near the crack edge with the crack velocity (Yoffe 1951) or other similar changes, such as in the principal stresses; cf. page 343. Such theories would therefore predict that branching occurs when a certain velocity is reached.

Clark and Irwin (1966), Congleton and Petch (1967), Anthony et al. (1970), Ramulu et al. (1983) suggested that the branching event seems to be stress intensity controlled rather than velocity controlled; see also Ramulu and Kobayashi (1985). This view recognizes that the stress intensity factor may not be uniquely related to the velocity, a view supported by, for instance, the experimental results by Kobayashi and Dally (1977) on an epoxy resin. They show acceleration to a constant velocity, but successful branching only after a fairly large amount of continued crack propagation with this velocity. During this period, the stress intensity factor increases by almost a factor of two. Attempted branching was observed in the constant velocity region well before the onset of successful branching. This is shown schematically in Fig. 9.1.1.

The theory of a correlation between branching and a critical value of the stress intensity factor has been found to be at variance with certain experimental results. Arakawa and Takahashi (1991a,b) used specimens loaded so that a crack accelerates to a maximum velocity and then decelerates. The stress intensity factor increases during acceleration and even during the initial phase of deceleration. Branching may occur during the deceleration phase. In several experiments, Arakawa and Takahashi (1991b) found branching after considerable deceleration, in one case (for Homalite-911) from about 510 m/s maximum velocity to about 440 m/s, and in another case (for Araldite D) from about 380 m/s to about 300 m/s. During this phase the stress intensity factor increased by a factor of about 1.25 and 2.6, respectively. However, the correlation between branching and the stress intensity factor was not found to be satisfactory. Rather, as Arakawa and Takahashi (1991b) found, a critical value of a combination between stress intensity factor and crack velocity gives a better correlation, interestingly enough the same combination as one that gives a good correlation with the surface roughness (Arakawa and Takahashi 1991a). The correlation is good both for a large range of velocities, including branching during acceleration as well as deceleration. The combination is

$$\frac{(K_I)^2 - (K_I^{arrest})^2}{E}\dot{a} \qquad (9.2.2)$$

where K_I^{arrest} is the stress intensity factor for an arresting crack and E is the dynamically determined modulus of elasticity. The expression does not seem to have a simple physical explanation. It may be noted that the energy flux into the crack edge in the plane stress approximation differs from K_I^2/E through the velocity dependent factor Y_I given by (6.2.51).

Arakawa and Takahashi (1991b) noted a tendency for the number of branches to increase with the velocity of the mother crack. For Homalite-911, the number was found to be two for velocities below about 500 m/s and three above about 500 m/s, but with some overlap. For Araldite D the number was found to be two from about 300–360 m/s, three from about 320–430 m/s and more than four in a few cases at about and just below 430 m/s, thus with overlapping regions.

Simulations of branching

The numerical simulations by Johnson (1992b,c, 1993) were able to simulate branching. Actually, this outcome was not anticipated: the main reason for the simulations was to test the idea (Broberg 1979a) that the cell model would lead to a loss of the significance of an intrinsic length parameter and the consequential loss of a unique relation between stress intensity factor and crack velocity in the high velocity range. The expected smooth increase of the process region with the crack velocity on a microscale did not occur, but rather a somewhat irregular development, in some simulations following a periodic pattern with attempted and finally successful branching.

The cells used in Johnson (1992b, 1993) were all identical. In Johnson (1992c), the cell parameters were varied according to a Gaussian distribution. Crack propagation simulations were performed for cases that previously (with identical cells) had shown a smooth development of the process region without branching. The result was that, even for a variation of the parameters with a standard deviation as small as 1 per cent, a clearly less smooth development of the process region was obtained as well as crack branching.

The development of a somewhat periodic pattern in some simulations, appears to be in accord with experimental results and an important part of the micro-structural mechanism leading to attempted and successful branching. However, the band patterns described by, e.g., Sharon and Fineberg (1996) cannot, of course, be reproduced in two-dimensional simulations.

Branching was also found to occur in the simulations by Abraham et al. (1994, 1997), Nakano et al. (1995), Xu and Needleman (1994, 1995,1996) and Needleman (1997). In these simulations, branching appears to be clearly associated with the increase of the ratio σ_x/σ_y beyond unity at higher crack velocities. The prescribed cleavage directions in the model by Xu and Needleman allow branches to develop along these directions, but also to change direction stepwise during continued crack growth.

9.3 The maximum crack velocity

Mode I

The notion of a specific energy dissipation that is independent of the crack velocity, or at least a bounded function of the crack velocity, would imply acceleration toward the Rayleigh wave velocity for certain body and loading geometries, such as an expanding crack in a large plate, subjected to remote loading (Broberg 1960). The experimental results by, for instance, Schardin (1950, 1959) that only about half the Rayleigh wave velocity could be obtained was therefore unexplained until Paxson and Lucas (1973) discovered that the specific energy dissipation in PMMA seemed to increase by as much as a factor of about 50 when the crack velocity increased from a small value to somewhat less than $0.7c_R$. The maximum crack velocity then seemed to be connected with a steep increase of the specific energy disspation with velocity. Other results, for instance those by Kobayashi and Dally (1977), which clearly indicated that the stress intensity factor continued to increase under a constant maximum velocity (cf. Fig. 9.1.1), gave further support to this theory. A first attempt to relate this steep increase to dynamic micro-structural events was made be Broberg (1979a) on the basis of a discussion of the implications of a cell model of the process region; see pages 630ff.

The results by Ravi-Chandar (1982) and Ravi-Chandar and Knauss (1984c) showed convincingly that the maximum constant velocity, that is known to be reached during crack propagation in a variety of materials, is not a material property. These results could be simulated by using the cell mode (Johnson 1992b,c, 1993), which gave support to the theory that the steep increase of the specific energy dissipation with the crack velocity depends on dynamic micro-structural events in the process region, leading to a loss of an intrinsic length parameter at high crack velocities; see page 631.

The loss of an intrinsic length parameter can explain why there is no unique relation between the specific energy dissipation and the crack velocity, but it cannot explain why cracks accelerate to a constant velocity. Consider a crack in a large plate subjected to remote loading. The energy flux to each crack edge increases with increasing crack length, but decreases with increasing velocity, cf. (9.1.2)-(9.1.3). Thus, the balance between energy flux and specific energy dissipation may prevail in different ways. One extreme is that the crack velocity increases, whereas the specific energy dissipation stays constant. The other is that the crack velocity stays constant whereas the specific energy dissipation increases. During crack acceleration, the balance is achieved by choices between these two extremes, but eventually the second extreme is chosen. There does not seem to be a simple thermodynamical explanation of this fact.

It may be observed that a self-similar solution for an expanding crack with constant velocity, as discussed in Section 6.9, also holds for a non-linear continuum, if no significant rate effects are involved. Now, after the loss of an intrinsic length parameter, the process region as well as the plastic region may be considered as a continuum, and constant velocity propagation is then a solution. On the other hand, constant velocity propagation has been observed for cases other than those involving self-similarity, and it has indeed been found to occur under decreasing as well as increasing stress intensity factor. The preference for constant velocity propagation is thus not well understood.

Now, when different experimental conditions give different constant velocities, one may ask whether there are certain conditions that yield the maximum possible crack velocity in a given material. In this context, the result obtained by Johnson (1992b), that the maximum constant velocity in one series of simulations was not obtained for the highest loads in the series (cf. page 636) is of some interest. Note that the highest loads also imply the highest stress intensity factors in Johnson's simulations. One may speculate about whether a shielding effect plays a part for this result and perhaps also for the very occurrence of constant velocities. Such a shielding effect would imply increasing difficulty for macroscopic stresses and strains to reach the innermost cells with increasing size of the process region.

Modes II and III

The process region character of a shear band in modes II and III, implies that the Barenblatt model appears to be very suitable for analysis of dynamic crack propagation in modes II and III. Only mode II will be discussed here: it is assumed that rapid mode III growth is rather similar in modes II and III. One essential difference, of course, is that only one velocity region, velocities below the S wave velocity, need to be considered for mode III, whereas there are three distinct regions to be considered for mode II: the sub-Rayleigh region, the super-Rayleigh region below the S wave velocity, and the intersonic region, between the S and P wave velocities.

The strong localization of the process region to a shear band and its coupling to thermal softening, seems to act against the widening of the region with the crack velocity that is found to occur for mode I (pages 630ff.). This would indicate that the strong increase of the specific energy dissipation with crack velocity would not be present for the shear modes. Then, there is a possibility that a mode III crack could accelerate to a velocity close to the S wave velocity in mode III and close to the P wave velocity in mode II, if the forbidden region between the Rayleigh and the S wave velocities could be bypassed. This was briefly discussed on pages 419.

The energy flux into the crack edge in small scale yielding has been determined previously for both sub-Rayleigh crack velocities by (6.2.42) and intersonic crack velocities by (6.3.72). Figure 9.3.1 shows this energy flux for all crack velocities. Note that the forbidden region only represents a narrow velocity interval. Disturbances during the sub-Rayleigh acceleration could lead to a bypass of this region, and so could a temporary delay of crack motion along some part of the crack front, caused, for instance, by an obstacle. After the delay, the crack velocity along that part would then accelerate to catch up, and perhaps also overshoot, the other parts of the crack front; cf. page 505. A further possibility consists of the development of a very pronounced stress peak, travelling with the S wave velocity ahead of a crack which is expanding with a high sub-Rayleigh velocity; cf. Fig. 6.9.11 and connected text.

If the velocity of an expanding mode II crack would, by any chance, exceed the S wave velocity, then the crack could accelerate toward the P wave velocity, if the specific energy dissipation does not increase too fast with increasing crack velocity. Note that the increasing crack length contributes an energy flux increase. Some discussion about the mechanism for transition through the forbidden region and acceleration toward the P wave velocity is given in Broberg (1995c). The possibility of intersonic crack propagation in mode II has got some support in recordings from the Imperial Valley Earthquake 1979 (Archuleta 1982, Scholtz 1990).

Numerical simulations of mode II crack propagation through the forbidden region

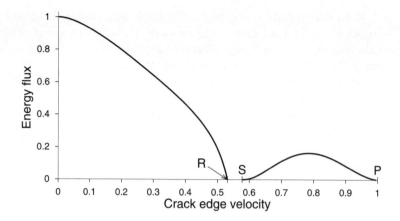

Fig. 9.3.1 Energy flux into the process region of a symmetrically expanding mode II crack in a large body, subjected to remote loading, as a function of the crack velocity. The region between the Rayleigh wave velocity (labelled R) and the S wave velocity (labelled S) is not available for crack propagation. The label P refers to the P wave velocity. The energy flux is calculated for constant velocity propagation, and it is normalized with respect to the energy flux for vanishingly small crack velocity and the same instantaneous length. The material constant $k^2 = 1/3$, corresponding to Poisson's ratio $\nu = 1/4$ for plane strain and $\nu = 1/3$ for plane stress.

to intersonic velocities have been made by e.g. Andrews (1976, 1985), Das and Aki (1977), Day (1982) and Johnson (1990, 1992a). In their simulations, the stress peak travelling with the S wave velocity appears to play a part for the transition from sub-Rayleigh to intersonic velocities.

9.4 Crack arrest

Situations in which a running crack may be arrested

The basic theory for crack arrest was given in Section 6.12, but the situations in which a running crack may be arrested were not identified. From experience, it has been found that four basic situations may be distinguished†:

1. The loading is such that the stress intensity factor decreases during crack propagation. An example is a double cantilever beam, subjected to loading under grip control; see Fig. 9.4.1. If, for instance, the crack starts propagating from a blunted notch, it may accelerate to a high velocity and then decelerate towards arrest. Another example is loading of such a short duration that crack propagation may be initiated but not sustained across the whole body. Also residual stresses may be included in this category of situations. An example is a segment of an earthquake fault along which the shear stresses are lower than in neighbouring segments, due to relaxation after a previous event.

---

† For simplicity, terms that are pertinent for small scale yielding are used here.

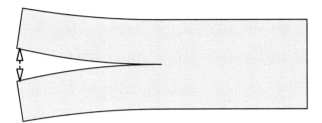

Fig. 9.4.1 Double cantilever beam. For fixed grip loading, the stress intensity factor for a running crack decreases in general with the crack length. Their relation depends upon the load magnitude and the acceleration-deceleration development.

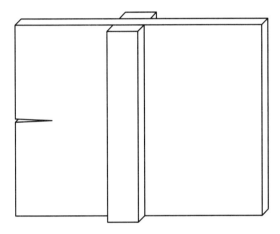

Fig. 9.4.2 Plate with a stiffener. The stiffener may, for instance, be glued, welded or riveted to the plate.

2. The fracture toughness increases along the prospective crack path. This may be due to a change of material in a plate, so that the crack crosses the interface between a less tough to a more tough material, or to a temperature gradient, so that the temperature, and thereby the fracture toughness, increases along the crack path.

3. The geometry changes, so that the length of the crack front increases during propagation. One example is a plate with varying thickness, for instance in the form of a stiffener (see Fig. 9.4.2).

4. The crack is running into a hole and may be arrested because the circumferential stress along the hole edge does not reach the level needed for initiating a new crack on the other side of the hole.

Conditions for crack arrest

Assume that small scale yielding prevails. Then, it is convenient to introduce the dynamic fracture toughness, $K_c^{dyn}(V)$, where V is the crack velocity. Its relation to the specific energy dissipation $\Gamma(V)$ is

$$K_{Ic}^{dyn} = 2\sqrt{\frac{(1-k^2)\mu\Gamma_I(V)}{Y_I(V/c_P)}} \quad \text{for mode I} \tag{9.4.1}$$

$$K_{IIc}^{dyn} = 2\sqrt{\frac{(1-k^2)\mu\Gamma_{II}(V)}{Y_{II}(V/c_P)}} \quad \text{for mode II} \tag{9.4.2}$$

$$K_{IIIc}^{dyn} = \sqrt{\frac{2\mu\Gamma_{III}(V)}{Y_{III}(V/c_S)}} \quad \text{for mode III} \tag{9.4.3}$$

These relations follow from (6.2.49), (6.2.42) and (6.4.14), respectively. The velocity dependence through the functions Y_I, Y_{II} and Y_{III} is negligible for low velocities: these functions differ, for instance, less than about 2 per cent from unity for $V/c_S < 0.15$.

Assume, for simplicity, that both $\Gamma(V)$ and $K_c^{dyn}(V)$ are single-valued and monotone increasing below the maximum crack velocity considered. The condition for crack growth is then $\mathcal{G} \geq \Gamma(0)$ or $K \geq K_c^{dyn}(0)$, where $K_c^{dyn}(0)$ is the static fracture toughness, which has been denoted K_{SS} or K_c previously.

Re-initiation

After a running crack has been arrested, there is still kinetic energy in the body, gradually dissipating under back and forth wave motion. Thus, the stress intensity factor at the arrested crack edge will oscillate until it finally reaches a static value. During the oscillations, it might happen that the stress intensity factor overshoots the static fracture toughness. Re-initiation of crack propagation may then occur after some delay, if this overshoot is sufficiently high and persistent.

Examples of re-initiation are found in Ravi-Chandar and Knauss (1984a) for Homalite 100 and in Naus et al. (1989) for steel ASME SA 533 B. The latter refers to 1000 mm wide plate tests at the Oak Ridge National Laboratory, USA. In one test, the crack was running along a temperature gradient, starting with about 200 mm length at the ambient crack edge temperature $-40°C$, being arrested first at about 460 mm and $+77°C$, then, after re-initiation about 1.6 ms later, arrested again at about 500 mm and $+88°C$, and finally, after a second re-initiation about 2 ms later or more, arrested at about 530 mm and $+95°C$. Crack propagation occurred by cleavage†.

Re-initiation occurs if the stress intensity factor after crack arrest overshoots the static fracture toughness during a sufficiently long time. The phenomenon is not basically different from that of initiation of unstable crack growth by the sudden application of a stress pulse. This has been studied by several authors, e.g. Shockey and Curran (1973), Shockey et al. (1983, 1986), Klepaczko (1984), Ravi-Chandar and Knauss (1984a), Kalthoff (1986), Prakash and Clifton (1992) and Rittel and Maigre (1996). In general, the results show that crack initiation occurs at the static fracture toughness (the value of a slowly increased stress intensity factor at onset of unstable crack growth) if the pulse duration is sufficiently long, for instance, longer than about

---

† Note that the term cleavage is used for fracture in polycrystalline materials even if it is only associated with a minor part of the total energy dissipation. The major part is usually due to plastic tearing of the ligaments between cleavage facets.

50μs for Homalite 100 (Ravi-Chandar and Knauss 1984a) and for a high strength steel (Kalthoff 1986). For shorter pulses, the stress intensity factor needed for crack initiation is in general higher, although some apparent anomalies may exist; cf Kalthoff (1986).

Several circumstances seem to influence the delay time for crack initiation, such as viscoplasticity, viscoelasticity, change of micro-separation mechanism and pulse shape. Consider, for simplicity, a linearly increasing mode I stress intensity factor. For a very slow increase, the stress intensity factor may develop as shown by Fig. 8.4.3. This implies that some nucleation and growth of micro-separation has occurred when the stress intensity factor reaches the value K_I^0 for onset of stable crack growth, when coalescences between micro-separations and the main crack also take place. Due to viscoplasticity and other delaying effects, a faster increase of the stress intensity factor would imply that onset of crack growth would not occur until this factor has reached a value beyond K_I^0, assuming the micro-separation mechanism to be essentially the same. For the same reason, onset of unstable crack growth will occur when the stress intensity factor has overshot K_{ISS} (K_{Ic}) for a faster increase of the stress intensity factor. The faster the increase, the higher the overshoot.

Consider now a pulsed stress intensity factor, for instance in the form of half a period of a sine pulse. It is obvious that the stress intensity factor close to the peak will contribute more to crack initiation than lower stress intensity factors, and below some level there will be no contribution at all. If the pulse amplitude is sufficiently high and the pulse duration is sufficiently long, onset of unstable crack growth will occur. It may occur before or after the peak is reached, depending on the amplitude. If it occurs, it would do so later or not at all for a triangular stress pulse with the same amplitude and duration as the sine pulse.

The oscillation of the stress intensity factor after crack arrest may contribute toward a re-initiation during periods when some critical level is exceeded. Re-initiation, if it occurs, is therefore not necessarily limited to the first time the oscillating stress intensity factor exceeds this value. In cases when the oscillations depend on wave reflections at the body boundaries rather than on varying stress wave loading, it is obvious that the oscillation period is essentially proportional to the length dimensions of the body. The probability of re-initiation may therefore be much higher for crack arrest in a large plate than in a smaller, scaled model. If the oscillations are sufficiently persistent and their period sufficiently long, then more than one re-initiation followed by arrest may occur, as demonstrated by the Oak Ridge tests (page 657).

The crack arrest stress intensity factor

So far it has been assumed that the dynamic fracture toughness is a single-valued and monotone increasing function of the velocity. However, as discussed in Section 9.1, this may not be the case. For some materials the relation might not be monotone increasing. If the relation looks like the one shown in the right part of Fig. 9.1.6 for crack deceleration, then the stress intensity factor may reach values that are lower than the value required for onset of unstable crack growth. Arrest would then occur at the minimum value, K_I^{arrest}, provided that the maximum velocity exceeds the velocity for which the minimum occurs. If the dynamic fracture toughness would be single-valued, then the crack arrest fracture toughness, K_I^{arrest}, would be a material constant, that would be lower than or equal to the static fracture toughness, depending on whether the dynamic fracture toughness possesses a minimum for a non-zero velocity or not.

If the relation between dynamic fracture toughness and crack velocity is not single-valued, then the minimum encountered during deceleration is history dependent, and no definite material property of a crack arrest fracture toughness can be found. This was indicated by Hoagland et al. (1977) on the basis of experimental experience. The mechanism for crack arrest, on the other hand, may be essentially the same as described for the single-valued case. Note also that besides the material and the loading history, also size effects are important. Thus, for instance, the development of shear lips seems to increase the likelihood of crack arrest (Machida et al. 1986).

The lack of a unique relation between dynamic fracture toughness and crack velocity may play an important part for crack arrest. The relation may be different, not only between different loading conditions or different geometries, but also between the acceleration phase and the deceleration phase in one and the same crack propagation event (Arakawa and Takahashi 1991a). These facts appear to rule out the possibility of finding a crack arrest stress fracture toughness that is a pure material property.

The experiments by Kalthoff et al. (1977), resulted in a crack arrest stress intensity factor that is almost 10 per cent lower than the static fracture toughness and about the same for different load histories. The experiments were performed for double cantilever beam Araldite B specimens subjected to fixed grip (wedge) loading. After arrest, the stresses in the specimen reverted to their static values after some oscillations. The static stress intensity factor for the arrested crack was considerably lower than the arrest stress intensity factor. The oscillations occurred essentially around the static stress intensity factor and there was no re-initiation of crack growth.

The role of the kinetic energy in crack arrest

Hahn et al. (1973) and Bergkvist (1974) discussed crack arrest from the perspective of kinetic energy. Hahn et al. used beam theory to calculate the kinetic energy in a double cantilever beam during crack growth. They found that kinetic energy could play an important part in the mechanism of crack arrest. Bergkvist, on the other hand, discussed the case of a crack in a large plate, so that wave reflections from the plate boundaries were insignificant. An illustration, using a simplified version of the theory by Bergkvist, will be made.

Consider a large plate with clamped ends, in which an edge crack is slowly opened from zero length to the length a_1. The energy dissipation during this manoeuvre is

$$D_{stat} = \int_0^{a_1} J_I(a) \mathrm{d}a \qquad (9.4.4)$$

where $J_I(a)$ is the J-integral for a stationary crack with length a.

Assume now, for simplicity, that the crack had reached the length a_1 by growing with constant edge velocity $\dot{a}$ from zero length. Then, the energy dissipation would have been

$$D_{dyn} = \int_0^{a_1} J_I(a) w(\dot{a}) \mathrm{d}a \qquad (9.4.5)$$

where $w(\dot{a})$ is a velocity dependent factor, analogous to $w_1(\dot{a})$ in (9.1.2) for a central crack in a large plate. The difference between D_{stat} and D_{dyn} equals the kinetic energy

$$T(a) = \int_0^{a_1} J_I(a)[1 - w(\dot{a})]\mathrm{d}a \qquad (9.4.6)$$

Suppose now that the crack meets a suddenly increased static fracture toughness, for instance by a change of material, an abrupt change of temperature or an increased plate thickness, such that a slowly moving crack would have been arrested at this discontinuity. If a crack running with velocity $\dot{a}$ is arrested at the obstacle, some part of the energy (9.4.6) is available for propagating the crack further. Bergkvist (1974) assumed, conservatively, that all kinetic energy would be available, and he studied the case when the fracture toughness only increased over a relatively short distance, such as caused by a narrow stiffener, so that there might be a possibility for continued crack propagation on the other side of the obstacle.

Again conservatively, Bergkvist (1974) assumed that the continued motion proceeds with a very low velocity, because this would be most energy-economic if the specific energy dissipation $\Gamma(\dot{a})$ increases with velocity. Thus, for $a > a_1$,

$$J_I(a)w(0) + p(a) = \Gamma_{obst}(0) \qquad (9.4.7)$$

where $\Gamma_{obst}(0)$ is the static fracture toughness of the obstacle, $w(0) = 1$ and $p(a)$ is the part of the available kinetic energy that is used to assist the energy flux to propagate the crack from the obstacle boundary to the position a inside the obstacle.

Obviously, the crack edge cannot reach $a = a_2$ if

$$\int_{a_1}^{a_2} p(a)\mathrm{d}a > T(a) \qquad (9.4.8)$$

i.e.,

$$\int_{a_1}^{a_2} [\Gamma_{obst}(0) - J(a)]\mathrm{d}a > \int_0^{a_1} J_I(a)[1 - w(\dot{a})]\mathrm{d}a \qquad (9.4.9)$$

In experiments, using plate stiffeners (cf. Fig. 9.4.2), Bergkvist (1974) found crack arrest in all cases when this conservative criterion was fulfilled, and also crack propagation through the stiffeners in cases when it was not fulfilled, except in a few cases close to the criterion limit.

The considerations by Bergkvist (1974) clearly showed the role of kinetic energy in carrying the crack beyond an obstacle that would arrest a crack moving slowly under the same load on the structure.

Appendices

A1 Formulae and general relations

Vector operations

$$Cartesian: \quad \text{div}\,\boldsymbol{\psi} = \frac{\partial \psi_x}{\partial x} + \frac{\partial \psi_y}{\partial y} + \frac{\partial \psi_z}{\partial z}$$

$$\text{grad}\,\phi = \frac{\partial \phi}{\partial x}\hat{x} + \frac{\partial \phi}{\partial y}\hat{y} + \frac{\partial \phi}{\partial z}\hat{z}$$

$$\text{curl}\,\boldsymbol{\psi} = \left(\frac{\partial \psi_z}{\partial y} - \frac{\partial \psi_y}{\partial z}\right)\hat{x} + \left(\frac{\partial \psi_x}{\partial z} - \frac{\partial \psi_z}{\partial x}\right)\hat{y} + \left(\frac{\partial \psi_y}{\partial x} - \frac{\partial \psi_x}{\partial y}\right)\hat{z}$$

$$\Delta \phi = \frac{\partial^2 \phi}{\partial x^2} + \frac{\partial^2 \phi}{\partial y^2} + \frac{\partial^2 \phi}{\partial z^2}$$

$$Cylindrical: \quad \text{div}\,\boldsymbol{\psi} = \frac{1}{r}\frac{\partial (r\psi_r)}{\partial r} + \frac{1}{r}\frac{\partial \psi_\varphi}{\partial \varphi} + \frac{\partial \psi_z}{\partial z}$$

$$\text{grad}\,\phi = \frac{\partial \phi}{\partial r}\hat{r} + \frac{1}{r}\frac{\partial \phi}{\partial \varphi}\hat{\varphi} + \frac{\partial \phi}{\partial z}\hat{z}$$

$$\text{curl}\,\boldsymbol{\psi} = \left(\frac{1}{r}\frac{\partial \psi_z}{\partial \varphi} - \frac{\partial \psi_\varphi}{\partial z}\right)\hat{r} + \left(\frac{\partial \psi_r}{\partial z} - \frac{\partial \psi_z}{\partial r}\right)\hat{\varphi} + \left[\frac{1}{r}\frac{\partial}{\partial r}(r\psi_\varphi) - \frac{1}{r}\frac{\partial \psi_r}{\partial \varphi}\right]\hat{z}$$

$$\Delta \phi = \frac{\partial^2 \phi}{\partial r^2} + \frac{1}{r}\frac{\partial \phi}{\partial r} + \frac{1}{r^2}\frac{\partial^2 \phi}{\partial \varphi^2} + \frac{\partial^2 \phi}{\partial z^2}$$

$$Spherical: \quad \text{div}\,\boldsymbol{\psi} = \frac{1}{r^2}\frac{\partial}{\partial r}(r^2 \psi_r) + \frac{1}{r\sin\theta}\frac{\partial \psi_\varphi}{\partial \varphi} + \frac{1}{r\sin\theta}\frac{\partial (\sin\theta\, \psi_\theta)}{\partial \theta}$$

$$\text{grad}\,\phi = \frac{\partial \phi}{\partial r}\hat{r} + \frac{1}{r\sin\theta}\frac{\partial \phi}{\partial \varphi}\hat{\varphi} + \frac{1}{r}\frac{\partial \phi}{\partial \theta}\hat{\theta}$$

$$\text{curl}\,\boldsymbol{\psi} = \frac{1}{r\sin\theta}\left[\frac{\partial \psi_\theta}{\partial \varphi} - \frac{\partial (\sin\theta\, \psi_\varphi)}{\partial \theta}\right]\hat{r} + \frac{1}{r}\left[\frac{\partial \psi_r}{\partial \theta} - \frac{\partial (r\psi_\theta)}{\partial r}\right]\hat{\varphi}$$
$$+ \frac{1}{r}\left[\frac{\partial (r\psi_\varphi)}{\partial r} - \frac{1}{\sin\theta}\frac{\partial \psi_r}{\partial \varphi}\right]\hat{\theta}$$

$$\Delta \phi = \frac{1}{r^2}\frac{\partial}{\partial r}\left(r^2 \frac{\partial \phi}{\partial r}\right) + \frac{1}{r^2 \sin^2\theta}\frac{\partial^2 \phi}{\partial \varphi^2} + \frac{1}{r^2 \sin\theta}\frac{\partial}{\partial \theta}\left(\sin\theta\, \frac{\partial \phi}{\partial \theta}\right)$$

Displacements and stresses expressed by displacement potentials

$$\boldsymbol{u} = \operatorname{grad} \phi + \operatorname{curl} \boldsymbol{\psi}$$

Cartesian:
$$u = \frac{\partial \phi}{\partial x} + \frac{\partial \psi_z}{\partial y} - \frac{\partial \psi_y}{\partial z}$$
$$v = \frac{\partial \phi}{\partial y} + \frac{\partial \psi_x}{\partial z} - \frac{\partial \psi_z}{\partial x}$$
$$w = \frac{\partial \phi}{\partial z} + \frac{\partial \psi_y}{\partial x} - \frac{\partial \psi_x}{\partial y}$$

For plane strain, $u_z = 0$, or for a thin plate treated with the plane stress approximation $\sigma_z = \tau_{zx} = \tau_{zy} = 0$, the potentials $\psi_x = \psi_y = 0$.

Cylindrical:
$$u_r = \frac{\partial \phi}{\partial r} + \frac{1}{r}\frac{\partial \psi_z}{\partial \varphi} - \frac{\partial \psi_\varphi}{\partial z}$$
$$u_\varphi = \frac{1}{r}\frac{\partial \phi}{\partial \varphi} + \frac{\partial \psi_r}{\partial z} - \frac{\partial \psi_z}{\partial r}$$
$$u_z = \frac{\partial \phi}{\partial z} + \frac{1}{r}\frac{\partial(r\psi_\varphi)}{\partial r} - \frac{1}{r}\frac{\partial \psi_r}{\partial \varphi}$$

For plane strain, $u_z = 0$, or for a thin plate treated with the plane stress approximation $\sigma_z = \tau_{zx} = \tau_{zy} = 0$, the potentials $\psi_r = \psi_\varphi = 0$.

Spherical (r, φ, θ):
$$u_r = \frac{\partial \phi}{\partial r} + \frac{1}{r \sin \theta}\left[\frac{\partial \psi_\theta}{\partial \varphi} - \frac{\partial}{\partial \theta}(\sin \theta \, \psi_\varphi)\right]$$
$$u_\varphi = \frac{1}{r \sin \theta}\frac{\partial \phi}{\partial \varphi} + \frac{1}{r}\frac{\partial \psi_r}{\partial \theta} - \frac{1}{r}\frac{\partial(r\psi_\theta)}{\partial r}$$
$$u_\theta = \frac{1}{r}\frac{\partial \phi}{\partial \theta} + \frac{1}{r}\frac{\partial(r\psi_\varphi)}{\partial r} - \frac{1}{r \sin \theta}\frac{\partial \psi_r}{\partial \varphi}$$

Cartesian:
$$\sigma_x = \frac{\mu}{k^2}\left[(1-2k^2)\Delta\phi + 2k^2\frac{\partial}{\partial x}\left(\frac{\partial \phi}{\partial x} + \frac{\partial \psi_z}{\partial y} - \frac{\partial \psi_y}{\partial z}\right)\right]$$
$$\sigma_y = \frac{\mu}{k^2}\left[(1-2k^2)\Delta\phi + 2k^2\frac{\partial}{\partial y}\left(\frac{\partial \phi}{\partial y} + \frac{\partial \psi_x}{\partial z} - \frac{\partial \psi_z}{\partial x}\right)\right]$$
$$\tau_{xy} = \mu\left[\frac{\partial}{\partial y}\left(2\frac{\partial \phi}{\partial x} + \frac{\partial \psi_z}{\partial y} - \frac{\partial \psi_y}{\partial z}\right) + \frac{\partial}{\partial x}\left(\frac{\partial \psi_x}{\partial z} - \frac{\partial \psi_z}{\partial x}\right)\right]$$
$$\tau_{yz} = \mu\left[\frac{\partial}{\partial z}\left(2\frac{\partial \phi}{\partial y} + \frac{\partial \psi_x}{\partial z} - \frac{\partial \psi_z}{\partial x}\right) + \frac{\partial}{\partial y}\left(\frac{\partial \psi_y}{\partial x} - \frac{\partial \psi_x}{\partial y}\right)\right]$$
$$\tau_{zx} = \mu\left[\frac{\partial}{\partial x}\left(2\frac{\partial \phi}{\partial z} + \frac{\partial \psi_y}{\partial x} - \frac{\partial \psi_x}{\partial y}\right) + \frac{\partial}{\partial z}\left(\frac{\partial \psi_z}{\partial y} - \frac{\partial \psi_y}{\partial z}\right)\right]$$

In general, $k^2 = (1-2\nu)/[2(1-\nu)]$. For plane strain, $u_z = 0$, or for a thin plate treated with the plane stress approximation $\sigma_z = \tau_{zx} = \tau_{zy} = 0$, the stresses are

$$\sigma_x = \frac{\mu}{k^2}\left[\frac{\partial^2\phi}{\partial x^2} + (1-2k^2)\frac{\partial^2\phi}{\partial y^2} + 2k^2\frac{\partial^2\psi_z}{\partial x\partial y}\right]$$

$$\sigma_y = \frac{\mu}{k^2}\left[(1-2k^2)\frac{\partial^2\phi}{\partial x^2} + \frac{\partial^2\phi}{\partial y^2} - 2k^2\frac{\partial^2\psi_z}{\partial x\partial y}\right]$$

$$\tau_{xy} = \mu\left(2\frac{\partial^2\phi}{\partial x\partial y} + \frac{\partial^2\psi_z}{\partial y^2} - \frac{\partial^2\psi_z}{\partial x^2}\right)$$

In the plane stress approximation for a thin plate, $k^2 = (1-\nu)/2$.

Cylindrical:

$$\sigma_r = \frac{\mu}{k^2}\left[(1-2k^2)\Delta\phi + 2k^2\left(\frac{\partial^2\phi}{\partial r^2} + \frac{1}{r}\frac{\partial^2\psi_z}{\partial r\partial\varphi} - \frac{1}{r^2}\frac{\partial\psi_z}{\partial\varphi} - \frac{\partial^2\psi_\varphi}{\partial r\partial z}\right)\right]$$

$$\sigma_\varphi = \frac{\mu}{k^2}\Big[(1-2k^2)\Delta\phi$$
$$+2k^2\left(\frac{1}{r}\frac{\partial\phi}{\partial r} + \frac{1}{r^2}\frac{\partial^2\phi}{\partial\varphi^2} + \frac{1}{r}\frac{\partial^2\psi_r}{\partial\varphi\partial z} - \frac{1}{r}\frac{\partial^2\psi_z}{\partial r\partial\varphi} + \frac{1}{r^2}\frac{\partial\psi_z}{\partial\varphi} - \frac{1}{r}\frac{\partial\psi_\varphi}{\partial z}\right)\Big]$$

$$\sigma_z = \frac{\mu}{k^2}\left[(1-2k^2)\Delta\phi + 2k^2\left(\frac{\partial^2\phi}{\partial z^2} + \frac{1}{r}\frac{\partial^2(r\psi_\varphi)}{\partial r\partial z} - \frac{1}{r}\frac{\partial^2\psi_r}{\partial\varphi\partial z}\right)\right]$$

$$\tau_{r\varphi} = \mu\Big[\frac{2}{r}\frac{\partial^2\phi}{\partial r\partial\varphi} - \frac{2}{r^2}\frac{\partial\phi}{\partial\varphi} + \frac{\partial^2\psi_r}{\partial r\partial z} - \frac{\partial^2\psi_z}{\partial r^2} - \frac{1}{r}\frac{\partial\psi_r}{\partial z} + \frac{1}{r}\frac{\partial\psi_z}{\partial r}$$
$$+ \frac{1}{r^2}\frac{\partial^2\psi_z}{\partial\varphi^2} - \frac{1}{r}\frac{\partial^2\psi_\varphi}{\partial\varphi\partial z}\Big]$$

$$\tau_{\varphi z} = \mu\left[\frac{2}{r}\frac{\partial^2\phi}{\partial\varphi\partial z} + \frac{1}{r}\frac{\partial^2(r\psi_\varphi)}{\partial\varphi\partial r} - \frac{1}{r^2}\frac{\partial^2\psi_r}{\partial\varphi^2} + \frac{\partial^2\psi_r}{\partial z^2} - \frac{\partial^2\psi_z}{\partial r\partial z}\right]$$

$$\tau_{zr} = \mu\Big[2\frac{\partial^2\phi}{\partial r\partial z} + \frac{1}{r}\frac{\partial^2\psi_z}{\partial\varphi\partial z} - \frac{\partial^2\psi_\varphi}{\partial z^2} + \frac{1}{r}\frac{\partial^2(r\psi_\varphi)}{\partial r^2} - \frac{1}{r^2}\frac{\partial(r\psi_\varphi)}{\partial r}$$
$$- \frac{1}{r}\frac{\partial\psi_r}{\partial\varphi\partial r} + \frac{1}{r^2}\frac{\partial\psi_r}{\partial\varphi}\Big]$$

For plane strain, $u_z = 0$, or for a thin plate treated with the plane stress approximation $\sigma_z = \tau_{zx} = \tau_{zy} = 0$, the potentials $\psi_r = \psi_\varphi = 0$. In the plane stress approximation for a thin plate, $k^2 = (1-\nu)/2$.

Spherical:

$$\sigma_r = \frac{\mu}{k^2}\Big\{(1-2k^2)\Delta\phi + 2k^2\Big[\frac{\partial^2\phi}{\partial r^2} + \frac{1}{r\sin\theta}\left(\frac{\partial^2\psi_\theta}{\partial r\partial\varphi} - \frac{\partial^2}{\partial r\partial\theta}(\sin\theta\,\psi_\varphi)\right)$$
$$-\frac{1}{r^2\sin\theta}\left(\frac{\partial\psi_\theta}{\partial\varphi} - \frac{\partial}{\partial\theta}(\sin\theta\,\psi_\varphi)\right)\Big]\Big\}$$

$$\sigma_\varphi = \frac{\mu}{k^2}\Big\{(1-2k^2)\Delta\phi + 2k^2\Big[\frac{1}{r^2\sin^2\theta}\frac{\partial^2\phi}{\partial\varphi^2} + \frac{1}{r^2\sin^2\theta}\frac{\partial^2\psi_r}{\partial\varphi\partial\theta}$$
$$-\frac{1}{r^2\sin\theta}\frac{\partial^2(r\psi_\theta)}{\partial r\partial\varphi} + \frac{\cot\theta}{r^2}\frac{\partial\phi}{\partial\theta} + \frac{\cot\theta}{r^2}\frac{\partial(r\psi_\varphi)}{\partial r}$$
$$-\frac{\cot\theta}{r^2\sin\theta}\frac{\partial\psi_r}{\partial\varphi} + \frac{1}{r}\frac{\partial\phi}{\partial r} + \frac{1}{r^2\sin\theta}\Big(\frac{\partial\psi_\theta}{\partial\varphi} - \frac{\partial}{\partial\theta}(\sin\theta\,\psi_\varphi)\Big)\Big]\Big\}$$

$$\sigma_\theta = \frac{\mu}{k^2}\Big\{(1-2k^2)\Delta\phi + 2k^2\Big[\frac{1}{r^2}\frac{\partial^2\phi}{\partial\theta^2} + \frac{1}{r^2}\frac{\partial^2(r\psi_\varphi)}{\partial r\partial\theta} - \frac{1}{r\sin\theta}\frac{\partial^2\psi_r}{\partial\varphi\partial\theta}$$
$$+\frac{\cos\theta}{r^2\sin^2\theta}\frac{\partial\psi_r}{\partial\varphi} + \frac{1}{r}\frac{\partial\phi}{\partial r} + \frac{1}{r^2\sin\theta}\Big(\frac{\partial\psi_\theta}{\partial\varphi} - \frac{\partial}{\partial\theta}(\sin\theta\,\psi_\varphi)\Big)\Big]\Big\}$$

$$\tau_{r\varphi} = \mu\Big\{\frac{2}{r\sin\theta}\frac{\partial^2\phi}{\partial r\partial\varphi} + \frac{1}{r^2\sin^2\theta}\Big[\frac{\partial^2\psi_\theta}{\partial\varphi^2} - \frac{\partial^2}{\partial\varphi\partial\theta}(\sin\theta\,\psi_\varphi)\Big]$$
$$-\frac{2}{r^2\sin\theta}\frac{\partial\phi}{\partial\varphi} + \frac{1}{r}\frac{\partial^2\psi_r}{\partial r\partial\theta} - \frac{2}{r^2}\frac{\partial\psi_r}{\partial\theta} - \frac{1}{r}\frac{\partial^2(r\psi_\theta)}{\partial r^2} + \frac{2}{r^2}\frac{\partial(r\psi_\theta)}{\partial r}\Big\}$$

$$\tau_{\varphi\theta} = \mu\Big\{\frac{2}{r^2\sin\theta}\frac{\partial^2\phi}{\partial\varphi\partial\theta} + \frac{1}{r^2}\frac{\partial^2\psi_r}{\partial\theta^2} - \frac{1}{r^2}\frac{\partial^2(r\psi_\theta)}{\partial r\partial\theta} - \frac{\cot\theta}{r^2}\frac{\partial\phi}{\partial\theta}$$
$$-\frac{\cot\theta}{r^2}\frac{\partial(r\psi_\varphi)}{\partial r} + \frac{\cot\theta}{r^2\sin\theta}\frac{\partial\psi_r}{\partial\varphi} + \frac{1}{r^2\sin\theta}\frac{\partial^2(r\psi_\varphi)}{\partial r\partial\varphi}$$
$$-\frac{1}{r^2\sin^2\theta}\frac{\partial^2\psi_r}{\partial\varphi^2}\Big\}$$

$$\tau_{\theta r} = \mu\Big\{\frac{2}{r}\frac{\partial^2\phi}{\partial r\partial\theta} - \frac{2}{r^2}\frac{\partial\phi}{\partial\theta} + \frac{1}{r}\frac{\partial^2(r\psi_\varphi)}{\partial r^2} - \frac{2}{r^2}\frac{\partial(r\psi_\varphi)}{\partial r} - \frac{1}{r\sin\theta}\frac{\partial^2\psi_r}{\partial r\partial\varphi}$$
$$+\frac{2}{r^2\sin\theta}\frac{\partial\psi_r}{\partial\varphi} + \frac{1}{r^2\sin\theta}\Big[\frac{\partial^2\psi_\theta}{\partial\varphi\partial\theta} - \frac{\partial^2}{\partial\theta^2}(\sin\theta\,\psi_\varphi)\Big]$$
$$-\frac{\cos\theta}{r^2\sin^2\theta}\Big[\frac{\partial\psi_\theta}{\partial\varphi} - \frac{\partial}{\partial\theta}(\sin\theta\,\psi_\varphi)\Big]\Big\}$$

Elastic constants

μ = modulus of rigidity, ν = Poisson's ratio, κ = modulus of compression, E = modulus of elasticity, and λ is one Lamé constant (the other is μ).

$$k^2 = \frac{1-2\nu}{2(1-\nu)} = \frac{3\kappa}{3\kappa+4\mu}, \quad \nu = \frac{1-2k^2}{2(1-k^2)} = \frac{3\kappa-2\mu}{2(3\kappa+\mu)}$$

$$\kappa = \frac{E}{3(1-2\nu)} = \frac{2(1+\nu)\mu}{3(1-2\nu)} = \frac{3-4k^2}{3k^2}\mu = \frac{1-k^2}{3k^2}E$$

$$\kappa + \frac{4\mu}{3} = \frac{2(1-\nu)\mu}{1-2\nu} = \frac{(1-\nu)E}{2(1+\nu)(1-2\nu)} = \frac{\mu}{k^2} = \frac{(1-k^2)E}{k^2(3-4k^2)}$$

$$E = \frac{3\kappa\mu}{\kappa+\mu/3}$$

$$\lambda = \kappa - 2\mu/3 = \frac{2\nu\mu}{1-2\nu} = \frac{1-2k^2}{k^2}\mu = \frac{\nu E}{(1+\nu)(1-2\nu)} = \frac{3(1-2k^2)}{3-4k^2}\kappa$$

The fundamental wave propagation velocities:
$c_S =$ velocity of S waves and $c_P =$ velocity of P waves.

$$c_S^2 = \frac{\mu}{\varrho}, \quad c_P = \frac{c_S}{k}, \quad c_P^2 = \frac{2(1-\nu)\mu}{(1-2\nu)\varrho} = \frac{(1-\nu)E}{(1+\nu)(1-2\nu)\varrho} = \frac{\kappa + 4\mu/3}{\varrho}$$

Thin plate, treated by the plane stress approximation:

$$k^2 = \frac{1-\nu}{2} = \frac{3\kappa + 4\mu}{4(3\kappa + \mu)}, \quad \nu = 1 - 2k^2 = \frac{3\kappa - 2\mu}{3\kappa + 4\mu}$$

$$c_S^2 = \frac{\mu}{\varrho}, \quad c_P = \frac{c_S}{k}, \quad c_P^2 = \frac{2\mu}{(1-\nu)\varrho} = \frac{E}{(1-\nu^2)\varrho} = \frac{3(1-2\nu)\kappa}{(1-\nu^2)\varrho}$$

A2 The first boundary value problem for the half-plane

In-plane problem

Given: the half-plane $y \leq 0$, subjected to static tractions $\sigma_y = \sigma_y^0(x)$ and $\tau_{xy} = \tau_{xy}^0(x)$ on $y = 0$. Sought: Stresses and displacements in the half-plane. This fundamental problem is described by Muskhelishvili (1953a), and its solution is shown here, because of its importance for several mode I and II crack problems. Also the corresponding anti-plane strain problem is treated.

Use will be made of the complex representations (4.4.9)-(4.4.10) for stresses and (4.4.12) for the displacements. Addition of (4.4.10) and (4.4.9) yields

$$\sigma_y + i\tau_{xy} = f'_-(z) + \overline{f'_-(z)} + \bar{z}f''_-(z) + g''_-(z) \tag{A2.1}$$

where the subscript minus is used to indicate that the two analytic functions are defined for the lower half-plane, only. It is, however, convenient to define also a function that is analytic in the upper half-plane:

$$f'_+(z) = -\overline{f'_-}(z) - z\overline{f''_-}(z) - \overline{g''_-}(z) \tag{A2.2}$$

The analyticity follows from the fact that if $F(z)$ is analytic in the lower half-plane, then $\overline{F}(z)$ is analytic in the upper half-plane (and *vice versa*). This is immediately obvious: in the vicinity of each point $z = z_0$ in the lower half-plane, the regularity of $F(z)$ implies that

$$F(z) = a_0 + a_1(z - z_0) + a_2(z - z_0)^2 + \cdots \tag{A2.3}$$

within some radius of convergence. Then, in the vicinity of the point $z = \overline{z_0}$ in the upper half-plane, the function

$$\overline{a_0} + \overline{a_1}(z - \overline{z_0}) + \overline{a_2}(z - \overline{z_0})^2 + \cdots = \overline{F}(z) \tag{A2.4}$$

is convergent within some (actually the same) radius. Hence, $\overline{F}(z)$ is regular at each point $z = \overline{z_0}$ in the upper half-plane.

Insertion of

$$g''_-(z) = -\overline{f'_+}(z) - f'_-(z) - zf''_-(z) \tag{A2.5}$$

A2 THE FIRST BOUNDARY VALUE PROBLEM FOR THE HALF-PLANE

into (A2.1) gives, after complex conjugation,

$$\sigma_y - i\tau_{xy} = f'_-(z) - \overline{f'_+(\bar{z})} + (z - \bar{z})\overline{f''_-(z)} \quad (A2.6)$$

and, hence, the boundary condition can be written in the form:

$$f'_+(x) - f'_-(x) = -\sigma_y^0 + i\tau_{xy}^0 \quad \text{for } y = 0 \quad (A2.7)$$

With $f(z)$ denoting the sectionally analytic function that equals $f_-(z)$ in the lower half-plane and $f_+(z)$ in the upper half-plane, Plemelj's formulae, page 677, give

$$f'(z) = \frac{1}{2\pi i}\int_{-\infty}^{+\infty}\frac{i\tau_{xy}^0(\xi) - \sigma_y^0(\xi)}{\xi - z}\,d\xi = \frac{1}{2\pi}\int_{-\infty}^{+\infty}\frac{\tau_{xy}^0(\xi) + i\sigma_y^0(\xi)}{\xi - z}\,d\xi \quad (A2.8)$$

because no additional polynomial has to be added if $\sigma_y - i\tau_{xy} \to 0$ as $|z| \to \infty$, $\Im z \leq 0$.

For simplicity, it is temporarily assumed that the load consists of normal and shear forces on the surface, concentrated to the line $x = 0$:

$$\sigma_y^0(x) = S_y\delta(x), \qquad \tau_{xy}^0(x) = T_{xy}\delta(x) \quad (A2.9)$$

where $\delta(x)$ is the Dirac delta function. Then,

$$f'_-(z) = -\frac{1}{2\pi}\cdot\frac{T_{xy} + iS_y}{z}, \qquad \overline{f'_+(\bar{z})} = -\frac{1}{2\pi}\cdot\frac{T_{xy} + iS_y}{\bar{z}} \quad (A2.10)$$

which, according to (A2.6) gives

$$\sigma_y - i\tau_{xy} = -\frac{2y}{\pi}\cdot\frac{(y - ix)(S_y y + T_{xy} x)}{(x^2 + y^2)^2} \quad (A2.11)$$

Further, according to (4.4.9),

$$\sigma_x + \sigma_y = 4\Re f'_-(z) = -\frac{2}{\pi}\cdot\frac{S_y y + T_{xy} x}{x^2 + y^2} \quad (A2.12)$$

Thus, the stresses are:

$$\sigma_x = -\frac{2x^2(S_y y + T_{xy} x)}{\pi(x^2 + y^2)^2}, \qquad \sigma_y = -\frac{2y^2(S_y y + T_{xy} x)}{\pi(x^2 + y^2)^2} \quad (A2.13)$$

$$\tau_{xy} = -\frac{2xy(S_y y + T_{xy} x)}{\pi(x^2 + y^2)^2} \quad (A2.14)$$

Note that $y \leq 0$. Before returning to distributed loads, the result will be given in cylindrical coordinates (r, φ, z) such that $\varphi = 0$ coincides with the negative y axis. Transformation gives:

$$\sigma_r = \frac{2(S_y \cos\varphi - T_{xy} \sin\varphi)}{\pi r}, \qquad \sigma_\varphi = \tau_{r\varphi} = 0 \quad (A2.15)$$

The fact that σ_φ and $\tau_{r\varphi}$ vanish shows that the solution also can be used for a wedge, $-\alpha \leq \varphi \leq \alpha$, after consideration of the equilibrium condition that σ_r shall balance the forces S_y and T_{xy} at the apex of the wedge. Thus, for the wedge it is found that

$$\sigma_r = \frac{2S_y \cos\varphi}{\pi r(2\alpha + \sin 2\alpha)} - \frac{2T_{xy}\sin\varphi}{\pi r(2\alpha - \sin 2\alpha)}, \qquad \sigma_\varphi = \tau_{r\varphi} = 0 \quad (A2.16)$$

Now, by returning to Cartesian coordinates and distributed loads,
$$S_y \to \sigma_y^0(x)dx, \qquad T_{xy} \to \tau_{xy}^0(x)dx \tag{A2.17}$$
the stresses are found to be
$$\sigma_x = -\frac{2}{\pi}\int_{-\infty}^{+\infty} \frac{[\sigma_y^0(\xi)y + \tau_{xy}^0(\xi)(x-\xi)](x-\xi)^2}{[(x-\xi)^2 + y^2]^2}d\xi \tag{A2.18}$$
$$\sigma_y = -\frac{2y^2}{\pi}\int_{-\infty}^{+\infty} \frac{\sigma_y^0(\xi)y + \tau_{xy}^0(\xi)(x-\xi)}{[(x-\xi)^2 + y^2]^2}d\xi \tag{A2.19}$$
$$\tau_{xy} = -\frac{2y}{\pi}\int_{-\infty}^{+\infty} \frac{[\sigma_y^0(\xi)y + \tau_{xy}^0(\xi)(x-\xi)](x-\xi)}{[(x-\xi)^2 + y^2]^2}d\xi \tag{A2.20}$$
for $y < 0$.

Using (4.4.12) and (A2.5), the displacement gradients are found to be
$$2\mu\left(\frac{\partial u}{\partial x} + i\frac{\partial v}{\partial x}\right)_- = \frac{1+k^2}{1-k^2}f'_-(z) + f'_+(\bar{z}) - (z-\bar{z})\overline{f''_-(z)} \tag{A2.21}$$
so that for $y = 0$, using (A2.8),
$$2\mu\left(\frac{\partial u}{\partial x} + i\frac{\partial v}{\partial x}\right)_- = \frac{1+k^2}{1-k^2}f'_-(x) + f'_+(x)$$
$$= \frac{i}{\pi(1-k^2)}\fint_{-\infty}^{+\infty} \frac{\sigma_y^0(\xi) - i\tau_{xy}^0(\xi)}{\xi - x}d\xi - \frac{k^2}{1-k^2}[\sigma_y^0(x) - i\tau_{xy}^0(x)] \tag{A2.22}$$
where the integral is to be taken in the sense of the Cauchy principal value.

Anti-plane problem

The corresponding anti-plane problem, where the shear stress $\tau_{yz} = \tau_{yz}^0(x)$ on the surface $y = 0$ of a half-plane $y \leq 0$, can be solved by using the complex potential representation (4.3.6),
$$\tau_{yz} + i\tau_{xz} = \mu\left(\frac{\partial w}{\partial y} + i\frac{\partial w}{\partial x}\right) = f'_-(z) \tag{A2.23}$$
where subscript minus indicates that $f_-(z)$ is defined for $\Im z \leq 0$. Define the function $f_+(z) = -\overline{f_-(\bar{z})}$, which is analytic for $\Im z > 0$. Then, for $y = 0$,
$$\tau_{yz} = \frac{1}{2}[f_-(x) + \overline{f_-(x)}] = \frac{1}{2}[f_-(x) - f_+(x)] = \tau_{yz}^0(x) \tag{A2.24}$$
i.e.,
$$f_+(x) - f_-(x) = -2\tau_{yz}^0(x) \tag{A2.25}$$
From this equation, the sectionally analytic function $f(z)$, which equals $f_-(z)$ in the lower and $f_+(z)$ in the upper half-plane, is determined by the Plemelj formulae; see page 677:
$$f(z) = -\frac{1}{\pi i}\int_{-\infty}^{+\infty} \frac{\tau_{yz}^0(\xi)}{\xi - z}d\xi \tag{A2.26}$$

No polynomial has to be added if $\tau_{yz} \to 0$ as $|z| \to \infty$. $\Im z < 0$ gives $f_-(z)$, $\Im z > 0$ gives $f_+(z)$. Letting $\Im z \to 0$ results in

$$f_\pm(x) = -\frac{1}{\pi i} \fint_{-\infty}^{+\infty} \frac{\tau_{yz}^0(\xi)}{\xi - x} d\xi \mp \tau_{yz}^0(x) \tag{A2.27}$$

Thus, from (A2.23), the stresses and the displacement gradients are (recall that $y < 0$)

$$\tau_{yz} = \mu \frac{\partial w}{\partial y} = -\frac{y}{\pi} \int_{-\infty}^{+\infty} \frac{\tau_{yz}^0(\xi)}{(x-\xi)^2 + y^2} d\xi \tag{A2.28}$$

$$\tau_{xz} = \mu \frac{\partial w}{\partial x} = -\frac{1}{\pi} \int_{-\infty}^{+\infty} \frac{(x-\xi)\tau_{yz}^0(\xi)}{(x-\xi)^2 + y^2} d\xi \tag{A2.29}$$

which for $y = 0$ default to

$$\tau_{yz} = \mu \frac{\partial w}{\partial y} = \tau_{yz}^0(x) \tag{A2.30}$$

$$\tau_{xz} = \mu \frac{\partial w}{\partial x} = \frac{1}{\pi} \fint_{-\infty}^{+\infty} \frac{\tau_{yz}^0(\xi)}{\xi - x} d\xi \tag{A2.31}$$

where the integral is to be taken in the sense of the Cauchy principal value.

A3 Some formulae related to the Airy stress function

Representation of stresses

Using cylindrical coordinates, the stresses are represented with the Airy stress function Φ by the expressions

$$\sigma_r = \frac{1}{r} \cdot \frac{\partial \Phi}{\partial r} + \frac{1}{r^2} \cdot \frac{\partial^2 \Phi}{\partial \varphi^2} \tag{A3.1}$$

$$\sigma_\varphi = \frac{\partial^2 \Phi}{\partial r^2} \tag{A3.2}$$

$$\tau_{r\varphi} = -\frac{\partial}{\partial r}\left(\frac{1}{r} \cdot \frac{\partial \Phi}{\partial \varphi}\right) \tag{A3.3}$$

and, assuming linear elasticity,

$$\sigma_z = \begin{cases} 0 & \text{for plane stress} \\ \nu(\sigma_r + \sigma_\varphi) & \text{for plane strain} \end{cases} \tag{A3.4}$$

Representation of strains and displacements at linear elasticity

Representations for the strains are obtained by using Hooke's law,

$$\epsilon_r = \frac{1}{4(1-k^2)\mu}[\sigma_r - (1-2k^2)\sigma_\varphi] \tag{A3.5}$$

$$\epsilon_\varphi = \frac{1}{4(1-k^2)\mu}[\sigma_\varphi - (1-2k^2)\sigma_r] \tag{A3.6}$$

$$\gamma_{r\varphi} = \frac{1}{\mu}\tau_{r\varphi} \tag{A3.7}$$

where k is related to Poisson's ratio ν by the equations

$$k^2 = \begin{cases} \dfrac{1-2\nu}{2(1-\nu)} & \text{for plane strain} \\ \dfrac{1-\nu}{2} & \text{for plane stress} \end{cases} \quad (A3.8)$$

Representations for the displacements u_r and u_φ can then be found by using the expressions

$$\frac{\partial u_r}{\partial r} = \epsilon_r, \quad \frac{1}{r} \cdot \frac{\partial u_\varphi}{\partial \varphi} + \frac{u_r}{r} = \epsilon_\varphi, \quad \frac{\partial u_\varphi}{\partial r} - \frac{u_\varphi}{r} + \frac{1}{r} \cdot \frac{\partial u_r}{\partial \varphi} = \gamma_{r\varphi} \quad (A3.9)$$

Equation and basic solutions

For an isotropic linearly elastic medium, the Airy stress function satisfies the biharmonic equation

$$\Delta\Delta\Phi = 0 \quad (A3.10)$$

The general solution can be expressed as:

$$\Phi(r,\varphi) = \Psi_0(r,\varphi) + \Psi_1(r,\varphi) r\cos\varphi + \Psi_2(r,\varphi) r\sin\varphi + \Psi_3(r,\varphi) r^2 \quad (A3.11)$$

where $\Psi_i(r,\varphi)$ are harmonic functions, i.e. solutions to the Laplace equation. Useful examples of solutions to the biharmonic equation are $\ln r$, r^2, $r^2 \ln r$, φ, $\ln r \cdot \varphi$, $r^2 \varphi$, $r^2 \ln r \cdot \varphi$, $r\varphi \cos\varphi$, $r\varphi \sin\varphi$, $r^s \cos(s\varphi)$, $r^{s+2}\cos(s\varphi)$, $r^s \sin(s\varphi)$, $r^{s+2}\sin(s\varphi)$, $r \ln r \cos\varphi$ and $r \ln r \sin\varphi$, where s is any real number.

Solutions for the vicinity of crack edges or sharp notches

Solutions in the form $\Phi = r^{s+2} F(\varphi)$ are sought. Insertion gives

$$F(\varphi) = A\cos(s\varphi) + B\cos[(s+2)\varphi] + C\sin(s\varphi) + D\sin[(s+2)\varphi] \quad (A3.12)$$

Then,

$$\sigma_r = r^s \left[(s+2)F + \frac{d^2 F}{d\varphi^2}\right] \quad (A3.13)$$

$$\sigma_\varphi = r^s(s+1)(s+2)F \quad (A3.14)$$

$$\tau_{r\varphi} = -r^s(s+1)\frac{dF}{d\varphi} \quad (A3.15)$$

$$\gamma_{r\varphi} = -\frac{r^s}{\mu}(s+1)\frac{dF}{d\varphi} \quad (A3.16)$$

$$\frac{\partial u_r}{\partial r} = \frac{r^s}{4(1-k^2)}\left\{(s+2)[2k^2(s+1)-s]F + \frac{d^2 F}{d\varphi^2}\right\} \quad (A3.17)$$

$$\frac{\partial u_\varphi}{\partial r} = \frac{r^s}{4(1-k^2)s}\left\{[2k^2 s(s+1) - 3(s+1)^2 - 1]\frac{dF}{d\varphi} - \frac{d^3 F}{d\varphi^3}\right\} \quad (A3.18)$$

A4 Analytic functions

Definition. Cauchy-Riemann equations

A general complex function (of x, y) is:

$$\phi(x, y) + i\psi(x, y) \tag{A4.1}$$

where ϕ and ψ are real functions. The function $\phi + i\psi$ possesses a value at each point z of the complex plane $z = x + iy$. Thus, formally,

$$\phi(x, y) + i\psi(x, y) = F(z) \tag{A4.2}$$

What happens if z is given the addition $dz = dx + i\,dy$? x will get the addition dx and y the addition dy. Thus, F will get the addition

$$dF = \frac{\partial \phi}{\partial x}dx + \frac{\partial \phi}{\partial y}dy + i\frac{\partial \psi}{\partial x}dx + i\frac{\partial \psi}{\partial y}dy = \left(\frac{\partial \phi}{\partial x} + i\frac{\partial \psi}{\partial x}\right)dx + \left(\frac{\partial \psi}{\partial y} - i\frac{\partial \phi}{\partial y}\right)i\,dy \tag{A4.3}$$

Two complex functions are obviously in general needed to determine dF, namely:

$$\frac{\partial \phi}{\partial x} + i\frac{\partial \psi}{\partial x} \quad \text{and} \quad \frac{\partial \psi}{\partial y} - i\frac{\partial \phi}{\partial y} \tag{A4.4}$$

But *if* these functions are equal – and only then – it is possible to write

$$dF = \left(\frac{\partial \phi}{\partial x} + i\frac{\partial \psi}{\partial x}\right)(dx + idy) = \left(\frac{\partial \phi}{\partial x} + i\frac{\partial \psi}{\partial x}\right)dz \tag{A4.5}$$

so that, formally $dF = F'(z)dz$, where

$$F'(z) = \frac{\partial \phi}{\partial x} + i\frac{\partial \psi}{\partial x} \tag{A4.6}$$

is the *derivative* of $F(z)$.

The condition under which the functions are equal,

$$\frac{\partial \phi}{\partial x} + i\frac{\partial \psi}{\partial x} = \frac{\partial \psi}{\partial y} - i\frac{\partial \phi}{\partial y} \tag{A4.7}$$

can be partitioned into one real and one imaginary part:

$$\boxed{\begin{aligned} \frac{\partial \phi}{\partial x} &= \frac{\partial \psi}{\partial y} \\ \frac{\partial \phi}{\partial y} &= -\frac{\partial \psi}{\partial x} \end{aligned}} \tag{A4.8}$$

These equations are called the *Cauchy-Riemann differential equations*. The functions $F(z) = \phi(x, y) + i\psi(x, y)$, for which these equations are satisfied, are called *analytic functions*. They constitute a very important subclass of the wider class of complex functions, and are simply *functions of z which can be differentiated with respect to z*. The proof can be reversed to show that the Cauchy-Riemann differential equations constitute a necessary condition for this property.

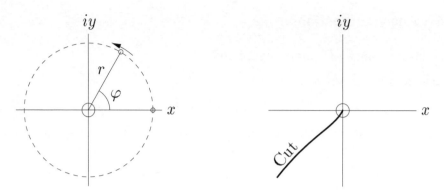

Fig. A4.1 Walking around a branch point.   **Fig. A4.2** Possible branch cut for $z^{1/2}$.

From the Cauchy-Riemann equations, it follows that

$$\boxed{\begin{aligned}\Delta\phi &= 0\\ \Delta\psi &= 0\end{aligned}} \qquad (A4.9)$$

for an analytic function. ϕ and ψ (real and imaginary part of an analytic function) thus satisfy the Laplace differential equation. Such functions are called *harmonic functions*.

Singular points

A function may be analytic in a certain region except for a finite or infinite number of *singular points*. Such singularities may be either poles or branch points.

A singularity $z = a$ is called a *pole* if, in the vicinity of a,

$$f(z) \approx \frac{A}{(z-a)^n} \qquad (A4.10)$$

where n is a positive integer. If $n = 1$ it is called a *simple pole*.

A singularity $z = a$ is called a *branch point* if the function is not single-valued when its argument encircles the point. Examples are given by

$$f(z) \approx A(z-a)^\alpha, \quad \alpha \text{ not an integer} \qquad (A4.11)$$
$$f(z) \approx A\ln(z-a) \qquad (A4.12)$$

Consider the specific example, $f(z) = z^{1/2}$. Prescribe that $z^{1/2} = \sqrt{x}\,(>0)$ on the positive real axis†. Then,

$$z^{1/2} = (re^{i\varphi})^{1/2} = \sqrt{r}e^{i\varphi/2} \qquad (A4.13)$$

$\varphi=0$ gives $z^{1/2} = \sqrt{r} > 0$. Now, walk around $z = 0$ which implies that φ increases from zero to 2π; see Fig. A4.1. One returns to the point $z = r$, but $z^{1/2} = -\sqrt{r} < 0$, violating the prescription. If one walks around once again, one comes back to $z^{1/2} = \sqrt{r} > 0$. Thus, $f(z) = z^{1/2}$ is not single-valued. It is, however, possible to arrange it so that no ambiguity persists.

---

† In the present work the square root sign is only used to denote a real and positive quantity.

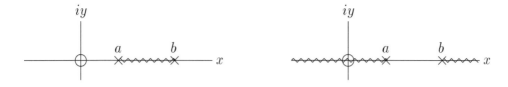

Fig. A4.3 Example of branch cut.

Fig. A4.4 Another possibility than in the previous figure.

This is done by making a *branch cut* along some path from $z = 0$ to infinity, see Fig. A4.2, and prescribing that the cut may not be traversed. Then, $f(z)$ becomes single-valued. Note that the prescription $z^{1/2} = \sqrt{x}\,(>0)$ on the positive real axis implies that a choice of function branch is made. In this case there is another possibility: to choose the branch for which $z^{1/2} = -\sqrt{x}\,(<0)$ on the positive real axis.

Another example is

$$f(z) = (z-a)^{1/2}(z-b)^{1/2} \tag{A4.14}$$

where a and b are real and $b > a$. Make a branch cut, for instance straight, from a to b, see Fig. A4.3, and choose the branch for which $f(z)$ equals $\sqrt{(x-a)(x-b)}$ for $x > b$, $y = 0$ (or, equivalently, $f(z) \to z$ as $z \to \infty$). The other possibility is $-\sqrt{(x-a)(x-b)}$ for $x > b$, $y = 0$ (or, equivalently, $f(z) \to -z$ as $z \to \infty$). I prefer to use a zig-zag line to mark a straight branch cut to avoid confusion, particularly if it coincides with part of a coordinate axis. Note that, as in the previous example, $f(z)$ becomes single-valued, if one does not cross the cut. In this case, φ increases with 2π when one walks around b and then further with 2π when one walks around a.

The cut does not necessarily need to go between a and b. Alternatively two cuts can be made, according to Fig. A4.4 with suitable choice of the branch. Also in this case $f(z)$ becomes single-valued, of course, but takes on different values (different sign) at some places than for a cut that goes from a to b.

A third example is $f(z) = \ln(z-a)$ with a straight branch cut from $z = a$ to $z = -\infty$. Choose the branch so that $f(z) = \ln|x - a|$ for $x > a$, $y = 0$ (the other possibilities are $\ln|x-a| + 2n\pi i$, n integer). Writing $f(z) = \ln(|z-a|e^{i\varphi}) = \ln|z-a| + i\varphi$, shows the difference $f(x+0) - f(x-0) = 2\pi i$ for $x < 0$, i.e. a jump across the branch cut.

For a function that is analytic in the whole plane except at isolated singularities, the singularities and the branch cuts are very important. In fact they determine the function uniquely if it tends toward zero at infinity, and if it does not it is determined by the singularities and the branch cuts, together with its behaviour at infinity.

In the following, some important properties of analytic functions will be shown.

Power series representation. Analytic continuation

If z_0 is a point in the region where $F(z)$ is analytic and $F(z_0) = a_0$, then

$$F(z) = a_0 + a_1(z - z_0) + a_2(z - z_0)^2 + a_3(z - z_0)^3 + \cdots \tag{A4.15}$$

for $|z - z_0| < R$, where R is the radius of convergence.

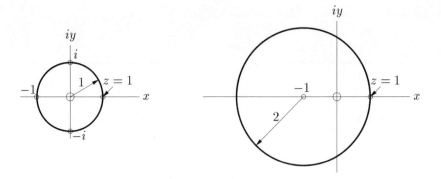

Fig. A4.5 Region of convergence for two power series.

This is an ordinary power series (Taylor series). Example:

$$F(z) = \frac{1}{1-z} = 1 + z + z^2 + z^3 + \cdots \quad \text{for } |z| < 1 \qquad (A4.16)$$

The radius $R = 1$ because the point $z = 1$ is a *singular point*. ($F(z)$ is not analytic at this point.)

For another choice, $z_0 = -1$, see Fig. A4.5,

$$F(z) = \frac{1}{2} + \frac{1}{2^2}(z+1) + \frac{1}{2^3}(z+1)^2 + \cdots \quad \text{for } |z+1| < 2 \qquad (A4.17)$$

The result, $R = 2$, is (still) due to the fact that $z = 1$ is a singular point.

Note that the expression $F_1(z) = 1/(1-z)$ defines a function which is analytic in the *whole plane* except at $z = 1$. But the expression $F_2(z) = 1 + z + z^2 + z^3 \cdots$ defines a function which is analytic only in $|z| < 1$. Now, $F_2(z) \equiv F_1(z)$ in $|z| < 1$. Then, is it possible to extend the region of analyticity for $F_2(z)$ to the same region as for $F_1(z)$ (i.e. to the whole plane except at the point $z = 1$)?

It is. This is called *analytic continuation*. One procedure is the following:

Assume that $F_2(z)$ is defined in $|z| < 1$, see Fig. A4.6. Develop $F_2(z)$ in a Taylor series around, say, $z = -1/2$:

$$F_2(z) = \frac{2}{3} + \left(\frac{2}{3}\right)^2 \left(z + \frac{1}{2}\right) + \left(\frac{2}{3}\right)^3 \left(z + \frac{1}{2}\right)^2 + \left(\frac{2}{3}\right)^4 \left(z + \frac{1}{2}\right)^3 + \cdots =$$

$$= \frac{2}{3} \cdot \frac{1}{1 - \frac{2}{3}(z+1/2)} = \frac{1}{1-z} \qquad (A4.18)$$

The series converges if the ratio

$$\left|\frac{2}{3}\left(z + \frac{1}{2}\right)\right| < 1, \quad \text{i.e.,} \quad \left|z + \frac{1}{2}\right| < \frac{3}{2} \qquad (A4.19)$$

Through this procedure $F_2(z)$ became extended over the lightly shadowed region in Fig. A4.7. The procedure can be repeated through development around, for instance, $z = i$ etc., until $F_2(z)$ has been extended over the whole plane, except for the singular point $z = 1$.

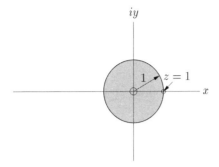

Fig. A4.6 Original region of definition.   **Fig. A4.7** Analytic continuation.

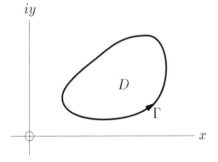 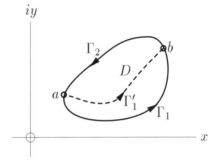

Fig. A4.8 Closed curve in the z plane.   **Fig. A4.9** Different paths of integration.

Connection with Fourier series

If $z_0 = 0$ and $F(z)$ is analytic for $|z| \leq 1$ then

$$F(z) = a_0 + a_1 z + a_2 z^2 + a_3 z^3 + \cdots, \quad |z| \leq 1 \tag{A4.20}$$

Now, put $z = re^{i\varphi}$. Then,

$$F(z) = a_0 + a_1 r e^{i\varphi} + a_2 r^2 e^{2i\varphi} + a_3 r^3 e^{3i\varphi} + \cdots \tag{A4.21}$$

This is a *Fourier series*! In particular, for $|z| = 1$, i.e. $r = 1$:

$$F(z) = a_0 + a_1 e^{i\varphi} + a_2 e^{2i\varphi} + a_3 e^{3i\varphi} + \cdots \tag{A4.22}$$

$$z = e^{i\varphi}, \quad 1/z = e^{-i\varphi} = \bar{z} \tag{A4.23}$$

Cauchy's integral theorem

Let Γ be a closed curve, and D the region inside Γ, see Fig. A4.8. Positive direction when walking along the curve is chosen so that D is on the left side of Γ. If $F(z)$ is analytic in D and on Γ then Cauchy's integral theorem states that

$$\boxed{\int_\Gamma F(z)\,\mathrm{d}z = 0} \tag{A4.24}$$

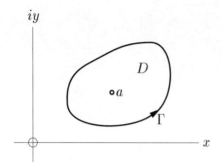

Fig. A4.10 Path of integration encircling the point a.

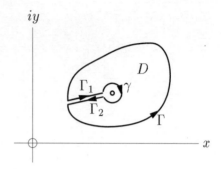

Fig. A4.11 Shrinking the path of integration.

One immediate interpretation of Cauchy's integral theorem is that (see Fig. A4.9):

$$\int_{\Gamma_1} F(z)\mathrm{d}z = -\int_{\Gamma_2} F(z)\mathrm{d}z, \quad \int_{\Gamma_1'} F(z)\mathrm{d}z = -\int_{\Gamma_2} F(z)\mathrm{d}z \qquad (A4.25)$$

$$\int_{\Gamma_1'} F(z)\mathrm{d}z = \int_{\Gamma_1} F(z)\mathrm{d}z = \int_a^b F(z)\mathrm{d}z \qquad (A4.26)$$

i.e., $\int_a^b F(z)\mathrm{d}z$ is determined by a and b, not by the orientation of the path between! Compare potential theory!

Cauchy's integral formula

Assume that $f(z)$ is analytic in D, including the boundary Γ, see Fig. A4.10. Consider the so-called Cauchy integral

$$I = \int_\Gamma \frac{f(\zeta)}{\zeta - a}\mathrm{d}\zeta \qquad (A4.27)$$

where ζ are points on Γ and a an arbitrary point inside, but not on, Γ. Because $f(z)$ is analytic in D and $(z-a)^{-1}$ is analytic in D except for $z = a$, it is possible to "shrink" the path of integration. Thus, see Fig. A4.11,

$$\int_{\Gamma+\Gamma_1-\gamma+\Gamma_2} \frac{f(\zeta)}{\zeta - a}\mathrm{d}\zeta = 0 \qquad (A4.28)$$

because the boundary encloses a region in which $f(z)/(z-a)$ is analytic. But $\int_{\Gamma_1+\Gamma_2} = 0$. Thus,

$$I = \int_\Gamma \frac{f(\zeta)}{\zeta - a}\mathrm{d}\zeta = \int_\gamma \frac{f(\zeta)}{\zeta - a}\mathrm{d}\zeta \qquad (A4.29)$$

i.e., the path of integration can be shrunk to γ.

Choose γ circular around a, with radius r. Then,

$$\zeta - a = re^{i\varphi} \tag{A4.30}$$

$$f(z) = f(a) + \frac{1}{1!}f'(a)(z-a) + \frac{1}{2!}f''(a)(z-a)^2 + \cdots \tag{A4.31}$$

Thus,

$$I = \int_\gamma \frac{f(a)}{\zeta - a} d\zeta + \frac{1}{1!}\int_\gamma f'(a) d\zeta + \frac{1}{2!}\int_\gamma f''(a)(\zeta - a) d\zeta + \cdots$$

$$= f(a)\int_0^{2\pi} \frac{1}{r}e^{-i\varphi} rie^{i\varphi} d\varphi + 0 + 0 + \cdots = 2\pi i f(a) \tag{A4.32}$$

i.e.,

$$\int_\Gamma \frac{f(\zeta)}{\zeta - a} d\zeta = 2\pi i f(a) \tag{A4.33}$$

But a is an arbitrary point inside Γ. Put $a = z$:

$$\boxed{f(z) = \frac{1}{2\pi i}\int_\Gamma \frac{f(\zeta)}{\zeta - z} d\zeta} \tag{A4.34}$$

This is Cauchy's integral formula. Thus, a function $f(z)$ which is analytic inside D can be represented with the aid of the Cauchy integral, which uses only values of $f(z)$ on the boundary (Γ). The formula also shows that a function that is analytic inside a contour Γ is completely determined by its values on Γ.

In the same way it can be shown that

$$\frac{1}{2\pi i}\int_\Gamma \frac{f(\zeta)}{(\zeta - z)^2} d\zeta = f'(z) \tag{A4.35}$$

and, in general

$$\frac{n!}{2\pi i}\int_\Gamma \frac{f(\zeta)}{(\zeta - z)^{n+1}} d\zeta = f^{(n)}(z) \tag{A4.36}$$

Thus, as seen from (A4.36), the right member of

$$f(z) = \frac{1}{2\pi i}\int_\Gamma \frac{f(\zeta)}{\zeta - z} d\zeta \tag{A4.37}$$

may be differentiated inside the integral sign.

Assume now that a is an arbitrary point outside Γ. Then, obviously, the Cauchy integral vanishes:

$$\int_\Gamma \frac{f(\zeta)}{\zeta - a} d\zeta = 0 \tag{A4.38}$$

An important consequence of Cauchy's formula follows from the generalization to integration of a function $f(z)$, analytic inside Γ, except for a number of poles, of which those at $z = a_1, a_2, \ldots a_n$ are simple poles. The path can be shrunk to small residual circular paths around each pole. Integration around these paths gives non-zero contribution from the simple poles, so that $\int_\Gamma f(\zeta) d\zeta = 2\pi i(a_1 + a_2 + \cdots a_n)$. This forms the basis for integral calculation by *residue calculus*.

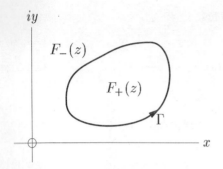

Fig. A4.12 Sectionally analytic function.

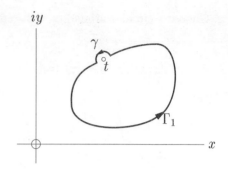

Fig. A4.13 Deformation of the path of integration as $z \to t$ from the inside.

Another consequence is that a function that is analytic inside Γ is completely determined by its values on Γ.

Plemelj's formulae

Assume now that $\varphi(\zeta)$ is a complex function given only on the boundary Γ. $\varphi(\zeta)$ may be discontinuous at some points ζ, i.e. either the real or the imaginary part (or both) may be discontinuous. Then, it is possible to show that

$$F(z) = \frac{1}{2\pi i} \int_\Gamma \frac{\varphi(\zeta)}{\zeta - z} d\zeta, \quad z \text{ not on } \Gamma \qquad (A4.39)$$

is an analytic function, or, rather, two analytic functions (see Fig. A4.12):

$$F_+(z) = \frac{1}{2\pi i} \int_\Gamma \frac{\varphi(\zeta)}{\zeta - z} d\zeta, \quad z \text{ inside } \Gamma \qquad (A4.40)$$

$$F_-(z) = \frac{1}{2\pi i} \int_\Gamma \frac{\varphi(\zeta)}{\zeta - z} d\zeta, \quad z \text{ outside } \Gamma \qquad (A4.41)$$

Let t be a point on Γ, at which $\varphi(\zeta)$ is continuous. Determine $F_+(z)$ and $F_-(z)$ when $z \to t$. By making an infinitesimally small half-circular indentation of Γ, the path of integration is prevented from going through the point t, and it may be considered as consisting of two parts, Γ_1 and γ; see Fig. A4.13. Thus, as $z \to t$,

$$F_+(z) \to \frac{1}{2\pi i} \int_{\Gamma_1} \frac{\phi(\zeta)}{\zeta - t} d\zeta + \frac{1}{2\pi i} \int_\gamma \frac{\varphi(t)}{\zeta - t} d\zeta$$

$$= \frac{1}{2\pi i} \int_{\Gamma_1} \frac{\varphi(\zeta)}{\zeta - t} d\zeta + \frac{\varphi(t)}{2\pi i} \pi i = F_+(t) \qquad (A4.42)$$

where Γ_1 is shown in Fig. A4.13. Thus,

$$F_+(t) = \frac{1}{2\pi i} \oint_\Gamma \frac{\varphi(\zeta)}{\zeta - t} d\zeta + \frac{\varphi(t)}{2} \qquad (A4.43)$$

where C on the integral sign denotes the Cauchy principal value, i.e. the integral along

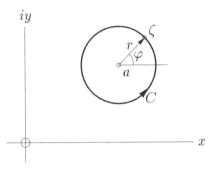

Fig. A4.14 The radius of the path of integration can be chosen arbitrarily large.

Γ, excluding the infinitesimally short half-circular arc with centre at the pole t on Γ. In the same way, it follows that

$$F_-(t) = \frac{1}{2\pi i} \oint_\Gamma \frac{\varphi(\zeta)}{\zeta - t} d\zeta - \frac{\varphi(t)}{2} \tag{A4.44}$$

Equations (A4.43) and (A4.44) give the *Plemelj formulae*†:

$$\boxed{F_+(t) + F_-(t) = \frac{1}{\pi i} \oint_\Gamma \frac{\varphi(\zeta)}{\zeta - t} d\zeta} \tag{A4.45}$$

$$\boxed{F_+(t) - F_-(t) = \varphi(t)} \tag{A4.46}$$

Liouville's theorem

Assume that $f(z)$ is bounded and analytic in the *whole* plane. Then, Liouville's theorem states that $f(z) = constant$

Proof: Consider an arbitrary point $z = a$; see Fig. A4.14. Equation (A4.35) gives

$$f'(a) = \frac{1}{2\pi i} \int_C \frac{f(\zeta)}{(\zeta - a)^2} d\zeta \tag{A4.47}$$

Because $f(z)$ is bounded there exists a constant M such that

$$|f(z)| \leq M \tag{A4.48}$$

Hence

$$f'(a) \leq \frac{1}{2\pi} \int_0^{2\pi} \frac{M}{r} d\varphi = \frac{M}{r} \tag{A4.49}$$

But r can be chosen arbitrarily large, which implies that $f'(a) = 0$. Because a is an arbitrary point it follows that $f'(z) = 0$, i.e., $f(z) = constant$.

Corollary. The Fundamental Theorem of algebra: The equation $P(z) = 0$ possesses at least one root, if $P(z)$ is a polynomial of at least degree 1. To prove this theorem,

---

† These formulae are also called the Sokhotski-Plemelj formulae, reflecting studies of Cauchy integrals made by Yu.V. Sokhotski, St Petersburg, in his doctor's thesis (1873).

assume that $P(z)\neq 0$ in the whole plane. Then, $1/P(z)$ is bounded and analytic in the whole plane. Thus, $P(z)$ is a constant. But this contradicts the assumption. Thus, $P(z)=0$ has at least one root.

Decomposition of analytic functions

Certain applications, particularly the Wiener-Hopf technique, require decomposition of functions into parts that are analytic in specified regions, such as the regions to the left and to the right of a specified vertical line in the complex plane. Here, the partition of a function $F(z)$ into two parts, $F_+(z)$ and $F_-(z)$, that are regular for $\Re z > a$ and $\Re z < b$, $a < b$, respectively, and such that $F_+(z) + F_-(z) = F(z)$, will be considered. It is assumed that $F(z)$ is regular in the strip $a \leq \Re z \leq b$ and that $F(z)/z^\alpha \to 0$ as $|z| \to \infty$ if $\alpha > 0$. Then, the required partition is obtained by

$$F_+(z) = -\frac{1}{2\pi i} \int_{a-i\infty}^{a+i\infty} \frac{F(\zeta)d\zeta}{\zeta - z}, \quad \Re z > a \tag{A4.50}$$

$$F_-(z) = +\frac{1}{2\pi i} \int_{b-i\infty}^{b+i\infty} \frac{F(\zeta)d\zeta}{\zeta - z}, \quad \Re z < b \tag{A4.51}$$

whereupon $F_+(z)$ for $\Re z \leq a$ and $F_-(z)$ for $\Re z \geq b$, if required, follow from analytic continuation.

For the proof, it is first observed that $F_+(z)$ is analytic for $\Re z > a$, because $1/(\zeta-z)$ is an analytic function of z for all ζ on the integration path, except when z is situated on this path. Similarly, $F_-(z)$ is analytic for $\Re z < b$. An example is $F(z) = 1/(z-c)$, $c > b$, for which the integrations give $F_+(z) = 0$ for $z > a$, $F_-(z) = 1/(z-c)$ for $z < b$.

Second, it is shown that $F_+(z) + F_-(z) = F(z)$ for z inside the strip $a < \Re z < b$. Addition gives

$$F_+(z) + F_-(z) = \frac{1}{2\pi i}\left(\int_{a+i\infty}^{a-i\infty} + \int_{b-i\infty}^{b+i\infty}\right) \frac{F(\zeta)d\zeta}{\zeta - z} \tag{A4.52}$$

Note the different path directions, which enable closing of the path to a rectangle – integration over the short rectangle sides gives no contribution. But then Cauchy's integral formula applies, showing that the integral equals $F(z)$ for z inside the strip. It is then possible to use analytic continuation to extend the definitions of $F_\pm(z)$, thereby also extending the relation $F_+(z)+F_-(z) = F(z)$ to the whole plane. This continuation is usually obvious; in the example $F_+(z) = 0$ for $z \leq a$ and $F_-(z) = 1/(z-c)$ for $z \geq b$.

The partition procedure holds even if the strip of analyticity degenerates to a line. Actually, it is often convenient to work with a line, even if a finite width strip of analyticity exists.

Factorization, i.e. writing $F(z) = F_+(z) \cdot F_-(z)$ or $F(z) = F_+(z)/F_-(z)$, is obtained in the same manner, by working with $\ln[F(z)] = \ln[F_+(z)] \pm \ln[F_-(z)]$.

A5 Laplace transforms

One-sided Laplace transforms

Original functions $f(t)$ and $g(t)$, defined for $t > 0$, are considered. Their one-sided Laplace transforms are denoted by $F(p)$ and $G(p)$. The Laplace transform of $f(t)$ is here defined as

$$\mathcal{L}_{pt}[f(t)] = p \int_0^\infty e^{-p\tau} f(\tau) \mathrm{d}\tau = F(p) \tag{A5.1}$$

so that $f(t)$ and $F(p)$ will be of the same dimension. $F(p)$ is analytic in $\Re p \geq \Re p_0$, the region of convergence of the definition integral.

Let a be a positive constant. Then,

$$\mathcal{L}_{pt}[f(at)] = p \int_0^\infty e^{-p\tau} f(a\tau) \mathrm{d}\tau = \frac{p}{a} \int_0^\infty e^{-(p/a)\cdot a\tau} f(a\tau) \mathrm{d}(a\tau) \tag{A5.2}$$

Thus,

$$\mathcal{L}_{pt}[f(at)] = F(p/a), \quad a > 0 \tag{A5.3}$$

Now, write

$$f(t) = \mathcal{L}_{pt}^{-1}[F(p)] = \mathcal{L}_{pt}^{-1}\left[\int_0^\infty p e^{-p\tau} f(\tau) \mathrm{d}\tau\right] \tag{A5.4}$$

where $\mathcal{L}^{-1}$ denotes the inverse Laplace transform, and compare with

$$f(t) = \int_0^\infty \delta(t-\tau) f(\tau) \mathrm{d}\tau \tag{A5.5}$$

where $\delta(t)$ is the Dirac delta function. Then, obviously,

$$\mathcal{L}_{pt}^{-1}[p e^{-p\tau}] = \delta(t-\tau) \tag{A5.6}$$

and, in particular,

$$\mathcal{L}_{pt}^{-1}[p] = \delta(t) \tag{A5.7}$$

Furthermore,

$$\frac{1}{p} F(p) G(p) = p \int_0^\infty e^{-p\tau} f(\tau) \mathrm{d}\tau \int_0^\infty e^{-p\tau_0} g(\tau_0) \mathrm{d}\tau_0$$
$$= \int_0^\infty \int_0^\infty p e^{-p(\tau+\tau_0)} f(\tau) g(\tau_0) \mathrm{d}\tau_0 \, \mathrm{d}\tau \tag{A5.8}$$

and thus, according to (A5.6):

$$\mathcal{L}_{pt}^{-1}\left[\frac{1}{p} F(p) G(p)\right] = \int_0^\infty f(\tau) \int_0^\infty \delta(t-\tau-\tau_0) g(\tau_0) \mathrm{d}\tau_0 \, \mathrm{d}\tau \tag{A5.9}$$

The inner integral vanishes if $t - \tau < 0$ and equals $g(t - \tau)$ if $t - \tau > 0$. Thus, the Laplace transform of a convolution integral, expressed in inverse form, is found to be

$$\boxed{\mathcal{L}_{pt}^{-1}\left[\frac{1}{p}F(p)G(p)\right] = \int_0^t f(\tau)g(t-\tau)d\tau} \qquad (A5.10)$$

Now, put

$$g(t) = \delta(t - \alpha) \qquad (A5.11)$$

i.e. according to (A5.6):

$$G(p) = pe^{-\alpha p} \qquad (A5.12)$$

Then, from (A5.10),

$$\mathcal{L}_{pt}^{-1}[F(p)e^{-\alpha p}] = \int_0^t f(\tau)\delta(t - \alpha - \tau)d\tau \qquad (A5.13)$$

i.e.

$$\boxed{\mathcal{L}_{pt}^{-1}[F(p)e^{-\alpha p}] = f(t-\alpha)U(t-\alpha)} \qquad (A5.14)$$

Alternatively, put

$$g(t) = U(t) \qquad (A5.15)$$

Then,

$$G(p) = p\int_0^\infty e^{-p\tau}d\tau = 1 \qquad (A5.16)$$

so that the Laplace transform of an integral, in inverse form, is found to be

$$\boxed{\mathcal{L}_{pt}^{-1}\left[\frac{1}{p}F(p)\right] = \int_0^t f(\tau)U(t-\tau)d\tau = \int_0^t f(\tau)d\tau} \qquad (A5.17)$$

Putting

$$f(t) = s'(t) \qquad (A5.18)$$

gives

$$\mathcal{L}_{pt}[s(t) - s(0)] = \frac{1}{p}F(p) \qquad (A5.19)$$

$$\mathcal{L}_{pt}[s(t)] = \frac{1}{p}F(p) + s(0) = S(p) \qquad (A5.20)$$

so that the Laplace transform of a derivative is found to be

$$\boxed{\mathcal{L}_{pt}[s'(t)] = F(p) = p[S(p) - s(0)]} \qquad (A5.21)$$

The Laplace transform of $f(t)$ multiplied by e^{at} is

$$p\int_0^\infty e^{-p\tau}e^{a\tau}f(\tau)d\tau = \frac{p}{p-a}(p-a)\int_0^\infty e^{-(p-a)\tau}f(\tau)d\tau \qquad (A5.22)$$

Thus,
$$\mathcal{L}_{pt}[e^{at}f(t)] = \frac{p}{p-a}F(p-a) \tag{A5.23}$$

One-sided Laplace transforms of "well behaved" originals are generally regular in the right p-half-plane, i.e., the definition integral converges if $\Re p > 0$. If the original increases as $\exp(ct)$ as $t \to \infty$, then they are regular to the right of $\Re p = c$.

From a theorem by Lerch (1903), it follows that a one-sided Laplace transform is uniquely determined if it is known at the points $p = p_0 + nq_0$, $(n = 1, 2, 3, \ldots)$ in its region of analyticity, where p_0 and q_0 are constants. Thus, it is possible to make the often very useful restriction to real and positive values of p in the region of analyticity.

One-sided Laplace transforms can be inverted by the formula

$$f(t) = \int_{c-i\infty}^{c+i\infty} e^{pt} \frac{F(p)}{p} dp \tag{A5.24}$$

where c must be situated inside the region of analyticity of $F(p)$.

Two-sided Laplace transforms

Two-sided Laplace transforms must be used if the original is not identically zero for $t < 0$. They are here defined as

$$\mathcal{L}_{pt} f(t) = p \int_{-\infty}^{+\infty} e^{-p\tau} f(\tau) d\tau = F(p) \tag{A5.25}$$

The region of convergence is limited to a vertical strip in the p-plane, often containing $\Re p = 0$, sometimes degenerated to only $\Re p = 0$.

Obviously (A5.3), (A5.6) and (A5.7) are still valid (though without limitation to positive t). Relation (A5.3) can be complemented with the relation

$$\mathcal{L}_{pt}[f(at)] = -F(p/a), \quad a < 0 \tag{A5.26}$$

and the Laplace transform of a convolution integral, in inverse form, is found to be

$$\mathcal{L}_{pt}^{-1}\left[\frac{1}{p}F(p)G(p)\right] = \int_{-\infty}^{+\infty} f(\tau)g(t-\tau)d\tau \tag{A5.27}$$

The displacement theorem (A5.14) is valid unchanged, but (A5.17) has to be exchanged by

$$\mathcal{L}_{pt}^{-1}\left[\frac{1}{p}F(p)\right] = \int_{-\infty}^{t} f(\tau)d\tau \tag{A5.28}$$

The two-sided Laplace transform of a derivative is

$$\mathcal{L}_{pt}[s'(t)] = pS(p) \tag{A5.29}$$

and the theorem (A5.23) remains unchanged. Finally, the inversion formula (A5.24) remains unchanged.

Asymptotic relations

It is often possible to make immediate conclusions about the asymptotic behaviour of the original when its Laplace transform is known and *vice versa*. For one-sided originals, relations exist between the original as $t \to +0$ and the transform as $p \to +\infty$ or between the original as $t \to +\infty$ and the transform as $p \to +0$. These relations are called *Abel theorems* when the asymptotic behaviour of the original is known, and *Tauber theorems* when the asymptotic behaviour of the Laplace transform is known.

The simplest form of Abel and Tauber theorems concerns cases when constant asymptotic limits exist (as opposed to functions of t or p). Then, the theorems tell that

$$f(\infty) = F(0), \qquad f(0) = F(\infty) \tag{A5.30}$$

Both relations can be understood after making the substitution $pt = u$ in the definition integral for one-sided originals, rendering

$$F(p) = \int_0^\infty e^{-u} f\left(\frac{u}{p}\right) du \tag{A5.31}$$

In more general cases, Abel theorems can be looked upon as possibilities to find the asymptotic behaviour of the transform by inverting an asymptotic expression for the original, which in general is much simpler than a full inversion. More specifically:

$$\lim_{t \to +0} \frac{f(t)}{t^\nu} \Longrightarrow \lim_{p \to +\infty} \frac{p^\nu F(p)}{\Gamma(1+\nu)}, \quad \nu > -1 \tag{A5.32}$$

$$\lim_{t \to +\infty} \frac{f(t)}{t^\nu} \Longrightarrow \lim_{p \to +0} \frac{p^\nu F(p)}{\Gamma(1+\nu)}, \quad \nu > -1 \tag{A5.33}$$

where $\Gamma(\cdot)$ is the gamma function. It is assumed that the limits in the left members exist, and that the strip of convergence for $F(p)$ contains the imaginary axis in the second relation.

Tauber theorems are basically reversed Abel theorems; however, such reversals are not always permitted. One particularly useful relation is given by van der Pol and Bremmer (1959):

$$\lim_{p \to +0} p^\nu F(p) \Longrightarrow \lim_{t \to +\infty} \Gamma(2+\nu) \frac{\int_0^t f(s) ds}{t^{1+\nu}}, \quad \nu \geq -1 \tag{A5.34}$$

It is valid under the rather general condition that a real constant C can be found such that $Ct^\nu + f(t) \geq 0$ for $t > 0$. For the more restricted case that $f(t)$ is monotonic for $t > 0$, the relation can be written as a direct reversal of the corresponding Abel Theorem:

$$\lim_{p \to +0} p^\nu F(p) \Longrightarrow \lim_{t \to +\infty} \Gamma(1+\nu) \frac{f(t)}{t^\nu}, \quad \nu \geq -1 \tag{A5.35}$$

Further discussions about Abel and Tauber theorems are found in the excellent book on Laplace transforms by van der Pol and Bremmer (1959).

A6 Mellin transforms

Let $f(r)$ be defined for $r > 0$. Its Mellin transform, $F(s)$, is defined by the relation

$$F(s) = \int_0^\infty r^s f(r) dr \tag{A6.1}$$

which is convergent in a strip $a < \Re(s) < b$ in the complex s-plane†. $F(s)$ is then analytic in the strip. The inversion of $F(s)$ is

$$f(r) = \frac{1}{2\pi i} \int_{c-i\infty}^{c+i\infty} r^{-s-1} F(s) ds, \quad a < c < b \tag{A6.2}$$

By partial integration it is found that the Mellin transform for the derivative $f'(r)$ equals

$$\int_0^\infty r^s f'(r) dr = -sF(s-1) \tag{A6.3}$$

and that the Mellin transform for the integral $\int_0^r f(\rho) d\rho$ equals

$$\int_0^\infty r^s \int_0^r f(\rho) d\rho \, dr = -\frac{1}{s+1} F(s+1) \tag{A6.4}$$

Similar formulae are found for higher order derivatives, etc. One useful example is the Mellin transform of $rf(r)$,

$$\int_0^\infty r^{s+1} f(r) dr = -(s+1) F(s) \tag{A6.5}$$

A7 Hankel transforms

Original functions $f(r)$, defined for $0 \leq r < \infty$ are studied. The complete Hankel transform F of $f(r)$ can be defined in a few different forms, which, however, in principle are equivalent. The choice made here is the definition

$$\boxed{F = F(\xi) = \int_0^\infty r f(r) J_\nu(\xi r) dr} \tag{A7.1}$$

where J_ν is a Bessel function of order ν. Note that this definition implies that F does not possess the same dimension as f. The Hankel transform variable ξ obviously is of dimension $1/r$.

A simple inversion formula exists:

$$\boxed{f(r) = \int_0^\infty \xi F(\xi) J_\nu(r\xi) d\xi} \tag{A7.2}$$

---

† In this definition the variable choice made by Sneddon (1942) is used, where the exponent s in the definition integral replaces the exponent $s-1$ originally suggested for the transform and still often used.

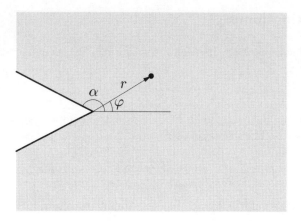

Fig. A8.1 Notch and cylindrical coordinates.

Example:
$$f(r) = e^{-ar}, \quad \nu = 0$$
$$F = \int_0^\infty r e^{-ar} J_0(\xi r) \mathrm{d}r = a(a^2 + \xi^2)^{-3/2} \tag{A7.3}$$

The integral is found in integral tables. Inversion gives

$$f = a \int_0^\infty \xi (a^2 + \xi^2)^{-3/2} J_0(r\xi) \mathrm{d}\xi = e^{-ar} \tag{A7.4}$$

as expected. Also this integral is found in integral tables.

Hankel transforms do not necessarily need to contain the Bessel function J. Even other Bessel functions can be used, for instance the Hankel function H, as suggested by the name of the transform.

A8 Stresses near a sharp notch

A sharp notch with internal angle $2\pi - 2\alpha$, $\alpha > \pi/2$, is subjected to some remote symmetrical (mode I) loading, whereas the notch faces are traction free. Plane stress or strain is assumed. A cylindrical coordinate system (r, φ) is introduced with the origin at the notch root and $\varphi = 0$ coinciding with the symmetry plane inside the body; see Fig. A8.1. The stresses in the notch edge vicinity are sought. This problem was solved by Williams (1952).

Stresses and strains are represented by an Airy stress function Φ, which satisfies the biharmonic equation

$$\Delta\Delta\Phi = 0 \tag{A8.1}$$

Solutions in the form $\Phi = r^{s+2} F(\varphi)$ are given in (A3.12). The part which is symmetrical in φ is

$$\Phi = A r^{s+2} \cos(s\varphi) + B r^{s+2} \cos[(s+2)\varphi] \tag{A8.2}$$

A8 STRESSES NEAR A SHARP NOTCH

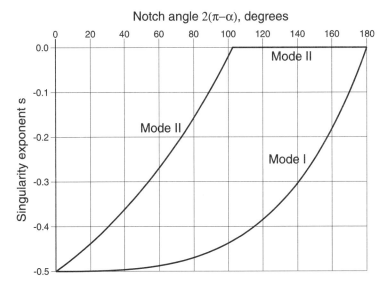

Fig. A8.2 Exponents s for symmetric (mode I) and anti-symmetric (mode II) loading.

The conditions of traction free notch faces give:

$$\sigma_\varphi = \frac{\partial^2 \Phi}{\partial r^2} = 0 \quad \text{for } \varphi = \alpha \tag{A8.3}$$

$$\tau_{r\varphi} = -\frac{\partial}{\partial r}\left(\frac{1}{r} \cdot \frac{\partial \Phi}{\partial \varphi}\right) = 0 \quad \text{for } \varphi = \alpha \tag{A8.4}$$

These expressions lead to the system of equations

$$A\cos(s\alpha) + B\cos[(s+2)\alpha] = 0 \tag{A8.5}$$
$$As\sin(s\alpha) + B(s+2)\sin[(s+2)\alpha] = 0 \tag{A8.6}$$

with non-trivial solutions if

$$(s+1)\sin(2\alpha) + \sin[2(s+1)\alpha] = 0 \tag{A8.7}$$

giving

$$\frac{B}{A} = -\frac{s\sin(s\alpha)}{(s+2)\sin[(s+2)\alpha]} \tag{A8.8}$$

which, after insertion into (A8.2) results in

$$\Phi = Cr^{s+2}\{(s+2)\cos(s\varphi)\sin[(s+2)\alpha] - s\cos[(s+2)\varphi]\sin(s\alpha)\} \tag{A8.9}$$

where C is a constant. The dominating stress-strain field in the vicinity of the notch root is given by the smallest value of $s \geq -1/2$ that satisfies (A8.7). Any stress or strain in this field can then be determined, for instance the normal stress on the plane

$\varphi = 0$:

$$\sigma_\varphi = \frac{\partial^2 \Phi}{\partial r^2} = (s+1)(s+2)Cr^s\{(s+2)\cos(s\varphi)\cos[(s+2)\alpha]$$
$$- s\cos[(s+2)\varphi]\sin(s\alpha)\} \quad (A8.10)$$

For anti-symmetric (mode II) loading, the anti-symmetric part of solution (A3.12) should be used, giving

$$(s+1)\sin(2\alpha) - \sin[2(s+1)\alpha] = 0 \quad (A8.11)$$
$$\Phi = Cr^{s+2}\{\sin(s\varphi)\sin[(s+2)\alpha] - \sin[(s+2)\varphi]\sin(s\alpha)\} \quad (A8.12)$$

instead of equations (A8.7)-(A8.9):

The exponents s for symmetric and anti-symmetric loading are shown as functions of the notch angle $2(\pi - \alpha)$ in Fig. A8.2. For anti-symmetric loading, there is no singularity for $2(\pi - \alpha) > \gamma$, where $\gamma \approx 102.55°$ is a solution of the equation $\tan\gamma = \gamma$.

A9 Basic relations for stress waves in elastic solids

Basic equations for motion. The two fundamental wave types

The equation of motion in a continuum is given by the balance between forces and inertia, so that

$$\sigma_{ji,j} + \rho f_i = \rho \frac{\partial^2 u_i}{\partial t^2} \quad (A9.1)$$

where ρ is the density of the solid and f_i is the body force per unit mass, which, however, is omitted for simplicity in the continuation. For an isotropic linearly elastic solid, Hooke's law may be written in the form, cf. (4.15.1),

$$\sigma_{ij} = \mu(u_{j,i} + u_{i,j}) + \left(\kappa - \frac{2\mu}{3}\right)u_{k,k}\delta_{ji} \quad (A9.2)$$

where μ and κ are the moduli of rigidity and compression, respectively. Hence,

$$\sigma_{ij,j} = \mu(u_{j,ij} + u_{i,jj}) + \left(\kappa - \frac{2\mu}{3}\right)u_{k,kj}\delta_{ji} \quad (A9.3)$$

Insertion into (A9.1), noting that

$$u_{k,kj}\delta_{ji} = u_{k,ki} = u_{j,ji} = u_{j,ij} \quad (A9.4)$$

leads to the equations of motion with the displacement as the only dependent variable:

$$\left(\kappa + \frac{\mu}{3}\right)u_{j,ij} + \mu u_{i,jj} = \rho \frac{\partial^2 u_i}{\partial t^2} \quad (A9.5)$$

The second order derivatives of the displacement can be expressed in vector notation by letting

$$u_{j,ij} \to \text{grad div } \boldsymbol{u}, \quad u_{j,ij} - u_{i,jj} \to \text{curl curl } \boldsymbol{u} \quad (A9.6)$$

A9 BASIC RELATIONS FOR STRESS WAVES IN ELASTIC SOLIDS

Then, the equation of motion can be written in the form

$$\boxed{c_P^2 \operatorname{grad} \operatorname{div} \boldsymbol{u} - c_S^2 \operatorname{curl} \operatorname{curl} \boldsymbol{u} = \frac{\partial^2 \boldsymbol{u}}{\partial t^2}} \qquad (A9.7)$$

where

$$c_P^2 = \frac{\kappa + 4\mu/3}{\rho}, \quad c_S^2 = \frac{\mu}{\rho} \qquad (A9.8)$$

This way of writing has the merit of freeing the equation of motion from a particular choice of coordinate system – it has been written in an *invariant form*. Thereby also physical interpretations are made easier. The expression $\operatorname{div} \boldsymbol{u}$ is the dilatation, the volume increase per unit original volume, $\operatorname{grad} \boldsymbol{u}$ expresses the displacement "slope", the maximum displacement increase per unit length (the strain) and the direction in which it occurs, and $\operatorname{curl} \boldsymbol{u}$ represents a rotation, the maximum angular change (which equals $1/2|\operatorname{curl} \boldsymbol{u}|$) and the direction in which it occurs.

Note that dilatation during wave motion in a solid is always accompanied by shear. A wave motion is due to *differences* between adjacent elements of the dilatation ($\operatorname{grad} \operatorname{div} \boldsymbol{u} \neq 0$) and/or of the rotation ($\operatorname{curl} \operatorname{curl} \boldsymbol{u} \neq 0$). Neither of these differences can occur without shear, like hydro-static compression or rigid body motion. This is also reflected in the fact that the coefficients c_P and c_S both contain μ.

It is always possible to decompose a differentiable displacement $\boldsymbol{u}$ into a sum of two components,

$$\boldsymbol{u} = \boldsymbol{u}_P + \boldsymbol{u}_S \qquad (A9.9)$$

so that

$$\operatorname{curl} \boldsymbol{u}_P = 0, \quad \operatorname{div} \boldsymbol{u}_S = 0 \qquad (A9.10)$$

Thus, the displacement $\boldsymbol{u}$ is decomposed into one *irrotational* part, $\boldsymbol{u}_P$, and one *equivoluminal* part, $\boldsymbol{u}_S$. This is Helmholtz' decomposition theorem; see e.g. Morse and Feshbach (1953), pp. 52-53. A shortened version of the proof assumes that it is possible to write $\boldsymbol{u}$ in the form $\boldsymbol{u} = \Delta \boldsymbol{U}$, i.e. that a vector function $\boldsymbol{U}$ can be found for any given vector function $\boldsymbol{u}$, so that this representation is possible. Then, the Helmholtz' decomposition follows from the identity

$$\Delta \boldsymbol{U} \equiv \operatorname{grad} \operatorname{div} \boldsymbol{U} - \operatorname{curl} \operatorname{curl} \boldsymbol{U} \qquad (A9.11)$$

if $\operatorname{grad} \operatorname{div} \boldsymbol{U}$ is identified with $\boldsymbol{u}_P$ and $-\operatorname{curl} \operatorname{curl} \boldsymbol{U}$ with $\boldsymbol{u}_S$, because $\operatorname{curl} \operatorname{grad} \equiv 0$ and $\operatorname{div} \operatorname{curl} \equiv 0$. Also, an extremely useful representation of the displacement with two potential functions, one scalar, ϕ, and one vectorial, $\boldsymbol{\psi}$, follows, simply by putting $\operatorname{div} \boldsymbol{U} = \phi$ and $-\operatorname{curl} \boldsymbol{U} = \boldsymbol{\psi}$:

$$\boxed{\begin{aligned} \boldsymbol{u}_P &= \operatorname{grad} \phi \\ \boldsymbol{u}_S &= \operatorname{curl} \boldsymbol{\psi} \end{aligned}} \qquad (A9.12)$$

where $\boldsymbol{\psi}$ is a divergence free vector, because $\operatorname{div} \operatorname{curl} \boldsymbol{U} \equiv 0$. This representation can be inserted into the equation of motion (A9.7) to yield

$$\operatorname{grad}\left[c_P^2 \operatorname{div} \operatorname{grad} \phi - \frac{\partial^2 \phi}{\partial t^2}\right] - \operatorname{curl}\left[c_S^2 \operatorname{curl} \operatorname{curl} \boldsymbol{\psi} + \frac{\partial^2 \boldsymbol{\psi}}{\partial t^2}\right] = 0 \qquad (A9.13)$$

Note that the representation (A9.12) is not unique: any function ϕ_0 that satisfies the condition grad $\phi_0 = 0$ can be added to ϕ, and any function $\boldsymbol{\psi}_0$ that satisfies the condition curl $\boldsymbol{\psi}_0 = 0$ can be added to $\boldsymbol{\psi}$. This property can be used to demonstrate that the equation of motion (A9.13) can be split up into two parts, one for ϕ and one for $\boldsymbol{\psi}$. Thus, among all functions ϕ and $\boldsymbol{\psi}$ that satisfy (A9.13), it is sufficient to select those which satisfy the vanishing of the expressions within the brackets [ ] separately, so that the equation of motion finally reads:

$$\boxed{\begin{aligned} c_P^2 \,\text{div grad}\, \phi &= \frac{\partial^2 \phi}{\partial t^2} \\ -c_S^2 \,\text{curl curl}\, \boldsymbol{\psi} &= \frac{\partial^2 \boldsymbol{\psi}}{\partial t^2}, \quad \text{div}\, \boldsymbol{\psi} = 0 \end{aligned}} \tag{A9.14}$$

The displacement $\boldsymbol{u}$ is found as

$$\boldsymbol{u} = \text{grad}\, \phi + \text{curl}\, \boldsymbol{\psi} \tag{A9.15}$$

From the displacements, the stresses are found, after differentiations and application of Hooke's law. Results for Cartesian, cylindrical and spherical coordinate systems are shown in A1, where expressions for certain vector operations are also given.

Note that div grad $\phi \equiv \Delta \phi$, so that the first of equations (A9.14) may be written as

$$\Delta \phi = \frac{1}{c_P^2} \cdot \frac{\partial^2 \phi}{\partial t^2} \tag{A9.16}$$

The equation for $\boldsymbol{\psi}$ may be written in a similar way, by introducing a vector Laplacian, $\boldsymbol{\Delta} = \text{grad div} - \text{curl curl}$, but there seems to be little advantage in doing so. Cartesian coordinates permit simple equations in component form, with use of the ordinary Laplacian:

$$\Delta \psi_x = \frac{1}{c_S^2} \cdot \frac{\partial^2 \psi_x}{\partial t^2}, \quad \Delta \psi_y = \frac{1}{c_S^2} \cdot \frac{\partial^2 \psi_y}{\partial t^2}, \quad \Delta \psi_z = \frac{1}{c_S^2} \cdot \frac{\partial^2 \psi_z}{\partial t^2} \tag{A9.17}$$

For general coordinates, the corresponding equations may be rather complicated, but in many common contexts, for instance in plane or rotationally symmetric cases, they can be written in the same simple form as (A9.16).

If, for in-plane deformation, the coordinate system x, y, z is oriented so that stresses and strains do not depend on z, then the equations of motion may be written in the form

$$\left.\begin{aligned} \Delta \phi &= \frac{1}{c_P^2} \cdot \frac{\partial^2 \phi}{\partial t^2} \\ \Delta \psi &= \frac{1}{c_S^2} \cdot \frac{\partial^2 \psi}{\partial t^2} \end{aligned}\right\} \tag{A9.18}$$

where $\phi = \phi(x, y, t)$, $\psi = \psi_z = \psi(x, y, t)$ and

$$u = u(x, y, t) = \frac{\partial \phi}{\partial x} + \frac{\partial \psi}{\partial y} \tag{A9.19}$$

$$v = v(x, y, t) = \frac{\partial \phi}{\partial y} - \frac{\partial \psi}{\partial x} \tag{A9.20}$$

If, for rotational symmetry, the coordinate system r, φ, z is oriented so that stresses and strains do not depend on φ, then the equations of motion may be written in the form

$$\left.\begin{array}{r}\Delta \phi = \dfrac{1}{c_P^2} \cdot \dfrac{\partial^2 \phi}{\partial t^2} \\[2mm] \Delta \psi - \dfrac{\psi}{r^2} = \dfrac{1}{c_S^2} \cdot \dfrac{\partial^2 \psi}{\partial t^2}\end{array}\right\} \quad (A9.21)$$

where $\phi = \phi(r, z, t)$, $\psi = \psi_\varphi = \psi(r, z, t)$ and

$$u_r = u_r(r, z, t) = \frac{\partial \phi}{\partial r} - \frac{\partial \psi}{\partial z} \quad (A9.22)$$

$$u_z = u_z(r, z, t) = \frac{\partial \phi}{\partial z} + \frac{1}{r} \cdot \frac{\partial (r\psi)}{\partial r} \quad (A9.23)$$

If, for anti-plane deformation, the coordinate system x, y, z is oriented so that the displacement vector points in the z direction, then the equations of motion may be written in the form

$$\boxed{\Delta w = \frac{1}{c_S^2} \cdot \frac{\partial^2 w}{\partial t^2}} \quad (A9.24)$$

where $w = w(x, y, t)$ and

$$\tau_{xz} = \mu \partial w / \partial x, \quad \tau_{yz} = \mu \partial w / \partial y \quad (A9.25)$$

Waves from an embedded point source

For spherical symmetry, equation (A9.16) can be written in the particularly simple form,

$$\frac{\partial^2 (r\phi)}{\partial r^2} = \frac{1}{c_P^2} \cdot \frac{\partial^2 (r\phi)}{\partial t^2} \quad (A9.26)$$

with the general solution

$$r\phi = f(r - c_P t) + g(r + c_P t) \quad (A9.27)$$

which implies outgoing and incoming waves from the origin, travelling with propagation velocity c_P. If the waves are caused by an excitation at the origin, i.e. a *point source*, incoming waves must be discarded. This is generally referred to as the *radiation condition*. Note that also $\boldsymbol{u}_P = \operatorname{grad} \phi$ will involve only waves travelling with velocity c_P. Moreover, in a spherical coordinate system r, φ, θ, the displacement components $u_\varphi = u_\theta = 0$, and

$$u_r = \frac{1}{r} f''(r - c_P t) \quad (A9.28)$$

where $'$ indicates differentiation with respect to $r - c_P t$.

To find the response to a point source consisting of a rotation at the origin, is somewhat more complicated. One way is to study the waves going out from a rigid sphere, oscillating around its North-South Pole axis, and then letting its radius shrink indefinitely. This is equivalent to inspecting the solution at sufficiently large distances from the sphere to be the result of a point source. The only non-vanishing displacement component is u_φ, which implies that the second of equations (A9.14) takes the form

$$\frac{\partial^2(ru_\varphi)}{\partial r^2} + \frac{1}{r^2} \cdot \frac{\partial}{\partial \theta}\left[\frac{1}{\sin\theta} \cdot \frac{\partial(\sin\theta\, ru_\varphi)}{\partial\theta}\right] = \frac{1}{c_S^2} \cdot \frac{\partial^2(ru_\varphi)}{\partial t^2} \qquad (A9.29)$$

For harmonic oscillations with angular frequency ω, the result turns out to be

$$u_\varphi \to C \sin\theta \sin\left[\omega\left(t - \frac{r}{c_S}\right)\right] \quad \text{as} \quad \frac{a\omega}{c_S} \to 0 \text{ and } \frac{\omega^2 a r}{c_S^2} \to \infty \qquad (A9.30)$$

where C is a constant and a is the radius of the sphere. The other displacement components, u_r and u_θ, are zero. The result shows that a rotational point source creates a wave, which is travelling outwards with propagation velocity c_S.

Obviously, waves from a point source in an infinte solid can be either irrotational, travelling with velocity c_P, or equivoluminal, travelling with velocity c_S. These are *the two fundamental wave types* and *the two fundamental wave velocities*.

Wave guides

Although there are only two fundamental wave types in isotropic, linearly elastic media, it is, in certain cases, possible to discern clearly waves travelling with propagation velocities other than c_S and c_P. This occurs when the waves are confined by some kind of a wave guide to a certain path, i.e., they cannot propagate freely. Irrotational and equivoluminal waves then interact, creating a new wave type with lower propagation velocity than c_P – obviously it can never be higher. A well known example is wave propagation along a rod, where waves with propagation velocity $\sqrt{E/\varrho} < c_P$ appear if the wavelength or pulse length is much larger than the rod diameter. In addition, pure equivoluminal waves can appear, travelling, as they should, with velocity c_S. They are transversal and are realized as torsional waves.

Why isn't it possible to obtain irrotational waves in a rod? The answer is simply that it *is* possible, for instance by impact at a rod end, but their amplitude decreases very rapidly, and they become virtually impossible to detect already after a distance of a few rod diameters from the end.

Another type of waveguide, similar in action to a rod, is a thin plate, where the wave velocity is effectively limited to the maximum velocity in the plane stress approximation, lower than c_P but higher than the maximum effective velocity in a rod. A further example is simply a free surface, which is guiding waves with a velocity slightly lower than c_S, the Rayleigh wave velocity c_R, page 694.

Waves in plane strain or plane stress

For a *plane wave*, the displacement vector may be written as

$$u_i = f(x_k g_k - ct)d_i \qquad (A9.31)$$

where c is the propagation velocity, g_i is the unit vector ($g_i g_i = 1$) in the propagation direction and d_i the unit vector in the direction of the displacement vector. With use

A9 BASIC RELATIONS FOR STRESS WAVES IN ELASTIC SOLIDS

of c_P and c_S, the equation of motion (A9.5) may be written as

$$(c_P^2 - c_S^2)u_{i,ji} + c_S^2 u_{j,ii} = \frac{\partial^2 u_j}{\partial t^2} \tag{A9.32}$$

Here, differentiations of the displacement vector gives

$$u_{i,j} = f'\delta_{jk}g_k d_i = f' g_j d_i, \quad u_{i,ji} = f'' g_j g_i d_i, \quad u_{j,ii} = c^2 f'' d_j \tag{A9.33}$$

whereupon insertion into (A9.5) results in

$$f'' \cdot [(c_P^2 - c_S^2)g_j g_i d_i + (c_S^2 - c^2)g_j d_j] = 0 \tag{A9.34}$$

Divide both members by f'' and multiply by g_j:

$$(c_P^2 - c_S^2)g_i d_i + (c_S^2 - c^2)g_j g_j = 0 \tag{A9.35}$$

i.e.,

$$\boxed{(c_P^2 - c_S^2)g_i d_i = 0} \tag{A9.36}$$

Thus, either $c^2 = c_P^2$ or $g_i d_i = 0$. The first possibility implies, according to (A9.34), that $g_j g_i d_i - d_j = 0$, which, after multiplication by d_j leads to

$$g_j d_j g_i d_i - d_j d_j = 0 \tag{A9.37}$$

which gives $(g_i d_i)^2 = 1$, or

$$g_i d_i = \pm 1 \tag{A9.38}$$

This implies that the propagation and displacement directions coincide or are exactly opposed. This is, by definition, *a longitudinal wave*.

The other possibility, $g_i d_i = 0$, implies that propagation and displacement direction are perpendicular to each other, which is the definition of *transverse waves*. Insertion of $g_i d_i = 0$ into (A9.34), shows that the propagation velocity is $c = c_S$.

The result is that there are two kinds of plane waves, longitudinal with propagation velocity c_P, and transverse with velocity c_S. Of course, the longitudinal waves are irrotational and the transverse waves are equivoluminal.

It was tacitly assumed that the medium is infinite. However, plane transverse waves can also appear in a plate.

In longitudinal plane waves, the strain is zero in a direction perpendicular to the propagation direction. Thus, plane strain prevails. Plane strain waves can also appear in some vicinity behind a plane front of an irrotational wave in a finite body, as long as there are no disturbances from the body boundaries. Approximately, plane strain can also prevail close to a non-plane front, for instance a spherical front.

In-plane longitudinal wave motion in a plate approaches plane strain in the plate interior, if the wavelength or pulse length is much smaller than the plate thickness.

Plane stress is approached in the plate interior at predominantly in-plane wave motion, provided that the wavelength or pulse length is much larger than the plate thickness. Such a motion is, however, neither logitudinal nor irrotational, because it implies thickness changes of the plate at the wave passage. Nevertheless, the formulae obtained for irrotational – and thus also for the special case of longitudinal – waves can be applied with almost cosmetic changes. With the x_3 axis normal to the plate surfaces, the plane stress approximation implies that

$$\left.\begin{array}{l}\sigma_{31} = \mu(u_{3,1} + u_{1,3}) = 0 \\ \sigma_{32} = \mu(u_{3,2} + u_{2,3}) = 0 \\ \sigma_{33} = 2\mu u_{3,3} + \left(\kappa - \dfrac{2\mu}{3} u_{k,k}\right) = 0\end{array}\right\} \quad (A9.39)$$

Let $\alpha, \beta = 1, 2$. Then, from (A9.5), for in-plane motion,

$$\left(\kappa + \frac{\mu}{3}\right)u_{\beta,\alpha\beta} + \mu u_{\alpha,\beta\beta} + S_\alpha = \rho\frac{\partial^2 u_\alpha}{\partial t^2} \quad (A9.40)$$

where

$$S_\alpha = \left(\kappa + \frac{\mu}{3}\right)u_{3,\alpha 3} + \mu u_{\alpha,33} = \left(\kappa - \frac{2\mu}{3}\right)u_{3,\alpha 3} \quad (A9.41)$$

where the first two of relations (A9.39) was used. With the aid of the third of relations (A9.39) this may be written as

$$S_\alpha = \frac{3\kappa\mu}{\kappa + 4\mu/3}u_{\beta,\alpha\beta} \quad (A9.42)$$

so that

$$\frac{3\kappa\mu}{\kappa + 4\mu/3}u_{\beta,\alpha\beta} + \mu u_{\alpha,\beta\beta} = \rho\frac{\partial^2 u_\alpha}{\partial t^2} \quad (A9.43)$$

which, by the notation changes

$$u_{\beta,\alpha\beta} \to \operatorname{grad} \operatorname{div} \boldsymbol{u}, \quad u_{\beta,\alpha\beta} - u_{\alpha,\beta\beta} \to \operatorname{curl} \operatorname{curl} \boldsymbol{u} \quad (A9.44)$$

where $\boldsymbol{u}$ is the in-plane displacement vector, may be written in the invariant form

$$(c_P^{pss})^2 \operatorname{grad} \operatorname{div} \boldsymbol{u} - c_S^2 \operatorname{curl} \operatorname{curl} \boldsymbol{u} = \frac{\partial^2 \boldsymbol{u}}{\partial t^2} \quad (A9.45)$$

where

$$(c_P^{pss})^2 = \frac{4\mu(\kappa + \mu/3)}{(\kappa + 4\mu/3)\rho} \quad (A9.46)$$

This is virtually the same equation of motion as (A9.7), the difference being only that the constant $(c_P^{pss})^2 \neq c_P^2$.

With this difference taken into account, all previous formulae for plane strain are valid also for the plane stress approximation. In particular, two fundamental wave types appear, one with velocity c_P^{pss} and the other with velocity c_S. However, for simplicity, the superscript pss is left out in the present work when there is no risk of confusing plane stress conditions with plane strain.

It is possible to formulate more accurate equations of motion for waves in a plate than in the plane stress approximation, but still only including the two in-plane displacement components, u and v. They are based on the assumption that the transverse strain, ϵ_z, is constant across the plate thickness, $2h$, and, by using Hamilton's principle,

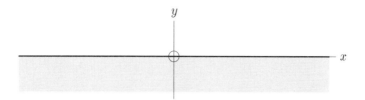

Fig. A9.1 Coordinate choice.

they are found to be (Broberg 1995b):

$$\Delta\phi - \frac{1}{(c_P^{pss})^2} \cdot \frac{\partial^2\phi}{\partial t^2} - \frac{(1-\nu)h^2}{6}\left(\Delta - \frac{1}{c_S^2}\cdot\frac{\partial^2}{\partial t^2}\right)\left(\Delta\phi - \frac{1}{c_P^2}\cdot\frac{\partial^2\phi}{\partial t^2}\right) = 0 \quad (A9.47)$$

$$\Delta\psi - \frac{1}{c_S^2}\cdot\frac{\partial^2\psi}{\partial t^2} = 0 \quad (A9.48)$$

The in-plane displacements are given as $u = \partial\phi/\partial x + \partial\psi/\partial y$ and $v = \partial\phi/\partial y - \partial\psi/\partial x$, whereas the transverse strain

$$\epsilon_z = \frac{1-\nu}{\nu}\left(\frac{1}{c_P^2}\cdot\frac{\partial^2\phi}{\partial t^2} - \Delta\phi\right) \quad (A9.49)$$

Similar equations of motion were given by Kane and Mindlin (1956). As a consequence of the approximate character, the condition of traction free plate surfaces cannot be satified – this would require a parabolic variation of ϵ_z over the plate thickness, and would probably lead to unwieldy equations of motion, still approximate, of course.

Surface waves – Rayleigh waves

The simplest waveguide is a free surface. This waveguide was studied by Rayleigh (1885) and the waves are called Rayleigh waves.

Regard the plane surface of a semi-infinite body (see Fig. A9.1). Do waves exist that are propagated along the surface? Assume that they do. Then, they should be expressible in terms of ϕ and $\psi_z = \psi$ as

$$\begin{cases} \phi = f(y)S(t - x/c) \\ \psi = g(y)T(t - x/c) \end{cases} \quad (A9.50)$$

Insertion into the first one of the equations (A9.18) results in

$$\frac{f''}{(c^{-2} - c_P^{-2})f} = \frac{1}{S}\cdot\frac{\partial^2 S}{\partial t^2} = \omega^2 \quad (A9.51)$$

where ω^2 does not depend on either y (because the middle member does not) or of $t - x/c$ (because the first member does not). A solution is

$$\phi = fS = Ae^{\omega(c^{-2} - c_P^{-2})^{1/2}y}\cos[\omega(t - x/c)] \quad (A9.52)$$

where A is a constant. The branch is chosen so that $(c^{-2} - c_P^{-2})^{1/2} = \sqrt{c^{-2} - c_P^{-2}}$

for $|c^{-1}| < c_P^{-1}$. The general solution contains one additional term which, however, becomes unbounded when $y \to -\infty$ and therefore must be discarded for physical reasons. By suitable choice of origin, a term with sine instead of cosine can be avoided without loss of generality.

In the same way,

$$\psi = gT = Ce^{\omega(c^{-2}-c_S^{-2})^{1/2}y}\sin[\omega(t-x/c)] \qquad (A9.53)$$

Actually, a term with cosine instead of sine should be added, but such a term would disappear after use of the boundary conditions. These are

$$\sigma_y = \tau_{yx} = 0 \text{ for } y = 0 \qquad (A9.54)$$

In Appendix A1, expressions for σ_y and τ_{xy} are given. Insertion of (A9.52) and (A9.53) into these expressions and use of the boundary conditions for $y = 0$ gives

$$\begin{cases} A(2k^2c^{-2} - c_P^{-2}) + 2k^2C(c^{-2} - c_S^{-2})^{1/2}c^{-1} = 0 \\ 2A(c^{-2} - c_P^{-2})^{1/2}c^{-1} + C(2c^{-2} - c_S^{-2}) \quad = 0 \end{cases} \qquad (A9.55)$$

The system possesses non-trivial solutions if

$$\boxed{4k^3(1-k_R^2)^{1/2}(k^2-k_R^2)^{1/2} - (2k^2-k_R^2)^2 = 0} \qquad (A9.56)$$

where the notation $c/c_P = k_R > 0$ has been used. From this equation, the propagation velocity $c_R = k_R c_P$ of Rayleigh waves (surface waves) can be determined.

Equation (A9.56) gives the zeros of the so-called *Rayleigh function*

$$\boxed{R(v) = 4k^3(1-v^2)^{1/2}(k^2-v^2)^{1/2} - (2k^2-v^2)^2} \qquad (A9.57)$$

where branch cuts are made along the portions $(-1,-k)$ and $(k,1)$ of the real axis in the complex v plane, and the branch is chosen so that $R(v)$ is non-negative on the real axis close to $v = 0$. Then, it is obvious that $0 < k_R < k$, because $R(v)$ is negative when $v \to k - 0$.

Function $R(v)$ often appears in the treatment of dynamic problems, in which free surfaces play a part. It can be normalized in different ways, no standard form exists. The form chosen here implies that

$$R(v) \to 2k^2(1-k^2)v^2 \text{ as } v \to 0 \qquad (A9.58)$$
$$R(v) > 0 \text{ for } 0 < v < k_R, \quad R(v) < 0 \text{ for } k_R < v \leq k \qquad (A9.59)$$

$R(v)$ has four zeros:

$$v = 0 \text{ (double zero)}, \quad v = \pm k_R \qquad (A9.60)$$

An associated function is

$$C(v) = 4k^3(1-v^2)^{1/2}(k^2-v^2)^{1/2} + (2k^2-v^2)^2 \qquad (A9.61)$$

with the same branch cuts as for $R(v)$ and the branch chosen so that $C(v) = 8k^4$ for $v = 0$. It also possesses four zeros,

$$v = \pm v_1, \quad v = \pm v_2 \qquad (A9.62)$$

If $k < k_0 \approx 0.5670$ ($k_0^2 \approx 0.3215$), then these zeros are complex and thus come in conjugated pairs, so that $v_2 = \overline{v_1}$, and the real part of v_1^2 and v_2^2 is positive. If $k > k_0$ the zeros are real, and then $|v_1| > 1$, $|v_2| > 1$. $|v_1| = |v_2| \approx 1.1495$ for $k = k_0$.

The four zeros of $R(v)$ together with the four zeros of $C(v)$ constitute the eight zeros of the function

$$R(v)C(v) = v^2[16k^6(1-k^2) - 8k^4(3-2k^2)v^2 + 8k^2v^4 - v^6] \quad (A9.63)$$

Note that k_R, v_1 and v_2 are obtained by solving a third degree equation. However, the handy expression

$$1 - k_R/k \approx 0.135/(3 - 4k^2) \quad (A9.64)$$

gives k_R/k to within $\pm 0.5\%$. For $k \to 0$ the relative error is less than 0.4 pro mille: the actual value is $k_R/k \approx 0.955313$.

Waves in linearly elastic anisotropic media

The constitutive relation for linearly elastic anisotropic media, see (4.14.6), is

$$\sigma_{ij} = C_{ijkl} u_{k,l} \quad (A9.65)$$

Insertion into the equations of motion

$$\sigma_{ij,j} = \varrho \frac{\partial^2 u_i}{\partial t^2} \quad (A9.66)$$

gives the wave equation for linearly elastic anisotropic media:

$$\boxed{C_{ijkl} u_{k,lj} = \varrho \frac{\partial^2 u_i}{\partial t^2}} \quad (A9.67)$$

Consider now the special case of plane waves by seeking solutions in the form

$$u_k = u_0 d_k f(t - g_m x_m/c) \quad (A9.68)$$

where d_k and g_m are unit vectors:

$$d_i d_i = g_i g_i = 1 \quad (A9.69)$$

The ansatz (A9.68) signifies a plane wave with propagation direction g_m. The velocity c is unknown, whereas g_m is considered as given. The vector d_k shows the particle velocity direction.

For insertion into (A9.67), the following derivatives are needed:

$$u_{k,l} = -u_0 d_k c^{-1} g_l f', \quad u_{k,lj} = u_0 d_k c^{-2} g_l g_j f'' \quad (A9.70)$$

$$\frac{\partial^2 u_i}{\partial t^2} = u_0 d_i f'' = u_0 \delta_{ik} d_k f'' \quad (A9.71)$$

Here, δ_{ik} is Kronecker's delta which equals unity when its indices are equal and otherwise equals zero. Insertion gives

$$[C_{ijkl} g_l g_j - c^2 \varrho \delta_{ik}] d_k = 0 \quad (A9.72)$$

The equation possesses non-trivial solutions for d_k, only if

$$\det[C_{ijkl} g_l g_j - c^2 \varrho \delta_{ik}] = 0 \quad (A9.73)$$

Put
$$C_{ijkl}g_l g_j = \Gamma_{ik} \tag{A9.74}$$

where Γ_{ik} is called the "Christoffel stiffness". The reason for the term "stiffness" can be understood if comparison is made between

$$\det[c^2 \delta_{ik} - \Gamma_{ik}/\varrho] = 0 \tag{A9.75}$$

and the relation

$$c^2 - \frac{\mu}{\varrho} = 0 \tag{A9.76}$$

for isotropic, dilatation free waves.

Equation (A9.75) is obviously the secular equation for Γ_{ik}/ϱ, and the eigenvalues are c^2. The secular equation is of third degree, and Γ_{ik} is positive definite and symmetric: thus there are three real eigenvalues – three phase velocities, c_1, c_2 and c_3. The associated directions d_I, d_{II} and d_{III} are found by insertion of c_1, c_2 and c_3, respectively, in (A9.72) and use of (A9.69).

If the propagation velocity of the plane wave coincides with the g_i direction, then there are three possible phase velocities: c_I, c_{II} and c_{III}. The velocity c_I is associated with a particle velocity in the direction $d_I^{(i)}$, the velocities c_{II} and c_{III} are associated with particle velocities in the directions $d_{II}^{(i)}$ and $d_{III}^{(i)}$. The unit vectors $d_I^{(i)}$, $d_{II}^{(i)}$ and $d_{III}^{(i)}$ are orthogonal, but not neccessarily parallel or perpendicular to the propagation direction g_i. Thus, in the general anisotropic case, plane waves with constant phase velocity are neither longitudinal nor transversal, but some mixture between these two.

A10 Formulae related to path-independent integrals

Energy flow to the process region

Consider a crack edge moving slowly in a plate subjected to in-plane loading. A cylindrical surface S_Γ perpendicular to the plate surfaces and encircling the process region is assumed to be positioned as shown by Fig. A10.1. In addition to S_Γ, another surface $S_{\Gamma pr}$, perpendicular to the plate surfaces, is introduced, drawn outside the process region and its wake, but open-ended at its rear, where it meets S_Γ. It is assumed to be positioned as close to the process region and its wake as possible, but the component in the x_1 direction of its outwards normal must be non-negative, i.e., $S_{\Gamma pr}$ must be non-concave outwards. A coordinate system (s, x_3), where s is the arc length along S_Γ or $S_{\Gamma pr}$ at constant x_3-level, will be used to indicate positions on the surface S_Γ or $S_{\Gamma pr}$. Arrows in Fig. A10.1 show positive directions for s. S_Γ may or may not cut through the wake of the process region, depending on how far back it is drawn.

The crack length a is assumed to be measured to the crack edge intersection with the mid-plane of the plate (note that the crack edge may be curved). The flow of energy into the region bounded by S_Γ due to an increase da of the crack length is

$$\Phi\, da \approx \oint_{S_\Gamma} n_j \sigma_{ji} \frac{\partial u_i}{\partial a} da\, ds\, dx_3 \qquad (i=1,2,3;\ j=1,2) \tag{A10.1}$$

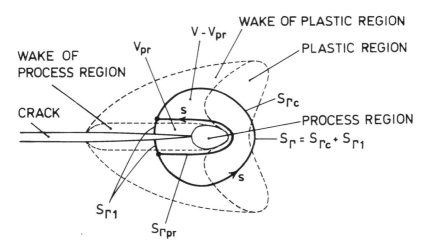

Fig. A10.1 Cylindrical surfaces, perpendicular to the plate surfaces and the volume $V - V_{pr}$. S_Γ must end behind the process region.

where σ_{ji} is to be put to zero along the parts of S_Γ that cut through the wakes of the process region, where concepts like stress and strain don't apply. This implies neglect of the energy flow through these two parts, which is justified because energy *changes* very near the crack faces behind the edge are certainly small compared to other energy terms involved. The displacement u_i is a function of x_1, x_2, x_3 and the crack length a. Due to development of a thumb-nail effect the advance of the crack edge might vary with x_3, but a and (consequently) da do not.

The increase of energy inside S_Γ consists of two distinguishable parts. One is the increase of energy in the process region and the other is the increase of stress-strain energy in the continuum part:

$$h(dD_{pr})_{av} + \int_{V-V_{pr}} \frac{\partial W}{\partial a} da\, dV + \int_{S_{\Gamma pr}} W da\, dx_2\, dx_3 \quad \text{(A10.2)}$$

where h is the plate thickness, D_{pr} is the energy dissipation in the process region per unit of plate thickness, V is the volume inside S_Γ and V_{pr} is the part of this volume that is inside $S_{\Gamma pr} + S_{\Gamma 1}$. Subscript av refers to the average over the plate thickness. The last integral is the energy loss in the continuum when the process region extends the distance da forwards; see Fig. A10.2. The positive sign depends on the direction of integration: note that the direction of s along $S_{\Gamma pr}$ is such that an increase of s implies a decrease (or no change) of x_2. Changes of the recoverable energy in the process region and its wake are neglected.

The flow of energy through S_Γ must equal the energy increase inside S_Γ so that

$$\left(\frac{dD_{pr}}{da}\right)_{av} \approx \int_{\Gamma_c} \left(n_j \sigma_{ji} \frac{\partial u_i}{\partial a}\right)_{av} ds - \int_{A-A_{pr}} \frac{\partial W_{av}}{\partial a} dA - \int_{\Gamma_{pr}} W_{av} dx_2 \quad \text{(A10.3)}$$

where $j \neq 3$, Γ_c and Γ_{pr} are the intersections between $S_{\Gamma c}$ respectively $S_{\Gamma pr}$ and the plate mid-plane, and $A - A_{pr}$ is the area between Γ_c and Γ_{pr}.

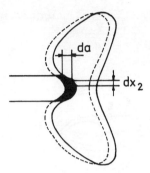

Fig. A10.2 Forwards motion of the process region contributes a loss of energy in the continuum in front.

Now, the area integral in (A10.3) is written as

$$\int_{A-A_{pr}} \frac{\partial W_{av}}{\partial x_1} dA - \int_{A-A_{pr}} \left[\frac{\partial W_{av}}{\partial a} + \frac{\partial W_{av}}{\partial x_1}\right] dA$$

$$= \int_{\Gamma_c+\Gamma_{pr}} n_1 W_{av} \, ds - \int_{A-A_{pr}} \left[\frac{\partial W_{av}}{\partial a} + \frac{\partial W_{av}}{\partial x_1}\right] dA \quad (A10.4)$$

where Green's theorem was used to transform the area integral to a line integral.

Finally, combination of Eqs (A10.3) and (A10.4) gives:

$$\left(\frac{dD_{pr}}{da}\right)_{av} \approx \int_{\Gamma_c} [W_{av} dx_2 + (n_j \sigma_{ji} \partial u_i/\partial a)_{av} \, ds]$$

$$- \int_{A-A_{pr}} \left[\frac{\partial W_{av}}{\partial a} + \frac{\partial W_{av}}{\partial x_1}\right] dA \quad (i=1,2,3;\ j=1,2) \quad (A10.5)$$

Relations between P- or J-integrals and load-displacement records

Equation (3.2.26),

$$hP(\delta, a) = -\int_0^\delta \frac{\partial Q(\delta', a)}{\partial a} d\delta' \quad (A10.6)$$

expresses the P-integral as a function of the displacement δ under the load Q on the boundary of a plate with thickness h and containing an edge crack with length a. Mode I or II is assumed. Differentiation with respect to δ gives

$$h\frac{\partial P(\delta, a)}{\partial \delta} = -\frac{\partial Q(\delta, a)}{\partial a} \quad (A10.7)$$

and, hence, by integration,

$$Q(\delta, a) = Q(\delta, 0) - h\int_0^a \frac{\partial P(\delta, a')}{\partial \delta} da' \quad (A10.8)$$

Here, the P-integral was regarded as a function of δ and a. Similar relations may be obtained for $P(Q,a)$. To this end, consider the work performed by the outer load until it reaches a value Q. It may be written as

$$\int_0^\delta Q(\delta', a)\mathrm{d}\delta' = Q(\delta, a) \cdot \delta - \int_0^Q \delta(Q', a)\mathrm{d}Q' \tag{A10.9}$$

because both members equal the area under the $Q - \delta$-curve, Fig. 3.2.7. δ is the displacement reached at load Q. In the same way, the work at crack length $a + \mathrm{d}a$ under the same load Q is found to be

$$\int_0^\delta \left[Q(\delta', a) + \frac{\partial Q(\delta', a)}{\partial a}\mathrm{d}a\right]\mathrm{d}\delta' + Q(\delta, a)\mathrm{d}\delta$$
$$= Q(\delta, a) \cdot (\delta + \mathrm{d}\delta) - \int_0^Q \left[\delta(Q', a) + \frac{\partial \delta(Q', a)}{\partial a}\mathrm{d}a\right]\mathrm{d}Q' \tag{A10.10}$$

where the displacement under the load Q is $\delta + \mathrm{d}\delta$. The difference in work between the case with crack length a and the case with crack length $a + \mathrm{d}a$ is

$$\mathrm{d}A = \int_0^Q \frac{\partial \delta(Q', a)}{\partial a}\mathrm{d}a\,\mathrm{d}Q' - Q(\delta, a)\mathrm{d}\delta \tag{A10.11}$$

Then, the difference in potential energy is

$$\mathrm{d}\Pi = -Q(\delta, a)\mathrm{d}\delta + \mathrm{d}A \tag{A10.12}$$

and it follows from relation (3.2.28), $hP = -\mathrm{d}\Pi/\mathrm{d}a$, that

$$hP(Q, a) = \int_0^Q \frac{\partial \delta(Q', a)}{\partial a}\mathrm{d}Q' \tag{A10.13}$$

Differentiation with respect to Q gives

$$h\frac{\partial P(Q, a)}{\partial Q} = \frac{\partial \delta(Q, a)}{\partial a} \tag{A10.14}$$

and, hence, by integration,

$$\delta(Q, a) = \delta(Q, 0) + h\int_0^a \frac{\partial P(Q, a')}{\partial Q}\mathrm{d}a' \tag{A10.15}$$

The equalities (3.2.26), (A10.8), (A10.13) and (A10.15) are useful for calculations of the P-integral when the load-displacement relation is known or *vice versa*. They may also be used to find $P(\delta, a)$ when $P(Q, a)$ is known or *vice versa*. If plane stress or strain prevails in the main part of the plate, then the expressions are valid for the J-integral as well as for the P-integral.

Load-displacement curves $Q = Q(\delta, a)$ for different values of a, the length of the stationary crack, may be obtained experimentally or theoretically. In the latter case, almost invariably, a state of plane stress, or sometimes plane strain, is assumed throughout the plate, so that P may be exchanged by J. As an example, the case of a 3PB specimen with height W, thickness B and length L is considered. Tada et al. (1985)

give the following expression for the stress intensity factor (see Section 3.3):

$$K_I = \frac{QL}{BW^{3/2}} \cdot \frac{3(a/W)^{1/2}}{2(1+2a/W)(1-a/W)^{3/2}}$$
$$\times \left[1.99 - \frac{a}{W}\left(1 - \frac{a}{W}\right)\left(2.15 - 3.93\frac{a}{W} + 2.7\frac{a^2}{W^2}\right)\right] \quad \text{(A10.16)}$$

This expression is valid for either plane stress or plane strain, but for test specimens that are not thicker than about half their height, plane stress prevails except in a close vicinity of the crack edge, so that plane stress may be assumed. Then, according to equation (3.5.14),

$$J(Q,a) = \frac{K_I^2}{E} \quad \text{(A10.17)}$$

The displacement $\delta(Q,a)$ is now obtained from (A10.15), after determination of $\delta(Q,0)$ from beam theory. But from $\delta = \delta(Q,a)$, the expression $Q = Q(\delta,a)$ is found, whereupon $J(\delta,a)$ follows from (3.2.26).

Even though an edge crack was studied, equalities (3.2.26), (A10.8), (A10.13) and (A10.15) are valid also for internal (double-ended) cracks, but some caution is advisable in applications, because two P- or J-integrals, one for each crack edge, are obtained. In order to avoid confusion, consider an x-axis along the crack, with the crack edges situated at $x = b$ and $x = c$, where $c - b = a$. Then, $Q = Q(\delta, b, c)$, etc. Let P_c be the P-integral enclosing crack edge c. The previous formulae may then be written as

$$hP_c(\delta, b, c) = -\int_0^\delta \frac{\partial Q(\delta', b, c)}{\partial c} d\delta' \quad \text{(A10.18)}$$

$$Q(\delta, b, c) = Q(\delta, b, b) - h\int_0^a \frac{\partial P_c(\delta, b, c)}{\partial \delta} dc \quad \text{(A10.19)}$$

$$hP_c(Q, b, c) = \int_0^Q \frac{\partial \delta(Q', b, c)}{\partial c} dQ' \quad \text{(A10.20)}$$

$$\delta(Q, b, c) = \delta(Q, b, b) + h\int_0^a \frac{\partial P_c(Q, b, c)}{\partial Q} dc \quad \text{(A10.21)}$$

In the case of symmetry with respect to a plane normal to the crack and to the plate surfaces, these equalities default to the following ones, after a change in notation, so that the crack length is called $2a$:

$$hP(\delta, a) = -\frac{1}{2}\int_0^\delta \frac{\partial Q(\delta', a)}{\partial a} d\delta' \quad \text{(A10.22)}$$

$$Q(\delta, a) = Q(\delta, 0) - 2h\int_0^a \frac{\partial P_i(\delta, a')}{\partial \delta} da' \quad \text{(A10.23)}$$

$$hP(Q, a) = \frac{1}{2}\int_0^Q \frac{\partial \delta(Q', a)}{\partial a} dQ' \quad \text{(A10.24)}$$

$$\delta(Q, a) = \delta(Q, 0) + 2h\int_0^a \frac{\partial P(Q, a')}{\partial Q} da' \quad \text{(A10.25)}$$

References

Abraham, F.F., Brodbeck, D., Rafey, R.A. and Rudge, W.E., 1994. Instability dynamics of fracture: A computer simulation investigation. *Phys. Rev. Letters*, **73**, 272-275.

Abraham, F.F., Brodbeck, D., Rudge, W.E. and Xu, X., 1997. A molecular dynamics investigation of rapid fracture mechanics. *J. Mech. Phys. Solids*, **45**, 1595-1619.

Achenbach, J.D., 1970a. Crack propagation generated by a horizontally polarized shear wave. *J. Mech. Phys. Solids*, **18**, 245-259.

Achenbach, J.D., 1970b. Brittle and ductile extension of a finite crack by a horizontally polarized shear wave. *Int. J. Eng. Sci.*, **8**, 947-966.

Achenbach, J.D., 1973. *Wave Propagation in Elastic Solids*, North-Holland, Amsterdam.

Achenbach, J.D. and Bazant, Z.P., 1975. Elastodynamic near-tip stress and displacement fields for rapidly propagating cracks in orthotropic media. *J. Appl. Mech.*, **42**, 183-189.

Achenbach, J.D. and Gautesen, A.K., 1977. Elastodynamic stress intensity factors for a semi-infinite crack. *J. Appl. Mech.*, **44**, 243-249.

Achenbach, J.D. and Li, Z.-L., 1985. Plastic deformation near a rapidly propagating crack tip. In *Fundamentals of Deformation and Fracture*, Eshelby Memorial Symposium, edited by B.A. Bilby, K.J. Miller and J.R. Willis. Cambridge Univ. Press, Cambridge, 217-231.

Achenbach, J.D. and Nuismer, R., 1971. Fracture generated by a dilatational wave. *Int. J. Fract. Mech.*, **7**, 77-88.

Achenbach, J.D. and Tolikas, P.K., 1977. Elastodynamic effects on crack arrest. In *Fast Fracture and Crack Arrest, ASTM STP 627*, edited by G.T. Hahn and M.F. Kanninen, American Society for Testing and Materials, 59-76.

Aifantis, E.C., 1984. On the microstructural origin of certain inelastic models. *Trans. ASME J. Engng Mater. Tech.*, **106**, 326-330.

Al-Ani, A.M. and Hancock, J.W., 1991. J-dominance of short cracks in tension and bending. *J. Mech. Phys. Solids*, **39**, 23-43.

Amazigo, J.C., 1974. Fully plastic crack in an infinite body under anti-plane shear. *Int. J. Solids Structures*, **10**, 1003-1015.

Amazigo, J.C., 1975. Fully plastic center-cracked strip under anti-plane shear. *Int. J. Solids Structures*, **11**, 1291-1299.

Amazigo, J.C. and Hutchinson, J.W., 1977. Crack-tip fields in steady crack growth with linear strain hardening. *J. Mech. Phys. Solids*, **25**, 81-97.

Anderson, T.L., 1995. *Fracture Mechanics: Fundamentals and Applications*, second edition. CRC Press, Boca Raton, Ann Arbor, London, Tokyo.

Andersson, H., 1973a. A finite-element representation of stable crack growth. *J. Mech. Phys. Solids*, **21**, 337-356.

Andersson, H., 1973b. The steadily growing, elastic-plastic crack tip in a finite element treatment. *Int. J. Fract.*, **9**, 231-233.

Andersson, H., 1974. Finite element treatment of a uniformly moving elastic-plastic crack tip. *J. Mech. Phys. Solids*, **22**, 285-308.

Andersson, H., 1975. Finite element methods applied to problems of moving cracks. In *Computational Fracture Mechanics*. The American Society of Mechanical Engineers, New York, 185-198.

Andersson, H., 1977. Analysis of a model for void growth and coalescence ahead of a moving crack tip. *J. Mech. Phys. Solids*, **25**, 217-233.

Andersson, H. and Bergkvist, H., 1970. Analysis of a non-linear crack model. *J. Mech. Phys. Solids*, **18**, 1-28.

Andrews, D.J., 1976. Rupture velocity at plane strain shear cracks. *J. geophys. Res.*, **81**, 5679-5687.

Andrews, D.J., 1985. Dynamic plane-strain rupture with a slip-weakening friction law calculated by a boundary-integral method. *Bull. seism. Soc. America*, **75**, 1-22.

Andrews, E.H., 1959. Stress waves and fracture surfaces. *J. Appl. Phys.*, **30**, 740-743.

Ang, D.D., 1958. *Some Radiation Problems in Elastodynamics*. Dissertation, California Institute of Technology, Pasadena, USA.

Anthony, S.R., Chubb, J.P. and Congleton, J., 1970. The crack-branching velocity. *Phil. Mag.*, **16**, 1201-1216.

Aoki, S., Kishimoto, K., Takeya, A. and Sakata, M., 1984. Effects of microvoids on crack blunting and initiation in ductile materials. *Int. J. Fract.*, **24**, 267-278.

Aoki, S., Nonoyama, Y. and Amaya, K., 1995. Boundary element study on the optical method of caustic for measuring fast crack propagation toughness K_{ID}. *Int. J. Fract.*, **71**, 379-390.

Arakawa, K. and Takahashi, K., 1991a. Relationships between fracture parameters and fracture surface roughness of brittle polymers. *Int. J. Fract.*, **48**, 103-114.

Arakawa, K. and Takahashi, K., 1991b. Branching of a fast crack in polymers. *Int. J. Fract.*, **48**, 245-254.

Arawas, N. and Blazo, D.A., 1991. Higher order terms in asymptotic elastoplastic mode-III crack tip solutions. *Acta Mechanica*, **90**, 130-153.

Archuleta, R.J., 1982. Analysis of near source static and dynamic measurements from the 1979 Imperial Valley earthquake. *Bull. seism. Soc. America*, **72**, 1927-1956.

Asaro, R.J., 1979. Geometrical effects in the inhomogeneous deformation of ductile single crystals. *Acta Metallurgica*, **27**, 445-453.

Ashby, M.F., Blunt, F.J. and Bannister, M., 1989. Flow characteristics of highly constrained metal wires. *Acta Metallurgica*, **37**, 1847-1857.

Atkinson. C., 1965. The propagation of a brittle crack in anisotropic material. *Int. J. Engng Sci.*, **3**, 77-91.

Atkinson, C., 1967. A simple model of relaxed expanding crack. *Arkiv för Fysik*, **35**, 469-476.

Atkinson, C., 1968. On the axially symmetric expanding boundary value problems in classical elasticity. *Int. J. Engng Sci.*, **6**, 27-35.

Atkinson, C., 1977. Dynamic crack problems in dissimilar media. In *Mechanics of Fracture, Vol. IV: Elastodynamic Problems*, edited by G.C. Sih, Leyden: Noordhoff, 213-248.

Atkinson, C., 1979. A note on some dynamic crack problems in linear viscoelasticity. *Archives of Mechanics (Archiwum Mechaniki Stosowanej)*, **31**, 829-849.

Atkinson, C. and Bourne, J.P., 1989. Stress singularities in viscoelastic media. *Quart. J. Mech. Appl. Math.*, **42**, 385-412.

Atkinson, C. and Bourne, J.P., 1990. Stress singularities in angular sections of viscoelastic media. *Int. J. Engng Sci.*, **28**, 615-630.

Atkinson, C. and Coleman, C.J., 1977. On some steady-state moving problems in the linear theory of viscoelasticity. *J. Inst. Maths Applics*, **20**, 85-106.

Atkinson, C. and Craster, R.V., 1995. Theoretical aspects of fracture mechanics. *Prog. Aerospace Sci.*, 31, 1-83.

Atkinson, C. and Eshelby, J.D., 1968. The flow of energy into the tip of a moving crack. *Int. J. Fract. Mech.*, **4**, 3-8.

Atkinson, C. and Popelar, C., 1979. Antiplane dynamic crack propagation in a viscoelastic strip. *J. Mech. Phys. Solids*, **27**, 431-439.

Atkinson, C. and Thiercelin, M., 1993. The interaction between the well bore and pressure-induced fractures.*Int. J. Fract.*, **59**, 23-40.

Atluri, S.N., Kobayashi, A.S. and Nakagaki, M., 1975. An assumed displacement hybrid finite element model for linear fracture mechanics. *Int. J. Fract.*, **11**, 257-271.

Baker, B.R., 1962. Dynamic stresses created by a moving crack. *J. Appl. Mech.*, **29**, 449-458.

Ball, J.M., 1977. Convexity conditions and existence theorems in nonlinear elasticity. *Arch. Rational Mech. Anal.*, **63**, 337-403.

Banks-Sills, Leslie, 1993. Weight functions for interface cracks. *Int. J. Fract.*, **60**, 89-95.

Barenblatt, G.I., 1959a. The formation of equilibrium cracks during brittle fracture. General ideas and hypotheses. Axially-symmetric cracks. *J. Appl. Math. Mech.*, **23**, 622-636. English translation from *PMM*, **23** (1959), 434-444.

Barenblatt, G.I., 1959b. Equilibrium cracks formed during brittle fracture. Rectilinear cracks in plane plates. *J. Appl. Math. Mech.*, **23**, 1009-1029. English translation from *PMM*, **23** (1959), 706-721.

Barenblatt, G.I., 1959c. Concerning equilibrium cracks forming during brittle fracture. The instability of isolated cracks, relationship with energetic theories. *J. Appl. Math. Mech.*, **23**, 1273-1282. English translation from *PMM*, **23** (1959), 893-900.

Barenblatt, G.I., 1993. Micromechanics of fracture. In Theoretical and Applied Mechanics 1992, edited by S.R. Bodner, J. Singer, A. Solan and Z. Hashin. Elsevier Science Publishers B.V., Amsterdam, 25-52.

Barenblatt, G.I. and Botvina, L.R., 1983. Self-similar nature of fatigue failure: Damage accumulation. *Mechanics of Solids*, **18**, 160-164. Translation from *Izv. Akad. Nauk SSSR: Mekhanika Tverdogo Tela*, **18**, (1983), 88-92.

Barenblatt, G.I. and Cherepanov, G.P., 1960. On the wedging of brittle bodies. *J. Appl. Math. Mech.*, **23**, 993-1014. English translation from *PMM*, **24** (1960), 667-682.

Barenblatt, G.I., Salganik, R.L. and Cherepanov, G.P., 1962. On the nonsteady motion of cracks. *J. Appl. Math. Mech.*, **26**, 469-477). English translation from *PMM*, **26** (1962), 328-334.

Barnett, D.M. and Asaro, R.J., 1972. The fracture mechanics of slit-like cracks in anisotropic elastic media. *J. Mech. Phys. Solids*, **20**, 353-366.

Barnett, D.M. and Lothe, J., 1973. Synthesis of the sextic and the integral formulations of dislocations, Green's functions and surface waves in anisotropic elastic solids. *Phys. Norv.*, **7**, 13-19.

Bassani, J.L. and Qu, J., 1989. Finite crack on bimaterial and bicrystal interfaces. *J. Mech. Phys. Solids*, **37**, 435-453.

Bažant, Z.P., Belytschko, T. and Chang, T.P., 1984. Continuum theory for strain softening. *J. Engng Mech., ASCE*, **115**, 755-767.

Bažant, Z.P. and Estenssoro, L.F., 1979. Surface singularity and crack propagation. *Int. J. Solids Structures*, **15**, 405-426.

Bažant, Z.P. and Estenssoro, L.F., 1980. Addendum to the paper "Surface singularity and crack propagation". *Int. J. Solids Structures*, **16**, 479-481.

Bažant, Z.P. and Lin, F.B., 1988. Non-local yield limit degradation. *Int. J. Numer. Methods Engng*, **26**, 1805-1823.

Bažant, Z.P. and Pfeiffer, P.A., 1986. Shear fracture tests of concrete. *Materials and Structures*, **19**, 111-121.

Beatty, M.F. and Hayes, M.A., 1992a. Deformations of an elastic, internally constrained material. Part 1: Homogeneous deformations. *J. Elasticity*, **29**, 1-84.

Beatty, M.F. and Hayes, M.A., 1992b. Deformations of an elastic, internally constrained material. Part 2: Nonhomogeneous deformations. *Quart. J. Mech. Appl. Math.*, **45**, 663-709.

Begley. J.A. and Landes, J.D., 1972. The J-integral as a fracture criterion. *Fracture Toughness* (Proceedings of the 1971 National Symposium on Fracture Mechanics, Part II), A.S.T.M., S.T.P. **514**, 1-23.

Bell, J.F., 1985. Contemporary perspectives in finite strain plasticity. *Int. J. Plasticity*, **1**, 3-27.

Bell, J.F., 1989. Experiments on the kinetics of large plastic strain in ordered solids. *Int. J. Solids Structures*, **25**, 267-278.

Beltz, G.E., Rice, J.R., Shih, C.F. and Xia, L., 1996. A self-consistent model for cleavage in the presence of plastic flow. *Acta Mat.*, **44**, 3943-3954.

Bentham, J.P., 1977. State of stress at the vertex of a quarter-infinite crack in a half-space. *Int. J. Solids Structures*, **13**, 479-492.

Bentham, J.P., 1980. The quarter-infinite crack in a half-space; alternative and additional solutions. *Int. J. Solids Structures*, **16**, 119-130.
Bergkvist, H., 1973. The motion of a brittle crack. *J. Mech. Phys. Solids*, **21**, 229-239.
Bergkvist, H., 1974. Crack arrest in thin sheets. *J. Mech. Phys. Solids*, **22**, 491-502.
Bergkvist, H., 1977. An investigation of axisymmetric crack propagation. *Fast Fracture and Crack Arrest, ASTM STP 627*, edited by G.T. Hahn and M.F. Kanninen, American Society for Testing and Materials, 321-335.
Bergkvist, H. and Andersson, H., 1972. Plastic deformation of a stably growing crack tip. *Int. J. Fract. Mech.*, **8**, 139-156.
Bergkvist, H. and Guex-Le, L.H., 1977. J-integral related quantities in axisymmetric cases. *Int. J. Fract.*, **13**, 556-558.
Bergkvist, H. and Guex, L., 1979. Curved crack propagation. *Int. J. Fract.*, **15**, 429-441.
Betegón, C. and Hancock, J.W., 1991. Two-parameter characterization of elastic-plastic crack-tip fields. *J. Appl. Mech.*, **58**, 104-110.
Bever, M.D., Holdt, D.L. and Titchner, A.L., 1973. The stored energy of cold work. *Progr. Materials Science*, **17**, 1-190.
Bhattacharjee, D. and Knott, J.F., 1993. Micromechanisms of fracture in steel tested under mixed-mode loading above the transition temperature. In *Mixed-Mode Fatigue and Fracture*, edited by H.P. Rossmanith and K.J. Miller, Mechanical Engineering Publications, London, 99-109.
Bhattacharjee, D. and Knott, J.F., 1994. Ductile fracture in HY100 steel under mixed mode I/mode II loading. *Acta Metall. Mater.*, **42**, 1747-1754.
Bilby, B.A. and Cardew, G.E., 1975. The crack with a kinked tip. *Int. J. Fract.*, **11**, 708-712.
Bilby, B.A., Cardew, G.E., Goldthorpe, M.R. and Howard, I.C., 1986. A finite element investigation of the effect of specimen geometry on the fields of stress and strain at the tip of stationary cracks. In *Size Effects in Fracture*, Institution of Mechanical Engineers, London, 37-46.
Bilby, B.A., Cottrell, A.H., Smith, E. and Swinden, K.H., 1964. Plastic yielding from from sharp notches. *Proc. Roy. Soc. (London)*, **A279**, 1-9.
Bilby, B.A., Cottrell, A.H. and Swinden, K.H., 1963. The spread of plastic yield from a notch. *Proc. Roy. Soc. (London)*, **A272**, 304-314.
Bishlinghoff, R.L., Mar, J.W. and Pian, T.H.H., 1990. *Statics of Deformable Solids*. Dover Publications, Inc., New York.
Bodner, S.R. and Partom, Y., 1975. Constitutive equations for elastic-viscoplastic strain hardening materials. *J. Appl. Mech.*, **42**, 385-389.
Boley, B.A. and Weiner, J.H., 1960. *Theory of Thermal Stresses*, J. Wiley and Sons, New York, N.Y.
Borodich, F.M., 1997. Some fractal models of fracture. *J. Mech. Phys. Solids*, **45**, 239-259.
Boulanger, Ph. and Hayes, M., 1993. *Bivectors and Waves in Mechanics and Optics*, Chapman & Hall, London.
Brace, W.F. and Bombolakis, E.G., 1963. A note on brittle crack growth in compression. *J. Geophys. Res.*, **68**, 3709-3713.
Brickstad, B., 1983. A viscoplastic analysis of rapid crack propagation experiments in steel. *J. Mech. Phys. Solids*, **31**, 307-327.
Broberg, K.B., 1959. A problem on stress waves in an infinite elastic plate. *Transactions Roy. Inst. Techn., Stockholm, Sweden*, **139**, 1-27.
Broberg, K.B., 1960. The propagation of a brittle crack. *Arkiv för Fysik*, **18**, 159-192.
Broberg, K.B., 1964. On the speed of a brittle crack. *J. Appl. Mech.*, **31**, 546-547.
Broberg, K.B., 1967. Discussion of fracture from the energy point of view. In *Recent Progress in Applied Mechanics*, edited by B. Broberg, J. Hult and F. Niordson. Almqvist and Wiksell, Stockholm, 125-151.
Broberg, K.B., 1968. Critical review of some theories in fracture mechanics. *Int. J. Fract. Mech.*, **4**, 11-18.
Broberg, K.B., 1971. Crack-growth criteria and non-linear fracture mechanics. *J. Mech. Phys. Solids*, **19**, 407-418.

Broberg, K.B., 1973a. Discussion of initial and subsequent crack growth. *Engng Fract. Mech.*, **5**, 1031-1035.

Broberg, K.B., 1973b. On dynamic crack propagation in elastic-plastic media. In *Dynamic Crack Propagation*, edited by G.C. Sih, Noordhoff International Publishing, Leyden, 461-499.

Broberg, K.B., 1974. The importance of stable crack extension in linear and non-linear fracture mechanics. In *Prospects of Fracture Mechanics*, edited by G.C. Sih, H.C. van Elst and D. Broek, Noordhoff International Publishing, Leyden, 125-137.

Broberg, K.B., 1975. On stable crack growth. *J. Mech. Phys. Solids* **23**, 215-237.

Broberg, K.B., 1976. On determination of the J-integral for paths close to a crack-tip in cases of axial symmetry. GALCIT SM Report 76-9, Graduate Aeronautical Laboratories, California Institute of Technology, Pasadena, California.

Broberg, K.B., 1978. On transient sliding motion. *Geophys. J. R. astr. Soc.*, **52**, 397-432.

Broberg, K.B., 1979a. On the behaviour of the process region at a fast running crack tip. In *High Velocity Deformation of Solids*, edited by K. Kawata and J. Shioiri, Springer-Verlag, Berlin Heidelberg, 182-194.

Broberg, K.B., 1979b. Mathematical methods in fracture mechanics. In *Trends in Applications of Pure Mathematics to Mechanics*, vol. II, edited by H. Zorski, Pitman Publ. Ltd, London, 57-78.

Broberg, K.B., 1982. The foundations of fracture mechanics. *Engng Fract. Mech.*, **16**, 497-515.

Broberg, K.B., 1983a. New approaches in fracture mechanics. In *Mechanical Behaviour of Materials*, edited by J. Carlsson and N.G. Ohlson, Pergamon Press, vol. 2, 927-934.

Broberg, K.B., 1983b. On crack paths. In *Workshop on Dynamic Fracture*, California Institute of Technology, Pasadena, California, 140-155.

Broberg, K.B., 1985. What happens at fast crack growth? In *Fundamentals of Deformation and Fracture*, Eshelby Memorial Symposium, edited by B.A. Bilby, K.J. Miller and J.R. Willis. Cambridge Univ. Press, Cambridge, 233-242.

Broberg, K.B., 1987a. A path-independent integral for plates. *J. Appl. Mech.*, **54**, 458-459.

Broberg, K.B., 1987b. On crack paths. *Engng Fract. Mech.*, **28**, 663-679.

Broberg, K.B., 1989a. The near-tip field at high crack velocities. *Int. J. Fracture*, **39**, 1-13.

Broberg, K.B., 1989b. Implications of viscoplasticity at rapid crack growth. In *Advances in Plasticity 1989*, Pergamon Press, Oxford, 709-712.

Broberg, K.B., 1992. Fracture mechanics - *theoria* or *tekhne*? *Int. J. Fract.*, **57**, 85-99.

Broberg, K.B., 1994. Intersonic bilateral slip. *Geophys. J. International*, **119**, 706-714.

Broberg, K.B., 1995a. Critical review of some methods in nonlinear fracture mechanics. *Engng Fract. Mech.*, **50**, 157-164.

Broberg, K.B., 1995b. Influence of lateral inertia on crack propagation in plates. *Int. J. Solids Structures*, **32**, 2457-2480.

Broberg, K.B., 1995c. Intersonic mode II crack expansion. *Archives of Mechanics (Archiwum Mechaniki Stosowanej)*, **47**, 859-871.

Brock, L.M. and Achenbach, J.D., 1973. Extension of an interface flaw under the influence of transient waves. *Int. J. Solids Structures*, **9**, 53-67.

Broek, D., 1968. Some considerations on slow crack growth. *Int. J. Fract. Mech.*, **4**, 19-34.

Broek, D., 1971. A study on ductile fracture. *Nationaal Lucht- en Raumtevaartlaboratorium*, The Netherlands, NLR TR 71021 U.

Broek, D., 1972. Some contributions of electron fractography to the theory of fracture. *Nationaal Lucht- en Raumtevaartlaboratorium*, The Netherlands, NLR TR 72029 U.

Broek, D., 1982. *Elementary Engineering Fracture Mechanics*, 3rd ed. Martinus Nijhoff Publ., The Hague.

Brook, R.J., 1991. *Concise Encyclopedia of Advanced Ceramic Materials*, Pergamon Press, Oxford.

Brown, W.F., Jr and Srawley, J.E., 1967. *Plane Strain Crack Toughness Testing of High Strength Metallic Materials. ASTM STP 410*.

Budiansky, B., Amazigo, J.C. and Evans, A.G., 1988. Small-scale crack bridging and the fracture toughness of particulate-reinforced ceramics. *J. Mech. Phys. Solids*, **36**, 167-187.

Budiansky, B. and O'Connell, R.J., 1975. Elastic moduli of a cracked solid. *Int. J. Solids Structures*, **12**, 81-97.

Budiansky, B. and Rice, J.R., 1973. Conservation laws and energy release rates. *J. Applied Mech.*, **26**, 201-203.

Bueckner, H.F., 1970. A novel principle for the computation of stress intensity factors. *Z. Angew. Math. Mech.*, **50**, 529-546.

Bui, H.D., 1974. Dual path independent integrals in the boundary-value problems of cracks. *Engng Fract. Mech.*, **6**, 287-296.

Burridge, R., 1973. Admissible speeds for plane-strain self-similar shear cracks with friction but lacking cohesion. *Geophys. J. Roy. astr. Soc.*, **35**, 439-455.

Burridge, R., 1976. An influence function for the intensity factor in tensile fracture. *Int. J. Eng. Sci.*, **14**, 725-734.

Burridge, R., Conn, G. and Freund, L.B., 1979. The stability of rapid mode II shear crack with finite cohesive traction. *J. geophys. Res.*, **84**, 2210-2222.

Burridge, R. and Willis, J.R., 1969. The self-similar problem of the expanding elliptical crack in an anisotropic solid. *Proc. Cambridge Phil. Soc.*, **66**, 443-468.

Burton, W.S., Sinclair, G.B., Solecki, J.S. and Swedlow, J.L., 1984. On the implications for LEFM of the three-dimensional aspects in some crack/surface intersection problems. *Int. J. Fract.*, **25**, 3-32.

Cagniard, L., 1939. *Reflexion et refraction des ondes seismiques progressives*, Gauthier-Villars, Paris.

Carlsson, A.J., 1962. Experimental studies of brittle fracture propagation. *Transactions Roy. Inst. Techn., Stockholm, Sweden*, **189**.

Carlsson, A.J., 1963. On the mechanism of brittle fracture propagation. Branching, fracture surface appearance, and the influence of stress waves. *Transactions Roy. Inst. Techn., Stockholm, Sweden*, **205**.

Carlsson, J., 1974. Path independent integrals in fracture mechanics and their relation to variational principles. In *Prospects of Fracture Mechanics*, edited by G.C. Sih, H.C. van Elst and D. Broek, Noordhoff International Publishing, Leyden, 139-157.

Carothers, S.D., 1920. Plane strain: the direct determination of stress. *Proc. Roy. Soc. (London)*, **A97**, 110-123.

Champion, C.R., 1988. The stress intensity factor history for an advancing crack under three-dimensional loading. *Int. J. Solids Structures*, **24**, 285-300.

Chang, Y.W. and Asaro, R.J., 1980. Lattice rotations and localized shearing in single crystals. *Archives of Mechanics (Archiwum Mechaniki Stosowanej)*, **32**, 369-388.

Chen, E.P. and Sih, G.C., 1977. Transient response of cracks to impact loads. In *Elastodynamic Crack Problems*, edited by G.C. Sih, Noordhoff International Publishing, Leyden, 1-58.

Chen, X.-Y. and Drugan, W.J., 1991. Plane strain elastic-ideally plastic crack fields for mode I quasistatic growth at large-scale yielding — II. Global analytical solutions for finite geometries. *J. Mech. Phys. Solids*, **39**, 895-925.

Cherepanov, G.P., 1962. The stress state in a heterogeneous plate with slits (in Russian). *Izvestia AN SSSR, OTN, Mekhan. i Mashin.*, **1**, 131-137. See: Cherepanov (1979).

Cherepanov, G.P., 1967. Crack propagation in continuous media. *J. Appl. Math. Mech.*, **31**, 503-512. English translation from *PMM*, **31** (1967), 476-488.

Cherepanov, G.P., 1979. *Mechanics of Brittle Fracture*. McGraw-Hill, New York.

Cherepanov, G.P. and Afanasev, E.F., 1974. Some dynamic problems of the theory of elasticity — A review. *Int. J. Engng Sci.*, **12**, 665-690.

Chitaley, A.D. and McClintock, F.A., 1971. Elastic-plastic mechanisms of steady crack growth under antiplane conditions. *J. Mech. Phys. Solids*, **19**, 147-163.

Christensen, R.M., 1982. *Theory of Viscoelasticity: An Introduction*, Academic Press, New York.

Chu, C.C. and Needleman, A., 1980. Void nucleation effects in biaxially stretched sheets. *J. Engng Mater. Technol.*, **102**, 249-256.

Clark, A.B.J. and Irwin, G.R., 1966. Crack-propagation behaviours. *Experimental Mechanics*, **6**, 321-330.

Clark, D.S. and Wood, D.S., 1949. The time delay for the initiation of plastic deformation at rapidly applied constant stress. *Proc. Am. Soc. Test. Mat.*, **49**, 717-737.

Clifton, R.J., 1990. High strain rate behaviour of metals. *Appl. Mech. Review*, **43**, S9-S22.

Cochard, A.P. and Rice, J.R., 1997. A spectral method for numerical elastodynamic fracture analysis without spatial replication of the rupture event. *J. Mech. Phys. Solids*, **45**, 1393-1418.

Comninou, M., 1977. The interface crack. *J. Appl. Mech.*, **44**, 631-636.

Comninou, M., 1978. The interface crack in a shear field. *J. Appl. Mech.*, **45**, 287-290.

Congleton, J., 1973. Practical applications of crack-branching measurements. In *Dynamic Crack Propagation*, edited by G.C. Sih, Noordhoff International Publishing, Leyden, 427-438.

Congleton, J. and Petch, N.J., 1967. Crack-branching. *Phil. Mag.*, **16**, 749-760.

Cook, T.S. and Erdogan, F., 1972. Stresses in bonded materials with a crack perpendicular to the interface. *Int. J. Engng Sci.*, **10**, 677-697.

Cotterell, B., 1977. Plane stress ductile fracture. In *Fracture Mechanics and Technology*, edited by G.C. Sih and C.L. Chow, Vol. 2, Sijthoff and Noordhoff International Publishers, The Netherlands, 785-795.

Cotterell, B. and Reddel, J.K., 1977. The essential work of plane stress fracture. *Int. J. Fract.*, **13**, 267-277.

Cotterell, B. and Rice, J.R., 1980. Slightly curved or kinked cracks. *Int. J. Fract.*, **16**, 155-169.

Craggs, J.W., 1960. On the propagation of a crack in an elastic-brittle material. *J. Mech. Phys. Solids*, **8**, 66-75.

Craggs, J.W., 1963. Fracture criteria for use in continuum mechanics. In *Fracture of Solids*, edited by D.C. Drucker and J.J. Gilman, Interscience Publishers, N.Y., 51-63.

Craggs, J.W., 1966. The growth of a disk shaped crack. *Int. J. Engng Sci.*, **4**, 113-124.

Curran, D.R., Seaman, L. and Shockey, D.A., 1987. Dynamic failure of solids. *Physics Reports*, **147**, 253-338.

Currie, P.K., Hayes, M.A. and O'Leary, P.M., 1977. Viscoelastic Rayleigh waves. *Quart. Appl. Math.*, **35**, 35-53.

Curry, D.A. and Knott, J.F., 1979. The effect of microstructure on the cleavage fracture toughness of quenched and tempered steels. *Metal Science*, **13**, 341-345.

Dahlberg, L., Nilsson, F. and Brickstad, B., 1980. Influence of specimen geometry on crack propagation and arrest toughness. In *Crack Arrest Methodoloy and Applications, ASTM STP 711*, edited by G.T. Hahn and M.F. Kanninen. American Society for Testing and Materials, Philadelphia, 89-108.

Dally, J.W., Fourney, W.L. and Irwin, G.R., 1985. On the uniqueness of the stress intensity factor – crack velocity relationship. *Int. J. Fract.*, **27**, 159-168.

Das, S. and Aki, K., 1977. A numerical study of two-dimensional spontaneous rupture propagation. *Geophys. J. R. astr. Soc.*, **50**, 643-668.

Day, S.M., 1982. Three-dimensional simulation of spontaneous rupture: the effect of non-uniform prestress. *Bull. Seism. Soc. Am.*, **72**, 1881-1902.

Dean, R.H. and Hutchinson, J.W., 1980. Quasi-static steady growth in small-scale yielding. In *Proc. Twelfth National Symposium on Fracture Mechanics, May 1979, St Louis, MO, ASTM-STP 700. American Society for Testing and Materials*, Philadelphia, 383-405.

de Borst, R. and Mühlhaus, H.-B., 1992. Gradient-dependent plasticity: Formulation and algorithmic aspects. *Int. J. Numer. Methods Engng*, **35**, 521-539.

de Hoop, A.T., 1961. A modification of Cagniard's method for solving seismic pulse problems. *Applied Scientific Research*, **B8**, 349-356.

Deng, H. and Nemat-Nasser, S., 1994. Dynamic damage evolution of solids in compression: microcracking, plastic flow, and brittle-ductile transition. *J. Engng Mater. Tech.*, **116**, 286-289.

Deng, X. and Rosakis, A.J., 1991. Dynamic crack propagation in elastic-perfectly plastic solids under plane stress conditions. *J. Mech. Phys. Solids*, **39**, 683-722.

Deng, X. and Rosakis, A.J., 1992. A finite element investigation of quasi-static and dynamic asymptotic crack-tip fields in hardening elastic-plastic solids under plane stress. *Int. J. Fract.*, **57**, 291-308.

Dickinson, J.T., 1990. Fracto-emission. In *Non-Destructive Testing of Fibre-Reinforced Plastics Composites*, edited by J. Summerscales. Elsevier, London, Vol. 2, 429-482.

Dieter, G.E., 1988. *Mechanical Metallurgy*. McGraw-Hill Book Company, London.

Drucker, D.C., 1951. A more fundamental approach to plastic stress-strain relations. In *Proceedings of the 1st National Congress for Applied Mechanics*, ASME, New York, 487-491.

Drucker, D.C., 1959. A definition of stable inelastic material. *J. Appl. Mech.*, **26**, 101-106.

Drucker, D.C. and Rice, J.R., 1970. Plastic deformation in brittle and ductile fracture. *Engng Fract. Mech.*, **1**, 577-602.

Drugan, W.J. and Chen, X.-Y., 1989. Plane strain elastic-ideally plastic crack fields for mode I quasistatic growth at large scale yielding — I. A new family of analytical solutions. *J. Mech. Phys. Solids*, **37**, 1-26.

Drugan, W.J. and Miao, Y., 1992. Influence of porosity on plane strain tensile crack-tip stress fields in elastic-plastic materials. Part I. *J. Appl. Mech.*, **59**, 559-567.

Drugan, W.J., Rice, J.R. and Sham, T.L., 1982. Asymptotic analysis of growing plane strain tensile cracks in elastic-ideally plastic solids. *J. Mech. Phys. Solids*, **30**, 447-473.

Drugan, W.J. and Rice, J.R., 1984. Restrictions on quasi-statically moving surfaces of strong discontinuity in elastic-plastic solids. In *Mechanics of Material Behaviour*, edited by G.J. Dvorak and R.T. Shield. Elsevier Science Publishers B.V., Amsterdam, 59-73.

Drugan, W.J. and Shen, Y., 1987. Restrictions on dynamically propagating surfaces of strong discontinuity in elastic-plastic solids. *J. Mech. Phys. Solids*, **35**, 771-787.

Du, Z.-Z. and Hancock, J.W., 1991. The effect of non-singular stresses on crack-tip constraint. *J. Mech. Phys. Solids*, **39**, 555-567.

Dugdale, D.S., 1960. Yielding of steel sheets containing slits. *J. Mech. Phys. Solids*, **8**, 100-104.

Dulaney, E.N. and Brace, W.F., 1960. Velocity behaviour of a growing crack. *J. Appl. Phys.*, **31**, 2233-2236.

Dunayevsky, V. and Achenbach, J.D., 1982a. Boundary layer phenomenon in the plastic zone near a rapidly propagating crack tip. *Int. J. Solids Structures*, **18**, 1-12.

Dunayevsky, V. and Achenbach, J.D., 1982b. Radial nonuniformity of the stress fields near a moving crack tip in a material with linear strain hardening. *J. Appl. Mech.*, **49**, 646-649.

Dunders, J., 1969. Edge-bonded dissimilar orthogonal wedges under normal and shear loading. *J. Appl. Mech.*, **36**, 650-652.

Dunders, J. and Comninou, M., 1979. Some consequences of the inequality conditions in contact and crack problems. *J. Elasticity*, **9**, 71-82.

Dunders, J. and Gautesen, A.K., 1988. An opportunistic analysis of an interface crack. *Int. J. Fract.*, **36**, 151-159.

Elssner, G., Korn, D. and Rühle, M., 1994. The influence of impurities on fracture energy of UHV diffusion bonded metal-ceramic bicrystals. *Scripta metall. mater.*, **31**, 1037-1042.

England, A.H., 1965. A crack between dissimilar media. *J. Appl. Mech.*, **32**, 400-402.

Erdogan, F., 1963. Stress distribution in a non-homogeneous elastic plane with cracks. *J. Appl. Mech.*, **30**, 232-236.

Erdogan, F., 1965. Stress distribution in bonded dissimilar materials with cracks. *J. Appl. Mech.*, **32**, 403-410.

Erdogan, F., 1968. Crack-propagation theories. In *Fracture: An Advanced Treatise*, edited by H. Liebowitz. Academic Press, New York, Vol. II, 497-590.

Erdogan, F. and Arin, K., 1975. Halfplane and a strip with an arbitrarily located crack. *Int. J. Fract.*, **11**, 191-204.

Erdogan, F. and Biricikoglu, V., 1973. Two bonded half-planes with a crack going through the interface. *Int. J. Engng Sci.*, **11**, 745-766.

Erdogan, F. and Gupta, G.D., 1972. On the numerical solution of singular integral equations. *Quart. Appl. Math.*, **30**, 525-534.

Erdogan, F. and Joseph, P.F., 1989. Toughening of ceramics through crack bridging by ductile particles. *J. Am. Soc.*, **72**, 262-270.

Erdogan, F. and Kibler, J.J., 1969. Cylindrical and spherical shells with cracks. *Int. J. Fract. Mech.*, **5**, 229-237.

Erdogan, F. and Sih, G.C., 1963. On the crack extension in plates under plane loading and transverse shear. *J. Basic Engng*, **85**, 519-523.

Erdogan, F. and Wu, B., 1993. Interface crack problems in layered orthotropic materials. *J. Mech. Phys. Solids*, 889-917.

Eringen, A.C., 1981. On nonlocal plasticity. *Int. J. Eng. Sci.*, **19**, 1461-1474.

Eringen, A.C., 1983. Theories of nonlocal plasticity. *Int. J. Eng. Sci.*, **21**, 741-751.

Eshelby, J.D., 1949. Uniformly moving dislocations *Proc. Phys. Soc.*, **A 62**, 307-314.

Eshelby, J.D., 1951. The force on an elastic singularity. *Phil. Trans. Roy. Soc. (London)*, **A244**, 87-112.

Eshelby, J.D., 1957. The determination of the elastic field of an ellipsoidal inclusion and related problems. *Proc. Roy. Soc. (London)*, **A241**, 376-396.

Eshelby, J.D., 1969. The elastic field of a crack extending nonuniformly under general anti-plane loading. *J. Mech. Phys. Solids*, **17**, 177-199.

Eshelby, J.D., 1974. The calculation of energy release rates. In *Prospects of Fracture Mechanics*, edited by G.C. Sih, H.C. van Elst and D. Broek, Noordhoff International Publishing, Leyden, 69-84.

Eshelby, J.D., Read, W.T. and Shockley, W., 1953. Anisotropic elasticity with some applications to dislocation theory. *Acta Metallurgica*, **1**, 251-259.

Eubanks, R.A. and Sternberg, E., 1956. On the completeness of the Boussinesq-Papkovich stress functions. *J. Rat. Mech. Anal.*, **5**, 735-746.

Evans, A.G. and Cannon, R.M., 1986. Toughening of brittle solids by martensitic transformation. *Acta Metallurgica*, **34**, 761-800.

Evans, A.G. and Faber, K.T., 1980. Toughening of ceramics by circumferential microcracking. *J. Am. Ceram. Soc.*, **64**, 394-398.

Evans, A.G. and McMeeking, R.M., 1986. Toughening of ceramics by strong reinforcements. *Acta Metallurgica*, **34**, 2235-2441.

Fabrikant, V.I., 1989. *Applications of Potential Theory in Mechanics*, Kluwer Academic Publishers, Dordrecht/Boston/London.

Faleskog, J. and Shih, C.F., 1997. Micromechanics of Coalescence – I. Synergistic effects of elasticity, plastic yielding and multi-size-scale voids. *J. Mech. Phys. Solids*, **45**, 21-50.

Fett, T., Tilscher, M. and Munz, D., 1997. Weight functions for cracks near the interface of a bimaterial joint, and applications to thermal stresses. *Engng Fract. Mech.*, **56**, 87-100.

Field, J.E., 1971. Brittle fracture: its study and application. *Contemporary Physics*, **12**, 1-31.

Field, F.A. and Baker, B.R., 1962. Crack propagation under shear displacements. *J. Appl. Mech.*, **29**, 436-437.

Fillipov, A.F., 1956. Some problems of elastic wave diffraction (in Russian). *PMM*, **20**, 688-703.

Fineberg, J., Gross, S.P., Marder, M. and Swinney, H.L., 1991. Instability in dynamic fracture. *Phys. Rev. Lett.*, **67**, 457-460.

Fineberg, J., Gross, S.P., Marder, M. and Swinney, H.L., 1992. Instability in the propagation of fast cracks. *Phys. Rev.*, **B45**, 5146-5154.

Fleck, N.A., 1991. Brittle fracture due to an array of microcracks. *Proc. Roy. Soc. (London)*. **A432**, 55-76.

Fleck, N.A. and Hutchinson, J.W., 1993. A phenomenological theory for strain gradient effects in plasticity. *J. Mech. Phys. Solids*, **41**, 1825-1857.

Fleck, N.A. and Hutchinson, J.W., 1997. Strain gradient plasticity. In *Advances in Applied Mechanics*, Vol. **33**, edited by J.W. Hutchinson and T.Y. Wu. Academic Press, New York, 295-361.

Fleck, N.A., Muller, G.M., Ashby, M.F. and Hutchinson, J.W., 1994. Strain gradient plasticity: Theory and experiment. *Acta metall. mater.*, **42**, 475-487.

Flitman, L.M., 1963. Waves generated by sudden crack in a continuous elastic medium. *Applied Mathematics and Mechanics*, **27**, 938-953. English translation from *PMM*, **27** (1963), 618-628.

Fok, V.A., 1944. O nekotorykh integral'nykh uravneniiakh matematicheskoi fiziki (On some integrals of the equations of mathematical physics). *Matem. sb.*, **14**, No. 1-2.

Folias, E.S., 1965a. A finite line crack in a pressurized cylindrical shell. *Int. J. Fract. Mech.*, **1**, 20-46.

Folias, E.S., 1965b. An axial crack in a pressurized cylindrical shell. *Int. J. Fract. Mech.*, **1**, 104-113.

Fontana, M.G., 1987. *Corrosion Engineering*. McGraw-Hill Book Company, Singapore.

Freudenthal, A.M., 1968. Statistical approach to brittle fracture. In *Fracture: An Advanced Treatise*, edited by H. Liebowitz. Academic Press, New York, Vol. II, 591-619.

Freund, L.B., 1971. The oblique refection of a Rayleigh wave from a crack tip. *Int. J. Solids Structures*, **7**, 1199-1210.

Freund, L.B., 1972a. Crack propagation in an elastic solid subjected to general loading – I. Constant rate of extension. *J. Mech. Phys. Solids*, **20**, 129-140.

Freund, L.B., 1972b. Crack propagation in an elastic solid subjected to general loading – II. Nonuniform rate of extension. *J. Mech. Phys. Solids*, **20**, 141-152.

Freund, L.B., 1973. Crack propagation in an elastic solid subjected to general loading – III. Stress wave loading. *J. Mech. Phys. Solids*, **21**, 47-61.

Freund, L.B., 1974a. Crack propagation in an elastic solid subjected to general loading – IV. Obliquely incident stress pulse. *J. Mech. Phys. Solids*, **22**, 137-146.

Freund, L.B., 1974b. The stress intensity factor due to normal impact loading of the faces of a crack. *Int. J. Engng Sci.*, **12**, 179-189.

Freund, L.B., 1976. The analysis of elastodynamic crack tip stress fields. In *Mechanics today*, vol. 3, edited by S. Nemat-Nasser, Pergamon, New York and London, 55-91.

Freund, L.B., 1978. Stress intensity factor calculations based on a conservation integral. *Int. J. Solids Structures*, **14**, 241-250

Freund, L.B., 1979. The mechanics of dynamic shear crack propagation. *J. geophys. Res.*, **84**, 2199-2209.

Freund, L.B., 1983. Some theoretical results on the dependence of the dynamic stress intensity factor on crack tip speed. In *Workshop on Dynamic Fracture*, California Institute of Technology, Pasadena, California, 129-136.

Freund, L.B., 1987. The stress intensity factor history due to three-dimensional transient loading on the faces of a crack. *J. Mech. Phys. Solids*, **35**, 61-72.

Freund, L.B., 1990. *Dynamic Fracture Mechanics*, Cambridge University Press, Cambridge.

Freund, L.B. and Clifton, R.J., 1974. On the uniqueness of elastodynamic solutions for running cracks. *J. Elasticity*, **4**, 293-299.

Freund, L.B. and Douglas, A.S., 1982. The influence of inertia on elastic-plastic antiplane shear crack growth. *J. Mech. Phys. Solids*, **30**, 59-74.

Freund, L.B. and Hutchinson, J.W., 1985. High strain-rate crack growth in rate-dependent plastic solids. *J. Mech. Phys. Solids*, **33**, 169-191.

Freund, L.B. and Rice, J.R., 1974. On the determination of elastodynamic crack tip stress fields. *Int. J. Solids Structures*, **10**, 411-417.

Freund, L.B. and Rosakis, A.J., 1992. The structure of the near-tip field during transient elastodynamic crack growth. *J. Mech. Phys. Solids*, **40**, 699-719.

Friedman, M.M., 1949. Diffraction of plane elastic waves by a semi-infinite traction free crack (in Russian). *Dokl. Akad. Nauk, SSSR*, **66**, 21-24.

Fuller, K.N.G., Fox, P.G. and Field, J.E., 1975. The temperature rise at the tip of fast running cracks in glassy polymers. *Proc. Roy. Soc. (London)*, **A341**, 537-557.

Gallagher, D., 1995. A crack at the apex of a V-shaped notch. *Int. J. Fract.*, **69**, R81-R87.

Gao, H., 1992. Weight function method for interface cracks in anisotropic bimaterials. *Int. J. Fract.*, **56**, 139-158.

Gao, H., 1993. Surface roughening and branching instabilities in dynamic fracture. *J. Mech. Phys. Solids*, **41**, 457-486.

Gao, H. and Rice, J.R., 1986. Shear stress intensity factors for a planar crack with a slightly curved front. *J. Appl. Mech.*, **53**, 774-778.

Gao, H. and Rice, J.R., 1989. A first-order perturbation analysis of crack trapping by arrays of obstacles. *J. Appl. Mech.*, **56**, 828-836.

Gao, Y.C., 1980. Elastic-plastic field at the tip of a crack growing steadily in perfectly plastic medium (in Chinese). *Acta Mechanica Sinica*, **1**, 48-56.

Gao, Y.C. and Nemat-Nasser, S., 1983. Near-tip dynamic fields for a crack advancing in a power-law elastic-plastic material: Modes I, II and III. *Mechanics of Materials*, **2**, 305-317.

Gao, Y.C. and Shi, Z.F., 1995. Large strain field near an interface crack tip. *Int. J. Fract.*, **69**, 269-279.

Gautesen, A.K., 1992. The interface crack in a tension field: an eigenvalue problem for the gap. *Int. J. Fract.*, **55**, 261-271.

Gautesen, A.K., 1993. The interface crack under combined loading. *Int. J. Fract.*, **60**, 349-361.

Gautesen, A.K. and Dunders, J., 1987a. On the solution to a Cauchy principal value integral equation which arises in fracture mechanics. *Siam J. Appl. Math.*, **47**, 109-116.

Gautesen, A.K. and Dunders, J., 1987b. The interface crack in a tension field. *J. Appl. Mech.*, **54**, 93-98.

Geubelle, P.H. and Knauss, W.G., 1994a. Finite strains at the tip of a crack in a sheet of hyperelastic material: I. Homogeneous case. *J. Elasticity*, **35**, 61-98.

Geubelle, P.H. and Knauss, W.G., 1994b. Finite strains at the tip of a crack in a sheet of hyperelastic material: II. Special bimetal cases. *J. Elasticity*, **35**, 99-137.

Geubelle, P.H. and Knauss, W.G., 1994c. Finite strains at the tip of a crack in a sheet of hyperelastic material: III. General bimetal case. *J. Elasticity*, **35**, 139-174.

Geubelle, P.H. and Rice, J.R., 1995. A spectral method for three-dimensional elastodynamic fracture processes. *J. Mech. Phys. Solids*, **43**, 1791-1824.

Ghahremani, F., Hutchinson, J.W. and Tvergaard, V., 1990. Three-dimensional effects in microcrack nucleation in brittle polymers. *J. Am. Ceram. Soc.*, 73, 1548-1554.

Gibbs, J.W., 1881. *Elements of Vector Analysis*, Privately printed. Published 1961 in *Scientific Papers*, Vol. 2, Dover, New York, 17-80.

Gilat, A., 1985. A viscoplastic theory with anisotropic hardening and its application to prssure-shear plate impact experiments. *J. Appl. Mech.*, **52**, 629-633.

Golden, J.M. and Graham, G.A.C., 1988. *Boundary Value Problems in Linear Viscoelasticity*, Springer-Verlag, Berlin Heidelberg.

Golden, J.M. and Graham, G.A.C., 1990. Energy balance criteria for viscoelastic fracture. *Quart. Appl. Math.*, **48**, 401-413.

Goldman, N.L. and Hutchinson, J.W., 1975. Fully plastic crack problems: The center-cracked strip under plane strain. *Int. J. Solids Structures*, **11**, 575-591.

Gol'dshtein, R.V., 1966. On steady motion of a crack along a straight-line boundary between two joined materials (in Russian). *Mekhanika Tverdogo Tela*, **1**, 94-102.

Gol'dshtein, R.V., 1967. On surface waves in jointed elastic materials. *Appl. Math. and Mech.*, **31**, 496-502.

Gol'dshtein, R.V. and Salganik, R.L., 1970. Plane problem of cuvilinear cracks in an elastic solid (in Russian). *Mekhanika Tverdogo Tela*, **5**, 69-82.

Gol'dstein, R.V. and Salganik, R.L., 1974. Brittle fracture of solids with arbitrary cracks. *Int. J. Fract.*, **10**, 507-523.

Gol'dshtein, R.V. and Savova, L.N., 1972. On the determination of opening stress intensity factors for a smooth curvilinear crack in an elastic plane (in Russian). *Mekhanika Tverdogo Tela*, **7**, 69-82.

Gradshteyn, I.S. and Ryzhik, I.M., 1980. *Table of Integrals, Series, and Products*, translated and edited by A. Jeffrey, Academic Press, New York.

Graham, G.A.C., 1968. The correspondence principle of linear viscoelasticity theory for mixed boundary value problems involving time-dependent boundary regions. *Quart. Appl. Math.*, **26**, 167-174.

Graham, G.A.C. and Sabin, G.C.W., 1973. The correspondence principle of linear viscoelasticity for problems that involve time-dependent regions. *Int. J. Eng. Sci.*, **11**, 123-140.

Green, A.E. and Hundy, B.B., 1956. Initial plastic yielding in notch bend tests. *J. Mech. Phys. Solids*, **4**, 128-145.

Green, A.E. and Sneddon, I.N., 1950. The distribution of stresses in the neighbourhood of a flat elliptical crack in an elastic solid. *Proc. Cambridge Phil. Soc.*, **46**, 159-163.

Green, A.E. and Zerna, W., 1954. *Theoretical Elasticity.* Oxford (Clarendon) Press, Oxford.
Green, A.K. and Pratt, P.L., 1974. Measurement of the dynamic fracture toughness of polymethylmethacrylate by high-speed photography. *Engng Fract. Mech.*, **6**, 71-80.
Griffith, A.A., 1920. The phenomena of rupture and flow in solids, *Phil. Trans. Roy. Soc. (London)*, **A221**, 163-198.
Gross, D., Heimer, St. and Hohe, J., 1997. In *IUTAM Symposium on Nonlinear Analysis of Fracture*, edited by J.R. Willis. Kluwer Academic Publishers, The Netherlands, 221-230.
Gudmundson, P., 1989. Validity of crack tip solutions for plastic materials. In *Advances in Fracture Research*, edited by K. Salama, K.Ravi-Chandar, D.M.R. Taplin and P.Rama Rao. Pergamon Press, New York, 315-322.
Gudmundson, P., 1990. Anisotropic microcrack nucleation in brittle materials. *J. Mech. Phys. Solids*, **38**, 531-551.
Gudmundson, P. and Östlund, S., 1990. Stress singularity at the free surface of a dynamically growing crack. *J. Appl. Mech.*, **57**, 112-116.
Günther, W., 1962. Über einige Randintegrale der Elastomechanik. *Abh. Braunschw. Wiss. Ges.*, **14**, 53-72.
Gunnars, J., Ståhle, P. and Wang, T.C., 1997. On crack path stability in a layered material. *Computational Mechanics*, **19**, 545-552.
Guo, Q.X., Li, Z.L. and Li, K.R., 1988. Dynamic effects on the near crack-line fields for crack growth in an elastic-perfectly plastic material. *Int. J. Fract.*, **36**, 71-81.
Gurson, A.L., 1977. Continuum theory of ductile rupture by void nucleation and growth: Part I – Yield criteria and flow rules for porous ductile media. *J. Engng Mater. Technol.*, **99**, 2-15.
Hadamard, J., 1903. *Leçons sur la Propagation des Ondes et les Équations de l'Hydrodynamique*, Libraire Scientific, Herman, Paris.
Hahn, G.T., Hoagland, R.G., Kanninen, M.F. and Rosenfield, A.R., 1973. A preliminary study of fast fracture and arrest in the DCB test specimen. In *Dynamic Crack Propagation*, edited by G.C. Sih, Noordhoff International Publishing, Leyden, 649-662.
Hahn, H.G., 1976. *Bruchmechanik.* B.G. Teubner, Stuttgart.
Hancock, J.W. and Brown, D.K., 1983. On the role of strain and stress state in ductile fracture. *J. Mech. Phys. Solids*, **31**, 1-24.
Hancock, J.W. and Mackenzie, A.C., 1976. On the mechanics of ductile fracture in high-strength steels subjected to multi-axial stress-states. *J.Mech. Phys. Solids*, **24**, 147-169.
Hart, E.W., 1980 A theory for stable crack extension rate in ductile materials. *Int. J. Solids Structures*, **16**, 807-823.
Hayashi, K. and Nemat-Nasser, S., 1981. On branched interface cracks. *J. Appl. Mech.*, **48**, 529-533.
He, M.Y., Bartlett, A., Evans, A.G. and Hutchinson, J.W., 1991. Kinking of a crack out of an interface: Role of plane stress, *J. Am. Ceram. Soc.*, **74**, 767-771.
He, M.Y., Evans, G.A. and Hutchinson, J.W., 1996. Interface cracking phenomena in constrained metal layers. *Acta Materialia*, **44**, 2963-2971.
He, M.Y. and Hutchinson, J.W., 1989a. Kinking of a crack out of an interface, *J. Appl. Mech.*, **56**, 270-278.
He, M.Y. and Hutchinson, J.W., 1989b. Crack deflection at an interface between dissimilar elastic materials. *Int. J. Solids Structures*, **25**, 1053-1067.
Hellan, K., 1984. *Introduction to Fracture Mechanics.* McGraw-Hill, New York.
Herrmann, J.M., 1989. An asymptotic analysis of finite deformation near the tip of an interface crack. *J. Elasticity*, **21**, 227-269.
Herrmann, J.M., 1992. An asymptotic analysis of finite deformation near the tip of an interface crack: Part II. *J. Elasticity*, **29**, 203-241.
Herrmann, K.P. and Hauck, T., 1995. Thermisch beanspruchte Grenzflächenrisse in elastoplastischen Medien. *27. Vortragsveranstaltung des DVM-Arbeitskreises Bruchvorgänge, Köln-Porz, 14.-15. Februar 1995, DVM Bruchmechanik von Verbundwerkstoffen und Soffverbunden*, 343-351.

Herrmann, K.P. and Hauck, T., 1997. Numerical modelling of interface cracking in thermomechanically loaded elastoplastic materials. *Computational Mechanics*, **20**, 41-45.

Herrmann, K.P. and Noe, A., 1995. Dynamic interface crack propagation and related problems of caustics. *Archives of Mechanics (Archiwum Mechaniki Stosowanej)*, **47**, 915-956.

Hertzberg, R.W., 1983. *Deformation and Fracture Mechanics of Engineering Materials*, J. Wiley, New York.

Hickerson Jr, J.P., 1977. Experimental confirmation of the J integral as a thin section fracture criterion. *Engng Fracture Mech.*, **9**, 75-85.

Hill, R., 1948. A theory of the yielding and plastic flow of anisotropic materials. *Proc. Roy. Soc. (London)*, **A193**, 281-291.

Hill, R., 1950. *The Mathematical Theory of Plasticity*, Clarendon, Oxford.

Hill, R., 1962. Acceleration waves in solids *J. Mech. Phys. Solids*, **10**, 1-10.

Hillerborg, A., Modéer, M. and Petersson, P.E., 1976. Analysis of crack formation and crack growth in concrete by means of fracture mechanics and finite elements. *Cement and Concrete Reasearch*, **6**, 773-782.

Hoagland, R.G., Rosenfield, A.R., Gehlen, P.C. and Hahn, T., 1977. A crack arrest measuring procedure for K_{Im}, K_{ID}, and K_{Ia} properties. In *Fast Fracture and Crack Arrest, ASTM STP 627*, edited by G.T. Hahn and M.F. Kanninen. American Society for Testing and Materials, Philadelphia, 1977, 177-202.

Hoek, E. and Bienawski, Z.T., 1965. Brittle fracture propagation in rock under compression. *Int. J. Fract. Mech.*, **1**, 137-155.

Hoenig, A., 1979. Elastic moduli of a non-randomly cracked body. *Int. J. Solids Structures*, **15**, 137-154.

Hopkinson, B., 1913. Discussion to paper by C.E. Inglis (see Inglis 1913), *Trans. Inst. Naval Architects*, **55**, 232-234.

Hori, S. and Nemat-Nasser, S., 1987. Interacting micro-cracks near the tip in the process zone of a main crack. *J. Mech. Phys. Solids*, **35**, 601-629.

Horii, H. and Nemat-Nasser, S., 1983. Overall moduli of solids with microcracks load-induced anisotropy. *J. Mech. Phys. Solids*, **31**, 155-171.

Horii, H. and Nemat-Nasser, S., 1985. Compression-induced microcrack growth in brittle solids: Axial splitting and shear failure. *J. Geophys. Res.*, **90**, 3105-3125.

Horii, H. and Nemat-Nasser, S., 1986. Brittle fracture in compression: Splitting, faulting and brittle-ductile transition. *Phil. Trans. Roy. Soc. (London)*, **A319**, 337-374.

Howard, I.C., 1981. A method of estimating biaxial fatigue growth rates. *Fatigue of Engineering Materials and Structures*, **3**, 265-270.

Huang, Y., 1989. Accurate dilation rates for spherical voids in triaxial stress fields. Report Mech-155, Division of Applied Sciences, Harvard University, Cambridge, USA.

Huang, Y., Hutchinson, J.W. and Tvergaard, V., 1991. Cavitation instabilities in elastic-plastic solids. *J. mech. Phys. Solids*, **39**, 223-241.

Huang, Y., Liu, C. and Rosakis, A.J., 1996. Transonic crack growth along a bimaterial interface: An investigation of the structure of near-tip fields. *Int. J. Solids Structures*, **33**, 2625-2645.

Huang, Y., Wang, W., Liu, C. and Rosakis, A.J., 1998. Intersonic crack growth in bimaterial interfaces: an investigation of crack contact. *J. Mech. Phys. Solids*, **46**, 2233-2259.

Huber, M.T., 1904. Właściwa praca odkształcenia jako miara wytężenia materyału. *Czasopismo techniczne, Lemberg*, **22**, I-IV: 38-40, V-VII: 49-50, VIII-X: 61-62, XI-XII: 80-81.

Hui, C.Y. and Riedel, H., 1981. The asymptotic stress and strain field near the tip of a growing crack under creep conditions. *Int. J. Fract.*, **17**, 409-425.

Hult, J.A., 1957. Fatigue crack propagation in torsion. *J. Mech. Phys. Solids*, **6**, 47-52.

Hult, J.A. and McClintock, F.A., 1956. Elastic-plastic stress and strain distribution around sharp notches under repeated shear. *Ninth International Congress for Applied Mechanics*, Brussels, **8**, 51-58.

Hutchinson, J.W., 1968a. Singular behavior at the end of a tensile crack in a hardening material. *J. Mech. Phys. Solids*, **16**, 13-31.

Hutchinson, J.W., 1968b. Plastic stress and strain fields at a crack tip. *J. Mech. Phys. Solids*, **16**, 337-347.

Hutchinson, J.W., 1987. Crack tip shielding by micro-cracking in brittle solids. *Acta Metallurgica*, **35**, 1605-1619.

Hutchinson, J.W., Mear, M.E. and Rice, J.R., 1987. Crack paralleling an interface between dissimilar materials. *J. Appl. Mech.*, **54**, 828-832.

Hutchinson, J.W. and Suo, Z., 1991. Mixed mode cracking in layered materials. In *Advances in Applied Mechanics*, Vol. **29**, edited by J.W. Hutchinson and T.Y. Wu. Academic Press, 63-191.

Hwu, C., 1991. Collinear cracks in anisotropic bodies. *Int. J. Fract.*, **52**, 239-256.

Inglis, C.E., 1913. Stresses in a plate due to the presence of cracks and sharp notches. *Trans. Inst. Naval Architects*, **55**, 219-241.

Irwin, G.R., 1957. Analysis of stresses and strains near the end of a crack traversing a plate. *J. Appl. Mech.*, **24**, 361-364.

Irwin, G.R., 1960. In *Structural Mechanics: Proceedings of the 1st Symposium on Naval Structural Mechanics*, edited by J.N. Goodier and N.J. Hoff, Pergamon Press, New York, 557-591.

Irwin, G.R. and Wells, A.A., 1965. A continuum-mechanics view of crack propagation. *Metallurgical Reviews*, **10**, 223-270.

Isida, M. and Noguchi, H., 1992. Stress intensity factors at tips of branched cracks under various loadings. *Int. J. Fract.*, **54**, 293-316.

Isida, M., Noguchi, H. and Yoshida, T., 1984. Tension and bending of finite thickness plates with a semi-elliptical surface crack. *Int. J. Fract.*, **26**, 157-188.

Johnson, E., 1990. On the initiation of unidirectional slip. *Geophys. J. Int.*, **101**, 125-132.

Johnson, E., 1992a. The influence of the lithospheric thickness on bilateral slip. *Geophys. J. Int.*, **108**, 151-160.

Johnson, E., 1992b. Process region changes for rapidly propagating cracks. *Int. J. Fract.*, **55**, 47-63.

Johnson, E., 1992c. Process region influence on crack branching. *Int. J. Fract.*, **57**, R27-R29.

Johnson, E., 1993. Process region influence on energy release rate and crack tip velocity during rapid crack propagation. *Int. J. Fract.*, **61**, 183-187.

Johnson, J.W. and Holloway, D.G., 1966. On the shape and size of the fracture zones on glass fracture surfaces. *Phil. Mag.*, **14**, 731-743.

Kachanov, M., 1986. On crack-microcrack interactions. *Int. J. Fract.*, **30**, R65-R72.

Kachanov, M., 1987. Elastic solids with many cracks: A simple method of analysis. *Int. J. Solids Structures*, **23**, 23-43.

Kachanov, M. and Laures, J.-P., 1989. Three-dimensional problems of strongly interacting arbitrarily located penny-shaped cracks. *Int. J. Fract.*, **41**, 289-313.

Kallivayalil, J.A., Hui, C.Y. and Zehnder, A.T., 1996. Method for thermo-mechanical analysis of steady-state dynamic crack growth. *Int. J. Solids Structures*, **13**, 1967-1889.

Kallivayalil, J.A. and Zehnder, A.T., 1994. Measurement of the temperature field induced by dynamic crack growth in Beta-C titanium. *Int. J. Fract.*, **66**, 99-120.

Kalthoff, J.F., 1973. On the propagation direction of bifurcated cracks. In *Dynamic Crack Propagation*, edited by G.C. Sih, Noordhoff International Publishing, Leyden, 449-458.

Kalthoff, J.F., 1983. On some current problems in experimental fracture mechanics. In *Workshop on Dynamic Fracture*, California Institute of Technology, Pasadena, California, 11-35.

Kalthoff, J.F., 1986. Fracture behaviour under high rates of loading. *Engng Fract. Mech.*, **23**, 289-298.

Kalthoff, J.F., 1988. Shadow optical analysis of dynamic shear fracture. *Optical Engng*, **27**, 835-840.

Kalthoff, J.F., 1990. Transition in the failure behaviour of dynamically shear loaded cracks. *Appl. Mech. Rev.*, **43**, S247-S250.

Kalthoff, J.F., Beinert, J. and Winkler, S., 1977. Measurements of dynamic stress intensity factors for fast running and arresting cracks in double-cantilever-beam specimens. In

Fast Fracture and Crack Arrest, ASTM STP 627, edited by G.T. Hahn and M.F. Kanninen. American Society for Testing and Materials, 161-176.

Kalthoff, J.F., Beinert, J., Winkler, S. and Klemm, W., 1980. Experimental analysis of dynamic effects in different crack arrest test specimens. In *Crack Arrest Methodology and Applications. ASTM STP 711*, edited by G.T. Hahn and M.F. Kanninen. American Society for Testing and Materials, Philadelphia, 109-127.

Kane, T.R. and Mindlin, R.D., 1956. High frequency extensional vibration of plates. *J. Appl. Mech.*, **23**, 277-283.

Kanninen, M.F. and O'Donoghue, P.E., 1995. Research challenges arising from current and potential applications of dynamic fracture mechanics to the integrity of engineering structures. *Int. J. Solids Structures*, **32**, 2423-2445.

Kanninen, M.F. and Popelar, C.H., 1985. *Advanced Fracture Mechanics*. Oxford University Press, Oxford.

Kanwal, R.P., 1971. *Linear Integral Equations, Theory and Technique*, Academic Press, New York.

Kamei, A. and Yokobori, T., 1974. Some results on stress intensity factors of the cracks and/or slip band system. *Rep. Res. Inst. Strength Fract. Materials*, Tohoku Univerity, **10**, 30-93.

Karihaloo, B.L., 1995. *Fracture Mechanics & Structural Concrete*. Longman Scientific & Technical. Essex, England.

Kassir, M.K. and Sih, G.C., 1966. Three-dimensional stress distribution around an elliptical crack under arbitrary loadings. *J. Appl. Mech.*, **33**, 601-611.

Kassir, M.K. and Sih, G.C., 1973. Application of Papkovich-Neuber potentials to a crack problem. *Int. J. Solids Structures*, **9**, 643-654.

Kassir, M.K. and Sih, G.C., 1975. *Mechanics of Fracture 2 – Three-dimensional Crack Problems*, Noordhoff International Publishing, The Netherlands.

Kerkhof, F., 1970. *Bruchvorgänge in Gläsern*. Verlag der Deutschen Glastechnischen Gesellschaft, Frankfurt.

Kerkhof, F., 1973. Wave fractographic investigations of brittle fracture dynamics. In *Dynamic Crack Propagation*, edited by G.C. Sih, Noordhoff International Publishing, Leyden, 1-35.

Khrapkov, A.A., 1971. The first basic problem for a notch at the apex of an infinite wedge. *Int. J. Fract.*, **7**, 373-382.

Kirk, M.T., Koppenhoefer, K.C. and Shih, C.F., 1993. Effect on constraint on specimen dimensions needed to obtain structurally relevant toughness measures. In *Constraint Effects in Fracture, ASTM STP 1171*, edited by E.M. Hackett, K.-H. Schwalbe and R.H. Dodds, American Society for Testing and Materials, Philadelphia, 79-103.

Kishimoto, K., Aoki, S. and Sakata, M., 1980. On the path independent integral-$\hat{J}$. *Engng Fract. Mech.*, **13**, 841-850.

Kishimoto, K., Aoki, S. and Sakata, M., 1982. Use of $\hat{J}$-integral in dynamic analysis of cracked linear viscoelastic solids by finite element method. *J. Appl. Mech.*, **49**, 75-80.

Klepaczko, J.R., 1984. Loading rate spectra for fracture initiation in metals. *Theoretical and Applied Fracture Mechanics*, **1**, 181-191.

Knauss, W.G., 1970a. Delayed failure – The Griffith problem for linearly viscoelastic materials. *Int. J. Solids Structures*, **6**, 995-1009.

Knauss, W.G., 1970b. An observation of crack propagation in anti-plane shear. *Int. J. Fract. Mech.*, **6**, 183-187.

Knauss, W.G., 1973. The mechanics of polymer fracture. *Appl. Mech. Reviews*, **26**, 1-17.

Knauss, W.G., 1974. On the steady propagation of a crack in a viscoelastic sheet: Experiments and analysis. In *Deformation and Fracture of High Polymers*, edited by H.H. Kausch, J.A. Hassell and R.I. Jaffee. Plenum Press, New York, 501-541.

Knauss, W.G. and Dietmann, H., 1970. Crack propagation under variable load histories in linearly viscoelastic solids. *Int. J. Engng Sci.*, **8**, 643-656.

Knott, J.F., 1979. *Fundamentals of Fracture Mechanics*. Butterworth, London.

Knott, J.F., 1994. The science and engineering of fracture. In *Advances in Fracture Resistance and Structural Integrity*, edited by V.V. Panasyuk, D.M.R. Taplin, M.C. Pandy, O.Ye. Andreykiv, R.O. Ritchie, J.F. Knott and P. Rama Rao, Pergamon, 13-49.

Knowles, J.K., 1977. The finite anti-plane shear field near the tip of a crack for a class of incompressible elastic solids. *Int. J. Fracture*, **13**, 611-639.

Knowles, J.K., 1981. A nonlinear effect in mode II crack problems. *Engng Fract. Mech.*, **15**, 469-476.

Knowles and Sternberg, E., 1972. On a class of conservation laws in linearized and finite elastostatics. *Arch. Rat. Mech. Anal.*, **44**, 187-211.

Knowles, J.K. and Sternberg, E., 1973. An asymptotic finite-deformation analysis of the elastostatic field near the tip of a crack. *J. Elasticity*, **3**, 67-107.

Knowles, J.K. and Sternberg, E., 1974. Finite-deformation analysis of the elastostatic field near the tip of a crack. *J. Elasticity*, **4**, 201-233.

Knowles, J.K. and Sternberg, E., 1983. Large deformation near a tip of an interface-crack between two Neo-Hookean sheets. *J. Elasticity*, **13**, 257-293.

Kobayashi, A.S. and Mall, S., 1978. Dynamic fracture toughness of Homalite-100. *Experimental Mechanics*, **18**, 11-18.

Kobayashi, T. and Dally, J.W., 1977. Relation between crack velocity and the stress intensity factor in birefringent polymers. In *Fast Fracture and Crack Arrest, ASTM STP 627*, edited by G.T. Hahn and M.F. Kanninen. American Society for Testing and Materials, 257-273.

Koiter, W.T., 1956a. On the flexural rigidity of a beam weakened by transverse saw cuts. I. *Proc. Kon. Ned. Ak. Wet.*, **B59**, 354-364.

Koiter, W.T., 1956b. On the flexural rigidity of a beam weakened by transverse saw cuts. II. *Proc. Kon. Ned. Ak. Wet.*, **B59**, 365-374.

Koiter, W.T., 1959. An infinite row of collinear cracks in an infinite elastic sheet. *Ingenieur-Archiv*, **28**, 168-172.

Koiter, W.T., 1965. Rectangular tensile sheet with symmetric edge cracks. *J. Appl. Mech.*, **32**, 237.

Kolosoff, G., 1914. Über einige Eigenschaften des ebenen Problems der Elastizitätstheorie. *Zeitschrift für Mathemetik und Physik*, **62**, 384-409.

Kolosov, G., 1909. *Doctoral dissertation*, Dorpat, Estonia. See: Kolosoff (1914).

Koskinen, M.F., 1963. Elastic-plastic deformation of a single grooved flat plate under longitudinal shear. *Transactions, Am. Soc. Mechanical Engrs (Journal of Basic Engineering)*, **85**, 585-594.

Kostrov, B.V., 1964a. The axi-symmetric problem of propagation of a tensile crack. *Applied Mathematics and Mechanics*, **28**, 793-803. English translation from *PMM*, **28** (1964), 644-652.

Kostrov, B.V., 1964b. Self-similar problems of propagation of shear cracks. *Applied Mathematics and Mechanics*, **28**, 1077-1087. English translation from *PMM*, **28** (1964), 889-898.

Kostrov, B.V., 1966. Unsteady propagation of longitudinal shear cracks. *Applied Mathematics and Mechanics*, **30**, 1241-1248. English translation from *PMM*, **30** (1966), 1042-1049.

Kostrov, B.V., 1975. On the crack propagation with variable velocity. *Int. J. Fract.*, **11**, 47-56.

Kostrov, B.V. and Nikitin, L.V., 1970. Some general problems of mechanics of brittle fracture. (In Russian). *Archives of Mechanics (Archiwum Mechaniki Stosowanej)*, **22**, 749-775.

Krafft, J.M. Sullivan, A.M. and Boyle, R.W., 1961. Influence of speed of deformation on strength properties in the post lower yield stress-strain curve of mild steel. *Proc. Crack Propagation Symp. College of Aeronautics, Cranfield, England*, **1**, 8-26.

Kubo, S., 1982. An application of M-integral to cracks in dissimilar elastic materials. *Int. J. Fract.*, **20**, R27-R30.

Kuna, M. and Sun, D.Z., 1996. Three-dimensional cell model analyses of void growth in ductile metals. *Int. J. Fract.*, **81**, 235-258.

Kuo, M.K., 1993. Stress intensity factors for a semi-infinite plane crack under a pair of point forces. *J. Elasticity*, **30**, 197-209.

Kuo, M.K. and Chen, T.Y., 1992. The Wiener-Hopf technique in elastodynamic crack problems with characteristic lengths in loading. *Engng Fract. Mech.*, **42**, 805-813.

Lam, P.S. and Freund, L.B., 1985. Analysis of dynamic crack growth of a tensile crack in an elastic-plastic material. *J. Mech. Phys. Solids*, **33**, 153-167.
Lamb, H., 1904. On the propagation of tremors over the surface of an elastic solid. *Phil. Trans. Roy. Soc. (London)*, **A203**, 1-42.
Lambros, J. and Rosakis, A.J., 1995. Shear dominated transonic crack growth in bimaterials – Part I: Experimental observations. *J. Mech. Phys. Solids*, **43**, 169-188.
Larsson, S.G. and Carlsson, A.J., 1973. Influence of non-singular stress terms and specimen geometry on small-scale yielding at crack tips in elastic-plastic materials. *J. Mech. Phys. Solids*, **21**, 263-277.
Laures, J.-P. and Kachanov, M., 1991. Three-dimensional interactions of a crack front with arrays of penny-shaped microcracks. *Int. J. Fract.*, **48**, 255-279.
Lawn, B.R., 1993. *Fracture of Brittle Solids*. Second Edition†. Cambridge University Press, Cambridge.
Le, K.Ch., 1992. On the singular elastostatic field induced by a crack in a Hadamard material. *Quart. J. Mech. Appl. Math.*, **45**, 101-117.
Le, K.Ch. and Stumpf, H., 1993. The singular elastostatic field due to a crack in rubberlike materials. *J. Elasticity*, **32**, 183-222.
Lebedev, N.N., 1949. On the expansion of an arbitrary function in an integral with respect to cylinder functions of imaginary order and argument. *PMM* (in Russian), **13**, (1949), 466-476.
Lee, Y.J. and Freund, L.B., 1990. Fracture initiation due to asymmetric impact loading of an edge cracked plate. *J. Appl. Mech.*, **57**, 104-111.
Leevers, P.S. and Radon, J.C., 1982. Inherent stress biaxiality in various fracture specimen geometries. *Int. J. Fract.*, **19**, 311-325.
Leighton, J.T., Champion, C.R. and Freund, L.B., 1987. Asymptotic analysis of steady dynamic crack growth in an elastic-plastic material. *J. Mech. Phys. Solids*, **35**, 541-563.
Lekhnitskii, S.G., 1950. *Theory of Elasticity of an Anisotropic Body*, Gostekhizdat, Moscow (in Russian). English translations published by Holden-Day, San Francisco, (1963), and by Mir Publ., Moscow, (1981).
Leonov, M.Ya and Panasyuk, V.V., 1959. Growth of the minutest cracks in a brittle body. In Ukrainian. *Prikhladnaya Mekhanika*, **5**, 391-401.
Lerch, M., 1903. Sur un point de la théorie des fonctions génératrices d'Abel. *Acta Mathematica*, **27**, 339-351.
Levita, G., 1996. Effects of thickness on the specific essential work of fracture in rigid PVC. *Polymer Engineering and Science*, **36**, 2534-2541.
Li, W., Deng, X. and Rosakis, A.J., 1996. Determination of temperature field around a rapidly moving crack-tip in an elastic-plastic solid. *Int. J. Heat Mass transfer*, **39**, 677-690
Li, Xiang-Ping and Liu, Chun-Tu, 1994. Three-dimensional transient wave response in a cracked elastic solid. *Engng Fract. Mech.*, **48**, 545-552.
Li, Xiang Ping and Liu, Chun Tu, 1995. Elastodynamic stress-intensity factors for a semi-infinite crack under 3-D combined mode loading. *Int. J. Fract.*, **69**, 319-339.
Li, Yaochen and Wang, Ziqiang, 1986. Higher-order asymptotic field of tensile plane strain nonlinear crack problems. *Scientia sinica*, **A29**, 941-955.
Liebowitz, H., editor, 1968-1972. *Fracture: An Advanced Treatise*, Vol. I-VII. Academic Press, New York.
Liu, C., Lambros, J. and Rosakis, A.J., 1993. Highly transient elastodynamic crack growth in a bimaterial interface: Higher order asymptotic analysis and optical experiment. *J. Mech. Phys. Solids*, **41**, 1887-1954.
Liu, C. and Rosakis, A.J., 1994. On the higher order asymptotic analysis of a non-uniformly propagating dynamic crack along an arbitrary path. *J. Elasticity*, **35**, 27-60.
Liu, C., Rosakis, A.J. and Freund, L.B., 1993. The interpretation of optical caustics in the presence of dynamic non-uniform crack-tip motion histories: A study based on a higher order transient crack-tip expansion. *Int. J. Solids Structures*, **30**, 875-897.
Lo, K.K., 1978. Analysis of branched cracks. *J. Appl. Mech.*, **45**, 797-802.

---

† The first edition, 1975, was written jointly by B.R. Lawn and T.R. Wilshaw.

Lo, K.K., 1982. Elastic-plastic field at the tip of a propagating shear crack. *Quart. Appl. Math.*, **40**, 27-36.

Lo, K.K., 1983. Dynamic crack-tip fields in rate-sensitive solids. *J. Mech. Phys. Solids*, **31**, 287-305.

Lu, J., Dhumne, A. and Ravi-Chandar, K., 1995. Dynamic fracture under pressure and shear. *Archives of Mechanics (Archiwum Mechaniki Stosowanej)*, **47**, 971-984.

Lu, M.-C. and Erdogan, F., 1983a. Stress intensity factors in two bonded elastic layers containing cracks perpendicular to and on the interface—I. Analysis. *Engng Fract. Mech.*, **18**, 491-506.

Lu, M.-C. and Erdogan, F., 1983b. Stress intensity factors in two bonded elastic layers containing cracks perpendicular to and on the interface—II. Solution and results. *Engng Fract. Mech.*, **18**, 507-528.

Lundström, A. and Tryding, J., 1991. A low cycle fatigue criterion based on non-linear fracture mechanics. *Engng Fracture Mech.*, **39**, 769-781.

Ma, C.C. and Freund, L.B., 1986. The extent of the stress intensity factor field during crack growth under dynamic loading conditions. *J. Appl. Mech.*, **53**, 303-310.

Maccagno, T.M. and Knott, J.F., 1991. The low temperature brittle fracture behaviour of steel in mixed modes I and II. *Engng Fract. Mech.*, **38**, 111-128.

Machida, S., Yoshinari, H. and Kanazawa, T., 1986. Some recent experimental work in Japan on fast fracture and crack arrest. *Engng Fract. Mech.*, **23**, 251-264.

Mai, Y.-W. and Cotterell, B., 1986. On the essential work of ductile fracture in polymers. *Int. J. Fract.*, **32**, 105-125.

Malyushev, B.M. and Salganik, R.L., 1965. The strength of adhesive joints using the theory of cracks. *Int. J. Fract.*, **1**, 114-128.

Mandelbrot, B.B., Passoja, D.E. and Paullay, A.J., 1984. Fractal character of fracture surfaces of metals. *Nature*, **308**, 721-722.

Marchand, A. and Duffy, J., 1988. An experimental study of the formation process of adiabatic shear bands in a structural steel. *J. Mech. Phys. Solids*, **36**, 251-283.

Mason, J.J. and Rosakis, A.J., 1993. On the dependence of the dynamic crack tip temperauture fields in metals upon crack tip velocity and material properties. *Mechanics of Materials*, **16**, 337-350.

Mason, J.J., Rosakis, A.J. and Ravichandran, G., 1994. On the strain and strain-rate dependence of the fraction of plastic work converted into heat: An experimental study using high speed infrared detectors and the Kolsky bar. *Mechanics of Materials*, **17**, 135-145.

Mataga, P.A., 1989. Deformation of crack-bridging ductile reinforcements in toughened brittle materials. *Acta Metallurgica*, **37**, 3349-3359.

Mataga, P.A., Freund, L.B. and Hutchinson, J.W., 1987. Crack tip plasticity in dynamic fracture. *J. Phys. Chem. Solids*, **48**, 985-1005.

Maue, A.W., 1953. Die Beugung elastischer Wellen an der Halbebene. *Zeitschrift für angewandte Mathematik und Mechanik*, **33**, 1-10.

Maue, A.W., 1954. Die Entspannungswelle bei plötzlichem Einschnitt eines gespannten elastischen Körpers. *Zeitschrift für angewandte Mathematik und Mechanik*, **34**, 1-12.

McClintock, F.A., 1958. Ductile fracture instability in shear. *J. Appl. Mech.*, **25**, 582-588.

McClintock, F.A., 1963. On the plasticity of the growth of fatigue cracks. In *Fracture of Solids*, edited by D.C. Drucker and J.J. Gilman, Interscience Publ., New York, 65-102.

McClintock, F.A., 1966. Ductile fracture. In *Mechanical Behaviour of Materials*, edited by F.A. McClintock and A.S. Argon. Addison-Wesley Publ. Company, Reading, Massachusetts, 518-545.

McClintock, F.A., 1968a. A criterion for ductile fracture by growth of holes. *J. Appl. Mech.*, **4**, 363-371.

McClintock, F.A., 1968b. Plasticity aspects of fracture. In *Fracture: An Advanced Treatise*, edited by. H. Liebowitz. Academic Press, New York, Vol. III, 47-225.

McClintock, F.A. and Irwin, G.R., 1965. Plasticity aspects of fracture mechanics. In *Fracture Toughness Testing and Its Applications, ASTM STP 381*, 84-113.

McClintock, F.A. and Sukhatme, S.P., 1960. Travelling cracks in elastic materials under longitudinal shear. *J. Mech. Phys. Solids*, **8**, 187-193.

McCrum, N.G., Buckley, C.P. and Bucknall, C.B., 1988. *Principles of Polymer Engineering*, Oxford University Press Inc., New York.
McMeeking, R.M., 1977. Finite deformation analysis of crack-tip opening in elastic-plastic materials and implications for fracture. *J. Mech Phys. Solids*, **25**, 357-381.
McMeeking, R.M. and Evans, A.G., 1982. Mechanics of transformation toughening in brittle solids. *J. Am. Ceramic Soc.*, **65**, 242-246.
Meade, K.P. and Keer, L.M., 1984. On the problem of a pair of point forces applied to the faces of a semi-infinite plane crack. *Journal of Elasticity*, **14**, 3-14.
Mecholsky, J.J., Passoja, D.E. and Feinberg-Ringel, K.S., 1989. Quantitative analysis of brittle fracture profiles using fractal geometry. *J. Am. Ceramic Soc.*, **72**, 60-65.
Melin, Solveig, 1983. Why do cracks avoid each other? *Int. J. Fract.*, **23**, 37-45.
Melin, Solveig, 1986. When does a crack grow under mode II conditions? *Int. J. Fract.*, **30**, 103-114.
Melin, Solveig, 1989. Why are crack paths in concrete and mortar different from those in PMMA? *Materials and Structures*, **22**, 23-27.
Melin, Solveig, 1991. On the directional stability of wedging. *Int. J. Fract.*, **50**, 293-300.
Melin, Solveig, 1992. Directional stability of an originally straight crack. *Int. J. Fract.*, **53**, 121-128.
Melin, Solveig, 1994. Accurate data for stress intensity factors at infinitesimal kinks. *J. Appl. Mech.*, **61**, 467-470.
Mercier, S. and Molinari, A., 1998. Steady-state shear band propagation under dynamic loading conditions. *J. Mech. Phys. Solids*, **46**, 1463-1495.
Miao, Y. and Drugan, J.W., 1993. Influence of porosity on plane strain tensile crack-tip stress fields in elastic materials. *J. Appl. Mech.*, **60**, 883-889.
Miao, Y. and Drugan, J.W., 1995. Asymptotic analysis of growing crack stress/deformation fields in porous ductile metals and implications for stable crack growth. *Int. J. Fract.*, **72**, 69-96.
Miller, K.J., 1977. Fatigue under complex stress. *Metal Science*, **11**, 432-438.
Mises, R. von, 1913. Mechanik der festen Körper im plastisch-deformablem Zustand. *Nachrichten der Königlicher Gesellschaft der Wissenschaften, Göttingen, Math. phys. Klasse*, **K1**, 582-592.
Mises, R. von, 1928. Mechanik der plastischen Formänderung von Kristallen. *Zeitschrift für Angewandte Mathematik und Mechanik*, **8**, 161-185.
Misnaevsky Jr, L.L., 1997. Methods of the theory of complex systems in modelling of fracture: A brief review. *Engng Fract. Mech.*, **56**, 47-56.
Møller, C.V. and Karihaloo, B.L., 1995. Crack front trapping in transformation-toughened ceramics. *Int. J. Fract.*, **72**, 171-181.
Mooney, M., 1940. A theory of large elastic deformation *J. Appl. Phys.*, **11**, 582-592.
Moran, B. and Shih, C.F., 1987. A general treatment of crack tip contour integrals. *Int. J. Fract.*, **35**, 295-310.
Morrisey, J.W. and Rice, J.R., 1998. Crack front waves. *J. Mech. Phys. Solids*, **46**, 467-487.
Morse, P. and Feshbach, H., 1953. *Methods of Theoretical Physics*, McGraw-Hill, New York.
Mott, N.F., 1948. Fracture of metals: Theoretical considerations. – *Engineering*, **165**, 16-18.
Movchan, A.B. and Willis, J.R., 1995. Dynamic weight functions for a moving crack. II. Shear loading. *J. Mech. Phys. Solids*, **43**, 1369-1383.
Mühlhaus, H.-B., 1994. Shear banding in fourth order gradient plasticity. *J. Mech. Behavior Materials*, **5**, 141-153.
Mühlhaus, H.-B. and Aifantis, E.C., 1991. A variational principle for gradient plasticity. *Int. J. Solids Structures*, **28**, 845-857.
Muskhelishvili, N.I., 1953a. *Some Basic Problems in the Mathematical Theory of Elasticity*, translated from Russian by J.R.M. Radok, Noordhoff, Groningen, Holland.
Muskhelishvili, N.I., 1953b. *Singular Integral Equations*, translated from Russian edition, GITTL, Moscow 1946, by J.R.M. Radok, Noordhoff, Groningen, Holland.
Nakano, A., Kalia, R.K. and Vashishta, P., 1995. Dynamics and morphology of brittle cracks: A molecular-dynamics study of silicon nitride. *Phys. Rev. Letters*, **75**, 3138-3141.
Narasimhan, R. Rosakis, A.J. and Moran, B., 1992. A three-dimensional numerical investiga-

tion of fracture initiation by ductile failure mechanisms in a 4340 steel. *Int. J. Fract.*, **56**, 1-24.

Naus, D.J., Keeney-Walker, J., Bass, B.R., Iskander, S.K., Fields, R.J., deWit, R. and Low, S.R., 1989. *SEN Wide-Plate Crack-Arrest Tests Using A 533 Grade B Class 1 Material: WP-CE Test Series*, NRC Publications, Washington DC.

Needleman, A., 1989. Dynamic shear band development in plane strain. *J. Appl. Mech.*, **56**, 1-9.

Needleman, A., 1997. Numerical modeling of crack growth under dynamic loading conditions. *Computational Mechanics*, **19**, 463-469.

Needleman, A. and Rice, J.R., 1978. Limits to ductility set by plastic flow localization. In *Mechanics of Sheet Forming*, edited by D.P. Kostinen and N.-M. Wang. Plenum Press, New York, 1978, 237-267.

Needleman, A., Tvergaard, V. and Hutchinson, J.W., 1992. Void growth in plastic solids. In *Topics in Fracture and Fatigue*, edited by A.S. Argon. Springer Verlag, New York, 145-178.

Needleman, A. and Tvergaard, V., 1995. Analysis of a brittle–ductile transition under dynamic shear loading. *Int. J. Solids Structures*, **32**, 2571-2590.

Neimitz, A., 1995. Crack growth equations. In *Constitutive Relation in High/Very High Strain Rates*, edited by K. Kawata and J. Shioiri, Springer-Verlag, Tokyo, 92-104.

Nemat-Nasser, S., 1995. Fracturing in anisotropic solids. In *IUTAM Symposium on Inhomogeneity and Nonlinearity in Solid Mechanics*, edited by D.F. Parker and A.H. England. Kluwer Academic Publishers, The Netherlands, 249-262.

Nemat-Nasser, S., 1997. Tensile fracturing in dynamic compression. In *Advances in Fracture Research. Proceedings of the Ninth Conference on Fracture*. Pergamon, Oxford, Vol. **5**, 2299-2308.

Nemat-Nasser, S. and Chang, S.-N., 1990. Compression-induced high strain rate void collapse, tensile cracking, and recrystallization in ductile single and polycrystals. *Mech. Mater.*, **10**, 1-17.

Nemat-Nasser, S. and Deng, H., 1994. Strain-rate effect on brittle fracture in compression. *Acta Metall. Mater.*, **42**, 1013-1024.

Nemat-Nasser, S. and Hori, M., 1987a. Toughening by partial or full bridging of cracks in ceramics and fiber reinforced composites. *Mech. Mater.*, **6**, 245-267.

Nemat-Nasser, S. and Hori, M., 1987b. Void collapse and void growth in crystalline solids. *J. Appl. Phys.*, **62**, 2746-2757.

Nemat-Nasser, S. and Horii, H., 1982. Compression-induced nonplanar crack extension with application to splitting, exfoliation and rockburst. *J. Geophys. Res.*, **87**, 6805-6821.

Nemat-Nasser, S. and Ni, L., 1995. A duality principle and correspondence relations in elasticity. *Int. J. Solids Structures*, **32**, 467-472.

Neuber, H., 1961. Theory of stress concentration for shear-strained prismatical bodies with arbitrary nonlinear stress-strain law. *J. Appl. Mech.*, **28**, 544-550.

Ni, L. and Nemat-Nasser, S., 1991. Interface cracks in anisotropic dissimilar materials: an analytic solution. *J. Mech. Phys. Solids*, **39**, 113-144.

Ni, L. and Nemat-Nasser, S., 1996. General duality principle in elasticity. *Mech. Mat.*, **24**, 87-123.

Nilsson, F., 1972. Dynamic stress intensity factors for finite strip problems. *Int. J. Fract. Mech.*, **8**, 403-411. See also Nilsson (1973b).

Nilsson, F., 1973a. A path-independent integral for transient crack problems. *Int. J. Solids Structures*, **9**, 1107-1115.

Nilsson, F., 1973b. Erratum to "Dynamic stress intensity factors for finite strip problems". *Int. J. Fract.*, **9**, 477.

Nilsson, F., 1973c. A transient crack problem for an infinite strip under anti-plane shear. In *Dynamic Crack Propagation*, edited by G.C. Sih, Noordhoff International Publishing, Leyden, 543-551.

Nilsson, F., 1974a. A note on the stress singularity at a nonuniformly moving crack tip. *J. Elasticity*, **4**, 73-75.

Nilsson, F., 1974b. Crack propagation experiments on strip specimens. *Engng Fract. Mech.*, **6**, 397-403.

Nilsson, F., 1977a. A suddenly stopping crack in an infinite strip under tearing action. In *Fast Fracture and Crack Arrest, ASTM STP 627*, edited by G.T. Hahn and M.F. Kanninen, American Society for Testing and Materials, 77-91.

Nilsson, F., 1977b. Sudden arrest of steadily moving cracks. In *Proc. Int. Conf. on "Dynamic Fracture Toughness"*, Cambridge: The Welding Institute, 249-257.

Nilsson, F., 1977c. Steady mode III crack propagation followed by non-steady growth. *Int. J. Solids Structures*, **13**, 543-548.

Nilsson, F., 1977d. Steady crack propagation followed by non-steady growth – mode I solution. *Int. J. Solids Structures*, **13**, 1133-1139.

Noble, B., 1958. *Methods Based on the Wiener-Hopf Technique*, Pergamon Press, London.

Noether, E., 1918. Invariante Variationsprobleme. *Nachrichten der Königlicher Gesellschaft der Wissenschaften, Göttingen*, 235-257.

Nye, J.F., 1957. *Physical Properties of Crystals: their Representation by Tensors and Matrices*, Oxford: Clarendon Press.

Obreimoff, J.D., 1930. The splitting of mica. *Proc. Roy. Soc. (London)*, **A127**, 290-297.

O'Dowd, N.P. and Shih, C.F., 1991. Family of crack-tip fields characterized by a triaxiality parameter — I. Structure of fields. *J. Mech. Phys. Solids*, **39**, 989-1015.

O'Dowd, N.P. and Shih, C.F., 1992. Family of crack-tip fields characterized by a triaxiality parameter — II. Fracture applications. *J. Mech. Phys. Solids*, **40**, 939-963.

O'Dowd, N.P. and Shih, C.F., 1994. Two-parameter fracture mechanics: Theory and applications. In *Fracture Mechanics: Twenty-Fourth Volume, ASTM STP 1207*, edited by J.D. Landes, D.E. McCabe and J.A.M. Boulet, American Society for Testing and Materials, 21-47.

O'Dowd, N.P., Shih, C.F. and Dodds, R.H., 1994. The role of geometry and crack growth on constraint and implications for ductile/brittle fracture. In *Constraint Effects in Fracture: Theory and Applications, ASTM STP 1244*, edited by M. Kirk and A. Bakker, American Society for Testing and Materials, Philadelphia, 134-159.

Ogden, R.W., 1972a. Large deformation isotropic elasticity: on the correlation of theory and experiment for incompressible rubberlike solids. *Proc. Roy. Soc. (London)*, **A326**, 565-584.

Ogden, R.W., 1972b. Large deformation isotropic elasticity: on the correlation of theory and experiment for compressible rubberlike solids. *Proc. Roy. Soc. (London)*, **A328**, 567-583.

Okubu, P.G. and Aki, K., 1987. Fractal geometry in the San Andreas fault system. *J. Geophs. Res.*, **92**, 345-355.

Orowan, E., 1945. Notch brittleness and the strength of metals. *Trans. Inst. Engrs Shipbuilders*, Scotland, **89**, 165-215.

Orowan, E., 1952. Fundamentals of brittle behaviour in metals. In *Fatigue and Fracture of Metals*, edited by W.M. Murray. John Wiley, New York, 139-154.

Östlund, S., 1990. On numerical modeling and fracture criteria of dynamic visco-plastic crack growth. *Int. J. Fract.*, **44**, 283-299.

Östlund, S., 1991. Large scale yielding for dynamic crack growth in a strip geometry. *Int. J. Fract.*, **49**, 219-237.

Östlund, S. and Gudmundson, P., 1988. Asymptotic crack tip fields for dynamic fracture of linear strain-hardening solids. *Int. J. Solids Structures*, **24**, 1141-1158.

Ouchterlony, F., 1978. Some stress intensity factors for selfsimilar cracks derived from path-independent integrals. *J. Elasticity*, **8**, 259-271.

Ouchterlony, F., 1980. Symmetric cracking of a wedge by transverse displacements. *J. Elasticity*, **10**, 215-223.

Pan, J. and Shih, C.F., 1986. Plane-strain crack-tip fields for power-law hardening orthotropic materials. *Mechanics of Materials*, **5**, 299-316.

Pan, J. and Shih, C.F., 1988. Plane-stress crack-tip fields for power-law hardening orthotropic materials. *Int. J. Fract.*, **37**, 171-195.

Pan, J. and Shih, C.F., 1992. Elastic-plastic analysis of combined mode I, II and III crack fields under small-scale yielding conditions. *Int. J. Solids Structures*, **29**, 2795-2814.

Panasyuk, V.V., 1960. On the theory of spreading of cracks during deformation of a brittle body (in Ukrainian). *Dopovidi Akademii Nauk Ukrainskoi RSR*, 1185-1188.

Papkovich, P.F., Vyrazhenie obshchego integrals osnovnykh uravnenii teorii uprugosti cherez garmonicheskie funktsii. (Expression of the general integral of the basic equations of the theory of elasticity in terms of harmonic functions.) *Izv. Akad. Nauk SSSR, ser. fiz.-met.*, No. 10, 1932.

Pärletun, L.G., 1979. Determination of the growth of branched cracks by numerical methods. *Engng Fract. Mech.*, **11**, 343-358.

Paxson, T.L. and Lucas, R.A., 1973. An experimental investigation of the velocity characteristics of a fixed boundary fracture model. In *Proceedings of an International Conference on Dynamic Crack Propagation*, edited by G.C. Sih. Noordhoff International Publishing, Leyden, 415-426.

Perrin, G., Rice, J.R. and Zheng, G., 1995. Self-healing slip pulse on a frictional surface. *J. Mech. Phys. solids*, **43**, 1461-1495.

Perzyna, P., 1963. The constitutive equations for rate sensitive plastic materials. *Quart. J. Appl. Math.*, **20**, 321-332.

Perzyna, P., 1966. Fundamental problems in viscoplasticity. In *Advances in Applied Mechanics*, Vol. **9**, Academic Press, New York, 243-377.

Pineau, A., 1981. Review of fracture mechanisms and a local approach to predicting crack resistance in low strength steels. In *Advances in Fracture Research (ICF5 1981)*, edited by D. Francois, Pergamon Press, 533-577.

Pipkin, A.C., 1972. *Lectures on Viscoelasticity Theory*. George Allen & Unwin Ltd., London. Springer-Verlag, New York, Heidelberg, Berlin.

Pol, van der, B. and Bremmer, H., 1959. *Operational Calculus Based on the Two-sided Laplace Integral*. Cambridge University Press, Cambridge.

Ponte Castañeda, P., 1986. Asymptotic fields of a perfectly-plastic, plane-stress mode II growing crack. *J. Appl. Mech.*, **53**, 831-833.

Ponte Castañeda, P., 1987a. Plastic stress intensity factors in steady crack growth. *J. Appl. Mech.*, **54**, 379-387.

Ponte Castañeda, P., 1987b. Asymptotic fields in steady crack growth with linear strain-hardening. *J. Mech. Phys. Solids*, **35**, 227-268.

Popelar, C.H. and Atkinson, C., 1980. Dynamic crack propagation in a viscoelastic strip. *J. Mech. Phys. Solids*, abf 28, 79-93.

Prakash, V. and Clifton, R.J., 1992. Experimental and analytical investigations of dynamic fracture under conditions of plane strain. In *Fracture Mechanics: Twenty-Second Symposium (Volume I), ASTM STP 1131*, edited by H.A. Ernst, A. Saxena and D.L. McDowell. American Society for Testing and Materials, Philadelphia, 412-444.

Predeleanu, M., 1992. Modeling dynamic strain localization in inelastic solids. *Journal of Materials Processing Technology*, **32**, 75-89.

Qu, J. and Bassani, J.L., 1989. Cracks on bimaterial and bicrystal interfaces. *J. Mech. Phys. Solids*, **37**, 417-433.

Radok, J.R.M., 1956. On the solution of problems of dynamic plane elasticity. *Quart. Appl. Math.*, **14**, 289-298.

Ramanathan, S. and Fischer, D.S., 1997. Dynamics and instabilities of planar tensile cracks in heterogeneous media. *Phys. Rev. Letter*, **79**, 877-880.

Ramirez, J.-C., 1987. The three-dimensional stress intensity factor due to the motion of a load on the faces of a crack. *Quart. Appl. Math.*, **45**, 361-375.

Ramulu, M. and Kobayashi, A.S., 1985. Mechanics of curving and branching – a dynamic fracture analysis. *Int. J. Fract.*, **27**, 187-201.

Ramulu, M., Kobayashi, A.S., Kang, B.S.J. and Barker, D.B., 1983. Further studies on dynamic crack branching. *Experimental Mechanics*, **23**, 431-437.

Ravera, R.J. and Sih, G.C., 1969. Transient analysis of stress waves around cracks under transient strain. *J. Acoust. Soc. Amer.*, **47**, 875-880.

Ravi-Chandar, K., 1982. An experimental investigation into the mechanics of dynamic fracture. *Ph.D. Thesis*. California Institute of Technology, Pasadena, California†.

Ravi-Chandar, K., 1995. On the failure mode transitions in polycarbonate under dynamic mixed-mode loading. *Int. J. Solids Structures*, **32**, 925-938.

Ravi-Chandar, K., 1998. Dynamic fracture of nominally brittle materials. *Int. J. Fract.*, **90**, 83-102.

Ravi-Chandar, K. and Knauss, W.G., 1984a. An experimental investigation into dynamic fracture: I Crack initiation and arrest. *Int. J. Fract.*, **25**, 247-262.

Ravi-Chandar, K. and Knauss, W.G., 1984b. An experimental investigation into dynamic fracture: II Microstructural aspects, *Int. J. Fract.*, *Int. J. Fract.*, **26**, 65-80.

Ravi-Chandar, K. and Knauss, W.G., 1984c. An experimental investigation into dynamic fracture: III On steady-state crack propagation and crack branching, *Int. J. Fract.*, **26**, 141-154.

Ravi-Chandar, K., Lu, J., Yang, B. and Zhu, Z., 1998. Failure mode transitions in polymers under high strain rate loading. Submitted for publication.

Ravi-Chandar, K. and Yang, B., 1997. On the role of microcracks in the dynamic fracture of brittle materials. *J. Mech. Phys. Solids*, **45**, 535-563.

Rayleigh, J.W.S., 1885. On waves propagated along the plane surface of an elastic solid. *Proc. London math. Soc.*, **17**, 4-11.

Reimanis, I.E., Dalgleish, B.J. and Evans, A.G., 1991. Fracture resistance of a model metal-ceramic interface. *Acta Metallurgica et Materialia*, **39**, 3133-3141.

Rice, J.R., 1966. Contained plastic deformation near cracks and notches under longitudinal shear. *Int. J. Fract. Mech.*, **2**, 426-447.

Rice, J.R., 1967. Stresses due to a sharp notch in a work-hardening elastic-plastic material loaded by longitudinal shear. *J. Appl. Mech.*, **34**, 287-298.

Rice, J.R., 1968a. A path independent integral and the approximate analysis of strain concentration by notches and cracks. *J. Appl. Mech.*, **35**, 379-386.

Rice, J.R., 1968b. Mathematical analysis in the mechanics of fracture. In *Fracture: An Advanced Treatise*, edited by H. Liebowitz. Academic Press, New York, Vol. II, 191-311.

Rice, J.R., 1972. Some remarks on elastic crack tip stress fields. *Int. J. Solids Structures*, **8**, 751-758.

Rice, J.R., 1977. The localization of plastic deformation. In *Theoretical and Applied Mechanics*, Proc. 14th Int. Congr. Theoret. Appl. Mech., edited by W.T. Koiter. North-Holland Publ. Co., Amsterdam, 207-220.

Rice, J.R., 1978. Thermodynamics of the quasi-static growth of Griffith cracks. *J. Mech. Phys. Solids*, **26**, 61-78.

Rice, J.R., 1980. The mechanics of earthquake rupture. In *Physics of the Earth's Interior* (Proceedings of the International School of Physics "Enrico Fermi", Course 78, 1979), edited by A.M. Dziewonski and F. Boschi, Italian Physical Society. North-Holland Publ. Co., Amsterdam, 555-649.

Rice, J.R., 1982. Elastic-plastic crack growth. In *Mechanics of Solids*, edited by H.G. Hopkins and M.J. Sewell, Pergamon Press, Oxford and New York, 539-562.

Rice, J.R., 1985a. Conserved integrals and energetic forces. In *Fundamentals of Deformation and Fracture*, Eshelby Memorial Symposium, edited by B.A. Bilby, K.J. Miller and J.R. Willis. Cambridge Univ. Press, Cambridge, 33-56.

Rice, J.R., 1985b. First-order variations in elastic fields due to variation in location of a planar crack front. *J.Appl. Mech.*, **52**, 561-579.

Rice, J.R., 1987. Tensile crack tip fields in elastic-ideally plastic crystals. *Mechanics of Materials*, **6**, 317-335.

Rice, J.R., 1988a. Two general integrals of singular crack tip deformation fields. *J. Elasticity*, **20**, 131-142.

Rice, J.R., 1988b. Elastic fracture concepts for interfacial cracks. *J. Appl. Mech.*, **55**, 98-103.

Rice, J.R., 1989. Weight function theory for three-dimensional elastic crack analysis. In *Fracture Mechanics Perspectives and Directions (Twentieth Symposium), ASTM STP 1020*,

---

† The essential results of this thesis are covered in Ravi-Chandar and Knauss (1984a,b,c).

edited by R.P. Wei and R.P. Gangloff. American Society for Testing and Materials, Philadelphia, 29-57.

Rice, J.R., 1992. Dislocation nucleation from a crack tip: An analysis based on the Peierls concept. *J. Mech. Phys. Solids*, **40**, 239-271.

Rice, J.R. and Beltz, G.E., 1994. An activation energy for dislocation nucleation from a crack tip. *J. Mech. Phys. Solids*, **42**, 333-360.

Rice, J.R., Drugan, W.J. and Sham, T.-L., 1980. Elastic-plastic analysis of growing cracks. In *Fracture Mechanics: 12th Conference, ASTM STP 700*, American Society for Testing and Materials, Philadelphia, 189-219.

Rice, J.R. and Johnson, M.A., 1970. The role of large crack tip geometry changes in plane strain fracture. In *Inelastic Behaviour of Solids*, edited by M.F. Kanninen et al, McGraw-Hill Publ.

Rice, J.R. and Levy, N., 1969. Local heating by plastic deformation at a crack tip. In *Physics of Strength and Plasticity*, edited by A.S. Argon. M.I.T. Press Cambridge, Massachusetts, and London, England.

Rice, J.R., Paris, P.C. and Merkle, J.G., 1973. Some further results of J integral analysis and estimates. In *Progress in Flaw Growth and Fracture Toughness Testing. ASTM STP 536*, 231-2459.

Rice, J.R. and Rosengren, G.F., 1968. Plane strain deformation near a crack tip in a power law hardening material. *J. Mech. Phys. Solids*, **16**, 1-12.

Rice, J.R. and Sih, G.C., 1965. Plane problems of cracks in dissimilar media. *J.Appl. Mech.*, **32**, 418-423.

Rice, J.R. and Thomson, R., 1974. Ductile versus brittle behaviour of crystals. *Phil. Mag.*, **29**, 73-97.

Rice, J.R. and Tracey, D.M., 1969. On the ductile enlargement of voids in triaxial stress fields. *J. Mech. Phys. Solids*, **17**, 201-217.

Richard, H.A., 1982. Bruchvorhersagen bei Überlagerter Normal- und Scherbeanspruchung. Dissertation, Universität Karlsruhe.

Riedel, H., 1987. *Fracture at High Temperatures*, Springer-Verlag, Berlin, Heidelberg.

Ritchie, R.O., Knott, J.F. and Rice, J.R., 1973. On the relationship between critical tensile stress and fracture toughness in mild steel. *J. Mech. Phys. Solids*, **21**, 395-410.

Rittel, D., 1998a. Experimental investigation of dynamic failure mode transitions. In *Non-linear Singularities in Deformation and Flow. Proceedings of the IUTAM Conference on Non-linear Singularities in Deformation and Flow, Haifa 1997*, edited by D. Durban and R.A Pearson, Kluwer Academic Publ., Dordrecht/Boston/London, 181-192.

Rittel, D., 1998b. Experimental investigation of transient thermoelastic effects in dynamic fracture. *Int. J. Solids Structures*, **35**, 2959-2973.

Rittel, D., Levin, R. and Maigre, H., 1997. The influence of mode-mixity on dynamic failure mode transitions in polycarbonate. *Journal de Physique IV France. Colloque C3, Supplément au Journal de Physique III d'août 1997*, **7**, C3-861–C3-866.

Rittel, D. and Maigre, H., 1996. An investigation of dynamic crack initiation in PMMA. *Mechanics of Materials*, **23**, 229-239.

Rivlin, R.S., 1947. Torsion of a rubber cylinder. *J. Appl. Phys.*, **18**, 444-449.

Rivlin, R.S., 1948. Large elastic deformations of isotropic materials. I. Fundamental concepts. *Phil. Trans. Roy. Soc. (London)*, A240, 459-490.

Rivlin, R.S. and Thomas, A.G., 1952. Rupture of rubber. I. Characteristic energy for tearing. *J. Pol. Sci.*, **10**, 251-318.

Roesler, J.W., 1956. Brittle fractures near equilibrium. *Proc. Phys. Soc.*, **B69**, 981-992 with photo on plate between pages 1068 and 1069.

Rogers, H.C., 1960. The tensile fracture of ductile metals. *Trans. Metallurgical Society of AIME*, **218**, 498-506.

Rosakis, A.J., Duffy, J. and Freund, L.B., 1984. The determination of dynamic fracture toughness of AISI 4340 steel by the shadow spot method. *J. Mech. Phys. Solids*, **32**, 443-460.

Rosakis, A.J., Liu, C. and Freund, L.B., 1991. A note on the asymptotic stress field of a non-uniformly propagating dynamic crack. *Int. J. Fract.*, **50**, R39-R45.

Rosakis, A.J., Mason. J.J. and Ravichandran, G., 1992. The conversion of plastic work to heat around a dynamically propagating crack in metals. *J. Mech. Behaviour Materials*, **4**, 375-385.

Rosakis, A.J. and Ravi-Chandar, K., 1986. On crack-tip stress state: an experimental evaluation of three-dimensional effects. *Int. J. Solids and Structures*, **22**, 121-134.

Rosakis, A.J., Ravichandran, G. and Zhou, M., 1997. Dynamically growing shear bands in metals: A study of transient temperature and deformation fields. In *IUTAM Symposium on Nonlinear Analysis of Fracture*, edited by J.R. Willis. Kluwer Academic Publishers, The Netherlands, 141-150.

Rosakis, A.J. and Zehnder, A.T., 1985. On the dynamic fracture of structural metals. *Int. J. Fract.*, **27**, 169-186.

Rose, L.R.F., 1976a. Recent theoretical and experimental results on fast brittle fracture. *Int. J. Fract.*, **12**, 799-813.

Rose, L.R.F., 1976b. On the initial motion of a brittle crack. *Int. J. Fract.*, **12**, 829-841.

Rose, L.R.F., 1986. Micro-crack interaction with a main crack. *Int. J. Fract.*, **31**, 233-242.

Rose, L.R.F., 1987. Crack reinforcement by distributed springs. *J. Mech. Phys. Solids*, **35**, 383-405.

Rossikhin, Y.A. and Shitikova, M.V., 1997. Applications of fractional calculus to dynamic problems of linear and nonlinear hereditary mechanics of solids. *Applied Mechanics Reviews*, **50**, 15-67.

Rossmanith, H.P., 1983. How 'mixed' is dynamic mixed-mode crack propagation? – A dynamic photoelastic study. *J. Mech. Phys. Solids*, **31**, 251-260.

Rubinstein, A.A., 1986. Macrocrack-microcrack interactions. *J. Appl. Mech.*, **53**, 505-510.

Rubinstein, A.A. and Wang Peng, 1998. The fracture toughness of a particulate-reinforced brittle matrix. *J. Mech. Phys. Solids*, **46**, 1139-1154.

Rudnicki, J.W. and Rice, J.R., 1975. Conditions for the localization of deformation in pressure-sensitive materials. *J. Mech. Phys. Solids*, **23**, 371-394.

Ruggieri, C. and Dodds Jr, R.H., 1996. A transferability model for brittle fracture including constraint and ductile tearing effects: a probabilistic approach. *Int. J. Fract.*, **79**, 309-340.

Ruggieri, C., Panontin, T.L. and Doods Jr, R.H., 1996. Numerical modeling of ductile crack growth in 3-D using computational cell elements. *Int. J. Fract.*, **82**, 67-95.

Sakata, M., Aoki, S. and Ishii, K., 1978. J-integral approach to fracture of rotating disk. *Trans. ASME J. Engng Mat. Tech.*, **100**, 128-133.

Sanders, J.L., 1960. On the Griffith-Irwin fracture theory. *J. Appl. Mech.*, **27**, 352-353.

Schapery, R.A., 1975. A theory of crack initiation and growth in viscoelastic media. I. Theoretical development. *Int. J. Fract.*, **11**, 141-159.

Schardin, H., 1950. Ergebnisse der kinematographischen Untersuchung des Glasbruchvorganges. *Glastechnische Berichte*, **23**, 1-10; 67-79; 325-336.

Schardin, H., 1959. Velocity effects in fracture. In *Fracture*, edited by B.L. Averbach, D.K. Felbeck, G.T. Hahn and D.A. Thomas, John Wiley & Sons, New York, 297-329.

Scholz, C.H., 1990. *The Mechanics of Earthquakes and Faulting*, Cambridge University Press, Cambridge, UK.

Sham, T.-L., 1991. The determination of the elastic T-term using higher order weight functions. *Int. J. Fract.*, **48**, 81-102.

Shand, E.B., 1958. *Glass Engineering Handbook*, McGraw-Hill Book Co. New York.

Sharma, S.M. and Aravas, N., 1991. Determination of higher-order terms in asymptotic elastoplastic crack problems. *J. Mech. Phys. Solids*, **39**, 1043-1072.

Sharon, E. and Fineberg, J., 1996. Micro-branching instability and the dynamic fracture of brittle materials. *Phys. Rev.*, B **54**, 7128-7149.

Sharon, E., Gross, S.P. and Fineberg, J., 1995. Local crack branching as a mechanism for instability in dynamic fracture. *Phys. Rev. Lett.*, **74**, 5096-5099.

Sharon, E., Gross, S.P. and Fineberg, J., 1996. Energy dissipation in dynamic fracture. *Phys. Rev. Lett.*, **76**, 2117-2120.

Shen, Y and Drugan, W.J., 1990. Constraints on moving strong discontinuity surfaces in dynamic plane-stress or plane-strain deformations of stable elastic-ideally plastic materials. *J. Appl. Mech.*, **57**, 569-576.

Shih, C.F., 1974. Small-scale yielding analysis of mixed-mode plane strain crack problems. In *Fracture Analysis, ASTM STP 560*, American Society for Testing and Materials, 187-210.

Shih, C.F., 1981. Relationships between the J-integral and the crack opening displacement for stationary and extending cracks. *J. Mech. Phys. Solids*, **29**, 305-326.

Shih, C.F., Cheng, L., Faleskog, J. and Gao, X., 1997. A cell model for ductile fracture with applications to the transition regime. In *Proceedings from ICF9 1997, Sydney*, Vol. **4**, Pergamon Press, Oxford, 1935-1946.

Shih, C.F., O'Dowd, N.P. and Kirk, M.T., 1993. A framework for quantifying crack tip constraint. In *Constraint Effects in Fracture, ASTM STP 1171*, edited by E.M. Hackett, K.-H. Schwalbe and R.H. Dodds, American Society for Testing and Materials, Philadelphia, 2-20.

Shih, C.F. and Xia, L., 1994. Modeling crack growth resistance using computational cells with micro-structurally-based length scales. In *Constraint Effects in Fracture: Theory and Applications, ASTM STP 1244*, edited by M. Kirk and A. Bakker, American Society for Testing and Materials, Philadelphia, 163-190.

Shioya, T. and Zhou, F., 1995. Dynamic fracture toughness and crack propagation in brittle material. In *Constitutive Relation in High/Very High Strain Rates*, edited by K. Kawata and J. Shioiri, Springer-Verlag, Tokyo, 105-112.

Shockey, D.A. and Curran, D.R., 1973. A method for measuring K_{Ic} at very high strain rates. In *Progress in Flaw Growth and Fracture Toughness Testing. ASTM STP 536*, American Society for Testing and Materials, 297-311.

Shockey, D.A., Erlich, D.C., Kalthoff, J.F. and Homma, H., 1986. Short-pulse fracture mechanics. *Engng Fract. Mech.*, **23**, 311-319.

Shockey, D.A., Kalthoff, J.F. and Erlich, D.C., 1983. Short pulse fracture mechanics. In *Workshop on Dynamic Fracture*, California Institute of Technology, Pasadena, California, 57-71.

Shum, K.M. and Hutchinson, J.W., 1990. On toughening of microcracks. *Mechanics of materials*, **9**, 83-91.

Sih, G.C., 1968. Some elasto-dynamic problems of cracks. *Int. J. Fract. Mech.*, **4**, 51-68.

Sih, G.C. and Embley, G.T., 1972. Impact response of a finite crack in plane extension. *Int. J. Solids Structures*, **8**, 977-993.

Sih, G.C., Embley, G.T. and Ravera, R.S., 1972. Impact response of a finite crack in plane extension. *Int. J. Solids Structures*, **8**, 977-993.

Sih, G.C. and Liebowitz, H., 1968. Mathematical theories of brittle fracture. In *Fracture: An Advanced Treatise*, edited by H. Liebowitz. Academic Press, New York, Vol. II, 67-190.

Sills, L.B. and Benveniste, Y., 1981. Steady state propagation of a mode III interface crack between dissimilar viscoelastic media. *Int. J. Engng Sci.*, **19**, 1255-1268.

Sjöberg, F. and Ståhle, P., 1992. On the autonomy of the process region. *Int. J. Fracture*, **54**, 1-20.

Slepyan, L.I., 1973. Deformation at the edge of a growing crack. *Mechanics of Solids*, **8**, 125-134. Translation from *Izv. Akad. Nauk SSSR: Mekhanika Tverdogo Tela*, **8** (1973), 139-148.

Slepyan, L.I., 1974. Growing crack during plane deformation in an elastic-plastic body. *Mechanics of Solids*, **9**, 46-55. Translation from *Izv. Akad. Nauk SSSR: Mekhanika Tverdogo Tela*, **9** (1974), 57-67.

Slepyan, L.I., 1976. Crack dynamics in an elastic-plastic body. *Mechanics of Solids*, **11**, 126-134. Translation from *Izv. Akad. Nauk SSSR: Mekhanika Tverdogo Tela*, **11** 1976, 145-153.

Slepyan, L.I., 1993. Principle of maximum energy dissipation rate in crack dynamics. *J. Mech. Phys. Solids*, **41**, 1019-1033.

Smirnov, V.I., 1964. *A Course of Higher Mathematics*, **III**, Addison-Wesley, Reading, Mass.

Smirnov, V.I. and Sobolev, S.L., 1932. Novyi metod resheniia ploskoi zadachi uprugikh kolebanii. (New method for solving the planar problem of elastic oscillations.) *Trudy seismich. in-ta Akad. Nauk SSSR*, No. 20, **5**.

Sneddon, I.N., 1942. *Fourier Transforms*, McGraw-Hill, New York.

Sneddon, I.N., 1946. The distribution of stress in the neighbourhood of a crack in an elastic solid. *Proc. Roy. Soc. (London)*, **A187**, 229-260.

Sneddon, I.N., 1958. Note on a paper by J.R.M. Radok. *Quart. Appl. Math.*, **16**, 197.

Sneddon, I.N., 1972. *The Use of Integral Transforms*, McGraw-Hill Book Company, New York.

Sorensen, E.P., 1978. A finite element investigation of stable crack growth in anti-plane shear. *Int. J. Fract.*, **14**, 485-500.

Ståhle, P., 1983. On the small scale fracture mechanics. *Int. J. Fracture*, **22**, 203-216.

Ståhle, P., 1985. Process region characteristics and stable crack growth. *Report LUTFD2-/(TFHF-3019) from Division of Solid Mechanics*, Lund Institute of Technology, Lund, Sweden.

Ståhle, P., 1986. On stable crack growth at large scale of yielding. *Computational Mechanics '86*. Proc. Int. Conf. Computational Mech., May 25-29. 1986. Tokyo, Springer-Verlag, V-365-370.

Ståhle, P., 1989. Crack-tip field at steady crack growth and vanishing linear strain hardening. In *Proc. of ICF7*, edited by K. Salama, K. Ravichandar, D.M.R. Taplin and P. Rama Rao, Pergamon Press New York, 455-462.

Ståhle, P., 1993. Dynamic crack tip fields at steady growth and vanishing strain hardening. *J. Mech. Phys. Solids*, **41**, 919-936.

Stam, G. and van der Giessen, E., 1996a. Crack growth in non-homogeneous transformable ceramics. Part I: Constrained straight cracks. *Int. J. Fract.*, **79**, 249-271.

Stam, G. and van der Giessen, E., 1996b. Crack growth in non-homogeneous transformable ceramics. Part II: Crack deflection. *Int. J. Fract.*, **79**, 273-293.

Steenbrink, A.C., van der Giessen, E. and Wu, P.D., 1997. Void growth in glassy polymers. *J. Mech. Phys. Solids*, **45**, 405-437.

Stephenson, R.A., 1982. The equilibrium field near the tip of a crack for finite plane strain of incompressible elastic materials. *J. Elasticity*, **12**, 65-99.

Stoneley, R., 1924. Elastic waves at the surface of separation of two solids. *Proc. Roy. Soc. (London)*, **A106**, 416-428.

Stören, S. and Rice, J.R., 1975. Localized necking in thin sheets. *J. Mech. Phys. Solids*, **23**, 421-441.

Stroh, A.N., 1958. Dislocations and cracks in anisotropic elasticity. *Philosophical Magazine*, **3**, 645-646.

Strömberg, L. and Ristinmaa, M., 1996. FE-formulation of a nonlocal plasticity theory. *Comp. Meth. in Appl. Mech. and Engng*, **136**, 127-144.

Sun, C.T., Achenbach, J.D and Herrmann, G., 1968. Continuum theory for a laminated medium. *J. Appl. Mech.*, **35**, 467-475.

Sun, C.T. and Jih, C.J., 1987. On strain energy release rates for interfacial cracks in bi-metal media. *Engng Fract. Mech.*, **28**, 13-20.

Sun, C.T. and Li, S., 1988. Three-dimensional effective elastic constants for thick laminates. *Journal of Composite Materials*, **22**, 629-639.

Sun, C.T. and Wu, X.X., 1996. On the J-integral in periodically layered composites. *Int. J. Fracture*, **78**, 89-100.

Suo, Z., 1990. Singularities, interfaces and cracks in dissimilar anisotropic media. *Proc. Roy. Soc. (London)*, **A427**, 331-358.

Suo, Z., Ortiz, M. and Needleman, A., 1992. Stability of solids with interfaces. *J. Mech. Phys. Solids*, **40**, 613-640.

Suo, Z., Shih, C.F. and Varias, A.G., 1993. A theory for cleavage cracking in the presence of plastic flow. *Acta Metallurgica et Materialia*, **41**, 1551-1557.

Tada, H., Paris, P.C. and Irwin, G.R., 1985. *The Stress Analysis of Cracks Handbook*, Del Research Corporation, St. Louis, Missouri, USA.

Takahashi, K. and Arakawa, K., 1987. Dependence of crack acceleration on the dynamic stress-intensity factor in polymers. *Experimental Mechanics*, **27**, 195-200.

Takahashi, K. and Kido, M., 1995. Velocity dependent dynamic fracture toughness of Araldite B simultaneously determined by caustic and photoelastic methods. In *Constitutive Relation in High/Very High Strain Rates*, edited by K. Kawata and J. Shioiri, Springer-Verlag, Tokyo, 121-128.

Tarantino, A.M., 1997. Nonlinear fracture mechanics for an elastic Bell material. *Quart. J. Mech. Appl. Math.*, **50**, 436-456.

Taylor, G.I. and Quinney, M.A., 1934. The latent energy remaining in a metal after cold working. *Proc. Roy. Soc. (London)*, **A143**, 307-326.

Thomas, T.Y., 1961. *Plastic Flow and Fracture in Solids*, Academic Press, New York.

Thomason, P.F., 1990. *Ductile Fracture of Metals*, Pergamon Press, Oxford.

Thomson, R.M., 1983. Fracture. In *Physical Metallurgy*, edited by R.W. Cahn and P. Hansen. North-Holland Publishing, Vol. II, 1487-1551.

Theocaris, P.S., 1977. Asymmetric branching of cracks. *J. Appl. Mech.*, **44**, 611-618.

Timoshenko, S.P., 1983. *History of Strength of Materials*, Dover Publications, Inc., New York.

Ting, T.C.T., 1992. Barnett-Lothe tensors and their associated tensors for monoclinic materials with symmetry plane at $x_3 = 0$. *J. Elasticity*, **27**, 143-165.

Ting, T.C.T., 1996. *Anisotropic Elasticity: Theory and Applications*, Oxford University Press, New York.

Tipper, C.F., 1949. The fracture of metals. *Metallurgia*, **39**, 133-137.

Tryding, J., 1991 Crack-growth-rate estimation on a three-point bend specimen during low-cycle fatigue. *Experimental Mechanics*, **31**, March 1991, 42-46.

Tvergaard, V., 1981. Influence of voids on shear band instabilities under plane strain conditions. *Int. J. Fract.*, **17**, 389-407.

Tvergaard, V., 1982a. Ductile fracture by cavity nucleation between larger voids. *J. Mech. Phys. Solids*, **30**, 265-286.

Tvergaard, V., 1982b. On localization in ductile materials comtaining spherical voids. *Int. J. Fract.*, **18**, 237-252.

Tvergaard, V., 1987. Effect of yield surface curvature and void nucleation on plastic flow localization. *J. Mech. Phys. Solids*, **35**, 43-60.

Tvergaard, V., 1995. Cavity growth in ductile particles bridging a brittle matrix crack. *Int. J. Fract.*, **72**, 277-292.

Tvergaard, V. and Hutchinson, J.W., 1992. On the relation between crack growth resistance and fracture process parameters in elastic-plastic solids. *J. Mech. Phys. Solids*, **40**, 1377-1397.

Tvergaard, V. and Hutchinson, J.W., 1993. Effect of initial void shape on the occurrence of cavitation instabilities in elastic-plastic solids. *J. Appl. Mech.*, **60**, 807-812.

Tvergaard, V. and Hutchinson, J.W., 1994. Effect of T-stress on mode I crack growth resistance in a ductile solid. *Int. J. Solids Structures*, **31**, 823-833.

Tvergaard, V. and Needleman, A., 1988. An analysis of the temperature and rate dependence of Charpy V-notch energies for a high nitrogen steel. *Int. J. Fract.*, **37**, 197-215.

Tvergaard, V. and Needleman, A., 1993. An analysis of the brittle–ductile transition in dynamic crack growth. *Int. J. Fract.*, **59**, 53-67.

Tylecote, R.F., 1976. *A History of Metallurgy*, The Metals Society, London.

Tylecote, R.F., 1987. *The Early History of Metallurgy in Europe*, Longman, London.

Uflyand, Y.S., 1965. *Survey of Articles on the Application of Integral Transforms in the Theory of Elasticity*. North Carolina State University, Dept. of Applied Mathematics Research Group, File No. PSR-24/6.

Underwood, E.E. and Banerji, K., 1986. Fractals in fractography. *Mater. Sci. Engng*, **80**, 1-14.

van der Pol, B. and Bremmer, H., 1959. *Operational Calculus Based on the Two-sided Laplace Integral*. Cambridge University Press, Cambridge.

Varias, A.G. and Shih, C.F., 1993. Quasi-static crack advance under a range of constraints—steady-state fields based on a characteristic length. *J. Mech. Phys. Solids*, **41**, 835-861.

Varias, A.G. and Shih, C.F., 1994. Dynamic steady crack growth in elastic-plastic solids – propagation of strong discontinuities. *J. Mech. Phys. Solids*, **42**, 1817-1848.

Varias, A.G., Suo, Z. and Shih, C.F., 1991. Ductile failure of a constrained metal foil. *J. Mech. Phys. Solids*, **39**, 963-986.

Varias, A.G., Suo, Z. and Shih, C.F., 1992. Mode mixity effect on the damage of a constrained ductile layer. *J. Mech. Phys. Solids*, **40**, 485-509.

Vitek, V., 1977. Plane strain stress intensity factors for branched cracks. *Int. J. Fract.*, **13**, 481-501.

Vitvisky, P.M. and Leonov, M.Y., 1961. On fracture of a plate containing a crack (in Ukranian). *Prikhladna Mekhanika*, **5**, 516-520.

von Mises, R., 1913. Mechanik der festen Körper im plastisch-deformablem Zustand *Nachr. d. Kgl. Ges. d. Wissensch. Göttingen, Math. phys. Klasse*, **K1**, 582-592.

von Mises, R., 1928. Mechanik der plastischen Formänderung von Kristallen. *Zeitschrift für Angewandte Mathematik und Mechanik*, **8**, 161-185.

Wallin, K., 1993. Statistical aspects of constraint with emphasis to testing and analysis of laboratory specimens in the transition region. In *Constraint Effects in Fracture, ASTM STP 1171*, edited by E.M. Hackett, K.-H. Schwalbe and R.H. Dodds, American Society for Testing and Materials, Philadelphia, 264-268.

Wallin, K., Saario, T. and Törrönen, K., 1984. Statistical model for carbide induced brittle fracture in steel. *Metal Science*, **18**, 13-16.

Wallner, H., 1939. Linienstrukturen an Bruchflächen. *Z. Physik*, **114**, 368-378.

Walton, J.R., 1982. On the steady-state propagation of an anti-plane shear crack in an infinite linearly viscoelastic body. *Quart. Appl. Math.*, **40**, 37-52.

Walton, J.R., 1985. The dynamic, steady-state propagation of an anti-plane shear crack in a general linearly viscoelastic layer. *J. Appl. Mech.*, **52**, 853-856.

Walton, J.R., 1987. The dynamic energy release rate for a steadily propagating antiplane shear crack in a linearly viscoelastic body. *J. Appl. Mech.*, **54**, 635-641.

Ward, G.N., 1955. *Linearized Theory of Steady Highspeed Flow*, Cambridge, London.

Washabaugh, P.D. and Knauss, W.G., 1993. Non-steady, periodic behavior in the dynamic fracture of PMMA. *Int. J. Fract.*, **59**, 189-197.

Washabaugh, P.D. and Knauss, W.G., 1994. A reconciliation of dynamic crack velocity and Rayleigh wave speed in isotropic brittle solids. *Int. J. Fract.*, **65**, 97-114.

Weertman, J., 1963. Dislocations moving uniformly on the interface between isotropic media of different elastic properties. *J. Mech. Phys. Solids*, **11**, 197-204.

Wei, Y. and Hutchinson, J.W., 1997. Steady-state crack growth and work of fracture for solids characterized by strain gradient plasticity. . *J. Mech. Phys. Solids*, **45**, 1253-1273.

Weibull, W., 1938. Investigations into strength properties of brittle materials. *Proceedings of the Royal Swedish Institute for Engineering Research*, No. 149.

Weibull, W., 1939a. A statistical theory of the strength of materials. *Proceedings of the Royal Swedish Institute for Engineering Research*, No. 151.

Weibull, W., 1939b. The phenomenon of rupture in solids. *Proceedings of the Royal Swedish Institute for Engineering Research*, No. 153.

Weibull, W., 1951. A statistical distribution function of wide applicability. *J. Appl. Mech.*, **18**, 293-297.

Weichert, R. and Schönert, K., 1978. Heat generation at the tip of a moving crack. *J. Mech. Phys. Solids*, **26**, 151-161.

Westergaard, H.M., 1939. Bearing pressures and cracks. *J. Appl. Mech.*, **6**, A49-A53.

Wiener, N. and Hopf, E., 1931. Über eine Klasse singulärer Integralgleichungen. *S. B. Preuss. Acad. Wiss.*, 691-706.

Wihlborg, G., 1976. High velocity tensile tests on a mild steel. *Report from the Division of Solid Mechanics*, Lund Institute of Technology, Lund, Sweden.

Williams, J.G. and Ewing, P.D., 1972. Fracture under complex stress – the angled crack problem. *Int. J. Fract.*, **8**, 441-446.

Williams, J.G. and Ewing, P.D., 1974. Authors' closure to discussion be Sih and Kipp on "Fracture under complex stress – the angled crack problem" by J.G. Williams and P.D. Ewing. *Int. J. Fract.*, **10**, 265.

Williams, M.L., 1952. Stress singularities resulting from various boundary conditions in angular corners of plates in extension. *J. Appl. Mech.*, **19**, 526-528.

Williams, M.L., 1957. On the stress distribution at the base of a stationary crack. *J. Appl. Mech.*, **24**, 109-113.

Williams, M.L., 1959. The stress around a fault or crack in dissimilar media. *Bull. seism. Soc. America*, **49**, 199-204.

Williams, M.L., 1963. The fracture of viscoelastic material. In *Fracture of Solids*, edited by D.C. Drucker and J.J. Gilman, Interscience Publishers, N.Y., 157-188.

Williams, M.L., 1965. Initiation and growth of viscoelastic fracture. *Int. J. Fract. Mech.*, **1**, 292-310.

Willis, J.R., 1967. Crack propagation in viscoelastic media. *J. Mech. Phys. Solids*, **15**, 229-240.

Willis, J.R., 1968. The stress field around an elliptical crack in an anisotropic elastic medium. *Int. J. Engng Sci.*, **6**, 253-263.

Willis, J.R., 1971. Fracture mechanics of interfacial cracks. *J. Mech. Phys. Solids*, **19**, 353-368.

Willis, J.R., 1973. Self-similar problems in elastodynamics. *Phil. Trans. Roy. Soc. (London)*, **A274**, 435-491.

Willis, J.R. and Movchan, A.B., 1995. Dynamic weight functions for a moving crack. I. Mode I loading. *J. Mech. Phys. Solids*, **43**, 319-341.

Willis, J.R. and Movchan, A.B., 1997. Three-dimensional dynamic perturbation of a propagating crack. *J. Mech. Phys. Solids*, **45**, 591-610.

Wnuk, M.P., 1971. Subcritical growth of fracture (inelastic fatigue). *Int. J. Fract. Mech.*, **7**, 383-407.

Wnuk, M.P., 1973. Prior to failure extension of flaws in a rate sensitive Tresca solid. In *Progress in Flaw Growth and Fracture Toughness Testing. ASTM STP 536*, American Society for Testing and Materials, 64-75.

Wnuk, M.P., 1974. Quasi-static extension of a tensile crack contained in a viscoelastic-plastic solid. *J. Appl. Mech.*, **42**, 234-242.

Wnuk, M.P. and Knauss, W.G., 1970. Delayed fracture in viscoelastic-plastic solids. *J. Appl. Mech.*, **38**, 483-488.

Wong, F.S. and Shield, R.T., 1969. Large plane deformations of thin elastic sheets of neo-Hookean material. *Zeitschrift für angwandte Mathematik und Physik*, **20**, 176-199.

Wright, T.W. and Walter, J.W., 1987. On stress collapse in adiabatic shear bands. *J. Mech. Phys. Solids*, **35**, 701-720.

Wu, J. and May, Y.-W., 1996. The essential fracture work concept for toughness measurement of ductile polymers. *Polymer Engineering and Science*, **36**, 2275-2288.

Wu, K.-C., 1989. On the crack-tip fields of a dynamically propagating crack in an anisotropic solid. *Int. J. Fract.*, **41**, 253-266.

Wu, K.-C., 1991. Explicit crack-tip fields of an expanding interface crack in an anisotropic bimaterial. *Int. J. Solids Structures*, **27**, 455-466.

Wylie, C.R. and Barrett, L.C., 1985. *Advanced Engineering Mathematics*, McGraw-Hill Co., Singapore.

Xia, L. and Shih, C.F., 1995a. Ductile crack growth – I. A numerical study using computational cells with microstructurally-based length scales. *J. Mech. Phys. Solids*, **43**, 233-259.

Xia, L. and Shih, C.F., 1995b. Ductile crack growth – II. Void nucleation and geometry effects on macroscopic fracture behaviour. *J. Mech. Phys. Solids*, **43**, 1953-1981.

Xia, L. and Shih, C.F., 1996. Ductile crack growth – III. Transition to cleavage fracture incorporating statitistics. *J. Mech. Phys. Solids*, **43**, 1953-1981.

Xia, L., Shih, C.F. and Hutchinson, J.W., 1995. A computational approach to ductile crack growth under large scale yielding conditions. *J. Mech. Phys. Solids*, **43**, 389-413.

Xia, Z.C. and Hutchinson, J.W., 1996. Crack tip fields in strain gradient plasticity. *J. Mech. Phys. Solids*, **44**, 1621-1648.

Xu, X.-P. and Needleman, A., 1994. Numerical simulations of fast crack growth in brittle solids. *J. Mech. Phys. Solids*, **42**, 1397-1434.

Xu, X.-P. and Needleman, A., 1995. Numerical simulations of dynamic interfacial crack growth allowing for crack growth away from the bond line. *Int. J. Fract.*, **74**, 253-275.

Xu, X.-P. and Needleman, A., 1996. Numerical simulations of dynamic crack growth along an interface. *Int. J. Fract.*, **74**, 289-324.
Xu, Y., Blume, J.A. and Shih, C.F., 1993. An interface crack between an orthotropic thin film and a substrate. *Int. J. Fract.*, **63**, 369-381.
Yang, W. and Freund, L.B., 1985. Transverse shear effects for through-cracks in an elastic plate. *Int. J. Solids and Structures*, **21**, 977-994.
Yang, W., Suo, Z. and Shih, C.F., 1991. Mechanics of dynamic debonding. *Proc. Roy. Soc. (London)*, **A433**, 679-697.
Yiantai, H. and Xinghua, Z., 1996. Collinear cracks in an anisotropic medium. *Int. J. Fract.*, **76**, 207-219.
Yoffe, E.H., 1951. The moving Griffith crack. *Phil. Mag.*, **42**, 739-750.
Yokobori, T., 1965. *The Strength, Fracture and Fatigue of Materials*, Noordhoff, Groningen, The Netherlands. Translated from Japanese original (1955).
Yokobori, T., 1968. Criteria for nearly brittle fracture. *Int. J. Fract. Mech.*, **4**, 179-187.
Yokobori, T. and Ichikawa, M., 1965. The interaction of two collinear dislocation cracks with special reference to brittle fracture strength of metals. *Rep. Res. Inst. Strength Fract. Materials*, Tohoku Univerity, **1**, 47-57.
Yokobori, T., Ichikawa, M. and Ohashi, M., 1965a. The interaction between an elastic crack and a slip band with special reference to brittle fracture strength of materials. *Rep. Res. Inst. Strength Fract. Materials*, Tohoku Univerity, **1**, 69-78.
Yokobori, T., Ohashi, M. and Ichikawa, M., 1965b. The interaction of two collinear asymmetrical cracks. *Rep. Res. Inst. Strength Fract. Materials*, Tohoku Univerity, **1**, 33-39.
Yokobori, T., Uozumi, M. and Ichikawa, M., 1971. Interaction between non-coplanar parallel staggered elastic cracks. *Rep. Res. Inst. Strength Fract. Materials*, Tohoku Univerity, **7**, 25-47.
Yu, H. and Yang,W., 1995. Mechanics of transonic debonding of a bimaterial interface: the in-plane case. *J. Mech. Phys. Solids*, **43**, 207-232.
Zak, A.R. and Williams, M.L., 1963. Crack point singularities at a bi-material interface. *J. Appl. Mech.*, **30**, 142-143.
Zbib, H. and Aifantis, E.C., 1988. On the structure and width of shear bands. *Scipta Metallurgica*, **22**, 703-708.
Zehnder, A.T. and Rosakis, A.J., 1990. Dynamic fracture initiation and propagation in 4340 steel under impact loading. *Int. J. Fract.*, **43**, 271-285.
Zehnder, A.T. and Rosakis, A.J., 1991. On the temperature distribution at the vicinity of dynamically propagating cracks in steel. *J. Mech. Phys. Solids*, **39**, 385-415.
Zener, C. and Hollomon, J.H., 1944. Effect of strain rate upon plastic flow of steel. *Phil. Mag.*, **15**, 22-32.
Zhou, M., Clifton, R.J. and Needleman, A., 1992. Shear bandformation in a W-Ni-Fe alloy under plate impact. *J. Mech. Phys. Solids*, **44**, 981-1006.
Zhou, M., Ravichandran, G. and Rosakis, A.J., 1996a. Dynamically propagating shear bands in impact-loaded prenotched plates: II – Numerical simulations *J. Mech. Phys. Solids*, **44**, 1007-1032.
Zhou, M., Rosakis, A.J. and Ravichandran, G., 1996b. Dynamically propagating shear bands in impact-loaded prenotched plates: I – Experimental investigations of temperature signatures and propagation speed. *J. Mech. Phys. Solids*, **44**, 981-1006.
Zhu, W. and Li, K., 1991. The dynamic near crack-line fields for mode II crack growth in an elastic perfectly-plastic solid. *Int. J. Fract.*, **52**, 307-317.

Index

3PB specimen, 82

α lines, 279
Abel theorems, 374, 455
Abelian integral equation, 96, 162, 200, 294
acoustic emission, 33, 573
adiabatic shear bands, 642
adiabatic temperature change, 536
Airy stress function, 77, 667, 684
    for anisotropic solids, 212
    biharmonic equation, 668
    for interface crack problems, 186
    representation of stresses, 667
    solutions for crack and notch edges, 668
amorphous structure, 11
analytic continuation, 672
analytic functions, 669–671
    branch cut, 671
    branch point, 670
    Cauchy's integral formula, 675
    Cauchy's integral theorem, 673
    decomposition, 678
    factorization, 678
    as Fourier series, 673
    Liouville's theorem, 677
    Plemelj's formulae, 677
    pole, 670
    as power series, 672
    sectionally analytic, 676
    simple pole, 670
    singularities, 670
Andersson's void growth model, 548
anisotropy, 206–236
    Airy stress function, 218
    anti-plane strain, 210
        complex potential representation of stresses and displacements, 211
        correspondence with isotropic cases, 211
        crack problems, 215
        energy flux into crack edge, 216
        equilibrium equations, 210
    central crack in a large plate
        asymptotic stresses and displacements, 222
        crack face loading, 223
        energy flux into the crack edge, 222
        stresses and displacements, 221
    contracted notations, 208
    degenerate materials, 206
    elastic compliances, 207
        coordinate transformations, 209
    elastic stiffnesses, 207
        coordinate transformations, 209
    equations of motion, 695
    general duality principle, 229
    generalized Hooke's law, 206, 207
    generalized plane strain, 207, 214, 216
        central crack in a large plate, 219
    hexagonal materials, 206
    interface cracks, 226
        definition of stress intensity factors, 232
        displacement jump, 232, 236
        energy flux into the crack edge, 233
        non-oscillatory solutions, 228
        oscillatory solutions, 230
        remote loading, 234
    monoclinic materials, 206
    orthotropic materials, 206
    path-independent integrals, 215
    plane cases, 210–215
    plane strain, 212
        use of complex potentials, 213
    plane stress, 214
        correspondence with plane strain, 214
    plane waves, 695
    steady state dynamic crack propagation, 360–364
    Stroh formalism, 216–236
        Barnett-Lothe tensors, 221, 225
        compact form, 218
        solution procedures, 219
    symmetry planes, 206
    transversely isotropic materials, 206
    use of symmetry modes, 215

wave equation, 695
weight functions, 215
anti-plane mode, *see also* mode III, 46
arc crack, *see* circular arc crack
arrays of cracks, 140
   periodic, 140
arrest of cracks, *see* crack arrest
assembly of sectors
   slowly moving cracks
      mode III, 303
associated flow rule, 248
ASTM convention for LEFM, 581
   relation to the T-stress, 581
asymptotic crack face slip
   mode III
      infinitesimally small scale yielding, 76
asymptotic displacements
   anisotropy, 222
   for cracks in non-linearly elastic media, 85
   dynamic cracks
      modes I and II, 336–341
   elastoplastic dynamic crack propagation
      mode III, 513
   mode III
      infinitesimally small scale yielding, 75, 76
   viscoplasticity, 640
asymptotic elastic crack face displacements
   compact form, 83
asymptotic fields
   non-linear elasticity, 84
asymptotic strains
   elastic-perfectly plastic dynamic crack propagation
      mode I, 530
      mode II, 535
   elastoplastic dynamic crack propagation
      mode III, 513
   mode I
      slip line theory, 282
   slowly moving cracks
      mode I, 320–325
      mode II, 325–327
      mode III, 300, 303
   slowly moving mode I and II cracks
      in centered fan sectors, 312
      in elastically deforming sectors, 314
   viscoplasticity, 638–640
asymptotic stress-strain fields
   region of dominance
      dependence on strain hardening, 276
asymptotic stresses
   anisotropy, 222
   for cracks in elastic media
      compact form, 83
      universal character, 83
   for cracks in nonlinearly elastic media, 85
   dynamic crack propagation in strain hardening material
      region of validity as function of crack velocity, 521
      vanishingly small crack velocity, 521
   dynamic cracks
      higher order terms, 343
      modes I and II, 336–341
   elastoplastic dynamic crack propagation
      mode III, 512
   mode I
      infinitesimally small scale yielding, 79
      slip line theory, 280
   mode II
      dynamic crack propagation, 352
      infinitesimally small scale yielding, 80
   mode III
      infinitesimally small scale yielding, 75, 76
   at notch edges, 684
   slowly moving cracks
      mode I, 320–325
      mode II, 325–327
      mode III, 300, 303
   viscoplasticity, 640
atomic bonds
   rupture, 6
autonomy, 3, 31, 39, 74
   before crack growth, 574
   class, 575, 588
   disappearance at large scale yielding, 268
   extension beyond small scale yielding
      at onset of fracture, 592
   illustration of its restriction to small scale yielding, 268
   large scale yielding and crack growth, 588
   near the edge of a mode III crack
      dependence on strain hardening, 276
axial splitting, 610

β lines, 279
back stress, 251
Barenblatt model, *see* Barenblatt region
Barenblatt region, 86, 92, 95–98, 119, 150, 244, 295, 345, 348, 421, 423, 504, 558, 564, 637, 653
Barnett-Lothe tensors, 221, 225
Bauschinger effect, 251
Bell's constraint, 84
Bessel functions, 683
   modified
      imaginary argument, 195
biaxiality parameter, 82
biharmonic equation, 77, 187, 668
   for notches, 684

solutions, 668
bimaterial
  anisotropy, 226
bimaterial combinations, 185
bimetals
  cracks perpendicular to the interface, 186
bipolar coordinates, 100
bivectors, 230
blunting, 29
  leading to plastic collapse, 558
Bodner-Partom constitutive equation, 638
body-centered cubic metals, 563
boundary collocation, 111
boundary conditions, 45
  continuity conditions, 45
boundary layer, 513
boundary layer approach, 119
  modified, 120
boundary value problems, 45, 239
  mixed, 106, 331
bounded stress-strain energy, 75
branch cut, 671
branch point, 670
branching, 38, 173, 625, 645–651
  angle, 173
  attempted, 625, 647, 648
  conditions for survival of the shorter
      branch, 647
  hypothesis of required energy flux, 645
  macroscopic criteria, 650
  number of branches, 651
  numerical simulations, 651
  relation to asymptotic stresses, 341
  relation to fracture surface patterns, 649
  symmetry, 647
bridging of cracks, see crack bridging
brittle fracture, 25
brittle materials, 597
brittle-ductile transition, 15, 611
building codes, 544
Burger's vector, 153

Cagniard's method, 401
canonical crack problem
  central crack in a large plate, 132–138
    energy relations in mode I, 136
    mode I, 132
    mode I crack shape, 134
    mode I stress intensity factor, 134
    mode I stresses and displacements, 133
    mode II counterpart, 135
    mode III counterpart, 135
carbide particles, 600
Cauchy principal value, 151
Cauchy's integral formula, 675
Cauchy's integral theorem, 673

Cauchy-Riemann equations, 103, 105, 265, 669
  in polar coordinates, 144
cavitation, 549–552
  deviation from spherical symmetry, 552
  in metal foils between ceramics, 549
  in metals, 9
  from spherical hole
    basic mechanism, 550
cavities, 5
cell model, 15–18
  cohesion, 16
  decohesion, 16
  for formation of micro-crack clusters, 571
centered fan characteristics, 257
centered fan sector
  dynamic crack propagation
    mode I, 527
    mode III, 512
  elastic-perfectly plastic dynamic crack
      propagation
    mode II, 532
  modes I and II, 310
  slip line theory
    mode I, 279
    mode II, 296
  slowly moving cracks
    mode I, 320
    mode II, 325
    mode III, 299
  slowly moving mode I and II cracks
    asymptotic strains, 312
central crack in a large plate, see canonical
    crack problem
ceramics
  cluster of micro-cracks, 569
  crack bridges
    energy flux, 567
  crack bridging
    fracture toughness, 566
  grain size, 12
  micro-crack shielding
    toughening ratio, 571
  particulate-reinforced, 565
  polycrystalline, 12
    elastic anisotropy mismatch, 569
    thermal expansion mismatch, 569
  profuse micro-cracking, 569
  toughening mechanisms, 565–572
    crack bridging, 565
    micro-crack shielding, 569
    toughening ratio, 568
    transformation toughening, 569
Chaplygin transformations, 398
characteristic lines, 255
  slowly moving cracks

mode III, 301
Charpy tests, 15
Christoffel stiffness, 696
circular arc crack, 142
  stress intensity factors, 147
circular cracks
  dynamic expansion, 425–428
    crack opening, 425
    energy flux, 426
    shear loading, 427
    stress intensity factor, 426
circular cylinder
  with longitudinal crack, 99
cleavage, 25, *see also* competition between cleavage and plastic flow
  along metal/ceramic interface, 550, 563
    different scales, 564
    size dependence, 550
  in ceramics, 11
  in ferritic steels, 599
  in metals, 9
  Obreimoff's experiments on mica, 583
  in the presence of plastic flow, 563
cleavage *versus* void growth
  dependence on the T-stress, 586
clip-gauge, 29
CMOD, 29
COD, 29, 575
codes, 544
cohesion modulus, 3, 97
  mode II, 152
    dynamic, 347
cohesive regions
  dynamic mode II crack expansion, 414
cohesive state, 17
cohesive strength
  opening mode, 18
  sliding mode, 18
compact tension specimen, 82
competition between cleavage and plastic flow, 562
  ductile-brittle transition, 563
  influence of thermal fluctuations, 563
  Rice-Thomson theory, 562
competition between decohesion and plastic flow
  process region share of total energy dissipation, 561
complex potentials, 103–126
  anti-plane strain, 103–111
    boundary conditions, 104
    polar coordinates, 105
    representation of stresses and displacements, 103
  conformal mapping, 126–132
    onto a half-plane, 128

    onto the unit circle, 130
    rational mapping functions, 132
    Schwartz-Christoffel transformation, 132
  in-plane strain, 111–126
    polar coordinates, 125
    rigid body displacement, 113
  mode III, 105
    finite body, 110
    infinite body, 106–110
complex stress intensity factors
  in anisotropic materials, 231
  for interface cracks, 180
compliance, 33, *see also* anisotropy, elastic compliances
  of loading device, 590
    influence on onset of fracture, 591
  measurement, 33
compliance function, 240
composites, 13
conductivity (heat), 537
configuration stability
  straight crack front, 204
conformal mapping, *see also* complex potentials, 104
  mode I
    elliptic hole, 130
  mode III
    large scale yielding, 265
    semi-infinite crack in strip, 127
    onto the unit semi-circle, 265
conservation laws, 49
constant stress sector
  dynamic crack propagation
    mode I, 527
    mode III, 512
  elastic-perfectly plastic dynamic crack propagation
    mode II, 532
  slip line theory
    mode I, 279
    mode II, 296
  slowly moving cracks
    mode I, 320
    mode II, 325
    mode III, 299
    modes I and II, 310
constant velocity crack expansion, 398–428
constant velocity phase
  dynamic crack growth, 625
constitutive equations
  anisotropic elasticity, 206
  isotropic elasticity, 236
  perfect plasticity
    mode III, 510
  plasticity, 248–254

viscoelasticity, 236
viscoplasticity, 637
constraint to plastic flow, 579
cooling
   thermoelastic, 536, 540
coordinate-invariant form
   equations of motion, 687
   Laplace equation, 69
correspondence principle
   viscoelasticity, 238, 365
correspondence static-dynamic crack
      propagation, 334
   mode III, 356
Coulomb friction, 345, 423
   dynamic mode II crack expansion, 414
crack
   definition, 1
crack arrest, 456–469, 654–659
   branching, 658
   conditions, 655
   different situations, 654
   of expanding crack, 468
   face loaded mode III cracks
      after motion from unloaded state, 457
      radiation of static crack face
         displacement, 461
      radiation of static stress along the
         symmetry plane, 462
   mode II, 643
   mode III
      after motion from equilibrium state, 462
      radiation of static stress field, 359
   modes I and II, 463
      radiation of the static crack face
         displacement, 464
      radiation of the static stress, 464
      stress intensity factor, 463
   re-initiation, 656
   from steady crack propagation, 467
   stress intensity factor, 657
   in welds, 36
crack arrest after stress pulse loading
   incident P wave, 482
   incident SH pulse, 481
   incident SV wave, 483
crack bridging, 565
   continuous distribution of springs, 566
   energy flux, 567
   fracture toughness, 566
   inner stress intensity factor, 566
   length of bridging region, 568
   toughening ratio, 568
crack detection, 544
crack edge equation of motion, 628–629
crack expansion with constant velocity,
   398–428

circular cracks, 425–428
   crack opening, 425
   energy flux, 426
   shear loading, 427
   stress intensity factor, 426
elliptic cracks
   shear loading, 428
energy flux, 421
intersonic mode II, 419–422
   crack opening, 421
mode I, 402–414
   crack opening, 406
   stress intensity factor, 409
   stresses and displacements, 408–412
   stresses and displacements flux, 412
mode II, 414–422
   energy flux, 415
   one trailing edge, 418
   stress intensity factor, 415
   stress peak travelling with the S wave
      speed, 416
mode III, 422–425
   energy flux, 425
   stress intensity factor, 424
crack expansion with nonsymmetric
   velocities
mode II, 417–418
   ratio between stress intensity factors,
      417
crack face loading, see face loaded cracks
crack face opening
   mode I
      infinitesimally small scale yielding, 79
crack face slip, 76
   mode II
      infinitesimally small scale yielding, 80
   mode III
      slowly oving cracks, 306
      strain hardening, 273
crack from the apex of a V-shaped notch,
   171
crack front waves, 500
crack growth, see also stable crack growth,
   unstable crack growth and dynamic
   crack propagation
in compression, 608
direction, see direction of crack growth
onset
   crack edge slip criterion, 270
   critical strain criterion, 270
   local condition, 30
   prediction, 574–579
unstable, 8
crack in plates
   mode I
      influence of transverse shear, 205

crack mouth opening displacement, 29
crack opening displacement, 29
  strain hardening, 290
crack paths, see direction of crack growth
crack propagation velocities
  plane stress approximation, 692
crack propagation velocity
  irregular, 626
  upper limit, 36, 652–654
    mode I, 338, 652
    mode III, 356
    modes I and II, 653
      Washabaugh and Knauss' experiments, 635
crack tip equation of motion, see crack edge equation of motion
crack trapping
  in ceramics, 571
Craggs' problem, 344
crazing
  periodic pattern, 626
  in polymers, 12
creep, 2
crystalline structure, 11
CT specimen, 82
CTOA, 575
CTOD, 575
curious velocity at intersonic crack propagation, 353
curved path dynamic crack propagation
  modes I and II, 344
curved shells, 191

$\delta(\cdot)$, see Dirac's delta function
decohesion, 5
  competition with plastic flow, 19, 556
  without plastic flow, 558
    dependence on T-stress, 559
decohesive state, 17
decomposition of analytic functions, 678
"deformation J", 591
deformation rate tensor, see rate of deformation tensor
deformation therory in plasticity, 254
delay time
  for crack growth, 447, 474–481
    anti-plane strain, 474–478
    in-plane deformation, 478–481
  crack re-initiation, 657
  kink formation, 644
  shear banding, 643
deposited slip, see slipping region
deviatoric stress, 250
diamond-cubics, 563
diffusivity (heat), 537
dilational waves, see P waves

dimples, 10
Dirac's delta function, 89
direction of crack growth, 27, see also directional stability, 603–614
  in compression, 608
    influence of hydrostatic stress, 608
  condition, 174
  criteria, 603–605
  in a layer on top of a substrate, 607
  microscopic, 603
  mixed mode II/III, 603
  periodic array of cracks, 606
  shear banding
    influence of rate of loading, 603
  small scale yielding, 605–608
    comparison with large scale yielding, 607
directional stability, 174
  cracks in a layer on top of a substrate, 608
  definition, 175
  during wedging, 613
  meeting of coplanar cracks, 605
  of originally straight crack, 174
disentangling of molecules
  in polymers, 5, 12
dislocation emission
  spontaneous, 563
    dependence on lattice structure, 563
dislocation slip
  in thin metal sheets, 6
dislocations
  Burger's vector, 153
  continuous arrays, 153
  emission, 548
    from crack edges, 562
  geometrically necessary, 549
  statistically stored, 549
  strength, 153
  stress fields, 153
displacement potential functions, 332
displacement potentials, 687
  formulae, 661
displacement rate, 312
dissipative regions, 27
distortional waves, see S waves
double cantilever, 81
Drucker's postulate, 554
DT tests, 15
dual integral equations, 160–163
ductile fracture, 25
ductile-brittle transition, 563
  connected with shear banding, 644
  in ferritic steels, 599
  statistical theory, 599
Dugdale model, see Leonov-Panasyuk-Dugdale model

Duhamel integral, 436
Dunders' parameters, 187
   dynamic, 383
dynamic crack propagation, 624–645
   cell model, 629
      numerical simulations, 634
   constant velocity phase, 625, 652
      in numerical simulations, 635
   minimum energy argument, 632
   modes II and III
      shear banding, 641–645
   morphology of micro-separations
      dependence on crack velocity, 633
   relation between stress intensity factor
      and crack velocity
      influence of T-stress, 627
   shielding effects by peripheral
      micro-separations, 633
   similarity argument, 631
   simulations
      cell model, 634
      decohesive boundaries between finite
         elements, 636
      molecular dynamics, 636
      super-Rayleigh mode I velocities, 637
   steady state problems, 330–397
   strain rate sensitivity, 634
   in a strip, 375
   summary of features, 627
   viscoplasticity, 637–641
      asymptotic displacements, 640
      asymptotic strains, 638–640
      asymptotic stresses, 640
      elastic crack edge enclave, 640
      energy flux into a crack edge, 640
dynamic-tear tests, 15

edge crack
   as continuous distribution of dislocations,
      155
   mode I, 164
   mode III, 109
   stress intensity factor, 159
effective plastic strain, 249
effective stress, 249
   Huber-von Mises material, 253
   Tresca material, 254
eigenfunction expansions, 99–103
elastic anisotropy mismatch, 10, 177
elastic cell
   dislocation free crack edge vicinity, 564
elastic constants, 663
elastic-perfectly plastic dynamic crack
   propagation
   mode I
      assembly of sectors, 527

asymptotic solutions, 525–531
asymptotic strains, 530
centered fan sectors, 527
constant stress sectors, 527
governing equations, 525
mode II
   assembly of sectors, 532
   asymptotic solutions, 531–536
   asymptotic strains, 535
   centered fan sectors, 532
   constant stress sectors, 532
   governing equations, 531
   logarithmic singularity, 535
mode III, 510–522
   an apparent paradox, 513
   assembly of sectors, 513
   asymptotic displacement, 513
   asymptotic solutions, 511
   asymptotic strains in primary plastic
      region, 513
   asymptotic stresses in primary plastic
      region, 512
   centered fan sector, 512
   constant stress sector, 512
   constitutive equations, 510
   equation of motion for active plastic
      region, 510
   logarithmic singularity, 513
   resolution of the paradox, 519
modes I and II
   asymptotic solutions, 522–536
   constitutive equations, 522
   equations of motion, 523
elastic-plastic cracks, 247–327
elastic-viscoelastic correspondence principle,
   238, 365
elastically deforming sector
   slowly moving cracks
      mode I, 323
      mode II, 325
      mode III, 298
elastoplastic crack dynamics, 509–536
   asymptotic solutions, 509
   full-field solutions, 509
   plastic wave front, 509
elastoplastic dynamic crack propagation, *see
   also* elastic-perfectly plastic dynamic
   crack propagation
   mode III
      strain hardening, 521
electrochemical processes at crack edges, 9
elliptic cracks, 191
   dynamic expansion
      shear loading, 428
embedded point source
   stress wave radiation, 689

embedment of the process region, 32, 556
  dependence on cohesive strength and
      T-stress, 560
  dependence on the T-stress, 586
emission of dislocations, 548
emission of particles during crack growth,
    573
energy dissipation
  small scale yielding
    dependence on crack velocity, 624
    lack of unique relation with crack
        velocity, 625
    specific, 628
energy dissipation in process region
  mode II dynamic crack propagation
    relation to total energy dissipation, 348
energy flux, 55
  dynamic cracks
    mode I, 337
    mode II, 336
  expressed by P- and J-integrals, 57
  interface cracks, 181
  intersonic interface crack propagation
    in-plane deformation, 390
  intersonic mode I crack propagation
    attempt to calculate, 355
  into active dissipative regions, 56
  into process region, 56
  Irwin's derivation, 88
  long strip configuration
    dynamic mode I crack propagation, 375
  mode I
    infinitesimally small scale yielding, 88
  mode II
    infinitesimally small scale yielding, 88
    intersonic crack propagation, 353
  mode III
    dynamic crack propagation, 358
    infinitesimally small scale yielding, 88
  for straight crack growth under mixed
      mode loading, 88
  subsonic interface crack propagation
    in-plane deformation, 385
energy release, 36
energy release rate, see energy flux
energy sink, 86
engineering fracture mechanics, 614–623
environmental assisted cracking, 9
equations of motion
  anisotropy, 695
  linear isotropy, 686
    anti-plane deformation, 689
    in Cartesian coordinates, 688
    in displacement potentials, 688
    in displacements, 687
    in-plane deformation, 688

invariant form, 687
    rotational symmetry, 689
  perfect plasticity
    mode III, 510
    modes I and II, 523
  plane stress approximation, 692
    for plates, 693
  viscoelasticity
    mode III, 365
equivoluminal part of vector, 687
equivoluminal waves, 329
essential work of fracture, 588
evolution law, 251
expanding crack
  constant velocity, 398–428
  orthotropic material, 399

face centered cubic metals, 563
face loaded cracks
  general time-dependent loading
    mode I, 451
  general time-independent load
    stress intensity factor, 450
  general time-independent loading
    mode III, 435
    modes I and II, 450
  mode I crack motion from equilibrium
      state
    stress intensity factor, 450
  mode II crack motion from equilibrium
      state
    stress intensity factor, 451
  mode III, 428–439
  mode III crack motion from equilibrium
      state, 436
  modes I and II, 439–456
  opposite concentrated forces, 61
  self-similar time-dependent loading
    mode I stress intensity factor, 456
    mode III, 438
    modes I and II, 451
  sudden motion of a mode I crack, 440–448
    interpenetration, 444
    moving pair of opposed forces, 449
    special cases, 445
    stress intensity factor, 444
    wave pattern, 443
  sudden motion of a mode II crack
    stress intensity factor, 448
    wave pattern, 449
  sudden motion of a mode III crack,
      428–434
    moving pair of opposed crack face
        forces, 434
    special cases, 431
    stress intensity factor, 431

wave pattern, 431
  superposition of elementary solutions, 434
factorization of analytic functions, 166, 678
factorization of matrix functions, 170
failure mode transition, 644
fatigue
  high-cycle, 2
  low-cycle, 2
faulting, 610
ferritic steels, 600
FFT (Fast Fourier Transform), 504
final stretch criterion, 575
first boundary value problem
  for half-plane, 664
flaw, 6
flow rule, 248
  associated, 248
flux, see energy flux
forbidden velocity region, 653
Fourier analysis, 70, 102
Fourier inversion, 195
Fourier series, 70
FPZ (Fracture Process Zone), see process region
fractal models for cracks, 572
  box counting method, 573
  Richardson plot, 573
  slit-island technique, 573
fracto-emission, 573
  acoustic emission, 573
fractography, 25
fracture
  definition, 1
  intergranular, 12
  in metals, 9
  prediction, 580–595
  transgranular, 9, 12
fracture mechanics, 614–623
  dynamic processes, 624–659
  general principles, 616
  objective, 1
  philosophy, 616
fracture process zone, see process region
fracture surfaces
  dimples, 10, 25
  facets, 11, 25
  fatigue, 25
  fibrous, 25
  fractal representation, 573
  hackle, 26, 626
  high crack velocities, 26
  mirror, 26, 626
  mist, 26, 626
  river pattern, 25
  shear banding, 644
  surface roughness

  correlation with stress intensity factor, 626
  dependence on acceleration/deceleration, 626
  velocity-dependent periodic pattern, 626
fracture toughness, 581
  in ceramics
    crack bridging, 566
  dependence on plate thickness, 594
Fredholm equation, 157, 158
  numerical solution, 159
Freund's fundamental solution
  mode I, 435, 450
  mode II, 450
friction coefficient
  dynamic, 346
fully plastic solutions, 277

Galilean transformation, 245, 332, 360, 366, 377, 391, 492, 510, 522, 539
Gauss integration with Legendre functions, 159
general duality principle, 229
generalized Hooke's law, 206, 207
geometrically necessary dislocations, 549
glass fibres
  Griffith's experiments, 596
glass plates
  statistical strength, 599
glassy moduli, 237
global criterion
  for onset of fracture
    large scale yielding, 588
    under general loading control, 590
    under grip control, 590
    under load control, 589
global instability, 35, 580
global versus local criteria, 580
grain boundaries
  cavitation, 5, 8
Green's theorem, 50, 64, 91, 94
Griffith's theory of fracture, 582
grip control, 16
Gurson model, 39, 249

hackle, see fracture surfaces
Hadamard materials, 84
half-plane
  first boundary value problem, 664
Hankel transforms, 161, 683
  inversion, 683
hardening
  isotropic, 251
  kinematic, 251
harmonic functions, 69, 670

healing, 330
  energy-neutral, 124, 152, 334
    dynamic mode III propagation, 358
  mode II, 152
  mode III, 109
    energy neutral, 109
heat conduction
  governing equation, 537
  in and from a layer, 538
  near a moving crack edge, 539–543
heat conduction equation
  for half-plane problems, 538
heat conductivity, 537
heat diffusivity, 537
heat generation by irreversible deformation, 536
Heaviside function, *see* unit step function
Helmholtz' decomposition, 687
hereditary integral, 237
Hermitian matrix, 230
higher order terms, *see also* T-stress
  mode I
    infinitesimally small scale yielding, 80
Hilbert problem, 106, 116, 122, 128, 145, 154, 157, 179, 228, 235, 350, 378, 383, 388, 405, 419, 458
  homogeneous, 107
hodograph transform, 258
  compatibility equation, 260
  elastic-perfectly plastic dynamic crack propagation
    mode III, 514
  equilibrium equation, 260
  inversion, 262, 267, 273
  potential function, 260
Hooke's law, 236
  generalized, 206
HRR singularity
  mode I, 285
  mode II, 297
Huber-von Mises material, 250
Huber-von Mises yield condition, 250
Hult-McClintock problem, 260
Hutchinson-Rice-Rosengren singularity, *see* HRR singularity
hydrogen embrittlement, 35
hydrostatic stress, 250
hyperelasticity, 84

in-plane modes, *see also* mode I *and* mode II, 46
incompressibility, 251
incremental strain theory, 254
infinitesimally small scale yielding, 260
  definition, 68
  mode I

asymptotic stresses, 79
crack face opening, 79
higher order terms, 80
mode II
  asymptotic stresses, 80
  crack face slip, 80
mode III
  asymptotic displacements, 75, 76
  asymptotic stresses, 75, 76
  crack face slip, 76
two different approaches, 99
influence functions, *see* weight functions
instability
  in continuum
    dependence on constitutive relations, 555
    opening mode, 555
    shearing mode, 555
  under load control, 552
    relation to cohesion-decohesion curve, 553
integral equations, 148–160
  from arrays of dislocations, 152–160
    finite regions, 159
  from dislocation array
    in finite and semi-infinite regions, 155
  dual, 160–163
  from elementary loads, 148–152
interface cracks, 177–190
  anisotropy, 226
    complex stress intensity factors, 231
  anti-plane strain, 185
  complex stress intensity factor, 180
  definition, 183
  contact regions, 183
  cracks kinking out of the interface, 190
  cracks penetrating the interface, 190
  cracks perpendicular to the interface, 186
    interface stresses, 189
    singularity exponent, 189
  cracks reaching the interface under oblique angle, 190
  dynamic crack propagation, 377–397
  energy flux, 181
  estimate of interpenetration extension, 182
  in-plane deformation, 178–185
  interpenetration of crack faces, 181
  neo-Hookean materials, 183
  oscillating interface displacements, 181
  oscillating interface stresses, 181
  subsonic crack propagation
    anti-plane strain, 377
    in-plane deformation, 379
  viscoelasticity
    dynamic mode III cracks, 371

interface stresses
  dynamic subsonic crack propagation
    in-plane deformation, 384
International Congress of Fracture, 3
interpenetration of crack faces, 177
  under superposition schemes, 439, 440, 501
intersonic crack velocity
  definition, 331
intersonic interface crack propagation
  in-plane deformation, 386
intersonic mode II crack propagation
  with crack face friction, 348–356
  observations from earthquakes, 624
intrinsic length parameter, 8
  in amorphous ceramics, 12
  in crystalline ceramics, 12
  disappearence at high crack velocities, 630
  in metals
    micro-crack domination, 11
    void domination, 10
  in polymers, 12
inverse square root terms
  dominance, 72
inverse square-root terms
  annular region of dominance, 72
    relative error, 72
  mode I
    control of the crack edge state, 81
    relative error in control of the crack edge state, 81
  mode III
    control of the crack edge state, 71
    relative error in control of crack edge state, 72
irreversible deformation energy
  fraction dissipated as heat, 536
irrotational part of vector, 687
irrotational waves, 329
Irwin's concept of energy release rate, 583
Irwin's theory of fracture, 583
isotropic hardening, 251

J_2 flow theory, 250
  multiaxial stresses, 253
"J-dominance", 579
J-integral, 3, 48, 52–53
  anti-plane shearing, 63
  axial symmetry, 63
  derived from potential energy, 60
  estimates, 59
  extension to plastic region, 53
  interpretation as energy flux, 57
  mode I
    relation to stress intensity factor, 88
  mode II
    relation to stress intensity factor, 88
  mode III, 63
    relation to stress intensity factor, 77
  path near crack edge, 53
  relation to displacement records, 698
  remote path, 53
J-N method, 615
  discussion, 622
  procedure for prediction of fracture, 618
  relation to the T-stress, 617
  testing procedure, 617
J-resistance curves
  engineering applications, 615
  large scale yielding, 588
  relation to energy flux, 586
  small scale yielding, 584
    dependence on the T-stress, 586
jump conditions at sector boundaries, 316

K-resistance curves, 584
Kalthoff's shear banding experiment, 642
kernels of nucleation, see nucleation kernels
kinematic conditions
  at sector boundaries, 316
kinematic hardening, 251
kinking, 37, 171–173
  of cracks under compressive loading, 608
  dynamic cracks
    mode II, 343
  infinitesimal, 171
  from shear banding, 643
  stress intensity factors, 172
  suppressed by hydrostatic pressure, 609
Kontorovich-Lebedev transform, 195
Kronecker's delta, 83

L-integral, 49
Lagrangian extrapolation, 159
Laplace equation, 69, 103
Laplace operator, 69
Laplace transformation, 366
Laplace transforms, 679–682
  Abel theorems, 682
  asymptotic relations, 682
  double, 399
  one-sided, 679
    inversion, 681
  Tauber theorems, 682
  two-sided, 681
    strip of convergence, 681
large scale yielding, 40
  deep cracks, 592
  shallow cracks, 592
Lebedev inversion, 195
LEFM, 3
  convention, 581

engineering applications, 614
Legendre functions, 159
Lekhnitskii-Stroh formalism, 215, 216, 360
Leonov-Panasyuk-Dugdale model, 24, 559
   mode I, 292
   mode II, 298
lift-off, 637
line load
   shear
      moving, 332
linear elastic fracture mechanics, 3
   convention, 581
   engineering applications, 614
linear thermal expansion coefficient, 540
Liouville's theorem, 107, 116, 167, 180, 227, 455, 677
load control, 16
loading
   cyclic, 1
   monotone, 1
   repeated, 1
local criterion
   for onset of crack growth, 30
   for onset of crack growth, 580
local instability
   continuum model, 552
localization
   plastic flow
      insufficiency of continuum model, 555
   plastic flow in shear
      loss of ellipticity, 556
   shear banding, 556
logarithmic singularity
   slowly moving cracks
      mode I, 324
      mode III, 302
logarithmic spiral, 282
long-time moduli, 237
longitudinal wave
   definition, 691
low-cycle fatigue, 617
lower shelf, 15

M-integral, 49, 61
   for determination of stress intensity factors, 61
Macdonald function, 195
macroscopic strains, 18
macroscopic stress, 18
martensite transformation, 569
matching elastic and plastic solutions, 277
matrix
   Hermitian, 230, 235
   principal minors, 208
matrix function
   factorization, 170

maximum crack velocity
   relation to the local Rayleigh wave velocity, 629
maximum possible crack velocity, 652
Maxwell-Betti reprocity theorem, 92
McClintock's void growth model, 545
mechanical condition at sector boundaries, 316
Mellin transforms, 164, 170, 683
   inversion, 683
   strip of convergence, 683
mesoscale, 562
metal foil between ceramics
   cleavage along interface, 563
metal/ceramic interface
   debond nucleations, 565
micro-crack clusters
   cell model approach, 571
   continuum approach, 569
   discrete approach, 571
   in rocks, 610–612
   under compression, 610–612
      axial splitting, 610
      brittle-ductile transition, 611
      faulting, 610
micro-crack shielding, 569
   toughening ratio, 571
micro-cracks, 5
   in ceramics, 569–572
   definition, 1
   formation
      influence of triaxiality, 8
   in metals, 9
micro-separations, 5
   in ceramics, 11, 569–572
   in composites, 13
   definition, 1
   influence of triaxiality, 9
   in metals, 9–11
   morphology
      dependence on crack velocity, 633
   nucleation, 6
   in polymers, 12
minimum energy argument, 632
mirror, see fracture surfaces
mist, see fracture surfaces
mixed boundary value problem, see boundary value problem
mixed mode crack growth, 2, 603
mixed mode loading, 550, 603
mode I, 2
   asymptotic elastic stresses, 79
   crack face opening, 79
   definition, 47
   displacements, 47
   higher order terms, 80

stresses, 47
mode II, 2
    asymptotic elastic stresses, 80
    crack face slip, 80
    definition, 47
    displacements, 47
mode III, 2
    asymptotic elastic crack face slip, 76
    asymptotic elastic displacements, 75, 76
    asymptotic elastic stresses, 76
    asymptotic stresses, 75
    definition, 47
    displacements, 47
    stress-strain fields outside the dissipative regions, 70
    stresses, 47
mode III cracks
    strain hardening
        infinitesimally small scale yielding, 271
molecular dynamics, 636
monoclinic materials, 206
Mooney-Rivlin materials, 84
moving line load, 332

nanoscale, 562
NDT, 15
necking
    in biaxially loaded sheets, 556
    near a crack edge, 13, 296
    to a sheet boundary, 15
    in uniaxially loaded sheets, 556
negated stress
    used in superposition schemes, 438
neo-Hookean materials, 84
    generalized, 84
net section plastic flow, 40
net section yielding, 54
nil ductility temperature, 15
Nilsson integral, 67, 488
    interpretation, 90
    viscoelastic fields, 67
niobium, 564
non-centered fan
    slip line theory
        mode I, 284
non-constant velocity crack propagation, 456–469
    after stress wave loading
        mode III, 483
        modes I and II, 485
    modes I and II, 464
    sudden change of crack speed
        modes I and II, 467
non-linear elastic materials, 84
    Bell's constraint, 84
    incompressibility, 84

non-local plasticity, 548, 549, 556
normality rule, 248
normative rules, 544
notches
    asymptotic stresses, 684
nucleation kernels
    dominating, 6

Obreimoff's experiments on mica, 583
Ogden-Ball materials, 84
onset of crack growth
    COA (CTOA) criterion, 575
    COD (CTOD) criterion, 575
    final stretch criterion, 575
    J-integral criterion, 575
    local criterion, 580
    mode I large scale yielding, 578
        J-Q-criterion, 578
    mode I small scale yielding, 576
        influence of the T-stress, 576
    mode II small scale yielding, 576
        influnce of compressive stresses, 576
    mode III small scale yielding, 575
        influence of compressive stresses, 575
    modes I and II large scale yielding, 578
        J-integral criterion, 578
    prediction, 574–579
    stress intensity factor criterion, 575
onset of fracture
    influence of loading device compliance, 591
    large scale yielding
        global criterion, 588
    small scale yielding, 581
opening mode, 2, 7, *see also* mode I
Orowan's modification of Griffith's theory, 583
orthorhombic materials, 206
orthotropic materials, 206, 209
oscillating interface stresses and displacements, 181

P waves, 329
P-integral
    derived from potential energy, 60
    estimates, 59
    extension to plastic region, 53
    interpretation as energy flux, 57
    non-closed path, 51
    path near crack edge, 53
    remote path, 53
Papkovich's representation of displacements, 111, 161, 192
particle populations
    in ferritic steels, 600
particulate-reinforced ceramics, 565

path-area integral
  axisymmetric deformation, 66
  dynamic cases, 67
    allowing for inelastic strains, 67
    allowing for viscoelasticity, 67
path-independence of J-integral
  net section yielding, 593
  well contained plastic region, 592
path-independent integrals, 48–67
  complementary integrals, 49
  J-integral, 52–53
  L-integral, 49
  M-integral, 61–63
  P-integral, 49–52
  for plates, 49–52
  stationary crack in dynamic field, 66
penny shaped crack, 160
  dynamic expansion, 425–428
    crack opening, 425
    energy flux, 426
    shear loading, 427
    stress intensity factor, 426
perfect plasticity, 252
  general relations, 254
  mode III
    crack edge slip, 258
    crack face slip, 263
    J-integral path independence, 258
    large scale yielding, 263, 264
    slowly moving cracks, 298
  mode III crack
    strain singularity, 257
    stresses and strains, 257
  stationary mode III crack
    infinitesimally small scale yielding, 260
    straight characteristic lines, 255
periodic arrays of cracks, see arrays of cracks, periodic
persistent wave
  propagating along a crack front, 505
Perzyna constitutive equation, 637
phase transition
  shear banding, 644
piecewise linear hardening
  tangent modulus, 252
P-integral
  relation to displacement records, 698
plane stress approximation, 692
plane waves
  anisotropy, 695
plastic collapse, 2, 558, 622
  in polymers, 12
plastic flow, see also competition between cleavage and plastic flow
  competition with decohesion, 556
  without decohesion

  dependence on T-stress, 557
  without decohesion, 556
    dependence on strain hardening, 557
  flow rule, 248
  high constraint, 579
  low constraint, 579
plastic incompressibility, 251
plastic potential, 248
plastic region, 27, 39–44
  during unstable crack growth
    size and shape, 43
  mode I
    Leonov-Panasyuk-Dugdale model, 292
    size and shape, 290
  mode II
    size and shape, 297
  at moving cracks, 43
    principal strain directions, 43
  primary, 44
  principal strain directions, 42
  secondary, 27
  shape, 41
    anti-plane shear mode, 42
    mode II, 42
  size, 39
  size and shape dependence on the
    T-stress, 577
  size increase during crack growth, 584
  at stationary cracks, 39
  strains, 40
    peculiar distribution, 42
  wake, 27
  well contained, 592
plasticity, 248–254
  back stress, 251
    updating, 251
  deformation theory, 254
  dynamic crack propagation, see
    elastoplastic crack dynamics and
    elastic-perfectly plastic dynamic
    crack propagation
  effective plastic strain, 249
  effective stress, 249
  field near the crack edge
    Q-parameter, 287
  flow rule, 248
  highly constrained flow
    cavitation, 549
  incremental strain theory, 254
  large strain theory
    stresses near the crack edge, 287
  mode I
    plane stress, 289
    stationary cracks, 278–296
    T-stress influence, 278
  mode II

stationary cracks, 296–298
mode III
  crack edge slip, 263
  stationary cracks, 254–278
normality rule, 248
perfect, 252
piece-wise linear hardening, 252
power-law strain hardening, 252
reference stress, 249
rigid-perfectly plastic, 279
slowly moving cracks
  mode III, 298–308
small strain theory, 248
strain hardening
  mode III cracks, 271
total strain theory, 254
Plemelj's formulae, 107, 157, 179, 442, 677
pole, 670
pop-in, 36
potential energy, 60
potential functions
  dynamic problems
    representation of stresses and displacements, 332
power-law strain hardening, 252, 271
Prandtl slip line field, 279, 323
Prandtl-Reuss flow rule, 250
prediction of fracture, 580–595
  Griffith's criterion, 582
  Irwin's criteria
    critical energy release rate, 583
    critical stress intensity factor, 582
  Orowan's criterion, 582
prediction of onset of crack growth, 574–579
pre-existing cracks, 27
pressure vessel codes, 544
primary plastic region, 44
  mode III, 298
  strains
    slowly moving mode III cracks, 300
  stresses
    slowly moving mode III cracks, 300
  wake, 44
principal minors, 208
probability of fracture, 597
process region, 5–26
  in ceramics, 11
  in composites, 13
  in concrete, 12
  definition, 18
  dimensions, 20–22
  embedment, 32
  experimental observations, 25
  in geological materials, 12
  geometry related, 14
  inertia effects, 21

length to height ratio, 20
  in concrete, 21
  at earthquake slip, 21
in metals, 9–11
models, 22–24
  Barenblatt model, 23
  box model, 23
  cell model, 23
  Leonov-Panasyuk-Dugdale model, 24
  point model, 24
as necking, 13
in polymers, 12
significance for slowly moving cracks, 327
size dependence on the T-stress, 577
size increase during crack growth, 584
size increase with crack velocity, 630
viscosity effects, 21
wake, 27, 44
process zone, see process region
propagation velocities
  elastic waves, 664
  plane waves
    anisotropy, 696
  Rayleigh waves, 694
  two fundamental, 664
  waves in thin plate, 664
proportional loading, 42, 53, 254, 270
  necessary and sufficient conditions, 271

Q-parameter, 287
  definition, 289
  small scale yielding
    connection to T-stress, 289
quasi-cleavage, 25

radial loading, see proportional loading
Ramberg-Osgood material, 252
rapidly expanding crack
  crack shape, 408
  stresses, 409
rate of deformation tensor, 311
Rayleigh function, 333, 694
  zeros, 694
Rayleigh velocity, 333, 694
  approximate expression, 695
Rayleigh waves
  propagation velocity, 694
re-initiation after crack arrest, 656
relaxation, 238
residual stresses, 177
  caused by elastic anisotropy mismatch, 569
  caused by martensite transformation, 569
  caused by thermal expansion mismatch, 569
resistance to crack growth, 302

rhombic materials, 206
Rice-Thomson theory, 562, 564
Rice-Tracey void growth model, 547
river pattern, 25
rocks
 igneous, 12
 under compressive loads
  axial splitting, 610
  brittle-ductile transition, 611
  faulting, 610
rotational waves, *see* S waves
rubbery moduli, 237

S waves, 330
sapphire, 563
scale of yielding, *see* small scale yielding *and* large scale yielding
 representation, 71, 72
scales of material processes
 macro, meso, micro and nano, 562
SCC, 9
secondary plastic region, 27, 44
 mode II, 325
 mode III, 298
 modes I and II, 320
 strains
  slowly moving mode III cracks, 303
 stresses
  slowly moving mode III cracks, 303
self-similar dynamic problems, 398
 Cagniard's method, 401
 double Laplace transform approach, 399
  inversion formulae, 401
  inversion procedure, 399
 expansion of circular cracks, 425–428
  crack opening, 425
  shear loading, 427
  stress intensity factor, 426
 expansion of elliptic cracks
  shear loading, 428
 homogeneous character, 399
 intersonic mode II crack expansion, 419–422
  crack opening, 421
  energy flux, 421
 mode II crack expansion with one trailing edge
  deposited slip, 418
 nonsymmetric mode II crack expansion, 417–418
  ratio between stress intensity factors, 417
 symmetric and nonsymmetric mode II crack expansion, 414–422
 symmetric mode I crack expansion, 402–414
  crack opening, 406
  energy flux, 412
  stress intensity factor, 409
  stresses and displacements, 408–412
 symmetric mode II crack expansion
  energy flux, 415
  stress intensity factor, 415
  stress peak travelling with the S wave speed, 416
 symmetric mode III crack expansion, 422–425
  energy flux, 425
  stress intensity factor, 424
semi-elliptic crack, 191
SENB specimen, 82
shear banding, 11, 556, 641–645
 arrest, 643
 heat generation, 536
 phase transition, 644
 temperature rise, 536, 641
shear lips, 14
shear waves, *see* S waves
shearing mode, 2, *see also* mode II *and* mode III, 7
shielding
 by cluster of micro-cracks near a crack edge, 569
 of energy flux in dynamic crack growth, 653
short-time moduli, 237
similarity argument, 631
simple pole, 670
single edge notched bend specimen, 82
singular stresses
 mode I, 78
singularity
 along a plane
  subsonic/supersonic interface crack propagation, 393
 analytic functions, 670
 logarithmic
  slowly moving mode III cracks, 302
 slowly moving mode I cracks, 324
 in stresses, 76
 in stress-strain energy density, 76
slanting edge crack, 169
sliding mode, 2, *see also* mode II *and* mode III, 7
slip line theory
 blunted crack edge
  maximum shear strain, 284
  centered fan sectors, 279, 281
  constant stress sectors, 279, 281
 large geometry changes
  non-centered fan, 284
 mode I, 279

asymptotic strains, 282
asymptotic stresses, 280
consideration of the T-stress, 285
crack blunting, 282
large geometry changes, 282
mode II, 296
stresses, 279
slipping region
expanding dynamic mode II crack with one trailing edge
deposited slip, 418
mode II
deposited slip, 124, 152
mode II, dynamic
deposited slip, 347
mode III, 106
deposited slip, 106
non-uniformly deposited slip, 358
slowly moving cracks
assembly of sectors
mode III, 303
matching conditions at sector boundaries, 316
jump conditions, 316
kinematic conditions, 316
mechanical conditions, 316
mode I, 320–325
mode II, 150, 325–327
mode III
asymptotic strains, 300, 303
asymptotic stresses, 300, 303
perfect plasticity, 298
modes I and II, 308–327
assembly of sectors, 320
strain hardening
mode III, 307
mode III angular stress-strain distribution, 308
mode III radial stress-strain distribution, 308
small scale yielding, 24
definition, 67
stress-strain fields, 67
small strain theory, 248
smooth closing, 152
at energy-neutral healing, 334
mode I
condition, 96, 97, 117
mode II
condition, 151
mode III, 109
viscoelasticity, 245
smooth opening, *see* smooth closing
Sneddon-Radok equations, 334
soft loading machine, 580

Sokhotski-Plemelj formulae, *see* Plemelj's formulae
specific energy dissipation, 628
specific heat, 537
spherulites, 12
St Venant's theory of torsion, 101
stability
under load control
condition, 552
stability criterion
for a continuum, 553
Drucker's postulate, 554
stable crack growth, 27, 31–34
definition, 31
large scale yielding
amount at onset of fracture, 589
measurement, 33
onset, 30–31
local condition, 30
screening of energy flux, 587
viscoplastic effects, 31
standard
testing procedures, 544
standard linear solid, 238
standards, 544
star shaped crack, 132
statistical theory of fracture, 596–603
statistically stored dislocations, 549
steady crack edge motion, 57
steady mode II crack propagation, 330–336
steady mode III crack propagation, 356–360
steady state, 54
definition, 54
steady state approximation, 54
steady state motion, 54
steady state problems
dynamic crack propagation, 330–397
stiff loading machine, 580
strain gradient plasticity, 548, 549, 556
strain hardening
crack opening displacement, 290
HRR singularity
mode I, 285
mode II, 286
mode III
large scale yielding, 273
small scale yielding, 271
strain softening, 641
stress continuity
mode I
condition, 117
mode II
condition, 151
stress corrosion cracking, 9
stress intensity factor, 74
complex

for interface cracks, 180
constant velocity crack expansion
  mode I, 409
defined by the crack face displacements, 84
distinction between mode II crack growth and mode II loading, 609
long strip configuration
  dynamic mode I crack propagation, 375
  dynamic mode III crack propagation, 376
mode I
  definition, 78
  relation to J-integral, 88
  sign convention, 78
mode II
  definition, 80
  relation to J-integral, 88
  sign convention, 80
mode III
  definition, 75
  relation to J-integral, 77
  sign convention, 75
stress invariants, 250
stress wave loading of cracks, *see also* three-dimensional dynamic crack problems, oblique incidence of P wave, 469–491
  incident P wave, 473
    crack growth after delay time, 478
    crack in a strip, 488
    crack of finite length, 487
  incident SH wave
    crack growth after delay time, 474
    crack moving after impact, 471
    crack of finite length, 486
    on stationary crack, 470
  incident SV wave, 472
    crack growth after delay time, 480
    crack of finite length, 487
stress-strain energy, 36
stress-strain energy density, 75
stretched zone, 29
strip
  long strip configuration, 36
    dynamic mode I crack propagation, 375
    dynamic mode III crack propagation, 376
  mode III crack, 127
strip of convergence
  for Mellin transforms, 683
  for two-sided Laplace transforms, 681
Stroh formalism, 216–236
sub-Rayleigh mode II crack propagation
  with crack face friction, 345–348
  with cohesive regions, 345–356

subsonic crack velocity
  definition, 331
subsonic/intersonic interface crack propagation
  in-plane deformation, 394
subsonic/supersonic interface crack propagation
  anti-plane deformation, 390
sudden change of crack speed
  modes I and II, 467
sulphide particles, 600
surface crack
  semi-elliptic, 191
surface energy, 582, 583
surface roughness, *see* fracture surfaces
symmetry modes, 46

T-stress, 19
  definition, 80
  implications for cleavage *versus* void growth, 586
  implications for decohesion, 559
  implications for plastic flow, 557
  implications for plastic region, 291, 577, 585
  implications for process region, 585
  implications for resistance curves, 586
tangent modulus, 252
tearing, 330
  mode III, 109
temperature change
  adiabatic, 536
  maximum temperature
    dependence on crack velocity, 543
  at moving crack edges, 536–543
    maximum temperature, 542
temperature distribution
  near a moving crack edge, 539–543
temperature rise
  connection with the cell model, 536
  in shear bands, 641
tensile cracking without tensile loading, 612
thermal expansion coefficient, 540
thermal expansion mismatch, 10, 177, 569
thermal shrinking, 7
thermal softening, 642
thermoelastic cooling, 536, 540
thermoelastic stresses, 36
three-dimensional crack growth
  influence of plate thickness, 594
three-dimensional dynamic crack problems, 491–508
  crack face loading
    semi-infinite cracks, 499–501
    wave front perturbation, 500
    weight functions, 500

general loading
  coupled modes II and III planar cracks, 506
  mode I planar cracks, 501
  planar cracks, 501–508
  spectral representations, 501
moving cracks
  crack front waves, 500, 505
  elimination of ghost events in spectral analysis, 508
  oblique incidence of P wave, 491–499
  mode I stress intensity factor, 498
  mode III stress intensity factor, 499
three-dimensional static crack problems, 190–206
configuration stability
  straight crack front, 204
cracks with slightly curved front, 201–205
curved shells, 191
elliptic cracks, 191
semi-elliptic crack, 191
semi-infinite crack
  crack face loading, 192–201
  opposed concentrated normal crack face forces, 193–196
  opposed concentrated tangential crack face forces, 196–201
semi-infinite crack with slightly curved front, 203
stress singularity at the corner between a crack and a plate surface, 205
three point bend specimen, 82
thumbnail effect, 14, 30, 33, 36, 53
torsionally loaded circular cylinder, *see* twisted bar
total strain theory in plasticity, 254
toughening mechanisms
  in ceramics, 565–572
toughening ratio, 568
  micro-crack shielding, 571
  transformation toughening, 569
  tetragonal-to-monoclinic transformation, 569
transverse waves
  definition, 691
transversely isotropic materials, 206
transversely orthotropic materials, 209
Tresca material, 250
trilogy theory, experiment and practice, 544
tunnelling, *see* thumbnail effect
twisted bar
  with longitudinal crack, 99
  St Venant's theory of torsion, 101
  stress function, 101

$U(\cdot)$, *see* unit step function

unit step function, 86
universal angular asymptotic stress distribution, 83
unstable crack growth, 27, 36–38
  global condition, 35
  Griffith's criterion, 582
  grip control, 36
  influence of loading rate, 35
  load control, 36
  onset, 35–36
  prediction, 580–595
  viscoplasticity, 35
upper shelf, 15

vector formulae, 660
vertices
  of yield surface, 555
viscoelastic fluid, 237
viscoelastic solid, 237
viscoelasticity, 236–246
  anti-plane strain
    glassy modulus, 366
    intersonic region, 366
    rubbery modulus, 366
    subsonic region, 366
    wave propagation velocities, 366
  boundary value problems
    mixed, 239
  central crack in a large plate
    suddenly applied displacement, 242
    suddenly applied load, 239
    time dependent crack face load, 242
  compliance function, 240
  constitutive equations, 236
  correspondence principle, 239
  crack meeting an interface, 244
  elastic-viscoelastic correspondence principle, 238–243
  energy dissipation
    mode III subsonic crack propagation, 373
  energy flux into the crack edge
    mode III subsonic crack propagation, 373
  extended correspondence principle, 239
  glassy moduli, 237
  hereditary integral, 237
  interface cracks
    dynamic mode III propagation, 371
  long-time moduli, 237
  models, 238
    fractional derivative models, 238
    fractional integral models, 238
    Kelvin materials, 238
    Maxwell materials, 238
    power law model, 238

spectrum models, 238
standard linear solid, 238
rather general material models
  mode III subsonic crack propagation, 371
relaxation, 238
rubbery moduli, 237
short-time moduli, 237
slowly moving cracks
  energy flow into a cohesive region, 244
  smooth closing, 245
slowly propagating crack, 243
  stress intensity factor, 244
standard linear solid
  mode III dynamic crack propagation, 367
  mode III intersonic crack propagation, 371
  mode III subsonic crack propagation, 368
stationary cracks, 239
  suddenly applied displacement, 242
  suddenly applied load, 239
  time dependent crack face load, 242
steady mode III crack propagation
  semi-infinite crack, 365
steady state dynamic crack propagation, 365–375
time dependent crack face load
  stress intensity factor, 242
viscoplasticity
  asymptotic displacements, 640
  asymptotic strains, 638–640
  asymptotic stresses, 640
  constitutive equations, 637
  dynamic crack propagation, 637–641
  elastic crack edge enclave, 640
  energy flux, 640
void growth, 7, 545–552
  Andersson's model, 548
  cylindrical hole, 545
  hole radius expansion, 547
  McClintock's model, 545
  Rice-Tracey model, 547
  small voids
    strain gradient or non-local plasticity, 548
  spherical hole, 547
    decohesion relation, 548
    representative cells, 548
void sheet, 11, 556
voids, 5
  coalescence, 7
  collapse under compression, 612
  growth, see void growth
  in metals, 9

nucleation
  strain-controlled, 600
  stress-controlled, 600
von Mises yield condition, see Huber-von Mises yield condition
Voronoi cells, 16

wake
  of plastic region, 27
    mode I, 323
    mode II, 325
    mode III, 298
  of plastic region
    asymptotic stresses in mode III, 306
  of process region, 27
Wallner lines, 500, 505
wave equation
  anisotropy, 695
wave guides, 690
wave propagation
  plane strain, 690
  plane stress, 690
wave propagation velocities
  expressions, 664
waves from embedded point source, 689
weakest link concept, see Weibull's statistical theory of fracture
Weibull distribution, 596, 598
Weibull's statistical theory of fracture, 596–603
weight functions, 90–95
  dynamic loading, 94
  moving cracks, 95
  three-dimensional
    moving cracks, 95, 500
    stationary cracks, 93
welds, 36
  arrest in base material, 36
  crack initiation in heat affected zone, 36
Westergaard method, 114–125
  boundary conditions, 114
  mode I, 115–121
    crack face shear stresses, 120
    finite body, 121
    infinite body, 116–120
    potential relations, 116
    semi-infinite crack, 118
  mode II, 121–125
    finite body, 125
    infinite body, 122–125
  representation of stresses and displacements, 114
  rigid body displacement, 115
Wiener-Hopf equations, 163, 166, 369, 452
  coupled, 170
Wiener-Hopf technique, 163–171, 366, 451

factorization, 166
matrix formulation, 170, 500
mode I edge crack, 164

yield condition, 248
yield function, 248
yield surface, 248
   vertices, 555
Yoffe functions, 338
Yoffe problem, 330–345
   mode II, 331–336
      intersonic, 355